U0252445

观赏植物育种

Ornamental Crops

〔比〕J. V. 霍伦布鲁克　主编

王继华　李绅崇　李　帆　主译

科 学 出 版 社

北　京

图字：01-2021-1781

内 容 简 介

观赏植物在全球范围内具有重要的经济价值和意义，不仅丰富和满足了人们的精神生活与审美需求，还为社会发展提供了强有力的经济支撑。随着我国社会经济的全面发展和人们生活水平的提高，原有的花卉品种已不能满足人们对美的精神需求，迫切需要培育出更多新、奇、异、罕品种。因而，观赏植物育种愈发显示出其重要性。通过传统的育种方法，如选择育种、杂交育种已培育出了一大批花卉新品种，这仍是今后花卉育种的主要手段，但日益成熟的生物技术，尤其是分子遗传学和基因工程技术为观赏植物育种提供了全新的思路，展现出广阔的发展前景和市场需求。

本书以现代植物育种技术理论和主要观赏植物的育种方法为重点，全面并详细地阐述了观赏植物育种的最新研究进展，为相关研究提供了翔实的第一手科研资料，可使从事观赏植物育种研究的科研人员、高校师生全面与充分了解世界观赏植物育种的前沿研究，也可为观赏植物生产者提供观赏植物育种和生产的关键技术指导。

图书在版编目（CIP）数据

观赏植物育种/（比）J. V. 霍伦布鲁克主编；王继华，李绅崇，李帆主译. —北京：科学出版社，2021.11

书名原文：Ornamental Crops

ISBN 978-7-03-069178-1

Ⅰ. ①观… Ⅱ. ①J… ②王… ③李… ④李… Ⅲ. ①观赏植物–育种 Ⅳ. ①S680.3

中国版本图书馆 CIP 数据核字（2021）第 111386 号

责任编辑：李秀伟 白 雪 闫小敏 / 责任校对：严 娜
责任印制：吴兆东 / 封面设计：刘新新

科学出版社 出版
北京东黄城根北街 16 号
邮政编码：100717
http://www.sciencep.com

北京虎彩文化传播有限公司 印刷
科学出版社发行 各地新华书店经销
*
2021 年 11 月第 一 版 开本：787×1092 1/16
2022 年 10 月第二次印刷 印张：45 1/4
字数：1 080 000

定价：468.00 元

(如有印装质量问题，我社负责调换)

译者分工及名单

第 1 章	邹　凌　王继华
第 2 章	邹　凌　王继华
第 3 章	马璐琳　王继华
第 4 章	蹇洪英
第 5 章	张艺萍　王继华
第 6 章	曹　桦
第 7 章	李　涵
第 8 章	阮继伟
第 9 章	李淑斌
第 10 章	段　青
第 11 章	张春英　解玮佳
第 12 章	彭绿春
第 13 章	张　露
第 14 章	卢珍红
第 15 章	莫锡君
第 16 章	李　帆　耿怀婷　王继华
第 17 章	金春莲　李　帆　李绅崇
第 18 章	段　青
第 19 章	曹　桦
第 20 章	崔光芬
第 21 章	欧阳琳　李　帆
第 22 章	阮继伟
第 23 章	李　涵
第 24 章	蔡艳飞
第 25 章	贾文杰
第 26 章	解玮佳　张春英
第 27 章	邱显钦　张　颢　李绅崇
第 28 章	杜文文
第 29 章	晏慧君　李绅崇

中 文 版 序

中国花卉产业经过三十多年的快速发展，取得了全球瞩目的成就，中国已成为世界花卉生产面积最大的国家，但与发达国家相比仍有较大差距。我国主要鲜切花和商品盆花品种大多从国外引进，新品种数量少、缺乏突破性、育种效率低一直是我国花卉产业发展的瓶颈之一。加快品种创制和技术创新是我国从花卉大国到花卉强国的必经之路。2021 年 7 月 9 日，中央全面深化改革委员会第二十次会议审议通过的《种业振兴行动方案》明确了实现种业科技自立自强的目标，提出了种业振兴的指导思想、重点任务和保障措施等，为打好我国种业翻身仗、推动我国由种业大国向种业强国迈进指明了方向。

花卉种业链的前端是育种，它是产业发展的关键。我国花卉育种优势明显。第一，中国是世界植物资源最丰富的国家之一，维管植物有 29 650 种，其中特有植物 15 000～16 000 种，占总数的 50%左右，具有观赏价值的有 7000 多种，蕴藏着独特的花色、花型、花香及抗性等优异基因，是名副其实的“植物王国”和“世界花园”，这是花卉新品种创制的重要基础。第二，我国有近百个花卉育种研究团队或课题组，近三十年来收集了大量的花卉资源，开展了多年的育种研究，积累了一定的经验，为培育新品种打下了重要基础。因此，了解国际花卉育种的前沿技术，高效地创制更多更优的新品种是当前花卉育种领域迫切的任务。

《观赏植物育种》（*Ornamental Crops*）一书由比利时农业与渔业研究所 Johan Van Huylenbroeck 博士和全球知名观赏植物研究人员共同撰写，涵盖了世界主要花卉植物的育种研究内容，汇集了当前观赏植物育种的前沿技术，是花卉育种理论、技术与经验的总结。该书是植物育种领域的大型系列专著《植物育种手册》（*Handbook of Plant Breeding*）中的观赏植物育种专集（第 11 卷）。《植物育种手册》由加拿大作物育种家 Istvan Rajcan 等组织全球植物育种工作者编写，前 10 卷包含了主要农作物、园艺植物等百余种植物育种方面的技术和成就。该系列专著出版十多年以来，广受育种工作者的欢迎。《观赏植物育种》则是围绕全球主要商品花卉育种的基础和技术问题，阐述了花卉育种的重要理论、进展和技术要点。该书包含二十九章，内容丰富。前九章涵盖了生物多样性在观赏植物育种中的作用，花器官遗传基础，花色形成调控机制及育种、倍性育种、突变育种、分子育种、抗病育种等方面，后二十章包含了月季、菊花、杜鹃花、香石竹、非洲菊、百合、蝴蝶兰、郁金香、报春花、花毛茛、补血草、铁筷子、伽蓝菜等 20 类主要商品花卉的育种。该专著的特点是科学性强、技术实用、内容广泛：既有主要切花，也有重要盆花；既有观叶植物，也有观花植物；既有草本花卉，也有温带木本观赏植物；既有传统育种，也有分子育种。这是一本有重要实用价值的花卉育种参考书。该书原版为英文版，很多遗传育种专业词汇对花卉育种工作者来说较为陌生，阅读并理解内容有一定难度。为了让我国广大花卉遗传育种工作者了解国外花卉育种进展、借鉴国外先进技术、加快我国花卉育种创新步伐，云南省农业科学院副院长王继华、花卉研究所副所长李绅崇及李帆博士等克服困难，组织花卉研究所和国家观赏园艺工程技术研究中心近30名科研人员对该书进行了翻译。总的来说，翻译忠实原著、语言顺畅、术语准确。该书的翻译倾注了主译和全体翻译人员的心血，凝聚了花卉研究所科研人员的智慧。作为花卉界的同行，借此作序之际代表广大读者向他们的辛勤付出表示由衷的感谢。

目前，基因组学与分子遗传学的技术积累与快速发展为观赏植物育种提供了全新的思路和方法，能否抓住前沿生物技术迭代重构全球花卉种业格局的历史机遇是建设花卉强国的关键，相信该书的翻译出版能为中国花卉产业的发展注入新的动力。

张启翔　教授

2021年11月于北京林业大学

原 书 序

观赏植物为人们的生活环境空间带来了活力，它们赋予了生活色彩，使人们身心愉悦，并对人类的健康产生积极的影响。同时，观赏植物改善了我们环境的空气质量，降低了环境中有害物质的浓度，并增加了生物多样性。但是，如果没有植物育种家的参与，观赏植物不可能在全球范围内被如此成功地推广应用。他们不断地培育新的品种，且孜孜不倦地努力满足种植者和消费者不断变化的需求。由于观赏植物分类繁多（鲜切花、盆栽花卉、多年生植物、灌木和木本植物）、繁育方法不同（种子、扦插、组培、鳞茎等），且植物种类广泛，因此观赏植物的育种不是容易的，其大规模的繁育仅局限于几个属的植物。大多数观赏植物是由小规模专业化的育种项目培育的，有些甚至是由业余爱好者培育的。由于对观赏植物育种知识的缺乏阻碍了许多观赏植物育种公司的研发，因此观赏植物研究与产业需求之间产生了巨大的鸿沟。为弥合这一鸿沟，我们编撰了本书。

在不断变化的世界中，观赏植物育种的主要目标是不变的：开发具有优良观赏性状的花卉新品种（花色、花型、花香、瓶插期长）、观叶和植物习性方面的创新育种。其中，抗性方面的育种研究是一项巨大的挑战，其对观赏植物产业的可持续发展具有重要作用，在未来的产业中具有巨大的潜力。

在几乎所有观赏植物的育种中，杂交育种仍然是创制新品种最重要的方法。同时，种间杂交和染色体加倍被广泛用于增加遗传多样性。生物学基础知识的最新研究也为观赏植物育种提供了新的机遇。分子遗传学和相关生物工程技术有望在一些重要经济观赏植物中获得突破性进展。这将彻底改变对遗传育种中目标性状的理解和认识，并提高观赏植物的育种效率。

本书汇集了这一研究领域杰出育种者和研究者的最新研究进展与成果。该领域的其他观赏植物育种者、研究人员、老师、学生和植物种植者均可学到高度相关的知识。在本书的 29 章中，前 9 章着重介绍了观赏植物育种中特定的性状或常用的育种技术，其余 20 章则专注于一种特定的园艺作物。本书介绍了与园艺作物有关的最新育种知识、主要的育种成就、当前的育种目标、育种方法和技术等信息。本书选择了广泛的切花植物、盆栽植物、球宿根植物、园林植物和木质观赏植物，以涵盖最重要的经济物种和一些小众物种。

本书是为观赏植物育种做出贡献的杰出育种者和研究人员知识与经验的结晶。他们中的许多人都是我的同事和好朋友。为此，我感谢他们所付出的时间、意愿和专业知识。

同时，我还要感谢许多在幕后工作的人们，是他们使这项工作成为可能。最后，我衷心感谢家人的支持。

Johan Van Huylenbroeck

于比利时梅勒

原书贡献者

Zaiton Ahmad Malaysian Nuclear Agency, Kajang, Selangor, Malaysia

Junji Amano Laboratory of Plant Cell Technology, Graduate School of Horticulture, Chiba University, Matsudo, Chiba, Japan

Paul Arens Wageningen University & Research, Wageningen, The Netherlands

Sakinah Ariffin Malaysian Nuclear Agency, Kajang, Selangor, Malaysia

Rodrigo Barba-Gonzalez Centro de Investigación y Asistencia en Tecnología y Diseño del Estado de Jalisco A.C, Guadalajara, Mexico

Suzanne Rodrigues Bento Laboratoire Reproduction et Développement des Plantes, Univ Lyon, ENS de Lyon, UCB Lyon 1, CNRS, INRA, Lyon, France

Paul E. Berry University of Michigan, EEB Department, Ann Arbor, MI, USA

Margherita Beruto Regional Institute for Floriculture (IRF), Sanremo (IM), Italy

Krishna Bhattarai University of Florida, IFAS, Department of Environmental Horticulture, Gulf Coast Research and Education Center, Wimauma, FL, USA

Mark P. Bridgen Cornell University, Riverhead, NY, USA

Andy Van den Broeck Denis-Plants bvba, Beervelde-Lochristi, Belgium

Evelien Calsyn Flanders Research Institute for Agriculture, Fisheries and Food (ILVO), Plant Sciences Unit, Applied Genetics and Breeding, Melle, Belgium

Pierre Chambrier Laboratoire Reproduction et Développement des Plantes, Univ Lyon, ENS de Lyon, UCB Lyon 1, CNRS, INRA, Lyon, France

Stephen Chandler School of Applied Sciences, RMIT University, Bundoora, VIC, Australia

Hong-Hwa Chen Department of Life Sciences, National Cheng Kung University, Tainan, Taiwan

Orchid Research and Development Center, National Cheng Kung University, Tainan, Taiwan

Wen-Huei Chen Orchid Research and Development Center, National Cheng Kung University, Tainan, Taiwan

Zhe Chen Yunnan Institute of Forest Inventory and Planning, Kunming, China

Mathilde Chopy Laboratoire Reproduction et Développement des Plantes, Univ Lyon, ENS de Lyon, UCB Lyon 1, CNRS, INRA, Lyon, France

Lívia Lopes Coelho Department of Plant and Environmental Sciences, University of Copenhagen, Tåstrup, Denmark

Ryan Contreras Oregon State University, Department of Horticulture, Corvallis, OR, USA

Mario G. R. T. de Cooker Ohé en Laak, The Netherlands

Enrico Costanzo Laboratoire Reproduction et Développement des Plantes, Univ Lyon, ENS de Lyon, UCB Lyon 1, CNRS, INRA, Lyon, France

Zhanao Deng University of Florida, IFAS, Department of Environmental Horticulture, Gulf Coast Research and Education Center, Wimauma, FL, USA

René Denis Denis-Plants bvba, Beervelde-Lochristi, Belgium

Emmy Dhooghe Flanders Research Institute for Agriculture, Fisheries and Food (ILVO), Plant Sciences Unit, Applied Genetics and Breeding, Melle, Belgium

Tom Eeckhaut Flanders Research Institute for Agriculture, Fisheries and Food (ILVO), Plant Sciences Unit, Applied Genetics and Breeding, Melle, Belgium

Keith Funnell The New Zealand Institute for Plant & Food Research Ltd, Palmerston North, New Zealand

Geert van Geest Deliflor Chrysanten B.V, Maasdijk, The Netherlands

Edwin J. Goulding Ipswich, UK

Affrida Abu Hassan Malaysian Nuclear Agency, Kajang, Selangor, Malaysia

Mai Hayashi Laboratory of Plant Cell Technology, Graduate School of Horticulture, Chiba University, Matsudo, Chiba, Japan

Josefine Nymark Hegelund Department of Plant and Environmental Sciences, University of Copenhagen, Tåstrup, Denmark

Stan C. Hokanson University of Minnesota, Department of Horticultural Science, St. Paul, MN, USA

Wouter Van Houtven Flanders Research Institute for Agriculture, Fisheries and Food (ILVO), Plant Sciences Unit, Applied Genetics and Breeding, Melle, Belgium

Chia-Chi Hsu Department of Life Sciences, National Cheng Kung University, Tainan, Taiwan

Johan Van Huylenbroeck Flanders Research Institute for Agriculture, Fisheries and Food (ILVO), Plant Sciences Unit, Applied Genetics and Breeding, Melle, Belgium

Rusli Ibrahim Malaysian Nuclear Agency, Kajang, Selangor, Malaysia

Mayuko Inari-Ikeda Department of Nutrition, School of Health and Nutrition, Tokai Gakuen University, Tenpaku, Nagoya, Aichi, Japan

Maja Dibbern Kaaber Department of Plant and Environmental Sciences, University of Copenhagen, Tåstrup, Denmark

Juntaro Kato Department of Biology, Aichi University of Education, Kariya, Aichi, Japan

Ellen De Keyser Flanders Research Institute for Agriculture, Fisheries and Food (ILVO), Plant Sciences Unit, Applied Genetics and Breeding, Melle, Belgium

Hyoung Tae Kim Department of Horticultural Science, Kyungpook National University, Daegu, South Korea

Nobuo Kobayashi Faculty of Life and Environmental Science, Shimane University, Matsue, Japan

Stephen L. Krebs The Holden Arboretum, Kirtland, OH, USA

Marie-Christine Van Labeke Ghent University, Department Plants and Crops, Ghent, Belgium

Katrijn Van Laere Flanders Research Institute for Agriculture, Fisheries and Food (ILVO), Plant Sciences Unit, Applied Genetics and Breeding, Melle, Belgium

Leen Leus Flanders Research Institute for Agriculture, Fisheries and Food (ILVO), Plant Sciences Unit, Applied Genetics and Breeding, Melle, Belgium

Ki-Byung Lim Department of Horticultural Science, Kyungpook National University, Daegu, South Korea

Bo Long School of Life Sciences, Yunnan University, Kunming, China

Chunlin Long College of Life and Environmental Sciences, Minzu University of China, Beijing, China

Kunming Institute of Botany, Chinese Academy of Sciences, Kunming, China

Henrik Lütken Department of Plant and Environmental Sciences, University of Copenhagen, Tåstrup, Denmark

Kathryn Kuligowska Mackenzie Department of Plant and Environmental Sciences, University of Copenhagen, Tåstrup, Denmark

Agnieszka Marasek-Ciolakowska Research Institute of Horticulture, Skierniewice, Poland

Heiko Mibus Hochschule Geisenheim University, Department of Urban Horticulture and Ornamental Plant Research, Geisenheim, Germany

Masahiro Mii Laboratory of Plant Cell Technology, Graduate School of Horticulture, Chiba University, Matsudo, Chiba, Japan

Center for Environment, Health and Field Sciences, Chiba University, Kashiwa, Chiba, Japan

Patrice Morel Laboratoire Reproduction et Dévelopement des Plantes, Univ Lyon, ENS de Lyon, UCB Lyon 1, CNRS, INRA, Lyon, France

Ed Morgan The New Zealand Institute for Plant & Food Research Ltd, Palmerston North, New Zealand

Renate Müller Department of Plant and Environmental Sciences, University of Copenhagen, Tåstrup, Denmark

Naonobu Noda Institute of Vegetable and Floriculture Science, National Agriculture and Food Research Organization, Tsukuba, Ibaraki, Japan

Peter Oenings Heuger, Glandorf, Germany

Hiroaki Ohashi Faculty of Agriculture, Ehime University, Matsuyama, Ehime, Japan

Naoko Okitsu Research Institute, Suntory Global Innovation Center Ltd., Kyoto, Japan

Takashi Onozaki Institute of Vegetable and Floriculture Science, NARO (NIVFS), Tsukuba, Japan

Teresa Orlikowska Research Institute of Horticulture, Skierniewice, Poland

Thierry Van Paemel Het Wilgenbroek, Oostkamp, Belgium

Małgorzata Podwyszyńska Research Institute of Horticulture, Skierniewice, Poland

Mario Rabaglio Biancheri Creations, Camporosso (IM), Italy

Elizanilda Ramalho do Rêgo Research Productivity, Centro de Ciências Agrárias, Universidade Federal da Paraíba – CCA-UFPB, Areia, Brazil

Mailson Monteiro do Rêgo Research Productivity, Centro de Ciências Agrárias, Universidade Federal da Paraíba – CCA-UFPB, Areia, Brazil

Jan De Riek Flanders Research Institute for Agriculture, Fisheries and Food (ILVO), Plant Sciences Unit, Applied Genetics and Breeding, Melle, Belgium

Shakinah Salleh Malaysian Nuclear Agency, Kajang, Selangor, Malaysia

Arwa Shahin Wageningen University & Research, Wageningen, The Netherlands

Marinus J. M. Smulders Wageningen University & Research, Wageningen, The Netherlands

Dariusz Sochacki Warsaw University of Life Sciences – SGGW, Departments of Ornamental Plants, Warsaw, Poland

Jaap Spaargaren Ingenieursbureau, Aalsmeer, The Netherlands

Julia Sparke Boehringer Ingelheim Pharma GmbH & Co. KG, Ingelheim am Rhein, Germany

Roman Szymański Horticulture Farm Roman Szymański, Poznań, Poland

Yoshikazu Tanaka Research Institute, Suntory Global Innovation Center Ltd., Kyoto, Japan

Jaap M. Van Tuyl Wageningen University & Research, Wageningen, The Netherlands

Michiel Vandenbussche Laboratoire Reproduction et Développement des Plantes, Univ Lyon, ENS de Lyon, UCB Lyon 1, CNRS, INRA, Lyon, France

Jeroen Van der Veken Flanders Research Institute for Agriculture, Fisheries and Food (ILVO), Plant Sciences Unit, Applied Genetics and Breeding, Melle, Belgium

Serena Viglione Regional Institute for Floriculture (IRF), Sanremo (IM), Italy

Jan H. Waldenmaier Herpen, The Netherlands

Traud Winkelmann Institute of Horticultural Production Systems, Woody Plant and Propagation Physiology, Leibniz Universität Hannover, Hannover, Germany

Ying Zhou Xiangxi Tujia and Miao Autonomous Prefecture Forest Resources Monitoring Center, Hunan, China

主 编 简 介

Johan Van Huylenbroeck 博士是比利时农业与渔业研究所（ILVO）植物育种家。除参与实际的育种工作，他还与他人协作完成了许多观赏植物方向的科研项目。自 2006 年起，Johan Van Huylenbroeck 博士担任研究所应用育种和遗传学小组的首席研究员，该研究团队由 20 名科研人员和 30 名技术人员组成，其研究领域主要集中在：①研发高效杂交和选择策略；②培育可持续利用的农学和园艺品种；③将 DNA 标记辅助育种技术用于数量和质量性状的选育；④基于分子标记的遗传多样性分析和鉴定。他的科研团队与产业密切合作，开展了杜鹃花、庭院月季和木本观赏植物等的育种研究。这些观赏植物的主要育种目标为提高植株抗病性、通过种间杂交丰富遗传多样性和培育紧凑型植物。

目　录

第 1 章　生物多样性与植物保护在观赏植物育种中的作用

Chunlin Long, Zhe Chen, Ying Zhou, and Bo Long

摘要　本章简要介绍了世界植物的生物多样性。根据生物多样性信息分析，估计地球上有 452 科 333 500～390 900 种维管植物。观赏植物及其野生近缘种的物种数量估计为 85 000～99 000。保护观赏植物的策略有 4 种，即就地保护、迁地保护、可持续利用和法律制度建设。《生物多样性公约》和其他国际法律制度，连同国家或地方法律和条例，正在将保护生物多样性和可持续利用生物资源纳入主流。《关于获取遗传资源和公正公平分享其利用所产生惠利的名古屋议定书》将深刻影响世界各地观赏植物的收集、育种及其商业化。基因组测序等新技术加快了我们对植物遗传多样性的认识，将促进观赏植物的育种和发展。

关键词　世界植物物种；观赏植物；保护；《名古屋议定书》

1.1　世界植物的生物多样性

生物多样性信息学是以信息科学和技术为手段，研究生物和其他相关研究而产生数据与信息的学科，正在迅速发展。许多国家和全球生物多样性信息学项目已经实施或已经实现（Ma 2014）。其中大多数已在生物多样性社区和其他组织之间传播与共享。在这些组织或网站中，包括物种 2000（COL-Species 2000）（Roskov et al. 2017）、全球生物多样性信息设施（GBIF）、国际植物名称索引（IPNI）、生命条码（CBOL）、国家生物技术信息中心（NCBI）、生命大百科全书（EOL）、生命之树（TOL）、美国农业部植物和生物多样性遗产图书馆（BHL）等，极具代表性。大数据时代的到来为通过互联网或数据库整合数据和信息提供了解决方案。这使人们能够以一种简单易行的方式分析和挖掘生物多样性信息。众所周知，根据植物的物种比例和科学家估计的数量，记载的植物比动物要多。近年来，通过大数据分析触及了世界植物的轮廓。

C. Long (✉)
College of Life and Environmental Sciences, Minzu University of China, Beijing, China
Kunming Institute of Botany, Chinese Academy of Sciences，Kunming, China
e-mail: long@mail.kib.ac.cn; long.chunin@muc.edu.cn
Z. Chen
Yunnan Institute of Forest Inventory and Planning, Kunming, China
Y. Zhou
Xiangxi Tujia and Miao Autonomous Prefecture Forest Resources Monitoring Center, Hunan, China
B. Long
School of Life Sciences, Yunnan University, Kunming, China

尽管大多数国家在不断发现新的植物物种时并未最终确定植物群，但可以从数据库和文献中获得与世界植物生物多样性有关的大量数据和信息。例如，乌干达的植物信息可以通过诸如 Hamilton 等 2016 年的出版著作和其他文献获得。从互联网特别是生物多样性信息学获得的数据，使科学家有可能在全球范围内估算植物的生物多样性。最近的一项统计分析显示，世界上有 600 多个生物多样性信息学项目被启动（Ma 2014；Wang et al. 2010）。表 1.1 列出了一些重要的植物生物多样性信息学网站或项目。皇家植物园为记录全球植物多样性做出了巨大贡献，如实施了国际植物名称索引（IPNI）、植物名录（TPL）、世界植物名录（WCSP），以及其他一些项目。密苏里植物园是另一个杰出的组织，它为丰富许多国家的植物多样性做出了贡献。

表 1.1　世界植物生物多样性信息学网站或项目名录

网站或项目名称（简写）	核心数据	网站
International Plant Names Index（IPNI）	植物物种名称	www.ipni.org
Species 2000 and ITIS Catalogue of Life（COL）	植物、动物、真菌和其他生物	www.catalogueoflife. org/col
Global Biodiversity Information Facility（GBIF）	植物标本	www.gbif.org/
The Plant List（TPL）	维管植物和苔藓植物	www.theplantlist.org/
The World Checklist of Selected Plant Families（WCSP）	来自 22 个国家/地区的 173 种种子植物	apps.kew.org/wcsp/
Barcoding of Life（CBOL）	DNA 条形码，主要为植物的条形码	www.barcodinglife. org/
Tree of Life（TOL）	生物的系统发育数据	tolweb.org/tree/
USDA PLANTS	植物种类信息	plants.usda.gov/
Encyclopedia of Life（EOL）	物种信息	www.eol.org/
Flora of China	中国植物志	flora.huh.harvard.edu/ china/
Flora brasiliensis-The Project	巴西植物志	florabrasiliensis.cria. org.br
Flora of North America	北美植物志	floranorthamerica.org/
Convention on International Trade in Endangered Species of Wild Fauna and Flora（CITES）	野生动植物国际贸易公约	www.cites.org/
2000 IUCN Red List of Threatened Species	濒危植物物种	www.iucnredlist.org/
Botanic Gardens Conservation International（BGCI）	植物保存和公众参与	www.bgci.org/
The Flower Expert	鲜花和园艺	www.theflowerexpert. com/

注：修改自 Ma（2014）

在这里，我们将仅就维管植物的审美潜力进行介绍。根据 *COL-Species 2000* 和《生物物种目录》（COL）的综合生物分类信息系统（ITIS），我们已对 336 369 种维管植物物种进行了分类（Roskov et al. 2017）。分类学家已知的维管植物物种数约为 333 500（表 1.2）。然而，根据《2016 年世界植物状况》，其他科学家认为，这一数字应该为 390 900，其中 369 400 种是开花植物或被子植物（RBG Kew 2016）。无论是 COL 的估计还是邱园（RBG Kew）中的保存陈列，都更新了我们对植物种类数量的了解，这远远高于传统教科书[如 Judd 等（2008）]中所报道的仅 26 万种维管植物。

表 1.2　世界维管植物种类

纲	数目	已知物种的估计数量	COL 所包括的百分数（%）	该纲内园艺植物示例
Cycadopsida	353	317	100	*Cycas revoluta*，*Zamia amazonum*
Equisetopsida	38	40	95	*Equisetum arvense*
Ginkgoopsida	1	1	100	*Gingko biloba*
Gnetopsida	112	112	100	*Gnetum montanum*，*Welwitschia mirabilis*
Liliopsida	74 230	72 926	100	*Roystonea regia*，*Tulipa gesneriana*
Lycopodiopsida	1 393	1 330	100	*Lycopodium japonicum*
Magnoliopsida	247 825	246 366	100	*Nelumbo nucifera*，*Rosa chinensis*
Marattiopsida	133	140	95	*Angiopteris fokiensis*
Pinopsida	615	615	100	*Araucaria cunninghamii*，*Podocarpus macrophyllus*
Polypodiopsida	11 530	11 530	100	*Cibotium arachnoideum*，*Adiantum flabellatum*
Psilotopsida	139	123	100	*Psilotum nudum*
总量（平均）	336 369	333 500	99	

资料来源：Roskov 等（2017），Catalogue of Life-Species（2000）

基于 DNA 测序技术的分子证据克服了植物形态特征的局限性，在植物物种鉴定中得到了广泛的应用。自 20 世纪 90 年代以来，被子植物系统发育组（APG）开始了各国植物学家之间的国际合作，建立被子植物系统发育学，并探索世界植物多样性。2016 年，他们发现了 416 个开花植物科。到目前为止，科学家已经在全世界发现了 452 个维管植物科，其中 12 个科是裸子植物，24 个是蕨类植物、木贼类植物和石松类植物（RBG Kew 2017）。

科学家将 17 个国家确定为世界上生物多样性最丰富的国家，特别关注其特有物种。这些国家包括澳大利亚、巴西、中国、哥伦比亚、刚果、厄瓜多尔、印度、印度尼西亚、马达加斯加、马来西亚、墨西哥、巴布亚新几内亚、秘鲁、菲律宾、南非、美国和委内瑞拉。这些物种高度多样化的国家非鱼类脊椎动物物种至少占所有的 2/3，高等植物物种占所有的 3/4（UNEP-WCMC 2014）。在这些生物多样性最丰富的国家中，巴西、中国、印度尼西亚、墨西哥和哥伦比亚是拥有最多维管植物种类的 5 个国家，分别是 32 364 种、29 650 种、29 375 种、25 036 种和 24 500 种（Li and Miao 2016）。这些物种高度多样化的国家也是维管植物新物种的主要贡献者。根据 2015 年新发现的植物物种统计，最多的国家是巴西、澳大利亚和中国，分别有 2220 种、1648 种和 1537 种。在其他国家如哥伦比亚、厄瓜多尔、墨西哥、秘鲁、马来西亚、南非和印度尼西亚等，也发现了许多新的植物物种（RBG Kew 2016）。

地球上有 14 种陆地生物群落或栖息地类型（Olson and Dinerstein 1998；Olson et al. 2001）。分别是：①红树林（亚热带和热带，被盐水淹没）；②热带和亚热带湿润阔叶林（热带和亚热带，湿润）；③热带和亚热带干燥阔叶林（热带和亚热带，半湿润）；④热带和亚热带针叶林（热带和亚热带，半湿润）；⑤温带阔叶和混交林（温带，潮湿）；⑥温带针叶林（温带，湿润至半湿润）；⑦北部森林/针叶林（亚热带，潮湿）；⑧热带和亚热带草原，热带稀树草原和灌丛（热带和亚热带，半干旱）；⑨温带草场，稀树草原和灌丛（温带，半干旱）；⑩沼泽和稀树草原（温和的热带，被淡水或微咸水淹没）；⑪山地和灌丛（高山或山地气候）；⑫苔原（北极）；⑬地中海森林，林地，灌丛或硬叶森林（温带，半湿润至半干旱，冬季降雨）；⑭沙漠和旱生灌丛（温带至热带，干旱）。这些不同的生物群落和植被类型为植物提供了多种栖息地。

在遗传多样性水平上，通过分子标记和 DNA 测序技术获得许多新的研究成果，揭示了植物的多样性。由于 DNA 测序的成本持续下降，更多的植物物种可以在线获得组装好的全基因组序列草案。截至 2017 年 1 月，获得全基因组序列的植物达到 225 种。它们主要是农作物（水稻、玉米、小麦、大麦、土豆等），其次是模式种、近缘种以及作物野生近缘种（RBG Kew 2016）。目前，已经对一些观赏植物进行了测序，如梅花（*Prunus mume*）（Zhang et al. 2012）。就在最近，位于中国南方深圳的全球最大的基因组测序中心——华大基因研究所（BGI），在 2017 年 7 月下旬举行的第 19 届国际植物学大会上发布了 10kb 植物基因组计划。根据该计划，BGI 及其合作者将在 2022 年底前完成 10 000 种植物的测序。在这个项目中，将选择更多的观赏植物进行测序。

1.2 观赏植物的多样性

世界上有多少种观赏植物，这仍然是一个有待回答的问题。根据荷兰阿斯米尔花卉拍卖市场记录，20 世纪 90 年代商业观赏植物的总分类单元（种和品种）约为 1600 种，英国皇家园艺学会编写的《园林植物百科全书》收录了约 15 500 种园林植物，涵盖新品种和外来品种（Brickell 2008）。Bailey 和 Bailey（1976）报道了 23 979 个园艺植物分类单元（科、属、种），这是观赏植物分类单元数量最多的。

以所有物种为分类单元，Bailey 和 Bailey（1976）报道的园艺植物仅占世界维管植物的 7.13%（COL）或 6.13%（Kew）。根据我们的调查，这些比例可能不准确（Long et al. 2015）。以中国为例，约 6000 种具有园艺价值的维管植物被普遍接受（Chen 2000），占中国维管植物总种数的 20.24%。一些作者认为，超过 15 000 种（Xue 2005）或 10 000 种（Zhu et al. 2007）的维管植物具有园林和园艺的潜力。最近的一项评估显示，中国本土的观赏植物和园林植物大约有 7500 种（Huang 2011）。根据观赏植物的美学特征，我们选取了 7500 种植物作为观赏植物，占中国植物区系总数的 25.29%。按照该比例，世界上应该有 85 067（COL）～98 858（Kew）种观赏植物。这些数字涵盖了人工栽培的观赏植物野生近缘种。

许多种类的维管植物是众所周知的观赏植物。例如，几乎所有杜鹃花品种都可能是园林植物，或被作为观赏植物进行栽培。该属约 1000 种，在亚洲、欧洲和北美已经培育了成千上万的新品种。仙人掌科属于热带和亚热带的一个科，有 1000 多种，大多数作为室内或温室的观赏植物被栽培。荚蒾属（五福花科）的成员也是不错的园林植物，有 200～250 个种（Dirr 2008；Long et al. 2015）。其他来自许多不同群体的成员，如槟榔科、秋海棠科、凤梨科、鸢尾科、木兰科、兰科、罗汉松科、山茶科、花烛属、醉鱼草属、菊属、铁线莲属、苏铁属、金丝桃属、凤仙花属、百合属、木犀属、芍药属、喜林芋属、李属、蔷薇属、紫丁香属和郁金香属，通常也作为观赏植物进行栽培。

观赏植物的栖息地多种多样，从热带到温带、从低地到高山、从潮湿地区到干旱地区，以及从陆地环境到水生环境。观赏植物的生活方式也非常多样。大多数针叶树和许多被子植物是乔木植物，在热带地区以南洋杉和棕榈树为代表，在亚热带地区以肉桂（Zhou and Yan 2015）和木兰为代表，在温带地区则以榆树和胡杨为代表。肉质植物现在作为室内装饰品很受欢迎，特别是芦荟属、酒瓶兰属、仙人掌科、青锁龙属、大戟属、伽蓝菜属、虎尾兰和景天属植物。在温暖的城市地区许多附生植物也是受喜爱的，如菠萝、天竺葵、石斛、蝴蝶兰和万带兰。藤本植物包括凌霄花、铁线莲、常春藤、金银花、爬山虎、使君子、崖角藤、月季、葡萄和紫藤等，被广泛应用于室内和花园的装饰（Chen et al. 2013）。大多数观赏植物是草本植物和灌木，用于切花、盆景、室内装饰、地被植物、园林景观及路边绿化美化等方面。

为了观赏，大量新品种已经被培育出来，如菊花（20 000～30 000 个品种）、梅花（400 多个品种）和牡丹（约 2000 个品种）（Long et al. 2015）。莲花是亚洲国家一种具有重要文化价值的观赏性植物。在中国、印度以及其他一些国家，已经培育出 200 多个品种，被当作水生观赏植物和食物，并用于文化传播。部分观赏植物如梅花（Zhang et al. 2012）、蝴蝶兰（Cai et al. 2015）及月季（Foucher et al. 2015）的全基因组序列已经发表。随着 DNA 测序成本的不断降低和大量基因组计划的实施（例如，深圳市兰科植物保护研究中心和华大基因研究院于 2017 年 7 月启动的兰科植物基因组计划，由中国科学院上海生命科学研究院和合作伙伴于 2017 年 8 月发起的旋花科植物基因组计划），大量观赏植物基因组数据将很快公布。

1.3　观赏植物的养护

自工业革命以来，全球范围内保护植物生物多样性的生态系统发生了翻天覆地的变化。过去 10 年间，土地覆盖发生了很大变化。在地球上的 14 种生物群落中，植被都受到了土地覆盖变化的影响。在 14 种植被类型中，红树林的变化是最大的，2012 年变化率达到 26%，而沙漠和旱生灌木丛变化最小，为 7%（RBG Kew 2016）。随着土地覆盖的变化，各种植被类型的观赏植物都受到了威胁。许多野生观赏物种都已濒临灭绝，如芦荟属、苏铁属、棒槌树属、兜兰属和瓶子草属。

几个世纪以来，人类活动导致生物有机体在世界各地迁徙。植物、动物和微生物，从它们的生长范围迁移到其他地方，可能会成为入侵物种。入侵植物作为生物多样性丧失的重要驱动因素之一，其对环境和经济的影响日益受到人们的关注。据估计，入侵物种的治理成本接近世界经济的 5%（RBG Kew 2016）。观赏植物的迁移和引进，一直被认为是外来物种入侵的主要驱动因素。有时，观赏植物本身也会成为有害的入侵物种。最著名的例子可能就是多叶羽扇豆、加拿大一枝黄花、万寿菊、凤眼莲和狐尾藻。它们因外形美丽被引进，但在许多国家造成了生态灾难，并威胁到其他观赏植物。观赏植物的引进也将病虫害带入当地环境，这将对包括观赏植物在内的生物多样性构成威胁。

保护生物多样性的一般策略包括：①就地保护；②迁地保护；③可持续利用；④法

律制度建设。重要植物区（IPA）标准体系提供了务实而科学严谨的数据集传递手段，使国家或地区范围的保护优先化成为可能，并为全球优先化系统做出了重要贡献（Darbyshire et al. 2017）。科学家在全球范围内已经确定了 1771 个 IPA，并且 IPA 已被正式确认为《生物多样性公约》全球植物保护战略目标 5 下的一个原位保护工具。应该优先保护这些 IPA，因为它们是动植物（包括观赏植物）最重要的栖息地。不幸的是，目前很少有有效的保护活动。即使在欧洲，某些 IPA 也没有法律保护或有效的管理计划，因此，相当数量的物种受到了迫在眉睫的威胁（RBG Kew 2016）。自然保护区（或保护地、国家公园）是生物多样性就地保护的重要类型，有时可能与 IPA 重叠。一些自然保护区，特别是发展中国家的自然保护区，由于缺乏资金支持、专业技术人员、能力或设施以及法律制度的保障而无法实施有效的保护。圣树林和圣地，可能是生物多样性就地保护系统中最便宜但有效的。许多动植物受到宗教、传统文化或习惯法的保护，而没有成本或成本很低。例如，西双版纳神林的基诺族保护了许多观赏植物，包括紫金牛属、石斛属、榕树属、山茱萸属、罗汉果属等（Long and Zhou 2001）。在传统农业生态系统中，观赏植物的养护是一种重要的就地保护或农场保护方法。厄瓜多尔东南部 7 个村庄的土著居民，在其传统家庭花园中种植了 10 种本地棕榈树。其中大部分是观赏植物，以及作为食品、建筑、药品和手工工艺品原料使用（Byg and Balslev 2006；Long et al. 2017）。

由于土地覆盖的变化和其他因素，迁地保护对于动植物至关重要。观赏植物可以保存在人类控制或管理的环境中，而植物园就是最传统的迁地保护方法。根据国际植物园保护组织（BGCI）（http://www.bgci.org/）的统计资料，世界上大约有 3300 个植物园（树木园）。据估计，世界上有 1/3 的维管植物已经保存在植物园中，其中有 10 000 种是珍稀濒危植物（Sharrock 2012）。自 20 世纪 90 年代以来，种质库已被广泛接受用于植物迁地保存。英国皇家植物园千年种子库（MSB-RBG Kew）和中国科学院昆明植物研究所野生物种种质库（GBWS-KIB）是最大的野生植物保存库。截至目前，来自 189 个国家和地区的 82 896 份（37 770 个品种）野生植物资源在 MSB-RBG Kew 的标本库中被收存，而在 GBWS-KIB 的标本库中有 179 570 份（16 554 种）野生植物资源被收存。著名的农作物种子库和野生亲缘关系种子库有美国国家遗传资源计划（NGRP-USDA）、中国农业科学院作物科学研究所作物种质资源中心、俄罗斯植物遗传资源研究所种子库、北欧基因库的斯瓦尔巴种子库（挪威 SSB-NGB），分别保存了 508 994 份、412 038 份、322 238 份和 860 000 份农作物种子（Liu 2015）。斯瓦尔巴种子库拥有 450 万份种子/样本，是全球约 1700 个种子库的备份。科学家提出了“3E”原则来收集和保存种子库中的种质资源。三个“E”指的是濒危（endangered）、地方性（endemic）和经济性（economic）。该原则有助于确定种子库中收集和保存植物的优先顺序（Huang and Long 2011）。MSB-RBG Kew 和 GBWS-KIB 采用了这一原理，并保存了种子、离体材料和 DNA 材料。

位于美国俄亥俄州哥伦布的观赏植物种质中心（OPGC），是一个迁地保护草本观赏植物的机构。OPGC 成立于 2001 年，致力于保护和开发具有美学价值的植物种质资源。OPGC 拥有约 200 属草本观赏植物的约 3200 份种质，应要求可将其分发给全世界的研究者、育种者和教育者（https://opgc.osu.edu/）。在 2016 年，中国发布了《国家花卉种质资源中心名单》。这 34 个活藏品和离体库包括 22 组观赏植物，即莲属、观赏蕨类、睡莲、石蒜科、百合科、山茶花属、木瓜属、菊花、兰属、石斛属、非洲菊、玉簪属、鸢尾属、紫薇属、木犀属、芍药属、蝴蝶兰属、梅花、杜鹃花属、中国月季、郁金香属和几种凝灰岩草。中国国家种质中心已保存了 27 000 多种观赏植物，而这些收藏的观赏植物资源是育种和园艺行业的宝贵研究材料。

根据联合国粮食及农业组织的数据，世界经济的 40%直接或间接地依赖于生物资源的利用。可持续利用是生物多样性保护的最终途径。如果可持续地使用一个物种或品种，它将极大地吸引公众的兴趣和关注。来自中国的例子可能为观赏植物的可持续利用提供了案例。云南拟单性木兰是一种珍稀濒危植物，由于野生数量非常有限，已被列入《中国物种红色名录》。它是中国西南部特有的美丽乔木树种。技术人员已经开发出用于园林及街道绿化的云南拟单性木兰苗木生产栽培技术，现如今已在市区和当地乡镇流行，减少了对野生种群资源的采挖。另一个例子是地涌金莲的利用，其是一种分布狭窄的地方性特有物种，野生种群很少。当地民族开发了与其可持续利用有关的技术，如繁殖、育种、栽培、疾病控制和可持续收获。因此，基于它们的可持续利用，这些濒危物种得到了有效的保护（Long et al. 2015）。

《生物多样性公约》（CBD）于 1992 年 6 月 5 日在联合国环境与发展会议（巴西里约“地球峰会”）开始签署。截止到 2004 年 2 月，已有 188 个缔约方（国家）在公约上签字，意味着在生物多样性保护、可持续和公平利用遗传资源方面取得了巨大进步（https://www.cbd.int/convention/）。《生物多样性公约》是保护包括植物在内的生物多样性最重要的法律文件。

《濒危野生动植物种国际贸易公约》（CITES）在以农业、园艺和木材工业为主的植物世界贸易中一直发挥着重要作用。根据希思罗机场（Heathrow Airport）2015 年的统计，65%的非法没收植物是观赏植物（42%的兰科植物、12%的蝴蝶亚属植物、6%的芦荟属植物和 5%的仙人掌属植物）（RBG Kew 2016）。

《关于获取遗传资源和公正公平分享其利用所产生惠利的名古屋议定书》（《名古屋议定书》）于 2014 年 10 月生效，可能对观赏植物产业产生巨大影响，涉及方面包括：①在当事方事先知情同意的情况下收集遗传资源和相关传统知识；②从育种到商业化；③从贸易到最终产品的利润分享。《获取和惠益分享》（ABS）是一项国际协议，旨在公平、公正地分享利用遗传资源所产生的惠益。到目前为止，已有 92 个缔约方（国家）签署了《名古屋议定书》（https://www.cbd.int/abs/）。这与传统的获取和利用观赏植物遗传资源进行育种而获利的方式有很大的不同，它将改善观赏植物的保护以及相关传统知识与公平利益的分享现状。

“2011—2020 年全球植物保护战略”作为《生物多样性公约》项目之一，正在与“生物多样性战略计划”、“全球植物保护伙伴关系”以及一系列其他项目一起实施，以了解、保护和可持续利用世界植物多样性。到 2020 年，该战略将实现以下 5 个目标：①充分了解、记录和认可植物多样性；②紧急有效地保护植物多样性；③以可持续和公平的方式利用植物多样性；④促进对植物多样性、其在可持续生计中的作用以及地球上所有生命的重要性的教育和认识；⑤发展实施该战略所需的能力和公众参与（https://www.cbd.

int/gspc/）。该战略正在促进世界观赏植物的保护、繁殖和可持续利用。

除国际公约外，在国家、省、地、县、社区层面上也已建立或将启动植物保护法律制度。与观赏植物有关的活动，应当遵守这些条例。

1.4 结　论

近年来，随着生物多样性信息学的发展和大数据的利用，人们有可能了解世界植物生物多样性的现状。根据COL、GBIF、IPNI、floras以及其他生物多样性数据库，世界维管植物估计有333 500～390 900种，隶属于452科。维管植物多样性最丰富的国家分别是巴西、中国、印度尼西亚、墨西哥和哥伦比亚，分别有32 364种、29 650种、29 375种、25 036种和24 500种。地球上的各种植被类型以及14种生物群落为维管植物提供了多样的生态环境，形成了复杂的生态系统。新技术加速了我们对植物遗传多样性的理解。目前，225种植物的全基因组已完成测序。在不久的将来，越来越多的观赏植物将会被测序。

在物种水平上，尽管Bailey和Bailey在1976年已报道了23 979个园艺植物分类群（科、属、种），但全球观赏植物物种的数量仍然是一个问号。我们估计世界上有85 000～99 000种观赏植物。在某些情况下，几乎全科或全属的所有成员都被用作装饰品，如槟榔科、仙人掌科、喜林芋属、蔷薇属等均为其代表。许多观赏植物的品种繁多。例如，已经培育出20 000～30 000个菊花新品种。

保护观赏植物的遗传资源是观赏植物育种和未来发展的关键。由于土地覆盖变化、入侵物种和其他因素，许多观赏植物的野生近缘种已濒临灭绝。观赏植物可以通过就地保护、迁地保护、可持续利用及法律制度建设等途径进行保护。自然保护区、种子库和种质中心，在观赏植物的保护中发挥着重要作用。在法律制度的支持下，观赏植物的可持续利用将是保护的最佳解决方案之一。《生物多样性公约》、《濒危野生动植物种国际贸易公约》以及全球、区域、国家和地方各级的其他法律制度将确保观赏植物的保护与可持续利用。ABS协议可能是观赏植物的一个里程碑，它与传统的收集、繁殖、商品化方式有很大的不同。

本章译者：

邹　凌，王继华
云南省农业科学院花卉研究所，国家观赏园艺工程技术研究中心，云南省花卉育种重点实验室，昆明 650200

参 考 文 献

Bailey LH, Bailey EZ (1976) Hortus third: a concise dictionary of plants cultivated in the United States and Canada. Macmillan Publishers, London
Brickell C (2008) Encyclopedia of garden plants, 3rd edn. Dorling Kindersley, London
Byg A, Balslev H (2006) Palms in indigenous and settler communities in southeastern Ecuador: farmers' perceptions and cultivation practices. Agrofor Syst 67:147–158

Cai J, Liu X, Vanneste K, Proost S, Tsai WC, Liu KW, Chen LJ, He Y, Xu Q, Bian C, Zheng Z, Sun F, Liu W, Hsiao YY, Pan ZJ, Hsu CC, Yang YP, Hsu YC, Chuang YC, Dievart A, Dufayard JF, Xu X, Wang JY, Wang J, Xiao XJ, Zhao XM, Du R, Zhang GQ, Wang MN, Su YY, Xie GC, Liu GH, Li LQ, LQ LYB, Chen HH, Van de Peer Y, Liu ZJ (2015) The genome sequence of the orchid *Phalaenopsis equestris*. Nat Genet 47:65–72

Chen JY (2000) Classification system for Chinese flower cultivars. China Forestry Press, Beijing

Chen HB, Zhang FJ, Ruan ZP, Chen RS (2013) Ornamental climbing plants. Huazhong University of Science & Technology Press, Wuhan

Darbyshire I, Anderson S, Asatryan A, Byfield A, Cheek M, Clubbe C, Ghrabi Z, Harris T, Heatubun CD, Kalema J, Magassouba S, McCarthy B, Milliken W, de Montmollin B, Lughadha EN, Onana J, Saïdou D, Sârbu A, Shrestha KK, Radford EA (2017) Important plant areas: revised selection criteria for a global approach to plant conservation. Biodivers Conserv 26(8):1767–1800

Dirr MA (2008) Manual of woody landscape plants, 8th edn. Stipes Publishing, Champaign

Foucher F, Hibrand-Saint Oyant L, Hamama L, Sakr S, Nybom H, Baudino S, Caissard JP, Byrne DM, Smulder JMS, Desnoyé B, Debener T, Bruneau A, De Riek J, Matsumoto S, Torres A, Millan T, Amaya I, Yamada K, Wincker P, Zamir D, Gouzy J, Sargent D, Bendahmane M, Raymond O, Vergne P, Dubois A, Just J (2015) Towards the rose genome sequence and its use in research and breeding. Acta Hortic 1064:167–175

Hamilton AC, Karamura D, Kakudidi E (2016) History and conservation of wild and cultivated plant diversity in Uganda: forest species and banana varieties as case studies. Plant Divers 38(1):23–24

Huang HW (2011) Plant diversity and conservation in China: planning a strategic bioresource for a sustainable future. Botanical Journal of the Linnean Society 166: 282–300.

Huang TC, Long CL (2011) Priorities for genetic resource collection and preservation of wild gymnosperms in Yunnan: an analysis based on the “3E” principle. Biodivers Sci 19(3):319–326

Judd WS, Campbell CS, Kellogg EA, Stevens PF, Donoghue ML (2008) Plant systematics: a phylogenetic approach, 3rd edn. Sinauer Associates, Inc, Sunderland

Li CX, Miao XY (2016) Notes on the rank of China in the world in terms of higher plant diversity. Biodivers Sci 24(6):725–727

Liu X (2015) The science report on biological germplasm resources in China. Science Press, Beijing

Long CL, Zhou YL (2001) Indigenous community forest management in Jinuo people’s swidden agroecosystems in SW China. Biodivers Conserv 10(5):756–768

Long CL, Ni YN, Long B, Zhang XB, Xin T (2015) Biodiversity of Chinese ornamentals. Acta Hortic 1087:209–220

Long CL, Long B, Bai YJ, Lei QY, Li JQ, Liu B (2017) Indigenous people’s ornamentals for future gardens. Acta Hortic 1167:17–22

Ma KP (2014) Rapid development of biodiversity informatics in China. Biodivers Sci 22(3):251–252

Olson DM, Dinerstein E (1998) The global 200: a representation approach to conserving the Earth’s most biologically valuable ecoregions. Conserv Biol 12:502–515

Olson DM, Dinerstein E, Wikramanayake ED, Burgess ND, Powell GVN, Underwood EC, D’Amico JA, Itoua I, Strand HE, Morrison JC, Loucks CJ, Allnutt TF, Ricketts TH, Kura Y, Lamoreux JF, Wettengel WW, Hedao P, Kassem KR (2001) Terrestrial ecoregions of the world: a new map of life on Earth. Bioscience 51(11):933–938

RBG Kew (2016) The state of the world’s plants report – 2016. Royal Botanic Gardens, Kew

RBG Kew (2017) The state of the world’s plants report – 2017. Royal Botanic Gardens, Kew

Roskov Y, Abucay L, Orrell T, Nicolson D, Bailly N, Kirk PM, Bourgoin T, DeWalt RE, Decock W, De Wever A, Nieukerken E van, Zarucchi J, Penev L (eds) (2017) Species 2000 & ITIS Catalogue of Life, 26th July 2017. Digital resource at www.catalogueoflife.org/col. Species 2000: Naturalis, Leiden. ISSN 2405-8858

Sharrock S (2012) Global strategy for plant conservation: a guide to the GSPC, all the targets, objectives and facts. BGCI, London

The Angiosperm Phylogeny Group (2016) An update of the Angiosperm Phylogeny Group classification for the orders and families of flowering plants: APG IV. Bot J Linn Soc 181(1):1–20

UNEP-WCMC (2014) Biodiversity A-Z Website: www.biodiversitya-z.org, UNEP-WCMC, Cambridge. 20th Aug 2017

Wang LS, Chen B, Ji LQ, Ma KP (2010) Progress in biodiversity informatics. Biodivers Sci 18(5):429–443

Xue DY (2005) Status quo and protection of bio-genetic resources in China. China Environmental Science Press, Beijing
Zhang QX, Chen WB, Sun LD, Zhao FY, Huang BQ, Yang WR, Tao Y, Wang J, Yuan ZQ, Fan GY, Xing Z, Han CL, Pan HT, Zhong X, Shi WF, Liang XM, DL D, Sun FM, ZD X, Hao RJ, Lü T, Lü YM, Zheng ZQ, Sun M, Luo L, Cai M, Gao YK, Wang JY, Yin Y, Xu X, Cheng TR, Wang J (2012) The genome of *Prunus mume*. Nat Commun 3:1318
Zhou Y, Yan WD (2015) Conservation and application of camphor tree (*Cinnamomum camphora*) in China: ethnobotany and genetic resources. Genet Resour Crop Evol 63(6):1049–1061
Zhu TP, Liu L, Zhu M (2007) Plant resources of China. Science Press, Beijing

第2章 花器官特征的遗传基础及其在观赏植物育种中的应用

Mathilde Chopy, Patrice Morel, Enrico Costanzo, Suzanne Rodrigues Bento, Pierre Chambrier, and Michiel Vandenbussche

摘要 *Petunia hybrida*（矮牵牛）是世界上最受欢迎的花坛植物之一。同时，矮牵牛作为科学研究的模式物种已有数十年的历史，主要用来研究涵盖花器官发育在内的各种生长和发育过程。在本章，我们以矮牵牛为例，系统地阐述了不同花器官特征的遗传基础，并系统地解释了花器官的同一性及其突变体在植物育种中的潜力。尽管 B 和 C 类编码植物花朵器官的基因在分子水平上非常保守，具有广泛的适用性，但是不同物种在重复同源基因对之间的冗余度与亚功能化/专业化程度方面可能存在着显著的差异。这是不同植物基因组起源复杂的直接结果，这些植物基因组是由全基因组经大规模或者小规模复制产生的，通常导致部分遗传物质的冗余。经典常用的基因筛选只能揭示非冗余遗传物质的功能，这大概是许多观赏植物尚未充分利用花器官同一性突变体作为育种目标的主要原因。由于这种问题和现象的存在，在本章我们讨论了不同的育种策略。

关键词 矮牵牛（*Petunia hybrida*）；花发育；同源突变；ABC 模型；观赏园艺；花器官特征

2.1 引 言

当我们在花园或野外环顾四周时，是否能意识到我们不经意偶然地参与了选美大赛？我们所看到的的确是一场选美大赛，这场激烈的选美大赛已经进行了数百万年，关乎参赛者的生死。从生物学的角度来看，植物通过美丽的花朵吸引授粉者并与授粉者互动来提高繁殖效率。为了实现这一目标，花的结构出现了许多变化，证明了植物进化和适应的能力，演化出了大概由 350 000 个植物物种构成的陆地植物生态系统。虽然大多数植物物种需要通过提供物质奖励（如花蜜）来吸引授粉者，但从分类学的角度来看，有一类通常称为“观赏园艺的植物”可能已达到花进化过程的最终步骤，因为这些植物成功地吸引了一个新的授粉者。这一类授粉者甚至不需要物质的奖励，仅仅只是由于花朵本身的美丽，这个新的授粉者就是我们人类，人类进一步扩大了观赏植物的栖息地范围，这种范围的扩大规模空前。

M. Chopy · P. Morel · E. Costanzo · S. Rodrigues Bento · P. Chambrier · M. Vandenbussche (✉) Laboratoire Reproduction et Développement des Plantes, Univ Lyon, ENS de Lyon, UCB Lyon 1, CNRS, INRA, Lyon, France
e-mail: michiel.vandenbussche@ens-lyon.fr

本章系统地阐述了不同花卉器官特征的分子水平的基础，并解释了这些基础研究知识对于观赏植物育种的意义。矮牵牛不同花卉器官特征调控的分子机制研究始于 20 多年前，在这期间矮牵牛既为研究花卉发育的模型植物，又是比较流行的观赏植物，其不同花卉器官特异突变体一直是观赏植物研究的热点。

2.2 矮牵牛作为观赏物种及其研究模式

矮牵牛在世界范围内是一种流行的花坛植物，经过近两个世纪的密集育种工作产生了不同的品种，这些品种具有令人印象深刻的多样性的花色和形态。矮牵牛原产于南美，属于茄科（Solanaceae）植物，茄科涵盖了主要的粮食作物（马铃薯、番茄、胡椒、茄子、烟草）以及观赏植物（如 *Petunia*、*Calibrachoa*、*Datura*、*Schizanthus* 和其他属植物）（Sarkinen et al. 2013）。杂合程度高的矮牵牛（*P. hybrida*）和许多实验室产生的品系均由蛾授粉的白色 *P. axillaris* 与蜜蜂授粉的 *P. integrifolia* 之间杂交而来，从进化的角度来看，和包括 *P. inflata* 在内的几种紫罗兰花色的种密切相关（图 2.1）（Stehmann et al. 2009; Segatto et al. 2014）。这些杂交品种是在 19 世纪初，由欧洲的植物学家选育的（Sink 1984）。

图 2.1 野生矮牵牛（*Petunia*）。左：依靠蜜蜂授粉的 *P. inflata*。右：依靠蛾授粉的 *P. axillaris*

P. hybrida 作为遗传学基础研究的模型植物有悠久的历史（Vandenbussche et al. 2016）。在关于矮牵牛的网站（http://flower.ens-lyon.fr/）上可以找到更多有关将矮牵牛作为模型植物的研究团队的信息。毫无疑问，矮牵牛研究的最新里程碑是 *P. axillaris* 与 *P. inflata* 基因组序列的发布（Bombarely et al. 2016），极大地促进了矮牵牛的基因组学研究。Gerats 和 Strommer（2009）的综述详细地阐述了矮牵牛作为一种观赏园艺植物在遗传和育种学中的进展。

2.3　花器官特征的遗传基础

尽管花朵的结构具有很大的形态多样性，但是几乎所有开花植物的花朵都具有相似的基本结构。花朵中都含有功能不同的花器官，从外到内分别是萼片、花瓣、雄蕊和雌蕊。值得注意的是，在拟南芥（*Arabidopsis thaliana*）和金鱼草（*Antirrhinum majus*）中都发现了花的突变体，为一种花器官被另一种花器官替代，这种突变通常发生在同一朵花的两个相邻轮生体中。这种类型的突变体，称为花同源突变体，可以分为不同的类别即 A、B、C 类，具体取决于受影响的花器官。A 类突变体为萼片突变为雌蕊或叶状的结构，还有花瓣转化为雄蕊状的结构。B 类突变体为花瓣突变为萼片，雄蕊突变为雌蕊。C 类突变体为雄蕊突变为花瓣，雌蕊突变为萼片。基于这些花的同源突变体和不同突变之间的遗传相互作用，建立了一个简单的遗传模型，阐述了不同花朵的器官如何发育成为特异的结构，这个花朵发育的模型称为 ABC 模型（Coen and Meyerowitz 1991；Bowman et al. 2012）。依据该模型，分别由 A、B 和 C 类不同的同源基因单独起作用或者相互作用发育成为独特的花器官（图 2.2）。A 类基因单独作用产生第一轮生体中的萼片，A 类和 B 类基因共同作用产生第二轮生体中的花瓣，B 类和 C 类基因共同作用产生第三轮生体中的雄蕊，C 类基因单独作用产生第四轮生体中的雌蕊。A 类和 C 类基因互相作用会抑制彼此的活性。当 C 类基因突变失去相应的功能时，会导致 A 类基因在第三和第四轮生体中得到表达，相应的，当 A 类基因突变失去相应的功能时，C 类基因在第一和第二轮生体中得到表达。早在 20 世纪 90 年代初，有一系列针对 *Arabidopsis* 和 *Antirrhinum* 中同源的 B 与 C 类基因的研究，结果表明它们均负责编码 MADS-box 转录

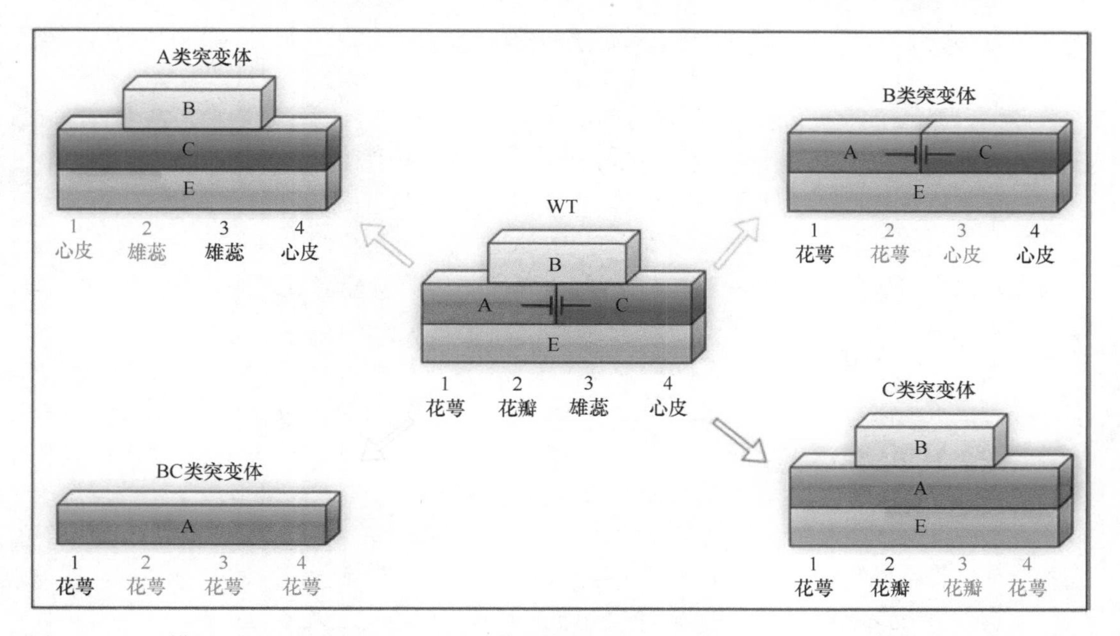

图 2.2　ABC 模型（图正中）与由 A、B、C 单独突变与 B、C 同时突变（图中四个角）导致的花内同源突变

因子（Sommer et al. 1990；Yanofsky et al. 1990；Jack et al. 1992；Trobner et al. 1992；Bradley et al. 1993；Goto and Meyerowitz 1994；Davies et al. 1999）。最初仅在 *Arabidopsis* 中清楚地阐述了 A 类基因的功能，主要的 A 类基因 *AP2*（*APETALA2*）属于植物特异性 AP2 / EREBP（对植物激素乙烯响应的结合蛋白）转录因子家族（Jofuku et al. 1994），另一个研究的结果表明另一类 A 类基因 *AP1*（*APETALA1*）也负责编码 MADS-box 转录因子（Mandel et al. 1992）。转录因子是一类有特殊调节作用的蛋白质，能够与 DNA 调节因子结合，通过结合与否，它们能够开启或关闭大量不同基因的表达，这样这些特殊的蛋白质可以控制和调节一些重要的生化反应，如花发育过程中的重要步骤。

而后，因为在 *Arabidopsis* 中发现了 *sep1 sep2 sep3* 三重突变体花是由负责编码 SEP（SEPALLATA）MADS-box 转录因子的 E 类基因变异导致的，这个发现对现有的 ABC 模型是重要的补充（Pelaz et al. 2000）；其突变体具体表现为所有花的器官被萼片代替，与 BC 双突变体的表型非常相似（图 2.2），因此我们可以得出结论，B、C 类基因编码的 MADS-box 转录因子需要 E 类基因编码的功能蛋白才能发挥其功能。

基于 MADS-box 转录因子能够彼此作用并有步骤地结合形成高阶的复合物，当前的花卉发育模型被形象地比作四重奏模型（Theissen and Saedler 2001），在每一个轮生体，都会形成不同的 MADS-box 转录因子复合物，并通过激活或抑制每个轮生体中特定的靶基因组，来调控不同花朵器官与不同花朵组织的形成。例如，由 B、C、E 类基因编码的 MADS-box 转录因子所组成的转录因子复合物会激活或抑制大量不同的基因，促进花朵第三个轮生体中雄蕊的发育。

2.4 ABCE 模型是否适用于所有开花植物？一个简单的模型，在现实中可能更复杂

ABC 模型由于通俗易懂，在本科的课程中诸如遗传学、发育生物学、花发育的遗传基础方面发挥了重要的作用（Bowman et al. 2012）。由于其通俗易懂以及人们对植物物种遗传多样性的低估，在某种程度上人们认为这个模型可能适用于大多数开花物种。

从观赏植物育种的角度来看，如果这个模型的通用性很广将非常有利于筛选品种，因为它可以直接以突变体作为亲本来选育新的花卉性状（如重瓣花的表型，一种非常理想的花卉性状）。

然而，研究结果证明不同开花植物物种的花朵发育可能要比这个模型复杂得多。例如，对 *P. hybrida*（Cartolano et al. 2007；Morel et al. 2017）和 *A. majus*（Keck et al. 2003；Cartolano et al. 2007）的 A 类基因进行深度研究表明，与 *Arabidopsis* 相比，以上两种植物的 A 类基因具有明显的差异。此外，有一个研究的结果表明，在矮牵牛不同的同源表型体中，其花朵的发育并不遵循经典的 ABC 花朵发育模型（Morel et al. 2017）。这些结果表明，尽管不同物种的花朵都具有由萼片、花瓣、雄蕊和雌蕊组成的基本结构，但蔷薇科下物种 *Arabidopsis* 与菊科下物种 *P. hybrida* 和 *A. majus* 在进化过程中，并没有演化出相同的控制开花的分子机制。由于 A 类基因突变在观赏园艺植物育种不占有重要的地

位，因此我们将在本书其他地方详细讨论这些结果。

第二个主要凸显花发育复杂性的方面在于不同植物基因组的起源复杂，在进化过程中，通过全基因组水平不同规模的变化，不同的物种得到了重塑。我们知道，一旦某些基因被复制，在进一步进化的过程中，可能会发生不同的情况（Force et al. 1999；Prince and Pickett 2002）。例如，在基因复制后，其中一个拷贝可能会丢失。但也可能两个基因拷贝同时存在，并可能相互独立地演化。这有可能会导致其中的一个拷贝演化出新功能，或者导致亚功能化，即两个基因中的一个或两个仅执行部分原始功能。然而，通常情况下，两个基因拷贝都可能保留其大部分功能（很大一部分），从而导致（部分）遗传冗余。因为这些基因功能的多样化过程本质上是随机发生的，所以它们在不同植物谱系中可能导致截然不同的结果。显然，这种现象也适用于影响花朵发育的基因，这将极大地影响不同的育种策略及其可行性。我们在以下各节中对此进行了说明，将在分子水平上更详细地比较 *Arabidopsis*、*Antirrhinum* 和 *Petunia* 花的 C、B、E 同源基因的功能。

2.4.1　花的 C 类功能基因

花的 C 类功能基因与雄蕊和雌蕊的发育有关，它们的突变能导致雄蕊转变为花瓣，而雌蕊转变为萼片样的器官。雄蕊转变成花瓣会让花朵显现重瓣的性状，这是一种非常理想的观赏性状，所以了解 C 谱系基因的进化和功能多样性非常重要。

C 谱系基因编码 MADS-box 转录因子，由于 C 谱系基因在被子植物进化过程的早期发生了复制，因此在大多数被子植物中通常存在一个以上的拷贝，到目前为止发现了两种类型的 C 谱系基因，称为 *euAG* 和 *PLE*-亚谱系（Becker and Theissen 2003；Kramer et al. 2004；Zahn et al. 2006）（图 2.3b）。在 *Arabidopsis* 中，*AG*（*AGAMOUS*）基因（*euAG* 型）与 C 类功能基因（Yanofsky et al. 1990），以及之后发现的 *PLE* 亚谱系基因 *SHATTERPROOF1* 和 *2*（*SHP1/2*）在胚珠、雌蕊内侧组织、果实的发育中具有高度的亚功能化作用（Liljegren et al. 2000；Pinyopich et al. 2003；Colombo et al. 2010）。在 *Antirrhinum* 中 C 功能则由 *PLE*（*PLENA*）基因编码，其 *euAG* 基因 *FAR*（*FARINELLI*）在雄蕊发育过程中具有高度的亚功能化作用（Bradley et al. 1993；Davies et al. 1999）。因此，我们可以看到被子植物祖先的 C 谱系基因发生了复制以后，C 谱系基因亚功能化在植物进化的过程中出现截然不同的轨迹，从而产生了 *Arabidopsis* 和 *Antirrhinum* 这两类植物。

然而，当我们分析 *Petunia* 中的 *euAG* 基因 *PMADS3* 和 *PLE* 基因 *FBP6*（花结合蛋白 6）的功能时，再次发现了不同的现象，即 *PMADS3* 和 *FBP6* 都为经典的 C 类功能基因（Heijmans et al. 2012），表明 *Petunia* 的 *euAG* 和 *PLE* 亚谱系基因的亚功能化作用与 *Arabidopsis* 或 *Antirrhinum* 例子相比，没有显著的差异。因此，如果我们想在 *Petunia* 得到重瓣表型，*PMADS3* 和 *FBP6* 基因都必须发生突变（图 2.3c）。

显然，多个 C 谱系基因的存在让我们获得基于 C 功能突变的新的重瓣品种的育种程序复杂化。在启动育种程序之前，我们应该首先得到 *PLENA* 和 *euAG* 基因的序列，然后通过原位杂交后进行有深度的表达分析。我们应该搞清楚这个花卉物种的 C 功能是否仅由一个 C 功能基因（如 *Arabidopsis* 或 *Antirrhinum*）编码，还是由两个 C

功能基因（如 *Petunia*）编码，或由更多的基因编码。根据结果，我们应采取不同的育种策略（详见下文）。

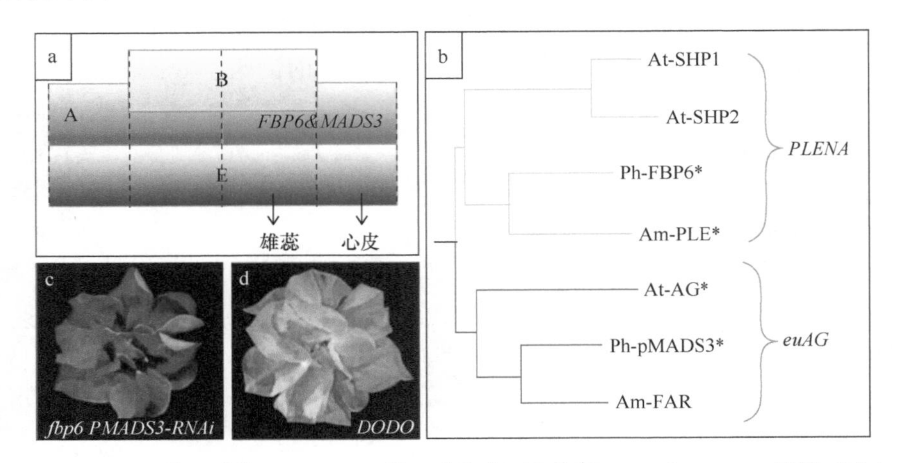

图 2.3 *Petunia hybrida* 的 C 功能。(a)*Petunia* 的 C 功能由两个基因 *FBP6* 和 *MADS3* 编码。(b)*Arabidopsis thaliana*（At）、*Antirrhinum majus*（Am）和 *Petunia hybrida*（Ph）的 C 谱系基因的系统发育树。表现出完全 C 功能活性的基因用*标注，花括号代表两个主要的 C 亚功能系，即 *PLENA* 和 *euAG*。（c）通过 RNA 干扰技术干扰 *fbp6 PMADS3* 基因表达的花。（d）遗传呈显性的重瓣性状

有趣的是，大自然似乎为我们找到了另一种方法来制造重瓣的花朵，这个方法和 C 功能基因的突变与否无关。在包括 *Petunia* 在内的许多观赏物种中，人们用经典的遗传筛选方法已选育出了商业化的重瓣品种。在这些重瓣品种中有可能发生了对应的 C 功能基因的突变，但是导致重瓣性状的还有其他的一些原因，并且通常是遗传显性的，如 *Petunia*，从表面上看，这种商业化的 *DODO*（*DOMINANT DOUBLE*）花表型（图 2.3d）看起来与 *fbp6 PMADS3*-RNAi 系（图 2.3c）的表型非常相似，即花瓣在第二轮生体中生长，但区别细节主要体现在：①*DODO* 仍存在类似花药的结构，这些结构仍可以产生可用于杂交的花粉。②*DODO* 花的中心仍然可以找到残留的雌蕊组织。③*DODO* 第二轮生体的花瓣数依然增加。然而，控制这种重瓣显性表型的突变仍然是一个没有解开的谜团，解开这个谜团仍然是当今观赏园艺科学的研究热点。

2.4.2 花的 B 类功能基因

花的 B 类功能基因与雄蕊和花瓣的发育有关，它们的突变导致花瓣转变为萼片，雄蕊转变为雌蕊。在 At 和 Am 中，B 功能由一对 MADS-box 蛋白编码，对于 At，由 AP3（APETALA3）和 PI（PISTILLATA）蛋白编码；对于 Am，由 DEF（DEFICIENS）和 GLO（GLOBOSA）蛋白编码；这些蛋白质为专一性强的异二聚体（Sommer et al. 1990；Jack et al. 1992；Trobner et al. 1992；Goto and Meyerowitz 1994）（图 2.4a，b）。这种专一性强的异二聚体的出现意味着只有在两种蛋白质均具有完全的功能且能相互结合时，花的 B 功能才具有活性。因此，*AP3*、*PI*、*DEF* 或 *GLO* 中的任意一个发生突变都会导致 B 功能的完全丧失。虽然 At 和 Am 为在进化树上相距较远的重要的双子叶植物物种，

但是两者具有非常相似的 B 功能组织，所以人们最初认为这个简单的遗传模型在很多双子叶植物中具有普遍的适用性。

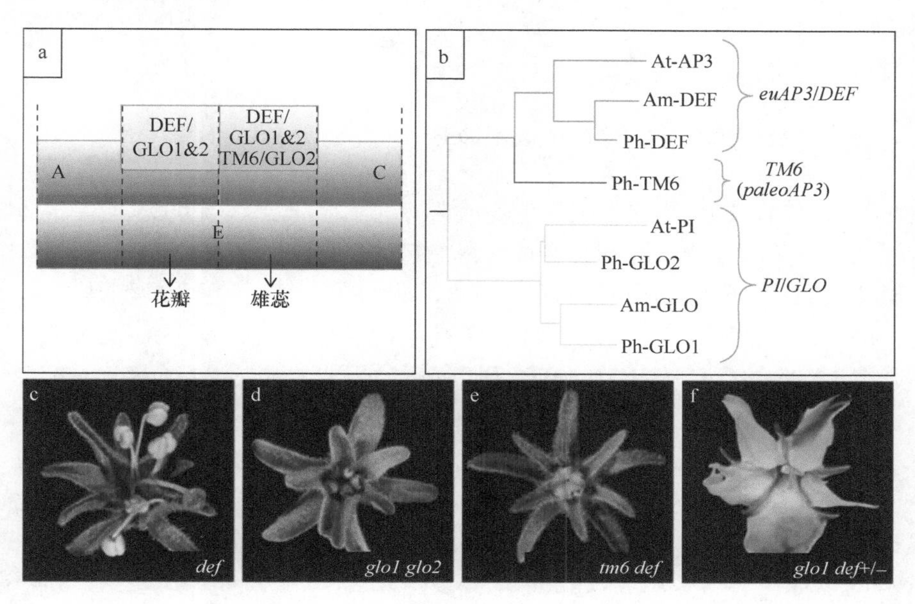

图 2.4　*Petunia hybrida* 的 B 功能。(a) DEF/GLO1 和 DEF/GLO2 异二聚体与花瓣和雄蕊发育的协同作用示意图。TM6/GLO2 异二聚体只与雄蕊的发育有关。(b) *Arabidopsis thaliana*（At）、*Antirrhinum majus*（Am）和 *Petunia hybrida*（Ph）的 B 类基因的系统发育树。花括号代表 B 类功能基因的三个谱系。(c～f) *Petunia hybrida* 的 B 类基因的突变体（*def*、*glo1 glo2*、*tm6 def*、*glo1 def*+/–代表不同的基因型）

但是，更多的研究在更多的双子叶植物物种中分析了 *DEF*/*AP3* 基因后发现了两个平行而存在差异的 *DEF*/*AP3* 谱系，称为 *euAP3* 和 *TM6*（*paleoAP3*）基因谱系，它们是由处于双子叶植物进化树基部的物种在早期发生了该基因的复制事件引起的（Kramer et al. 1998；Vandenbussche et al. 2003a）。At 和 Am 在进化过程中有可能丢失了 *TM6* 基因，但是大多数双子叶植物物种同时具有 *euAP3* 和 *TM6* 基因。因此，与 At 和 Am 相比，这些物种中花朵 B 功能的表达在基因水平上可能更为复杂。研究证明在 *Petunia* 中，花朵 B 功能的表达确实要更复杂一些，它同时具有 *euAP3* 和 *TM6* 基因（图 2.4a，b）。我们的研究表明 *PhTM6* 作为 B 类功能基因与 *PhDEF*（*euAP3* 基因）一起促进雄蕊的发育，但与花瓣的发育无关（Rijpkema et al. 2006）。因此，*Petunia* 中 *euAP3* 同源基因 *PhDEF* 的突变（图 2.4c）会导致花瓣转变为萼片，但是雄蕊的发育不受影响（van der Krol et al. 1993）。此外，*GLO*/*PI* 谱系中的一个基因复制事件在 *Petunia* 中产生了两个类似 *GLO*/*PI* 的基因：我们的研究表明 *PhGLO1* 和 *PhGLO2* 在很大程度上是多余的。因此，只有 *phglo1 phglo2* 双重突变体才会导致花朵 B 功能的完全丧失（Vandenbussche et al. 2004），表现型与 *phdef phtm6* 双重突变体相同（Rijpkema et al. 2006）（图 2.4d，e）。

因为 B 类突变体会导致花瓣和雄蕊完全缺失，所以即便完全开花的 B 突变体可能看起来仍然像营养生长。因此乍一看，B 类基因的突变可能对观赏花卉的育种者来说并不重要。但是，在 *Petunia* 等由多个基因编码 B 功能的物种中，通过量化地进行基因水平的剂量操作可能会产生一些中间表型，可能产生一些具有观赏价值的变异体。例如，*Petunia* 的 *phglo1 phdef* +/–表型的花具有精致的花瓣，但是花瓣的主静脉明显地转变成萼片状组织（图 2.4f）。这使这个特定的花朵具有独特的外观，对观赏花卉育种者来说，可能具有潜在的价值。

2.4.3 花的 E 类功能或 *SEPALLATA* 功能基因

SEP（SEPALLATA）蛋白是被子植物所特有的（Zahn et al. 2005），它是 MADS-box 转录因子的另一个亚家族，在花卉器官发育中起重要作用。在 *Arabidopsis* 中，SEP 家族由 4 个成员组成，分别为 sep1、sep2、sep3、sep4（图 2.5b）。因为它们是高度冗余的，所以它们的功能只能通过反向遗传学方法来揭示，因此其机制比 ABC 类突变体发现的时间要晚得多。在 *Arabidopsis* 中，一个研究首先发现了 *sep1 sep2 sep3* 三重突变体的花瓣、雄蕊、雌蕊转变为萼片（Pelaz et al. 2000），而 *sep1 sep2 sep3 sep4* 四重突变体的花只产生了叶状组织（Ditta et al. 2004）。所以 *SEP* 基因很有可能是所有花器官发育过程中所必需的，这些基因通过桥梁作用来促使 B 和 C 功能基因的产物形成高阶的复合物并最终发挥该有的作用（Honma and Goto 2001；Theissen and Saedler 2001；Immink et al. 2009；Melzer et al. 2009）。这些发现将 SEP 或 E 功能添加到经典 ABC 花卉发育模型中。然而值得注意的是，关于 *Antirrhinum* 的 *SEP* 基因研究结果还没有确定性的进展，这可能是由于 *Antirrhinum* 的 *SEP* 基因存在广泛的遗传冗余。最近，在 *Petunia*、水稻和玉米中发现，与 *SEP* 基因联系最紧密的 *AGL6* 亚家族成员具有类似 *SEP* 基因（Becker and Theissen 2003）的功能（Ohmori et al. 2009；Rijpkema et al. 2009；Thompson et al. 2009）。这为花朵 SEP 功能的研究增加了复杂性或者说是冗余性。

E 类功能的丧失会导致所有的花器官转化为萼片或叶状组织，因此与 B 功能突变体一样，这类突变体对观赏园艺似乎没有太大的意义。然而，尽管 *Petunia* 含有 6 个 *SEP* 基因和一个 *AGL6* 基因（图 2.5a，b），但与 *Arabidopsis* 中的类似基因相比，其中一些基因的功能似乎没有那么丰富。我们研究发现，*Petunia* 中 *SEP3* 的直系同源物 *FBP2*（花朵结合蛋白 2）可以防止花瓣边缘组织转变为萼片组织（Vandenbussche et al. 2003b）。所以 *fbp2* 单突变体展现出带有绿色边缘的花朵，从而产生非常具有观赏性的表型（图 2.5c）。此外，我们发现该表型可以通过 *SBP1/2* 的同源基因 *FBP5* 的突变进一步增强（Vandenbussche et al. 2003b）。数年来，具有绿色边缘的 *Petunia* 品种已在市场上出售（称为“green picotee”），这个品种可能源自将在野生物种 *Petunia* 中发现的 *fbp2* 进行转座子插入（Matsubara et al. 2008）。

尽管很难预测其他物种中的 *SEP* 基因在什么条件下是完全冗余的，但 *Petunia* 的 *fbp2* 突变体表明，作为一种新的表型变异来源，筛选 *sep* 突变体将可能有助于观赏花卉的育种工作。

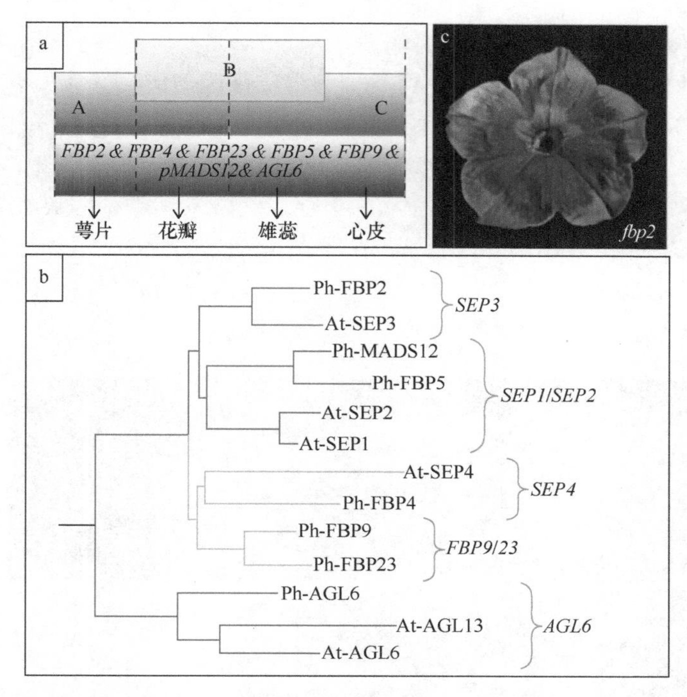

图 2.5　*Petunia hybrida* 的 E 功能。（a）*P. hybrida* 含有 6 个 *SEP* 基因和 1 个具有 SEP 功能的 *AGL6* 基因。（b）*Arabidopsis thaliana*（At）和 *Petunia hybrida*（Ph）的 E 类基因的系统发育树（Malcomber and Kellogg 2005），花括号代表了 5 个 E 类功能基因谱系。（c）*fbp2* 突变体的花朵性状

2.5　花瓣形态变化

开花植物物种的花瓣形状令人印象深刻（Galliot et al. 2006；Irish 2006），但对这种形态多样性背后的遗传学机制仍知之甚少。尽管本段中讨论的 *Petunia* 突变体不能归类为花朵器官的特征突变体，但这些突变的确会很大程度地影响花瓣的形态，这对观赏花卉育种者来说是有吸引力的。*Petunia* 花中的 5 个花瓣完全融合在一起，形成了一个花瓣管（图 2.1，图 2.6a）。因为花瓣管在花的中心包裹了花的生殖器官和蜜腺，所以花瓣管成为进入蜜腺的屏障。不同 *Petunia* 种类的花瓣管和管径变化与不同授粉媒介相关性的研究（Wijsman 1983；Ando 2001；Stuurman et al. 2004；Hoballah et al. 2007）表明：花瓣融合导致了 *Petunia* 不同授粉体系的进化。为了深入了解花瓣融合的机制，我们分析了两个隐性突变体，分别为 *maw*（*maewest*）和 *chsu*（*chorip-etala suzanne*），其中一些花瓣融合被部分阻断。我们发现 *MAW* 编码一个同源的转录因子（Vandenbussche et al. 2009），属于 WOX（与 WUSCHEL 相关的同源盒）转录因子家族（Mayer et al. 1998；Haecker et al. 2004）。类似的功能由 *A. thaliana* 中的 *WOX1* 和 *PRS*（*PRESSED FLOWER*）冗余部分编码，这些结果表明：*MAW*/*WOX1*/*PRS* 基因的功能非常保守，在调节花朵的侧向器官发育中起着相同的作用（Vandenbussche et al. 2009）。这些发现被进一步证实，因为在其他几个物种中，*WOX1* 突变体都在花瓣发育中显示出缺陷（详见综述 Costanzo et al. 2014）。有趣的是，有一种新的由 Thompson & Morgan 公司近年来选育的 *Petunia hybrida* 品种‘Sparklers’的花瓣形态缺陷性状与 *maw* 突变体的性状非常相似（图 2.6d）。我们发现‘Sparklers’突变体与 *maw* 突变体为等位基因的突变（未发表数据），这说明了 *maw/wox1* 突变体对于观赏花卉育种有潜在的用途。此外，*Petunia* 的 *chsu* 突变（目前在分子水平上已经被阐述）将能够与编辑 *maw* 结合进一步人为地改变花瓣的形态（图 2.6c）。我们最近的研究结果表明：*Petunia* 的 *AP2* 转录因子基因 *ROB1*、*ROB2*、*ROB3*（分别为 *B-FUNCTION-1*、*B-FUNCTION-2*、*B-FUNCTION-3* 的阻遏物）突变也极大地改变了花瓣的形状（Morel et al. 2017），与野生型 WT 的圆形花瓣相比，产生了一个类似五角星状的花瓣（图 2.6a，e）。

图 2.6 *Petunia* 的花瓣形态的突变。（a～e）野生型（WT）以及各种花瓣的突变体（图中的白色字体代表突变体的基因型）

2.6 以花器官特性突变体作为育种目标

我们希望通过 *Petunia* 的例子，证明花卉同源突变可以为观赏花卉育种提供新颖性，并且这些突变不限于 C 功能的丧失，众所周知后者是与重瓣表型相关的。

然而，在我们更详细地比较 *Petunia*、*A. thaliana* 和 *A. majus* 这三者花器官发育的遗传学机制时发现，很明显，尽管花同源的 B、C 和 E 功能在分子水平上很保守，但是在 B、C 和 E 谱系基因的重复拷贝之间，冗余度/亚功能化/专业化程度和基因损失程度存在重要差异。因此，一个物种中单个基因编码的功能可能在另一个物种中由两个（或多个）基因冗余编码，反之亦然。

因此，我们应根据物种的特定遗传背景来调整育种策略。长期以来，观赏植物育种中新花型的筛选仅限于经典的正向遗传筛选，其中含有具有吸引力的表型，这些表型要么源自自然的遗传变异，要么源于随机诱变的群体，如通过类似 EMS（甲磺酸乙酯）的辐射诱变。但是，经典的正向遗传筛选只能发现非冗余基因编码的功能（单个隐性或显性等位基因）。加之同源基因之间的遗传冗余，这些都可能是在许多观赏植物中我们尚未尝试使用花器官同一性突变体作为育种目标的主要原因。

如今，随着现代基因组育种技术的出现，冗余功能也可以相对容易地解决。例如，使用 TILLING 方法（Wang et al. 2012），我们可以利用由 EMS 诱变的群体对目的基因中的突变等位基因进行反向遗传鉴定，然后可以将其组合起来以创建双重或更高阶的突变体。近来，精确的基因编辑技术如 CRISPR-Cas9 技术已成功地被应用于多种植物物种（Lozano-Juste and Cutler 2014），其具有彻底改变植物育种方法的潜力。因为该技术允许同时靶向多个同源基因，所以它特别适合敲除冗余的功能基因。但是，CRISPR-Cas9 系统目前仅适用基于农杆菌方法进行转化的植物物种，而且目前尚不清楚（尤其在欧洲）CRISPR-Cas9 技术在植物育种将在多大程度上可以被使用，因为我们可以预见 CRISPR-Cas9 技术将受到未来相关 GMO 法规的限制。

本章译者：

邹　凌，王继华

云南省农业科学院花卉研究所，国家观赏园艺工程技术研究中心，云南省花卉育种重点实验室，昆明 650200

参 考 文 献

Ando T (2001) Reproductive isolation in a native population of Petunia sensu Jussieu (Solanaceae). Ann Bot 88:403–413

Becker A, Theissen G (2003) The major clades of MADS-box genes and their role in the development and evolution of flowering plants. Mol Phylogenet Evol 29:464–489

Bombarely A, Moser M, Amrad A, Bao M, Bapaume L, Barry CS, Bliek M, Boersma MR, Borghi L, Bruggmann R et al (2016) Insight into the evolution of the Solanaceae from the parental genomes of Petunia hybrida. Nat Plants 2:16074

Bowman JL, Smyth DR, Meyerowitz EM (2012) The ABC model of flower development: then and now. Development 139:4095–4098

Bradley D, Carpenter R, Sommer H, Hartley N, Coen E (1993) Complementary floral homeotic phenotypes result from opposite orientations of a transposon at the plena locus of Antirrhinum. Cell 72:85–95

Cartolano M, Castillo R, Efremova N, Kuckenberg M, Zethof J, Gerats T, Schwarz-Sommer Z, Vandenbussche M (2007) A conserved microRNA module exerts homeotic control over *Petunia hybrida* and *Antirrhinum majus* floral organ identity. Nat Genet 39:901–905

Coen ES, Meyerowitz EM (1991) The war of the whorls: genetic interactions controlling flower development. Nature 353:31–37

Colombo M, Brambilla V, Marcheselli R, Caporali E, Kater MM, Colombo L (2010) A new role for the SHATTERPROOF genes during Arabidopsis gynoecium development. Dev Biol 337:294–302

Costanzo E, Trehin C, Vandenbussche M (2014) The role of WOX genes in flower development. Ann Bot 114:1545–1553

Davies B, Motte P, Keck E, Saedler H, Sommer H, Schwarz-Sommer Z (1999) PLENA and FARINELLI: redundancy and regulatory interactions between two Antirrhinum MADS-box factors controlling flower development. EMBO J 18:4023–4034

Ditta G, Pinyopich A, Robles P, Pelaz S, Yanofsky MF (2004) The SEP4 gene of Arabidopsis thaliana functions in floral organ and meristem identity. Curr Biol 14:1935–1940

Force A, Lynch M, Pickett FB, Amores A, Yan YL, Postlethwait J (1999) Preservation of duplicate genes by complementary, degenerative mutations. Genetics 151:1531–1545

Galliot C, Stuurman J, Kuhlemeier C (2006) The genetic dissection of floral pollination syndromes. Curr Opin Plant Biol 9:78–82

Gerats T, Strommer J (2009) Petunia: evolutionary, developmental and physiological genetics. In: Gerats T, Strommer J (eds) Petunia. Springer, New York, pp 1–433

Goto K, Meyerowitz EM (1994) Function and regulation of the Arabidopsis floral homeotic gene PISTILLATA. Genes Dev 8:1548–1560

Haecker A, Gross-Hardt R, Geiges B, Sarkar A, Breuninger H, Herrmann M, Laux T (2004) Expression dynamics of WOX genes mark cell fate decisions during early embryonic patterning in Arabidopsis thaliana. Development 131:657–668

Heijmans K, Ament K, Rijpkema AS, Zethof J, Wolters-Arts M, Gerats T, Vandenbussche M (2012) Redefining C and D in the Petunia ABC. Plant Cell 24:2305–2317

Hoballah ME, Gubitz T, Stuurman J, Broger L, Barone M, Mandel T, Dell'Olivo A, Arnold M, Kuhlemeier C (2007) Single gene-mediated shift in pollinator attraction in Petunia. Plant Cell 19:779–790

Honma T, Goto K (2001) Complexes of MADS-box proteins are sufficient to convert leaves into floral organs. Nature 409:525–529

Immink RG, Tonaco IA, de Folter S, Shchennikova A, van Dijk AD, Busscher-Lange J, Borst JW, Angenent GC (2009) SEPALLATA3: the 'glue' for MADS box transcription factor complex formation. Genome Biol 10:R24

Irish VF (2006) Duplication, diversification, and comparative genetics of angiosperm MADS Box genes. Adv Bot Res 44:129–161

Jack T, Brockman LL, Meyerowitz EM (1992) The homeotic gene *APETALA3* of Arabidopsis thaliana encodes a MADS box and is expressed in petals and stamens. Cell 68:683–697

Jofuku KD, Boer B, Montagu MV, Okamuro JK (1994) Control of Arabidopsis flower and seed development by the homeotic gene APETALA2. Plant Cell 6:1211–1225

Keck E, McSteen P, Carpenter R, Coen E (2003) Separation of genetic functions controlling organ identity in flowers. EMBO J 22:1058–1066

Kramer EM, Dorit RL, Irish VF (1998) Molecular evolution of genes controlling petal and stamen development: duplication and divergence within the APETALA3 and PISTILLATA MADS-box gene lineages. Genetics 149:765–783

Kramer EM, Jaramillo MA, Di Stilio VS (2004) Patterns of gene duplication and functional evolution during the diversification of the AGAMOUS subfamily of MADS box genes in angiosperms. Genetics 166:1011–1023

Liljegren SJ, Ditta GS, Eshed Y, Savidge B, Bowman JL, Yanofsky MF (2000) *SHATTERPROOF* MADS-box genes control seed dispersal in *Arabidopsis*. Nature 404:766–770

Lozano-Juste J, Cutler SR (2014) Plant genome engineering in full bloom. Trends Plant Sci 19:284–287

Malcomber ST, Kellogg EA (2005) SEPALLATA gene diversification: brave new whorls. Trends Plant Sci 10:427–435

Mandel MA, Gustafson-Brown C, Savidge B, Yanofsky MF (1992) Molecular characterization of the Arabidopsis floral homeotic gene APETALA1. Nature 360:273–277

Matsubara K, Shimamura K, Kodama H, Kokubun H, Watanabe H, Basualdo IL, Ando T (2008) Green corolla segments in a wild Petunia species caused by a mutation in FBP2, a SEPALLATA-like MADS box gene. Planta 228:401–409

Mayer KF, Schoof H, Haecker A, Lenhard M, Jurgens G, Laux T (1998) Role of WUSCHEL in regulating stem cell fate in the Arabidopsis shoot meristem. Cell 95:805–815

Melzer R, Verelst W, Theissen G (2009) The class E floral homeotic protein SEPALLATA3 is sufficient to loop DNA in 'floral quartet'-like complexes in vitro. Nucleic Acids Res 37:144–157

Morel P, Heijmans K, Rozier F, Zethof J, Chamot S, Bento SR, Vialette-Guiraud A, Chambrier P, Trehin C, Vandenbussche M (2017) Divergence of the Floral A-Function between an Asterid and a Rosid Species. Plant Cell 29:1605–1621

Ohmori S, Kimizu M, Sugita M, Miyao A, Hirochika H, Uchida E, Nagato Y, Yoshida H (2009) MOSAIC FLORAL ORGANS1, an AGL6-like MADS box gene, regulates floral organ identity and meristem fate in rice. Plant Cell 21:3008–3025

Pelaz S, Ditta GS, Baumann E, Wisman E, Yanofsky MF (2000) B and C floral organ identity functions require SEPALLATA MADS-box genes. Nature 405:200–203

Pinyopich A, Ditta GS, Savidge B, Liljegren SJ, Baumann E, Wisman E, Yanofsky MF (2003) Assesing the redundancy of MADS-box genes during carpel and ovule development. Nature 424:85–88

Prince VE, Pickett FB (2002) Splitting pairs: the diverging fates of duplicated genes. Nat Rev Genet 3:827–837

Rijpkema AS, Royaert S, Zethof J, van der Weerden G, Gerats T, Vandenbussche M (2006) Analysis of the Petunia TM6 MADS box gene reveals functional divergence within the DEF/AP3 lineage. Plant Cell 18:1819–1832

Rijpkema AS, Zethof J, Gerats T, Vandenbussche M (2009) The petunia AGL6 gene has a SEPALLATA-like function in floral patterning. Plant J 60:1–9

Sarkinen T, Bohs L, Olmstead RG, Knapp S (2013) A phylogenetic framework for evolutionary study of the nightshades (Solanaceae): a dated 1000-tip tree. BMC Evol Biol 13:214

Segatto AL, Ramos-Fregonezi AM, Bonatto SL, Freitas LB (2014) Molecular insights into the purple-flowered ancestor of garden petunias. Am J Bot 101:119–127

Sink KC (1984) Taxonomy. In: Sink KC (ed) Petunia: monographs on theoretical and applied genetics, vol 9. Springer, Berlin, pp 3–9

Sommer H, Beltran JP, Huijser P, Pape H, Lonnig WE, Saedler H, Schwarz-Sommer Z (1990) Deficiens, a homeotic gene involved in the control of flower morphogenesis in Antirrhinum majus: the protein shows homology to transcription factors. EMBO J 9:605–613

Stehmann JR, Lorenz-Lemke AP, Freitas LB, Semir J (2009) The genus Petunia. In: Gerats T, Strommer J (eds) Petunia. Springer, New York, pp 1–28

Stuurman J, Hoballah ME, Broger L, Moore J, Basten C, Kuhlemeier C (2004) Dissection of floral pollination syndromes in Petunia. Genetics 168:1585–1599

Theissen G, Saedler H (2001) Plant biology. Floral quartets. Nature 409:469–471

Thompson BE, Bartling L, Whipple C, Hall DH, Sakai H, Schmidt R, Hake S (2009) Bearded-ear encodes a MADS box transcription factor critical for maize floral development. Plant Cell 21:2578–2590

Trobner W, Ramirez L, Motte P, Hue I, Huijser P, Lonnig WE, Saedler H, Sommer H, Schwarz-Sommer Z (1992) GLOBOSA: a homeotic gene which interacts with DEFICIENS in the control of Antirrhinum floral organogenesis. EMBO J 11:4693–4704

van der Krol AR, Brunelle A, Tsuchimoto S, Chua NH (1993) Functional analysis of petunia floral homeotic MADS box gene pMADS1. Genes Dev 7:1214–1228

Vandenbussche M, Theissen G, Van de Peer Y, Gerats T (2003a) Structural diversification and neo-functionalization during floral MADS-box gene evolution by C-terminal frameshift mutations. Nucleic Acids Res 31:4401–4409

Vandenbussche M, Zethof J, Souer E, Koes R, Tornielli GB, Pezzotti M, Ferrario S, Angenent GC, Gerats T (2003b) Toward the analysis of the petunia MADS box gene family by reverse and forward transposon insertion mutagenesis approaches: B, C, and D floral organ identity functions require SEPALLATA-like MADS box genes in petunia. Plant Cell 15:2680–2693

Vandenbussche M, Zethof J, Royaert S, Weterings K, Gerats T (2004) The duplicated B-class heterodimer model: whorl-specific effects and complex genetic interactions in Petunia hybrida flower development. Plant Cell 16:741–754

Vandenbussche M, Horstman A, Zethof J, Koes R, Rijpkema AS, Gerats T (2009) Differential recruitment of WOX transcription factors for lateral development and organ fusion in Petunia and Arabidopsis. Plant Cell 21:2269–2283

Vandenbussche M, Chambrier P, Rodrigues Bento S, Morel P (2016) Petunia, your next super-model? Front Plant Sci 7:72

Wang TL, Uauy C, Robson F, Till B (2012) TILLING in extremis. Plant Biotechnol J 10:761–772

Wijsman HJW (1983) On the interrelationships of certain species of Petunia II.Experimental data: crosses between different taxa. Acta Botanica Neerlandica 32:1–128

Yanofsky MF, Ma H, Bowman JL, Drews GN, Feldmann KA, Meyerowitz EM (1990) The protein encoded by the Arabidopsis homeotic gene agamous resembles transcription factors. Nature 346:35–39

Zahn LM, Kong H, Leebens-Mack JH, Kim S, Soltis PS, Landherr LL, Soltis DE, Depamphilis CW, Ma H (2005) The evolution of the SEPALLATA subfamily of MADS-box genes: a preangiosperm origin with multiple duplications throughout angiosperm history. Genetics 169:2209–2223

Zahn LM, Leebens-Mack JH, Arrington JM, Hu Y, Landherr LL, dePamphilis CW, Becker A, Theissen G, Ma H (2006) Conservation and divergence in the AGAMOUS subfamily of MADS-box genes: evidence of independent sub- and neofunctionalization events. Evol Dev 8:30–45

第 3 章　花色及其基因改造工程

Naoko Okitsu, Naonobu Noda, Stephen Chandler, and Yoshikazu Tanaka

摘要　花色主要由类黄酮化合物和花色苷化合物有色亚类的成分决定。花朵中的类黄酮物质通常是特定的，由于遗传的局限性，一个物种的花色是有限的。通过表达外源基因来改造类黄酮的生物合成途径，使得通过杂交或突变育种来获得一个物种新的花色品种成为可能。根据在矮牵牛和蓝猪耳等一些模式植物上的研究结果，已成功建立了通过基因工程改变花色的一般策略。外源基因的高效表达可以通过启动子、翻译增强子、编码区序列和终止子的最佳组合来实现。除了表达外源基因外，竞争代谢途径的下调以及利用花色合成突变体也是必要的因素。在考虑合适遗传背景的同时，选择具有较高市场地位和价值的观赏植物作为受体也很重要。每个目标物种应建立一套高效的遗传转化体系。技术技能和充裕资金也是获得转基因植物商业化许可证的必要条件。通过表达外源矮牵牛、三色堇和风铃草等的类黄酮 3′,5′-羟化酶（flavonoid 3′,5′-hydroxylase，F3′5′H）基因，已开发出了紫色的香石竹、月季和菊花等花卉，其中转基因香石竹和月季品种已商业化。除了 F3′5′H 外，外源的花色苷 3′,5′-葡萄糖基转移酶（anthocyanin 3′,5′-glucosyltransferase，A3′5′GT）基因在菊花中表达与菊花中的内源黄酮物质发生复合作用，从而使菊花产生纯蓝花色。因外源玉米二氢黄酮醇 4 还原酶（dihydroflavonol 4-reductase，DFR）基因的表达，从而在体内生成并积累了天竺葵素的橙色矮牵牛已在世界范围内广泛种植。尽管这是一种无意释放的转基因生物，但也提供了一个很好的例子，表明基因工程和杂交育种技术相结合可以产生高度受欢迎的商业品种。

关键词　花色苷；香石竹；菊花；类黄酮；转基因生物；矮牵牛；月季

3.1　引　　言

花色是花卉的一个重要消费性状，新奇的花色一直以来都是育种工作者追求的育种目标。杂交育种和诱变育种已成功用于产生新的花色。例如，黄色栽培月季（*Rosa hybrida*）就是用一个波斯野生的开黄花的异味蔷薇（*R. foetida*）不断杂交而育成的。但

N. Okitsu·Y. Tanaka (✉)
Research Institute, Suntory Global Innovation Center Ltd., Kyoto, Japan
e-mail: Yoshikazu_Tanaka@suntory.co.jp
N. Noda
Institute of Vegetable and Floriculture Science, National Agriculture and Food Research Organization, Tsukuba, Ibaraki, Japan
S. Chandler
School of Applied Sciences, RMIT University, Bundoora, VIC, Australia

育种者常被物种有限的基因库所限制。基因工程技术将育种者从这种限制中解放出来，第一个通过基因工程进行花色修饰的植物——砖红/橙色矮牵牛在 30 年前就已有报道。这种转基因矮牵牛通过表达一个外源玉米 DFR 基因从而在体内积累了天竺葵素花色苷（非转基因矮牵牛植株不含这种天竺葵素花色苷）（Meyer et al. 1987）。自此以后，许多经过花色改造修饰的花卉产生了，如花色为紫罗兰色的香石竹和月季，这种花色是其他香石竹和月季品种所没有的。由于花色修饰领域的研究已被包括本章作者在内的多位研究者（Nishihara and Nakatsuka 2011；Tanaka and Brugliera 2013；Sasaki and Nakayama 2015；Tanaka and Ohmiya 2008）评论过，在这里我们仅简要介绍描述与花色相关的类黄酮生物合成途径的研究进展，重点介绍近年来具有重要经济意义的花色改造方法及实用的花色修饰策略。

3.2　类黄酮生物合成途径

3.2.1　花色相关常规途径

类黄酮化合物是一类具有 C6-C3-C6 结构的苯丙醇类化合物的总称。呈现橙色、红色、洋红色、紫色和蓝色的类黄酮化合物的有色类别称为花色苷。类黄酮生物合成途径（图 3.1）是最具特色的生物合成途径之一。很多研究者之前已对该途径的生物化学和演化过程做过综述性的论述（Rausher 2006；Tanaka et al. 2008；Forkmann and Heller 1999），此处仅简要介绍该途径中与花色有关的化合物和酶。查耳酮合酶（chalcone synthase，CHS）是类黄酮化合物生物合成过程中的第一个酶，该酶将 1 个 *p*-香豆酰基分子和 3 个丙二酰辅酶 A 分子催化聚合成 2′,4,4′,6′-四羟基查耳酮（2′,4,4′,6′- tetrahydroxychalcone，THC）。THC 是一种不稳定的化合物，可自发异构化为柚皮素，但在体内可立体异构化为（2*S*）-柚皮苷。当通过依赖于 UDP-葡萄糖的葡萄糖基转移酶（GT）的作用，将 THC 2′-葡萄糖基化时，就会合成像香石竹中一样呈淡黄色的 THC 2′-葡萄糖苷。而在黄色的金鱼草中，THC 在细胞溶质中被 4′-葡萄糖基化，然后被液泡中的金鱼草素合酶（aureusidin synthase，AS）（一种多酚氧化酶）异构化，生成金鱼草素 6-*O*-葡萄糖苷和 bracteatin 6-*O*-葡萄糖苷。柚皮素是一种黄烷酮物质，在黄烷酮 3-羟化酶（flavanone 3-hydroxylase，F3H）的作用下，其被催化转变为二氢山奈酚。柚皮素也可被黄酮合酶（flavone synthase，FNS）催化成芹黄素（一种黄酮物质）。FNS 有两种类型：可溶性的 2-酮戊二酸依赖型 FNS I 和更常见的细胞色素 P450 型 FNS II。二氢黄酮醇在 DFR 和花色素合酶（anthocyanidin synthase，ANS）的作用下转化为有色的复合前体物质——花色素。由于某些物种的 DFR 具有严格的底物（二氢黄酮醇）特异性，因此 DFR 通常是改造该途径的靶点。花色素很容易被花色素 3GT（或称为类黄酮 3GT）糖基化（通常在 3 位点）。二氢黄酮醇也可被 FLS 代谢为黄酮醇。黄酮醇和黄酮被糖基化后通常输送到液泡中（见下文）。编码上述酶的基因已被用来改变各种植物的花色。

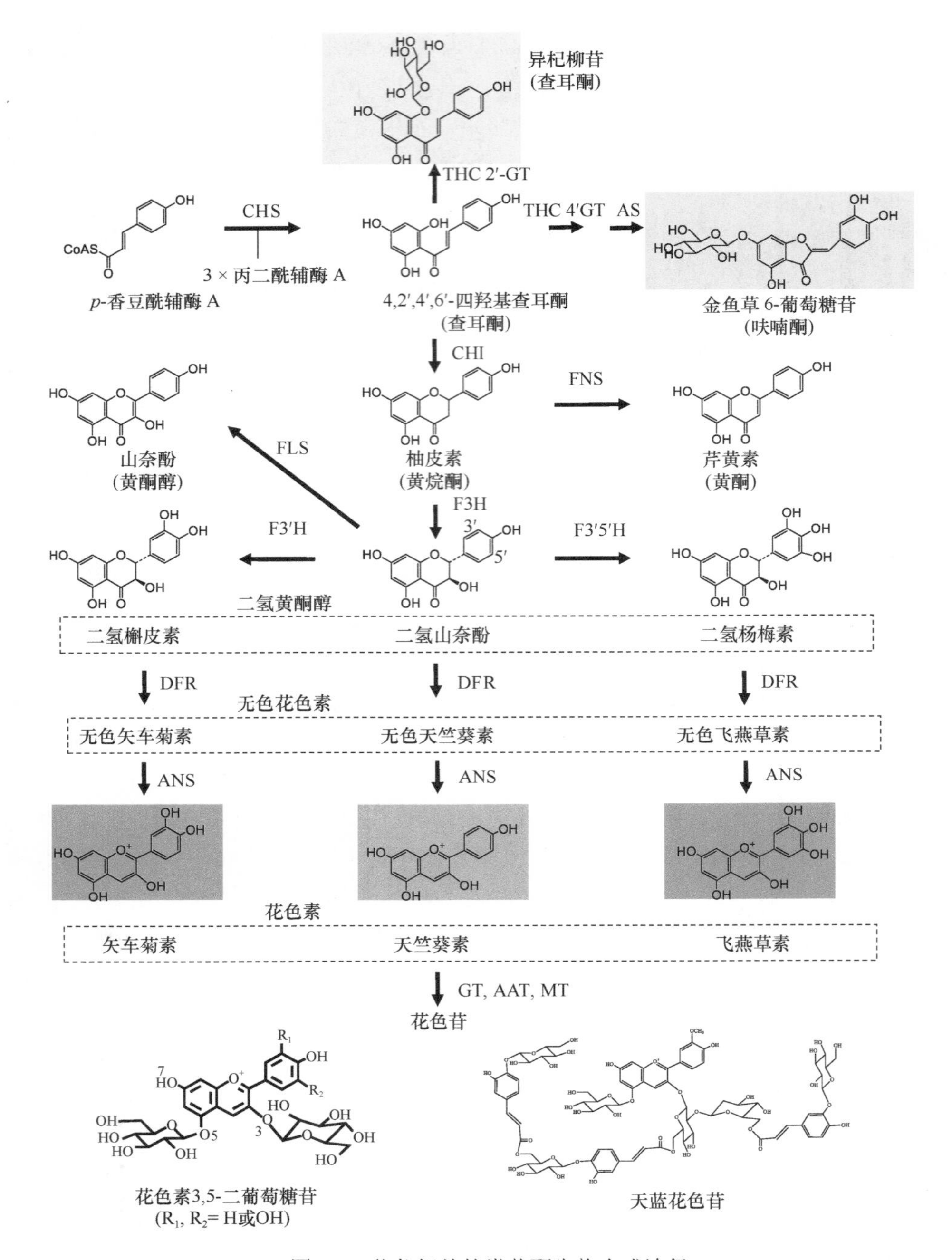

图 3.1 花色相关的类黄酮生物合成途径

花色素 3-葡萄糖苷通常是第一个合成的花色苷，并以物种特异性的方式被进一步修饰。牵牛花中的花色素 3,5-二葡萄糖苷和天蓝花色苷也在图中列出。请注意 FNS 有两种类型，即 GT 和 AAT。更多信息请参见文字。CHS：查耳酮合酶，CHI：查耳酮异构酶，F3H：黄烷酮 3-羟化酶，F3′H：类黄酮 3′-羟化酶，F3′5′H：类黄酮 3′,5′-羟化酶，DFR：二氢黄酮醇 4-还原酶，ANS：花色素合酶，GT：葡萄糖基转移酶，AAT：花色苷酰基转移酶，MT：甲基转移酶，AS：金鱼草素合酶，FLS：黄酮醇合酶，FNS：黄酮合酶

花色在很大程度上取决于花色苷的结构，特别是 B 环上羟基的数目。蓝/紫色花倾向于积累飞燕草素花色苷，橙色/深红色花则是积累天竺葵素花色苷。F3′H 和 F3′5′H 催

化黄烷酮或二氢黄酮醇的羟基化过程确证了这一点。F3′H 和 F3′5′H 也催化黄酮醇与黄酮的羟基化过程，但该过程对花色无直接影响。*F3′H* 和 *F3′5′H* 是修改花色的主要目标基因，如下所述。花色苷可被甲基、糖基和酰基修饰。经多个芳香酰基修饰的花色苷会更蓝、更稳定，如龙胆、蝶豆和瓜叶菊等。Yoshida 等（2009）对蓝色花色的化学机制已做过综述性介绍。

除了花色苷的结构外，花色还取决于液泡的 pH（中性或高 pH 环境下颜色更蓝），因为花色苷和类黄酮存在于液泡中。与花色苷共存的类黄酮物质（如黄酮和黄酮醇等）以及金属离子如 Al^{3+}、Fe^{3+}和 Mg^{2+}等是影响花色的其他因素（Yoshida et al. 2009）。鸭跖草和矢车菊中就存在由花色苷、黄酮/黄酮醇和金属离子组成的大型络合物（Takeda 2006；Yoshida et al. 2009）。

类黄酮生物合成途径研究得已较详细，也取得了一些新的发现，这些发现可能有助于改变花色。下面将对这些进展情况进行介绍。

3.2.2　花色苷修饰

根据种类不同，花色苷可以被糖基、脂肪族酰基（如乙酰基、丙二酰基、苹果酰基）、芳香酰基（如 *p*-香豆酰基、咖啡酰基、*p*-羟基苯甲酰基）和/或甲基等修饰。这些修饰由特定的酶完成，这些酶一般严格按照类黄酮的修饰位置来发挥作用，并根据其特异性形成不同的系统发育簇。在这些修饰酶中，UDP-糖基依赖型的 GT 酶以及 BAHD 型酰基辅酶 A 依赖型的花色苷酰基转移酶（anthocyanin acyltransferase，AAT）自 20 世纪末就开始了详细的研究。GT 和 AAT 这 2 种酶都是水溶性酶，存在于细胞溶质中（Sasaki and Nakayama 2015；Tanaka et al. 2008）。花色苷被这些修饰酶按照特定顺序如在蝶豆（Kogawa et al. 2007b）和半边莲（Hsu et al. 2017）中那样进行修饰。

最近，依赖于酰基葡萄糖（即有机酸的 1-*O*-*β*-D-葡萄糖酯）的花色苷 GT 酶和酰基转移酶也被分离（Sasaki and Nakayama 2015）。这类 GT 酶首次是在香石竹（花色素-3-*O*-葡萄糖苷的 5-*O*-葡萄糖基化）和飞燕草（花色素-3-*O*-葡萄糖苷的 7-*O*-葡萄糖基化）（Matsuba et al. 2010）中被研究报道的。接着分别在百子莲（花色素 3-*O*-葡萄糖苷的 7-*O*-葡萄糖基化）（Miyahara et al. 2012）、拟南芥（花色苷 4-香豆酸部分的糖基化）（Miyahara et al. 2013）、飞燕草（7-多糖基化）（Nishizaki et al. 2013）和风铃草（花色素 3-芸香苷的 7-糖基化）（Miyahara et al. 2014）中被研究报道。这些 GT 酶都属于糖苷水解酶家族 1（GH 1）成员，定位于液泡中。

酰基葡萄糖依赖型的 AAT 基因已从蝶豆（*p*-香豆酰基从 *p*-香豆酰葡萄糖上转移到花色苷 3′位上的葡萄糖部分）（Noda et al. 2006）、香石竹（1-*O*-苹果酰-*β*-D-葡萄糖依赖型 1-*O*-苹果酰葡萄糖：天竺葵素 3-葡萄糖苷-6″-*O*-苹果酰转移酶）（Umemoto et al. 2014）以及拟南芥和飞燕草（花色苷的 *p*-羟基苯甲酰化）（Nishizaki et al. 2013）等植物中分离得到。这些 AAT 酶是丝氨酸羧肽酶类似物（serine carboxypeptidase-like，SCPL）并定位于液泡中。特别有趣的是，*p*-羟基苯甲酰葡萄糖是 GH1 类 GT 和 SCPL 类 AAT 的“双方供体”（Nishizaki et al. 2013）。

催化酰基葡萄糖合成的UDP-葡萄糖依赖型GT酶已在飞燕草（Nishizaki et al. 2014）和拟南芥中被鉴定。在转基因植物中，这种GT酶的共表达可能是通过GH1类GT和SCPL类AAT对花色苷进行有效修饰的必要条件。可能还需优化它们的“液泡分类信号”，以便它们能在异源植物中表达。

S-腺苷甲硫氨酸依赖型的花色苷甲基转移酶（anthocyanin methyltransferase）基因，已在葡萄（Lucker et al. 2010；Hugueney et al. 2009）、仙客来（Akita et al. 2011）、矮牵牛（Provenzano et al. 2014）和蓝猪耳（Nakamura et al. 2015）等植物中被研究。蓝猪耳的花色苷甲基转移酶基因与三色堇的*F3'5'H*基因已在转基因月季中共表达。由锦葵素所产生的花色比单独积累飞燕草素的花色更鲜红（Nakamura et al. 2015）。花色苷甲基化对仙客来的花色有影响。用离子束照射原本积累锦葵素3,5-*O*-二葡萄糖苷的紫色仙客来，结果发现了一个积累飞燕草素3,5-*O*-二葡萄糖苷的紫红色突变株（Kondo et al. 2009）。

3.2.3 黄酮C-糖基化

黄酮物质尤其是黄酮*C*-葡萄糖苷，如日本庭院鸢尾玉蝉花（*Iris ensata* Thunb.）中的异牡荆苷（芹黄素6-*C*-葡萄糖苷）（Yabuya et al. 1997），是一种很强的辅色素。黄酮*O*-葡萄糖苷是由GT介导的黄酮骨架直接糖基化合成的。导入*FNS*基因后，天然不含黄酮的月季体内便会积累黄酮*O*-葡萄糖苷（结果未发表）。在水稻中发现了长期以来一直是个谜的黄酮*C*-葡萄糖苷的生物合成途径。在水稻中，P450酶介导的由黄烷酮合成的2-羟基黄烷酮开链形式被GT酶催化糖基化，再通过脱水酶脱水生成黄酮*C*-葡萄糖苷（Brazier-Hicks et al. 2009）。最近，有研究者发现龙胆叶中含有异荭草素（木犀草素6-*C*-葡萄糖苷），并分离了可将葡萄糖直接催化转移至黄酮C-6位置上的GT酶基因（Sasaki et al. 2015）。该基因的表达以及异荭草素6-*C*-葡萄糖苷在花中积累较高而在叶中积累较少，表明异荭草素对龙胆花的蓝色形成未起主要作用。这些酶尤其是龙胆6*C*GT酶，可以用作转基因花中积累黄酮*C*-葡萄糖苷的分子工具。

3.2.4 蛋白质-蛋白质互作

生物合成途径中的代谢酶形成一种称为“metabolon”的蛋白质复合物（Winkel 2004）。类黄酮生物合成酶之间存在蛋白质-蛋白质相互作用的直接证据已在拟南芥中被验证（Burbulis and Winkel-Shirley 1999），其中酵母双杂交实验显示了CHS、CHI和DFR的相互作用，亲和色谱和免疫沉淀测定显示了CHS、CHI与F3H的相互作用。最近，进一步令人信服的结果支持了“metabolon”的形成。通过分裂泛素化膜酵母双杂交系统和烟草叶片细胞生物分子荧光互补实验证明了大豆2-羟基异黄酮合酶（P450）与CHS或CHI存在相互作用。更有趣的是，这种互作产生的荧光信号在细胞内呈现出网络分布模式，类似于荧光蛋白标记的GmIFS1（大豆异黄酮合酶）在ER的定位荧光模式（Waki et al. 2016）。在金鱼草中存在CHI和DFR与FNS相互作用（Fujino et al. 2018）。

利用突变体和转座子标记技术，从日本牵牛花（*Ipomoea nil*）中分离了一系列参与类黄酮生物合成的酶和蛋白质。从一个浅色突变体中鉴定出一个类黄酮生成增强蛋白

（enhancer protein for a flavonoid production，EFP）的基因。EFP 属于一类查耳酮异构酶相关 IV 型酶。在日本牵牛花中，*EFP* 基因突变导致了花色苷和黄酮醇的显著减少（减少约 1/4）。在矮牵牛和蓝猪耳中也分离到 *EFP* 的同源基因。这些 *EFP* 基因在转基因植物中的下调表达导致花色苷和黄酮醇（矮牵牛中）或黄酮（蓝猪耳中）减少 70%～80%，同时花色变淡。EFP 在类黄酮生物合成过程中的作用尚不清楚。然而，有人认为 EFP 可能不是 CHI、转录因子或花色苷转运体，EFP 与 CHS 和/或相关产物的互作证明 EFP 起着增强子的作用，尽管其增强子作用的生化机制尚待研究。EFP 在植物包括小立碗藓（*Physcomitrella patens*）中普遍存在。有趣的是，小立碗藓含有 EFP 但没有 CHI，表明可能在进化上 EFP 出现的时间要比 CHI 早（Morita et al. 2014）。

当类黄酮生物合成相关蛋白的蛋白质-蛋白质互作具有特异性时，内源蛋白与外源蛋白的兼容性可以作为修饰花色的工具，然而现在还不可行。

3.2.5　液泡运输

大多数类黄酮物质，包括花色苷都是在细胞溶质中合成的，再被运输到液泡中。植物细胞采用多种机制来转运类黄酮物质，如①谷胱甘肽 *S*-转移酶/多药耐药相关蛋白（multidrug resistance-associated protein，MRP）介导的转运，②多药和有毒化合物外排（multidrug and toxic compound extrusion，MATE）转运体介导的膜转运，③囊泡转运，这些已被详细介绍描述（Zhao 2015）。这些机制并不是单独存在的，而是在植物细胞中相互共存。

谷胱甘肽 *S*-转移酶（GST）在类黄酮转运中的作用已被很好地证实，如在玉米突变体 Bz2、矮牵牛 AN9、拟南芥 TT19、葡萄和其他植物等中（Zhao 2015）。尚不清楚 GST 在生化上如何参与花色苷向液泡的转运过程。GST 本身不运输类黄酮，也不与类黄酮形成复合物（Zhao 2015）。MRP 是一种 C 型 ATP 结合蛋白，已被证明在葡萄中转运锦葵素 3-葡萄糖苷依赖型的游离谷胱甘肽（Francisco et al. 2013）。

MATE 转运蛋白已被证明在蒺藜苜蓿中转运表儿茶素 3-*O*-葡萄糖苷和花色苷苹果酸盐，而在拟南芥中则是转运原花色素前体（Zhao and Dixon 2009）。

花色苷和原花色素通过高尔基体中的囊泡转运进入液泡。拟南芥 GFS9（TT9）是一种位于高尔基体的外周膜蛋白，通过与液泡的膜融合来促进液泡发育。GFS9 是囊泡介导的蛋白质和原花色素、黄酮醇等植物化学物质转运系统的重要组成部分（Ichino et al. 2014）。

如果这些转运系统对花色苷或类黄酮结构具有特异性，为目标物种设计转运系统可能是积累所需类黄酮的一种策略。

3.2.6　液泡中花色苷的分布

花色苷并不总是均匀分布在液泡中。花色苷的聚集或浓缩形式，即花色苷液泡内含物（anthocyanin vacuolar inclusion，AVI）常在各种不同植物中被观测到（Markham et al. 2000）。AVI 的出现赋予花瓣一种独特的色调。蓝色月季品种如‘Rhapsody in Blue’花瓣含有由矢车菊素 3,5-*O*-二葡萄糖苷组成的 AVI（Gonnet 2003）。在积累花色苷的拟南芥幼苗中，AVI 的形成与矢车菊素 3-*O*-葡萄糖苷的含量有关，5GT 的突变增加了 AVI 的数量（Pourcel et al. 2010）。结果表明，液泡膜吞噬细胞质中的花色苷聚集体，然后由液泡产生的单个膜包围聚集体，导致液泡腔中出现 AVI。拟南芥中矢车菊素 3-*O*-葡萄糖苷的增加和 GST（TT19）的缺乏促进了 AVI 的形成（Chanoca et al. 2015）。该途径被认为是花色苷向液泡转运的一种机制（Chanoca et al. 2015）。

香石竹花瓣通常积累苹果酰天竺葵素或矢车菊素葡萄糖苷。表明具有金属光泽颜色花朵的香石竹品系含有高度密集的由不含苹果酰基的花色苷组成的 AVI，并且产生 AVI 的原因是花色苷苹果酰基转移酶基因失活（Okamura et al. 2013）。

在葡萄培养细胞中，AVI 富含 *p*-香豆酰化花色苷（Conn et al. 2003）。从葡萄细胞培养物中分离纯化的 AVI 表明它们是致密的、高度有机的结构，该结构带有一个膜外壳标示性的脂质部件。纯化的 AVI 含有长链单宁（Conn et al. 2010）。最近，利用通过表达金鱼草转录因子积累大量花青素的转基因烟草发现，花色苷的芳香酰化是形成 AVI 的必要条件（Kallam et al. 2017）。AVI 及其形成机制可能存在多样性。

3.2.7 液泡 pH 调控

花色苷的颜色在很大程度上取决于 pH，因此液泡 pH 可影响花色，但花色苷在较高 pH 下稳定性较差。仙客来主要积累锦葵素 3,5-*O*-二葡萄糖苷，过去没有蓝色品种，经过长期培育现已有多个蓝色品种。这是由于一个隐性突变引起液泡 pH 升高而产生的（Takamura et al. 2011），但这些品种的花色容易褪色。

尽管通过两个质子泵（V-ATPase 和焦磷酸酶）的作用，常使液泡的 pH 普遍呈现弱酸性，但已证明植物细胞还含有其他的离子转运/交换机制。牵牛花花瓣通过表达 K^+/H^+ 交换器，将液泡 pH 碱性化至 7.7，从而使其花色在开放时从紫色变为蓝色（Yoshida et al. 2005；Fukada-Tanaka et al. 2000）。高度修饰后的牵牛花花色苷（含有 6 个葡萄糖基和 3 个芳香族酰基部分的芍药素）在如此高的 pH 下可保持稳定的颜色。

对矮牵牛进行广泛的遗传和生化分析发现它含有一个 3A 族的质膜 ATPase（P_{3A}-ATPase）同源物（由 PH5 编码）（Verweij et al. 2008）和一个定位在“tonoplant”上的 P_{3B}-ATPase 同源物（由 PH1 编码）（Li et al. 2016）。PH1 本身不具有 H^+转运活性，但 PH1 与 PH5 相互作用可增强其转运活性。二者都是使矮牵牛花瓣液泡过度酸化以获得紫色而非蓝紫色的必要因素。已证实，功能性的 PH1 和 PH5 同源物分布在不同的被子植物中。调节它们的表达有望改变花的颜色（Li et al. 2016）。特别的是，上调交换器以及下调 PH5 或 PH1 是获得蓝色花色的一个诱人方法。然而，据我们所知还没有成功的先例。

绿玉藤的青蓝色花色源自一个异常高的表皮细胞液泡 pH（为 7.9）条件下的锦葵素 3,5-*O*-二葡萄糖苷和异牡荆素 7-*O*-葡萄糖苷（二者物质的量为 1∶9）。这种高 pH 的分子机制尚未被研究。

3.3　基因工程改变花色

3.3.1　模式植物花色改变策略

矮牵牛在花色遗传研究中应用广泛，易于转化，又是一种重要的经济植物，因此被用作改变花色的模式植物。许多类黄酮生物合成途径中的相关基因都是从矮牵牛中首次分离出来的，它们的功能已在矮牵牛的互补或抑制实验中得到证实（Tanaka and Ohmiya 2008）。1987 年，外源玉米 *DFR* 基因在转基因矮牵牛花中表达产生了天竺葵素，从此之后开始了矮牵牛花色的改良工作（Meyeret al. 1987）（这部分内容将在第 3.3.2 节中详细描述）。接着，通过 *CHS-A* 基因的反义（van der Krol et al. 1990b）和共抑制（van der Krol et al. 1990a；Napoli et al. 1990）使花色苷合成量下调。烟草、高杯花（类似矮牵牛的一种茄科植物）和蓝猪耳（Aida，2008）也被用作改变花色的模式植物。

在试图将蓝猪耳的蓝色花色（源于飞燕草素花色苷）改变为粉红色（源于天竺葵素花色苷）时，通过下调蓝猪耳内源 *F3'5'H* 和 *F3'H* 基因只会使其花色变淡，而月季或天竺葵表达外源 *DFR* 基因可使其花色变深，且天竺葵 *DFR* 比月季 *DFR* 更有效（Nakamura et al. 2010）。

值得注意的是，化合物通常以多种途径代谢。例如，二氢黄酮醇是花色苷和无色的黄酮醇的前体物质，而且花色苷的含量与黄酮醇的含量呈负相关。外源月季 *FLS* 基因在矮牵牛中表达可提高矮牵牛黄酮的含量，而外源矮牵牛 *FLS* 基因的过度转录则降低其黄酮的含量，这可能是由共抑制现象所致（Tsuda et al. 2004）。高杯花积累飞燕草素花色苷和黄酮醇。高杯花 *F3'5'H* 基因的下调导致其花色苷合成量减少，但黄酮醇合成量的增加又弥补了这种减少，最终使其花色变淡（Ueyama et al. 2006）。而内源 *FLS* 基因的下调以及外源月季 *DFR* 基因的表达可导致产生天竺葵素而使花色变为粉红色（Takeda et al. 2010）。同样的，*F3'5'H*、*FLS* 和花色素 3-芸香苷 *AT* 基因的下调可导致矮牵牛中的矢车菊糖苷积聚，从而使矮牵牛的花色由洋红色变为红色（Tsuda et al. 2004）。*F3'H* 和 *FLS* 基因的下调表达以及外源非洲菊 *DFR* 基因的表达导致烟草中天竺葵素的积累，从而使烟草花色表现为深红色（Nakatsuka et al. 2007）。通过黄酮积累、花色苷减少已使烟草产生了浅色花色（Nakatsuka et al. 2006）。通过这种策略可使龙胆和鸢尾产生红色花。

内源基因表达下调最初是通过反义和共抑制技术实现的，后来则是通过 RNAi（双链 RNA 的转录），RNAi 效率在蓝猪耳中比反义或共抑制更好（Nakamura et al. 2006）。利用 RNAi 获得的表型并不总是稳定的，特别是在自然环境条件下（Tanaka et al. 2010），因此 RNAi 有可能在不久的将来被基因编辑技术所取代。

花色苷生物合成相关基因的转录被由 R2R3 型 Myb（矮牵牛 An2）、bHLH 蛋白（basic helix-loop-helix protein）（矮牵牛 An1）和 WD40（矮牵牛 An11）组成的复合物以协作的方式调控（Koes et al. 2005）。已从许多花卉中分离了相应的基因，R2R3 型 Myb 或 bHLH 蛋白在异源植物中的组成型表达导致异位花色苷或类黄酮的积累。在这里仅举几个例子：外源拟南芥 *PAP1*（生成花色苷色素 1）基因的表达会使烟草的花色变深（He et al.

2017），外源金鱼草 R3R3 型 *Myb*（Rosea 1 和 Delila）基因的表达会使番茄产生紫色果实，并积累花色苷和内源性类胡萝卜素（Butelli et al. 2008），月季中相应基因的表达会使其叶色变红（Gion et al. 2012）。

3.3.2 产生天竺葵素的橙色矮牵牛

因为矮牵牛的 DFR 不能催化二氢山奈酚生成天竺葵素，所以矮牵牛没有橙色或砖红色的花色，利用花椰菜花叶病毒 35S（CaMV35S）启动子使外源玉米 *DFR* 基因在缺少 *F3′H* 和 *F3′5′H* 基因的矮牵牛中表达，会使矮牵牛积累天竺葵素从而产生砖红花的花色（Meyer et al. 1987）。但产生的花色并不总是很稳定，有时会出现带有斑点的或白色的花。花色的稳定性与整合到矮牵牛基因组中的外源 *DFR* 基因的拷贝数以及 CaMV35S 启动子的甲基化程度有关（Linn et al. 1990）。对 30 000 株含有单拷贝外源 *DFR* 基因的矮牵牛植株进行田间试验发现，CaMV35S 启动子甲基化会引起表型不稳定。季节也会影响花色颜色深度（Meyer et al. 1992）。之后的研究发现，非洲菊 *DFR* 比玉米 *DFR* 在矮牵牛中表现更稳定，产生的花色也更深，这表明基因来源是获得合适表型需要考虑的一个因素（Elomaa et al. 1995）。

表达外源玉米 *DFR* 基因的矮牵牛被纳入矮牵牛育种工程后，其花色的不稳定性问题被解决，并获得了具有深橙色花色并且栽培性良好的 F4 系矮牵牛（Oud et al. 1995）。然而，这种矮牵牛本不该商业化应用。芬兰食品安全局（Finnish Food Safety Authority，芬兰语简称 EVIRA）在 2017 年 4 月 27 日发现了表达外源玉米 *DFR* 基因的橙色矮牵牛，并要求将其从种子和种苗销售中剔除。美国和澳大利亚政府以及日本农林水产省（MAFF）紧随其后。这种转玉米 *DFR* 基因的矮牵牛后代被用于培育橙色矮牵牛商业品种（图 3.2a）（Bashandy and Teeri 2017）。很多专业的种子和种苗公司出售这些转基因矮牵牛品种，却不知道它们是转基因植物。这种花色的矮牵牛很新颖且很流行。数十个这种花色的矮牵牛品种的数百万种子和植株已被销往世界各地。荷兰基因改良委员会（COGEM）审查了包括这种产生天竺葵素的矮牵牛在内的矮牵牛育种工作后得出结论，这种转基因矮牵牛的释放不大可能会影响生物多样性和人类健康（COGEM 2017）。日本 MAFF 和美国的权威机构评估后也指出这种转基因矮牵牛不会影响日本与美国的生物多样性。在中国，*CHS* 基因被抑制的转基因白色矮牵牛已商业化（COGEM 2017）。

尽管已下达了将已释放的转基因矮牵牛恢复至以前或者直接摧毁掉的命令，但由于许多业余育种家是在家里进行矮牵牛杂交工作，并且当 F3′H、F3′5′H 和 FLS 有活性时，来自玉米 *DFR* 基因的表型是潜在存在的，因此不可能恢复所有的已释放的转基因矮牵牛植株。实际上，一些转基因矮牵牛品种的花色并非为橙色。尽管从基因改造的监管角度来看，这次释放是个意外事件，但从某种意义上说，这次意外释放也是一次大规模的实地实验。同时，这也是一个很好的例子，说明基因工程和育种计划结合可以产生在商业上成功的品种。尽管与食品相关的转基因报道通常会得到公众和大众媒体的负面回应，但这次矮牵牛事件几乎没有主流媒体提到。

图 3.2　转基因花卉。(a) 积累天竺葵素的橙色矮牵牛；(b) 积累飞燕草素的香石竹‘Moon’系列（标准，喷雾）的瓶插寿命测试及其生产过程，这些转基因植物与非转基因香石竹一样进行繁殖和生产；(c) 日本销售的积累飞燕草素的转基因月季，研究温室里的月季和商业种植者生产的月季切花；(d) 非转基因菊花（IR）与转基因菊花的花色比较，转基因菊花可以积累飞燕草素，其花色比非转基因菊花（IR）的花色表现得更蓝

3.3.3　紫罗兰色的香石竹‘Moon’

由于 *F3'5'H* 基因的缺失，香石竹、月季和菊花不能产生飞燕草素花色苷，因此缺乏蓝紫色的花色。可生成飞燕草素的紫罗兰色转基因香石竹自 1996 年开始商业化。它们的遗传结构及转基因结果在之前已被详细介绍过（Tanaka and Brugliera 2013）。矮牵牛 *F3'5'H* 基因在原本产生天竺葵素的香石竹中过表达可使其生成一定的飞燕草素，但花色变化不明显。为了避免香石竹内源 *DFR*（生成天竺葵素）与外源 *F3'5'H* 之间的竞争，一种缺失 *DFR* 基因的 midi 型白色香石竹被选作进行基因转化的材料，矮牵牛 *F3'5'H* 和 *DFR* 的 cDNA（通过构建 pCGP1470 载体）被转化到该品种香石竹中，产生的转基因香石竹因只积累飞燕草素而显示出淡紫色的花色，这个香石竹品种（Florigene®Moondust™）

是世界上第一个商业化的花卉作物，接着是一个导入三色堇 *F3'5'H* 和矮牵牛 *DFR*（通过 pCGP1991 载体）、花色呈深紫色的香石竹品种（Florigene®Moonshadow™）。这些香石竹品种的单朵花在商业上价值不大，被后来更优良的转基因品种所取代。新的转基因香石竹品种有很长的茎秆，而且即使在经过长途国际运输后也能保持较长的瓶插寿命。

利用香石竹普通品种进行遗传转化培育了 Florigene®Moonaqua™、Florigene® Moonlite™、Florigene®Moonshade™和 Florigene®Moonvista™（花色最暗的品种）等一系列蓝色花色的香石竹转基因品种。这些品种间蓝色花色的深浅程度不同是由各自花瓣中飞燕草素的含量不同所造成的。有趣的是，Florigene®Moonaqua™和 Florigene® Moonvista™使用了同样的转化载体 pCGP1991，但其花色表现不同，这表明同一转基因结果的后代表型可能会存在一定差异，因此需要保障足够多的转基因植物数量才有可能筛选出理想的表型植株。

在接下来的香石竹花色转基因育种工作中，通过 RNAi 下调内源 *DFR* 基因的表达，同时表达两个外源 *F3'5'H* 基因，或增强 *F3'5'H* 基因的电子传递而培育出的多头型紫蓝色香石竹品种（Florigene®Moonberry™、Florigene®MoonPearl™、Florigene®Moonique™ 和 Florigene®Moonvelvet™）（图 3.2b）已经被商业化推广应用。由于是转基因植物，因此必须获得相关国家或地区的生产和销售许可证才可以推广销售。许可证是在证明这些转基因香石竹释放不太可能会影响生物多样性、环境或人类健康之后才颁发的。获得许可证的过程取决于各国家政府，有时需要数年（如第 3.12 节所述）。

值得注意的是，即便在这些转基因香石竹中产生了同样的飞燕草素花色苷，但其蓝色深浅程度还会因受体品种不同而存在一定差异。品种‘Moonshadow’的花色之所以表现得最蓝，是因为 midi 型香石竹体内含有的一种黄酮 *C*-葡萄糖苷可作为辅色素使其花色变深（Fukui 2003）。

3.3.4 蓝紫色月季

在来源于矮牵牛、三色堇、龙胆、蝶豆、薰衣草、马鞭草以及其他少数植物的 *F3'5'H* 基因中（其中矮牵牛和三色堇各有 2 个 *F3'5'H* 基因），只有三色堇的 2 个 *F3'5'H* 基因导入月季后会产生大量的飞燕草素从而引起花色发生变化。转基因植株中生成的飞燕草素水平取决于转基因受体，因为外源酶 F3'5'H 与内源酶如 F3'H、FLS 和 DFR 等之间的竞争因受体不同而存在差异。三色堇 *F3'5'H* 基因的组成型表达在少数月季品种中可合成大量的飞燕草素（占总花色素总量的 90%以上）。花色变化取决于亲本品种（Katsumoto et al. 2007）。从一个新颖的蓝色花色品种中筛选出了两个相似的转基因月季品种。这两种转基因月季在日本的商业化应用已得到了批准，因为研究表明它们都是 L1 嵌合体（只有 L1 细胞层含有转入的基因），在自然条件下不太可能发生从现代月季到日本野生月季的基因漂移（Nakamura et al. 2011a，2011b）。其中一个转基因月季品种从 2009 年已开始在日本国内销售（图 3.2c）。但因为这个品种的瓶插寿命较短而限制了它在其他国家的销售。培育瓶插寿命较长的转基因月季品种将更有利于得到全球商业化应用。通过外源三色堇 *F3'5'H* 基因和矮牵牛或鸢尾 *DFR* 基因在月季中联合过表达，再利用 RNA 干扰（RNAi）技术使月季本身的 *DFR* 基因下调表达，实现了更有意义的花色改造工程，通过这种方式培育的月季体内能生成几乎 100%的飞燕草素，尽管其花色依然为紫色，但来源于品种‘Lavande’的转基因月季已表现出月季中最蓝的花色（Katsumoto et al. 2007）。然而，这些转基因月季由于不明原因无法正常生长，因此无法被商业化推广应用。

3.3.5　紫色及蓝色菊花

日本皇室徽章常用菊花图案做装饰，菊花也是日本的国花。其一直以来都是进行蓝色花色改造的目标物种，但菊花成功的蓝色花色改造工作比香石竹和月季都要晚，因为外源基因在菊花中较难表达，且仅有很少的启动子在菊花中表现出很好作用（Aida et al. 2004）。

商业多头菊花品种中内源 *F3′H* 的表达，结合类黄酮生物合成相关基因启动子（如月季 *CHS* 基因启动子）驱动三色堇 *F3′5′H* 的表达，使其体内积累了含量高达 80%的飞燕草素，最终转基因菊花花色显著变蓝（Brugliera et al. 2013）。其中一些品系通过田间试验来评价它们的商业性状（图 3.2d）。然而由于它们的蓝色花色还不够新颖以至于不能商业化，因为利用传统育种技术已培育出一个花色较蓝（相对于非转基因香石竹和月季而言）的菊花品种。

Noda 等（2013）也将外源 *F3′5′H* 基因在菊花中进行表达。研究了不同类黄酮生物合成相关基因的启动子对三色堇 *F3′5′H* 基因在菊花花瓣中产生飞燕草素的影响。菊花 *F3H* 基因启动子的效果明显好于月季 *CHS*、三色堇 *CHS* 及月季 *DFR* 基因的启动子。在菊花 *F3H* 基因启动子的驱动下，风铃草 *F3′5′H* 基因（*CamF3′5′H*）相比瓜叶菊、马鞭草、三色堇及洋桔梗的 *F3′5′H* 基因可合成更多的飞燕草素。另外，将烟草、拟南芥和水稻等的乙醇脱氢酶基因的 5′-非翻译区作为翻译增强因子，引入导入的外源基因起始密码子的 5′端，也可增加转基因菊花植株中飞燕草素的含量。随着飞燕草素的积累并几乎达到菊花花瓣花色苷总量的 100%，此时菊花花色变得更蓝（Noda et al. 2013）。以上研究表明，对基因表达的物种特异性进行优化是获得理想表型的必要条件。

单独生成飞燕草素可以将花色改变为紫色或紫蓝色，但不能变为蓝色，还需要采取额外的策略来设计菊花的蓝色花色。Noda 等（2017）在这方面取得了重大突破并已成功培育出真正的蓝色菊花。这一突破是菊花花瓣中芳香聚酰化花色苷积累研究工作的一部分。

聚酰化花色苷的分子内辅助成色作用是蓝色花色形成的机制之一（Yoshida et al. 2009；Sasaki and Nakayama 2015）。蝶豆（*Clitoria ternatea*）蓝色花瓣积累的聚酰化花色苷称为 ternatin，如 ternatin A1（图 3.3a）（Saito et al. 1985；Terahara et al. 1990）。如果 ternatin 的 3′和 5 位置上缺失聚酰化葡萄糖基，则会相应地积累飞燕草素 3-(6″-丙二酰)葡萄糖苷，导致花色从蓝色变为淡紫色（kazuma et al. 2003）。UDP-葡萄糖依赖型的花色苷 3′,5′-*O*-GT 催化的 3′-和 5′-*O*-葡萄糖基化反应是蓝色蝶豆中 ternatin 生物合成的第一步（Kogawa et al. 2007a）。与淡紫色蝶豆一样，表达 *F3′5′H* 的紫色转基因菊花中也能产生飞燕草素 3-(6″-丙二酰)葡萄糖苷。这表明，如果飞燕草素的 3′和 5′位置被葡萄糖基化并进一步被芳香酰基修饰，菊花的花色将会从紫色或紫蓝色变为蓝色。合成最简单的蓝色 ternatin D3 和 3′,5′-二- *p*-香豆酰糖基化的飞燕草素 3-(6″-丙二酰)葡萄糖苷（Terahara et al. 1998）的候选基因是 UDP-葡萄糖依赖型花色苷 3′,5′-*O*-GT 基因（*CtA3′5′GT*）、UDP-葡萄糖依赖型羟基肉桂酸 1-*O*-葡萄糖基转移酶（hydroxycinnamate 1-*O*-glucosyltransferase）基因以及 1-*O*-酰基葡萄糖依赖型花色苷 3′,5′-酰基转移酶（anthocyanin 3′,5′-acyltransferase）基因（Noda et al. 2006）。

这些基因从蝶豆中分离出来后再导入菊花中，同 *CamF3′5′H* 一起合成 ternatin D3（图 3.3a），由于分子内的辅助呈色作用，在菊花中，由 ternatin A1 前体积累的 ternatin D3 生成的花色要比飞燕草素 3-(6″-丙二酰)葡萄糖苷蓝得多，因此其产生的转基因菊花呈现出很好的正蓝色（图 3.3b）。花色苷分析意外地表明，被积累的是 ternatin D3 而不是非 *p*-香豆素化（non-*p*-coumaroylated）的 ternatin C5（图 3.3c）。这个结果在只转化表达了 *CtA3′5′GT* 和 *CamF3′5′H* 基因的不同菊花中可重复再现。已在 10 多个不同花型如莲座型、蜂窝型和托桂型等育种系和栽培种中获得了蓝色花色（Noda et al. 2017）。

a

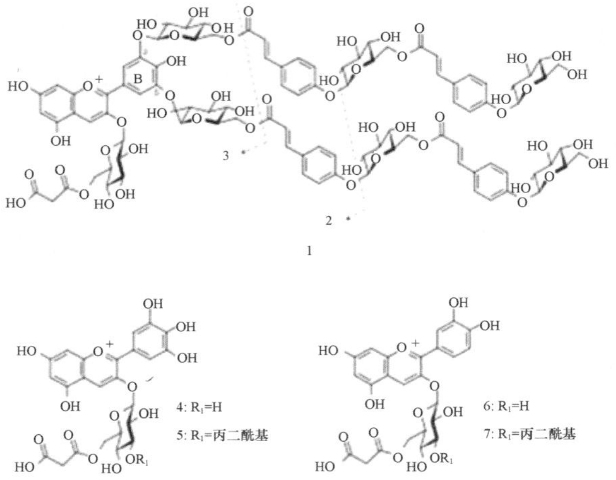

1

4: R_1=H
5: R_1=丙二酰基

6: R_1=H
7: R_1=丙二酰基

b

1

2

c

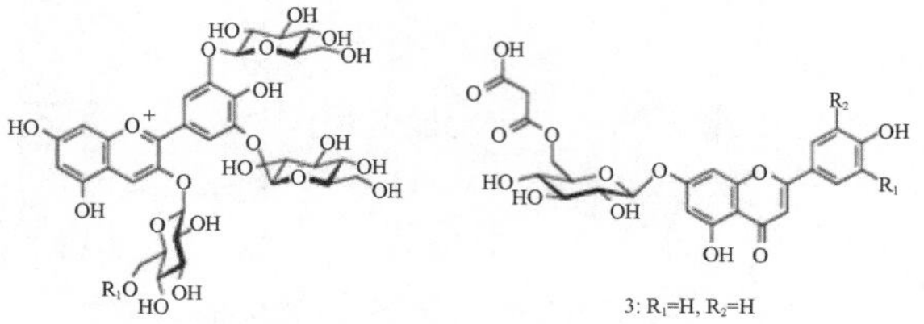

1: R_1=丙二酰基

2: R_1=H

3: R_1=H, R_2=H

4: R_1=OH, R_2=H

5: R_1=OH, R_2=OH

图 3.3　蓝色菊花的产生。（a）蝶豆（*Clitoria ternatea*）蓝色花瓣中的 ternatin A1（1）、ternatin D3（2）、ternatin C5（3），表达 F3′5′H 的转基因紫色菊花中的飞燕草素 3-(6″-丙二酰)葡萄糖苷（4）、飞燕草素 3-(3″,6″-二聚丙二酰)葡萄糖苷（5），野生型菊花花瓣中的矢车菊素 3-(6″-丙二酰)葡萄糖苷（6）和矢车菊素 3-(3″,6″-二聚丙二酰)葡萄糖苷（7）的结构图。（b）转基因紫色和蓝色菊花：（1）积累飞燕草素 3-(丙二酰)葡萄糖苷的紫色菊花呈现出与英国皇家园艺学会（RHSCC）比色卡上 83B 对应的紫蓝色；（2）积累飞燕草素 3-(丙二酰)葡萄糖苷 3′,5′-二葡萄糖的蓝色菊花呈现出对应 95C 的蓝紫色。（c）转基因菊花蓝色花色的分子间辅助成色。辅助成色反应是主要花色苷 ternatin C5 [飞燕草素 3- (6″-丙二酰)葡萄糖苷 3′,5′-二葡萄糖苷]（1）、ternatin C5 前体（飞燕草素 3,3′,5′-三葡萄糖苷）（2）和黄酮 7-(6″-丙二酰)葡萄糖苷、芹黄素 7-(6″-丙二酰)葡萄糖苷（3）、木犀草素 7-(6″-丙二酰)葡萄糖苷（4）、五羟黄酮 7-(6″-丙二酰)葡萄糖苷（5）联合作用的结果

在酸性条件下，蓝色菊花中的主要花色苷 ternatin C5 比飞燕草素 3-(6″-丙二酰)葡萄糖苷和矢车菊素 3-(6″-丙二酰)葡萄糖苷更红。在 pH 5.6 的弱酸性菊花花瓣中，ternatin C5 则呈现紫蓝色而非蓝色。不考虑菊花的蓝色，在缓冲溶液中菊花花瓣在可见光区的最大吸收波长（λ_{vismax}）要比 ternatin C5 的 λ_{vismax} 长。Cross-TLC（cross thin-layer chromatography）分析表明，ternatin C5 是通过与菊花蓝色花瓣中黄酮 7-(6″-丙二酰)葡萄糖苷之间发生辅助成色反应而形成蓝色的。进一步的类黄酮化合物和重组分析表明，内源的木犀草素 7-(6″-丙二酰)葡萄糖苷、芹黄素 7-(6″-丙二酰)葡萄糖苷及其经 F3′5′H 催化生成的衍生物五羟黄酮 7-(6″-丙二酰)萄糖苷等都是花色苷的强辅色素（图 3.3c）（Noda et al. 2017）。

结果表明，被认为是创造蓝色花色途径的金属复合物和聚酰化花色苷的形成，对某些物种的蓝色花色形成可能不是必需的。花色苷的合成方法，如 ternatin C5 与花瓣中的辅色素相互作用形成蓝色的过程，将有助于在不同观赏花卉中生成蓝色。

3.3.6　蓝色、红色及黑色花色的其他植物

通过鸭跖草（*Commelina communis*）*F3′5′H* 的表达，已产生了蓝色的大丽花和蝴蝶兰，该基因在转基因植株体内的表达效率比矮牵牛 *F3′5′H* 更高（Mii 2012）。该研究的具体细节尚未公布。蓝色蝴蝶兰在日本的一些花展上被公开展出。蓝色大丽花和蓝色蝴蝶兰的花色比之前分别在第 3.3 及 3.4 节描述过的转基因月季与转基因香石竹的花色更蓝。这可能是因为在飞燕草素存在的条件下，大丽花和蝴蝶兰相比月季与香石竹具有更适宜使其变蓝的细胞环境。CaMV35 启动子驱动三色堇 *F3′5′H* 在非洲菊中表达，使非

洲菊积累了含量约 50%的飞燕草素，花色也发生了变化（Florigene 公司，结果未发表）。Qi 等（2013）将蝴蝶兰 *F3'5'H* 基因和风信子（*Hyacinthus*）*DFR* 基因在粉色百合‘sorbonne’的花被中进行了瞬时共表达，发现转基因细胞呈现紫色。

仙客来 *F3'5'H* 基因经反义技术抑制后，其花色变浅，花色苷从以飞燕草素花色苷为主变成了以矢车菊素花色苷为主，从而使其花色从紫色变为红色或粉色（Boase et al. 2010）。龙胆花的蓝色源自以聚酰化飞燕草素为主的花色苷，尤其是龙胆翠雀花素（5,3′-*di*-(6-*O*-caffeoyl β-D-glucosyl)-3-*O*-(β-D-glucosyl)delphinidin）。通过 RNAi 技术分别抑制龙胆的 *CHS* 和 *ANS* 表达，产生了白色和浅色花色的龙胆，而抑制 *F3'5'H* 则产生淡粉色花色的龙胆（Nakatsuka et al. 2008）。抑制 *5,3'-AAT* 导致龙胆的花色变浅（Nakatsuka et al. 2010）。假如龙胆的遗传背景使其更适宜产生蓝色花色，那么要想使其产生鲜红色的花色还是具有挑战性的。

深紫和深红（即所谓的黑色）色的花常会吸引消费者的兴趣。通过杂交和诱变育种，在多种观赏植物中培育出了黑色和棕色的花色品种。呈现黑色有时是因为积累了高浓度的花色苷。大丽花呈现黑色是由 *FNS II* 基因（*DvFNS*）转录沉默导致花色苷增加而黄酮物质减少而导致的（Thill et al. 2012；Deguchi et al. 2013）。

除了呈红色的花色苷之外，黄色的类胡萝卜素和绿色的叶绿素等积累也会使花瓣呈现一定的黑色。植物体内类胡萝卜素和叶绿素的代谢过程已做过详细的综述（Tanaka and Tanaka 2006；Ohmiya 2013；Yuan et al. 2015；Nisar et al. 2015）。

3.3.7 橙酮及类胡萝卜素调控的黄色和红色

在金鱼草中由 THC 合成橙酮葡萄糖苷需经过 2 个酶催化：细胞溶质 THC 4′-GT（Ono et al. 2006）和液泡金鱼草素合酶（aureusidin synthase）（Nakayama et al. 2000）。这些基因在蓝色蓝猪耳中表达导致金鱼草素 6-葡萄糖苷的积累，但是由于蓝猪耳体内花色苷的存在，看不到黄色。进一步下调 *F3H* 或 *DFR* 基因的表达则会生成淡黄色（Ono et al. 2006）。该表型在温室中通常会稳定存在，但在田间试验下则表现不稳定（Tanaka et al. 2010）。

淡红色的类胡萝卜素物质虾青素和酮类胡萝卜素也是植物代谢工程的研究对象。类胡萝卜素合成途径中 7 个基因在新铁炮百合（*Lilium* × *formolongi*）中表达使其体内生成酮类胡萝卜素，愈伤组织和叶片表现出深橙色（Azadi et al. 2010）。前人已对类胡萝卜素改造工程进行过描述（Giuliano et al. 2000，2008）。类胡萝卜素异构酶基因突变而产生的橙色金盏花中生成并积累呈红色的 5-顺式-类胡萝卜素（Kishimoto and Ohmiya 2012）。

大多数类胡萝卜素是以羟基化类胡萝卜素、叶黄素和/或叶黄素环氧化物（游离脂肪酸酯化）的形式在花瓣发育的色素母细胞中积累（Ohmiya 2011）。但花瓣中酯化步骤和高积累机制尚不明确。淡黄色番茄花中的叶黄素酯化相关基因 *PALE YELLOW PETAL 1* 已被报道（Ariizumi et al. 2014）。在很多花中，叶绿素和类胡萝卜素随着花瓣的发育而被分解代谢，同时伴随着花色苷或甜菜碱的生物合成而形成鲜艳青蓝色的花色。通过控制与菊花开白花相关的类胡萝卜素裂解双加氧酶（carotenoid cleavage dioxygenase，CmCCD4a）基因，可以实现菊花白花及黄花的分子育种（Ohmiya et al. 2006；Ohmiya et al. 2009）。有研究者对香石竹花瓣中叶绿素积累和降解相关基因进行了研究（Ohmiya 2014）。

为了获得目标花色，不仅需调控类黄酮生物合成相关基因，还需要调控参与类胡萝卜素和叶绿素代谢的基因。

3.3.8　甜菜碱调控的黄色和红色

甜菜碱和花色苷在被子植物中不能同时存在。在石竹目植物（石竹科、粟米草科、蓬粟草科除外）中，甜菜色素、洋红色甜菜红素和黄色甜菜黄素等是负责植物花色的主要色素（Brockington et al. 2015）。

一些主要的花卉作物缺乏黄色花色。甜菜黄素是一种黄色的甜菜碱，含有一个源自酪氨酸的氮原子，它可使仙人掌、紫茉莉等植物开出深黄色的花。甜菜碱在设计改变花色中的应用还是非常有限的，但也取得了一定突破性的进展：香菇（*Lentinula edodes*）的酪氨酸酶基因和紫茉莉的 DOPA 4,5-双加氧酶基因共表达，在烟草 BY2 和拟南芥 T87 的培养细胞中成功地生成了甜菜碱。由于甜菜黄素的积累，转基因烟草 BY2 细胞呈亮黄色（Nakatsuka et al. 2013）

甜菜红素是一种红色的甜菜碱，在一些植物中呈现亮红色到洋红色。红甜菜 *CYP76AD1* 基因和大花马齿苋（*Portulaca grandiflora*）*DODA1*（*PGDODA*）基因在烟草属（*Nicotiana*）、茄属（*Solanum*）和矮牵牛属（*Petunia*）等植物中共表达生成了洋红色甜菜红素。*CYP76AD6*（而不是 *CYP76AD1*）与 *PgDODA* 共表达只会导致黄色甜菜黄素的积累。*CYP76AD1*、*CYP76AD6* 和 *PgDODA* 共表达通过甜菜红素和甜菜黄素的积累而生成橙色色素（Polturak et al. 2016）。这些策略通过黄酮和花色苷的积累可能有望产生更深的黄色与红色花色。

3.3.9　荧光花

一些花卉通过诸如甜菜黄素之类的代谢物发出荧光（Gandia Herrero et al. 2005）。水母的荧光蛋白如绿色荧光蛋白（green fluorescent protein，GFP）导入花卉中也会产生荧光花卉（Mercuri et al. 2002）。来源于一种海洋浮游生物 *Chiridius poppei* 的黄绿色荧光蛋白（CpYGFP），通过利用热激蛋白终止子的优化序列和拟南芥乙醇脱氢酶基因的 5′-非翻译区，成功地在蓝猪耳中高效表达。通过刺激和滤波器可以看到这种强烈荧光（Sasaki et al. 2014）。这种蛋白质高效表达策略应该适用于改变类黄酮化合物生物合成途径，如用在菊花中生成飞燕草素。

3.3.10 花色改良的一般策略和最新技术

想要得到期望的花色必须积累特定的花色苷。要做到这一点，第一步应在该花色苷生物合成途径的分支点过度表达合适的基因（如 *F3'5'H*）。基因的表达水平主要依赖其转录效率，而转录效率又受启动子和终止子调控。翻译效率也很重要，但比转录调控作用更难理解。在起始密码子前插入一个公认的翻译增强子序列已被证明是有效的。选择一个合适的编码序列也很重要，如上述的 *F3'5'H* 和 *DFR* 等基因。

由于导入的外源基因会与内源基因之间发生竞争作用，因此最好选择缺乏该竞争途径的突变株或者通过 RNAi 以及最新基因编辑技术如 CRISPR/Cas9 来下调表达该途径中的内源竞争基因。基因编辑技术可能会永久敲除存在干扰的基因。

通过 CRISPR/Cas9 使用两个单链向导 RNA（single guide RNA）对转基因菊花植株中基因进行定点突变编辑，成功地将一个六倍体菊花中的 *CpYGFP* 基因进行了修饰敲除而使转基因菊花植株失去了荧光（Kishi-Kaboshi et al. 2017）。在多倍性较高的无性繁殖植物中，对于那些已在某些位点进行过突变诱导的植物，为了在多个位点引入突变，促进其腋芽生长是一种有效的方法（Kishi-Kaboshi et al. 2017）。

基因编辑技术尤其是 CRISPR/Cas9，近年来已引起世人极大关注。基因编辑植物监管的法律框架仍未确定，尽管被编辑过的植物与变异植物并没有太大区别，但欧盟仍持谨慎态度。如果基因编辑植物的监管负担远低于使用以往技术产生的转基因生物的监管负担，那么基因编辑技术将会得到蓬勃发展。

花朵的着色模式取决于花瓣发育过程中类黄酮生物合成过程中 R2R3- MYB 转录因子的表达，并且可以通过调节同源异型盒基因的表达而使其发生改变（Davies et al. 2012）。融合抑制基因沉默技术（chimeric REpressor gene-silencing technology，CRES-T）通过表达融合了强转录抑制域（SRDX）的转录因子使植物产生一个显著的负表型（Hirats et al. 2003）。CRES-T 对验证多余转录因子功能以及修饰观赏作物性状来说都是一种有效的手段。在蓝猪耳（*Torenia fournieri*）（Narumi et al. 2011；Shikata et al. 2011）、仙客来（*Cyclamen percisum*）（Tanaka et al. 2011，2013）以及其他一些观赏植物中，通过融合了 MADS-box 或 TCP（TEOSINTE BRANCHED1、CYCLOIDEA 和 PCF 家族）的 SRDX 的表达来改变花的形态及花器官的形状。通过改变花形以及使用不同的花器官特异性启动子，在蓝猪耳中可以诱导产生不同的色素沉着模式（Sasaki et al. 2016；Kasajima et al. 2017）。通过转录因子 GTMYB3-SRDX 的表达，会在花的花瓣中产生花边（Nakatsuka et al. 2011）。表型信息可以通过 FioreDB 数据库访问查询（http://www.cres-t.org/fiore/public_db/，Mitsuda et al. 2011）。研究表明，矮牵牛的花边和流星品种的天然双色花是由 *CHS-A* 基因的空间抑制作用产生的。这些品种的矮牵牛具有相同的由两个拷贝 *CHS-A* 基因 *PhCHS-A1* 和 *PhCHS-A2* 头尾串联而组成的小干扰 RNA（siRNA）产生位点。两个拷贝 *CHS-A* 外显子 2 区的 siRNA 引起 *CHS-A* 降解以及花色苷减少（Morita et al. 2012）。

3.3.11　花卉的遗传转化

无论选择何种基因修饰策略来改变花色，最终的实施都依赖遗传转化体系的成功应用。观赏植物中许多重要的切花、盆花以及花坛植物等的遗传转化体系已经公布，此处不一一列出。最近的报道说明月季、香石竹、菊花、百合、矮牵牛、天竺葵以及一些兰花种类等都是易于进行遗传转化的物种，这些结果令人欣慰（Noman et al. 2017；Azadi et al. 2016；Koetle et al. 2015；Milošević et al. 2015）。我们试图总结在为目标作物制定遗传转化方案时应考虑的关键点，而不是某一物种的转化细节。以下概述的要点是整体产品开发计划的一部分，这对商业化的监管审批有一定作用。观赏植物遗传转化体系的详细优化过程与其他植物一样，即对农杆菌菌株、共培养或基因枪条件、再生激素处理、培养基和培养环境等因素进行系统评估。受篇幅所限这里不对这些因素进行讨论。

进行遗传转化时植物基因型（品种或种类）的选择可能起最重要的决定作用。如之前所述，植物基因型的筛选可能会受含有一定类黄酮物质、变异或细胞 pH 以及金属离子等背景条件的实验所需植物品种的限制，如果根据这些条件进行植物种类选择的话可能又会存在一些问题，因为观赏植物可能因品种不同而再生及遗传转化能力存在很大差异。为了克服可能在制定任何一种植物的遗传转化方案时都会遇到的这种品种间差异问题，最明智的做法就是在初始的再生和遗传转化研究中尽可能选择多种不同的基因型作为实验材料。对于一些很难进行遗传转化的观赏植物（如一些单子叶和木本观赏植物种类），可先选择一个能与该目标品种进行杂交的易转化品种来进行遗传转化，然后通过杂交将外源基因从易转化品种转移到难转化的目标品种中。这种方法对一些品种可以进行遗传转化，但转化效率很低的品种也可能有效。因为用这种方法可以扩大前期遗传转化个体数，从而增大从不受无性系变异影响的再生混合体中筛选出目标性状的概率。

使用筛选标记时，筛选标记基因的选择非常重要。原因有二，首先：选择的筛选标记在杀灭曲线（kill curve）实验中可显示清晰并可重复的半数致死剂量（LD_{50}）数值。如果杀灭曲线不是指数曲线或终点不清晰，则会增加选择过程中出现偏离的机会。筛选标记基因一般在抗生素抗性基因和抗除草剂基因之间选择，这就带来第二个问题，就是尽管抗除草剂标记基因筛选效果可能会更好，但是会引起其他的监管制度问题，因为很多国家的监管机构都想了解除草剂用量增加可能带来的潜在影响。如果企图通过最终的转基因植株中抗除草剂筛选标记基因的表达来达到控制杂草的目的，则还需另外注册登记。开发人员还应该意识到某些除草剂抗性基因是受到知识产权保护的，无论其表型是否会被利用。

用于转化和再生的外植体材料在遗传转化方案制定过程中的最终确定，从最有可能再生的植物组织部位开始着手。在观赏植物中，幼嫩或者新生组织细胞具有更好的再生能力。例如，香石竹茎尖分生组织周围的细胞最易遗传转化（Lu et al. 1991）。木本观赏植物茎分生组织由于产生多酚和单宁类物质而比较难转化，因此愈伤组织或悬浮培养物可能是进行遗传转化和再生的首选或唯一的细胞来源（Katsumoto et al. 2007）。在观赏植物产业中，植物的洁净度是极其重要的，没有被列为无病原体和无病毒的植物材料是不能进行国际贸易的。因此在这里强烈建议，在外植体类型确定后，用于遗传转化的材料应来源于在无病环境下生长的索引植物。如果体外培养材料可以作为遗传转化材料的来源，那么这些脱毒材料的保存和管理则相对更加容易。

通过不定芽或体细胞胚的再生方式通常与外植体的选择直接相关，因为最直接的外植体（如茎、叶或其他组织外植体）再生方法通常是通过不定芽形成，而来自细胞培养的再生通常是通过体细胞胚发生的。有必要从以下两个方面对再生途径进行详细了解。首先，不定芽的形成尤其是通过中间愈伤组织阶段形成时，在很大程度可能会发生体细胞无性系变异。在再生观赏植物中，体细胞无性系变异通常表现为矮化、早花、叶片或花瓣白化以及花色变异等。假如需要产生更多商业上有用的转基因植物的话，体细胞无性系变异的发生还是具有重要意义的。另外，尽管体细胞胚的再生更容易由单细胞产生，但也可能由借助于不定芽或体细胞胚的多细胞再生而产生。多细胞再生也导致转基因植物存在转基因和非转基因细胞形成嵌合体的可能，如月季一样（Nakamura et al. 2011a）。尽管嵌合体植物的分子解析比较困难，但从监管的角度来看还是具有一定潜在益处的。这是因为假如嵌合体是周缘嵌合体，那么导入的外源基因在花粉中可能不会表达，这样就限制了转基因植物中外源基因漂移的发生。从监管要求的角度可能需要提供可证明转基因植物是嵌合体的原位杂交结果等实验数据。

观赏植物进行转基因的方法可能主要是通过农杆菌共培养或基因枪轰击（Azadi et al. 2016；Noman et al. 2017）。尽管转基因方法的选择可能因某些因素（如农杆菌通常难以侵染单子叶植物）而受到限制，但这些方法仍然不失为产生转基因植物的重要方法。开发方应意识到，选择农杆菌侵染在美国可能会触发不同的监管评估途径，而在其他国家，可能要通过实验证明（如通过 PCR 技术检测）所有残留的细菌都已从最终的转基因植物中去除干净。而选择基因枪法时，对插入基因拷贝数进行早期筛选是一个很重要的过程。无论采用哪种转基因方法都是如此，但对于基因枪法可能会增加转基因过程中出现多个插入位点的频率。多插入位点更难进行分子及生物信息分析和解析，因此一般选择单插入位点的转基因植株进行商品化应用。另外，从更长远的角度看，如果转基因植物用于后期的杂交育种，单插入位点的转基因植物对简化监管流程也更有利。

3.3.12 转基因花卉及其监管问题

转基因观赏植物的实验工作及环境释放程序由一些国家的及国际的法规制定。这些法规制定了处理、使用及清除这些转基因植物的规则，然而它们在欧洲以外的国家基本没有一个统一的标准。在国际上，《生物多样性公约卡塔赫纳生物安全议定书》（Cartagena Biosafety Protocol）（https://www.cbd.int/doc/legal/cartagena-protocol-en.pdf）为改性活生物体（living modified organisms，LMO，类似于转基因生物体的术语 GMO，genetically modified organism）的国际贸易提供了规则，包括为释放到环境中的所有转基因植株分配唯一的识别码。遵守相应法律规则至关重要，因为最终需要获得必要的许可和批准是产品开发过程的关键部分。在某些情况下，如需要进行详细的分子分析或比较实验的情况下，监管数据的收集可能需大量研究费用。正如最近发生在欧洲、美国、日本和澳大利亚等地的撤销与销毁未经批准的转基因橙色矮牵牛变种案例显示的那样，不遵守法规会造成非常有害的影响。

一般而言，转基因观赏植物的监管方式与转基因粮食作物的监管方式相同，只是不需要专门针对食品或动物饲料安全性评估的法规。但如果观赏植物有可能被用作食物或者是食物上的装饰用配菜如观赏辣椒，则需要对其进行食品安全性评估。对转基因生物进行食品和饲料安全性评估需要进行广泛的测试，而对于观赏植物而言这样的花费过于高昂。

另外，一般在大多数情况下，一个转基因生物需要大规模地（即商业性）释放或进口时，与将要进口的最终产品（如切花）相比，一个国家对转基因生物所需的信息量要更大。在前一种情况下，可能需要实验数据提供证据来证明转基因生物与其亲本几乎没有什么形态上的差异。监管机构则主要根据此信息来评估是否有增加基因漂移的潜在风险。

如之前所述，国际上对转基因生物的监管缺乏一个的统一标准，一些管辖区根据转基因生物的表型进行监管，而另外一些管辖区则是通过产生转基因生物的过程对其进行监管。尽管英语已被欧盟所接受，但在大多数情况下，转基因生物的批准申请还需根据管辖地而翻译成当地语言。在欧洲和日本，需对每个转基因生物进行非常详细的分子检测，而在其他国家并非如此。一些国家并不需要对进口的转基因切花进行详细的风险评估。这些立法上的差异很重要，因为许多观赏植物在进行国际贸易时，它们在一个国家生产，又出口到另一个国家。因此，在产品开发过程中必须有规划地收集一定的监管信息以满足可能种植或交易转基因生物的所有地区的需要。

在改变修饰花色时，主要表型除了筛选标记基因表达引起的变化外，还包括花中色素分布的改变。这可能是由物种本身自然生成的色素的平衡和浓度发生变化引起的，也可能是由生成了物种本身不存在的色素引起的。对这两种情况下色素分布变化潜在影响的评估表明，这种变化不会对环境造成特定的风险，也不会通过潜在的毒性或过敏源而对人类或动物的健康构成任何潜在威胁。这是因为至少到目前为止，经基因修饰后花色改变的观赏植物生成的是在自然界广泛分布的花色苷和类胡萝卜素物质。含有花色苷、类胡萝卜素、甜菜碱和黄酮醇等色素物质的、可能被用作花色修饰对象的观赏植物已具有悠久的安全使用历史。例如，经花色修饰过的香石竹和月季会产生原本它们不可自然生成的飞燕草素花色苷。而飞燕草素花色苷在其他观赏植物中广泛天然存在，在一些常见的粮食作物中也自然存在（Chandler et al. 2013；Nakamura et al. 2011a，2011b）。

在评估一种新的转基因观赏植物释放后带来的潜在影响时，需要提供一些能够证明无论在基因插入、新蛋白质表达，还是表型改变方面都不存在主要或次要影响的信息。如上所述，重点应通过亲本植物和转基因植物之间的比较来评估基因漂移可能增强的潜力。基因漂移增强的潜力很大程度上取决于物种及其生长的环境，一个环境中的观赏植物在另一个环境中有可能成为有害杂草。可以测量的性状包括如花粉和种子的产生及活力等影响生殖力的性状。植物扩散的可能性也可以通过实验来评估。特别是在欧洲，生物信息学分析是监管过程的一个重要部分，而进行生物信息学分析必须要得到转基因过程中所有插入基因的全长序列以及插入基因侧翼的基因组序列信息，分析其可能编码的所有可读框。

法规监管的成本并非微不足道，除了进行实验和分析的直接费用外，一些国家或地区还会收取一些诸如翻译等额外的费用。在欧洲，提交并验证一份鉴定协议需花费数万欧元。必须仔细评估任何一种新的转基因观赏品种的潜在商业效益，并与产品开发及监管申请所需的花费进行权衡。

3.4 结　　论

既然大多数的类黄酮生物合成相关基因已被分离并解析，并且一些主要花卉作物的遗传转化体系已建立，因此通过改变类黄酮物质结构来获得理想的花色是可行的。然而，要在花瓣中有目的地大量或者以最佳的比例（即花色苷和辅色素的比例）积累一种化合物并非一件易事。经多个糖基和酰基高度修饰的花色苷很难在异源植物中合成。通过改变寄主植物的液泡 pH 或金属离子浓度而不致其生长减缓来优化寄主植物的遗传背景仍然是一个挑战。了解目标物种的全部情况可能为解决这些困难提供了一定线索，而基因组测序可能是获得物种全部情况的第一步。花卉作物的基因组研究已落后于其他植物。迄今为止，只有少数花卉作物完成了基因组测序工作，如香石竹（Yagi et al. 2014）、2种矮牵牛（Bombarely et al. 2016）、日本牵牛花（Hoshino et al. 2016）以及蔷薇（*Rosa multiflora*）（Nakamura et al. 2018）。将最佳的基因表达系统与优化后的启动子、增强子和终止子相结合，以及使用最新的基因编辑技术，将促进花卉作物花色修饰研究领域的发展。

致谢： Yuko Fukui 绘制图 3.1。

本章译者：
马璐琳，王继华
云南省农业科学院花卉研究所，国家观赏园艺工程技术研究中心，云南省花卉育种重点实验室，昆明 650200

参 考 文 献

Aida R (2008) *Torenia fournieri* (torenia) as a model plant for transgenic studies. Plant Biotechnol 25(6):541–545. https://doi.org/10.5511/plantbiotechnology.25.541

Aida R, Ohira K, Tanaka Y, Yoshida K, Kishimoto S, Shibata M, Omiya A (2004) Efficient transgene expression in chrysanthemum, *Dendranthema grandiflorum* (Ramat.) Kitamura, by using the promoter of a gene for chrysanthemum chlorophyll-a/b-binding protein. Breeding Sci 54(1):51–58. https://doi.org/10.1270/jsbbs.54.51

Akita Y, Kitamura S, Hase Y, Narumi I, Ishizaka H, Kondo E, Kameari N, Nakayama M, Tanikawa N, Morita Y, Tanaka A (2011) Isolation and characterization of the fragment cyclamen *O*-methyltransferase involved in flower coloration. Planta 234(6):1127–1136. https://doi.org/10.1007/s00425-011-1466-0

Ariizumi T, Kishimoto S, Kakami R, Maoka T, Hirakawa H, Suzuki Y, Ozeki Y, Shirasawa K, Bernillon S, Okabe Y, Moing A, Asamizu E, Rothan C, Ohmiya A, Ezura H (2014) Identification of the carotenoid modifying gene *PALE YELLOW PETAL 1* as an essential factor in xanthophyll esterification and yellow flower pigmentation in tomato (*Solanum lycopersicum*). Plant J 79(3):453–465. https://doi.org/10.1111/tpj.12570

Azadi P, Otang NV, Chin DP, Nakamura I, Fujisawa M, Harada H, Misawa N, Mii M (2010) Metabolic engineering of *Lilium* × *formolongi* using multiple genes of the carotenoid biosynthesis pathway. Plant Biotechnol Rep 4(4):269–280. https://doi.org/10.1007/s11816-010-0147-y

Azadi P, Bagheri H, Nalousi AM, Nazari F, Chandler SF (2016) Current status and biotechnological advances in genetic engineering of ornamental plants. Biotechnol Adv 34(6):1073–1090. https://doi.org/10.1016/j.biotechadv.2016.06.006

Bashandy H, Teeri TH (2017) Genetically engineered orange petunias on the market. bioRxiv 142810. https://doi.org/10.1101/142810

Boase MR, Lewis DH, Davies KM, Marshall GB, Patel D, Schwinn KE, Deroles SC (2010) Isolation and antisense suppression of *flavonoid 3′, 5′-hydroxylase* modifies flower pigments and colour in cyclamen. BMC Plant Biol 10:107. https://doi.org/10.1186/1471-2229-10-107

Bombarely A, Moser M, Amrad A, Bao M, Bapaume L, Barry CS, Bliek M, Boersma MR, Borghi L, Bruggmann R, Bucher M, D'Agostino N, Davies K, Druege U, Dudareva N, Egea-Cortines M, Delledonne M, Fernandez-Pozo N, Franken P, Grandont L, Heslop-Harrison JS, Hintzsche J, Johns M, Koes R, Lv X, Lyons E, Malla D, Martinoia E, Mattson NS, Morel P, Mueller LA, Muhlemann J, Nouri E, Passeri V, Pezzotti M, Qi Q, Reinhardt D, Rich M, Richert-Poggeler KR, Robbins TP, Schatz MC, Schranz ME, Schuurink RC, Schwarzacher T, Spelt K, Tang H, Urbanus SL, Vandenbussche M, Vijverberg K, Villarino GH, Warner RM, Weiss J, Yue Z, Zethof J, Quattrocchio F, Sims TL, Kuhlemeier C (2016) Insight into the evolution of the Solanaceae from the parental genomes of *Petunia hybrida*. Nat Plants 2(6):16074. https://doi.org/10.1038/nplants.2016.74

Brazier-Hicks M, Evans KM, Gershater MC, Puschmann H, Steel PG, Edwards R (2009) The *C*-glycosylation of flavonoids in cereals. J Biol Chem 284(27):17926–17934. https://doi.org/10.1074/jbc.M109.009258

Brockington SF, Yang Y, Gandia-Herrero F, Covshoff S, Hibberd JM, Sage RF, Wong GKS, Moore MJ, Smith SA (2015) Lineage-specific gene radiations underlie the evolution of novel betalain pigmentation in Caryophyllales. New Phytol 207(4):1170–1180. https://doi.org/10.1111/nph.13441

Brugliera F, Tao GQ, Tems U, Kalc G, Mouradova E, Price K, Stevenson K, Nakamura N, Stacey I, Katsumoto Y, Tanaka Y, Mason JG (2013) Violet/blue chrysanthemums--metabolic engineering of the anthocyanin biosynthetic pathway results in novel petal colors. Plant Cell Physiol 54(10):1696–1710. https://doi.org/10.1093/pcp/pct110

Burbulis IE, Winkel-Shirley B (1999) Interactions among enzymes of the *Arabidopsis* flavonoid biosynthetic pathway. Proc Natl Acad Sci U S A 96(22):12929–12934

Butelli E, Titta L, Giorgio M, Mock HP, Matros A, Peterek S, Schijlen EG, Hall RD, Bovy AG, Luo J, Martin C (2008) Enrichment of tomato fruit with health-promoting anthocyanins by expression of select transcription factors. Nat Biotechnol 26(11):1301–1308. https://doi.org/10.1038/nbt.1506

Chandler SF, Senior M, Nakamura N, Tsuda S, Tanaka Y (2013) Expression of flavonoid 3′,5′-hydroxylase and acetolactate synthase genes in transgenic carnation: assessing the safety of a nonfood plant. J Agric Food Chem 61(48):11711–11720. https://doi.org/10.1021/jf4004384

Chanoca A, Kovinich N, Burkel B, Stecha S, Bohorquez-Restrepo A, Ueda T, Eliceiri KW, Grotewold E, Otegui MS (2015) Anthocyanin vacuolar inclusions form by a microautophagy mechanism. Plant Cell 27(9):2545–2559. https://doi.org/10.1105/tpc.15.00589

COGEM (2017) Unauthorised GM garden petunia varieties with orange flowers COGEM advice CGM/170522-0422May 2017 https://www.cogem.net/index.cfm/en/publications?order=relevance&q=petunia&category=advice&from=30-09-1998&to=28-04-2018&sc=fullcontent

Conn S, Zhang W, Franco C (2003) Anthocyanic vacuolar inclusions (AVIs) selectively bind acylated anthocyanins in *Vitis vinifera* L. (grapevine) suspension culture. Biotechnol Lett

25(11):835–839. https://doi.org/10.1023/A:1024028603089

Conn S, Franco C, Zhang W (2010) Characterization of anthocyanic vacuolar inclusions in *Vitis vinifera* L. cell suspension cultures. Planta 231(6):1343–1360. https://doi.org/10.1007/s00425-010-1139-4

Davies KM, Albert NW, Schwinn KE (2012) From landing lights to mimicry: the molecular regulation of flower colouration and mechanisms for pigmentation patterning. Funct Plant Biol 39(8):619–638. https://doi.org/10.1071/FP12195

Deguchi A, Ohno S, Hosokawa M, Tatsuzawa F, Doi M (2013) Endogenous post-transcriptional gene silencing of flavone synthase resulting in high accumulation of anthocyanins in black dahlia cultivars. Planta 237(5):1325–1335. https://doi.org/10.1007/s00425-013-1848-6

Elomaa P, Helariutta Y, Griesbach RJ, Kotilainen M, Seppanen P, Teeri TH (1995) Transgene inactivation in *Petunia hybrida* is influenced by the properties of the foreign gene. Mol Gen Genet 248(6):649–656. https://doi.org/10.1007/BF02191704

Forkmann G, Heller W (1999) Biosynthesis of flavonoids. In: Sankawa U (ed) Polyketides and other secondary metabolites including fatty acid and their derivatives, Comprehensive natural products chemistry, vol 1. Elsevier, Amsterdam, pp 713–748

Francisco RM, Regalado A, Ageorges A, Burla BJ, Bassin B, Eisenach C, Zarrouk O, Vialet S, Marlin T, Chaves MM, Martinoia E, Nagy R (2013) ABCC1, an ATP binding cassette protein from grape berry, transports anthocyanidin 3-*O*-glucosides. Plant Cell 25(5):1840–1854. https://doi.org/10.1105/tpc.112.102152

Fujino N, Tenma N, Waki T, Ito K, Komatsuzaki Y, Sugiyama K, Yamazaki T, Yoshida S, Hatayama M, Yamashita S, Tanaka Y, Motohashi R, Denessiouk K, Takahashi S, Nakayama T (2018) Physical interactions among flavonoid enzymes in snapdragon and torenia reveal the diversity in the flavonoid metabolon organization of different plant species. Plant J. 94(2):372–392. https://doi.org/10.1111/tpj.13864

Fukada-Tanaka S, Inagaki Y, Yamaguchi T, Saito N, Iida S (2000) Colouring-enhancing protein in blue petals. Nature 407:581. https://doi.org/10.1038/35036683

Fukui Y, Tanaka Y, Kusumi T, Iwashita T, Nomoto K (2003) A rationale for the shift in colour towards blue in transgenic carnation flowers expressing the flavonoid 3′,5′-hydroxylase gene. Phytochemistry 63(1):15–23. https://doi.org/10.1016/S0031-9422(02)00684-2

Gandia-Herrero F, Escribano J, Garcia-Carmona F (2005) Betaxanthins as pigments responsible for visible fluorescence in flowers. Planta 222(4):586–593. https://doi.org/10.1007/s00425-005-0004-3

Gion K, Suzuri R, Ishiguro K, Katsumoto Y, Tsuda S, Tanaka Y, Mouradova E, Brugliera F, Chandler S (2012) Genetic engineering of floricultural crops: modification of flower colour, flowering and shape. Acta Hortic 953:209–216. https://doi.org/10.17660/ActaHortic.2012.953.29

Giuliano G, Aquilani R, Dharmapuri S (2000) Metabolic engineering of plant carotenoids. Trends Plant Sci 5(10):406–409

Giuliano G, Tavazza R, Diretto G, Beyer P, Taylor MA (2008) Metabolic engineering of carotenoid biosynthesis in plants. Trends Biotechnol 26(3):139–145. https://doi.org/10.1016/j.tibtech.2007.12.003

Gonnet JF (2003) Origin of the color of Cv. Rhapsody in blue rose and some other so-called "blue" roses. J Agric Food Chem 51(17):4990–4994. https://doi.org/10.1021/jf0343276

He X, Li Y, Lawson D, Xie DY (2017) Metabolic engineering of anthocyanins in dark tobacco varieties. Physiol Plant 159(1):2–12. https://doi.org/10.1111/ppl.12475

Hiratsu K, Matsui K, Koyama T, Ohme-Takagi M (2003) Dominant repression of target genes by chimeric repressors that include the EAR motif, a repression domain, in *Arabidopsis*. Plant J 34(5):733–739. https://doi.org/10.1046/j.1365-313X.2003.01759.x

Hoshino A, Jayakumar V, Nitasaka E, Toyoda A, Noguchi H, Itoh T, Shin IT, Minakuchi Y, Koda Y, Nagano AJ, Yasugi M, Honjo MN, Kudoh H, Seki M, Kamiya A, Shiraki T, PA-Ohoo C, Asamizu E, Nishide H, Tanaka S, Park KI, Morita Y, Yokoyama K, Uchiyama I, Tanaka Y, Tabata S, Shinozaki K, Hayashizaki Y, Kohara Y, Suzuki Y, Sugano S, Fujiyama A, Iida S, Sakakibara Y (2016) Genome sequence and analysis of the Japanese morning glory *Ipomoea nil*. Nat Commun 7:13295. https://doi.org/10.1038/ncomms13295

Hsu YH, Tagami T, Matsunaga K, Okuyama M, Suzuki T, Noda N, Suzuki M, Shimura H (2017) Functional characterization of UDP-rhamnose-dependent rhamnosyltransferase involved in anthocyanin modification, a key enzyme determining blue coloration in *Lobelia erinus*. Plant J 89(2):325–337. https://doi.org/10.1111/tpj.13387

Hugueney P, Provenzano S, Verries C, Ferrandino A, Meudec E, Batelli G, Merdinoglu D, Cheynier V, Schubert A, Ageorges A (2009) A novel cation-dependent *O*-methyltransferase involved in anthocyanin methylation in grapevine. Plant Physiol 150(4):2057–2070. https://doi.org/10.1104/pp.109.140376

Ichino T, Fuji K, Ueda H, Takahashi H, Koumoto Y, Takagi J, Tamura K, Sasaki R, Aoki K, Shimada T, Hara-Nishimura I (2014) GFS9/TT9 contributes to intracellular membrane trafficking and flavonoid accumulation in *Arabidopsis thaliana*. Plant J 80(3):410–423. https://doi.org/10.1111/tpj.12637

Kallam K, Appelhagen I, Luo J, Albert N, Zhang H, Deroles S, Hill L, Findlay K, Andersen OM, Davies K, Martin C (2017) Aromatic decoration determines the formation of anthocyanic vacuolar inclusions. Curr Biol 27(7):945–957. https://doi.org/10.1016/j.cub.2017.02.027

Kasajima I, Ohtsubo N, Sasaki K (2017) Combination of *Cyclamen persicum* mill. Floral gene promoters and chimeric repressors for the modification of ornamental traits in *Torenia fournieri* Lind. Hortic Res 4:17008. https://doi.org/10.1038/hortres.2017.8

Katsumoto Y, Fukuchi-Mizutani M, Fukui Y, Brugliera F, Holton TA, Karan M, Nakamura N, Yonekura-Sakakibara K, Togami J, Pigeaire A, Tao GQ, Nehra NS, Lu CY, Dyson BK, Tsuda S, Ashikari T, Kusumi T, Mason JG, Tanaka Y (2007) Engineering of the rose flavonoid biosynthetic pathway successfully generated blue-hued flowers accumulating delphinidin. Plant Cell Physiol 48(11):1589–1600. https://doi.org/10.1093/pcp/pcm131

Kazuma K, Noda N, Suzuki M (2003) Flavonoid composition related to petal color in different lines of *Clitoria ternatea*. Phytochemistry 64(6):1133–1139. https://doi.org/10.1016/S0031-9422(03)00504-1

Kishi-Kaboshi M, Aida R, Sasaki K (2017) Generation of gene-edited *Chrysanthemum morifolium* using multicopy transgenes as targets and markers. Plant Cell Physiol 58(2):216–226. https://doi.org/10.1093/pcp/pcw222

Kishimoto S, Ohmiya A (2012) Carotenoid isomerase is key determinant of petal color of *Calendula officinalis*. J Biol Chem 287(1):276–285. https://doi.org/10.1074/jbc.M111.300301

Koes R, Verweij W, Quattrocchio F (2005) Flavonoids: a colorful model for the regulation and evolution of biochemical pathways. Trends Plant Sci 10(5):236–242. https://doi.org/10.1016/j.tplants.2005.03.002

Koetle MJ, Finnie JF, Balázs E, Van Staden J (2015) A review on factors affecting the *Agrobacterium*-mediated genetic transformation in ornamental monocotyledonous geophytes. S Afr J Bot 98:37–44. https://doi.org/10.1016/j.sajb.2015.02.001

Kogawa K, Kato N, Kazuma K, Noda N, Suzuki M (2007a) Purification and characterization of UDP-glucose: anthocyanin 3′,5′-*O*-glucosyltransferase from *Clitoria ternatea*. Planta 226(6):1501–1509. https://doi.org/10.1007/s00425-007-0584-1

Kogawa K, Kazuma K, Kato N, Noda N, Suzuki M (2007b) Biosynthesis of malonylated flavonoid glycosides on the basis of malonyltransferase activity in the petals of *Clitoria ternatea*. J Plant Physiol 164(7):886–894. https://doi.org/10.1016/j.jplph.2006.05.006

Kondo E, Nakayama M, Kameari N, Tanikawa N, Morita Y, Akita Y, Hase Y, Tanaka A, Ishizaka H (2009) Red-purple flower due to delphinidin 3, 5-diglucoside, a novel pigment for *Cyclamen* spp., generated by ion-beam irradiation. Plant Biotechnol 26(5):565–569. https://doi.org/10.5511/plantbiotechnology.26.565

Li Y, Provenzano S, Bliek M, Spelt C, Appelhagen I, Machado de Faria L, Verweij W, Schubert A, Sagasser M, Seidel T, Weisshaar B, Koes R, Quattrocchio F (2016) Evolution of tonoplast P-ATPase transporters involved in vacuolar acidification. New Phytol 211(3):1092–1107. https://doi.org/10.1111/nph.14008

Linn F, Heidmann I, Saedler H, Meyer P (1990) Epigenetic changes in the expression of the maize A1 gene in *Petunia hybrida*: role of numbers of integrated gene copies and state of methylation. Mol Gen Genet 222(2–3):329–336. https://doi.org/10.1007/BF00633837

Lu CY, Nugent G, Wardley-Richardson T, Chandler SF, Young R, Dalling MJ (1991) *Agrobacterium*-mediated transformation of carnation (*Dianthus caryophyllus* L.). Nat Biotechnol 9:864–868. https://doi.org/10.1038/nbt0991-864

Lucker J, Martens S, Lund ST (2010) Characterization of a *Vitis vinifera* cv. Cabernet sauvignon 3′,5′-*O*-methyltransferase showing strong preference for anthocyanins and glycosylated flavonols. Phytochemistry 71(13):1474–1484. https://doi.org/10.1016/j.phytochem.2010.05.027

Markham KR, Gould KS, Winefield CS, Mitchell KA, Bloor SJ, Boase MR (2000) Anthocyanic vacuolar inclusions-their nature and significance in flower colouration. Phytochemistry 55(4):327–336. https://doi.org/10.1016/S0031-9422(00)00246-6

Matsuba Y, Sasaki N, Tera M, Okamura M, Abe Y, Okamoto E, Nakamura H, Funabashi H, Takatsu M, Saito M, Matsuoka H, Nagasawa K, Ozeki Y (2010) A novel glucosylation reaction on anthocyanins catalyzed by acyl-glucose-dependent glucosyltransferase in the petals of carnation and delphinium. Plant Cell 22(10):3374–3389. https://doi.org/10.1105/tpc.110.077487

Mercuri A, Sacchetti A, De Benedetti L, Schiva T, Alberti S (2002) Green fluorescent flowers. Plant Sci 162(5):647–654. https://doi.org/10.1016/S0168-9452(02)00044-4

Meyer P, Heidemann I, Forkmann G, Saedler H (1987) A new petunia flower colour generated by transformation of a mutant with a maze gene. Nature 330:677–678. https://doi.org/10.1038/330677a0

Meyer P, Linn F, Heidmann I, Meyer H, Niedenhof I, Saedler H (1992) Endogenous and environmental factors influence 35S promoter methylation of a maize A1 gene construct in transgenic petunia and its colour phenotype. Mol Gen Genet 231(3):345–352. https://doi.org/10.1007/BF00292701

Mii M (2012) Ornamental plant breeding through interspecific hybridization, somatic hybridization and genetic transformation. Acta Hortic 953:43–54

Milošević S, Cingel A, Subotic A (2015) *Agrobacterium*-mediated transformation of ornamental species: a review. Genetika 47(3):1149–1164

Mitsuda N, Takiguchi Y, Shikata M, Sage-Ono K, Ono M, Sasaki K, Yamaguchi H, Narumi T, Tanaka Y, Sugiyama M, Yamamura T, Terakawa T, Gion K, Suzuri R, Tanaka Y, Nakatsuka T, Kimura S, Nishihara M, Sakai T, Endo-Onodera R, Saitoh K, Isuzugawa K, Oshima Y, Koyama T, Ikeda M, Narukawa M, Matsui K, Nakata M, Ohtsubo N, Ohme-Takagi M (2011) The new FioreDB database provides comprehensive information on plant transcription factors and phenotypes induced by CRES-T in ornamental and model plants. Plant Biotechnol 28(2):123–130. https://doi.org/10.5511/plantbiotechnology.11.0106a

Miyahara T, Takahashi M, Ozeki Y, Sasaki N (2012) Isolation of an acyl-glucose-dependent anthocyanin 7-*O*-glucosyltransferase from the monocot *Agapanthus africanus*. J Plant Physiol 169(13):1321–1326. https://doi.org/10.1016/j.jplph.2012.05.004

Miyahara T, Sakiyama R, Ozeki Y, Sasaki N (2013) Acyl-glucose-dependent glucosyltransferase catalyzes the final step of anthocyanin formation in *Arabidopsis*. J Plant Physiol 170(6):619–624. https://doi.org/10.1016/j.jplph.2012.12.001

Miyahara T, Tani T, Takahashi M, Nishizaki Y, Ozeki Y, Sasaki N (2014) Isolation of anthocyanin 7-*O*-glucosyltransferase from Canterbury bells (*Campanula medium*). Plant Biotechnol 31(5):555–559. https://doi.org/10.5511/plantbiotechnology.14.0908a

Morita Y, Saito R, Ban Y, Tanikawa N, Kuchitsu K, Ando T, Yoshikawa M, Habu Y, Ozeki Y, Nakayama M (2012) Tandemly arranged chalcone synthase a genes contribute to the spatially regulated expression of siRNA and the natural bicolor floral phenotype in *Petunia hybrida*. Plant J 70(5):739–749. https://doi.org/10.1111/j.1365-313X.2012.04908.x

Morita Y, Takagi K, Fukuchi-Mizutani M, Ishiguro K, Tanaka Y, Nitasaka E, Nakayama M, Saito N, Kagami T, Hoshino A, Iida S (2014) A chalcone isomerase-like protein enhances flavonoid production and flower pigmentation. Plant J 78(2):294–304. https://doi.org/10.1111/tpj.12469

Nakamura N, Fukuchi-Mizutani M, Suzuki K, Miyazaki K, Tanaka Y (2006) RNAi suppression of the anthocyanidin synthase gene in *Torenia hybrida* yields white flowers with higher frequency and better stability than antisense and sense suppression. Plant Biotechnol 23(1):13–18. https://doi.org/10.5511/plantbiotechnology.23.13

Nakamura N, Fukuchi-Mizutani M, Fukui Y, Ishiguro K, Suzuki K, Tanaka Y (2010) Generation of pink flower varieties from blue *Torenia hybrida* by redirection of the flavonoid pathway from delphinidin to pelargonidin. Plant Biotechnol 27(5):375–383. https://doi.org/10.5511/plantbiotechnology.10.0610a

Nakamura N, Fukuchi-Mizutani M, Katsumoto Y, Togami J, Senior M, Matsuda Y, Furuichi K, Yoshimoto M, Matsunaga A, Ishiguro K, Tanaka Y (2011a) Environmental risk assessment and field performance of rose (*Rosa* × *hybrida*) genetically modified for delphinidin production. Plant Biotechnol 28(2):251–261. https://doi.org/10.5511/plantbiotechnology.11.0113a

Nakamura N, Tems U, Fukuchi-Mizutani M, Chandler S, Matsuda Y, Takeuchi S, Matsumoto S, Tanaka Y (2011b) Molecular based evidence for a lack of gene-flow between *Rosa* × *hybrida* and wild *Rosa* species in Japan. Plant Biotechnol 28(2):245–250. https://doi.org/10.5511/plantbiotechnology.10.1217a

Nakamura N, Katsumoto Y, Brugliera F, Demelis L, Nakajima D, Suzuki H, Tanaka Y (2015) Flower color modification in *Rosa hybrida* by expressing the *S*-adenosylmethionine: anthocyanin 3′,5′-*O*-methyltransferase gene from *Torenia hybrida*. Plant Biotechnol 32(2):109–117. https://doi.org/10.5511/plantbiotechnology.15.0205a

Nakamura N, Hirakawa H, Sato S, Otagaki S, Matsumoto S, Tabata S, Tanaka Y (2018) Genome structure of *Rosa multiflora*, a wild ancestor of cultivated roses. DNA Research 25(2):113–121. https://doi.org/10.1093/dnares/dsx042

Nakatsuka T, Nishihara M, Mishiba K, Yamamura S (2006) Heterologous expression of two gentian cytochrome P450 genes can modulate the intensity of flower pigmentation in transgenic tobacco plants. Mol Breed 17(2):91–99. https://doi.org/10.1007/s11032-005-2520-z

Nakatsuka T, Abe Y, Kakizaki Y, Yamamura S, Nishihara M (2007) Production of red-flowered plants by genetic engineering of multiple flavonoid biosynthetic genes. Plant Cell Rep 26(11):1951–1959. https://doi.org/10.1007/s00299-007-0401-0

Nakatsuka T, Mishiba K, Abe Y, Kubota A, Kakizaki Y, Yamamura S, Nishihara M (2008) Flower color modification of gentian plants by RNAi-mediated gene silencing. Plant Biotechnol 25(1):61–68. https://doi.org/10.5511/plantbiotechnology.25.61

Nakatsuka T, Mishiba K, Kubota A, Abe Y, Yamamura S, Nakamura N, Tanaka Y, Nishihara M (2010) Genetic engineering of novel flower colour by suppression of anthocyanin modification genes in gentian. J Plant Physiol 167(3):231–237. https://doi.org/10.1016/j.jplph.2009.08.007

Nakatsuka T, Saito M, Yamada E, Nishihara M (2011) Production of picotee-type flowers in Japanese gentian by CRES-T. Plant Biotechnol 28(2):173–180. https://doi.org/10.5511/plantbiotechnology.10.1101b

Nakatsuka T, Yamada E, Takahashi H, Imamura T, Suzuki M, Ozeki Y, Tsujimura I, Saito M, Sakamoto Y, Sasaki N, Nishihara M (2013) Genetic engineering of yellow betalain pigments beyond the species barrier. Sci Rep 3:1970. https://doi.org/10.1038/srep01970

Nakayama T, Yonekura-Sakakibara K, Sato T, Kikuchi S, Fukui Y, Fukuchi-Mizutani M, Ueda T, Nakao M, Tanaka Y, Kusumi T, Nishino T (2000) Aureusidin synthase: a polyphenol oxidase homolog responsible for flower coloration. Science 290(5494):1163–1166. https://doi.org/10.1126/science.290.5494.1163

Napoli C, Lemieux C, Jorgensen R (1990) Introduction of a chimeric chalcone synthase gene into petunia results in reversible co-suppression of homologous genes in trans. Plant Cell 2(4):279–289. https://doi.org/10.1105/tpc.2.4.279

Narumi T, Aida R, Koyama T, Yamaguchi H, Sasaki K, Shikata M, Nakayama M, Ohme-Takagi M, Ohtsubo N (2011) *Arabidopsis* chimeric TCP3 repressor produces novel floral traits in *Torenia fournieri* and *Chrysanthemum morifolium*. Plant Biotechnol 28(2):131–140. https://doi.org/10.5511/plantbiotechnology.11.0121a

Nisar N, Li L, Lu S, Khin NC, Pogson BJ (2015) Carotenoid metabolism in plants. Mol Plant 8(1):68–82. https://doi.org/10.1016/j.molp.2014.12.007

Nishihara M, Nakatsuka T (2011) Genetic engineering of flavonoid pigments to modify flower color in floricultural plants. Biotechnol Lett 33(3):433–441. https://doi.org/10.1007/s10529-010-0461-z

Nishizaki Y, Yasunaga M, Okamoto E, Okamoto M, Hirose Y, Yamaguchi M, Ozeki Y, Sasaki N (2013) *p*-Hydroxybenzoyl-glucose is a zwitter donor for the biosynthesis of 7-polyacylated anthocyanin in *Delphinium*. Plant Cell 25(10):4150–4165. https://doi.org/10.1105/tpc.113.113167

Nishizaki Y, Sasaki N, Yasunaga M, Miyahara T, Okamoto E, Okamoto M, Hirose Y, Ozeki Y (2014) Identification of the glucosyltransferase gene that supplies the *p*-hydroxybenzoyl-glucose for 7-polyacylation of anthocyanin in delphinium. J Exp Bot 65(9):2495–2506. https://doi.org/10.1093/jxb/eru134

Noda N, Kazuma K, Sasaki T, Kogawa K, Suzuki M (2006) Molecular cloning of 1-*O*-acylglucose dependent anthocyanin aromatic acyltransferase in ternatin biosynthesis of butterfly pea (*Clitoria ternatea*). Plant Cell Physiol 47(S109). https://doi.org/10.14841/jspp.2006.0.341.0

Noda N, Aida R, Kishimoto S, Ishiguro K, Fukuchi-Mizutani M, Tanaka Y, Ohmiya A (2013) Genetic engineering of novel bluer-colored chrysanthemums produced by accumulation of delphinidin-based anthocyanins. Plant Cell Physiol 54(10):1684–1695. https://doi.org/10.1093/pcp/pct111

Noda N, Yoshioka S, Kishimoto S, Nakayama M, Douzono M, Tanaka Y, Aida R (2017) Generation of blue chrysanthemums by anthocyanin B-ring hydroxylation and glucosylation and its coloration mechanism. Sci Adv in press 3(7):e1602785. https://doi.org/10.1126/sciadv.1602785

Noman A, Aqeel M, Deng J, Khalid N, Sanaullah T, Shuilin H (2017) Biotechnological advancements for improving floral attributes in ornamental plants. Front Plant Sci 8:530. https://doi.org/10.3389/fpls.2017.00530

Ohmiya A (2011) Diversity of carotenoid composition in flower petals. Japan Agric Res Q 45(2):163–171. https://doi.org/10.6090/jarq.45.163
Ohmiya A (2013) Qualitative and quantitative control of carotenoid accumulation in flower petals. Sci Hortic 163:10–19. https://doi.org/10.1016/j.scienta.2013.06.018
Ohmiya A, Kishimoto S, Aida R, Yoshioka S, Sumitomo K (2006) Carotenoid cleavage dioxygenase (CmCCD4a) contributes to white color formation in chrysanthemum petals. Plant Physiol 142(3):1193–1201. https://doi.org/10.1104/pp.106.087130
Ohmiya A, Sumitomo K, Aida R (2009) "Yellow Jimba": suppression of carotenoid cleavage dioxygenase (CmCCD4a) expression turns white chrysanthemum petals yellow. J Japan Soc Hort Sci 78(4):450–455. https://doi.org/10.2503/jjshs178.450
Ohmiya A, Hirashima M, Yagi M, Tanase K, Yamamizo C (2014) Identification of genes associated with chlorophyll accumulation in flower petals. PLoS One 9:e113738. https://doi.org/10.1371/journal.pone.0113738
Okamura M, Nakayama M, Umemoto N, Cano EA, Hase Y, Nishizaki Y, Sasaki N, Ozeki Y (2013) Crossbreeding of a metallic color carnation and diversification of the peculiar coloration by ion-beam irradiation. Euphytica 191(1):45–56. https://doi.org/10.1007/s10681-012-0859-x
Ono E, Fukuchi-Mizutani M, Nakamura N, Fukui Y, Yonekura-Sakakibara K, Yamaguchi M, Nakayama T, Tanaka T, Kusumi T, Tanaka Y (2006) Yellow flowers generated by expression of the aurone biosynthetic pathway. Proc Natl Acad Sci U S A 103(29):11075–11080. https://doi.org/10.1073/pnas.0604246103
Oud JSN, Schneiders H, Kool AJ, van Grinsven MQJM (1995) Breeding of transgenic orange *Petunia hybrida* varieties. Euphytica 84:175–181. https://doi.org/10.1007/978-94-011-0357-2_49
Polturak G, Breitel D, Grossman N, Sarrion-Perdigones A, Weithorn E, Pliner M, Orzaez D, Granell A, Rogachev I, Aharoni A (2016) Elucidation of the first committed step in betalain biosynthesis enables the heterologous engineering of betalain pigments in plants. New Phytol 210(1):269–283. https://doi.org/10.1111/nph.13796
Pourcel L, Irani NG, Lu Y, Riedl K, Schwartz S, Grotewold E (2010) The formation of anthocyanic vacuolar inclusions in *Arabidopsis thaliana* and implications for the sequestration of anthocyanin pigments. Mol Plant 3(1):78–90. https://doi.org/10.1093/mp/ssp071
Provenzano S, Spelt C, Hosokawa S, Nakamura N, Brugliera F, Demelis L, Geerke DP, Schubert A, Tanaka Y, Quattrocchio F, Koes R (2014) Genetics and evolution of anthocyanin methylation. Plant Physiol 165(3):962–977. https://doi.org/10.1104/pp.113.234526
Qi Y, Lou Q, Quan Y, Liu Y, Wang Y (2013) Flower-specific expression of the *Phalaenopsis* flavonoid 3′,5′-hydoxylase modifies flower color pigmentation in *Petunia* and *Lilium*. Plant Cell Tissue Organ Cult 115(2):263–273. https://doi.org/10.1007/s11240-013-0359-2
Rausher MD (2006) The evolution of flavonoids and their genes. In: Grotewold E (ed) The science of flavonoids. Springer, New York, pp 177–212
Saito N, Abe K, Honda T, Timberlake CF, Bridle P (1985) Acylated delphinidin glucosides and flavonols from *Clitoria ternatea*. Phytochemistry 24(7):1583–1586
Sasaki N, Nakayama T (2015) Achievements and perspectives in biochemistry concerning anthocyanin modification for blue flower coloration. Plant Cell Physiol 56(1):28–40. https://doi.org/10.1093/pcp/pcu097
Sasaki K, Kato K, Mishima H, Furuichi M, Waga I, Takane K, Yamaguchi H, Ohtsubo N (2014) Generation of fluorescent flowers exhibiting strong fluorescence by combination of fluorescent protein from marine plankton and recent genetic tools in *Torenia fournieri* Lind. Plant Biotechnol 31(4):309–318. https://doi.org/10.5511/plantbiotechnology.14.0907a
Sasaki N, Nishizaki Y, Yamada E, Tatsuzawa F, Nakatsuka T, Takahashi H, Nishihara M (2015) Identification of the glucosyltransferase that mediates direct flavone *C*-glucosylation in *Gentiana triflora*. FEBS Lett 589(1):182–187. https://doi.org/10.1016/j.febslet.2014.11.045
Sasaki K, Yamaguchi H, Kasajima I, Narumi T, Ohtsubo N (2016) Generation of novel floral traits using a combination of floral organ-specific promoters and a chimeric repressor in *Torenia fournieri* Lind. Plant Cell Physiol 57(6):1319–1331. https://doi.org/10.1093/pcp/pcw081
Shikata M, Narumi T, Yamaguchi H, Sasaki K, Aida R, Oshima Y, Takiguchi Y, Ohme-Takagi M, Mitsuda N, Ohtsubo N (2011) Efficient production of novel floral traits in torenia by collective transformation with chimeric repressors of *Arabidopsis* transcription factors. Plant Biotechnol 28(2):189–199. https://doi.org/10.5511/plantbiotechnology.10.1216a
Takamura T, Mizuoka Y, Kage T (2011) Mechanism of blue coloration in cyclamen (in Japanese). Hort Res (Japan) S2:281
Takeda K (2006) Blue metal complex pigments involved in blue flower color. Proc Jpn Acad Ser B Phys Biol Sci 82(4):142–154. https://doi.org/10.2183/pjab.82.142

Takeda K, Fujii A, Senda Y, Iwashina T (2010) Greenish blue flower colour of Strongylodon macrobotrys. Biochem Syst Ecol 38(4):630–633. https://doi.org/10.1016/j.bse.2010.07.014

Tanaka Y, Brugliera F (2013) Flower colour and cytochromes P450. Philos Trans R Soc Lond B Biol Sci 368(1612):20120432. https://doi.org/10.1098/rstb.2012.0432

Tanaka Y, Ohmiya A (2008) Seeing is believing : engineering anthocyanin and carotenoid biosynthetic pathways. Curr Opin Biotechnol 19(23):190–197. https://doi.org/10.1016/j.copbio.2008.02.015

Tanaka A, Tanaka R (2006) Chlorophyll metabolism. Curr Opin Plant Biol 9:248–255. https://doi.org/10.1016/j.pbi.2006.03.011

Tanaka Y, Sasaki N, Ohmiya A (2008) Plant pigments for coloration: anthocyanins, betalains and carotenoids. Plant J 54(4):733–749. https://doi.org/10.1111/j.1365-313X.2008.03447.x

Tanaka Y, Brugliera F, Kalc G, Senior M, Dyson B, Nakamura N, Katsumoto Y, Chandler S (2010) Flower color modification by engineering of the flavonoid biosynthetic pathway: practical perspectives. Biosci Biotechnol Biochem 74(9):1760–1769. https://doi.org/10.1271/bbb.100358

Tanaka Y, Yamamura T, Oshima Y, Mitsuda N, Koyama T, Ohme-Takagi M, Terakawa T (2011) Creating ruffled flower petals in *Cyclamen persicum* by expression of the chimeric cyclamen TCP repressor. Plant Biotechnol 28(2):141–147. https://doi.org/10.5511/plantbiotechnology.10.1227a

Tanaka Y, Oshima Y, Yamamura T, Sugiyama M, Mitsuda N, Ohtsubo N, Ohme-Takagi M, Terakawa T (2013) Multi-petal cyclamen flowers produced by *AGAMOUS* chimeric repressor expression. Sci Rep 3:2641. https://doi.org/10.1038/srep02641

Terahara N, Saito N, Honda T, Toki K, Osajima Y (1990) Structure of ternatin A1, the largest ternatin in the major blue anthocyanins from *Clitoria ternatea* flowers. Tetrahedron Lett 31(20):2921–2924. https://doi.org/10.1016/0040-4039(90)80185-O

Terahara N, Toki K, Saito N, Honda T, Matsui T, Osajima Y (1998) Eight new anthocyanins, Ternatins C1–C5 and D3 and Preternatins A3 and C4 from young *Clitoria ternatea* flowers. J Nat Prod 61(11):1361–1367. https://doi.org/10.1021/np980160c

Thill J, Miosic S, Ahmed R, Schlangen K, Muster G, Stich K, Halbwirth H (2012) 'Le Rouge et le Noir': a decline in flavone formation correlates with the rare color of black dahlia (*Dahlia variabilis* hort.) flowers. BMC Plant Biol 12:225. https://doi.org/10.1186/1471-2229-12-225

Tsuda S, Fukui Y, Nakamura N, Katsumoto Y, Yonekura-Sakakibara K, Fukuchi-Mizutani M, Ohira K, Ueyama Y, Ohkawa H, Holton TA, Kusumi T, Tanaka Y (2004) Flower color modification of *Petunia hybrida* commercial varieties by metabolic engineering. Plant Biotechnol 21(5):377–386. https://doi.org/10.5511/plantbiotechnology.21.377

Ueyama Y, Katsumoto Y, Fukui Y, Fukuchi-Mizutani M, Ohkawa H, Kusumi T, Iwashita T, Tanaka Y (2006) Molecular characterization of the flavonoid biosynthetic pathway and flower color modification of *Nierembergia* sp. Plant Biotechnol 23(1):19–24. https://doi.org/10.5511/plantbiotechnology.23.19

Umemoto N, Abe Y, Cano EA, Okumyra M, Sasaki N, Yoshida S (2014) Carnation serine carboxypeptidase-like acyltransferase is important for anthocyanin malyltransferase activity and formation of anthocyanic vacuolar inclusions. Paper presented at the XXVIIth international conference on polyphenols (ICP2014), Nagoya, Japan

van der Krol AR, Mur LA, Beld M, Mol JN, Stuitje AR (1990a) Flavonoid genes in petunia: addition of a limited number of gene copies may lead to a suppression of gene expression. Plant Cell 2(4):291–299. https://doi.org/10.1105/tpc.2.4.291

van der Krol AR, Mur LA, de Lange P, Mol JN, Stuitje AR (1990b) Inhibition of flower pigmentation by antisense CHS genes: promoter and minimal sequence requirements for the antisense effect. Plant Mol Biol 14(4):457–466. https://doi.org/10.1007/BF00027492

Verweij W, Spelt C, Di Sansebastiano GP, Vermeer J, Reale L, Ferranti F, Koes R, Quattrocchio F (2008) An H^+ P-ATPase on the tonoplast determines vacuolar pH and flower colour. Nat Cell Biol 10(12):1456–1462. https://doi.org/10.1038/ncb1805

Waki T, Yoo D, Fujino N, Mameda R, Denessiouk K, Yamashita S, Motohashi R, Akashi T, Aoki T, Ayabe S, Takahashi S, Nakayama T (2016) Identification of protein-protein interactions of isoflavonoid biosynthetic enzymes with 2-hydroxyisoflavanone synthase in soybean (*Glycine max* (L.) Merr.). Biochem Biophys Res Commun 469(3):546–551. https://doi.org/10.1016/j.bbrc.2015.12.038

Winkel BSJ (2004) Metabolic channeling in plants. Annu Rev Plant Biol 55:85–107. https://doi.org/10.1146/annurev.arplant.55.031903.141714

Yabuya T, Nakamura M, Iwashina T, Yamaguchi M, Takehara T (1997) Anthocyanin-flavone copigmentaion in blusih purple flowers of Japanese garden iris (*Iris ensata* Thunb.). Euphytica

98(3):163–167. https://doi.org/10.1023/A:1003152813333

Yagi M, Kosugi S, Hirakawa H, Ohmiya A, Tanase K, Harada T, Kishimoto K, Nakayama M, Ichimura K, Onozaki T, Yamaguchi H, Sasaki N, Miyahara T, Nishizaki Y, Ozeki Y, Nakamura N, Suzuki T, Tanaka Y, Sato S, Shirasawa K, Isobe S, Miyamura Y, Watanabe A, Nakayama S, Kishida Y, Kohara M, Tabata S (2014) Sequence analysis of the genome of carnation (*Dianthus caryophyllus* L.). DNA Res 21(3):231–241. https://doi.org/10.1093/dnares/dst053

Yoshida K, Kawachi M, Mori M, Maeshima M, Kondo M, Nishimura M, Kondo T (2005) The involvement of tonoplast proton pumps and $Na^+(K^+)/H^+$ exchangers in the change of petal color during flower opening of morning glory, *Ipomoea tricolor* cv. Heavenly Blue. Plant Cell Physiol 46(3):407–415. https://doi.org/10.1093/pcp/pci057

Yoshida K, Mori M, Kondo T (2009) Blue flower color development by anthocyanins: from chemical structure to cell physiology. Nat Prod Rep 26(7):857–974. https://doi.org/10.1039/b800165k

Yuan H, Zhang J, Nageswaran D, Li L (2015) Carotenoid metabolism and regulation in horticultural crops. Hortic Res 2:15036. https://doi.org/10.1038/hortres.2015.36

Zhao J (2015) Flavonoid transport mechanisms: how to go, and with whom. Trends Plant Sci 20(9):576–585. https://doi.org/10.1016/j.tplants.2015.06.007

Zhao J, Dixon RA (2009) MATE transporters facilitate vacuolar uptake of epicatechin 3′-*O*-glucoside for proanthocyanidin biosynthesis in *Medicago truncatula* and *Arabidopsis*. Plant Cell 21(8):2323–2340. https://doi.org/10.1105/tpc.109.067819

第 4 章 货架期和瓶插寿命的育种与遗传研究

Heiko Mibus

摘要 大部分切花是在中美洲或东非生产，但在欧洲、美国和亚洲销售。由于生产日益集约化，盆花的运输距离也越来越远。因此，耐贮藏、运输和展销的品质对栽培观赏植物来说是一个很重要的性状。为了延长观赏寿命，人们采用了各种各样的可能并没有效果但非常昂贵的采后处理方法。因此，选育观赏寿命较长的新品种是最具可持续性的策略。

大多数杂交育种分析显示观赏寿命性状的遗传性高，这为通过有效选择来逐步改良观赏寿命性状奠定了基础。在香石竹和矮牵牛上主要通过分子遗传分析详细解析了观赏寿命性状及其相关基因。这些结果描述了观赏寿命性状的遗传特征，但在育种项目中只在一定程度上可用。通过对衰老和脱落的分子机制研究，尤其是对乙烯效应的发现，已用转基因的方法成功地提升了观赏植物的观赏寿命。基于消费者对转基因植物的接受度有限，对观赏寿命性状进行传统育种和筛选将是开发新的观赏寿命长的品种的决定性手段。可利用分子标记辅助选择（marker-assisted selection，MAS）来提升筛选技术，在一些观赏植物（如月季、菊花）的观赏寿命性状筛选中已经建立了这种技术。通过有效利用第二代测序技术（NGS），有可能开发出更多的能在育种和筛选中利用的分子标记。

关键词 切花；乙烯；寿命；观赏植物；采后处理；盆栽植物；衰老

4.1 引　　言

切花和盆花的采后寿命是指从采收（或停止生产）开始到失去观赏价值的这一段时间。该性状在切花上称为瓶插期或瓶插寿命，在盆花上被为货架期。决定该时间跨度的生理特征是花的开放和闭合、花朵和叶片的颜色变化、植株器官（如花瓣和叶片）的衰老、叶片和花朵的“弯颈”，以及植株器官的脱落。这些生理过程在植株或切花体内同时或相继进行。此外，每一个观赏种类在某种情况下，甚至每个品种都有自己不同的萎蔫模式。因此，植物不同物种间自然的观赏寿命相差很大，如从牵牛花（*Ipomoea*）的几小时到蝴蝶兰（*Phalaenopsis*）的多个月不等（Ashman and Schoen 1994）。在已观察到的观赏植物中还有这些因素的生理反应的组合。叶片黄化或花朵衰老如不与它们的脱落相关，会对观赏寿命性状有负面影响。因此，在评价观赏植物的采后观赏寿命时，必

H. Mibus (✉)
Hochschule Geisenheim University, Department of Urban Horticulture and Ornamental Plant Research, Geisenheim, Germany
e-mail: heiko.mibus-schoppe@hs-gm.de

须考虑复杂的生理反应的总体情况。

基于成本原因，欧洲、美国和日本等主要市场的大部分切花是在中美洲或者东非等地进行生产的（CBI 2016；Hübner 2015；Riisgaard and Hammer 2011）。盆花通常则仍然在销售的区域进行生产。然而，由于专业化、集约化、生产商的规模以及杂货零售商的销量增加，盆花也正在向更远的距离运输（Hübner 2015；Rikken 2010）。因此，尤其是耐贮存、运输和展售的性状是产出的切花与盆花的重要特征。

为提高切花和盆花的观赏寿命，人们采用了各种采后处理方法。开始是对采收的切花和盆花在运输前进行冷藏，使用适宜运输的包装材料，以及控制贮存、运输和销售过程中的温度与空气成分（如 CO_2、O_2 和乙烯）（Nowak and Rudnicki 1990；Reid 2004；Scariot et al. 2014）。为防止病原菌，还使用了可对设施设备、贮存间和瓶插水进行消毒的试剂（Macnish et al. 2010b；van Doorn et al. 1990；van Doorn and de Witte 1997）。为了防止因衰老和脱落诱导产生的植物激素乙烯发挥作用，用无毒性的 1-MCP（1-甲基环丙烯，一种新型乙烯受体）来代替包含了具重金属毒性的硫代硫酸银试剂（Blankenship and Dole 2003）。

然而，这些采后处理措施的收效甚微并且很昂贵，因此培育不需这些处理的新切花或盆花品种是最可持续的策略。

通常来说，所有观赏植物的观赏寿命的高变异性在品种筛选过程中即可表现出来，使得在育种中通过目标性状的选择来提升观赏寿命成为可能。此外，与新种质杂交和种内杂交育种表明即使在现存品系的基因池中观赏寿命性状仍可被改良。

基于瓶插寿命性状的复杂性，不仅其表型分析通常很昂贵，而且其遗传是多基因控制的，因此很复杂。尽管在性状筛选和澄清其遗传特性上存在上述困难，但近年来在一些重要的观赏植物[如盆栽月季、切花月季、香石竹和伽蓝菜（长寿花）]上通过长期的育种而取得胜利还是有可能的。

4.2 改良观赏寿命性状的育种策略及选择

4.2.1 种内及种间杂交改良观赏寿命性状

利用具有典型特征的杂交亲本组合获得的种内或种间杂种，其后代的观赏寿命性状表现出高度的变异性，因此可以通过遗传和育种进行改良。

改良凤仙花（*Impatiens walleriana*）观赏寿命所用有效育种策略的一个好例子是对259 个商业品种自交系进行分析，并有针对性地用于创制新杂合子（Howard et al. 2012）。在对亚洲百合的花朵寿命进行遗传分析时，研究了 10 个栽培品种及其后代群体。亲本间、后代间及后代内的衍生材料间都表现出高度明显的变异性，表明针对性的育种和筛选可能是高度有效的（van der Meulen-Muisers et al. 1999）。利用两个具有不同观赏寿命的蝴蝶兰基因型进行正反杂交培育 F_1 代杂种并鉴定其观赏寿命性状已成为可能。对分别具有 34 个和 63 个个体的两个 F_1 后代群体进行开花持续时间检测，结果表现出显著的母性效应（Vo et al. 2015）。

对于香石竹（*Dianthus caryophyllus* L.）来说，在一个育种项目中对其连续 4 代进行分析，结果表明瓶插寿命性状在一些后代中可以提升 4 倍。对不同个体的生理指标检测表明，其与内源乙烯的生成高度相关（Onozaki et al. 2006）。利用种内和种间杂交对香石竹可育性的遗传特征进行综合分析，研究了石竹（*Dianthus chinensis*）、须苞石竹（*Dianthus barbatus*）和瞿麦（*Dianthus superbus*）及其后代的花粉活力、可授性、结实率及花衰老过程（Fu et al. 2011）。

通过育种来改变花朵观赏寿命的好例子是松红梅属（*Leptospermum*）的种间杂交。在一个用松红梅（*L. scoparium*）38 个品种和同属的其他 16 个野生种进行杂交的育种项目中，种间杂交后代的瓶插寿命与所有的松红梅物种相比都得到显著提高（Bicknell 1995）。

种内和种间杂交育种的主要缺点是不能直接用于对现有品种进行改良。多数情况下，在选育其他重要的园艺性状时常需采取耗时的回交和筛选。这个问题一定程度上可以采用分子标记辅助选择（MAS）来解决。然而，只有开发出可靠并验证了与花观赏寿命相关的标记后才可行。

4.2.2　分子标记辅助选择

4.2.2.1　形态和生理标记

过去 20 年虽然已开发出一些用于重要观赏植物育种筛选的分子标记（Boxriker et al. 2017b；Carvalho et al. 2015；Spiller et al. 2010；Su et al. 2017；Zhang et al. 2013），但目前在观赏植物的实际育种应用中形态标记仍然经常用于观赏寿命性状的筛选。

特别重要的是花或花瓣的形态性状。花瓣表面的结构和厚度可影响呼吸速率并与衰老有关（Goodwin et al. 2003；Kawarada et al. 2013；Kitamura and Ueno 2015）。具较厚且坚固组织结构的花瓣能更好地抵抗灰霉菌（*Botrytis cinerea*）孢子的入侵，因此这个形态性状可以忍耐或抵抗灰霉病侵染（Hammer and Evensen 1994；Hazendonk et al. 1995；Pie and Brouwer 1993）。归因于这种对真菌病原菌忍耐性的增强，切花月季的耐运输和贮存能力以及瓶插寿命也表现出增强（Hammer and Evensen 1994；Pie and Brouwer 1993）。叶片的形态也会影响切花和盆花的观赏寿命。因此，具较厚角质和多毛的叶片通过蒸腾作用阻止水分的流失并阻碍对热和辐射胁迫产生反应（Fanourakis et al. 2011，2012，2013a，2016；Urban et al. 2002；Welker and Furuya 1994；Xue et al. 2017）。通常来说，切花月季品种的高需水性也与气孔的数量或延迟功能相关（Carvalho et al. 2015）。花枝细弱导致花头弯曲，从而降低瓶插寿命。同样，花型也能强烈地影响耐运输和贮存能力以及相应的瓶插寿命。因此，大花和饱满的花会导致花枝更明显的弯头，也就是切花月季中的“弯颈”现象（Kohl 1968；Matsushima et al. 2012；Tanigawa et al. 1999；Torre et al. 2001）。然而，饱满的花朵也有正面的影响，如高度饱满的花朵对乙烯的反应降低，这样的长寿花品种更容易销售，因为花朵不会快速凋谢。

花色在检测采收、贮存和运输过程中形成的机械损伤中起着重要作用。花粉数量减少直到完全的花粉不育可阻止传粉，从而延迟萎蔫发生（Jones and Woodson 1997；

Onozaki et al. 2006；Thongkum et al. 2015）。在一些观赏植物物种中，传粉与乙烯的产生相关，从而诱导衰老（Chapin and Jones 2007；Jones and Woodson 1997；Onozaki et al. 2006；Pech et al. 1987；Thongkum et al. 2015）。尽管百合品种‘Brindisi’授粉后乙烯产量增加，但不能表明是由乙烯引起了衰老和脱落（Pacifici et al. 2015）。

有报道讨论了切花尤其是月季的瓶插寿命与花香之间的相关性，且与花瓣结构有联系，但这些文章没有详细地研究和分析这些相关性（Barletta 1995；Spiller et al. 2010）。只有一个比较 12 个月季品种的研究（Borda et al. 2011）探究了两个性状之间的可能相关性。结果确凿地表明无法证明花香和瓶插寿命之间存在相关性（Borda et al. 2011）。到目前为止，尚未确定是否存在特定基因型的月季品种其强烈的香气和瓶插寿命降低之间具有明确的相关性。这种特定的相关性背后潜在的生理和分子机制也尚未被研究。

切花的耐贮存和运输能力以及随后的采后处理行为也取决于初生和次生成分。例如，切花花瓣和叶片的可溶性碳水化合物含量能影响瓶插寿命，根据不同的基因型或较高或较低（Ichimura et al. 2005）。因推测可溶性碳水化合物含量的增加可能会延长瓶插寿命，故研究了各种月季品种花瓣中可溶性碳水化合物的含量。尽管一些研究表明了确实存在这种联系（Ichimura et al. 2005；Nabigol et al. 2010），但争论仍在持续。例如，Marissen（2001）的研究发现 4 个月季品种的花瓣中可溶性碳水化合物的含量与瓶插寿命之间没有相关性。然而，遗传因素可能能解释这些品种间的差异，虽然并未被研究过，且现在对于月季来说，遗传因素的影响仍然存在较大的争议。

4.2.2.2 分子标记

观赏寿命性状生理连接体的分子遗传知识让我们能鉴定相关候选基因并通过单核苷酸多态性（SNP）或 SRAP 标记在育种过程中进行应用。这方面的例子与编码乙烯合成酶或乙烯信号转导因子的基因相关。这些基于候选基因进行关联分析研究的主要困难是大多数观赏寿命性状是多基因遗传的，导致其在育种过程中用于后代筛选时受遗传背景的深层次影响而难以实施。尽管如此，将由候选基因序列变异衍生来的分子标记整合到连锁图谱中从而将其整合到数量性状基因座（QTL）标记筛选中是可能的。通过这一策略，可以提升 QTL 标记和所要筛选性状之间的关联性（Carvalho et al. 2015；Spiller et al. 2010；Zhang et al. 2013）。

只有几个已发表的研究将分子标记辅助选择（MAS）应用到了观赏寿命性状的筛选中。用 QTL 标记进行切花月季气孔功能性状筛选，从而获得更好的耐运输、贮存能力以及更长的瓶插寿命（Carvalho et al. 2015）。在该分析中，将 108 个四倍体 F_1 代个体栽培在相对湿度高（85%）的环境下，然后确定叶片的相对含水量（RWC）和切花的瓶插寿命。通过应用 3 个已检测到的 QTL 标记（两个主效标记和一个微效标记），RWC 的变化率达 32%。这些 QTL 标记的辅助将会促进未来育种过程中切花瓶插寿命性状的筛选（Carvalho et al. 2015）。

此外，对于菊花（*Chrysanthemum morifolium*）的开花时间已用 SRAP 标记进行了遗传分析（Zhang et al. 2011），对花芽形成性状（Zhang et al. 2013）和耐潮湿性状（Su et al.

2017）也进行了 QTL 标记分析。

近年来，通过对一些重要的经济观赏植物物种进行第二代测序（next-generation sequencing，NGS）得到了大量的序列数据，对开发 SNP 标记做出了贡献并将持续做出贡献。因此，Boxriker 等（2017b）用转录组测序（RNA-Seq）和 cDNA 末端规模化分析（massive analysis of cDNA end，MACE）的新方法检测到许多 SNP 标记并将其应用到香石竹（*Dianthus caryophyllus* L.）的瓶插寿命研究中。他们用群分法（bulked segregant analysis，BSA）对 500 个基因型的瓶插寿命性状进行了表型分析。不过，这些筛选出的 SNP 标记仅能在两种类型中的其中一类香石竹的瓶插寿命性状筛选中进行有效应用，该研究也表明该筛选方法的可转移性是有限的（Boxriker et al. 2017b）。在红马蹄莲（*Zantedeschia rehmannii* Engl.）上，通过 Illumina® HiSeq™ 2000 测序后的转录组分析检测到近 10 000 个 SSR（single sequence repeat）标记及 7000 多个 SNP 标记，将在未来用于重要性状包括观赏寿命的筛选（Wei et al. 2016）。

利用这些新的可用技术，可产生许多能在育种和筛选过程中应用的分子标记。然而，把通过这种方法建立的分子标记辅助选择（MAS）技术转移到多种遗传类型的难度，以及对瓶插寿命性状进行表型鉴定以建立筛选方法的高成本都将在未来持续存在（详见 4.3 节）。

4.2.3　生物技术

基本上，自 20 世纪 90 年代以来利用转基因技术改良观赏植物的瓶插寿命具有较大的潜力，且已成功地得到应用（Azadi et al. 2016；Noman et al. 2017）。许多经济观赏植物已建立了转基因体系；目前主要是基于农杆菌介导的转化，只在少数兰科植物上采用了基因枪法（Azadi et al. 2016；Lutken et al. 2012；Noman et al. 2017）。

当前，绝大多数改良观赏植物观赏寿命的研究集中在对乙烯合成和乙烯感知的影响上。因此，通过反义转化将 ACC 氧化酶和 ACC 合成酶等主效酶基因沉默，目的是减少内源乙烯的产生。这种转化策略已在香石竹、矮牵牛、秋海棠和蝴蝶草（*Torenia*）上成功地实施（Aida 1998；Chen et al. 2004；Einset 1996；Iwazaki et al. 2004；Savin et al. 1995）。所有的转基因植株均表现出更好的与乙烯产生减少相关的观赏寿命。乙烯反应的转基因修饰是通过转化源自拟南芥的乙烯受体突变子（*etr1-1*）来进行的，产生对乙烯不敏感的显性表达（Bovy et al. 1999；Mibus et al. 2009；Raffeiner et al. 2009；Sanikhani et al. 2008；Sriskandarajah et al. 2007；Winkelmann et al. 2016）。在矮牵牛中，乙烯信号链中的加效基因 *PhEIN2* 和 *PhEIN3* 被用在转基因技术中，从而减少对乙烯的反应，就可以延迟衰老的开始（Ciardi et al. 2003；Shibuya et al. 2004）。通过沉默具有 ERF 转录因子功能的 *PhHDZip* 基因，*PhACS* 和 *PhACO* 的表达会降低，从而使花的寿命增加 20%（Chang et al. 2014）。

除了上述乙烯效应的影响，对在细胞分裂素合成中具有重要作用的 IPT 基因进行遗传转化并促进其表达的方法，也可推迟月季叶片衰老开始的时间（Chang et al. 2003；Khodakovskaya et al. 2005，2009；Zakizadeh et al. 2013）。

所有采用组成型启动子的转基因实验都以产生不良的副作用而结束，因此采用了组织特异性表达的和诱导型启动子。此类包括*fbp*2、*FS*19 和 *FS*26（Bovy et al. 1999；Satoh et al. 2008）等花特异性表达启动子，*pSAG*-12（Zakizadeh et al. 2013）等衰老促进启动子，或者类固醇促进启动子（Wang et al. 2013）。

考虑到修饰乙烯合成和乙烯感知的知识与多种应用实验的先进性，最近出现的新基因编辑技术，特别是 CRISPR/Cas9，为改良观赏植物的观赏寿命性状提供了极大的潜力。至今还未见在观赏植物中应用这种新技术来获得更好的观赏寿命结果的研究报道。然而，已在几种观赏植物中建立了该技术（Subburaj et al. 2016；Zhang et al. 2016），现在也开始了几个项目，那么其结果将在未来几年内知晓（Kemp et al. 2017）。

最近几十年的研究结果表明转基因策略可以显著改良观赏寿命，因此表现出巨大的潜力。然而，观赏植物虽然具有丰富的物种多样性，但导致单一个体物种的经济相关性低，这阻碍了这些技术更广泛的应用。较高的成本不仅来自有效利用这些方法，而且来自专利和许可成本以及较高的审批门槛。此外，消费者对遗传修饰植物的接受度对商业转基因品种来说至关重要。即便如此，最近揭露的在过去 10 年大量出售时未被发现的转基因矮牵牛非法市场并未引发消费者的抱怨（Bashandy and Teeri 2017）。

4.3 观赏寿命性状育种中筛选方法的评价和分级

一般来说，盆栽开花植物和切花大多是在可控制的温室条件下生产的。这种受保护的环境使植物快速生长和对基质、肥料、温度、湿度和光照等生长参数进行优化控制成为可能。因此，生产条件决定性地影响了观赏植物的产品质量。考虑到这些方面，对于新培育的品种来说，育种公司与生产者应紧密合作从而将这些生产因素整合到观赏寿命性状的筛选中。在重要经济切花和盆花的许多研究中都调查了生产、采收和贮存因素的影响。Fanourakis 等（2013b）在综述中总结和讨论了在切花月季上的这些研究。因为生产、采收和贮存因素对瓶插寿命有非常大的影响，遗传影响就有可能被这些因素的影响覆盖。因此，观赏寿命性状成功筛选所要求的生产及实验设计很重要（Poorter et al. 2012）。

4.3.1 观赏寿命性状筛选的方法

除了统一的生产条件，成功完成观赏寿命性状的表型分析需要适合的标准化测量和评价方法。大多数育种者和生产者都有开展这种类型评价的采后处理室或气候箱。Reid 和 Kofranek（1980）发表了瓶插寿命测试的第一个推荐标准。他们建议的温度为 18℃±2℃、相对湿度为 60%～70%、光照强度为 13.5μmol/(m^2·s) PAR（光合有效辐射）以及 12h 的光周期。空气中的乙烯浓度必须低于 0.05ppm[①]，CO_2 浓度为 200～400ppm。然而，比较调控观赏植物和切花观赏寿命的实际条件与评价条件时发现，除了温度，在实际中有很多条件是不一样的。在测试瓶插寿命时，光照强度从 5.3μmol/(m^2·s)（Boxriker et al. 2017b）

① 1ppm=10^{-6}

到非常高的 140μmol/(m^2·s)（Jin et al. 2006）都有。光周期则从短到 8h（Särkkä and Rita 1997）到长至 16h（Buanong et al. 2005）直至持续不停光照（Bayleyegn et al. 2012）都有。同样，瓶插寿命测试中的相对湿度记录从非常低的 20%～40%（Särkkä and Eriksson 2003）到非常高的 80%±20%（Ahmad et al. 2011）都有。在瓶插寿命测试中没有空气组成成分的标准测量方法，因此这方面几乎没有细节记载。在育种过程中要想成功获得筛选目标，需要对瓶插寿命测试设置一致的标准的环境参数。此外，仍具瓶插寿命的表象以及观赏价值的评价大多情况下是主观的，且在不同的文章中分歧很大（Fanourakis et al. 2013b）。既然观赏植物是通过零售商店或者大的批发市场，如欧洲的 Flora-Holland 交易，许多育种公司使用这些经销商发布的指导方针如评价卡（VBN 2017）来规范相应产品的质量和观赏寿命。

在一篇对香石竹瓶插寿命性状筛选的文章中，开发并检验了一种在产量和瓶插寿命测试方面都有效的随机实验设计（Boxriker et al. 2017b）。这些结果表明，对产量方面进行有效统计评价采用随机区组（行列）设计就足够了。如果没有位置影响或者梯度影响，甚至采用 α 设计进行评价也可行。在第二阶段，也就是瓶插寿命测试，证明完全随机设计比区组设计更有效，因此建议采用这种方法（Boxriker et al. 2017a）。

4.3.2　预估观赏寿命的方法和预测模型

综上所述，由于观赏寿命测试非常复杂和成本高，近年来开发并测试了一些方法，旨在有可能预测后期的观赏寿命。许多研究对后代进行表型分析时，常对其他参数如花朵大小、花瓣数量或者叶片大小进行详细说明并与观赏寿命性状相关联（Díaz et al. 2017；van der Meulen-Muisers and van Oeveren 1996，1997；Weber et al. 2005）（见 4.2.2 节）。

另一种可预测观赏寿命的方法是利用无损检测法。通过近红外光谱（near-infrared spectroscopy，NIR）测量，使确定切花枝中的碳水化合物（糖）及其他物质的状态以及后期的耐贮存能力成为可能（Lohr et al. 2017）。使用热成像摄像机测量切花月季的呼吸速率，在多重回归分析（MRA）的辅助下可以预报后期的瓶插寿命（In et al. 2016）。

然而，利用生产、运输和贮存（如通过数据记录仪收集的）中产生的数据通过数学模型计算也可预测瓶插寿命。对于切花月季来说，除了环境、生产和贮存数据，再加上形态和生理数据，已被用于开发神经网络（neural network，NN）模型。结果表明这个新模型比传统的统计学方法能输出一个更精确的瓶插寿命预报（In et al. 2009）。

Tromp 等（2012）成功开发了一个在切花月季上用从运输到抵达期间的温度-时间之和来预测瓶插寿命的模型。然而，该模型只能在 2～6℃做出精确的预测，因此仅在非常有限的范围内适用。基于同样的数据还设计了用于预测盆栽蝴蝶兰（*Phalaenopsis*）和安祖花（*Anthurium*）货架期的模型。这些模型在被作为可信的预报工具之前都必须进行进一步验证（Tromp et al. 2017）。

4.4 盆栽植物观赏寿命改良的遗传特性及育种

盆栽开花植物和观叶植物是在温室中最佳条件下栽培的，因此大型生产单位的生产条件可以依据每个基因型或品种进行调节。与切花相比，盆栽植物的优势是整体植株离开了生产地，因此在理论上来说具有长期的生命力。即便如此，贮存、运输和销售中植株内部的生物过程是十分活跃的，意味着碳水化合物要从上部的植物器官中转运到根系，从而通过叶片或花的脱落导致采后损失。因此，盆栽植物观赏寿命的最大问题包括花、花蕾及叶片的脱落，黄叶，叶和花褪色，花瓣衰老或花朵凋谢（van Doorn and Stead 1997；Woltering and van Doorn 1988）。与切花不同，盆栽植物通常不在国外的产地进行生产。然而，由于生产商的专业化和集约化以及零售商店销量的增加，即使盆栽植物也正经历长距离的运输（Rijswick 2015；Rikken 2010）。这就使得提高盆栽植物对贮存、运输和销售的适应性正成为近年来越来越重要的育种目标。

4.4.1 盆栽植物观赏寿命的遗传基础和育种

虽然许多盆栽植物的育种目标是改良观赏寿命，但这也只在一些重要的经济观赏植物的选育中才是既定步骤。目前只在少数观赏植物物种中有追踪改良观赏寿命性状育种目标的结果发表（表 4.1）。

表 4.1 提高盆栽植物货架期的主要育种方法

物种	基因型（*n*）或后代（*n*）	采后处理评估	配合力	筛选方法	参考文献
矮牵牛	自交系（11） 选择系（4） 6 个 F_1 后代群体（10） F_2 代（F_1 代自交获得）（35） 12 个回交后代群体（25）	花的观赏寿命	SCA = 1.75（m.s.） GCA = 3.81（m.s.）	离体花的寿命测定	Krahl and Randle 1999
凤仙花	自交系（259） 选择系（12） F_1 代杂合子（66）	花的观赏寿命和脱落、花瓣衰老	SCA=83.60（m.s.） GCA=2.28（m.s.）	花观赏寿命的平均值	Howard et al. 2012
蝴蝶兰	正反交的 F_1 代（34 和 63）	花蕾脱落和花的观赏寿命		花观赏寿命的平均值	Vo et al. 2015
月季	正反交的 F_1 代（233）	花蕾、花和叶片的脱落		乙烯敏感性的平均值	Ahmadi et al. 2009

注：GCA（general combining ability）：一般配合力，SCA（specific combining ability）：特殊配合力，m.s.（mean square）：均方，*n*（number of investigated genotypes）：研究的基因型数量

盆栽月季在欧洲有很大的市场关联性（Flora-Holland 2016）。由于它们可以全年生产、销售和使用，常被暴露在各种不同的环境条件下。它们在欧洲一些区域（丹麦和荷兰）的高度专业化公司中生产，因此必须耐受较长时间的运输。盆栽月季的主要育种目标是提高生产特性、调控花色和花型以及延长观赏寿命。最近几十年，盆栽月季品种的观赏寿命已在育种项目中通过目标性状选择逐步提升。尽管如此，即使在今天，具有特定性状如不一样的花色或香味的盆栽月季品种比标准花色的盆栽月季表现出较短的观赏寿命（Ahmadi et al. 2009；Müller et al. 1998，2000a，2000b，2001）。首先，深入的盆

栽月季研究在 20 年前开始，结果表明观赏寿命高度变异（Müller et al. 1998，1999，2000a）。对 Poulsen Roser A/S 公司‘天堂’系列 9 个品种和 Kordes Söhne GmbH & Co KG 公司‘Kordana’系列 6 个品种 20 天内的采后表现进行研究，同时研究内源乙烯产生及外源乙烯反应对叶片、花蕾和花脱落的影响。观赏寿命最短的品种在 12 天后即表现出完全失去采后质量；观赏寿命最长的品种在同样时间内则仅失去了 10%的采后质量（Müller et al. 1998）。其他研究也表明了这种观赏寿命存在较大的差异以及其与乙烯效应之间的紧密相关性。对由 Kordes Söhne GmbH & Co KG 公司‘Kordana’系列的两个品种通过正反交所构建的两个 F_1 后代群体进行分析，表明与两个亲本相比，一些筛选出的后代的观赏寿命得到了显著提高（Ahmadi et al. 2009）。比较在观赏寿命性状上具有最大差异的基因型对外源乙烯的反应，并测量内源乙烯的产生量，结果表明乙烯效应和反应与观赏寿命之间有高度的相关性（Ahmadi et al. 2009；Müller et al. 1998，2000a，2000b，2001）。对乙烯合成（*ASC* 和 *ACO*）和乙烯感受（*ETR*、*CTR*、*EIN*3）基因的转录分析也表明乙烯效应和盆栽月季的观赏寿命之间存在紧密关系（Müller and Stummann 2003；Müller et al. 2000b，2002，2003）。这种乙烯信号产生基因的转录与植株乙烯敏感性状之间的关联在 F_1 代的基因型分析中得到确认（Ahmadi et al. 2009）。由于观赏寿命仍然很重要，尽管近年来才有改良的盆栽月季品种，但观赏寿命长的基因型和品种已在育种项目中得到应用以获得新品种。不过，在对 F_1 代种子苗进行首次筛选时，观赏寿命的表型分析很难实施，因此该性状的变异性可能到进行观赏寿命测试时已丢失了。

目前，从经济上来说欧洲市场上最重要的盆栽花卉是蝴蝶兰（Flora-Holland 2016）。虽然蝴蝶兰早已拥有可延续几个月的花期，但也可能会发生花朵衰老、花芽和叶片的早熟脱落。这主要是由运输和销售过程中的外源乙烯或经胁迫产生的内源乙烯对花蕾和花产生危害所诱导（Favero et al. 2016；Sun et al. 2009）。受传粉诱导而产生的乙烯，同样也促进蝴蝶兰的自然衰老（Bui and O’Neill 1998；Halevy et al. 1996；Porat et al. 1994，1995）。已有报道表明乙烯在其他兰科植物如石斛（*Dendrobium*）的传粉中也以同样的方式产生影响（Ketsa and Luangsuwalai 1996；Phetsirikoon et al. 2012，2016）。研究表明除乙烯的作用外，蝴蝶兰中由衰老诱导的乙酰辅酶 A 氧化酶[Acyl-CoA oxidase（PAC01）]也会起作用（Do and Huang 1997），花瓣中碳水化合物的浓度和花的观赏寿命间也有相关性（Hou et al. 2011）。尽管有对观赏寿命性状进行筛选的育种项目，但方法和策略都未被育种者公开发表。所以，目前只有一篇发表的研究，介绍了由两个具有不同观赏寿命的蝴蝶兰基因型构建的正反交组合，并描述了 F_1 代杂合子的观赏寿命性状（Vo et al. 2015），检测了这两个分别由 34 个和 63 个 F_1 代个体构成的群体的花观赏寿命。由于仅有两个基因型被用作亲本，不可能分析其配合力。然而，明显的母系影响证据意味着通过选择亲本来进行有针对性的育种是可行的（Vo et al. 2015）。

除了盆栽植物，地被植物和阳台植物代表了另外一些重要的经济观赏植物（Flora-Holland 2016）。对于地被和阳台类型的观赏植物如天竺葵、矮牵牛、菊花和凤仙花的栽培来说，可以确定乙烯对观赏寿命的影响（van Doorn 2002；Woltering and van Doorn 1988）。矮牵牛常被用作采用生物技术澄清花朵衰老机制的模式植物。接下来，我们将介绍和探究提升地被和阳台植物观赏寿命的传统育种方法。

通过凤仙花（*Impatiens walleriana*）自交系的双列杂交研究了花朵观赏寿命的遗传特性（Howard et al. 2012）。通过分析 259 个自交系的花朵寿命和脱落，表明花朵的寿命为 3.3 天±0.4 天至 15.8 天±2.5 天。用 12 个筛选出的自交系（6 个花朵寿命长，6 个花朵寿命短）产生了 66 个杂交系，计算了一般配合力和特殊配合力。结果表明，一般配合力比特殊配合力高 37 倍，因此加性遗传在花朵寿命的遗传中有着更重要的作用。结合凤仙花观赏寿命性状的变异，表明有针对性的育种是可行的（Howard et al. 2012）。

通过 4 个基因型（2 个花朵寿命长，2 个花朵寿命短）的双列杂交研究了矮牵牛花朵观赏寿命的遗传特性。用这种方法构建了 F_1 代群体，分析了其一般配合力和特殊配合力，通过自交和回交可以再产生 F_2 代并进行分析。用这种方法，可以表明 F_1 代群体的花朵观赏性状是显著的或非显著的加性基因效应。另外，对 F_2 代和回交群体的分析却显示出意外偏差，需要进一步通过研究来确定（Krahl and Randle 1999）。

为了改良园艺性状，产生了三倍体和四倍体的芙蓉葵（*Hibiscus moscheutos* L.），一部分植株的花粉失活。这种对授粉的阻止导致了开花时间显著延长（Li and Ruter 2017）。

总之，目前已发表的结果表明已检测了的盆栽植物的后代群体在观赏寿命上表现出非常广泛的变异性（表 4.1）。这使得在育种中逐渐改良观赏寿命成为可能。然而，只有少数研究细致地分析了育种项目中观赏寿命的遗传特性，将来的项目应更集中在这方面。

4.4.2 通过生物技术改良盆栽植物的观赏寿命

基于花朵和叶片衰老、花蕾和花朵脱落的基础生理与遗传机制，最近几十年来利用生物技术辅助改良了一些盆栽植物的观赏寿命性状。绝大多数都是以乙烯合成和乙烯效应的分子机制为基础的。因此，矮牵牛、秋海棠和蝴蝶草的乙烯合成（Aida 1998；Aida et al. 1998；Einset and Kopperud 1995；Huang et al. 2007），以及乙烯对矮牵牛、风铃草（*Campanula*）、长寿花（*Kalanchoe*）、文心兰（*Oncidium*）和布加兰（*Burrageara*）的影响（Gubrium et al. 2000；Raffeiner et al. 2009；Sanikhani et al. 2008；Sriskandarajah et al. 2007；Wang et al. 2013；Winkelmann et al. 2016）等受转基因策略的抑制。

由于矮牵牛是研究花和叶衰老的分子特征的模式植物，大多数发表的关于花卉观赏寿命转基因修饰的文章也集中于这种植物。乙烯在矮牵牛花的衰老中起着关键作用这一观点早已通过授粉过程受 ACC 合成酶基因（乙烯合成中的关键基因）调控这一证据所证实（Lindstrom et al. 1999）。因此，可通过由病毒介导的基因沉默法（VIGS）对 ACC 合成酶基因进行短暂沉默（Chen et al. 2004），以及转化从花椰菜中克隆的反义 *BoACS1*（花椰菜 ACC 合成酶基因）和反义 *BoACO1*（花椰菜 ACC 氧化酶基因）序列（Huang et al. 2007）来减少乙烯的合成，以达到改良花朵观赏寿命的目的。为了产生乙烯不敏感性状，开展了多个来源于拟南芥（*Arabidopsis*）的乙烯受体基因显性突变子的遗传转化研究。Gubrium 等（2000）采用了 35S 组成型启动子，获得的转基因矮牵牛表现出更长的花朵观赏寿命，但也伴随着额外的不良性状，如根系发育力降低、花粉活力减弱、种子成熟度和种子质量变差（Clevenger et al. 2004）。利用这些乙烯不敏感转基因矮牵牛，使鉴定并描述参与花衰老的其他基因如核酸酶（*PhNUC1*）基因成为可能（Langston et al. 2005）。

为防止转 *etr1-1* 导致的乙烯不敏感所带来的副作用，通过针对性地应用地塞米松诱导启动子（dexamethasone-induced promoter）来激活 *etr1-1*（Wang et al. 2013）。为研究矮牵牛的花朵衰老，用遗传转化方法修饰乙烯信号链的 *PhEIN2* 和 *PhEIN3* 基因，使植株对乙烯的反应减弱并获得与之相关的衰老延缓性状成为可能（Ciardi et al. 2003；Shibuya et al. 2004）。通过沉默转录因子 *PhHDZip*，*PhACS* 和 *PhACO* 基因的表达降低，从而使花的观赏寿命提高 20%（Chang et al. 2014）。更进一步，用转基因方法研究了另一个通过影响矮牵牛乙烯合成从而影响花朵衰老的转录因子 *PhFBH*4（Yin et al. 2015）。

除了这些对乙烯效应的直接干扰，通过转基因方法还影响了参与衰老的其他植物激素。以衰老诱导启动子 SAG12 来转化细胞分裂素合成基因、异戊烯基转移酶基因（isopentenyl transferase，IPT），可显著提高矮牵牛花的观赏寿命，转基因植株的花观赏寿命可延长 5～10 天（Chang et al. 2003）。为防止在黑暗中贮存所造成的叶绿素降解和黄化等损害，将 *IPT* 基因结合到冷诱导启动子上并转化到矮牵牛中，用该方法，使通过冷诱导导致的细胞分裂素增加来降低转基因矮牵牛的叶片受损程度成为可能（Khodakovskaya et al. 2005）。

前面描述的 *SAG12:IPT* 组合也应用到了盆栽月季中以提高观赏寿命。乙烯处理后的转基因盆栽月季在黑暗中的细胞分裂素浓度增加了一倍且延缓了叶片的衰老（Zakizadeh et al. 2013）。

更前面提到的源自拟南芥的突变显性乙烯受体（*etr1-1*），与源自矮牵牛花的特异性表达启动子——*FLOWER BINDING PROTEIN* 1（*fbp*1）结合在一起最先被应用到香石竹中，其转化株系表现出显著延长的花朵观赏寿命（Bovy et al. 1999）。过去 20 年，这种组合被成功应用到一些盆栽植物如风铃草（Sriskandarajah et al. 2007）、长寿花（Sanikhani et al. 2008）、文心兰（Raffeiner et al. 2009）和布加兰（Winkelmann et al. 2016）的遗传转化中。结果表明 *etr1-1* 基因在花朵中特异性大量表达导致组织对乙烯不敏感，从而显著提升盆栽植物的花朵观赏寿命。

在长寿花（*Kalanchoe blossfeldiana*）中，应用野生型毛根农杆菌（*Agrobacterium rhizogenes*）转化 *rol*（根位点）基因，获得的转基因系表现出比对照植株更长的花朵观赏寿命，花对乙烯的敏感性更低，产生该表型的原因可能是 ABA 的浓度降低（Christensen and Müller 2009）。

目前已开展的转基因方法应用研究尤其表明针对观赏寿命的遗传改良是可行的。然而，结果也阐明基因型会对性状如何产生实质性的影响。此外，所有的研究也表明采用组成型启动子来调控目标基因会产生不良的作用。一些研究用特异性表达启动子可以较大程度地解决该问题。尽管如此，未来用新的基因编辑方法将某个目标基因敲除掉或是将其修饰，需对这些基因的自然启动子的调控进行精确的研究。

4.5　切花瓶插寿命改良的遗传特性和育种

出于成本考虑，欧洲、美国、日本等主要市场的大部分切花在国外如肯尼亚和哥伦比亚生产（CBI 2016；Hübner 2015；Rijswick 2015）。因此，耐贮存和运输能力是在这些地方生产的切花的至关重要的性状。近年来，由于在食品零售商店中销售的切花比例已上升到市场份额的40%以上（Hübner 2015；Rikken 2010），这就要求所有提供的切花在瓶插寿命上具有高度均一性和可靠性。通过新的营销结构，在切花生产、贮存、运输和销售中引进了完善的质量管理体系。生产后和供应链中每一步的瓶插寿命检测都有重要作用（Fanourakis et al. 2013b）。近几十年来，很明显只有特定的品种才能较好地适应在东非或中美洲等地生产且运输和销售后仍有较好的瓶插寿命。对于新切花育种项目中的育种者来说，这些性状的筛选正变得越来越重要。

4.5.1 切花瓶插寿命性状的育种和筛选

在经济上最重要的切花是月季（Flora-Holland 2016）。因此，大部分瓶插寿命测试发生在切花月季的生产、贮存、运输和销售的质量控制中。切花月季的瓶插寿命性状高度复杂，依赖于生长、采收、贮存、运输和销售中的许多环境条件（Fanourakis et al. 2013b）。此外，由于其四倍体基因组（Carvalho et al. 2015），切花月季的遗传和育种都非常具有挑战性。

通过比较在同一条件下生产的品种，许多研究评价鉴定了切花月季瓶插寿命的遗传影响（Fanourakis et al. 2012；Marissen 2001；Mortensen and Gislerod 1999；Särkkä 2002）。尽管表明基因型变异对瓶插寿命具有强烈的影响，然而遗传影响产生的原因至今并未被完全了解（Fanourakis et al. 2013a）。如本章第2.1节中已讨论过的一样，人们总是假定切花月季瓶插寿命和香味之间具有相关性（Barletta 1995；Bent 2007；Spiller et al. 2010）。至今这种相关性也没能被文献所证明，对12个切花月季品种的比较研究证明这种相关性是不存在的（Borda et al. 2011）。盆栽月季及一些切花月季品种中内源乙烯的影响导致内源乙烯的自催化诱导，可能引起花朵衰老和花朵脱落（Mor et al. 1989；Müller et al. 2000b；Xue et al. 2008b）。然而，在一些研究中也鉴定出施用乙烯后一些品种的瓶插寿命并未受到影响（Borda et al. 2011；Macnish et al. 2010a）。虽然大多数切花月季品种的瓶插寿命缩短，但内源乙烯产生或乙烯合成调控与植株对乙烯敏感性之间并无相关性（Borda et al. 2011）。除了对观赏寿命的影响，对乙烯生物合成相关基因（*RhACS* 和 *RhACO*）及乙烯转导相关基因（*RhETR*、*RhCTR*1、*RhEIN*3）的转录组分析也表明切花月季中乙烯能调控花朵的开放（Ma et al. 2006；Tan et al. 2006；Xue et al. 2008a）。

延长切花月季品种尤其是干法运输后的观赏寿命的关键是供水、再水合及蒸腾调控。因此，需要更深入地研究不同切花月季品种较好水分供应的关键特征。于是，品种的瓶插寿命长可能与更好的气孔管理（Carvalho et al. 2015；Fanourakis et al. 2012；Mortensen and Gislerod 1999；Torre et al. 2003）、再水合中更好的水分吸收（Hu et al. 1998；Spinarova et al. 2007；van Doorn 2012；van Doorn and Suiro 1996）以及木质部更优化的水分运输（Tijskens et al. 1996；van Doorn and Reid 1995）相关。通过对四倍体的月季 F_1 代群体气孔功能和 QTL 标记进行遗传分析，表明相对含水量（RWC）的变异可达32%（Carvalho et al. 2015）。

在石竹属中，花瓣的衰老受乙烯诱导，因此切花香石竹的瓶插寿命也受到直接影响

（Woodson et al. 1992）。通过对乙烯合成基因 *DcACS* 和 *DcACO* 的表达分析表明授粉后的自然衰老过程刺激了内源乙烯的产生和乙烯的自催化（Park et al. 1992；tenHave and Woltering 1997；Woltering et al. 1993；Woodson et al. 1992）。品种比较研究表明如果香石竹品种中尤其是 *DcACS*1 基因的表达减弱，则花的瓶插寿命较长（Nukui et al. 2004；Tanase et al. 2008，2011，2013，2015）。从 6 个商业香石竹品种开始，一个育种项目采用传统的杂交育种通过筛选两代后成功将香石竹的瓶插寿命平均提升了 3.6～15 天（Onozaki et al. 2001）。随之立即对 39 个品系的内源乙烯产生量和乙烯敏感程度进行测量，从而筛选乙烯产生量显著较低和乙烯反应延迟的基因型（Onozaki et al. 2001）。从这个多年的育种项目中筛选和评价鉴定到的基因型瓶插寿命长、内源乙烯产生量非常低，且对乙烯高度不敏感。即便像其他基因型一样，使用乙烯和 ACC 使乙烯产生量增加，显示乙烯自催化仍然具有活性，但并不会促进其花瓣衰老（Onozaki et al. 2015；Tanase et al. 2008）。由于并不是所有品种在花朵观赏寿命上的差异都能用内源乙烯产生或乙烯敏感性来解释，一些研究对具不同瓶插寿命的香石竹品种进行了衰老诱导基因的表达分析（Itzhaki et al. 1994；Tanase et al. 2013，2015）。结果表明许多其他因子在调控香石竹乙烯合成、乙烯反应以及花朵衰老过程中起作用。一些研究显示花瓣中的蔗糖浓度对乙烯合成有显著影响（Hoeberichts et al. 2007；Pun et al. 2016）。花瓣的萎蔫大概受细胞成分降解所激活的水解酶及其调控因子，如脂肪酶（Lip）（Hong et al. 2000；Kim et al. 1999a）、半胱氨酸蛋白酶（cysteine proteinase，CP）（Jones et al. 1995）以及半胱氨酸蛋白酶抑制子（CPIn）（Kim et al. 1999b；Sugawara et al. 2002）的触发。因此，有可能在不同基因型中研究多种受衰老影响的表达模式。在花朵衰老过程中及乙烯处理后，*DcCP1*、*DcbGal*（β-半乳糖苷酶，β-galactosidase）、*DcGST1*（谷胱甘肽转移酶，glutathione-*S*-transferase）以及 *DcLip*（Hong et al. 2000；Kim et al. 1999a；Tanase et al. 2015；Verlinden et al. 2002）等基因的转录水平增加。相反，在这些条件下 *DcCPIn* 基因的转录水平降低（Sugawara et al. 2002；Tanase et al. 2013，2015）。近年来第二代测序技术（NGS）的应用产生了大量的序列数据，从中可以获得新的候选基因和新的分子标记（Tanase et al. 2012；Villacorta-Martin et al. 2015；Yagi 2015；Yagi et al. 2014，2017）。在 Boxriker 等（2017b）最近发表的一个研究中，cDNA 末端规模化分析（MACE-massive analysis of cDNA end）等新方法被直接用于香石竹瓶插寿命性状育种目标的筛选。用群分法（bulked segregant analysis，BSA）对 500 个具 SNP 标记的基因型进行了瓶插寿命表型分析。在该研究中，近年来已证实的导致香石竹花瓣衰老的分子遗传机制在具不同瓶插寿命的基因型中再次被大量确证（Boxriker et al. 2017b）。

除了月季和香石竹这两种在瓶插寿命性状研究中作为模式植物的切花物种外，在其他切花物种上也开展了遗传分析工作，简要介绍如下。

菊花的采后寿命有差异的原因多种多样，如叶片黄化（Satoh et al. 2008）、叶片萎蔫（van Doorn and Cruz 2000；van Meeteren 1992）、花萎蔫（Adachi et al. 2000），以及盘状小花的失绿（van Geest et al. 2016）。为了遗传分类和育种，一个育种项目从对 44 个菊花品种进行瓶插寿命性状的表型分析开始，创制了一个有 381 个基因型的双亲本群体（van Geest et al. 2016）。计算品种组间（$n = 44$）瓶插寿命的皮尔逊相关系数（Pearson's correlation coefficient）得到的 r 值为 0.7（$p \leqslant 0.0001$），亚群体间（$n = 145$）的 $r = 0.67$（$p < 0.0001$）。此外，还证明了其具中等偏高的遗传力（$h^2 = 0.73$），从而打算继续该育种项目，且将来在筛选花朵衰老性状时应用 QTL 标记（van Geest et al. 2016）。

在亚洲百合（*Lilium* L.）花寿命的遗传分析中，对 63 个百合品种进行鉴定筛选时发现极显著的高变异性（$p < 0.001$）（van der Meulen-Muisers et al. 1999）。在接下来的育种项目中，利用筛选出的 10 个品种得到了 45 个后代群体并进行了分析（van der Meulen-Muisers et al. 1999）。对于瓶插寿命性状，两个亲本间、后代群体之间以及后代群体内都存在极显著差异（$p < 0.001$）。基于个体植株水平计算的广义遗传力达到了 $H^2 = 0.79$，因此可以设想对瓶插寿命较长的百合个体进行筛选应是非常有效的。更进一步，一般配合力（GCA）的影响极显著（$p < 0.001$），而预测的狭义遗传力 $h^2 = 0.74$ 也很高，（van der Meulen-Muisers et al. 1999）。总之，通过对百合瓶插寿命性状遗传力的大规模研究，表明实施有针对性的育种和筛选项目可得到非常好的效果。

改良金鱼草（*Antirrhinum majus* L.）瓶插寿命的育种项目从对两个具有不同瓶插寿命的自交系进行杂交获得 F_1 代群体开始。接下来通过对每个后代群体进行自交创制了 F_2 代至 F_5 代群体，明确了瓶插寿命具有高达 0.79～0.81（±0.06）的遗传力（Martin and Stimart 2005；Weber et al. 2005）。

非洲菊（*Gerbera hybrida* Hort.）瓶插寿命性状的育种研究在早期就进行了（Harding et al. 1981）。这些研究分析了 98 个亲本无性系，预测其广义遗传力为 $H^2 = 0.36$～0.46。检测的 3 个后代群体的瓶插寿命的狭义遗传力低至 $h^2 = 0$、0.24 和 0.38（Harding et al. 1981）。在这个最先开始的非洲菊育种项目中，开发了一个可对瓶插寿命进行有效筛选的选择指数（Harding et al. 1987）。另一个对非洲菊（*Gerbera jamesonii*）12 个亲本系和 77 个后代群体进行杂交分析的项目表明针对瓶插寿命的双列杂交的 GCA 值之间有显著的差异（De Jong and Garretsen 1985）。对一个非洲菊（*Gerbera* × *hybrid* Hort）5 × 5 双列杂交的分析表明瓶插寿命性状的广义遗传力低至 $H^2 = 0.28$，狭义遗传力为 $h^2 = 0.28$（Wernett et al. 1996a，1996b）。基于这些结果推测其瓶插寿命性状具加性基因效应，而一般基因型变异和加性遗传变异之间的差异很轻微（Wernett et al. 1996a，1996b）。

至今发表的结果表明已研究的切花品种的瓶插寿命性状在其后代群体中表现出较大的变异性（表 4.2）。这使得在将来的育种中逐渐改良瓶插寿命性状成为可能。

4.5.2 生物技术改良“瓶插寿命”的例子

所有在经济上比较重要的切花（六出花、花烛、香石竹、菊花、非洲菊、唐菖蒲、百合、洋桔梗和月季）都已建立了遗传转化体系（Azadi et al. 2016）。然而，延长瓶插寿命的生物技术方法仅在菊花（Narumi et al. 2005）和香石竹上实施过（Bovy et al. 1999；Kinouchi et al. 2006）。

在菊花的采后行为中，叶片黄化作用明显（Satoh et al. 2008）（见第 4.5.1 节）。由于这种叶片黄化在少数品种中受乙烯诱导（Doi et al. 2003），人们尝试通过转化修饰过的乙烯受体（*mDG-ERS1*）来降低叶片的乙烯敏感性。受体的序列改变后，具有诱导

表 4.2　改良切花瓶插寿命性状的主要育种方法

物种	基因型（n）	采后处理评价	遗传力（h^2，H^2）	筛选	参考文献
金鱼草	品种，F_1至F_5代自交系	花朵脱落、衰老和瓶插寿命	$h^2=0.79$～0.81	瓶插寿命平均值	Martin and Stimart 2005；Weber et al. 2005
菊花	品种、育种系（44）、F_1代群体（381）、F_1次级后代群体（145）	盘状小花失绿、碳水化合物含量	$h^2=0.734$ $h^2=0.755$	瓶插寿命平均值、颜色及碳水化合物渗出物的相对电导率测量、QTL 标记（准备中）	Van Geest et al. 2016
非洲菊	亲本无性系（98）、5×5 双列杂交	花朵衰老、“弯颈”	$H^2=0.36$ 和 0.46 $h^2=0$、0.24 和 0.38 $H^2=0.28$ $h^2=0.28$	瓶插寿命平均值	Harding et al. 1981；Wernett et al. 1996a，1996b
香石竹	品种选育项目，有几个 Fn F_1（500）后代群体	花瓣和花朵衰老、群分法（BSA）		瓶插寿命平均值和乙烯敏感性 SNP、瓶插寿命平均值、QTL 标记（准备中）	Onozaki et al. 2001，2011，2015；Boxriker et al. 2017b
百合	63 个百合品种 45 个后代	花朵寿命	$H^2=0.79$ $h^2=0.74$	瓶插寿命平均值、观赏寿命	van der Meulen-Muisers et al. 1999
月季	F_1代群体（108）	相对含水量（RWC）	F_1代中的 RWC 变异为 7%～62%	瓶插寿命平均值、相对含水量，两个主效 QTL 标记和 1 个次效标记(32%)	Carvalho et al. 2015

注：h^2（narrow-sense heritability）：狭义遗传力；H^2（broad-sense heritability）：广义遗传力；n（number of investigated genotypes）：所研究的基因型的数量

乙烯不敏感基因如拟南芥中的 *etr1-1* 显性表达的效果。为了防止产生副作用，将烟草伸长因子 11α（*EF1*α）基因的启动子与 *mDG-ERS1* 融合在一起后进行转化。获得的转基因系叶片显示出预期的乙烯敏感性降低（Narumi et al. 2005）。

对于香石竹来说，由于乙烯合成和乙烯反应直接影响瓶插寿命（如第 4.5 节所述），可通过转基因方法来修饰乙烯合成和乙烯反应。只有这样，才能通过采用编码 ACC 氧化酶和 ACC 合成酶基因反义序列的策略使乙烯合成降低，从而使转基因系的瓶插寿命翻倍（Chandler and Tanaka 2007；Kinouchi et al. 2006；Savin et al. 1995）。为减少乙烯反应，同时转化源自拟南芥的乙烯受体基因突变子（*etr1-1*）与组成型启动子 35S 和花特异性表达启动子 *PhFbp1*（见第 4.4.2 节）。施用乙烯后香石竹转基因系不再发生花朵衰老反应。而且，转基因香石竹产生更少的乙烯这一结果，可证明内源乙烯产生的自催化受到了抑制（Bovy et al. 1999）。

因此，对切花来说应用生物技术方法可改良观赏寿命。正如已在第 4.4.2 节描述的盆栽植物一样，不良作用可以通过使用组织特异性表达启动子来避免。例如，采用新的基因编辑技术可以修饰具显著特征的乙烯反应基因，因此，最大的挑战是对这些目标基因的自然启动子进行调控和修饰。

4.6　影响采后行为的病害

采后行为受到栽培过程中许多病虫害的影响。如果侵染比较明显，则该盆栽植物或切花就不能销售或采收。然而，如果害虫、真菌病害、细菌或病毒病是潜伏的未被发现，这将降低观赏寿命。因此，对承受生物因素的能力进行选择也常常会提高观赏寿命。

采收前、贮存或运输过程中受坏死性病原菌侵染是切花和盆栽植物采后损失的最大原因。至今为止，观赏植物最重要的病原菌是真菌型病原菌——灰霉菌（*Botrytis cinerea* Pers. Ex Fr）（Dik and Wubben 2004；Gleason and Helland 2003；Williamson et al. 1995，2007）。虽然这种病原菌导致了较高的采后损失，但对于最重要的切花物种（如月季、香石竹和菊花）来说，对其的抗性遗传学知之甚少。在拟南芥中病原菌和寄主植物间互作的可能解释是茉莉酸和乙烯信号的调控。产生亚麻荠素（Camalexin）可诱导对灰霉菌的抗性（Williamson et al. 2007）。

通过品种比较研究了对灰霉病易感的切花月季基因型差异以及特定品种的灰霉病发生频率（Hammer and Evensen 1994；Hazendonk et al. 1995；Vrind 2005）。较低的易感性可以用如灰霉菌菌丝体在某些品种花瓣上的穿透和生长能力降低来解释（Pie and Brouwer 1993）。因此，推测花瓣的角质层厚度通过减慢真菌菌丝体的穿透速度来降低易感性。然而在品种鉴定筛选中不可能将角质层厚度与灰霉病的易感性相关联（Hammer and Evensen 1994），也无法表明月季的抗氧化剂浓度和灰霉病抗性之间有关联（Friedman et al. 2010）。

灰霉病在非洲菊中也会导致瓶插寿命的显著降低从而产生采后损失（Kerssiers 1993）。最近对非洲菊的灰霉病抗性进行了 QTL 分析（Fu et al. 2017b），部分源自亲本的新 SNP 标记被用于对两个 F_1 代群体进行分析（Fu et al. 2016）。用整个花序、盘状小花的基部以及舌状小花的基部三个不同部位对灰霉病易感性进行了表型鉴定。以这种方式可检测到 20 个 QTL，较多的 QTL 数量和性状的高度变异性证明了应答灰霉病侵染的抗性的复杂遗传机制（Fu et al. 2017a）。

对 62 个天竺葵（*Pelargonium hortorum* L. H. Bail）品种和育种系的比较研究成功表明了易感性存在基因型差异，并从中筛选出 2 个抗性品种。然而，这种花序的易感性并不与叶片的易感性相关联。基于这些结果，推测二倍体品种比四倍体品种耐灰霉病的能力更强（Uchneat et al. 1999）。

概括来说，至今发表的结果表明这些切花的灰霉病抗性机制非常复杂。另外，这些结果也暗示着这些抗性机制只能被解释为是通过复杂的多基因遗传的，这就使得将来通过灰霉病抗性育种来筛选观赏寿命这一关键性状很困难。

4.7 结论和展望

生产、采收和贮存因素对观赏寿命性状的影响很大，以至于在建立新的育种项目时必须考虑这些影响。由此可见，对于科学研究和园艺实践来说，实验设计对观赏寿命性状的产生和评价都是非常重要的（Fanourakis et al. 2013b；Poorter et al. 2012）。

现有的已发表的结果表明研究过的大部分群体在观赏寿命性状上都表现出较高的变异性（表 4.1，表 4.2）。这为在育种项目中逐渐改良观赏寿命性状开辟了道路。正如许多杂交育种研究中显示的那样，观赏寿命性状的高度可遗传性使在育种中对其进行有效选择成为可能。然而，只有少数研究对育种项目中的观赏寿命性状进行了细致描述和相关基因的鉴定。对香石竹和矮牵牛这两个研究观赏寿命与花朵衰老的模式植物进行分子遗传学研究获得了观赏寿命性状分子生理机制尤其是乙烯效应的详细结果。然而，在观赏寿命性状选育的育种实践中这些见解只能有条件地适用。目前已开展的转基因技术研究表明观赏寿命性状的定向改良是可行的。得益于新的基因编辑技术（如CRISPR/Cas9），将来某些特定的目标基因可以被敲除或修饰。然而在那种情况下，对这些目标基因自然启动子的调控进行精确鉴定及随之的修饰将是较大的挑战。由于消费者对遗传修饰植物的接受度有限以及其在大多数国家获得批准的成本较高，需要怀疑地看待转基因品种的商业潜力。

因此，观赏寿命性状的传统育种和筛选将仍然是培育新观赏品种的决定性手段。种内和种间杂交的缺点是通过这种方法得到的后代通常不能直接应用在新品种的筛选中，意味着大多数情况下必须要采用更耗时的回交和筛选来获得其他重要的园艺性状。这个问题在某种程度上可以通过分子标记辅助选择（MAS）来解决。尽管如此，这也只在观赏寿命性状的可信标记被开发并验证后才可行。采用第二代测序技术（NGS）可以产生无数的分子标记并在育种和筛选过程中加以应用。将来，随着那些特征不太鲜明的观赏植物品种的基因组覆盖率越来越高，会有更多可用的分子标记。然而，将以这种方式建立的分子标记辅助选择技术转移到多个遗传后代还存在困难。目前，建立这种新的选育技术需要进行的观赏寿命表型分析的成本仍然很高，可通过采用新的分析方法（如红外光谱测量 NIRS）和对观赏寿命进行有效的标准化检测来进行优化。

本章译者：

蹇洪英

云南省农业科学院花卉研究所，国家观赏园艺工程技术研究中心，云南省花卉育种重点实验室，昆明 650200

参 考 文 献

Adachi M, Kawabata S, Sakiyama R (2000) Effects of temperature and stem length on changes in carbohydrate content in summer-grown cut chrysanthemums during development and senescence. Postharvest Biol Technol 20:63–70. https://doi.org/10.1016/s0925-5214(00)00106-x

Ahmad I, Joyce DC, Faragher JD (2011) Physical stem-end treatment effects on cut rose and acacia vase life and water relations. Postharvest Biol Technol 59:258–264. https://doi.org/10.1016/j.postharvbio.2010.11.001

Ahmadi N, Mibus H, Serek M (2009) Characterization of ethylene-induced organ abscission in F1 breeding lines of miniature roses (*Rosa hybrida* L.). Postharvest Biol Technol 52:260–266. https://doi.org/10.1016/j.postharvbio.2008.12.010

Aida R (1998) Gene silencing in transgenic *Torenia* and its applications for breeding. J Jpn Soci Hortic Sci 67:1200–1202

Aida R, Yoshida T, Ichimura K et al (1998) Extension of flower longevity in transgenic torenia plants incorporating ACC oxidase transgene. Plant Sci 138:91–101. https://doi.org/10.1016/s0168-9452(98)00139-3

Ashman T-J, Schoen DJ (1994) How long should flowers live? Nature 371:788–791

Azadi P, Bagheri H, Nalousi AM et al (2016) Current status and biotechnological advances in genetic engineering of ornamental plants. Biotechnol Adv 34:1073–1090. https://doi.org/10.1016/j.biotechadv.2016.06.006

Barletta A (1995) Scent makes a comeback. Flora Cul Int January: 23–25

Bashandy H, Teeri TH (2017) Genetically engineered orange petunias on the market. Planta 246:277–280. https://doi.org/10.1007/s00425-017-2722-8

Bayleyegn A, Tesfaye B, Workneh TS (2012) Effects of pulsing solution, packaging material and passive refrigeration storage system on vase life and quality of cut rose flowers. Afr J Biotechnol 11:3800–3809

Bent E (2007) Fragrance is unpredictable, but breeders undeterred. Flora Cult Int September:32–33

Bicknell R (1995) Breeding cut flower cultivars of Leptospermum using interspecific hybridisation. N. Z. J Crop Hortic Sci 23:415–421

Blankenship SM, Dole JM (2003) 1-methylcyclopropene: a review. Postharvest Biol Technol 28:1–25. https://doi.org/10.1016/s0925-5214(02)00246-6

Borda AM, Clark DG, Huber DJ et al (2011) Effects of ethylene on volatile emission and fragrance in cut roses: the relationship between fragrance and vase life. Postharvest Biol Technol 59:245–252. https://doi.org/10.1016/j.postharvbio.2010.09.008

Bovy AG, Angenent GC, Dons HJM et al (1999) Heterologous expression of the Arabidopsis etr1-1 allele inhibits the senescence of carnation flowers. Mol Breed 5:301–308. https://doi.org/10.1023/a:1009617804359

Boxriker M, Boehm R, Mohring J et al (2017a) Efficient statistical design in two-phase experiments on vase life in carnations (*Dianthus caryophyllus* L.). Postharvest Biol Technol 128:161–168. https://doi.org/10.1016/j.postharvbio.2016.12.003

Boxriker M, Boehm R, Krezdorn N et al (2017b) Comparative transcriptome analysis of vase life and carnation type in Dianthus caryophyllus L. Sci Hortic 217:61–72. https://doi.org/10.1016/j.scienta.2017.01.015

Buanong M, Mibus H, Sisler EC et al (2005) Efficacy of new inhibitors of ethylene perception in improvement of display quality of miniature potted roses (*Rosa hybrida* L.). Plant Growth Regul 47:29–38. https://doi.org/10.1007/s10725-005-1768-y

Bui AQ, O'Neill SD (1998) Three 1-aminocyclopropane-1-carboxylate synthase genes regulated by primary and secondary pollination signals in orchid flowers. Plant Physiol 116:419–428. https://doi.org/10.1104/pp.116.1.419

Carvalho DRA, Koning-Boucoiran CFS, Fanourakis D et al (2015) QTL analysis for stomatal functioning in tetraploid *Rosa* x *hybrida* grown at high relative air humidity and its implications on postharvest longevity. Mol Breed 35:172. https://doi.org/10.1007/s11032-015-0365-7

CBI (2016) CBI Trade Statistics 2016: Cut Flowers and Foliage http://www.cbi.eu/sites/default/files/market_information/researches/trade-statistics-cut-flowers-foliage-2016.pdf

Chandler S, Tanaka Y (2007) Genetic modification in floriculture. Crit Rev Plant Sci 26:169–197. https://doi.org/10.1080/07352680701429381

Chang HS, Jones ML, Banowetz GM et al (2003) Overproduction of cytokinins in petunia flowers transformed with P(SAG12)-IPT delays corolla senescence and decreases sensitivity to ethylene. Plant Physiol 132:2174–2183. https://doi.org/10.1104/pp.103.023945

Chang X, Donnelly L, Sun D et al (2014) A *Petunia* Homeodomain-Leucine Zipper Protein, PhHD-Zip, Plays an Important Role in Flower Senescence. PLoS One 9(2):e88320. https://doi.org/10.1371/journal.pone.0088320

Chapin L, Jones M (2007) Nutrient remobilization during pollination-induced corolla senescene in *Petunia*. Acta Hortic 55:181–190

Chen JC, Jiang CZ, Gookin TE et al (2004) Chalcone synthase as a reporter in virus-induced gene silencing studies of flower senescence. Plant Mol Biol 55:521–530. https://doi.org/10.1007/s11103-004-0590-7

Christensen B, Müller R (2009) *Kalanchoe blossfeldiana* transformed with rol genes exhibits improved postharvest performance and increased ethylene tolerance. Postharvest Biol Technol 51:399–406. https://doi.org/10.1016/j.postharvbio.2008.08.010

Ciardi J, Barry K, Shibuya K et al (2003) Increased flower longevity in petunia through manipulation of ethylene signaling genes. In: Vendrell MKH, Pech JC et al (eds) Biology and biotechnology of the plant hormone ethylene III vol 349 Book Series: Nato Science Series, Sub-Series I: Life And Behavioural Sciences, pp 370–372

Clevenger DJ, Barrett JE, Klee HJ et al (2004) Factors affecting seed production in transgenic ethylene-insensitive petunias. J Am Soc Hortic Sci 129:401–406

De Jong J, Garretsen F (1985) Genetic analysis of cut flower longevity in gerbera. Euphytica 34:779–784 van der Meulen-Muisers JJM, van Oeveren JC (1996) Influence of variation in plant characteristics caused by bulb weight on inflorescence and individual flower longevity of Asiatic hybrid lilies after harvest. J Am Soc Hortic Sci 121:33–36

van der Meulen-Muisers JJM, van Oeveren JC (1997) Influence of bulb stock origin, inflorescence harvest stage and postharvest evaluation conditions on cut flower longevity of Asiatic hybrid lilies. J Am Soc Hortic Sci 122:368–372

van der Meulen-Muisers JJM, van Oeveren JC, Jansen J et al (1999) Genetic analysis of postharvest flower longevity in Asiatic hybrid lilies. Euphytica 107:149–157

Díaz JMS, Jimenez-Becker S, Jamilena M (2017) A screening test for the determination of cut flower longevity and ethylene sensitivity of carnation. Hortic Sci 44:14–20. https://doi.org/10.17221/134/2015-hortsci

Dik AJ, Wubben JP (2004) Epidemiology of Botrytis cinerea diseases in greenhouses. Botrytis: Biology, Pathology and Control. Springer, Dordrecht, pp 319–333

Do YY, Huang PL (1997) Characterization of a pollination-related cDNA from *Phalaenopsis* encoding a protein which is homologous to human peroxisomal Acyl-CoA oxidase. Arch Biochem Biophys 344:295–300. https://doi.org/10.1006/abbi.1997.0212

Doi M, Nakagawa Y, Watabe S et al (2003) Ethylene-induced leaf yellowing in cut chrysanthemums (*Dendranthema grandiflora* Kitamura). J Jpn Soc Hortic Sci 72:533–535

van Doorn WG (2002) Effect of ethylene on flower abscission: a survey. Ann Bot 89:689–693. https://doi.org/10.1093/aob/mcf124

van Doorn WG (2012) Water relations of cut flowers. An update. Hortic Rev 40:55–106

van Doorn WG, Cruz P (2000) Evidence for a wounding-induced xylem occlusion in stems of cut chrysanthemum flowers. Postharvest Biol Technol 19:73–83. https://doi.org/10.1016/s0925-5214(00)00069-7

van Doorn WG, de Witte Y (1997) Sources of the bacteria involved in vascular occlusion of cut rose flowers. J Am Soc Hortic Sci 122:263–266

van Doorn WG, Reid MS (1995) Vascular occlusion in stems of cut flowers exposed to air – Role of Xylem anatomy and rates of transpiration. Physiol Plant 93:624–629. https://doi.org/10.1034/j.1399-3054.1995.930407.x

van Doorn WG, Stead AD (1997) Abscission of flowers and floral parts. J Exp Bot 48:821–837. https://doi.org/10.1093/jxb/48.4.821

van Doorn WG, Suiro V (1996) Relationship between cavitation and water uptake in rose stems. Physiol Plant 96:305–311. https://doi.org/10.1034/j.1399-3054.1996.960221.x

van Doorn W, De Witte Y, Perik R (1990) Effect of antimicrobial compounds on the number of bacteria in stems of cut rose flowers. J Appl Bacteriol 68:117–122

Einset JW (1996) Differential expression of antisense in regenerated tobacco plants transformed with an antisense version of a tomato ACC oxidase gene. Plant Cell Tissue Org Cult 46:137–141. https://doi.org/10.1007/bf00034847

Einset JW, Kopperud C (1995) Antisense ethylene genes for begonia flowers. Acta Hortic 405:190–196

Fanourakis D, Carvalho SMP, Almeida DPF et al (2011) Avoiding high relative air humidity during critical stages of leaf ontogeny is decisive for stomatal functioning. Physiol Plant 142:274–286. https://doi.org/10.1111/j.1399-3054.2011.01475.x

Fanourakis D, Carvalho SMP, Almeida DPF et al (2012) Postharvest water relations in cut rose cultivars with contrasting sensitivity to high relative air humidity during growth. Postharvest Biol Technol 64:64–73. https://doi.org/10.1016/j.postharvbio.2011.09.016

Fanourakis D, Heuvelink E, Carvalho SMP (2013a) A comprehensive analysis of the physiological and anatomical components involved in higher water loss rates after leaf development at high humidity. J Plant Physiol 170:890–898. https://doi.org/10.1016/j.jplph.2013.01.013

Fanourakis D, Pieruschka R, Savvides A et al (2013b) Sources of vase life variation in cut roses: a review. Postharvest Biol Technol 78:1–15. https://doi.org/10.1016/j.postharvbio.2012.12.001

Fanourakis D, Giday H, Li T et al (2016) Antitranspirant compounds alleviate the mild-desiccation-induced reduction of vase life in cut roses. Postharvest Biol Technol 117:110–117. https://doi.org/10.1016/j.postharvbio.2016.02.007

Favero BT, Poimenopoulou E, Himmelboe M et al (2016) Efficiency of 1-methylcyclopropene (1-MCP) treatment after ethylene exposure of mini-Phalaenopsis. Sci Hortic 211:53–59. https://doi.org/10.1016/j.scienta.2016.08.010

Flora-Holland R (2016) Facts and Figures 2016 The Netherlands. doi:http://annualreport.royalfloraholland.com/#/feiten-en-cijfers/kamerplanten?_k=nld2i9

Friedman H, Agami O, Vinokur Y et al (2010) Characterization of yield, sensitivity to *Botrytis cinerea* and antioxidant content of several rose species suitable for edible flowers. Sci Hortic 123:395–401. https://doi.org/10.1016/j.scienta.2009.09.019

Fu XP, Zhang JJ, Li F et al (2011) Effects of genotype and stigma development stage on seed set following intra- and inter-specific hybridization of *Dianthus* spp. Sci Hortic 128:490–498. https://doi.org/10.1016/j.scienta.2011.02.021

Fu Y, Esselink GD, Visser RGF et al (2016) Transcriptome Analysis of *Gerbera hybrida* Including in silico Confirmation of Defense Genes Found. Front Plant Sci 7:247. https://doi.org/10.3389/fpls.2016.00247

Fu YQ, Van Silfhout A, Shahin A et al (2017a) Genetic mapping and QTL analysis of *Botrytis* resistance in *Gerbera hybrida*. Mol Breed 37:13. https://doi.org/10.1007/s11032-017-0648-2

Fu YQ, van Silfhout A, Shahin A et al (2017b) Genetic mapping and QTL analysis of *Botrytis* resistance in *Gerbera hybrida*. Mol Breed 37(2):13. https://doi.org/10.1007/s11032-016-0617-1

van Geest G, Choi YH, Arens P et al (2016) Genotypic differences in metabolomic changes during storage induced-degreening of chrysanthemum disk florets. Postharvest Biol Technol 115:48–59. https://doi.org/10.1016/j.postharvbio.2015.12.008

Gleason ML, Helland S (2003) *Botrytis*. In: Roberts AV et al (eds) Encyclopedia of rose science. Elsevier Academic Press, Amsterdam, pp 144–148

Goodwin SM, Kolosova N, Kish CM et al (2003) Cuticle characteristics and volatile emissions of petals in *Antirrhinum majus*. Physiol Plant 117:435–443. https://doi.org/10.1034/j.1399-3054.2003.00047.x

Gubrium EK, Clevenger DJ, Clark DG et al (2000) Reproduction and horticultural performance of transgenic ethylene-insensitive petunias. J Am Soc Hortic Sci 125:277–281

Halevy AH, Porat R, Spiegelstein H et al (1996) Short-chain saturated fatty acids in the regulation of pollination-induced ethylene sensitivity of *Phalaenopsis* flowers. Physiol Plant 97:469–474

Hammer PE, Evensen KB (1994) Differences between Rose cultivars in subsceptibility to infection by *Botrytis cinerea*. Phytopathology 84:1305–1312. https://doi.org/10.1094/Phyto-84-1305

Harding J, Byrne T, Nelson R (1981) Heritability of cut-flower vase longevity in *Gerbera*. Euphytica 30:653–657

Harding J, Byrne T, Drennan D (1987) The use of a selection index to improve gerbera cut flowers. Acta Hortic 205:57–64

Hazendonk H, ten Hoope M, van der Wurff T (1995) Methods to test rose cultivars on their susceptibility to *Botrytis cinerea* during the postharvest stage. Acta Hortic 405:39–45

Hoeberichts FA, van Doorn WG, Vorst O et al (2007) Sucrose prevents up-regulation of senescence-associated genes in carnation petals. J Exp Bot 58:2873–2885. https://doi.org/10.1093/jxb/erm076

Hong YW, Wang TW, Hudak KA et al (2000) An ethylene-induced cDNA encoding a lipase expressed at the onset of senescence. Proc Natl Acad Sci USA 97:8717–8722. https://doi.org/10.1073/pnas.140213697

Hou JY, Miller WB, Chang YCA (2011) Effects of simulated dark shipping on the carbohydrate status and post-shipping performance of *Phalaenopsis*. J Am Soc Hortic Sci 136:364–371

Howard NP, Stimart D, de Leon N et al (2012) Diallel analysis of floral longevity in *Impatiens walleriana*. J Am Soc Hortic Sci 137:47–50

Hu YX, Doi M, Imanishi H (1998) Competitive water relations between leaves and flower bud during transport of cut roses. J Jpn Soc Hortic Sci 67:532–536

Huang L-C, Lai UL, Yang S-F et al (2007) Delayed flower senescence of *Petunia hybrida* plants transformed with antisense broccoli ACC synthase and ACC oxidase genes. Postharvest Biol Technol 46:47–53. https://doi.org/10.1016/j.postharvbio.2007.03.015

Hübner S (2015) International Statistics Flowers and Plants 2015. Statistical Yearbook of AIPH and Union Fleurs 63:15–22

Ichimura K, Kishimoto M, Norikoshi R et al (2005) Soluble carbohydrates and variation in vase-life of cut rose cultivars 'Delilah' and 'Sonia'. J Hortic Sci Biotechnol 80:280–286

In BC, Inamoto K, Doi M (2009) A neural network technique to develop a vase life prediction model of cut roses. Postharvest Biol Technol 52:273–278. https://doi.org/10.1016/j.postharvbio.2009.01.001

In BC, Inamoto K, Doi M et al (2016) Using thermography to estimate leaf transpiration rates in cut roses for the development of vase life prediction models. Hortic Environ Biotechnol 57:53–60. https://doi.org/10.1007/s13580-016-0117-6

Itzhaki H, Maxson JM, Woodson WR (1994) An ethylene responsive enhancer element is involved in the senescence-related expresssion of the carnation Glutathion-S-Transferase (GSTI) gene. Proc Natl Acad Sci U S A 91:8925–8929. https://doi.org/10.1073/pnas.91.19.8925

Iwazaki Y, Kosugi Y, Waki K et al (2004) Generation and ethylene production of transgenic carnations harboring ACC synthase cDNA in sense or antisense orientation. J Appl Hortic 6:67–71

Jin JS, Shan NW, Ma N et al (2006) Regulation of ascorbate peroxidase at the transcript level is involved in tolerance to postharvest water deficit stress in the cut rose (*Rosa hybrida* L.) cv. Samantha. Postharvest Biol Technol 40:236–243. https://doi.org/10.1016/j.postharbio.2006.01.014

Jones ML, Woodson WR (1997) Pollination-induced ethylene in carnation – role of stylar ethylene in corolla senescence. Plant Physiol 115:205–212

Jones ML, Larsen PB, Woodson WR (1995) Ethylene regulated expression of a carnation cysteine proteinase during flower petal senescence. Plant Mol Biol 28:505–512. https://doi.org/10.1007/bf00020397

Kawarada M, Nomura Y, Harada T et al (2013) Cloning and expression of cDNAs for biosynthesis of very-long-chain fatty acids, the precursors for Cuticular Wax Formation, in Carnation (Dianthus caryophyllus L.). Petals. J Jpn Soc Hortic Sci 82:161–169

Kemp O, Favero BT, Hegelund JN et al (2017) Modification of ethylene sensitivity in ornamentel plants using CRISPR/Cas9. Acta Hortic 1167:271–280

Kerssiers A (1993) Influence of environmental-conditions on dispersal of *Botrytis cinerea* conidia and on postharvest infection of gerbera flowers under glass. Plant Pathol 42:754–762

Ketsa S, Luangsuwalai K (1996) The relationship between 1-aminocyclopropane-1-carboxylic acid content in pollinia, ethylene production and senescence of pollinated *Dendrobium* orchid flowers. Postharvest Biol Technol 8:57–64. https://doi.org/10.1016/0925-5214(95)00053-4

Khodakovskaya M, Li Y, Li JS et al (2005) Effects of cor15a-IPT gene expression on leaf senescence in transgenic *Petunia x hybrida* and *Dendranthema x grandiflorum*. J Exp Bot 56:1165–1175. https://doi.org/10.1093/jxb/eri109

Khodakovskaya M, Vankova R, Malbeck J et al (2009) Enhancement of flowering and branching phenotype in chrysanthemum by expression of ipt under the control of a 0.821 kb fragment of the LEACO1 gene promoter. Plant Cell Rep 28:1351–1362. https://doi.org/10.1007/s00299-009-0735-x

Kim JY, Chung YS, Ok SH et al (1999a) Characterization of the full-length sequences of phospholipase A(2) induced during flower development. Biochim Biophys Acta Gene Struct and Expr 1489:389–392. https://doi.org/10.1016/s0167-4781(99)00193-1

Kim JY, Chung YS, Paek KH et al (1999b) Isolation and characterization of a cDNA encoding the cysteine proteinase inhibitor, induced upon flower maturation in carnation using suppression subtractive hybridization. Mol Cells 9:392–397

Kinouchi T, Endo R, Yamashita A et al (2006) Transformation of carnation with genes related to ethylene production and perception: towards generation of potted carnations with a longer display time. Plant Cell Tissue Org Cult 86:27–35. https://doi.org/10.1007/s11240-006-9093-3

Kitamura Y, Ueno S (2015) Inhibition of transpiration from the inflorescence extends the vase life of cut hydrangea flowers. Hortic J 84:156–160. https://doi.org/10.2503/hortj.MI-016

Kohl HC (1968) Gerberas: their culture and commercial possibilities. S Flor Nursery 28:24–26

Krahl KH, Randle WM (1999) Genetics of floral longevity in petunia. Hortscience 34:339–340

Langston BJ, Bai S, Jones ML (2005) Increases in DNA fragmentation and induction of a senescence-specific nuclease are delayed during corolla senescence in ethylene-insensitive (etr1-1) transgenic petunias. J Exp Bot 56:15–23. https://doi.org/10.1093/jxb/eri002

Li Z, Ruter JM (2017) Development and evaluation of diploid and polyploid *Hibiscus moscheutos*. Hortscience 52:676–681. https://doi.org/10.21273/hortsci11630-16

Lindstrom JT, Lei CH, Jones ML et al (1999) Accumulation of 1-aminocyclopropane-1-carboxylic acid (ACC) in petunia pollen is associated with expression of a pollen-specific ACC synthase late in development. J Am Soc Hortic Sci 124:145–151

Lohr D, Tillmann P, Druege U et al (2017) Non-destructive determination of carbohydrate reserves in leaves of ornamental cuttings by near-infrared spectroscopy (NIRS) as a key indicator for quality assessments. Biosyst Eng 158:51–63. https://doi.org/10.1016/j.biosystemseng.2017.03.005

Lutken H, Clarke JL, Müller R (2012) Genetic engineering and sustainable production of ornamentals: current status and future directions. Plant Cell Rep 31:1141–1157. https://doi.org/10.1007/s00299-012-1265-5

Ma N, Tan H, Liu X et al (2006) Transcriptional regulation of ethylene receptor and CTR genes involved in ethylene-induced flower opening in cut rose (*Rosa hybrida*) cv. Samantha. J Exp Bot 57:2763–2773. https://doi.org/10.1093/jxb/erl033
Macnish AJ, Leonard RT, Borda AM et al (2010a) Genotypic Variation in the Postharvest Performance and Ethylene Sensitivity of Cut Rose Flowers. Hortscience 45:790–796
Macnish AJ, Morris KL, de Theije A et al (2010b) Sodium hypochlorite: A promising agent for reducing Botrytis cinerea infection on rose flowers. Postharvest Biol Technol 58:262–267. https://doi.org/10.1016/j.postharvbio.2010.07.014
Marissen N (2001) Effects of pre-harvest light intensity and temperature on carbohydrate levels and vase life of cut roses. Acta Hortic 543:191–197
Martin WJ, Stimart DP (2005) Genetic analysis of advanced populations in Antirrhinum majus L. with special reference to cut flower postharvest longevity. J Am Soc Hortic Sci 130:434–441
Matsushima U, Hilger A, Graf W et al (2012) Calcium oxalate crystal distribution in rose peduncles: Non-invasive analysis by synchrotron X-ray micro-tomography. Postharvest Biol Technol 72:27–34. https://doi.org/10.1016/j.postharvbio.2012.04.013
van Meeteren U (1992) Role of air-embolism and low water temperature in water-balance of cut chrysanthemum flowers. Sci Hortic 51:275–284
Mibus H, Sriskandarajah S, Serek M (2009) Genetically modified flowering potted plants with reduced ethylene sensitivity. Acta Hortic 847:75–79
Mor Y, Johnson F, Faragher JD (1989) Preserving the quality of cold-stored rose flowers with ethylene antagonists. Hortic Sci 24:640–641
Mortensen LM, Gislerod HR (1999) Influence of air humidity and lighting period on growth, vase life and water relations of 14 rose cultivars. Sci Hortic 82:289–298. https://doi.org/10.1016/s0304-4238(99)00062-x
Müller R, Stummann BM (2003) Genetic regulation of ethylene perception and signal transduction related to flower senescence. J Food Agric Environ 1:87–94
Müller R, Andersen AS, Serek M (1998) Differences in display life of miniature potted roses (*Rosa hybrida* L.). Sci Hortic 76:59–71. https://doi.org/10.1016/s0304-4238(98)00132-0
Müller R, Stummann BM, Andersen AS et al (1999) Involvement of ABA in postharvest life of miniature potted roses. Plant Growth Regul 29:143–150. https://doi.org/10.1023/a:1006237311350
Müller R, Sisler EC, Serek M (2000a) Stress induced ethylene production, ethylene binding, and the response to the ethylene action inhibitor 1-MCP in miniature roses. Sci Hortic 83:51–59. https://doi.org/10.1016/s0304-4238(99)00099-0
Müller R, Lind-Iversen S, Stummann BM et al (2000b) Expression of genes for ethylene biosynthetic enzymes and an ethylene receptor in senescing flowers of miniature potted roses. J Hortic Sci Biotech 75:12–18
Müller R, Stummann BM, Andersen AS (2001) Comparison of postharvest properties of closely related miniature rose cultivars (*Rosa hybrida* L.). Sci Hortic 91:325–338. https://doi.org/10.1016/s0304-4238(01)00252-7
Müller R, Owen CA, Xue ZT et al (2002) Characterization of two CTR-like protein kinases in Rosa hybrida and their expression during flower senescence and in response to ethylene. J Exp Bot 53:1223–1225. https://doi.org/10.1093/jexbot/53.371.1223
Müller R, Owen CA, Xue ZT et al (2003) The transcription factor EIN3 is constitutively expressed in miniature roses with differences in postharvest life. J Hortic Sci Biotechnol 78:10–14
Nabigol A, Naderi R, Mostofi Y (2010) Variation in vase life of cut rose cultivars and soluble carbohydrates content. Acta Hortic 858:199–204
Narumi T, Aida R, Ohmiya A et al (2005) Transformation of chrysanthemum with mutated ethylene receptor genes: mDG-ERS1 transgenes conferring reduced ethylene sensitivity and characterization of the transformants. Postharvest Biol Technol 37:101–110. https://doi.org/10.1016/j.postharvbio.2005.04.008
tenHave A, Woltering EJ (1997) Ethylene biosynthetic genes are differentially expressed during carnation (D*ianthus caryophyllus* L) flower senescence. Plant Mol Biol 34:89–97. https://doi.org/10.1023/a:1005894703444
Noman A, Aqeel M, Deng JM et al (2017) Biotechnological advancements for improving floral attributes in ornamental plants. Front Plant Sci 8:15. https://doi.org/10.3389/fpls.2017.00530
Nowak J, Rudnicki RM (1990) Post harvest handling and storage of cut flowers, florist greens, and potted plants. Tiber Press, Portland, p 210
Nukui H, Kudo S, Yamashita A et al (2004) Repressed ethylene production in the gynoecium of long-lasting flowers of the carnation 'White Candle': role of the gynoecium in carnation flower senescence. J Exp Bot 55:641–650. https://doi.org/10.1093/jxb/erh081

Onozaki T, Ikeda H, Yamaguchi T (2001) Genetic improvement of vase life of carnation flowers by crossing and selection. Sci Hortic 87:107–120. https://doi.org/10.1016/s0304-4238(00)00167-9
Onozaki T, Tanikawa N, Yagi M et al (2006) Breeding of carnations (*Dianthus caryophyllus* L.) for long vase life and rapid decrease in ethylene sensitivity of flowers after anthesis. J Jpn Soc Hortic Sci 75:256–263. https://doi.org/10.2503/jjshs.75.256
Onozaki T, Yagi M, Tanase K et al (2011) Crossings and selections for six generations based on flower vase life to create lines with ethylene resistance or ultra-long vase life in carnations (*Dianthus caryophyllus* L.). J Jpn Soc Hortic Sci 80:486–498
Onozaki T, Yagi M, Tanase K (2015) Selection of carnation line 806-46b with both ultra-long vase life and ethylene resistance. Hortic J 84:58–68. https://doi.org/10.2503/hortj.MI-011
Pacifici S, Prisa D, Burchi G et al (2015) Pollination increases ethylene production in *Lilium hybrida* cv. Brindisi flowers but does not affect the time to tepal senescence or tepal abscission. J Plant Physiol 173:116–119. https://doi.org/10.1016/j.jplph.2014.08.014
Park K, Drory A, Woodson W (1992) Molecular cloning of an 1-Aminocyclopropane-1-Carboxylate Synthase from senescing Carnation flowers petals. Plant Mol Biol 18:377–386
Pech J, Latche A, Larrigaudiere C et al (1987) Control of early ethylene synthesis in pollinated petunia flowers. Plant Physiol Biochem 25:431–437
Phetsirikoon S, Ketsa S, van Doorn WG (2012) Chilling injury in *Dendrobium* inflorescences is alleviated by 1-MCP treatment. Postharvest Biol Technol 67:144–153. https://doi.org/10.1016/j.postharvbio.2011.12.016
Phetsirikoon S, Paull RE, Chen N et al (2016) Increased hydrolase gene expression and hydrolase activity in the abscission zone involved in chilling-induced abscission of *Dendrobium* flowers. Postharvest Biol Technol 117:217–229. https://doi.org/10.1016/j.postharvbio.2016.03.002
Pie K, Brouwer Y (1993) Susceptibility of cut rose flower cultivars to infections by different isolates of *Botrytis cinerea*. J Phytopathol 137:233–244
Poorter H, Fiorani F, Stitt M et al (2012) The art of growing plants for experimental purposes: a practical guide for the plant biologist Review. Funct Plant Biol 39:821–838. https://doi.org/10.1071/fp12028
Porat R, Borochov A, Halevy AH et al (1994) Pollination-induced senescence of *Phalaenopsis* petals- The wilting process, ethylene production and sensitivity to ethylene. Plant Growth Regul 15:129–136. https://doi.org/10.1007/bf00024102
Porat R, Halevy AH, Serek M et al (1995) An increase in ethylene sensitivity following pollination is the initial event triggering an increase in ethylene production and enhance senescence of *Phalaenopsis* orchid flower. Physiol Plant 93:778–784. https://doi.org/10.1034/j.1399-3054.1995.930429.x
Pun UK, Yamada T, Azuma M et al (2016) Effect of sucrose on sensitivity to ethylene and enzyme activities and gene expression involved in ethylene biosynthesis in cut carnations. Postharvest Biol Technol 121:151–158. https://doi.org/10.1016/j.postharvbio.2016.08.001
Raffeiner B, Serek M, Winkelmann T (2009) *Agrobacterium tumefaciens*-mediated transformation of *Oncidium* and *Odontoglossum* orchid species with the ethylene receptor mutant gene etr1-1. Plant Cell Tissue Org Cult 98:125–134. https://doi.org/10.1007/s11240-009-9545-7
Reid MS (2004) Handling of cut flowers for air transport IATA perishable cargo manual – flowers. http://ucce.ucdavis.edu/files/datastore/234-1373.pdf
Reid MS, Kofranek AM (1980) Recommendations for standardized vase life evaluations. Acta Hortic 113:171–173
Riisgaard L, Hammer N (2011) Prospects for labour in global value chains: labour standards in the cut flower and banana industries. Br J Ind Relat 49:168–190. https://doi.org/10.1111/j.1467-8543.2009.00744.x
Rijswick V (2015) World Floriculture Map 2015 Rabobank Industry Note #475
Rikken M (2010) The European Market for Fair and Sustainable Flowers and Plants BTC, Belgian development agency doi:http://proverde.nl/Documents/ProVerde%20-%20The%20European%20Market%20for%20Fair%20and%20Sustainable%20Flowers%20and%20Plants.pdf?x15400
Sanikhani M, Mibus H, Stummann BM et al (2008) *Kalanchoe blossfeldiana* plants expressing the Arabidopsis etr1-1 allele show reduced ethylene sensitivity. Plant Cell Rep 27:729–737. https://doi.org/10.1007/s00299-007-0493-6
Särkkä LE (2002) Effects of rest period length and forcing temperature on yield, quality and vase life of cv. Mercedes roses. Acta Agriculturae Scand Sect B-Soil Plant Sci 52:36–42. https://doi.org/10.1080/090647102320260026

Särkkä LE, Eriksson C (2003) Effects of bending and harvesting height combinations on cut rose yield in a dense plantation with high intensity lighting. Sci Hortic 98:433–447. https://doi.org/10.1016/s0304-4238(03)00065-7

Särkkä L, Rita H (1997) Significance of plant type and age, shoot characteristics and yield on the vase life of cut roses grown in winter. Acta Agriculturae Scand Sect B-Soil Plant Sci 47:118–123. https://doi.org/10.1080/09064719709362449

Satoh S, Watanabe M, Chisaka K et al (2008) Suppressed leaf senescence in chrysanthemum transformed with a mutated ethylene receptor gene mDG-ERS1(etr1-4). J Plant Biol 51:424–427

Savin KW, Baudinette SC, Graham MW et al (1995) Antisense ACC oxidase RNA delays Carnation petal senescence. Hortscience 30:970–972

Scariot V, Paradiso R, Rogers H et al (2014) Ethylene control in cut flowers: classical and innovative approaches. Postharvest Biol Technol 97:83–92. https://doi.org/10.1016/j.postharvbio.2014.06.010

Shibuya K, Barry KG, Ciardi JA et al (2004) The central role of PhEIN2 in ethylene responses throughout plant development in petunia. Plant Physiol 136:2900–2912. https://doi.org/10.1104/pp.104.046979

Spiller M, Berger RG, Debener T (2010) Genetic dissection of scent metabolic profiles in diploid rose populations. Theor Appl Genet 120:1461–1471. https://doi.org/10.1007/s00122-010-1268-y

Spinarova S, Hendriks L, Steinbacher F et al (2007) Cavitation and transpiration profiles of cut roses under water stress. Eur J Hortic Sci 72:113–118

Sriskandarajah S, Mibus H, Serek M (2007) Transgenic *Campanula carpatica* plants with reduced ethylene sensitivity. Plant Cell Rep 26:805–813. https://doi.org/10.1007/s00299-006-0291-6

Su JS, Zhang F, Yang XC et al (2017) Combining ability, heterosis, genetic distance and their inter-correlations for waterlogging tolerance traits in chrysanthemum. Euphytica 213:42. https://doi.org/10.1007/s10681-017-1837-0

Subburaj S, Chung SJ, Lee C et al (2016) Site-directed mutagenesis in *Petunia x hybrida* protoplast system using direct delivery of purified recombinant Cas9 ribonucleoproteins. Plant Cell Rep 35:1535–1544. https://doi.org/10.1007/s00299-016-1937-7

Sugawara H, Shibuya K, Yoshioka T et al (2002) Is a cysteine proteinase inhibitor involved in the regulation of petal wilting in senescing carnation (*Dianthus caryophyllus* L.) flowers? J Exp Bot 53:407–413. https://doi.org/10.1093/jexbot/53.368.407

Sun Y, Christensen B, Liu F et al (2009) Effects of ethylene and 1-MCP (1-methylcyclopropene) on bud and flower drop in mini *Phalaenopsis* cultivars. Plant Growth Regul 59:83–91. https://doi.org/10.1007/s10725-009-9391-y

Tan H, Liu XH, Ma N et al (2006) Ethylene-influenced flower opening and expression of genes encoding Etrs, Ctrs, and Ein3s in two cut rose cultivars. Postharvest Biol Technol 40:97–105. https://doi.org/10.1016/j.postharvbio.2006.01.007

Tanase K, Onozaki T, Satoh S et al (2008) Differential expression levels of ethylene biosynthetic pathway genes during senescence of long-lived carnation cultivars. Postharvest Biol Technol 47:210–217. https://doi.org/10.1016/j.postharvbio.2007.06.023

Tanase K, Onozaki T, Satoh S et al (2011) Effect of age on the auto-catalytic ethylene production and the expression of ethylene biosynthetic gene Dc-ACS1 in petals of long-life carnations. Jpn Agric Res Q 45:107–116

Tanase K, Nishitani C, Hirakawa H et al (2012) Transcriptome analysis of carnation (*Dianthus caryophyllus* L.) based on next-generation sequencing technology. BMC Genomics 13:292. https://doi.org/10.1186/1471-2164-13-292

Tanase K, Otsu S, Satoh S et al (2013) Expression and regulation of senescence-related genes in carnation flowers with low ethylene production during senescence. J Jpn Soc Hortic Sci 82:179–187

Tanase K, Otsu S, Satoh S et al (2015) Expression levels of ethylene biosynthetic genes and senescence-related genes in carnation (*Dianthus caryophyllus* L.) with ultra-long-life flowers. Sci Hortic 183:31–38. https://doi.org/10.1016/j.scienta.2014.11.025

Tanigawa T, Kobayashi Y, Matsui H et al (1999) Histological observations on crooked neck, its degree and rate of development among clonal lines of chrysanthemum cv. Shuhonochikara. J Jpn Soc Hortic Sci 68:655–660

Thongkum M, Burns P, Bhunchoth A et al (2015) Ethylene and pollination decrease transcript abundance of an ethylene receptor gene in *Dendrobium* petals. J Plant Physiol 176:96–100. https://doi.org/10.1016/j.jplph.2014.12.008

Tijskens LMM, Sloof M, Wilkinson EC et al (1996) A model of the effects of temperature and time on the acceptability of potted plants stored in darkness. Postharvest Biol Technol 8:293–305. https://doi.org/10.1016/0925-5214(96)00008-7

Torre S, Fjeld T, Gislerod HR (2001) Effects of air humidity and K/Ca ratio in the nutrient supply on growth and postharvest characteristics of cut roses. Sci Hortic 90:291–304. https://doi.org/10.1016/s0304-4238(01)00230-8

Torre S, Fjeld T, Gislerod HR et al (2003) Leaf anatomy and stomatal morphology of greenhouse roses grown at moderate or high air humidity. J Am Soc Hortic Sci 128:598–602

Tromp S-O, van der Sman RGM, Vollebregt HM et al (2012) On the prediction of the remaining vase life of cut roses. Postharvest Biol Technol 70:42–50. https://doi.org/10.1016/j.postharvbio.2012.04.003

Tromp SO, Harkema H, Hogeveen E et al (2017) On the validation of improved quality-decay models of potted plants. Postharvest Biol Technol 123:119–127. https://doi.org/10.1016/j.postharvbio.2016.09.008

Uchneat MS, Spicer K, Craig R (1999) Differential response to floral infection by *Botrytis cinerea* within the genus *Pelargonium*. Hortscience 34:718–720

Urban L, Six S, Barthelemy L et al (2002) Effect of elevated CO2 on leaf water relations, water balance and senescence of cut roses. J Plant Physiol 159:717–723. https://doi.org/10.1078/0176-1617-0602

VBN (2017) Evaluation cards for cut flowers

Verlinden S, Boatright J, Woodson WR (2002) Changes in ethylene responsiveness of senescence-related genes during carnation flower development. Physiol Plant 116:503–511. https://doi.org/10.1034/j.1399-3054.2002.1160409.x

Villacorta-Martin C, Gonzalez FFC, de Haan J et al (2015) Whole transcriptome profiling of the vernalization process in *Lilium longiflorum* (cultivar White Heaven) bulbs. BMC Genomics 16:550. https://doi.org/10.1186/s12864-015-1675-1

Vo TC, Mun J-H, Yu H-J et al (2015) Phenotypic analysis of parents and their reciprocal F-1 hybrids in *Phalaenopsis*. Hortic Environ Biotechnol 56:612–617. https://doi.org/10.1007/s13580-015-0063-8

Vrind T (2005) The *Botrytis* problem in figures. Acta Hortic 669:99–102

Wang H, Stier G, Lin J et al (2013) Transcriptome changes associated with delayed flower senescence on transgenic *Petunia* by inducing expression of etr1-1, a mutant ethylene receptor. PLoS One 8(7):e65800. https://doi.org/10.1371/journal.pone.0065800

Weber JA, Martin WJ, Stimart DP (2005) Genetics of postharvest longevity and quality traits in late generation crosses of *Antirrhinum majus* L. J Am Soc Hortic Sci 130:694–699

Wei Z, Sun Z, Cui B et al (2016) Transcriptome analysis of colored calla lily (*Zantedeschia rehmannii* Engl.) by Illumina sequencing: de novo assembly, annotation and EST-SSR marker development. Peerj 4:e2378. https://doi.org/10.7717/peerj.2378

Welker OA, Furuya S (1994) Surface-structure of leaves in heat tolerant plants. J Agron Crop Sci-Zeitschrift Fur Acker Und Pflanzenbau 173:279–288. https://doi.org/10.1111/j.1439-037X.1994.tb00565.x

Wernett HC, Sheehan TJ, Wilfret GJ et al (1996a) Postharvest longevity of cut-flower *Gerbera* .1. Response to selection for vase life components. J Am Soc Hortic Sci 121:216–221

Wernett MC, Wilfret GJ, Sheehan TJ et al (1996b) Postharvest longevity of cut-flower *Gerbera* .2. Heritability of vase life. J Am Soc Hortic Sci 121:222–224

Williamson B, Duncan GH, Harrison JG et al (1995) Effect of humidity on infection of rose petals by dry inoculated conidia of *Botytis cinerea*. Mycol Res 99:1303–1310. https://doi.org/10.1016/s0953-7562(09)81212-4

Williamson B, Tudzynsk B, Tudzynski P et al (2007) Botrytis cinerea: the cause of grey mould disease. Mol Plant Pathol 8:561–580. https://doi.org/10.1111/j.1364-3703.2007.00417.x

Winkelmann T, Warwas M, Raffeiner B et al (2016) Improved Postharvest Quality of Inflorescences of fbp1::etr1-1 Transgenic *Burrageara* 'Stefan Isler Lava Flow. J Plant Growth Regul 35:390–400. https://doi.org/10.1007/s00344-015-9545-2

Woltering E, van Doorn WGJ (1988) Role of ethylene in senescence of petals morphological and taxonomical relationships. J Exp Bot 39:1605–1616

Woltering E, Van Hout M, Somhorst D et al (1993) Roles of pollination and short-chain saturated fatty acids in flower senescence. Plant Growth Regul 2:1–10

Woodson W, Park K, Drory A et al (1992) Expression of ethylene biosynthetic pathway transcripts in senescing carnation flowers. Plant Physiol 99:526–532

Xue J, Li Y, Tan H et al (2008a) Expression of ethylene biosynthetic and receptor genes in rose floral tissues during ethylene-enhanced flower opening. J Exp Bot 59:2161–2169. https://doi.org/10.1093/jxb/em078

Xue JQ, Li YH, Tan H et al (2008b) Expression of ethylene biosynthetic and receptor genes in rose floral tissues during ethylene-enhanced flower opening. J Exp Bot 59:2161–2169. https://doi.org/10.1093/jxb/em078

Xue DW, Zhang XQ, Lu XL et al (2017) Molecular and Evolutionary Mechanisms of Cuticular Wax for Plant Drought Tolerance. Front Plant Sci 8:621. https://doi.org/10.3389/fpls.2017.00621

Yagi M (2015) Recent progress in genomic analysis of ornamental plants, with a focus on carnation. Hortic J 84:3–13. https://doi.org/10.2503/hortj.MI-IRO1

Yagi M, Kosugi S, Hirakawa H et al (2014) Sequence analysis of the genome of carnation (*Dianthus caryophyllus* L.). DNA Res 21:231–241. https://doi.org/10.1093/dnares/dst053

Yagi M, Shirasawa K, Waki T et al (2017) Construction of an SSR and RAD marker-based genetic linkage map for carnation (*Dianthus caryophyllus* L.). Plant Mol Biol Report 35:110–117. https://doi.org/10.1007/s11105-016-1010-2

Yin J, Chang XX, Kasuga T et al (2015) A basic helix-loop-helix transcription factor, PhFBH4, regulates flower senescence by modulating ethylene biosynthesis pathway in petunia. Hortic Res 2:15059. https://doi.org/10.1038/hortres.2015.59

Zakizadeh H, Lutken H, Sriskandarajah S et al (2013) Transformation of miniature potted rose (*Rosa hybrida* cv. Linda) with P (SAG12) -ipt gene delays leaf senescence and enhances resistance to exogenous ethylene. Plant Cell Rep 32:195–205. https://doi.org/10.1007/s00299-012-1354-5

Zhang F, Chen S, Chen F et al (2011) SRAP-based mapping and QTL detection for inflorescence-related traits in chrysanthemum (*Dendranthema morifolium*). Mol Breed 27:11–23. https://doi.org/10.1007/s11032-010-9409-1

Zhang F, Chen S, Jiang J et al (2013) Genetic Mapping of Quantitative Trait Loci Underlying Flowering Time in Chrysanthemum (*Chrysanthemum morifolium*). PLoS One 8(12):e83023. https://doi.org/10.1371/journal.pone.0083023

Zhang B, Yang X, Yang CP et al (2016) Exploiting the CRISPR/Cas9 System for Targeted Genome Mutagenesis in *Petunia*. Sci Rep 6:8. https://doi.org/10.1038/srep20315

第5章 观赏植物抗病育种

Leen Leus

摘要 许多物种经驯化成为观赏的切花、盆栽及园林植物。物种的多样性与栽培和售后期间由植物病原物引起的各种各样的问题是密切相关的。因此，提高抗病性通常是观赏植物育种者优先考虑的因素。对于粮食作物，已经在大力发展植物病理学研究和育种。筛选亲本和后代植物的生物测定方法、生物技术手段或其他提高抗病性的方法应用可获得更具抗性的品种。但是，对于观赏植物，这些技术和资源仅得到有限的应用。

本章概述了观赏植物抗病育种的研究进展。列举的具体观赏植物实例主要基于科学文献。我们还介绍了商业育种者的实践经验和他们应用于育种实践的方法。具体的植物病原物案例包括红掌（*Anthurium andreanum*）上的地毯草黄单胞菌花叶万年青致病变种（*Xanthomonas axonopodis* pv. *dieffenbachiae*），杜鹃花（*Rhododendron simsii*）上的疫霉（*Phytophthora plurivora*）和丽赤壳属病菌（*Calonectria pauciramosa*），黄杨上的丽赤壳属2个真菌（*Calonectria pseudonaviculata* 和 *C. henricotiae*），香石竹（*Dianthus caryophyllus*）上的尖孢镰刀菌香石竹专化型（*Fusarium oxysporum* f. sp. *dianthi*），菊花（*Chrysanthemum* × *morifolium*）上的白锈病菌（*Puccinia horiana*），百合（*Lilium*）上的尖孢镰刀菌百合专化型。

关键词 生物测定；育种者经验；抗病性；植物病原物；筛选

5.1 引　　言

许多观赏植物作为盆栽、切花或园林植物在栽培和售后过程中会受到许多不同微生物的危害。类病毒和病毒、植原体、细菌、卵菌及真菌均可危害观赏植物。由于新出现病害、病原物的变化及寄主范围的变化，粮食作物及观赏植物的植物病理学问题也在不断出现。新病原物数量的增加受气候变化、植物材料的国际贸易等影响。与作物化学保护产品的短期效应相比，提高抗病性的育种是一种可持续的作物保护方法。植物的抗病性研究是很有价值的，因为可以减少对其他控制策略的需求，也减少了种植者必需的工作。

与其他农作物相比，观赏植物的抗病育种仍不发达，有几个原因，一是观赏植物的抗病育种受到大量观赏物种和病原物种类的多样性以及其特定的生活方式、寄主范围等的限制。二是观赏植物育种主要是一个直观的过程，而不是基于特定的选择方案。通常

L. Leus (✉)
Flanders Research Institute for Agriculture, Fisheries and Food (ILVO), Plant Sciences Unit, Applied Genetics and Breeding, Melle, Belgium
e-mail: leen.leus@ilvo.vlaanderen.be

仅在育种后期才观察到植物的抗性（Debener 2009）。三是可用的资源在小型育种公司和育种方案中应用较少。而低经济价值的作物和世代周期进一步限制了资源的利用。只有少数的公司和研究人员投资于抗病育种工作。在较广泛的学科领域有可共享的资源和信息，如常见的蔬菜育种，但在观赏植物中运用相当有限（Arens et al. 2012）。四是观赏植物育种的局限性还包括通常由多倍体引起的复杂遗传学。首先涉及的常常是挖掘特征新奇且具有市场价值的观赏植物。育种目标通常取决于经济重要性。因此，抗病性的需求还取决于病原物的经济影响。新品种在抗病性提高的同时，应具有所需的品质和产量特性。特别是在观赏植物中，在抗性育种文献中介绍的例子并不一定是正在进行的育种工作，抗病性的需求通常大于实际应用（Debener 2009）。

无论如何，可以通过常规育种（包括生物测定），应用生物技术手段或分子标记辅助选择及引入新的育种技术来显著提高观赏植物的抗病性。

与病害相比，观赏植物抗虫育种仍处于起步阶段。只有一些较少的例子，如在杜鹃花上进行抗螨虫的筛选（Luypaert et al. 2014）。

因此，在本章，我们主要聚焦于真菌和细菌性病害。讨论了可用的不同的生物测定方法，并详细阐述了一些具体的观赏植物上的病原物及抗性筛选工作，包括育种者的观点。

5.2 病原物生活方式和宿主特异性

观赏植物上病原物的多样性丰富：有些病原物是高度特异性的，称为宿主特异性，而另一些病原物则有很广的寄主范围。对于育种者来说，了解特定病原物和介导宿主特异性或生活方式的因素是必不可少的，了解植物感病或抗病的机制有助于培育出具有抗病性的作物。观赏植物和病害的多样性及病原物特定的寄主范围和生活方式使得抗病育种变得复杂。此处提供了病原物生活方式和观赏植物寄主范围的一些例子。

根据病原物以植物为食的方式，可以将它们分为三类：腐生型、寄生型和半寄生型。腐生型病原物在定殖之前或期间杀死植物细胞。寄生型病原物通常是专性寄生物，没有植物就无法生长。半寄生型病原物以寄生型开始，但在后期变成腐生型的。寄生型通常是寄主特异性的，而腐生型通常具有更广泛的寄主范围，但有些是寄主特异性的（Höfte 2015）。

在卵菌纲疫霉属（*Phytophthora*）中，如在桤木上的 *P. alni* 或橡树上的 *P. quercina* 就是寄主特异性的（Jung et al. 1999）。疫霉属的其他真菌寄主范围很广：*P. ramorum* 有 100 多种寄主植物（Grünwald et al. 2012），*P. cinnamomi* 甚至可以感染数千种寄主植物（Hardham 2005）。疫霉属真菌都能在许多木本观赏植物上发现，如山茶、杜鹃花和荚蒾。一些疫霉属病原物可能是腐生型的，而另一些可能是半寄生型的：有时它们的行为在其生命周期中会发生变化，或者它们的行为不仅取决于其所属物种，还取决于其寄主植物（Fawke et al. 2015）。

真菌葡萄孢属中的植物病原物也是如此。灰霉病的病原物灰葡萄孢（*Botrytis cinerea*）是一种腐生型真菌的教科书级示例，该真菌在真双子叶植物（包括许多观赏植物种）中具有非常广的寄主范围（Jarvis 1977）。然而，有些葡萄孢属物种的寄主范围非常狭窄，包括一些观赏植物上的特异性葡萄孢，如郁金香上的 *B. tulipa*，百合上的 *B. eliptica*，唐菖蒲上的 *B. gladiolorum*，番红花上的 *B. croci*，金盏花上的 *B. calthae*，花毛茛上的 *B. ranunculi*，天竺葵上的 *B. pelargonii*，芍药上的 *B. paeoniae*，风信子上的 *B. hyacinthi*，水仙鳞茎上的 *B. narcissicola* 和水仙叶上的 *B. polyblastis*，雪莲花上的 *B. galanthina* 及鸢尾上的 *B. convuluta*（Hennebert 1973）。葡萄孢属的进化已被 Staats 等（2004）的分子系统发育研究所证实。值得注意的是，并不是所有的葡萄孢属都是腐生型的，有些具有腐生型生活方式（Van Kan et al. 2014）。

病毒、类病毒以及真菌病原物如白粉病菌和锈病菌是寄生型植物病原物的示例。许多植物物种，包括观赏植物，都容易发生白粉病。尽管如此，致病菌要么是严格寄主特异性的，要么具有非常狭窄的寄主范围。例如，在月季上，白粉病由真菌 *Podosphaera pannosa* 引起；在杜鹃花上，该病是由 *Erysiphe azalea* 引起的；在紫丁香上，则是由 *Microsphaera penicillata* 引起的等。锈病也是如此：菊花上的白锈病是由 *Puccinia horiana* 引起的，月季上的锈病是由 *Phragmidium* spp.引起的等。

像 *Xylella fastidiosa* 这样的毁灭性细菌实际上是半寄生型的。当细菌转换到腐生阶段时才会发病，而在其半寄生阶段可能很长一段时间内都不会发病（Armijo et al. 2016）。

5.3　观赏植物的病害控制

在园艺中，可以通过集约且高度可控的栽培方法，即配备有水、肥和气候控制系统的温室来避免非生物胁迫。园艺中的生物胁迫可以应用化学药品来成功控制。然而今天，在观赏植物的保护方面显然存在紧张关系。一方面，针对农药使用有更严格的规定，有害生物综合治理（IPM）措施的义务（欧盟指令 2009/128/EC）以及消费者对环境问题认识的提高，如关注“授粉友好型”植物上的化学残留物（Lentola et al. 2017），导致了在观赏植物贸易中对残留物有更多的规范和要求。另一方面，观赏植物市场成功的主要标准是植物的审美价值，因此，病虫害造成的所有损害都会降低植物的价值。对昆虫、螨虫或其他生物的零容忍度通常用于商业最终产品，这阻碍了天敌在 IPM 策略中的应用。同样，许多不同的农作物和生长系统使 IPM 在观赏植物中的充分应用具有挑战性（Skirvin et al. 2002）。因此，化学保护仍然是获得顶级产品的首选解决方案。尽管如此，让专业人员和最终用户减少农药使用的政府政策是必需的，而且消费者对环境友好型园艺产品的认识和需求越来越高。

与农作物相比，观赏产品的附加值更高。然而，由于耕种面积通常很小，观赏植物保护方面的创新受到限制。从开阔的田间苗圃到温室栽培的盆栽植物和切花，植物种类和生长系统的高度差异给必要的植物毒性测试增加了困难。因此，农作物保护产品的生产者很少去申请专门应用于观赏植物的产品注册。为了克服这些问题，一些国家允许如第三方（官方机构、科学机构或咨询机构）可以要求获得次要作物的授权。

与植物病理学有关的法律也可能干扰观赏植物的贸易或栽培。通常，禁止特定种类的观赏植物的种植和贸易。例如，在蔷薇科（Rosaceae）中，已知诸如栒子属（*Cotoneaster*）、花楸属（*Sorbus*）、红果树属（*Stranvaesia*）、山楂属（*Crataegus*）、榅桲属（*Cydonia*）、火棘属（*Pyracantha*）易感染解淀粉欧文氏菌（*Erwinia amylovora*）(Zeller 1979)。在一些国家，观赏性蔷薇灌木的种植很容易被采取植物检疫措施，因为采取此项措施可以保护果树免遭同一病原物的侵害，从而避免果树上发生火疫病。

植物检疫证书旨在通过植物贸易来限制植物病原物的传播（欧盟指令 2000/29/EC）。但是，这也对特定观赏植物的栽培和贸易产生影响。众所周知的例子是栎树猝死病菌（*Phytophthora ramorum*）和 *Xylella fastidiosa*。迄今为止，对于 *Xylella fastidiosa*，在法国暴发时都会感染观赏植物桃金娘叶远志（*Polygala myrtifolia*），虽然已确认该细菌还以其他 19 种植物为其寄主，但其中大部分是观赏植物（EU Fact Sheet 2015）。实施特定的检疫措施可避免新病虫害的传播。

病虫害也影响消费者对特定观赏植物产品的偏好。英国的一项调查明确表明，由真菌 *Ophiostoma ulmi* 引起的榆树病害和由 *Chalara fraxinea* 引起的白蜡树病害的问题已为公众所熟知（Jepson and Arakelyan 2017）。这些树种的种植及其在园林中的应用容易出现上述植物病理学问题。在欧洲大部分地区，黄杨木（一种流行的园林植物）的种植正遭遇由真菌（*Calonectria pseudonaviculata*）和黄杨木蛾（*Cydalima perspectalis*）为害问题导致的销售突然下滑的困扰。园林设计师建议消费者从他们的花园中清除黄杨木，并常常建议他们改用一种与其外观相似的灌木冬青（*Ilex crenata*）。

5.4 抗病机制

植物病原物的攻击会触发植物的免疫反应。Jones 和 Dangl（2006）的拉链模型是植物对病原物侵袭产生的反应最受人们广泛接受的解释。在第一阶段，病原体相关分子模式（plant pathogen-associated molecular pattern，PAMP）通过植物模式识别受体（pattern recognition receptor，PRR）识别病原物的保守结构。保守结构的一个例子是几乎所有鞭毛菌中都存在细菌鞭毛蛋白（Boller and Felix 2009）。这种对保守结构的识别导致了第一步防御反应的激活：病原体相关分子模式激发的免疫（pattern-triggered immunity，PTI）反应，通常可以成功地抵御病原物。PTI 反应包括胼胝质沉积、细胞壁变化以及防御相关蛋白和植保素的积累（Oßwald et al. 2014）。一些病原物分泌毒性因子，从而抑制病原物相关分子模式激发的免疫反应，导致植物产生效应因子激活的感病性（effector-triggered susceptibility，ETS）。该阶段的植物反应是细胞程序性死亡（programmed cell death，PCD）和过敏反应（hypersensitivity response，HR），这些都是基于效应因子激活的免疫反应（ETI）。已知这些机制对（半）寄生型致病菌最有效（Jones and Dangl 2006）。当受到昆虫和腐生型病原物的攻击后，激发了基于损伤相关分子模式（damage-associated molecular pattern，DAMP）的机制。这种类型的损伤会导致各种宿主防御反应，类似于 PAMP 确发的反应（Galletti et al. 2009）。

由 ETI 和 PTI 诱导的信号转导途径的下游，涉及三种植物胁迫激素：水杨酸（SA）、茉莉酸（JA）及乙烯（ET），是由生物胁迫反应诱导的胁迫激素。当对寄生型和半寄生型病原物产生抗性时，会开启 SA 途径。腐生型病原物和虫害会刺激 JA 与 ET 途径。植物防御信号转导途径也涉及其他激素如脱落酸（ABA）、生长素、赤霉素（GA）、细胞分裂素（CK）、油菜素内酯（BR）和肽激素（Bari and Jones 2009）。对防御网络激素调节的更多了解可能会为抗病育种开辟新的前景。育种者对植物免疫反应的激活很感兴趣。通过应用特定激发子或植物激活剂上调抗病相关基因在抗病性方面具有优势。触发免疫反应将导致病程相关蛋白的形成、细胞壁增强、激素调节的变化等（Höfte 2015）。

植物与病原物之间的某些特定相互作用已得到充分研究，并用于改良作物的抗病性。在农作物中，抗病基因（*R* 基因）和与抗病相关的不同 *R* 基因或数量性状基因座（QTL）组合的品种的应用得到了发展。聚合 *R* 基因和 QTL 被认为是抗性育种的一种较好策略（Pilet-Nayel et al. 2017）。然而，在观赏植物育种中，这些应用尚不发达。

另一种方法是基于病原物与宿主之间兼容性的丧失。根据 *S* 基因的原理，抗性本身不是渗入的，而是感病性的丧失。这种基于 *S* 基因失活的抗性被认为是更广谱和更持久的，因此具有非常好的育种潜力（Pavan et al. 2010）。根据感染的阶段，可以识别出三类 *S* 基因：早期病原物侵入、宿主防御调节和病原物在宿主中继续侵染（Van Schie and Takken 2014）。基于 *S* 基因的抗病性的教科书级示例是在大麦中发现的霉病位点抗性基因 *O*（*MLO*）（Jorgensen 1992）。在发现大麦的广谱白粉病抗性是基于 *mlo* 基因功能丧失后，还在拟南芥、番茄和豌豆等其他植物中研究了该基因与白粉病之间的关系。功能丧失的隐性等位基因（*mlo*）可以介导对白粉病的广谱抗性，并在育种上很有前途（Van Schie and Takken 2014）。许多观赏植物容易感染白粉病；因此，*mlo* 功能丧失可为观赏植物抗白粉病的研究提供新思路。第一次的尝试是在月季中，已从一种大型月季的 EST 序列中分离出 4 个 *MLO* 同源基因并进行了鉴定（Kaufmann et al. 2012）。

植物病原物，特别是细菌的基因组测序使得对病原物的多样性和进化以及细菌生活方式与致病性有了更深入的了解（Arnold and Jackson 2011）。现在，植物病原物正受到大规模遗传方法的影响；突变体的收集用于提供基于功能的毒性基因信息。将来，不仅在病害管理策略中，而且在宿主抗性育种中，病原物效应物的鉴定将变得越来越重要（Motaung et al. 2017）。

5.5　抗 病 育 种

5.5.1　常规抗病育种

抗性育种始于 100 多年前。1900 年左右的第一批报道涉及棉花的枯萎病（Orton 1918）和亚麻的枯萎病（Bolley 1901）。在这些研究中，抗病基因型的选择是基于基因型对病原物的暴露程度。100 年后，仍没有通用的抗性育种方法。通常，抗病育种工作费力、费时且昂贵。

在观赏植物中，关于抗病育种的例子很少。因此，在某些情况下，当化学农药无法维持作物的正常生长时，就需要其他的解决方案。病害的经济影响与对抗病的迫切需要有关。例如，在由 *Calonectria* spp.引起黄杨木的巨大损失之后，启动育种计划的唯一原因就是提高抗病性（Van Laere et al. 2015）。

抗病育种中选用的方法取决于作物和病原物的类型。一种常规方法是循环选择，定义为旨在进行反复的选择。可以通过连续选择如聚合抗性基因来实施。经过几代后就会出现不同的特征。另一种方法是对每个特征（包括抗病性）应用独立的淘汰标准。这样，只有在每种性状（包括抗病性）阈值水平以上的植株才能保留。在指标选择方面，对基因型的综合表现进行了评价。一个基因型表现优异的性状被认为是对表现较差的性状的补偿。根据抗病性在选择中作为不同性状组合的价值，其可以在几代间得到改善。然而，抗病育种可能会产生副作用，如在抗病和作物产量之间进行权衡的适应度代价（Vyska et al. 2016）。病原物的重要性及其流行和危害界定了特定农作物抗病性状的重要性。

抗病育种在重要经济作物中很常见，如小麦、水稻、马铃薯、番茄等。培育出的品种能抗不同的病原物。抗病育种成功的关键因素之一是获得的遗传变异且抗性能在后代中遗传。抗病育种需要一个包含抗性来源的基因库，这种来源通常包括野生种或遗传上不同的种质。在苹果中，在遗传资源中心保存的古老品种中发现了抗病基因（Kellerhals et al. 2012）。另一个著名的例子是番茄。自 1917 年以来，就开始在番茄的野生近缘种中进行近亲繁殖。最近，对番茄基因组的分子分析揭示了野生近缘种遗传物质渗入的历史过程和性状遗传背景，其中最重要的一项是引入抗病性（Menda et al. 2014）。

在观赏植物中，不经常报道（再）利用野生远缘种和种间杂交来定向引入抗病性。例如，在月季中，在几个月季种中发现了对几种病害的抗性。但是，到目前为止，野生种质的利用仅应用于庭院月季，在庭院月季中，月季类型和倍性水平会发生很大变化，并且由于种植者和消费者的需求，抗病性已经作为优先考虑的性状。在切花月季中，狭窄的遗传背景、需要对不同世代回交的需要以及四倍体是在育种中引入较好抗性月季品种的一个障碍。在最近被引入花园作为切花应用的 *Anigozanthos* spp.（袋鼠爪）中，由 *Puccinia haemodori* 引起的锈病和由 *Alternaria anigozanthi* 引起的墨斑病的种间杂种被用来培育抗病品种（Growns 2015）。在其他情况下，抗病性是种间杂交的副效应。在绣球花育种中已经看到了这一点。*Hydrangea angustifolia* 与商业上重要的 *H. macrophylla* 杂交后产生的杂种对白粉病具有更好的抗性，这就是以导入其他更理想的性状为主要育种目标的一种副效应（Kardos et al. 2009）。

需要有效的筛选技术来选择最佳的父母本和优良的后代。因此，表型鉴定是筛选植物抗病性的基础。在过去的 10 年中，通过传感器开发和高性能计算技术的进步，在高通量表型分析方法的开发方面取得了巨大的进步（Shakoor et al. 2017）。同样，观赏植物育种尚未利用高通量表型技术。

选择筛选分析和生物测定需要病理学方面的知识，特别是病原菌群体内的变异可以影响有效的筛选。因此，测试方法必须可控且可重复，并应反映植物对病原物种群内多样性的抗性表现。

生物测定是测试候选品种并实施轮回选择过程的常用方法。荷兰组织 NAK-tuinbouw（Netherlands Inspection Service for Horticulture，荷兰园艺检验局）提供抗病育种的抗性测试商业服务。对于蔬菜，可以提供 12 种作物上 44 种植物病原物和昆虫的测试服务。对于某些病原物的许多致病型可从表 5.1 中获得。NAKtuinbouw 拥有 70 种从植物病害样本中收集到的病原物分离物。例如，可以测试莴苣基因型对 15 种 *Bremia lactucae* 致病型的抗性。也可给在自己的设施中进行生物测定的育种者提供病原物（NAKtuinbouw 2018a）。

表 5.1　观赏植物商业化抗病筛选的生物测定方法一览表

植物种	病原菌	提供测定的机构
六出花（*Alstroemeria* sp.）	六出花花叶病毒	NAKtuinbouw（荷兰）
菊花（*Chrysanthemum* × *morifolium*）	白锈病菌（*Puccinia horiana*）	ILVO（比利时）、NAKtuinbouw（荷兰）
菊花（*Chrysanthemum* × *morifolium*）	尖孢镰刀菌菊花专化型（*Fusarium oxysporum* f. sp. *chrysanthemi*）	NAKtuinbouw（荷兰）
菊花（*Chrysanthemum* × *morifolium*）	大丽轮枝菌（*Verticillium dahlia*）	NAKtuinbouw（荷兰）
香石竹（*Dianthus caryophyllus*）	尖孢镰刀菌香石竹专化型（*Fusarium oxysporum* f. sp. *dianthi*）	NAKtuinbouw（荷兰）
小苍兰（*Freesia hybrida*）	尖孢镰刀菌唐菖蒲专化型（*Fusarium oxysporum* f. sp. *gladioli*）	NAKtuinbouw（荷兰）
月季（*Rosa*）	蔷薇双壳菌（*Diplocarpon rosae*）	INRA（法国）

该机构还提供了少量观赏植物的抗病性测试：可以在 4 种植物中测试 6 种病原菌唯一筛选方法（NAKtuinbouw 2018b）。概述了一些 NAKtuinbouw 和其他机构提供的筛选与测试的几种观赏植物病害种类，见表 5.1。育种公司与研究机构之间在特定测试方面存在合作。例如，位于法国昂热的 INRA 提供了一种商业生物测定方法，采用了在法国收集的 10 个黑斑病菌分离株，用于测试玫瑰插条（Hibrand Saint-Oyant，个人交流）。

文献中描述了用于观赏植物的几种生物测定方法。5.6 节讨论了用于重要观赏植物的不同测试方法。测试由育种公司或有时由商业服务实验室进行。育种者的主要关注点在商业方面；种植者也是如此。因此，每个抗性增强的候选品种都应满足市场的所有其他需求。寄主、病原物和环境之间的相互作用通常很复杂，可能使植物的抗病性评估变得复杂甚至抵消抗病性。在实践中，经常观察到生物测定的可重复性受环境变化或其他（非）生物胁迫的影响。测试可以在实验室和生长室中进行，也可以在正常的生长条件下甚至在组织培养中进行（van den Bulk 1991）。实验室测试可提供更好的标准化和更好的受控环境。它们通常在离体的叶片或叶盘上进行，如天竺葵属（*Pelargonium* spp.）对灰霉病（*Botrytis*）的抗性（Uchneat et al. 1999）或大叶绣球对白粉病（*Erysiphe polygoni*）的抗性（Li et al. 2009）。组织培养中的再生可用作选择机制，如耐镰刀菌的香石竹（Thakur et al. 2002）和百合（Zhang et al. 2014b）的筛选。组织培养测试是在添加了尖孢镰刀菌培养滤液的培养基中进行的。在尖孢镰刀菌滤液存在的情况下，从两种作物的再生植株中都筛选到了对病原菌抗性明显提高的植株。

5.5.2　生物技术在抗病育种中的应用

分子标记辅助选择（MAS）和与抗病性基因连锁的 QTL 已开发并应用于许多农作物及病原物中。MAS 在植物育种中的应用存在以下问题：①性状难以表达、成本高或周期长；②性状取决于发育阶段或特定环境；③回交选择隐性等位基因或④聚合基因（Xu and Crouch 2008)。抗病性往往满足上述 4 个条件。但是，总的来说，MAS 在育种计划中的应用落后于预期，问题在于复杂的性状、与表型选择相比较高的成本、QTL 定位的高通量精确表型分析、有用的生信工具、与环境互作的基因型和上位性。因此，在许多农作物中，QTL 缺乏有效的 MAS 验证是一个问题（Neale and Kremer 2011）。尽管如此，在多倍体作物中也可以找到一些例子。在马铃薯中，使用分子标记辅助选择筛选抗病毒和抗线虫基因（Ortega and Lopez-Vizcon 2012）。同样在小麦中， MAS 已成功应用于实际育种中，以选择小麦的抗锈基因、抗眼斑病基因和抗枯萎病（*Fusarium*）基因。在大麦中，MAS 应用于选择抗大麦黄化花叶病毒和基于 *mlo* 基因选择抗大麦白粉病的抗病育种（Miedaner and Korzun 2012）。在观赏植物中，选择抗病性分子标记应用有限。Arens 等（2012）综述了观赏植物的研究进展，主要列举了郁金香对三种病原物即尖孢镰刀菌（*Fusarium oxysporum*）、郁金香碎裂病毒、灰霉菌（*Botrytis tulipae*）的抗性及 SNP 标记的应用。同样在多倍体中，已经开发了应用于抗病性的标记，如在月季中（Koning-Boucoiran et al. 2012），已经开发出抗月季不同病原菌的抗性标记和 QTL，但在月季育种实践中并不常用。Debener 和 Byrne（2014）综述了不同的新技术，如第二代测序（NGS）方法和靶向突变方法。

受不同基因影响的复杂性状对于传统的 MAS 来说太困难了。但是现在基因组选择使用所有的标记数据进行性能预测，从而获得更准确的预测。系统方法将改进和廉价的“基因组”技术与高通量表型分析相结合。这种方法将有助于研究植物和病原菌代谢组的相互作用及其生理学机制，并将这些数据与环境变化关联起来。这些将最终改善病害管理策略（Leach et al. 2014）。这里的挑战是要将实验室知识转化为田间条件并应用于商业育种。越来越多的测序基因组可供使用，包括几种观赏植物，如蝴蝶兰（*Phalaenopsis*）（Cai et al. 2015）。

基因改良（GM）已被发现可用于抗病育种：值得注意的是，北美洲已批准的转基因田间试验总数中约有 10%旨在提高抗病性。这个比例已经保持 15 年不变（Collinge et al. 2010）。在允许转基因作物种植的国家，确实找到了转基因品种进入市场的途径。马铃薯就是一个很好的例子，马铃薯的某些品种中积累了对马铃薯晚疫病菌（*Phytophthora infestans*）具有抗性的基因（Zhu et al. 2012）。在某些地方，法律的限制也阻碍了转基因方法在观赏植物上的应用。尽管法律上有严格的限制，但仍有许多文献介绍了有关观赏植物的基因转化，其中大多数适用于切花。与许多其他技术相反，大量的基因工程技术应用于改良许多不同观赏植物的抗病性。最近，发表了两篇有关切花的基因工程（Sharma and Messar 2017）和普通观赏植物的基因工程（Azadi et al. 2016）方面的综述，其中包括抗病性的遗传转化工作。许多观赏植物组织培养方案的可用性可能解释了其在遗传转化研究中的受欢迎程度。迄今为止，尚无抗病性提高的转基因观赏植物商品化生产。

遗传改良中最流行的观赏植物是菊花，主要是在病虫害的抗性方面。在菊花中，基于对 $hpaG_{Xoo}$ 基因（Xu et al. 2010，2011）和 *Prunus mumei* 中 *PGIP*（polygalacturonase-inhibiting protein）基因过表达的研究基础（Miao et al. 2012），获得对 *Alternaria tenuissima* 的抗病性。通过将水稻几丁质酶基因（*CHI-II*）整合到黄瓜花叶病毒（cucumber mosaic virus，CMV）的外壳蛋白（cp）基因中（Sen et al. 2013）可获得对由针壳孢属（*Septoria*）

引起的叶斑病的抗性（Kumar et al. 2012），通过将核衣壳基因整合可获得对番茄斑萎病毒（TSWV）的抗性（Sherman et al. 1998），基于水稻几丁质酶（Takatsua et al. 1999）和 *N*-甲基转移酶基因（*CaXMT1*、*CaMXMT1* 和 *CaDXMT1*）可获得对灰霉病菌（*Botrytis cinerea*）的抗性（Kim et al. 2011）。利用转基因技术还实现了对昆虫的抗性。基于对苏云金芽孢杆菌（*Bacillus thuringiensis*，Bt）的抗性（*Cry1Ab* 基因）引入了对毛毛虫（如烟青虫和斜纹叶蛾）的抗性（van Wordragen et al. 1993；Shinoyama et al. 2003，2008，2012）。通过将 *Lycoris longituba* 凝集素（*LLA*）基因转化到菊花中获得对蚜虫的抗性（He et al. 2009）。同样在菊花中，通过苏云金芽孢杆菌 *Cry1Ab* 基因的改良和病原物（*Sarcophaga peregrine*）的 *Sarcotoxin I1* 基因改良获得对昆虫的综合抗性，从而对毛毛虫和白锈病产生很强的抗性（Shinoyama et al. 2015）。

在月季中，转基因方法已用于改良抗病性，但所有方法仅导致增加部分抗性。通过引入不同的基因可增强黑斑病抗性（Marchant et al. 1998；Dohm et al. 2001，2002）。据报道，在引入抗菌蛋白 AceAMP1（Li et al. 2003）和水稻几丁质酶基因（Pourhosseini et al. 2013）后，白粉病抗性增强。

在非洲菊中，通过导入 *NP* 基因（核蛋白基因）获得了对 TSWV 有抗性的植株（Korbin 2006）。在一品红中，病毒衍生的发夹 RNA 结构对一品红花叶病毒（Poinsettia mosaic virus，PnMV）具有抗性（Clarke et al. 2008）。在唐菖蒲中，应用转基因方法引入了对大豆黄化花叶病毒（BYMV）和 CMV 的抗性。由于对 BYMV 的抗性不稳定，因此结果各不相同（Kamo et al. 2005）。对 CMV 的抗性取决于病毒亚组对植物的侵染力（Kamo et al. 2010，2012）。Kamo 等（2015，2016）还介绍了唐菖蒲对尖孢镰刀菌（*Fusarium oxysporum*）的抗性。

在百合中，通过过表达水稻几丁质酶 10（*RCH10*）基因开发了对灰葡萄孢（*Botrytis cinerea*）抗性增强的转基因植株（De Cáceres González et al. 2015）。通过应用有缺陷的 CMV 复制酶基因来改良百合对 CMV 的抗性（Azadi et al. 2011）。Vieira 等（2015）通过 *OcIΔD86* 的过表达获得了抗根结线虫（*Pratylenchus penetrans*）的植株。

在蝴蝶兰中，通过 CymMV 外壳蛋白（CP）基因渗入获得了对大花蕙兰花叶病毒（Cymbidium mosaic virus，CymMV）的抗性（Liao et al. 2004）。将具有 CymMV CP 的蝴蝶兰植株用甜椒铁氧还蛋白的 cDNA 重新转化，可获得对病毒和胡萝卜杆菌 *Pectobacterium carotovora* 的抗性（Chan et al. 2005）。

在纽约紫菀（“紫色紫菀”）中，在引入 Ace-AMP1 后可提高其对白粉病的抗性（Mork 2011）。

新的育种技术（NBT）在研究和应用育种方面都有很高的期望。应用 NBT 提高抗病性受到越来越多的关注：这些技术包括锌指核酸内切酶、类似转录激活子的效应核酸酶（TALEN）和 CRISPR/Cas9 技术。CRISPR/Cas9 等基因组编辑技术目前获得了最多的关注。CRISPR/Cas9 的作用方式是使基因失活、突变或插入，与其他技术相比，其精度要高得多。像转基因技术一样，政府现在必须决定是否将使用这种技术产生的栽培品种投放市场。NBT 在提高抗病性方面的应用潜力与对植物-病原菌互作的了解取得了巨大进展有关，但是这种知识仍主要应用于实验室和生长室测试的模型系统中。对不同作物和病原物组合的病原物效应知之甚少。因此，用于病害控制的模式触发和效应触发免疫（PTI 和 ETI）的调节是一个新兴的研究领域。观赏植物遗传转化方面的大量经验提高了人们对 NBT 应用的期望，尤其是由于预期的法律限制较少。Xiong 等（2015）综述了基因组编辑技术在园艺作物育种（包括观赏植物）中应用的潜力。*MLO* 基因的应用引起了人们的关注；一个有前景的例子是通过禁用 *MLO* 基因获得抗白粉病植株的可能性（Rispail and Rubiales 2016）。

5.6 观赏植物抗病育种：筛选案例

在这一部分中，我们将叙述部分植物病原物案例的抗性育种现状。因此，我们将文献中的科学知识与育种者在商业育种中的观点和经验相结合。这些来自育种实践的信息可以更好地指导如何在公司进行抗病育种研究。

5.6.1 红掌上的地毯草黄单胞菌花叶万年青致病变种（*Xanthomonas axonopodis* pv. *dieffenbachiae*）

1983 年，整个加勒比（Caribbean）地区的红掌生产被细菌性疫病毁灭（Anaïs et al. 2000）。从那时起，地毯草黄单胞菌花叶万年青致病变种（*Xanthomonas axonopodis* pv. *dieffenbachiae*，Xad；原名 *Xanthomonas campestris* pv. *dieffenbachiae*），大多数生产红掌盆栽和切花的国家都有报道（Anaïs et al. 2000；Elibox and Umaharan 2008a）。常见控制措施的无效性使抗病育种成为继续生产红掌的唯一途径（Anaïs et al. 2000；Elibox and Umaharan 2008a）。植物的症状分两个阶段显示。第一阶段是通过叶片感染发生的，在叶片背面有不规则形状的水渍状斑点，叶片反面黄化，佛焰苞也黄化。随后，斑点变成棕色或黑色，并可能迅速发展到系统性阶段，从而导致植株死亡（Elibox and Umaharan 2008a）。

在病害发展的两个阶段都可进行生物测定。在第一阶段，评价叶片对细菌性侵染的抗性（Elibox and Umaharan 2008a），而在第二阶段，则要评价细菌的系统性扩展（Elibox and Umaharan 2008b）。在两个时期的红掌中都发现了抗病性，并已经研究了其在红掌中的遗传（Elibox and Umaharan 2007，2008b，2010）。在叶片水平，抗病性决定了叶片损伤程度，其是一种定量遗传，具有很强的加性遗传效应。系统水平上的感病性是由相互作用的两个等位基因间主要基因决定的，这些基因是疫病感染导致植株死亡的原因（Elibox and Umaharan 2010）。在红掌中发现了更抗病的种质（Anaïs et al. 2000；Elibox and Umaharan 2010）。*A. andreanum* 和 *A. antioquiense* 的种间杂交增强了对疫病的抗性。不幸的是，这些杂种中没有预期的典型心形佛焰苞。

应用类似抗菌肽 Shiva 1 的裂解肽转基因增强了对 Xad 的抗病性（Kuehnle et al. 2004a，2004b）。但是，存在植物基因型效应，在某些情况下表达水平过低。最近，Teixeira da Silva 等（2015）综述了红掌的分子生物学方法和遗传转化技术，包括抗病性。

育种者的观点：Anthura（荷兰）

"过去，Anthura 与外部合作伙伴合作进行了抗黄单胞菌的生物测定。不幸的是，独立测试之间的差异很大；这可能是由生长条件与其他生物学因素的相互作用所致。我们注意到不同季节的天气条件可能会影响温室中的感染水平。即使在温室内条件保持稳定的情况下，在晴朗和干燥的天气与在多云和阴雨的天气条件下进行的生物测定也会得出不同的结果。

从幼苗到开花，已经测试了不同生长期植株的性能。尽管在基因型之间发现了明显的感病性差异，但该公司决定不继续使用生物测定法。主要原因是无法与外部合作伙伴进行生物测定。对于这种类型的病害，抗病育种非常复杂。尽管如此，这种病害仍在给种植者带来麻烦，特别是在热带地区。需要一种有效的生物测定方法，并为有效筛选植物材料提供更多的可能性。"

5.6.2　杜鹃花上的多主寄生疫霉（*Phytophthora plurivora*）和丽赤壳属病菌（*Calonectria pauciramosa*）

盆栽杜鹃花（*Rhododendron simsii* hybrids）作为开花盆栽植物培育，在植物的温室栽培期间，多种病原菌可引发病害。最重要的病原菌是疫霉属（*Phytophthora* spp.）病菌和丽赤壳属病菌（*Calonectria pauciramosa*）。

已知杜鹃花（*Rhododendron*）上有许多疫霉属（*Phytophthora*）病菌的发生，如 *P. cactorum*、*P. ramorum*、*P. plurivora* 和 *P. pini*（Rytkönen et al. 2012）。在温室栽培的盆栽杜鹃花上最常出现的种是 *P. plurivora* 和 *P. multivora*（之前称为 *P. citricola*）。疫霉病菌侵染的症状是叶片变色，然后落叶（Backhaus 1994b）。在细枝的横切面，可以观察到疫霉病菌的典型症状褐化变色。已知对疫霉菌的感病性取决于杜鹃花的基因型（Van Huylenbroeck et al. 2015）。

杜鹃花上丽赤壳属（*Calonectria*）病害的典型症状是根和茎腐烂。当植物在不利条件下生长时，真菌会造成更多的伤害（Backhaus 1994a）。潮湿的土壤条件将有利于病害的发展。

杜鹃花抗病育种工作做得越来越多。已经开发了两种病害的生物测试方法，并应用于候选品种抗病性评价。主要采用的是传统的离体叶片法（De Keyser et al. 2008）。现在，一种使用插条的改良方法具有更好的可重复性：将 10 周龄生根的候选品种插条与已知抗病性或感病性的品种一起进行测试（Van Huylenbroeck et al. 2015）。对于两种病原菌，在测试中都使用了混合分离株。疫霉病菌的接种是切割插条的尖端创造伤口后在其表面用游动孢子进行接种。对于丽赤壳属病菌，是将孢子悬浮液倒在植物-土壤的交界面上。接种后 1 个月进行疫霉病菌感病性评价。对于丽赤壳属病菌，接种后 4 个月对插条进行评价，因为通常需要很长时间才能出现症状。测试之间的差异被认为是由气候因素和插条质量的差异所致（Van Huylenbroeck et al. 2015）。

育种者的观点：Evelien Calsyn-ILVO（比利时）

"对丽赤壳属（*Calonectria*）和疫霉病菌的生物测定每年作为 ILVO 杜鹃花筛选计划的规定部分进行。为了抵消测试年之间的差异，杜鹃花基因型在至少 3 年内进行重复测试。疫霉病菌接种后的结果大部分为出现黑色或白色斑点。如丽赤壳属病菌接种的预期，测试内和测试间的变异性更高。对于丽赤壳属病菌，真菌的毒性取决于寄主的生长条件。该测试需要几个月的时间才能完成。

生物测定法的缺点是所需的插条数量。因此，只能在随后的选择阶段进行生物测定。

根据生物测定结果，选择了抗病性较好的植株，但到目前为止，在实际生长条件下的抗病性尚不清楚，需要进一步研究。也没有关于抗性遗传背景的信息。尚未尝试有针对性的杂交组合来提高抗病性。”

5.6.3 香石竹上的尖孢镰刀菌香石竹专化型（*Fusarium oxysporum* f. sp. *dianthi*）

对由尖孢镰刀菌香石竹专化型（*Fusarium oxysporum* f. sp. *dianthi*）引起的枯萎病进行抗性育种是香石竹（*Dianthus caryophyllus*）育种的重要挑战。香石竹中维管束萎蔫可能由不同的病原菌引起。荷兰的 Proefstation voor Bloemisterij（1984）报道了该病害。在荷兰，关于枯萎病的研究始于 1939 年。在本研究开始时，瓶霉菌是主要原因；后来发现由假单胞菌（*Pseudomonas caryophylli*）和 *Dickeya* spp.病菌引起细菌型枯萎病。尖孢镰刀菌（*Fusarium oxysporum*）和氧化镰刀菌（*F. redolens*）分别发现于 1949 年和 1966 年。自 20 世纪 70 年代以来，不再有细菌感染的报道，但是镰刀菌感染的报道增加了。因此，研究的重点转变为镰刀菌（*Fusarium*）(Proefstation voor Bloemisterij 于 1984 年在荷兰报道)。香石竹维管束萎蔫的症状是由木质部堵塞引起的。不同的病原菌可以引起相同的症状，并且可以同时发生。病茎的横切面在木质部处呈现出褐化变色。症状通常仅出现在植物的一侧。其他症状是叶片黄化。两种病害通常都会导致植株死亡。

利用对瓶霉属病菌（*Phialophora cinerescens*）有抗性的香石竹庭院品种种子生长产生的庭院香石竹品种进行单头大花型和多头型香石竹的抗病育种。通过筛选幼苗群体中的抗性植株，在三代中获得了抗性植株。在此阶段改变了育种程序，不再对幼苗进行抗病性测试，而是对已经用于品质和生产力性状筛选的无性繁殖植物进行测试（Sparnaaij and Demmink 1976）。

尖孢镰刀菌香石竹专化型（*Fusarium oxysporum* f. sp. *dianthi*）的抗病育种始于 20 世纪 60 年代。20 年后，第一个抗病品种投放市场（Mitteau 1987）。尖孢镰刀菌香石竹专化型具有很好的特征：已知 8 个小种，其中 2 个小种普遍存在（Demmink et al. 1989; Prados-Ligero et al. 2007），并且在世界各地均有发生（Prados-Ligero et al. 2007）。寄主响应系统也叙述过。环境条件对生物测定结果的影响阻碍了香石竹的抗病性选择（Ben-Yephet et al. 1993，1996）。尽管如此，通过田间试验和人工接种已获得了高抗尖孢镰刀菌的品种。目前尚不清楚香石竹如何遗传对镰刀菌所有小种的抗性(Prados-Ligero et al. 2007)，因为不同的作者得出了不同的作用方式。例如，在对 1 号小种的抗性研究中，提出了单基因显性抗性（Demmink et al. 1989）。对 2 号小种的抗性描述为水平抗性和多基因抗性（Prados-Ligero et al. 2007）。Baayen 等（1991）叙述了对真菌不同小种的抗病性与植物抗毒素的产生有关。尽管在田间条件下的抗病性可能与人工接种有所不同（Prados-Ligero et al. 2007），但已报道了 6 个小种的抗性基因型。

使用转基因方法，来自粘质沙雷氏菌（*Serratia marcescens*）的细菌几丁质酶基因被用于获得抗镰刀菌枯萎病的香石竹品种（Brugliera et al. 2000）。

育种者的观点：Frans van der Helm-Hilverdakooij（荷兰）

“在香石竹中，全球最重要的病原菌是尖孢镰刀菌。多年前，瓶霉属（*Phialophora*）病菌也是一个问题，但随着定向抗病育种情况发生了变化。在过去，像荷兰 IVT 这样的研究所一直在寻找抗这两种病菌的父母本，并在抗病育种方面取得了长足进展。产生的基因型应用于进一步的育种，且仍然是当前育种中抗镰刀菌的主要材料。当前品种对瓶霉属病菌的抗性可能降低了。

镰刀菌仍然很重要，并且研究已经持续了很长时间。在某些时候，像哥伦比亚和肯尼亚这样的国家引入无土栽培后，对土传病害的关注减少了。当溴甲烷禁用且土壤蒸汽消毒并不总是可行时，对镰刀菌的关注再次加强。

育种者应用生物测定法筛选对镰刀菌的抗病性。目前的测试几乎与 20 世纪 90 年代初的测试相同，仅有一些小改动。例如，除了常规测试外，现在还使用了更快的筛选测试方法。随着时间的流逝，出现了更多的致病型；目前描述了超过 25 种致病型。其中一些是罕见的，最普遍的是 2 号、1 号、4 号和 8 号小种。特别是 2 号小种用于育种公司和服务实验室的筛选。

由于注意到重复之间存在重要差异，因此在不同的时间点对品种进行 3 次测试。极端反应通常很明显，但是可以发现抗病性的变化，特别是在具水平抗性的植株中。一系列的标准品种始终采用一个对照品种。生物测定可提供有用的结果：大多数的结果与栽培中的观察基本一致。在实践中仅有几个品种表现更好。

由于生物测定为表型鉴定提供了良好的工具，因此在镰刀菌抗病性筛选中采用分子方法是可行的。日本研究人员开始使用标志物。在 20 世纪 90 年代的日本，针对伯克霍尔德氏菌（*Burkholderia*）开发了一种很好的生物测定方法，分子标记的开发部分成功。该病主要是日本人的问题，但是随着日本商业香石竹育种的减少，人们对该技术发展的关注减少了。

已知对重要的镰刀菌 2 号小种的抗性是多基因控制的，估计约 5 个基因占变异的 90%。对致病型 1 号和 8 号小种的抗病性可以通过简单的单基因方式进行控制。基于标记的方法可以在苗期进行选择。由于所需的成本高和大规模应用难，当前的生物测定法不能在苗期使用。

在新品种中抗镰刀菌的附加价值是很重要的。在抗其他病虫害方面可以获得更多的进展，但是有用的生物测定方法开发缓慢且成本高。因为栽培面积减少，在公共资助的研究中就没有太多的资金投入。与其他观赏植物相比，用于香石竹的公共基金减少，对其发展和地位形成负面影响。作物保护仍然是研究中非常重要的主题。观赏植物中许多不同的小种类使育种公司很难开发诸如生物测定类筛选技术，且很难获得投资回报。蔬菜作物生物测定技术的进步使人们对观赏植物寄予厚望，但很难将生物测定应用于观赏植物。”

5.6.4 黄杨木上的 *Calonectria pseudonaviculata* 和 *C. henricotiae*

黄杨木枯萎病是由 *Calonectria pseudonaviculata*（syns. *Cylindrocladium pseudonaviculatum* 和 *C. buxicola*）或 *Calonectria henricotiae* 引起的。尽管描述了黄杨木上丽赤壳属（*Calonectria*）的两个不同种，但到目前为止，欧洲仅报道了 *C. henricotiae* 这个种（Gehesquière et al. 2016）。在黄杨木中首次发现这种病原菌是在 20 世纪 90 年代的英国和新西兰（Henricot and Culham 2002；Ridley 1998）。黄杨木枯萎病的地理范围已迅速扩展到全球温带地区的 21 个以上国家（Gehesquière 2014）。

该病在黄杨属植物（*Buxus* spp.）上引起叶片出现斑点和茎病变。受感染的植株会迅速落叶并表现为顶梢枯死。这种病害主要在黄杨木苗圃、园林业、历史上的黄杨木花园及野生黄杨木种群中引起了关注。在苗圃中，化学控制是有效的，但在花园和公共绿地中则应避免使用。

黄杨属是一个传统上只有很少病虫害问题的属，但出现这种“新”病原菌后，使得世界各地不同研究小组对黄杨木基因型的这种病菌感病性测试方法进行研究。最近的许多文献描述了黄杨木基因型对丽赤壳属病菌（*Calonectria pseudonaviculata*）的感病性差异（Ganci et al. 2013；Gehesequière et al. 2012；Henricot et al. 2008；LaMondia 2015；Shishkoff et al. 2015）。在所有手稿中均使用了人工接种。大多数用离体的茎或插条进行测试（Guo and Olsen 2015；Henricot et al. 2008；La Mondia 2015；Shishkoff et al. 2015）。所有作者都描述了黄杨木基因型之间感病性存在巨大差异。

黄杨木大约有 70 个种，但是大多数商业基因型都属于 *B. sempervirens* 和 *B. microphylla*。种植者已经注意到，与属 *B. microphylla* 遗传簇的基因型相比，商业上最流行的 *B. sempervirens* 栽培品种更感病。尽管主要在感病性水平上存在差异，但从未在所测试的黄杨属（*Buxus*）基因型中观察到免疫（Gehesquière 2014）。

观察到种间育种的渗入主要从 *B. microphylla*、*B. sinica*、*B. harlandii*、*B. bodinieri* 及 *B. balearica* 渗入商业上更受欢迎的 *B. sempervirens* 遗传背景中（Van Laere et al. 2011）。在育种过程中使用人工接种进行抗病性测试。

育种者的观点：Didier Hermans-Herplant（比利时）

“基于对更强抗病性黄杨木基因型的需求，开始了育种计划，其唯一目的是降低黄杨木对枯萎病的感病性。生物测定法用于测试黄杨木对丽赤壳属（*Calonectria*）病菌的抗病性。在生物测定中获得的结果通常与在苗圃中观察到的抗病性相当。然而，在某些情况下，*Buxus sempervirens* 的品系在生物测定中表现良好，但在栽培过程中表现出更高的感病性。这可能是由接种浓度和环境条件存在差异造成的。在苗期选择过程中使用生物测定，随后在高侵染率的田间苗圃中进行第二次抗病选择，并使用有利于病害传播的感病植株。我们不使用生物测定法来选择父母本，因为大多数有关父母本性状的信息都源于栽培期间的经验。幼苗的生物测定由外部合作伙伴进行。尤其是在育苗场进行育种的情况下，该公司没有合适的基础设施和人员来以受控方式进行生物测定。我们的苗圃从黄杨木的园林绿化和生产开始，直到最近 10 年才开始增加育种这项工作。这种演变使我们能够确定某些种所需或有用的性状。与研究合作伙伴一起，我们能够获得资金；

这种合作为我们公司的育种工作获得了额外的价值。例如，植物学信息与黄杨属（*Buxus*）的分类学研究相结合，得出了潜在特定亲本植株抗病育种潜力的有价值信息。可以开发改良的品种，我们希望在 2019 年推销首个抗丽赤壳属（*Calonectria*）病菌的黄杨木品种。”

5.6.5　菊花上的白锈病菌（*Puccinia horiana*）

由白锈病菌（*Puccinia horiana*）引起的菊花白锈病是菊花（*Chrysanthemum* × *morifolium*）的一个主要问题。病原菌最初来自日本，已传播到世界上所有菊花生长的地区。它在许多国家/地区（如欧盟的插条）处于检疫状态。当检测到病原菌存在时，必须销毁幼苗（CABI/EPPO 2017）。

自 1975 年以来就已经开发出了抗白锈病的菊花。在 20 世纪 80 年代已经调查了其抗病性模式（De Jong and Rademaker 1986）。但是，当时没有使用致病型。基因库中存在极好的抗性基因，但是出于三个原因，在育种公司中很难进行真正的抗性育种。首先，病原菌的检疫地位要求育种公司安装所需的生物安全设施并据此进行操作，这既费钱又费力。其次，在世界范围内，病原菌群体内有许多致病型（生理小种）（De Backer et al. 2011）。由于新品种通常是面向全球市场生产的，因此这意味着抗病性筛选需要包括病原菌的特异分离物，以确保这些品种对所有已知的致病型病菌具有抗性。最后，白锈病菌（*Puccinia horiana*）是专性寄生的，不会产生生存结构。这需要通过将其定期转移到新植株上来维持相关测试分离株的致病性，增加了测试所需的成本和人力。因此，商业育种公司倾向于将抗菊花白锈病的筛选测试外包。例如，ILVO（比利时）可维持相关分离株的致病性，并为商业育种公司提供候选品种的筛选测试服务（表 5.1）。*Chrysanthemum* × *morifolium* 中抗白锈病菌（*Puccinia horiana*）系统的抗性基因依据基因与基因的相互作用来起作用，当存在抗性时，通常是抗性基因完全起作用的。

育种者的观点：Wim Declercq-Paraty Breeding/Gediflora（比利时）

“抗白锈病在菊花中很重要，因为该病原菌的检疫地位及已注册的杀菌剂数量减少。零售连锁店对其残留量提出了更多要求。允许使用的植物保护产品更少，并且可用产品的效力越来越差。预防性治疗仍是有用的，但几乎不可能实现治疗性应用，这可能部分是由于出现了对某些使用的杀菌剂更具耐受性的病原菌。现在已知白锈病菌有几种致病型。不同致病型病菌的检疫地位和发生增加了标准化生物测定的必要性。

在发展抗性品种的同时，仍在引进许多新的感病品种。因此，抗白锈病被视为新品种的额外优点。除繁殖效率外，商业方面的性状对育种者仍然最重要。尽管如此，在选择过程中仍会评价抗白锈病性。虽然在这方面已经取得进展，但进展缓慢。

在育种实践中，使用抗性基因型不能保证后代具有抗病性。不管对白锈病的抗性如何，都要将不具有其他特性的后代淘汰。

病害的发生取决于国家、地区和栽培方法。由于控制了叶片湿度，温室种植的菊花不易感病。因此，育种区域和系统都会影响抗病品种的选育。在南欧，由于气候温暖和干燥，病害发生的可能性较低。在某些国家，如美国，没有发现白锈病。然而，在比利时和荷兰，由于凉爽、潮湿的气候，病害发生的可能性更高。在病害发生可能性更高的地区，由于自然感染的可能性，在整个选择周期中可以更容易筛选出可室外生长的盆栽抗病性菊花。对于在温室中种植的切花和盆栽菊花，较低的病害发生可能阻碍了抗病性的选择。”

5.6.6 百合上的尖孢镰刀菌百合专化型（*Fusarium oxysporum* f. sp. *lilii*）

由镰刀菌引起的严重损失发生在百合小鳞茎和一年生种球阶段。定期使用杀菌剂可防止种球腐烂。开发了用于选择不同栽培阶段亚洲百合的筛选方法：早期选择的规模实验和最终检验的规模种球实验（Straathof and Löffler 1994）。Bakhshaie 等（2016）综述了百合在生物技术方面的进展，包括抗病性。20 多年前，Straathof 等（1997）发表了与尖孢镰刀菌抗性基因连锁的 RAPD 标记。采用了 150 个亚洲百合后代的回交群体。3 个标记可解释发现的镰刀菌抗性的 24%。

基于 251 个 AFLP 标记，开发了 100 个后代的遗传图谱。在不同的连锁群中发现了 4 个抗镰刀菌的 QTL。在该图谱上也确定了一个郁金香碎裂病毒的抗性基因（Van Heusden et al. 2002）。Shahin 等（2011）通过添加阵列技术（DArT）和核苷酸结合位点（NBS）分析改进了该图谱，绘制了 1 个与百合斑驳病毒（LMoV）抗性相关的位点和 6 个与抗镰刀菌相关的推定 QTL。一个单一的 QTL 解释了抗病性变异的 25%。

Liu 等（2011）在不同倍性水平的东方百合中发现了对镰刀菌的抗性。同样在该研究中，发现具有较高皂苷含量的突变体表现出较高的抗病性。Rao 等（2014）研究了岷江百合（*Lilium regale*）和尖孢镰刀菌（*F. oxysporum*）之间不亲和的百合与菌互作的基因表达谱。利用 EST 的寡核苷酸微阵列分析研究感病的东方百合杂种和尖孢镰刀菌（*F. oxysporum*）的亲和性相互作用。岷江百合（*Lilium regale*）对植物致病菌和干旱具有天然抗性。因此，该百合种被认为是筛选非生物和生物胁迫抗性基因的来源。已经从岷江百合（*Lilium regale*）中分离出谷胱甘肽-*S*-转移酶基因（*LrGSTU5*）（Han et al. 2016）、基础的亮氨酸拉链基因（*LrbZIP1*）（Zhang et al. 2014a）和萌发素类蛋白基因（*LrGLP1*）（Zhang et al. 2017），并描述为尖孢镰刀菌（*Fusarium oxysporu*m）抗性基因。

育种者的观点：Stefan van der Heijden-Royal Van Zanten（荷兰）

“目前，镰刀菌抗性在百合育种者的优先考虑列表中并不排在前面，因此在育种计划的早期阶段并未专门针对该病害。由于种球实践栽培经验的提高，抗病性品种的附加值很低。种植者在一个地块上种植不同的品种，只有几行抗尖孢镰刀菌品种无法克服由真菌引起的问题。种植者主要通过轮作或土壤消毒来解决镰刀菌的问题。在育种过程的最后阶段，在生物测定中用混合接种物测试抗病性。品种之间存在差异，但与旧品种相比，新品种总体上未表现出较大的改良。百合的育种周期相当长（12～15 年），部分原因是繁殖种球需要时间。除镰刀菌外，病毒问题正变得越来越重要，主要问题是由昆虫在繁殖期间和贮藏期间传播病毒引起的。因此，在百合育种和繁殖中通常会启动抗病毒的筛选。预计这些问题对于将来的育种会变得重要。

我们公司已在育种计划中投资应用分子标记。我们相信特定标记的应用将显著改善百合的抗病育种。对多种致病菌的抗病性将成为育种过程早期选择阶段筛选的特征之一。我们还期望这将对百合的镰刀菌抗性产生积极影响。”

5.7 展　　望

使用抗性品种是控制植物病原菌最有效、最持久的方法。简便而适当的筛选方法有助于创建更多的抗病性观赏植物。不同观赏植物育种者分享的经验表明，生物测定法是普遍使用的方法，对于某些作物，它是筛选和选择的一个重要工具。这些测试的可重复性和有效性至关重要，但是，对于某些植物-病原菌组合而言，可能会出现问题。在许多观赏植物中，通过育种提高了抗病性。这些成功的例子将有助于在其他观赏植物上开发类似的计划，并有望激发育种者对在其他作物上培育抗其他病害品种的热情。

例如，与其他 IPM 手段结合应用时，抗病性的小幅提高可能对种植者产生积极影响。较慢较低程度的病害发展可以增加通过采取其他措施来对抗病原物的成功率。

在观赏植物中，可以获得大量关于生物技术方法特别是遗传转化方面的文献。然而，主要的法律限制了市场化转基因品种的数量。抗病分子标记已经在某些观赏植物中开发。尚不清楚是否有任何商业育种公司将这种标记应用于抗病性。通常，与分子方法相比，普通方法对植物性状筛选的成本更高。

基于高通量测序的新进展已用于其他农作物。现在，这些应用为观赏植物育种铺平了道路。需要更多关于植物-病原菌相互作用的知识来研究每种植物上的病原菌，植物种类的广泛差异，阻碍了观赏植物抗病育种的进展。然而，对模式植物和重要经济作物的基本了解与可用的基础知识都越来越多。

开发具有较好抗性植物的新技术正在进行中。新的育种技术可能会在没有法律限制的情况下取代遗传转化，但是在将这些技术应用于观赏植物之前，仍有许多工作要做。

最后，成本效益将决定特定观赏植物对病原菌抗性的重要性。可以预期，减少农药使用的立法只会增加种植者和消费者对抗病植物的需求，并将迫使育种者大量投资于抗病性的研究和开发。

致谢：作者感谢 Miriam Levenson、Kurt Heungens 和 Johan Van Huylenbroeck 对手稿结构与需要核实的常见问题的有益评论以及对英语的纠正。还要感谢 Kurt Heungens 在某些植物病理学问题方面提供的有益帮助。作者还想对育种公司和观赏植物育种者表示感谢，他们提供了有价值的信息和抗病育种方面的相关经验。

本章译者：

张艺萍，王继华

云南省农业科学院花卉研究所，国家观赏园艺工程技术研究中心，云南省花卉育种重点实验室，昆明 650200

参考文献

Anaïs G, Darrasse A, Prior P (2000) Breeding anthuriums (*Anthurium andreanum* L.) for resistance to bacterial blight caused by *Xanthomonas campestris* pv. *dieffenbachiae*. Acta Hortic 508:135–140

Arens P, Bijman P, Tang N, Shahin A, van Tuyl JM (2012) Mapping of disease resistance in ornamentals: a long haul. Acta Hortic 953:231–237

Armijo G, Schlechter R, Agurto M, Muñoz D, Nuñez C, Arce-Johnson P (2016) Grapevine pathogenic microorganisms: understanding infection strategies and host response scenarios. Front Plant Sci 7:382

Arnold DL, Jackson RW (2011) Bacterial genomes: evolution of pathogenicity. Curr Opin Plant Biol 14:385–391

Azadi P, Otang NV, Supaporn H, Khan RS, Chin DP, Nakamura I, Mii M (2011) Increased resistance to cucumber mosaic virus (CMV) in *Lilium* transformed with a defective CMV replicate gene. Biotechnol Lett 33:1249–1255

Azadi P, Bagheri H, Nalousi AM, Nazari F, Chandler SF (2016) Current status and biotechnological advances in genetic engineering of ornamental plants. Biotechnol Adv 34:1073–1090

Baayen RP, Sparnaaij LD, Jansen J, Niemann GJ (1991) Inheritance of resistance in carnation against *Fusarium oxysporum* f.sp. *dianthi* races 1 and 2, in relation to resistance components. Neth J Plant Pathol 97:73–86

Backhaus GF (1994a) *Cylindrocladium scoparium* causing wilt disease in *Rhododendron* and azalea. Acta Hortic 364:163–166

Backhaus GF (1994b) *Phytophthora citricola* (Sawada) – cause of an important shoot rot of *Rhododendron* and azalea. Acta Hortic 364:145–154

Bakhshaie M, Khosravi S, Azadi P, Bagheri H, Van Tuyl JM (2016) Biotechnological advances in *Lilium*. Plant Cell Rep 35:1799–1826

Bari R, Jones JD (2009) Role of plant hormones in plant defence responses. Plant Mol Biol 69:473–488

Ben-Yephet Y, Reuven M, Mor Y (1993) Selection methods for determining resistance of carnation cultivars to *Fusarium oxysporum* f.sp. *dianthi*. Plant Pathol 42:517–521

Ben-Yephet Y, Reuven M, Zveibil A, Shtienberg D (1996) Effects of abiotic variables on the response of carnation cultivars to *Fusarium oxysporum* f.sp.*dianthi*. Plant Pathol 45:98–105

Boller T, Felix G (2009) A renaissance of elicitors: perception of microbe-associated molecular patterns and danger signals by pattern-recognition receptors. Annu Rev Plant Biol 60:379–406

Bolley HL (1901) Flax wilt and flax sick soil. Norht Dakota Experiment Station Bulletin 50

Brugliera F, Kalc-Wright G, Hyland C, Webb L, Herbert S, Sheehan B, Mason JG (2000) Improvement of *Fusarium* wilt tolerance in carnations expressing chitinase. Int Plant Mol Biol Rep 18:522–529

CABI/EPPO (2017) n° 236 https://www.cabi.org/isc/datasheet/45806. Accessed 04 Mar 2018

Cai J, Liu X, Vanneste K, Proost S, Tsai WC, Liu KW, Chen LJ, He Y, Xu Q, Bian C, Zheng ZJ, Sun FM, Liu WQ, Hsiao YY, Pan ZJ, Hsu CC, Yang YP, Hsu YC, Chuang YC, Dievart A, Dufayard JF, Xu X, Wang JY, Wang J, Xiao XJ, Zhao XM, Du R, Zhang GQ, Wang MN, Su YY, Xie GC, Liu GH, Li LQ, Huang LQ, Luo YB, Chen HH, Van de Peer Y, Liu ZJ (2015) The genome sequence of the orchid *Phalaenopsis equestris*. Nat Genet 47:65–72

Chan YL, Lin KH, Sanjaya, Liao LJ, Chen WH, Chan MT (2005) Gene stacking in Phalaenopsis orchid enhances dual tolerance to pathogen attack. Transgenic Res 14:279–288

Clarke JL, Spetz C, Haugslien S, Xing S, Dees MW, Moe R, Blystad DR (2008) Agrobacterium tumefaciens-mediated transformation of poinsettia, *Euphorbia pulcherrima*, with virus-derived hairpin RNA constructs confers resistance to Poinsettia mosaic virus. Plant Cell Rep 27:1027–1038

Collinge DB, Jorgensen HJL, Lund OS, Lyngkjaer MF (2010) Engineering pathogen resistance in crop plants: current trends and future prospects. Annu Rev Phytopathol 48:269–291

De Backer M, Alaei Shah Anar Vanar H, Van Bockstaele E, Roldàn-Ruiz I, van der Lee T, Maes M, Heungens K (2011) Identification and characterization of pathotypes in Puccinia horiana, a rust pathogen of Chrysanthemum x morifolium. Eur J Plant Pathol 130:325–338

De Cáceres González FFN, Davey MR, Sanchez EC, Wilson ZA (2015) Conferred resistance to *Botrytis cinerea* in *Lilium* by overexpression of the *RCH10* chitinase gene. Plant Cell Rep 34:1201–1209

De Jong J, Rademaker W (1986) The reaction of *Chrysanthemum* cultivars to *Puccinia horiana* and the inheritance of resistance. Euphytica 3:945–952

De Keyser E, De Riek J, Heungens K (2008) Development of supporting techniques for pot azalea (*Rhododendron simsii* hybrids) breeding focused on plant quality, disease resistance and enlargement of the assortment. Acta Hortic 766:361–366

Debener T (2009) Current strategies and future prospects of resistance breeding in ornamentals. Acta Hortic 836:125–130

Debener T, Byrne D (2014) Disease resistance breeding in rose: current status and potential of biotechnological tools. Plant Sci 228:107–117

Demmink JF, Baayen RP, Sparnaaij LD (1989) Evaluation of the virulence of races 1, 2 and 4 of *Fusarium oxysporum* f. sp. *dianthi* in carnation. Euphytica 42:55–63

Dohm A, Ludwig C, Schilling D, Debener T (2001) Transformation of roses with genes for antifungal proteins. Acta Hortic 547:27–33

Dohm A, Ludwig C, Schilling D, Debener T (2002) Transformation of roses with genes for antifungal proteins to reduce their susceptibility to fungal diseases. Acta Hortic 572:105–111

Elibox W, Umaharan P (2007) The inheritance of systematic resistance to the bacterial blight pathogen (*Xanthomonas axonopodis* pv. *dieffenbachiae*) in *Anthurium andreanum*. Sci Hortic 115:76–81

Elibox W, Umaharan P (2008a) A quantitative screening method for the detection of foliar resistance to *Xanthomonas axonopodis* pv. *dieffenbachiae* in anthurium. Eur J Plant Pathol 121:35–42

Elibox W, Umaharan P (2008b) Genetic basis of resistance to systemic infection by *Xanthomonas axonopodis* pv. *dieffenbachiae* in *Anthurium*. Phytopathology 98:421–426

Elibox W, Umaharan P (2010) Inheritance of resistance to foliar infection by *Xanthomonas axonopodis* pv. *dieffenbachiae* in *Anthurium*. Plant Dis 94:1243–1247

EU Fact Sheet (2015) Emergency control measures by species. https://ec.europa.eu/food/plant/plant_health_biosecurity/legislation/emergency_measures/xylella-fastidiosa_en. Accessed 04 Mar 2018

Fawke S, Doumane M, Schornack S (2015) Oomycete interactions with plants: infection strategies and resistance principles. Microbiol Mol Biol Rev 79:263–279

Galletti R, De Lorenzo G, Ferrari S (2009) Host-derived signals activate plant innate immunity. Plant Signal Behav 4:33–34

Ganci M, Benson DM, Ivors K (2013) Susceptibility of commercial boxwood varieties to boxwood blight (boxwood cultivars with tolerance to box blight). North Carolina Coop. Ext., Plant Pathology, Raleigh. https://plantpathology.ces.ncsu.edu/wp-content/uploads/2013/05/final-Cult-trials-summary-2013.pdf?fwd=no. Accessed 04 Mar 2018

Gehesquiere B (2014) *Cylindrocladium buxicola* nom. cons. prop. (syn. *Calonectria pseudonaviculata*) on *Buxus*: Molecular characterization, epidemiology, host resistance and fungicide control. PhD dissertation, Ghent University

Gehesquière B, Rys F, Maes M, Gobin B, Van Huylenbroweck J, Heungens K (2012) Genotypic and phenotypic variation in *Cylindrocladium buxicola*. (Abstr) Commun Agric Appl Biol Sci 77:95–96

Gehesquière B, Crouch JA, Marra RE, Van Poucke K, Rys F, Maes M, Gobin B, Höfte M, Heungens K (2016) Characterization and taxonomic reassessment of the box blight pathogen *Calonectria pseudonaviculata*, introducing *Calonectria henricotiae* sp. nov. Plant Pathol 65:37–52

Growns DJ (2015) Phenotypic recurrent selection for disease tolerance in *Anigozanthos* spp. L. Acta Hortic 1097:101–106

Grünwald NJ, Garbelotto M, Goss EM, Heungens K, Prospero S (2012) Emergence of the sudden oak death pathogen Phytophthora ramorum. Trends Microbiol 20:131–138

Guo Y, Olsen RT (2015) Effective bioassays for evaluating boxwood blight susceptibility using detached stem inoculations. Hortic Sci 50:268–271

Han Q, Chen R, Yang Y, Cui X, Ge F, Chen C, Liu D (2016) A glutathione S-transferase gene from Lilium regale Wilson confers transgenic tobacco resistance to Fusarium oxysporum. Sci Hortic 198:370–378

Hardham (2005) Mol Plant Pathol 6:589–604

He JP, Chen FD, Chen SM, Fang WM, Miao HB, Luo HL (2009) Transformation of *Lycoris longituba* agglutinin gene to cut chrysanthemum and identification of aphid resistance in the transgenic plants. Acta Bot Boreal Occident Sin 29:2318–2325

Hennebert GL (1973) *Botrytis* and *Botrytis*-like genera. Persoonia 7:183–204

Henricot B, Culham A (2002) *Cylindrocladium buxicola*, a new species affecting *Buxus* spp., and its phylogenetic status. Mycologia 94:980–997

Henricot B, Gorton C, Denton G, Denton J (2008) Studies on the control of *Cylindrocladium buxicola* using fungicides and host resistance. Plant Dis 92:1273–1279

Höfte M (2015) Basal and induced disease resistance mechanisms in ornamentals. Acta Hortic 1087:473–478

Jarvis WR (1977) *Botryotinia* and *Botrytis* species: taxonomy, physiology and pathogenicity, Monograph no. 15. Canadian Department of Agriculture, Ottawa

Jepson P, Arakelyan I (2017) Exploring public perceptions of solutions to tree diseases in the UK: implications for policy-makers. Environ Sci Pol 76:70–77

Jones JDG, Dangl JL (2006) The plant immune system. Nature 444:323–329

Jorgensen IH (1992) Discovery, characterization and exploitation of *Mlo* powdery mildew resistance in barley. Euphytica 63:141–152

Jung T, Cooke DEL, Blaschke H, Duncan JM, Oßwald W (1999) *Phytophthora quercina* sp. nov., causing root rot of European oaks. Mycol Res 103:785–798

Kamo K, Gera A, Cohen J, Hammond J, Blowers A, Smith F, Van Eck J (2005) Transgenic gladiolus plants transformed with the bean yellow mosaic virus coat-protein gene in either sense or antisense orientation. Plant Cell Rep 23:654–663

Kamo K, Jordan R, Guaragna MA, Hsu HT, Ueng P (2010) Resistance to Cucumber mosaic virus in Gladiolus plants transformed with either a defective replicase or coat protein subgroup II gene from Cucumber mosaic virus. Plant Cell Rep 29:695–704

Kamo K, Aebig J, Guaragna MA, James C, Hsu HT, Jordan R (2012) Gladiolus plants transformed with single-chain variable fragment antibodies to Cucumber mosaic virus. Plant Cell Tissue Organ Cult 110:13–21

Kamo K, Lakshman D, Bauchan G, Rajasekaran K, Cary J, Jaynes J (2015) Expression of a synthetic antimicrobial peptide, D4E1, in *Gladiolus* plants for resistance to *Fusarium oxysporum* f. sp *gladioli*. Plant Cell Tissue Organ Cult 121:459–467

Kamo K, Lakshman D, Pandey R, Guaragna MA, Okubara P, Rajasekaran K, Cary J, Jordan R (2016) Resistance to *Fusarium oxysporum* f. sp *gladioli* in transgenic *Gladiolus* plants expressing either a bacterial chloroperoxidase or fungal chitinase genes. Plant Cell Tissue Organ Cult 124:541–553

Kardos JH, Robacker CD, Dirr MA, Rinehart TA (2009) Production and verification of *Hydrangea macrophylla* × *H. angustipetala* hybrids. Hortscience 44:1534–1537

Kaufmann H, Qiu X, Wehmeyer J, Debener T (2012) Isolation, molecular characterization, and mapping of four rose *MLO* orthologs. Front Plant Sci 3:244

Kellerhals M, Szalatnay D, Hunziker K, Duffy B, Nybom H, Ahmadi-Afzadi M, Höfer M, Richter K, Lateur M (2012) European pome fruit genetic resources evaluated for disease resistance. Trees 26:179–189

Kim YS, Lim S, Yoda H, Choi YE, Sano H (2011) Simultaneous activation of salicylate production and fungal resistance in transgenic *Chrysanthemum* producing caffeine. Plant Signal Behav 6:409–412

Koning-Boucoiran CF, Gitonga VW, Yan Z, Dolstra O, van der Linden CG, van der Schoot J, Uenk GE, Verlinden K, Smulders MJ, Krens FA, Maliepaard C (2012) The mode of inheritance in tetraploid cut roses. Theor Appl Genet 125:591–607

Korbin M (2006) Assessment of gerbera plants genetically modified with TSWV nucleocapsid gene. J Fruit Ornam Plant Res 14:85–93

Kuehnle AR, Fujii T, Chen FC, Alvarez A, Sugii N, Fukui R, Aragon SL (2004a) Peptide biocides for engineering bacterial blight tolerance and susceptibility in cut flower anthurium. Hortscience 39:1327–1331

Kuehnle AR, Fujii T, Mudalige R, Alvarez A (2004b) Gene and genome mélange in breeding of *Anthurium* and *Dendrobium* orchid. Acta Hortic 651:115–122

Kumar S, Raj SK, Sharma AK, Varma HN (2012) Genetic transformation and development of cucumber mosaic virus resistant transgenic plants of *Chrysanthemum morifolium* cv. Kundan. Sci Hortic 134:40–45

LaMondia JA (2015) Management of *Calonectria pseudonaviculata* in boxwood with fungicides and less susceptible host species and varieties. Plant Dis 99:363–369

Leach JE, Leung H, Tisserat NA (2014) Plant disease and resistance. In: Van Alfen NK (ed) Encyclopedia of agriculture and food systems, vol 4. New York, pp 360–374

Lentola A, David A, Abdul-Sasa A, Tapparo A, Goulson D, Hill EM (2017) Ornamental plants on sale to the public are a significant source of pesticide residues with implications for the health of pollinating insects. Environ Pollut 228:297–304

Li X, Gasic K, Cammue B, Broekaert W, Korban SS (2003) Transgenic rose lines harboring an antimicrobial protein gene, Ace-AMP1, demonstrate enhanced resistance to powdery mildew (*Sphaerotheca pannosa*). Planta 218:226–232

Li Y, Trigiano R, Reed S, Rinehart T, Spiers J (2009) Assessment of resistance components of bigleaf hydrangeas (*Hydrangea macrophylla*) to *Erysiphe polygoni* in vitro. Can J Plant Pathol 31:348–355

Liao LJ, Pan IC, Chan YL, Hsu YH, Chen WH, Chan MT (2004) Transgene silencing in Phalaenopsis expressing the coat protein of Cymbidium Mosaic Virus is a manifestation of RNA-mediated resistance. Mol Breed 13:229–242

Liu WL, Wu LF, Wu HZ, Zheng SX, Wang JH, Liu FH (2011) Correlation of saponin content and *Fusarium* resistance in hybrids from different ploidy levels of *Lilium* oriental. Sci Hortic 129:849–853

Luypaert G, Van Huylenbroeck J, De Riek J, De Clercq P (2014) Screening for broad mite susceptibility in *Rhododendron simsii* hybrids. J Plant Dis Protect 121:260–269

Marchant R, Davey MR, Lucas JA, Lamb CJ, Dixon RA, Power JB (1998) Expression of a chitinase transgene in rose (Rosa hybrida L.) reduces development of blackspot disease (*Diplocarpon rosae* Wolf). Mol Breed 4:187–194

Menda N, Strickler SR, Edwards JD, Bombarely A, Dunham DM, Martin GB, Mejia L, Hutton SF, Havey MJ, Maxwell DP, Mueller LA (2014) Analysis of wild-species introgressions in tomato inbreds uncovers ancestral origins. BMC Plant Biol 14:287

Miao H, Jiang B, Chen S, Zhang S, Chen F, Fang W, Teng N, Guan Z (2010) Isolation of a gibberellin 20-oxidasec DNA from and characterization of its expression in chrysanthemum. Plant Breed 129:707–714

Miedaner T, Korzun V (2012) Marker-assisted selection for disease resistance in wheat and barley breeding. Phytopathology 102:560–566

Mitteau Y (1987) Breeding of new carnations resistant to *Fusarium oxysporum*. Acta Hortic 216:359–366

Mork EK (2011) Disease resistance in ornamental plants – transformation of Symphyotrichum novi-belgii with powdery mildew resistance genes. PhD dissertation, Aarhus University

Motaung TE, Saitoh H, Tsilo TJ (2017) Large-scale molecular genetic analysis in plant-pathogenic fungi: a decade of genome-wide functional analysis. Mol Plant Pathol 18:754–764

NAKtuinbouw (2018a) https://www.naktuinbouw.com/test/resistance-tests-vegetable-crop-varieties?_ga=2.100665095.724202317.1519139659-93668408.1515697636. Accessed 04 Mar 2018

NAKtuinbouw (2018b) https://www.naktuinbouw.com/test/resistance-tests-ornamental-crop-varieties. Accessed 04 Mar 2018

Neale DB, Kremer A (2011) Forest tree genomics: growing resources and applications. Nat Rev Genet 12:111–122

Ortega F, Lopez-Vizcon C (2012) Application of molecular marker-assisted selection (MAS) for disease resistance in a practical potato breeding programme. Potato Res 55:1–13

Orton WA (1918) Breeding for disease resistance in plants. Am J Bot 5:279–283

Oßwald W, Fleischmann F, Rigling D, Coelho AC, Cravador A, Diez J, Dalio RJ, Horta Jung M, Pfanz H, Robin C, Sipos G, Solla A, Cech T, Chambery A, Diamandis S, Hansen E, Jung T, Orlikowski LB, Parke J, Prospero S, Werres S (2014) Strategies of attack and defence in woody plant–*Phytophthora* interactions. For Pathol 44:169–190

Pavan S, Jacobsen E, Visser RGF, Bai Y (2010) Loss of susceptibility as a novel breeding strategy for durable and broad-spectrum resistance. Mol Breed 25:1–12

Pilet-Nayel ML, Moury B, Caffier V, Montarry J, Kerlan MC, Fournet S, Durel CE, Delourme R (2017) Quantitative resistance to plant pathogens in pyramiding strategies for durable crop protection. Front Plant Sci 8:1838

Pourhosseini L, Kermani MJ, Habashi AA, Khalighi A (2013) Efficiency of direct and indirect shoot organogenesis in different genotypes of *Rosa hybrida*. Plant Cell Tissue Organ Cult 112:101–108

Prados-Ligero AM, Basallote-Urba MJ, Lopze-Herrera CJ, Melero-Vara JM (2007) Evaluation of susceptibility of carnation cultivars to fusarium wilt and determination of *Fusarium oxysporum* f.sp. *dianthi* races in Southwest Spain. Hortscience 42:596–599

Proefstation voor de Bloemisterij in Nederland (1984) Overzicht van het onderzoek over schimmelvaatziekten bij anjers in de periode 1939-1983. Report 23

Rao J, Liu D, Zhang N, He H, Ge F, Chen C (2014) Differential gene expression in incompatible interaction between *Lilium regale* Wilson and *Fusarium oxysporum* f. sp. *lilii* revealed by combined SSH and microarray analysis. Mol Biol 48:802–812

Ridley G (1998) New plant fungus found in Auckland box hedges (*Buxus*). Forest Health News 77:1

Rispail N, Rubiales D (2016) Genome-wide identification and comparison of legume *MLO* gene family. Sci Rep 6:32673

Rytkönen A, Lilja A, Vercauteren A, Sirkiä A, Parikka P, Soukainen M, Hantula J (2012) Identity and potential pathogenicity of *Phytophthora* species found on symptomatic *Rhododendron* plants in a Finnish nursery. Can J Plant Pathol 34:255–267

Sen S, Kumar S, Ghani M, Thakur M (2013) *Agrobacterium* mediated genetic transformation of chrysanthemum (*Dendranthema grandiflora* Tzvelev) with rice chitinase gene for improved resistance against *Septoria obesa*. Plant Pathol J 12:1–10

Shahin A, Arens P, Van Heusden AW, Van Der Linden G, Van Kaauwen M, Khan N, Schouten HJ, Van De Weg WE, Visser RGF, Van Tuyl JM (2011) Genetic mapping in *Lilium*: mapping of major genes and quantitative trait loci for several ornamental traits and disease resistances. Plant Breed 130:372–382

Shakoor N, Lee S, Mockler TC (2017) High throughput phenotyping to accelerate crop breeding and monitoring of diseases in the field. Curr Opin Plant Biol 38:184–192

Sharma R, Messar Y (2017) Transgenics in ornamental crops: creating novelties in economically important cut flowers. Curr Sci 113:43–52

Sherman JM, Moyer JW, Daub ME (1998) Tomato spotted wilt virus resistance in chrysanthemum expressing the viral nucleocapsid gene. Plant Dis 82:407–414

Shinoyama H, Mochizuki A, Komano M, Nomura Y, Nagai T (2003) Insect resistance in transgenic chrysanthemum (*Dendranthema* x *grandiflorum* (Ramat) Kitamura) by the introduction of a modified δ-endotoxin gene of *Bacillus thuringiensis*. Breed Sci 53:359–367

Shinoyama H, Mochizuki A, Nomura Y, Kamada H (2008) Environmental risk assessment of genetically modified chrysanthemums containing a modified cry1Ab gene from *Bacillus thuringiensis*. Plant Biotechnol 25:17–29

Shinoyama H, Sano T, Saito M, Ezura H, Aida R, Nomura Y et al (2012) Induction of male sterility in transgenic chrysanthemums (*Chrysanthemum morifolium* Ramat.) by expression of a mutated ethylene receptor gene, Cm-ETR1/H69A, and the stability of this sterility at varying growth temperatures. Mol Breed 29:285–295

Shinoyama H, Mitsuhara I, Ichikawa H, Kato K, Mochizuki A (2015) Transgenic chrysanthemums (Chrysanthemum morifolium Ramat.) carrying both insect and disease resistance. Acta Hortic 1087:485–497

Shishkoff N, Daughtrey M, Aker S, Olsen RT (2015) Evaluating boxwood susceptibility to *Calonectria pseudonaviculata* using cuttings from the National Boxwood Collection. Plant Health Prog 16:11–15

Skirvin DJ, De Courcy Williams ME, Fenlon JS, Sunderland KD (2002) Modelling the effects of plant species on biocontrol effectiveness in ornamental nursery crops. J Appl Ecol 39:469–480

Sparnaaij LD, Demmink JF (1976) Breeding for resistance to *Phialophora cinerescens* (Wr.) Van Beyma in glasshouse carnations (*Dianthus caryophyllus* L.). Euphytica 25:329–338

Staats M, van Baarlen P, van Kan JAL (2004) Molecular phylogeny of the plant pathogenic genus *Botrytis* and the evolution of host specificity. Mol Biol Evol 22:333–346

Straathof TP, Löffler HJM (1994) Screening for Fusarium resistance in seedling populations of Asiatic hybrid lily. Euphytica 78:43–51

Straathof TP, Löffler HJM, Linfield CA, Roebroeck EJA (1997) Breeding for resistance to Fusarium oxysporum in flower bulbs. Acta Hortic 430:477–486

Takatsua Y, Nishizawa Y, Hibi T, Akutsu K (1999) Transgenic chrysanthemum (*Dendranthema grandiflorum* (Ramat.) Kitamura) expressing a rice chitinase gene shows enhanced resistance to gray mold (*Botrytis cinerea*). Sci Hortic 82:113–123

Teixeira da Silva JA, Dobránszki J, Zeng S, Winarto B, Lennon AM, Jaufeerally-Fakim Y, Christopher DA (2015) Genetic transformation and molecular research in Anthurium: progress and prospects. Plant Cell Tissue Organ Cult 123:205–219

Thakur M, Sharma D, Sharma S (2002) In vitro selection and regeneration of carnation (*Dianthus caryophyllus* L.) plants resistant to culture filtrate of *Fusarium oxysporum* f.sp. *dianthi*. Plant Cell Rep 20:825–828

Uchneat MS, Zhigilei A, Craig R (1999) Differential response to foliar infection with *Botrytis cinerea* within the genus *Pelargonium*. J Am Soc Hortic Sci 124:76–80
van den Bulk RW (1991) Application of cell and tissue culture and in vitro selection for disease resistance breeding. Euphytica 56:269–285
Van Heusden AW, Jongerius MC, Van Tuyl JM, Straathof TP, Mes JJ (2002) Molecular assisted breeding for disease resistance in lily. Acta Hortic 572:131–138
Van Huylenbroeck J, Calsyn E, De Keyser E, Luypaert G (2015) Breeding for biotic stress resistance in *Rhododendron simsii*. Acta Hortic 1104:375–379
van Kan JAL, Shaw MW, Grant-Downton RT (2014) *Botrytis* species: relentless necrotrophic thugs or endophytes gone rogue? Mol Plant Pathol 15:957–961
Van Laere K, Hermans D, Leus L, Van Huylenbroeck J (2011) Genetic relationships in European and Asiatic *Buxus* species based on AFLP markers, genome sizes and chromosome numbers. Plant Syst Evol 293:1–11
Van Laere K, Hermans D, Leus L, Van Huylenbroeck J (2015) Interspecific hybridisation within *Buxus* spp. Sci Hortic 185:139–144
van Schie CCN, Takken FLW (2014) Susceptibility genes 101: how to be a good host. Annu Rev Phytopathol 52:551–581
van Wordragen MF, Honée G, Dons HJM (1993) Insect resistant chrysanthemum calluses by introduction of a *Bacillus thuringiensis* crystal protein gene. Transgenic Res 2:170–180
Vieira P, Wantoch S, Lilley CJ, Chitwood DJ, Atkinson HJ, Kamo K (2015) Expression of a cystatin transgene can confer resistance to root lesion nematodes in *Lilium longiflorum* 'Nellie White'. Transgenic Res 24:421–432
Vyska M, Cunniffe N, Gilligan C (2016) Trade-off between disease resistance and crop yield: a landscape-scale mathematical modelling perspective. J R Soc Interface 13:20160451
Xiong JS, Ding J, Li Y (2015) Genome-editing technologies and their potential application in horticultural crop breeding. Hortic Res 2:15019
Xu Y, Crouch JH (2008) Marker-assisted selection in plant breeding: from publication to practice. Crop Sci 48:391–407
Xu G, Chen S, Chen F (2010) Transgenic chrysanthemum plants expressing a harpin (Xoo) gene demonstrate induced resistance to alternaria leaf spot and accelerated development. Russ J Plant Physiol 57:548–553
Xu G, Liu Y, Chen S, Chen F (2011) Potential structural and biochemical mechanisms of compositae wild species resistance to *Alternaria tenuissima*. Russ J Plant Physiol 58:491–497
Zeller W (1979) Resistance and resistance breeding in ornamentals. EPPO Bull 9:35–44
Zhang NN, Liu DQ, Zheng W, He H, Ji B, Han Q, Ge F, Chen CY (2014a) A bZIP transcription factor, LrbZIP1, is involved in *Lilium regale* Wilson defense repsonses against *Fusarium oxysporym* f.sp. *lilii*. Genes Genom 36:789–798
Zhang YP, Jiang S, Qu SP, Yang XM, Wang XN, Ma LL, Wu LL, He YQ, Wang JH (2014b) In vitro selection for Fusarium resistant oriental lily mutants using culture filtrate of the fungal agent. Acta Hortic 1027:205–212
Zhang N, Guan R, Yang Y, Bai Z, Ge F, Liu D (2017) Isolation and characterization of a *Fusarium oxysporum*-resistant gene LrGLP1 from *Lilium regale* Wilson. In Vitro Cell Dev Biol Plant 53:461–468
Zhu S, Li Y, Vossen JH, Visser RGF, Jacobsen E (2012) Functional stacking of three resistance genes against *Phytophthora infestans* in potato. Transgenic Res 21:89–99

第6章　植物组织培养技术

Mark P. Bridgen, Wouter Van Houtven, and Tom Eeckhaut

摘要　几种组织培养技术在植物中可产生遗传变异。这些离体培养技术已成功应用于多个物种，并已用于创建抗病虫害和耐水胁迫的植物品种，这些品种表现出更高的生长率和产量。传统植物育种通过如挽救杂交后代的胚培养等技术得到发展。原生质体融合和体细胞克隆变异被用于开发具有更多遗传多样性或新的遗传变异性的植株。一旦育成新品种，微繁殖技术可以为商业市场进行有效的大规模克隆繁殖。如果没有植物组织培养技术的参与，近代基因工程是不可能发展的。

关键词　胚培养；微繁殖；植物组织培养；原生质体；体细胞克隆变异

6.1　引　　言

植物组织培养是在可控的环境条件下，即温度、湿度、光照和营养物质均能提供适宜的生长环境，在培养基上进行植物细胞、组织、器官、胚胎、原生质体或种子的无菌培养。成功的植物组织培养得益于科学知识和实践经验。而实践这些方法和技术的次数越多，使用这些技术的成功率就越高。植物组织培养是一门科学，因需要经过培训和相应的知识结构才能了解预期结果是什么以及如何达到预期的结果。

组织培养是一种简单而强大的生物技术，可让研究人员在无菌环境中控制植物生长，是一种对植物育种特别有用的方法。植物组织培养有许多用途，包括快速克隆繁殖或微繁殖；抗病性筛选；用于产生单倍体植物的花药、花粉培养；用于产生多倍体的原生质体培养；胚培养；植物次生代谢物的生产；细胞选择和突变；植物解剖学基础研究，发育和营养方面的研究；植物生物技术；体细胞胚发生；人工种子的合成；愈伤组织培养和花卉栽培（Bridgen 1994b；Kyte et al. 2013）。

Gottlieb Haberlandt 在 1902 年提出组织培养的假设，该假设指出植物具有细胞全能性，因此能够由单个细胞再生成新植物（Haberlandt 1902）。组织培养发展的第一个阶段是在 1922 年，Knudson 在培养基中成功萌发了兰花种子并获得植株（Knudson 1922）。该培养基后来被称为 KC 培养基，通常用于兰花种子的萌发。组织培养发展的第二个阶段是在 1929 年，Went（Went 1929）发现了促使根生成的物质。1948 年，Skoog 和 Tsui

M. P. Bridgen (✉)
Cornell University, Riverhead, NY, USA
e-mail: mpb27@cornell.edu
W. Van Houtven · T. Eeckhaut
Flanders Research Institute for Agriculture, Fisheries and Food (ILVO), Plant Sciences Unit, Applied Genetics and Breeding, Melle, Belgium

发现了诱导芽形成的激动素，这一突破性发现使植物组织培养中的繁殖成为可能（Skoog and Tsui 1948）。后来，Skoog 和 Miller 指出了生长素与细胞分裂素比例的重要性，以及该比例对芽诱导、根和愈伤组织形成的影响（Skoog and Miller 1957）。Toshio Murashige 在 1974 年对植物的微繁殖及其培养步骤进行了概述（Murashige 1974）。

植物的组织培养是植物在人工培养基上生长，该培养基为植物提供了生长所需的所有营养元素以及控制植物生长的调节剂。培养基提供植物所需的营养元素，包括大量元素和微量元素。植物生长调节剂，如细胞分裂素、生长素、赤霉素、脱落酸及其他生长调节剂被添加到培养基中，以调节和控制组织培养的植物或器官的生长。

蔗糖作为碳源常被添加到培养基中以替代植物正常情况下通过光合作用所固定的碳。在某些情况下也会添加如维生素、氨基酸等其他有机化合物及椰子汁以提供植物生长所需的营养物质（George et al. 2008b）。然而，在实际操作过程中，因为椰子汁确切的成分尚不清楚，在组织培养中使用可能会影响实验的可重复性，所以其使用受到局限。但仍有人在实验中偏向使用椰子汁。

组织培养中最常用的培养基是 Murashige 和 Skoog 介绍的培养基（Murashige and Skoog 1962），通常简称为 MS 培养基。MS 培养基最初是为了培养烟草愈伤组织而研发的。因植物的基本组成成分各不相同，MS 培养基可能不是其他植物的最佳培养基。但是，MS 培养基中的盐类物质已成功用于其他植物的组织培养。MS 培养基成分可以适当调整以适用于其他植物生长。

植物组织培养也可以在液体培养基或半固体培养基中开展。琼脂和吉兰糖胶是最常用的固化剂（George et al. 2008b）。但是，琼脂也有多种替代品，如合成聚合物、藻酸盐和胶凝剂等（Kyte et al. 2013）。

组织培养技术在园艺、育种、制药和生物技术领域有许多用途（Pasqual et al. 2014）。除了使用微繁殖技术获得大量克隆外，还使用无菌可控的组织培养环境来保存植物种质资源、生产植物次生代谢物和生产脱毒种苗。本章不着重讨论植物组织培养的所有用途，而着重于在植物育种的应用：微繁殖、胚培养、体细胞克隆变异和原生质体。本书其他章详细讨论了染色体加倍、单倍体诱导和诱变等其他应用。

6.2　微　繁　殖

当植物经过无性繁殖后，单个植物的所有后代都称为克隆。这意味着每个后代的基因构成与所有其他后代以及亲本的基因构成相同。“克隆”在组织培养中是指经过培养繁殖大量与亲本基因型相同的植物的过程（Kyte et al. 2013）。因为可以迅速繁殖新品种并推广到园艺市场，因此组培繁殖过程对于观赏植物的育种非常有用。

在国际上，微繁殖有 5 个阶段（0～4）（Debergh and Maene 1981）。第一次对微繁殖描述时，仅确定了三个阶段：建立、繁殖和移栽（Murashige 1974）。后来 Anderson 在 1980 年对微繁殖的阶段进行了重新定义（Anderson 1980），一共 5 个阶段，包括外植体的选择、培养体系的建立、增殖、移植前准备/生根和移栽。由于每个阶段都有明确的目标，因此新添加的阶段对于微繁殖体系建立是必要的。

6.2.1 第 0 阶段

第 0 阶段即微繁殖的起始阶段，即外植体的选择。此阶段有以下几个要求。在第 0 阶段，为了确保将来微繁殖的成功，必须对母体或供体植物进行仔细检查（Read 1992）。必须确保外植体种或者品种的准确性，因为这是组织培养的开始。如果外植体选择错误，则繁殖出来的所有克隆都是错的。另外，外植体必须健康且无病虫害，并且处于适当的成熟度。因为克隆患病植株会使病原物与克隆一起繁殖，因此外植体必须健康。为了确保灭菌的成功，最好选择表面比较干净的植株作为外植体，这可以通过在温室或其他清洁设施中种植外植体来实现。最后，应根据植株的成熟度选择外植体，因为与植物的成熟部分相比，植株幼苗或幼嫩组织在组织培养中通常更能获得成功。

6.2.2 第 1 阶段

当外植体满足第 0 阶段的所有要求后，即开始第 1 阶段：培养体系的建立。这一阶段有两个目的：成功对外植体灭菌，确保灭菌后外植体的成活率。灭菌流程取决于外植体的类型。外植体可以是种子、插条、块茎、根、花药、花粉甚至叶子。选择外植体后，必须选择适当的灭菌过程来确保灭菌的成功率以及外植体的成活率。对于相对坚韧的外植体如种子，可以选择稍强烈的灭菌处理，对于幼嫩的外植体如软组织切片，必须选择更精细的灭菌手段。在组织培养的第 1 阶段，通常情况下将茎段作为外植体。常规灭菌流程是在流水下冲洗外植体，然后用次氯酸钠或漂白剂溶液进行灭菌。在第 1 阶段结束时，外植体将不含任何细菌及真菌孢子。灭菌后，将外植体在培养基上培养几天，以观察是否有明显的污染。如果不存在污染，则可进入第 2 阶段（George et al. 2008a）。

6.2.3 第 2 阶段

第 2 阶段被认为是增殖或繁殖阶段，其目的是生产繁殖体，使其能够长成新植物。在第 2 阶段，使用高浓度的细胞分裂素诱导外植体产生新芽（Trigiano and Gray 2005）。以下细胞分裂素可用于诱导芽，如 6-苄基腺嘌呤（6-BA）、激动素、玉米素和 6-(γ,γ-二甲基烯丙基氨基)嘌呤（2iP）。但植物种类不同，细胞分裂素诱导效率也不同。另外，还需要考虑植株顶端优势的影响（Cline 1991）。顶端优势可能会降低细胞分裂素的效果；因此，可能需要去除顶芽来促进最大数量芽的生长。因为顶芽会产生植物生长素，所以它可能在第 2 阶段影响培养基中细胞分裂素与生长素的比例，目前已有证据表明植物内源激素的比例会影响外植体的生长（Skoog and Miller 1957），为了在第 2 阶段最大程度地提高芽的数量，细胞分裂素的最佳浓度需要通过实验来确定。

6.2.4 第 3 阶段

第 3 阶段是移植前准备即生根阶段。在此阶段，已繁殖的芽将在体外被诱导生根。

用于体外诱导生根的激素称为生长素。生长素有吲哚-3-丁酸（IBA）、萘乙酸（NAA）和吲哚-3-乙酸（IAA）。研究表明，与天然生长素 IAA 相比，人工合成生长素（IBA 和 NAA）在诱导生根方面更为有效（Zimmerman and Wilcoxon 1935）。外植体在离体条件下生根有以下几个原因。通常，用于微繁殖的植物是那些难以通过传统技术繁殖并且难以生根的植物，根诱导或离体生根加速了生根阶段，并确保了克隆体的存活率。此外，生根的植株在第 4 阶段（移栽阶段）更易于成活，因为生根的植株能够吸收水分，从而减轻水分流失所带来的影响。价值高的植物也可在第 3 阶段进行体外生根。但在某些情况下，如易于生根的植物，第 3 阶段可跳过，茎段直接在第 4 阶段进行生根。

6.2.5　第 4 阶段

微繁殖的第 5 个也是最后一个阶段是第 4 阶段，即离体培养植物的过渡移栽。第 4 阶段的目的是将离体繁殖的外植体从培养基上移栽出去。移栽后，通常将它们在高湿度的环境中放置几天，再将它们移到温室（George et al. 2008a）。

以下几个因素影响第 4 阶段的成功（Preece and Sutter 1991）。第一个因素是失水。离体培养的植物生长在高湿度的人工环境中，高湿度导致植物表皮缺乏蜡质层，因此当将其放置在湿度较低的环境中时，它们容易脱水（Sutter and Langhans 1979）。为了使植株适应环境，它们需要在离开培养基之前发育出叶片蜡质层。为此，需要逐渐降低生长环境的湿度，以使植物形成蜡质层。

第 4 阶段需要考虑的第二个因素是植物离体培养时添加的蔗糖的作用。如果第 3 阶段植物生长在含有琼脂的培养基上，则需将琼脂从根部洗掉，再移植到合适的生根培养基中。已有研究表明，离体培养改变了植物的生理构造，离体培养植物的光合作用机制没有正常生长植株那样完善（Desjardins et al. 1995）。经离体培养的植物叶绿素密度较低，气孔无功能或反应缓慢。因为植株一直在使用蔗糖作为碳源，不需要光合作用的满负荷运转，所以组织培养出来的植物光合作用机制较弱。为了激活这一机制，必须在不加蔗糖的情况下生长，才能促使植物自行进行光合作用。

在第 4 阶段需要考虑的最后一个因素是光照强度。组织培养的植物在光照强度非常低的人造光下生长。组织培养出来的植物需要保护其免受全日照。因为全日照不仅会产生光线压力，还会产生水分压力，所以将组织培养的植物逐渐暴露在阳光下之前，应将其置于阴凉处过渡。

6.3　胚　培　养

植物胚培养，也称为胚挽救或合子胚培养，是一种有价值的植物体外育种工具。它是最早使用的体外育种技术之一，并已被证明对育种者具有极大的价值。在植物育种中主要用于种间杂交（Norstog 1979）。

胚培养是在无菌的营养培养基上分离合子胚并使其生长，目的是获得有活力的植物。发育停滞或流产的胚胎中保留着杂交后代的全部基因组，当提供适当的营养物质时，杂种可能恢复正常生长。对于植物育种者而言，胚培养相对容易，只需要简单的营养培养即可缩短育种周期。

种间和属间杂交通常可能导致种子败育，胚培养被用于挽救杂交胚胎。可以通过挽救脆弱的、不成熟的杂交胚并保持其正常生长来获得重要的杂交后代。也能挽救因没有细胞核而完全不能自然发育的胚胎。该技术还可用于拯救无核三倍体胚胎、培养单倍体、通过消除发芽抑制剂打破种子休眠和鉴定种子活力。胚培养技术也有助于了解胚胎的形态发生和早熟发芽（Bridgen 1994a）。

胚培养依赖于无损伤地分离胚胎并诱导胚胎生长，最后形成幼苗。胚培养被用于挽救通常会流产或不会经历个体发育的胚胎。由于必须进行繁琐的解剖以及可能需要复杂的营养培养基，因此该过程很困难。胚培养的成功在很大程度上取决于胚胎被分离时的发育阶段（Bridgen 1994a）。

胚培养非常容易获得无菌合子外植体，因为在大多数情况下，胚胎即胚珠，位于无菌环境中。因此，不需要对胚进行灭菌，而是对整个子房进行灭菌，然后在无菌环境下从周围组织中取出胚胎。由于胚胎通常受到周围组织的良好保护，因此灭菌过程可以很严格。常用95%乙醇或次氯酸钠（漂白剂）对胚珠周围的子房进行灭菌。胚胎的无菌培养通常很容易建立。但如果子房受损，则需要直接灭菌胚胎，这是一个困难的过程，因为胚胎通常会因灭菌而受损。

大的胚胎很容易分离。但是，未成熟或小的胚胎需要使用无菌解剖工具和解剖镜进行分离。当切开种皮并去除脆弱的胚性组织后，胚胎很容易受损。同样，一旦将胚从胚乳中取出，胚会很快失水。如果胚在形成球状体之前被剥离，则胚培养会很困难，所以建议胚培养时可以培养整个胚珠甚至整个子房（Monnier 1995；Sharma et al. 1996；Van Tuyl and De Jeu 1997）。

植物胚培养的成功取决于所用培养基，该培养基应该可使胚继续发育和生长。为了使年幼的胚胎发育，培养基的成分通常更为复杂（Haslam and Yeung 2011）。胚胎离体培养所需的营养随胚胎发育阶段的不同而变化。依赖胚乳和周围组织的年幼胚胎生长需要更复杂的营养培养基，并且可能需要比成熟胚胎更高的渗透压。用于培养年幼胚胎的培养基可能需要添加维生素、氨基酸和生长激素。在胚培养早期，可以将天然提取物如番茄汁和椰子汁加入培养基中。为了增加渗透压还可以在培养基中加入高浓度的蔗糖溶液（Sharma et al. 1996；Van Creij et al. 2000）。

成熟胚胎可以在仅包含 Knop 矿物盐和 2.5%～5%蔗糖的半固体培养基上正常生长，因为此培养基能够提供糖类和营养物质。用于胚培养的一些常见的培养基通常为优化后的 MS 培养基（1962）和 B5 培养基（Gamborg et al. 1968），其通过降低糖的含量来控制渗透压。

植物生长调节剂通常在胚培养中不发挥重要作用。已有证据表明外源生长素并非植物胚芽生长所必需的。细胞分裂素作为唯一激素时促进幼胚生长的作用无效或轻微。然而，当细胞分裂素与一些生长素结合使用时，则可促进胚胎的生长和分化。有证据表明生长素和细胞分裂素可用于胚培养，以促进愈伤组织的形成。如果胚胎发育过程中有生理上的休眠过程，赤霉素可以克服休眠并刺激其萌发。

6.4　体细胞克隆变异

当离体培养植物时，因为需要的是克隆而不是突变，所以应避免植物和植物细胞的突变。然而，植物育种者可能希望通过诱变物质使未分化的组织发生变异，从而开发新品种。体细胞克隆变异通过使用组织培养技术来产生遗传和表观遗传多样性变异。产生的变异可能是预先存在的或诱发的；一些植物比其他植物更容易发生突变。在离体培养中诱发突变的诱因为各种胁迫因素，如伤口、灭菌剂的使用、未分化细胞的培养、高浓度植物激素（如生长素和细胞分裂素）、糖、光照条件和高湿度等。长时间的继代培养，尤其是愈伤组织的培养，以及某些基因型的使用也往往会导致变异的增加。氧化应激和活性氧的产生被认为在诱导遗传变异中起重要作用（Krishna et al. 2016）。

体细胞克隆变异的频率和性质取决于外植体来源与再生方式。通常，用于再生的组织分化程度越高、越成熟获得变异的机会就越高。高度分化的组织（如叶、茎和根）要比原先存在分生组织（如腋芽和芽尖）的外植体产生更多的变异。不定芽再生导致较高的体细胞克隆变异率，但通过腋芽产生的植物通常不会发生变异。通常，在经历愈伤组织阶段的培养体系中，突变率会增加。体外离体培养的时间越长，获得体细胞克隆变异的机会就越高。因此，建议定期开始新的培养，以避免离体培养系统中不必要的体细胞克隆变异（Jain et al. 1998；Krishna et al. 2016）。

利用离体培养产生的体细胞克隆变异已用于常规观赏植物的育种。Jain 等（1998）和 Krishna 等（2016）概述了几种在观赏植物离体培养中观察到的体细胞克隆变异的形态变化，报道了株型、叶型、花色、花型和叶色的变化。除了形态学变化外，有时还观察到细胞学变化，如倍性水平的改变和非整倍体植株的产生以及生理生化变化（Kaeppler et al. 2000）。除了会对植物形态产生影响外，体细胞克隆变异还可以用于选择对生物和非生物胁迫具有抗性的基因型。通过观赏植物的细胞悬浮培养，可以选择耐受低温或高盐胁迫的基因型。利用真菌毒素或培养液作为选择剂可能有助于选择对病原物具有抗性的基因型（Jain et al. 1998；Krishna et al. 2016）。

利用体细胞克隆变异作为育种工具的缺点是对基因型的依赖，缺乏用于识别突变体的标记，以及遗传变异恢复的风险。应该记住的是，体细胞遗传变异并不总是不可逆的正向变异。对于不存在变异的无性繁殖植物，体细胞克隆变异是植物育种者的一种工具，但是，通过种子繁殖产生的遗传变异仍然可以产生大量成功的变异。

组织培养技术经常与化学试剂和物理手段结合使用，或与有丝分裂抑制剂结合以实现染色体改变，从而诱导遗传突变。定量和定性的性状突变、序列修饰和核型改变都是可能的。这些技术将在本书的不同章节介绍。

6.5　原 生 质 体

体细胞杂交或原生质体融合被用来克服有性生殖过程中遇到的困难，是一种很有前途的育种方法。原生质体可以在不影响核基因组的情况下替代种间细胞质。原生质体融合可以克服品种间、远缘植物种间或性功能障碍品种间的杂交不亲和性。通过这项技术可以转移不同的特性，如抗病性、抗除草剂和抗逆性等（Johnson and Veilleux 2001；Tapingkae et al. 2012）。原生质体是指去除细胞壁的植物细胞。从理论上讲，原生质体是全能的，这意味着其可以去分化并重新进入细胞周期，即经过有丝分裂，长成愈伤组织，再分化成不同的器官。原生质体融合避开了有性杂交中合子前和合子后的障碍，是一种实用的育种工具。该技术的瓶颈通常是融合的原生质体再生成植株（Eeckhaut et al. 2013）。

原生质体融合的研究目前已在农业和观赏植物上开展。在 Mii（2001）的综述中，列出了观赏植物从原生质体到植株再生的方案。第一次观赏植物的原生质体融合发生在烟草（*Nicotiana*）、矮牵牛（*Petunia*）和曼陀罗（*Datura*）之间，因为这些原生质体很容易再生。该列表中还包括萱草（*Hemerocallis*）、石竹（*Dianthus*）、蔷薇（*Rosa*）、高杯花（*Nierembergia*）、补血草（*Limonium*）、洋桔梗（*Eustoma*）、非洲堇（*Saintpaulia*）和报春花（*Primula*）。其他原生质体研究包括金光菊（*Rudbeckia*）（Al-Atabee and Power 1987）、亮叶忍冬（*Lonicera nitida*）（Ochatt 1991）、朱槿（*Hibiscus rosa-sinensis*）、苏云金缕梅（*Lavatera thuringiaca*）（Vazquez-Thello et al. 1996）和胡杨（*Populus euphratica*）（Kang et al. 1996）。种间体细胞杂种已在石竹（*Dianthus*）、矮牵牛（*Petunia*）、鸢尾（*Iris*）和报春花（*Primula*）中制成（do Valle Rego and Tadeu de Faria 2001；Mii 2001）。但是，如须苞石竹（*Dianthus barbatus*）和满天星（*Gypsophila paniculata*）之间的种内杂交也是可能的（Nakano et al. 1996）。

Eeckhaut 等（2013）回顾了 2004～2013 年植物的原生质体研究进展。不同的观赏植物开展了原生质体研究，包括切花，如六出花（*Alstroemeria* spp.）和日本百合（*Lilium japonicum*）等；盆栽植物，如火鹤花（*Anthurium scherzerianum*）、小花矮牵牛（*Calibrachoa* spp.）、仙客来（*Cyclamen* spp.）、圣诞伽蓝菜（*Kalanchoe blossfeldiana*）等；木本观赏植物，如黄檗（*Phellodendron amurense*）、刺槐（*Robinia pseudoacacia*）和欧洲野榆（*Ulmus minor*）等；以及园林植物，如针叶石竹（*Dianthus acicularis*）、中亚秦艽（*Gentiana kurroo*）、向日葵（*Helianthus annuus*）和贯叶连翘（*Hypericum perforatum*）等。其中大约 70%植物的原生质体能够再生成植株。其他 30%则形成愈伤组织或芽。

后来，以下植物的原生质体再生获得了成功，包括天竺葵（*Pelargonium* spp.）（Klocke et al. 2013）、莉黛柏蕊（*Lilium ledebourrii*）（Tahami et al. 2014）、美国树（*Ulmus americana*）（Jones et al. 2015）、龙胆（*Gentiana macrophylla*）（Hu et al. 2015）和合欢（*Albizia julibrissin*）（Rahmani et al. 2016）。最近，在玉兰（*Magnolia denudata*）白玉兰‘黄河’（*Magnolia acuminata*‘Yellow River’）（Shen et al. 2017）、中亚秦艽（*Gentiana kurroo*）和十字龙胆（*Gentiana cruciata*）以及十字龙胆（*Gentiana cruciata*）和西藏秦艽（*Gentiana tibetica*）（Tomiczak et al. 2015）、不同的绣球（*Hydrangea* spp.）之间成功进行了体细胞融合（Kästner et al. 2017）。

在以下章节中简要概述了进行体细胞融合的不同技术。本节将介绍原生质体的分离、融合和再生，通常需要优化参数，从而最大程度地提高成功率。目的是为每种技术的不同策略和技术选择提供见解，并总结其典型的缺陷，因此，本节中提及的内容被认为具有指导意义。

6.5.1 原生质体分离

在原生质体分离过程中，通常使用酶解或机械方法去除细胞壁，仅保留细胞膜。由于机械方法去除细胞壁后原生质体的数量和质量较低，因此最优选的是酶解分离（George et al. 2008a）。原生质体分离的成功取决于许多因素。

首先，植物材料的来源是分离成功与否的关键。尽管单个物种中可能存在基因型效应，但最重要的因素是供体材料类型。原生质体可以从各种组织获得，如愈伤组织、分生组织、叶肉、细胞悬浮培养物、芽和花。离体培养时首先应对供体材料进行灭菌，但大多数时候都使用组培材料（Chabane et al. 2007）。就原生质体分离效率而言，叶肉是理想的材料，但在许多农作物尤其是单子叶植物中，叶肉的再生能力可能较低。

其次，酶溶液的组成成分影响分离效率和原生质体质量。细胞壁的三个主要成分是纤维素、半纤维素和果胶。纤维素和半纤维素负责细胞壁的硬度，而果胶则将相邻的细胞连在一起（do Valle Rego and Tadeu de Faria 2001）。因此，去除细胞壁的酶液应由果胶酶和纤维素酶组成。前者使细胞分离，后者通过分解细胞壁的纤维素和半纤维素而使细胞壁破裂（George et al. 2008a）。常用的纤维素酶分别是纤维素酶 Onozuka R-10 和纤维素酶，浓度分别为 0.2～3mg/L 和 0.1～6mg/L。普通果胶酶分别是离析酶 R10 和果胶酶 Y23（do Valle Rego and Tadeu de Faria 2001），浓度分别为 0.3～1mg/L 和 0.01～0.6mg/L。也可使用如 Rhozyme 或 MKC 半纤维素酶（Kästner et al. 2017）。也可以添加 0.05～0.5mg/L 浓度的崩溃酶，它是具有纤维素酶和果胶酶活性的酶的混合物。然后根据酶特性优化培养基的 pH。

由于原生质体缺乏细胞壁的保护，因此必须使用渗透剂来防止细胞破裂。通常将 0.4～0.7mol/L 的甘露醇或山梨糖醇添加到悬浮培养液、酶液混合物以及用于纯化和培养原生质体的培养基中。为了获得不同密度的原生质体悬浮液，这些糖醇可以分别用糖或盐代替。在用酶液替代细胞培养基之前，通常将具有划痕或切片的植物材料在添加有盐、糖或糖醇的预溶酶溶液中孵育 1h 左右。通过增加溶酶前溶液的渗透性，进而增加酶的渗透率。样品与酶溶液在温度为 22～27℃的黑暗中孵育 4～18h 以溶解细胞壁。具体的培养条件取决于酶的类型。通常是将植物材料放在转速为 10～30r/min 的旋转摇床上培养。

酶促孵育后通过旋转式离心机旋转收集纯化原生质体。为了防止原生质体破损，离心速度应保持较低（约 100 × *g*）。纯化通常是基于原生质体培养液之间的浓度差来实现的（Johnson and Veilleux 2001；Chabane et al. 2007；George et al. 2008a）。图 6.1 显示了从植物到原生质体再到植株再生的过程。

6.5.2 原生质体融合

原生质体融合可以分为两种不同类型。对称融合包含了父母双方完整的核与细胞质遗传物质。不对称融合是在融合之前使至少一个亲本的核或细胞质失活后进行的。染色体可以通过对原生质体进行辐射（紫外线、X 射线或 γ 射线）而失活。γ 射线可以使用 50～150Gy 的剂量，254nm 的 UV 辐射的通常剂量是 $50J/m^2$ 或更高，持续照射 2～10min。或者可以在与微原生质体融合后通过添加附加系来转移完整的染色体。也可以通过灭活

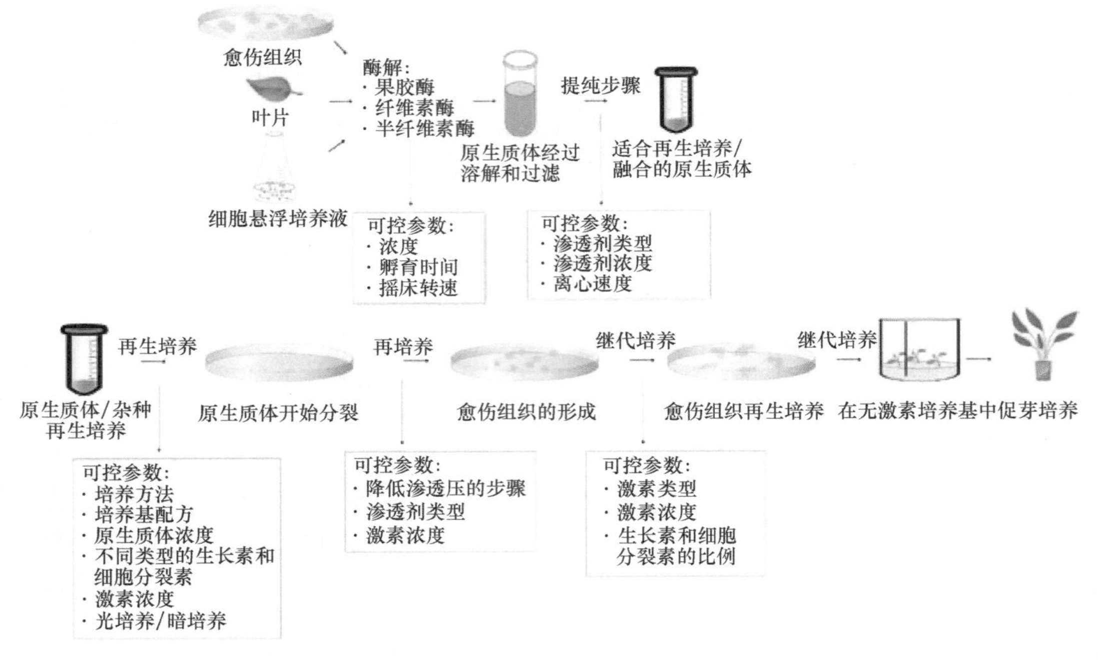

图 6.1　确保原生质体分离、纯化和再生成功的参数

的 IOA（碘乙酸盐）处理来完全失活线粒体 DNA 而不破坏核 DNA。这提供了将一个亲本的细胞核遗传信息与另一亲本的胞质遗传信息相结合的可能（Tapingkae et al. 2012）。一般将原生质体在 3～15mmol/L IOA 中于室温或更冷的环境中孵育 10～20min。在这两种原生质体融合类型中都可以看到染色体丢失、不想要的基因渗入和不育。因此，在对称融合之后也可能出现不对称杂种。悬浮液中的原生质体由于表面带负电荷而相互排斥，融合很少自发发生，需借助化学或物理方法触发。化学融合和电融合的选择取决于物种（Eeckhaut et al. 2013）。

6.5.2.1　物理融合

物理融合或电融合是使用电流融合原生质体。原生质体被交流电产生的电场极化，在两个极之间排列形成一条链，通过施加短暂的直流电使相邻细胞的细胞膜发生可逆性破损，然后再次施加交流电，使细胞完成融合（Tomiczak et al. 2015）。典型的对准电流介于 75～100V/cm，持续 30～60s。用于电击的直流电通常在 1000～2500V/cm，施加带有一个或两个脉冲的 40μs 电压。原生质体融合密度为 1 000 000 个/mL；融合双方原生质体的比例通常为 1∶1。常用的融合溶液为 0.5mol/L 甘露醇，它是一种低渗溶液，在电融合过程中用于保持原生质体的稳定。

6.5.2.2　化学融合

钙离子和聚乙二醇（PEG）均可诱导化学融合。大多数情况下用 20%～50% PEG（MW 3350～6000）进行处理以诱导原生质体聚集。这种聚集克服了原生质体本身带负电的质膜而能够融合。通常在碱性溶液中，Ca^{2+}（0.05～0.25mol/L）也可以诱导融合（Tomiczak et al. 2015)，孵育时间通常在 15～30min（Evan et al. 1983；Constabel 1984；Navrátilová 2004）。PEG 和 Ca^{2+}处理通常结合使用。一般情况下，PEG 用 Ca^{2+}溶液洗脱，以进一步支持相邻原生质体的融合。需要对 PEG 类型、孵育时间和 Ca^{2+}溶液进行相应调整才能应用于不同的植物物种。图 6.2 显示了 PEG 介导融合的示意图。如果 PEG 浓度太高或孵育时间太长，就会发生大量凝集，导致原生质体变形。如果孵育条件不是最佳条件，则会发生聚集，从而妨碍融合。与电融合相似，化学融合时双方原生质体通常以 1∶1 的比例混合，原生质体融合密度通常约为 1 000 000 个/mL。

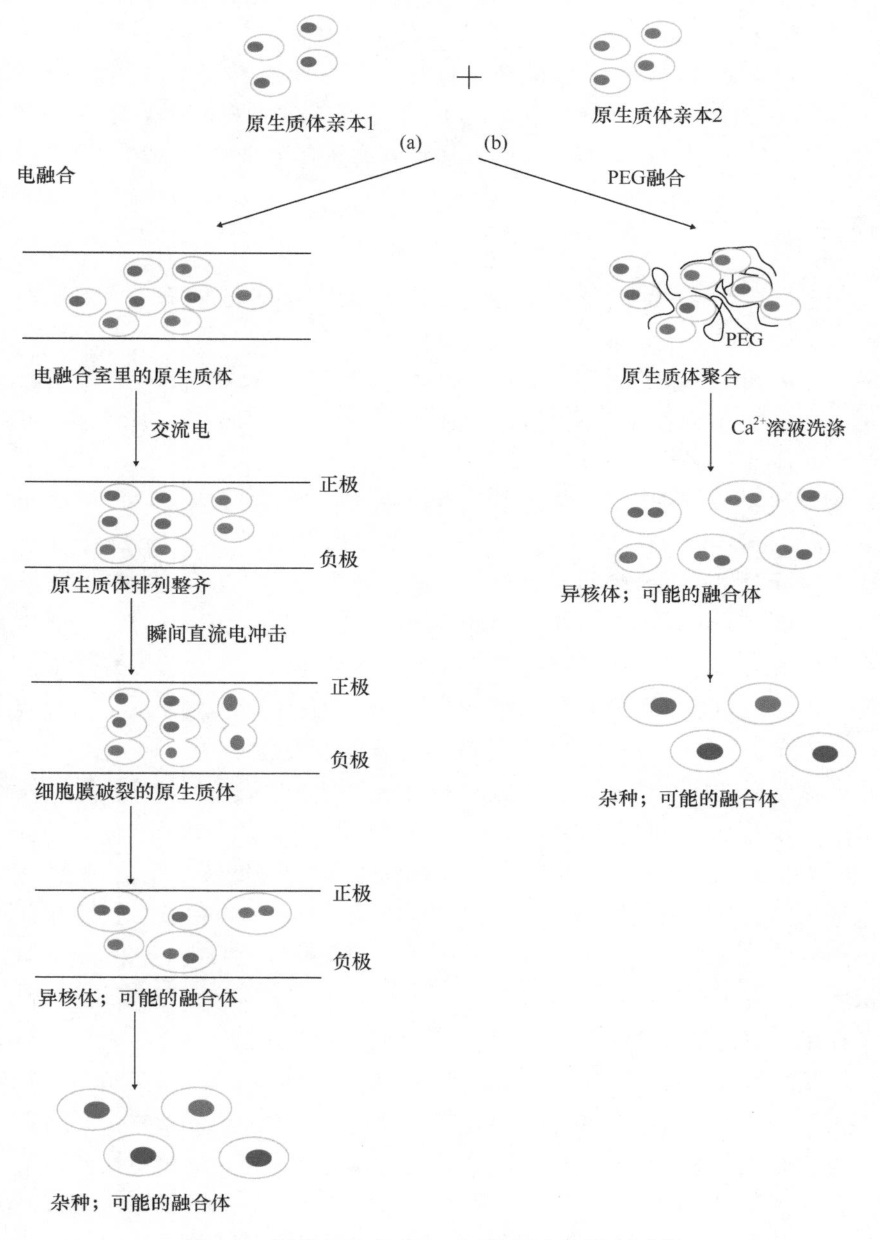

图 6.2　电融合（a）和 PEG 融合（b）的示意图

物理融合比化学融合需要更少的洗涤步骤。但 PEG 具有细胞毒性。另外，化学融合易于使用，无须购买昂贵的设备。

6.5.3 微观评估

为了确定融合后原生质体的活力，可用二乙酸荧光素（FDA）或 EB 或酚藏花红染料染色进行检测。FDA 能通过细胞膜，经细胞内酯酶分解产生荧光素，积累在细胞中发出荧光。活的原生质体不能渗透染料，而死的原生质体则可以渗透并着色（do Valle Rego and Tadeu de Faria 2001）。

如果原始原生质体群体被差异染色或具有独特的形态，如在绿色叶肉原生质体和从黄化组织分离的无色原生质体融合的情况下，融合效率的微观评估是可行的。在杂种中，父母双方都有特征，则可通过双重染色或通过形态学参数来鉴定杂种。

6.5.4 原生质体再生

成功融合后，必须从杂交细胞中再生出植株。许多因素影响再生（图 6.1）。亲缘关系较近的杂交后代通常比远缘杂交后代更容易再生（Tapingkae et al. 2012）。有效的再生方案还应与分离和融合步骤兼容（Duquenne et al. 2007）。常见方法是使用先前开发的未融合原生质体再生方案来诱导杂交原生质体再生（Tomiczak et al. 2015）。也可以通过诱导愈伤组织或体细胞胚发生来实现再生。

原生质体到植株再生的不同培养基配方很容易获得，如 MS 培养基（Murashige and Skoog 1962）、Gamborg 培养基（Gamborg 1968）和 KM 培养基（Kao and Michayluk 1975）。通常情况下需要对培养基配方进行修改。因糖能被代谢并迅速去除，则可通过添加甘露醇、山梨糖醇或糖（0.4～0.7mol/L）调节渗透压（Davey et al. 2010）。典型的再生培养基需要补充生长素 NAA、细胞分裂素 6-BA 和维生素等。最适合的培养基是具有基因型依赖性的，融合后的细胞可能与亲本对培养基的适应性不一样。根据植物种类不同，原生质体培养应在温度为 22～25℃的黑暗或明亮条件下进行。

主要的原生质体培养方法是在半凝固的培养基中进行液体培养。原生质体培养密度为 100 000 个/mL。在细胞壁恢复后，细胞分裂并形成微愈伤组织，其进一步生长成愈伤组织。典型的操作过程是每周换新的培养基，逐渐降低渗透液浓度直到愈伤组织长大。愈伤组织通常转移到含有植物激素的半固体或固体培养基中。生长素和细胞分裂素的比例通常逐渐降低以促进芽的形成。定期的培养基更换对防止酚类的积累至关重要。芽形成后，植物可以逐渐适应环境。

为了在半凝固的培养基中建立培养体系，应首先将原生质体悬浮液与含有藻酸盐或琼脂糖等的胶凝剂按 1∶1 混合。在最高 40℃的温度下，将原生质体嵌入终浓度为 2～6g/L 的琼脂糖胶凝剂的小液滴中。然后将它们吸移到大培养皿中，并在凝固后添加液体培养基。与纯液体培养相比，此培养基的更换很容易，因为只需要更换液体培养基，原生质体不会受到干扰。此外，由于已经固定了原生质体，因此可以对其进行监测。无论是液体、嵌入琼脂糖还是嵌入其他固化剂，最佳的培养体系都取决于植物特性。

6.6 结　　论

成功的植物发育取决于遗传多样性，然后是基因型选择和评估。在过去的 50 年中，由于技术进步，植物组织培养已成为一种常规的技术。现在，体外培养技术通过增加植物特性的多样性和开发优质植物，提供了新的有效方法来加速基因选择。微繁殖、胚培养、原生质体的分离和利用以及体外诱变是观赏植物育种者可以使用的育种手段。

本章译者：

曹　桦

云南省农业科学院花卉研究所，国家观赏园艺工程技术研究中心，云南省花卉育种重点实验室，昆明 650200

参 考 文 献

Al-Atabee JS, Power JB (1987) Plant regeneration form protoplasts of *Dimorphotheca* and *Rudbeckia*. Plant Cell Rep 6:414–416

Anderson WC (1980) Mass propagation by tissue culture. In: Principles and techniques. Agricultural research results, United States Department of Agriculture, pp 1–9

Bridgen MP (1994a) A review of plant embryo culture. Hortscience 29:1243–1246

Bridgen MP (1994b) Tissue culture's potential for introducing new plants. Proceedings of the International Plant Propagators' Society 44:595–601

Chabane D, Assani A, Bouguedoura N, Haïcour R, Ducreux G (2007) Induction of callus formation from difficile date palm protoplasts by means of nurse culture. CR Biol 330:392–401

Cline MG (1991) Apical dominance. Bot Rev 57:318–358

Constabel F (1984) Fusion of protoplasts by polyethylene glycol (PEG). In: Vasil IK (ed) Cell culture and somatic cell genetics of plants: volume 1: laboratory procedures and their applications. Academic Press Inc, Orlando, Florida, pp 414–422

Davey RM, Anthony P, Patel D, Power JB (2010) Plant protoplasts: isolation, culture and plant regeneration. In: Davey RM, Anthony P (eds) Plant cell culture: essential methods. Wiley-Blackwell, West Sussex, UK, pp 153–174

Debergh PC, Maene LJ (1981) A scheme for commercial propagation of ornamental plants by tissue culture. Sci Hortic 14(4):335–345

Desjardins Y, Hdider C, De Riek J (1995) Carbon nutrition *in vitro*- regulation and manipulation of carbon assimilation in micropropagated systems. In: George EF, Hall MA, De Klerk GJ (eds) Automation and environmental control in plant tissue culture. Springer, Dordrecht, pp 441–471

do Valle Rego L, Tadeu de Faria R (2001) Tissue culture in ornamental plant breeding: a review. Crop Breed Appl Biotechnol 1:283–300

Duquenne B, Eeckhaut T, Werbrouck S, Van Huylenbroeck J (2007) Effect of enzyme concentrations on protoplast isolation and protoplast culture of *Spathiphyllum* and *Anthurium*. Plant Cell Tissue Org Cult 91(2):165–173

Eeckhaut T, Lakshmanan PS, Deryckere D, Van Bockstaele E, Van Huylenbroeck J (2013) Progress in plant protoplast research. Planta 238:991–1003

Evans DA, Bravo JE, Gleba YY (1983) Somatic hybridization: fusion methods, recovery of hybrids, and genetic analysis. Int Rev Cytol 16:143–159

Gamborg OL, Miller RA, Ojima K (1968) Nutrient requirements of suspension cultures of soybean root cells. Exp Cell Res 50:151–158

George EF, Hall MA, De Klerk G-J (2008a) Plant tissue culture procedure – background. In: George EF, Hall MA, De Klerk GJ (eds) Plant propagation by tissue culture. Vol. 1, The background, 3rd edn. Springer, Dordrecht, pp 1–28

George EF, Hall MA, De Klerk G-J (2008b) The component of plant tissue culture media II: organic additions, osmotic and pH effects, and support systems. In: George EF, Hall MA, De Klerk GJ (eds) Plant propagation by tissue culture, 3rd edn. Springer, Dordrecht, pp 115–173

Haberlandt G (1902) Culturversuche mit isolierten pflanzenzellen, sitz-ungsb. Akad. D. wissensch. Mathermatusch-naturwissenschaftlicher c169
Haslam TM, Yeung EC (2011) Zygotic embryo culture: an overview. In: Thorpe TA, Yeung EC (eds) Plant embryo culture: methods and protocols, Methods in molecular biology, vol 710. Humana Press, Totowa, pp 3–15
Hu X, Yin Y, He T (2015) Plant regeneration from protoplasts of *Gentiana macrophylla* pall. Using agar-pool culture. Plant Cell Tissue Organ Cult 121(2):345–351
Jain SM, Buiatti M, Gimelli F, Saccardo F (1998) Somaclonal variation in improving ornamental plants. In: Jain SM, Brar DS, Ahloowalia BS (eds) Somaclonal variation and induced mutations in crop improvement, Current plant science and biotechnology in agriculture, vol 32. Springer, Dordrecht, pp 81–104
Johnson AA, Veilleux RE (2001) Somatic hybridization and application in plant breeding. Plant Breed Rev 20:167–225
Jones AM, Shukla MR, Biswas GC, Saxena PK (2015) Protoplast-to-plant regeneration of American elm (*Ulmus americana*). Protoplasma 252:925–931
Kaeppler SM, Kaeppler HF, Rhee Y (2000) Epigenetic aspects of somaclonal variation in plants. Plant Mol Biol 43:179–188
Kang J-M, Kojima K, Ide Y, Sasaki S (1996) Plantlet regeneration from leaf protoplasts of *Populus euphratica*. J For Res 1:99–102
Kao KN, Michayluk MR (1975) Nutritional requirements for growth of *Vicia hajastana* cells and protoplasts at a very low population density in liquid media. Planta 126:105–110
Kästner U, Klocke E, Abel S (2017) Regeneration of protoplasts after somatic hybridization of *Hydrangea*. Plant Cell Tissue Organ Cult 129:359–373
Klocke E, Weinzierl K, Abel S (2013) Occurrence of endophytes during Pelargonium protoplast culture. In: Schneider C (ed) Endophytes for plant protection: the state of the art. DPG Verlag, Germany, pp 94–99
Knudson L (1922) Nonsymbiotic germination of orchid seeds. Bot Gaz 73:1–25
Krishna H, Alizadeh M, Singh D, Singh U, Chauhan N, Eftekhari M, Sadh RK (2016) Somaclonal variations and their applications in horticultural crops improvement. 3. Biotech 6(1):54. https://doi.org/10.1007/s13205-016-0389-7
Kyte L, Klyen J, Scoggins H, Bridgen M (2013) Plants from test tubes: an introduction to micropropagation, 4th edn. Timber Press, Portland, p 250
Mii M (2001) Somatic hybridization in ornamental species. The Korean breeding society conference: Introduction pp 20–22
Monnier M (1995) Culture of zygotic embryos. In: Thorpe TA (ed) In vitro embryogenesis in plants. Kluwer Academic Publishers, Dordrecht, pp 117–153
Murashige T (1974) Plant propagation through tissue cultures. Annu Rev Plant Physiol 25:135–166
Murashige T, Skoog F (1962) A revised medium for rapid growth and bio assays with tobacco tissue cultures. Physiol Plant 15:473–497
Nakano M, Hoshino Y, Mii M (1996) Intergeneric somatic hybrid plantlets between *Dianthus barbatus* and *Gypsophila paniculata* obtained by electrofusion. Theor Appl Genet 92(2):170–172
Navrátilová B (2004) Protoplast cultures and protoplast fusion focused on Brassicaceae – a review. Hortic Sci 31(4):140–157
Norstog K (1979) Embryo culture as a tool in the study of comparative and developmental morphology. In: Sharp WR, Larsen PO, Paddock EF, Raghavan V (eds) Plant cell and tissue culture. Ohio State University Press, Columbus, pp 179–202
Ochatt SJ (1991) Requirements for plant regeneration from protoplasts of the shrubby ornamental honeysuckle, *Lonicera nitida* cv. Maigrun. Plant Cell Tissue Organ Cult 25:161–167
Pasqual M, Soares JD, Rodrigues FA (2014) Tissue culture application for the genetic improvement of plants. Biotechnology and Plant Breeding: applications and approaches for developing improved cultivars 225:157–199
Preece JE, Sutter EG (1991) Acclimatization of micropropagated plants to the greenhouse and field. In: Debergh PC, Zimmerman RH (eds) Micropropagation. Kluwer Academic Publisher/Springer, Dordrecht, pp 71–93
Rahmani MS, Shabanian N, Pijut PM (2016) Protoplast isolation and genetically true-to-type plant regeneration from leaf- and callus-derived protoplasts of *Albizia julibrissin*. Plant Cell Tissue Organ Cult 127:475–488
Read PE (1992) Environmental and hormonal effects in micropropagation. In: Kurata K, Kozai T (eds) Transplant production systems. Kluwer Academic Publisher/Springer, Dordrecht, pp 231–246

Sharma D, Kaur R, Kumar K (1996) Embryo rescue in plants – a review. Euphytica 89:325–337
Shen Y, Meng D, McGrouther K, Zhang J, Cheng L (2017) Efficient isolation of *Magnolia* protoplasts and the application to subcellular localization of MdeHSF1. Plant Methods 13:44–54
Skoog F, Miller CO (1957) Chemical regulation of growth and organ formation in plant tissues cultured. Vitro Symp Soc Exp Biol 11:118–131
Skoog F, Tsui C (1948) Chemical control of growth and bud formation in tobacco stem segments and callus cultured in vitro. Forestry 45:197–210
Sutter E, Langhans RW (1979) Epicuticular wax formation on carnation plantlets regenerated from shoot tip culture. J Am Soc Hort Sci 104:493–496
Tahami SK, Chamani E, Zare N (2014) Plant regeneration from protoplasts of *Lilium ledebourrii* (Baker) Boiss. J Agric Sci Technol 16:1133–1144
Tapingkae T, Zulkarnain Z, Kawaguchi M, Ikeda T, Taji A (2012) Somatic (asexual) procedures (haploids, protoplasts, cell selection) and their applications. In: Altman A, Hasegawa PM (eds) Plant biotechnology and agriculture: prospects for the 21st century. Academic Press, Oxford, pp 141–162
Tomiczak K, Mikula A, Rybczynski JJ (2015) Protoplast culture and somatic cell hybridization of gentians. In: Rybczynski JJ, Davey MR, Mikula A (eds) The Gentianaceae – volume 2: biotechnology and applications. Springer, Heidelberg/New York/Dordrecht/London, pp 163–186
Trigiano RN, Gray DJ (2005) Plant development and biotechnology. CRC Press LLC, Boca Raton, p 348
Van Creij MGM, Kerckhoffs DMFJ, De Bruijn SM, Vreugdenhil D, Van Tuyl JM (2000) The effect of medium composition on ovary-slice culture and ovule culture in intraspecific *Tulipa gesneriana* crosses. Plant Cell Tissue Organ Cult 60:61–67
Van Tuyl J, De Jeu M (1997) Methods for overcoming interspecific crossing barriers. In: Shivanna K, Sawhney V (eds) Pollen biotechnology for crop production and improvement. University Press, Cambridge, UK, pp 273–292
Vazquez-Thello A, Yang L, Hidaka M, Uozumi T (1996) Inherited chilling tolerance in somatic hybrids of transgenic *Hibiscus rosa-sinensis* x transgenic *Lavatera thuringiaca* selected by double-antibiotic resistance. Plant Cell Rep 15:506–511
Went FW (1929) On a substance, causing root formation. Proceedings Royal Academy, Amsterdam xxxii:35–39
Zimmerman PW, Wilcoxon F (1935) Several chemical growth substances which cause initiation of root and other responses in plants. Contrib Boyce Thompson Inst 7:209–229

第 7 章　观赏植物倍性育种

Tom Eeckhaut, Jeroen Van der Veken, Emmy Dhooghe, Leen Leus,
Katrijn Van Laere, and Johan Van Huylenbroeck

摘要　倍性育种常用于培育观赏植物新品种、克服杂交障碍及创造同源品系。本章总结了三种具有较强操作性的体外加倍方法，即通过化学诱变剂导致有丝分裂受阻产生加倍、通过 2*n* 配子减数分裂产生加倍、通过雄性或雌性配子再生或有性杂交技术获得单倍体。本章将介绍三种诱变技术的最新进展，总结相关技术在观赏植物育种应用上的优、缺点。在有丝分裂多倍体化技术方面，本章对比了有丝分裂抑制剂对分生组织短期及长期处理的不同影响，N_2O 及温度对配子体形成的影响；在单倍体诱导技术中，比较了孤雌生殖、小孢子胚胎发生概率和雌激素与雄激素生成比的关系。同时，本章还概述了该技术领域的最新研究进展。另外，阐述了 2*n* 配子形成的机制、CENH3 修饰和环境的互作关系。尽管我们无法提供详细的资料，但我们尽量总结了倍性育种的相关资料。本章提供了对不同诱导方法的见解，并与实际应用进行全面结合。

关键词　2*n* 配子；单倍体；离体培养；多倍体；再生

7.1　引　　言

观赏植物种类繁多，育种目标也各不同。总的来说，观赏植物的育种目标主要集中在观赏性状上，如植株形态、花色、花形、较长的开花期、香味等。同时集中在其他经济指标上，如产量和生物与非生物胁迫耐受性等。倍性育种是通过对植物基因组中染色体组数（*x*）处理开展种质创新的主要技术之一。由于倍性育种通常会导致形态发生变化，因此可作为观赏植物育种的主要手段。Dhooghede 等（2011）总结，多倍体诱导通常会引起植物变化，如叶片增厚、叶宽与叶长比增加、叶色变深、茎增粗、株型紧凑、花朵变大、花瓣数增多、花色变深以及开花时间延迟等。同时，总结了多倍体诱导后给植物带来的负面影响，如白化病、畸形、不育或植株易折等。

多倍体化是植物进化的主要驱动力，也是形成新物种的主要途径之一。自然界中所有被子植物在进化过程中至少经历过一次多倍体化历程（Soltis and Soltis 2009），但其不应被视为基因进化唯一的、最适合的途径（Husband et al. 2013）。多倍体的产生通常与植物的适应性增强有直接关系，在恶劣环境中，多倍体化为植株的生长适应性提

T. Eeckhaut (✉)· J. Van der Veken · E. Dhooghe · L. Leus · K. Van Laere. J. Van Huylenbroeck
Flanders Research Institute for Agriculture, Fisheries and Food (ILVO), Plant Sciences Unit, Applied Genetics and Breeding, Melle, Belgium
e-mail: tom.eeckhaut@ilvo.vlaanderen.be

供了较大的支持，使其在新的环境中更易于生存（Te Beest et al. 2012）。例如，生长于安第斯山脉地区的伏氏凤梨（*Fosterella*）具有较强的抗逆性。一般来说，年平均温度差异较大、温度较低的区域更易于发现多倍体（Paule et al. 2017）。同时，在观赏植物育种中，多倍性与植物对非生物胁迫的耐受程度有直接的关系。例如，菊花（*Dendranthema*）的抗旱性、抗冻性、抗盐碱性都较强（Liu et al. 2011）；白鹤芋（*Spathiphyllum*）的抗旱性较强（Eeckhaut et al. 2004；Van Laere et al. 2011）。杂交月季'Yesterday' × *Rosa wichurana* 的四倍体与二倍体相比具有较高的耐旱性。一些多倍体杂交种对干旱胁迫有较高的耐受性，而有一些杂交种的耐受性就较差；但在某些情况下，耐旱性不受倍性水平的影响（未发表）。

在自然界中，多倍体是通过形成的未减数（$2n$）配子自发形成的。因此，植物育种家通常会在自然界的物种中寻找自发产生 $2n$ 配子的基因型，然后利用这些基因型的种子或花粉亲本进行三倍体或四倍体的育种研究。同时，我们可以通过温度（高温、低温）、抗有丝分裂剂或其他化学物质处理来诱导 $2n$ 配子的形成。例如，经证明，秋海棠（*Begonia*）的 $2n$ 配子是在自然界中自发形成的（Dewitte et al. 2009）。当然，也可利用化学物质或一氧化二氮诱导形成 $2n$ 配子（Dewitte et al. 2010）。大多数情况下，多倍体诱导的方法仍然以利用化学物质处理体细胞阻止有丝分裂为主。这些化学物质通过干扰纺锤体形成所需的微管蛋白聚合阻碍纺锤体形成，导致染色体不能向细胞两极分开，从而引起细胞内染色体加倍。目前，该技术已经相当成熟。与大多数农作物和蔬菜利用种子进行繁殖不同，观赏植物一般通过无性方式进行繁育，只要提供一个相对稳定的、标准的环境，就能够通过化学诱导方法实现染色体的加倍。

单倍体诱导是创建纯合双单倍体基因的桥梁。纯合双单倍体可作为 F_1 杂交后代的亲本。事实上，每一个植物都是由单个单倍体细胞（花粉、花药或小孢子培养物）或卵细胞（胚珠培养物）再生而来的。因此，可以结合种间或属间杂交、体外胚挽救来获得单倍体。由于通过 F_1 种子可以快速获得目标基因型，因此 F_1 杂种在育种上是极其重要的。在可控环境中，这些 F_1 株系的特性是完全可预测的。如果没有 F_1 的培育，观赏特性的稳定则只能通过扦插、组织培育或其他劳动密集型的营养繁殖来保证。所以，单倍体诱导是观赏植物 F_1 育种过程中的重要中间环节，一些单倍体本身也具有较高的商业开发价值，如天竺葵（*Pelargonium*）品种'Kleine Liebling'（Daker 1966）。F_1 杂种可商用的观赏物种种类繁多，包括金鱼草（*Antirrhinum*）、秋海棠（*Begonia*）、翠雀（*Delphinium*）、石竹（*Dianthus*）、老鹳草（*Geranium*）、向日葵（*Helianthus*）、洋桔梗（lisianthus）、天竺葵（*Pelargonium*）、矮牵牛（*Petunia*）、报春花（*Primula*）、堇菜（*Viola*）等。虽然这些物种可以通过自交产生多数纯合系，相比之下，单倍体诱导作为桥梁的方法将更加有效。

为了保障多倍体及单倍体诱导后再生植株的倍性水平，可以采用以下几种方法。直接计数法可以直接确定染色体数目，但劳动强度大；流式细胞法（FCM）是一种快速、可靠、高通量的计数工具，可通过间接计数方法来完成染色体数目的确定（Eeckhaut et al. 2005；Doležel et al. 2007；Leus et al. 2009）。流式细胞法只需要少量的植物组织作为检测材料，在任何环境下均可完成检测。该方法是使用刀片把植物材料切碎（Galbraith et al.

1983）或利用磨珠破碎（Roberts 2007）植物组织释放细胞核，然后对核 DNA 进行染色，在流式细胞仪中进行流体动力聚焦，使核对齐，在激光、UV 灯、LED 的光源下激发荧光核，再进行检测。荧光核激发量与每个细胞（2C）的 DNA 含量有直接的关系。FCM 一次运行即可计数成千上万个细胞。FCM 可对同个属种样品与已知倍性水平对照的荧光反应量进行对比。在物种之间，该比较结果仅与基因组大小有关，而与倍性水平无关，因为 2*n*/2C 值取决于染色体大小和数目的差异。

现在，本章将从三个方面进行阐述。首先介绍有丝分裂多倍体化、减数分裂多倍体化和单倍体在观赏植物领域的最新进展；还将通过 Dhooghe 等（2011）以及 Ferrie 和 Caswell（2011）的综述介绍有丝分裂多倍体化和单倍体诱导的相关进展；减数分裂多倍体化在观赏植物上应用的研究结果概述。其次，总结多倍体和单倍体诱导的不同技术方法及其步骤与优缺点。最后，将简要讨论植物倍性育种科学应用前景。

7.2 有丝分裂多倍体化

7.2.1 方法与应用

有丝分裂多倍体化一般通过使用抗有丝分裂剂使染色体加倍。Dhooghe 等（2011）对植物的有丝分裂体外多倍体化诱导进行了综述，涵盖了 63 篇研究论文，其中 30 篇是关于观赏植物的，包含了 15 篇木本观赏植物及 15 篇草本观赏植物的研究。作者总结认为，由于抗有丝分裂剂的种类、浓度、处理时间、培养基质、外植体类型与处理结果有直接的关系，实验设计显得尤为重要。在综述所列举的观赏植物研究中，大量使用了氟乐灵（trifluralin）、甲基胺草磷（amiprophos methyl）和氨磺乐灵（oryzalin）等（20～30 种诱导剂）。但开发研究更多诱导剂及降低诱导嵌合率仍是重点方向。

自发表该综述以来，观赏植物在多倍体诱导方面取得了较大的进展（表 7.1）。共计有 36 种外植体组合已成功获得多倍体。除了湖北紫荆（*Cercis glabra*）节间小叶和紫锥菊（*Echinacea purpurea*）叶与根外，其他物种的所有外植体均可被诱导为多倍体。表 7.1 列出了最有效的处理方法。在 9 种组合中，结节/节段/切口通常被用作外植体材料；芽、芽分生组织被成功诱导了 6 次；种子、幼苗和分生组织被成功诱导 9 次。较少见的材料是原球茎（4 次）、叶片（3 次）；除此之外，还有愈伤组织、原生质体、花茎、叶柄和鳞片（均为 1 次）。

18 个诱导组合结果表明，秋水仙碱是产生多倍体效果最好的诱导剂。在 14 个组合中，诱导效果最好的是氨磺乐灵。其他抗有丝分裂剂的使用频率较低，氟乐灵和甲基草胺磷被用于多倍体诱导试验 3 次，有 2 次为公开发表。在特定情况下，多倍体化是从原生质体开始自发发生的，无须添加外源抗有丝分裂剂。

最成功的 MI（抗有丝分裂剂）诱导方法是在分生组织部位进行滴液处理。在表 7.1 中有 3 种植物使用了这种方法。具体操作为使用高浓度抗有丝分裂剂溶液连续数天施用于枝条或幼苗的分生组织上（图 7.1a），液滴也可以制成半固化（Schulze and Contreras 2017）。另一种常见的方法是在相对较短的时间内将外植体浸泡于含有高浓度抗有丝分

表 7.1　自 Dhooghe 等（2011）以来观赏植物多倍体化的研究进展

种类[a]	外植体[b,c]	诱导剂（mg/L）[c, d]	诱导时间[c]	诱导方法[e]	参考文献
木本观赏植物					
南鼠刺科					
紫红南鼠刺（*Escallonia rosea*）	N	ORY 52*; TRI 50; COL 400/800	2d/3d*	SL	Denaeghel et al. 2015
粉红南鼠刺（*Escallonia rubra*）	N	ORY 17/52/87; TRI 17/50/84 或 ORY 0.3/1.7/3.5; TRI 0.3*/1.7/3.4	2d/3d/4d 6*w/8w/10w	SL CS	
豆科					
湖北紫荆（*Cercis glabra*）	STM*	ORY 52	6/12/24/48/96*h 1/2/4/8/16d	SL SS	Nadler et al. 2012
	NS*	ORY 52	6/12/24/48/96*h 1/2/4/8/16d	SL SS	
	IC	ORY 52	12h	SL	
	IC	ORY 52	5/10/15/20/25/30/35/40d	SS	
唇形科					
薰衣草（*Lavandula* × *intermedia*）	STM	COL 1000	3d	D	Urwin 2014
	C	COL 1000	6h	SL	
	N	COL 1000	6h	SL	
穗花牡荆（*Vitex agnus-castus*）	S	COL 500*/1 000; ORY 50/100; TRI 50/100	36h	SL	Ari et al. 2015
木犀科					
日本女贞(*Ligustrum japonicum*)	SLM	COL 1 000/2 000*/3 000	3d	D	Fetouh et al. 2016
车前草科					
Hebe ‘Oratia Beauty’	N	COL 200/400; ORY 4/20/100*	48h	SL	Gallone et al. 2014
蔷薇科					
卢李梅（*Prunus lusitanica*）	SLM	ORY 43*; COL 2 000/ 4 000/8 000	10d	D	Schulze and Contreras 2017
杨柳科					
杨树（*Populus* spp.）	L	COL 20/30*/40	2d/3d*/4d	SL	Xu et al. 2016
双子叶草本观赏植物					
菊科					
松果菊（*Echinacea purpurea*）	L PT* R	COL 100	15d/22d/30d/37d*/47d	CS	Li et al. 2016
非洲菊（*Gerbera jamesonii*）	ST	COL 100/500/1 000*/ 5 000/10 000	2h/4h/8h*	SL	Gantait et al. 2011
香金光菊（*Rudbeckia subtomentosa*）	STM	ORY 10	5d	SL	Oates et al. 2012
黄雏菊（*Rudbeckia subtomentosa* × *Rudbeckia hirta*）	STM	ORY 10	5d	SL	
十字花科					
香雪球（*Lobularia maritima*）	S	COL 500/1 000/2 000*	6h/12h*/24h/36h	SL	Huang et al. 2015
	SLM*	COL 500*/1 000/2 000	48h	棉球	
石竹科					
剪红纱花（*Lychnis senno*）	NS	COL 100/500/1 000; ORY 10*/20/50; APM 1/5/10	24h	SL	Nonaka et al. 2011
大花剪秋罗（*Lychnis fulgens*）	NS	ORY 10	24h	SL	

续表

种类 [a]	外植体 [b,c]	诱导剂（mg/L）[c,d]	诱导时间 [c]	诱导方法 [e]	参考文献
狭叶剪秋罗（*Lychnis sieboldii*）	NS	ORY 10	24h	SL	
龙胆科					
斜升秦艽（*Gentiana decumbens*）	P	Spontaneous			Tomiczak et al. 2015
锦葵科					
芙蓉葵（*Hibiscus moscheutos*）	SL	ORY35*/43/52μmo/L; COL 250/500/1 000	6h*/12h/24h	SL	Li and Ruter 2017
毛茛科					
金莲花（*Trollius chinensis*）	S	COL 500/1 000*/2 000	12h/24h*/36h/48h	SL	Zhang et al. 2016
茄科					
矮牵牛（*Petunia hybrida*）	L	COL 200	15d*/30d	CS	Regalado et al. 2017
马鞭草科					
Glandularia peruviana × scrobiculata	NS	COL 10/100*	24h/48h*	SL	Gonzalez et al. 2015
单子叶草本观赏植物					
天南星科					
彩叶芋（*Caladium × hortulanum*）	L	COL 500/1 000/2 000*	2d/4d*/6d	SL	Cai et al. 2015
鸢尾科					
黄金臭藏红花（*Crocosmia aurea*）	S	COL 0.1* 或 COL 10	16h 3d	SL SL	Hannweg et al. 2013
百合科					
杂交百合（*Lilium × formolongi× Oriental hybrid*）	BSS	COL 500*/1 000	18h/24h*/36h	SL	Zhang et al. 2017
郁金香（*Tulipa gesneriana*）	FS	ORY 5/10; APM 5/10*	7*/14d	SS	Podwyszynska 2012
兰科					
石斛（*Dendrobium* 'Gatton Sun Ray'）	PLB	ORY 5*/10/20	3d/6d*	SL	Miguel and Leonhardt 2011
美洲石斛（*Epidendrum* 'Helen's Pride'）	PLB	ORY 5/10/20*	3d/6d*	SL	
齿唇兰（*Odontioda* 'Emma Sander'）	PLB	ORY 5*/10/20	3d/6d*	SL	
蝴蝶兰（*Phalaenopsis bellina*）	PLB	ORY 5*/10/20	3d*/6d	SL	
柳枝稷（*Panicum virgatum*）	SL	APM 3*/6; COL 1 000/ 2 000	5h	SL	Yoon et al. 2016
姜科					
海南三七（*Kaempferia rotunda*）	ST	COL 2 000	2/4*/8/12d	SS	Soonthornkalump et al. 2014

a. 如果文章中未提及，则用品种名称代种名

b. BSS 基础鳞片，C 插条，FS 花茎，IC 节间愈伤组织，L 叶，N 茎段，NS 茎段数，P 原生质体，PLB 原球茎，PT 叶柄，R 根，S 种子，SL 幼苗，SLM 苗分生组织，ST 芽，STM 芽分生组织

c. 标有*的为最有效的处理方法

d. APM 甲基草胺磷，COL 秋水仙碱，ORY 氨磺乐灵，TRI 氟乐灵

e. CS 在固体介质中的缓慢处理，D 分生组织的液滴处理，SL 液体介质中的冲击处理，SS 固体介质中的冲击处理

裂剂的液体培养基中（图 7.1b），然后再转移到无抗有丝分裂剂的培养基上继续培育。在表 7.1 中有 26 种植物使用了该方法。第三种方法是在固体培养基中添加抗有丝分裂剂进行缓效处理（图 7.1c），在表 7.1 中有 5 种植物使用了该方法。在表 7.1 中还有一种植物使用了将秋水仙碱饱和棉球包裹在幼苗周围的技术，与标准液滴法相比，可使分生组织得到有效处理。

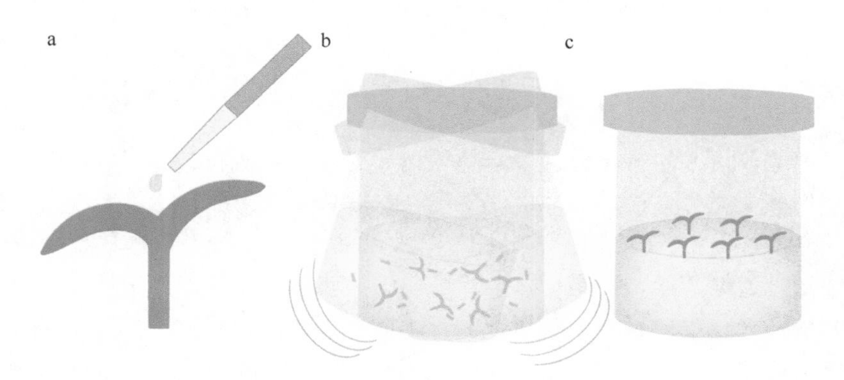

图 7.1　有丝分裂多倍体化不同的诱导方法：（a）用抗有丝分裂剂处理分生组织，连续 2～5 天，每天向幼苗或新梢的顶端滴施抗有丝分裂剂溶液；（b）浸泡处理，将外植体放入添加了较高浓度抗有丝分裂剂的液体培养基中，震荡处理 6h 至 12 天，之后，将外植体转移至无抗有丝分裂剂的培养基中继续培养；（c）缓释处理，将外植体放在补充有相对低浓度的抗有丝分裂剂的固体培养基上处理 15～70 天，然后再转移到无抗有丝分裂剂的培养基中继续培养

表 7.2 列出了上述每种方法的优缺点。多倍体诱导最常遇到的问题之一是嵌合体及混倍体的形成。在部分倍性嵌合体中，植物组织有的为多倍体，有的为非多倍体，也就是说植物组织由不同倍性水平的细胞嵌合而成；有时也会出现周缘嵌合体（即不同细胞层中有不同的倍性水平）。在分析单个植物区域或细胞层时，经常会把嵌合体误认为是稳定的四倍体或二倍体。

表 7.2　有丝分裂多倍体化不同系统的优缺点

技术方法	优点	缺点
分生组织处理	细胞处理单一，嵌合体少	基因表现不明显
缓效处理	死亡率低	诱导率低
脉冲处理	诱导率高	死亡率高

因此，多倍体需要全方位地进行鉴定，如在白鹤芋（*Spathiphyllum*）诱导过程中，就必须对再生体的多个位点和细胞层进行倍性鉴定（Vanstechelman et al. 2010）。当然，这些嵌合体是可以直接用于四倍体再生的，如紫锥菊（Li et al. 2016）的四倍体诱导。

浸泡处理可以在液体或固体介质中使用，在固体介质上多倍体诱导过程较为缓慢，液体介质则可以更好地分散抗有丝分裂剂。缩小外植体也可以改善抗有丝分裂剂的吸收效率。目前，在液体培养基添加抗有丝分裂剂进行处理是观赏植物多倍体诱导最常用的方法（表 7.1）。药剂剂量与外植体死亡率呈线性关系，当然，也可以通过增加外植体数量来提高诱导率。由于该方法需要多次移植材料再培养，因此，在诱导过程中会增加材料污染的风险。

多倍体诱导通常使用克隆繁殖苗作为诱导材料进行染色体加倍。同样，种子和幼苗也常用作多倍体诱导的材料。使用种子作为外植体进行诱导可以节约劳动力，但存在未知基因型取代多倍体基因型的可能。有时，在处理过程中还需要将种皮进行软化或破除，以提高抗有丝分裂剂的浸入效率（Hannweg et al. 2013）。另一种方法是直接利用种子萌发产生的分生组织为材料，直接进行组织培养，该方法可减少嵌合体的发生，但会增加劳动力。

7.2.2 展望

基于组织培养技术的有丝分裂多倍体化诱导已成为常规育种技术。育种家发现染色体的加倍与形态变化有直接的关系，如茎粗、叶片长宽比增加、叶色加深、生物和非生物胁迫耐受性增加、生长变紧凑等。我们认为，目前观赏植物多倍体育种取得的成果远远多于出版物报道的数量，主要是由于多倍体诱导通常是在育种公司进行，公司与公立或私立研究机构合作，成果往往是保密的，因此不会出现在科学文献中。

多倍体诱导也常用于杂交亲本倍性的匹配及三倍体株系的育种过程中。例如，木槿（*Hibiscus syriacus*）'DVPAzurri' Azurri®就是 *H. s.* 'Oiseau Bleu' 和其多倍体进行杂交后获得的新品种，其表现出生长势增强、花期延长和不育。有丝分裂多倍体化也能恢复杂交品种的繁殖力，在这种情况下，不同染色体组间的融合受到阻碍，并阻碍了正常的减数分裂。尽管杂交后代繁殖力也可以通过 F_2 代来恢复，但由于同源染色体的优先配对，通常无法实现基因渗入（Eeckhaut et al. 2006）。

因此，多倍体诱导技术仍然是繁育新品种很好的技术手段。相比于 Dhooghe 等（2011）的综述，目前，多倍体诱导技术在外植体选择及诱导剂类型方面均未发生较大变化。秋水仙碱（colchicine）仍然是应用最广的抗有丝分裂剂，而氨磺乐灵、氟乐灵和甲基草胺磷的使用依旧比较局限。氨磺乐灵对动物微管蛋白没有亲和力，对人体毒害较小。在表 7.1，大多数多倍体诱导仅使用 1 种试剂进行诱导，没有在同一物种上使用多种药物来做诱导率对比。但是，使用氨磺乐灵与秋水仙碱进行比较时，3/4 的物种在使用氨磺乐灵情况下诱导率更高。我们希望在未来几年中，会有新的观赏植物在多倍体诱导方面获得成功；同时，许多正在进行的研究还应进一步优化。尽管如此，多倍体诱导新技术及新诱导剂估计在近期很难会有大的突破。

7.3 减数分裂多倍体化

7.3.1 方法与应用

减数分裂多倍体化技术是通过未减数（$2n$）配子获得多倍体植株的技术。未减数配子的染色体数等于体细胞的染色体数。异常减数分裂（图 7.2a）能够产生未减数的配子（De Storme and Geelen 2011）。这些异常情况中最常见的是省略了减数分裂 I 期或 II 期，使纺锤体方向改变或联会发生异常。根据配子的染色体组成，可以分为两种主要类型：第一次减数分裂核复原形成的 $2n$ 雌配子（FDR）、第二次减数分裂核复原形成的 $2n$ 雌配子（SDR）。

SDR 配子减数分裂 II 期被省略，保持姐妹染色单体在一起（图 7.2b），因此，这些配子是相当纯合的。FDR 配子的非姐妹染色单体和着丝粒共定位，获得杂合配子体。FDR 配子其实就是缺失了一次减速分裂的产物，如果不发生配对及重组，则配子基因型与亲本相同。但是，当减数分裂 II 期出现异常纺锤体（二级分裂、三级分裂或融合分裂），形成的配子会发生配对及重组（图 7.2c）。

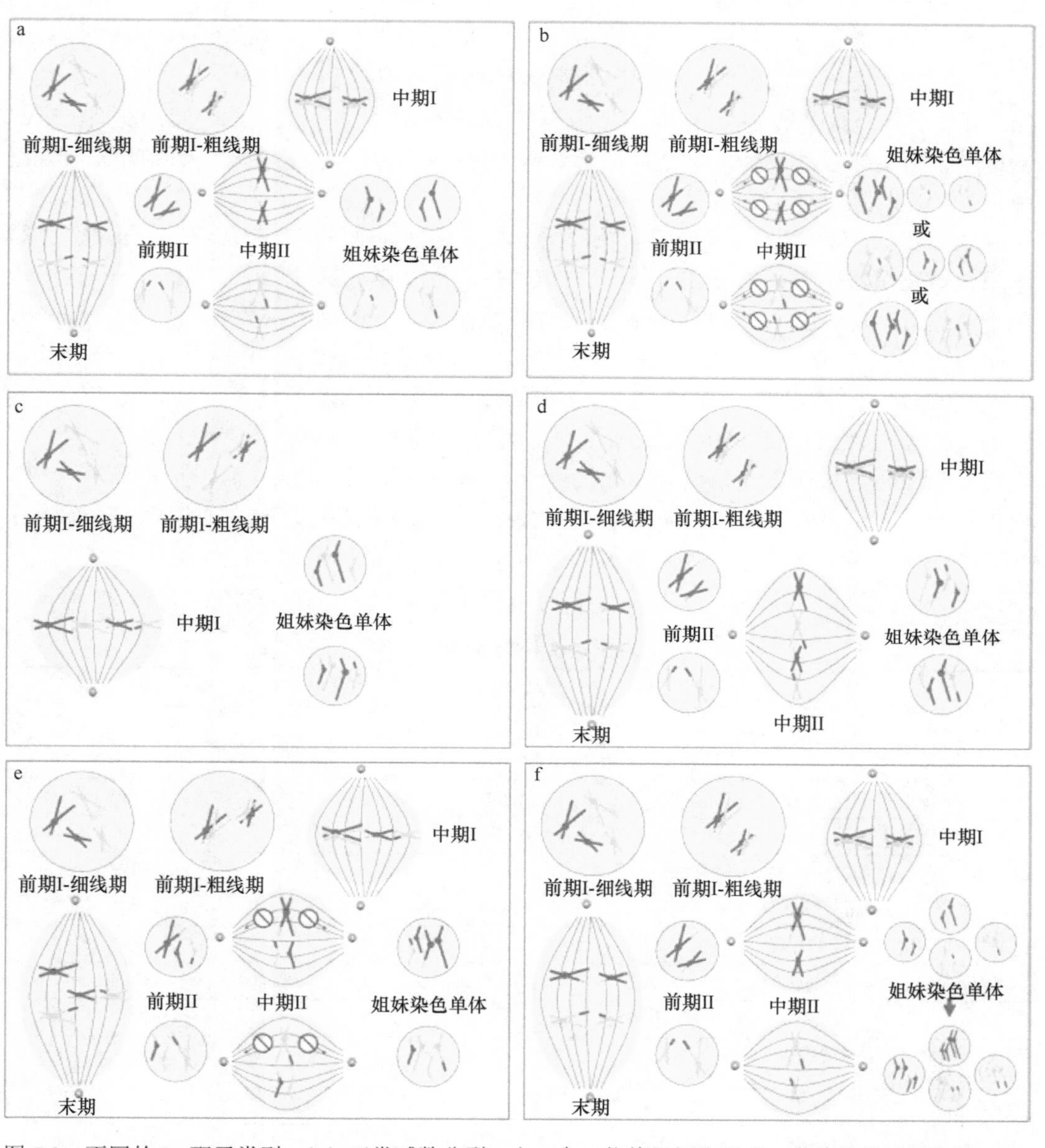

图 7.2　不同的 $2n$ 配子类型：（a）正常减数分裂，由一个二倍体母细胞形成 4 个单倍体子代细胞；（b）第二次减数分裂核复原形成的 $2n$ 雌配子（SDR），在分裂中期 II 中，纺锤体未形成，导致出现二倍体细胞，原染色体组的着丝粒在子细胞之间分布不均；（c）第一次减数分裂核复原形成的 $2n$ 雌配子（FDR），减数分裂受阻，染色体重组不发生；（d）FDR 等效，减数分裂 I 后获得的子代细胞在减数分裂 II 之前融合，第一次重组分裂后，子细胞包含原始染色体组的相等着丝粒数；（e）不确定的减数分裂重组，这是 SDR 和 FDR 之间的中间形式；（f）减数分裂后的恢复，减数分裂正常进行，二倍体子代细胞是纯合的

尽管细胞学异常发生在减数分裂 II 期，虽不同于传统 FDR 配子的异常，但它也被归为 FDR（Bretagnolle and Thompson 1995），这种配子还是保留了亲本的杂合性（Zamariola et al. 2014），该系统称为“FDR 模拟系统”（图 7.2d）。在文献中，术语“FDR”

也称为“FDR 类似物”，说明科学文献中 2*n* 配子的细胞学背景并不是很清楚。因此，SDR 与 FDR 未减数配子之间的区别不是基于细胞学的，而是基于遗传学的。不常见的 2*n* 配子类型是不确定减数分裂重组（IMR）配子，它是 SDR 和 FDR 配子之间的中间产物（图 7.2e）与减数分裂后恢复（PMR）配子（图 7.2e）。PMR 配子的产生基于规则减数分裂后染色体的加倍，因此是纯合配子。

Dewitte 等（2012）已对形成未减数配子的植物物种进行了完整的综述。表 7.3 总结了观赏植物领域中有关 2*n* 配子诱导的文献。大多数文章（21～35 篇文章）描述了 FDR 配子的形成，尽管并不清楚是由平行/融合纺锤体形成，还是由减数分裂 I 期遗漏导致 FDR 配子或其他类型的细胞学畸变形成。SDR、PMR 和 IMR 配子分别被描述了 2 次，在 11 篇文献中未明确配子类型。在 13 篇文献中，作者描述了 2*n* 配子对植物基因型的影响，或通过种间杂交（10 次），或多倍体育种（3 次），或单倍体育种（2 次）。此外，也有文献报道环境因素可以诱导形成 2*n* 配子。有 11 篇文献显示，抗有丝分裂剂（一氧化二氮、秋水仙碱、氟乐灵）可诱导产生 2*n* 配子。在某种情况下，咖啡因也可诱导 2*n* 配子的产生。有 4 个例子证明，变温处理可诱导 2*n* 配子。在其余的 6 篇文献中，未发现基因型的变化、环境因素与 2*n* 配子的形成有直接的联系。在文献中，共有 29 篇文章

表 7.3　观赏植物 2*n* 配子诱变

种类	2*n* 配子类型[a]	处理方式[b,c]	雄配子/雌配子[d]	参考文献
木本观赏植物				
醉鱼草科				
醉鱼草（*Buddleja lindleyana*）	NS	SPONT	M	Van Laere et al. 2009b
杜鹃花科				
欧石楠（*Calluna vulgaris*）	FDR	SPONT	F，M	Przybyla et al. 2014
高山杜鹃（*Rhododendron* spp.）	NS	POLY	M	Jones and Ranney 2009
绣球花科				
绣球花（*Hydrangea macrophylla*）	NS	SPONT	M	Alexander 2017
锦葵科				
木槿（*Hibiscus syriacus* × *H. paramutabilis*）	NS	IC	F	Van Laere et al. 2009a
蔷薇科				
月季杂交品种（*Rosa* hybrid）	FDR，SDR	HAP	F，M	Crespel et al. 2006
月季杂交品种（*Rosa* hybrid）	FDR[eq]	HAP	M	El Mokadem et al. 2002
玫瑰（*Rosa* spp.）	FDR[eq]	TEMP 30/33/36*℃	M	Pécrix et al. 2011
杨柳科				
响叶杨（*Populus adenopoda*）	FDR	TEMP 38*/41/44℃	F	Lu et al. 2013
毛白杨（*Populus tomentosa*）	FDR[eq]	SPONT	M	Zhang and Kang 2010
毛白杨（*Populus tomentosa*）	FDR	COL 5mg/L 24/36/48*/54/58/62h 花后诱导，3-5-7*处理	M	Zhao et al. 2017
杨树（*Populus* spp.）	PMR	TEMP 38*/41/44℃	F	Guo et al. 2017
马鞭草科				
马缨丹（*Lantana camara*）	FDR	SPONT	F	Czarnecki and Deng 2009

续表

种类	2*n* 配子类型[a]	处理方式[b,c]	雄配子/雌配子[d]	参考文献
双子叶草本观赏植物				
菊科				
欧蓍草（*Achillea borealis*）	NS	SPONT	M	Ramsey 2007
凤仙花科				
新几内亚凤仙（*Impatiens hawkeri* × *I. platypetala*）	FDR^{eq}	IC	M	Stephens 1998
秋海棠				
秋海棠（*Begonia*）	NS	N_2O，6bar（1bar=10^5Pa），48h TRI 3/33*/335mg/L	M	Dewitte et al. 2010
蝶形花科				
百脉根（*Lotus*）	FDR^{eq}	TEMP，28℃	M	Negri and Lemmi 1998
柳叶菜科				
灯笼海棠（*Fuchsia* spp.）	FDR	IC	M	Talluri 2011
报春花科				
报春花（*Primula denticulate* × *P. rosea*）	NS	IC，POLY	F	Hayashi et al. 2009
单子叶草本观赏植物				
六出花科				
六出花（*Alstroemeria* spp.）	FDR	IC	F，M	Ramanna et al. 2003
百合科				
百合（*Lilium*）	FDR	IC	M	Barba-Gonzalez et al. 2005
百合（*Lilium*）	FDR，IMR	N_2O，6bar，24*/48h	F，M	Barba-Gonzalez et al. 2006
百合（*Lilium*）	FDR	N_2O 24h	M	Kitamura et al. 2009
Lilium × *formolongi*	FDR	N_2O，6.1bar，48h	M	Sato et al. 2010
百合（*Lilium*）	IMR	IC	M	Lim et al. 2001
百合（*Lilium*）	SDR	IC	M	Lim et al. 2004
百合（*Lilium*）	FDR 或 PMR	咖啡因 0.3%	F，M	Lim et al. 2005
百合（*Lilium*）	NS	COL 0/0.2/0.5/1*/2mg/L	F	Wu et al. 2007
百合（*Lilium*）	NS	N_2O，6.1bar，24h	M	Akutsu et al. 2007
百合（*Lilium*）	FDR	N_2O，6bar，24/48*/72h	M	Luo et al. 2016
Lilium auratum × *L. henryi*	FDR	IC	M	Chung et al. 2013
Tulipa gesneriana × *T. fosteriana*	FDR	IC，POLY	F，M	Marasek-Ciolakowska et al. 2014
Tulipa gesneriana，*T. fosteriana*	NS	N_2O 24/48/24+24h*	M	Okazaki et al. 2005
兰科				
美丽蝴蝶兰（*Phalaenopsis amabilis*）	NS	N_2O 24*/48/68h	M	Wongprichahan et al. 2013
鹤望兰科				
鹤望兰（*Strelitzia reginae*）	FDR^{eq}	COL 0.5/1/2*mg/L	M	Xiao et al. 2007

a. FDR 第一次减数分裂核复原形成的 2*n* 配子，FDR^{eq} 等价于第一次减数分裂核复原形成的 2*n* 配子，IMR 不确定减数分裂重组形成的 2*n* 配子，NS 未指定，PMR 减数分裂后重组形成的 2*n* 配子，SDR 第一次减数分裂核复原形成的 2*n* 配子

b. COL 秋水仙碱处理，HAP 单倍体，IC 种间杂交，N_2O 一氧化二氮处理，POLY 多倍体，SPONT 自发形成，TEMP 温度处理，TRI 氟乐灵处理

c. 标记为*表示最为有效

d. F 雌，M 雄

对 2*n* 雄配子进行了阐述，12 篇文章对 2*n* 雌配子进行了阐述。减数分裂多倍体化的优点及缺点可见表 7.4。利用化合物如秋水仙碱、氨磺乐灵、氟乐灵及相关复合物干扰纺锤体形成是诱导 2*n* 配子最常规的方法。最合适的诱导时机是粗线期（Younis et al. 2014；Zhao et al. 2017）。秋水仙碱经常存在于球根植物中，具有毒性（Van Tuyl et al. 1992）。由于在作物中，组织层会保护分生组织，因此，可以通过高压强（±6bar）N_2O 处理植物组织来诱导 2*n* 配子。在高压条件下，N_2O 可以更有效地穿透组织，使用该方法最适合的阶段是中期 I（Dewitte et al. 2010）。实验设计时，可将植物组织暴露于具高压 N_2O 的特殊设备中 24～72h（图 7.3）。在百合属植物中有报道，咖啡因也可以诱导 2*n* 配子。虽然 2*n* 配子形成的机制已经有了 3 种假设，即阻止同源重组、抑制胞质分裂和减数分裂后恢复，但确切的作用方式尚不清楚（Lim et al . 2005；Younis et al. 2014）。

表 7.4 减数分裂多倍体化不同系统的优缺点

技术方法	优点	缺点
传统抑制剂处理	容易掌握	毒性高，组织穿透性差
N_2O 高压处理	毒性低 组织穿透性好	需要专用设备
温度处理	操作容易	使用局限性小

极端温度（高温或低温）会显著促进 2*n* 的配子形成（Pécrix et al. 2011；De Storme et al. 2012；De Storme and Mason 2014；Zhou et al. 2015）。但也有极端温度无法诱导 2*n* 配子形成的例子（Lokker et al. 2004），估计这与植物基因型有较大的关系。在基础研究层面上，科学家已揭示了涉及 2*n* 配子形成的几个基因（De Storme and Geelen 2011；De Storme and Mason 2014；Younis et al. 2014）。目前，已证明外源赤霉素对未减数花粉形成没有干扰作用（Liu et al. 2017）。显然，基因型是 2*n* 配子形成的决定性因素。

多倍体、单倍体和种间杂种也可以产生 2*n* 配子体（表 7.3）。可以假设，在新的基因组减数分裂过程中发生细胞畸变最终形成配子。尽管这个过程是一种反作用，但具有较广的育种潜力。大部分物种种间杂交后，正常配子配对受阻，形成的 2*n* 配子可以产生 F_2 代。这是因为杂交可以在很多物种中发生，所以 2*n* 配子比 F_1 种间杂种有丝分裂多倍体化后形成的正常配子更有优势，而在正常配子中，突变通常只发生在同一染色体组之间，而未减数配子能为种间杂种提供更有优势的多倍体化（Marasek-Ciolakowska et al. 2016）。

利用 2*n* 配子开展百合育种是 2*n* 配子在育种上应用的典型成功案例。Jaap Van Tuyl 的研究小组创建了不同杂种群（Longiflorum、Asiatic、Oriental 和 Trumpet），用作 2*n* 配子的来源，从而获得了三倍体种间杂种。通过该方法创造的百合品种占全球百合种植面积的 45%以上。多倍体水仙、郁金香和秋海棠品种也有一部分起源于 2*n* 配子（Marasek-Ciolakowska et al. 2016）。

7.3.2 展望

多倍体在植物界很普遍。这些自然形成的多倍体是形成 2*n* 配子的结果。自然条件下诱导的多倍体可能比其对应的二倍体具有更高的变异性（Harlan and De Wet 1975）。因此，

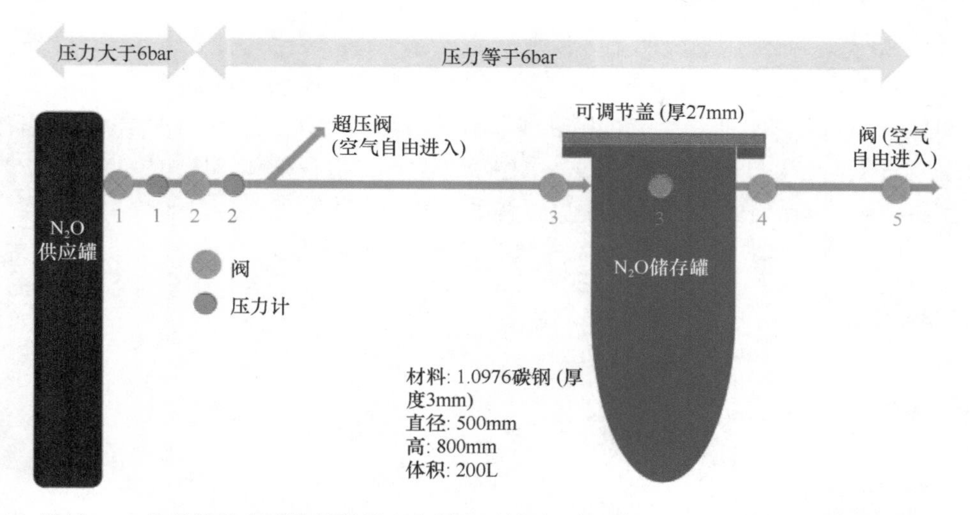

图 7.3　通过 N_2O 处理诱导未减数的配子。在程序开始时，打开阀 3 和 4。首先，关闭阀 2，并打开压力计 1 对气体管道的第一部分加压；缓慢打开阀 2，10s 后关闭阀 5，允许在管道和孵化器中 N_2O 代替空气；然后，通过进一步打开阀 2 并控制压力计 2 和 3，逐步增加培养容器中的压力；1h 后，压力稳定在 6bar；关闭阀 2，处理 24～48h。处理结束后，关闭阀 1，打开阀 5；之后，通过打开阀 2 释放阀 1 和 2 之间气体管道中的气体；该装置可承受高达 9bar 的压力，但配有安全阀以防止超过 6bar 的压力

多倍体化是许多重要农作物获得新品种的基础（Bretagnolle and Thompson 1995）。在一些经济作物（如小麦、马铃薯、咖啡、月季等）中，通过选择诸如产量、生物和非生物胁迫耐受性或观赏性等特征，2*n* 配子被发现，至今已发现多种类型（Bretagnolle and Thompson 1995），特别是在模式作物中取得了重大进展，但在许多情况下仍未阐明其合成的分子背景。

在拟南芥中已经鉴定出许多涉及 2*n* 配子形成的基因。例如，OSD1 基因产生 SDR 配子（d' Erfurth et al. 2009），而 AtPS1（d' Erfurth et al. 2008）和 JASON（JAS）（De Storme and Geelen 2011）突变体的纺锤体形成导致类似 FDR 配子形成。Brownfield（2015）研究了 atps1 和 JAS 突变体的胞质分裂。双子叶植物胞质分裂为同时型，形成减数分裂 I 和 II 的两个纺锤体没有分裂。在正常的减数分裂中，细胞质中的两个纺锤体应该分开，即阻止合并的机制（Brownfield and Köhler 2011）。Brownfield（2015）总结到，在 2*n* 配子形成过程中，在野生型植物中提供物理屏障的细胞器因 JAS 蛋白的丧失而被破坏。

对 2*n* 配子形成的环境-遗传相互作用的分子基础进行研究是未来研究的热门话题。环境的压力可能会触发 2*n* 配子的形成，这与多倍体物种形成是对不断变化的环境反应的理论相一致（Mason and Pires 2015）。在育种过程中，温度是主要的媒介，但其他胁迫因素（包括植物营养、虫食性、水分和疾病）也会影响 2*n* 配子发生（De Storme and Geelen 2014）。糖代谢、植物激素、表观遗传调控、活性氧（ROS）都与配子的形成有直接的关系，但分子调控机制仍然未知。也有观点认为在整个观赏植物生长过程中 2*n* 配子的变化具体取决于温度（Crespel et al. 2015；Zhou et al. 2015），如对月季进行热处理会诱发平行和三极纺锤体的形成（Pécrix et al. 2011）。但无法证明 AtPS1 或 JASON 玫瑰类似物相关蛋白的功能改变。为了完全阐明 2*n* 配子的发生，商业品种的任何转录谱或翻译后修饰方面的研究尚待取得更大进展。

优化 $2n$ 花粉/卵细胞形成的第一步是充分了解其遗传及生理背景，也可以为筛选特定的 $2n$ 配子类型铺平道路。在许多观赏作物中，可通过染色体重组诱导二倍体花粉。特定物种的杂交障碍可能是由倍性提高或非目标基因的转移造成的。未成熟花粉可产生微原生质体（Saito and Nakano 2002），原生质体融合使得在受体基因组中仅添加一个重组染色体，从而阻止了使用 $2n$ 配子后一系列回交的发生。然而，由于原生质体再生和微原生质体分离的要求，这种选择在技术上仍然具有挑战性。

7.4 单 倍 体

7.4.1 方法与应用

Ferrie 和 Caswell（2011）发表了有关在观赏物种中诱导单倍体的科学文献及综述，结果显示差异很大，而且培养基和培养条件因物种变化而产生较大的变化。在一个物种中，基因型对于实验方案也至关重要。该文献中共列出了 46 个属的观赏植物，大多通过一种或多种技术完成了单倍体的诱导。其中，最成熟的技术是花药培养（33 属）、小孢子培养（13 属）、辐照花粉（8 属）、胚珠/卵巢培养（6 属）。

表 7.5 总结了 10 篇最近发表的文献，均成功完成了观赏植物的单倍体诱导。列出使用花药培养获得单倍体 5 次，使用小孢子培养获得单倍体 2 次；在其中一份文献中，胚珠培养也获得成功。在另外两个文献中，通过辐射花粉或受精花粉可诱导雌核发育。

在 5 篇文献中，提到了通过温度处理获得单倍体的方法。在 4 篇文献中，使用冷处理（4℃）获得了单倍体；而马蹄莲小孢子则是在 32℃下获得的。如表 7.5 所示，各种设计方案的诱导率会随着物种的基因型变化而变化。同样，生长调节剂（Jia et al. 2014）和培养类型（Ari et al. 2016）也对单倍体的诱导具有决定性的作用。

将双单倍体育种技术作为诱导大麦、小麦和油菜籽等二倍体或假二倍体经济作物的 F_1 杂种的中间步骤具有较大的应用潜力（Thomas et al. 2003）。自然界中的一些观赏植物，如月季、菊花是多倍体，而单倍体植株仍然是杂合体，不适合用于 F_1 育种。由于观赏植物经济价值较小，因此，与粮食作物相比，其单倍体育种研究投入较为有限。此外，大部分观赏植物可以成本较低的方式进行营养繁殖，因此，单倍体诱导及 F_1 育种仅因兴趣而进行。在许多种子繁殖植物（如金鱼草、秋海棠、翠雀、石竹、老鹳草属、向日葵、洋桔梗、天竺葵属、矮牵牛、报春花和三色堇）中，纯合的亲本是通过自交形成杂交品种而制成的。所有单倍体育种技术在商业观赏育种中的实际应用有限，尽管技术可行性已在许多出版物中得到了证明（Ferrie and Caswell 2011）（表 7.5）。

图 7.4 和表 7.6 列出了单倍体诱导的最常见育种方法。小孢子培养技术具有较大的应用潜力。因为在培养过程中，没有体细胞的参与，只有单位体细胞被培养。因此，无论是自然发生的，还是由抗有丝分裂剂诱导获得的二倍体都是纯合的。通常，单核晚期或双核早期的小孢子最适合进行小孢子培养。标准程序是进行 DAPI 染色，随后通过显微镜观察将芽的大小与小孢子发育阶段联系起来。在诱导的头一天会通过温度胁迫或营养饥饿培养来诱导分裂。如果该技术有效，则可以形成胚胎，然后可以对其进行传代

表 7.5 Ferrie 和 Caswell（2011）关于观赏植物单倍体的进展概述

品种	外植体 a, b	处理方法 a, c	参考文献
双子叶植物			
菊科			
杭白菊（*Chrysanthemum morifolium*）	2 DAP	木茼蒿花粉混合授粉	Wang et al. 2014
Dendranthema × *grandiflorum*	花药，U	无	Gao et al. 2011
十字花科			
甘蓝（*Brassica oleracea* var. *acephala*）	单核发育阶段的胚珠（球形、心形、鱼雷形、子叶状*）	4℃，0/2*/5*/10d	Wang et al. 2011b
龙胆科			
龙胆（*Gentiana scabra*）		4℃，0/3/7/14d*	Doi et al. 2013
三花龙胆（*Gentiana triflora*）	胚珠形成前期	4℃，0/3/7*/14d	
Gentiana triflora × *G. scabra*	开花期	4℃，0/3/7/14d（GD）	
苦苣苔科			
非洲紫罗兰（*Saintpaulia ionantha*）	花药，U	无	Uno et al. 2016
透骨草科			
Mimulus aurantiacus	胚珠、种子 21*/28/35 DAP	授粉（辐射后的花粉）	Murovec and Bohanec 2013
报春花科			
小报春（*Primula forbesii*）	花药，LU，EB	无	Jia et al. 2014
茄科			
辣椒（*Capsicum annuum*）	花药，LU，EB	4℃，1 天	Ari et al. 2016
辣椒（*Capsicum annuum*）	花药，LU，EB	4℃，1/2/3/4d（GD）	Popova et al. 2016
单子叶植物			
天南星科			
马蹄莲（*Zantedeschia aethiopica*）	小孢子，MU/LU*/EB	4/25/32*℃	Wang et al. 2011a

a. *表示最有效的方法

b. DAP 授粉后的天数；EB 双核（双子叶植物）早期或双细胞（单子叶植物）早期发育阶段的小孢子；LU 单核（双子叶植物）晚期或单细胞（单子叶植物）晚期发育阶段的小孢子；MU 处于单细胞发育中期的小孢子；U 单细胞相中的小孢子

c. GD 表示最好的处理方法是基因依赖的

培养。但该技术的局限在于仅适用有限种类的科，如十字花科和禾本科，其中也包括较少量的观赏植物。

由于存在杂合组织，因此会有愈伤组织或器官被诱导形成，它们将与小孢子共同竞争营养物质，从而干扰单倍体的再生。因此，每一次再生分裂都有可能是杂合体，只能通过分子技术进行甄别。与小孢子培养相似，胁迫处理可诱导小孢子体的发育，而不是进一步的配子体发育。

在胚珠培养中，一般不施用外部胁迫。与花药培养相似，愈伤组织的形成是单倍体细胞向器官发生的中间步骤。但是，可诱导为单倍体的细胞数量是有限的（每胚珠 5～8 个细胞，不同的物种数量不同）。胚珠大小也与物种有直接的关系，因此，需要进行子房初步的培养，以便更容易地分离胚珠及后续培养。然而，二倍体的再生都可能来自体细胞组织，因此是杂合的。

另一种方法是利用辐射过的花粉或无法进行杂交的花粉通过授粉来诱导胚珠。尽管

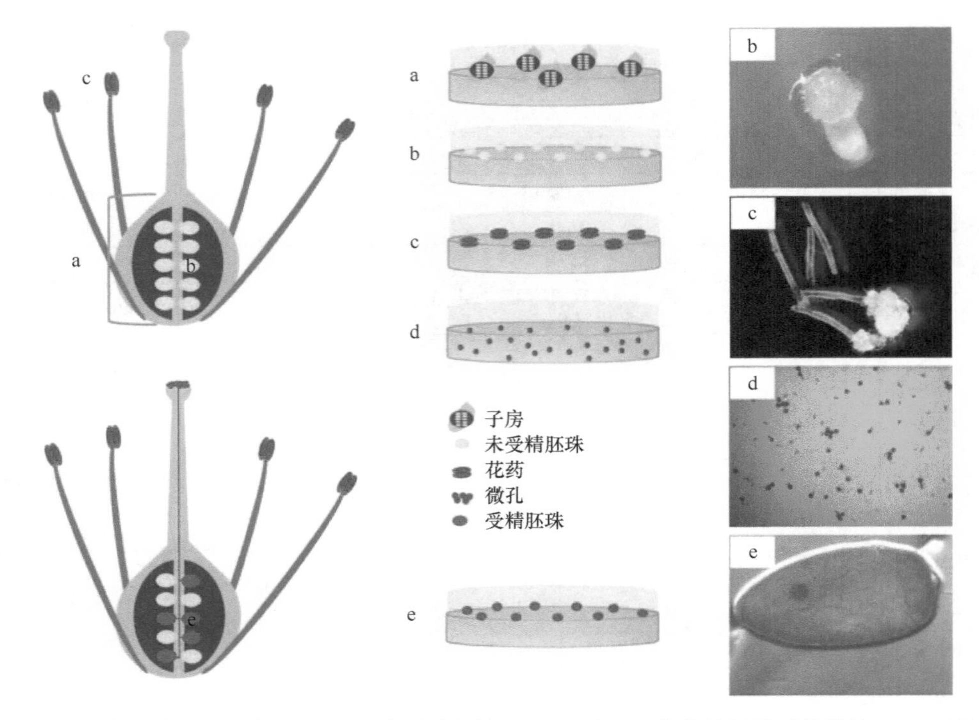

图 7.4 通过未受精花的器官（a～c）、未受精花的胚（d）或无功能花粉授粉后的雌核（e）诱导开花植物单倍体的育种策略：(a) 卵巢培养；(b) 胚珠培养（图片，愈伤组织间期）；(c) 花药培养（图片，愈伤组织间期）；(d) 小孢子培养；(e) 胚胎挽救（图片，具有球形胚的受精胚珠）。

表 7.6 单倍体不同系统的优缺点

技术方法	优点	缺点
胚珠培养	依赖性低	有二倍体细胞再生的可能性，获得的细胞较少
花药培养	使用的种类较多	有二倍体细胞再生的可能性，胁迫处理
小孢子培养	只有单细胞，很多细胞均适用	极少数品种可用胁迫处理
授粉	没有应力处理	劳动力成本高，效率低

这些花粉能够通过花粉管并能使胚珠受精，但合子的基因组不稳定，从而导致亲本基因组丢失。该技术称为雌核发育或雄核发育，具体取决于贡献单倍体中剩余染色体组的亲本，该技术也称为“Hordeum bulbosum”技术，常用于大麦授粉并诱导单倍体。授粉后，胚珠被转移到胚挽救培养基中继续进行胚培养。大麦种子亲本通常是去雄的，但需要筛选再生的二倍体，因为可能存在自花授粉或种间杂种。

无论采用何种技术，植物的基因型尤其重要。在诱导实验过程中，可以对植物激素、培养物类型或环境条件进行深入的筛选。

7.4.2 展望

目前也有利用修饰的 CENH3 基因进行单倍体诱导研究的。CENH3 是定位于着丝粒的组蛋白，位于活性着丝粒的动粒复合体的组装位点，真核生物几乎都有。Ravi 和 Chan

（2010）在拟南芥中证明了通过与 CENH3 GFP 尾部交换互补创建 cenh3 突变体来创建单倍体诱导系。功能减退型 CENH3 基因可诱导有丝分裂（纯合 cenh3 具有致命性）。减数分裂也被启用。但是，在将该变异体与野生型植物杂交后，合子中将存在两种 CENH3 类型，这是由于 GFP 尾交换 CENH3 的功能降低而产生“竞争情况”。这将阻碍早期胚胎产生后的下一个有丝分裂期间 GFP 尾交换母染色体的正确分离。有证据表明，在消除了配子中原有的 CENH3 组蛋白后，这些染色体的 CENH3 着丝粒重载受到削弱或延迟（Karimi-Ashtiyani et al. 2015）。这将导致着丝粒失活或细胞周期不同，最终引发染色体分裂后期落后。不依赖有丝分裂的染色体出芽也在间期发生（Ishii et al. 2016）。虽然野生型亲本的染色单体沿着纺锤丝向两极分开，但是诱导细胞系的染色单体可能会组装在缠绕的微核中，因此不会并入新形成的细胞中，而是会退化。相反，当只有一种 CENH3 类型存在“非竞争情况”时，两个染色体组的着丝粒装配是相同的，且常规细胞分裂不会停止。换句话说，也就是需要不同的 CENH3 类型来产生可以在单倍体植物中再生的单倍体细胞。这是由 CENH3 差异补充、着丝粒失活、染色体出芽还是其他因素（综合）引起的，这个有趣的问题很可能在不久的将来得到解决。

近年来，该技术已经变得更加完善。CENH3 的不同转基因修饰对诱导物的互补以及单倍体和非整倍体的形成有不同的作用（Kelliher et al. 2016）。但是非转基因方法也是可行的（Britt and Kuppu 2016）。化学诱导的单个氨基酸突变体可能已经显著影响 CENH3 的功能并引起单倍体形成（Karimi-Ashtiyani et al. 2015）。用相关物种的 CENH3 补充 cenh3 无效突变体也可以具有这种效果。这表明，经过稍微不同的 CENH3 组蛋白的组合，种间授粉可能会根据同一原理形成单倍体。

雌性和雄性基因型都可以是单倍体，这意味着该技术可以实现雌雄同体发育，也可以诱导雌核发育（Britt and Kuppu 2016）。此外，“功能失调的”CENH3 突变体不会传递至单倍体。CRISPR/Cas9 等基因组编辑技术具有巨大的潜力，我们预计它在未来几年内在经济作物和目前较难得物种方面取得许多突破。但是必须根据植物种类基因型的不同建立诱导再生技术。此外，在单个基因组中可能会出现功能不明的 CENH3 的多个拷贝，因此可能需要多个基因组编辑周期。

7.5　结　论

倍性育种是观赏植物中成熟的育种技术，育种者较易掌握。该技术可导致未减数配子的形成，也可以获得比有丝分裂多倍体化后杂合度更高的四倍体。但是，$2n$ 配子形成的分子背景是不完整的，因此育种者尚不能激发其全部潜能。当配子为异性杂合配子时，倍性育种技术在观赏作物中的应用将会增加。例如，使用 $2n$ 配子恢复种间杂种可成为有丝分裂多倍体化的有用替代方法。单倍体育种技术使得本不可以进行基因交换的植物进行基因交换成为可能。已证明，单倍体育种可创建许多物种的 F_1 杂种。尽管传统技术仍是单倍体育种的主要手段，但 CENH3 修饰技术仍然具有很大潜力，因此，只要法律允许该技术应用到商业中，便可以成为观赏植物育种的又一创新工具。

本章译者：

李 涵

云南省农业科学院花卉研究所，国家观赏园艺工程技术研究中心，云南省花卉育种重点实验室，昆明 650200

参 考 文 献

Akutsu M, Kitamura S, Toda R, Miyajima I, Okazaki K (2007) Production of 2n pollen of Asiatic hybrid lilies by nitrous oxide treatments. Euphytica 155:143–152

Alexander L (2017) Production of triploid *Hydrangea macrophylla* via unreduced gamete breeding. Hortscience 52:221–224

Ari E, Djapo H, Mutlu N, Gurbuz E, Karaguzel O (2015) Creation of variation through gamma irradiation and polyploidization in *Vitex agnus-castus* L. Sci Hortic 195:74–81

Ari E, Yildirim T, Mutlu N, Buyukalaca S, Gokmen U, Akman E (2016) Comparison of different androgenesis protocols for doubled haploid plant production in ornamental pepper (*Capsicum annuum* L.). Turk J Biol 40:944–954

Barba-Gonzalez R, Lim K, Ramanna M, Visser R, Van Tuyl M (2005) Occurrence of 2n gametes in the F1 hybrids of oriental x Asiatic lilies (*Lilium*): relevance to intergenomic recombination and backcrossing. Euphytica 143:67–73

Barba-Gonzalez R, Miller C, Ramanna M, Van Tuyl J (2006) Nitrous oxide N_2O induces 2n gametes in sterile F1 hybrids of oriental x Asiatic lilies (*Lilium*) and leads to intergenomic recombination. Euphytica 148:303–309

Bretagnolle F, Thompson J (1995) Gametes with the somatic chromosome number: mechanisms of their formation and role in the evolution of autopolyploid plants. New Phytol 129:1–22

Britt A, Kuppu S (2016) Cenh3: an emerging player in haploid induction technology. Front Plant Sci 7:357

Brownfield L, Köhler C (2011) Unreduced gamete formation in plants: mechanisms and prospects. J Exp Bot 62:1659–1668

Brownfield L, Yi J, Jiang H, Minina E, Twell D, Kohler C (2015) Organelles maintain spindle position in plant meiosis. Nat Commun 6:6492

Cai X, Cao Z, Xu S, Deng Z (2015) Induction, regeneration and characterization of tetraploids and variants in 'tapestry' caladium. PCTOC 120:689–700

Chung M, Chung J, Ramanna M, Van Tuyl J, Lim K (2013) Production of polyploids and unreduced gametes in *Lilium auratum* × *L. henryi* hybrid. Int J Biol Sci 9:693–701

Crespel L, Ricci S, Gudin S (2006) The production of 2n pollen in rose. Euphytica 151:155–164

Crespel L, Le Bras C, Relion D, Roman H, Morel P (2015) Effect of high temperature on the production of 2n pollen grains in diploid roses and obtaining tetraploids via unilateral polyploidization. Plant Breed 134:356–364

Czarnecki D, Deng Z (2009) Occurrence of unreduced female gametes leads to sexual polyploidization in *Lantana*. J Am Soc Hortic Sci 134:560–566

d'Erfurth I, Jolivet S, Froger N, Catrice O, Novatchkova M, Simon M, Jenczewski E, Mercier R (2008) Mutations in AtPS1 (*Arabidopsis thaliana* parallel spindle 1) lead to the production of diploid pollen grains. PLoS Genet 4:e1000274

d'Erfurth I, Jolivet S, Froger N, Catrice O, Novatchkova M, Mercier R (2009) Turning meiosis into mitosis. PLoS Biol 7:e1000124

Daker M (1966) 'Kleine Liebling', a haploid cultivar of *Pelargonium*. Nature 211:549–550

De Storme N, Geelen D (2011) The *Arabidopsis* mutant *jason* produces unreduced first division restitution male gametes through a parallel/fused spindle mechanism in meiosis II. Plant Physiol 155:1403–1415

De Storme N, Geelen D (2014) The impact of environmental stress on male reproductive development in plants: biological processes and molecular mechanisms. Plant Cell Environ 37:1–18

De Storme N, Mason A (2014) Plant speciation through chromosome instability and ploidy change: cellular mechanisms, molecular factors and evolutionary relevance. Curr Plant Biol 1:10–33

De Storme N, Copenhaver GP, Geelen D (2012) Production of diploid male gametes in *Arabidopsis* by cold-induced destabilization of postmeiotic radial microtubule arrays. Plant Physiol 160:1808–1826

Denaeghel H, Van Laere K, Leus L, Van Huylenbroeck J, Van Labeke MC (2015) Induction of

tetraploids in *Escallonia* spp. Acta Hortic 1087:453–458
Dewitte A, Eeckhaut T, Van Huylenbroeck J, Van Bockstaele E (2009) Occurrence of viable unreduced pollen in a *Begonia* collection. Euphytica 168:81–94
Dewitte A, Eeckhaut T, Van Huylenbroeck J (2010) Induction of unreduced pollen by trifluralin and N_2O treatments. Euphytica 171:283–293
Dewitte A, Van Laere K, Van Huylenbroeck J (2012) In: Abdurakhmonov I (ed) Use of 2n-gametes in plant breeding. Plant breeding. Intech, Rijeka, pp 59–86
Dhooghe E, Van Laere K, Eeckhaut T, Leus L, Van Huylenbroeck J (2011) Mitotic chromosome doubling of plant tissues in vitro. PCTOC 104:359–373
Doi H, Hoshi N, Yamada E, Yokoi S, Nishihara M, Hikage T, Takahata Y (2013) Efficient haploid and doubled haploid production from unfertilized ovule culture of gentians (Gentiana spp.). Breed Sci 63:400–406
Doležel J, Greilhuber J, Suda J (2007) Flow cytometry with plants: an overview. In: Dolezel J, Greilhuber J, Suda J (eds) Flow cytometry with plant cells. Wiley, Weinheim, pp 41–65
Eeckhaut T, Werbrouck S, Leus L, Van Bockstaele E, Debergh P (2004) Chemically induced polyploidization of *Spathiphyllum wallisii* Regel through somatic embryogenesis. PCTOC 78:241–246
Eeckhaut T, Leus L, Van Huylenbroeck J (2005) Exploitation of flow cytometry for ornamental breeding. Acta Physiol Plant 27:743–750
Eeckhaut T, Van Laere K, De Riek J, Van Huylenbroeck J (2006) Overcoming interspecific barriers in ornamental plant breeding. In: Teixeira da Silva J (ed) Floriculture, ornamental and plant biotechnology: advances and topical issues, 1st edn. Global Science Books, London, pp 540–551
El Mokadem H, Crespel L, Meynet J, Gudin S (2002) The occurrence of 2n pollen and the origin of sexual polyploids in dihaploid roses (*Rosa* hybrid L.). Euphytica 125:169–177
Ferrie A, Caswell K (2011) Review of doubled haploidy methodologies in ornamental species. Propagation of ornamental plants 11:63–77
Fetouh M, Kareem A, Knox G, Wilson S, Deng Z (2016) Induction, identification, and characterization of tetraploids in Japanese privet (*Ligustrum japonicum*). Hortscience 51:1371–1377
Galbraith D, Harkins K, Maddox J, Ayres N, Sharma D, Firoozabady E (1983) Rapid flow cytometric analysis of the cell cycle in intact plant tissues. Science 220:1049–1051
Gallone A, Hunter A, Douglas C (2014) Polyploid induction in vitro using cochicine and oryzalin on hebe ‘Oratia beauty’: production and characterization of the vegetative traits. Sci Hortic 179:59–66
Gantait S, Mandal N, Bhattacharyya S, Kanti Das P (2011) Induction and identification of tetraploids using in vitro colchicine treatment of Gerbera jamesonii Bolus cv. Sciella. PCTOC 106:485–493
Gao Y, Chen B, Zhang J (2011) Anther culture of garden *Chrysanthemum*. Acta Hortic 923:103–110
Gonzalez R, Iannicelli J, Coviella A, Bugallo V, Bologna P, Pitta-Alvarez S, Escandon A (2015) A protocol for the in vitro propagation and polyploidization of an interspecific hybrid of *Glandularia* (*G. peruviana* x *G. scrobiculata*). Sci Hortic 184:46–54
Guo L, Xu W, Zhang Y, Zhang J, Wei Z (2017) Inducing triploids and tetraploids with high temperatures in *Populus* sect. Tacamahaca. Plant Cell Rep 36:313–326
Hannweg K, Sippel A, Bertling I (2013) A simple and effective method for the micropropagation and in vitro induction of polyploidy and the effect on floral characteristics of the South African iris, *Crocosmia aurea*. S Afr J Bot 88:367–372
Harlan J, De Wet J (1975) On Ö. Winge and a prayer: the origins of polyploidy. Bot Rev 41(4):361–390. ISSN 0006-8101
Hayashi M, Kato J, Ohashi H, Mii M (2009) Unreduced 3x gamete formation of allotriploid hybrid derived from the cross of *Primula denticulata* (4x) x *P. rosea* (2x) as a causal factor for producing pentaploid hybrids in the backcross with pollen of tetraploid P. denticulate. Euphytica 169:123–131
Huang R, Liu D, Zhao M, Li Z, Li M, Sui S (2015) Artificially induced polyploidization in *Lobularia maritima* (L.) Desv. and its effect on morphological traits. Hortscience 50:636–639
Husband B, Baldwin S, Suda J (2013) The incidence of polyploidy in natural plant populations: major patterns and evolutionary processes. In: Leitch I, Greilhuber J, Dolezel J, Wendel J (eds) Plant genome diversity, Physical structure, behaviour and evolution of plant genomes, vol 2. Springer, Wien, pp 255–276
Ishii T, Karimi-Ashtiyani R, Houben A (2016) Haploidization via chromosome elimination: means

and mechanisms. Annu Rev Plant Biol 67:421–438
Jia Y, Zhang Q, Pan H, Wang S, Liu Q, Sun L (2014) Callus induction and haploid plant regeneration from baby primrose (*Primula forbesii* Franch) anther culture. Sci Hortic 176:273–281
Jones J, Ranney T (2009) Fertility of neopolyploid *Rhododendron* and occurrence of unreduced gametes in triploid cultivars. J Amer Rhododendron Soc 63:131–135
Karimi-Ashtiyani R, Ishii T, Niessen M, Stein N, Heckmann S, Gurushidze M, Mohammad Banaei-Moghaddam A, Fuchs J, Schubert V, Koch K, Weiss O, Demidov D, Schmidt K, Kumlehn J, Houben A (2015) Point mutation impairs centromeric CENH3 loading and induces haploid plants. PNAS 36:112111–111216
Kelliher T, Starr D, Wang W, McCuiston J, Zhong H, Nuccio M, Martin B (2016) Maternal haploids are preferentially induced by *CENH3-tail-swap* transgenic complementation in maize. Front Plant Sci 7:414
Kitamura S, Akutsu M, Okazaki K (2009) Mechanisms of action of nitrous oxide gas applied as a polyploidizing agent during meiosis in lilies. Sex Plant Reprod 22:9–14
Leus L, Van Laere K, Dewitte A, Van Huylenbroeck J (2009) Flow cytometry for plant breeding. Acta Hortic 836:221–226
Li Z, Ruter J (2017) Development and evaluation of diploid and polyploid *Hibiscus moscheutos*. Hortscience 52:676–681
Li Q, Yang Y, Wu H (2016) In vitro segregation of tetraploid and octoploid plantlets from colchicine-induced ploidy chimeras in *Echinacea purpurea* L. Hortscience 51:549–577
Lim K, Ramanna M, De Jong J, Jacobsen E, Van Tuyl J (2001) Indeterminate meiotic restitution (IMR): a novel type of meiotic nuclear restitution mechanism detected in interspecific lily hybrids by GISH. Theor Appl Genet 103:219–230
Lim K, Shen T, Barba-Gonzalez R, Ramanna M, Van Tuyl J (2004) Occurrence of SDR 2n-gametes in *Lilium* hybrids. Breed Sci 54:13–18
Lim K, Barba-Gonzalez R, Zhou S, Ramanna M, Van Tuyl J (2005) Meiotic polyploidization with homologous recombination induced by caffeine treatment in interspecific lily hybrids. Korean J Genet 27:219–226
Liu S, Chen S, Chen Y, Guan Z, Yin D, Chen F (2011) In vitro induced tetraploid of *Dendranthema nankingense* (Nakai) Tzvel. shows an improved level of abiotic stress tolerance. Sci Hortic 127:411–419
Liu B, De Storme N, Geelen D (2017) Gibberellin induces diploid pollen formation by interfering with meiotic cytokinesis. Plant Physiol 173:338–353
Lokker A, Barba-Gonzalez R, Lim K, Ramanna M, Van Tuyl J (2004) Genotypic and environmental variation in production of 2n-gametes of oriental x Asiatic lily hybrids. Acta Hortic 673:453–456
Lu M, Zhang P, Kang X (2013) Induction of 2n female gametes in *Populus adenopoda* maxim by high temperature exposure during female gametophyte development. Breed Sci 63:96–103
Luo J, Arens P, Niu L, Van Tuyl J (2016) Induction of viable 2n pollen in sterile oriental x Trumpet *Lilium* hybrids. J Hortic Sci Biotechnol 91:258–263
Marasek-Ciolakowska A, Xie S, Arens P, Van Tuyl J (2014) Ploidy manipulation and introgression breeding in Darwin hybrid tulips. Euphytica 198:389–400
Marasek-Ciolakowska A, Arens P, Van Tuyl J (2016) The role of polyploidization and interspecific hybridization in the breeding of ornamental crops. In: Mason A (ed) The breeding of ornamentals. CRC Press, Taylor & Francis Group, Boca raton. Chapter 7, pp 159–181
Mason A, Pires C (2015) Unreduced gametes: meiotic mishap or evolutionary mechanism? Trends Genet 31:1
Miguel T, Leonhardt K (2011) In vitro polyploidy induction of orchids using oryzalin. Sci Hortic 130:314–319
Murovec J, Bohanec B (2013) Haploid induction in *Mimulus aurantiacus* Curtis obtained by pollination with gamma irradiated pollen. Sci Hortic 162:218–225
Nadler J, Pooler M, Olsen R, Coleman G (2012) In vitro induction of polyploidy in *Cercis glabra* Pamp. Sci Hortic 148:126–130
Negri V, Lemmi G (1998) Effect of selection and temperature stress on the production of 2n gametes in *Lotus tenuis*. Plant Breed 117:345–349
Nonaka T, Oka E, Asano M, Kuwayama S, Tasaki H, Han D, Godo T, Nakano M (2011) Chromosome doubling of *Lychnis* spp. by in vitro spindle toxin treatment of nodal segments. Sci Hortic 129:832–839
Oates K, Ranney T, Touchell D (2012) Influence of induced polyploidy on fertility and morphol-

ogy of *Rudbeckia* species and hybrids. Hortscience 47:1217–1221
Okazaki K, Kurimoto K, Miyajima I, Enami A, Mizuochi H, Matsumoto Y, Ohya H (2005) Induction of 2n pollen in tulips by arresting the meiotic process with nitrous oxide gas. Euphytica 143:101–114
Paule J, Wagner N, Weising K, Zizka G (2017) Ecological range shift in the polyploid members of the South American genus *Fosterella* (Bromeliaceae). Ann Bot 120:233–243
Pécrix Y, Rallo G, Folzer H, Cigna M, Gudin S, Le Bris M (2011) Polyploidization mechanisms: temperature environment can induce diploid gamete formation in *Rosa* sp. J Exp Bot 62:3587–3597
Podwyszynska M (2012) In vitro tetraploid induction in tulip (*Tulipa gesneriana* L.). Acta Hortic 961:391–396
Popova T, Grozeva S, Todorova V, Stankova G, Anachkov N, Rodeva V (2016) Effects of low temperature, genotype and culture media on in vitro androgenic answer of pepper (*Capsicum annuum* L.). Acta Physiol Plant 38:273
Przybyla A, Behrend A, Bornhake C, Hohe A (2014) Breeding of polyploidy heather (*Calluna vulgaris*). Euphytica 199:273–282
Ramanna M, Kuipers A, Jacobsen E (2003) Occurrence of numerically unreduced (2n) gametes in *Alstroemeria* interspecific hybrids and their significance for sexual polyploidisation. Euphytica 133:95–106
Ramsey J (2007) Unreduced gametes and neopolyploids in natural populations of *Achillea borealis*. Heredity 98:143–150
Ravi M, Chan S (2010) Haploid plants produced by centromere-mediated genome elimination. Nature 464:615–619
Regalado J, Carmona-Martin E, Querol V, Velez C, Encina C, Pitta-Alvarez S (2017) Production of compact petunias through polyploidization. PCTOC 129:61–71
Roberts A (2007) The use of bead beating to prepare suspensions of nuclei for flow cytometry from fresh leaves, herbarium leaves, petals and pollen. Cytometry Part A 71:1039–1044
Saito H, Nakano M (2002) Isolation and characterization of gametic microprotoplasts from developing microspores of *Lilium longiflorum* for partial genome transfer in the liliaceous ornamentals. Sex Plant Reprod 15:179–185
Sato T, Miyoshi K, Okazaki K (2010) Induction of 2n gametes and 4n embryo in *Lilium* (*Lilium* x *formolongi* Hort.) by nitrous oxide gas treatment. Acta Hortic 855:243–248
Schulze J, Contreras R (2017) In vivo chromosome doubling of *Prunus lusitanica* and preliminary morphological observations. Hortscience 52:332–337
Soltis P, Soltis D (2009) The role of hybridization in plant speciation. Annu Rev Plant Biol 60:561–588
Soonthornkalump S, Chuenboonngarm N, Soonthornchainaksaeng P, Jenjittikul T, Thammasiri K (2014) Effect of colchicine incubation time on tetraploid induction of *Kaempferia rotunda*. Acta Hortic 1025:89–92
Stephens L (1998) Formation of unreduced pollen by an *impatiens Hawkeri* x *Platypetala* interspecific hybrid. Hereditas 128:251–255
Talluri R (2011) Gametes with somatic chromosome number and their significance in interspecific hybridization in *Fuchsia*. Biol Plant 55:596–600
Te Beest M, Le Roux J, Richardson D, Brysting A, Suda J, Kubesova M, Pysek P (2012) The more the better? The role of polyploidy in facilitating plant invasions. Ann Bot 109:19–45
Thomas W, Forster B, Gertsson B (2003) Doubled haploids in breeding. In: Maluszynski M, Kasha K, Forster B, Szarejko I (eds) Doubled haploid production in crop plants: a manual. Kluwer Academic Publishers, Dordrecht, pp 337–349
Tomiczak K, Mikula A, Sliwinska E, Rybczynski J (2015) Autotetraploid plant regeneration by indirect somatic embryogenesis from leaf mesophyll protoplasts of diploid *Gentiana decumbens* L.f. In vitro cell dev boil 51:350–359
Uno Y, Koda-Katayama H, Kobayashi H (2016) Application of anther culture for efficient haploid production in the genus Saintpaulia. PCTOC 125:241–248
Urwin N (2014) Generation and characterization of colchicine-induced polyploidy *Lavandula* x *intermedia*. Euphytica 197:331–339
Van Laere K, Dewitte A, Van Huylenbroeck J, Van Bockstaele E (2009a) Evidence for the occurrence of unreduced gametes in interspecific hybrids of *Hibiscus*. J Hort Sci Biotech 84:240–247
Van Laere K, Leus L, Van Huylenbroeck J, Van Bockstaele E (2009b) Interspecific hybridisation and genome size analysis in Buddleja. Euphytica 166:445–456
Van Laere K, França S, Vansteenkiste H, Van Huylenbroeck J, Steppe K, Van Labeke M (2011)

Influence of ploidy level on morphology, growth and drought susceptibility in *Spathiphyllum wallisii*. Acta Physiologia Plantarum 33:1149–1156
Van Tuyl J, Meijer B, van Diën M (1992) The use of oryzalin as an alternative for colchicine in in vitro chromosome doubling of *Lilium* and *Nerine*. Acta Hort 352:625–630
Vanstechelman I, Eeckhaut T, Van Huylenbroeck J, Van Labeke M-C (2010) Histogenic analysis of chemically induced mixoploids in *Spathiphyllum wallisii*. Euphytica 174:61–72
Wang S, Li X, Yang L, Wu H, Zheng S, Zhang X, Zhang L (2011a) Microspore culture of *Zantedeschia aethiopica*: the role of monosaccharides in sporophytic development. Afr J Biotechnol 10:10287–10292
Wang Y, Tong Y, Li Y, Zhang Y, Zhang J, Feng J, Feng H (2011b) High frequency plant regeneration from microspore-derived embryos of ornamental kale (*Brassica oleracea* L. var. *acephala*). Sci Hortic 130:296–302
Wang H, Dong B, Jiang J, Fang W, Guan Z, Liao Y, Chen S, Chen F (2014) Characterisation of in vitro haploid and doubled haploid *Chrysanthemum morifolium* plants via unfertilized ovule culture for phenotypical traits and DNA methylation pattern. Front Plant Sci 5:738
Wongprichachan P, Huang KL, Hsu ST, Chou YM, Liu TY, Okubo H (2013) Induction of polyploid *Phalaenopsis amabilis* by N_2O treatment. J Fac Agri Kyushu Univ 58:33–36
Wu H, Zheng S, He Y, Yan G, Bi Y, Zhu Y (2007) Diploid female gametes induced by colchicines in oriental lilies. Sci Hortic 114:50–53
Xiao Y, Zheng S, Long C, Zheng L, Guan W, Zhao Y (2007) Initial study on 2n-gametes induction of *Strelitzia reginae*. J Yunnan Agri Univ 22:475–479
Xu C, Huang Z, Liao T, Li Y, Kang X (2016) In vitro tetraploid plants regeneration from leaf explants of multiple genotypes in *Populus*. PCTOC 125:1–9
Yoon S, Aucar S, Hernlem B, Edme S, Palmer N, Sarath G, Mitchell R, Blumwald E, Tobias C (2016) Generation of octaploid switchgrass by seedling treatment with mitotic inhibitors. Bioenergy Res 10:344. https://doi.org/10.1007/s12155-016-9795-2
Younis A, Hwang Y, Lim K (2014) Exploitation of induced 2n-gametes for plant breeding. Plant Cell Rep 33:215–233
Zamariola L, Tiang C, De Storme N, Pawlowski W, Geelen D (2014) Chromosome segregation in plant meiosis. Front Plant Sci 5:279
Zhang Z, Kang X (2010) Cytological characteristics of numerically unreduced pollen production in *Populus tomentosa* Carr. Euphytica 173:151–159
Zhang Q, Zhang F, Li B, Zhang L, Shi H (2016) Production of tetraploid plants of *Trollius chinensis* Bunge induced by colchicine. Czech J Genet Plant Breed 52:34–38
Zhang X, Cao Q, Jia G (2017) A protocol for fertility restoration of F1 hybrid derived from *Lilium* × *formolongi* 'Raizan 3' × oriental hybrid 'Sorbonne'. PCTOC 129:375–386
Zhao C, Tian M, Li Y, Zhang P (2017) Slow-growing pollen-tube of colchicine-induced 2n pollen responsible for low triploid production rate in *Populus*. Euphytica 213:94
Zhou X, Mo X, Gui M, Wu X, Jiang Y, Ma L, Shi Z, Luo Y, Tang W (2015) Cytological, molecular mechanisms and temperature stress regulating production of diploid male gametes in *Dianthus caryophyllus* L. Plant Physiol Biochem 97:255–263

第 8 章 观赏园艺植物突变育种

Rusli Ibrahim, Zaiton Ahmad, Shakinah Salleh, Affrida Abu Hassan,
and Sakinah Ariffin

摘要 突变诱导技术是近 30 年来一直被用于观赏园艺植物育种的一种有效工具。因为表型特征的变化，如花的颜色、形状和大小，叶片中的叶绿素变异以及生长习性可以很容易检测，突变育种在观赏植物中应用更成功。另外，许多观赏植物基因杂合度高的特性提高了突变率。因为突变是在单细胞中诱导的，所以化学和物理诱变剂对多细胞的诱导将表现为嵌合体。事实证明，使用不定芽技术进行体外培养是避免嵌合体产生的最有效方法之一。使用 X 射线和伽马射线诱变已经成功地在不同的观赏植物中产生了许多新品种，这些品种已经商业化。事实证明，采用适当的诱变策略，如将体外培养技术与慢性伽马射线辐照相结合，是一种有效的诱变方法，可在短时间内产生新的有前景的观赏植物突变体。在过去的 20 年中，离子束已成为一种改善观赏植物的有效且独特的诱变剂，与 X 射线和伽马射线相比，离子束辐照产生更高的突变率。目前，观赏植物改良研究的兴趣已转向分子育种和基因工程，但两者各有优缺点。突变育种仍然是创造遗传变异的一种有吸引力的方法，并且已成为许多无性繁殖观赏植物的常规育种技术。

关键词 不定芽；嵌合；慢性辐照；遗传变异；突变体；观赏植物育种

8.1 引 言

诱变处理已诱导了广泛的遗传变异，并应用于植物育种和作物改良计划（Schaart et al. 2016）。物理诱变是优先选择的诱变方法，因为它是一种方便、易于重现和用户友好的方法。由于各种原因，突变育种在观赏植物以及许多其他无性繁殖园艺植物中应用特别成功。首先，通常不难选择具有可直接感知特征的突变，包括花的形状、大小或颜色。其次，许多品种是杂合的，通过突变和杂交容易扩大变异。最后，体外或体内繁殖方法容易成功繁殖突变体，该突变体随后可被识别（Broertjes 1968a）。

突变育种是一种改良作物的成熟方法，并且在观赏植物许多新的花色/形突变体的开发中发挥了重要作用（Broertjes 1966）。观赏植物仅出于其美学价值而生产。因此，改善诸如花色、寿命和形态与植株形态等质量属性以及创造新的突变体具有重要经济意义。通过杂交和选择的常规育种已经产生了几种无性繁殖的优良基因型，这些基因型具

R. Ibrahim (✉) · Z. Ahmad · S. Salleh · A. A. Hassan · S. Ariffin
Malaysian Nuclear Agency, Kajang, Selangor, Malaysia

有理想的特征，如颜色、形状、瓶插寿命以及对病虫害的抗性。但是，传统育种的缺点之一是任何单个物种的基因库都很有限。

原则上，观赏植物的育种与其他作物的育种没有区别，但是育种目标通常与粮食作物不同。许多观赏植物是“新的”应用种，育种历史很短，新颖性是观赏植物育种筛选的重要考虑因素（Broertjes and Van Harten 1988）。观赏植物是应用诱变技术的理想材料，因为诱变处理后很容易发现许多重要经济性状变化，如花的特性或生长习性。此外，它们中的许多是杂合的，并且通常以营养方式繁殖，这使得早在 M_1 代就可以检测、选择和保存突变体。在无性繁殖作物中进行诱变育种的主要优点是能够改变一个本来就很突出的品种的一个或几个特征，而又不改变基因型的其余部分（通常是独特部分）（Broertjes and Van Harten 1988）。

突变体大多以嵌合体形式出现（Wolf 1996；Chakrabarty et al. 1999；Mandal et al. 2000a；Misra et al. 2004）。嵌合体的分离也可以在突变以后的阶段实现。例如，在再生植物开花后观察突变的花瓣区域，利用体外培养实现。在开始实验之前，必须考虑不同技术的假设机会和局限性以及不同再生途径产生的后果。使用源自单细胞的不定芽或体细胞胚可以解决与嵌合体形成有关的问题。Broertjes 等（1976）首次提出使用辐照诱导菊花中的突变。根据 Broertjes 等（1968）的理论，在体内和体外利用叶片再生的植物的不定分生组织最终来自单个初始细胞。因此，不定芽技术在辐照育种中的使用使再生植物，包括突变植物，由遗传上相同的组织组成。这样，避免了嵌合体的产生，可以大大缩短繁殖过程。

观赏植物第一个人工诱导的商业突变品种是‘Faraday’，一个郁金香的花色突变品种。继两位研究人员于 1936 年尝试 X 射线诱变育种（Broertjes and Van Harten 1988）后，‘Faraday’由经 X 射线处理的‘Fantasy’中筛选而来，由 W. E. de Mol 于 1949 年在荷兰发布。郁金香中第二个花色突变品种‘Estella Rijnveld’由同一位研究人员于 1954 年发布。商业突变品种的另一个早期著名的例子是香石竹（*Dianthus caryophyllus*）‘White Sim No.1’，是由伽马射线辐照生根的插条产生，于 1962 年由 Mehlquist 引入美国。

使用化学物质和辐照进行育种已成功地在不同类型的观赏植物中选育了大量的突变新品种，其中大多数是花朵颜色的变化。在过去的 30 年中，书籍和评论报道了数百种观赏植物商业突变新品种的释放（Broertjes and Van Harten 1978，1988；Schum and Preil 1998；Van Harten 1998；Maluszynski et al. 2000；Ahloowalia et al. 2004；Jain 2006；Datta 2009a；Nagatomi and Degi 2009；Pathirana 2011）。位于奥地利维也纳的国际原子能机构（IAEA）的一系列文献，列出了正式发布的突变体品种的详细清单。联合国粮食及农业组织/国际原子能机构（FAO/IAEA）突变品种数据库的最新信息显示，迄今为止，全世界已选育超过 3245 种植物突变体品种，总共 705 种观赏植物突变体品种已正式注册（FAO/IAEA 2017）。此类诱导突变新品种的实际数量比报道的数量大得多，部分原因是许多育种者不愿透露其新品种的育种细节以及所采用的方法。

物理诱变剂（如 X 射线和 γ 射线）和化学诱变剂在观赏植物中诱导理想突变的能力已从大量通过直接突变育种开发的新品种中得到了充分证明。有关在全球范围内开发和发布的诱变衍生品种的影响力可参见许多详细综述（Ahloowalia et al. 2004；Lagoda

2009；Kharkwal and Shu 2009；Jain 2010）。

根据 FAO/IAEA 突变品种数据库 2017 年最新的数据，诱导的菊花突变品种占官方注册的观赏植物新品种的比例最大，在 37%以上，其次是玫瑰（16%）（FAO/IAEA 2017）。菊花是非常流行的切花和经济价值高的观赏盆栽植物。在全球花卉市场上，它是仅次于玫瑰的第二大切花（Kumar et al. 2006），切花的经济回报主要取决于花朵的质量（Kaul et al. 2011）。其染色体倍性高、杂合度高以及自交不亲和现象使得采用传统育种方法有性杂交选育比较困难（Minano et al. 2009），而适合于采用突变育种。目前所有的菊花新品种都是自发突变、杂交、诱导突变和选择的结果。通过诱导突变和幼苗选择已经开发出许多商业品种（Broertjes and Van Harten 1988；Datta 1988；Data et al. 2001；Schum 2003；Zalewska et al. 2007）。

电离辐照育种已被广泛用于改良适应性强的植物品种的一种或几种重要性状（Kumar et al. 2006；Jain and Spencer 2006）。与急性伽马射线照射相比，将慢性伽马射线照射与体外培养相结合可在菊花中产生更高的花突变率和更宽的色谱突变（Nagatomi et al. 1993；Nagatomi et al. 1996a；Nagatomi and Degi 2009）。使用高频离子束还成功地诱导出菊花的新型花色和形状突变体。此外，使用离子束对菊花的花瓣辐照获得了较高的复杂花色突变率（Nagatomi et al. 1997；Ueno et al. 2003）。因此，离子束已经发展成为一种新的独特的辐照诱变方法，可以在菊花和其他观赏植物中产生有用的突变（Okamura et al. 2003），如仙客来（Sugiyama et al. 2008b）、大丽花（Hamatani et al. 2001）、飞燕草（Chinone et al. 2008）、香石竹（Sugiyama et al. 2008a）、补血草（Chinone et al. 2008）、兰花（Affrida et al. 2008）、蓝眼菊（Iizuka et al. 2008）、矮牵牛（Miyazaki et al. 2002）、玫瑰（Hara et al. 2003；Yamaguchi et al. 2003）、蝴蝶草（Miyazaki et al. 2006；Sasaki et al. 2008）和马鞭草（Kanaya et al. 2008）。

8.2 诱变技术

8.2.1 辐照诱变

无论是自发的还是诱发的，突变通常是染色体大规模缺失、倒位或易位的结果，或者是由 DNA 中的点突变（一种遗传材料发生单一位点插入或缺失的突变类型）所致。物理诱变剂通常会导致染色体变化和大片段 DNA 缺失，而化学诱变剂通常会导致点突变。

使用物理辐照诱变剂（主要是电子束）进行辐照诱变可将自然突变率提高 1000～1×10^6 倍，并已被广泛用于诱导可遗传突变。超过 70%的通过诱导选育的作物新品种是通过物理诱变手段获得的。最初物理辐照诱变主要使用的是 X 射线，但是随着许多发展中国家通过 IAEA 获得了这些放射源，60Co 和 137Cs 等放射源的伽马射线使用越来越普遍。也可以选择使用来自核反应堆的快中子，这种辐照服务由位于维也纳的联合国粮食及农业组织/国际原子能机构提供。电离辐照会导致染色体断裂，从而使 DNA 链发生交联。在电离辐照中，伽马射线使用广泛，因为它们具有良好的穿透力和精确的剂量测定法，同时对整个植株或花粉等器官辐照后造成的伤害较小（Liew et al. 2008）。伽马射线可以深入细胞内并与原子或分子相互作用以产生自由基。这些自由基会根据辐照水平破坏或修饰植物的重要成分（Wi et al. 2005）。

伽马射线的 LET（线性能量转移）较低，这可能导致相对更多的基因突变和缺失。对已被伽马射线照射的水稻的核苷酸序列进行分析表明，最常见的突变是缺失，尤其是较小的缺失（1～16bp），只有少数显示出大的缺失（9.4～129.7kb）、单碱基取代和倒位（Morita et al. 2009）。同时，Yamaguchi 等（2008）调查了总辐照剂量和剂量率对花朵颜色与核 DNA 含量（菊花中辐照损伤的指标）的影响。他们发现，低剂量率下的高总剂量在诱导辐照损伤较少的花色突变体方面更为有效。

结构突变是染色体断裂及重排的结果。电离辐照主要导致结构突变（Van Harten 1998；FAO/IAEA 1977；Kumar and Yadav 2010）。可以将此类突变分为 4 类：删除或缺陷、重复、倒位和易位。电离辐照带来的染色体畸变中约有 90%是缺失，并且常常是致命的（Van Harten 1998）。物理诱变剂通常会导致染色体变化和大片段的 DNA 缺失，而化学诱变剂通常会诱导点突变。但是，某些缺失会阻止生化合成途径。例如，如果突变发生在导致有毒代谢物产生的途径中，则这种缺失可能会产生有用的无毒植物产品。

同时，在伽马射线处理中，处理后植物 M_1 幼苗生长率减少 30%～50%或成活率达到 40%～60%时的辐射量通常被认为是有效的伽马射线辐照量标准（Yamaguchi 2013）。剂量率也是伽马射线辐照需考虑的重要因素，其作用已通过评估各种反应进行了研究，包括致死性（Broertjes 1968b；Killion and Constantin 1971；Sripichtt et al. 1988）、生长（Bottino et al. 1975；Killion and Constantin 1971；Killion et al. 1971）和生育力（Killion and Constantin 1971）。根据剂量率的差异，γ 射线处理分为两类：急性和慢性辐照。在急性辐照下，植物材料以高剂量率辐照伽马射线几分钟至一天的时间；而在慢性辐照下，植物材料以低剂量率辐照伽马射线数周甚至更长。据推测，慢性辐照对无性繁殖作物进行突变育种更有效，因为获得的突变体受到的辐照损害较小，并可直接用作新品种。

目前，有 4 个慢性伽马辐照设施正在积极地进行植物突变育种，即日本辐照育种研究所的伽马辐照场，这是世界上最大的用于辐照诱导植物育种的室外设施；泰国曼谷 Kasetsart 大学核技术研究中心（NTRC）的伽马辐照室；马来西亚核研究机构的伽马辐照温室；韩国原子能研究所（KAERI）的伽马辐照光培养室（图 8.1）。

通常，急性伽马射线处理的剂量处于最佳水平，并且仅处理一次。慢性伽马辐照处理的剂量低，分次辐照，可在数分钟、数小时甚至数月内完成。通常，与相同总剂量的急性伽马照射处理相比，慢性伽马辐照处理导致的突变频率更低（Van Harten 1998）。然而，Nagatomi（2003）进行了一项菊花诱变研究发现，暴露于急性伽马辐照仅产生非常有限的花色变异谱，而暴露于慢性伽马射线辐照产生的突变体具有更宽的色谱变异范围。

8.2.2 离子束

在过去的 20 年中，离子束辐照已成为一种用于改善观赏植物的有效且独特的诱变剂。离子束是包括质子和氦在内的带电粒子，较重的带电粒子被诸如回旋加速器之类的加速器加速，与 γ 射线或 X 射线相比蓄能更高。与低 LET 辐照（如 γ 射线或 X 射线）

图 8.1　慢性（长期）伽马辐照设施：(a) 日本辐照育种研究所的伽马辐照场（由 Hiroshi Degi 博士提供）；(b) 马来西亚核研究机构的伽马辐照温室；(c) 泰国曼谷 Kasetsart 大学核技术研究中心（NTRC）的伽马辐照室（由 Peeranuch Sompok 博士提供）；(d) 韩国原子能研究所（KAERI）的伽马辐照光培养室（由 Si-Yong Kang 博士提供）

相比，其具有更高的线性能量转移（LET）、致死性和使细胞失活的相对生物学有效性（RBE）（Tanaka et al. 1997；Hase et al. 2002；Shikazono et al. 2002；Blakely 1992；Lett 1992）。离子束还具有将高能量聚焦在目标部位上的潜力，因此，离子束可引起高水平的诱变作用（Yamaguchi et al. 2010）。值得注意的是，离子束主要导致单链或双链 DNA 断裂，端基受损，其导致的损伤可修复性很低（Goodhead 1995）。

与低 LET 辐照相比，离子束会产生更多的 DNA 损伤和双链断裂（Ward 1994）或较大的 DNA 片段改变，如倒位、易位和大的缺失，而不是点突变（Tanaka 2009）。根据 Tanaka（1999）的研究，获得的拟南芥突变体中有一半表现出小的突变，如碱基变化和涉及几个碱基的小缺失；另一半显示出较大的 DNA 改变，如倒位、易位和缺失。这些结果表明离子束照射具有广谱和高频率的突变。

表型常被用来研究突变率。对于菊花的花色，由碳离子（LET = 122keV/μm，剂量 = 5～20Gy）辐照诱导的样品花瓣和叶子的突变率分别为 16%和 7%，大约是伽马射线突变率的一半（Tanaka et al. 2010）。香石竹中花朵颜色和形状的突变率分别为 2.8%、2.3%和 1.3%，分别由碳离子（LET = 122keV/μm）、γ 射线和 X 射线辐照诱导（Ueno et al. 2004）。通过用 20Gy 氖离子（LET = 64keV/μm）照射水稻，可获得高频率（11.6%）的叶绿素缺陷型突变体，包括白化病（Abe et al. 2002）。由离子束引起的玫瑰花的突变率也曾被研究过（Yamaguchi et al. 2003）。

离子束的突变率高于伽马射线的突变率，尤其是在碳离子的情况下。非常有趣的是，在所有测试的辐照中，突变产量，即在子代中产生叶绿素突变体的 M_1 植物数量除以辐照后播种的 M_1 种子数量，在存活曲线的肩剂量附近出现峰值。尽管长期以来人们一直认为 LD_{50} 是诱导突变体的最佳剂量，但在任何辐照下几乎不影响存活率的肩剂量足以有效地获得突变体（Tanaka et al. 2010）。与 γ 射线（Fujii et al. 1966；Mei et al. 1994）、X 射线（Hirono et al. 1970；Yang and Tobias 1979）和电子（Shikazono et al. 2003）相比，离子束显示出更高的突变诱导作用。同样在菊花（Yamaguchi et al. 2010）和玫瑰（Yamaguchi 2003）中，使用不影响枝条再生或存活的低剂量离子束辐照，也获得了突变体。

此外，Ueno 等（2004）指出，低剂量辐照对于诱变选育几乎没有腋芽的‘Aladdin’菊花品种至关重要。我们还认为，低剂量辐照对于获得单点突变体至关重要，特别是对于无性繁殖作物而言，因为这样获得的突变体可直接用作新品种。因此，与 γ 射线相比，离子束被认为更适合诱发点突变（Yamaguchi 2013）。

日本原子能机构（JAEA）是开发离子束辐照设备的先驱之一。图 8.2 显示了它在高崎 TIARA 的设施（用于高剂量辐照的高崎离子加速器），适用于生物技术和材料科学研究（Magori 2010）。该设施使用通过方位角变化场（AVF）回旋加速器加速的碳离子辐照植物材料。就诱变而言，相比其他类型的离子束（100MeV 和 320MeV），220MeV 碳离子束似乎是最合适的，因为它具有高突变率且染色体损伤率低（Yamaguchi 2013）。辐照处理对突变率和核 DNA 含量的影响根据离子束的类型而有所不同（Yamaguchi 2010）。

图 8.2 日本高崎 TIARA 的离子束设施（用于高剂量辐照的高崎离子加速器）（由日本高崎国立量子与放射科学技术研究所、量子束科学研究所研究理事会、计划和促进办公室主任 Atsushi Tanaka 博士提供）

离子束辐照已成功地诱导了各种观赏植物突变株，如大丽花（Hamatani et al. 2001）、香石竹（Okamura et al. 2003）、玫瑰（Hara et al. 2003；Yamaguchi et al. 2003）、蓝眼菊（Iizuka et al. 2008）、仙客来（Sugiyama et al. 2008b）、飞燕草（Chinone et al. 2008）、兰

花（Affrida et al. 2008）、石竹（Sugiyama et al. 2008a）和菊花（Watanabe et al. 2008；Zaiton et al. 2014）。根据 Nagatomi 等（1998），通过离子束辐照可以产生通过伽马射线辐照不能获得的菊花特定花色突变体。类似的具有不同花色的新突变品种（Nagatomi et al. 2003）或侧芽较少突变种（Ueno et al. 2005）也被报道。

8.2.3　化学诱变

突变可以由烷基诱变剂化学诱导产生。化学诱变剂的烷基与 DNA 反应，可能会改变核苷酸序列并引起点突变，但这些化合物几乎不会引起染色体突变（Broertjes and Van Harten 1988）。相反，细胞对电离辐照的吸收会使染色体发生结构畸变（FAO/IAEA 1977）。已有的文献已经讨论了主要农作物的化学诱变剂的分类、诱变处理方法、诱变后处理和处理后筛选（Van Harten 1998；Medina et al. 2004；FAO/IAEA 1977，2011）。在没有辐照设施的情况下，化学诱变剂可以考虑用于诱导突变。在某些情况下，化学诱变剂的效率比物理诱变剂要高（Jacobs 2005；Rego and Faria 2001）。

尽管有大量的诱变剂，但只有少数在植物中进行了测试。其中，只有极有限的烷基诱变剂小类在植物诱变实验和植物突变育种中得到广泛应用。通过化学诱变剂获得的 IAEA 突变品种数据库（FAO/IAEA 2017）报告的注册新突变植物品种有 80%以上是由烷基诱变剂诱导的。其中，三种化合物很重要：甲基磺酸乙酯（EMS）、1-甲基-1-亚硝基脲（MNU）和 1-乙基-1-亚硝基脲（ENU），占这些品种的 64%。其他化学诱变剂包括 Fang（2011）报道的乙撑亚胺（EI）、硫酸二甲酯（DMS）、硫酸二乙酯（DES）、秋水仙碱和 NaN_3（叠氮化钠）。EMS 是最常用的化学诱变剂。它导致高频率的基因突变和低频率的染色体畸变（Lai et al. 2004；Van Harten 1998）。EMS 已发展成为高通量筛选突变群体的诱变剂，如正在发展的 TILLING 群体（McCallum et al. 2000a，2000b；Till et al. 2007）。

化学诱变可以用于所有类型的植物材料（如种子、幼苗和体外培养的细胞），但最常用的植物材料是种子。用于营养繁殖的各种形式的植物繁殖体，如鳞茎、块茎、球茎、根茎和外植体，如插条、接穗或离体培养的组织，如叶和茎秆、花药、愈伤组织、细胞培养物、小孢子、胚珠等繁殖体可采用化学诱变处理。

通常，一般的突变育种采用化学诱变方法。通常使用谱系方法，将 M_1 植物单独收获并分别播种为 M_2 家族；以及批量方法，所有 M_1 植物都进行批量收获并播种 M_2 后代。第一种方法可以在 M_1 和 M_2 世代之间建立更好的相关性，并且可以更好地比较不同诱变剂的生物学效应。EMS 是一种潜在的致癌物，因此其制备和处理必须在化学通风橱内进行。可以将组织培养的外植体（如叶切片）浸入 0、0.2%、0.4%和 0.6% EMS 溶液中 0min、30min、60min、120min 和 240min。处理后，通常在 4 周后记录外植体存活率，在 4 周和 8 周后观察外植体生芽率。

通过用 EMS 处理非洲紫罗兰叶片外植体进行突变诱导研究表明，在总共 1838 株植株中，鉴定出 10 个突变体，其中 4 个突变体是叶片斑驳嵌合体，其余 6 个在花色或花瓣边缘上表现出变异（Fang 2011）。嵌合体的发生表明非洲紫罗兰不定芽的形成可能是由多细胞起源而不是由 Broertjes 和 Van Harten（1988）报道的单细胞起源引起的。为了稳定突变的部分，可能需要进行反复多次的营养繁殖。根据这些结果，建议将来可以使用 0.4%的 EMS 处理 30～120min 来对非洲紫罗兰进行体外诱变操作，以获得新的突变品种。诱变后的非洲紫罗兰发生嵌合体现象已见不少报道。Sparrow 等（1960）观察到由辐照非洲紫罗兰插条再生的 154 个突变植物（0.7%）中有一个是嵌合的。利用秋水仙碱对非洲紫罗兰叶插枝叶柄处理，在 29 株（3.5%）中产生了一株嵌合体植物（Arisumi and Frazier 1968）。

化学诱变剂已应用于多种植物的突变诱导。Rodrigo 等（2004）通过 EMS 处理获得了具有各种花瓣颜色（粉红色、浅粉红、青铜色、白色、黄色和浅橙色）的菊花突变体。然而，化学诱变剂向营养组织的低渗透性是化学诱变中的主要问题，因为这可能导致突变率低和难以重复实验（Van Harten 1998）。但是，可以通过进行体外诱变来解决这个问题，在体外诱变中，与体内诱变相比，据报道突变发生得更均匀（Constantin 1984）。通过化学诱变诱导正式注册的突变品种的详细清单在 FAO/IAEA 突变品种数据库（FAO/IAEA 2017）中列出。

8.3 影响突变率的因素

观赏植物突变育种成功与否取决于许多因素，如选择正确的基因型、使用的诱变剂、外植体类型、放射敏感性测试和最佳剂量、材料的处理以及筛选有前途的突变体。这方面相关的有用信息可参照相关综述（FAO/IAEA 1977；Velmurugan et al. 2010；Shu 2009；Suprasanna and Nakagawa 2012）。

8.3.1 基因型选择

与纯合材料（AA 或 aa）相比，使用杂合起始材料（Aa）进行突变工作更为实用。这是因为诱发的突变大多是隐性的；所以，它们只能在具有原始基因型 Aa 且显性等位基因发生突变的植物中表达。从隐性到显性的突变发生率非常低，即可能少于所有突变事件的 5%（Suprasanna and Nakagawa 2012）。

选择特定的目标原始品种是诱变育种重要的第一步。对高性能原始品种的恰当选择以及对突变目标性状的恰当选择将最终决定突变育种的成功或失败。当选择成功的品种用物理诱变剂（如 X 射线或 γ 射线）诱变或化学诱变剂诱变时，诱变剂处理通常仅导致单个或简单性状的修饰，并维持原始品种其他较好和实用的农艺性状（Broertjes 1968a）。

在大多数情况下，以观赏植物的流行品种作为育种材料的目的是，通过创建种质库中不存在的新等位基因，扩大其遗传变异性，以供进一步选择，或者诱导新的遗传性状，从而可以直接用作商业品种。以无性繁殖观赏植物，如菊花、玫瑰或香石竹为例，诱变育种很可能是在这些品种中诱导与鉴定出消费者与种植者更喜欢的花朵颜色及形状突变。

8.3.2 诱变剂和诱变材料的选择

化学诱变剂和物理诱变剂均已用于农作物的突变诱导，但超过 90%已释放的突变品种是通过物理诱变剂诱导无性繁殖作物开发的（FAO/IAEA 2017）。所有类型的电离辐照，如 X 射线、γ 射线、快中子和离子束，都可以有效地诱发突变。但是，实际上只有 X 射线和伽马射线以及最近使用的重离子束最常用于观赏植物的突变诱导。

诱变剂的选择和使用方法是突变育种获得成功的重要因素。物理诱变剂（如伽马射线和 X 射线）或化学诱变剂如甲基磺酸乙酯（EMS）已广泛用于诱导各种农作物突变（Van Harten 1998）。

也有许多报道指出化学诱变的效果较差，这是因为化学物质的吸收差和渗透性差。大型材料，如扦插苗、鳞茎、茎和接穗，很难用化学诱变剂得到重复性好的处理效果。虽化学诱变剂可以应用于组织培养，但仍需要通过超滤方法对诱变剂溶液进行灭菌。

在无性繁殖观赏植物中，可以通过体内或体外技术进行突变诱导。体内技术通常利用诸如块茎、鳞茎、根茎、走茎、茎秆或叶插条等作为目标材料。体外技术利用愈伤组织、细胞悬浮培养物、原生质体、体细胞胚和不定芽等形式的细胞，或利用微型扦插、叶或花蕾等形式的组织作为诱导材料。由于突变是在单细胞中诱导形成的，因此多细胞结构中的突变将导致嵌合体的出现。非嵌合突变体可以通过无性繁殖突变体（部分）来获得。如果是茎秆扦插，建议至少进行 4～5 次削减扦插（反复用新的再生芽）的程序。如果使用腋芽或离体芽，则建议反复切取新的再生芽并继代培养 4～5 次（$M1_1V_1$～$M1_1V_5$）以上。通过这样做，将有助于最小化嵌合体的存在概率，并且可以获得具有期望性状的非嵌合突变体。

Broertjes 和 Verboom（1974）用伽马射线处理了六出花二倍体与三倍体品种的幼嫩根茎，并检测到稳定的突变体。对三倍体秋海棠植株上的叶片照射后，可以获得更高的突变频率以及稳定的突变体（Doorenbos and Karper 1975；Doorenbos 1973）。类似的研究结果可见 Zalewska 等（2010）的报道；此外，Jerzy 和 Zalewska（1996）在菊花中进行研究也有类似的报道，他们通过用 X 射线照射活体叶外植体和用伽马射线照射离体微型扦插幼苗进行研究。之前，Nagatomi 等（1996a）通过伽马场慢性伽马射线辐照生根的扦插条来证明菊花诱变的有效性。花瓣花朵的颜色突变率最高，其次是由愈伤组织再生的花芽和叶子。使用慢性伽马射线处理离体材料的突变率比处理活体扦插材料高 10 倍，并且产生了非嵌合突变体（Nagatomi et al. 1996a）。Nagatomi（1997，2003）研究表明，由花瓣再生植物花朵颜色的突变率高于经离子束和伽马射线处理叶片再生的植株。可能是因为花瓣细胞中的花色基因比叶细胞中的对应基因更活跃，所以花瓣再生植物的突变率更高。

8.3.3 放射敏感性试验

许多观赏植物如菊花、玫瑰、秋海棠、香石竹和叶子花（Srivastava et al. 2002）对电离辐照的放射敏感性已经被测定。用于辐照叶子花茎秆扦插苗的伽马射线的合适剂量已标准化为250～1250拉德（Datta 1992）。突变率通常随诱变剂量的增加而线性增加，但存活率和再生能力随剂量的增加而降低。主要农作物种子和芽的辐照剂量与条件已被研究优化并公布（Van Harten 1998；Medina et al. 2004；Datta 2009b；FAO/IAEA 1977，2011）。对于新的或次要作物种类，由于研究信息的缺乏，需要通过实验得出这些新的或次要作物品种的辐照剂量和条件。通常建议使用的辐照剂量是：与未处理的对照组相比，该剂量下处理植株的生长和形态发生性能降低50%（GR 50）（FAO/IAEA 1977，2011）。

使用辐照的育种计划的第一步是评估辐照剂量，即放射敏感性测试。在进行实际的突变诱导之前，建议对要使用的材料或外植体对伽马射线的敏感性进行初步研究。实际上，放射敏感性的差异不仅可以通过品种或物种之间固有的遗传差异来解释，还可以通过生理差异来解释。最适剂量的选择通常基于辐照处理导致的生长量减少程度或存活率降低程度（Ahloowalia and Maluszynski 2001）。初步研究的目的是评估导致生长量减少50%的剂量。50%致死剂量（LD_{50}）是指枝条的增殖率降低50%的辐照剂量（Van Harten 1998）。根据所测试的材料，依据处理后第一个继代培养中获得的再生枝或增殖的数量确定生长量，即表明根尖和腋生分生组织的存活率（Ahloowalia 1998）。由于大多数组织培养材料（如体外芽、腋芽或愈伤组织）对放射线非常敏感，因此100～120Gy的高剂量对生长具有致命性。

放射敏感性测试有助于人们对电离辐照的作用机制了解。确定任何作物的放射敏感性是开展大规模诱变的前提。研究表明，各种生物、物理和化学因素，如基因型、组织年龄等，都可能改变辐照对植物的影响（Sparrow et al. 1968；Datta 2006）。例如，Nagatomi等（1996a）和Zalewska等（2010）报道菊花的放射敏感性机制是基因型依赖性的。确定诱导体细胞突变的最佳辐照剂量是非常重要的。突变率随基因型和剂量而变化，一些基因型不那么敏感，而另一些基因型更敏感，一些对辐照有抵抗力。评估最适剂量的常用程序是了解和调查组织和/或细胞对给定诱变剂的敏感性，通过绘制基于剂量和存活率的敏感性曲线来进行测量。杀死约50%的已处理材料的剂量称为LD_{50}，可从敏感性曲线获得。放射敏感性随物种和基因型、植物和器官的生理状况以及诱变处理之前和之后对辐照物质的处理而变化。

8.3.4 确定最佳处理剂量

确定LD_{50}值后，最有效的剂量或最佳的处理剂量便容易选择确定。特定育种项目中使用的实际剂量要根据特定植物材料、其遗传学和生理的经验来确定，目的是最大程度地挽救有用的突变体。Heinze和Schmidt（1995）建议，实验的起初剂量范围应为$LD_{50}\pm10\%$。低于LD_{50}的剂量有利于处理后的植物恢复，这意味着发生突变的可能性较小，而使用更高剂量则增加了诱导更多突变（阳性或阴性）的可能性。因此，为了安全起见，也为了诱导更多的阳性突变，建议选择三种剂量：低于LD_{50} 10%、LD_{50}以及高于LD_{50} 10%。确定诱导突变的最佳辐照剂量对诱变育种项目非常重要。一些研究已经确定了一些最常见的观赏物种的工作剂量（Datta 2009a）。突变率随照射材料的种类和照射剂量而变化。

一旦确定了诱导突变的剂量，就需要更多信息来决定将使用多少外植体进行辐照。根据所需性状的突变率，可计算出要辐照的外植体数量。单个性状的突变率预计为 0.1%～1.0%（Predieri and Zimmerman 2001）。Predieri 和 Zimmerman（2001）也报道，对于不同植物品种的许多性状，突变率范围为 0.14%～1.93%。因此，为了安全起见，建议将预期突变率设置为 0.5%，这样，如果照射 1000 株植物将有 5 个个体具有所需的性状。但是，如果用于辐照的外植体数量有限（少于 500 个），将限制成功选择具有所需性状的突变体的可能性。我们建议至少要有 1000 棵受辐照植物，以便能够成功选择目标突变性状。

Ibrahim 等（1998）利用 2.5～10kR 剂量 X 射线照射叶片外植体，分析了三个现代月季基因型的放射敏感性。结果表明，从 2.5kR 到 10kR 处理剂量，观察到芽的再生速率降低（从 47%到 0）。也研究了 X 射线辐照对三个现代月季基因型叶片离体培养萌发的不定芽再生的影响，选择 0.5kR 和 1.5kR 的 X 射线剂量，并观察处理后基因型之间的差异。在实验的三个基因型中，RUI 317 株系的不定芽再生率最高（Ibrahim et al. 1998）。Dao 等（2006）开发了一种在菊花中进行体外 γ 射线诱变的方案，发现愈伤组织的致死剂量约为 5.0kR。用 1.0kR 和 3.0kR 的 γ 射线处理突变种群产生了丰富的变异，包括花瓣数量和颜色，而对照种群则未观察到变异。对于离子束诱变的研究，菊花（Yamaguchi et al. 2010）和玫瑰（Yamaguchi et al. 2003）的最佳处理方法已被确定，并获得了突变体。人们普遍认为，低剂量辐照对于获得单点突变体至关重要，特别是对于无性繁殖作物的辐照育种，因为这样获得的突变体可直接作为新品种应用（Yamaguchi 2013）。

8.3.5　产生稳定突变体的体外方法

在无性繁殖观赏植物的常规突变育种中，用化学或物理诱变剂对多细胞组织（如茎或有根插枝、块茎、鳞茎、茎或根茎）进行诱变处理会产生嵌合现象。在嵌合组织中，突变细胞与正常细胞一起存在。在随后的细胞分裂过程中，突变的细胞与周围的正常细胞竞争生长或存活，这称为二倍期选择（Datta et al. 2005）。在这种情况下，可以通过无性繁殖突变体组织（细胞）来获得非嵌合突变体。原则上，突变育种涉及突变诱导、分离和选择有用突变的过程。固定单个植物水平上表型多样突变体的有效方法很多，包括不定芽技术、连续修剪、嫁接和剪枝技术以及体外培养技术。通常需要几个营养世代之后才能产生稳定的突变体。因此，体外诱变的概念被提出，这为诱导和固定更多数量的突变体提供了新的可能。该技术的主要优点是克服了嵌合体的形成。诱导突变与体外培养技术相结合已被常规用于观赏性改良，并且已被证明是增加具有经济潜力品种遗传变异性的有效方法。

与常规繁殖手段（如插条和种子）相比，诸如微繁殖和体细胞胚发生等体外（组培）繁殖技术为大规模繁殖无病、遗传上统一的植物提供了一种有效的方法。营养繁殖观赏植物的突变育种包括突变诱导、分离和选择有用突变的过程。针对在单个植物水平上表型多样的突变体，固定有效突变的方法包括不定芽技术、连续修剪、嫁接和剪枝技术以及体外培养技术。通常，需要数个营养世代之后才能产生稳定突变体（Suprasanna et al. 2012）。

体外（组培）技术在辐照育种中的使用使再生植物，包括突变植物，由在遗传上均质的组织组成（Broertjes and Keen 1980）。这样就避免了嵌合体的产生，从而大大缩短了育种过程。Zalewska 和 Jerzy（1997）使用体外（组培）技术，通过用 X 射线辐照‘Red Nero’菊花的叶片外植体，产生了矮化突变体，即具有较小花序的品种‘Mini Nero’，其有作为盆栽植物的潜力。然而，使用体外（组培）技术，由经 γ 射线辐照的菊花品种‘Richmond’叶片外植体再生的不定芽，产生了多种花色突变体（Jerzy and Zalewska 1996）。突变率为 2.4%～11.6%，且所有突变体均为稳定的、非周缘型和扇区型嵌合体。Datta 等（2005）用急性伽马射线辐照 4 个菊花品种的小花，在体外培养的植株中观察到体细胞突变产生的颜色和小花形状变异株。在所有变异植物中均未检测到嵌合突变，突变率为 10%～20%。

类似于 Jerzy 和 Zalewska（1996）与 Datta 等（2005）所应用的方法，Nagatomi 等（1996a）使用急性和慢性伽马射线，建立了一种有效的诱导菊花品种‘Taihei’花色突变的方法。使用急性伽马射线辐照方法：从未经辐照的开花植物上切下花瓣、花蕾和叶外植体。这些外植体首先进行体外培养，然后用急性伽马射线辐照，再由愈伤组织再生植物。使用慢性伽马射线辐照方法：首先在伽马射线场辐照生根的插条，直到开花。切下诸如花瓣、花蕾和叶外植体并进行体外培养，由愈伤组织再生植物。观察到所有再生的突变体都是非嵌合的，并且使用慢性伽马射线辐照的花色突变率更高（花瓣 38.7%，花蕾 37.5%，叶 13.8%），同时比获得的花色色谱具有更宽的范围。通过急性伽马射线辐照的花色突变率分别为：花瓣 29.7%，花蕾 12.0%，叶子 12.9%。这些发现表明，体外培养和电离辐照是诱变的有效方法，可以在短时间内产生新的潜在的观赏植物突变体。

体外（组培）条件下的不定芽技术是突变育种的重要工具（Broertjes et al. 1976；Zalewska and Jerzy 1997）。该技术使获得商业上有吸引力的稳定突变体成为可能。据推测，菊花的许多品种是周缘型嵌合体，其中一个细胞层（主要是外部的）在遗传上不同于其他细胞层。嵌合现象可能是与遗传变异产生有关的重要现象，特别是在突变育种计划的早期阶段（Van Harten 1998）。根据 Broertjes 和 Keen（1980）的理论，植物由叶片经体内和体外再生的不定分生组织最终来自单个初始细胞。因此，不定芽技术在辐照育种中的应用使再生植株，包括突变植株，由在遗传上均质的组织组成。这样，既避免了嵌合体的产生，又可以大大缩短繁殖过程。

通过由叶外植体再生，在第二代无性繁殖（M_1V_2）中，具有一致花序变化的突变体比例显著增加。例如，在菊花中，Broertjes 等（1980）估计，在 X 射线辐照后，对于特定品种 1～2 年产生的花色突变体数量相当于自然状态下 10～20 年自发出现的变异体数量。由于在体内条件下，不定芽的再生是一个耗时的过程，并且并非所有菊花品种都具有不定器官发生的能力，因此体外培养是突变育种更好的解决方案。在非洲菊中，对体外生长的叶片进行 X 射线或伽马射线处理，随后产生不定芽，共培育了 1250 株 M_1V_2 植株，其中分离出 187 个突变体，仅 6 个被确认是稳定突变体（Laneri et al. 1990）。

一种通过由菊花花瓣直接再生来克服嵌合组织出现的技术已经被开发出来（Mandal et al. 2000b；Datta et al. 2005）。这种分离嵌合体和建立稳定突变体的方法有两个步骤：首先，对舌状小花（复合花外围环上的小花）进行体内诱变处理；其次，用突变的部位进行植株体外再生。此方法代表了创建稳定突变体的快速方法（M_1V_2）。在唐菖蒲中，通常在春季用大约 75Gy 的伽马射线辐照休眠的球茎。当扇区型嵌合体 M_1V_1（诱变后第一个无性繁殖植物）植物通过球茎繁殖后，得到的 M_1V_2 几乎完全由稳定突变体或稳定的正常植物组成（Banerji et al. 1994）。

8.3.6　突变体的筛选

突变育种程序的最后一步是鉴定和筛选所需的突变体。该步骤应在非嵌合植物上进行。与涉及体内芽处理的方法相比，体外培养可提供更多的植物材料供筛选，而且这些材料由突变的少数几个甚至一个细胞发育而来（Maluszynski et al. 1995）。这有利于减少获得嵌合植物的风险，而提高获得具有所需特征的突变株系以及筛选出生长受抑制植株的可能性（Predieri and Zimmerman 2001）。进行体外（组培）筛选可处理大量植株群体，避免了植株群体数量不足的问题。对于观赏植物，花色或植物形态突变的选择非常容易；可以直接在温室或露天实验地里完成。如果要采用连续修剪技术，反复扦插新生的芽几次即可获得具所需性状的稳定突变植株。

根据植物种类的不同，有用突变体的筛选需要（照射后）3～4 年的时间。利用这些突变体，可以通过营养繁殖（克隆）的手段扩大植株群体用于进一步评估。通过这种方式，已经获得了一些有价值的突变体（Van Harten 1998）。目视筛选突变表型变异是最有效的选择方法。它可以有效地识别常见的性状和特征，如颜色变化、植物形态、早熟性、对病虫害的抵抗力等。选择通常从 M_1V_2 开始，并在 M_1V_3 或 M_1V_4 世代中继续进行确认。因为在同质 M_1V_2 群体中鉴定到所需突变体的可能性非常低，所以有必要在 M_1V_3 代或某些情况下在 M_1V_4 代中进行筛选。就诸如郁金香之类的观赏植物而言，选择至少要持续 4 年，以使突变的细胞长出来，并产生稳定突变的鳞茎或至少更大的可见扇区型突变，这些扇区型突变需要更长的时间才能表现出来。

8.4　观赏植物花色突变谱的诱导

诱导新花色是突变育种的主要目标之一。通过诱变处理诱导观赏植物花色变化的研究已见报道（Mandal et al. 2000a，2000b）。Schum 和 Preil（1998）报道，观赏植物诱导突变的记录中有 55%与花色的变化有关，而 15%与花的形态有关。如上所述，突变大多以嵌合体形式出现（Wolf 1996；Chakrabarty et al. 1999；Mandal et al. 2000a；Misra et al. 2004）。当整个分枝都发生突变时，可以通过常规方法分离突变枝条。但是，当花的一个部分发生突变时，通过常规手段很难分离出来（Datta et al. 2001）。由于菊花的许多品种是周缘型嵌合体，因此菊花通常被认为是体细胞克隆变异频繁的植物（Dowrick and Bayoumi 1966；Shibata et al. 1998；Minano et al. 2009）。但是，外植体的来源和性质可能会影响变异的发生（Zalewska et al. 2007）。正如许多报道表明的那样，对于菊花的突变育种，选择适当的基因型将有助于创造新的花色突变体。

在菊花中，观察到粉红色的花朵颜色将是使用电离辐照（如 X 射线和伽马射线）产生其他花朵颜色的最佳起始材料（Nagatomi et al. 2000）。Broertjes 等（1966）早些时候就观察到了这一点，他对粉红色品种‘霍滕森玫瑰’（Hortensien rose）的扦插生根插条进行辐照后获得了几乎所有现有的菊花花色。这是菊花育种的“标准做法”，事实上，杂交育种群体内粉红色花朵幼苗总是会被仔细研究，因为众所周知，稍后可以通过诱导突变而容易地改善或改变花朵的颜色。Schum 和 Preil（1998）报道，在粉红色的菊花品种暴露于诱变因子后，突变率最高，而诱变效果最差的初始基因型是具黄色花的植物。

Lamseejan 等（2000）证明，通过用急性（高剂量）伽马射线照射处理来自菊花浅紫色株系（多头型）舌状小花的愈伤组织，可以观察到窄的颜色突变谱：深紫色、浅紫色和黄色。但是，Nagatomi 等（2000）用慢性（低剂量）伽马射线辐照处理粉红色基因型‘Taihei’后，产生了许多颜色的突变体，包括白色、浅粉红色、深粉红色、橙色、黄色、古铜色和条纹状。此外，粉红色的基因型容易芽变出各种颜色，其中包括菊花中能观察到的大多数颜色（Schum and Preil 1998）。因此，Lamseejan 等（2003）使用的菊花的原始紫色基因型不适合作为花颜色突变诱导材料，因为它产生很少的花色突变。

为了获得广泛的颜色变化，有必要选择适当的初始基因型进行辐照处理（Schum and Preil 1998；Lamseejan et al. 2003；Nagatomi et al. 2000）。最近，重离子或离子束已被用于诱变选育具有商业价值的复色或新颖颜色的新突变体或品种，特别是在菊花（Nagatomi et al. 1996c；Yamaguchi 2013）、香石竹（Okamura et al. 2003）、玫瑰（Yamaguchi et al. 2003）、仙客来（Sugiyama et al. 2008b）、矮牵牛（Okamura et al. 2006）和蝴蝶草属（Miyazaki et al. 2006；Sasaki et al. 2008）。这是由于与低线性能量转移（LET）辐照（如 X 射线和伽马射线）相比，离子束具有较高的线性能量转移，因此具有更大的生物学诱变效应（Tanaka et al. 2010）。所以，离子束有望产生更高的突变率和新型突变体。诱变剂和外植体的选择是突变诱导成功与否的最重要因素之一。

8.5 观赏植物商业突变品种的开发

许多观赏植物突变体是由物理和化学诱变剂诱导的，并已作为新品种正式发布（FAO/IAEA 2017）。它们中的大多数，特别是在市场上具有较大经济价值的切花和盆栽植物已经商业化；而其他装饰性和园林植物等许多植物在园艺行业中被认为具有观赏价值。观赏植物的诱变育种计划着重于改善各种植株性状以提高观赏价值，包括花的颜色、大小和形状以及生产质量。创建具有新颖颜色的新花形是大多数观赏植物的主要育种目标。

通过诱导突变开发并商业化的一些观赏品种包括马齿苋（Gupta 1979）、秋海棠（Matsubara et al. 1975；Mikkeksen et al. 1975）、菊花、矮牵牛、风信子（Broertjes 1966；Broertjes and Van Harten 1988）、香石竹（Simard et al. 1992；Cassels and Walsh 1993）、叶子花（Gupta and Nath 1977；Datta and Banerji 1990，1994）和大丽花（Das et al. 1974；Dube et al. 1980）。迄今为止，诱导的菊花突变品种占比最多，为 37%以上，其次是玫瑰，为 16%（FAO/IAEA 2017）。具有花色和形态的变化被认为是已释放的观赏植物品种的最显著特征。在荷兰花卉市场上，菊花突变体品种占近 40%，并且通过 X 射线或伽马射线进行突变育种获得了约 236 个正式宣布的商业突变体（Teixeira da Silva 2004）。通过突变技术已经产生了许多具有新的、理想性状的种质（Chatterjee et al. 2006；Jain and

Spencer 2006；Jain 2010；Micke et al. 1990）。

在国际原子能机构支持的先驱项目中，泰国 Kasetsart 大学与当地种植者合作，成功选育了 50 多个新的突变品种，包括美人蕉（37）、菊花（6）、马齿苋（10）和沙漠玫瑰（2）。一些通过慢性（低剂量）伽马射线辐照诱变的具新颜色的美人蕉突变体品种具有独特的花色，吸引了花店老板和种植者，并且已经用于商业生产（图 8.3）。

图 8.3　慢性伽马射线辐照诱导而来的具有独特花色的美人蕉新突变体。左侧的第一列是原始的亲本品种，第二列和第三列是突变植株。a 是亲本植株‘GISC 1’、突变植株‘Narippawaj’和‘Pink Peeranuch’；b 是亲本植株‘GISC 2’、突变植株‘Red Ridthee’和‘Sumin’；c 是亲本植株‘GISC 24’、突变植株‘Mattana’和‘Orange Siranut’；d 是亲本植株‘GISC 12’和突变植株‘Yellow Arunee’[由泰国曼谷 Kasetsart 大学理学院核技术研究中心（NTRC）负责人 Peeranuch Jompuk 博士提供]

马来西亚原子能研究机构也利用急性和慢性伽马射线辐照成功开发了观赏植物的新突变体品种，包括具有新花色与花形的兰花、木槿、矮牵牛、百合和美人蕉，具有叶斑的朱蕉和金露花，开花期延长的时钟花。兰花的改良育种仍然是一项艰巨的任务，因为它们的不亲和性和不育性，并且大多数品种需要很长时间才能育成，这就是没有更多新的兰花突变品种发表的原因。用经 35Gy 的急性伽马射线辐照分生组织产生的原球茎，成功地产生了一些有前途的突变品种，这些品种具有新的花色和形状（图 8.4）。对于木槿，用 20Gy 的急性伽马射线辐照扦插植株的方法，也选育并注册了具有独特且有趣花朵颜色和形状的新品种（图 8.5）。

图 8.4　急性伽马辐照诱导选育的兰花新突变品种。上排从左到右依次是原始的石斛兰亲本‘Sonia’（左），颜色为红紫色，突变品种‘Keena Hieng Ding’（中），花瓣宽阔，突变品种‘Keena Pearl’（右），白色和紫色边缘。下排从左到右依次是突变品种‘Keena Ahmad Sobri’（左），具有均匀的菱形紫色花瓣，突变品种‘Keena Pastel’（中），具有柔和的紫色阴影花瓣，突变品种‘Keena Radiant’（右），具有淡紫色的花朵（由马来西亚原子能研究机构提供）

在韩国原子能研究所（KAERI），通过用 100Gy 的伽马射线辐照花色为粉红色的芙蓉花原始亲本种子，开发了 4 个颜色突变体。通过突变改良，具有蓝紫色花色和较大花型的品种‘Changhae’在 2006 年被批准为新品种（图 8.6）。另一个有趣的突变品种‘Ggoma’于 2006 年作为新品种发布，其具有矮化和小型花等主要特性，可能用作盆景或盆栽植物。据报道，这些植物品种权已于 2009 年转让给名为 Supro 的私人公司进行商业化推广（个人通讯）。

针对菊花 *Chrysanthemum morifolium* 品种‘Taihei’，Nagatomi 等（1996a）研究表明，使用辐照剂量为 20～100Gy 的急性伽马射线辐照花瓣、芽和叶的突变率相对较低，并且花朵的色谱非常有限。相比之下，使用剂量为 25～150Gy 的伽马射线长期辐照获得

图 8.5　急性伽马射线辐照产生的木槿花色突变体。(a) 从左到右依次是原始的亲本材料（左），花瓣为粉红色，突变品种 'Siti Hasmah RedShine'（中），突变品种 'Siti Hasmah PinkBeauty'（右）；(b) 从左到右依次是原始亲本品种（左），具有醒目的红色，突变品种 'Nori'（中），具有多层红色花瓣，突变植株（右），具有紧凑的卷曲花瓣（由马来西亚原子能研究机构提供）

高突变率，并具有多种颜色突变体，包括白色、浅粉红色、深粉红色、橙色、黄色、青铜色和条纹状。这表明长期伽马射线辐照对产生具有更宽花色谱和更高突变率的突变更为有效。从 3689 株植物中获得了总共 550 个（14.91%）突变体，结果 6 个有前途的切花突变株系已被登记为新品种（图 8.7）。

Zalewska 等（2010）报道了利用不同菊花（*Chrysanthemum* × *grandiflorum*）品种进行 X 射线和伽马射线诱变研究工作。用 25Gy 剂量的 X 射线辐照处理菊花品种 'Red Nero' 植株的叶片，产生了矮化突变株系 'Mini Nero'，具有作为盆栽应用的潜能。使用体外（组培）技术，用 15Gy 的伽马射线辐照处理菊花品种 'Richmond' 的微型扦插苗，从继代培养产生的不定芽中，成功筛选出 21 个具有独特花色的株系，与原始品种（'Richmond'）典型的紫粉红色显著不同。在这个研究中，Zalewska 等（2010）采用急性（短时）伽马射线照射能够产生各种有前途的颜色突变体，如金甜菜色、金棕色、粉红色、紫金色、红棕色、浅橙色、黄色和白色。这表明，接受辐照的不同品种和植株材料对不同电离辐照剂敏感，其都对菊花花色突变率产生影响。

韩国原子能研究所（KAERI）在多头型菊花品种 'Kitam' 中也观察到了类似的高突变率和多种颜色突变体。以 30～50Gy 剂量的急性（短时）伽马射线照射幼嫩的离体（组培）植物成功产生了一些新的有前途的颜色突变体，并且这些突变体已被注册为新品种（图 8.8）（FAO/IAEA 2017）。

图 8.6 伽马射线诱导的木槿花新突变品种。上排从左至右依次是突变品种‘Dasom’，红色花朵，突变品种‘Changhae’，蓝紫色花朵。中排从左至右依次是突变品种‘Daegoang’，粉色花朵；突变品种‘Seonnyo’，白色花朵。下排是一种新的粉色花突变品种‘Ggoma’，经伽马射线诱导突变为矮生小花品种[由韩国原子能研究所(KAERI)先进辐照技术研究所(ARTI)辐照育种研究小组首席研究员 Si-Yong Kang 博士提供]

自 1990 年日本原子能机构（JAEA）和 RIKEN 通过回旋加速器在日本建立离子束设施以来，许多新颖和优良的花色突变观赏植物新品种被发表并实现商业化推广应用，包括菊花（Nagatomi et al. 1996b，2003）、香石竹（Okamura et al. 2003）、玫瑰（Yamaguchi et al. 2003）、仙客来（Sugiyama et al. 2008b）、矮牵牛（Okamura et al. 2006）、蝴蝶草（Miyazaki et al. 2006；Sasaki et al. 2008）和马鞭草（Kanaya et al. 2008）。通过选择尽可能低的辐照剂量，诱变选择出的这些品种仅有一个性状改变，没有其他有害的特征，如畸形的叶子和细胞中 DNA 含量减少。这表明离子束可以诱导只有一个点（性状）改变的突变，称为“单点育种”，这是之前的诱变剂很难取得的效果（Ueno et al. 2003）。

如 Nagatomi 等（1996a）报道，伽马射线辐照处理菊花品种‘Taihei’产生的花色突变体有浅粉红色和少数深粉红色。但是，由离子束诱导的花色突变体不仅有浅粉色和深粉色，还有橙色、白色、黄色、复色以及白色和黄色的条纹状。Tanaka 等（2010）报道，随着离子束剂量的增加，复色花突变体的数量也增加。通过伽马射线辐照，没有获得复色和条纹状突变体的案例（图 8.9）。

图 8.7　由具粉红色花朵的原始品种‘Taihei’（1995 年注册）通过长期伽马射线辐照诱导选育出的不同颜色的 Ramat 系列菊花品种。上排从左到右依次是白色突变品种‘Haeno Hatsuyuki’，橙色突变品种‘Haeno Kirameki’和深粉红色突变品种‘Haeno Kurenai’。下排从左到右依次是浅粉红色突变品种‘Haeno Mirayabi’，青铜色突变品种‘Haeno Yuugure’和亮黄色突变品种‘Haeno Kagayaki’（由日本国家农业和食品研究组织作物科学研究所放射育种研究所所长 Hiroshi Kato 博士提供）

图 8.8　伽马射线辐照处理菊花品种‘Kitam’诱导选育的 6 个新品种。上排从左到右依次是突变品种‘ARTI-Rollypop’、‘ARTI-Purple Lady’和‘ARTI -Rising Sun’。下排从左到右依次是突变品种‘ARTI-Dark Chocolate’、‘ARTI-Red Star’和‘ARTI-Queen’[由韩国原子能研究所（KAERI）先进辐照技术研究所（ARTI）辐照育种研究小组首席研究员 Si-Yong Kang 博士提供]

图 8.9　第一批由离子束诱导的具有不同花色的菊花突变新品种并于 1998 年注册。上排从左到右依次是原始亲本品种‘Taihei’、粉色突变品种‘Ion No Koki’。下排从左到右依次是橙色突变品种‘Ion No Seiko’和青铜色突变品种‘Ion No Maho’（由日本高崎国立量子与放射科学技术研究所、量子束科学研究理事会、研究计划和促进办公室主任 Atsushi Tanaka 博士提供）

Kirin Agribio 公司和日本原子能机构（JAEA）的合作项目通过离子束辐照选育了香石竹新花色品种。带有褶皱花瓣、樱桃粉红色花朵的多头型香石竹品种‘Vital’作为原始材料，通过伽马射线、X 射线辐照或甲基磺酸乙酯（EMS）处理获得花朵颜色突变体，如粉红色、红色或条纹状；而通过碳离子（LET＝122keV/μm）辐照获得的突变体色谱更为丰富，包括粉红色、浅粉红色、浅橙色、红色、黄色以及复色和条纹类型（图 8.10）。此外，除花朵颜色外，还通过碳离子辐照诱导出许多不同的圆形花瓣，这表明与其他诱变剂相比，离子束可以高频诱导出新颖的花朵颜色和花朵形状（个人通讯）。

图 8.10　离子束诱导选育的香石竹新品种。左上角的樱桃粉红色花朵是原始的亲本品种‘Vital’，其他则是诱导产生的花色和花形突变体（由日本高崎国立量子与放射科学技术研究所、量子束科学研究理事会、研究计划和促进办公室主任 Atsushi Tanaka 博士提供）

Nagayoshi（2003）和 Ueno 等（2003）报道，使用离子束诱变选育的商业上最成功的花卉品种是菊花品种‘Aladdin’和‘Aladdin 2’，这 2 个品种由起始品种‘Jimba’诱导而来，主要特征是侧生花芽较少（图 8.11）。据报道，每年销售超过 3000 万枝切花，并应邀授权日本 35 个以上营销合作社销售（Tanaka et al. 2010）。

图 8.11　左侧的原始菊花品种‘Jimba’显示出过多的腋芽。中间是离子束诱导新品种‘Aladdin’，具有腋芽较少的特性。右边是离子束诱导新品种‘Aladdin 2’，其腋芽减少，由‘Aladdin’诱导而来，较‘Aladdin’早开花（由日本高崎国立量子与放射科学技术研究所、量子束科学研究理事会、研究计划和促进办公室主任 Atsushi Tanaka 博士提供）

对于非洲菊和仙客来说，通过离子束辐照诱导也选育了一些商业品种。JAEA 离子束技术公司、农业技术中心和花卉种植者之间的区域产学研组织合作成功选育了非洲菊新品种（图 8.12）。通过离子束辐照非洲菊黄色花品种的叶子组织培养物，成功诱导出各种新的花色突变体，包括橙色和白色；其中最著名的品种‘Vient Flamingo’花朵浅粉

图 8.12　离子束诱导选育的非洲菊和仙客来商业品种。上排最左侧是原始品种，后面依次是花色突变品种，有柔和的粉红色、橙色和白色。下排左侧为紫红色的原始香气品种‘Kaori-no-mai’，右侧是新的香气品种，其颜色为红色（由日本高崎国立量子与放射科学技术研究所、量子束科学研究理事会、研究计划和促进办公室主任 Atsushi Tanaka 博士提供）

色，耐高温和低温，开花期长（1～6 月）。对于仙客来说，由日本 Saitama 县选育的原始品种‘Kaori-no-mai’的花色是紫红色，其含有锦葵素，具独特的香味；通过离子束辐照诱导产生的突变体含有飞燕草素，有新的香味。

8.6 突变育种的前景

突变是无性繁殖植物（如观赏植物）遗传变异的主要来源。通常，诱导变异与自发突变产生的变异没有什么不同。直接使用诱导突变对于观赏植物的育种很重要，特别是当育种目标是改善一个或两个性状但又保持成熟品种中剩余的其他性状。直接使用突变植株是植物育种非常有价值的一种补充手段。与杂交育种过程相比，该技术的主要优点是获得具有改良性状新品种所需的时间较短。此外，诱导的突变体具有进一步遗传改良的潜力。

由于各种原因，突变育种在观赏植物以及其他无性繁殖植物中特别成功。首先，选择包括花色、形状或花大小等直接可察觉的突变体比较容易。其次，许多观赏品种是杂合的，可以通过突变和杂交扩大变异。而且，体内或体外（组培）繁殖方法结合侧芽技术有利于某些需要较长时间才能识别出来的突变体的产生。

具有理想性状的观赏植物的繁殖和改良可以通过创造新颖变异来实现，这有利于花卉产业的发展。遗传变异可以通过对叶片、插条和体外（组培）生长的植物材料进行物理与化学诱变处理来诱导。辐照处理在已报道的几种观赏植物，如菊花、玫瑰、兰花等中，诱导新的突变体非常有效，此手段用于开发观赏植物新突变体潜力较大，包括引入和本地观赏植物。离子束已被证明是比 X 射线和伽马射线更好的物理诱变剂，在针对菊花、香石竹、非洲菊的研究中可以诱导出具更宽花色谱的突变体。

随着分子遗传学和基因组学的最新发展，诱导突变的科学发展已从基础研究和应用于育种发展到基于先进基因组学的技术。新的和更具体的技术非常有前途，如允许对植物进行相对便宜和快速的基因组测序的高通量测序。诸如 TILLING（靶向基因组的诱导局部病变）、锌指核酸酶介导的诱变以及核酸酶的使用等方法是潜在的筛选技术，可以帮助在观赏植物中产生靶向突变，描绘出基因功能并助力品种改良。此外，诱导产生的观赏植物突变在下一步遗传改良中有很高的应用潜力。

本章译者：
阮继伟
云南省农业科学院花卉研究所，国家观赏园艺工程技术研究中心，云南省花卉育种重点实验室，昆明 650200

参 考 文 献

Abe T, Matsuyama T, Sekido S, Yamaguchi I, Yoshida S, Kameya T (2002) Chlorophyll-deficient mutants of rice demonstrated the deletion of a DNA fragment by heavy-ion irradiation. J Radiat Res 43(Suppl):S157–S161
Affrida AH, Sakinah A, Zaiton A, Mohd Nazir B, Tanaka A, Narumi I, Oono Y, Hase Y (2008) Mutation induction in orchids using ion beams. JAEA Takasaki Annu Rep 2007:61
Ahloowalia BS (1998) *In-vitro* techniques and mutagenesis for the improvement of vegetatively

propagated plants. In: Jain SM, Brar DS, Ahloowalia BS (eds) Somaclonal variation and induced mutations in crop improvement. Kluwer Academic Publishers, Dordrecht, pp 293–309

Ahloowalia BS, Maluszynski M (2001) Induced mutations – a new paradigm in plant breeding. Euphytica 118:167–173

Ahloowalia BS, Maluszynski M, Nichterlein K (2004) Global impact of mutation-derived varieties. Euphytica 135(2):187–204

Arisumi T, Frazier LC (1968) Cytological and morphological evidence for the single-cell origin of vegetatively propagated shoots in thirteen species of *Saintpaulia* treated with colchicines. Proc Am Soc Hortic Sci 93:679–685

Banerji BK, Datta SK, Sharma SC (1994) Gamma irradiation studies on gladiolus cv. White Friendship. J Nucl Agric Biol 23(3):127–133

Blakely EA (1992) Cell inactivation by heavy charged particles. Radiat Environ Biophys 31:181–196

Bottino PJ, Sparrow AH, Schwemmer SS, Thompson KH (1975) Interrelation of exposure and exposure rate in germinating seeds of barley and its concurrence with dose-rate theory. Radiat Bot 15:17–27

Broertjes C (1966) Mutation breeding of chrysanthemums. Euphytica 15(2):156–162

Broertjes C (1968a) Mutation breeding of vegetatively propagated crops. In: 5th Congress of the European Association for Research on Plant Breeding, 30.9-2.10, Milano, pp 139–165

Broertjes C (1968b) Dose-rate effects in *Saintpaulia*. In: Mutations in plant breeding II. Proceedings of a panel, Vienna, 11–15 September 1967, Jointly organized by the IAEA and FAO, IAEA Vienna, pp 63–71

Broertjes C, Van Harten AM (1978) Application of mutation breeding methods in the improvement in vegetatively propagated crops. Elsevier Scientific Publishing Company, Amsterdam., 353pp

Broertjes C, Keen A (1980) Adventitious shoots: do they develop from one cell? Euphytica 29:73–87

Broertjes C, Van Harten AM (1988) Applied mutation breeding for vegetatively propagated crops. Elsevier Scientific Publishing Co, Amsterdam. 345pp

Broertjes C, Verboom H (1974) Mutation breeding of *Alstroemeria*. Euphytica 23:39–44

Broertjes C, Haccius B, Weidlich S (1968) Adventitious bud formation on isolated leaves and its significance for mutation breeding. Euphytica 22:415–423

Broertjes C, Roest S, Bokelmann GS (1976) Mutation breeding of *Chrysanthemum morifolium* Ramat. using *in vivo* and *in vitro* adventitious bud techniques. Euphytica 25:11–19

Broetrjes C, Koene P, Van Veen JWH (1980) A mutant of a mutant of a mutant of a...: Irradiation of progressive radiation-induced mutants in a mutation-breeding programme with *Chrysanthemum morifolium* Ramat. Euphytica 29:525–530

Cassels AC, Walsh PC (1993) Diplontic selection as a positive factor in determining the fitness of mutants of *Dianthus* 'Mystere' derived from X-irradiation of nodes in *in vitro* culture. Euphytica 70:167–174

Chakrabarty D, Mandal AKA, Datta SK (1999) Management of chimera through direct shoot regeneration from florets of chrysanthemum (*Chrysanthemum morifolium* Ramat.). J Hortic Sci Biotechnol 74:293–296

Chatterjee J, Mandal AKA, Ranade SA, Teixeira da Silva JA, Datta SK (2006) Molecular systematics in *Chrysanthemum x grandiflorum* (Ramat.) Kitamura. Sci Hortic 110:373–378

Chinone S, Tokuhiro K, Nakatsubo K, Hase Y, Narumi I (2008) Mutation induction on *Delphinium* and *Limonium sinuatum* irradiated with ion beams. JAEA Takasaki Annu Rep 2007:68

Constantin MJ (1984) Potential of *in vitro* mutation breeding for the improvement of vegetatively propagated crop plants, pp 59–77. International Atomic Energy Agency, Vienna, Austria. Induced mutations for crop improvement in Latin America, Vienna, Austria

Dao TB, Nguyen PD, Do QM, Vu TH, Le TL, Nguyen TKL, Nguyen HD, Nguyen XL (2006) In vitro mutagenesis of chrysanthemum for breeding. Plant Mutat Rpt 1:26–27

Das PK, Ghosh P, Dube S, Dhua SP (1974) Induction of somatic mutations in some vegetatively propagated ornamentals by gamma radiation. Technology (Coimbatore, India) 11(2&3):185–188

Datta SK (1988) Chrysanthemum varieties evolved by induced mutations at Botanical Research Institute, Lucknow. Chrysanthemum 44(1):72–75

Datta SK (1992) Mutation studies on double bracted Bougainvillea at National Botanical Research Institute (NBRI), Lucknow, India. Mutat Breed Newsl 39(January):8–9

Datta SK (2006) Parameters for detecting effects of ionizing radiations on plants. In: Tripathi RD, Kulshreshtha K, Agrawal M, Ahmad KJ, Varshney CK, Krupa SV, Pushpangadan P (eds) Plant

responses to environmental stress. International Book Distributing Cp, Lucknow, pp 257–265
Datta SK (2009a) A report on 36 years of practical work on crop improvement through in induced mutagenesis. In: Shu QY (ed) Induced plant mutations in the genomic era. Joint FAO/IAEA Division of Nuclear Techniques in Food and Agriculture, International Atomic Energy Agency, Vienna, pp 253–256
Datta SK (2009b) Role of classical mutagenesis for development of new ornamental varieties. In: Shu QY (ed) Induced plant mutations in the genomics era. Proceedings of an international joint FAO/IAEA symposium. International Atomic Energy Agency, Vienna, Austria, pp 300–2
Datta SK, Banerji BK (1990) "Los Banos Variegata" – new double bracted chlorophyll variegated bougainvillea induced by gamma rays. J Nucl Agric Biol 19(2):134–136
Datta SK, Banerji BK (1994) 'Mahara variegata' – a new mutant of bougainvillea. J Nucl Agric Biol 23(2):114–116
Datta SK, Chakrabarty D, Mandal AKA (2001) Gamma ray induced genetic variation and their manipulation through tissue culture. Plant Breed 120:91–92
Datta SK, Misra P, Mandal AKA (2005) In vitro mutagenesis – a quick method for establishment of solid mutant in chrysanthemum. Curr Sci 88(1):155–158
Doorenbos J (1973) Breeding 'Elatior' begonia (*Begonia x hiemalis* Fotsch). Acta Hortic 31:127–131
Doorenbos J, Karper JJ (1975) X-ray induced mutations in *Begonia x Hiemalis*. Euphytica 24(1):13–19
Dowrick GJ, Bayoumi AE (1966) The induction of mutations in chrysanthemum using X- and gamma radiation. Euphytica 15:204–210
Dube S, Das PK, Dev AK, Bid NN (1980) Varietal improvement of *Dahlia* by gamma irradiation. Indian J Hortic 37(1):82–87
Fang JY (2011) In vitro mutation induction of Saintpaulia using ethyl methane sulfonate. Hortscience 46(7):981–984
FAO/IAEA (1977) Manual on mutation breeding second edition. Technical reports series No. 119. International Atomic Energy Agency, Vienna, 288pp
FAO/IAEA (2011) Mutation induction for breeding. International Atomic Energy Agency, Vienna. URL: http://mvgs.iaea.org/LaboratoryProtocols.aspx. Accessed 10 May 2011
FAO/IAEA (2017) FAO/IAEA mutant variety database of the joint FAO/IAEA division of nuclear techniques in food and agriculture. Available online: http://mvd.iaea.org/. Accessed March 2017
Fujii T, Ikenaga M, Lyman JT (1966) Radiation effects on *Arabidopsis thaliana*. II. Killing and mutagenic efficiencies of heavy ionizing particles. Radiat Bot 6:297–306
Goodhead DT (1995) Molecular and cell models of biological effects of heavy ion radiation. Radiat Environ Biophys 34:67–72
Gupta MN (1979) Improvement of some ornamental plants by induced somatic mutations at National Botanical Research Institute. In: Proceedings of symposium on the role of induced mutations in crop improvement, Sept. 10–13, Department of Genetics, Osmania University, Hyderabad, pp 75–92
Gupta MN, Nath P (1977) Mutation breeding in bougainvillea II. Further experiments with varieties 'Partha' and 'President Rosevilles Delight'. J Nucl Agric Biol 6:122–124
Hamatani M, Iitsuka Y, Abe T, Miyoshi K, Yamamot M, Yoshida S (2001) Mutant flowers of dahlia (*Dahlia pinnata* Cav.) induced by heavy-ion beams. RIKEN Accel Prog Rep 34:169
Hara Y, Abe T, Sakamoto K, Miyazawa Y, Yoshida S (2003) Effects of heavy-ion beam irradiation in rose (*Rosa Hybrid* cv. 'Bridal Fantasy'). RIKEN Accel Prog Rep 36:135
Hase Y, Yamaguchi M, Inoue M, Tanaka A (2002) Reduction of survival and induction of chromosome aberrations in tobacco irradiated by carbon ions with different LETs. Int J Radiat Biol 78:789–806
Heinze B, Schmidt J (1995) Mutation work with somatic embryogenesis in woody plants. In: Jain SM, Gupta K, Newton J (eds) Somatic embryogenesis in woody plants, vol 1. Kluwer Academic Publishers, Dordrecht, pp 379–IAEA 1970398
Hirono Y, Smith HH, Lyman JT, Thompson KH, Baum JW (1970) Relative biological effectiveness of heavy ions in producing mutations, tumors and growth inhibition in the crucifer plant, *Arabidopsis*. Radiat Res 44:204–223
Ibrahim R, Mondelaers W, Debergh PC (1998) Effects of X-irradiation on adventitious buds regeneration from *in vitro* leaf explants of *Rosa hybrid*. Plant Cell Tissue Organ Cult 54:37–44
Iizuka M, Yoshihara R, Hase Y (2008) Development of commercial variety of *Osteospermum* by a stepwise mutagenesis by ion beam irradiation. JAEA Takasaki Annu Rep 2007:65

Jacobs M (2005) Comparaison de l'action mutagen d'agents alkylants et des radiations gamma chez Arabidopsis thaliana. Radiat Bot 9:251–268

Jain SM (2006) Mutation-assisted breeding for improving ornamental plants. In: Proceedings of the 22nd international Eucarpia symposium section ornamentals: breeding for beauty. Issue 714, pp 85–98

Jain SM (2010) Mutagenesis in crop improvement under the climate change. Rom Biotechnol Lett 15(2):88–106

Jain SM, Spencer MM (2006) Biotechnology and mutagenesis in improving ornamental plants. In: Teixeira da Silva JA (ed) Floriculture, ornamental and plant biotechnology: AdVances and topical issues, vol 1. Global Science Books, Ilseworth, UK, pp 589–600

Jerzy M, Zalewska M (1996) Polish varieties of *Dendranthema grandiflora* Tzvelev and *Gerbera jamesonii* Bolus bred *in vitro* by induced mutations. Mutat Breed Newsl 42:19

Kanaya T, Saito H, Hayashi Y, Fukunishi N, Ryuto H, Miyazaki K, Kusumi T, Abe T, Suzuki K (2008) Heavy-ion beam-induced sterile mutants of verbena (*Verbena hybrida*) with an improved flowering habit. Plant Biotechnol 25:91–96

Kaul A, Kumar S, Thakur M, Ghani M (2011) Gamma ray induced *in-vitro* mutations in flower colour in *Dendranthema grandiflora* Tzelev. Floricult Ornament Bio-Technol 5:71–73

Kharkwal MC, Shu QY (2009) The role of induced mutations in world food security. In: Shu QY (ed) Induced plant mutations in the genomics era. Food and Agriculture Organization of the United Nations, Rome, pp 33–38

Killion DD, Constantin MJ (1971) Acute gamma irradiation of the wheat plant: effects of exposure, exposure rate, and developmental stage on survival, height, and grain yield. Radiat Bot 11:367–373

Killion DD, Constantin MJ, Siemer EG (1971) Acute gamma irradiation of the soybean plant: effects of exposure, exposure rate and developmental stage on growth and yield. Radiat Bot 11:225–232

Kumar G, Yadav RS (2010) Induced intergenomic chromosomal rearrangements in *Sesamum indicum* L. Cytologia 75(2):157–162

Kumar S, Prasad KV, Choudhary ML (2006) Detection of genetic variability among chrysanthemum radiomutants using RAPD markers. Curr Sci 90:1108–1113

Lagoda PJL (2009) Networking and fostering cooperation in plant genetics and breeding. Role of the joint FAO/IAEA division. In: Shu QY (ed) Induced plant mutations in the genomics era. Food and Agriculture Organization of the United Nations, Rome, pp 27–30

Lai YP, Huang J, Li J, Wu ZR (2004) A new approach to random mutagenesis *in vitro*. Biotechnol Bioeng 86(6):622–627

Lamseejan S, Jompuk P, Wongpiyasatid A, Deeseepan S, Kwanthammachart P (2000) Gamma-rays induced morphological changes in chrysanthemum (*Chrysanthemum morifolium*). Kasetsart J (Nat Sci) 34:417–422

Lamseejan S, Jompuk P, Deeseepan S (2003) Improvement of chrysanthemum var. 'Taihei' through *in vitro* induced mutation with chronic and acute gamma rays. J Nucl Soc Thail 2003(4):2–13

Laneri U, Franconi R, Altavista P (1990) Somatic mutagenesis of *Gerbera jamesonii* hybrid: irradiation and *in vitro* culture. Acta Hortic 28:395–402

Lett JT (1992) Damage to cellular DNA from particulate radiations, the efficacy of its processing and the radiosesitivity of mammalian cells. Radiat Environ Biophys 31:257–277

Liew OW, Ching P, Chong J, Li B, Asundi AK (2008) Signature optical cues: emerging technologies for monitoring plant health. Sensors 8:3205–3239

Magori S, Tanaka A, Kawaguchi M (2010) Physical induced mutations: ion beam mutagenesis. In: Kahl G, Meksem K (eds) The handbook of plant mutation screening. Wiley-VCH Verlag GmbH & Co, Weinheim

Maluszynski M, Ahloowalia BS, Sigurbjörnsson B (1995) Application of *in vivo* and *in vitro* mutation techniques for crop improvement. Euphytica 85:303–315

Maluszynski M, Nichterlein K, Zanten V, Ahloowalia BS (2000) Officially released mutant varieties – the FAO/IAEA database. Mutat Breed Rev 12:1–84

Mandal AKA, Chakrabarty D, Datta SK (2000a) Application of *in vitro* technique in mutation breeding of chrysanthemum. Plant Cell Tissue Organ Cult 60:33–38

Mandal AKA, Chakrabarty D, Datta SK (2000b) *In vitro* isolation of solid novel flower colour mutants from induced chimeric ray florets of chrysanthemum. Euphytica 114:9–12

Matsubara H, Suda H, Sawada Y, Nouchi I (1975) An-ozone-sensitive strain from the mutants induced by gamma ray irradiation of the Begonia rex, variety Winter Queen. Agric Hortic 50(6):811–812

McCallum CM, Comai L, Greene EA, Henikoff S (2000a) Targeted screening for induced mutations. Nat Biotechnol 18(4):455–457

McCallum CM, Comai L, Greene EA, Henikoff S (2000b) Targeting induced local lesions in genomes (TILLING) for plant functional genomics. Plant Physiol 123(2):439–442

Medina FIS, Amano E, Tano S. (2004) FNCA mutation breeding manual. Forum for Nuclear Cooperation in Asia. URL: http://www.fnca.mext.go.jp/english/mb/mbm/e_mbm.html. Accessed 10 May 2011

Mei M, Deng H, Lu Y, Zhuang C, Liu Z, Qiu Q, Qiu Y, Yang TC (1994) Mutagenic effects of heavy ion radiation in plants. Adv Space Res 14:363–372

Micke A, Donini B, Maluszynski N (1990) Induced mutations for crop improvement. In: Mutation breeding review (FAO/IAEA), no. 7, Vienna (Austria), IAEA, p 41

Mikkeksen JC, Ryan J, Constantin MJ (1975) Mutation breeding of Rieger's Elatior begonias. Am Hortic 54(3):18–21

Minano HS, Gonzalez-Benino ME, Martic C (2009) Molecular characterization and analysis of somaclonal variation in chrysanthemum varieties using RAPD markers. Sci Hortic 122:238–243

Misra P, Datta SK, Chakbarty D (2004) Mutation in flower colour and shape of *Chrysanthemum morifolium* induced by gamma irradiation. Biol Plant 47:153–156

Miyazaki K, Suzuki K, Abe T, Katsumoto Y, Yoshida S, Kusumi T (2002) Isolation of variegated mutants of *Petunia* hybrid using heavy-ion beam irradiation. RIKEN Accel Prog Rep 35:130

Miyazaki K, Suzuki K, Iwaki K, Kusumi T, Abe T, Yoshida S, Fukui H (2006) Flower pigment mutations induced by heavy ion beam irradiation in an interspecific hybrid of *Torenia*. Plant Biotechnol 23:163–167

Morita R, Kusaba M, Iida S, Yamaguchi H, Nishio T, Nishimura M (2009) Molecular characterization of mutations induced by gamma irradiation in rice. Genes Genet Syst 84:361–370

Nagatomi S (2003) Development of flower mutation breeding through ion beam irradiation. Res J Food Agric 26:33–38

Nagatomi S, Degi K (2009) Mutation breeding of Chrysanthemum by gamma field irradiation and *in vitro* culture. In: Shu QY (ed) Induced plant mutations in genomic era. Food and Agriculture Organization of the United Nations, Rome, pp 258–261

Nagatomi S, Degi K, Yagaguchi M, Miyahira E, Skamoto M, Takaesu K (1993) Six mutant varieties of different flower colour induced by floral organ culture of chronically irradiated chrysanthemum plants. Tech News-Inst Radiat Breed 43:1–2

Nagatomi S, Miyahira E, Degi K (1996a) Combined effect of gamma irradiation methods and in vitro explants sources on mutation induction of flower colour. Gamma field symposia No. 35, Institute of Radiation Breeding, NIAR, MAFF, Japan

Nagatomi S, Miyahira E, Degi K (1996b) Induction of flower mutation comparing with chronic and acute gamma irradiation using tissue culture technique in *Chrysanthemum morifolium*. Ramat. Acta Hortic 508:69–73

Nagatomi S, Tanaka A, Kato A, Watanabe H, Tano S (1996c) Mutation induction of chrysanthemum plants regenerated from *in vitro* cultured explants irradiated with C ion beam. TIARA Annu Rep 5:50–52

Nagatomi S, Tanaka A, Tano S, Watanabe H (1997) Chrysanthemum mutants regenerated from in vitro explants irradiated with 12C5+ ion beam. Technical News of Institute of Radiation Breeding, No. 60

Nagatomi S, Tanaka A, Watanabe H, Tano S (1998) Enlargement of potential chimera on Chrysanthemum mutants regenerated from 12C5+ ion beam irradiated explants. JAERI-Rev 97-015:48–50

Nagatomi S, Miyahira E, Degi K (2000) Combined effect of gamma irradiation methods *in vitro* explant sources on mutation induction of flower colour in *Chrysanthemum morifolium* Ramat. Gamma field symposia, No. 35, 1996 Institute of Radiation Breeding, NIAR, MAFF, Japan

Nagatomi S, Watanabe H, Tanaka A, Yamaguchi H, Degi K, Morishita T (2003) Six mutant varieties induced by ion beams in chrysanthemum. Institute of Radiation Breeding Technical News 65

Nagayoshi S (2003) Radiation breeding in Kagoshima prefecture – breeding of 'chrysanthemum with a few axillary flower buds' by ion beams. Radiat Ind 98:10–16

Okamura M, Yasuno N, Ohtsuka K, Tanaka A, Shikazono N, Hase Y (2003) Wide variety of flower-colour and –shape mutants regenerated from leaf cultures irradiated with ion beams. Nucl Instrum Methods Phys Res B206:574–578

Okamura M, Tanaka A, Momose M, Umemoto N, Teixeira da Silva JA, Toguri T (2006) Advances of mutagenesis in flowers and their industrialization. In: da Silva JAT (d). Floriculture, ornamental and plant biotechnology Vol. I. Global Science Books, Isleworth, 619–628

Pathirana R (2011) Plant mutation breeding in agriculture. Perspectives in Agriculture, Veterinary Science, Nutrition and Natural Resources, CAB 6, No. 032
Predieri S, Zimmerman RH (2001) Pear mutagenesis: *in vitro* treatment with gamma-rays and field selection for productivity and fruit traits. Euphytica 3:217–227
Rego LV, Faria RT (2001) Tissue culture in ornamental plant breeding review. Crop Breed Appl Biotechnol 1:285–300
Rodrigo RL, Alvis HA, Neto TA (2004) *In vitro* mutation of chrysanthemum (*Dendranthema grandiflora* Tzvelev) with ethyl methane sulphonate (EMS) in immature floral pedicels. Plant Cell Tissue Organ Cult 77:103–106
Sasaki K, Aida R, Niki T, Yamaguchi H, Narumi T, Nishijima T, Hayashi Y, Ryuto H, Fukunishi N, Abe T, Ohtsubo N (2008) High-efficiency improvement of transgenic torenia flowers by ion beam irradiation. Plant Biotechnol 25:81–89
Schaart JG, Van de Wiel CCM, Lotz LAP, Smulders MJM (2016) Opportunities for products of new plant breeding techniques. Trends Plant Sci 21:438–448. https://doi.org/10.1016/j.tplants.2015.11.006
Schum AR (2003) Mutation breeding in ornamentals: and efficient breeding method. Acta Hortic 612:47–60
Schum A, Preil W (1998) Induced mutations in ornamental plants. In: Jain SM, Brar S, Ahloowli BS (eds) Somaclonal variation and induced mutations in crop improvement. Kluwer Academic Publishers, Dordrecht, pp 333–366
Shibata M, Kishimoto S, Hirai M, Aida R, Ikeda I (1998) Analysis of the periclinal chimeric structure of chrysanthemum sports by random amplified polymorphic DNA. Acta Hortic 454:347–353
Shikazono N, Tanaka A, Kitayama S, Watanabe H, Tano S (2002) LET dependence of lethality in *Arabidopsis thaliana* irradiated by heavy ions. Radiat Environ Biophys 41:159–162
Shikazono N, Yokota Y, Kitamura S, Suzuki C, Watanabe H, Tano S, Tanaka A (2003) Mutation rate and novel tt mutants of Arabidopsis thaliana induced by carbon ions. Genetics 163:1449–1455
Shu QY (2009) Turning plant mutation breeding into a new era: molecular mutation breeding. In: Shu QY (ed) Induced plant mutations in the genomic era. Food and Agriculture Organization of the United Nations, Rome, pp 425–427
Simard MH, Michaux-Ferriera N, Silvy A (1992) Variations of carnation (Dianthus caryophllus L.) obtained by organogenesis from irradiated petals. Plant Cell Tissue Organ Cult 29:37–42
Sparrow AH, Sparrow RC, Schairer LA (1960) The use of x-rays to induce somatic mutations in Saintpaulia. Afr Violet Mag 13:32–37
Sparrow AH, Rogers AF, Susan SS (1968) Radiosensitivity studies with woody plants. I. Acute gamma irradiation survival data for 28 species and prediction for 190 species. Radiat Bot 8:149–186
Sripichtt P, Nawata E, Shigenaga S (1988) The effects of exposure dose and dose rate of gamma radiation on *in vitro* shoot-forming capacity of cotyledon explants in red pepper. Jpn J Breed 38:27–34
Srivastava R, Datta SK, Sharma SC, Roy RK (2002) Gamma rays induced genetic variability in Bougainvillea. J Nucl Agric Biol 31(1):28–36
Sugiyama M, Hayashi Y, Fukunishi N, Ryuto H, Terakawa T, Abe T (2008a) Development of flower colour mutant of *Dianthus chinensis* var. semperflorens by heavy-ion beam irradiation. RIKEN Accel Prog Rep 41:229
Sugiyama M, Saito H, Ichida H, Hayashi Y, Ryuto H, Fukunishi N, Terakawa T, Abe T (2008b) Biological effects of heavy-ion beam irradiation on cyclamen. Plant Biotechnol 25:101–104
Suprasanna P, Nakagawa H (2012) Mutation breeding of vegetatively propagated crops. In: Shu QY, Forster B, Nakagawa H (eds) Plant mutation breeding and biotechnology. Joint FAO/IAEA Programme, Nuclear Techniques in Food and Agriculture, CABI, Wallingford, UK, pp 347–369
Suprasanna P, Jain SM, Ochatt SJ, Kulkarni VM, Predieri S (2012) Applications of *in vitro* techniques in mutation breeding of vegetatively propagated crop. In: Shu QY, Forster B, Nakagawa H (eds) Plant mutation breeding and biotechnology. Joint FAO/IAEA Programme, Nuclear Techniques in Food and Agriculture, CABI, Wallingford, UK, pp 371–385
Tanaka A (1999) Mutation induction by ion beams in Arabidopsis. Gamma Field Symp 38:19–27
Tanaka A (2009) Establishment of ion beam technology for breeding. In: Shu QY (ed) Induced plant mutations in the genomics era. Food and Agriculture Organization of the United Nations, Rome, pp 216–219
Tanaka A, Tano S, Chantes T, Yokota Y, Shikazano N, Watanabe H (1997) A new Arabidopsis

mutant induced by ion beams affect flavonoid synthesis with spotted pigmentation in testa. Genes Genet Syst 72:141–148
Tanaka A, Shikazono N, Hase Y (2010) Studies on biological effects of ion beams on lethality, molecular nature of mutation, mutation rate, and spectrum of mutation phenotype for mutation breeding in higher plants. J Radiat Res 51:223–233
Teixeira da Silva JA (2004) Ornamental chrysanthemum: improvement by biotechnology. Plant Cell Tissue Organ Cult 79:1–8
Till BJ, Cooper J, Tai TH, Colowit P, Greene EA, Henikoff S et al (2007) Discovery of chemically induced mutations in rice by TILLING. BMC Plant Biol 7:19
Ueno K, Nagayoshi S, Hase Y, Shikazono N, Tanaka A. 2003. Effects of ion beam irradiation on the mutation induction from chrysanthemum leaf disc culture. TIARA Annu Rep 2002. JAERI-Rev 2003-033: 52–54
Ueno K, Nagayoshi S, Hase Y, Shikazono N, Tanaka A (2004) Additional improvement of chrysanthemum using ion beam re-irradiation. TIARA Annu Rep 2003. JAERI Rev 2004-025: 53–55
Ueno K, Shirao T, Nagayoshi S, Hase Y, Tanaka A (2005) Additional improvement of *Chrysanthemum* using ion beam re-irradiation. TIARA Annu Rep 2004:60
Van Harten AM (1998) Mutation breeding: theory and practical applications. Cambridge University Press, Cambridge, UK
Velmurugan M, Rajamani P, Paramaguru R, Gnanam JR, Bapu K, Harisudan C, Hemalatha P (2010) *In vitro* mutation in horticultural crops. Agric Rev 31(1):63–67
Ward JF (1994) The complexity of DNA damage: relevance to biological consequences. Int J Radiat Biol 66:427–432
Watanabe H, Toyota T, Emoto K, Yoshimatsu S, Hase Y, Kamisoyama S (2008) Mutation breeding of a new chrysanthemum variety by irradiation of ion beams to ‘Jinba’. JAEA Takasaki Annu Rep 2007:81
Wi SG, Chung BY, Kim JH, Baek MH, Yang DH, Lee JW, Kim JS (2005) Ultrastructure changes of cell organelles in Arabidopsis stem after gamma irradiation. J Plant Biol 48(2):195–200
Wolf K (1996) RAPD analysis of sporting and chimerism in chrysanthemum. Euphytica 89:159–164
Yamaguchi H (2013) Characteristics of ion beams as mutagens for mutation breeding in rice and chrysanthemums. JARC 47(4):339–346. http://www.jircas.affrc.go.jp
Yamaguchi H, Nagatomi S, Morishita T, Degi K, Tanaka A, Shikazono N, Hase S (2003) Mutation induced with ion beam irradiation in rose. Nucl Instrum Methods Phys Res B206:561–564
Yamaguchi H, Shimizu A, Degi K, Morishita T (2008) Effects of dose and dose rate of gamma ray irradiation on mutation induction and nuclear DNA content in chrysanthemum. Breed Sci 58:331–335
Yamaguchi H, Shimizu A, Hase Y, Tanaka A, Shikazono N, Degi K, Morishita T (2010) Effect of ion beam irradiation on mutation induction and nuclear DNA content in chrysanthemum. Breed Sci 60:398–404
Yang TC, Tobias CA (1979) Potential use of heavy-ion radiation in crop improvement. Gamma Field Symp 18:141–154
Zaiton A, Affrida AH, Shakinah S, Nurul Hidayah M, Nozawa S, Narumi I, Hase Y, Oono Y (2014) Development of new *Chrysanthemum morifolium* Pink mutants through ion beam irradiation. JAEA Rev 2014-050:108
Zalewska M, Jerzy M (1997) Mutation spectrum in *Dendranthema grandiflora* Tzvelev after *in vivo* and *in vitro* regeneration of plants from irradiated leaves. Acta Hortic 447:615–618
Zalewska M, Lema-Ruminska J, Miller N (2007) *In vitro* propagation using adventitious buds technique as a source of new variability in chrysanthemum. Sci Hortic 113:70–73
Zalewska M. Miler N, Tymoszuk A, Drzewiecka B, Winiecki J (2010) Results of mutation breeding activity on *Chrysanthemum* × *grandiflorum* (Ramat.) Kitam. In Poland, EJPAU 13(4), 27. Available Online: http://www.ejpau.media.pl/volume13/issue4/art-27.html

第9章　观赏植物分子育种技术新进展

Marinus J. M. Smulders and Paul Arens

摘要　由于观赏植物在发现研究、标记开发、遗传图谱构建和性状定位等遗传育种研究方面进展缓慢，加上遗传背景复杂、远缘杂交、长的童期、多倍体化以及较大的基因组等，遗传和分子育种工具就农业与园艺作物而言发展相对落后。

过去10年，观赏植物分子工具的发展潜能得到了显著提升：①基于基因组和转录组测序，第二代测序技术可以规模化产出单核苷酸多态性标记（SNP）；②高效和自动化SNP基因分型系统；③不同倍性基因分型数据分析方法和软件的开发发掘性状与标记的关联，以及开发分子标记辅助育种技术的工具。较低价格、少数标记对大量的样本进行基因分型变得可能。近几年的主要挑战是利用这些技术加速观赏植物育种。

越来越多观赏植物及其近缘种基因组序列的公布，使分子标记辅助育种到性状变异基因的发掘变为现实。功能等位基因变异信息有可能考虑定向的生物合成与代谢途径，如不同花色和香气的结合。

新的植物育种技术（又称精确育种技术）增加了新的可能性去指导育种过程。尤其是基因编辑技术或基因组编辑技术（CRISPR/Cas）可用于增加功能变异，但是在观赏植物中应用具有很大的挑战性，需要可利用的候选基因序列信息、现成的转化和再生体系等。

关键词　单核苷酸多态性标记；数量性状基因座；分子标记辅助育种；DNA信息育种；功能变异；植物育种

9.1　引　　言

由于众多因素，观赏植物遗传和分子育种手段的发展已经落后于主要农作物与园艺作物。这些因素包括主要农作物和园艺作物在发掘研究、标记开发、遗传图谱构建和基因定位等方面的发展是相对更快的。而众多观赏植物远缘杂交且进行无性繁殖，具有较高水平的遗传多样性，具有较长的童期，大多是多倍体而具较大的基因组。另外，观赏植物育种过程中，表型是重要的目标性状，性状的评估不需要分子标记，因此这方面的投资是非常有限的。目前，这一现象正逐步得到改变，育种者开始关注难以评估及多基因调控性状。例如抗病性、茎生长、开花时间及花的大小等（Smulders et al. 2012）。

在过去的10年，基于以下三个方面的发展，分子工具的发展潜能得到显著提高：①第二代测序技术的进步，使得基于基因组及转录组高通量开发单核苷酸多态性标记变得相对

M. J. M. Smulders (✉) · P. Arens
Wageningen University & Research, Wageningen, The Netherlands
e-mail: rene.smulders@wur.nl

容易；②开发了高效且自动化的基因分型系统；③即便是多倍体，数据分析方法和软件已经具备发掘性状与标记间关联以及开发分子标记辅助育种技术的功能。因此，对于大群体，利用少数基因位点进行低成本分型变为现实。未来几年的挑战是将分子工具应用到观赏植物育种中，这些分子工具包括：开发标记、标记和性状间关联以及应用于加速育种。因为植物保护药剂的使用压力逐日倍增，实际应用标记很多可能和抗病性相关。抗病能力评估是比较困难和昂贵的，除非有 DNA 标记的支持。例如，球根作物抗镰刀菌。然而番茄抗镰刀菌是被绑定在 NBS-LRR 型抗病基因控制的。观赏植物球根百合和郁金香抗镰刀菌是多基因控制性状，最少有 6 个数量性状基因座（QTL）参与调控（Shahin et al. 2010；Tang et al. 2015）。和这些位点联合遗传的标记可以用于筛选后代，筛选可以在幼苗阶段进行，这样可以加快育种进程尤其是对于具有较长童期的球根观赏植物（Smulders et al. 2012）。非常遗憾的是，由于较大的 QTL 区间以及非常高的遗传多样性，很难做到将 QTL 转化为可靠的标记应用于 MAS。

随着观赏植物种类及其近缘种基因组测序的完成，从 QTL 区间分析及标记开发到候选基因定位。功能等位基因变异信息不仅有可能定向考虑如不同花色和香气的结合的生物合成与代谢途径，而且可以开发广谱抗病寄主（Fu et al. 2016）。

最终，新的植物育种技术（又称精确育种技术）有可能直接指导观赏植物育种过程。尤其是基于 CRISPR/Cas 系统的基因编辑技术有助于为育种者增加遗传变异。但是，该技术的应用也是有挑战的，不仅需要考虑细胞生物学，因为有效的转化体系特别是再生体系是基本要求，而且可利用的作物基因序列信息对于目标设计是必需的。

在本章节，我们将重点阐述相关进展和明确观赏植物育种可能的方法。在适当的地方我们将参考有用的文献。

9.2 DNA 标记

DNA 标记是指可以在杂交中进行遗传追踪的任何 DNA 序列差异。部分可能统计上和表型性状相关联。在分子标记辅助育种这些标记可以作为性状（单个主效基因或多个数量性状位点）的替代。

单核苷酸多态性（SNP）标记是指 DNA 序列上单个核苷酸的差异，它们是遗传变异中最广泛应用的标记，因为 SNP 标记在基因组中广泛存在并且可以进行自动探测，这与需要人工计数的传统标记是完全相反的。因此我们认为，即便是传统标记包含更多的遗传信息，但 SNP 标记是最终选择（如 SSR 标记在单个位点常常具有多个等位基因而 SNP 标记只有 2 个等位基因，这对于多倍体作物来说是一个限制，因为多倍体作物单个基因位点往往有多个等位基因）。

为了鉴定育种作物的 SNP 标记，可以对两个或者更多的个体进行比较。较低深度的测序对于可靠 SNP 标记的探测和其后续在渐渗育种中应用通常已经足够。然而，大多观赏植物属于远交作物，所以是杂合体。在杂合体中，即便是单个二倍体基因型其所具有的两条染色体也是不同的，因此可以开发单个植株不同姐妹染色单体间 SNP 标记。分析两个基因型（如杂交群体的两个亲本间）SNP 标记时，获得的 SNP 标记既包括单个基因型内的又包括两个基因型间的标记。对于开发更多基因型间 SNP 标记而言，深度测序是开发可靠 SNP 标记、区别序列错误和旁系同源等位基因间序列差异所必需的，如转录组测序（Shahin et al. 2012a；Koning-Boucoiran et al. 2015；Van Geest et al. 2017b）。

9.3　高通量测序的发展

自首个人类基因组计划完成以来，测序技术得到了突飞猛进的发展（Schmutz et al. 2004）。得益于第二代测序技术的发展，DNA 测序已经变得更快速和经济。测序技术的开发日新月异，更新、更便宜的技术开发了新的应用领域。这些应用不仅包括因为技术昂贵尚未开展的应用，也包括一些原来应用其他技术的新领域，或者是不存在和是不可能的。

Goodwin 等（2016）阐述了不同技术在增加产出和短 reads 以及准确率间的平衡。随着产出的增加，开销急剧降低，单个人类基因组测序纯的花费目前仅为 1000 美元，而首个人类基因组计划的支出为 1 亿美元（https://www.genome. gov/27565109/the-cost-of-sequencing-a-human-genome/）。但是，测序只是开销的一部分，数据的储存和分析也需要支出（Muir et al. 2016）。当我们所讨论的便宜主要考虑的是 reads 产出支出的降低，往往这方面的降低超过计算机存储费用的降低。幸运的是，伴随着新的测序技术的出现，基因组组装、注释和变型探测等方面的信息分析水平得到了显著的提高，这将显著促进杂合或大基因组的测序，也使进行观赏植物物种高质量基因组测序变得可行。

开发 SNP 标记，利用 Illumina 技术进行短 reads 测序是目前最便宜的手段。基于 PacBio 或 MinION 技术长 reads 测序（1000～10 000bp）约需要 10 倍的花销。这些长片段读数常常结合短片段读数来进行高质量基因组的组装、RNA 拼接变异体分析或大范围的单倍型分析。

开发 SNP 标记，仅仅需要测序小部分基因组，但是需要测序基因组的相同区间以便获得可靠的 SNP 标记，避免不同旁系同源基因在同一个基因组内比对，也就是发生错位比对。其中的策略是对特殊候选基因进行 PCR 扩增。如果研究目的是获得全基因组范围内的标记，对于小基因组作物来说，进行高通量测序技术是相对容易的。3Gb 左右基因组大小的物种可以测序少数基因组，然后进行两两比对或者建立两个序列池进行 SNP 标记开发，如 BSA 分析途径。对于较大基因组而言，全基因组重测序还是相对昂贵的，但 BSA 分析途径是可行的。

降低 DNA 复杂程度的简单方法是，将感兴趣的 DNA 限定在基因组的一小部分区间，测序单个或几个组织 mRNA 生成的 cDNA，即通常所说的 RNA 测序。RNA 测序可以产出庞大数量的转录本，加之充足的测序深度可确保高质量 SNP 标记的鉴定（Shahin et al. 2012a；Kim et al. 2014）。该方法的优点主要体现在：①SNP 标记位于有丝分裂期间染色体发生重组的基因区域；②基因内部很少有阻碍基因分型的旁翼 SNP 出现；③对于建立孤儿作物基因组资源是适宜的（如候选基因途径），因为不需要更多已知的信息。仅仅需要从单个或多个组织提取 RNA，合成 DNA，再完成第二代测序即可。毫无疑问的是，目前许多物种通过 RNA 测序研究已经开发了数千万个 SNP 标记，其中观赏植物包括六出花（*Alstroemeria*）、秋海棠（*Begonia*）、贝母（*Caladium*）、菊花（*Chrysanthemum*）、非洲菊（*Gerbera*）、百合（*Lilium*）、郁金香（*Tulipa*）、梅（*Prunus mume*）、蝴蝶兰（*Phalaenopsis*）、蔷薇（*Rosa*）、蝇子草属（*Silene*）、马蹄莲属（*Zantedeschia*）。另外一种策略是降低基因组复杂性，并已经在菊花中应用 DArT 测序（捕获目标序列的杂交）和 SLAF 测序（特定位点扩增片段测序）（Chong et al. 2016）。

观赏植物由于高杂合度组装短 reads 有可能是困难的，目前绝大部分组装利用 Trinity。鉴定 SNP 的软件已经开发出来，可以鉴定真实或假阳性 SNP（如 QualitySNPng）（Nijveen et al. 2013）。假阳性 SNP 可能由于测序错误，或者由基因组旁系同源基因比对产生，无论是二倍体还是多倍体都是个问题。该问题可以通过测序一个群体或种质资源库中的多个基因型得到一定程度的消除（Shahin et al. 2012b；Van Geest et al. 2017b）。

9.4 开发基因分型技术

对于观赏植物来说，首要的基因分型是全基因组测序，但对于育种而言，代表部分基因组的标记就能满足需求。对于 SNP 标记基因分型，根据 SNP 标记数量发展了多种方法。

数千个标记可以获得高密度的基因组覆盖度，较为经济的方案是采用 SNP 芯片技术。目前有两个系统可以采用，分别为 Illumina（如玉米、水稻、马铃薯和苹果等）和 Affymetrix（如苹果、草莓、蔷薇和菊花等）提供的技术支持，个性化的微阵列也可以进行预定。Illumina 公司最小的样本数为 1200 个基因型，Affymetrix 公司最小的样本数为 488 个基因型。

SNP 芯片已经广泛应用于遗传图谱构建和 QTL 定位等，如蔷薇（Vukosavljev et al. 2016；Bourke et al. 2017）、菊花（Van Geest et al. 2017a，2017b）等。蔷薇 SNP 芯片还应用于关联作图研究（Schulz et al. 2016；Nguyen et al. 2017）。

并不是所有的应用都需要如此大数量的标记。如果用于鉴定，且等位基因频率是平衡的，几百个 SNP 标记已经足够。一旦 QTL 区间确定下来，很少数量的 SNP 关联标记就能追踪后代的对应基因组区间。渐渗育种依靠染色体筛选，对于负选择而言，每条染色体有 4 个分布理想的标记就足够了。

当需要数百个 SNP 标记对大的群体进行分型时，KASP（Koning-Boucoiran et al. 2012；Holdsworth and Mazourek 2015）或者 Fluidigm（Jung et al. 2017）技术是相对经济的选择。扩增子测序最近已经发展起来，联合 KASP 标记和 SNP 基因芯片是非常经济的。

9.5 方法和软件开发

在每个 SNP 位点，SNP 基因分型芯片输出两种等位基因类型，对于每个标记，杂交探针输出两种不同荧光信号 A 和 B，对应 2 种等位基因 A 和 B。在二倍体中，从 SNP 芯片或者 KASP 芯片得到的信号必须分为三个基因型等级（纯合位点 AA 和 BB 以及杂合位点 AB）。芯片公司的软件系统可以自动完成信号的收集，即便如此，建议对整个自动程序进行仔细检查，尤其是关键的标记。对于高杂合度的作物而言，相对多的无效等位基因可能出现。例如，等位基因因为旁翼序列 SNP 而没有杂交成功导致未获得任何一种信号。完成记录后部分信息仍然可以应用（Tang et al. 2015）。为了便于探测和自动重新计数，特定的工具已经开发出来（Di Guardo et al. 2015）。随后可以利用这些数据通过 Joinmap 软件（版本 6 可以处理更多的 SNP 标记）或 MSDmap 软件（Preedy and Hackett 2016）进行遗传图谱构建。

对于多倍体作物而言，首先需要根据等位基因数量对 SNP 信号进行归类。FitTetra 针对四倍体和 FitPloy 针对所有的倍性类型已经开发应用。随后，遗传图谱构建可以在 ploymapR 中完成，该软件利用 MSDmap 对标记进行排序。在 ploymapR 中，蔷薇和菊花分别作为四倍体和六倍体的例子进行整合图谱的构建。利用整合图谱，每个后代基于血缘的身份作为连锁群可能性可以进行评估。对于四倍体，特定的 IBD 评估软件已经开发出来。对于其他倍性水平，基于最大信息标记的鉴定途径可以进行连锁评估。

当鉴定 QLTs，标记或基因组区间与相关性状的关联，或者鉴定标记和优势等位基因的关联。从根本上来说，需要给育种者提供易于接受的界面给予所需步骤的方案。对于二倍体远交物种来说，最近几年在这些方面取得了进展。软件和工具的进展都快于多倍体物种。谱系重建的方法，多育种群体 QTL 分析，通过谱系种质资源定位 SNP 标记，指定单倍型等软件都已经可以应用，如图 9.1 所示（月季育种信息管理系统）。

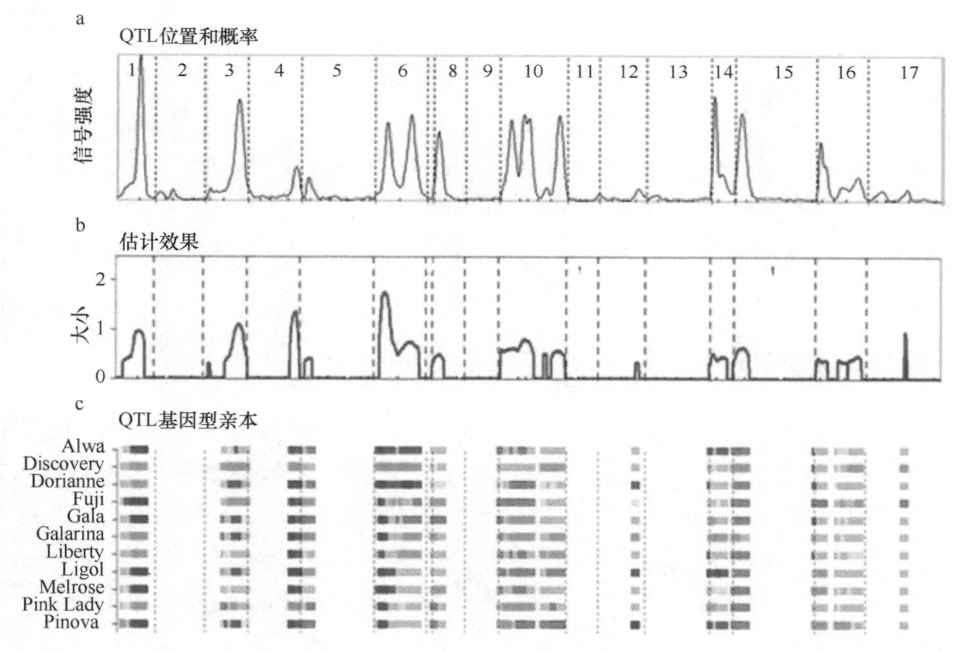

图 9.1 FlexQTL™（www.flexqtl.nl）软件 27 个远交同胞群体 QTL 分析定位 1 个园艺性状的输出案例。（a）QTL 的概率和在基因组上的位置。（b）观察尺度上的影响评估。（c）部分亲本相关染色体评估的 QTL 基因型，染色体被垂直的线分开。标记的位置在水平轴用垂直符号显示出来。QTL 基因型直观地用水平的红色、绿色和蓝色条块展示。红色表示 QQ，绿色表示 Qq，蓝色标识 qq，因此假设 QTL 是双等位基因。Q 和 q 分别与高表型值与低表型值关联。除此以外，QTL 基因型信息可能支持和良好结合的亲本选择。QTL 概率通过波峰面积而不是波峰的高度，数量统计采用贝叶斯因子。本研究的统计分析中，标记的密度相对于群体大小是受限制的。因此，利用 SNP 芯片增加标记的密度很可能显著提升 QTL 位置的精度和准确度，缩小 QTL 区间（也就是波峰的宽度）。多亲本群体的概念和软件的运行由 Bink 等（2014）验证。直到 2017 年底，已经有 17 篇应用该软件的学术论文发表

9.6 如何使用标记

Peace（2017）曾制定了将基因组研究产出应用于常规育种的系统，被他称为 DNA 信息育种（DNA-informed breeding）。该育种系统相当于分子标记辅助育种，而且包含了运用共享的中性标记或者 DNA 序列、亲本测定和身份鉴定。另外，DNA 信息育种因为不需要去解释什么是标记、标记如何工作等问题更容易交流，他认为有以下 5 个步骤。

1）确立利用 DNA 信息的育种请求。

2）改变工具以适应当地育种机械。

3）确定高效的应用计划。

4）访问提供 DNA 诊断的有效服务机构。

5）获得 DNA 信息育种经验。

DNA 信息可以被用于一系列的目的，我们列出一些不完全但是有可能的应用。

9.6.1 品种保护

作为育种者，DNA 信息对于侵权的确认是非常有帮助的。对某些作物，如月季和蝴蝶兰，DNA 标记已经用于 DUS 测试通过比较筛选品种，表型相似的品种可能是突变体，从原生品种衍生出来的一系列突变体。分子标记可以准确区分杂交种子品种基于唯一的基因型，然而突变品种的标记序列是具有相同的。11 个具有丰富多态性的 microsatellite 位点可以完全鉴定 700 个除了突变的月季品种（Smulders et al. 2009）。100 个或更多的 SNP 标记应该具有相似或者更好的分辨率。

9.6.2 育种系统中的身份检测

在许多大田作物中，亲本及筛选的后代必须开展一年一度的检测，繁殖公司确定可能的材料混杂及出现的错误标签。这样的检测是非常有必要且经济的，因为这样的错误是很难完全避免的，造成的损失也可能是巨大的。对种质资源材料大规模繁殖这样的检测也是经济的。检测需要相同数量的标记以便对品种区分鉴定。基因型分析也可以为校正亲本的后代提供信息。当我们研究一个庭院月季群体时，发现部分后代是母本的自交结果（Vukosavljev et al. 2016）。非典型（不同的父本或者错误的标签）后代也可以快捷地鉴定出来。

9.6.3 种质资源遗传结构

DNA 信息也可以提供所用种质资源遗传结构，如检验品种和筛选的历史谱系记录。出现育种标识，如基因组区间某一标记等位基因高频率出现，说明是育种和筛选的目标，即便性状基础本身可能还不清楚。最后 DNA 信息还提示可用的遗传多样性。对于蔷薇属而言，这些信息可以增加将新性状从二倍体引入四倍体切花月季的效率。另外，DNA 信息使材料在生产中维持遗传多样性成为可能。对于应用繁殖的树木、灌木等多年生物种尤其重要。一些改良的品种需要保存起来，因为这些遗传多样性可能能忍受将来的全球变暖或其他变化。

9.6.4 实生苗筛选

分子标记显而易见的应用是在实生苗筛选中进行性状预估。这对于等位基因联合不可见的表型是有用的。也就是说，如 1 个杂交组合的目的是在后代中产出纯合抗性植株，或者是在亲本中增加抗病基因，而后代遗传了 2 个抗性等位基因和 1 个抗性等位基因的都具有抗性，所以它们不能基于表型进行区分。其他的应用可能是性状非常昂贵不容易进行评估，或者性状评估只能在早期的繁殖阶段完成。

为优化实生苗筛选，筛选要尽可能地提早，减少在舍弃后代上花费的时间和空间。实生苗很小且移栽之前有一个非常短的窗口期。方便的形式是播种在 8 × 12 孔的穴盘，配套相对应的 96 孔微升平板用于实生苗叶片打孔。打孔获取的材料可以自己分析或者送给专业的服务公司开展 DNA 提取和标记分析。可以根据分析结果直接筛选实生苗。唯一的要求是盘子必须是编号和定向的。采用的生产时间是关键，结果信息必须在实生苗移栽前获得。

9.6.5 亲本筛选

相比实生苗筛选，亲本筛选常常可能更有效及更易于实施。在亲本筛选育种系统中，所有可能的亲本遗传组成信息都被用于优化亲本组合。这使得育种者可以设计更好的杂交组合。亲本筛选可以基于全基因组数据，利用基因组筛选来优化遗传信息利用。也可以基于亲本的家谱信息，因为 FlexQTL 已经对多种木本水果种类进行了详细说明。

亲本筛选同样可以应用于单个性状的筛选，如少数几个主效基因控制的 1 个重要 QTL。有时选择 1 个理想的亲本或者排除某一特定的遗传背景，可以避免在候选实生苗中筛选到希望排除的性状。

一个例子是利用标记确认两个独立遗传资源分离所得的性状是否具有共同的起源。在多倍体作物中是尤其有意义的，不同 QTL 的效能很难从表型上加以区分，因为开展大规模基因型和环境互作性状的表型实验并非易事。

亲本筛选的另一个目标，尽管非常有限的育种可能性，基于 DNA 信息时其常被用于加速引入供者基因到重复遗传背景，在观赏植物育种中被用于重要单个显性基因，如抗病基因、重复开花、化型突变体如重瓣花基因。接下来筛选包含感兴趣性状的引入片段，贯穿于基因组的标记被用于筛选回交后代中含供体基因组的其他个体，因为品种基因组背景可能恢复。在远交多倍体中也是可能的，连续的回交通常不是和单个植株，但是植株来自相同的栽培植物基因库。

9.6.6 减数分裂和偏分离

染色体配对的不规则导致了性状的分离，导致了杂交后代出现期望的等位基因频率、最优的杂交组合方向以及可开展遗传分析。基于 SNP 标记在全同胞家系中的共分离，可以对减数分裂进行检查。例如，在 1 个四倍体杂交组合中，四体染色体配对是遵循一定规则的，但是在1个亲本的一个区域是二体配对的（Bourke et al. 2017）。因此，后代1号染色体并不是全部理想地来自母本的等位基因结合。Van Geest等（2017b）证实菊花是六倍体，表现为多倍体的遗传。双倍的减少事件提供了快速筛选某一具有优良性状的特定染色体片段的可能性。

9.7 基因组测序及候选基因途径

下一步利用和性状关联的标记进行替代，可以尝试放大和评估这些性状基因的功能变异。生物合成和调控通路功能等位基因变异的信息让改变这些通路的方向成为可能，如不同花色和香气的结合。

9.7.1 第一个观赏植物基因组及其发展

越来越多的作物完成全基因组测序，包括了部分观赏植物种类，如香石竹（Yagi et al. 2014）、黄花九轮草（*Primula veris*）（Nowak et al. 2015）、桃红蝴蝶兰（*Phalaenopsis equestris*）（Cai et al. 2015）、2个矮牵牛（*Petunia hybrida*）亲本基因组（Bombarely et al. 2016）、梅（*Prunus mume*）（Zhang et al. 2012）、蔷薇属（*Rosa*）（Hibrand Saint-Oyant et al. 2018），结合最近的长片段测序技术和高准确度的短reads测序，庞大的菊花基因组已经在Wageningen（荷兰）根进行测序和评估，二倍体菊花的测序已经由中国和韩国的科学家团队进行，同时，利用长片段测序技术获得的郁金香基因组已经公布信息。

基因组信息使得研究从QTL区间关联标记开发转移到该区间候选基因的变异解析。其他方面是利用观赏植物基因来剖析近亲物种的同线性关系，聚焦和鉴定候选基因。例如，Schulz（2016）和Nguyen等（2017）利用高密度的SNP基因芯片开展了栽培四倍体庭院月季全基因组关联分析研究，鉴定了与花瓣花青素及类胡萝卜素含量、叶柄定向芽再生能力相关联的基因组区间。将关联标记锚定到近缘种草莓基因组上，大部分是共线性关系（Gar et al. 2011；Vukosavljev et al. 2016；Bourke et al. 2017）。这些标记定位的基因组区间包含相关性状已知候选基因，这些基因可以在月季中进行进一步功能验证研究和等位基因变异研究以及研究它们对性状的影响。

9.7.2 混合样本分组分析

另外一种可能是利用混合样本分组测序从一个杂交后代两组中筛选不同的标记或基因，如抗病和感病后代。这些后代连同亲本进行全基因组测序分析。基因组中和抗病性关联的区间及存在的基因可以被鉴定出来。利用该方法Hawkins等（2016）研究发现，1个MYB基因内的SNP和草莓果的黄色相关联。

9.7.3 候选基因途径

在目标性状候选基因可以找到的情况下，候选基因途径可以直接检测某一基因中

SNP 标记和目标性状的关联，如基于其他物种的研究，利用不同物种基因编码序列的保守性，或者同一基因家族基因高度相似。在 QTL 区间较大且包括数百个基因的情况下，候选基因途径是有效的策略。例如，Fu 等（2017）定位了非洲菊的抗葡萄孢菌病基因，探测到了 20 个 QTL，还通过转录组数据收集了 29 个候选基因的清单，利用其他物种已知的抗葡萄孢菌病基因，基于同源性开发了 SNP 标记。对这些 SNP 标记利用基因特定高分辨率熔点进行定位，其中 7 个基因和 QTL 共定位（Fu et al. 2016）。对其中 2 个基因进行了测验，发现葡萄孢菌感染后表达上调。利用 VIGS 沉默基因后，感染的病斑显著变小，说明这些基因是可能的治病基因。

Kaufmann 等（2012）根据其他双子叶植物 MLO 基因参与植物与白粉病互作，分离和鉴定了 4 个蔷薇属 MLO 同源基因。因此，它们被认为是功能 MLO 基因的候选基因，这些基因的等位基因在月季中可能具有抗病性。豌豆、番茄、黄瓜、苹果以及许多作物，单个 MLO 基因隐性突变体可以完全抵抗白粉病（Berg et al. 2017）。

研究参与休眠和开花时间等重要性状的调控通路获得了足够的重视，因为对于感兴趣的观赏植物来说，通过 RNA 测序可以较为容易地获得这些通路的候选基因。Leeggangers 等（2013）描述了开花时间和球根物种郁金香与百合营养繁殖能力候选基因的可能应用（Moreno-Pachon et al. 2016）。

需要说明的是，候选基因途径常常出现假阳性。很多基因在某一特定阶段或处理后可能会上调和下调，因此验证步骤不能缺少。

9.8　精确育种技术

新的植物育种技术（又称精确育种技术）是将一系列的技术添加到育种手段中（Schaart et al. 2016；Van de Wiel et al. 2017），同样可以应用到观赏植物育种中。尤其是基于 CRISPR/Cas 定向诱变（基因组编辑或基因编辑）或者其他可设计的位点定向核酸酶可以扩展育种者可用的遗传变异库。这些可设计核酸酶可以使基因组精确位点双链断裂。有时细胞修护断裂的双链会导致错误的发生，在基因组这个位置出现小的删除或插入的植物可以筛选出来。从生物学的观点来看，是一种目标突变形式。基因编辑以非常快的速度取得成功，部分由于技术简单，也由于第二代测序技术使得产出需要的序列信息非常快速。例如，最近的文献大多是敲除基因（如颜色合成基因），但是突变元件在基因启动子区域也是有可能的，可据此来改变基因的时间和空间表达模式。这方面的发掘将是非常有意义的，因为许多不同表型是由基因表达时间长短不同引起的。Rodriguez 等（2017）开展了系统筛选三个转录因子基因启动子元件突变组合对于番茄株型是关键的，该研究获得了新的遗传变异，包括新颖的表型性状。

基因编辑是潜在进行定向修饰或创制的非转基因途径。例如，月季隐性抗病是由于 MLO 基因一个新的突变（Debener and Byrne 2014），利用基因编辑技术可以在单个细胞内对该基因的所有拷贝实施诱变。因此其对于诱导多倍体突变尤其适合。有许多隐性删除破坏基因导致新功能的例子，如众所周知的作物驯化和传统的突变育种（同时利用辐射和化学诱变）（Van de Wiel et al. 2017）。

开展基因编辑目前主要存在三个方面的挑战：调控性状基因的知识信息；可用的基因组信息或者完整的一系列转录组信息，特定基因的标靶设计需要引导 RNA；最后但同样重要的是，有效的转化特别是再生体系，这对完成突变的细胞再生为植株是必不可少的。再生体系是最大的瓶颈，因为许多植物通过引导 RNA 瞬时表达 Cas9 蛋白是容易获得突变的，如在原生质体或未成熟胚胎。但是由原生质体再生观赏植物植株的再生体系是稀少的。因此，基因编辑将复兴细胞生物学和组织培养研究。

9.9 结　论

这是激动人心的时代。本章提到的技术将为未来的观赏植物育种提供新的方向。尤其是测序和基因分型方法已为开发加速育种的工具做好准备，对于缺乏基因组信息、杂合或多倍体物种，可能需要较长的时间。

对于观赏植物育种公司而言，引入 DNA 信息育种就得开始考虑：①所关心的性状有需要继续解决的问题（如生产者需要抗病品种，但是通过常规杂交和筛选项目很难取得进展）；②可以最大限度地减少筛选后代的费用（如目前育种最后的评估可以提前到早期对性状进行筛选）；③可以不需要通过杂交分裂阶段更好地优化品种改良。

下一步最好是联络一个研究团队（大学、研究所或者在某些情况下可以是一个专业的育种研究公司），因为在绝大多数情况下，在开展 DNA 信息育种前必须开展发掘与性状基因相关联标记的研究。在最初的步骤中，关注植物材料或者群体对于发掘研究是必需的。因为这些产出需要较高的关于真实亲本、群体大小以及性状分离的专业背景。育种公司进行 DNA 信息育种需要在公司内部聘用或培养熟悉 DNA 信息育种和操作步骤的人员，在起始阶段就关注的作物直接和育种者讨论与沟通育种步骤。

致谢：该项目部分获得 KB-24-002-017、TKI-U 多倍体项目 BO-26.03-002-001 和 BO-26.03-009-004 的支持。对参与多倍体项目公司的支持表示衷心感谢。

本章译者：

李淑斌

云南省农业科学院花卉研究所，昆明 650200

参 考 文 献

Arens P, Bijman P, Tang N, Shahin A, van Tuyl JM (2012) Mapping of disease resistance in ornamentals: a long haul. Acta Hortic (ISHS) 953:231–238. https://doi.org/10.17660/ActaHortic.2012.953.32

Berg JA, Appiano M, Bijsterbosch G et al (2017) Functional characterization of cucumber (*Cucumis sativus* L.) Clade V MLO genes. BMC Plant Biol 17:80. https://doi.org/10.1186/s12870-017-1029-z

Bink MCAM, Jansen J, Madduri M et al (2014) Bayesian QTL analyses using pedigreed families of an outcrossing species, with application to fruit firmness in apple. Theor Appl Genet 127:1073–1090. https://doi.org/10.1007/s00122-014-2281-3

Bombarely A, Moser M, Amrad A et al (2016) Insight into the evolution of the Solanaceae

from the parental genomes of *Petunia hybrida*. Nat Plants 2:16074. https://doi.org/10.1038/nplants.2016.74
Bourke PM, Arens P, Voorrips RE et al (2017) Partial preferential chromosome pairing is genotype dependent in tetraploid rose. Plant J 90:330–343. https://doi.org/10.1111/tpj.13496
Bourke P, van Geest G, Voorrips RE et al (2018) polymapR: linkage analysis and genetic map construction from F1 populations of outcrossing polyploids. Bioinformatics, bty371. https://doi.org/10.1093/bioinformatics/bty371
Cai J, Liu X, Vanneste K et al (2015) The genome sequence of the orchid *Phalaenopsis equestris*. Nat Genet 47:65–72. https://doi.org/10.1038/ng.3149
Cao Z, Deng Z (2017) De novo assembly, annotation, and characterization of root transcriptomes of three *Caladium* cultivars with a focus on necrotrophic pathogen resistance/defense-related genes. Int J Mol Sci 18:712. https://doi.org/10.3390/ijms18040712
Casimiro-Soriguer I, Narbona E, Buide ML, del Valle JC, Whittall JB (2016) Transcriptome and biochemical analysis of a flower color polymorphism in *Silene littorea* (Caryophyllaceae). Front Plant Sci 7:204. https://doi.org/10.3389/fpls.2016.00204
Chong X, Zhang F, Wu Y et al (2016) A SNP-enabled assessment of genetic diversity, evolutionary relationships and the identification of candidate genes in *Chrysanthemum*. Genome Biol Evol 8:3661–3671. https://doi.org/10.1093/gbe/evw270
Debener T, Byrne DH (2014) Disease resistance breeding in rose: current status and potential of biotechnological tools. Plant Sci 228:107–117. https://doi.org/10.1016/j.plantsci.2014.04.005
Di Guardo M, Micheletti D, Bianco L et al (2015) ASSIsT: an automatic SNP scoring tool for in- and outbreeding species. Bioinformatics 31:3873–3874. https://doi.org/10.1093/bioinformatics/btv446
Fu Y, Esselink GD, Visser RGF, van Tuyl JM, Arens P (2016) Transcriptome analysis of *Gerbera hybrida* including in silico confirmation of defense genes found. Front Plant Sci 7:247. https://doi.org/10.3389/fpls.2016.00247
Fu Y, van Silfhout A, Shahin A et al (2017) Genetic mapping and QTL analysis of Botrytis resistance in *Gerbera hybrida*. Mol Breed 37:13. https://doi.org/10.1007/s11032-016-0617-1
Gar O, Sargent DJ, Tsai C-J et al (2011) An autotetraploid linkage map of rose (*Rosa hybrida*) validated using the strawberry (*Fragaria vesca*) genome sequence. PLoS One 6:e20463. https://doi.org/10.1371/journal.pone.0020463
Goodwin S, McPherson JD, McCombie WR (2016) Coming of age: ten years of next-generation sequencing technologies. Nat Genet 17:333–351. https://doi.org/10.1038/nrg.2016.49
Hawkins C, Caruana J, Schiksnis E, Liua Z (2016) Genome-scale DNA variant analysis and functional validation of a SNP underlying yellow fruit color in wild strawberry. Sci Rep 6:29017. https://doi.org/10.1038/srep29017
Hibrand Saint-Oyant L, Ruttink T, Hamama L et al (2018) A high-quality sequence of *Rosa chinensis* to elucidate genome structure and ornamental traits. In: bioRxiv. https://doi.org/10.1101/254102
Holdsworth WL, Mazourek M (2015) Development of user-friendly markers for the pvr1 and Bs3 disease resistance genes in pepper. Mol Breed 35:28
Huang J, Lin C, Cheng T et al (2016) The genome and transcriptome of *Phalaenopsis* yield insights into floral organ development and flowering regulation. PeerJ 4:e2017. https://doi.org/10.7717/peerj.2017
Jung H-J, Veerappan K, Natarajan S et al (2017) A system for distinguishing octoploid strawberry cultivars using high-throughput SNP genotyping. Trop. Plant Biol 10:68–76. https://doi.org/10.1007/s12042-017-9185-8
Kaufmann H, Qiu X, Wehmeyer J, Debener T (2012) Isolation, molecular characterization, and mapping of four rose MLO orthologs. Front Plant Sci 3:244. https://doi.org/10.3389/fpls.2012.00244
Kim JE, Oh SK, Lee JH, Lee BM, Jo SH (2014) Genome-wide SNP calling using next generation sequencing data in tomato. Mol Cells 37:36–42. https://doi.org/10.14348/molcells.2014.2241
Koning-Boucoiran CFS, Smulders MJM, Krens FA, Esselink GD, Maliepaard C (2012) SNP genotyping in tetraploid roses. Acta Hortic (ISHS) 953:351–356. https://doi.org/10.17660/ActaHortic.2012.953.49
Koning-Boucoiran CFS, Esselink GD, Vukosavljev M et al (2015) Using RNA-Seq to assemble a rose transcriptome with more than 13,000 full-length expressed genes and to develop the WagRhSNP 68k Axiom SNP array for rose (*Rosa* L.). Front Plant Sci 6:249. https://doi.org/10.3389/fpls.2015.00249
Leeggangers HACF, Moreno-Pachon N, Gude H, Immink RGH (2013) Transfer of knowledge about flowering and vegetative propagation from model species to bulbous plants. Int J Dev

Biol 57:611–620. https://doi.org/10.1387/ijdb.130238ri

Leeggangers HACF, Nijveen H, Nadal Bigas J, Hilhorst HWM, Immink RGH (2017) Molecular regulation of temperature-dependent floral induction in *Tulipa gesneriana*. Plant Physiol 173:1904–1919. https://doi.org/10.1104/pp.16.01758

Moreno-Pachon NM, Leeggangers HACF, Nijveen H et al (2016) Elucidating and mining the *Tulipa* and *Lilium* transcriptomes. Plant Mol Biol 92:249–261. https://doi.org/10.1007/s11103-016-0508-1

Muir P, Li S, Lou S et al (2016) The real cost of sequencing: scaling computation to keep pace with data generation. Genome Biol 17:53. https://doi.org/10.1186/s13059-016-0917-0

Nguyen THN, Schulz D, Winkelmann T, Debener T (2017) Genetic dissection of adventitious shoot regeneration in roses by employing genome-wide association studies. Plant Cell Rep 36:1493–1505. https://doi.org/10.1007/s00299-017-2170-8

Nijveen H, van Kaauwen M, Esselink DG, Hoegen B, Vosman B (2013) QualitySNPng: a user-friendly SNP detection and visualization tool. Nucleic Acids Res 41:W587–W590. https://doi.org/10.1093/nar/gkt333

Nowak MD, Russo G, Schlapbach R et al (2015) The draft genome of *Primula veris* yields insights into the molecular basis of heterostyly. Genome Biol 16:12. https://doi.org/10.1186/s13059-014-0567-z

Peace CP (2017) DNA-informed breeding of rosaceous crops: promises, progress and prospects. Hortic Res 4:17006. https://doi.org/10.1038/hortres.2017.6

Preedy KF, Hackett CA (2016) A rapid marker ordering approach for high-density genetic linkage maps in experimental autotetraploid populations using multidimensional scaling. Theor Appl Genet 129:2117–2132. https://doi.org/10.1007/s00122-016-2761-8

Rodríguez-Leal D, Lemmon ZH, Man J, Bartlett ME, Lippman ZB (2017) Engineering quantitative trait variation for crop improvement by genome editing. Cell 171:470–480.e8. https://doi.org/10.1016/j.cell.2017.08.030

Schaart JG, van de Wiel CCM, Lotz LAP, Smulders MJM (2016) Opportunities for products of new plant breeding techniques. Trends Plant Sci 21:438–449. https://doi.org/10.1016/j.tplants.2015.11.006

Schmutz J, Wheeler J, Grimwood J et al (2004) Quality assessment of the human genome sequence. Nature 429:365–368

Schulz DF, Schott RT, Voorrips RE, Smulders MJM, Linde M, Debener T (2016) Genome-wide association analysis of the anthocyanin and carotenoid contents of rose petals. Front Plant Sci 7:1798. https://doi.org/10.3389/fpls.2016.01798

Shahin A, Arens P, Van Heusden AW et al (2010) Genetic mapping in *Lilium*: mapping of major genes and quantitative trait loci for several ornamental traits and disease resistances. Plant Breed 130:372–382. https://doi.org/10.1111/j.1439-0523.2010.01812.x

Shahin A, van Gurp T, Peters SA, Visser RGF, van Tuyl JM, Arens P (2012a) SNP markers retrieval for a non-model species: a practical approach. BMC Res Notes 5:79. https://doi.org/10.1186/1756-0500-5-79

Shahin A, van Kaauwen M, Esselink D et al (2012b) Generation and analysis of expressed sequence tags in the extreme large genomes Lilium and Tulipa. BMC Genomics 13:640. https://doi.org/10.1186/1471-2164-13-640

Smulders MJM, Esselink D, Voorrips RE, Vosman B (2009) Analysis of a database of DNA profiles of 734 hybrid tea rose varieties. Acta Hortic 836:169–174. http://www.actahort.org/books/836/836_24.htm

Smulders MJM, Vukosavljev M, Shahin A, van de Weg WE, Arens P (2012) High throughput marker development and application in horticultural crops. Acta Hortic (ISHS) 961:547–551. https://doi.org/10.17660/ActaHortic.2012.961.72

Smulders MJM, Voorrips RE, Esselink GD et al (2015) Development of the WagRhSNP Axiom SNP array based on sequences from tetraploid cut roses and garden roses. Acta Hortic 1064:177–184. https://doi.org/10.17660/ActaHortic.2015.1064.20

Tang N, van der Lee T, Shahin A et al (2015) Genetic mapping of resistance to *Fusarium oxysporum* f. sp. tulipae in tulip. Mol Breed 35:122. https://doi.org/10.1007/s11032-015-0316-3

Van de Weg E, Di Guardo M, Jänsch M et al (2018) Epistatic fire blight resistance QTL alleles in the apple cultivar 'Enterprise' and selection X-6398 discovered and characterized through pedigree-informed analysis. Mol Breed 38:5. https://doi.org/10.1007/s11032-017-0755-0

Van de Wiel CCM, Schaart JG, Lotz LAP, Smulders MJM (2017) New traits in crops produced by genome editing techniques based on deletions. Plant Biotechnol Rep 11:1–8. https://doi.org/10.1007/s11816-017-0425-z

Van Geest G, Bourke PM, Voorrips RE et al (2017a) An ultra-dense integrated linkage map for hexaploid chrysanthemum enables multi-allelic QTL analysis. Theor Appl Genet 130:2527–2541. https://doi.org/10.1007/s00122-017-2974-5

Van Geest G, Voorrips RE, Esselink D, Post A, Visser RGF, Arens P (2017b) Conclusive evidence for hexasomic inheritance in chrysanthemum based on analysis of a 183k SNP array. BMC Genomics 18:585. https://doi.org/10.1186/s12864-017-4003-0

Voorrips RE, Gort G, Vosman B (2011) Genotype calling in tetraploid species from bi-allelic marker data using mixture models. BMC Bioinformatics 12:172

Voorrips RE, Bink MCAM, Kruisselbrink JW et al (2016) PediHaplotyper: software for consistent assignment of marker haplotypes in pedigrees. Mol Breed 36:119. https://doi.org/10.1007/s11032-016-0539-y

Vukosavljev M, Arens P, Voorrips RE et al (2016) High-density SNP-based genetic maps for the parents of an outcrossed and a selfed tetraploid garden rose cross, inferred from admixed progeny using the 68k rose SNP array. Hortic Res 3:16052. https://doi.org/10.1038/hortres.2016.52

Wei Z, Sun Z, Cui B, Zhang Q, Xiong M, Wang X, Zhou D (2016) Transcriptome analysis of colored calla lily (*Zantedeschia rehmannii* Engl.) by Illumina sequencing: de novo assembly, annotation and EST-SSR marker development. PeerJ 4:e2378. https://doi.org/10.7717/peerj.2378

Yagi M, Kosugi S, Hirakawa H et al (2014) Sequence analysis of the genome of carnation (*Dianthus caryophyllus* L.). DNA Res 21:231–241. https://doi.org/10.1093/dnares/dst053

Zhang Q, Chen W, Sun L et al (2012) The genome of *Prunus mume*. Nat Commun 3:1318. https://doi.org/10.1038/ncomms2290

Zhang J, Zhao K, Hou D et al (2017) Genome-wide discovery of DNA polymorphisms in Mei (*Prunus mume* Sieb. et zucc.), an ornamental woody plant, with contrasting tree architecture and their functional relevance for weeping trait. Plant Mol Biol Report 35:37–46. https://doi.org/10.1007/s11105-016-1000-4

Zheng C, Boer MP, Van Eeuwijk FA (2015) Reconstruction of genome ancestry blocks in multiparental populations. Genetics 200:1073–1087. https://doi.org/10.1534/genetics.115.177873

第 10 章　六　出　花

Mark P. Bridgen

摘要　六出花属是南美一类非常迷人的花卉，颜色多种多样，可作切花、盆栽或庭院种植。六出花俗称印加百合，易种植。单子叶植物从地下根茎发出营养枝和花枝。通过调控植物的光照和温度，尤其是根部的光照和温度，很容易诱导植株开花。通过常规的种内和种间杂交，特别是将离体培养技术应用于育种中时，就可以制定出一个有效的六出花育种方案。该属是一类自交亲和、具有同步雄蕊先熟花序的草本植物。花药先于雌蕊成熟并开裂释放花粉。当花粉散开且没活性之后，雌蕊的三分柱头才会发生反应并接受花粉。六出花自花授粉很容易产生种子。然而，六出花种间杂交却很少能产生种子，因此胚培养是成功培育杂种植株的必要条件。通过组织培养，新植株很容易在离体条件下快速繁殖。虽然种间杂种通常是不育的，但可通过使用秋水仙碱等化学处理使染色体数目加倍来改善和恢复其育性。

关键词　胚培养；印加百合；六出花；快繁；组织培养；育种

10.1　引　　言

六出花又名印加百合、秘鲁百合，隶属六出花科，以其艳丽而持久的切花而闻名，颜色多彩多样，有紫红色、淡紫色、红色、粉色、白色、橙色、桃红色和黄色以及双色。切花的瓶插寿命很长，可持续 2 周甚至更长时间。20 世纪 70 年代，六出花在欧洲和美国开始流行，主要用作切花。近年来，它作为盆栽植物（Olate et al. 2000）和庭院花卉而流行。

六出花的原生种主要分布于智利和巴西，但其他南美洲国家也有分布（Bridgen et al. 2000；Sanso et al. 2005）。其原产地的生境多种多样，包括安第斯山脉雪线、海洋海岸线、高原森林和沙漠。尽管六出花种类繁多，但几乎没有商业价值，因为该植物在一年中有一段时间处于休眠状态，花朵通常很小而显得微不足道，并且采后状态也不理想。由于种间和种内的变异、野外采集原生种的机会（Bridgen 2001），以及种间杂交的可能性，因此培育出了可以作为切花、盆栽植物和庭院花卉等各种用途的六出花新品种（Bridgen 2006b）。新的六出花品种是多年来通过种间和种内杂交育种以及辐射诱变育种等技术获得的结果（Bridgen et al. 2009；Broertje and Verboom 1974）。育种家对该植物的育种目标包括长势旺盛、周年连续开花、花朵芳香以及花色新颖等。

M. P. Bridgen (✉)
Cornell University, Riverhead, NY, USA
e-mail: mpb27@cornell.edu

10.2　生长与培养

六出花是一种人们喜欢栽种的植物，特别是专业种植者，部分原因是它在凉爽的温度下生长最好，花朵的产量很高，并且具有一个一旦开始开花就会不断开花的习性。当生长在露地时，这些植物在比美国气候带 5 更加温暖的地区以多年生草本植物的形式生长。最近的育种计划引入了新的栽培品种，这些品种可在比美国气候带 6 更冷的地区生长（Bridgen 1997）。在寒冷地区，六出花被视为一年生植物或柔嫩的多年生草本植物，它在阳光充足的情况下生长最佳，但在夏季炎热的地区，尤其是在炎热的下午，适当遮阴才有利于其生长。

根茎植物在排水良好的有机基质中生长最好。为了获得最佳的长势和花枝产量，根茎应浅栽，生长点仅需 7～10cm 深；如果作为盆栽植物种植，则只需距表面 2.5～3cm 深即可（Bridgen 1993，1997）。浅栽的开花量比深栽的要多。在生长季节，植物需要充足的水分才能生长良好。充足的养分同样会促进植株生长，氮肥能促进开花和增加产量，最好是施用 400ppm 的氮肥（Elliott et al. 1993；Smith et al. 1998）。

六出花属于草本地下芽植物，地下根茎会产生花芽和营养芽两种类型的芽（Bridgen 1993）。在满足开花条件后，地下根茎就会不断产生新的花茎。六出花的开花调控过程首先需要低温条件，其次需要长光照。不同品种对诱导开花的低温有不同要求。这些温度可能在 10～17℃（50～63°F）变化；所需的低温持续时间也会因其基因型不同而不同。一旦开花，植株就会源源不断地分化花芽，直到土壤温度达到 18～21℃（65～70°F）以上并持续较长时间时才会停止分化（Bridgen 1997；Bridgen and Bartok 1990）。

10.3　育 种 规 程

通过常规的种内和种间杂交以及离体技术，我们可以制定出一个有效的六出花育种方案。对于六出花，育种家的育种目标是长势旺盛、周年连续开花、花朵芳香及花色新颖。由于这些植物是通过营养繁殖得到的，因此其受到专利保护（Bridgen 2006a，2012）。

六出花属是一类自交亲和、具有同步雄蕊先熟花序的草本植物（Harder and Aizen 2004）。花药在雌蕊接受花粉之前就成熟开裂并释放花粉。当花粉散开且没活性之后，雌蕊的三分柱头才会发生反应并接受花粉。这种同步雌雄蕊异熟的进化目的是减少同株植物自花授粉的发生。对于传统的六出花育种者来说，这种结构特征具有很大的优势，因为在异花授粉时不需要去雄。

当六出花自花授粉时，通常会结实并产生种子。对于智利和巴西的原生植物来说，这是一个正常的有性繁殖过程。然而，当进行种间杂交时，却很少能产生种子。一般母本植株会产生种子，但是这些种子在授粉后大约 2 周内就会败育（Ishikawa et al. 2001）。离体胚挽救已经被证明是六出花成功进行种间杂交的必要条件（Bridgen 1994；

Lu and Bridgen 1996，1997）。授粉后 10～14d 采集子房，置于 95%乙醇中燃烧；子房燃烧过程可重复两次，以保证表面无菌。然后，从杂种子房中取出无菌胚珠，接种至每升含有 25%无机盐和维生素 D 的 Murashige 与 Skoog（MS）培养基上（Murashige and Skoog 1962），蔗糖 7.5g，琼脂 7.0g。在 18～20℃的黑暗条件下培养直到胚萌发。萌发后，幼苗可以在相同的培养基上生长，温度 18℃，光照 18h/d。4～8 周后，幼苗可作为再生植株进行快繁。

六出花的组培快繁非常容易。杂种植株可以在每升含 30g 蔗糖和 2mg 6-苄基腺嘌呤（6-BA）的无胶凝剂（液体）的 MS 完全培养基上进行快速繁殖（Bridgen et al. 1991；Chiari and Bridgen 2000）。在继代培养过程中，通过茎尖分生培养可使六出花的增殖率达到 200%（Chiari and Bridgen 2002）。每 4 周对植物进行一次继代培养，以保持活跃生长，这一点很重要。液体培养基可以产生优质植株，只要植物没有完全浸没，就无须摇动。

由于亲本基因型的差异，通过胚挽救获得的种间杂种往往不育。为了提高和恢复这些杂种的育性，可以通过化学处理使染色体数目翻倍，从而获得四倍体或其他多倍体（Lu and Bridgen 1997）。由于同源染色体配对发生在减数分裂时期，这些多倍体植物的基因组平衡性和育性得到改善，并产生有活力的花粉粒和卵细胞。

秋水仙碱是一种抗有丝分裂剂，已成功应用于离体培养中，从而获得了六出花多倍体植株并恢复了其育性。秋水仙碱的浓度为 0.2%～0.6%，处理时间为 6～24h。使用时，秋水仙碱应在 2%二甲基亚砜中现配现用并过滤灭菌。在无菌条件下将带有 2～3 个健壮幼嫩芽的无菌外植体完全浸入含有 50mL 秋水仙碱溶液的玻璃培养皿中。培养皿用塑料膜包裹后置于摇床上，室温条件下，转速 60r/min，处理结束后，用无菌去离子水漂洗外植体 4～5 次，去除秋水仙碱残留物，每次漂洗应持续 25～35min。然后在无菌条件下将外植体接种至增殖培养基进行离体增殖和生根培养。秋水仙碱处理过的植株大约有 50%的成活率（Bridgen et al. 2009）。

多倍体植株可采用根尖细胞的染色体计数和气孔大小的测量来鉴别，秋水仙碱处理之后，二倍体植株的染色体数目（$2n = 2x = 16$）会加倍至四倍体（$2n = 4x = 32$）。一些植株是同时含有二倍体和四倍体组织的嵌合体。从形态上看，秋水仙碱诱导的六出花四倍体植株与二倍体植株相比，花更大，茎更高、更大、更绿，叶更厚。四倍体植株的花粉母细胞染色体配对正常，然而，在后期 I 可能会出现染色体桥和落后染色体等不规则的减数分裂行为。需要注意的是，秋水仙碱诱导的四倍体植株在染色体加倍后育性并没有得到改善，仍然保持完全不育（Bridgen et al. 2009）。

巴西六出花和智利六出花的种间杂交培育出了具有新型花色、花形的新杂种，同时也证明了培育具有综合性状如耐寒性强、株形紧凑的盆栽植物和花香浓郁等新杂种的可能性。印加百合的首个香型品种于 1997 年引进，研究表明控制花香的基因可能是核基因，可以通过花粉传递（Aros et al. 2012，2015）。香型品种‘Sweet Laura’（图 10.1）是智利六出花 *Alstroemeria aurea* 和巴西六出花 *A. caryophyllaea* 杂交而成的（Bridgen et al. 1997）。

图 10.1 六出花‘Sweet Laura’——首个香型杂交品种

10.4 结 论

植物组织培养技术，如胚培养、体细胞胚胎发生、诱变和快繁等，已被用于六出花属的经典育种。由于许多种间杂交不能产生有活力的种子和后代，因此胚培养技术特别重要。从胚珠中取出未成熟的胚置于无菌的营养培养基上进行离体培养，胚就可以存活并生长。自发的体细胞胚胎发生和秋水仙碱诱导突变可以产生新的植株。一旦开发出一种新型植株，就可以利用组培快繁来缩短其进入商业市场的时间。

本章译者：

段 青

云南省农业科学院花卉研究所，昆明 650200

参 考 文 献

Aros D, Gonzalex V, Allemann RK, Müller CT, Rosati C, Rogers HJ (2012) Volatile emissions of scented *Alstroemeria* genotypes are dominated by terpenes, and a myrcene synthase gene is highly expressed in scented *Alstroemeria* flowers. J Exp Bot 63:2739–2752

Aros D, Spadafora N, Venturi M, Núñez-Lillo G, Meneses C, Methven L, Müller CT, Rogers H (2015) Floral scent evaluation of segregating lines of *Alstroemeria caryophyllaea*. Sci Hortic 185(30):183–192

Bridgen MP (1993) *Alstroemeria*. In: De Hertogh AA, Le Nard M (eds) The physiology of flower bulbs: a comprehensive treatise on the physiology and utilization of ornamental flowering bulbous and tuberous plants. Elsevier Publishing, Amsterdam, London, New York, Tokyo, pp 201–209

Bridgen MP (1994) A review of plant embryo culture. HortScience 29:1243–1246

Bridgen MP (1997) *Alstroemeria*. In: Ball V (ed) Ball Red Book, 16th edn. Geo. J. Ball Publishing Co., W. Chicago, IL, pp 341–348

Bridgen M (2001) A nondestructive harvesting technique for the collection of native geophyte plant species. Herbertia 56:51–60

Bridgen MP (2006a) New Plant Introductions: *Alstroemeria* Mauve Majesty PPAF. Comb Proc Int Plant Propag Soc 56:398–399

Bridgen MP (2006b) Using traditional and biotechnological breeding for new plant development. Comb Proc Int Plant Propag Soc 56:307–310

Bridgen MP (2012) *Alstroemeria* ‘Tangerine Tango.’ U.S. Patent 22,701. Filed May 11, 2010, and issued May 1, 2012

Bridgen MP, Bartok J (1990) Evaluation of a growing medium cooling system and its effects on

the flowering of *Alstroemeria*. HortSci 25:1592–1594
Bridgen MP, King JJ, Pedersen C, Smith MA, Winski PJ (1991) Micropropagation of *Alstroemeria* hybrids. Proc Int Plant Propag Soc 41:380–384
Bridgen MP, Lu C, Neuroth M (1997) *Alstroemeria* 'Sweet Laura.' U.S. Patent 10,030. Filed December 15, 1995, and issued September 16, 1997
Bridgen MP, Olate E, Schiappacasse F (2000) Flowering geophytes from Chile. Acta Hortic 570:75–80
Bridgen MP, Kollman EE, Lu C (2009) Interspecific hybridization of *Alstroemeria* for the development of new, ornamental plants. Acta Hortic 836:73–78
Broertje C, Verboom H (1974) Mutation breeding of *Alstroemeria*. Euphytica 23:39–44
Chiari A, Bridgen M (2000) Rhizome splitting: a new micropropagation technique to increase *in vitro* propagule yield in *Alstroemeria*. J Plant Cell Org Tiss Cult 62:39–46
Chiari A, Bridgen MP (2002) Meristem culture and virus eradication in *Alstroemeria*. Plant Cell Tiss Org Cult 68:49–55
Elliott GC, Smith MA, Bridgen MP (1993) Growth response of *Alstroemeria* 'Parigo Pink' to phosphate supply *in vitro*. Plant Cell Tiss Org Cult 32:199 204
Harder LD, Aizen MA (2004) The functional significance of synchronous protandry in *Alstroemeria aurea*. Funct Ecol 18:467–474
Ishikawa T, Takayama T, Ishizaka H, Ishikawa K, Mii M (2001) Production of interspecific hybrids between *Alstroemeria pelegrina* L. var. rosea and *A. magenta* Bayer by ovule culture. Euphytica 118:19–27
Lu C, Bridgen MP (1996) Effects of genotype, culture medium, and embryo developmental stage of the *in vitro* responses from ovule cultures of interspecific hybrids of *Alstroemeria*. Plant Sci 116:205–212
Lu C, Bridgen MP (1997) Chromosome doubling and fertility study of *Alstroemeria aurea* X *A. caryophyllaea*. Euphytica 94:75–81
Murashige T, Skoog F (1962) A revised medium for rapid growth and bioassays with tobacco tissue cultures. Physiol Plant 15:473–497
Olate E, Ly D, Elliott G, Bridgen MP (2000) Influence of the timing of propagation and cold storage on the growth and development of *Alstroemeria* pot plants. Proc Int Plant Propag Soc 50:379–391
Sanso AM, de Assis MC, Xifreda CC (2005) *Alstroemeria*: a charming genus. Acta Hortic 683:63–77
Smith MA, Elliott GC, Bridgen MP (1998) Calcium and nitrogen fertilization of *Alstroemeria* for cut flower production. HortSci 33:55–59

第11章 杜 鹃 花

Jan De Riek, Ellen De Keyser, Evelien Calsyn, Tom Eeckhaut,
Johan Van Huylenbroeck, and Nobuo Kobayashi

摘要 通常认为比利时盆栽杜鹃花亲本的遗传基础相对狭窄，其主要是由植物采集者引种到植物园和私人爱好者从东亚地区收集的杜鹃花种类杂交培育而来。日本和中国的部分杜鹃花传统品种栽培历史起码有400多年。比利时盆栽杜鹃花始于约200年前，第一个比利时培育并商业推广的杜鹃花品种为'范德·克鲁森夫人'（'Madame Van der Cruyssen'）。系统发育学和分子标记研究表明，应该存在一个由映山红亚属的多个种类参与培育而成的过渡类型。该类型也曾被亚洲和欧洲的育种者利用，而且证明其是一个良好的遗传变异和有趣株型的来源。直到今天，常规杂交育种和芽变选种仍然是品种改良的主要方式。盆栽杜鹃花育种史上的重大突破有以下几点：①品种实现了自根苗的培育，不再受限于嫁接技术；②早花（秋季开花）品种的成功培育，以品种'赫尔穆特·沃格尔'（'Hellmut Vogel'）为代表；③成品盆栽观赏期延长。育种目标是根据大规模生产目标和消费者意愿而定的。消费者总是希望看到新的品种，而且需求变化很快，如从小型花到大型的重瓣花，从彩色花到白色花等。直到今天，花朵特征仍然是最重要的选择性状。除此之外，叶片的形状和颜色、叶表面的光泽度、是否存在表皮毛、生长势、植物株型、早花性状与成品苗的质量（包括花期、没有褐色的芽鳞等）都是重要的育种目标。最近，针对生物抗性的育种受到关注，特别是对某些真菌疾病的抵抗力已成为主要的选择标准。此外，早期候选品种已经在商业环境条件下进行了测试，以评估它们对特定培养条件的适应性反应，如修剪、生长调节剂，以及胁迫环境对植物生长发育的影响。

关键词 杜鹃花；盆栽杜鹃花；育种；分类；生物胁迫；花色；植物株型；遗传关系

11.1 比利时盆栽杜鹃花介绍

11.1.1 历史遗传资源研究

通常认为比利时盆栽杜鹃花亲本的遗传基础相对狭窄，其主要是由植物采集者引种到植物园和私人爱好者从东亚地区收集的杜鹃花种类杂交培育而来。比利时杜鹃花的盆

J. De Riek (✉) · E. De Keyser · E. Calsyn ·T. Eeckhaut · J.Van Huylenbroeck
Flanders Research Institute for Agriculture, Fisheries and Food (ILVO), Plant Sciences Unit, Applied Genetics and Breeding, Melle, Belgium
e-mail: jan.deriek@ilvo.vlaanderen.be
N.Kobayashi
Faculty of Life and Environmental Science, Shimane University, Matsue, Japan

栽始于约 200 年前。但是，Galle（1987）认为，“比利时盆栽杜鹃花的起源从 19 世纪初就存在来源混淆，1808 年映山红（*Rhododendron simsii*）被当作皋月杜鹃（*Azalea indica*）引入英国”。Scheerlinck 等（1938）报道映山红被引进比利时后不久，在 1818 年，就与真正的皋月杜鹃（*R. indicum*）混为一谈，被认为是同一种杜鹃花，这种混淆一直持续到今天。Galle（1987）谈到，“同时，杜鹃花品种‘白色尹迪卡’（‘Indica Alba’）或‘毛白’（‘Mucronatum’）在 1819 年从中国的花园引种到英国，随后在 1824 年杜鹃花品种‘凤凰’（‘Phoeniceum’）也被引种至英国。1833 年，几个不同变型的皋月杜鹃均以 *A. indica lateritia* 的名字到达了英格兰，随意地归入皋月杜鹃变种类群，作为重要的亲本”。利用这些最早的杜鹃花种质培育了早期的英国皋月杜鹃杂交种，其成为 1840 年前后欧洲流行的室内观赏花卉和温室植物。然而，Galle（1987）指出，“比利时杜鹃的主要亲本来源于 Robert Fortune 从上海苗圃收集的 3 个映山红单株，在 1851 年映山红被送到英国，也给予了不正确的名字“尹迪卡”（“Indica”）。1854 年，这 3 个映山红单株进入比利时，开启了伟大的杜鹃花育种和培育项目”。杜鹃花新品种的选育工作由此开始了，主要培育国家是比利时和英国，还有法国和德国。杂交工作达到了惊人的水平并产生了大量的杂交类群。早期引种至欧洲的种源混淆和后来不可忽略的未知中国或日本种源的加入使比利时盆栽杜鹃花的起源蒙上了一层神秘的面纱。

11.1.2 比利时盆栽杜鹃花起源的分子生物学研究

20 世纪 90 年代兴起的 DNA 测序和分子标记指纹图谱技术为深入研究杜鹃花野生和栽培类群的遗传多样性与类群内及类群间的基因型提供了便利。首先，分类学家研究了常绿杜鹃花野生种的种间关系和系统发育。在研究的基础上，不同学者修订了杜鹃花分类系统（Spethmann 1987；Chamberlain and Rae 1990；Cullen 1991；Cox 1995，1997）。几位学者公开发表了应用 *rbcL*、ITS 或 *matK* 序列技术的研究结果（Kron 1993，1997；Kurashige et al. 1998）。应用分类学的经典基因测序技术不能区分比利时杜鹃花杂种类型，只有应用分辨率较高的分子标记（AFLP、SSR 或 iSSR）才可以成功建立品种与杂交类群间的系统关系。

11.1.3 比利时盆栽杜鹃花的品种选育

常规杂交育种在杜鹃花品种改良方面取得了重要进展。1867 年推出了第一个商业品种‘范德·克鲁森夫人’（图 11.1）。由此激发了杜鹃花育种的热情（van Trier 2012）。自从 1867 年品种‘范德·克鲁森夫人’取得商业成功后，又推出了两个重要的栽培品种：一个是‘帕特里克夫人’（1880 年），花期早（圣诞节前开花），粉红色重瓣花；另一个品种是‘赫尔穆特·沃格尔’（1967 年），花期更早，在秋季就能开花（图 11.1），该品种还有一个优点是能够自根生长，无须嫁接，并可以更精确地调控花期（van Trier 2012）。

目前，盆栽杜鹃花育种主要在比利时和德国开展。在比利时，两个育种项目主导了市场方向：一个为私有园艺育种公司（Hortibreed）主导，一个为公共研究机构法兰德斯农业、渔业和粮食研究院（Flanders Research Institute for Agriculture，Fisheries and Food，ILVO）主导。在品种研发和市场营销方面，ILVO 与名为阿扎诺巴（AZANOVA，www.azanova.be）

图 11.1　历史上重要的杜鹃花品种：(a)'范德·克鲁森夫人'('Madame Van der Cruyssen')(1867 年)；(b)'帕特里克夫人'('Madame Petrick')(1880 年)；(c)'赫尔穆特·沃格尔'('Hellmut Vogel')(1967 年)

的种植者合作社协作。少量的品种由比利时种植者培育而成。在德国，一直只有两个重要的育种家分别为 Rannacher 和 Stahnke-Dettmer。从历史育种成绩上看，德国育种者相对于比利时育种者更加成功，在比利时市场上，47%的杜鹃花品种由德国育种者培育而成，而比利时育种者培育的品种为 38%（Heursel 1999）。

培育满足消费者需求的新颖品种是所有观赏植物育种者的目标。由于培育一个杜鹃花新品种需要 10～15 年的时间（Heursel 1999），因此常出现这样一种情况，当育种者成功培育了满足当时消费者需求的新品种时，而现在消费者对其新颖性的渴望已经消失。因为人们的需求总是在变化。品种'赫尔穆特·沃格尔'及其芽变系列控制了杜鹃花消费市场多年，占据了市场 60%以上的份额。该品种系列成功的原因可归结为良好的植物品质（易于生产栽培）与其芽变系列品种具有多种花色，而且花期早。因此，培育一个特性足够优良能够替代'赫尔穆特·沃格尔'的新品种成为许多育种者的梦想。

11.1.4　比利时盆栽杜鹃花的栽培技术

盆栽杜鹃花的现代品种均是采用扦插无性繁殖方式。根据盆的大小，在一个盆内扦插 1～5 根插条，以酸性泥炭（pH = 4.5）为扦插基质。在幼苗营养生长过程中，需要多次摘心，促进腋芽生长，最后达到所需要的冠幅和球形的株型。在夏季，将植物放在室外的容器苗培育场地。在最后一次摘心后，为了控制营养生长和启动生殖生长，需要使用植物生长调节剂。在开花之前，杜鹃花需要一段冷凉时期以打破花芽的休眠。打破休眠的最佳温度是 7℃。根据需冷量，杜鹃花可分为超早花（8 月 15 日始花）、早花（12 月 1 日始花）、中花（1 月 15 日始花）和晚花（2 月 15 日始花）品种。最后，杜鹃花盆栽在加温温室内催花，冬季需要补充光照。通过对栽培品种选择，生产时间合理安排，配合使用植物生长调节剂，冷藏处理和催花期间适当补光，可以实现盆栽杜鹃花的周年生产（Heursel 1999；Christiaens 2014）。

11.2　分类学和系统发育

11.2.1　杜鹃花属于杜鹃花属（不包括一些特殊的种类）

杜鹃花科杜鹃花属以其美丽和丰富多样的植被类型而闻名，其包括了 1000 多种植物。该属植物种类多，分类学家依据形态特征（如花、叶、表皮毛等）又进行了属下分类，分为 8 个亚属（表 11.1）。4 个最重要的亚属分别是映山红亚属[除了轮生叶杜鹃花组外的常绿杜鹃花]、羊踯躅亚属（落叶杜鹃花）、杜鹃花亚属（有鳞杜鹃花）和常绿杜鹃花亚属（无鳞杜鹃花）（Chamberlain et al. 1996）。常绿杜鹃花亚属的种类常绿，叶片大，不具鳞片。杜鹃花亚属的种类叶片较小，具鳞片，通常是常绿的，也有少量半落叶种类。杜鹃花亚属和常绿杜鹃花亚属在园艺上统称为高山杜鹃（rhododendron），映山红亚属、羊踯躅亚属和马银花亚属则统称为杜鹃花（azalea）。

表 11.1　杜鹃花属（Chamberlain et al. 1996）

亚属	组	含有种类
杜鹃花亚属	杜鹃花组	27 个亚组，211 个种，叶片有鳞
	类越橘杜鹃花组	7 个亚组，310 个种
	髯花杜鹃花组	21 个种
常绿杜鹃花亚属	常绿杜鹃花组	24 个亚组，302 个种，叶片无鳞，园林用杂种具大花
羊踯躅亚属	羊踯躅组	23 个种，落叶杜鹃
	Rhodora	2 个种
	沼生杜鹃花组	1 个种
	十花药杜鹃花组	4 个种
映山红亚属	映山红组	94 个种，常绿杜鹃花
	轮生叶杜鹃花	23 个种，落叶杜鹃花
马银花亚属	马银花组	11 个种
	长蕊杜鹃花组	19 个种
叶状苞亚属	—	2 个种
异蕊杜鹃花亚属	—	1 个种
纯白杜鹃花亚属	—	1 个种

由于杜鹃花具有重要的观赏价值，许多均对其亚属的分类进行了修订和论述（Spethmann 1987；Chamberlain and Rae 1990；Cullen 1991；Cox 1995，1997；Goetsch et al. 2005；Kron and Powell 2009）。最近发表的文献均参考了由基因测序获得的系统发育数据（最新的杜鹃花分类就是基于系统发育学的数据，有关内容参见本书中的 Krebs 分类法）。杜鹃花在地球上已经生存了 5000 多万年（Irving and Hebda 1993）。亚洲是其最大的自然资源基因中心，原产 300 多种杜鹃花种类。其分布范围从尼泊尔到喜马拉雅山脉，一直到缅甸北部及中国西南部的云南和四川省（Leach 1961）。现在杜鹃花在整个北半球大陆均有分布。大多数杜鹃花分布于中国和喜马拉雅山，一些热带种类（*Vireya*）分布在马来西亚和印度尼西亚。杜鹃花属植物在东北亚、高加索和北美洲地区也有分布。从其分布上看，杜鹃花属植物适应性相当广泛，但是最适宜其生长的土壤 pH 为 4.5～6.0（Galle 1987）。

杜鹃花分为落叶和常绿种类。落叶杜鹃花通常株型较高、直立，相对于常绿杜鹃花分枝少，在冬季清晰可见其茎秆和枝条的形状；大多数种类的花朵具有相对较长而狭窄的花筒管；花色有白色、不同深浅的黄色、橙色、偏黄的粉红色、红色、粉红色、紫色和紫粉色（从浅到深）（Voss 2001）。典型的常绿杜鹃花是分枝较多的灌木，株高与冠幅相当，覆盖着或深或浅的绿色叶片，多为光滑、全缘叶。冬季有些品种的叶子呈现或深或浅的紫红色。虽然常绿杜鹃花整年都具有良好的株型和叶片，但人们还是最关注其春季开花时的状态。杜鹃花栽培品种丰富，花型有单瓣、半重瓣和重瓣，花瓣有纯色、斑纹、条纹的扇形或半圆形，花色有白色、淡橙色、红色、粉红色、紫色和紫粉色等。Kron 和 Powell（2009）根据叶片、嫩枝和花冠的特性把映山红亚属分为轮生叶杜鹃花组和映山红组。分类的分子生物学证据来自 30 种映山红亚属植物和 5 种 *Menziesia* 植物的 3 个叶绿体基因序列（*matK*、*ndhF* 和 *trnS-trnG*）与两个核基因序列（*nrITS* 和 *rpb2* 的第三个内含子）片段。基于总证据分析的简约法、贝叶斯推断法和相似性分析用于评估映山红亚属内植物的单系统关系及采集样本之间的物种关系，尤其着重阐述了大武杜鹃（*Rhododendron tashiroi*）的分类地位。研究结果支持大武杜鹃为轮生叶杜鹃花组分支的成员。分子证据同时支持皋月杜鹃、*R. tsusiophyllum*、*R. tschonoskii* 和 *R. serpyllifolium* 归入映山红亚属，这几种杜鹃花以前没有证据表明其与映山红亚属系统关系密切。

常绿杜鹃花育种的一个主要困难是映山红亚属系统关系较近的种类缺乏黄色、橙色和蓝色花。历史上（从 1800 年开始），根特的园艺家培育了著名的抗性根特杂种。这些落叶杜鹃花杂种与欧洲的黄花杜鹃（*R. luteum*）、亚洲的羊踯躅（*R. molle*）和一些北美洲羊踯躅亚属的种类系统关系较近。这些杂种与最近推出的羊踯躅杂种群，如 Knap Hill、Exbury 和 Mollis 等杂交类群均具有醒目的黄色及橙色花朵。许多盆栽杜鹃花种植者希望将这种花特性转移到常绿杜鹃花中。但是，杜鹃花亚属和羊踯躅亚属的系统发育研究者 Kron（1993，2002）着重提醒想利用羊踯躅亚属植物杂交的育种者：这个看似非常古老的杜鹃花谱系，近来开始变得多样化，尤其在北美洲地区。有趣的是，相对于映山红亚属，羊踯躅亚属（落叶杜鹃花）与高山杜鹃（杜鹃花亚属和常绿杜鹃花亚属）亲缘关系更近。Kron 认为落叶羊踯躅类杜鹃花代表了杜鹃花属中较老和较原始的谱系，而常绿杜鹃花（evergreen azalea）是较为进化的类型，与常绿高山杜鹃（evergreen rhododendron）之间的亲缘关系更近。

11.2.2 杂交类群的命名

在欧洲，根据主要亲本不同，常绿杜鹃花（evergreen azalea）至少可分为 4 个品种类群（Heursel 1999）：比利时盆栽杜鹃群（映山红杂种类群，以前被错误地称为皋月杜鹃）、平户杜鹃群（火红杜鹃杂种类群）、久留米杜鹃群（为九州杜鹃 *R. kiusianum* Makino var. *kiusianum* 与 var. *sataense* 和山杜鹃 *R. kaempferi* Planch.的杂交类群）和皋月杜鹃群[皋月杜鹃和 *R. eriocarpum* (Hayata) Nakai 的突变体或杂种类群]。比利时盆栽杜鹃花的种植者经常将平户杜鹃、久留米杜鹃和皋月杜鹃称为“日本杜鹃花”，这种叫法很普遍，主要由于其地理起源。在西欧和北欧，区分日本杜鹃花的明显特征是其在冬季的耐寒性。

据推测，至少有 4 个不同的杜鹃花种类参与了现代映山红杂种群的培育（表 11.2）（Heursel 1999；van Trier 2012）。

表 11.2　4 个最重要的祖先对现代比利时盆栽杜鹃花的贡献

特性	皋月杜鹃	映山红	火红杜鹃	白花杜鹃
起源	日本	中国	日本	日本
株型	球形	直立	直立	直立
花大小	小花	中花	大花	大花
花色	玫红色	红色	深红色	白色带紫
开花时间	晚	早	较早	中

摘自 Heursel（1999）

• 映山红（*Rhododendron simsii* Planch.）

该物种被认为是主要祖先，原产于中国的丘陵地区（长江流域），为常绿灌木，广泛分布于中国亚热带海拔 500～1200m 的山地（Li et al. 2012），日本琉球群岛、中国台湾、缅甸、老挝和泰国也有分布。映山红不耐寒，因为它的自然栖息地温度在 8.6～20.2℃，花是典型的红色（Heursel 1999）。

• 皋月杜鹃[*Rhododendron indicum* (L.) Sweet]

该物种原产于日本本州、九州和屋久诸岛（Heursel 1999）。以前其曾与映山红混淆。这两个种可以根据花药数量区分。皋月杜鹃有 5 个花药，而映山红有 10 个（有时是 8 个）（Bean 1980）。两种植物的花都是胭脂红色（Heursel 1999）。

• 火红杜鹃（*Rhododendron scabrum* G. Don）

该物种在中国东海和菲律宾海岸的琉球群岛（日本）上有栽培。其原产于亚热带气候区，不耐寒，花为深红色（Yamazaki 1996）。

• 白花杜鹃（*Rhododendron* × *mucronatum* G. Don）

该物种（图 11.2）起源于日本，被认为是 *R. ripense* Makino 和 *R. macrosepalum* Maxim 之间的杂交种。该杂交类群有很多品种，如‘白龙圭’（‘Shiroryukyu’）、‘白如町’（‘Shiromanyō’）、‘琉球隔膜’（‘Ryukyushibori’）、‘松风’（‘Minenomatsukaze’）等（Yamazaki 1996）。白花杜鹃品种‘白龙圭’开白色花，是最有代表性的品种。历史书籍《金寿樱》（*Kinshū-makura*）（1692 年）也提到了这种大型白色花朵。该品种在日本和中国地区很常见，因为容易扦插繁殖。该品种由中国引入英国，更名为‘诺迪亚娜’（‘Noordtiana’）。‘白龙圭’开白花，但携带紫色花的遗传信息，可能是育种中使用最多的亲本材料。白花杜鹃是耐寒的（Heursel 1999）（图 11.2）。

将映山红花期早和易于产生芽变的能力引入现代盆栽杜鹃花。芽变即植物个体或器官自发性地突然出现的超出正常变异范围的性状改变，即基因突变，通常是体细胞突变（Pratt 1980）。大花基因则来自火红杜鹃和白花杜鹃；而且白花杜鹃同时带有紫色花的基因。球形株型和胭脂红色源自皋月杜鹃（表 11.2）（Heursel 1999）。

日本和中国有着不同的分类系统。日本杜鹃花育种历史悠久，形成了几个重要的园艺杂种群，主要亲本为杜鹃花原生种类，而不是比利时盆栽杜鹃花（表 11.2，图 11.3）。在日本原产的 50 种杜鹃花中（Yamazaki 1996），几个常绿杜鹃花种类，如山杜鹃、*R. macrosepalum*、皋月杜鹃、*R. ripense* 等，都具有很高的观赏价值。自 Edo 时代（1603～1867 年）以来，园艺家从这些杜鹃花自然种群和杂交群体中选育了数百个品种（Kobayashi 2013）。

图 11.2 白花杜鹃（*R.* × *mucronatum*）。起源于日本，推测是 *R. ripense* Makino 和 *R. macrosepalum* Maxim. 的种间杂种。插图摘自 *Kinshū-makura*（1692 年）

杜鹃花品种群，如江户雾岛、久留米、琉球、平户和皋月等品种群，均选育自以上杜鹃花的自然杂交和人工杂交群体的。第一本杜鹃花的专著是 *Kinshū-makura*，1692 年由 Ito 主编，其中介绍了 337 个杜鹃花品种和皋月杜鹃杂交种。该专著的英文版名为 *A Brocade Pillow: Azaleas of Old Japan*（Ito and Creech 1984）。这本专著里描述的许多品种现在仍有栽培，有着特异且引人注目的花朵和叶片特征。

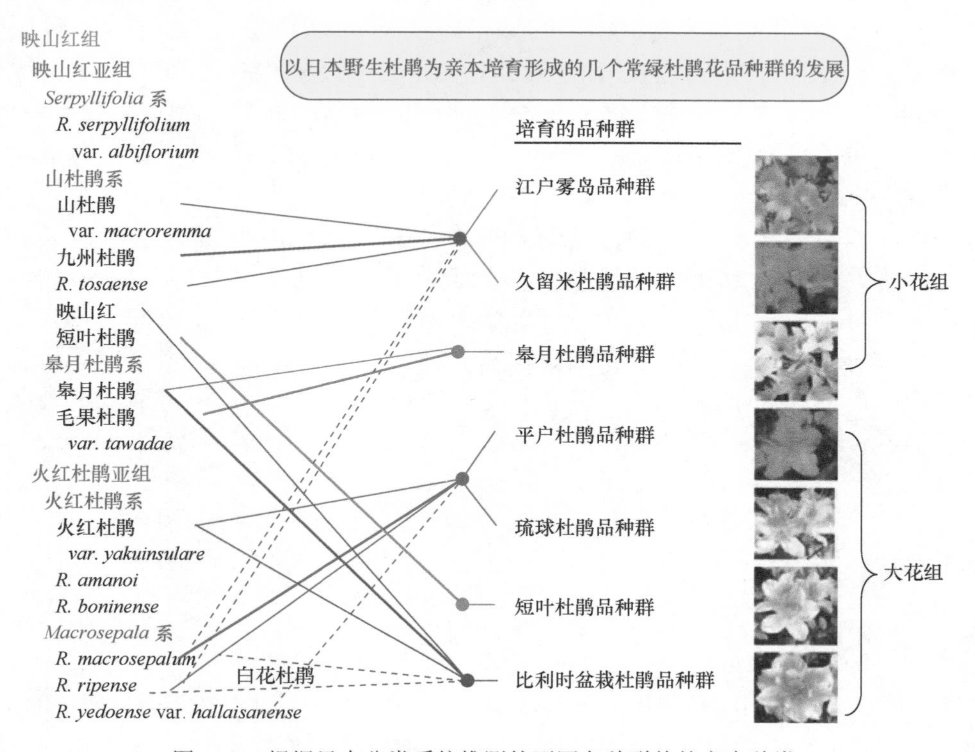

图 11.3 根据日本分类系统推测的不同杂种群体的亲本种类

在中国，Huang 和 Qiang（1984）根据表型性状和品种来源对杜鹃花进行了分类，分为 4 种类型：①东洋鹃，具有小叶、套瓣花的品种；②西洋鹃，为重瓣、半重瓣的大花品种；③毛鹃，单瓣大花，茎和叶密布细短柔毛的品种；④夏鹃，初夏开花的品种。这种模糊分类不利于实际应用，常导致不同品种使用相同名称或同一品种有多个名称（Zhou et al. 2013）。

11.3 遗传资源和育种基因库的特征与保育

11.3.1 西欧与亚洲联系的重建

1963～1997 年，Jozef Heursel 博士在比利时梅勒的 ILVO 负责比利时盆栽杜鹃花的育种项目。他积极寻找日本和中国的杜鹃花新品种与野生物种，以期拓宽西欧育种者可用的基因库。他是第一个重建欧洲和亚洲之间联系的人，先是与日本，然后与中国。他因在比利时培育日本杜鹃花杂交种而闻名，首次成功培育了适合室内栽培的日本杜鹃花（典型的户外种植的中型灌木）和比利时杜鹃花之间的杂交种。后来通过双方合作，他从中国科学院昆明植物研究所得到了不同野生种群杜鹃花的种子。这些野生杜鹃花种类大多数属于映山红亚属，其中 8 个是映山红。同时，他还引种了栽培种类，如锦绣杜鹃（*R.* × *pulchrum*）和白花杜鹃（*R.* × *mucronatum*）。如今，ILVO 拥有世界最大的比利时盆栽杜鹃花活植物种质资源库。该资源库中大约有 350 个映山红杂种与 250 个其他杜鹃花种类和变种。为了保护最重要和最有价值的遗传资源，ILVO 正在尝试对核心种质进行冷冻保存。核心种质包括 65 种资源（Van Huylenbroeck and Calsyn 2011）。至今，其中 60 种已成功启动了外植体芽诱导和组培扩繁。在冷冻保存方面，使用了 Verleysen 等（2005）开发的包埋脱水技术。迄今为止，ILVO 已成功冻存了 12 个品种（每个品种保存 3 个独立样品）。

ILVO、意大利和日本研究人员之间的联系是在 21 世纪初建立的，当时都灵大学希望挖掘坐落于意大利北部的历史园林中古老观赏植物品种的文化遗产特征。过去，杜鹃花从欧洲不同国家来到意大利，主要栽培在马焦雷湖地区（Lake Maggiore area，意大利北部），当地花卉种植者由此开始对其杂交和选择，建立了一个重要的杜鹃花基因库，在当地被归为三个不同的类群：①尹迪卡杜鹃群（Indica），以大花为特征的品种；②阿迈纳杜鹃群（Amoena），以极小、紫色花为特征的品种；③日本杜鹃群（Japonica），在形态特征上介于以上两者之间的品种（Scariot 2006）。

Heursel 建立的研究联系成为比利时、意大利、日本和中国的研究组之间持续密切合作的起点。

11.3.2 当前 ILVO 育种基因库中的遗传相似性

AFLP 分子标记的应用（De Riek et al. 1999）及平均 Jaccard 遗传相似性矩阵的构建，均使用主坐标分析方法建立一个育种基因库的二维分类图（图 11.4）。根据这些数据，盆栽杜鹃花可以清楚地与野生杜鹃花分开，它们与久留米杜鹃群聚在一起。平户杜鹃与火红杜鹃聚为一组。久留米杜鹃和平户杜鹃均与它们最重要的祖先聚在一起（分别是九州杜鹃和火红杜鹃）。

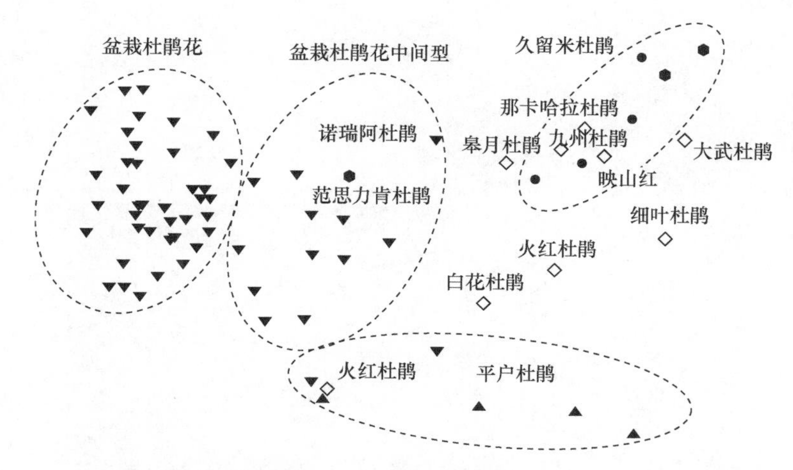

图 11.4 基于 AFLP 数据的育种者基因库的 PCO 图（改编自 De Riek et al. 1999）

De Riek 等（1999）研究表明映山红杂种可分为两个亚组：一组为比利时盆栽杜鹃花的所谓的原始类型（图 11.4）（盆栽杜鹃花）。植株呈球形，叶片深绿色，花胭脂红色、重瓣且花期早。该组的代表品种是‘帕特里克夫人’（‘Madame Petrick’）（1880 年）、‘保罗 • 沙姆’（‘Paul Schame’）（1890 年）、‘安布罗萨那’（‘Ambrosiana’）（1948 年）和‘莱因霍尔德•安布罗休斯’（‘Reinhold Ambrosius’）（1930 年）。在‘赫尔穆特 • 沃格尔’（‘Hellmut Vogel’）（1967 年）品种推出之前，以上杜鹃花品种栽培最为广泛。该亚组品种构成了当前商业杜鹃花品种的遗传基础。另一组的品种构成则较松散，包括一些中间类型（intermediate pot azalea）（图 11.4）。它们常被描述为开花晚，花单瓣或半重瓣，其中有些是老品种，如‘克里斯汀’（‘Coelestine’）（不出名）、‘沃尔特斯教授’（‘Professor Wolters’）（1871 年）、‘利奥波德 • 阿斯特里德’（‘Leopold-Astrid’）（1923 年）或‘禁酒’（‘Tempérance’）（1890 年）。其中一些是由杂交产生的中间类型。尖塔形诺娅杜鹃‘凯布斯’（‘Cheops’）是 *R. noriakianum* 的杂交后代；品种‘拉拉’（‘Lara’）和‘米思初’（‘Mistral’）是平户杜鹃与映山红的杂交后代；‘范思力肯主任’（‘Directeur Van Slycken’）是久留米杜鹃和映山红的杂交后代。‘克里斯汀’，一个非常古老的起源不明的品种，可能是一种杂交中间类型。它的球形株型、单瓣花和晚花期等特性与九州杜鹃和久留米杜鹃相似。该特性的部分性状在其后代‘格拉泽 10 号’（‘Glaser Nr. 10’）、‘多萝西 •吉斯’（‘Dorothy Gish’）、‘翠鸟’（‘Kingfisher’）、‘海克斯’（‘Hexe’）和‘欧原子’（‘Euratom’）套瓣花品种中仍有体现（图 11.5）。品种‘麒麟’（‘Kirin’ ）和‘雷克斯’（‘Rex’ ）是典型的久留米杜鹃花。品种‘克里斯汀’是一个突变品种。

品种‘禁酒’（‘Tempérance’）和‘詹姆斯 • 贝尔顿’（‘James Belton’）表现出了映山红杂种的典型特征：开花较晚，具典型木本植物的直立特性。‘禁酒’开不常见的淡紫色至蓝色花，‘詹姆斯 • 贝尔顿’（‘James Belton’）叶片小，绿色，具毛（De Riek et al. 1999）。

图 11.5　套瓣花，典型久留米杜鹃花

11.3.3　AFLP 分子标记在意大利常绿杜鹃花分类中的应用

如上所述，位于马焦雷湖地区的苗圃和历史园林拥有出色的花卉种质资源。然而至今未见关于其遗传多样性信息的报道。Scariot（2006）根据形态和分子数据进行特性描述，以阐明意大利杜鹃起源、分类和多样性信息。采用 AFLP 分子指纹图谱技术鉴定了 93 个杜鹃花种质的遗传信息[48 个尹迪卡杜鹃（Indica）品种，41 个日本杜鹃（Japonica）品种，4 个阿迈纳杜鹃（Amoena）品种]（Scariot 2006；De Riek et al. 2008）。尹迪卡杜鹃类群似乎来自一个称为“南部皋月杜鹃杂种”的群体，1840 年左右在比利时由常绿杜鹃花杂交种培育而来（被称为比利时皋月杜鹃杂种），或者来源于皋月杜鹃，该种杜鹃花早在 1680 年在荷兰以另一个植物学名报道过（Lee 1958）。此外，林奈首次在《植物种志》（*Species Plantarum*）（1753 年）中将杜鹃花归为皋月杜鹃（*Azalea indica*）。日本杜鹃类群名称的由来可能与九州杜鹃几个变种之一有关，在其名称里有“var. *japonicum*”或“forma *japonicum*”（Chamberlain and Rae 1990）。最后，阿迈纳类群应该属于日本杜鹃类群，因为它与久留米（Kurume）群体的‘愉悦’（‘Amoenum’）品种有关。为了提供有助于分类和系谱重建的信息，在数据分析中应用了一系列杜鹃花资源作为参考（Scariot et al. 2007）。首先，使用系统分析法（UPGMA 聚类和 PCO）研究了意大利基因库内以及意大利基因库和参考基因库之间的 DNA 序列相关性。

在 AFLP 分子标记数据聚类分析中（图 11.6），大多数尹迪卡杜鹃类群品种的样本主要采自古老庭院和公园，与日本杜鹃类群品种各自聚为一组，该样本主要采自苗圃。第三个本地类群阿迈纳杜鹃（Amoena）聚为单独的一支，与日本杜鹃（Japoinica）关系较近（Scariot 2006）。基于 AFLP 分子标记和形态数据的 PCO 分析结果与聚类分析的结果相似。尹迪卡杜鹃类群品种特征是花大，单瓣，紫色或白色花（较少见）和 9 个雄蕊。日本杜鹃类群品种特征为花白色到粉蓝色和 5 个雄蕊。

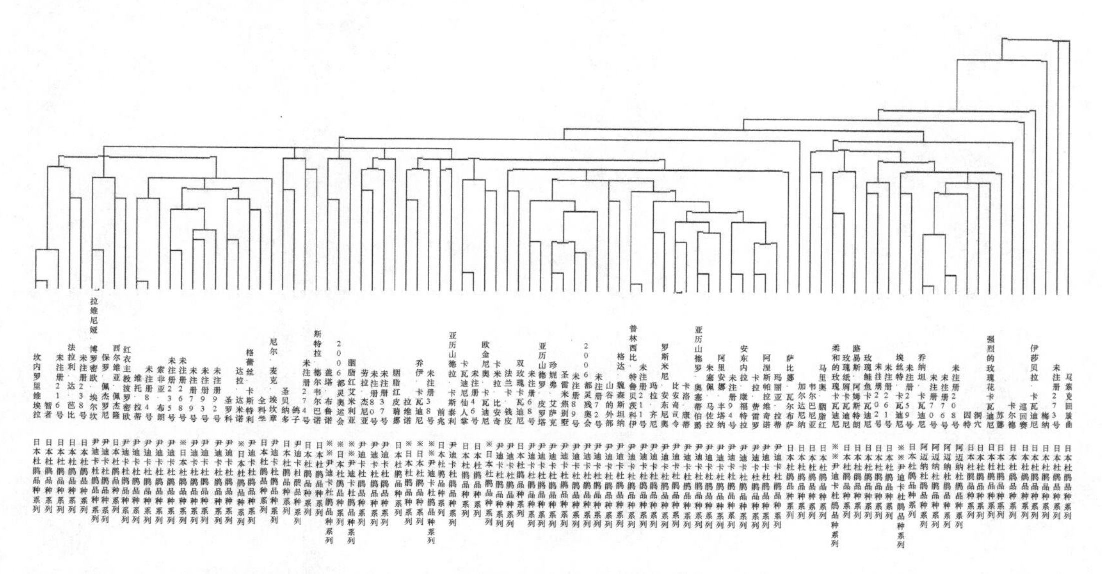

图 11.6 马焦雷湖地区（意大利北部）杜鹃花品种 AFLP 分子标记聚类图，标注了与当地园艺组的隶属关系（Scariot，2006）。尹迪卡（A）和日本杜鹃花 Japonica（B）组在图中已标。意大利品种已在皇家园艺协会的国际杜鹃花注册处注册。没有注册名称的品种标为‘Unito’（= Università di Torino）。※表示尹迪卡杜鹃组中包含的品种是根据 AFLP 数据进行测试分配的（software Doh）。※※表示日本杜鹃组中包含的品种是根据 AFLP 数据进行测试分配的（software Doh）

在下一步的聚类分析中，采用平均 Jaccard 相似性和赋值运算分析马焦雷湖地区杜鹃花品种与参考种质之间的遗传关系。有一些例外的情况，尹迪卡类群倾向与平户杜鹃和比利时杜鹃聚在一起，日本杜鹃和阿迈纳杜鹃类群与久留米及皋月杜鹃聚在一起。在两个园艺类群的所有植物之间进行成对比较后，平均杰卡相似度的计算结果表明：①尹迪卡杜鹃类群明显与比利时杜鹃和平户杜鹃杂交种有关；②阿迈纳杜鹃类群很可能来自品种‘愉悦’（‘Amoenum’），因此与久留米杜鹃杂交种关系更近；③日本杜鹃类群，来源多样，不是单一族系，与阿迈纳杜鹃类群、久留米杜鹃和皋月杜鹃杂交种具有较高的相似性（Scariot et al. 2007）。

根据同一个园艺类群内品种间相似程度和最相似种类，可以勾画出几个意大利种质的起源相似图（主要由形态学数据支持），平均 Jaccard 系数高（≥0.7）的意大利种质可能是芽变品种[如‘卡拉·费雷罗’（‘Carla Ferrero’）/‘阿格尼丝·派勒’（‘Agnese Pallavicino’）和 69 号（未注册）/‘圣雷米吉欧别墅’（‘Villa San Remigio’）或实生苗选育而来，如‘坎内罗里维埃拉’（‘Cannero Riviera’）/‘萨宾采拉’（‘Sabioncella’）和 207 号/94 号（为注册）]。同样，平均 Jaccard 系数在 0.5～0.7 应该表明意大利种质间有直接或间接亲缘关系（Scariot et al. 2007）。

通常 AFLP 技术结合形态学特征可以帮助阐明常绿杜鹃花基因资源的起源和分类。尹迪卡杜鹃类群遗传上接近平户杜鹃和比利时杜鹃类群，似乎是特意从比利时杜鹃类群中分离出来的一支，而日本杜鹃类群则来源比较广泛，与其他杂种杜鹃类群相关。所有数据均表明，日本杜鹃类群不应被归为一个单独或界限分明的类群。意大利阿迈纳杜鹃群体无疑是起源于久留米杜鹃类群（Scariot 2006；De Riek et al. 2008）。根据以上信息应当建立一个新分类体系，为将来种质保护和育种项目战略计划的制定奠定基础。

11.3.4　日本常绿杜鹃花品种与野生物种的遗传关系

在日本，特别是九州（Kyusyu）岛（日本岛南部），野生杜鹃花资源丰富，分布广泛。这些杜鹃花属于映山红亚属，是常绿杜鹃花的祖先。九州杜鹃（图 11.7）原产于九州火山高海拔地区，而山杜鹃则是九州至整个日本山脚下最常见的种类（Yamazaki 1996）。在雾岛山区的九州杜鹃和山杜鹃种间自然杂交基因渗入研究中，采用叶绿体DNA的 PCR-RFLP 标记检测到 rDNA 16S 序列 *Hha*I 限制酶切片段的以上两个物种的特异条带。花色多样的这个种间杂交群体，由许多具有九州杜鹃或山杜鹃遗传特征的个体组成。这表明九州杜鹃和山杜鹃是两个明显不同的物种，它们的自然杂交群体单株含有两个物种之一的叶绿体 DNA（cpDNA）（Kobayashi et al. 2000）。使用相同的 cpDNA 分子标记检测表明：云仙山地区表型多样的种群是九州杜鹃和山杜鹃种间杂交群体。在阿苏山脉（Aso Mountains）、Kujyu 山脉和周围山脉分布的种群，根据形态学特征分析认为九州杜鹃具有山杜鹃的叶绿体 DNA（cpDNA），因而推测山杜鹃和九州杜鹃之间存在细胞质渗入（Kobayashi et al. 2007）。AFLP 分析已经揭示了以上种群内部和种群之间的遗传变异。在九州杜鹃和山杜鹃中检测到一些特异条带，并应用 AFLP 多态性鉴定了每个种群的遗传构成（Handa et al. 2002）。对大花品种群的另一个祖先，即分布在日本山阴（San-in）

图 11.7　九州岛火山高海拔地区特有的九州杜鹃（*R. kiusianum*）

和四国（Shikoku）地区的 *R. ripense* Makino 自然种群，作为有潜在价值的育种材料进行了研究。采用形态特征和叶绿体 DNA 多态性评价了其遗传多样性。相对于大约 40 年前的植被特征，研究认为山阴地区的 *R. ripense* 野生生境因修建水坝和河道改造受到干扰。该地区的 *R. ripense* 花冠直径约 6cm，并在花冠特征和萼片长度上表现出局部形态多样性。与四国地区的 *R. ripense* 花朵相比，它们的花冠直径更大，颜色更深，叶子更宽。而且当地遗传多样性研究也检测到在叶绿体 DNA 的 *trnW-trnP* 内含子中存在序列突变（Kobayashi et al. 2008）。

11.3.5　中国品种和杂交群体

在中国，估计约有 300 个杜鹃花品种栽培在各地苗圃中，但是关于品种本身或品种之间关系的报道很少。有文献应用 AFLP 分子标记对 Huang 和 Qiang（1984）传统分类中东洋鹃、西洋鹃、毛鹃和夏鹃品种以及日本杜鹃、一些没有名称的品种与野生种类进行了亲缘关系分析（Zhou et al. 2013）。三组 AFLP 引物扩增产生了 408 个多态性标记，应用 NJ 树和 PCO 分析法对多态性数据进行了分析。根据分析结果，并结合开花时期和叶片形态，作者认为中国的夏季群可能属于皋月杂种群，西洋鹃很可能属于比利时杂种，或者说品种起源于比利时杜鹃。但是，在该研究中以没有文献记载的欧洲品种作为样本。所以，很难推断中国西洋鹃群体是否为本土映山红杂种群，该杂种群被认为是所有比利时大花型杂交品种的起源，或者是来自日本和欧洲且起源于映山红品种群的第二个分化中心。貌似可信的东洋鹃和久留米杂种群之间的关系有点复杂。大部分东洋鹃品种聚集在一个单独的分支中，其小叶和套瓣小花为典型的久留米杜鹃形态特征。其余品种与毛鹃品种的来源需要进一步的研究证据和参考资料。中国杜鹃花品种的命名有些混乱，一个品种有时会有两个名字。中国品种（可能还有来自国外的品种）之间已经发生相互杂交和与当地杂种间产生了基因混合。Xu 等（2016）将一组杜鹃花品种分为春花型、夏花型和春夏花型，得到了类似于 Zhou 等（2013）的研究结果。AFLP 和 iSSR 分子标记明显地将夏花型品种归为一类，在聚类中同时考虑了叶片形态与开花习性。由于样本的品种数量较少，因此该研究结论的可采纳性低于 Zhou 等（2013）学者的研究。

11.4　比利时盆栽杜鹃花及相关常绿杜鹃花的育种

11.4.1　育种目标

比利时盆栽杜鹃花育种史上的几个重大突破：①品种实现了自根苗的培育，不再受限于嫁接技术；②早花（秋季开花）品种的成功培育，以品种‘赫尔穆特·沃格尔’为代表；③成品盆栽观赏期延长。传统上，育种目标依据大规模生产目标和消费者意愿而改变。近 40 年来，杜鹃花育种目标一直在改变，如从小型花到大型重瓣花，从彩色花到白色花等。消费者总是希望看到新的品种，而且需求会变化很快。

直到今天，花朵特征仍然是最重要的选择目标，主要包括花色、花型、花大小、花形、色斑和花蕾颜色。此外，叶片形状和颜色、叶表面光泽、是否被毛、生长势、株型、早花性状与成品花质量（如花期长短、没有褐色芽鳞）等都是育种关注的重要性状。最近，植物抵抗生物因子的抗性育种受到关注，特别是对某些真菌疾病的抗性已成为主要的选择标准。此外，早期候选品种已经开始在商业环境条件下测试，以评估它们对特定培养条件的适应性反应，如修剪、生长调节剂，以及胁迫环境对植物生长发育的影响。

未来，杜鹃花育种的主要挑战之一可能是选择需要较少植物生长调节剂的品种。一个有趣的观点是，常绿杜鹃花的其他潜在育种目标（包括非生物胁迫、香味）和达到这些育种目标的可用遗传资源应该由 Kobayashi 在 De Riek 等（2008）研究基础上提出。

11.4.2 杂交

在现有类群中进行常规杂交和选择仍是杜鹃花新品种培育的主要方式。阻碍杂交进程的因素有多种，如早花和晚花品种的花期不遇或者不育花尤其是重瓣花缺少雄蕊或心皮畸形，均是限制杂交的重要因子。杜鹃花是异花传粉植物，也可以自花授粉，但会导致近交衰退，如结实率低和后代生活力下降。

直到 20 世纪 90 年代，盆栽杜鹃花常规育种的一般规则仍是从现有类群中选择优良亲本配对杂交，而且亲本的选择不是十分严谨。从 90 年代开始，制定了更加严格的育种计划，需要改良的目标性状明确。按照计划实施育种操作，需要大约 10 年的时间才能推出一个新品种（图 11.8）。

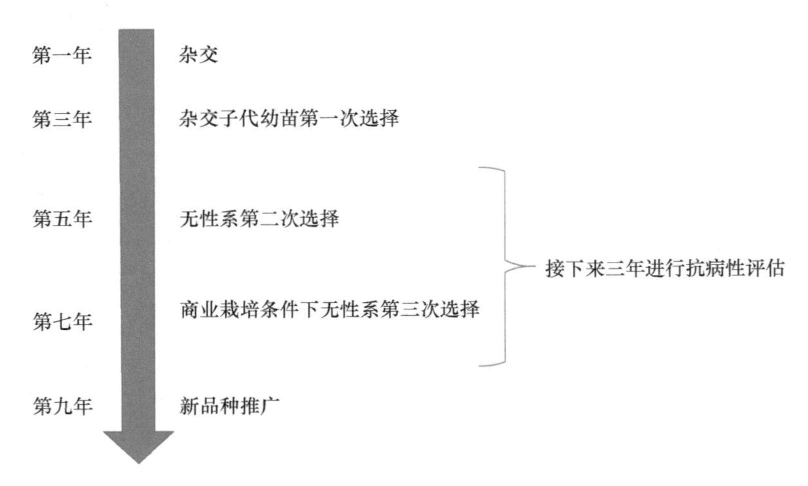

图 11.8　ILVO 商业盆栽杜鹃花育种计划示意图

11.4.3 种间杂交

在杜鹃花属植物中，同一个亚属物种间杂交通常易于成功，不同亚属的种间杂交存在明显的杂交障碍（Heyting 1970）。杜鹃花亚属（主要是类越橘杜鹃花组）与映山红亚属种间杂交研究表明，杂交不亲和性可以出现在从授粉到开花的任一阶段（Rouse 1993）。

Preil 和 Ebbinghaus（1985）在单系杂交中使用了来自多个亚属的物种，但成功率很低。然而，杜鹃花和高山杜鹃类群的种间杂交也不是不可能的（Bowers 1936）。Salley 和 Greer（1992）曾报道了获得了腋花杜鹃（*R. racemosum*）与常绿杜鹃花种类之间的杂交后代。映山红亚属和羊踯躅亚属种间三向杂交得到了杂交苗，子代获得了羊踯躅亚属的黄色花基因，变成（半）落叶且生长势差（Ureshino et al. 1998）。为了培育新花色的映山红杂种，如黄色、橙色、蓝色花，Eeckhaut（2003）启动了一个开拓性的种间杂交育种计划，使用了常绿杜鹃花亚属、羊踯躅亚属和杜鹃花亚属的物种或杂种作为亲本。针对合子形成前障碍，对亲本倍性和花粉管生长进行了检测与研究；针对合子形成后障碍，开展了胚胎和胚珠培养。映山红亚属种类为母本和羊踯躅亚属为父本的杂交苗因白化病死去，而在反交情况下，花粉在羊踯躅亚属种类的柱头上生长停滞。类似的单向杂交不亲和性在类越橘杜鹃花组与映山红亚属、常绿杜鹃花亚属与映山红亚属的种间杂交中也有表现。以上两个杂交组合中，当映山红亚属种类作为父本（花粉供体）时能获得杂种。杜鹃花亚属种间杂交可以得到杂种后代，尤其是常绿杜鹃花亚属或杜鹃花亚属与映山红亚属的种间杂种。如前所述，花园里栽培的杜鹃花与常绿杜鹃花亲缘关系更近（Kron 2000）。这可能正好解释了使用杜鹃花亚属或常绿杜鹃花亚属以及羊踯躅亚属的种类进行杂交可以获得一定的成功率。获得的杂交子代的特征可以用 AFLP 和 SSR 标记检测。尽管杂交结果令人鼓舞，但期望获得新花色杜鹃花品种的最终目标仍未实现，主要是因为很难获得 F_2 子代和大多数杂种的生长势较差。

映山红杂种与南湖大山杜鹃（*R. noriakianum*，原产于台湾北部）的种间杂交成功获得了杂种（Heursel 1999）。尽管南湖大山杜鹃观赏价值不高，但它具有自然直立生长习性。在 F_2 代群体中，可以选择具有自然金字塔树形和引人注目的花朵、花色的单株，从中选择培育成功的新品种商品名为‘诺利阿杂种’（‘Noria hybrids’）。种间杂交育种成功的另一个例子是将原产中国的茶绒杜鹃（*R. rufulum*）香味引入杜鹃花基因库中（Van Huylenbroeck et al. 2018，刊印中）。Kobayashi 等（2008）也成功地获得了不同亚属的种间杂种，生长势旺盛且花有香味，该杂种选育自那克哈杜鹃（*R. nakaharae*）或者其杂种（映山红亚属）和落叶具有花香的杜鹃花 *R. arborescens* 或 *R. viscosum*（羊踯躅亚属）之间的互交子代群体。存活的杂交子代具有双亲的核糖体 DNA（nrDNA）、母本的线粒体 DNA（mtDNA），叶绿体 DNA（cpDNA）则来自落叶杜鹃，与杂交组合无关。以上结果表明，落叶杜鹃花的叶绿体基因组与常绿杜鹃花的核基因组在杂种后代中是共融的。关于这些亚属间杂种后代的鉴定，基于核内转录间隔区（ITS）和叶绿体 *matK-trnK* 区的序列开发了新型多重 PCR 标记（Mizuta et al. 2008）。这些多重 PCR 标记可以轻松、快速地区分常绿和落叶杜鹃花，以及有活性杂种的 DNA 遗传模式。分子标记辅助育种可用于研究杜鹃花广泛的交叉杂交中复杂的核基因组和细胞器基因组杂交机制。

11.4.4　其他育种方法

杜鹃花新品种培育的另一种重要方式是芽变选种。据估计，盆栽杜鹃花栽培品种中有 52%来自芽变（Horn 2002）。‘赫尔穆特·沃格尔’是一个开花很早的杜鹃花品种，于 1967 年育成，以其易于产生芽变而闻名。近几十年来，已经从其芽变中选择培育了 30 多个品种。芽变是基因表达过程中发生的一种体细胞突变，可形成表型上具明显差异的芽（De Schepper et al. 2003）。芽变可自发产生，也可以用外源激素诱导外植体（Samyn et al. 2002）或用低剂量 X 射线辐射产生（De Loose 1979）。

在杜鹃花品种培育中，很少有育种者诱导多倍体。因此，几乎所有的商业品种都是二倍体。使用抗有丝分裂剂，虽然经常会诱导出多倍体嵌合体，但纯四倍体也是可以得到的（Eeckhaut 2003）。第二种获得四倍体的方法是以嵌合体中四倍体花瓣边缘组织为外植体进行培养再生（De Schepper et al. 2004）。花朵芽变呈现多色，花瓣宽大（>7mm），不同颜色的花瓣边缘被证明是四倍体，而其余花组织是二倍体（图 11.9）（De Schepper et al. 2001b）。四倍体杜鹃花有异常的花形，花瓣更厚，特别是在花中心。其他“典型”多倍体特征是育性降低（花药发育不良和开花异常），叶片变脆，长宽比明显降低和叶基增厚（Eeckhaut et al. 2006）。至今，多倍体杜鹃花的商业价值看来仍很有限。

图 11.9 （a）阔白边花瓣芽变‘马塞拉’（‘Marcella’）；（b）窄白边花瓣芽变‘格达·基森’（‘Gerda Keessen’）

采用农杆菌介导法对常绿杜鹃花进行遗传转化，获得了可育的转基因植株（Mertens et al. 2000）。这项研究始于 20 世纪 90 年代，但由于转化效率低和欧盟消费者对转基因生物的抗拒而很快终止。

11.5 主要目标性状介绍

11.5.1 花色

对于所有开花植物，花朵特征，尤其是花朵颜色是最重要的育种目标。杜鹃花花色丰富，具有从紫色、胭脂红、红色、粉红色到白色系列花色。类黄酮物质是杜鹃花花色的主要着色成分（De Loose 1969）。类黄酮由一定的分支途径合成，可生成可着色的色素（花青素）和无色的黄酮醇。根据 RHS 花色图谱，常绿杜鹃花野生种类和品种的花色分为红色与紫色组。依据红色组的色差指标 a^*和 b^*值，如红色花的山杜鹃，仅包含花青素系列色素，因此呈现相近的花色分布。另外，紫色组花色，如紫色花的九州杜鹃和 *R. ripense*，同时含有花青素和飞燕草素系列色素，因此呈现的花色更加丰富。根据 HPLC 分析，红色组样品含有 2～4 个主要花色苷类型，而紫色组样品具有 2～6 个类型。

久留米杜鹃品种群的主要花色苷类型，与其祖先野生物种九州杜鹃和山杜鹃花瓣中的花色苷类型一致，呈现丰富的花色变异（Mizuta et al. 2009）。锦绣杜鹃及其红色花芽变品种‘大村崎’（‘Oomurasaki’）的花瓣色素分析表明，其含有花青素和飞燕草素系列的花色苷，而红色花芽变品种仅含有花青素系列色素。*F3'5'H* 基因转录水平较低导致红色花芽变品种的花瓣中缺乏飞燕草素系列色素的积累（Mizuta et al. 2010）。

杜鹃花花色分离由三个主效基因（*P*、*W* 和 *Q*）决定，且遵循孟德尔遗传规律（Heursel and Horn 1977），但是该分离规律不能阐明粉红色花的成因，推测粉色的形成与基因表达量有关。此外，杜鹃花花瓣有时出现花斑，中心和边缘具有不同的颜色，推测花斑是由转座子活动引起的（De Schepper et al. 2001a）。芽变造成的花色和花斑改变在杜鹃花中很常见。

作为杜鹃花花色新品种培育的工具，从锦绣杜鹃品种‘大村崎’的花瓣中分离了参与类黄酮生物合成途径的 8 个结构基因的 cDNA 序列：查耳酮合酶（*CHS*）、查耳酮异构酶（*CHI*）、黄酮 3-羟化酶（*F3H*）、黄酮 3'-羟化酶（*F3'H*）、类黄酮 3', 5'-羟化酶（*F3'5'H*）、二氢黄酮醇还原酶（*DFR*）、花色素苷合成酶（*ANS*）和黄酮合成酶（*FLS*），并对这 8 个基因在杜鹃花花发育过程中的表达模式进行了研究。在花瓣发育期中 *F3'5'H* 基因表达量最高，*F3'5'H* 基因与杜鹃花花瓣着色密切相关（Nakatsuka et al. 2008）。

De Keyser 等（2013a，2013b）对杜鹃花花色呈色机制进行了深入研究。采用简单的表型分类结合图像分析对两种粉红色花杜鹃花（‘西玛’和实生苗‘98-13-4’）的后代花色进行了分析。可见花色分为 4 类：胭脂红、砖红、淡粉红色（包括有斑点的花朵）和白色。根据图像分析结果，每种特定花色可以表示为 RGB 值[*R*/(*G*+*B*)、*R*/*G* 和 *R*/*B*]（图 11.10）。*R*/(*G*+*B*)值是一个良好的花色度量值，白花的 *R*/(*G*+*B*)值接近 0.5，而彩色花的值显著更高，*R*/*B* 值在白花和彩色花之间有显著差异（粉色花除外），可以看作共着色

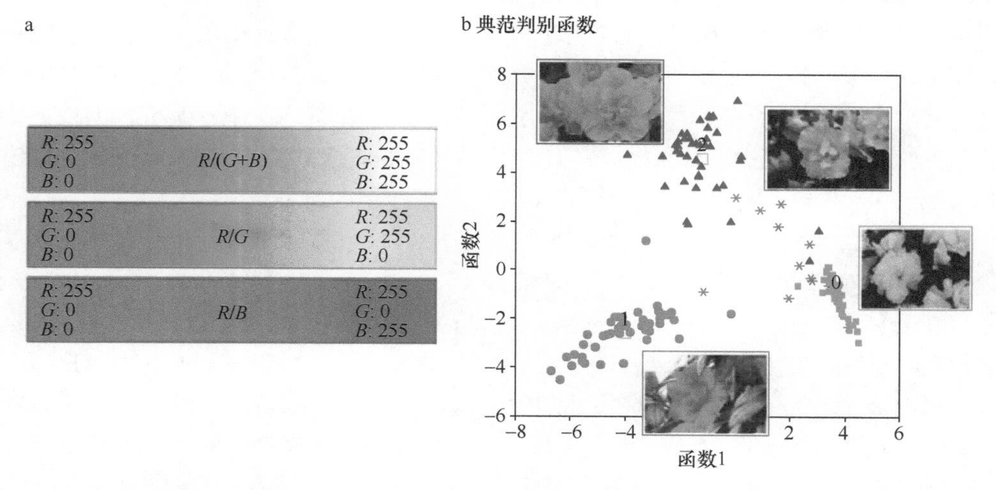

图 11.10　花朵颜色分析：(a) RGB 参数的范围，*R* 值均设为 255，*G* 和 *B*（上面方框）、*G*（中间方框）、*B*（下面方框）值在 0～255；(b) 2D 散点图，为胭脂红色（三角形）、红色（圆形）和白色（正方形）花的 RGB 与 HSV 值分析图。粉红色和有花斑的花（星）散布在其他组之间，靠近与其相近的类群（根据 De Keyser et al. 2013b 绘制）

基因的表达，相对于红色花，其胭脂红色花的表达量更高。用 *R/G* 值变化不能轻易区分颜色组，但可以检测红色的浓淡。

根据 RGB 值可把粉色和有花斑的花色与其他花色组区分开来（图 11.10），可以识别两种不同的粉色：胭脂粉色（近胭脂红色）和红粉色（近红色）。这种粉花色的差异在亲本花色中也有体现：'西玛'（'Sima'）的花色属于胭脂粉色，而父本'98-13-4'为红粉色。图像分析把这两种颜色归为有类黄酮醇和无类黄酮醇的粉色。子代中红花色的出现进一步肯定了花色苷基因表达量影响花色呈现的推论。假如粉色是由显性等位基因控制，后代群体中应出现较多的粉色花。在花色组中花色苷含量的变化使粉色与红色（胭脂红色）界限不清楚，图像分析则更加精确。图像分析产生的系列数据也可以用于构建 QTL 花色图谱。两个主要花色 QTL 图谱已经构建完成（DeKeyser et al. 2013b）。QTL 位点与 Heursel 和 Horn（1977）确定的 *Q* 及 *W* 基因位点相吻合。但是没有发现粉色的主要调控 QTL 位点，可能由于群体中粉色花出现的频率较低。然而，两个次要的表示花色强度的 QTL 位点可以更恰当地描绘为粉色基因区域。但不幸的是，在这些 QTL 位点中没有备选基因。以上研究结果表明，粉色不是由生物合成中基因突变所致，而是由于液泡中色素的数量积累较少。

最近，使用柯尼卡美能达 CM700D 色差仪的 CIEL * *a* * *b* *色彩指数进行杜鹃花色的测定，这样可以区分不同的花色组。而且，RHS 颜色代码都可以转换为 CIEL * *a* * *b* *值，这样育种者可以用色差仪快速、客观地测定花色，从而确定申请植物新品种保护所需的 RHS 代码。

11.5.2 花型

商业杜鹃花栽培品种大多具有单瓣花、套瓣花和重瓣花。ILVO 最近培育出了 AIKO® 杜鹃花，具有特殊的山茶花形状的花朵（图 11.11）。这种高度重瓣的花使得该系列品种的花期大为延长。日本在江户时代就开始从野生种群中选育花型突变单株培育杜鹃花新品种。花型突变包括套瓣花型、似花瓣的持久花型、萼片瓣化花型；雌雄蕊演变的蜘蛛状花型；雄蕊瓣化或演变为细长叶状的花型等（图 11.11，图 11.12）（Kobayashi et al. 2010；Kobayashi 2013）。有些品种的花型组合了以上不同花型的特征。

花器官构造可用 ABC 模型诠释，该模型由 MADS 基因组成。为了把这些特殊花型品种用于杜鹃花育种，已经有学者对 MADS 相关基因及其遗传特性进行了研究。从常绿杜鹃花品种中，已经分离了主要的 MADS 基因：APETALA1/FRUITFULL 型（A 类）、APETALA3 型和 PISTILLATA 型（B 类）、AGAMOUS 型（C 类）和 SEPALATA 型（E 类）。使用同源突变品种已经对这些基因的表达模式进行了分析。与野生型相比，相应基因的表达模式在突变的花器官中发生了变化。花型特征的改变是由同源基因突变引起的。利用后代杂交对这些突变的遗传模式进行了研究，并在套瓣花和其他花型突变中开发了表型选择 DNA 标记（Cheon et al. 2016，2017a，2017b；Tasaki et al. 2014，2015）。在 AIKO®杜鹃花中对这些基因也进行了研究，高度重瓣花型似乎与一个 B 类（APETALA3）和 C 类基因（AGAMOUS）的差异有关。

图 11.11 花型突变的积累效应。左上方是茶花形状；右下角是长花期的绿色花朵

图 11.12 蜘蛛形花突变

11.5.3 株型

植物株型主要由枝条分枝状况、株高和花序形态等因素决定。杜鹃花的株型似乎是多结构的，即没有清晰的树干，所有枝条粗细相似。在许多种类中，杜鹃花生长素信号

来源于主枝芽顶端，可调控侧芽的向外生长并抑制侧枝发育（Brown et al. 1967；Cline 1991）。摘心可减轻主芽的抑制作用（Schmitz and Theres 2005）。在杜鹃花培育中，通过在固定的发育阶段摘心来调控侧芽生长以促使其形成需要的球形株型。增加摘心步骤时，摘心后形成的芽数减少（Heursel and Volckaert 1980）。育种者主要选择株型优良和早期分枝的幼苗。选择其他株型的品种，就像 Noria 杂种群具有尖塔形树型，为满足市场需求则无须修剪。

迄今为止，尚未见到关于杜鹃花中涉及植物株型基因的信息。确定植物株型 QTL 图谱的第一步，需要记录三个绘图种群“A × B”、“C × D”和“E × F”的幼苗分枝数、枝条长、每枝叶片的数量以及植物总面积（De Keyser et al. 2010）。采集第一次和第三次摘心后幼苗分枝数据。第三次摘心后，每个枝条的叶数和枝条长度也具有决定作用（图 11.13）。在第三次摘心后 9 个月使用图像分析法对植物总面积进行测量。

图 11.13 植物株型结构。（a）通过测量第一次摘心（P1）和 5 个分枝第三次摘心（P3-1～P3-5）后形成的枝条数预估植物分枝习性与生长势，在每个分枝上形成的最长枝条，其分枝长和叶数量是确定的；（b）直立型盆栽杜鹃花；（c）低矮丛生型杜鹃花

根据杜鹃花 UPOV 测试指南（www.upov.org），定义了三类生长习性的株型：直立型、宽阔丛生型和扁平丛生型。我们把亲本植株的株型分为丛生型（A、C 和 D）和直立型（B、E 和 F），各 3 个级别。直立型以分枝较少为特征，而丛生型以分枝较多为特征。

所有分枝特性的测量值在类群内有分离，而类群之间差异显著。在种群“C × D”中，第三次摘心后平均形成了 3.4 个新枝，而种群“E × F”和“A × B”分别形成 4.2 个和 4.6 个新枝（表 11.3）。种群“C × D”父母本均倾向于自然球形株型，第一次摘心后产生的新枝数量最多，但是在第三次摘心后，形成的新枝数量相对于其他种群明显减少。Heursel 和 Volckaert（1980）也描述了杜鹃花经过频繁地摘心后新形成枝条数量减少的状况，这明显不是“A × B”和“E × F”种群应该出现的现象，应该描述为是具有直立生长特性的种群或者亲本中至少有一个属于直立生长型。第一次摘心后，形成较少的新枝数应是直立型株型的好兆头。种群“A × B”是丛生型和直立型亲本间的杂交群体，预期会有中间型的株型出现。但是，该种群新枝总数量（第一次和第三次摘心后分枝总数）为 14.3，低于种群“E × F”的 16.0 个。该结果表明直立型是一种显性性状。当用植物面积比较时，得出的结论是相似的。种群“C × D”平均植物面积显著高于种群“E × F”。

表 11.3　植物株型特征的平均值

	种群	均值	标准差	最小值	最大值
第一次摘心后新枝数	A × B	3.1[a]	2.0	0	20
	C × D	5.6[c]	2.6	1	15
	E × F	3.8[b]	1.8	0	11
第三次摘心后新枝数	A × B	4.6[c]	1.7	1	8.0
	C × D	3.4[a]	1.2	2.0	5.5
	E × F	4.2[b]	1.5	2.4	8.0
植物面积（cm^2）	A × B	887.7[a,b]	320.4	265.3	1900.2
	C × D	950.5[b]	288.9	281.9	1986.4
	E × F	844.6[a]	294.6	212.0	1887.7

第一次和第三次摘心后新枝数（Kruskal-Wallis，$p<0.05$）与植物面积（Tukey HSD，$p<0.05$）在群体之间的差异显著性用 a、b、c 表示

只有种群“E × F”的植物面积与第一次摘心后形成的新枝数相关。相关性分析表明，植物面积主要由枝条长度决定，而与分枝方式相关性不大。在种群“A × B”中，第一次摘心后形成的新枝数量与第三次摘心后的相关参数一致。但是，在其他种群中，则不存在这种相关性。在两个种群中，新枝数似乎也与叶片数相关。当着生更多叶片时，可能意味着枝条具有生长的潜力。

11.5.4　叶片形态

在杜鹃花品种选育中，叶片大小、颜色和光泽度均是重要的性状，因为盆花是在花芽开放前上市的，需要依靠叶片吸引消费者。但是，关于杜鹃花叶片特性的遗传信息很少。De Keyser 等（2013b）曾使用经典参数和椭圆傅里叶描述子描述了叶片形态。两种方法 QTL 作图的输入值均是来自绘图群体的 4 株不相关的杜鹃花。由于以提高未来育种效率为目的，多种群一致的 QTL 相对于次要种群特的 QTL 更加具有吸引力。QTL 作图揭示了杜鹃花叶片形态性状遗传网络的复杂性。一些与叶片形状和大小相关的主要 QTL 位点可以确定，主要与叶片长宽比有关，在整个作图种群中具有一致性。这些 QTL 位点是将来性状选择分析的最吸引力的标记，但是仍需要更多的研究来确定最有价值的 QTL，以方便将来在 MAS 中进行叶片形态的选择。在叶色方面，QTL 图谱也给出了一些遗传信息。叶子的颜色由三种色素决定：叶绿素、类胡萝卜素和花青素。在绿色叶片中，叶绿素占主导作用。叶子颜色可以用 RGB 值定量（De Keyser et al. 2013b）。虽然基因型差异导致不同类群间叶色很难区分，但图像分析仍然能够为 QTL 分析生成有价值的参数。在所有类群中仅发现了一个主要的 QTL 位点，一些次要的种群特异性 QTL 作为补充。所幸，叶色呈现是由少量基因调控的，似乎并不是一个很复杂的性状。

11.5.5　香味

关于杜鹃花香味的报道较少。在 20 世纪 80 年代，从映山红与一个平户杜鹃品种‘和

平无光’（‘Heiwa-no-hikari’）杂交后代中选育了几个有香味的品种，分别为‘拉拉’（‘Lara’）、‘米斯特拉尔’（‘Mistral’）和‘马克·范·埃特维尔德夫人’（‘Mevrouw Marc Van Eetvelde’）（Heursel 1999）。但是这几个品种生长过于旺盛，限制了它们的商业用途。大约 15 年前，ILVO 启动了一项育种计划，把茶绒杜鹃的香味引入映山红杂种中，得到 F_1 代杂种后，开始了广泛而复杂的回交育种。预计仍需要一至两代回交才能选育出商业用香味品种（Van Huylenbroeck et al. 2018，刊印中）。Kobayashi 等（2008）报道了常绿杜鹃花那克哈杜鹃（*R. nakaharae*）或其杂种（映山红亚属）和落叶具香味的杜鹃花树形杜鹃（*R. arborescens*）或 *R. viscosum*（羊踯躅亚属）杂交成功的范例。

11.5.6 抗生物胁迫

在过去 10 年中，生物胁迫抗性成为杜鹃花育种的重要目标。特别是盆花生产过程中感染柠檬疫霉菌（*Phytophthora plurivora*）和丽赤壳属病原菌（*Calonectria pauciramosa*，原名 *Cylindrocladium* spp.）所致的两种真菌病害，目前已纳入抗病育种项目。针对这两种真菌开展了生物测试实验，目的就是选择对这两类病害抗性较强的候选品种（Van Huylenbroeck et al. 2015）。该生物实验采用 10 周龄的生根插条。在接种方式上，可以用疫霉菌活性孢子（*Phytophthora citricola*）悬浮液喷洒直至覆盖整个叶片表面，或者用丽赤壳属（*Calonectria*）病原菌孢子溶液浸泡处理茎基部。接种 3～4 周后，两种真菌的危害情况可以用肉眼清楚地观察到。Van Huylenbroeck 等（2015）概述了 ILVO 育种项目连续 4 年（2010～2013 年）开展的抗病筛选结果。结果表明，用柠檬疫霉菌或丽赤壳属菌株接种生根的插条是测试杜鹃花种质对病菌敏感性的适当方法。与以前使用离体叶片进行生物测试方法相比（De Keyser et al. 2008），生根插条测试法提供了更加可靠和稳定的数据，这在长达数年的不同筛选实验中均得到了证实。在多年的测试中，大多数品种或品系均得到了稳定的测试数据，其中包含了众所周知的抗病或易感品种，这种测试稳定性让不同测试实验的筛选结果可以互相比较，同时为种质对所有病害的易感性或耐受性提供了很好的指示。多年来观测到的评分结果变化可能是由外部气候因素、插条质量不同等原因造成的。为此，选择方案中的抗病测试要在几年内重复进行（图 11.8）。

最近，由于一些广谱性农药的禁用，茶黄螨（*Polyphagotarsonemus latus*）危害成为杜鹃花栽培的重要问题。该害虫以幼嫩的顶芽为食，可导致花芽和顶生叶畸形（图 11.14），受损叶片严重发育不良和变硬、边缘卷曲，叶下表面呈褐色或红色。Luypaert 等（2014）选择了 32 个杜鹃花品种（主要是映山红杂种），测试它们对茶黄螨的敏感性。在温室中人工接种螨虫，对杜鹃花每个顶芽上接种螨虫数量进行计数，然后统计危害率。尽管测试结果表明危害率和接种螨虫数量有显著相关性，但以单个品种为观测对象时，危害率和接种螨虫数量间的直接相关性并不显著。在检测的品种中，‘艾琳’（‘Elien’）和‘米斯特拉尔’表现出对该螨虫的完全抗性（Luypaert et al. 2014）。

后来，试验选择了对茶黄螨敏感性不同的 6 个基因型品种开展了进一步研究：6 个品种分别为‘北极光’（‘Nordlicht’）、‘艾琳’（‘Elien’）、‘爱子粉’（‘Aiko Pink’）、‘米歇尔·玛丽’（‘Michelle Marie’）、‘杰拉德·金特夫人’（‘Mevrouw Gerard Kint’）和‘萨

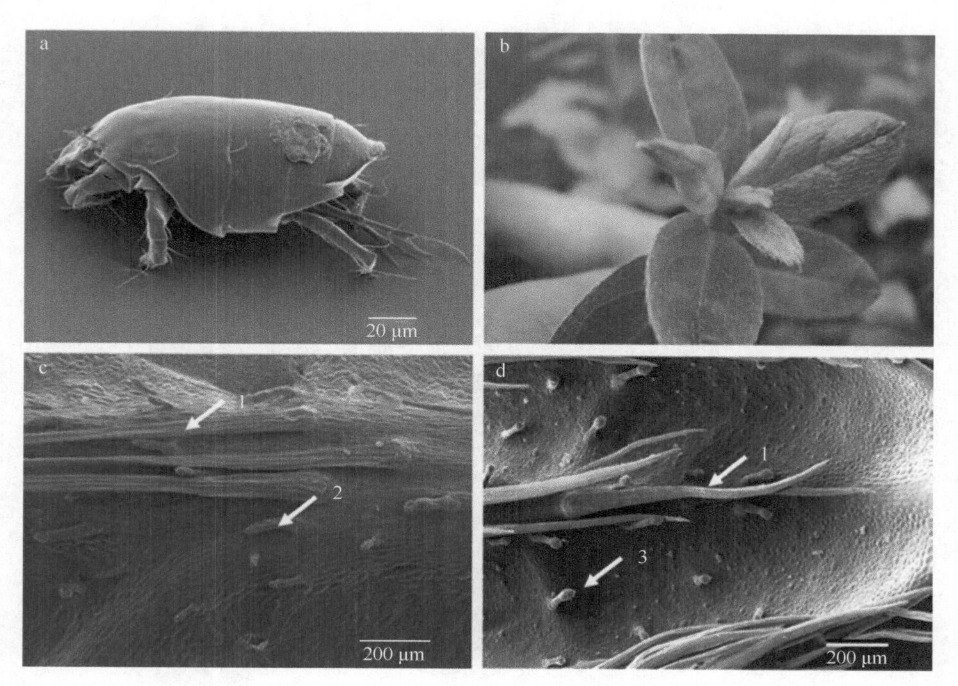

图 11.14　杜鹃花上茶黄螨（*Polyphagotarsonemus latus*）。（a）成年雌虫；（b）受损叶片；（c）映山红品种‘北极光’（*R. simsii*‘Nordlicht’）的表皮毛状体，（1）长毛状体和（2）短毛状体；（d）映山红品种‘艾琳’（*R. simsii*‘Elien’）表皮毛状体，（1）长毛状体和（3）带有黏性末端的短毛状体

河森斯特恩’（‘Sachsenstern’）。试验结果表明，不同基因型对该螨虫感染的反应差异明显。接种 6 周后，‘艾琳’几乎没有危害症状，植株上也没有检测到螨虫存在，而‘爱子粉’与‘北极光’危害率很高和螨虫数量很多（平均 50 头螨虫以上）。同样条件下，品种‘米歇尔·玛丽’螨虫危害率也很高，而植株上检测到的螨虫数量很少。品种‘萨河森斯特恩’和‘杰拉德·金特夫人’都受到了中等程度的损害，但几乎没有检测到螨虫。为了探究以上杜鹃花品种对该类螨虫的抗性，从植物形态特征和防御途径两个方面进行了研究。电子显微镜扫描结果揭示，不同形态的表皮毛与抗螨虫程度有关，抗性品种‘艾琳’和‘米斯特拉尔’着生典型的叶表面腺体毛（Luypaert et al. 2015）。螨虫可能被这些腺毛产生的黏性和/或有毒分泌物困住。而品种‘马克·范·埃特维尔德’（‘Marc Van Eetvelde’）也着生叶表腺体毛，但容易受螨虫危害。值得注意的是这些品种的培育亲本，品种‘米斯特拉尔’和‘马克·范·埃特维尔德’的亲本之一是品种‘和平无光’，而品种‘艾琳’是从‘米斯特拉尔’芽变中选育而来，因此它也与品种‘和平无光’有关。这种可观测到的抗性也许可追溯到平户杜鹃类群（Hirado azalea）。筛选这种典型腺体毛可以成为抗性品种培育的一种策略，由于腺体毛着生在植物营养器官上，因此可以在杂交后快速选择具有该性状的植株。但是作为观赏植物，腺体毛产生黏性分泌物是其劣势，不是市场需要的特性。因而市场对培育具有高密度腺体毛的杜鹃花品种兴趣不大。

以品种‘北极光’和‘艾琳’为材料开展的 RNA-Seq 实验结果表明，茉莉酸（JA）和水杨酸（SA）途径参与了茶黄螨的侵染过程。实验开发了一组有效的标记基因，可用于鉴定映山红杂种生理学过程中以上途径的作用。结果表明，在映山红基因型中，螨虫侵染后可诱发 JA 反应。基因表达分析和植物激素测定结果也表明，敏感基因型品种‘北极光’在螨虫侵染后产生 SA，而在抗性基因型品种‘艾琳’中则未检测到 SA，这表明 SA 水平越高则该品种对螨虫侵染的敏感性越高（Luypaert 2015）。然而，目前仍需要进一步研究阐明以上两种途径对螨虫适应性的机制。

11.5.7 非生物抗逆性

在盆栽杜鹃花中，非生物抗逆性不是主要育种目标，因为在整个盆栽期间，影响杜鹃花生长的气候因子、水分和营养条件都是人工控制的。然而，当杜鹃花在园林中栽培应用时（如在日本），非生物胁迫抗性就变得更为重要。杜鹃花栽培的一个重要限制因子是其适宜土壤 pH 在 4.5～6.0。在钙质土壤中，经常出现严重的缺铁失绿症状和生长抑制现象。Demasi（2015）曾筛选了几个常绿杜鹃花种类，对缺铁失绿症具有良好的抗性。5 个缺铁不敏感基因型种类[即‘尤科’、*R. macrosepalum* 品种‘花车’、火红杜鹃、锦绣杜鹃品种‘大村崎’（‘Oomurasaki’）和‘赛意乌默’（‘Sen-e-oomurasaki’）]在实验中到确认，这些种类可以在石灰质土壤地区栽培和应用于园林，或作为培育抗缺铁失绿症品种的育种材料。

11.6 结　论

作为最早使用 AFLP 标记的研究组之一，De Riek 等（1999）利用 ILVO 育种基因库对比利时盆栽杜鹃花的起源进行了深入研究。后来，亚洲小组使用 AFLP 与 SSR 标记研究了当地可收集到的栽培杂交种、自然种群和野生种间杂种类型的遗传变异。在比较分析比利时盆栽杜鹃花与中国野生映山红种群和其他杜鹃花种类的亲缘关系时，盆栽杜鹃花并没有像其命名和推测的历史起源那样与在映山红种群聚在一起（De Riek et al. 2008）。这一点也许可以用历史上分类的混淆来解释。然而，另一个更容易被接受的解释是，实际上来自日本和中国的栽培种与野生种间以及种内的杂交可能已经形成了第二代分化基因库，该基因库已经在地理起源地被驯化和引入亚洲园艺栽培。即使在中国（Huang and Qiang 1984）这一事实也得到了证实，大量的植物材料频繁交换，而且几乎可以肯定的是，伴随着从日本进口的栽培品种。同时此第二代基因库已经被证明是盆栽杜鹃花育种的新遗传变异的宝贵资源。在欧洲，由于在历史园林中大量引入了不同类群的常绿杜鹃花，专业和业余育种者也创造了新型遗传变异的第二代基因库。最典型的例子之一是意大利北部湖泊周围的花园和苗圃，当地的园艺师已经创造了一个独特的、“半自然”的常绿杜鹃花园艺种群。根据分子标记指纹图谱揭示的遗传一致性，应该存在一个由映山红亚属许多种类参与形成的遗传过渡类型。过去该过渡类型已经被亚洲和欧洲的育种者利用，并且被证明仍是遗传变异和有趣株型的良好来源。

到目前为止，杜鹃花育种几乎完全基于常规杂交和容易产生花卉芽变的特性。种间杂交和其他育种方法对商业品种培育的影响较小。最近，针对主要真菌病害的生物测试实验已纳入商业育种计划。但是遗传背景知识仅限于比较简单的性状，这些性状由少数的基因调控，如花型和花色。然而，更加复杂的调控植物质量性状的候选基因正在利用杜鹃花转录组第二代测序技术进行分离。

本章译者：

张春英[1]，解玮佳[2]

1. 上海植物园，上海城市植物资源开发应用工程技术研究中心，上海 200231
2. 云南省农业科学院花卉研究所，国家观赏园艺工程技术研究中心，云南省花卉育种重点实验室，昆明 650200

参 考 文 献

Bean WJ (1980) Trees and shrubs, hardy in the British Isles, vol. III. John Murray, London

Bowers C (1936) Rhododendrons and azaleas. The MacMillan Company, New York. 525pp

Brown CL, McAlpine RG, Kormanik PP (1967) Apical dominance and form in woody plants: a reappraisal. Am J Bot 54:153–162

Chamberlain DF, Rae SJ (1990) A revision of Rhododendron IV subgenus Tsutsusi. Edinb J Bot 47:89–200

Chamberlain DF, Hyam G, Argent G, Fairweather G, Walter K (1996) The genus Rhododendron. Its classification & synonymy. Royal Botanical Garden, Edinburgh

Cheon KS, Nakatsuka A, Kobayashi N (2016) Mutant PI/GLO homolog confers the hose-in-hose flower phenotype in Kurume azaleas. Hort J 85:380–387

Cheon KS, Nakatsuka A, Tasaki K, Kobayashi N (2017a) Floral morphology and MADS gene expression in double-flowered Japanese Evergreen azalea. Hort J 86:269–276

Cheon KS, Nakatsuka A, Gobara Y, Kobayashi N (2017b) Mutant RoPI-1 allele-based marker development for selection of the hose-in-hose flower phenotype in Rhododendron obtusum cultivars Euphytica. Hort J 213:1–8. https://doi.org/10.1007/s10681-016-1808-x

Christiaens A (2014) Factors affecting flower development and quality in Rhododendron simsii. PhD thesis, Ghent University, Ghent, Belgium, 145pp

Cline MG (1991) Apical dominance. Bot Rev 57:318–358

Cox PA, Cox KNE (1995) Encyclopedia of Rhododendron hybrids. Batsford Ltd, London

Cox PA, Cox KNE (1997) Encyclopedia of Rhododendron species. Glendoick Publishing, Perth

Cullen J (1991) The logic of Rhododendron classification. Rhododendrons (1992). R Hortic Soc Lond 44:14–19

De Keyser E, Pauwels E, Heungens K, De Riek J (2008) Development of supporting techniques for pot azalea (Rhododendron simsii hybrids) breeding focused on plant quality, disease resistance and enlargement of the assortment. Acta Hortic 766:361–366

De Keyser E, Shu QY, Van Bockstaele E, De Riek J (2010) Multipoint-likelihood maximization mapping on 4 segregating populations to achieve an integrated framework map for QTL analysis in pot azalea (Rhododendron simsii hybrids). BMC Mol Biol 11:1

De Keyser E, Desmet L, Van Bockstaele E, De Riek J (2013a) How to perform RT-qPCR accurately in plant species? A case study on flower color gene expression in an azalea (Rhododendron simsii hybrids) mapping population. BMC Mol Biol 14:13

De Keyser E, Lootens P, Van Bockstaele E, De Riek J (2013b) Image analysis for QTL mapping of flower color and leaf characteristics in pot azalea (Rhododendron simsii hybrids). Euphytica 189:445–460

De Loose R (1969) The flower pigments of the Belgian hybrids of Rhododendron simsii and other species and varieties from Rhododendron subseries obtusum. Phytochemistry 88:253–259

De Loose R (1979) Characterisation of Rhododendron simsii planch. Cultivars by flavonoid and isozyme markers. Sci Hortic 11:175–182

De Riek J, Dendauw J, Mertens M, De Loose M, Heursel J, Van Bockstaele E (1999) Validation of criteria for the selection of AFLP markers to assess the genetic variation of a breeders' collection of evergreen azaleas. Theor Appl Genet 99:1155–1165

De Riek J, Scariot V, Eeckhaut T, De Keyser E, Kobayashi N, Handa T (2008) The potential of molecular analysis and interspecific hybridization for azalea phylogenetic research. In: Sharma AK, Sharma A (eds) Plant genome: biodiversity and evolution, vol. 1, part E: phanerogams – angiosperm. Science Publishers, Enfield, NH

De Schepper S, Debergh P, Van Bockstaele E, De Loose M (2001a) Molecular characterisation of flower color genes in azalea sports (Rhododendron simsii hybrids). Acta Hortic 552:143–150
De Schepper S, Leus L, Mertens M, Debergh P, Van Bockstaele E, De Loose M (2001b) Somatic polyploidy and its consequences for flower coloration and flower morphology in azalea. Plant Cell Rep 20:583–590
De Schepper S, Debergh P, Van Bockstaele E, De Loose M, Gerats A, Depicker A (2003) Genetic and epigenetic aspects of somaclonal variation: flower color bud sports in azalea, a case study. S Afr J Bot 69:117–128
De Schepper S, Leus L, Eeckhaut T, Van Bockstaele E, Debergh P, De Loose M (2004) Somatic polyploid petals: regeneration offers new roads for breeding Belgian pot azaleas. Plant Cell Tissue Org Cult 76:183–188
Demasi S (2015) Iron deficiency tolerance in evergreen azaleas (Rhododendron spp.). PhD thesis, University Torino, Italy, 121pp
Eeckhaut T (2003) Ploidy breeding and interspecific hybridization in Spathiphyllum and woody ornamentals. PhD thesis, Ghent University, Ghent, Belgium, 126pp
Eeckhaut T, Van Huylenbroeck J, De Schepper S, Van Labeke MC (2006) Breeding for polyploidy in Belgian azalea (Rhododendron simsii hybrids). Acta Hortic 714:113–118
Galle F (1987) Azaleas. Timber Press, Portland, Oregon
Goetsch L, Eckert AJ, Hall BD, Hoot SB (2005) The molecular systematics of Rhododendron (Ericaceae): a phylogeny based upon RPB2 gene sequences. Syst Bot 30(3):616–626
Handa T, Eto J, Kita K, Kobayashi N (2002) Genetic diversity of Japanese wild evergreen azaleas in Kyusyu (south main island of Japan) characterized by AFLP. Acta Hortic 572:159–162
Heursel J (1999) Azalea's: oorsprong, veredeling en cultivars. Tielt, Lannoo
Heursel J, Horn W (1977) A hypothesis on the inheritance of flower colors and flavonoids in Rhododendron simsii planch. Zeitschrift für Pflanzenzüchtung 79:238–249
Heursel J, Volckaert E (1980) Azaleateelt. Ministerie van Landbouw, Brussel, Belgium, 160p
Heyting J (1970) Hybrids between elepidote and lepidote rhododendrons. Q Bull Am Rhodod Soc 24:97–98
Horn W (2002) Breeding methods and breeding research. In: Vainstein A (ed) Breeding for ornamentals: classical and molecular approaches. Kluwer Academic Publishers, Dordrecht, The Netherlands, pp 47–84
Huang MR, Qiang HL (1984) Rhododendron. China Forestry Publishing House, Beijing, China
Irving E, Hebda R (1993) Concerning the origin and distribution of rhododendrons. J Am Rhodod Soc 47(3):139–162
Ito I, Creech JL (1984) A brocade pillow: azaleas of old Japan. Weatherhill, New York and Tokyo
Kobayashi N (2013) Evaluation and application of evergreen azalea resources of Japan. Acta Hortic 990:213–219
Kobayashi N, Handa T, Yoshimura K, Tsumura Y, Arisumi K, Takayanagi K (2000) Evidence for introgressive hybridization based on chloroplast DNA polymorphisms and morphological variation in wild evergreen azalea populations in the Kirishima Mountains, Japan. Edinb J Bot 57:209–219
Kobayashi N, Handa T, Miyajima I, Arisumi K, Takayanagi K (2007) Introgressive hybridization between Rhododendron kiusianum and R. kaempferi (Ericaceae) in Kyushu, Japan based on chloroplast DNA markers. Edinb J Bot 64:283–293
Kobayashi N, Mizuta D, Nakatsuka A, Akabane M (2008) Attaining intersubgeneric hybrids in fragrant azalea breeding and the inheritance of organelle DNA. Euphytica 159:67–72
Kobayashi N, Ishihara M, Ohtani M, Cheon KS, Mizuta D, Tasaki K, Nakatsuka A (2010) Evaluation and application of the long-lasting flower trait (misome-sho) of azalea cultivars. Acta Hortic 855:165–168
Kron KA (1993) A revision of Rhododendron section Pentanthera. Edinb J Bot 50:249–363
Kron KA (1997) Phylogenetic relationships of Rhododendroideae (Ericaceae). Am J Bot 84:973–980
Kron K (2000) Evolutionary relationships of azaleas and rhododendrons. In: Jacobs R (ed) Jaarboek van de Belgische Dendrologische vereniging, Belgische Dendrologie Belge BDB, Bonheiden, Belgium, pp 30–40
Kron KA (2002) Phylogenetic relationships and major clades of Rhododendron (Rhodoreae, Ericoideae, Ericaceae). In: Argent G, McFarlane M (eds) Rhododendrons in horticulture and science. Royal Botanic Garden, Edinburgh, pp 79–85
Kron KA, Powell E (2009) Molecular systematics of Rhododendron subgenus Tsutsusi (Rhodoreae, Ericoideae, Ericaceae). Edinb J Bot 66(1):81–95

Kurashige Y, Mine M, Kobayashi N, Handa T, Takayanagi K, Yukawa T (1998) Investigation of sectional relationships in the genus Rhododendron (Ericaceae) based on matK sequences. J Jpn Bot 73:143–154

Leach D (1961) Rhododendrons of the world. Charles Scribner's Sons, New York

Lee FP (1958) The azalea book. Van Nostrand Company Inc., Princeton, NJ. American Horticultural Society

Li Y, Yan HF, Ge XJ (2012) Phylogeographic analysis and environmental niche modeling of widespread shrub *Rhododendron simsii* in China reveals multiple glacial refugia during the last glacial maximum. J Syst Evol 50:362–373

Luypaert G (2015) The broad mite, Polyphagotarsonemus latus, and its interactions with pot azalea, Rhododendron simsii hybrid. PhD thesis, Ghent University, Ghent, Belgium, 191pp

Luypaert G, Van Huylenbroeck J, De Riek J, De Clercq P (2014) Screening for broad mite susceptibility in Rhododendron simsii. J Plant Dis Prot 121:260–269

Luypaert G, Van Huylenbroeck J, De Riek J, Mechant E, Pauwels E, De Clercq P (2015) Opportunities to breed for broad mite resistance in Rhododendron simsii hybrids. Acta Hortic 1087:479–484

Mertens M, Heursel J, Van Bockstaele E, De Loose M (2000) Inheritance of foreign genes in transgenic azalea plants generated by Agrobacterium-mediated transformation. Acta Hortic 521:127–132

Mizuta D, Nakatsuka A, Kobayashi N (2008) Development of multiplex PCR markers to distinguish evergreen and deciduous azaleas. Plant Breed 127:533–535

Mizuta D, Ban T, Miyajima I, Nakatsuka A, Kobayashi N (2009) Comparison of flower color with anthocyanin composition patterns in evergreen azalea. Sci Hortic 122:594–602

Mizuta D, Nakatsuka A, Miyajima I, Ban T, Kobayashi N (2010) Pigment composition patterns and expression analysis of flavonoid biosynthesis genes in the petals of evergreen azalea 'Oomurasaki' and its red flower sport. Plant Breed 129:558–562

Nakatsuka A, Mizuta D, Kii Y, Miyajima I, Kobayashi N (2008) Isolation and expression analysis of flavonoid biosynthesis genes in evergreen azalea. Sci Hortic 118:314–320

Pratt C (1980) Somatic selection & chimeras. In: Moore JN, Janick J (eds) Methods in fruit breeding. Purdue University Press, West Lafayette, IN, pp 172–185

Preil W, Ebbinghaus R (1985) Bastardierungen von Topfazaleen (Rhododendron simsii) mit anderen Rhododendron-Arten. Rhododendron und immergrüne Laubgehölze, Jahrbuch, pp 85–92

Rouse J (1993) Inter- and intraspecific pollinations involving Rhododendron species. J Am Rhodod Soc 47:23–45

Samyn G, De Schepper S, Van Bockstaele E (2002) Adventitious shoot regeneration and appearance of sports in several azalea cultivars. Plant Cell Tissue Org Cult 70:223–227

Scariot V (2006) The DNA-typing of ornamental plants: Evergreen azaleas (Rhododendron spp.) and old garden roses (Rosa spp.). PhD thesis, University of Torino, Turin, Italy

Scariot V, Handa T, De Riek J (2007) A contribution to the classification of evergreen azalea cultivars located in the Lake Maggiore area (Italy) by means of AFLP markers. Euphytica 158:47–66

Scheerlinck H, Delbeke V, Hendriks WJ et al (1938) Tuinbouw encyclopedie. De Sikkel, Antwerpen, Belgium

Schmitz G, Theres K (2005) Shoot and inflorescence branching. Curr Opin Plant Biol 8:506–511

Spethmann W (1987) A new infrageneric classification and phylogenetic trends in the genus Rhododendron (Ericaceae). Plant Syst Evol 157:9–31

Salley H, Greer H (1992) Rhododendron hybrids, 2nd edn. Timber Press, Portland, Oregon. 344pp

Tasaki K, Nakatsuka A, Cheon KS, Kobayashi N (2014) Expression of MADS-box genes in narrow-petaled cultivars of Rhododendron macrosepalum Maxim. J Jpn Soc Hortic Sci 83(1):52–58

Tasaki K, Nakatsuka A, Cheon K, Kobayashi N (2015) Inheritance of the narrow leaf mutation in traditional Japanese evergreen azaleas. Euphytica 206(3):649–656

Ureshino K, Miyajima I, Akabane M (1998) Effectiveness of three-way crossing for the breeding of yellow-flowered evergreen azalea. Euphytica 104:113–118

Van Huylenbroeck J, Calsyn E (2011) Cryopreservation of an azalea germplasm collection. Acta Hortic 908:489–493

Van Huylenbroeck J, Calsyn E, De Keyser E, Luypaert G (2015) Breeding for biotic stress resistance in Rhododendron simsii. Acta Hortic 1104:375–380

Van Huylenbroeck J, Calsyn E, De Keyser E, Eeckhaut T, De Riek J (2018) Use of genetic resources to develop new commercial Rhododendron simsii hybrids. Acta Hortic. (in press)

Van Trier H (2012) Gentse azalea. Stichting Kunstboek, Oostkamp

Verleysen H, Van Bockstaele E, Debergh P (2005) An encapsulation–dehydration protocol for cryopreservation of the azalea cultivar 'Nordlicht' (Rhododendron simsii Planch.). Sci Hortic 106:402–414

Voss D (2001) What is an azalea? J Am Rhodod Soc 55:188–192

Xu JJ, Zhao B, Shen HF, Huang WM, Yuan LX (2016) Assessment of genetic relationship among Rhododendron cultivars using amplified fragment length polymorphism and inter-simple sequence repeat markers. Genet Mol Res 15(3). https://doi.org/10.4238/gmr.15038467

Yamazaki T (1996) A revision of the genus Rhododendron in Japan, Taiwan, Korea, and Sakhalin. Tsumura Laboratory, Tokyo

Zhou H, Liao J, Xia YP, Tang YW (2013) Determination of genetic relationships between evergreen azalea cultivars in China using AFLP markers. J Zhejiang Univ Sci 14(4):299–308

第12章 五 彩 芋

Zhanao Deng

摘要 五彩芋因其鲜艳的叶色和多样的叶形，被广泛应用于盆栽种植和园林景观。其原产于南美洲热带地区，18 世纪中叶引入欧洲。在过去的 70 年中，佛罗里达州一直是五彩芋的主要生产地，提供了全球 95%以上的块茎。过去 160 年的育种过程中，已经培育出了众多五彩芋品种。杂交一直是五彩芋育种的主要方法，新品种选择主要集中在鲜艳的叶色、新颖的斑纹图案、多叶片和块茎高产能力等性状上。近期的育种目标则包括增强对腐霉根腐病、镰刀菌根腐病、细菌性枯萎病和根结线虫病的抗性，以及对晒伤和寒害的耐受性。有性杂交将继续在实现这些育种目标中发挥重要作用。对重要的叶和园艺性状的遗传模式的新认识，加之分子标记技术，有望提高五彩芋的育种效率。体细胞变异、倍性诱变和种间杂交都有创造新的颜色与斑纹图案，并提高抗病性的潜力。其他细胞和分子技术研究，诸如原生质体培养、体细胞杂交、遗传转化、基因组和转录组分析以及针对性的基因编辑，对改进五彩芋的观赏和园艺性状也至关重要。

关键词 天南星；五彩芋；育种；品种改良；细胞学；杂交；遗传；倍性诱变；体细胞变异；四倍体

12.1 引 言

五彩芋（*Caladium* Vent.）属天南星科（Araceae），因其鲜艳的叶色和多样的叶形被用作盆栽、点睛或镶边植物种植在景观中。该属原产于热带南美洲（Mayo et al. 1997）。18 世纪中叶，五彩芋被引入欧洲，并在 19 世纪成为欧洲流行的温室植物（Hayward 1950）。五彩芋育种在 19 世纪 60 年代初始于法国，并于 80 年代在巴西和英国继续发展（Hayward 1950；Wilfret 1993）。自 20 世纪初以来，五彩芋的繁殖主要在佛罗里达进行（Hayward 1950；Wilfret 1993）。1976 年，佛罗里达大学在佛罗里达州中部的墨西哥湾海岸研究中心启动了一个五彩芋育种项目（Wilfret 1992）。在过去的 40 年中，该育种项目育成了近 35 个新品种。佛罗里达州私营企业的五彩芋育种也非常活跃（Hayward 1950；Hartman 2008）。近年来，在其他几个国家（包括泰国、中国、韩国等）也报道了五彩芋的育种。考虑到许多五彩芋栽培种可能的杂交起源，Birdsey（1951）提出将 *Caladium* × *hortulanum* 作为阔心形叶五彩芋的种名。

Z. Deng (✉)
University of Florida, IFAS, Department of Environmental Horticulture, Gulf Coast Research and Education Center, Wimauma, FL, USA
e-mail: zdeng@ufl.edu

除非另有说明，否则在本章中使用的“caladium”或“caladiums”一词指的是 *C.* × *hortulanum* 中的栽培种和原种。

12.1.1 繁殖

五彩芋可以用种子繁殖，但是需要人工授粉，并且子代之间存在相当大的遗传变异（Hartman et al. 1972）。因此，在五彩芋的商业生产中不使用种子繁殖。直接由种子生产的五彩芋植株可能不带病毒（Hartman et al. 1972）。因此，种子繁殖可用于消除病毒，如芋花叶病毒。

绝大多数商业化生产的五彩芋都是通过切分块茎繁殖的。切小的块茎种植在五彩芋田中，为观赏植物市场生产新的块茎。全球超过 95%的五彩芋块茎产自佛罗里达州中部（Bell et al. 1998；Deng et al. 2008c）（图 12.1a）。在过去的 70 年中，佛罗里达州一直是五彩芋块茎的主要生产地。五彩芋块茎的田间生产需要一年的时间：3～5 月，播种种苗（分块或切块的块茎）；在温暖多雨的夏季精心管理植株；11 月至次年 2 月，收获新的块茎（植株）（Wilfret 1993）。佛罗里达州的五彩芋种植者，每年向美国和世界上其他 40 个国家/地区运送 7000 万至 1 亿个块茎，以用于盆花生产和景观打造。

通过分生组织、叶盘和叶柄段，五彩芋很容易进行组织培养（Hartman 1974；Cai and Deng 2016）。然而，组织培养尚未广泛用于五彩芋栽培种的商业繁殖。限制组织培养使用的因素包括相对较高的生产成本，生产中易出现异型或体细胞克隆变异（Ahmed et al. 2004；Cao et al. 2016a），并且从组培苗移栽到长成容器苗需要较长的生产周期。基于这些原因，组织培养在五彩芋中的主要用途是消除病原体并生产不带病的原种（Hartman 1974）。

12.1.2 细胞学

五彩芋染色体数的首次报道可以追溯到 Kurakubo（1940）。自那时以后，从 $2n = 19$（Pfitzer 1957）、26（Marchant 1971）、28（Sharma and Sarkar 1964；Marchant 1971）、30（Jones 1957; Pfitzer 1957; Sharma and Sarkar 1964; Marchant 1971）到 32（Marchant 1971），相继报道了五彩芋几个种和栽培种的染色体数。最近，Cao 等（2014）检测了属于 10 个五彩芋属的 39 个五彩芋种的染色体数，并估计了其基因组大小（核 DNA 含量）。检测的 6 个阔心形叶五彩芋栽培种（*C.* × *hortunalum*）的染色体数均为 $2n = 30$。有时在两个栽培品种‘Candidum’和‘Miss Muffet’的某些细胞中观察到一条额外的染色体。这条染色体几乎是圆形的，比其他染色体小，类似于其他植物中的 B 染色体。*C. bicolor* 的大部分种和 *C. schomburgkii* 的所有种都有 $2n = 30$ 条染色体。其他五彩芋种也存在多样化的体细胞染色体数目：$2n = 18$（*C. picturatum*）、20（*C. humboldtii*）、24（*C. steudneriifolium*）、26（*C. lindenii* 和 *C. praetermissum*）、34（*C. marmoratum*）和 38（*C. clavatum*）（表 12.1）。26 个五彩芋栽培品种的基因组大小为 8.8～9.9pg/2C，平均基因组大小为 9.2pg/2C（Cao et al. 2014）。其他五彩芋种的基因组大小为 3.0～9.8pg/2C（表 12.1）。

根据五彩芋体细胞染色体数目和基因组大小，可以将其分为 4 个细胞型组（CCG1～4）

图 12.1　（a）佛罗里达州中部块茎生产大田里的五彩芋；（b）两种叶形、多种颜色和叶斑的混合种植；（c）五彩芋的杂交实生苗，依据叶色和着色图案，种植在穴盘里，准备进行第一阶段的筛选；（d）全日照下进行第四阶段田间特性的重复筛选；（e）种植在方形容器（尺寸 11.5cm × 11.5cm）中的两株'Summer Pink'，均由一个五彩芋一级块茎（3.8～6.4cm）培育成；（f）直径 20cm 圆形容器中 4 个一级块茎，长成 1 株'Red Hot'

（Cao et al. 2014）。*Caladium lindenii*、*C. praetermissum* 和 *C. steudneriifolium* 属于 CCG-1 型；*C. picturatum* 和 *C. humboldtii* 属于 CCG-2 型；*C. clavatum* 和 *C. marmoratum* 属于 CCG-3 型；*C. bicolor*、*C.* × *hortulanum* 和 *C. schomburgkii* 属于 CCG-4 型（表 12.1）。似乎过去的基因组复制或四倍体化事件在 CCG-2 型物种的形成中起了重要作用。Cao 等（2014）假设 CCG-3 和 CCG-4 型中的五彩芋种是从 CCG-2 型的物种演化而来的；CCG-4 型物种的进化中染色体发生融合，CCG-3 型物种的形成中发生剧烈的染色体重排和 DNA 丢失。这些新的细胞学信息可能对指导将来的五彩芋种间杂交育种很有用。

表 12.1 10 种五彩芋的染色体数目、基因组大小和细胞型组

种	染色体数目（参考资料）	基因组大小（pg/2C）（Cao et al. 2014）	细胞型组（CCG）（Cao et al. 2014）	叶片类型
C. bicolor	2*n* = 30（Kurakubo 1940；Marchant 1971；Simmonds 1954；Ramachandran 1978；Subramanian and Munian 1988；Cao et al. 2014） *n* = 15（Pfitzer 1957；Ramachandran 1978）	9.1～9.8	4	阔心形
C. × *hortulanum*	2*n* = 30（Jones 1957；Marchant 1971） *n* = 15（Pfitzer 1957） 2*n* = 28，30，32（Sharma and Sarkar 1964） 2*n* = 26，28，30，32（Sarkar 1986） 2*n* = 30（Cao et al. 2014）	8.7～9.9	4	阔心形、披针形、带形
C. clavatum	2*n* = 38（Cao et al. 2014）	5.3～5.4	3	阔心形
C. humboldtii	2*n* = 19（Pfitzer 1957） 2*n* = 20（Cao et al. 2014）	3.8～4.1	2	阔心形
C. lindenii	*n* = 13（Pfitzer 1957） 2*n* = 28（Sharma 1970） 2*n* = 26（Cao et al. 2014）	3.0	1	
C. marmoratum	2*n* = 34（Cao et al. 2014）	5.6	3	阔心形
C. picturatum	2*n* = 18（Cao et al. 2014）	4.5～4.8	2	
C. praetermissum	2*n* = 28（Hetterscheid et al. 2009） 2*n* = 26（Cao et al. 2014）	4.1～4.2	1?	阔心形
C. schomburgkii	2*n* = 28（Jones 1957） 2*n* = 18，28，30（Sarkar 1986） 2*n* = 30（Cao et al. 2014）	9.0～9.2	4	披针形
C. steudneriifolium	2*n* = 24（Cao et al. 2014）	3.2～5.1	1	阔心形

12.1.3 分类学

专门从事天南星科分类的分类学家将芋属分为 7（Madison 1981）～17 种（Croat 1994）。分类学家之间的分歧主要集中在 4 个物种的分类上，即 *C. bicolor* Vent.、*C. marmoratum* Mathieu、*C. picturatum* C. Koch.和 *C. steudneriifolium*。Madison（1981）将 *C. marmoratum*、*C. picturatum* 和 *C. steudneriifolium* 合并为 *C. bicolor*。但因独特且长而窄的叶子，*C. picturatum* 被认为是一个独立的种（Croat 1994；Mayo et al. 1997；The Plant List 2013）。然而，关于 *C. marmoratum* 和 *C. steudneriifolium* 地位的争论仍然存在。*C. marmoratum* 被视为 *C. bicolor* 的一种变形（Hetterscheid et al. 2009）或 *C. bicolor* 的代名词（The Plant List 2013）。同样，*C. Steudneriifolium* 被认为是 *C. bicolor* 的代名词（The Plant List 2013）。最近，Cao 等（2014）表明，*C. marmoratum* 和 *C. steudneriifolium* 的染色体数彼此不同，并且它们在染色体数和基因组大小上也不同于 *C. bicolor*。染色体数目和基因组大小数据似乎支持 *C. bicolor*（2*n* = 30，9.1～9.8pg/2C）、*C. marmoratum*（2*n* = 34，5.6pg/2C）和 *C. steudneriifolium*（2*n* = 24，3.2～5.1pg/2C）是相互独立的物种，即使它们在叶片特征上具有相似性（Cao et al. 2014）（表 12.1）。

近年来，*C. clavatum* 和 *C. praetermissum* 被引入了芋属（Hetterscheid et al. 2009）。*C. clavatum* 是 *C. bicolor* var. *rubicundum* 或 *C. bicolor* var. *rubicunda* 的栽培种（Graf

1976）。由于其与 *C. bicolor* 在具药雄蕊、花柱、花序梗和块茎形态上明显不同，因此被作为一个新种（Hetterscheid et al. 2009）。*C. clavatum* 独特的基因组大小和染色体数支持了这种处理方法（Cao et al. 2014）。*C. praetermissum* 已被作为海芋‘Hilo Beauty’种植，它可能是 *C. marmoratum* 的斑点突变体（Hetterscheid et al. 2009）。*C. praetermissum* 的基因组比 *C. marmoratum* 少至少 30%，染色体数少了 8 条，这就支持了 *C. praetermissum* 的物种地位。以前，认为 *C. praetermissum* 的染色体数为 $2n = 28$（Hetterscheid et al. 2009）。Cao 等（2014）报道了 *C. praetermissum* 的染色体数为 $2n = 26$。

C. humboldtii 和 *C. bicolor* 有一些相同的叶片着色模式。Madison（1981）认为 *C. humboldtii* 可能是 *C. bicolor* 的一个染色体小种。分子标记分析揭示了 *C. humboldtii* 和 *C. bicolor* 的巨大差异，认为 *C. humboldtii* 是不同的物种（Loh et al. 2000；Deng et al. 2007）。染色体计数和基因组大小数据也支持 *C . humboldtii* 的物种地位（Cao et al. 2014）。Pfitzer（1957）报道了该物种有 $2n = 19$ 条染色体。Cao 等（2014）报道了 *C. humboldtii* 中 3 个种质染色体均为 $2n = 20$。

最近的分子标记分析提出了一个关于 *C. lindenii* 地位的有趣问题。该种曾经是黄体芋属（*Xanthosoma*）的一员，由于在花粉脱落模式上的相似性而转归到了五彩芋中（Madison 1981）。Loh 等（2000）注意到 *C. lindenii* 不具有任何黄体芋属的特异条带，因此它不属于黄体芋属。Deng 等（2007）表明，*C. lindenii* 与其他芋属物种之间的距离非常远。Gong 和 Deng（2011）报道 *C. lindenii* 在简单重复序列（SSR）标记位点上包含独特的等位基因。*C. lindenii* 和其他芋属物种之间如此低的遗传和形态相似性似乎暗示了需要将 *C. lindenii* 列为单独的属。

据推测，*C. bicolor* 和 *C. marmoratum* 是阔心形叶五彩芋栽培种（*C.* × *hortulanum*）的主要亲本（Hayward 1950；Wilfret 1993），而 *C. picturatum* 或 *C. schomburgkii* Schott 为一些五彩芋栽培品种提供了披针形叶性状（Hayward 1950；Graf 1976；Huxley et al. 1992；Wilfret 1993）。最近的染色体计数和基因组大小估计（Cao et al. 2014）以及以前的分子标记分析（Deng et al. 2007）数据支持 *C. bicolor* 和 *C. schomburgkii* 在 *C.* × *hortunalum* 栽培种发展中起主要作用，但没有指出 *C. picturatum* 和 *C. marmoratum* 在 *C.* × *hortulanum* 栽培种育种中有直接作用。

12.2 重要性状及育种目标

12.2.1 植株和叶片性状

容器和景观中五彩芋的观赏价值在很大程度上取决于其植株与叶的特征，包括植物的生长习性和株高，叶片数量、大小、类型、颜色、色斑图案和脉纹。现有的五彩芋品种在这些叶片特征上表现出显著的多样性（Wilfret 1993）（图 12.1b，e，f）。五彩芋品种具有三种明显的生长习性：顶芽优势、不完全顶芽优势和非顶芽优势（Wilfret and Hurner 1982）。“顶芽优势”的品种通常从块茎的中心眼（或芽）上长出一到两片非常大的叶子，并在 3～4 周后长出几片较小的叶子。“不完全顶芽优势”的品种从中心眼产生一到两片大叶子，从块茎的外围产生较小的叶子。“非顶芽优势”的品种可以从块茎上同时长出8～10个叶片。非顶芽优势生长习性的品种，可以在最短的时间内在容器中长出最丰满的植株（Wilfret and Hurner 1982），因此，更适于用作盆栽生产。

五彩芋栽培品种的单株叶片数量变化很大，从5个（‘Jubilee’）到41个（‘Rosalie’）不等（Wilfret and Hurner 1982）。在不同的品种中，成熟叶片的长度和宽度可以从几厘米到超过30cm不等（Wilfret and Hurner 1982）。广义上讲，五彩芋叶可分为三类：阔心形、披针形和带形（图12.1b，e，f）。阔心形叶（图12.1e）为心形（三角形或圆形、卵形），具有三条主脉，连接着盾状叶柄和两个明显的基部裂片，它们的连接长度超过主脉的1/5，并被短而狭窄的痕道分离。带形叶窄而呈条带状，有一条主脉，没有明显的基部裂片。披针形叶（图12.1f）介于阔心形叶和带形叶之间，宽矢状披针形至心状披针形，基部裂片不明显，如有裂片则被痕道大致隔开。叶的类型不仅直接影响五彩芋的被接受程度和销售，而且与其他重要特征（如植物生长习性、胁迫耐受性和块茎产量潜力）紧密相关。通常，带形或披针形植株矮得多，没有顶芽优势或具有不完全顶芽优势，大小相似的块茎长出的叶子更多，更耐日晒和低温，并且与阔心形叶植株相比，产生的块茎更小（Deng and Harbaugh 2008）。

五彩芋的叶色由叶脉（主脉、次脉和边缘脉）的颜色、脉间区域、底色、斑点和/或色块决定。五彩芋的叶片颜色和着色模式极为多样且引人入胜（图12.1b，e，f）。Wilfret和Hurner（1982）将五彩芋叶着色模式分为4个基本类别：白色网状；白色、粉红色或玫瑰色斑点；中央区域着色；主脉着色。在佛罗里达州商业种植的五彩芋品种中已观察到不同的类型以及这些类型的复合型。创造新颖的着色图案一直是五彩芋的主要育种目标。从美国佛罗里达和泰国引进的新品种丰富了五彩芋的着色模式（Deng and Harbaugh 2006c，2008；Deng 2012；Deng et al. 2013；Kosakul，个人交流）。

12.2.2 块茎特性和产量潜能

五彩芋品种在块茎大小、发枝习性、顶芽优势、镰刀菌茎腐病抗性和发芽时间方面差异很大。块茎大小、发枝习性和顶芽优势是影响盆栽五彩芋生产质量的最重要因素。根据五彩芋块茎的直径，可以将其分为5个级别（二级、一级、特大、巨大、超巨大型）。根据容器大小，选择合适的块茎尺寸非常重要。在当前市场上，对一级和特大块茎的需求比例比较高。顶芽优势是五彩芋植株生长习性的主要决定因素。顶芽优势强的五彩芋块茎发芽时会长出1～3个大而高的叶片，导致产生易倒伏、细高的劣质植株。为了使这些块茎在容器中生产高质量的完整植株，需要人工去除生长中心，这是一个费时的过程。五彩芋育种中希望选育具有较弱顶芽优势的品种。

五彩芋块茎产量通常表示为每个等级的块茎数量。可以依据块茎产量和不同等级块茎的相对经济价值计算出生产指数（PI）（Wilfret 1992）。该指数已用于比较五彩芋新品种的潜在经济价值。受园艺管理、天气条件、病害等的影响，每年五彩芋块茎产量、大小分布和生产指数可能会有很大不同。要获得五彩芋块茎产量潜力的良好评估和新品种潜在的生产价值，必须进行多年重复的田间试验。

12.2.3　抗病性

近年来，抗病性已成为五彩芋育种中最重要的性状之一（Deng et al. 2008c）。在过去的 15 年中，已经筛选出商业化的五彩芋品种，从而鉴定出了许多对真菌和细菌病原体以及一种线虫具有抗性的品种（表 12.2）。一些抗性品种在新品种培育中已被用作育种亲本。

表 12.2　最近鉴定的对两种真菌病害、一种细菌病害、一种线虫和冷害、日晒有抗性的五彩芋品种资源

品种	镰刀菌茎腐病	腐霉根腐病	细菌性枯萎病	根结线虫	冷害	日晒
Aaron	R	S	MS			T
Apple Blossom		MR				
Candidum	R	MR	HS	R	S	S
Candidum Junior	S	MR	MS	R	S	S
Carolyn Whorton	HS	S	MS		S	T
Etta Moore		MR				
Fannie Munson	HS	VS	MS	R		T
Florida Blizzard		MR	HS			S
Florida Red Ruffles		S	R		R	T
Florida Sweetheart	R	S	MS		S	T
Freida Hemple	HS	MR		R	S	T
Marie Moir		VS			R	
Miss Muffet	S	S		S	R	T
Mrs. Arno Nehrling		S	R			
Pink Beauty		S		R	S	S
Pink Gem		S	MS	R		T
Pink Symphony		VS	R	R		
Postman Joyner	S	VS	HS	R		T
Red Flash	MR	VS	HS		S	T
Rosebud	R	VS	HS	R		S
White Christmas	R	MR	HS	R	S	S
White Queen	S	VS	R	S	S	T
White Wing	MR	VS	MS	S		T

资料来源于 McSorley et al. 2004；Deng et al. 2005a，2005b；Dover et al. 2005；Deng and Harbaugh 2006a；Goktepe et al. 2007；Seijo et al. 2010；HS 高感，MR 中抗，MS 中感，R 抗病，S 易感或敏感，T 耐病，VS 十分易感

12.2.3.1　腐霉根腐病抗性

腐霉根腐病可发生在块茎生产田中的植株，以及容器植株生产或景观种植的植株上。腐霉菌（*Pythium myriotylum* Drechs）是五彩芋腐霉根腐病的最常见病原（Ridings and Hartman 1976）。在感染了腐霉菌的基质中播种的五彩芋块茎在 3～5 周后萌芽，植株生长量最多可减少 70%，新块茎的产量最多可减少 40%（Ridings and Hartman 1976）。当幼嫩五彩芋植株直接接种腐霉菌后，其根部多达 90%会腐烂，而在 2～3 周会失去多达 85%的叶片（Deng et al. 2005a，2005b）。根部的腐霉菌感染还导致叶片变色和出现坏死斑块，叶柄突起和枯萎，以及整片叶子塌陷。这些叶片症状最早可在接种腐霉菌后 3 天出现。接种

腐霉菌后，叶片损失与根腐病严重程度之间存在线性关系（Deng et al. 2005a，2005b）。

Deng 等（2005a）已经提出了腐霉菌接种和腐霉根腐病评估方案。该方案为在装满粗蛭石的小型塑料盘中种植幼嫩的五彩芋植株约 8 周后，将腐霉菌孢子悬浮液施用到根球上，之后接种的植株培养在 26～37℃的温度下，并不断用水浸泡根球，在接种后 2～3 周评估根腐和叶片损失的程度。使用该方法，Deng 等（2005a，2005b）评估了 42 个主要五彩芋商业品种对腐霉根腐病的抗性，并鉴定出 7 个品种（‘Apple Blossom’、‘Candidum’、‘Candidum Junior’、‘Etta Moore’、‘Florida Blizzard’、‘Freida Hemple’和‘White Christmas’）对腐霉根腐病具有中等抗性（表 12.2）。

12.2.3.2 细菌性枯萎病抗性

五彩芋细菌性枯萎病，也称黄单胞菌叶斑，是由地毯草黄单胞菌花叶万年青致病变种（*Xanthomonas axonopodis* pv. *dieffenbachiae*，Xad；原名 *X. campestris* pv. *dieffenbachiae*）引起的（Seijo et al. 2010）。这种病每年都可发生在热带和亚热带地区，那里在五彩芋的生长季节通常会出现温暖、潮湿和阴雨的天气。严重的叶子坏死和脱落，特别是在生产季节的早期发生时，会导致块茎尺寸（或等级）降低，从而造成商业块茎产量显著降低。坏死的叶斑还降低了苗圃和温室中五彩芋的销售性。虽然大多数五彩芋品种很容易患上这种疾病，但还有几个品种：‘Candidum Junior’、‘Carolyn Whorton’、‘Florida Red Ruffles’、‘Florida Sweetheart’、‘Mrs. Arno Nehrling’、‘Pink Symphony’和‘White Queen’，人工接种后，对这种疾病表现出良好的抗性（Seijo et al. 2010）（表 12.2）。有趣的是，‘Florida Red Ruffles’、‘Florida Sweetheart’像它们的亲本‘Candidum Junior’一样对这种疾病有抗性。这似乎暗示在某些五彩芋品种中观察到的细菌性枯萎病抗性可以从亲本传递给后代，因此对于开发新的抗性品种有用。‘White Queen’似乎与其他抗性品种不同，因为它在接种细菌后并未在叶片上显示出典型的坏死斑，但其看上去健康的叶片从下表面渗出了致病性黄单胞菌。‘White Queen’的抗性特征可能对培育新的抗性品种没有用，因为具有这种抗性的五彩芋植株可能会携带黄单胞菌病原，并成为感染邻近易感植株的病源。

12.2.3.3 根结线虫抗性

根结线虫（*Meloidogyne incognita*）可感染五彩芋的根和块茎。其严重感染五彩芋根部会阻碍植株生长并降低块茎产量。五彩芋生产中根结线虫的控制主要依赖于热水处理块茎和土壤熏蒸。近年来，寻找寄主植物抗性已作为控制根结线虫的另一种工具。McSorley 等（2004）和 Dover 等（2005）通过将培育的线虫幼虫接种在容器中生长的五彩芋植株周围的土壤中，评估了许多五彩芋品种对根结线虫的抗性。在这些评估中，不同五彩芋品种利于线虫种群发展的能力显著不同。几个品种对根结线虫有很高的抗性，在这些品种的根和/或根周围的土壤中没有或只有很少的线虫卵（表 12.2）（McSorley et al. 2004；Dover et al. 2005）。

12.2.3.4 镰刀菌茎腐病抗性

镰刀菌茎腐病是块茎收获后加工和贮藏期间最重要的疾病（Knauss 1975；McGovern 2004）。病原体是镰刀菌（*Fusarium solani*）。当环境条件适宜时，这种疾病会导致大量块茎迅速损失。在过去的几十年中，镰刀菌茎腐病已导致许多商业品种的块茎产量稳定下降，并导致商业生产中淘汰了许多品种（McGovern 2004）。Goketpe 等（2007）表明温度是影响真菌菌丝生长和引起块茎腐烂的最重要因素之一。较低的温度（13～18℃）更有利于真菌引起块茎腐烂。镰刀菌菌株引起块茎腐烂的侵袭性，或者说五彩芋品种对镰刀菌茎腐病的抗性方面存在相当大的差异（Goktepe et al. 2007）。在筛选出的 17 个五彩芋品种中，有 5 个品种（‘Aaron’、‘Candidum’、‘Florida Sweetheart’、‘Rosebud’和‘White Christmas’）被归类为抗镰刀菌茎腐病，2 个品种（‘Red Flash’、‘White Wing’）为中抗镰刀菌茎腐病（表 12.2）。以‘Candidum’作为育种亲本，已经开发并发布了中高抗镰刀菌茎腐病的新品种（‘Icicle’）。

12.2.4 抗逆性

12.2.4.1 低温和冷害耐受性

五彩芋是热带植物，对低温和冷害非常敏感。低于 15℃的温度会延迟五彩芋块茎发芽并损害叶片（Wilfret 1993）。这种低温敏感性限制了五彩芋在亚热带和温带地区景观中的使用，降低了植物的性能，并限制了五彩芋块茎的生产和运输。抗寒品种的开发将是有价值的，但是由于缺乏抗寒亲本而阻碍了这一尝试。Deng 和 Harbaugh（2006b）观察到五彩芋品种之间叶片对冷害的敏感性不同。‘Florida Red Ruffles’、‘Marie Moir’和‘Miss Muffet’这三个品种的叶片冷害损伤比其他 13 个品种小得多（表 12.2）。有趣的是，‘Florida Red Ruffles’的亲本（‘Candidum Junior’和‘Red Frill’）都对冷害非常敏感。这可能表明具有在五彩芋育种后代群体中鉴定超亲耐冷株系的潜力。‘Florida Red Ruffles’的抗寒性已遗传给‘Dr. Brent’——一个最近发布的新品种（Deng et al. 2008b）。

12.2.4.2 日晒耐受性

五彩芋叶组织对高强度的阳光敏感，尤其是在高温下。强烈的光线会使五彩芋叶颜色褪色，并使叶片上形成坏死斑（Wilfret 1993）。提高日晒耐受性已成为五彩芋育种的主要目标。现有商业品种对日晒的耐受性存在差异。阔心形叶品种中的‘Aaron’、‘Carolyn Whorton’、‘Freida Hemple’、‘Miss Muffet’和‘Red Flash’，和披针形叶品种中的‘Florida Red Ruffles’、‘Florida Sweetheart’、‘Pink Gem’和‘White Wing’比其他品种具有更好的耐日晒性。佛罗里达州的公共和私人五彩芋育种项目，引入了许多具有强耐日晒特性的新品种（Deng and Harbaugh 2006c；Hartman 2008；Deng et al. 2013）。

12.3 遗 传 模 式

12.3.1 叶片类型

五彩芋栽培品种的叶片大概分为三种类型：阔心形叶、披针形叶和带形叶。Wilfret（1986）提出，这些叶片类型受一个单一基因控制，其中一种纯合基因型是阔心形叶（图 12.1e），另一种纯合基因型是带形叶，而杂合基因型则为披针形叶（图 12.1f）。这种遗传模型得到了 Deng 和 Harbaugh（2006a）随后对杂交后代进行的大规模分离分析的支持。在此分离分析中，阔心形 × 阔心形杂交的后代表现出阔心形叶，多数阔心形 × 披针形的后代及其正反交后代中，阔心形：披针形按 1∶1 分离，而披针形 × 披针形杂交的后代显示三种类型的叶片（阔心形、披针形和带形 = 1∶2∶1）。这些分离比表明，一个位点具有两个共显性等位基因（*F* 和 *f*），这些等位基因控制杂交后代的叶片类型。五彩芋阔心形叶、披针形叶和带形叶的推断基因型分别为 *FF*、*Ff* 和 *ff*。其他基因、基因互作或基因的非随机组合可能参与了五彩芋叶类型的表达，因为在几个阔心形 × 披针形杂交中观察到偏斜的分离模式（Deng and Harbaugh 2006a）。

12.3.2 主脉颜色

五彩芋叶主脉的颜色可大致分为绿色、红色和白色（图 12.1b）。带有绿色脉纹的品种有 ‘Candidum’、‘Candidum Junior’ 和 ‘White Christmas’。红色的主脉存在于许多五彩芋品种中，其中包括 ‘Florida Fantasy’、‘Florida Red Ruffles’、‘Freida Hemple’ 和 ‘Red Flash’。主脉为白色的有 ‘Aaron’、‘Florida Moonlight’、‘Garden White’、‘Gingerland’ 和 ‘White Wing’。杂交后代分离分析一致表明，绿色主脉为隐性等位基因，红色主脉为显性等位基因（Wilfret 1983，1986；Deng and Harbaugh 2006a）。曾经有人认为白脉对红脉和绿脉是显性（Wilfret 1986），但是此模型在随后的分析中并未得到证明。Deng 和 Harbaugh（2006a）指出，白脉对绿脉是显性，而对红脉是隐性。具有三个等位基因（V^r、V^w 和 V^g）的单位点（*V*）控制着五彩芋品种的主脉颜色，并且具有以下主导顺序：V^r（红色）＞V^w（白色）＞V^g（绿色）。红脉品种 ‘Florida Fantasy’、‘Florida Red Ruffles’、‘Florida Sweetheart’ 和 ‘Red Flash’ 是杂合的，包含红色和绿色主脉等位基因（表 12.3）。白脉品种 ‘Aaron’ 是具有 V^wV^w 基因型的纯合体，而 ‘Florida Moonlight’ 是具有 V^wV^g 基因型的杂合体（表 12.3）。

表 12.3　五彩芋品种的叶片类型、主脉颜色、叶片底色、叶斑点、叶斑块、叶片皱褶表型和根据最近研究推断的基因型

品种	叶片类型		主脉颜色		叶片底色		叶斑点		叶斑块		叶片皱褶	
	表型	基因型	表型	基因型	表型	基因型	表现	基因型	表型	基因型	表型	基因型
Aaron	阔心形	*FF*	白	V^wV^w	绿	*lemlem*					无	*rlfrlf*
Candidum	阔心形	*FF*	绿	V^gV^g	绿	*lemlem*	无	*ss*	无	*bb*	无	*rlfrlf*
Candidum Junior	阔心形	*FF*	绿	V^gV^g								
Carolyn Whorton	阔心形	*FF*	红	V^rV^g					有	*Bb*	无	*rlfrlf*
Fannie Munson	阔心形	*FF*	红				无	*ss*	无	*bb*		
Fairytale Princess	披针形	*Ff*	红	V^rV^g	绿	*lemlem*	无	*ss*			无	*rlfrlf*
Florida Blizzard	阔心形	*FF*	白	V^wV^g			无	*ss*	有	*Bb*		
Florida Fantasy	阔心形	*FF*	红	V^rV^g							无	*rlfrlf*
Florida Irish Lace	披针形	*Ff*	绿	V^gV^g								
Florida Moonlight	阔心形	*FF*	白	V^wV^g	绿	*lemlem*					无	*rlfrlf*
Florida Red Ruffles	披针形	*Ff*	红	V^rV^g								
Florida Sweetheart	披针形	*Ff*	红	V^rV^g							无	*rlfrlf*
Freida Hemple	阔心形	*FF*	红	V^rV^g					无	*bb*		

续表

品种	叶片类型		主脉颜色		叶片底色		叶斑点		叶斑块		叶片皱褶	
	表型	基因型	表型	基因型	表型	基因型	表现	基因型	表型	基因型	表型	基因型
Gingerland	披针形	*Ff*	白	$V^{w}V^{g}$	绿	*lemlem*	有	*Ss*	无	*bb*	有	*RLFrlf*
Miss Muffet	阔心形	*FF*	白	$V^{w}V^{g}$	柠檬	*LEMlem*	有		无	*bb*	有	*RLFrlf*
Red Flash	阔心形	*FF*	红	$V^{r}V^{g}$	绿	*lemlem*	有				无	*rlfrlf*
Red Hot	披针形	*Ff*	红	$V^{r}V^{g}$			无	*ss*			无	*rlfrlf*
Rosebud	阔心形	*FF*	红	$V^{r}V^{w}$					无	*bb*		
Tapestry	阔心形	*FF*	红	$V^{r}V^{g}$							无	*rlfrlf*
White Christmas	阔心形	*FF*	绿	$V^{g}V^{g}$			无	*ss*	有	*Bb*		

资料来源：Wilfret 1983，1986；Deng and Harbaugh 2006a；Deng et al. 2008a；Deng and Harbaugh 2009；Cao et al. 2016b；Cao et al. 2017

12.3.3 叶片底色

五彩芋叶片的底色可大致分为绿色和柠檬色两类。柠檬色的五彩芋叶片在整个生长季节均为淡黄色。尽管可能存在绿色阴影，但大多数五彩芋商业品种的叶片底色为绿色。Cao 等（2017）表明，叶片的两种底色由具有两个等位基因 *LEM* 和 *lem* 的单个核基因控制，分别控制了显性的柠檬色和隐性的绿色叶片底色（表 12.3）。柠檬色的‘Miss Muffet’和一些育种品系具有杂合基因型 *LEMlem*。‘Miss Muffet’自交时，后代分离为 3（柠檬色）：1（绿色）。当‘Miss Muffet’与绿色亲本杂交时，它们的后代以 1（柠檬色）：1（绿色）的比例分离。绿色亲本的后代，包括‘Aaron’、‘Candidum’、‘Fairytale Princess’、‘Florida Moonlight’、‘Gingerland’和‘Red Flash’，均是绿叶。

12.3.4 叶斑点

许多重要的五彩芋品种在叶片上都表现出彩色斑点（图 12.1b）。这些品种之一‘Gingerland’，它的叶子上有许多砖红色斑点。‘Gingerland’的自交后代在有斑点或无斑点的性状上按 3∶1（斑点∶无斑点）分离（Deng et al. 2008a），不管无斑点品种在测交中是父系还是母系，‘Gingerland’与无斑点品种测交的后代以 1∶1 分离（斑点∶无斑点）。Deng 等（2008a）表明，‘Gingerland’中的叶斑点受单个核基因控制，该基因具有两个等位基因 *S* 和 *s*，分别控制斑点和无斑点性状（表 12.3）。

品种‘Painter's Palette’同一片叶子上有红色和白色斑点，表现出不同的叶斑模式。Zettler 和 Abo El-Nil（1979）提出了：一个位点具有等位基因 *R* 和 *W*，分别控制红色和白色斑点。Gager（1991）研究了‘Painter's Palette’与无斑点品种‘Aaron’或‘Florida Cardinal’杂交的后代，以及这些后代相互杂交的后代中叶斑点和斑点颜色的分离，并推测单个位点控制斑点的表达，有两个共显性等位基因，S_R 控制红色斑点，S_W 控制白色斑点，而一个隐性等位基因 *s* 控制无斑点，并且一个单独的基因位点控制着斑点的位置（独立的、接触边界的或重叠的）。但是，在许多杂交中，观察到的分离比与预期有所偏离，可能是由对基因型 S_R/s，特别是 S_R/S_R、S_W/S_W 和 S_R/S_W 产生致命影响的因素造成的。‘Painter's Palette’中叶斑的位点与‘Gingerland’中 *S* 基因位点之间的关系尚待确定。

12.3.5 叶斑块

几种流行的五彩芋栽培种，如‘Carolyn Whorton’和‘White Christmas’，在叶面上显示出形状不规则的彩色区域，这种着色模式称为叶片斑块（Deng and Harbaugh 2009）。斑块品种自交后代以 3（斑块）∶1（无斑块）的比例分离，而无斑块品种自交后代都是没有斑块的。Deng 和 Harbaugh（2009）提出了一个具有两个等位基因 *B* 和 *b* 的单核基因位点分别控制叶片有斑块与无斑块性状（表 12.3）。到目前为止测试过的斑块品种在该基因位点都是杂合的。在斑块品种和斑点品种‘Gingerland’的一些杂交中，斑块性状的分离偏向于无斑块的个体，这可能表明存在其他因素和/或基因互作影响叶片斑块性状。

12.3.6 皱褶叶

皱褶叶是两侧不规则凹陷或突起的、不均匀的叶片。具有这一特征的五彩芋品种有‘Miss Muffet’和‘Gingerland’（表 12.3）。Cao 等（2016b）揭示了一个具有两个等位基因 *RLF* 和 *rlf* 的单基因位点控制五彩芋中皱褶叶的存在与否。‘Miss Muffet’和‘Gingerland’在此基因位点具有杂合基因型（*RLFrlf*）（表 12.3）。当它们彼此杂交时，大约 3/4 的后代会形成具皱褶的叶子。而当两者其中之一与非皱叶品种杂交时，大约一半的后代表现为具皱褶叶子。这些皱叶品种可以成为有价值的育种亲本，用于开发具有皱叶的新品种。

12.3.7 叶性状的遗传连锁

叶片类型与主脉颜色（Deng and Harbaugh 2006a）、叶斑点（Deng et al. 2008a）、叶斑块（Deng and Harbaugh 2009）、皱褶叶（Cao et al. 2016b）和叶底色（Cao et al. 2017）的遗传的独立分类已经得到确认。由于这种独立的分类，以及主脉颜色的多等位基因系统与叶片类型等位基因的共显性，在五彩芋育种群体中观察到了叶片类型和主脉颜色之间的几种分离模式（表 12.4）。

在‘Gingerland’中发现叶斑与绿色的主脉紧密连锁（Deng et al. 2008a）。Deng 等（2008a）用两个基因型为 $V^g s//V^g s$ 的双隐性栽培品种（‘Candidum’和‘White Christmas’）作为测交亲本，估计‘Gingerland’中叶斑和绿脉之间的平均重组率约为 4.4%。在‘White Christmas’、‘Carolyn Whorton’和‘Florida Blizzard’等斑块品种中，观察到了叶片斑块和绿脉之间的遗传连锁。尽管这些品种中的叶片斑块位点 *B* 和脉色位点 *V* 之间的遗传距离仍有待确定，*B* 和 *V* 位点之间的连锁似乎比 *S*（斑点）和 *V* 位点之间的连锁更为紧密，因为在‘White Christmas’、‘Carolyn Whorton’和‘Florida Blizzard’之间的杂交中没有观察到双隐性重组体（Deng and Harbaugh 2009）。

叶皱褶位点 *RLF* 和叶底色位点 *LEM* 也与叶主脉颜色位点 *V* 及叶斑位点 *S* 紧密连锁（Cao et al. 2016b，2017）。Cao 等（2016b，2017）利用由‘Miss Muffet’（$LEM\ V^w s\ rlf//lem\ V^g S\ RLF$）和‘Candidum’（$lem\ V^g s\ rlf//lem\ V^g s\ rlf$）互交得到的两个群体的分离数据，绘制了 4 个基因位点的遗传连锁图谱，这些基因位点控制了五彩芋中的 4 个主要叶片性

表 12.4　五彩芋杂交后代叶片类型和主脉颜色的分离比

杂交组合（♀×♂）	阔心形叶（*FF*）			披针形叶（*Ff*）			带形叶（*ff*）		
	红	白	绿	红	白	绿	红	白	绿
Aaron × Florida Sweetheart	1	1		1	1				
Florida Sweetheart × Aaron	1	1		1	1				
Florida Sweetheart × White Christmas	1		1	1		1			
White Christmas × Florida Sweetheart	1		1	1		1			
Florida Irish Lace × Florida Fantasy	1		1	1		1			
Candidum × Florida Sweetheart	1		1	1		1			
Red Flash × Florida Sweetheart	3		1	3		1			
Florida Red Ruffles ×Florida Fantasy	3		1	3		1			
Florida Red Ruffles ×Florida Moonlight	2	1	1	2	1	1			
Florida Sweetheart × Florida Moonlight	2	1	1	2	1	1			
Florida Irish Lace × Florida Red Ruffles	1		1	2		2	1		1

资料来源：Deng and Harbaugh 2006a

状（主脉颜色、叶底色、斑点和皱纹）。连锁图的范围大约 15cM，主脉颜色位点 *V* 和叶斑位点 *S* 位于叶底色位点 *LEM* 及皱褶叶位点 *RLF* 之间。该遗传连锁谱表明，如果创造皱纹叶、主脉颜色和叶斑点的新组合是主要育种目标，则可能需要大的群体。

12.3.8　其他性状

已经进行了一些研究来了解其他重要性状的遗传模式，如块茎产量潜力、腐霉根腐病抗性和镰刀菌茎腐病抗性。初步数据似乎表明这些特征是数量性状。Deng 等（2005a，2005b）观察到腐霉根腐病严重程度在所评估的 42 个五彩芋品种中连续分布，这可能表明腐霉根腐病抗性是加性基因互作的数量性状。腐霉根腐病的几个抗性品种为阔心形叶和白脉。尚不清楚叶脉颜色、叶片类型和腐霉根腐抗性之间是否存在真正的关联。

12.4　育种方法和技术

12.4.1　杂交

五彩芋依靠繁殖系与现有品种之间的有性杂交来创造新品种（Wilfret 1993）。控制授粉、群体构建以及后代筛选和选择是五彩芋杂交育种的基本方法。五彩芋的许多生物学因素会阻碍其育种授粉中的种子生产和群体构建。五彩芋植物在自然条件下几乎不产生花序。例如，‘Freida Hemple’、‘Carolyn Whorton’和‘Candidum’平均每株植物开花 0.2～1.9 朵（Harbaugh and Wilfret 1979）。五彩芋开花（开裂）通常是零星的且不可预测的。在 23℃下，将块茎在 GA_3 中（250mg/L）浸种 8～16h 可刺激花朵的产生（Harbaugh and Wilfret 1979）。经 GA_3 处理的植株还可以更均匀地产生花朵。不同的五彩芋品种对这种处理的反应可能不同，开花所需天数也不同（Deng and Harbaugh 2004）。

五彩芋花粉寿命短且难以储存，在自然室温和湿度下放置 1 天，仅约 5%的花粉粒能在培养基中保持萌发能力。这可能是由于五彩芋花粉在成熟时是三核的，其中包含两个精子核和一个营养核（Grayum 1986）。这种短的花粉寿命，加上开花少，常常使得难以开展所希望的五彩芋杂交组合。将花粉储存在 4℃，可以在一定程度上减缓萌发能力的下降；贮藏 4 天后，约有 4%的花粉粒可以离体萌发并长出正常的花粉管（Deng and Harbaugh 2004）。这项技术可用于增加一个季节进行杂交的次数。离体花粉萌发可能是评估五彩芋花粉可育性的一种有价值的方法（Deng and Harbaugh 2004）。离体萌发的最佳蔗糖浓度约为 6.8%。

五彩芋花是雌蕊先熟，在开花前几天就具有可授性（Deng and Harbaugh 2004）。柱头可授性峰值可能仅持续约 24h（Job et al. 1979）。到花粉脱落时（通常在开花后 1 天），雌蕊花柱头的表面通常会变色、变软并被一层很黏的渗出液覆盖。

如果雌蕊花的柱头具有可授性，并有花粉，通过人工授粉可以容易地生产出五彩芋种子。通常在授粉后 5～6 周，五彩芋果实成熟。每个浆果里最多有 14 粒种子。一个授粉花序可以产生多达 1500 粒种子（Hartman et al. 1972；Deng and Harbaugh 2004）。新鲜种子在潮湿的泥炭或发芽室容易发芽。光线对于五彩芋种子发芽至关重要，持续 4～12h 的照明会使种子在最短的时间内达到最高的发芽率（Carpenter 1990）。种子发芽的最佳温度为 25～30℃，在这样的温度和持续光照下，大约 90%的种子可以在 2 周内发芽。温度保持在 15℃和 22%～52%的相对湿度可以将五彩芋种子储存 6 个月而发芽率不会显著减少（Carpenter 1990）。

从芽苗发芽后的 3 个月开始，五彩芋苗就开始显色。如果需要的话，后代选择可以在显色后的 2 或 3 个月内开始。在五彩芋的育种中，后代选择分 5 个阶段进行（Deng et al. 2011；Deng 2012）。第 1 阶段的选择在温室中穴盘里生长的单个幼苗上进行，当幼苗 4～6 个月大时，叶子已经显色（图 12.1c）。此阶段的主要目的是丢弃具有绿色叶子或不良着色与着色模式的后代。在这个阶段可能会丢弃约 50%的后代，但这并不罕见，剩余的后代被移栽到地里，并在充足的阳光下生长。第 2 阶段的选择是针对田间生长的单个株系进行的，选择标准包括植株生长习性、叶色、耐日晒性、细菌性枯萎病抗性等。若单个株系在第 2 年末形成块茎，无性繁殖（块茎切分）可在第 3 年春季开始。当田间有一定量由单株繁殖来的植株时，便开始第 3 阶段的选择。除上述植株和叶片特征外，块茎大小/重量也可包括在选择标准中。这个阶段可能会持续 1～4 年，取决于通过块茎分割五彩芋品系繁殖得有多快。当有足够的繁殖体时，五彩芋品系可进入第 4 阶段的选择，重复田间试验（图 12.1d）。这些试验目的是评估对照、意向新品系与现有商业品种的块茎产量潜力、植物性能和耐日晒性。通常，此类试验每 2～4 年重复一次。从这些试验中收集的数据将作为品种发布的基础。第 5 阶段的选择是评估选定的五彩芋品系用于盆栽的表现（图 12.1e）。所有生产的五彩芋块茎中，约有 70%用于盆栽生产。为了确保所发布的品种适合用作盆栽植物生产，第五阶段的选择或重复盆栽试验很重要。其他筛选或选择也可以进行，包括抗病筛选、田间试验、块茎商业生产田间试验以及品种真实性实验。在这个多年选择过程的每个阶段，商业品种和/或亲本品种通常被作为对照品种。

12.4.2　种间杂交

20 世纪 70 年代以前，关于五彩芋栽培种的起源或亲缘关系的文献很少，所以很难评估种间杂交在五彩芋育种史上的作用。

除了 *C. bicolor* 和 *C. schomburghii*，还有一些其他品种可能是五彩芋育种中有前途的育种亲本。*C. clavatum* 的植株特征是大叶子，有浅到深紫色斑点和主脉。*C. picturatum* 的叶子形状独特，具长而窄的叶片和由两个孔分开的裂片。*C. humboldtii* 是微型植物（株高＜25cm），在小的绿色叶子上有许多白色斑块，通常长＜10cm，宽＜5cm。引入部分这些叶和植株特征，可扩大 *C.*× *hortulanum* 品种的叶片类型、颜色和着色模式的范围。关于这些物种的育种，努力的第一步将是确定这些物种之间的亲和性。如果种间杂交可行并且可以产生种间杂种，则可以按照种内有性杂交后所使用的相同步骤进行后代筛选、选择和鉴定。

考虑到 *C.*× *hortulanum* 品种和 *C. clavatum*、*C. humboldtii* 或 *C. picturatum* 在染色体数量和基因组大小上有很大差异（Cao et al. 2014），这些物种之间种间杂交的后代（如果这些后代能够产生），部分可能高度不育。这种预期的不育性可能会阻碍有利于种间杂种获得的回交，因此，旨在从 *C. clavatum*、*C. humboldtii* 或 *C. picturatum* 引入品种特性到 *C.*× *hortunalum* 品种的种间杂交，需要在第一代集中精力选择好的种间杂种，而不是在后面的世代中选择。如果这些杂种表现理想，株型、叶或块茎性状好，则可以通过块茎切分或组织培养方法进行商业繁殖。

12.4.3　诱变育种

到目前为止，突变在五彩芋育种中的使用受到限制。有报道描述了由 γ 射线辐照引起的五彩芋突变率和突变谱，但是辐照创造的五彩芋品种还没有报道或没有被注册（Chu et al. 2000）。在五彩芋育种中不喜欢用这种方法的部分原因可能是，组织培养后五彩芋体细胞克隆变异率普遍高。Chu 和 Yazawa（2001）报道，‘Fire Chief’、‘Miss Muffet’和‘Tropicana’中 65%～80%再生植株的叶脉颜色从白色、红色或粉红色变为绿色。在其他五彩芋品种中，也观察到了体细胞无性繁殖的高频变异，包括‘Pink Cloud’、‘Rosalie’、‘White Queen’和‘WhiteWing’。观察到的变化性状包括叶片模式特征（形状、叶端、边缘和基部）、叶柄特征（连接和颜色）与叶片颜色特征（叶片、斑块、叶片边缘、主脉、次脉和汇合端的颜色）。每个变体个体具有一个或多个这些特征的变化（Thongpukdee et al. 2010）。总体而言，叶色突变率最高，其次是叶片模式变异。叶片类型突变的发生频率要低得多。

培养基中的植物生长调节剂对五彩芋体细胞克隆变异的发生产生显著影响。在植物生长素中，2,4-D 引起的变异比 NAA 多得多。Ahmed 等（2004）指出，当‘Pink Cloud’叶片外植体在含有 0.1～1.0mg/L 的 2,4-D 培养基上培养时，100%的再生植株表现出叶色变化，而当培养基中含有 0.1mg/L NAA 时，15%的再生植株会发生变化。与茎尖相比，叶片外植体似乎在组织培养中产生更多的体细胞克隆变异体（Ahmed et al. 2002）。与易于产生体细胞变异的五彩芋品种相比，‘Red Flash’相对稳定，并且体细胞变异少得多。但是，当在含 2,4-D 的培养基上培养成熟的叶片组织时，25%的‘Red Flash’再生植株在叶片形状、主脉颜色、斑点和叶缘颜色以及叶片大小方面表现出大的变化（Cao et al. 2016a）。已认可了 10 种类型的突变体。

直到现在，在细胞和分子水平上对五彩芋体细胞无性突变体还知之甚少。Cao 等（2016a）研究了‘Red Flash’的 24 个体细胞克隆变异体的核 DNA 含量、染色体数目和分子标记条带。一半突变体的核 DNA 含量减少 1.1%～5.4%，丢失了一条染色体；两个突变体的核 DNA 含量减少 5.4%～9.2%，丢失两条染色体；一个突变体的核 DNA 含量增加 95.0%，并具有 $2n = 58$ 条染色体。在 24 个突变体中，有 8 个在 *CaM1* 或 *CaM103* 标记位点丢失了两个等位基因之一，并且一个突变体在 *CaM103* 标记位点显示出等位基因大小变化（Cao et al. 2016a）。总体而言，五彩芋的体细胞克隆变异涉及多种细胞学和/或分子原因，而染色体数目变化可能是导致非整倍体和标记等位基因丢失频率最高的最常见原因。带有 *CaM1* 的 SSR 标记基因位点的染色体似乎非常不稳定，在组织培养过程中容易丢失（Cao et al. 2016a）。

来源于体细胞克隆变异或自然突变的五彩芋品种的数量仍然非常有限。‘Victoria’是自然发生的整株突变，从无性繁殖的‘Florida Sweetheart’群体中鉴定出来（Hartman 2007）。选择该品种的依据是其独特的叶片质地和叶片上的深红色。与‘Florida Sweetheart’的对照比较显示，植株和叶片的特性还发生了其他变化，包括更紧凑的株型，起伏更大，叶片更皱和叶柄较短。五彩芋突变体的其他变种包括在 *C. humboldtii* 中发现的两个：‘Marcel’和‘Snow White Park’。‘Marcel’是由 Marcel Lecoufle 的实验室在组织培养过程中发现突变体而得来的（Lecoufle 1981）。‘Snow White Park’是 *C. humboldtii* 在商业化种植中发现的一个整株突变体，其株型矮小，叶片更白（Park 2012）。

12.4.4 倍性育种

由于五彩芋的茎被压缩到地下块茎中，而分生组织被包裹在叶鞘内和土壤表层以下，因此很难使五彩芋的染色体加倍。即使可以在块茎的分生组织中诱导出四倍体细胞，也很难发育成稳定或纯合的四倍体。最近，Cai 等（2015）报道了有效的染色体离体加倍技术，通过预培养五彩芋叶盘，用秋水仙碱处理培养物并用倍性分析仪筛选再生植株。在一些处理中，多达 24%的再生植株表现为四倍体。四倍体的五彩芋植株表现为圆形和加厚的叶片，以及加厚的叶柄。一些明显的四倍体显示出不同的叶片形态。由于这种差异，Cai 等（2015）分析了外形变化明显的四倍体的核 DNA 含量和染色体数目。一些四倍体具有预期的 $2n = 4x = 60$ 条染色体，但是发现一个四倍体具有 4 条额外的染色体，两个明显的四倍体缺少 2 条或 4 条以下染色体。这些非整倍体的出现是出乎意料的，还有待进一步研究。一个潜在的原因可能是叶片组织中存在非整倍体细胞，秋水仙碱处理后其染色体加倍。在未经秋水仙碱处理的某些品种的再生植株中也观察到了高频的染色体加倍。已经尝试通过在四倍体和二倍体之间进行杂交来产生三倍体。利用诱导的四倍体，未来的五彩芋改良将扩展到四倍体或三倍体水平。

12.5　生物技术方法的发展和应用

12.5.1　体细胞杂交

通过融合和培养原生质体，已经在许多水果、蔬菜和大田作物中实现了体细胞杂交。这种方法对于将不同品种，甚至不同物种或属间的病虫害抗性性状相结合，创造新种质具有强大的作用（Johnson and Veilleux 2001）。该技术可能极大地促进将不同品种的镰刀菌茎腐病、腐霉根腐病和根结线虫的抗性组合到新的体细胞杂种中。Jing 和 Wang（1991）发表了有关五彩芋原生质体分离和培养以及植株再生的综述。从那时起，文献中关于该主题没有任何进一步的报道。来自其他植物的原生质体培养和融合实验的经验表明，诱导和建立胚性愈伤组织，并使用胚性悬浮细胞作为组织材料分离原生质体至关重要。这些领域的研究工作可能为将体细胞杂交技术应用于五彩芋育种奠定基础。

12.5.2　遗传转化

五彩芋的遗传转化已在许多报道中提到。农杆菌介导的转化率高达 10%（Li et al. 2005）。玉米花青素合成调控基因 *Lc* 在白脉品种‘Jackie Suthers’中的过表达，使转基因植株在叶、叶柄、块茎和根中积累花青素并使这些器官变红（Li et al. 2005）。以前，一种人工合成生长的激素基因是通过类似的方法引入五彩芋中的（Li et al. 1994）。这些研究表明，基于农杆菌的系统适用于五彩芋的遗传转化。

但在这些研究和未来的实验中生产的转基因五彩芋植株，似乎不太可能直接开发成新的品种。然而，这些研究中使用的转化技术可以用于提供基因编辑所需的元件，以在五彩芋中创造新的园艺和观赏性状。

12.5.3　分子标记

分子标记辅助选择（MAS）已被用于许多农艺和园艺作物的育种。这种工具在选择对根结线虫具有抗性的五彩芋品种方面非常有用。在许多植物-线虫的病理系统中，根被咬伤的严重程度与对线虫的敏感性密切相关，并且根不被咬伤已被普遍用作线虫抗性指标。在五彩芋中，已经观察到抗性品种和易感品种的根部都有咬伤，并且易感品种的根咬伤没有抗性品种严重（Dover et al. 2005；McSorley et al. 2004）。单独的根咬伤可能不是反映五彩芋对线虫抗性的可靠指标。抗线虫五彩芋的育种可能有赖于分子标记的开发和应用。基于标记的选择系统还可用于鉴定对其他疾病（如腐霉根腐病和镰刀菌茎腐病）具有抗性的育种系。

据报道，分子标记在五彩芋中的应用主要是在遗传多样性和种间关系的分析中（Loh

et al. 1999，2000；Deng et al. 2007）。通过对‘Florida Sweetheart’含有SSR的基因组DNA进行富集和测序，开发出五彩芋特异性SSR标记（Gong and Deng 2011）。这些SSR标记在五彩芋品种中具有高度多态性，且存在多个位点，在五彩芋不同物种中是可以通用的。它们已成为区分五彩芋栽培种、检测体细胞克隆变异（Cao et al. 2016a）、评估遗传多样性（Deng et al. 2007）以及了解五彩芋种与种间遗传关系的有力工具（Loh et al. 1999，2000；Deng et al. 2007；Gong and Deng 2011）。更多的分子标记和高通量标记分析系统，对于分子标记在五彩芋育种中的实际应用必不可少。

12.5.4 基因组信息的开发和应用

Cao 和 Deng（2017）使用 RNA-Seq 方法表征了三个五彩芋变种（‘Candidum’、‘Gingerland’和‘Muff Muffet’）的根转录组。从头组装了近1.7亿个过滤后的短reads，得到了大约13万个unigenes。研究中用到的三个五彩芋品种对腐霉根腐病的抗性水平不同。比较‘Candidum’（对腐霉根腐病有中等抵抗力）与‘Gingerland’和‘Miss Muffet’（均对腐霉根腐病极易感）的转录本，鉴定出有4518个unigenes只存在于‘Candidum’根转录组中。这些unigenes中的98个似乎与抗病性和防御反应有关，包括植物-病原相互作早期信号感知和信号转导，植物对坏死病原抵抗反应期间的激素调节，以及植物防御反应时具有抗菌活性的次生代谢产物产生。此外，在这些品种的转录组序列中鉴定出将近29 000个SSR位点和45 000个单核苷酸多态性（SNP）位点。这些SSR和SNP位点可以用于设计新的寡核苷酸引物与开发新的SSR和/或SNP标记的靶区域。随着五彩芋基因组和转录组测序数据的陆续出现，育种家可据此设计出高度特异性的向导RNAs和其他载体，用于精准育种或靶向基因编辑。

12.6 展　望

人们对种植观赏作物——五彩芋的兴趣大大增加，培育新的和改良的五彩芋栽培种可满足这种需要。最新获得的有关观赏和园艺性状的遗传模式知识，以及新开发的分子标记有望提高五彩芋育种的效率。作为一种可靠的育种方法，杂交将继续在五彩芋新品种的开发中发挥重要作用。同时，体细胞克隆变异、倍性育种和种间杂交可能对于创建新颖的颜色和着色图案，甚至提高作物产量和抗病性很有用。展望未来，关于细胞和分子技术的更多研究，如原生质体培养、体细胞杂交、遗传转化、基因组和转录组测序以及靶向基因编辑，对于继续改善五彩芋、创造理想的性状（如新颖的颜色和着色模式、耐除草剂与抗病）至关重要。

本章译者：

彭绿春
云南省农业科学院花卉研究所，国家观赏园艺工程技术研究中心，云南省花卉育种重点实验室，昆明 650200

参考文献

Ahmed EU, Hayashi T, Zhu Y, Hosokawa M, Yazawa S (2002) Lower incidence of variants in *Caladium bicolor* Ait. plants propagated by culture of explants from younger tissue. Sci Hortic 96:187–194

Ahmed EU, Hayashi T, Yazawa S (2004) Auxins increase the occurrence of leaf-color variants in caladium regenerated from leaf explants. Sci Hortic 100:153–159

Bell ML, Wilfret GJ, Devoll DA (1998) Survey of caladium tuber producers for acreage of cultivars grown. Proc Fla State Hortic Soc 111:32–34

Birdsey MR (1951) The cultivated aroids. Gillick Press, Berkeley

Cai X, Deng Z (2016) Thidiazuron promotes callus induction and proliferation in *Caladium ×hortulanum* Birdsey UF-4609. Propag Ornam Plants 16(3):90–97

Cai X, Cao Z, Xu S, Deng Z (2015) Induction, regeneration and characterization of tetraploids and variants in 'tapestry' caladium. Plant Cell Tissue Organ Cult 120:689–700

Cao Z, Deng Z (2017) De Novo assembly, annotation, and characterization of root transcriptomes of three caladium cultivars with a focus on necrotrophic pathogen resistance/defenses-related genes. Int J Mol Sci 18:712. https://doi.org/10.3390/ijms18040712

Cao Z, Deng Z, McLaughlin M (2014) Interspecific genome size and chromosome number variation shed new light on species classification and evolution in *Caladium*. J Am Soc Hortic Sci 139:449–459

Cao Z, Sui S, Cai X, Yang Q, Deng Z (2016a) Somaclonal variation in 'Red Flash' caladium: morphological, cytogenetic and molecular characterization. Plant Cell Tissue Organ Cult 126(2):269–279. https://doi.org/10.1007/s11240-016-0996-3

Cao Z, Sui Z, Yang Q, Deng Z (2016b) Inheritance of rugose leaf in caladium and genetic relationships with leaf shape, main vein color, and leaf spotting. J Am Soc Hortic Sci 141:527–534. https://doi.org/10.21273/JASHS03854-16

Cao Z, Sui Z, Yang Q, Deng Z (2017) A single gene controls leaf background color in caladium (Araceae) and is tightly linked to genes for leaf main vein color, spotting and rugosity. Hortic Res 4:16067. https://doi.org/10.1038/hortres.2016.67

Carpenter WJ (1990) Light and temperature govern germination and storage of caladium seed. HortScience 25:71–74

Chu Y, Yazawa S (2001) The variation and the hereditary stability on leaf character of plantlets regenerated from micropropagation in caladiums. J Chin Soc Hortic Sci 47:59–67

Chu Y, Hu T, Tsai YC, Chen JJ (2000) Effects of γ–rays irradiation and leaf blade culture on mutations in caladiums. J Chin Soc Hortic Soc 46:381–388

Croat T (1994) Taxonomic status of neotropical Araceae. Aroideana 17:33–60

Deng Z (2012) Fancy-leaved caladium varieties recently introduced by the UF/IFAS caladium breeding program. Proc Fla State Hortic Soc 125:307–311

Deng Z, Harbaugh BK (2004) Technique for in vitro pollen germination and short-term pollen storage in caladium. HortScience 39:365–367

Deng Z, Harbaugh BK (2006a) Independent inheritance of leaf shape and main vein color in caladium. J Am Soc Hortic Sci 131:53–58

Deng Z, Harbaugh BK (2006b) Evaluation of caladium cultivars for sensitivity to chilling. HortTechnology 16:172–176

Deng Z, Harbaugh BK (2006c) 'Garden White' – a large white fancy-leaved caladium for sunny landscapes and large containers. HortScience 41:840–842

Deng Z, Harbaugh BK (2008) Caladium breeding: progress in developing lance-leaved cultivars. Proc Fla State Hortic Soc 121:395–398

Deng Z, Harbaugh BK (2009) Leaf blotching in caladium (Araceae) is under simple genetic control and tightly linked to vein color. HortScience 44:40–43

Deng Z, Harbaugh BK, Kelly RO, Seijo T, McGovern RJ (2005a) Pythium root rot resistance in commercial caladium cultivars. HortScience 40:549–552

Deng Z, Harbaugh BK, Kelly RO, Seijo T, McGovern RJ (2005b) Screening for resistance to Pythium root rot among twenty-three caladium cultivars. HortTechnology 15:631–634

Deng Z, Goktepe F, Harbaugh BK (2007) Assessment of genetic diversity and relationships among caladium cultivars and species using molecular markers. J Am Soc Hortic Sci 132:219–229

Deng Z, Goktepe F, Harbaugh BK (2008a) Inheritance of leaf spots and their genetic relationships with leaf shape and vein color in caladium. J Am Soc Hortic Sci 133:78–83

Deng Z, Harbaugh BK, Peres NA (2008b) 'UF-404' – Dwarf, red caladium for container-forcing and sunny landscape. HortScience 43:1907–1910

Deng Z, Harbaugh BK, Schoellhorn RK, Andrew RC (2008c) 2003 Survey of the Florida caladium tuber production industry. http://edis.ifas.ufl.edu/ep258/

Deng Z, Peres NA, Harbaugh BK (2011) Improving disease resistance in caladium: progress and prospects. Acta Hort 886:69–76. https://doi.org/10.17660/ActaHortic.2011.886.7

Deng Z, Harbaugh BK, Peres NA (2013) UF 4412 and UF 4424 – lance-leaved caladium cultivars. HortScience 48:239–244

Dover KD, McSorley R, Wang K-H (2005) Resistance and tolerance of caladium cultivars to *Meloidogyne incognita*. Soil Crop Sci Soc Fla Proc 64:98–102

Gager CR (1991) Leaf spot color and *venation* pattern inheritance in *Caladium*. Dissertation, University of Florida

Goktepe F, Seijo T, Deng Z, Harbaugh BK, Peres NA (2007) Toward breeding for resistance to Fusarium tuber rot in caladium: inoculation techniques and sources of resistance. HortScience 42:1135–1139

Gong L, Deng Z (2011) Development and characterization of microsatellite markers for caladiums (*Caladium* Vent.). Plant Breed 130:591–595. https://doi.org/10.1111/j.1439-0523.2011.01863.x

Graf AB (1976) Exotica, series 3: pictorial cyclopedia of exotic plants from tropical and near-tropical regions, 9th edn. Roehrs, Rutherford

Grayum MH (1986) Pylogenetic implications of pollen nuclear number in *Araceae*. Plant Syst Evol 151:145–161

Harbaugh BK, Wilfret GJ (1979) Gibberellic acid (GA_3) stimulates flowering in *Caladium hortulanum* Birdsey. HortScience 14:72–73

Hartman RD (1974) Dasheen mosaic virus and other phytopathogens eliminated from caladium, taro, and cocoyam through culture of shoot tips. Phytopathology 64:237–240

Hartman RD (2007) Caladium plant named 'Victoria'. United States Plant Patent. U.S. PP18,087, 25 Sept 2007

Hartman RD (2008) Classic caladium sun series. GPN, Greenhouse Product News: 58

Hartman RD, Zettler FW, Knauss JF, Hawkins EM (1972) Seed propagation of caladium and dieffenbachia. Proc Fla State Hortic Soc 85:404–409

Hayward W (1950) Fancy-leaved caladiums. Plant Life 6:131–142

Hetterscheid WLA, Bogner J, Boos J (2009) Two new *Caladium* species (Araceae). Aroideana 32:126–131

Huxley A, Griffiths M, Levy M (1992) The new Royal Horticultural Society dictionary of gardening. MacMillan Press, London

Jing Z, Wang Z (1991) Plant regeneration from leaf callus protoplasts of *Caladium bicolor* Vent. Physiol Plant 82:A17

Job JS, Vijaya Bai K, Hrishi N (1979) Major factors limiting seed set in aroids. Paper presented at the 5th international symposium on tropical root and tuber crops, Visayas State College of Agriculture, Los Banos, Laguna, Philippines, 17–21 September 1979. http://www.istrc.org/images/Documents/Symposiums/Fifth/5th_symposium_proceedings_0065_697.pdf

Johnson AAT, Veilleux RE (2001) Somatic hybridization and application in plant breeding. Plant Breed Rev 20:167–225

Jones EJ (1957) Chromosome number and phylogenetic relationships in the Araceae. Dissertation, University of Virginia

Knauss JF (1975) Description and control of Fusarium tuber rot of caladium. Plant Dis Rep 59:975–979

Kurakubo Y (1940) Über die chromosomenzahlen von Araceae-Arten. Bot Zool 8:1492

Lecoufle M (1981) *Caladium humboldtii* and its cultivar 'Marcel'. Aroideana 4:114–115

Li BJ, Wang JF, Xu ZF, Xu YQ, Yu MZ, He X, Shen Y (1994) Integration and expression of human growth hormone gene in *Caladium bicolor*. Sci China B 37:280–285

Li SJ, Deng XM, Mao HZ, Hong Y (2005) Enhanced anthocyanin synthesis in foliage plant *Caladium bicolor*. Plant Cell Rep 23:716–720

Loh JP, Kiew R, Kee A, Gan LH, Gan Y-Y (1999) Amplified fragment length polymorphism (AFLP) provides molecular markers for the identification of *Caladium bicolor* cultivars. Ann Bot 84:155–161

Loh JP, Kiew R, Hay A, Kee A, Gan LH, Gan Y-Y (2000) Intergeneric and interspecific relationships in Araceae tribe Caladieae and development of molecular markers using amplified fragment length polymorphism (AFLP). Ann Bot 85:371–378

Madison M (1981) Notes on *Caladium* (Araceae) and its allies. Selbyana 5:342–277
Marchant CJ (1971) Chromosome variation in *Araceae*: II: Richardieae to Colocasieae. Kew Bull 25:47–56
Mayo SJ, Bogner J, Boyce PC (1997) The genera of Araceae. Royal Botanical Gardens, Kew
McGovern RJ (2004) Fighting Fusarium. Greenh Grow 22:146–150
McSorley R, Wang K-H, Frederick JJ (2004) Host suitability of caladium varieties to *Meloidogyne incognita*. Nematropica 34:97–101
Park SK (2012) Caladium plant named 'Snow White Park'. United States Plant Patent. US PP22,706, 1 May 2012
Pfitzer P (1957) Chromosomenzahlen von Araceen. Chromosoma 8:436–446
Ramachandran K (1978) Cytological studies on south Indian Araceae. Cytologia 43:289–303
Ridings WH, Hartman RD (1976) Pathogenicity of *Pythium myriotylum* and other species of *Pythium* to caladium derived from shoot-tip culture. Phytopahtology 66:704–709
Sarkar AK (1986) Mode of evolution in *Caladium bicolor* (Ait.) Vent. (Araceae). Biologisches Zentralblatt 105:621–639
Seijo TE, Peres NA, Deng Z (2010) Characterization of strains of *Xanthomonas axonopodis* pv. *dieffenbachiae* from bacterial blight of caladium and identification of sources of resistance for breeding improved cultivars. HortScience 45:220–224
Sharma AK (1970) Annual report, 1967–1968. Res. Bul. Univ. Calcutta (Cytogenetics Lab.) 2:1–50. (cited from Petersen, 1989)
Sharma AK, Sarkar AK (1964) Studies on the cytology of *Caladium bicolor* with special reference to the mode of speciation. Genetica Iberica 16:21–47
Simmonds NW (1954) Chromosome behavior in some tropical plants. Heredity 8:139–146
Subramanian D, Munian M (1988) Cytotaxonomical studies in south Indian Araceae. Cytologia 53:59–66
The Plant List (2013) Version 1.1. http://www.theplantlist.org/tpl1.1/search?q=caladium
Thongpukdee A, Thepsithar C, Chiensil P (2010) Somaclonal variation of *Caladium bicolor* (Ait.) Vent. 'Jao Ying' after *in vitro* culture propagation. Acta Hortic 855:281–287
Wilfret GJ (1983) Inheritance of vein color in caladium leaves. HortScience 18:610
Wilfret GJ (1986) Inheritance of leaf shape and color patterns in *Caladium* (Araceae). HortScience 21:750
Wilfret GJ (1992) 'Florida Cardinal' caladium. HortScience 27:1342–1344
Wilfret GJ (1993) Caladium. In: de Hertogh A, LeNard M (eds) The physiology of flower bulbs. Elsevier, Amsterdam, pp 239–247
Wilfret GJ, Hurner GT Jr (1982) A survey of caladium cultivars grown in Florida and their characteristics as potted plants. Proc Fla State Hortic Soc 95:190–194
Zettler FW, Abo El-Nil MM (1979) Mode of inheritance of foliage color in caladiums. J Hered 70:433–435

第 13 章 肖 竹 芋

Johan Van Huylenbroeck, Evelien Calsyn, Andy Van den Broeck,
René Denis, and Emmy Dhooghe

摘要 竹芋科（Marantaceae），特别是包含其中的肖竹芋属（*Calathea*）、锦竹芋属（*Ctenanthe*）、竹芋属（*Maranta*）、红里蕉属（*Stromanthe*）植物在热带和亚热带气候区通常被作为观赏植物，或是在温带地区作为室内盆栽。除了黄苞竹芋（*Calathea crocata*），大部分皆是因为它们叶片的可观赏性而被栽培应用。在 20 世纪 70～80 年代，除野生资源以外，仅通过组织培养进行无性繁殖产生体细胞无性系变异来形成新品种。在 21 世纪初，针对花卉植物的开发利用逐渐开始并盛行，种间杂交结合胚挽救技术造就了肖竹芋属植物新品种的亮点，如叶片观赏性好、花期更长。如今的育种目标则是株型紧凑、生产周期较短和花色新颖。属间染色体数目和基因组大小的不同阻碍了种间杂交育种，为了使种间和属间能混合杂交，目前的研究主要集中于多倍体化、选择性授粉技术和子房培养等方面。

关键词 肖竹芋属；竹芋科；种间杂交；育种；染色体数目；基因组大小

13.1 引 言

竹芋科是较为进化的单子叶植物门姜目植物（Kress 1995）。它包括 31 个属的约 550 个种的多年生草本和藤本植物（Andersson 1998）。其中，14 个属是在美国发现的，11 个属在非洲和马达加斯加发现，8 个属在亚洲发现，只有 2 个属在 2 个陆地板块出现（Prince and Kress 2006）。绝大部分种（>450 个种）位于美国的热带地区，它们大部分（约 300 种）属于肖竹芋属。竹芋科植物具有两个区别于其他科植物的特征：①叶片“C”形侧脉和平行交叉脉；②存在一个控制叶片运动的特殊细胞区域，即叶枕（Kennedy 2000）。在过去的 35 年里，Kennedy（1983，1995，2011）对许多新种和主要为肖竹芋属的品种进行过描述。Ley 和 Claßen-Bockhoff（2011，2013）对非洲物种进行了详细的研究。科内的分类和不同属间的系统进化关系几十年来一直具有极大的争议。1998 年，Andersson 提出在全世界范围内将竹芋科细分为 5 个非正式的类别。之后，依据 DNA 测序数据结合形态学描述针对竹芋科的进化又提出了新的观点（Andersson and Chase

J. Van Huylenbroeck (✉) · E. Calsyn · E. Dhooghe
Flanders Research Institute for Agriculture, Fisheries and Food (ILVO), Plant Sciences Unit, Applied Genetics and Breeding, Melle, Belgium
e-mail: Johan.vanhuylenbroeck@ilvo.vlaanderen.be
A. Van den Broeck · R. Denis
Denis-Plants bvba, Beervelde-Lochristi, Belgium

2001；Prince and Kress 2006；Ley and Claßen-Bockhoff 2011）。基于分子进化学方面的数据，Borchsenius 等（2012）赋予了肖竹芋属内植物新的遗传学边界，并将肖竹芋属植物归到了葛伯楠属（*Goeppertia*）。由于这种新的命名法至今没有在肖竹芋科植物的商业贸易领域中广泛使用，因此在本书中我们仍然将其作为肖竹芋属。

尽管竹芋科植物具有丰富的自然多样性，但只有有限数量的物种和品种可作为观赏植物。它们大多具有多彩或形态各异的叶子，因而才在商业生产中得以利用。色彩斑斓的叶子有白色、橙色或红色的斑点或条带，通常下表面为明亮的紫色。它们对弱光的耐受力是它们能成为室内景观的关键因素。肖竹芋属（葛伯楠属）、锦竹芋属、竹芋属、红里蕉属植物的代表在热带和亚热带气候区作为景观植物或者在温带地区作为室内盆栽植物。只有很少的物种可作为花卉进行售卖。黄苞竹芋（*Calathea crocata*）是最受欢迎的一种，其头状花序为短穗状花序，有明亮的橙黄色苞片。紫背天鹅绒竹芋（*Calathea warscewiczii*）有白色的苞片延伸到植株上部。除了这些品种，育种家还培育出了其他品种，如 ‘Mia’ 、 ‘Indri’ 和 Bicajoux®系列（图 13.1）。竹芋科盆栽的年产量为 800 万～1000 万株。

在一些地区，有些竹芋科植物是做成切花或切叶的。据报道，在夏威夷，箭羽竹芋（*C. insignis*）与金黄肖竹芋（*C. lutea*）分别在夏末至初冬和 1～3 月的盛花期进行季节性生产；圆柱竹芋（*C. cylindrica*）、响尾蛇竹芋（*C. crotalifera*）和一些肿节竹芋（*C. burle-marxii*）的各种变种则进行全年生产。

图 13.1　花部图片。（a）黄苞竹芋（*Calathea crocata*）；（b）紫背天鹅绒竹芋（*Calathea warscewiczii*）；（c）波浪竹芋（*Calathea rufibarba*）；（d）Bicajoux®系列 ‘Anaconda Pink’；（e）Bicajoux®系列 ‘Gekko Pink’；（f）Bicajoux®系列 ‘Mamba White’

从夏末到深秋，粉被多穗竹芋（*Pleiostachya pruinosa*）闻名于它的指形或麦穗形而在市场上流行。紫背竹芋（*Stromanthe sanguinea*）则是全年生产，5 月为高峰期（Criley 2015），有关其准确的生产量不详。

13.2 生殖生物学

13.2.1 花

竹芋科的花部结构很独特，使之很容易跟其他科植物区分开来（图 13.2）。虽然花朵有三个花瓣和三个离生萼片，但雄蕊结构不同，有一个兜状退化雄蕊、一个胼胝质退化雄蕊和一个或两个外退化雄蕊及一个仅一半花药是可育的雄蕊。花柱具弹性且柱头为杯状，花朵开放时在兜状退化雄蕊的张力下起到支撑作用。在传粉者来访期间，花柱被释放向前跃动并向内弯曲进而接受花粉（Kennedy 1978，2000）。竹芋科中这种不可逆转的花柱运动在很大程度上是以特殊组织结构为基础的。Pischtschan 和 Claßen-Bockhoff（2010）用采用组织切片的显微技术来研究竹芋科不同物种的花柱。另外，由于竹芋花柱的生长与封闭性兜状的退化雄蕊相关，因而这个运动的生物力学原理得以被研究（Pischtschan and Claßen-Bockhoff 2008）。

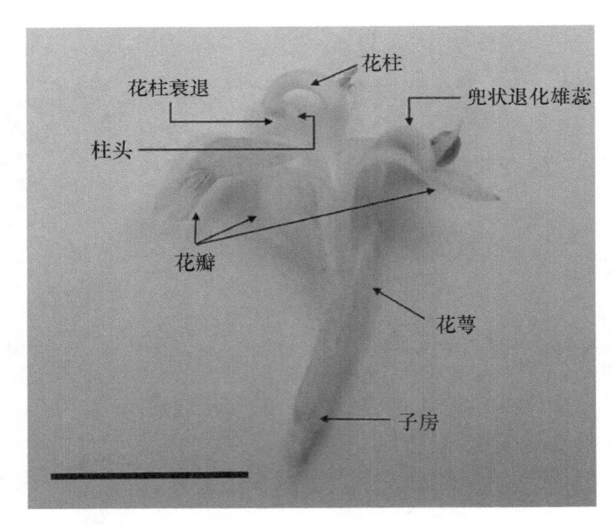

图 13.2　竹芋的花部形态结构（标尺 = 1 cm）

在花序中产生带苞片的不对称的花则更具有吸引力。这些花通常成为另一朵花的镜像。每个苞片对生一朵花或对应一个聚伞形花序（Kennedy 1978，2000）。

13.2.2 自然授粉

竹芋科的传粉机制被称为爆炸性的次级花粉展示（explosive secondary pollen presentation），并且被认为是一种关键的创新性适应机制，有助于提升物种的形成速率

（Kennedy 1978，2000）。在自然界中，花药中的花粉通常在开花前一晚散落，花柱的向上生长迫使花药中的花粉沉积在柱头背面的浅凹处。在开花期，柱头在清晨处于开放状态，并在兜状退化雄蕊的张力下保持原位。当传粉者进入花中寻找花蜜时，它便压向兜状退化雄蕊的附器，花柱向前咔嗒一声，柱头就接触到了传粉者访问过的花的花粉。同时，花朵自身的花粉则沉积在传粉者的身上（Kennedy 2000）。因为爆炸性的授粉机制是不可逆的，且花只开放一天，因此每一朵花就只有一次机会得以授粉和散落花粉（Kunze 1984；Kennedy 1978）。

在自然界中，竹芋科主要由蜜蜂授粉。花的形态和大小差异（花冠筒的长度与柱头的形状）通常用于适应不同的传粉者。Kennedy（2000）曾报道，与非洲和亚洲物种（平均 4.6mm）相比，美国物种的花粉管明显更长（平均 17.6mm）。前者主要适应长舌蜂的传粉，而后者则适应短舌蜂的传粉，有些物种在自然条件下也可由鸟类传粉。在黄苞竹芋中，观察到 4 种蜂鸟拜访植物并为花朵授粉（Nolasco et al. 2013）。所有的竹芋科植物都是自交亲和的。精巧的授粉机制阻止了自交。然而，当一个花序上的几朵花同时开放时，通常被认为是雌雄同株授粉。在竹芋科里约 8%的物种中，存在不同程度的自交（Kennedy 2000）。在自花授粉的情况下，授粉过程中花粉在花蕾内发生转移。花柱顶端和柱头方向的轻微改变可确保花粉管在伸长过程中进入柱头，然后柱头接受花粉开始萌发，从而促成自花授粉。在竹芋属的某些物种中，授粉后除了发生形态变化，还有明显的颜色变化。柱头的背部由最初的白色或黄色变为深棕色甚至黑色。Ley 和 Claßen-Bockhoff（2012）详细研究了竹芋科内存在的花朵多样性是否会影响繁育体系和传粉者的花粉放置位置。研究分析了 66 个不同物种的花朵，各物种在它们的基本花型上都类似，研究表明细微的形态变化会决定植物是异花授粉或是自花授粉，从而改变植物的繁育体系。

13.2.3 人工授粉

由于花的形态特殊，花粉和柱头存在一定的空间隔离。因为花柱运动是不可逆转的，除了自交授粉植物，一定的空间隔离可防止其他任何花粉随后进入柱头。这样的授粉机制对繁育来说是非常有利的，因此没有必要在授粉之前掐花，也没有必要在授粉之后进行套袋。如果进行人工授粉，可将兜状退化雄蕊拉向一边（在移除或移动胼胝质退化雄蕊后），然后用针把花粉从花粉盘上刮下来，再转移到另一朵花的柱头腔里（用于给另一朵花授粉）。在种间杂交中，人工授粉的成功率在很大程度上取决于物种本身和杂交组合。Ley 和 Claßen-Bockhoff（2013）报道，在自然条件下，对于非洲物种，可以通过人工授粉提高结实率。人工授粉后，自交授粉的平均结实率可达 18.5%，种间杂交可达到 12.9%。

每个果实有 1～3 粒种子。种子是弯生的，而且种子有一个外胚乳通道、一个孔盖、珠孔领和假种皮（Grootjen 1983）。竹芋科种子的外胚乳管是其最显著的解剖学特征，且具有很大的变异性（Benedict et al. 2016）。

13.3 育　　种

13.3.1 育种史

20 世纪 70 年代初期，开始商业化种植竹芋科植物。1975 年，仅有三个物种（*C. insignis*、*C. makoyana* 和 *C. roseopicta*）在美国得以普遍种植（Chao et al. 2005）。在欧洲，20 世纪 70～80 年代第一批用于商业生产的竹芋来自野外。在投向市场之前，这些收集来的植物都经过了观赏价值和生长适应性评价，评估其是否适合作为盆栽植物。组织培养规程的开发对竹芋科商业品种的大规模繁殖是一项重大突破（Dunston and Sutter 1984；Podwyszyńska 1997），这使得新品种可以更快地投放市场。但是，与大多数其他观赏植物相比，竹芋科植物的组织培养平均繁殖率仍然很低，内源性细菌污染是一个主要问题。

通过组织培养进行扩繁也会引起某些物种的无性系变异，如彩虹竹芋（*Calathea roseopicta*）。有些无性变异最终会形成新品种（图 13.3）。Chao 等（2005）的工作很好地证明了无性系变异是肖竹芋属（*Calathea*）植物新品种的重要来源。他们采用 AFLP 技术研究了商业种植的观赏性竹芋属植物之间的遗传关系，共包含了 15 个不同种的 34 个商业品系。16 个品系的种质背景来源于野生种，10 个品种来源于彩虹竹芋的无性系变异或由之芽变而来，仅有 3 个品种为杂交种。5 个品系来源不详。在这 3 个杂交种中，‘Picta Royale’是由彩虹竹芋（*C. roseopicta*）和美丽竹芋（*C. veitchiana*）种间杂交而成。‘Pink Aurora’是罗氏竹芋（*C. loesenerii*）和品种‘Asian Beauty’的杂交种，而第三个杂交种‘Corona’的亲本则未知。

图 13.3　（a）彩虹竹芋；（b）彩虹竹芋的无性系变异品种‘Angela’

Criley（2015）报道，2000 年市场中约有 25 种竹芋科的野生种或栽培种。更多的新品种是通过组织培养手段选择而得，有一些则是从收集的野生种中选择而来。

自 21 世纪初，竹芋科的育种和选择工作才真正开始。主要育种目标是在开花型品种基础上开发观叶型竹芋品种。Gregori Hambali 在印度尼西亚的茂物（Bogor）进行了一项育种计划。该育种计划的目的是创制适宜容器生产的、具理想开花习性和颜色的、具理想采后寿命的开花型竹芋新品种。2004～2005 年，市场对一些开花型竹芋品种（如‘Indri’和‘Mia’）进行了商业化引入。一些新发布的竹芋品种来源于 *Calathea loesenerii*

（父本）与 *C. roseopicta* ‘Eclipse’（母本）的杂交（Hambali 2005，2006）。从 2003 年开始，比利时 Denis-Plants bvba 公司与 ILVO 合作，发起了一项有关竹芋属植物的广泛性育种计划。其主要目的也是创造一个集新叶形和足够吸引人们眼球的花型于一身的开花型竹芋品种系列。通过此育种计划，产生了商标名为 Bicajoux®的一系列开花型竹芋新品种（图 13.1）（Van Huylenbroeck et al. 2012）。

13.3.2 育种目标

在竹芋属植物的培育和选择中，各不同特征都很重要。可将目标细分为花序或花朵特征（颜色、大小、花序数、寿命）、叶色、形态和模式、植物习性、生理以及对病虫害的耐受性。迫切需要株型紧凑且生产周期短的植物。此外，体外繁殖率是关系到是否将其引入市场的一个重要考虑因素。某些重要性状会受到消费者偏好的影响，花朵特征就是一个很好的例子。开花型竹芋的吸引力实际在于花序。花序中的单个花只保持开放一天，然后开始萎蔫变成棕色。消费者不喜欢变萎蔫的花序。通过育种，得到了一个形成不完全花且花朵仍隐藏于花序中的突变体（图 13.4）。

图 13.4 肖竹芋属杂交种的花序。（a）花；（b）隐藏于花序中的不完全花

13.3.3 育种方案

竹芋的育种存在一些实际的局限性：①一些物种的自然开花期可能会明显不同，对于某些物种而言，有必要进行开花诱导处理。母株的开花诱导处理关系到开花的一致性问题。例如，在黄苞竹芋中，可以通过在特定温度下结合短日照处理进行花诱导（Van Huylenbroeck and Debergh 1993）。在 18℃下进行为期 9 周的 10h 短日照处理，14～16 周后就能开花。一旦开花诱导成功，日照时间就变得不重要了。②种间杂交甚至属间杂交往往最可能产生人们所想要的性状。③竹芋科的体外培养耗时长，繁殖率低，延缓了新产品的上市速度。因此，通过胚挽救获得的幼苗可首先通过体外繁殖，然后一部分移植到温室中进行评估，其余部分进行体外保存。如此在选育过程早期就可以对候选品种母株进行选育，这样至少可以使新品种提前一年引入市场。图 13.5 所示为竹芋科植物的

育种方案。一般来说，一个新品种从杂交到进入市场的时间是4～5年。

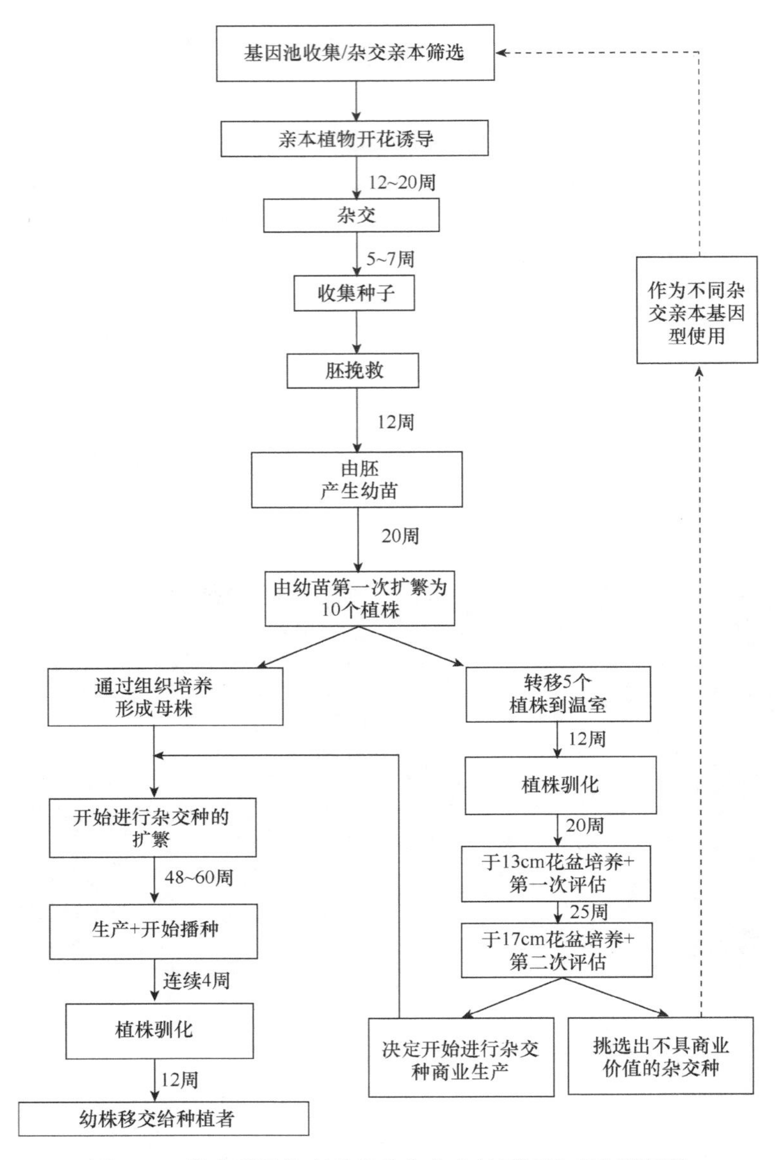

图 13.5　竹芋科植物商业品种生产中的不同选育步骤概览

13.4　种间杂交

13.4.1　细胞学研究

在竹芋科中，不同属和种的染色体数目与基因组大小各不相同。在染色体数目数据库（Rice et al. 2015）中可以找到肖竹芋属（*Calathea*）、竹芋属（*Maranta*）、红里蕉属（*Stromanthe*）、锦竹芋属（*Ctenanthe*）的42个物种的染色体数目。肖竹芋属的染色体基

数为 $x = 12$、13、14，竹芋属的为 $x = 12$、13，红里蕉属的为 $x = 11$（Mukhopadhyay and Sharma 1987），锦竹芋属的为 $x = 9$、10（Rice et al. 2015）。Mukhopadhyay 和 Sharma（1987）发表了最为广泛的细胞学研究。他们描述了 14 个物种和变种的核型。染色体分析表明，在肖竹芋属中，染色体数目的范围为 $2n = 24$、26、28，在竹芋属中为 $2n = 48$、52，而在红里蕉属中为 $2n = 44$。染色体长度为 25.86～65.18μm，但是观察到的差异与总染色体数或 DNA 含量没有直接关系。另外，多倍体不一定是染色体基数的倍数增加，也可能反映了染色体的结构变化（Mukhopadhyay and Sharma 1987）。

在我们的育种研究计划中，对根尖细胞的染色体进行了计数分析，并利用流式细胞技术对 12 种商业品种、栽培品种和杂种进行了基因组大小的测量（表 13.1，图 13.6）。

表 13.1 采用内标检测得到的和文献记载的有关锦竹芋属、红里蕉属和肖竹芋属及其种内杂交种中不同种的基因组大小与染色体数目

基因型	杂交亲本	内标[a]	基因组大小（pg/2C）[b]	染色体数目（$n = 6$～21）	文献记载的染色体数目	参考文献
Ct. burle-marxii		大豆	1.09 ± 0.001	18		
Str. sanguinea		大豆	1.45 ± 0.037	44	44	Mukhopadhyay and Sharma 1987
C. crocata		大豆	0.74 ± 0.013	26		
C. insignis		大豆	1.04 ± 0.016	28	28 / 22	Mukhopadhyay and Sharma 1987；Kumari et al. 2014
C. orbifolia		大豆	0.91 ± 0.048	28		
C. ornata		大豆	1.05 ± 0.024	28	28	Mukhopadhyay and Sharma 1987
C. loesenerii		豌豆	1.92 ± 0.060	26		
C. roseopicta		豌豆	1.79 ± 0.079	26	26	Venkatasubban 1946
C. roseopicta 'Tanja'		豌豆	1.90 ± 0.044	26		
C. rufibarba		大豆	1.03 ± 0.004	26		
C. warscewiczii		大豆	0.77 ± 0.018	26	26 / 24	Sharma and Bhattacharyya 1958；Kumari et al. 2014
C. 'Mia'	*C. loesenerii* × *C. roseopicta* 'Eclipse'	豌豆	1.64 ± 0.000	50%：26[c] 50%：27		
CH 148.2	*C. loesenerii* ×（*C. loesenerii* × *C. roseopicta*）	豌豆	1.82 ± 0.064	64%：26 36%：27		
1461 V	*C. loesenerii* ×（*C. loesenerii* × *C. roseopicta*）	豌豆	1.91 ± 0.038	75%：26 25%：27		
146FH	*C. loesenerii* ×（*C. loesenerii* × *C. roseopicta*）	豌豆	1.90 ± 0.042	50%：26 50%：27		
107 W6	*C. loesenerii* × *C. roseopicta*	豌豆	1.92 ± 0.076	60%：26 40%：27		
107	*C. loesenerii* × *C. roseopicta*	豌豆	2.74 ± 0.114	44%：52 11%：46 17%：45 5.5%：44 17%：42 5.5%：39		

a. 内参的基因组大小：大豆（*Glycine max* 'Polanka'）2C = 2.50pg，豌豆（*Pisum sativum* 'Ctirad'）2C = 9.09pg

b. 计算标准差的基因组大小（$n = 4$）

c. 某特定染色体数目的细胞百分比

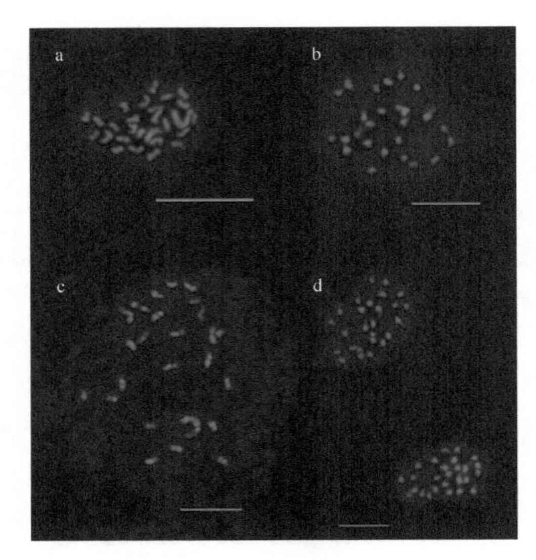

图 13.6 肖竹芋属的染色体计数。(a)黄苞竹芋(*C. crocata*)的 26 条染色体;(b)彩虹竹芋(*C. roseopicta*)的 26 条染色体;(c)波浪竹芋(*C. rufibarba*)的 26 条染色体;(d)罗氏竹芋(*C. loesenerii*)的 2 个细胞核及其 26 条染色体(标尺 = 10μm)

Mukhopadhyay 和 Sharma(1987)与 Hanson 等(1999)在紫背竹芋(*Stromanthe sanguinea*)中观察到体细胞的染色体数为 $2n = 44$。对于栉花竹芋(*Ctenanthe burlemarxii*),则有 18 条染色体。在紫背栉花竹芋(*Ctenanthe oppenheimiana*)中发现了类似的染色体数目(Sato 1948),而黄斑栉花竹芋(*Ctenanthe lubbersiana*)体细胞的染色体数为 $2n = 20$(Mahanthy 1970)。对于肖竹芋属,既有 $2n = 26$,也有 $2n = 28$(表 13.1)。这与早期已发布的肖竹芋属的数据相吻合。对于黄苞竹芋(*C. crocata*)($2n = 26$)、苹果竹芋(*C. orbifolia*)($2n = 28$)、波浪竹芋(*C. rufibarba*)($2n = 26$)和罗氏竹芋(*C. loesenerii*)($2n = 26$),为首次公开染色体数目。对于紫背天鹅绒竹芋(*C. warscewiczii*),我们发现其 $2n = 26$,这与 Sharma 和 Bhattacharyya(1958)针对 *Maranta picta*(*C. warscewiczii* 的代名词)发布的数据一致。然而,Kumari 等(2014)发表的结果是 $2n = 24$。对于箭羽竹芋(*C. insignis*)($2n = 28$),我们的研究结果与 Sharma 和 Mukhopadhyay(1984)发表的相符,而 Kumari 等(2014)报道的则是 22 条染色体。由此表明不同物种间的染色体大小有较高的差异。有关竹芋科植物基因组大小信息的文献报道有限。在已公布的 14 个竹芋科物种中基因组大小为 1C = 0.33～0.53pg,竹芋属为 1C = 0.38～0.55pg,红里蕉属为 1C = 0.68pg(Sharma and Mukhopadhay 1984; Bharathan et al. 1994; Hanson et al. 1999; Bennett and Leitch 2012)。我们有关紫背竹芋(*Stromanthe sanguinea*)、红羽竹芋(*Calathea ornate*)和箭羽竹芋(*C. insignis*)基因组大小的数据与 Hanson 等(1999)、Sharma 和 Mukhopadhay(1984)发表的数据相差无几。它们之间的微小差异可归结于评估方法的不同。

我们的数据是通过流式细胞技术得到的,而以前的作者则使用 Feulgen 显微分光光度法来估计核 DNA。肖竹芋属植物可以根据基因组大小区分为三类。第一类由 *Crocata crocata*、*C. orbifolia* 和 *C. warscewiczii* 组成,基因组大小在 2C = 0.74～0.91pg,与 Sharma 和 Mukhopadhay(1984)发表的其他肖竹芋属植物的数据一致。第二类由 *C. insignis*、*C. ornata*

和 *C. rufibarba* 组成，其 2C = 1.03～1.05pg，而第三类由 *C. loesenerii*、*C. roseopicta* 和 *C. roseopicta* ‘Tanja’ 组成，其 2C = 1.90～1.92pg（表 13.1）。染色体数目与基因组大小之间没有相关性，这表明不同肖竹芋属物种之间的染色体大小差异很大。染色体大小差异可能在一定程度上解释了某些种间杂交组合无法产生杂种的原因，尽管它们的染色体数目相等。

13.4.2 花粉萌发与受精前障碍

许多竹芋科植物花粉无法在人工萌发培养基上萌发。只能在渗透压经调节后的培养基中可观察到花粉萌发。可以使用 Cuevas 等（1994）描述的苯胺蓝染色方法在雌蕊中跟踪花粉管的生长。

在大多数种内组合中，在授粉后 2～4 天可观察到花粉管生长，而种间组合并非总是如此，表明可能存在合子形成前障碍（数据未显示）。此外，花粉和雌蕊的年龄对于授粉成功与否很重要：在某些组合中，幼龄花粉受精失败或幼龄雌蕊授粉失败（图 13.7）。为了加速花粉管的生长，我们也评价了在不同授粉时间进行各不同激素处理的效果（表 13.2）。紫背天鹅绒竹芋（*Calathea warscewiczii*）种内杂交的结果是：在授粉后 3h 用 100ppm GA_3 或 GA_3 和 2,4-D 各 50ppm 处理后能增加授粉成功率与结实率（表 13.2），这些结果可应用于种间杂交以提高成功率。

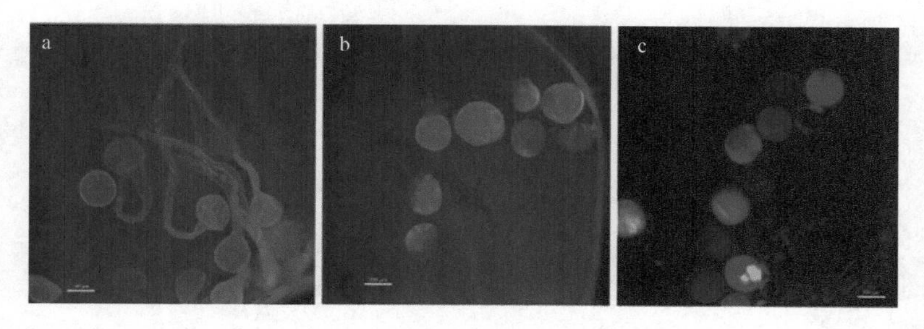

图 13.7 紫背天鹅绒竹芋（*C. warscewiczii*）的种内杂交种的苯胺蓝染色。（a）一朵开放的花朵与另一朵开放的花朵授粉杂交对照实验；（b）花朵开放前 2～3 天的花粉与另一朵开放花朵授粉杂交；（c）一朵开放花朵的花粉与另一朵开放前 2～3 天的花朵授粉杂交（标尺 = 100μm）

表 13.2 不同授粉方法对紫背天鹅绒竹芋（*C. warscewiczii*）种内杂交的结实率和成胚率的影响（数据基于 *n* 次的授粉实验）

激素	浓度（mg/L）	处理时间	数量（*n*）	胚珠膨大率 [a]（%）	结实率 [b]（%）	成胚率 [c]（%）
2,4-D	50	授粉后 3h	343	35.9	64.4	52.0
GA_3	50	授粉后 3h	340	36.2	68.3	47.9
2,4-D	100	授粉后 3h	65	63.1	48.7	32.6
GA_3	100	授粉后 3h	64	89.1	74.5	46.2
2,4-D+ GA_3	50	授粉后 3h	39	84.6	65.8	60.0
2,4-D+ GA_3	50	随后两天：第一天用 2,4-D 处理，第 2 天用 GA_3 处理	40	75.0	55.8	50.0
对照			161	50.9	64.4	56.4

a. 胚珠膨大率为活性胚珠与授粉数量的百分比，每一次授粉可能产生 3 粒种子（每个胚珠含 3 粒种子）

b. 结实率为形成的种子数量与（授粉数量 ×3）的百分比

c. 成胚率为含有胚的种子数量与所形成的种子总数的百分比

13.4.3 胚挽救

种间杂交很常见，通常需要对获得的杂交种子进行胚挽救。已开发了一种高效的实验方案（未发布）。根据母本不同，所得种子会在 8～11 周成熟。对授粉 4～5 周后收集到的胚进行胚挽救可获得最高的成株率。在胚胎非常小的早期阶段，无法进行有效的解剖；而在后期阶段，种皮变得太硬，解剖胚胎也变得困难。完成胚挽救后的大约 12 周，便可获得脱毒苗（图 13.8）。

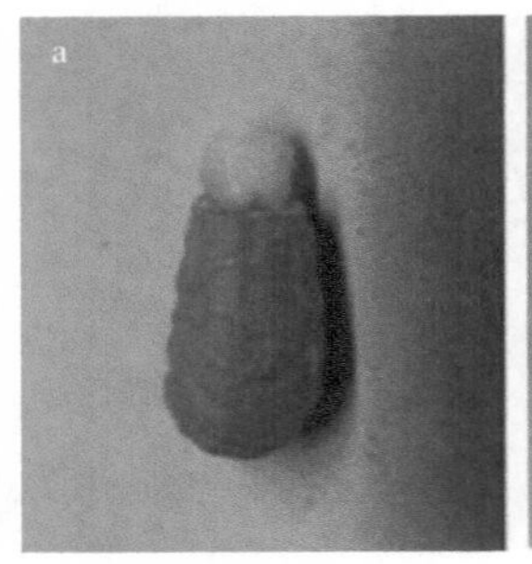
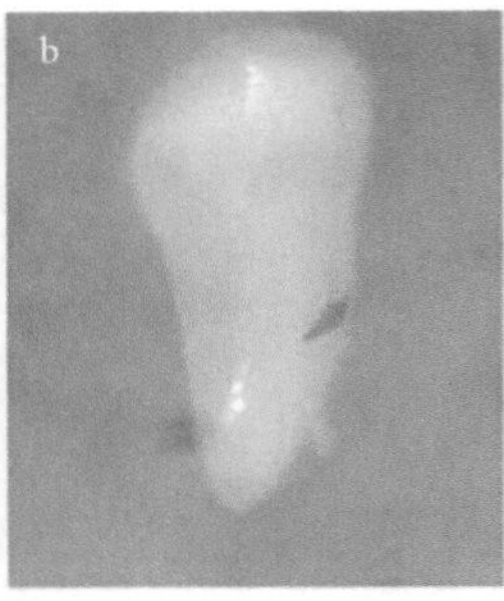
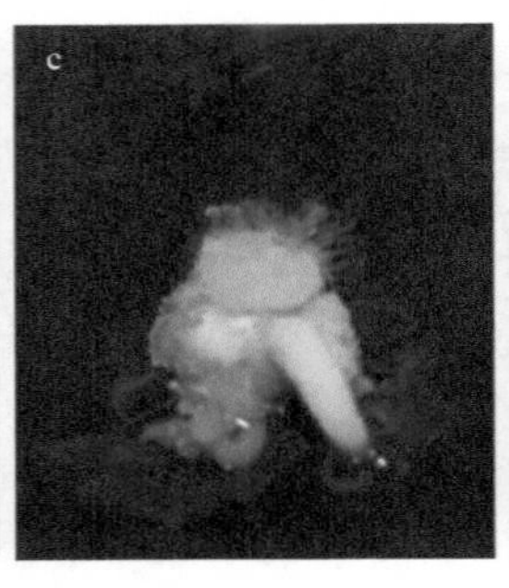
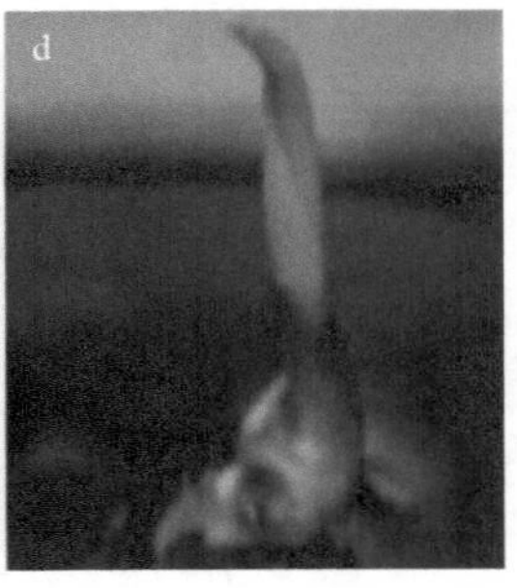

图 13.8 肖竹芋属（*Calathea*）植物的胚挽救。(a) 种子；(b) 分离胚；(c) 开始生长；(d) 成苗

13.4.4 种间育种

种间杂交育种结果表明，平均约有 7%的杂交组合可产生种子。然而，成功率在很大程度上取决于杂交组合本身，在 0～31%。在解剖胚中，通过胚挽救后大约有 4%能形成完整植株（Van Huylenbroeck et al. 2012）。这意味着平均每授粉 1000 朵花只能产生 3 株小苗。在种内杂交中，可获得 60%～75%的结实率（表 13.2）。引入黄色对于育种者来说是巨大的挑战，因此，尝试使用紫背天鹅绒竹芋（*Calathea warscewiczii*）、波浪竹芋（*Calathea rufibarba*）和黄苞竹芋（*Calathea crocata*）。这些杂交表明并非所有组合都能成功，并且观察到了一些杂交偏好（表 13.3）。例如，与波浪竹芋（*Calathea rufibarba*）相比，紫背天鹅绒竹芋（*Calathea warscewiczii*）似乎更适合作母本，而黄苞竹芋（*Calathea crocata*）作母本则不能成功结实。

表 13.3 紫背天鹅绒竹芋（*C. warscewiczii*）、波浪竹芋（*C. rufibarba*）和黄苞竹芋（*C. crocata*）的种间杂交成种率

母本	父本	成种率（%）
C. warscewiczii	*C. rufibarba*	5.14
C. warscewiczii	*C. crocata*	3.62
C. rufibarba	*C. warscewiczii*	0.28
C. rufibarba	*C. crocata*	1.08
C. crocata	*C. warscewiczii*	0
C. crocata	*C. rufibarba*	0

为了克服这些育种障碍，尝试体外授粉。通过直接为胚珠或子房授粉，可规避花柱这一重要障碍。初步结果表明，成功的关键因素是子房和花药的发育阶段、培养基、培养基更换频率以及花粉在子房上的位置。授粉的成功率可以在 2 或 3 周后通过观察体内子房的膨胀一致性的方式来确定。

种间杂交产生 F_1 子代，其杂种特征通过 AFLP 分析得到证实（Van Huylenbroeck et al. 2012）。杂种还表现出两种亲本的中间形态特征。除此之外，对一些种间杂种的染色体组成进行了详细研究（表 13.1）。所有分析过的杂种均是以罗氏竹芋（*C. loesenerii*）为母本，以彩虹竹芋（*C. roseopicta*）或罗氏竹芋（*C. loesenerii*）和彩虹竹芋（*C. roseopicta*）的杂交种为父本。除杂种 107 外，所有其他杂交种的基因组大小均在 2C＝1.82～1.92pg。基于亲本的基因组大小，杂交种的基因组大小与所期望的大小一样。然而，所有杂交种的染色体数目都是嵌合体，大约只含有 $2n=26$ 条染色体或 $2n=27$ 条染色体细胞的一半。品种 *C.*‘Mia’则拥有与 26 或 27 条染色体细胞相似的染色体结构（图 13.9）。杂种 107 可能是一种多倍体，其基因组大小为 2C ＝ 2.74pg，染色体数目为 52 的一半，而其他细胞的染色体数目则是 39～46 条。该杂种的自发多倍体化可能发生在胚挽救的早期。到目前为止，我们还没办法解释为什么在种间杂种的部分细胞中观察到非整倍性。

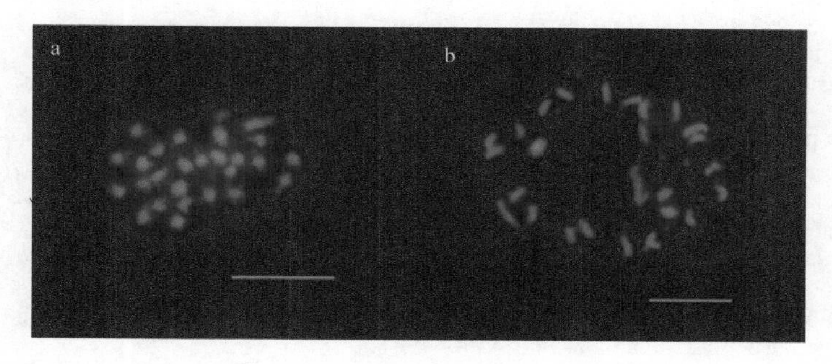

图 13.9 肖竹芋属‘Mia’品种的染色体计数（细胞核数量相等，但染色体数目不等）。(a) 27 条染色体；(b) 26 条染色体（标尺 ＝10μm）

13.5 多 倍 体 化

竹芋科内物种和属之间的染色体数目不同以及具有不同染色体数目的嵌合杂种细胞核的出现表明，染色体加倍可能有助于解决某些繁育障碍。迄今为止，还没有有效地用于竹芋科的多倍体化方案。为了获得杂种 146.1V 的多倍体植株，我们分别以三种不同的浓度测试了黄草消和秋水仙碱的处理效果。只有秋水仙碱[100mg/(mg·L)和 200mg/(mg·L)]诱导出一些混合倍体或四倍体芽。然而，这些多倍体小苗要么无法生存，要么随着时间的流逝变得不稳定，然后降至二倍体水平。可能处于植株中部的分生组织难以接触到有丝分裂抑制剂，或者倍性过高导致了诱导的分生组织死亡。

13.6 结　　论

长期以来，竹芋科新品种的主要来源为野生种或组织培养得到的体细胞无性系变种。在过去的 20 年里，育种才刚刚开始，而且仅集中在肖竹芋属的开花习性上，该目标只能通过种间杂交来实现。尽管产生杂交种的效率很低，并且存在各种杂交障碍，但吸引人的叶子与持久花序相结合的全新植物品种系列已然投放市场。

本章译者：

张　露

云南省农业科学院花卉研究所，国家观赏园艺工程技术研究中心，云南省花卉育种重点实验室，昆明 650200

参考文献

Andersson L (1998) Marantaceae. In: Kubitzki K (ed) The families and genera of vascular plants: 4. Springer-Verlag, Berlin/Heidelberg/New York, pp 278–293

Andersson L, Chase MW (2001) Phylogeny and classification of Marantaceae. Bot J Linn Soc 135:275–287

Benedict JC, Smith SY, Specht CD, Collinson ME, Leong-Škorničková J, Parkinson DY, Marone F (2016) Species diversity driven by morphological and ecological disparity: a case study of comparative seed morphology and anatomy across a large monocot order. AoB Plants 8:plw063

Bennett MD, Leitch IJ (2012) Plant DNA C-values database (release 6.0, Dec. 2012) http://www.kew.org/cvalues/

Bharathan G, Lambert G, Galbraith DW (1994) Nuclear DNA content of monocotyledons and related taxa. Am J Bot 81:381–386

Borchsenius F, Suárez Suárez LS, Prince LM (2012) Molecular phylogeny and redefined generic limits of *Calathea* (Marantaceae). Syst Bot 37:620–635

Chao C-CT, Devanand PS, Chen J (2005) AFLP analysis of genetic relationships among *Calathea* species and cultivars. Plant Sci 168:1459–1469

Criley RA (2015) Alpinia to Zingiber – Zingiberales in commercial floriculture. Acta Hortic 1104:435–454

Cuevas J, Rallo L, Rapoport HF, Lavee S, Klein I (1994) Staining procedure for the observation of olive pollen tube behavior. Acta Hortic 356:264–267

Dunston S, Sutter E (1984) In vitro propagation of prayer plants. Hortscience 19:511–512

Grootjen CJ (1983) Development of ovule and seed in Marantaceae. Acta Bot Neerl 32:69–86

Hambali GG (2005) *Calathea* plant named 'Indri' US Patent 16,028, Oct 11, 2005

Hambali GG (2006) *Calathea* plant named 'Mia' US Patent 16, 425, Apr 11, 2006

Hanson L, Leitch IJ, Bennett MD (1999) Unpublished values from the Jodrell Laboratory, Royal Botanic Gardens, Kew

Kennedy H (1978) Systematics and pollination of the 'closed-flowered' species of *Calathea* (Marantaceae). Univ Calif Publ Bot 71:1–90

Kennedy H (1983) A pattern mimic of *Calathea veitchiana* (Marantaceae) from Peru. Can J Bot 61:1429–1434

Kennedy H (1995) *Calathea ornata* and relatives, an ornate confusion. Acta Hortic 413:169–174

Kennedy H (2000) Diversification in pollination mechanisms in the Marantaceae. In: Wilson KI, Morrison DA (eds) Monocots: systematics and evolution. CSIRO Publishing, Collingwood, pp 335–343

Kennedy H (2011) Three new Costa Rican species of *Calathea* (Marantaceae) from montane wet forests. Novon 21:49–57

Kress WJ (1995) Phylogeny of the Zingiberanae: morphology and molecules. In: Ruddall PJ, Cribb PJ, Cutler DF, Humphries CJ (eds) Monocotyledons: systematics and evolution, vol 12.

Royal Botanical Gardens, Kew, pp 443–460
Kumari A, Lahiri K, Mukhopadhyay MJ, Mukhopadhyay S (2014) Genome analysis of species of Calathea utilizing chromosomal and nuclear DNA parameters. Nucleus 57:203–208
Kunze H (1984) Vergleichende Studien an Cannaceen- und Marantaceenblüten. Flora 175:301–318
Ley AC, Claßen-Bockhoff R (2011) Evolution in African Marantaceae – evidence from phylogenetic, ecological and morphological studies. Syst Bot 36:277–290
Ley AC, Claßen-Bockhoff R (2012) Floral synorganization and its influence on mechanical isolation and autogamy in Marantaceae. Bot J Linn Soc 168:300–322
Ley AC, Claßen-Bockhoff R (2013) Breeding system and fruit set in African Marantaceae. Flora 208:532–537
Mahanty HK (1970) A cytological study of the Zingiberales with special reference to their taxonomy. Cytologia 35:13–49
Mukhopadhyay S, Sharma AK (1987) Karyomorphological analysis of different species and varieties of *Calathea*, *Maranta* and *Stromanthe* of Marantaceae. Cytologia 52:821–831
Nolasco EC, Coelho AG, Machado CG (2013) Primeiro Registro de ornitofilia confirmada em *Calathea* (Marantacea). First verified record of ornithophily in *Calathea* (Marantaceae). Biosci J 29:1328–1338
Pischtschan E, Claßen-Bockhoff R (2008) Setting up tension in the style of Marantaceae. Plant Biol 10(4):441–450
Pischtschan E, Claßen-Bockhoff R (2010) Anatomic insights into the thigmonastic style tissue in Marantaceae. Plant Syst Evol 286:91–102
Podwyszyńska M (1997) Micropropagation of *Calathea ornata* Koern. Biol Plant 39:179–186
Prince LM, Kress WJ (2006) Phylogenetic relationship and classification in Marantaceae: insights from plastid DNA sequence data. Taxon 55:281–296
Rice A, Glick L, Abadi S, Einhorn M, Kopelman NM, Salman-Minkov A, Mayzel J, Chay O, Mayrose I (2015) The chromosome counts database (CCDB) – a community resource of plant chromosome numbers. New Phytol 206(1):19–26
Sato D (1948) Karyotype and systematics of Zingiberales. Jpn J Genet 23:44–45
Sharma AK, Bhattacharyya NK (1958) Inconstancy in chromosome complements in species of Maranta and Calathea. Proc Natl Inst Sci India 24B:101–117
Sharma AK, Mukhopadhyay S (1984) Feulgen microspectrophotometric estimation of nuclear DNA of species and varieties of three different genera of Marantaceae. Proc Indiana Acad Sci (Plant Sci) 93:337–347
Van Huylenbroeck J, Debergh P (1993) Year-round production of flowering *Calathea crocata*: influence of light and carbon dioxide. Hortscience 28:897–898
Van Huylenbroeck J, Calsyn E, Van den Broeck A, Denis R (2012) Breeding new flowering ornamentals: the Bicajoux® story. Acta Hortic 953:135–138
Venkatasubban KR (1946) A preliminary survey of chromosome numbers in scitamineÆ of Bentham and hooker. Proc Indiana Acad Sci Section B 23:281

第 14 章　菊　　花

Jaap Spaargaren and Geert van Geest

摘要　菊花是一个起源于杂交复合体的栽培植物，是东亚不同野生种之间的多重杂交产物。单核苷酸多态性分析结果表明菊花是六倍体植株，但是具有多变性。最早的“原菊”大约在 1600 年前起源于位于中国的主要基因中心。在接下来的几个世纪中，菊花传播至其他的亚洲国家，直到 1789 年才传播到西方国家。菊花在花卉交易市场的产值位列第二，仅次于月季。

菊花的野生种是珍贵的育种资源，本章将对这些野生种进行讨论。在欧洲，林奈是最早拥有野菊（*Chrysanthemum indicum*）标本的植物学家，他的野菊标本足足占据了两个标本室。1999 年，国际植物学大会批准了保留栽培菊花群体的植物学名为 *Chrysanthemum* L.的提案，所选的植物标本是林奈植物标本室的野菊（*Chrysanthemum indicum* L.）。据国际藻类、真菌和植物命名法规（ICN）的规定，1792 年 De Ramatuelle 首次提出将栽培菊花的拉丁文名定义为 *Chrysanthemum* 和种加词 *morifolium* 的组合。大量成功的试验结果表明，一些菊花的野生种是抗虫、抗真菌、抗特殊化学物质以及耐寒、耐高温、耐盐和耐干旱的资源。但是，有些物种也是白锈病的寄主植物。

本章也对菊花的选择育种技术进行了讨论，包括特性性状相关 DNA 标记的开发，以及 SNP 和 CRISPR/Cas9 的使用。总之，利用这些技术和结果并结合连锁图谱可共同鉴定调控特定性状的基因。

关键词　菊花；育种；DNA 标记；多倍体；选择型（lectotypification）；春黄菊族；体细胞遗传；属间；种间；同物种间；采后；独脚金内酯；突变；抗病性

14.1　生产和贸易

菊花是世界上主要的观赏作物之一，可以作为切花和盆栽植物。在全球范围内，亚洲国家的菊花供应量和需求量都是最高的，主要为了满足本地需求。这很难用产量或者价值这样的数据来体现，但是对于特定的产地，本地产品比出口产品更重要。切花菊生产面积较大的国家是中国和日本，其受保护的切花菊种植面积分别为 8475hm^2（2013 年）和 5230hm^2（2009 年），韩国为 700～800hm^2，以及中国台北为 728hm^2（2012 年）（AIPH

J. Spaargaren
Ingenieursbureau, Aalsmeer, The Netherlands
G. van Geest (✉)
Deliflor Chrysanten B.V, Maasdijk, The Netherlands
e-mail: geert.vangeest@wur.nl

2014 年）。这些国家对日本的菊花出口量相对适中。在印度，菊花这种观赏作物也广受欢迎，约有 19 000hm^2 未受保护的切花菊生产基地，生产的切花菊专门在国内市场销售；在泰国也是这种情况（2008 年约有 2199hm^2）。马来西亚和越南已经开始大幅度提高本国切花菊的生产能力，用于满足日本的需求。因为在日本有土地减少、人口老龄化等问题，切花菊的本地产量大量减少，造成进口需求量急剧提高。

与切花菊在亚洲的大量种植面积形成鲜明对比的是，荷兰瓦格宁根（2016 年，383hm^2）和哥伦比亚（2012 年，716hm^2）的切花菊种植面积相对较少。在全球范围内，切花菊主要销往包括俄罗斯在内的欧洲和美国市场。值得注意的是，在这些国家每公顷切花菊的生产力远高于亚洲国家每公顷切花菊的生产力。在西方国家中，仅有墨西哥在 2012 年开放了 2365hm^2 的土地用于生产切花菊，并且这些切花菊基本上是专门用于本地的供求。

很长时间以来，在切花的国际贸易中，切花菊的产值仅次于月季，高于香石竹。2014 年的联合国商品贸易统计结果显示，月季、切花菊和香石竹在世界出口中所占的份额总共为 918.8 万美元，占比分别为 37.6%、8.6%和 5.4%。2016 年，荷兰、立陶宛和哥伦比亚的菊花出口总份额为±75%。

物流和运输成本严重影响了切花菊的全球贸易。目前，切花菊在全球范围内的运输主要是通过航空运输，但是一些国家也利用可控集装箱（冷藏箱）进行海上运输，如哥伦比亚的海上运输比例正在显著增加。在切花菊的国际贸易中，哥伦比亚通过利用美国较低的航空成本和自由贸易协定（FTA）获得利润（Spaargaren 2015）。

14.2　起源、基因中心

14.2.1　基因中心

野生菊及其杂交种在中国、日本和韩国均有分布。然而，至今尚未发现栽培菊花（*C.* × *morifolium* Ramat.）真正的/最原始的野生种。大约在 16 世纪前，栽培菊花从自然杂交变种中经过人工选择而来，这些自然杂交种主要来源于野菊（*C. indicum* L.）（$2n = 18$，36）、甘菊[*C. lavandulifolium* (Fisch. Ex Trautv.) Makino]（$2n = 18$）、菊花脑（*C. nankingense* Hand.-Mazz.）（$2n = 18$）、小红菊（*C. chanetii* H. Lév.）（$2n = 54$）和紫花野菊（*C. zawadskii* Herbich）（$2n = 54$）。这种人工选择方式造成了栽培复合体在栽培菊花中广泛存在的现象（整个栽培群体呈现杂交复合体的形态）（Dai et al. 1998，2005；Chen 2004；Yang et al. 2006；Kim et al. 2009；Sun et al. 2010；Chen et al. 2012；Ma et al. 2016）。

最近的研究表明，即使是由具有不同形态花序和叶子的 6 个中国单头菊品种组成的一个小群体，似乎也有不同的祖先（Ma et al. 2016）。野菊（*C. indicum*）与其中的 3 个栽培品种聚在一起，紫花野菊（*C. zawadskii*）和菊花脑（*C. nankingense*）与其中的 2 个栽培品种聚在一起。毛华菊（*C. vestitum*）是一些品种公认的祖先品种。野菊（*C. indicum*）（$4x$）本身似乎是同源多倍体，其基因组深受二倍体甘菊（*C. lavandulifolium*）基因组的影响。分子标记和 SNP 的试验结果表明，日本的野生种对栽培菊花的选育具有重要的影响（Wang 2005；Chong et al.2016）。在后面的研究中，涉及了北野菊（*C. boreale*）、野路菊（日本）（*C. japonense*）、若狭滨菊（*C. makinoi* var. *wakasaense*）、萨摩野菊（*C. ornatum*）、小滨菊（*C. yezoense*）和那贺川野菊（*C. yoshinaganthum*）。此外，还有野生的中国种异色菊（*C. dichrum*）、甘菊（*C. lavandulifolium*）、菊花脑（*C. nankingense*）和毛华菊（*C. vestitum*）。

栽培菊花及其野生种的主要基因中心位于中国中部的安徽、河南、湖北和江西，均分布于30～40°N。就气候而言，虽然中国武汉山区秋季和冬季的阳光更为充足，但总体来说武汉山区与法国南部（43°N）相当。与19世纪上半叶的伦敦（51°N）和巴黎（49°N）地区不利于秋季开花型植物的气候相反，法国南部温和的气候对种子成熟和秋季开花类型的植物来说是相当适宜的。1832年，伦敦Watlington Hall的John Freestone实验室首次促成一些栽培菊花种子成熟（Genders 1971），这些种子可能来自早花类型。

与月季相比，现代菊花可以归类为对光有高要求的农作物（Spaargaren 2001）。它是从半阴的自然种进化而来的（Dai et al. 1995）。因此，栽培菊花的切花和插穗生产被转移到了全年光照充足的赤道附近的高海拔国家。

14.2.2 多倍体

种间杂交和多倍体化被认为是开花植物进化（物种形成）与驯化的最重要来源，如菊花（Van Tuyl and Lim 2003）。当整个基因组融合或者同源多倍体机制发生时就会产生多倍体化，如2*x*、4*x*、6*x*、8*x*、10*x*的紫花野菊和2*x*、4*x*、6*x*的野菊。在这些二倍体的野生种中存在一条同源多倍体化途径（Kim et al. 2009）。在第一次减数分裂初期，两个相邻的花粉母细胞（PMC）可能发生融合，融合频率为1.1%～1.3%。融合细胞（合子）（4*x*）同时具有两个PMC的染色体，随后作为单个大型PMC进行减数分裂并产生4个能够萌发的2*n*花粉粒。三倍体在回交时可能产生4*x*植株，自交时产生6*x*植株。目前，研究者已在*C. makinoi* Matsum. & Nakai和若狭滨菊的自然群体中发现三倍体（Tanaka 1959a，b）。减数分裂二倍体化导致2*n*或者未减数分裂花粉的形成广泛存在于包括栽培菊花（*C.* × *morifolium*）在内的各个物种中（Veilleux 1985；Bino et al.1990）。

14.2.3 在中国的实际应用

在亚洲，栽培菊花及其野生种以药用、驱虫剂、食用、茶用以及观赏性为目的种植、驯化和繁殖都十分广泛。观赏型的栽培菊花最早出现在中国的一些中心省份，随后逐渐传播到亚洲的其他地区，包括日本和韩国（Chen et al. 2012；Spaargaren 2015）。

关于栽培菊花（*C.* × *morifolium*）品种最早的记载来自诗人陶渊明（公元365～427年），他在自己的花园里种植了白色半重瓣的栽培菊花类型。陈俊愉教授将其视为观赏型栽培菊花在世界范围内出现的标志。在海拔100～2900m的中国主要基因中心区域仍然可以找到栽培菊花的原始种，它们通常生长于混合植物群落中。这些原始种与栽培菊花（*C.* × *morifolium* Ramat.）形态十分相近，但仍具有原始的性状特征。这些原始种包括具有白色到淡黄色花朵的天然种间杂种，形态特征介于白色毛华菊（*C. vestitum*）（包括阔叶毛华菊）和黄色野菊（*C. indicum* L.）之间。这些植物群落的分布区域相互重叠，这给种间杂种与其中一个亲本的重复回交提供了机会，从而促进了种间杂交和基因交流（基因渗入）。在位于河南的山上还发现了野菊（*C. indicum*）和甘菊（*C. lavandulifolium*）的群落。二倍体甘菊（*C. lavandulifolium*）和二倍体菱叶菊（*C. rhombifolium*）与六倍体毛华菊（*C. vestitum*）相邻生长，产生四倍体并最终形成六倍体类型（Chen et al. 1996）。类似的种间杂交在日本和韩国也有发现。有趣的是，研究表明，野菊（*C. indicum*）（4*x*）（AABB 基因组）是毛华菊（*C. vestitum*）（6*x*）（AABBCC）的主要供体（Dai et al. 2005）。

14.2.4　中药、杀虫剂

花和叶的化合物成分主要包含单萜与倍半萜（Teixeira da Silva 2004；Lawal et al. 2014）。这些化合物的定性和定量结果显示，不同的菊花栽培品种和野生种（含萜烯，如樟脑、桉树脑）对昆虫的抗药性、气味敏感度和药物活性是多种多样的。野菊（*C. indicum*）、甘菊（*C. lavandulifolium*）、毛华菊（*C. vestitum*）、野地菊（*C. japonense*）和委陵菊（*C. potentilloides*）是这些化合物的重要资源（Sun et al. 2010）。尤其是在亚洲国家，存在经过长期人工选育后用作茶、香料、调料和蔬菜的特殊栽培类型（Greger 1977；Uchio et al. 1981；Ito et al. 1990；Park and Kwon 1997；Lawal et al. 2014）。

菊属物种在民间医学中具有重要作用，最重要的是野菊（*C. indicum*），其次是甘菊（*C. lavandulifolium*）、栽培菊花（*C.* × *morifolium*）。这些菊属植物广泛应用于治疗普通感冒、头痛、头晕、心绞痛和高血压的药物中（Huang 1999）。

在古代，野菊（*C. indicum*）因其杀虫作用而闻名（Needham 1986）。它曾被用来熏蒸房屋、防昆虫以及控制蠹虫。从野菊提取的香精油具有驱蚊和熏蒸活性，作用类似于青蒿（Wang et al. 2006），但浓度低很多（Teixeira da Silva 2004；Shunying et al. 2005）。

菊属植物具有杀虫和许多其他作用主要是由于樟脑、桉叶素（具有强烈的樟脑气味）和茨醇的存在。它们在一些栽培菊花（*C.* × *morifolium*）品种中几乎是缺乏的（‘英雄’、‘紫安妮’和‘冲浪仔’）（Storer et al. 1993），但在尼日利亚等其他国家的一些栽培品种中含量相对较高（Lawal et al. 2014）。

茨醇具有药用功效，是许多日本香料和香水的重要配方与成分。茨醇在龙脑菊 *C. japonicum* 或 Ryunou-giku（日本的植物性药材）中含量较高。

14.2.5　传播到西方世界

尽管欧洲人对中国和日本植物的描述最早始于 16 世纪，但没有任何栽培菊花及其野生种的活体标本在欧洲存活（Fournier 1910；Spaargaren 2015），直到 1789 年布兰卡德船长首先在法国马赛市引种了栽培品种‘老紫色’（图 14.1）。栽培菊花在公元 400～

1800 年从中国传播到邻近国家，于 1789 年起向欧洲传播。日本的栽培品种从 1861 年开始传播到西方国家。1689 年，雅各布·布瑞恩描述了在荷兰花园中观察到的日本品种，该品种已消失（Spaargaren 2015）。在中国记录有 400 个原生栽培种，400～1800 个原生种通过中国移民和佛教僧侣、文化、帝国交往传播到邻国（日本）。1789 年皮埃尔・布兰卡（1787 年在澳门）设法将第一批中国栽培菊花品种的插穗带到法国马赛市。此后，许多其他国家也进行了效仿，尤其是英格兰。1798 年，第一个中国栽培菊花品种由 John Stevens 从英国传播到美国新泽西州。1882 年，来自英国 Cheltenham 的 Thomas Pockett 开始在澳大利亚墨尔本附近的维多利亚州种植英国的栽培菊花品种（Fournier 1910；Spaargaren 2015）。

图 14.1　印度菊，‘老紫色’（来源：柯蒂斯植物杂志，1796 年，327 版）

在 1860 年以前，中国丰富的种质资源仅有少部分传播到了西方。尽管如此，法国和英国的育种者运用熟练的栽培管理方法培育出了各种小花、多花、单头、绒球甚至夏花类型。1860/1861 年日本新种质的到来，成为西方菊花育种的转折点。迄今为止，西方菊花育种已发展并培育出了包括多头菊在内的多样化栽培品种（Dowrick 1953；Spaargaren 2015），育种的重点一直都是疏蕾。

14.3　菊花的命名法

一般而言，亚洲药典和诗中描述的是在野生环境中生长的野菊 *C. indicum* L.和种间杂种菊花 *C.* × *morifolium* Ramat.。但是，由于知识的缺乏，似乎在分类学历史上，经常将杂交种错误地命名为野菊 *C. indicum* L.。此外，在过去的 4 个世纪中，西方植物学家一直在为菊花的分类争论不休，于是就有了许多菊花的通用和专用名称。自 1999 年以来，亚洲菊科植物的属名一直是 *Dendranthema*。1993 年，国际植物学大会接受了将野菊命名为

C. indicum L.的提案（Humphries 1993；Jarvis et al. 1993；Trehane 1995；Nicolson 1999）。

根据国际藻类、真菌和植物命名法规（现在称为 ICN），栽培菊花被分类为菊花 *C.* × *morifolium* Ramat.。自 1999 年之后，基于实际情况，《国际栽培植物命名法》（ICNCP）也规定了菊花品种名称。另外，基于形态和韧性，可以将栽培品种分为几类，如将切花和盆栽菊花归类为 *indicum* 组。最后，可以添加一个附加的商标名称，无论该商标是否具有专利权，但该商标名称并不属于科学名称，如野菊 *C. indicum* 组 Eurobelle ‘EURO’（小型大写字母），其中 EURO 为商标名称。为了获得育种者的权利，将栽培菊花品种正式命名为 *C.* × *morifolium* ‘Eurobelle’，EURO 是商标名称。

一些品种或品系可以用另一个商标名称表示，这些商标名称有的受保护，有的不受保护。例如，在贸易中，根据国际藻类、真菌和植物命名法规，Santini MADIBA® LINDI YELLOW 被命名为 *C.* × *morifolium*‘Deklindi White’MADIBA LINDI YELLOW，而根据国际栽培植物命名法规，它则被命名为 *C. indicum* group‘Deklindi’MADIBA LINDI YELLOW。

14.4 属间繁殖

14.4.1 春黄菊族内的属间杂交

有许多重要的遗传资源（图 14.2 中的一些例子）可用于栽培菊花的遗传改良（见综述 Zhao et al. 2009；Sun et al. 2010）。由于这些种质资源相似度很高，因此大多数野生种都可以与菊花进行杂交（Tatarenko et al. 2011）。

因此，在杂交中应谨慎选择亲本的基因型（Fukai et al. 2000）。据报道，青蒿亚科之间的杂交比青蒿亚科与春黄菊亚族之间的杂交更成功。属间杂交种可作为多属杂交育种的桥梁，以实现将菊蒿（*Tanacetum vulgare*）中调控药用或杀虫效果等特定性状的基因导入。为了丰富菊花（*C.* × *morifolium*）基因库，可以使用杂交方法，成功地使用（*Chrysanthemum makinoi*）（蒿亚族）和菊蒿（*Tanacetum vulgare*）（春黄菊亚族）的杂交（Tanaka and Shimotomai 1968）或菊花脑（*Chrysanthemum nankingense*）和菊蒿（*Tanacetum vulgare*）的杂交（Wang et al. 2013）或者是蓬莱菊（*Chrysanthemum horaimontanum*）和菊蒿（*Tanacetum vulgare*）的杂交（Kondo et al. 1999；Abd El-Twab and Kondo 2001b）。Cumming 等（1939）成功地将滁虫菊（*Tanacetum cinerariifolium*）和菊花（*Chrysanthemum* × *morifolium*）进行了杂交。但是，该结果无法重复，这表明这种杂交可能不具有重要意义。

为了提高对昆虫的抗性，将北艾（*Artemisia vulgaris*）与菊花（*C.* × *morifolium*）杂交，尽管它们均属于同一亚族（蒿亚族）（Oberprieler et al. 2006），但只能通过胚挽救培养才能成功（Deng et al. 2010b）。体细胞杂种是由菊花（*C.* × *morifolium*）和大籽蒿（*Artemisia sieversiana*）杂交获得的，其对白锈病的抗性有所提高（Furuta et al. 2004）。

图 14.2　日本菊花野生种。资料来源：日本 Katsuhiko Sumitomo，NIFS。(a) 大岛野路菊 *Chrysanthemum crassum* (Kitam.) Kitam.（$2n = 10x$）；(b) 野菊 *Chrysanthemum indicum* L. var. *indicum*（$2n = 4x$）；(c) 野地菊 *Chrysanthemum japonense* Nakai（$2n = 6x$）；(d) 若狭滨菊 *Chrysanthemum makinoi* var. *wakasaense* Matsum. and Nakai（$2n = 2x$）；(e) 阴岐油菊 *Chrysanthemum okiense* Kitam.（$2n = 4x$）；(f) 萨摩野菊变种‘奥兰多’*Chrysanthemum ornatum* Hemsl. var. *ornatum*（$2n = 8x$）；(g) 太平洋菊花 *Chrysanthemum pacificum* Nakai（$2n = 10x$）；(h) 甘野菊 *Chrysanthemum seticuspe* (Maxim.) Hand.-Mazz；f. *boreale* H. Ohashi and Yonek；(i) 纪伊潮菊 *Chrysanthemum shiwogiku* Kitam.（$2n = 8x$）；(j) 那贺川野菊 *Chrysanthemum yoshinaganthum* Makino ex. Kitam.（$2n = 4x$）；(k) 小滨菊 *Chrysanthemum yezoense* Maek.（$2n = 10x$）；(l) 紫花野菊 *Chrysanthemum zawadskii* Herbich（$2n = 6x$）

目前已有很多青蒿属之间成功杂交的案例被报道：大岛野路菊（*Chrysanthemum crassum*）($2n = 10x = 90$)和多花亚菊(*Ajania myriantha*)($2n = 2x = 18$)(Tang et al. 2010)，菊花（*C.* × *morifolium*）与近缘属日本雏菊（*Nipponanthemum nipponicum*）和 *Leucanthemella linearis*（Tanaka and Shimotomai 1961；Abd El-Twab and Kondo 2001a，2006，2007a)，菊属和亚菊属（Shibata et al. 1988a，1988b；De Jong and Rademaker 1989)，菊属和太行菊属（Yang et al. 2010）等（Kondo et al. 2003；Anderson 2007）以及芙蓉菊属

和菊属（Tang et al. 2009）。

木茼蒿可以与茼蒿属杂交（Ohtsuka and Inaba 2008），两个种类都属于 Glebionidinae 亚族。

在人工杂交时应考虑野生种的利用有利有弊。一些栽培菊花品种的某些原始种为白锈病（*Puccinia*）的寄主植物，如野菊（*C. indicum*）、北野菊（*C. boreale*）、野地菊（*C. japonense*）、龙脑菊（*C. japonicum*）和紫花野菊（*C. zawadskii*），还有其他的如若狭滨菊（*C. makinoi* var. *wakasaense*）、纪伊潮菊（*C. shiwogiku*）和 *C. makinoi*（Punithalingam 1968）。具有抗性和耐寒性的野生种包括耐寒的野生种：银背菊（*C. argyrophyllum*）、小红菊（*C. chanetii*）、野菊（*C. indicum*）、蒙菊（*C. mongolicum*）、蒿（*Artemisa* spp.）、亚菊（*Ajania* spp.）以及太行菊属（*Opisthopappus*）；耐干旱、耐高温、耐盐渍土的野生种：芙蓉菊属（*Crossostephium*）；耐寒的野生种：异色菊（*C. dichrum*）；具有抗病性和抗虫性的野生种：野菊（*C. indicum*）、甘菊（*C. lavandulifolium*）、委陵菊（*C. potentilloide*）和毛华菊（*C. vestitum*）；耐高温和干旱的野生种：菊花脑（*C. nankingense*）；耐寒的野生种：小山菊（*C. oreastrum*）；以及耐涝的野生种：紫花野菊（*C. zawadskii*）（Sun et al. 2010）。

14.4.2 提高对蚜虫和干旱的抗性

利用胚挽救获得的蒿属和菊属杂交种对菊花蚜虫（*Macrosiphoniella sanborni*）与黑斑病的抗性增强，生根能力提高（Deng et al. 2010a）。蒿属植物中含有编码生物活性成分的基因，青蒿（*Artemisia annua*）中也含有这种成分。蚜虫感染后会造成蒿属植物中蚜虫报警信号素(*E*)-β-farnese 和蒿酮散发量急剧增加，这两种挥发物可能存在参与抵抗蚜虫的作用（Sun et al. 2015）。

通过三属杂交可以获得较好的抗蚜虫性和耐盐性，如两个属间 F_1 杂种杂交：*Chrysanthemum grandiflorum* 和北艾（*Artemisia vulgaris*）杂交以及大岛野路菊（*Chrysanthemum crissum*）和芙蓉菊（*Crossostephium chinense*）杂交。耐盐性的提高有赖于后一杂种的遗传物质（Deng et al. 2012）。最近研究发现，*DgWRKY4* 基因过表达可以增强菊属植物的耐盐性（Wang et al. 2017）。

研究表明，与菊花（*C.* × *morifolium*）杂交种相比，野菊（*C. indicum*）杂交种的抗性得到了显著的提高（Sun et al. 2010）。*CmWRKY10* 通过 ABA 介导的途径增强了菊花对干旱的耐受性（Fan et al. 2016；Jaffar et al. 2016）。

其他成功的属间杂交报道包括紊蒿[*Elachanthemum intricatum* (Franch.) Ling et Ling]和太行菊[*Opisthopappus taihangensis* (Y. Ling)]。依据 nrDNA ITS 序列和形态学特征发现，在这些杂交种中只有一种紊蒿的后代是真正的杂种（Zheng et al. 2013）。太行菊主要生长在太行山南麓的岩石上，是一种多年生草本植物，匍匐枝长达 10～30cm（Ling and Shih 1983；Li et al. 2009；Zhao et al .2009）。紊蒿株型独特，有许多从基部生长出的压缩茎，二倍体野生种的抗寒和抗旱性都很强，分布在中国宁夏、内蒙古、甘肃、青海和新疆的沙漠或荒漠草原（Ling and Shih 1983；Kondo et al. 1998；Zhao et al. 2009）。

14.4.3 匍匐或悬垂的生长习性

异色菊（*C. dichrum*）（父本）和太行菊（*Opisthopappus taihangensis*）（母本）可以成功杂交。其杂交种具有匍匐下垂的生长习性，适宜作为吊篮和地被植物。该栽培种（图 14.3）被称为‘太行银河’，具有白色的花瓣，花型简单，并表现出极高的种间杂种优势（Zhao et al. 2010）。

图 14.3 ‘太行银河’，异色菊×太行菊的属间杂交种（来源：赵惠恩）

14.4.4 舌状花的形态变异

喀什女蒿[*Hippolytia kaschgarica* (Krascheninnikov) Poljakov]和蒙托克菊[*Nipponanthemum nipponicum* (Franch. ex Maxim.) Kitam.]之间也可以进行属间杂交（Hong et al. 2015）。杂种植株的头状花序形态是两个亲本头状花序形态的混合。其头状花序基本上与母本植株（*Hippolytia*）头状花序相似，具有黄色（辐射对称）的管瓣舌状花，但是其头状花序外轮着生有 5～13 个白色的四基数或五基数的管瓣舌状花，更为独特且更大。与父本植株日本雏菊属（*Nipponanthemum*）的白色平瓣舌状花不同的是，杂交种的舌状花是管瓣的（五基数）。最近，同样的现象出现在灌木小甘菊（*Cancrinia maximowiczii* C. Winkl.）（管瓣）的属间杂种中以及楔叶菊（*C. naktongense* (Nakai) Tzvel.）× 菊花品种‘爱妃’（*C.* × *morifolium* ‘Aifen’）的种间杂交后代中（平瓣的，圆盘型的）（Wu et al.

2015）。后者可能被当作在春黄菊族 Anthemideae 中繁殖多基因杂种的桥梁。杂种后代新表型的分子调控机制可能是由 *CYCLOIDEA-like* 基因控制的，该基因主要具有调控花朵对称性和两侧对称（*CYC2*）的功能。作为 *CYC2* 亚家族 6 个成员之一，*CYC2c* 参与决定舌状花身份以及其后续的发育阶段。过表达 *CYC2c* 可以使舌状花的数目和花瓣片的长度显著增加（Huang et al. 2016）。

14.4.5　使用亚菊属培育小花型栽培品种

研究表明，亚菊属植物可以用于切花菊育种，如矶菊（*A. pacifica*）（Shibata et al. 1988b；De Jong and Rademaker 1989）和纪伊潮菊（*A. shiwogiku*）（Douzono and Ikeda 1998）。20 世纪 90 年代，日本冲绳县的杂交品种太平洋菊花（*Pacific* mum，菊属的回交 1 代 × 矶菊 *A. pacifica*）的切花生产非常成功（Fukai 2003）。在荷兰，利用这些杂交品种培育了小花型‘桑蒂尼’（‘Santini’）类型，见图 14.4。

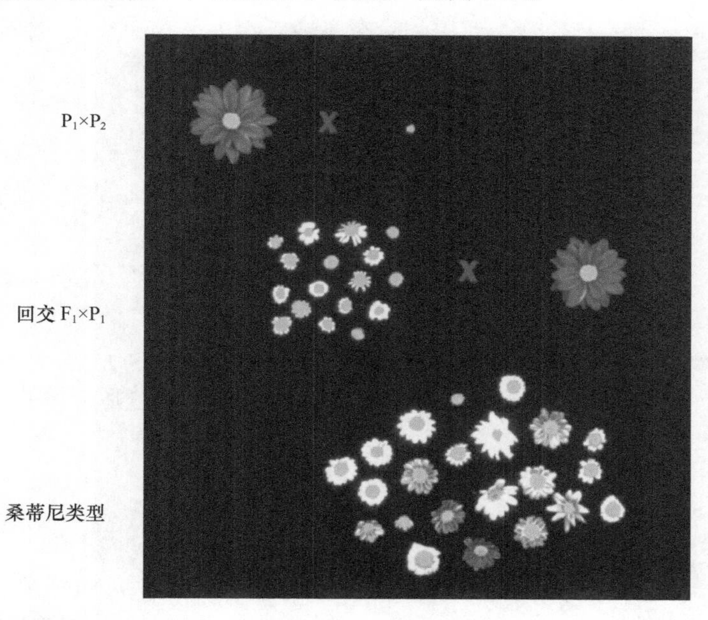

图 14.4　顶部：菊花的栽培品种 *Chrysanthemum* × *morifolium*‘Accent’和 *Chrysanthemum pacifum* 的杂交；中间：回交（BC1）；下部：小花类型的‘桑蒂尼’（来源：De Jong and Rademaker 1989）

此外，甘菊[*C. lavandulifolium* (Fisch. ex Trautv.) Ling et Shih]和疏齿亚菊[*A. remotipinna* (Hand.-Mazz.) Ling et Shih]，疏齿亚菊[*A. remotipinna* (Hand.-Mazz.) Ling et Shih]和小红菊[*C. chanetii* (Lévl.) Shih]之间也有杂交栽培种的产生（Kondo et al. 2003）。

许多菊属（*Chrysanthemum*）和亚菊属（*Ajania*）的天然与人工属间杂种的存在表明，应该将亚菊属（*Ajania*）、北极菊属（*Arctanthemum*）以及栎叶菊属（*Phaeostigma*）重新分类并将其归为菊属（Ohashi and Yonekura 2004）。除栎叶菊属（*Phaeostigma*）（未提及）外，种质资源信息网（GRIN）现在将上述种视为菊花的同义词。

14.5 种 间 育 种

14.5.1 菊花的种间杂交

菊属种类繁多，许多种间杂交可以产生（可育）后代（Anderson 2007）。不同种对当前盆栽、花坛、展览和切花类型的育种工作的相对贡献仍然未知。但是，仍然有一些野生材料被广泛用于当前的菊花育种工作中，如野菊（*C. indicum*）、甘菊（*C. lavandulifolium*）以及前文提到的一些野生种。

菊科植物均呈现自交不育，只能通过杂交繁育后代（Drewlow et al. 1973）。该属大约包括41个种，具有 $2n = 18$、36、54、72和90条染色体的物种，多倍体的形成和种间杂交产生了多种同源多倍体物种（Kadereit and Jeffrey 2007；Chen et al. 2012；Klie et al. 2014）。

根据倍性和核型的相似性，可以相对容易地获得人工种间杂种（Chen et al. 2012）。例如，日本的Dr. N. Shimotomai（1933）已获得野生菊花之间的杂交种（Kitamura 1937）。

中国的研究者已经培育了许多人工种间杂交品种，从而获得了真正的菊花 *Chrysanthemum* × *morifolium* 类型，并确定了其原始亲本（Chen et al. 2012）。1961年，通过人工杂交获得了真正的菊花 *C.* × *morifolium*，称为‘北京雏菊’。亲本是野菊变种 *C. indicum* var. *acutum* 和小红菊 *C. chanetii*。

菊花与10个野生种（*C. articum* ssp. *maekawanum*、甘菊 *C. lavandulifolium*、大岛野路菊 *C. crassum*、野菊 *C. indicum*、龙脑菊 *C. japonicum*、萨摩野菊 *C. ornatum*、足摺野路菊 *C. occidentali-japonense* var. *ashizuriense*、那贺川野菊 *C. yoshinaganthum*、小红菊 *C. chanetii*、矶菊 *A. pacifica*）之间的反交是可以通过仔细选择合适的亲本基因型来实现的（Fukai et al. 2000）。

14.5.2 耐寒性

自20世纪初以来，美国育种者一直忙于选择耐寒的庭院品种。明尼苏达大学在这个领域处于领先地位。他们引入了一些抗寒、自然呈球形、花朵繁密且耐冬季低温的多年生草本植物（Widmer 1978；Anderson and Ascher 2001；Anderson 2006；Anderson et al. 2008）。

在具有温带海洋性气候的荷兰，根据ICCNP规则，依照栽培品种不同程度的耐霜冻性可以将其划分为野菊（*C. indicum*）组、朝鲜菊花（*C. koreanum*）组、红菊花（*C. rubellum*）组、*C. weyrichii*（俄罗斯种）、小滨菊品种‘露丝’（*C. yezoense* ‘Roseum’）和紫花野菊变种‘拉特露伯’（*C. zawadskii* var. *latilobum*）。它们大多数属于野生菊组（*C.* × *morifolium* Ramat.），其次是来自紫花野菊亚种[*C. zawadskii* Herbich subsp. *coreanum* (Nakai) Y. N. Lee]，原产韩国]和乙女百合群（= 紫花野菊变种 *C. zawadskii* var. *latilobum*）（Hoffman 2005）。

小滨菊现在被重新分类为紫花野菊亚种。其他野生种也被用于培育抗寒性栽培品种，如北极菊（*C. arcticum*‘Astrid’）（可能是紫花野菊亚种，原产于韩国）、日本滨菊（*C. nipponicum*）和西伯利亚菊（*C. sibiricum* Turcz. ex DC.）（紫花野菊亚种）。

将菊花（*C.* × *morifolium*）和菊花脑（*C. nankingense*）（$2n = 2x = 18$）杂交成功获得了种间杂种（Cheng et al. 2009）。杂种是四倍体，并显示出比其菊花（*C.* × *morifolium*）亲本更好的耐寒性。在属间杂种中，该性状与游离脯氨酸的增加和丙二醛含量的减少（与氧化应激相关）有关，来源具有高度抗性的细裂亚菊（*Ajania przewalskii*）（Deng et al. 2011）。脯氨酸是一种天然氨基酸，可提高植物的抗性（Verbruggen and Hermans 2008）。显著的耐寒性是一种存在于某些菊属植物中的应激反应，用于适应寒冷的环境。

14.6　种间育种

14.6.1　传统育种

菊花主要利用外植体进行无性繁殖。因此，菊花的繁殖方法与其他进行无性繁殖的植物相似，如一些其他观赏植物、马铃薯和甘薯。在利用外植体进行无性繁殖的植物的育种方法中，生产种子只是为了改良基因型（图 14.4）。获得杂交种子后（图 14.5），种子生长为单个植株，选择其中长势较好的植株进一步用于繁殖。在第一轮播种之后，将

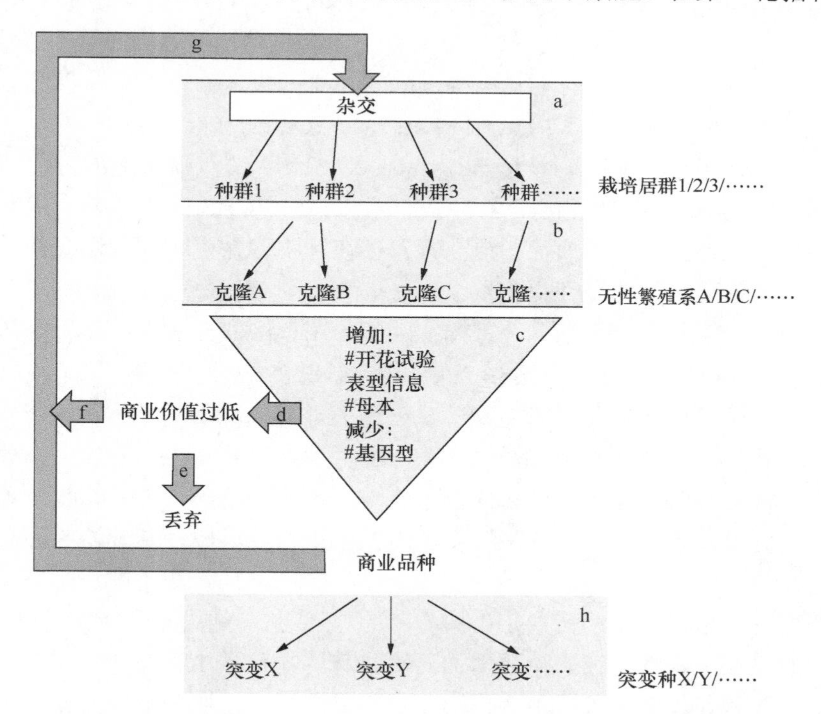

图 14.5　菊花育种方案（资料来源：Geert van Geest 2017）：首先将优良的植株材料进行杂交以获得父母本的综合性状（a），从这些杂交中收获种子并培育成植株（a），对这些植株进行第一次选择（b），这些植株均是通过无性繁殖方式获得的，随后这些无性繁殖系进入选择过程，在此过程中，要对它们进行多次评估（c），在每一轮评估中，都要决定是否将该植物作为候选的商业品种（d）

具有相同基因型的单株种植在同一块区域里，并对其各项性状进行评估。将这样的繁殖方式多次重复，就可以获得更多的表型数据，同时植株的表型也会变得更加优良，母本植株的数量也会增加。在任何一个繁殖周期里，育种者都可以针对某个基因型进行进一步的评估，并决定是将其丢弃还是用作杂交亲本。

如果某个基因型的无性繁殖系被培育为商业品种或接近成为商业品种的状态，那么可以通过突变育种诱导其花色和花型变异。获得此类突变体的优势是可以获得一系列性状高度相似的花朵。这种突变是自发产生的，或者是由伽马射线或 X 射线引起的。辐射诱导的突变体可能主要是通过染色体畸变产生的（Dowrick and El-Bayoumi 1966；Ichikawa et al. 1970；Yamaguchi et al. 2008）。

这种新获得的颜色或形状的遗传主要依赖于亲本的嵌合。花朵是由 L1 层和 L2 层组成的，但配子是由 L2 层形成的。L1 层的嵌合突变最为明显，因为它形成了舌状花的上部。然而，这种新获得的花色并不会遗传（Langton 1980）。

14.6.2 重要特征

14.6.2.1 花的特征

花色和花型可能是菊花最重要的性状。当前菊花栽培品种的主要颜色类群包括白色、黄色、橙色、粉红色、紫色、红色及绿色。这些颜色由类胡萝卜素（黄色）、花青素（紫色）和叶绿素（绿色）的不同组合、浓度及类型决定。花青素的形成受到多个能相互作用的显性等位基因调控（Hattori 1992；Miyake and Imai 1935；Van Geest et al. 2017a）。菊花中查耳酮合成酶（chalcone synthase，CHS）的沉默会造成白色花朵的产生（Courtney-Gutterson et al. 1994），在这一过程中，可能是 CHS 的表达减少或者查耳酮合成酶失活导致了花青素含量的变化。与花青素存在比不存在更重要的情况相反，类胡萝卜素的缺乏比类胡萝卜素的存在更重要（Hattori 1992；Miyake and Imai 1935）。类胡萝卜素的积累主要受类胡萝卜素裂解双加氧酶（CmCCD4a）的调节，其活性升高导致类胡萝卜素的缺乏（Ohmiya et al. 2006；Yoshioka et al. 2012）。绿色是菊花栽培品种中比较新奇的颜色，但其遗传机制尚未得到广泛的研究。

菊花栽培品种花型十分丰富（图 14.6）。由数百朵小花组成的。野生型的菊花外轮舌状花数目较少，其中三个融合的花瓣片合生成一个舌状的花瓣片，另外两个则保持不发育的状态。它们只有柱头并且不产生花粉。内轮由管状花组成，既有柱头又有花药，但花瓣很小。

花型的变异主要由三个因素决定：小花的类型、管状花的大小和舌状花花瓣的形态。在半重瓣型的菊花中，舌状花的轮数增加而管状花的轮数减少。在某些情况下，所有的小花均会变成舌状花，造成雄性不育，形成完全重瓣型。在桂瓣型的菊花中，管状花的 5 个融合花瓣会变大并且具有颜色。在管瓣型中，舌状花的花瓣完全融合形成管状的舌状花。这种管状类型由花瓣片的长度决定，管状的花瓣片长度为小花长度的一半时称为匙瓣。将这三个因素与小花的大小、长宽比的变化相结合，可以得到极为丰富的花型变异植株（图 14.7）。

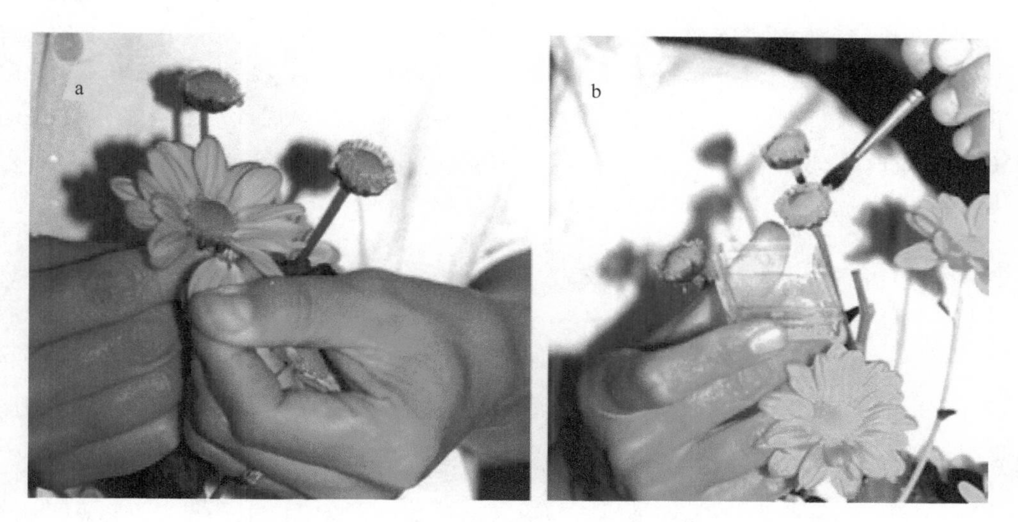

图 14.6　菊花杂交：（a）去除舌状花；（b）人工授粉

图 14.7　三种主要的花型变异及其组合：（a）野生型菊花花型；（b）重瓣型，其中所有管状花都变成了舌状花；（c）舌状花花瓣部分融合而形成匙瓣；（d）完全融合的舌状花，形成重瓣的管瓣型；（e）托桂型具有彩色且花瓣较大的管状花；（f）管瓣的舌状花结合大花瓣的管状花形成管桂型（spider-anemone）

14.6.2.2　植物形态与生长

植物的株型和花型具有重要的观赏价值。生长习性决定了这些价值的高低，与栽培的适应性相关。其中开花时间和生物量是重要的生长习性。

菊花栽培品种的生长习性多样，包括非常浓密的栽培类型，可以用作盆花或者地被植株；垂直生长且侧芽较少的栽培类型，可以用作单头切花。这一生长习性主要与分枝数量有关，可以根据分枝数进行定量区分，受到独脚金内酯合成途径的影响（Klie et al. 2016；Liang et al. 2010）。*BRC1*、*CCD7*、*CCD8* 和 *MAX2* 等位基因变异可能是导致这种遗传变异的基础（Klie et al. 2016）。

开花时间是栽培菊花的另一个重要特征，通常用开花的最短天数来表示。这个性状具有相当高的广义遗传力（De Jong 1984；Van Geest et al. 2017a），因此针对此性状的选择育种工作较为容易。开花所需时间较短的植株，其营养生长天数较短，在开花时间和植物生物量之间可能存在一种平衡关系。开花时间受多个基因位点调控（Van Geest et al. 2017a；Zhang et al. 2011b），但只有少数基因位点对性状有重要影响。

14.6.2.3 采后表现

观赏价值是切花的最重要特征，应在整个采后链和瓶插期间保持其观赏价值。因此，切花菊的采后表现非常重要。但是，由于包括萎蔫、色泽下降以及真菌或者细菌感染等多种潜在症状的存在，使得这一性状的遗传机制相对复杂（Van Geest et al. 2017b）。对于育种而言，研究这些采后出现的潜在症状并选育采后保鲜效果较好的切花菊品种具有重要的意义。例如，对管状花褪绿性状的选择，这一性状与管状花碳水化合物的缺乏有关（Van Geest et al. 2017b）。此外，对采后切花菊运输环境的评估是育种者面临的另一挑战。运输、储存或保存鲜花的技术正在高速变化并提高，因此针对某些特定性状的育种在将来可能会过时。

14.6.2.4 抗虫性

仅有少数害虫可以侵害菊花。但是抗虫性的遗传机制通常很复杂，大多数情况下它是受多种基因位点调控的。研究最多的害虫是蓟马、潜叶蝇、红蜘蛛和蚜虫。针对前三种害虫的抗性遗传物质成分似乎有一些相同的成分（Kos et al. 2014）。

西方的花蓟马（*Frankliniella occidentalis*）在许多地区是最常见且最为严重的菊花害虫。它也是一种广为人知的病毒载体。菊花的抗蓟马育种非常复杂，因为易感性与植物年龄、器官和环境高度相关。其中包括绿原酸（de Jager et al. 1996）和特定的异丁酰胺（Leiss et al. 2009）在内的代谢物对菊花的抗蓟马性具有重要作用。这类丰富的代谢物不仅有望成为抗性标记，也有助于帮助了解菊花对蓟马的抗性遗传。

菊花潜叶蝇（*Liriomiza trifolii*）严重侵害菊花的叶片，造成菊花叶片减少，进而降低菊花的观赏价值和产量。菊属植物中存在大量的抗潜叶蝇基因（De Jong and Rademaker 1991）。研究表明，抗潜叶蝇的基因主要来自矶菊（*Ajania pacifica*）（De Jong and Van De Vrie 1987；Zhao et al. 2012）。

二斑叶螨（*Tetranychus urticae*）为蜘蛛纲的一种，以叶子为食，能显著降低植株的产量。在后期阶段，它们会侵害花朵，降低观赏价值。菊花对二斑叶螨的抗性可能与酚醛化合物有关（Kielkiewicz and van de Vrie 1990）。但其潜在的机制仍需要进一步阐明。

在栽培菊花（*C.* × *morifolium*）中，对蚜虫（*Macrosiphoniella sanborni*）的抗性可能是通过多个基因座遗传而获得的（Wang et al. 2014a）。如前所述，除栽培菊花（*C.* × *morifolium*）的基因库之外，在北艾（*A. vulgaris*）中也发现了一种抗性基因（Deng et al. 2010a，2012；Zhu et al. 2014）。

14.6.2.5　抗病性

菊花中最主要的三种真菌病是镰刀菌病、黄萎病和柄锈菌病。尽管镰刀菌病和黄萎病是菊花中最广泛的真菌病，但几乎没有关于其土壤传播的相关文章发表。

对这些真菌的抗性可能是通过显性遗传的抗性等位基因获得的，如番茄、西瓜、胡萝卜、黄瓜和香蕉都是通过显性遗传而获得的相关抗性。对真菌病的治理方式通常是利用熏蒸或蒸土等方法来控制病害，这是严重破坏土壤生态系统和破坏能源需求的措施。因此，迄今为止，抗性育种是应对这些真菌病最可持续的策略。

菊花中研究最全面的与植物相互作用的病原体是堀柄锈菌（*Puccinia horiana*），存在几种不同易感基因型的病理类型（De Backer et al. 2011）。*P. horiana* 大部分是无性繁殖，但有明显的有性生殖证据（De Backer et al. 2013）。这种繁殖方式使得病原体更具多样性，并且可以适应不同的抗性等位基因。菊花对堀柄锈菌的抗性遗传以显性遗传为主（De Backer2012；De Jong and Rademaker 1986），一些基因型对多种病理类型具有抗性（De Backer et al. 2011）。虽然抗性育种相对简单，但引入抗性材料时应谨慎，因为抗性很容易被破坏。

14.7　菊 花 遗 传

栽培菊花的一个正常染色体组含有 9 条染色体，多为六倍体（$2n = 6x = 54$）。但也有很多关于栽培菊花为非整倍体的报道，染色体数目从 49 条到 67 条不等（Dowrick 1952，1953；Endo and Inada 1992；Li et al. 2009；Roxas et al. 1995；Zhang et al. 2013c）。但是，54 条染色体似乎是最常见和最稳定的结构。减数分裂时的染色体配对主要是二价的，但是确实也存在多价配对（Roxas et al. 1995；Gupta et al. 2013），表明发生了多体遗传。尽管在较早的报道中（Langton 1989）有一些有说服力的证据表明栽培菊花发生了多体遗传，但长期以来人们一直认为栽培菊花为双染色体遗传，这是根据其可能为多物种起源并且普遍存在二价染色体配对而推断的。最近对广泛的 DNA 分子标记分析的结果表明，菊花为六倍体遗传（Park et al. 2015；Van Geest et al. 2017c）。因此，最好将菊花描述为同源六倍体。在一个六倍体中，如果要使所有子代至少遗传某个等位基因的一个拷贝，那么就需要 4 个等位基因剂量。然而，在二倍体中，纯合体只需要一个亚基因组（即在一个亚基因组中有 2 个等位基因剂量）就可以至少将该等位基因的一个拷贝传给后代个体。多倍体遗传的另一个影响是，与二倍体遗传相比，其后代中可以获得更多的等位基因组合类型。当两个多倍体遗传的六倍体之间杂交时，每个位点可以获得 400 种可能的等位基因组合。对于二倍体来说，这个数字要小得多，仅为 64 种。这表明，与二倍体相比，在多倍体中性状分离可能更加广泛并且更加难以固定。

理想的等位基因应该可以独立于其起源而富集到菊花的基因库中（Klie et al. 2014）。

14.8 分子育种

14.8.1 栽培菊花（*C.* × *morifolium*）的育种资源

菊花基因库中的表型十分丰富。依生长习性，栽培菊花可以划分为两大类：直立型（具有单个直立茎）和匍匐型（具有丛生茎）。具有特定生长习性的群体之间存在一定程度的选择和隔离，并且不同的类型内存在一些特定的种群结构（Chong et al. 2016）。在欧洲和美洲，切花类型可分为多头型和单头型。本章中将多头型进一步细分为'Santini'、'regular'和'disbud'三种类型。与'regular'和'disbud'型相比，'Santini'型花枝更短，花朵数量更多，花朵体量更小，并且生长期更短。单头型的菊花通常为大花，在生长过程中会去除侧芽以获得单个大花。从遗传学上来讲，许多品种既可以作为单头类型又可以作为多头类型，因此'disbud'类型和'regular'类型之间的界限是相对模糊的。地被型大致可以分为多花型和盆花型，二者的区别是多花型的生长更为茂盛。此外，根据使用者利用的生长习性来看，展览型和更加多样化的花坛类型也可以划分到这个类群中。

目前，研究大多集中在栽培菊花的种群结构上。依照所调查的种质、DNA 标记的数量和类型，研究人员可以定义不同的组。根据 448 个扩增片段长度多态性（AFLP）位点的研究结果来看，欧洲种质资源中不存在盆花和切花类型（Klie et al. 2013）。随后，利用简单序列重复（SSR）技术进行基因分型的结果显示，一些主要的亚洲种质资源中存在一些地理区域特异性等位基因（Li et al. 2016；Zhang et al. 2013b）。最近研究表明，可以使用大量 SNP 标记（约 90 000）划分不同的组（Chong et al. 2016）。利用 DNA 标记区分菊花的不同用途或者来源时遇到的难题可能是由其高度杂合性和六倍体性引起的。由于六倍体的每个基因座可以携带 6 个等位基因，因此与二倍体相比，六倍体的每个个体具有更丰富的遗传变异。此外，许多等位基因都可以保留在种质中，只有少数特异性 SNP 可以用来定义特定群体的单倍型。微型卫星（或 SSR）标记可能是检测种群结构最有效的方式，因为它们的突变率通常更高并且更具有单倍体特异性。

14.8.2 新的育种工具

在观赏植物育种中相对较新的手段是利用 DNA 信息育种。在菊花中，与目标性状相关的 DNA 标记的开发伴随着重重障碍。第一个原因是菊花是多体性的六倍体（Van Geest et al. 2017c）。到目前为止，仍没有关于开发栽培菊花综合遗传连锁图谱的方法被报道。我们需要开发栽培菊花的综合遗传连锁图谱来鉴定与目的性状相关的基因座。接下来的章节中，我们描述了近年来菊花中利用 DNA 信息育种的进展。

14.8.2.1 连锁和数量性状基因定位

Zhang 等（2010）发表了第一个菊花遗传连锁图谱。此图谱是由双亲异交构建的，并且是基于单亲或者双亲的单一剂量标记。这使研究者可以使用二倍体物种的开发软件在同源染色体的水平上构建非整合型的连锁群。之后，研究者又对该图谱进行了改进，并用于检测花型性状、开花时间和持续期性状、生长习性以及叶片性状等的 QTL（Zhang et al. 2011a，2011b，2012a，2013）。这些公开的连锁图均是基于 SRAP、RAPD、ISSR 和 AFLP 标记，这些标记结果并不会因为实验室的不同而改变。到目前为止，研究者并未对这些遗传图谱进行整合。因此，QTL 所处的染色体位置是未知的，使得 QTL 无法在其他图谱群体中得到验证。最近，van Geest 等（2017a）发表了基于 SNP 标记的综合连锁图谱，这是实现菊花基因组研究标准化的第一步。在此连锁图上，检测到多个位点与花朵的颜色（粉色）、开花时间、采后表现和管状花数量等性状有关。

14.8.2.2　*相关研究*

尽管缺少参考基因组和连锁图谱，但仍有一些与性状相关的遗传关联的研究（Klie et al. 2016）。这些研究主要围绕植物形态建成（Klie et al. 2016）、植物和花的形态（Chong et al. 2016；Li et al. 2016）与抗涝性（Su et al. 2016）。这些研究发现了与某一类基因型性状相关的等位标记基因，但是它们在基因组中的位置未知。因此，很难在其他材料中验证或使用这些标记。Klie 等（2016）和 Chong 等（2016）基于候选基因的方法并结合这些基因的序列对这些可能的候选基因进行了分析，这些位点可以被整合到新的遗传图谱上或者是在基因组序列上被找到。研究还预测到位于 6 个基因范围内的 97 个 SNP 标记与舌状花类型、栽培类型和花型相关，但是相关性不显著。

14.8.2.3　*序列资源*

菊花和一些近缘野生种中有少许可以利用的序列资源。Xu 等（2013）基于 Illumina RNA 测序技术进行了菊花的转录组测序工作，用于研究菊花的耐水胁迫机制。Sasaki 等（2017）利用 454 RNA 测序技术对大量的组织进行了转录组测序。其他研究者也对菊花的 RNA 序列组装进行了一些研究，但是数据尚未公开。此外，一些原始序列可以从 Li 等（2014）对链格孢菌感染菊花叶片的研究，Van Geest 等（2017C）开发菊花 SNP 标记的研究以及 Liu 等（2016）对花型的研究中获得。目前尚无栽培菊花或者其近缘种的参考基因组被公布。参考基因组将有助于 DNA 标记的开发，并将为许多基于序列的基因分型方法提供大量有效信息，如参考基因组对于协助基因分型或者简化基因组测序（GBS 技术）非常有用。

14.8.3　诱变育种

菊花的诱变育种历史悠久，但主要是基于自发突变或辐射诱导的突变，这些突变通常主要是染色体畸变的结果。如果将这种材料用于育种，将会增加育种的复杂性。例如，经甲基磺酸乙酯（EMS）诱变获得的突变体，其表型变化通常不明显，但是会产生更为微妙的突变。因为这种突变通常是隐性的，并且很有可能被六倍体菊花中存在的其他 5 个等位基因所覆盖。靶向突变可能是在六倍体菊花育种中可使用的非常有趣的突变工具（Kishi-Kaboshi et al. 2017）。目前可以通过使用 CRISPR/Cas9 技术获得多个同源等位基因突变所诱导的突变体，如在小麦（Liang et al. 2017；Wang et al. 2014a，2014b）或亚麻荠属（Jiang et al. 2017；Morineau et al. 2017）植物中已获得相关突变体。因此，这也为开发新的隐性性状提供了可能，如突变一些疾病易感基因，从而解决一些以前通过传统育种方法难以改良的病害问题。

14.9 结　论

新技术将有可能推动传统的菊花育种工作朝着理想的方向加速发展。菊花育种工作高度依赖于对其遗传机制的深度解析，但是仍会受到菊花的杂合性、异源多倍体特性以及体细胞遗传模式的限制。

本章译者：

卢珍红

云南省农业科学院花卉研究所，国家观赏园艺工程技术研究中心，云南省花卉育种重点实验室，昆明 650200

参考文献

Abd El-Twab MH, Kondo K (2001a) Molecular cytogenetic identification of the parental genomes in the intergeneric hybrid between *Leucanthemella linearis* and *Nipponanthemum nipponicum* during meiosis and mitosis. Caryologia 54:109–114

Abd El-Twab MH, Kondo K (2001b) Genome territories of *Dendranthema horaimontana* in mitotic nuclei of F1 hybrid between *Dendranthema horaimontana* and *Tanacetum parthenium*. Chromosome Sci 5:63–71

Abd El-Twab MH, Kondo K (2006) Fluorescence in situ hybridization and genomic in situ hybridization to identify the parental genomes in the intergeneric hybrid between *Chrysanthemum japonicum* and *Nipponanthemum nipponicum*. Chromosome Bot 1:7–11

Abd El-Twab MH, Kondo K (2007a) FISH physical mapping of 5 s rDNA and telomere sequence repeats identified a peculiar chromosome mapping and mutation in *Leuchanthemella linearis* and *Nipponanthemum nipponicum* in *Chrysanthemum* sensu lato. Chromosome Bot 2:11–17

Abd El-Twab MH, Kondo K (2007b) Identification of parental chromosomes, intra-chromosomal changes and relationship of the artificial intergeneric hybrid between *Chrysanthemum horaimontanum* and *Tanacetum vulgare* by single color and simultaneous bicolor of FISH and GISH. Chromosome Bot 2:113–119

AIPH (2014) International statistics flowers and plants 2014, Reading, UK

Anderson NO (2006) *Chrysanthemum*. *Dendranthema grandiflora* Tzvelev. In: Anderson NO (ed) Flower breeding and genetics: issues, challenges, and opportunities for the 21st century. Springer, Dordrecht, pp 389–437

Anderson NO (ed) (2007) Flower breeding and genetics. Springer, Dordrecht. With chapter on chrysanthemum by Anderson

Anderson NO, Ascher PD (2001) Selection of day-neutral, heat-delay-insensitive *Dendranthema x grandiflora* genotypes. J Am Soc Hortic Sci 126(6):710–721

Anderson NO, Ascher PD, Gesick E (2008) Winter-hardy Mammoth series garden chrysanthemums ‘Red Daisy’, ‘White Daisy’, and ‘Coral Daisy’ sporting a shrub plant habit. Hortscience 43(3):648–654

Bino RJ, van Tuyl JM, de Vries JN (1990) Flow cytometric determination of relative nuclear DNA contents in bicellulate and tricellulate pollen. Ann Bot 65:3–8

Chen J-Y (2004) Ornamental plants. Webarticle on bio-diversity and resources of National CBD and Biosafety Office, China

Chen F, Chen P, Li H (1996) Genome analysis and their phylogenetic relationships of several wild species of *Dendranthema* in China. Acta Hortic Sin 23:67–72

Chen J-Y, Wang CY, Zhao HE, Zhou J (2012) The origin of garden Chrysanthemum, Beijing

Cheng X, Chen S, Chen F, Fang W, She L (2009) Interspecific hybrids between *Dendranthema morifolium* (Ramat.) Kitamura and *D. nankingense* (Nakai) Tzvel. achieved using ovary rescue and their cold tolerance characteristics. Euphytica 172:101–108

Chong X, Zhang F, Wu Y, Yang X, Zhao N, Wang H, Guan Z, Fang W, Chen F (2016) A SNP-enabled assessment of genetic diversity, evolutionary relationships and the identification of candidate genes in *Chrysanthemum*. Genome Biol Evol 8(12):3661–3671

Courtney-Gutterson N, Napoli C, Lemieux C, Morgan A, Firoozabady E, Robinson KE (1994) Modification of flower color in florist's chrysanthemum: production of a white-flowering variety through molecular genetics. Biotechnology 12(3):268–271

Cumming A (1939) Hardy chrysanthemums. Whittlesey House, New York

Dai SL, Zhong Y, Zhang XY (1995) Study on numerical taxonomy of some Chinese species of *Dendranthema* (DC) Des Moul. J Beijing For Univ 17(4):9-14–9-15

Dai SL, Chen J-Y, Li W-B (1998) Application of RAPD analysis in the study on the origin of Chinese cultivated chrysanthemum. Acta Bot Sin 11:1053–1059

Dai SL, Wang WK, Xu YX (2005) Phylogenetic relationship of *Dendranthema* (DC.) Des Moul. revealed by fluorescent in situ hybridization. J Integr Plant Biol 7:783–e791

De Backer M (2012) Characterization and detection of *Puccinia horiana* on chrysanthemum for resistance breeding and sustainable control. Thesis, Ghent University

De Backer M, Alaei H, Van Bockstaele E, Roldan-Ruiz I, Van der Lee T, Maes M, Heungens K (2011) Identification and characterization of pathotypes in *Puccinia horiana*, a rust pathogen of *Chrysanthemum x morifolium*. Eur J Plant Pathol 130:325–338

De Backer M, Bonants P, Pedley K, Maes M, Roldán-Ruiz I, Van Bockstaele E, Van der Lee T, Heungens K (2013) Genetic relationships in an international collection of *Puccinia horiana* isolates based on newly identified molecular markers and demonstration of recombination. Phytopathology 103(11):1169–1179

De Jager CM, Butôt RPT, Klinkhamer PGL, van der Meyden E (1996) The role of primary and secondary metabolites in chrysanthemum resistance to Franklienella occidentalis. J Chem Ecol 22:1987–1999

De Jong J (1984) Genetic analysis in Chrysanthemum morifolium. I. Flowering time and flower number at low and optimum temperature. Euphytica 33:455–463

De Jong J, Rademaker W (1986) The reaction of Chrysanthemum cultivars to Puccinia horiana and the inheritance of resistance. Euphytica 35:945–952

De Jong J, Rademaker W (1989) Interspecific hybrids between two Chrysanthemum species. Hortscience 24(2):370–372

De Jong J, Rademaker W (1991) Life history studies of the leafminer *Liriomyza trifolii* on susceptible and resistent cultivars of *Dendranthema grandiflora*. Euphytica 56:47–53

De Jong J, van de Vrie M (1987) Components of resistance to *Liriomyza trifolii* in *Chrysanthemum morifolium* and *Chrysanthemum pacificum*. Euphytica 36:719–724

Deng Y, Chen S, Chen F, Cheng X, Zhang F (2010a) The embryo rescue derived intergeneric hybrid between chrysanthemum and *Ajania przewalskii* shows enhanced cold tolerance. Plant Cell Rep 30(12):2177–2186

Deng Y, Chen S, Lu A, Chen F, Jang F, Guan Z, Teng N (2010b) Production and characterisation of the intergeneric hybrids between *Chrysanthemum morifolium* and *Artemisia vulgaris* exhibiting enhanced resistance to Chrysanthemum aphid (*Macrosiphoniella sanbourni*). Planta 231:693–703

Deng Y, Chen S, Chen F, Cheng X, Zhang F (2011) The embryo rescue derived intergeneric hybrid between chrysanthemum and *Ajania przewalskii* shows enhanced cold tolerance. Plant Cell Rep 30:2177–2186

Deng YM, Jiang JF, Chen S, Teng N, Song A, Guan Z, Fang W, Chen F (2012) Combination of multiple resistance traits from wild relative species in *Chrysanthemum* via trigeneric hybridization. PLoS One 7(8):e44337

Douzono M, Ikeda H (1998) All year round productivity of F1 and BC1 progenies between *Dendranthema grandiflorum* and *D. shiwogiku*. Acta Hortic 454:303–310

Dowrick GJ (1952) The chromosomes of *Chrysanthemum*, I: the species. Heredity 6:365–375

Dowrick GJ (1953) The chromosomes of *Chrysanthemum*, II:garden varieties. Heredity 7:59–72

Dowrick GJ, El-Bayoumi A (1966) The induction of mutations in *Chrysanthemum* using X- and gamma radiation. Euphytica 15:204–210

Drewlow LW, Ascher PD, Widmer RE (1973) Genetic studies of self incompatibility in the garden chrysanthemum, *Chrysanthemum morifolium* Ramat. Theor Appl Genet 43:1–5

Endo M, Inada I (1992) On the karyotypes of garden chrysanthemums, *Chrysanthemum morifolium* Ramat. J Jpn Soc Hortic Sci 61:413–420
Fan Q, Song A, Jiang J, Zhang T, Sun H, Wang Y et al (2016) CmWRKY1 enhances the dehydration tolerance of chrysanthemum through the regulation of ABA-associated genes. PLoS One 11:e015057210.1371
Fournier J (1910) Les voyages de P. Blancard. Bulletin de la Société de Géographie de Marseille:72–88, 205–225
Fukai S (2003) *Dendranthema* species as *Chrysanthemum* genetic resources. Acta Hortic 620:223–230
Fukai S, Zhang W, Goi M (2000) Cross compatibility between *Chrysanthemum* (*Dendranthema grandiflorum*) and *Dendranthema* species native to Japan. Acta Hortic 508:337–340
Furuta H, Shinoyama H, Nomura Y, Maeda M, Makara K (2004) Production of intergeneric somatic hybrids of chrysanthemum (*Dendranthema grandiflorum* Ramat.) and wormwood (*Artemisia sieversiana* J. F. Ehrh. ex. Willd) with rust (*Puccinia horiana* Henning) resistance by electrofusion of protoplasts. Plant Sci 166(3):695–702
Genders R (1971) Pelham's new gardening annual : new flowers, new vegetables, new ideas, The Gardening Book Club, Pelham, London
Greger H (1977) Anthemideae, critical review. In: Heywood VH, Harborne JB (eds) The biology and chemistry of the compositae. Academic Press, London
Gupta RC, Bala S, Sharma S, Kapoor M (2013) Cytomorphological studies in some species of *Chrysanthemum* L. (Asteraceae). Chromosome Bot 8(3):69–74
Hattori K (1992) Inheritance of Anthocyanin Pigmentation in Flower Color of Chrysanthemum. Japanese J Genet 67:253–258
Hoffman MHA (2005) List of names of perennials. Applied Plant Research, Netherlands
Hong G, Wu X, Liu Y, Xie F (2015) Intergeneric hybridization between *Hippolytia kaschgarica* (Krascheninnikov) Poljakov and *Nipponanthemum nipponicum* (Franch. ex Maxim.) Kitam. Genet Resour Crop Evol 62(2):255–263
Huang KC (1999) The pharmacology of Chinese herbs, second edn. CRC Press LLC, Florida
Huang D, Li X, Sun M, Zhang T, Pan H, Cheng T, Wang J, Zhan Q (2016) Identification and characterization of CYC-like genes in regulation of ray floret development in *Chrysanthemum morifolium*. Front Plant Sci 7:1633
Humphries CJ (1993) Lectotypification of *Chrysanthemum indicum*. In: Jarvis CE, Barrie FR, Allan DM, Reveal JL (eds) A list of Linnaean generic names and their types, Regnum Vegetabile, Koeltz scientific books, Oberreifenberg, Germany vol 127, p 33
Ichikawa S, Yamakawa KY, Sekiguchi F, Tatsuno T (1970) Variation in somatic chromosome number found in radiation-induced mutants of *Chrysanthemum morifolium* Hemsl. cv. Yellow Delaware and Delaware. Radiat Bot 10:557–562
Ito T, Tada S, Sato S (1990) Aroma constituents of edible chrysanthemum. J Fac Agric Iwate Univ 20:35–42
Jaffar AM, Song A, Faheem M, Chen S, Jiang J, Liu C et al (2016) Involvement of CmWRKY10 in drought tolerance of chrysanthemum through the ABA-signaling pathway. Int J Mol Sci 17(5):693
Jarvis CE, Barrie FR, Allan DM, Reveal JL (1993) A list of Linnaean generic names and their types. Regnum Veg 127:1–100
Jiang WZ, Henry IM, Lynagh PG, Comai L, Cahoon EB, Weeks DP (2017) Significant enhancement of fatty acid composition in seeds of the allohexaploid, Camelina sativa, using CRISPR/Cas9 gene editing. Plant Biotechnol J 15:648–657
Kadereit JW, Jeffrey C (2007) The families and genera of vascular plants. In: Kubitzki K (ed) Flowering Plants Eudicots, Asterales, vol VIII. Springer, Berlin/Heidelberg/New York
Kielkiewicz M, van de Vrie M (1990) Within-leaf differences in nutritive value and defence mechanism in chrysanthemum to the two-spotted spider mite (*Tetranychus urticae*). Exp Appl Acarol 10:33–43
Kim JS, Oginuma K, Tobe H (2009) Syncyte formation in the microsporangium of *Chrysanthemum* (Asteraceae):a pathway to infraspecific polyploidy. J Plant Res 122:439–444
Kishi-Kaboshi M, Aida R, Sasaki K (2017) Generation of gene-edited *Chrysanthemum morifolium* using multicopy transgenes as targets and markers. Plant Cell Physiol 58(1):216–226
Kitamura S (1937) Compositae Japonicae. Memoirs of the College of Science, Kyoto Imperial University, Series B, Vol. XV, No. 3, Art 9:1–350
Klie M, Menz I, Linde M, Debener T (2013) Lack of structure in the gene pool of the highly polyploid ornamental chrysanthemum. Mol Breed 32:339–348

Klie M, Schie S, Linde M, Debener T (2014) The type of ploidy of chrysanthemum is not black or white:a comparison of a molecular approach to published cytological methods. Front Plant Sci 5, 479:1–8

Klie M, Menz I, Linde M, Debener T (2016) Strigolactone pathway genes and plant architecture: association analysis and QTL detection for horticultural traits in chrysanthemum. Mol Gen Genomics 291:957–969

Kondo K, Tanaka R, Hizume M, Kokubugata G, Hong D, Ge S, Yang Q (1998) Cytogenetic studies on wild Chrysanthemum sensu lato in China VI. Karyomorphological characters of five species of Ajania and each one species of Brachanthemum, Dendranthema, Elachanthemum, Phaeostigma and Tanacetum in highlands of Gansu, Qinghai and Sichuan Province. J Jpn Bot 73:128–136

Kondo K, Abd El-Twab MH, Tanaka R (1999) Fluorescence in situ hybridization identifies reciprocal translocation of somatic chromosomes and origin of extra chromosome by an artificial, intergeneric hybrid between *Chrysanthemum japonica* × *Tanacetum vulgare*. Chromosome Sci 3:15–19

Kondo K, Abd El-Twab MH, Idesawa R, Kimura S, Tanaka R (2003) Genome phylogenetics in Chrysanthemum sensu lato. In: Sharma AK, Sharma A (eds) Plant genome: biodiversity and evolution, Phanerogams, vol 1A. Science Publisher, Plymouth, pp 117–200

Kos SP, Klinkhamer PGL, Leiss KA (2014) Cross-resistance of chrysanthemum to western flower thrips, celery leafminer, and two-spotted spider mite. Entomol Exp Appl 151(3):198–208

Langton FA (1980) Chimerical structure and carotenoid inheritance in *Chrysanthemum morifolium* Ramat. Euphytica 29:807–812

Langton FA (1989) Inheritance in *Chrysanthemum morifolium* Ramat. Heredity 62:419–423

Lawal OA, Ogunwande IA, Olorunloba OF, Opoku AR (2014) The essential oils of *Chrysanthemum morifolium* Ramat. from Nigeria. Am J Essent Oil Nat Prod 2(1):63–66

Leiss K, Maltese F, Choi YH, Verpoorte R, Klinkhamer PGL (2009) Identification of Chlorogenic acid as a resistance factor for Thrips in Chrysanthemum. Plant Physiol 150:1567–1575

Li J, Teng N, Chen F, Chen S, Sun C, Fang W (2009) Reproductive characteristics of *Opisthopappus taihangensis* (Y. Ling) C. Shih, an endangered Asteraceae species endemic to China. Sci Hortic 121:474–479

Li H, Chen S, Song A, Wang H, Fang W, Guan Z, Jiang J, Chen F (2014) RNA-Seq. derived identification of differential transcription in the chrysanthemum leaf following inoculation with *Alternaria tenuissima*. BMC Genomics 15:9–23

Li P, Zhang F, Chen S, Jiang J, Wang H, Su J, Fang W, Guan Z, Chen F (2016) Genetic diversity, population structure and association analysis in cut chrysanthemum (*Chrysanthemum morifolium* Ramat.). Mol Gen Genomics 291(3):1117–1125

Liang J, Zhao L, Challis R, Leyser O (2010) Strigolactone regulation of shoot branching in chrysanthemum (*Dendranthema grandiflorum*). J Exp Bot 61:3069–3078

Liang Z, Chen K, Li T, Zhang Y, Wang Y, Zhao Q, Liu J, Zhang H, Liu C, Ran Y, Gao C (2017) Efficient DNA-free genome editing of bread wheat using CRISPR/Cas9 ribonucleoprotein complexes. Nat Commun 8:14261

Ling R, Shih Z (1983) Anthemideae, Flora Republicae Popularis Sinicae, vol 76 (1). Science Press, Beijing:73–74, 97–98

Liu H, Sun M, Du D, Pan H, Cheng T, Wang J, Zhang Q, Gao Y (2016) Whole-transcriptome analysis of differentially expressed genes in the ray florets and disc florets of *Chrysanthemum morifolium*. BMC Genomics 17:389

Ma YP, Chen MM, Wei JX, Zhao L, Liu PL, Dai SL, Wen J (2016) Origin of Chrysanthemum cultivars, evidence from nuclear low-copy LFY gene sequences. Biochem Syst Ecol 65:129–136

Miyake K, Imai Y (1935) A chimerical strain with variegated flowers in *Chrysanthemum sinense*. Zeitschr f ind Abst- u Vererbgsl 68:300–302

Morineau C, Bellec Y, Tellier F, Gissot L, Kelemen Z, Nogué F, Faure J-D (2017) Selective gene dosage by CRISPR-Cas9 genome editing in hexaploid *Camelina sativa*. Plant Biotechnol J 15:729–739

Needham J (1986) Science and civilisation in China. Botany, Vol. VI.1. Cambridge University Press, Cambridge, UK

Nicolson DH (1999) Report of the General Committee:8. Proposal 1172 of 1998 accepted to conserve Chrysanthemum with a conserved type. This proposal by Trehane (1995) was accepted by the Committee for Spermatophyta in 1998, according to report 46 by Brummitt in Taxon 47:443–444

Oberprieler C, Vogt R, Watson LE (2006) Tribe Anthemideae Cass. In: Kadereit JW, Jeffrey C

(eds) The families and genera of vascular plants 8. Springer, Berlin, pp 342–374
Ohashi H, Yonekura K (2004) New combinations in Chrysanthemum (Compositae-anthemideae) of Asia with a list of Japanese species. J Jpn Bot 79:186–195
Ohmiya A, Kishimoto S, Aida R, Yoshioka S, Sumitomo K (2006) Carotenoid cleavage dioxygenase (CmCCD4a) contributes to white color formation in chrysanthemum petals. Plant Physiol 142:1193–1201
Ohtsuka H, Inaba Z (2008) Intergeneric hybridization of marguerite (*Argyranthemum frutescens*) with annual chrysanthemum (*Glebionis carinatum*) and crown daisy (*G. coronaria*) using ovule culture. Plant Biotechnol 25:535–539
Park NY, Kwon JH (1997) Chemical composition of petals of *Chrysanthemum* spp. J Food Sci Nutr 2(4):304–309
Park SK, Arens P, Esselink D, Lim JH, Shin HK (2015) Analysis of inheritance mode in chrysanthemum using EST-derived SSR markers. Sci Hortic 192:80–88
Punithalingam E (1968) *Puccinia horiana*. C MI Descriptions of Pathogenic Fungi and Bacteria No. 176. CAB International, Wallingford, UK
Roxas NJL, Tashiro Y, Miyazaki S, Isshiki S, Takeshita A (1995) Meiosis and pollen fertility in Higo chrysanthemum (*Dendranthema* × *grandiflorum* (Ramat.) Kitam). J Jpn Soc Hortic Sci 64(1):161–168
Sasaki K, Mitsuda N, Nashima K, Kishimoto K, Katayose Y, Kanamori H, Ohmiya A (2017) Generation of expressed sequence tags for discovery of genes responsible for floral traits of *Chrysanthemum morifolium* by next-generation sequencing technology. BMC Genomics 18:683–696
Shibata M, Amano M, Kawata J, Uda M (1988a) Breeding process and characteristics of Summer Queen, a spray chrysanthemum for summer production. Bull Nat Res Inst Veg Orn Plants Tea Ser A 2:245–255
Shibata M, Kawata J, Amano M, Kameno T, Yamagashi M, Toyoda T, Yamaguchi T, Okimura M, Uda M (1988b) Breeding process and characteristics of Moonlight, an interspecific hybrid between *Chrysanthemum morifolium* and *C. pacificum*. Bull Nat Res Inst Veg Orn Plants Tea Ser A 2:257–277
Shimotomai N (1933) Zur karyogenetik der Gattung Chrysanthemum. J.Sci. (Hiroshima Univ.) Ser.B. Div. 2. 2:1–98
Shunying Z, Yang Y, Huaidong Y, Yue Y, Guolin Z (2005) Chemical composition and antimicrobial activity of the essential oils of *Chrysanthemum indicum*. J Ethnopharmacol 96(1–2):151–158
Spaargaren JJ (2001) Supplemental lighting for greenhouse crops. Hortilux Schréder, P.L. Light Systems
Spaargaren JJ (2015) Origin and spreading of the cultivated chrysanthemum, Aalsmeer. ISBN:978-90-803929-2-2
Storer JR, Elmore JS, van Embden HF (1993) Airborne volatiles from the foliage of three cultivars of autumn flowering *Chrysanthemum*. Phytochemistry 34(6):1489–1492
Su J, Zhang F, Li P, Guan Z, Fang W, Chen F (2016) Genetic variation and association mapping of waterlogging tolerance in chrysanthemum. Planta 244:1241–1252
Sun CQ, Chen FD, Fang WM, Liu ZL, Teng NJ (2010) Advances in research on distant hybridization of *Chrysanthemum*. Sci Agric Sin 43(12):2508–2517
Sun H, Zhang F, Chen S, Guan Z, Jiang J, Fang W, Chen F (2015) Effects of aphid herbivory on volatile organic compounds of *Artemisia annua* and *Chrysanthemum morifolium*. Biochem Syst Ecol 60:225–233
Tanaka R (1959a) On the speciation and karyotype in diploid and tetraploid species of *Chrysanthemum*. II. Karyotype in *Chrysanthemum makinoi* (2n=18). J Sc Hiroshima Univ Series B Div 2 (Botany) 9:17–30
Tanaka R (1959b) On the speciation and karyotype in diploid and tetraploid species of *Chrysanthemum*. IV. Karyotype in *Chrysanthemum wakasaense* (2n=36). J Sc Hiroshima Univ Series B Div 2 (Botany) 9:41–57
Tanaka R, Shimotomai N (1961) Cytogenetic studies on the F1 hybrid of *Chrysanthemum lineare* × *Ch. Nipponicum*. Zeitschrift für Vererbungslehre 92:190–196
Tanaka R, Shimotomai N (1968) A cytogenetic study on the F1 hybrid of *Chrysanthemum makinoi* × *Ch. vulgare* (now *Tanacetum vulgare*). Cytologia 33:241–245
Tang F, Chen F, Chen S, Teng N, Fang W (2009) Intergeneric hybridization and relationship of genera within the tribe Anthemideae Cass. (I. *Dendranthema crassum* (Kitam.) Kitam. x *Crossostephium chinense* (L.) Makino). Euphytica 169(1):133–140
Tang FP, Chen SM, Deng YM, Chen FD (2010) Intergeneric hybridization between *Dendranthema crassum* and *Ajania myriantha*. Acta Hortic 855:267–272

Tatarenko E, Kondo K, Smirnov SV, Kucev M, Yang Q, Hong D, Ge S, Zhang D, Zhou S, Damdinsuren O, Abd El-Twab MH, Hizume M, Cao R, Vallès J, Motohashi T, Masuda Y (2011) Chromosome relationships among the *Chrysanthemum fruticulosum* complex. Chromosome Bot 6(3):61–66

Teixeira da Silva JA (2004) Mining the essential oils of the Anthemideae. Afr J Biotechnol 3(12):706–720

Trehane P (1995) Proposal to conserve *Chrysanthemum* L. with a conserved type (Compositae). Taxon 44:439–441

Tsao R, Attygalle AB, Schroeder FC, Marvin CH, McGarvey BD (2003) Isobutylamides of unsaturated fatty acids from *Chrysanthemum morifolium* associated with host-plant resistance against the western flower thrips. J Nat Prod 66(9):1229–1231

Tsao R, Marvin CH et al (2005) Evidence for an isobutylamide associated with host plant resistance to western flower thrips, Frankliniella occidentalis, in chrysanthemum. J Chem Ecol 31(1):103–110

Uchio Y, Tomosue K, Nakayama M, Yamammura A, Waki T (1981) Constituents of the essential oil from three tetraploid species of chrysanthemum. Phytochemistry 20(12):2691–2693

van Geest G (2017) Disentangling hexaploid genetics. Towards DNA-informed breeding for post-harvest performance in chrysanthemum. Thesis, WUR, Wageningen

van Geest G, Bourke PM, Voorrips RE et al (2017a) An ultra-dense integrated linkage map for hexaploid chrysanthemum enables multi-allelic QTL analysis. Theor Appl Genet 130:2527

van Geest G, Post A, Arens P, Visser RGF, van Meeteren U (2017b) Breeding for postharvest performance in chrysanthemum by selection against storage-induced degreening of disk florets. Postharvest Biol Technol 124:45–53

van Geest G, Voorrips RE, Esselink D, Post A, Visser RGF, Arens P (2017c) Conclusive evidence for hexasomic inheritance in chrysanthemum based on analysis of a 183 k SNP array. BMC Genomics 18(1):1471–2164

Van Tuyl JM, Lim K-B (2003) Interspecific hybridisation and polyploidisation as tools in ornamental plant breeding. Acta Hort. 612. ISHS 2003:13–22

Veilleux RE (1985) Diploid and polyploidy gametes in crop plants:mechanisms of formation and utilization in plant breeding. Plant Breed Rev 3:253–288

Verbruggen N, Hermans C (2008) Proline accumulation in plants: a review. Amino Acids 35(4):753–759

Wang C (2005) Chrysanthemum genetic diversity of germplasm evaluation and molecular genetic evolution of CDS on molecular markers. Doctoral dissertation of Beijing Forestry University, China

Wang J, Zhu F, Zhou XM, Niu CY, Lei CL (2006) Repellent and fumigant activity oil from *Artemisia vulgaris* to *Tribolium casteum* (Herbst) (Coleoptera:Tenebrionidae). J Stores Products Res 42(3):339–347

Wang H, Jiang J, Chen S, Fang W, Guan Z, Liao Y, Chen F (2013) Rapid genomic and transcriptomic alterations induced by wide hybridization: *Chrysanthemum nankingense* × *Tanacetum vulgare* and *C. crassum* × *Crossostephium chinense* (Asteraceae). BMC Genomics 14:902

Wang C, Zhang F, Guan Z, Chen S, Jiang J, Fang W, Chen F (2014a) Inheritance and molecular markers for aphid (*Macrosiphoniella sanborni*) resistance in chrysanthemum (*Chrysanthemum morifolium* Ramat.). Sci Hortic 180:220–226

Wang Y, Cheng X, Shan Q, Zhang Y, Liu J, Gao C, Qiu J (2014b) Simultaneous editing of three homoeoalleles in hexaploid bread wheat confers heritable resistance to powdery mildew. Nat Biotechnol 32:947–951

Wang K, Wu YH, Tian XQ, Bai ZY, Liang QY, Liu QL, Pan YZ, Zhang L, Jiang BB (2017) Overexpression of *DgWRKY4* enhances salt tolerance in Chrysanthemum seedlings. Front Plant Sci 13. https://doi.org/10.3389/fpls.2017.01592

Widmer RE (1978) Chrysanthemum named Minngopher. U.S. Plant Patent, No. 4,327. U.S. Patent Office, Washington, DC

Wu X, Hong G, Liu Y, Xie F, Liu Z, Liu W, Zhao H (2015) Possible intergeneric hybridization between *Cancrinia maximowiczii* C. Winkl. and *Chrysanthemum naktongense* (Nakai) Tzvel. × C. ×*morifolium* Ramat. ‘Aifen’. Genet Resour Crop Evol 62(2):293–301

Xu Y, Gao S, Yang Y, Huang M, Cheng L, Wei Q, Fei Z, Gao J, Hong B (2013) Transcriptome sequencing and whole genome expression profiling of chrysanthemum under dehydration stress. BMC Genomics 14(1):1–15

Yamaguchi H, Shimizu A, Degi K, Morishita T (2008) Effects of dose and dose rate of gamma ray irradiation on mutation induction and nuclear DNA content in chrysanthemum. Breed Sci

58:331–335
Yang W, Glover BJ, Rao GY, Yang J (2006) Molecular evidence for multiple polyploidization and lineage recombination in the *Chrysanthemum indicum* polyploid complex (Asteraceae). New Phytol 171(4):875–886
Yang D, Hu X, Liu Z, Zhao H (2010) Intergeneric hybridizations between Opisthopappus taihangensis and Chrysanthemum lavandulifolium. Sci Hortic 125:718–723
Yoshioka S, Aida R, Yamamizo C et al (2012) The carotenoid cleavage dioxygenase4 (CmCCD4a) gene family encodes a key regulator of petal color mutation in chrysanthemum. Euphytica 184:377
Zeven AC, Zhukovsky PM (1975) Dictionary of cultivated plants and their centres of diversity. Excluding ornamentals, forest trees and lower plants. Pudoc, Wageningen
Zhang F, Chen S, Chen F, Fang W, Li F (2010) A preliminary genetic linkage map of chrysanthemum (*Chrysanthemum morifolium*) cultivars using RAPD, ISSR and AFLP markers. Sci Hortic 125:422–428
Zhang F, Chen S, Chen F, Fang W, Chen Y, Li F (2011a) SRAP-based mapping and QTL detection for inflorescence-related traits in chrysanthemum (*Dendranthema morifolium*). Mol Breed 27:11–23
Zhang F, Chen S, Chen F, Fang W, Deng Y, Chang Q, Liu P (2011b) Genetic analysis and associated SRAP markers for flowering traits of chrysanthemum (*Chrysanthemum morifolium*). Euphytica 177:15–24
Zhang F, Jiang J, Chen S, Chen F, Fang W (2012a) Detection of quantitative trait loci for leaf traits in chrysanthemum. J Hortic Sci Biotechnol 87:613–618
Zhang F, Jiang J, Chen S, Chen F, Fang W (2012b) Mapping single-locus and epistatic quantitative trait loci for plant architectural traits in chrysanthemum. Mol Breed 30:1027–1036
Zhang F, Chen SM, Jiang JF, Guan ZY, Fang WM, Chen FD (2013a) Genetic mapping of quantitative trait loci underlying flowering time in chrysanthemum (*Chrysanthemum morifolium*). PLoS One 8(12):e83023
Zhang Y, Wang C, Ma HZ, Dai SL (2013b) Assessing the genetic diversity of chrysanthemum cultivars with microsatellites. J Am Soc Hortic Sci 138(6):479–486
Zhang Y, Zhu M, Dai S (2013c) Analysis of karyotype diversity of 40 Chinese chrysanthemum cultivars. J Syst Evol 51:335–352
Zhao HE, Liu ZH, Hu X, Yin JL, Li W, Rao GY, Zhang XH, Huang CL, Anderson N, Zhang QX, Chen JY (2009) Chrysanthemum genetic resources and related genera of Chrysanthemum collected in China. Genet Resour Crop Evol 56:937
Zhao HB, Chen F, Chen S, Wu G, Guo W (2010) Molecular phylogeny of *Chrysanthemum*, *Ajania* and its allies (Anthemideae, Asteraceae) as inferred from nuclear ribosomal ITS and chloroplast trnL-F IGS sequences. Plant Syst Evol 284:153–169
Zhao HB, Chen F, Tang F, Jiang J, Li C, Miao H, Chen F, Fang W, Guo W (2012) Morphological characteristics and chromosome behaviour in F1, F2 and BC1 progenies between *Chrysanthemum × morifolium* and *Ajania pacifica*. Russ J Genet 48(8):808–818
Zheng Y, Shen J, An YM, Zhang JQ, Rao GY (2013) Intergeneric hybridization between *Elachanthemum intricatum* (Franch.) Ling et Ling and *Opisthopappus taihangensis* (Y. Ling) C. Shih. Genet Resour Crop Evol 60(2):473–482
Zhu WY, Zhang ZF, Chen SM, Xu L, Wang L, Wang H, Qi X, Li H, Chen F (2014) Intergeneric hybrids between *Chrysanthemum morifolium* 'Nannongxiaoli' and *Artemisia vulgaris* 'Variegata' show enhanced resistance against both aphids and *Alternaria* leaf spot. Euphytica 197(3):399–408

第15章　石　　竹

Takashi Onozaki

摘要　香石竹（*Dianthus caryophyllus* L.）是世界上主要的花卉作物之一。石竹属是石竹科的一员，包括300多种一年生植物和多年生常绿植物。由于这个属的育种历史很悠久，现代香石竹品种背后的亲缘关系非常复杂。在本章中，总结了香石竹育种史和栽培品种的性状变迁，这些品种包括单头类型和多头类型的切花品种，以及矮小的盆栽品种。本章主要介绍了香石竹育种研究的最新进展，这些研究涉及花色、花型、突变、抗病性、瓶插寿命、种间杂交、花香和多倍体育种。其中着重介绍了白色品种的花色及色素组成的遗传，以及我们研究团队近30年来在以下几个方面的育种研究进展：抗细菌性枯萎病育种，延长瓶插寿命育种，利用具有较长瓶插寿命的香石竹选择系和长萼瞿麦变种的种间杂交研究（该野生的瞿麦变种原产于日本）。研究团队还总结了香石竹的基因组分析，包括使用高通量测序技术进行的大规模转录组测序。本章最后描述了遗传连锁图谱的构建、数量性状基因座（QTL）分析，以及基因组测序研究。

关键词　石竹；香石竹；抗病性育种；遗传连锁图谱；种间杂交；分子标记辅助选择；突变育种；瓶插寿命

15.1　引　　言

香石竹（*Dianthus caryophyllus* L.）跟菊花和月季一样，是世界上主要的花卉作物之一。它在全世界的温带地区可常年种植，尤其是在气候凉爽的高原，如哥伦比亚、中国和肯尼亚的部分地区，这些地区也就是香石竹主产区（Nimura et al. 2006b；Onozaki 2006）。石竹属是石竹科的一员，它包括300多种一年生植物和多年生常绿植物，常被称为“香石竹”或“石竹”（Itoh et al. 1989，Galbally and Galbally 1997）。石竹属的很多种，分布在整个欧洲和亚洲的部分地区，包括日本。少数几个种为非洲山区特有，而北极石竹仅分布在北美洲（Hamilton and Walters 1989）。

香石竹的育种历史悠久，产生了多种多样的品种。野生香石竹很可能起源于地中海地区（西班牙南部、意大利本土、撒丁岛、西西里岛和爱奥尼亚群岛），但是由于栽培历史悠久，难以确认其确切来源（Tutin 1964）。有几个石竹属的种，特别是香石竹，作为观赏植物，已经被种植了数百年，正是由于其悠久的育种历史，因此现代石竹栽培品

T. Onozaki (✉)
Institute of Vegetable and Floriculture Science, NARO (NIVFS), Tsukuba, Japan
e-mail: onozaki@affrc.go.jp

种呈现很高的杂合性。

石竹是人类历史中最古老的观赏植物之一（Halmagyi and Lambardi 2006）。过去，香石竹因其美丽，以及浓烈的香气让人联想到丁香（Kingman 1983），而被称为“加冕”“麝香石竹”或“神圣之花”。希腊植物学家 Theophrastus（公元前 372～前 288 年）命名的石竹（*Dianthus*），通常被作为香石竹的拉丁学名。该名称来自两个希腊词：*dios*（指宙斯神）和 *anthos*（指花）。因此，香石竹被称为“神圣之花”。该名称的石竹部分是指丁香（*Caryophyllus aromaticus*）的旧名称，因为它具有丁香的香味。古希腊人用石竹编织成运动员的冠冕，这种做法可能为“加冕花”名字的由来（Holley and Baker 1963）。

香石竹的品种改良始于 16 世纪，主要是在欧洲的花园中。在法国，路易十四（1643～1715 年）时期，香石竹是宫廷王室最喜欢的花（Halmagyi and Lambardi 2006）。能常年开花（即持续开花）的第一个现代香石竹——‘Atim’，是在 1835～1845 年，由法国里昂的 M. Dalmais 培育。他的育种至少有两个种参与，很可能是香石竹和中国石竹（Moyal-Ben Zvi and Vainstein 2007）。1852 年，持续开花品种的实生苗从里昂传到美国，被作为亲本材料，育成了许多美国优良品种（Holley and Baker 1963；Takeda 1989）。在随后的世纪中，持续开花的香石竹育种在美国已经很成熟了。然而，自 1960 年左右以来，香石竹商业育种的重心转移到了欧洲。

15.2 香石竹品种分类

当今的香石竹品种，根据植物形态、花大小和花型的表型差异，可分为三种主要类型（Vainstein et al. 1991）：单头切花品种、多头切花品种和盆栽品种。

单头类型，为一茎一花，花大。栽培上要摘去侧芽和侧生花蕾，以增加顶生花朵的大小。1938 年，美国育种家 William Sim 培育出一个红色品种——‘William Sim’。该品种是最著名的单头切花品种之一，为香石竹产业的发展做出了巨大贡献。由它产生的芽变品种达 300 多个，花色极为丰富，该品种群被称作 Sim 系品种。它们都具有生长势强、花大、茎强壮、持续开花等习性。在 20 世纪 50～90 年代近半个世纪的时间，世界各地广为种植。但是，西姆系品种有一个主要缺陷，对石竹镰刀孢菌枯萎病（*Fusarium oxysporum* f. sp. *dianthi*）和细菌性枯萎病（*Burkholderia caryophylli*）高度敏感。

大约自 1960 年以来，在法国和意大利里维埃拉地区，培育出了一种新型的单头香石竹品种，被称为“地中海系”。该地区主要的香石竹公司和育种公司有 Barberet & Blanc（后迁往西班牙）、Nobbio、Flavio Sapia、Baratta、La Villetta 和 Taroni。意大利圣雷莫的 Giacomo Nobbio 博士，是世界著名的香石竹育种者，为地中海系品种的培育做出了巨大贡献。地中海系品种抗石竹镰刀菌枯萎病，并具有出色的花形，良好的品质，花苞裂萼现象减少甚或消除，而这个问题在 Sim 系品种上多有发生。Sim 系品种已基本消失，目前的单头切花香石竹商业品种均为地中海系。

多头类型，为一茎多花，花小。第一个多头品种‘Exquisite’，是花瓣具有粉红色底淡紫色镶边的复色花品种。它是 1952 年由‘William Sim’分离鉴定出来的芽变品种，很快又在 1958 年育成一个芽变品种，花瓣为粉红色底白色边缘的‘Elegance’（Galbally and Galbally 1997）。此后，随着单头香石竹的发展，多头品种已经在许多国家广泛种植。荷兰 Hilverda Kooij 公司育成的‘巴巴拉’（‘Barbara’）和‘太子’（‘Tessino’），以及由这两个品种产生的花色多样的突变品种群，成为多头香石竹品种的领头羊。

第三种类型是矮小的盆栽品种，主要用作盆栽植物，也用于生产切花。盆栽香石竹矮小而花朵繁多。日本花卉育种公司即坂田种子公司（Sakata Seed Corporation），于 1970 年培育出了第一批可进行种子繁殖的盆栽香石竹‘Piccadilly Special Mix’（Onozaki 2016）。由他们改良的品种‘Juliet’赢得了 1975 年由全美精选颁发的铜质奖章，全美精选（AAS）是全球花卉竞赛的权威组织。取得这一胜利后，盆栽香石竹受到了世界各地观赏植物种植者越来越高的关注。随着这些品种的到来，在日本，盆栽香石竹作为母亲节（5 月的第二个星期日）的礼物被大量生产和销售（Onozaki 2016）。目前，早期的种子繁殖品种已被新的无性繁殖品种取代，矮化的盆栽香石竹已成为全球商业种植者的热门产品。

许多品种在世界各地已有种植，而新的品种不断被培育出来，它们具有许多吸引人的特征，诸如新颖的花色、花朵大小、改变了的花形，以及获得改良的花香、产量、抗病性和瓶插寿命等。

15.3　花 的 颜 色

花色是香石竹最重要的性状之一，因为它会强烈影响消费者购买鲜花时的选择倾向。在香石竹悠久的育种史中，育种人员已经开发出多种多样的花瓣颜色，包括红色、粉红色、黄色、橙色、棕色、白色和绿色。另外，有些花显示出边缘杂色、斑点或扇形条带（Okamura et al. 2003）。花色主要取决于花青素的化学结构、色素的存在与否，以及花瓣维管束 pH 等。在香石竹的红色、粉红色和黄色花朵中占主导作用的色素主要是花青素与查耳酮衍生物，与这些色素生物合成有关的大多数基因得到了鉴定，并且对其进行了详细研究（Britsch et al. 1993；Mato et al. 2000，2001；Yoshimoto et al. 2000；Itoh et al. 2002；Ogata et al. 2004；Abe et al. 2008；Matsuba et al. 2010；Momose et al. 2013）。

已知香石竹中有 4 种主要花青素（Bloor 1998；Nakayama et al. 2000）：天竺葵素 3-苹果糖苷（Pg3MG）、花青素 3-苹果糖苷（Cy3MG）、天竺葵素 3,5-环苹果酰二糖苷（Pg3,5cMdG）和花青素 3,5-环苹果酰二糖苷（Cy3,5cMdG）。这些花青素被糖基化，并且通常被苹果酸酰化。一个香石竹品种将特定的花色苷作为其主要花色苷。具有 Pg3MG、Cy3MG、Pg3,5cMdG 和 Cy3,5cMdG 的香石竹品种的花通常分别表现为红色、深红色、粉红色和紫色。

基因工程的最新进展突破了传统育种的局限性，使育种者可以添加新颖的蓝色。由于缺少编码类黄酮 3′,5′-羟化酶（*F3′5′H*）的基因，而该基因负责翠雀花色苷的生物合成，因此野生型香石竹不会积聚这些导致蓝色的色素，所以香石竹一直未出现过蓝色花。诸如‘月尘’和‘月影’等香石竹转基因品种，它们具有 *F3′5′H* 基因，花朵呈蓝色，已成功实现了商业化，现在，在世界各地的切花市场上有销售（Tanaka et al. 1998，Fukui et al. 2003）。有关蓝色花色遗传学的更多内容，详见 3.3 节。

15.3.1 纯白色香石竹品种及其色素成分的鉴定

尽管大多数香石竹品种都经过培育产生了强烈的着色，但是对于某些客户来说，希望没有这种着色。花色育种的一个重要目标是获得纯白色（无氰）花，以及更常见的彩色（青色）花（Mato et al. 2000）。许多种植者和研究人员都试图培育纯白色的品种，这些品种不会在花瓣中积聚大量可见的色素，因为纯白色的花朵可以使花束变亮，并且在商业用途方面有很大的潜在市场。要系统地培育无氰的花朵，必须弄清楚白色与可以阻止花色苷生物合成或改变颜色形成的合成后反应的因素之间的关系。

香石竹白花着色及其遗传的一些研究已有报道。Geissman 和 Mehlquist（1947）以及 Mehlquist 和 Geissman（1947）研究表明，花色素的形成涉及三个具有互补作用的基因 *Y*、*I* 和 *A*。他们假设白花的基因型为 *yyI_aa* 或 *yyI_A_* 或 *Y_I_aa*。Forkmann 和 Dangelmayr（1980）发现基因 *I* 调节查耳酮-黄烷酮异构酶的活性，并报道了具有隐性等位基因（*ii*）的基因型完全没有这种酶活性，并产生黄色花朵。Stich 等（1992）随后发现基因 *A* 调节二氢黄酮醇-4-还原酶的合成。隐性等位基因（*aa*）的存在，导致二氢黄酮醇-4-还原酶的缺乏，从而中断了二氢黄酮醇和黄烷-3,4-二醇之间的花青素生物合成途径。因此，*aa* 产生无氰的花。*Y* 基因似乎通过稀释效应影响类黄酮的浓度，从而导致白花。该基因可能在 *I* 和 *A* 之前起作用，但其作用的细节尚不清楚（图 15.1）。

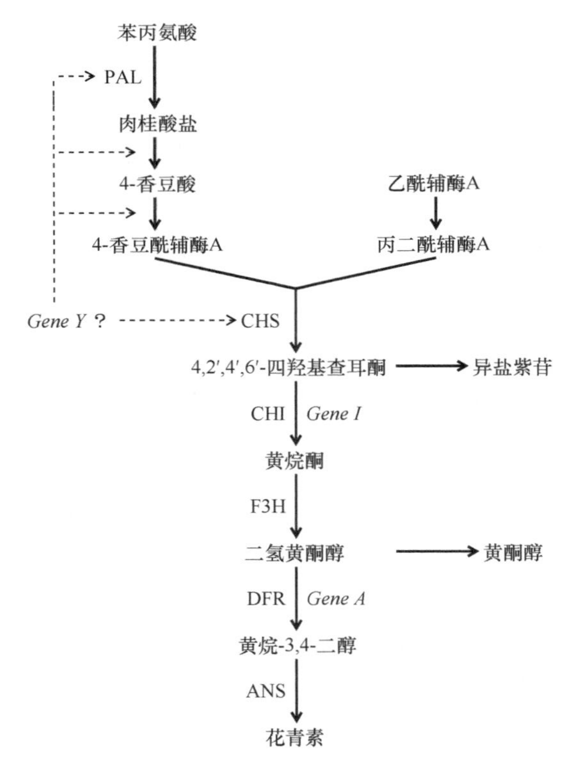

图 15.1 香石竹花中类黄酮化合物的生物合成途径。ANS，花青素合酶；CHI，查耳酮-黄烷酮异构酶；CHS，查耳酮合酶；DFR，二氢黄酮醇-4-还原酶；F3H，黄烷酮-3-羟化酶；PAL，苯丙氨酸解氨酶

利用颜色分析仪对13个白色香石竹品种的花色进行了测定，结果表明，13个品种之间的花色差异较大（Onozaki et al. 1999c）。三个品种的花瓣（‘White Mind’、‘Kaly’和‘White Barbara’）几乎是纯白色的，而其他十个品种有微妙但明显的颜色（乳白色）。用二维薄层色谱法分析了13个白色香石竹品种的色素组成，将其分为3类：缺类黄酮品种、柚皮素积累品种和山奈酚积累品种（表15.1）。几乎是纯白色的‘White Mind’，是1975年由‘U Conn Sim’的一个芽变株建立的（Yoshida et al. 2004），花瓣中缺乏类黄酮化合物。‘Kaly’和‘White Barbara’积累了大量的柚皮素衍生物。其他正常的白色品种都将山奈酚衍生物作为主要的类黄酮。Onozaki等（1999c）得出结论，‘White Mind’、‘Kaly’和‘White Barbara’几乎是纯白色的，因为它们缺少诸如山奈酚衍生物之类的类黄酮。我们的结果表明，在‘White Mind’中，类黄酮生物合成途径在查耳酮合成之前被阻断，而在‘Kaly’和‘White Barbara’中，黄酮生物合成途径在黄烷酮-3-羟化酶步骤被阻断（Onozaki et al. 1999c；Mato et al. 2000）。此外，在积累山奈酚的乳白色香石竹品种中，从紫外线、质谱、^{1}H和^{13}C核磁共振谱中鉴定出三种山奈酚苷（Iwashina et al. 2010）。

表15.1　13个白色香石竹品种花瓣中的有机酸、黄烷酮和黄酮醇

品种	花色	色素成分[a]		
		有机酸	黄烷酮（柚皮素）	黄酮醇（山奈酚）
山奈酚积累品种				
California White	乳白色	+	−	+
Sonnet Silky	乳白色	+	−	+
U Conn Sim	乳白色	+	−	+
Swan	乳白色	+	−	+
Albivette	乳白色	+	−	+
Florence	乳白色	+	−	+
Delphi	乳白色	+	−	+
Tobia	乳白色	+	−	+
Bagatel	乳白色	+	−	+
White Lucena	乳白色	+	−	+
柚皮素积累品种				
Kaly	纯白色	+	+	±
White Barbara	纯白色	+	+	±
缺黄酮品种				
White Mind	纯白色	+	−	−

a. +，有；−，无；±，数量很少

15.4　芽　　变

芽变是一种体细胞突变，它导致枝或个体发生突变，是培育新品种的重要手段，特别是对于无性繁殖植物（如观赏植物和果树）而言（Momose et al. 2013）。

芽变是由分生组织细胞在发育过程中的某一阶段突变引起的。由于香石竹的自然突变率很高（Holley and Baker 1963），芽变被广泛应用于培育香石竹新品种。例如，起源于‘William Sim’的Sim系品种群，就包括300多种具不同花瓣颜色的芽变品种。

香石竹品种具有高度杂合性，近交会引起衰退（Sato et al. 2000），因此，大多数具有商业意义的品种都是筛选出单株通过无性繁殖保持下来的无性系（Nimura et al. 2006a）。在这样的植物中，带有周缘嵌合体的分枝自发地产生芽突变体，这些突变体具有与原始植物不同的表型，并且经常被选择来培育新的品种（Yoshida et al. 2004）。花色的变异或杂色很容易被认为是一种新的表型，因此这种变异很容易被发现。花色变异之所以具有吸引力，是因为它们不会影响生长习性或植物形状等特性，从而使种植者能够有效地培育出保留亲本期望特性的新品种（Momose et al. 2013）。

转座基因元件在细胞分裂过程中如果移动到临界位置，就可以诱发芽突变。例如，黄色或橙色杂色花是查耳酮2′-葡萄糖苷合成和积累的结果，其产生需要转座因子*dTdic1*破坏查耳酮异构酶活性（Itoh et al. 2002，Okamura et al. 2006）。Momose等（2013）鉴定了编码香石竹F3′H的基因，并提供了证据表明将活性*hAT*型转座因子（*Tdic101*）插入*F3′H*会破坏该基因的功能，从而导致颜色从紫色变为深粉红色。

诱导突变已被用于扩大可用遗传资源的范围。包括化学处理以及伽马射线和X射线照射在内的几种技术，已被用于诱导各种农作物突变。日本在20世纪90年代开发的使用离子束辐照进行诱变已被证明可有效诱导对香石竹改良有用的突变。离子束辐照可以诱导多种颜色和形状的突变体（Okamura et al. 2003，2006）。例如，多头香石竹‘Vital’具有高产、瓶插寿命长以及抗镰刀菌枯萎病等特性。Okamura等（2003）通过离子束诱变‘Vital’，能够在2年内产生一系列颜色和形状突变体品种群，这些品种保留了亲本品种的优良特性。此外，Okamura等（2013）将杂交与离子束辐照结合，创造出一种由花青素液泡内含物引起的具有金属光泽的特殊颜色的香石竹。

15.5 重 瓣 花

香石竹的重要育种目标是产生单瓣花或重瓣花。单瓣花有5瓣，很容易区分；相反，重瓣花是由同源突变形成的。5个花瓣的单瓣香石竹品种在商业上已不再重要，而重瓣香石竹主导了大部分市场。100多年前，Saunders（1917）提出香石竹花表型是一种单基因性状，并将涉及的位点命名为‘*D*’。他假设隐性等位基因的纯合子（*dd*）产生单瓣花，而杂合子（*Dd*）产生重瓣花，而显性等位基因的纯合子（DD）产生带有花萼的超级重瓣花（Saunders 1917；Holley and Baker 1963）。Scovel等（1998）开发了显性随机扩增多态性DNA（RAPD）标记和与单瓣等位基因*d*连锁的限制性片段长度多态性标记，但没有发现*D*等位基因的标记。

我们选择了与控制瓣型的基因相关的RAPD标记，这些基因来自由127个单株构成的分离群体，该群体来自重瓣的香石竹‘Nou No.1’和单瓣的‘Pretty Favvare’杂交（Onozaki et al. 2006c）。我们通过使用696个引物进行大规模分离子分析确定了4个RAPD标记，这些标记与一个隐性基因相关联，该隐性基因控制单瓣花型，该基因来自一个石竹属的野生种（*Dianthus capitatus* ssp. *andrzejowskianus*）。我们成功地将与控制单瓣花型基因紧密连锁的 RAPD 标记 AT90-1000 转换为序列标记位点（STS）标记。连锁分析表明，该单一基因位于香石竹遗传连锁图中的第 16 组（Yagi et al. 2006）。但是，在 4 个单瓣香石竹品种中未检测到标记。该结果表明，该 STS 标记对源自 *D. capitatus* ssp. *andrzejowskianus* 的单个基因具有高度特异性。

Yagi 等（2014b）使用基于 SSR 标记的遗传连锁图绘制 *D85* 基因座，该 SSR 标记是由重瓣花第四代选择系 85-11 和单瓣花 'Pretty Favvare' 之间杂交产生的 F_2 后代的 91 个单株构建的（见第 15.7 节和图 15.6c），并鉴定了 4 个共分离 SSR 标记。以 SSR 标记为锚定位点的定位比较表明，*D85* 基因座的图谱位置对应于控制源自 *D. capitatus* ssp. *andrzejowskianus* 的单瓣花类型的单基因座。

Yagi 等（2014b）使用两个简单的重复序列（SSR）标记（CES1982 和 CES0212）估计了控制花型的 *D85* 基因座上每个品种的基因型，这两个标记与显性等位基因 *D* 紧密相连，确信后者负责重瓣花的表型。13 个多头香石竹，包括了纯合（*DD*）和杂合（*Dd*）的品种。有趣的是，在 *D85* 位点，11 个二倍体的单头品种和地中海系的 1 个品系均为纯合（*DD*），而 5 个 Sim 系品种均为杂合（*Dd*）。如前所述，Saunders（1917）提出 *DD* 基因型会产生“超重瓣”表型。但是，Yagi 等的研究表明，在 Sim 系之后才培育出来的地中海系品种，尽管具有 *DD* 基因型，但仍是正常的重瓣花类型。这些结果表明地中海系的 *DD* 品种具有抑制花器官异常发育并在这些遗传背景下产生正常花的机制。为了检验该假设并确定抑制异常花发育是如何进行的，将需要进一步的研究。

15.6 抗 病 性

香石竹的两种主要疾病严重影响了香石竹产业的经济：镰刀菌枯萎病[*F. oxysporum* Schlechtend: Fr. f. sp. *dianthi* (Prill. & Delacr.) W. C. Snyder & H. N. Hansen]，该病在世界范围内流行，而细菌性枯萎病[*B. caryophylli* (Burkholder) Yabuuchi, Kasako, Oyaizu, Yano, Hotta, Hashimoto, Ezaki and Arakawa；早期认为是 *Pseudomonas caryophylli* (Burkholder) Starr and Burkholder]，则主要分布于日本。对于香石竹种植者来说，这两种病害都是一个严重的问题。为了保护自己的花朵，在 20 世纪 70 年代，种植者开始使用无病插条、离地种植床，以及用蒸汽或化学药剂对土壤进行消毒，从而使两种疾病的发生率均下降了。然而，尽管采取了这些预防措施，仍然经常大面积暴发，导致严重的症状，这是因为较老的 Sim 系品种对这两种病害都非常敏感。

15.6.1 镰刀菌枯萎病（*F. oxysporum* f. sp. *dianthi*）

镰刀菌枯萎病是世界范围内香石竹最严重、最广泛和最具破坏力的病原体（Garibaldi and Gullino 2012）。受感染的植物容易枯萎，其下部叶片变黄变干，木质部组织变成褐色，在最严重的情况下，植物死亡。诸如 Sim 系这类易感品种，可能会严重减少。因此，对这种真菌性病害的抗性是香石竹育种中最重要的选择标准之一。多头香石竹和地中海系单头香石竹，现都已有抗镰刀菌枯萎病的品种。然而，由于对该病原体的抗性具有多基因和数量特性（Baayen et al. 1991），大多数商业品种只表现出部分抗性。

Garibaldi 和 Gullino（2012）报道，香石竹镰刀菌枯萎病的不同生理小种在 1975 年被首次描述。从那时起，生理小种变异得到广泛研究。到 20 世纪 80 年代末，根据枯萎病菌对不同品种的毒力差异，鉴定出 8 个枯萎病生理小种。第 2 生理小种的地理分布最广，种植香石竹的每个国家都有报道（Garibaldi and Gullino 2012）。在 20 世纪 90 年代，又报道了 3 个新的生理小种（9、10 和 11）（Baayen et al. 1997）。第 9 生理小种仅限于澳大利亚。在日本，Mizuno（1993）用 18 个分离物研究了来自国内香石竹种植区的镰刀菌分离物的生理小种分化；根据 4 个不同品种（‘Lena’、‘Pallas’、‘Elsy’和‘Novada’）的反应，证明所有分离物都属于第 2 生理小种。

在欧洲初步进行的抗性研究表明，‘Novada’、‘Lanathena’、‘Pallas’、‘Clair Avril’，‘Amapola’、‘Coralie’、‘Capello’、‘Diano’、‘Duca’、‘Edda’、‘Ficco’、‘Marlene’、‘Madrigal’、‘Ophelia’、‘Nero 90’、‘Monica’、‘Roland’、‘Sirio’和‘Tip-Top’对几种致病型都有抵抗力（Garibaldi and Gullino 1987）。Demmink 等（1989）揭示了尖孢镰刀菌不同生理小种的存在，并报道了 9 个多头品种中有两个品种（‘Novada’和‘Revada’）对生理小种 1、2 和 4 有抗性。在整个实验过程中，只有‘Novada’没有显症。石竹专化型尖孢镰刀菌在‘Novada’木质部的传播，受导管堵塞的限制，而在易感品种‘Early Sam’中，则没有发生这种限制（Ouellette et al. 1999）。

15.6.2 细菌枯萎病（*Burkholderia caryophylli*）

在日本，细菌性枯萎病是香石竹最重要和最具破坏性的疾病之一。它在日本温暖的地区造成严重的农作物损失，尤其是在夏季（Yamaguchi 1994）。细菌性枯萎病首先于 1941 年在美国华盛顿的斯波坎被报道（Jones 1941），并于 1964 年在日本神奈川县首次被记录下来（Tsuchiya et al. 1965）。美国的一项初步研究报道了 5 个老香石竹品种之间的易感性差异，发现‘杜兰戈’是免疫的（Thomas 1954）。随后，Nelson 和 Dickey（1963）对 21 个品种进行了测试，发现除‘Elegance’、‘Starlite’和‘Northland’外的所有品种都高度敏感。由于不到 50%的植物被感染，这三个品种具有一定的抗性。但是，‘Durango’、‘Elegance’、‘Starlite’和‘Northland’不再可用。Uematsu 等（1991）测试了日本的 126 个香石竹商业品种的抗性，发现它们都易感。

这种病菌一旦侵入作物，就很难控制。培育抗病品种似乎是克服该病的最佳策略。因此，日本 NARO 蔬菜与花卉科学研究所（NIVFS）于 1988 年启动了一项育种计划，以提高香石竹对细菌性枯萎病的抗性。

15.6.2.1 细菌性枯萎病的抗性育种

在我们的研究中，我们使用切根浸泡法筛选了 277 个栽培品种和 70 个石竹属野生材料对细菌性枯萎病的抗性，几乎所有品种都易感（Onozaki et al. 1999a）。但是，我们在 70 个野生材料中，鉴定出了一个高抗的野生种：*D. capitatus* Balbis ex DC. ssp.

andrzejowskianus Zapal.（图 15.2a）（Onozaki et al. 1999b）。我们通过种间杂交将这个种的抗性引入栽培的香石竹中（Onozaki et al. 1998），并开发了新的抗性品系‘Carnation Nou No. 1’（图 15.2b）（Onozaki et al. 2002）。‘Carnation Nou No. 1’是从与多头香石竹品种‘Super Gold’杂交的 F_1 后代中选出的，并于 1996 年在日本农林水产省注册（Onozaki et al. 2002）。尽管亲本品种 *D. capitatus* ssp. *andrzejowskianus* 仅在春末开花，但该系有持续开花的习性。另外，该品系的产量（每株切花数量）非常高。‘Carnation Nou No. 1’还能产生具有抗性的后代，表明其适合作为育种材料，将其抗性传递给其他栽培品种（Onozaki et al. 2002）。但是，较其他商业品种，它有一些缺点，如花径小、半重瓣（平均花瓣数为 18 枚）、切花品质较低。因此，在商业化生产之前，有必要通过回交做进一步的改良。

图 15.2　细菌性枯萎病的高抗野生种：*Dianthus capitatus* ssp. *andrzejowskianus*（a）和抗性品系‘Carnation Nou No. 1’（b）

15.6.2.2　与细菌性枯萎病抗性基因连锁的 STS 标记

我们使用切根浸泡方法筛选对细菌性枯萎病有抗性的后代（Onozaki et al. 1998，1999a，1999b）。然而，使用这种接种鉴定法确定育种材料的抗性需要 3 个多月的时间，并且这种鉴定方法需要大量的人力来筛选这些数量庞大的材料。DNA 标记为筛选抗性提供了强有力的替代工具，因而可以加快育种进程。此外，在幼小的种子实生苗阶段，开展分子标记辅助选择（MAS），可以大大减少所需的劳动力。

因此，我们用抗病的‘Carnation Nou No. 1’（参见 15.6.2.1）与感病的‘Pretty Favvare’（Onozaki et al. 2003）杂交产生的 134 个单株后代组成分离群体，开发细菌性枯萎病抗性基因的相关 RAPD 标记，这些杂交后代的抗性程度不同（参见 15.6.2.1 节）。平均发病率显示出连续分布，其值为 5.1%～100%（图 15.3）。整个群体的总体平均发病率为 60.4%±25.3%。‘Carnation Nou No. 1’的平均发病率是 14.0%，‘Pretty Favvare’的平均发病率是 100%（图 15.3）。我们共筛选了 505 个引物，获得了有助于筛选抗枯萎病品系的 RAPD 标记。通过大量的分离分析，我们鉴定了 8 个与主要抗性基因相关的标记，其中 WG44-1050 与抗性的相关性最大（Onozaki et al. 2004b）。通过数量性状基因座（QTL）分析，在 WG44-1050 附近定位了一个对抗性影响较大的基因座。RAPD 标记 WG44-1050 成功地转化为适合用于分子标记辅助选择的 STS 标记（STS-WG44）（Onozaki et al.

2004b)(图 15.4)。使用高抗的野生种(*D. capitatus* ssp. *andrzejowskiowski*)和抗性品系(‘Carnation Nou No. 1’)作为育种材料,该 STS 标记在抗细菌性枯萎病育种中是有用而可靠的选择标记。因此,在日本,STS-WG44 目前正被用作提高细菌性枯萎病抗性的实用育种方案中的选择标记。

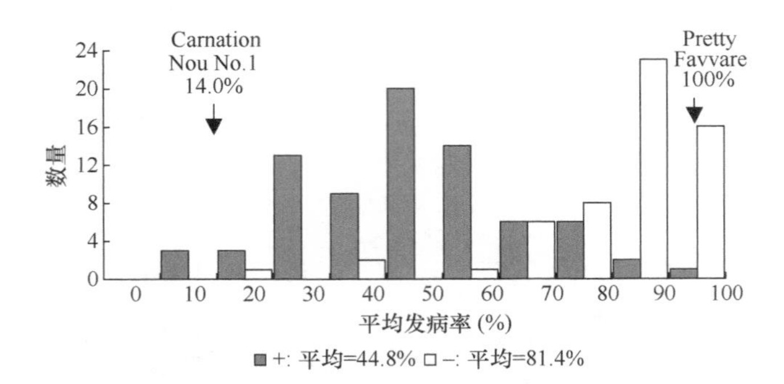

图 15.3 采用 RAPD 分析 WG44-1050 标记存在或不存在的香石竹品系平均发病率的频率分布。实心条,存在 WG44-1050;白色条,WG44-1050 缺失。箭头表示父母的平均值

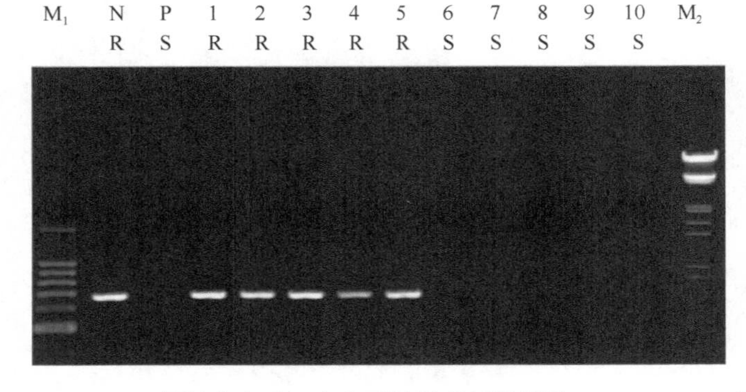

图 15.4 STS-WG-44 标记来自于一个主要抗性基因连锁的 WG44-1050。用 44-F80(5′-GGATTTTGTCACACCCTCCT-3′)和 44-R758(5′-TGCACAGAAACTTTTGAC-3′)两个引物进行 PCR 扩增。M_1,分子标记(100bp 的 DNA 梯段);M_2,分子标记(λ/*Eco*RⅠ+*Hin*dⅡ);N,‘Carnation Nou No. 1’(抗性亲本);P,‘Pretty Favvare’(易感亲本);1~10,分离个体;R,抗性;S,易感

15.6.2.3 细菌性枯萎病抗病品种‘Karen Rouge’的选育

2010 年,我们同时使用了抗性筛选和分子标记辅助选择(MAS)方法,开发了世界上第一个对细菌性枯萎病有高度抗性的实用香石竹品种‘Karen Rouge’(Yagi et al. 2010)。为开发出具有来自 *D. capitatus* ssp. *andrzejowskianus* 的抗性的实用品种,我们使用重复杂交和抗性筛选方法。

‘Karen Rouge’是 2004 年以来利用与一个主要抗性 QTL 连锁的 STS-WG44 标记,靠 MAS 技术开发出来的。在 6 次抗性测试中,‘Karen Rouge’的平均发病率为 7.1%,明显低于 87%的‘Francesco’和 97.1%的‘Nora’。‘Karen Rouge’是一种红色单头类型品种,花径为 7.5cm(图 15.5)。该品种是用我们开发出来具有较长瓶插寿命的‘Miracle Rouge’(参见第 15.7 节)和我们的抗性系 4AZ31-5 杂交的后代。每株切花总数为 4.5 支,

低于‘Francesco’的 5.0 支，但高于‘Nora’的 3.1。因此，我们长达 20 年的育种研究结果表明，分子标记辅助选择（MAS）是香石竹育种的有力工具。

图 15.5 ‘Karen Rouge’，抗细菌性枯萎病香石竹品种

15.7 瓶插寿命

切花的瓶插寿命，或花的寿命，是决定一个品种的品质和满足消费者喜好能力的最重要特征之一，由此可以刺激持续消费（Onozaki 2008）。因此，控制花瓣衰老和花衰老的能力是育种家非常感兴趣的（Zuker et al. 1998）。由于瓶插寿命是一个高度复杂的性状，在 20 世纪 90 年代以前，人们很少试图提高花卉的寿命。然而，现在，切花的寿命已经成为一个重要的品质因素，因为短命的花对消费者的吸引力有限，因此市场也有限。

香石竹的花对乙烯高度敏感（Woltering and van Doorn 1988）。完全开放的香石竹花朵暴露于乙烯中会诱导自身催化乙烯的产生和花瓣的枯萎（Halevy and Mayak 1981）。因此，乙烯是决定切花瓶插寿命的重要因素。

乙烯是一种重要的植物激素，参与植物生长发育的许多方面，包括花的衰老和脱落（Klee and Clark 2004）。香石竹花的衰老通常表现为乙烯产量的高峰和随后的下降（Mayak and Tirosh 1993）。乙烯产量的增加导致了花瓣的卷曲和随后的萎蔫（Halevy and Mayak 1981）。花瓣衰老是由雌蕊在自然衰老过程中产生的乙烯引发的（Satoh 2005，2011；Shibuya et al. 2000）。

尽管普通香石竹的瓶插寿命为 5～7 天，但通过采后化学处理，如硫代硫酸银，可显著延长其瓶插寿命（Veen 1979）。种植者通常会使用硫代硫酸银作为切花采后处理液。它的持久性和流动性允许非常短时的（脉冲）处理。但是，人们已经开始关注废弃溶液对环境的污染（Klee and Clark 2004），并且在一些国家禁止使用这种化学物质（Serek et al. 2007）。银离子很难从环境中去除，如果处理不当，高浓度的银离子会导致健康问题（Chandler 2007）。由于化学品的使用对环境和公众健康都造成了威胁，硫代硫酸银和许多其他目前使用的乙烯抑制剂可能很快就会被禁止（Scariot et al. 2014）。此外，这些化学品的购买和使用都很昂贵。因此，寻找延长瓶插寿命的替代方法是必要的。

15.7.1 使用杂交技术改善香石竹瓶插寿命

从基因上改善瓶插寿命是可取的，因为改进的品种将不需要化学处理。因此，NIVFS于 1992 年开始了一项研究育种计划，通过常规育种技术改善香石竹花的瓶插寿命。为了提高瓶插寿命，我们采用常规杂交技术，在 1992～2008 年，连续 7 代多次杂交选育出瓶插寿命长的优良后代（Onozaki et al. 2001，2006b，2011b）。我们选择了瓶插寿命差异较大的 6 个商业单头品种作为我们的初始育种材料（图 15.6a）：4 个地中海系品种（'Pallas'、'Sandrosa'、'Candy'和'Tanga'）与 2 个 Sim 系品种（'White Sim'和'Scania'）（Onozaki et al. 2001）。在 1992 年春季开始第一批杂交。

在育种过程中，我们选择了每一代瓶插寿命长的品系作为下一代的亲本（Onozaki et al. 2001，2006b，2011b）。没有选择乙烯产量高或敏感的植株。在下文中，我们用'第一代'替换了'父母一代'（这些先前研究中使用的术语），并相应地对后代重新编号。

杂交和选择使每一代的瓶插寿命显著改善，特别是许多选择系的乙烯产生显著减少（Onozaki et al. 2001，2006b，2011b）。各代瓶插寿命均呈连续正态分布（图 15.6b）。平均瓶插寿命从第一代的 7.4 天增加到第七代的 15.9 天，净增 8.5 天（图 15.6b）。杂交和选择的效果并没有保持不变，即改进程度因代而异。最显著的遗传改良效应发生在第四代和第五代之间，增加了 4.2 天。因此，通过使用常规杂交技术选育出了许多瓶插寿命长的香石竹品系。

15.7.2 长寿品种'Miracle Rouge'和'Miracle Symphony'的培育

2005 年，我们培育了两个品种'Miracle Rouge'和'Miracle Symphony'，在 23℃、70%相对湿度和 12h 光周期的标准条件下，通过基因测定，瓶插寿命为 17.7～20.7 天（是'White Sim'的 3.2～3.6 倍）（Onozaki et al. 2006a）。'Miracle Rouge'是从第四代中选出的红色单头品种（图 15.6c），'Miracle Symphony'是从第三代中选择出的白色底上有红色条纹的单头品种（图 15.6c）。两者都显示出较高的切花品质和商业生产所需的切花产量水平，此外，还有很长的瓶插寿命。

它们较长的瓶插寿命与雌蕊和花瓣中 1-氨基环丙烷-1-羧酸合酶及相应氧化酶（DC-ACS1、DC-ACS2 和 DC-ACO1）基因的低表达有关（Tanase et al. 2008），导致乙烯产量极低（Onozaki et al. 2006a）。这些结果表明，这些品种的乙烯生物合成途径在自然衰老过程中几乎完全被阻断，这一阻断是提高其瓶插寿命的原因。然而，它们的乙烯敏感性一般与正常的乙烯敏感 Sim 系品种相当（Onozaki et al. 2006a）。

15.7.3 香石竹花开花后乙烯敏感性的视频评估

我们通过使用延时录像系统（Onozaki et al. 2004a），开发了一种简单、准确地

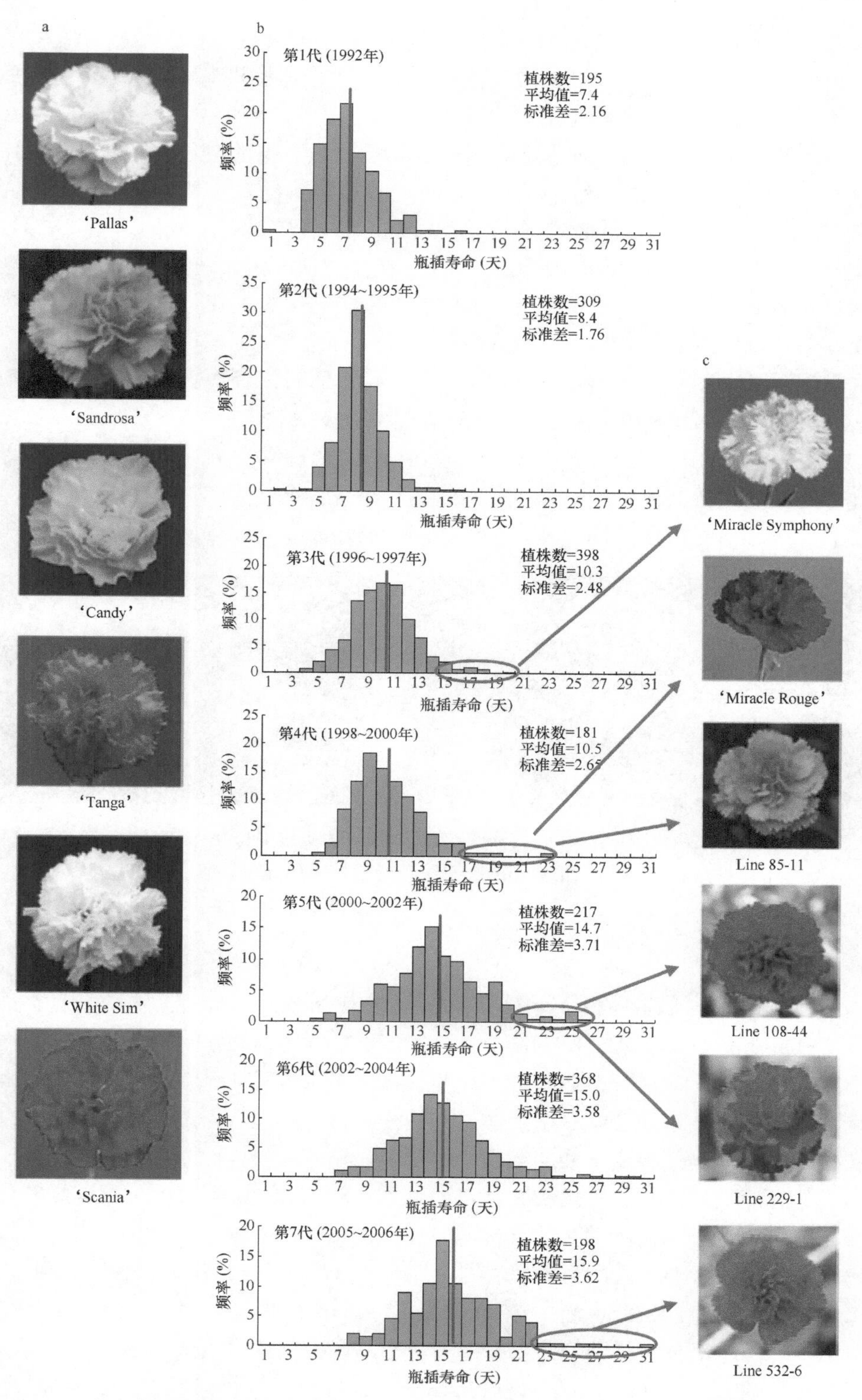

图 15.6 (a) 6 个亲本品种用作初始育种材料。(b) 第一代至第七代的花朵瓶插寿命的频率分布。竖线表示平均值。(c) 'Miracle Symphony'、'Miracle Rouge' 以及品系 85-11、108-44、229-1 和 532-6 的选育

评估香石竹花朵乙烯敏感性的方法，并得出结论，10μL/L 是使用新系统进行敏感性评估的最佳浓度。利用这一系统，育种家可以通过视频“快速推进”，直到枯萎症状变得明显，这比必须频繁地观察大量花朵更快、更有效。

我们还利用延时录像系统研究了花开后花朵乙烯敏感性的变化。视频记录系统可以检测到对乙烯反应的变化，这是以前的方法无法检测到的。在测试的 Sim 系品种中（‘White Sim’、‘Scania’、‘U Conn Sim’和‘Nora’），无论是早采收的切花（在第 0 天，外轮花瓣呈水平状），还是晚采收的切花（在开花后 3 或 6 天），开花后乙烯敏感性均随年龄增加而显著降低。这些结果清楚地表明，乙烯敏感性的下降与花的生理年龄增高有关（Onozaki et al. 2004a）。

借助延时录像系统（Onozaki et al. 2004a），我们证明传统杂交技术可以降低乙烯敏感性（Onozaki et al. 2008），并开发了 13 个抗乙烯选择系（图 15.7）（Onozaki et al. 2008，2011b）。但是，这些选择系的平均瓶插寿命在标准条件下为 7.6～15.2 天，不及两个对乙烯敏感的品种以及其他一些乙烯产量极低的品系（如编号为 532-6 的选择系），尽管这两个品种的瓶插寿命长（‘Miracle Rouge’和‘Miracle Symphony’）。

图 15.7　乙烯处理开始后（a）0h、（b）24h 和（c）48h，（左）004-17（高度敏感）、（中）234-52S（抗性）和（右）‘White Sim’（对照）对乙烯（10μL/L）的敏感性变化

15.7.4　超长瓶插寿命育种系的选育

在我们测试的 4 个品种和 18 个品系中，第七代品系 532-6（图 15.6c）具有最长的瓶插寿命（Onozaki et al. 2011b），2007 年为 32.7 天，2008 年为 27.8 天，分别相当于‘White Sim’的 5.4 倍和 4.6 倍（图 15.8a）。我们把这种罕见的超长寿命定义为“超长瓶插寿命”。其他选择系的花瓣从边缘开始变褐，然后慢慢变干褪色。然而，532-6 品系的则没有褐变（图 15.8b），而是慢慢变干和起皱。

为了选择瓶插寿命长、乙烯敏感性低的品系，我们在 2003～2010 年连续三代进行了多次杂交选育。2010 年，我们从 606-65S 和 609-63S 杂交产生的 50 个后代中，最终开发出了 806-46b 这个品系，兼有超长的瓶插寿命（图 15.9a）和抗乙烯性（图 15.9b）（Onozaki et al. 2015）。在标准条件下，806-46b 品系的平均瓶插寿命为 27.1 天（为‘White Sim’的 4.4 倍），是我们测试的 6 个品种和 7 个品系中瓶插寿命最长的。

532-6 和 806-46b 两个品系具有超长的瓶插寿命，但对外源乙烯的敏感性不同。806-46b 对 10μL/L 乙烯的反应时间为 21.8h，而‘White Sim’的反应时间为 5.8h。尽管

图 15.8　瓶插寿命超长的第七代品系 532-6 与两个对照品种‘Miracle Rouge’和‘White Sim’的瓶插寿命差异。(a) 和 (b) 分别在收获后 25 和 26 天拍摄。(a) 左，‘Miracle Rouge’；中，532-6 品系；右，‘White Sim’。(b) 左，‘Miracle Rouge’；右，532-6 品系

图 15.9　(a) 超长瓶插寿命的 806-46b 品系和 3 个对照品种（收获后 35 天）瓶插寿命的差异。切花保存在蒸馏水中，温度 23℃，相对湿度 70%，光周期 12h。从左到右为‘Miracle Rouge’、‘Miracle Rouge’、‘806-46b’品系和‘Karen Rouge’。(b) 806-46b 品系和‘Karen Rouge’之间的乙烯敏感性差异（10μL/L 乙烯处理后 12h）。左，乙烯敏感品种，‘Karen Rouge’；右，耐乙烯品系，806-46b

532-6 具有超长的瓶插寿命和衰老时产生的乙烯水平低，但它对外源乙烯的施用是敏感的（Onozaki et al. 2011b）。相比之下，806-46b 品系显示乙烯敏感性较低（图 15.9b）。这些结果表明，利用常规杂交技术，可以选育出瓶插寿命极长和乙烯抗性极强的品系。

806-46b 品系在衰老过程中花瓣边缘没有出现棕色变色（图 15.9a），这是其他低乙烯生产品系的典型特征。相反，由于水分流失，它的花瓣逐渐失去了表面的膨润度。这种罕见的衰老模式与 532-6 品系的相同（图 15.8b）。

806-46b 品系将有助于今后乙烯合成相关基因和外源乙烯反应的研究。

15.7.5　育种系乙烯产量与瓶插寿命的关系

本研究以作为育种初始材料的 6 个品种和从第一代到第七代选出的 123 个选择系为材料，对其瓶插寿命、自然衰老时乙烯产生量、乙烯处理后乙烯产生量（乙烯自催化生物合成）、乙烯处理反应时间（乙烯敏感性）和花径进行了研究（Onozaki et al. 2018）。我们发现这 5 个性状有很大的遗传变异性。我们观察到，瓶插寿命与自然衰老时乙烯产生量（$r = -0.88$，$p < 0.01$）（图 15.10）之间、瓶插寿命与乙烯处理后乙烯产生量（$r = -0.90$，

$p<0.01$）之间，以及瓶插寿命与花径（$r=-0.92$，$p<0.01$）之间，都存在显著的负相关。然而，花朵瓶插寿命与乙烯敏感性无显著相关性。因此，所选香石竹品系的瓶插寿命长与其乙烯产量的减少密切相关（Onozaki et al. 2018）。我们的结果表明，瓶插寿命的变化不是因乙烯敏感性的差异所致，而是因乙烯产生量的差异而导致的结果。花的大小是商品化生产中一个重要的栽培性状，由于杂交和针对切花瓶插寿命的选择，花的直径大大减小。这个问题必须在以后的研究中解决。

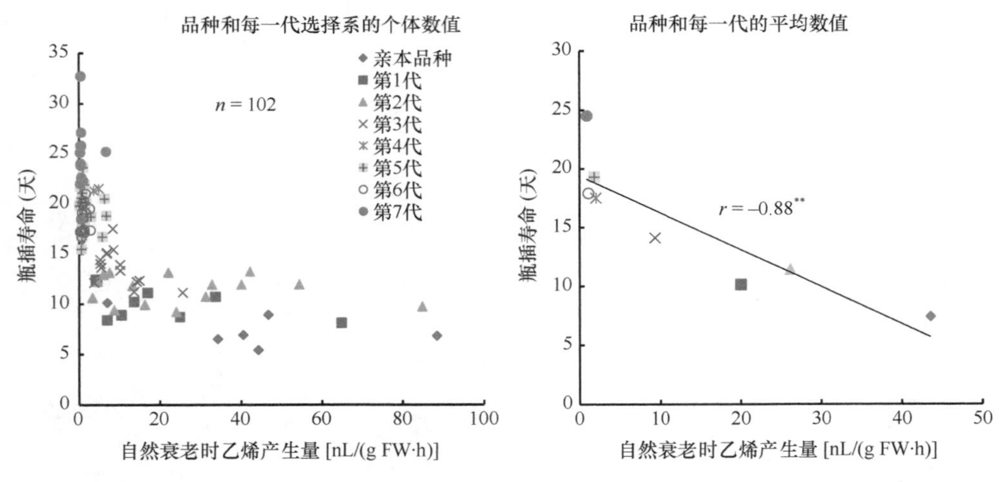

图 15.10　自然衰老时乙烯产量与瓶插寿命之间的关系

**：显著水平为 1%

15.7.6　延长瓶插寿命的多头品种‘Kane Ainou 1-go’的选育

为了培育多头类型品种，我们从 2006 年起与爱知县农业研究中心进行了联合研究。2015 年，我们利用第五代选择系 108-44（图 15.6c）作亲本（Hotta et al. 2016），开发了‘Kane-Ainou 1-go’，这是一种具有较长瓶插寿命的新型多头品种。108-44 是单头品种，具有长的瓶插寿命和低乙烯生产能力（Onozaki et al. 2006b）。‘Kane Ainou 1-go’的花在衰老过程中乙烯含量较低。在 25℃、60%相对湿度以及 12h 光周期下，瓶插寿命平均为 19.2～21.3 天（大约是多头香石竹‘Silhouette’的 3 倍）。该品种的年切花产量比对照品种高，开花早。这种植物有强壮的茎和大的淡粉色花（Hotta et al. 2016）。

15.8　种间杂交

种间杂交是在观赏植物中创造变异的最重要策略之一，因为它具有将双亲有用的性状（如良好的形态、抗病性和某些环境耐受性）组合在一起的潜力，而这是不能通过种内杂交获得的（Nimura et al. 2008a）。因此，要扩大遗传变异，种间杂交是最有用的策略之一（Nimura et al. 2008b）。石竹属植物包括许多重要的观赏植物，它们适应不同的环境条件，具有不同的生长习性和开花模式，如香石竹（*D. caryophyllus*）、石竹（*D. chinensis*）、常夏石竹（*D. plumarius*）和须苞石竹（*D. barbatus*）。该属的一个显著特征是，即使是在自然界和栽培中的不同种之间也可以相互杂交，而没有严重问题（Hamilton and Walters 1989；Sparnaaij and Koehorst-van Putten 1990）。历史上，种内杂交和种间杂交创造了许多香石竹与石竹属其他种的新品种（Fu et al. 2011）。该属的第一个杂交种早在 18 世纪初就已培育成功（Andersson-Kotto and Gairdner 1931）。

创制杂交种以引入新的理想特性，如早开花、抗细菌性枯萎病和耐热性（Sparnaaij and Koehorst-van Putten 1990；Onozaki 2006；Nimura et al. 2008a）。在欧洲，20 世纪 80 年代和 90 年代通过种间杂交培育了许多不同类型的品种（例如，'Micro'、'Diantini'、'Multiflora'、'Chinese' 和 'Gipsy'）（Sparnaaij and Demmink 1983）。在日本，通过种间杂交培育了用于切花生产的品种，如 'Angel' 型、'Sonnet' 型和 'Raffine' 型的品种（Onozaki 2016）。

几位日本研究者报道了种间杂交，如香石竹与 *D. deltoides*、*D. japonicus*、*D. knappii* 和 *D. superbus* 的杂交（Kanda 1992；Kanda et al. 1998）；涉及石竹 22 个种的杂交（Ohtsuka et al. 1995）；以及香石竹和 *D. capitatus* ssp. *andrzejowskianus*（Onozaki et al. 1998）、*D. japonicus*（Nimura et al. 2003）与 *D. xisensis*（Nimura et al. 2006a）的杂交种的培育。此外，通过原生质体融合，还对中国石竹和须苞石竹（Nakano and Mii 1993a），以及香石竹和中国石竹（Nakano and Mii 1993b）进行了体细胞杂交。

15.8.1　日本石竹属野生种的调查和收集

NIVFS 于 1989 年和 2000 年调查并收集了日本本土的野生石竹种（Yamaguchi et al. 1990；Onozaki 2001）。在日本，已经发现了 4 个野生种（*D. japonicus*、*D. kiusianus*、*D. shinanensis* 和 *D. superbus*）与两个变种（*D. superbus* var. *longicalycinus* 和 *D. superbus* var. *speciosus*）（Itoh et al. 1989）。1989 年，NIVFS 从宫崎县（Miyazaki）和长崎县（Nagasaki Prefectures）的海岸采集了 5 份 *D. japonicus* 样本，从鹿儿岛县（Kagoshima Prefecture）的海岸和路边悬崖采集了 5 份 *D. kiusianus* 样本。2000 年，从三重县（Mie Prefecture）海岸和北海道石狩海岸（Ishikari subpre-fecture）采集了 12 份 *D. superbus* 的长萼变种样本，从北海道鄂霍次克海沿岸采集了 21 份 *D. superbus* 样本。对材料进行无性繁殖，并评估其良好特性。我们将在下一节讨论这些特性。

15.8.2　瓶插寿命长的香石竹品系与 *D. superbus* var. *longicalycinus* 种间杂种的特性

对瓶插寿命长的香石竹品系和 *D. superbus* var. *longicalycinus* 进行种间杂交，以结合开花早、高产和瓶插寿命长等优良性状（图 15.11）（Onozaki et al. 2011a）。尽管这种杂交非常困难，但 22 个杂交组合中有 4 个结实，但所有的种子都不正常，表面呈皱褶的棕色。由于这种异常，发芽率很低。这种异常的结果是发芽率低。然而，在不需要离体培养的情况下，获得了 15 株 F_1 代实生苗。筛选出来的 9 个 F_1 品系均表现为雌雄可育，既可作母本，又可作父本。SSR 分析证明所有 9 个品系均为种间杂种。所选 F_1 或 BC1 系与瓶插寿命长的香石竹品系回交，结实率显著提高（提高 56%～75%），种子发芽率也较高（发芽率约 70%）。

图 15.11 种间杂种 4K38-15 的花特征及其回交后代

利用 *D. superbus* var. *longicalycinus* 选育早花后代是有效的。F_1 代表现出很早的开花习性。香石竹品系 229-1 开花天数为 189.7 天，F_1 代开花天数为 64.7～84.3 天。在 BC1 和 BC2 代中，还观察到了开花早的实生苗分离现象。我们的研究结果表明，瓶插寿命与开花早无关。尽管如此，9 个筛选出的 BC2 系仍然具有很长的瓶插寿命、开花早和产量高的特点，即使使用瓶插寿命很短的 *D. superbus* var. *longicalycinus* 作为育种材料（Onozaki et al. 2011a）。

15.9 香　气

香味是观赏花卉重要而迷人的商业特性，也是切花最重要的消费特性之一。然而，几十年来，由于观赏植物育种更关注于其他特性，主要是花色、花形、高产和抗病性，因此切花香气研究减少（Zuker et al. 1998）。

香石竹香料由苯类化合物、萜类化合物、脂肪酸衍生物和其他次要成分组成（Hudak and Thompson 1997；Clery et al. 1999；Schade et al. 2001；Zuker et al. 2002）。经典的香石竹有一种辛辣的丁香味，主要由苯类化合物引起（Clery et al. 1999）。然而，在大多数现代香石竹中，香气的种类在减少，强度也在减弱，苯甲酸甲酯已成为主要成分（Clery et al. 1999）。因此，通过增加其他香气物质的总量以及引入（或重新引入）其他香气物质，可以提高现代香石竹的商业价值。

最近的两份报告确定了几种石竹属野生种的化学成分（Jürgens et al. 2003；Kishimoto et al. 2011）。有些种有浓烈的或特有的香味。Kishimoto 等（2011）从 NIVFS 野生石竹基因资源中筛选出 10 个香型和 1 个弱香型种质，并根据其感官特性将其分为 4 组：药用香味（1 组）、类柑橘香味（2 组）、带花香味的绿叶香味（3 组）、几乎没有香味（4 组）。气相色谱-质谱分析表明，野生种的香气中含有 18 种主要的挥发性化合物，它们是苯类、萜类或脂肪酸衍生物。*Dianthus superbus* 具有较高的萜类化合物（β-罗勒烯和 β-石竹烯）含量。我们的结果与 Jürgens 等（2003）的报告一致。

石竹属野生种具有强烈或独特的香味，可能是改善香石竹香气的有用遗传资源。采用气相色谱-质谱技术，对香石竹与有香气的野生种的种间杂种的香气进行了研究，并评价了野生种提高香石竹香气的育种价值。4K38 新增加的 8 个品系（4K38-2、4K38-3、4K38-5、4K38-6、4K38-10、4K38-11、4K38-14 和 4K38-15）是第五代品系 229-1（图 15.6c）与 *D. superbus* var. *longicalycinus*（Onozaki et al. 2011a）的种间杂种。后者具有大量的 β-龙脑烯和 β-石竹烯。这些萜类化合物作为主要气味化合物被后代遗传到一些种间杂种中，这些杂种是这些种和缺少萜类化合物的香石竹的后代。这些结果表明，*D. superbus* var. *longicalycinus* 是一种很有前途的资源，可用于向缺乏萜类化合物的种质中添加萜类化合物，并增加香石竹花挥发物中苯类化合物的种类（Kishimoto et al. 2013）。

15.10　多　倍　体

石竹属的染色体数遵循基本染色体组 $x = 15$ 的简单倍性序列。最初已知三种不同的倍性水平（Carolin 1957）：二倍体（$2n = 2x = 30$）、四倍体（$2n = 4x = 60$）和六倍体（$2n = 6x = 90$）。我们用流式细胞仪对 99 份保存在 NIVFS 中的石竹种质进行了倍性水平测定，发现 8 份二倍体、6 份四倍体和 85 份六倍体（Ushio et al. 2002）。最近的研究表明，原产于伊比利亚半岛（即葡萄牙和西班牙）的 *Dianthus broteri* 种群包含 4 种不同的倍性水平，细胞类型有 $2x$、$4x$、$6x$ 和 $12x$（Balao et al. 2009，2010）。

香石竹通常是二倍体（$2n = 2x = 30$）植物（Carolin 1957）。多倍体，特别是三倍体和四倍体，因其花大、叶厚、茎粗、生长旺盛等特点，在观赏育种中被广泛应用。在香石竹中，用秋水仙碱处理叶腋芽（Sparnaaij 1979）是比较容易获得四倍体的。这些四倍体比原来的二倍体花大，但生长慢、茎短、开花晚、产量低。

我们使用流式细胞仪检测了 304 个在 NIVFS 保存的香石竹品种的多倍性，这些品种是为切花生产而培育的（Yagi et al. 2007）。我们发现 297 个品种是二倍体（$2n = 2x = 30$），3 个品种（‘Wiko’、‘Scarlet Bell’和‘piral Vivid Red’）是三倍体（$2n = 3x = 45$），4 个品种（‘salya’、‘Pink Roland’、‘Youkihi’和‘Sonnet saille’）是四倍体（$2n = 4x = 60$）。四倍体植株开花晚，降低了切花产量，一般不适合实际栽培。这或许可以解释为什么在香石竹中很少见商业四倍体品种。

最近，Agulló-Antón 等（2013）报道，两个单头香石竹品种‘Reina’和‘Roble’的 DNA 含量是他们研究的‘Master’等其他品种的两倍，表明这两个品种是四倍体。

此外，他们还报道了栽培香石竹的倍性、细胞大小和花瓣大小之间呈正相关。Zhou 等（2013）报道了四倍体和二倍体香石竹品种间杂交的生殖障碍。使用香石竹品种‘蝴蝶’（$2n = 4x = 60$）与两个育种系 NH10（$2n = 2x = 30$）和 NH14（$2n = 2x = 30$）进行了两个杂交。结果表明，受精前和受精后的障碍降低了这些杂交的成功率。最近，他们成功地在 $4x$ 和 $2x$ 香石竹材料之间进行了杂交，获得了 5 个三倍体杂交植株和 7 个四倍体杂交植株（Zhou et al. 2017）。

15.11 香石竹遗传连锁图谱及 QTL 分析

为了阐明构成重要农艺性状的遗传和生理机制，并将这些知识应用于实际育种，需要一些遗传和分子工具。遗传连锁图谱对于分析数量性状和质量性状都很重要，是为许多基因组分析提供框架的宝贵资源。我们通过使用 134 株具有不同细菌性枯萎病抗性的分离群体，基于 RAPD 和 SSR 标记构建了第一个香石竹遗传连锁图谱，分离群体来自抗性‘Carnation Nou No. 1’（图 15.2b）和感病‘Pretty Favvare’（Yagi et al. 2006）的杂交。利用 QTL 分析确定了抗病性位点（Yagi et al. 2006，2012a，2013）和花中花青素含量（Yagi et al. 2013）。此外，我们还开发了与细菌性枯萎病抗性（Onozaki et al. 2004b；Yagi et al. 2012a）和花型（Onozaki et al. 2006c；Yagi et al. 2014b）基因紧密相关的 DNA 标记。一个与细菌性枯萎病抗性相关的 STS 标记（STS-WG44）在日本正用于培育抗病商业品种。

Tanase 等（2012）在高通量测序仪（Roche 454 GS FLX）上进行了香石竹全转录组的分析（通过 RNA-Seq 方法），获得了 300 740 个单基因（37 844 个重叠群和 262 896 个单峰）。最近，Yagi 等（2017a）结合限制性位点相关 DNA（RAD）标记，构建了基于 SSR 标记的香石竹高密度连锁图谱。共有 2404 个（285 SSR 和 2119 RAD）标记被分为 15 个连锁群，跨距 971.5cm，平均标记间隔 0.4cm。该连锁图谱的构建使用了育种系 806-46b（见第 15.7.4 节），其具有超长的瓶插寿命。基于 SSR 和 RAD 标记的香石竹高密度连锁图谱为首次报道。

利用第四代品系 85-11（图 15.6c）和 806-46b（图 15.9）对瓶插寿命进行了 QTL 分析（Yagi et al. 2012b），这两个品系具有超长的瓶插寿命和抗乙烯性（Yagi et al. 2017b）。然而，与瓶插寿命紧密相关的 DNA 标记尚未开发出来。利用常规育种技术进行长瓶插寿命育种需要大量的时间和精力，因为需要对许多幼苗进行表型鉴定。通过将传统育种技术与瓶插寿命的分子遗传学知识相结合，我们将能够开发出更可靠的香石竹育种方法。

15.12 香石竹全基因组测序

目前可用的香石竹品种大多为二倍体，染色体数目为 $2n = 2x = 30$（Gatt et al. 1998；Yagi et al. 2007）。据报道，香石竹的核 DNA 含量在 1.23pg/2C 至 1.48pg/2C（Figueira et al. 1992；Nimura et al. 2003；Agulló-Antón et al. 2013），这表明香石竹的核基因组相对较

小，只为拟南芥（0.30pg/2C）的4倍（Arumuganathan and Earle 1991）。根据植物DNA C值数据库中的值（http://data.kew.org/cvalues/），香石竹的估计基因组大小（670 Mb；Agulló-Antón等，2013），与其他观赏花卉相比非常小，如月季（*Rosa hybrida*）（1.1Gb）、大头草（*Antirrhinum majus*）（1.5Gb）、矮牵牛（*Petunia hybrida*）（1.6Gb）、菊花（9.4Gb）、郁金香（26Gb）和长百合（34Gb）。因此，香石竹相对较小的二倍体基因组比其他染色体倍数水平较高或具有较大基因组大小的观赏植物更适合用于基因组分析。

NIVFS与Kazusa DNA研究所、东京农业科技大学和三得利全球创新中心进行了联合研究，在观赏植物中首次使用新一代多重测序平台组合，确定了地中海系红色香石竹主要品种'Francesco'的全基因组（Yagi et al. 2014a）。非冗余序列的总长度为569Mb，包含45 088个拼接序列，覆盖了622Mb香石竹基因组的91%（小于先前估计的670Mb）。香石竹基因组研究产生的信息和物质资源可以增进香石竹及相关植物的基础研究与应用研究。获得的有关基因组序列信息可在香石竹数据库网站上免费获取，该网站由Kazusa DNA研究所主办（http://carnation.kazusa.or.jp/）。

致谢：我衷心感谢Elsevier、Springer Nature和日本园艺科学学会（JSHS）对以下图片再版的授权。图15.1来源于Onozaki等（1999c）Scientia Horticulturae 82: 103-111。图15.3和图15.4来源于Onozaki等（2004b）Euphytica 138: 255-262。图15.2来源于Onozaki等（2002）Horticultural Research（Japan）1: 13-16。图15.8来源于Onozaki等（2011b）Journal of Japanese Society for Horticultural Science 80: 486-498。图15.9来源于Onozaki等（2015）The Horticulture Journal 84: 58-68。图15.10来源于Onozaki等（2018）The Horticulture Journal 87: 106-114。图15.11来源于Onozaki等（2011a）Horticultural Research（Japan）10: 161-172。

本章译者：
莫锡君
云南省农业科学院花卉研究所，国家观赏园艺工程技术研究中心，云南省花卉育种重点实验室，昆明 650200

参考文献

Abe Y, Tera M, Sasaki N, Okamura M, Umemoto N, Momose M, Kawahara N, Kamakura H, Goda Y, Nagasawa K, Ozeki Y (2008) Detection of 1-*O*-malylglucose: pelargonidin 3-*O*-glucose-6′′-*O*-malyltransferase activity in carnation (*Dianthus caryophyllus*). Biochem Biophys Res Commun 373:473–477

Agulló-Antón MA, Olmos E, Pérez-Pérez JM, Acosta M (2013) Evaluation of ploidy level and endoreduplication in carnation (*Dianthus* spp.). Plant Sci 201–202:1–11

Andersson-Kotto I, Gairdner AE (1931) Interspecific crosses in the genus *Dianthus*. Genetica 13:77–112

Arumuganathan K, Earle ED (1991) Nuclear DNA content of some important plant species. Plant Mol Biol Report 9:208–218

Baayen RP, Sparaaij LD, Jansen J, Niemann GJ (1991) Inheritance of resistance in carnation against *Fusarium oxysporum* f. sp. *dianthi* races 1 and 2, in relation to resistance components. Neth J Plant Pathol 97:73–86

Baayen RP, van Dreven F, Krijger MC, Waalwijk C (1997) Genetic diversity in *Fusarium oxysporum* f. sp. *dianthi* and *Fusarium redolens* f. sp. *dianthi*. Eur J Plant Pathol 103:395–408

Balao F, Casimiro-Soriguer R, Talavera M, Herrera J, Talavera S (2009) Distribution and diversity of cytotypes in *Dianthus broteri* as evidenced by genome size variations. Ann Bot 104:965–973

Balao F, Valente LM, Vargas P, Herrera J, Talavera S (2010) Radiative evolution of polyploid races of the Iberian carnation *Dianthus broteri* (Caryophyllaceae). New Phytol 187:542–551

Bloor SJ (1998) A macrocyclic anthocyanin from red/mauve carnation flowers. Phytochemistry 49:225–228

Britsch L, Dedio J, Saedler H, Forkmann G (1993) Molecular characterization of flavanone 3β-hydroxylases. Consensus sequence, comparison with related enzymes and the role of conserved histidine residues. Eur J Biochem 217:745–754

Carolin RC (1957) Cytological and hybridization studies in the genus *Dianthus*. New Phytol 56:81–97

Chandler S (2007) Practical lessons in the commercialization of genetically modified plants –long vase-life carnation. Acta Hortic 764:71–81

Clery RA, Owen NE, Chambers SF (1999) An investigation into the scent of carnations. J Essent Oil Res 11:355–359

Demmink JF, Baayen RP, Sparnaaij LD (1989) Evaluation of the virulence of races 1, 2 and 4 of *Fusarium oxysporum* f. sp. *dianthi* in carnation. Euphytica 42:55–63

Figueira A, Janick J, Goldsbrough P (1992) Genome size and DNA polymorphism in *Theobroma cacao*. J Am Soc Hortic Sci 117:673–677

Forkmann G, Dangelmayr B (1980) Genetic control of chalcone isomerase activity in flowers of *Dianthus caryophyllus*. Biochem Genet 18:519–527

Fu XP, Zhang JJ, Li F, Zhan PT, Bao MZ (2011) Effects of genotype and stigma development stage on seed set following intra- and inter-specific hybridization of *Dianthus* spp. Sci Hortic 128:490–498

Fukui Y, Tanaka Y, Kusumi T, Iwashita T, Nomoto K (2003) A rationale for the shift in colour towards blue in transgenic carnation flowers expressing the flavonoid 3′,5′-hydroxylase gene. Phytochemistry 63:15–23

Galbally J, Galbally E (1997) Carnation and pinks for garden and greenhouse. Timber Press, Portland, Oregon

Garibaldi A, Gullino ML (1987) *Fusarium* wilt of carnation: present situation, problems and perspectives. Acta Hortic 216:45–54

Garibaldi A, Gullino ML (2012) *Fusarium* wilt of carnation. In: Gullino ML, Katan J, Garibaldi A (eds) *Fusarium* wilts of greenhouse vegetable and ornamental crops. APS Press, St. Paul, MN, pp 191–198

Gatt MK, Hammett KRW, Markham KR, Murray BG (1998) Yellow pinks: interspecific hybridization between *Dianthus plumarius* and related species with yellow flowers. Sci Hortic 77:207–218

Geissman TA, Mehlquist GAL (1947) Inheritance in the carnation, *Dianthus caryophyllus*. IV. The chemistry of flower color variation, I. Genetics 32:410–433

Halevy AH, Mayak S (1981) Senescence and postharvest physiology of cut flowers, part 2. Hortic Rev 3:59–143

Halmagyi A, Lambardi M (2006) Cryopreservation of carnation (*Dianthus caryophyllus* L.). In: Teixeira da Silva JA (ed) Floriculture, ornamental and plant biotechnology. Advances and topical issues, vol 2. Global Science Books, Isleworth, pp 415–423

Hamilton RFL, Walters SM (1989) *Dianthus* Linnaeus. In: Walters SM, Alexander JCM, Brady A, Brickell CD, Cullen J, Green PS, Heywood VH, Matthews VA, Robson NKB, Yeo PF, Knees SG (eds) The European garden flora, vol 3. Cambridge University Press, Cambridge, pp 185–191

Holley WD, Baker R (1963) Carnation production. Wm.C. Brown Co-Inc, Dubuque, IA

Hotta M, Hattori H, Hirano T, Kume T, Okumura Y, Inubushi K, Inayoshi Y, Ninura M, Matsuno J, Onozaki T, Yagi M, Yamaguchi H, Yamaguchi N (2016) Breeding and characteristics of spray-type carnation "Kane Ainou 1 go" with long vase life. Res Bull Aichi Agric Res Ctr 48:63–71. (In Japanese with English abstract)

Hudak KA, Thompson JE (1997) Subcellular localization of secondary lipid metabolites including fragrance volatiles in carnation petals. Plant Physiol 114:705–713

Itoh A, Takeda Y, Tsukamoto Y, Tomino K (1989) *Dianthus* L. In: Tsukamoto Y (ed) The grand dictionary of horticulture, vol 3. Shogakukan, Tokyo, pp 455–462. (In Japanese)

Itoh Y, Higeta D, Suzuki A, Yoshida H, Ozeki Y (2002) Excision of transposable elements from the chalcone isomerase and dihydroflavonol 4-reductase genes may contribute to the variegation of the yellow-flowered carnation (*Dianthus caryophyllus*). Plant Cell Physiol 43:578–585

Iwashina T, Yamaguchi M, Nakayama M, Onozaki T, Yoshida H, Kawanobu S, Ono H, Okamura M (2010) Kaempferol glycosides in the flowers of carnation and their contribution to the creamy white flower color. Nat Prod Commun 5:1903–1906

Jones LK (1941) Bacterial wilt of carnations. Phytopathology 31:199

Jürgens A, Witt T, Gottsberger G (2003) Flower scent composition in *Dianthus* and *Saponaria* species (Caryophyllaceae) and its relevance for pollination biology and taxonomy. Biochem Syst Ecol 31:345–357

Kanda M (1992) Ovule culture for hybridization between carnation and the genus *Dianthus*. J Jpn Soc Hortic Sci 61(Suppl. 2):464–465. (In Japanese)

Kanda M, Horikawa T, Nakamura Y, Motoori S, Kotake H (1998) Breeding of carnation cultivars "Aqua Red" and "Aqua Yellow" through the interspecific hybridization between *Dianthus caryophyllus* and *D. superbus*. J Jpn Soc Hortic Sci 67(Suppl. 1):247

Kingman R (1983) The carnation industry in the United States. Acta Hortic 141:249–252

Kishimoto K, Nakayama M, Yagi M, Onozaki T, Oyama-Okubo N (2011) Evaluation of wild *Dianthus* species as genetic resources for fragrant carnation breeding based on their floral scent composition. J Jpn Soc Hortic Sci 80:175–181

Kishimoto K, Yagi M, Onozaki T, Yamaguchi H, Nakayama M, Oyama-Okubo N (2013) Analysis of scents emitted from flowers of interspecific hybrids between carnation and fragrant wild *Dianthus* species. J Jpn Soc Hortic Sci 82:145–153

Klee HJ, Clark DG (2004) Ethylene signal transduction in fruits and flowers. In: Davies PJ (ed) Plant hormones: biosynthesis, signal transduction, action, 3rd edn. Kluwer Academic Publishers, Dordrecht, pp 369–390

Mato M, Onozaki T, Ozeki Y, Higeta D, Itoh Y, Yoshimoto Y, Ikeda H, Yoshida H, Shibata M (2000) Flavonoid biosynthesis in white-flowered Sim carnations (*Dianthus caryophyllus*). Sci Hortic 84:333–347

Mato M, Onozaki T, Ozeki Y, Higeta D, Itoh Y, Hisamatsu T, Yoshida H, Shibata M (2001) Flavonoid biosynthesis in pink-flowered cultivars derived from "William Sim" carnation (*Dianthus caryophyllus*). J Jpn Soc Hortic Sci 70:315–319

Matsuba Y, Sasaki N, Tera M, Okamura M, Abe Y, Okamoto E, Nakamura H, Funabashi H, Takatsu M, Saito M, Matsuoka H, Nagasawa K, Ozeki Y (2010) A novel glucosylation reaction on anthocyanins catalyzed by acyl-glucose-dependent glucosyltransferase in the petals of carnation and delphinium. Plant Cell 22:3374–3389

Mayak S, Tirosh T (1993) Unusual ethylene-related behavior in senescing flowers of the carnation Sandrosa. Physiol Plant 88:420–426

Mehlquist GAL, Geissman TA (1947) Inheritance in the carnation, *Dianthus caryophyllus* III Inheritance of flower colour. Ann Mo Bot Gard 34:39–75

Mizuno H (1993) Race differentiation of *Fusarium oxysporum* f. sp. *dianthi* collected from carnation growing areas in Japan. Proc Kanto-Tosan Plant Prot Soc 40:157–159. (In Japanese with English abstract)

Momose M, Nakayama M, Itoh Y, Umemoto N, Toguri T, Ozeki Y (2013) An active hAT transposable element causing bud mutation of carnation by insertion into the flavonoid 3′-hydroxylase gene. Mol Gen Genomics 288:175–184

Moyal-Ben Zvi M, Vainstein A (2007) Carnation. In: Pua EC, Davey MR (eds) Biotechnology in agriculture and forestry. Transgenic crops VI, vol 61. Springer, Berlin, pp 241–252

Nakano M, Mii M (1993a) Somatic hybridization between *Dianthus chinensis* and *D. barbatus* through protoplast fusion. Theor Appl Genet 86:1–5

Nakano M, Mii M (1993b) Interspecific somatic hybridization in *Dianthus*: selection of hybrids by the use of iodoacetamide inactivation and regeneration ability. Plant Sci 88:203–208

Nakayama M, Koshioka M, Yoshida H, Kan Y, Fukui Y, Koike A, Yamaguchi M (2000) Cyclic malyl anthocyanins in *Dianthus caryophyllus*. Phytochemistry 55:937–939

Nelson PE, Dickey RS (1963) Reaction of twenty-one commercial carnation varieties to *Pseudomonas caryophylli*. Phytopathology 53:320–324

Nimura M, Kato J, Mii M, Morioka K (2003) Unilateral compatibility and genotypic difference in crossability in interspecific hybridization between *Dianthus caryophyllus* L. and *Dianthus japonicus* Thunb. Theor Appl Genet 106:1164–1170
Nimura M, Kato J, Mii M, Katoh T (2006a) Amphidiploids produced by natural chromosome-doubling in interspecific hybrids between *Dianthus* × *isensis* Hirahata *et* Kitam. and *D. japonicus* Thunb. J Hortic Sci Biotechnol 81:72–77
Nimura M, Kato J, Horaguchi H, Mii M, Sakai K, Katoh T (2006b) Induction of fertile amphidiploids induction by artificial chromosome-doubling in interspecific hybrid between *Dianthus caryophyllus* L. and *D. japonicus* Thunb. Breed Sci 56:303–310
Nimura M, Kato J, Mii M (2008a) Carnation improvement: Interspecific hybridization and polyploidization in carnation breeding. In: Teixeira da Silva JA (ed) Floriculture, ornamental and plant biotechnology. Advances and topical issues, vol 5. Global Science Books, Isleworth, pp 105–121
Nimura M, Kato J, Mii M, Ohishi K (2008b) Cross-compatibility and the polyploidy of progenies in reciprocal backcrosses between diploid carnation (*Dianthus caryophyllus* L.) and its amphidiploid with *Dianthus japonicus* Thunb. Sci Hortic 115:183–189
Ogata J, Itoh Y, Ishida M, Yoshida H, Ozeki Y (2004) Cloning and heterologous expression of cDNAs encoding flavonoid glucosyltransferases from *Dianthus caryophyllus*. Plant Biotechnol 21:367–375
Ohtsuka H, Horiuchi M, Inaba Z, Wakasawa H, Fukushima T (1995) Interspecific hybrids between carnation and *Dianthus* species by using embryo culture and their characteristics. Bull Shizuoka Agric Exp Stn 40:27–38. (in Japanese)
Okamura M, Yasuno N, Ohtsuka M, Tanaka A, Shikazono N, Hase Y (2003) Wide variety of flower-color and -shape mutants regenerated from leaf cultures irradiated with ion beams. Nucl Instrum Methods Phys Res B 206:574–578
Okamura M, Tanaka A, Momose M, Umemoto N, Teixeira da Silva JA, Toguri T (2006) Advances of mutagenesis in flowers and their industrialization. In: Teixeira da Silva JA (ed) Floriculture, ornamental and plant biotechnology. Advances and topical issues, vol 1. Global Science Books, Isleworth, pp 619–628
Okamura M, Nakayama M, Umemoto N, Cano EA, Hase Y, Nishizaki Y, Sasaki N, Ozeki Y (2013) Crossbreeding of a metallic color carnation and diversification of the peculiar coloration by ion-beam irradiation. Euphytica 191:45–56
Onozaki T (2001) Exploration and collection of *Dianthus superbus* var. *longicalycinus* and *D. superbus* in Mie prefecture and Hokkaido. Annu Rep Explor Introd Plant Genet Resour 17:49–54. (In Japanese with English abstract)
Onozaki T (2006) Carnation. In: JSHS (ed) Horticulture in Japan 2006. Shoukadoh Publication, Kyoto, pp 223–230
Onozaki T (2008) Improvement of flower vase life using cross-breeding techniques in carnation (*Dianthus caryophyllus* L.). JARQ 42:137–144
Onozaki T (2016) Carnation. In: Shibata M (ed) Japanese history on flower breeding. Yushokan, Tokyo, pp 31–62. (In Japanese)
Onozaki T, Ikeda H, Yamaguchi T, Himeno M (1998) Introduction of bacterial wilt (*Pseudomonas caryophylli)* resistance in *Dianthus* wild species to carnation. Acta Hortic 454:127–132
Onozaki T, Yamaguchi T, Himeno M, Ikeda H (1999a) Evaluation of 277 carnation cultivars for resistance to bacterial wilt (*Pseudomonas caryophylli*). J Jpn Soc Hortic Sci 68:546–550
Onozaki T, Yamaguchi T, Himeno M, Ikeda H (1999b) Evaluation of wild *Dianthus* accessions for resistance to bacterial wilt (*Pseudomonas caryophylli*). J Jpn Soc Hortic Sci 68:974–978
Onozaki T, Mato M, Shibata M, Ikeda H (1999c) Differences in flower color and pigment composition among white carnation (*Dianthus caryophyllus* L.) cultivars. Sci Hortic 82:103–111
Onozaki T, Ikeda H, Yamaguchi T (2001) Genetic improvement of vase life of carnation flowers by crossing and selection. Sci Hortic 87:107–120
Onozaki T, Ikeda H, Yamaguchi T, Himeno M, Amano M, Shibata M (2002) "Carnation Nou No.1", a carnation breeding line resistant to bacterial wilt (*Burkholderia caryophylli*). Hortic Res (Japan) 1:13–16. (In Japanese with English abstract)
Onozaki T, Kudo K, Funayama T, Ikeda H, Tanikawa N, Shibata M (2003) Identification of random amplified polymorphic DNA markers linked to bacterial wilt resistance in carnations. Acta Hortic 612:95–103
Onozaki T, Ikeda H, Shibata M (2004a) Video evaluation of ethylene sensitivity after anthesis in carnation (*Dianthus caryophyllus* L.) flowers. Sci Hortic 99:187–197

Onozaki T, Tanikawa N, Taneya M, Kudo K, Funayama T, Ikeda H, Shibata M (2004b) A RAPD-derived STS marker is linked to a bacterial wilt (*Burkholderia caryophylli*) resistance gene in carnation. Euphytica 138:255–262

Onozaki T, Ikeda H, Shibata M, Tanikawa N, Yagi M, Yamaguchi T, Amano M (2006a) Breeding process and characteristics of carnation Norin No. 1 "Miracle Rouge" and No. 2 "Miracle Symphony" with long vase life. Bull Natl Inst Flor Sci 5:1–16. (In Japanese with English abstract)

Onozaki T, Tanikawa N, Yagi M, Ikeda H, Sumitomo K, Shibata M (2006b) Breeding of carnations (*Dianthus caryophyllus* L.) for long vase life and rapid decrease in ethylene sensitivity of flowers after anthesis. J Jpn Soc Hortic Sci 75:256–263

Onozaki T, Yoshinari T, Yoshimura T, Yagi M, Yoshioka S, Taneya M, Shibata M (2006c) DNA markers linked to a recessive gene controlling single flower type derived from wild species, *Dianthus capitatus* ssp. *andrzejowskianus*. Hortic Res (Japan) 5:363–367. (In Japanese with English abstract)

Onozaki T, Yagi M, Shibata M (2008) Selection of ethylene-resistant carnations (*Dianthus caryophyllus* L.) by video recording system and their response to ethylene. Sci Hortic 116:205–212

Onozaki T, Yagi M, Fujita Y, Tanase K (2011a) Characteristics of interspecific hybrids between carnation (*Dianthus caryophyllus*) lines with long vase life and *D. superbus* var. *longicalycinus*, and their backcrossing progenies. Hortic Res (Japan) 10:161–172. (In Japanese with English abstract)

Onozaki T, Yagi M, Tanase K, Shibata M (2011b) Crossings and selections for six generations based on flower vase life to create lines with ethylene resistance or ultra-long vase life in carnations (*Dianthus caryophyllus* L.). J Jpn Soc Hortic Sci 80:486–498

Onozaki T, Yagi M, Tanase K (2015) Selection of carnation line 806-46b with both ultra-long vase life and ethylene resistance. Hortic J 84:58–68

Onozaki T, Yamada M, Yagi M, Tanase K, Shibata M (2018) Effects of crossing and selection for seven generations based on flower vase life in carnations (*Dianthus caryophyllus* L.), and the relationship between ethylene production and flower vase life in the breeding lines. Hortic J 87:106–114

Ouellette GB, Baayen RP, Simard M, Rioux D (1999) Ultrastructural and cytochemical study of colonization of xylem vessel elements of susceptible and resistant *Dianthus caryophyllus* by *Fusarium oxysporum* f. sp. *dianthi*. Can J Bot 77:644–663

Sato S, Katoh N, Yoshida H, Iwai S, Hagimori M (2000) Production of doubled haploid plants of carnation (*Dianthus caryophyllus* L.) by pseudofertilized ovule culture. Sci Hortic 83:301–310

Satoh S (2005) Induction of flower senescence by ethylene. In: Hashiba T (ed) Development in plant protection – bridging bioscience. SoftScience, Inc., Tokyo, pp 305–317. (In Japanese)

Satoh S (2011) Ethylene production and petal wilting during senescence of cut carnation (*Dianthus caryophyllus*) flowers and prolonging their vase life by genetic transformation. J Jpn Soc Hortic Sci 80:127–135

Saunders ER (1917) Studies in the inheritance of doubleness in flowers, II. *Meconopsis*, *Althaea* and *Dianthus*. J Genet 6:165–184

Scariot V, Paradiso R, Rogers H, De Pascale S (2014) Ethylene control in cut flowers: classical and innovative approaches. Postharvest Biol Technol 97:83–92

Schade F, Legge RL, Thompson JE (2001) Fragrance volatiles of developing and senescing carnation flowers. Phytochemistry 56:703–710

Scovel G, Ben-Meir H, Ovadis M, Itzhaki H, Vainstein A (1998) RAPD and RFLP markers tightly linked to the locus controlling carnation (*Dianthus caryophyllus*) flower type. Theor Appl Genet 96:117–122

Serek M, Sisler EC, Woltering EJ, Mibus H (2007) Chemical and molecular genetic strategies to block ethylene perception for increased flower life. Acta Hortic 755:163–169

Shibuya K, Yoshioka T, Hashiba T, Satoh S (2000) Role of the gynoecium in natural senescence of carnation (*Dianthus caryophyllus*) flowers. J Exp Bot 51:2067–2073

Sparnaaij LD (1979) Polyploidy in flower breeding. HortScience 14:496–499

Sparnaaij LD, Demmink JF (1983) Carnations of the future. Acta Hortic 141:17–22

Sparnaaij LD, Koehorst-van Putten HJJ (1990) Selection for early flowering in progenies of interspecific crosses of ten species in the genus *Dianthus*. Euphytica 50:211–220

Stich K, Eidenberger T, Wurst F, Forkmann G (1992) Enzymatic conversion of dihydroflavonols to flavan-3,4-diols using flower extracts of *Dianthus caryophyllus* L. (carnation). Planta 187:103–108

Takeda Y (1989) Horticultural history of carnation. In: Tsukamoto Y (ed) The grand dictionary of horticulture, vol 1. Shogakukan, Tokyo, pp 485–491. (In Japanese)

Tanaka Y, Tsuda S, Kusumi T (1998) Metabolic engineering to modify flower color. Plant Cell Physiol 39:1119–1226

Tanase K, Onozaki T, Satoh S, Shibata M, Ichimura K (2008) Differential expression levels of ethylene biosynthetic pathway genes during senescence of long-lived carnation cultivars. Postharvest Biol Technol 47:210–217

Tanase K, Nishitani C, Hirakawa H, Isobe S, Tabata S, Ohmiya A, Onozaki T (2012) Transcriptome analysis of carnation (*Dianthus caryophyllus* L.) based on next-generation sequencing technology. BMC Genomics 13:292

Thomas WD Jr (1954) The reaction of several carnation varieties to bacterial wilt. Phytopathology 44:713–715

Tsuchiya Y, Minakami T, Kagito T (1965) Bacterial wilt of carnation. Ann Phytopathol Soc Jpn 30(5):268. (In Japanese)

Tutin TG (1964) *Dianthus* L. In: Tutin TG, Heywood VH, Burges NA, Moore DM, Valentine DH, Walters SM, Webb DA (eds) Flora Europaea, vol 1. Cambridge University Press, Cambridge, pp 188–204

Uematsu S, Hosoya M, Sekiyama K (1991) Reaction of 126 commercial carnation varieties to *Pseudomonas caryophylli*. Proc Kanto-Tosan Plant Prot Soc 38:107–110. (In Japanese with English abstract)

Ushio A, Onozaki T, Shibata M (2002) Estimation of polyploidy levels in *Dianthus* germplasms by flow cytometry. Bull Natl Inst Flor Sci 2:21–26. (In Japanese with English abstract)

Vainstein A, Hillel J, Lavi U, Tzuri G (1991) Assessment of genetic relatedness in carnation by DNA fingerprint analysis. Euphytica 56:225–229

Veen H (1979) Effects of silver on ethylene synthesis and action in cut carnations. Planta 145:467–470

Woltering EJ, van Doorn WG (1988) Role of ethylene in senescence of petals — morphological and taxonomical relationships. J Exp Bot 39:1605–1616

Yagi M, Onozaki T, Taneya M, Watanabe H, Yoshimura T, Yoshinari T, Ochiai Y, Shibata M (2006) Construction of a genetic linkage map for the carnation by using RAPD and SSR markers and mapping quantitative trait loci (QTL) for resistance to bacterial wilt caused by *Burkholderia caryophylli*. J Jpn Soc Hortic Sci 75:166–172

Yagi M, Fujita Y, Yoshimura T, Onozaki T (2007) Comprehensive estimation of polyploidy level in carnation cultivars by flow cytometry. Bull Natl Inst Flor Sci 7:9–16. (In Japanese with English abstract)

Yagi M, Onozaki T, Ikeda H, Tanikawa N, Shibata M, Yamaguchi T, Tanase K, Sumitomo K, Amano M (2010) Breeding process and characteristics of carnation "Karen Rouge" with resistance to bacterial wilt. Bull Natl Inst Flor Sci 10:1–10. (In Japanese with English abstract)

Yagi M, Kimura T, Yamamoto T, Isobe S, Tabata S, Onozaki T (2012a) QTL analysis for resistance to bacterial wilt (*Burkholderia caryophylli*) in carnation (*Dianthus caryophyllus*) using an SSR-based genetic linkage map. Mol Breed 30:495–509

Yagi M, Yamamoto T, Kimura T, Isobe S, Tabata S, Onozaki T (2012b) QTL analysis for flower vase life in carnation. In: Book of abstracts – 24th International Eucarpia Symposium – Section ornamentals 129

Yagi M, Yamamoto T, Isobe S, Hirakawa H, Tabata S, Tanase K, Yamaguchi H, Onozaki T (2013) Construction of a reference genetic linkage map for carnation (*Dianthus caryophyllus* L.). BMC Genomics 14:734

Yagi M, Kosugi S, Hirakawa H, Ohmiya A, Tanase K, Harada T, Kishimoto K, Nakayama M, Ichimura K, Onozaki T, Yamaguchi H, Sasaki N, Miyahara T, Nishizaki Y, Ozeki Y, Nakamura N, Suzuki T, Tanaka Y, Sato S, Shirasawa K, Isobe S, Miyamura Y, Watanabe A, Nakayama S, Kishida Y, Kohara M, Tabata S (2014a) Sequence analysis of the genome of carnation (*Dianthus caryophyllus* L.). DNA Res 21:231–241

Yagi M, Yamamoto T, Isobe S, Tabata S, Hirakawa H, Yamaguchi H, Tanase K, Onozaki T (2014b) Identification of tightly linked SSR markers for flower type in carnation (*Dianthus caryophyllus* L.). Euphytica 198:175–183

Yagi M, Shirasawa K, Waki T, Kume T, Isobe S, Tanase K, Yamaguchi H (2017a) Construction of an SSR and RAD marker-based genetic linkage map for carnation (*Dianthus caryophyllus* L.). Plant Mol Biol Rep 35:110–117

Yagi M, Shirasawa K, Isobe S, Tanase K, Yamaguchi H (2017b) QTL analysis for flower vase life in carnation breeding line 806-46b. Hortic Res (Japan) 16(Suppl. 2):531. (In Japanese)

Yamaguchi T (1994) Carnation. In: Konishi K, Iwahori S, Kitagawa H, Yakuwa T (eds) Horticulture in Japan. Asakura Publishing Co. Ltd., Tokyo, pp 139–144

Yamaguchi T, Himeno M, Onozaki T, Shibata M (1990) Exploration and collection of wild *Dianthus* species in the south-western region in Japan. Annu Rep Explor Introd Plant Genet Resour 6:73–82. (In Japanese with English abstract)

Yoshida H, Akimoto H, Yamaguchi M, Shibata M, Habu Y, Iida S, Ozeki Y (2004) Alteration of methylation profiles in distinct cell lineages of the layers during vegetative propagation in carnation (*Dianthus caryophyllus*). Euphytica 135:247–253

Yoshimoto Y, Higeta D, Ito Y, Yoshida H, Hasebe M, Ozeki Y (2000) Isolation and characterization of a cDNA for phenylalanine ammonia-lyase (PAL) from *Dianthus caryophyllus* (carnation). Plant Biotechnol 17:325–329

Zhou X, Gui M, Zhao D, Chen M, Ju S, Li S, Lu Z, Mo X, Wang J (2013) Study on reproductive barriers in 4x-2x crosses in *Dianthus caryophyllus* L. Euphytica 189:471–483

Zhou X, Su Y, Yang X, Zhang Y, Li S, Gui M, Wang J (2017) The biological characters and polyploidy of progenies in hybridization in 4x-2x crosses in *Dianthus caryophyllus*. Euphytica 213:118

Zuker A, Tzfira T, Vainstein A (1998) Genetic engineering for cut-flower improvement. Biotechnol Adv 16:33–79

Zuker A, Tzfira T, Ben-Meir H, Ovadis M, Shklarman E, Itzhaki H, Forkmann G, Martens S, Neta-Sharir I, Weiss D, Vainstein A (2002) Modification of flower color and fragrance by antisense suppression of the flavanone 3-hydroxylase gene. Mol Breed 9:33–41

第16章 倒挂金钟

Mario G. R. T. de Cooker, Edwin J. Goulding,
Jan H. Waldenmaier, and Paul E. Berry

摘要 倒挂金钟属植物的品种杂交和培育开始于19世纪30年代，其中大部分是由早期的园艺家培育的，他们不愿意透露品种的来源，还有一部分是最近由业余种植者随机选择方法和材料进行培育的。通过DNA测序，我们现在对倒挂金钟属植物的进化及为何将其划分为12个组（每个组包含一个或多个密切相关的种）有了相当全面的了解。这可以指导我们未来杂交工作的品种选择，以及更好地了解过去培育的品种的杂交历程。倒挂金钟属的基本染色体数为11条（n = 11）。在本章中，我们将介绍倒挂金钟属不同种和栽培种的倍性水平，包括多倍体是如何通过人为化学处理培育的。通过对花青素色素代谢的研究，可以更好地了解花色的变化和寻找新的花色。倒挂金钟属的品种选育现在已经从耐寒性转向利用包括三叶和花朵较小及短日照（冬季开花）品种在内的更广泛的组间杂交。

关键词 倒挂金钟属；育种；种间杂交；花青素；流式细胞术；系统发育

16.1 进化与系统发育关系

倒挂金钟属是柳叶菜科一个独特的属，其浆果肉质饱满，大部分为两孔花粉。通过分子和形态学技术的结合（Berry et al. 2004），已经建立了该属的系统发育图谱，这极大地促进了将倒挂金钟属的杂交工作纳入生物学研究中（图 16.1）。根据这些研究结果可知，倒挂金钟属在大约 4000 万年前从柳叶菜科的近亲北方草本物种露珠草属中分离。然后，在大约 3000 万年之前，倒挂金钟属分化形成许多我们现在所看到的不同种类。倒挂金钟属最初的多样分化在地质时期就迅速发生了，以至于很难得出有把握的结论，即该属中的哪个组首先分化或分布在地球的哪个部分。两个现存的早期分化群体表明最初的分化可能发生在南半球，即巴西和巴塔哥尼亚的 *Quelusia* 组与目前仅存于新西兰的 *Skinnera* 和 *Procumbentes* 组（尽管我们通过对化石花粉的研究得知，倒挂金钟属也出现

M. G. R. T. de Cooker (✉)
Ohé en Laak, The Netherlands
E. J. Goulding
Ipswich, UK
J. H. Waldenmaier
Herpen, The Netherlands
P. E. Berry
University of Michigan, EEB Department, Ann Arbor, MI, USA
e-mail: peberry@umich.edu

在澳大利亚的渐新世和中新世)。另一种可能性为，倒挂金钟属最初的分化发生在中美洲地区，因为那里有许多组聚集。现在，倒挂金钟属物种最丰富的地方为热带的安第斯山脉，但是根据我们的分子生物学研究结果，这些物种可能是最近才形成的。我们已经鉴定了倒挂金钟属内的12个亚组。表16.1给出了这些组的名称、地理分布和物种数量。

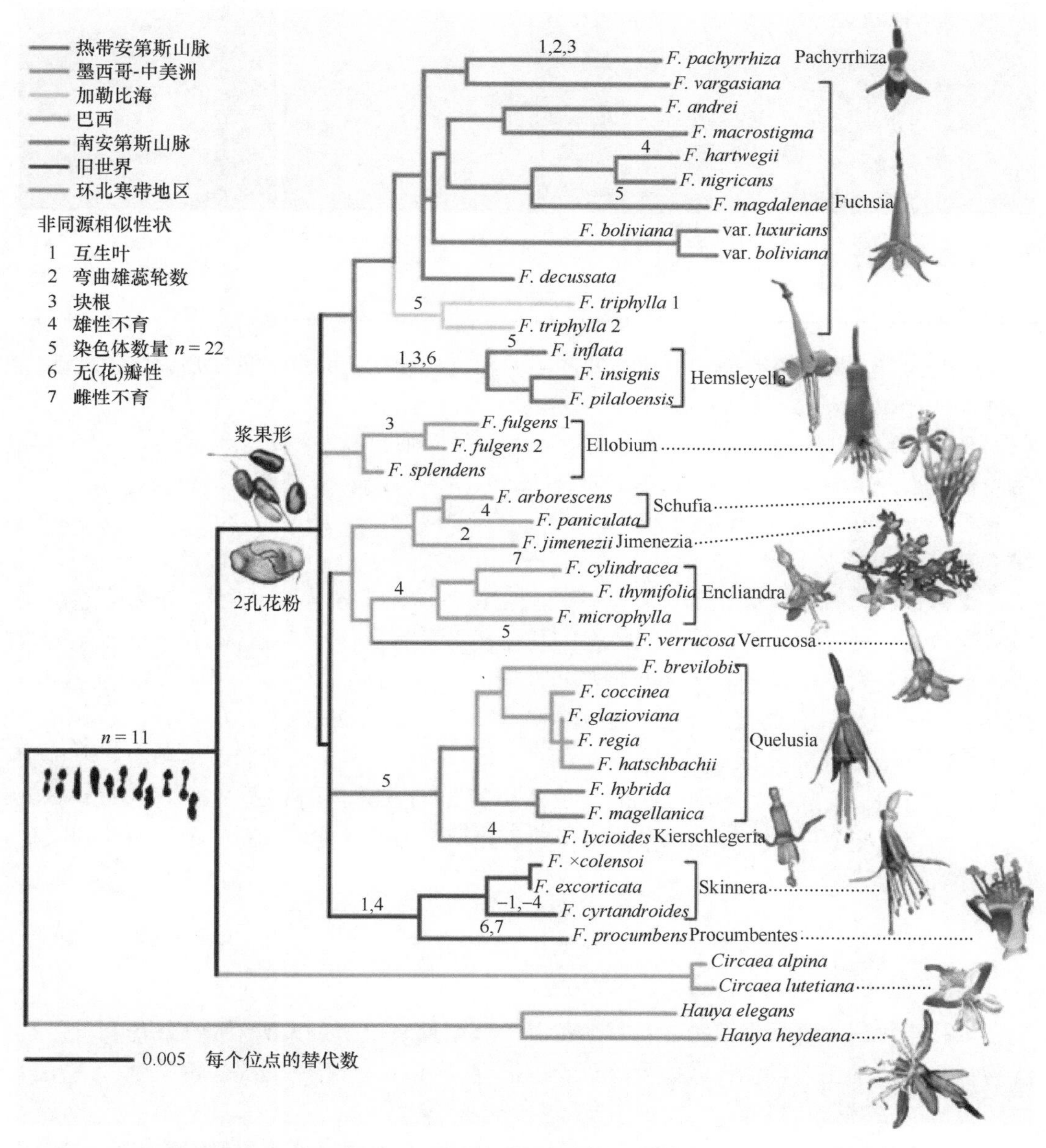

图16.1 通过对非编码叶绿体和细胞核DNA的分析得出的倒挂金钟属之间的关系图谱。树上单个分支的长度与沿着该分支的DNA序列中碱基变化的数量成正比（转载自Berry et al. 2004，获得了密苏里植物园的许可）

表 16.1 倒挂金钟属的地理分布和物种数量

分组	物种数	地域分布
Procumbentes	1	新西兰
Skinnera	3	新西兰
Quelusia	9，具有 3 个亚种	巴西东南部、阿根廷南部和智利
Kierschlegeria	1	智利
Encliandra	6，具有 11 个亚种	墨西哥、中美洲
Schufia	2，具有 2 个亚种	墨西哥、中美洲
Ellobium	3	墨西哥、中美洲
Jimenezia	1	中美洲（巴拿马、哥斯达黎加）
Verrucosa	1	北安第斯山脉（哥伦比亚和委内瑞拉）
Pachyrrhiza	1	中央安第斯山脉（秘鲁）
Fuchsia	65	热带安第斯山脉和加勒比海
Hemsleyella	15	热带安第斯山脉
总数：12	108	

一般而言，物种最容易与自己所属的组内种进行杂交，但育种者成功地将匍枝倒挂金钟（*Fuchsia procumbens*）等新西兰种与新世界的种杂交，这表明，尽管进行杂交有些困难但 3000 万年的物理分离并未阻断它们之间的杂交能力。毫无疑问的是大多数栽培品种均来源于 *Quelusia* 组，其中 *F. magellanica* 和 *F. regia* 可能是最主要的种。早期可能还将 *F. fulgens* 与倒挂金钟组（*Fuchsia*）整合形成了 *Ellobium* 组，该组在与 *F. triphylla* 和 *F. boliviana* 等重要物种的杂交中发挥了作用。

支持该属系统发育关系的其他一些特征是染色体数量（倍性水平）、花粉形态、花瓣缺失数量、是否有块茎、雄性或雌性的功能丧失（图 16.1）。尽管像 *Hemsleyella* 组的特征是完全没有花瓣，且其大多数种都具有块茎，但 *Procumbentes* 组同样没有花瓣，*Pachyrrhiza* 组也具有块茎。倒挂金钟属的倍性水平基本为 $2n = 2x = 22$，但 *Quelusia* 和 *Kierschlegeria* 组基本是四倍体。在某些种中，尤其是在许多不同的栽培品种中观察到更高的倍性水平。在倒挂金钟属中报道了多个种是四倍体，特别是三叶倒挂金钟（*F. triphylla*）。倒挂金钟属高倍性水平通常伴随着花粉形态的变化，花粉构型从两孔到三孔（更多的三角形晶粒）。

16.2 花和花粉形态

倒挂金钟属的种和栽培种在花朵颜色与形状上均表现出很大的差异。红色和橙色是花粉管与萼片的主要颜色。这种颜色对蜂鸟很有吸引力，而蜂鸟是美洲倒挂金钟属在自然环境中的主要授粉媒介。

在本地种中，*Fuchsia magellanica* 和南美洲 *Quelusia* 组的 8 个种都具有红色花粉管和萼片、具紫色花瓣的腋生花。南美洲 *Fuchsia* 组中的大部分物种在花型和花瓣排列上差异很大，从腋生花到浓密的圆锥状或总状花序，且通常相对于花粉管的长度花瓣很小。安第斯山脉的 *Hemsleyella* 组通常具有块茎，且所有种都完全没有花瓣；该组包括倒挂金钟属中具最长花朵的种，即 *Fuchsia apetala*，具有长达 15cm 的花粉管。一些具有最小花朵的种出现在墨西哥和中美洲地区；*Encliandra* 组中的种有微小的腋生花，而 *Schufia* 组的花则紧密地排列在圆锥花序中。

倒挂金钟属的花朵也可通过花瓣数量彼此区分。单瓣花在倒挂金钟的种和栽培种中普遍存在。4 个花瓣构成了花冠的基本结构。在栽培种中，某些植物的花瓣数为 4～80。增加染色体数目也具有增加重瓣的可能性。

多数物种是具有 22 条染色体的二倍体，花粉粒是外观呈柠檬状的 2 孔结构（图 16.2a）。原生四倍体物种，如 *F. lycioides* 和 *Quelusia* 组的许多种，倾向于具有三孔的花粉粒（图 16.2b）。在具有三孔或三孔以上花粉粒的栽培种中，倍性水平可能增加了（图 16.2c）。多倍体为杂交者提供了较大的灵活性，并增加了在更困难的杂交育种中成功的可能性。影响生育能力的环境因素已经得到了充分的研究，如温度和相对空气湿度水平（Goulding 2002）。从 2013 年 12 月的第 2 期开始，《倒挂金钟育种者计划》（*The Fuchsia Breeders Initiative*）在线发表了影响花粉育性的因素、所用设备和所涉及技术的详细分析。

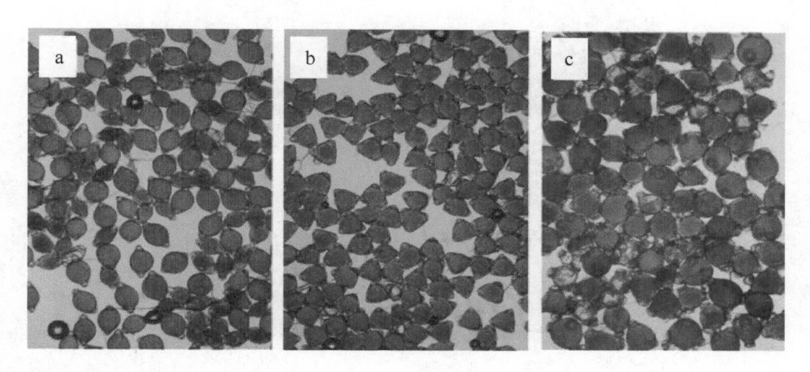

图 16.2　倒挂金钟花粉粒的示例：（a）两孔粒：如二倍体植物 *F. boliviana* var. *boliviana*；（b）三孔粒：如四倍体植物 *F. lycioides*；（c）花粉粒孔径数量和大小变化很大的品种：*F.* ‘Lye's Unique’，其可能是多倍体品种

16.3　栽培种历史

倒挂金钟属因其花期长、婉约优美和观赏价值高而被种植了数百年。18 世纪初，法国修道士兼植物学家 Charles Plumier 在 Hispaniola 岛上首次发现与记录了倒挂金钟这种植物。此后，花了一个多世纪的时间倒挂金钟才在欧洲确立了其园林植物的地位。在 19 世纪，多个新物种被发现且普遍存在，如 *F. coccinea*、*F. magellanica*、*F. regia* 和 *F. splendens* 等。

最早的杂交种和许多现代引种很可能是由 *Quelusia* 和 *Ellobium* 这两个组中的 4 个种种间杂交获得的。来源于 *Ellobium* 组的长筒倒挂金钟（*Fuchsia fulgens*）（图 16.3a）是特别有价值的亲本。1840 年，有了一个重要的育种突破，第一个具有紫色花瓣，但花粉管和萼片为白色的品种 *F.* ‘Venus Victrix’ 被推向市场。这引发了杂交主义者争相引进更多花色品种的风潮。

图 16.3 （a）长筒倒挂金钟（*Fuchsia fulgens*）；（b）*F.* ‘Lye’s Excelsior’（Lye 1886）；（c）*F.* ‘Gartenmeister Bonstedt’（Bonstedt 1905）；（d）*F.* ‘Cecile’（Whitfield 1981）

1850 年左右，第一个具有双花冠的品种被培育出。1870～1890 年，著名的英国杂交育种家 James Lye 培育了许多新品种，其中大多数具有相对较长的管状萼片，如品种 *F.* ‘Lye’s Excelsior’ 的萼片呈乳白色（图 16.3b）。花冠主要为单花冠，颜色从粉红色过渡到红色再到橙色不等。在 19 世纪末，经过德国杂交育种家和园艺师 Georg Bornemann 与 Carl Bonstedt 等的培育，源自三叶倒挂金钟（*F. triphylla*）的杂交品种开始流行。许多古老而珍贵的品种目前仍然存在，且在当前杂交育种中使用，其中就包括 *F.* ‘Göttingen’ 和 *F.* ‘Gartenmeister Bonstedt’（图 16.3c）。在近代，世界各地的杂交育种家对丰富倒挂金钟属品种的多样性做出了巨大贡献，其中美国倒挂金钟属协会（AFS）就已注册了超过一万个品种。*F.* ‘Cecile’ 就是众多品种中的一个例子（图 16.3d）。

尽管许多当地苗圃或收集家那里保存着数千种倒挂金钟属植物，但只有少数品种用于商业交易。在商业方面，数百万的插条在葡萄牙和肯尼亚等国用于生产，然后将无根插条空运到荷兰和德国等较北的国家。这些插条在那里被培育成生根的植株，并在 3～4 个月后推向市场。图片标签经常被用于展示花朵的样子，以为出售的迷你插条（mini-plug）和小扦插苗提供栽培信息。品牌推广在许多公司中越来越流行，如荷兰的 Hendriks Young Plants 公司，其通过 Bella®系列每年生产数百万株植物。在出售的产品中，较大的植物有时会培养到开花出售，还可以找到鞭状或标准的植株，甚至可以用大的容器栽种花卉出售给愿意支付高价的人。

16.4　倒挂金钟的种植协会

倒挂金钟广泛的大众吸引力从大量致力于推广该属的国家和地方协会上可见一斑。他们的重点是举办大型比赛或展览，在那里通常可以看到各种选育的植物。这些可能会无意间将公众的关注度限制于长日照的栽培品种上。那些大量但小众的物种及其衍生品主要是短日照植物，在很大程度上已被忽略或很少见。

美国倒挂金钟属协会成立于 1929 年，它成为负责全球倒挂金钟属新品种注册的组织。1938 年，英国倒挂金钟属协会成立，由于第二次世界大战，它主要致力于保存和推广最能在人们花园中生存的植物。荷兰倒挂金钟属花友会（The Dutch Circle of Fuchsia Friends，NKvF）于 1965 年成立后，迅速成为世界上最大的倒挂金钟属协会，到 20 世纪末已拥有 6000 多名会员。最后，欧洲倒挂金钟协会（Euro-Fuchsia）成立于 1982 年，其宗旨是加强跨国合作与联系，且每年举行一次会议。在过去的 20 年中，全球协会会员人数有所下降。这可能是因为倒挂金钟属植物在园艺上是一个小众市场，并且很少有年轻人对种植这些植物感兴趣。

在过去的 150 年中，许多关于倒挂金钟属的书籍以不同的语言在许多国家出版。大部分信息都是以品种图片和描述的形式出现，通常提供花色和花形图片，也有单花、双花和三叶杂种的信息。这些书籍中也会考虑倒挂金钟的种植方面，如关于生长习性和培育植物的不同方法（如创建标准）。许多书籍在某种程度上也涵盖了诸如病虫害之类的章节。

在过去的 20 年中，很多信息已经可以通过网络获取，尤其是大量的视频资料。大量关于品种杂交技术的建议和亲本信息也可获取。与大多数其他内容一样，由于不同网站的质量可能相差很大，因此需要谨慎选择。

通常，关于倒挂金钟属遗传学的技术信息很少，可能是因为大多数倒挂金钟的育种者都是预算较低的业余爱好者。在 21 世纪初期的短暂时间里，倒挂金钟国际研究组织（Fuchsia Research International）试图在物种和杂交方面填补这一空白，但该组织在 2008 年倒闭。新西兰的 R. Talluri 采用更系统的方法研究了倒挂金钟遗传学的一些方面（Talluri 2007，2012）。

最近，《倒挂金钟育种者计划》（*the Fuchsia Breeders Initiative*）开始以半年刊电子杂志提供倒挂金钟属的高质量信息。该杂志可为感兴趣的人们免费提供（ISSN：2214-7551，http://www.eurofuchsia.org）。

我们还设想了开发倒挂金钟物种和品种的专门收藏机构的可能性，可通过收取参观门票费用来维护一些不常见和稀有的种与杂交种。

16.5　育 种 目 标

16.5.1　非生物胁迫抗性

杂交学家青睐的领域之一是寻求耐寒性品种。这非常符合现代需求，即大多数园艺爱好者没有太多的时间来打理他们的植物，而且希望植物生长没有太多麻烦，理想的情况是整个夏天都开花。市场需求推动了这一趋势，除了在协会成员中寻找比赛获胜者外，直到最近该需求仍然具有显著的重要性。

耐寒性很难被定义。这不仅仅是在霜冻下存活的问题。的确，如果植物都不能开花，仅生存下来是毫无意义的。除了极寒以外，降水量和降水时间、光照强度和日照时间、土壤质地和肥力水平、温度的突然变化及其持续时间都有影响。

W. P. Wood 在 1950 年发表的《倒挂金钟属调查》中集中讨论了这个主题，因为战争年代人们对该主题的重视程度很高（Wood 1950）。随后，Tom Thorne 在《所有用途的倒挂金钟属》中尝试提供一种实用的方法来评估品种的抗寒性，即根据植物在英国冬季中的平均生存能力打分（1～3 分）（Thorne 1964）。最近，Van Veen（1992）在 *Winterharde Fuchsia's* 中尝试提供确切的答案。

这些年来进行了许多田间试验。这些活动主要涉及欧洲国家和组织，因此，仅为全球种植者提供了部分图片。那些能够承受零下温度的物种主要分布于 *Quelusia* 组。其中大多数花朵的颜色为红色和紫色。*Ellobium* 组的倒挂金钟（*F. splendens*）也被证明相对较为耐寒，它的一个后代 *F.* 'Speciosa' 在很大程度上保留了这一特性。

然而，这一特性并不是专营市场在目前杂交育种中唯一关注的。耐热和耐旱性是三叶倒挂金钟所具有的特性。植物在光照水平最低的几个月内开花将延长其在市场上的销售期，并提供广受欢迎的盆栽植物，普通大众也可以在家中种植。

16.5.2 生物胁迫抗性

病虫害防治对于倒挂金钟而言很重要，因为如果病虫害失控，会严重限制公众对种植这些植物和开发新品种的兴趣。倒挂金钟主要的病害是由灰葡萄孢（灰霉病）和锈菌引起的。前者在黑暗、潮湿或寒冷的温室等条件下暴发严重，通常是忽视了环境清洁导致的。后者往往是由风传播的，并且来自附近的共同宿主。防止上述病害的暴发和蔓延需要敏锐的观察。

倒挂金钟主要的虫害是由粉虱和红蜘蛛引起的，主要出现在温室中。如今，许多更为有效的化学品被禁止业余人员使用，自然捕食者往往是首选。使用它们的目的是限制暴发和破坏，而不是消灭所有害虫。任何与它们结合使用的化学物质都必须经过仔细评估，以便在虫害控制中获得所需的生态平衡。

但值得注意的是，某些杂交品种对病原体具有高度抗性，如 *Fuchsia glazioviana*（图 16.4），但到目前为止，在培育新的引进品种方面发掘抗性在很大程度上被忽略了。Triphylla 杂交种和 Encliandra 栽培品种也不容易感染病原体。

最近，倒挂金钟瘿螨（*Aculops fuchsiae*）的传播带来了另一个挑战，近几十年倒挂金钟的瘿螨已经从加利福尼亚迅速传播到了欧洲，特别是在法国的布列塔尼和英格兰南部。抗瘿螨可能成为将来杂交育种的主要目标。

图 16.4　*Fuchsia glazioviana*，一种对许多病虫害具有抵抗力的巴西物种

16.5.3　花色

倒挂金钟植物中存在的花色素是花青素（表 16.2）（Averett et al. 1986；Crowden et al. 1976；Jordheim et al. 2011），其中的一个例外是 *F. procumbens*，其花粉管呈黄色可能是因为存在类胡萝卜素。倒挂金钟属的主要花色是橙色，与色素的优势顺序相反（橙色→粉色/红色→蓝色/紫色）。这可能与美洲倒挂金钟主要由蜂鸟授粉有关。

表 16.2　倒挂金钟属的花青素

色素名称	紫外线下的纯色素颜色	倒挂金钟花朵中的色素颜色
天竺葵素-3-单葡糖苷	暗橙红色	橙色
天竺葵素-3,5-二葡糖苷	荧光黄色	亮黄色
矢车菊素-3-单葡糖苷	砖红色	红色
矢车菊素-3,5-二葡糖苷	亮红色	亮红色
芍药素-3-单葡糖苷	砖红色	粉色
芍药素-3,5-二葡糖苷	荧光粉色	亮粉色
飞燕草素-3-单葡糖苷	紫色	蓝色
飞燕草素-3,5-二葡糖苷	亮紫色	亮蓝色
矮牵牛素-3-单葡糖苷	紫色	紫丁香色
矮牵牛素-3,5-二葡糖苷	亮紫色	淡紫色
锦葵素-3-单葡糖苷	紫色	紫色
锦葵素-3,5-二葡糖苷	荧光蜡红色	亮紫色

在倒挂金钟属花色素的 3′端始终有一个葡萄糖分子。如果没有其他葡萄糖分子存在，它们则是 3-单葡糖苷。如果第二个葡萄糖分子存在于 5′端，它们则是 3,5-二葡糖苷。后者具有与单葡糖苷相同的颜色，但较亮。如果没有发生羟基化，产生的色素是天竺葵色素。其他倒挂金钟花的色素还可进一步羟基化和甲基化（图 16.5）。而飞燕草色素仅在某些倒挂金钟属花中发现一些痕迹。最近在倒挂金钟属 *Skinnera* 组的后代中发现了“新”的花色紫茄色，其来源于大量的锦葵素。有时花青素骨架中的 3′端被乙酰化。乙酰化的天竺葵素在紫外线下呈黄色，这也许是未来培育黄花倒挂金钟的另一种方法。

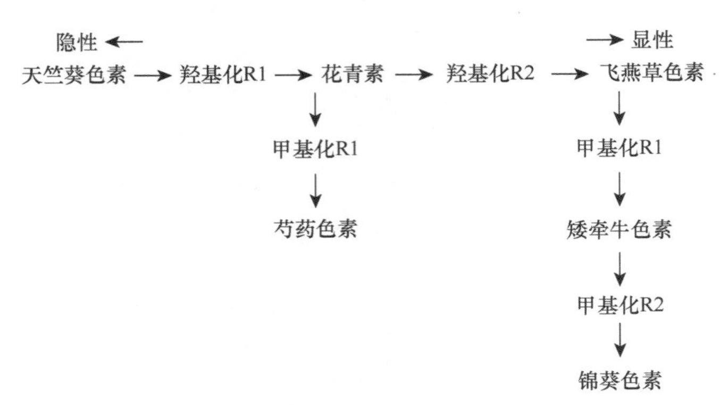

图 16.5　倒挂金钟属花色素中羟基化的优势顺序。甲基化的计数可能相同（锦葵色素是最主要的）

白花倒挂金钟中缺乏花青素合成所需的酶，色素的丢失可能发生在其合成的几个中间阶段。缺乏相同酶的白花杂交后代均为白花。如果它们含有不同的酶，其杂交 F_1 代花朵中会产生有颜色的后代，并可以根据亲本的基因型来预期。在全阳光下，白色花朵则显示出其遗传色彩的阴影，因为充分的阳光会激发花青素的产生。

羟基化的遗传调控通常是组织特异性的，在大多数情况下仅发生在花器官中。例如，与花瓣相比，萼片可能具有不同的花青素。即使遗传调控影响了整个植物的色素合成，基因的表达也可能在组织之间有所不同，如 *F.*‘Herps Saar’（图 16.6）。

图 16.6　*F.*‘Herps Saar’（Waldenmaier 2012）有紫茄色花瓣、白色带红色阴影的花粉管和萼片

对于大多数倒挂金钟的种和许多栽培品种来说，其花青素中色素的类型及其相对含量（相对于标准的锦葵色素-3,5-二葡糖苷）已经确定了（Waldenmaier，未公开结果，可在 http://members.home.nl/henkwaldenmaier/fuchsiapigments.pdf 获得）。

16.5.4 多花

花形和花朵数量是倒挂金钟属非常重要的特征。为了实现多花，在过去的几十年中，利用 *F. triphylla*（总状花序）和 *F. paniculata*（圆锥花序）的开花特性在多花品种的育种方面取得了长足的进步。最近，在品种 *F.*‘Spray’中引入了多花的特征（图 16.7）。该品种的花朵为大单瓣花，与已经存在的许多品种的花朵没有明显不同。但是，一旦开始开花，每个腋芽上就会产生大量花朵。这些花朵不是在总状花序或圆锥花序中产生的，而是在密集的分枝中产生的。该品种的生长尚有许多不足之处，但作为亲本，它在帮助开发其他品种方面显示出巨大的潜力。

图 16.7 *F.*‘Spray’

16.6 育 种

倒挂金钟通常是一种很容易实现杂交的植物，由高度可育的多倍体杂交种可以很容易地产生幼苗。种与种之间的杂交，特别是那些种间关系较远的种，杂交成功率要低得多，且在随后形成的后代幼苗中会出现一些问题。一般来说，在适当的栽培照料下，一年内可以从大多数杂交后代中获得幼苗，且在第二年可以开花。

在过去的几十年中，全世界数百个杂交育种家已经培育了几千个新品种。其中许多新品种已在美国倒挂金钟属协会注册。然而，利用这些有限的遗传资源进行育种增加了新品种表型相似的可能性。显然，很少有种植者真正意识到许多品种已经存在了，可能最意料之中的是新品种的名称没有重复的。近交繁殖也削弱了许多杂交种的抗性，使它们更容易受到病虫害的侵袭。新的遗传材料增加了提高 F_1 代抗性的可能性，也引入了以前没有的遗传特征。

利用倒挂金钟属其他品种向杂交种中引入新的特性展现出了巨大的前景。总共约有108种及几个亚种和许多本地变种可供选择。除了被广泛使用的品种，如 *F. magellanica*、*F. fulgens* 和 *F. triphylla* 外，很少有其他品种被广泛使用。获得许多不常见的品种是非常困难的。大多数品种掌握在少数热心的爱好者和收藏家手中，只有相对较少的一部分是可在市场上买到的。法国的 Chevreloup 植物园有收藏，在英国也有国家品种收藏。收藏最多的是荷兰倒挂金钟属花友会（NKvF），其收藏了数百种编号的品种、亚种、变种。其中大多数是由 Paul Berry 博士实地收集和鉴定的。

16.6.1 种间杂交

种间杂交种在野外被频繁发现（Berry 1989）。Talluri（2007）提供了许多种间杂交的信息。通常认为种间杂交父母亲本的亲缘关系密切，或者至少有适当的系统或过程使种子形成、发芽和生长。在杂交实验中，许多亲缘关系不密切的种在早期得到了似乎成功的结果，然而，经常出现的一种不幸的结果是“幼龄期死亡综合征”。

虽然 F_1 代有其自身的问题，但 F_2 杂交后代更难以成功。像倒挂金钟很多育种情况一样，也存在例外。两个二倍体 *Fuchsia fulgens* 与 *F. splendens* 进行杂交，产生了也是二倍体的 *F.*‘Speciosa’。父母亲本的亲缘关系密切，同属于 *Ellobium* 组。*Fuchsia*‘Speciosa’可以成功地与许多其他种和栽培品种杂交，但通常会产生不育的后代。

物种间的大部分杂交都是利用了有限的亲本。一个例子是‘Whiteknight's Ruby’品种，它源于 *F. triphylla* 和 *F. procumbens* 的杂交。前者是来自加勒比海地区的四倍体，后者是来自新西兰的二倍体。英国杂交育种家 John Wright 对实现利用更为广泛的基因库进行杂交育种具有重大影响。他的许多高度原创杂交品种都是在20世纪80年代完成的。

在荷兰，荷兰倒挂金钟属花友会（NKvF）在利用一些不常见品种进行杂交育种方面有着杰出的历史。其中最著名的是 Herman de Graaff，他致力于利用新西兰物种培育出紫茄色的花朵。后来，他研究了 *F. paniculata* 和 *F. triphylla* 间的杂交，并由此创建了‘Pantri’杂交种。

Henk Waldenmaier 还利用四倍体 *F. magdalenae* 和二倍体变种 *F. fulgens* var. *rubra grandiflora* 进行了开创性工作。他培育的品种‘B83-05’[五倍体栽培种，可能是由 *F. magdalenae* 在第一次减数分裂重组（FDR）或第二次减数分裂重组（SDR）过程引起的]，成为一系列引进橙色和长管状花表型性状的起点。其幼苗本身是可育的，不管是作为母株还是花粉亲本。更重要的是，下一代也被证明是可育的。

对上述系列实验和可考证研究工作做出重要贡献的还包括 Martin Beije、Jan de Boer、Egbert Dijkstra、Mia Goedman、Broer de Keijzer 和 Ronald Schwab 等荷兰杂交育种家。在德国，Lutz Bögeman 利用许多种来创制新品种。但遗憾的是它们中的大多数是不育的，因此在随后的杂交育种中很少被使用。在所有这些研究中，最重要的是实现遗传基因库的显著改变所需的大量工作。

16.6.2 多倍体化

16.6.2.1 染色体数目和基因组大小

倒挂金钟属中的大多数种是二倍体（$2n = 2x = 22$）。Goulding（2002）提供了倍性信息最全面的综述。在本土物种中，Berry（1989）指出除了主要的二倍体外，还存在四倍体和八倍体。四倍体受到许多杂交育种者的青睐，尤其会被用于与育种相关的工作中。

从 2000 年开始，荷兰倒挂金钟属花友会（NKvF）开始使用荧光染料 DAPI（4′,6-二脒基-2-苯基吲哚染色）进行流式细胞仪进行检测。该检测覆盖了大约 70%的现有物种、亚种和许多其他品种。一部分检测数据发表在《国际倒挂金钟属研究杂志》（Rosema 2001）。Talluri（2007）也发表了少数物种的检测数据。De Cooker（2016）最近获得了以碘化丙啶（PI）作为荧光染料和以豌豆作为内标的倍性检测结果（未发表）。不同来源物种在基于 PI 的测量中存在相对较大的差异，可能归因于所用物种材料的差异。通常，使用 DAPI 进行测量更为便宜，但获取的 DNA 值可靠性稍差。

流式细胞技术为杂交育种者提供了一套可用于鉴定亲本和后代的有用工具。种和品种经化学诱导产生的染色体加倍结果可以被精确确定。在倒挂金钟属中，不同物种单倍体（$n = 11$）染色体组的 DNA 值之间可能存在很大差异。通常，即使没有精确的亲缘关系，也可以对后代的可能基因组做出可靠的猜测。但是，在许多这种情况下，只有染色体数目才能提供基因组组成的必要证据，而实现这一目标的设备很大程度上超出了大多数业余杂交爱好者所能承受的范围。可以理解的是，随着杂交的进一步发展和染色体的重组，流式细胞技术在基因组分析中的可靠性就变得不那么精确了。

流式细胞技术在倒挂金钟属杂交中的实用性在 De Cooker（2016）的最新研究中展现，该研究测量了 *Fuchsia*‘Göttingen’及其许多杂交后代的基因组大小。*Fuchsia*‘Göttingen’是一种三叶杂交品种，是德国杂交育种家 Carl Bonstedt 于 1906 年从 *F. fulgens* 与 *F. triphylla* 的杂交后代中选育的。它仍然是现在最流行和最常见的三叶杂交品种之一，且经常用于倒挂金钟属的杂交育种中。

流式细胞仪测量结果表明，三叶倒挂金钟（*F. triphylla*）单套染色体的 DNA 值为 0.99pg，而 *F. fulgens* 的为 1.48pg。DNA 值的巨大差异使我们能够对 *F.*‘Göttingen’的基因组组成进行可靠的估计，因为只有有限数量的染色体组合才能解释检测到的 5.30pg 的 2C DNA 值。*F.*‘Göttingen’很可能源自一个三叶倒挂金钟（*F. triphylla*）未减数的配子与单倍体 *F. fulgens* 配子结合产生的 TTTTF 的基因组（De Cooker 2016）。最近已通过实际染色体计数确认了这一点（K. Van Laere，个人交流）。在已知幼苗亲本的情况下，通常只有有限数量的基因组组成可以解释组合后测量的 2C DNA 值和表型特性。橙/粉三叶幼苗（N 12-24）的 2C DNA 测得值为 6.83pg，与 TTTTFF 基因组 6.92pg 的计算值非常接近。当同一个 N 12-24 与 *F. juntasensis*（2C DNA 值为 5.04pg 的四倍体）杂交时，产生了 2C DNA 值为 5.90pg 的紫色幼苗，这与 TTFJJ 基因组具有 5.98pg 的 2C DNA 值相对应。这种特殊的基因组组成来源于 TTF 和 JJ 杂交的预期配子。

16.6.2.2　有丝分裂多倍体化

四倍体以上的多倍体具有多优点：表型结果不同；生长更加旺盛；枝叶较厚实粗壮；花朵更大。倒挂金钟多倍体可通过在有丝分裂过程中秋水仙碱或谷氨酰胺人工诱导生长点上快速分裂的细胞形成（图 16.8）。但没有证据表明人工诱导在过去广泛应用于该属植物中。混合倍性可能是此过程的不良结果。近年来，Waldenmaier 成功地在多个物种幼苗萌发的不同阶段培育了多倍体。

图 16.8　秋水仙碱处理之后的幼苗。注意茎的增厚

通过化学方法加倍染色体数目可以提高种间杂交的育性。生育能力通常会恢复，而三倍体可以加倍形成六倍体。这样的异源多倍体包含两组不同的基因组。因此，等位基因更多，杂合度也更高。有时会导致减数分裂异常，反而使得植物的育性降低。

倍性水平有一个最佳值，这与花的大小等特性有关。像八倍体所表现的那样，高倍性水平可能导致较小的花朵和营养繁殖难度的增加。多倍体杂交后代的繁殖能力差异很大。有时会形成杂种优势，即两个纯合亲本的后代比父母亲本中任何一个都更为健康强壮（图 16.9）。在某些情况下，杂种优势的影响是巨大的（图 16.10）。

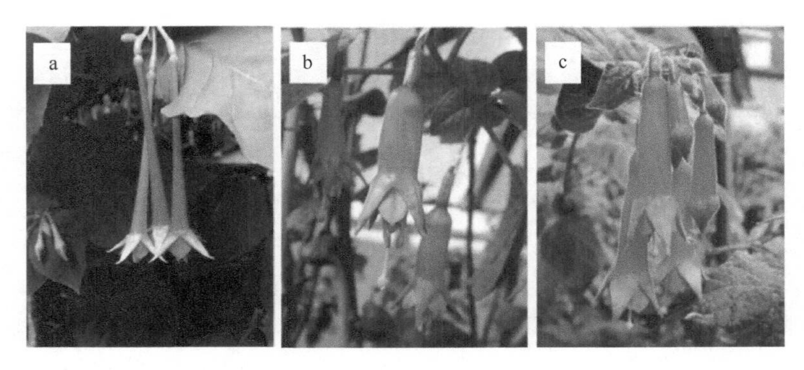

图 16.9　杂种优势：（a）四倍体植物 *Fuchsia fulgens* var. *gesneriana*；（b）四倍体植物 *F. splendens*；（c）四倍体 *F. splendens* 和四倍体 *Fuchsia fulgens* var. *gesneriana* 的杂交后代，其具有较大的杂种优势

图 16.10 四倍体 *F. paniculata*（左侧）与 *F. cylindracea* × *F. paniculate* 杂交后代（右侧）表现出的显著杂种优势

16.6.2.3 减数分裂多倍体化

在自然界中，多倍体通常是由发生减数分裂但染色体未分离形成的。一些物种经常形成未减数配子，而一些却基本上不产生。温度的急剧变化可能会增加第一次减数分裂重组（FDR）和第二次减数分裂重组（SDR）异常的发生率。这些重组异常可以利用显微镜观察花粉粒来鉴定，如花粉粒的直径比普通花粉粒至少增加了两倍。

多个倒挂金钟属植物似乎能够相当容易地形成未减数配子。这导致许多植株由未减数配子形成三倍体、四倍体或者五倍体基因组，这些多倍体要么是二倍体与二倍体杂交形成的三倍体基因组，要么是四倍体与二倍体杂交形成的四倍体或五倍体基因组。*F.* ‘Göttingen’之类的品种似乎是通过这种方式形成的，其 TTTTF 五倍体基因组必定是由三叶倒挂金钟未减数配子与 *F. fulgens* 正常单倍体配子组合形成的。

另一个例子是由 Henk Waldenmaier 培育的未上市的（B83-05）品种，其携带来自 *F. magdalenae* 未减数的四倍体配子和来自 *F. fulgens* 的单倍体配子，组成了 MMMMF 基因组。该品种高度可育，而且是 WALZ 系列中许多不寻常品种的亲本，包括第一个具有橙色重瓣花冠的 WALZ Mandoline 和许多其他具有长花管的品种。此外，在之后的几年里，该品种在其他杂交育种家培育的许多品种中发挥了重要作用。

这种由雄雌亲本产生的不同倍性水平的配子经常产生意想不到的后代表型。二倍体杂交组合 B83-05 的流式细胞检查结果表明其大部分配子为二倍体，少数为单倍体（11 条染色体）、三倍体（33 条染色体）和四倍体（44 条染色体）。这些 B83-05 配子的基因组可由 M（*F. magdalenae*）、F（*F. fulgens*）、MM、MMM、MMMM、MF 和 MMF 代表。相当一部分配子似乎是非整倍体（Waldenmaier，未发表）。

F. ‘Göttingen’ × *F.* ‘Our Ted’ 杂交组合的流式细胞测量结果表明，其杂交形成的五倍体变种极有可能拥有与 B83-05 相似的配子模式，从而导致杂交后代具有高度不可预测的特征。这种多倍体杂交的另一个复杂问题是由不均衡配子形成了非整倍体植株。研究表明，新多倍体样本自交后代很可能是非整倍体，只有约 60%的配子是整倍体（Ramsey and Schemske 2002）。这个数字在不同属植物中波动很大，且已经在拟南芥中进行了广泛的研究（Henry et al. 2005）。但是，到目前为止还未在倒挂金钟中进行系统的研究。

由此可以假设，类似的非整倍体植物可在新多倍体倒挂金钟品种的多倍体杂交中产生，如 *F.* ‘Göttingen’（五倍体）、*F.* ‘Our Ted’（五倍体）、*F.* ‘Wake The Harp’（五倍体）、*F.* ‘Strike the Viol’（六倍体）（De Cooker，由 Luca Comai 私下交流提供）。表观遗传效应可能进一步增加了新多倍体倒挂金钟杂交的复杂性和不可预测性。（*F.* ‘Göttingen’ × *F.* ‘Our Ted’） × *F. fulgens* var. *gesneriana* 的杂交结果表现出严重的由表观遗传诱导的基因沉默效应。从该杂交中得到一系列微型三叶倒挂金钟，其花管长度远低于预期的中间值（图 16.11）（De Cooker 2016）。这种影响也可能在幼苗存活的前几年造成遗传不稳定的问题。

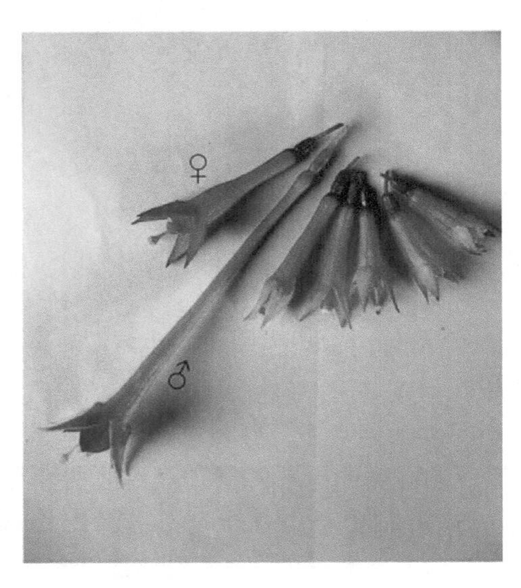

图 16.11　可能是表观遗传诱导的基因沉默效应：母本为 N 02-16 = *F.* ‘Göttingen’ × *F.* ‘Our Ted’；父本为 *F. fulgens* var. *gesneriana*

16.6.3　利用新的遗传资源进行育种

值得一提的是，倒挂金钟属的种类可大致分为由昆虫授粉的种和主要由蜂鸟授粉的种。这不是一个精确的划分，但是对于杂交育种者很有用。后者的花朵较长且较稀疏，最可能来自 *Fuchsia* 组和 *Hemsleyella* 组，且在温带气候下的云雾森林的原生栖息地中被发现。前者是倒挂金钟属中最大的组，花色展现出对鸟类强烈的吸引，但也最不适合在霜冻环境中生存。将这两组进行杂交是可能的，但并不总是那么容易。Goulding（2002）为有兴趣进一步研究此问题的人们提供了资源。

16.6.3.1 三叶

近年来，这种以前被低估的品种变得越来越重要。其部分原因是全球变暖，也因为房屋及其花园通常较小，因此庭院中盆栽种植脱颖而出；还有部分原因在于人们现在比以前在花园上花费的时间更少。更好的耐热性和抗旱性是这类品种的基本属性。

英国 Edwin Goulding、Brian Stannard 和 Brian Kimberley 等在早期开展了许多工作。Brian Kimberley 在育种研究中培育了双色三叶品种 *F.*'Frederick Woodward'（图 16.12a）。在荷兰，Hermann de Graaff 通过将新西兰 *F. triphylla* 和 *F. exorticata* 杂交，将紫茄色引入倒挂金钟的花色中。最早的白色花朵品种可能是 Drude Reimann 于 1983 年培育的 *F.* 'Challenge'，但其没有被推向市场，且迅速消亡了。第一个商业白色三叶品种是 *F.* 'Our Ted'，由 Gouldings Fuchsias 在 1983 年推出。

图 16.12 倒挂金钟属杂交后代：(a) 双色三叶品种 *F.* 'Frederick Woodward'（Kimberley 2005）；(b) 浅粉红色品种 *F. triphylla* 'Purcellian Elegancy'（De Cooker 2016）；(c) 彩叶品种 *F.* 'Sophie's Surprise'（Stannard 1992）；(d) Pantri 杂交种 *F.* 'Gerharda's Panache'（De Graaff 1996）；(e) Paniculate 杂交种 *F.* 'Wapenveld's Bloei'（Kamphuis 1991）；(f) Encliandra 杂交种 *F.* 'Straat Futami'（De Boer 2005）

荷兰杂交育种家 Hans van Aspert 在 2007 年和 2014 年分别推出了浅粉红色的三叶品种 *F.* 'Jaspers Triphywhite' 和几乎白色的品种 *F.* 'Phileine'。2010 年，另一位荷兰杂交育种家 Hans van der Post 培育了浅粉红色品种 *F. triphylla* 'HvdP'，其是 *F. triphylla* 'Herrenhausen' 的自交后代。尽管该品种没有花粉，但将它作为母本具有育性。由于它是一种长势较弱的植物，其并没有为一般大众提供，只是被少数杂交育种家使用。

在2008年，Paul Berry博士利用种子繁殖了幼苗，并命名为“PB7760＃xx”系列。这些植株被用于杂交育种中。从2013年起，De Cooker从*F. triphylla*‘HvdP’和*F. triphylla*‘PB7760#7’的杂交后代中选育了一系列近白色和粉红色的*F. triphylla*优株。在这些优株中，*F. triphylla*‘Purcellian Elegancy’（图16.12b）已经向公众出售。最后一组幼苗表现出很大的表型变异，可作为未来许多新品种培育的基础材料。即将问世的新品种是白色、双色杂合及茄色或紫罗兰色的品种。该群体还开展了叶片的改良。最初引进的是*F.*‘Sophie’s Surprise’（图16.12c），该植物由Brian Stannard发现并由Gouldings Fuchsias于1992年引入。随后，John Ridding在英格兰推广了一种更为强壮的品种*F.*‘Firecracker’。*F. triphylla*和*F. fulgens*变种之间进行杂交后，也可能存在一系列具有吸引力的叶片。还应该指出的是，尽管这类植物中的许多植物都具有浓密和直立生长的习性，但是相当多的植物同样能够在悬挂容器中生长良好。

16.6.3.2 “Pantris”和Paniculates

在*Quelusia*组和*Ellobium*组中，涉及同一物种的主流杂交品种相差甚远，Hermann de Graaff通过将*F. paniculata*与*F. triphylla*杂交获得了一些令人鼓舞的新成果，包括*F.*‘Gerharda’s Panache’。“Pantris”指的是来自*Schufia*组的圆锥花序品种和来自*Fuchsia*组的*F. triphylla*的杂交组合。圆锥花序品种可在圆锥花序末端产生大量的花朵，延长开花时间。其与三叶倒挂金钟的杂交后代中即可出现显著变化。就其自身而言，它们还需要引入更广泛的花朵颜色，也许还需要寻求更大的花朵。该系列最有价值的性状之一就是它的活力。在其原生境中，*F. paniculata*和来自同一*Schufia*组的*F. arborescens*一样，可以生长为大灌木和小型树木。*F. arborescens*品种中多花的杂交种有*F.*‘Miep Aalhuizen’和*F.*‘Wapenveld’s Bloei’）（图16.12e）。

16.6.3.3 “Encpans”

在这一点上，*Schufia*组和*Encliandra*组（“Encpans”）的种间杂交群体可能是本书所述杂交种中最不优秀的。应该指出的是，圆锥花序和伞形花序植物在其原生境中主要是由昆虫授粉。因此，它们往往会形成大量花朵，但花朵比许多其他组的种小。通过这些中美洲不同组之间的杂交，可以在F_1代中筛选到活力、花色和花型更广泛的植株。

16.6.3.4 Encliandras

对于小花型*Encliandra*组的植物，已经开始探索和开发其杂交遗传潜力的工作。杂交育种研究的一个实例是*F.*‘Straat Futami’（图16.12f）。该组的一个奇特表型是许多种和品种只有一种花蕊，主要是雌蕊。在Goulding最近几年未发表的实验研究中，这些植物的种子似乎比其两性植物的更容易获得。这一点在异交时尤其重要。

16.7 结 论

本章表明，目前倒挂金钟属杂交的趋势已大大扩展到了在杂交中使用不同种类和组别的范围。除南美洲外，中美洲已成为重要的资源地区。对耐寒性和竞争性植物的关注已转移到其他领域，如冬季开花、黄色和茄色等新颖的颜色。

随着流式细胞技术等新技术的普及和遗传分析成本的不断降低，我们有望在破译现有品种的起源、为改良和实验开发更多杂交种方面取得进一步发展。

本章译者：

李　帆[1]，耿怀婷[2]，王继华[1]

1. 云南省农业科学院花卉研究所，国家观赏园艺工程技术研究中心，云南省花卉育种重点实验室，昆明 650200

2. 云南大学资源植物研究院，昆明 650091

参考文献

Averett JE, Hahn WJ, Berry PE, Raven PH (1986) Flavonoids and flavonoid evolution in *Fuchsia* (Onagraceae). Am J Bot 73:1525–1534

Berry PE (1989) A systematic revision of the genus *Fuchsia* section *Quelusia* (Onagraceae). Ann Mo Bot Gard 76:532–584

Berry PE, Hahn WJ, Sytsma KJ, Hal JC, Mast A (2004) Phylogenetic relationships and biogeography of *Fuchsia* (Onagraceae) based on noncoding nuclear and chloroplast DNA data. Am J Bot 91:601–614

Crowden RK, Wright JR, Harborne JB (1976) Anthocyanins of *Fuchsia* (Onagraceae). Phytochemistry 16:400–402

De Cooker MGRT (2016) The Fuchsia Breeders Initiative. 7:11–15

Goulding EJ (ed) (2002) Fuchsias the complete guide. Batsford, London

Henry IM, Dilkes BP, Young K, Watson B, Wu H, Comai L (2005) Aneuploidy and genetic variation in the *Arabidopsis thaliana* triploid response. Genetics 170:1979–1988

Jordheim M, Lunder H, Skaar I, Andersen OM (2011) Anthocyanins from fuchsia flowers. Nat Prod Commun 6:35–40

Ramsey J, Schemske DW (2002) Neopolyploidy in flowering plants. Annu Rev Ecol Syst 33:589–639

Rosema G (2001) Flow cytometry, a useful tool in determining the genome formula of a hybrid. J Fuchsia Res Int 1(2):22–24

Thorne T (ed) (1964) Fuchsias for all purposes. Collingridge, London, pp 1–175

Talluri RS (2007) Interspecific hybridisation in *Fuchsia.* Ph.D. thesis, School of Biological Sciences, University of Auckland, 1–215

Talluri RS (2012) Interploidy interspecific hybridization in *Fuchsia*. J Genet 91:71–74

Van Veen G (1992) Winterharde Fuchsia's. Gottmer, Haarlem, pp 1–120

Wood WP (1950) A Fuchsia survey. Benn, London, pp 1–182

第 17 章 非 洲 菊

Zhanao Deng and Krishna Bhattarai

摘要 栽培非洲菊是全球花卉贸易中五大鲜切花之一，在多个国家被广泛应用于盆栽和园林种植。过去 50 年的非洲菊育种培育了大量的无性系品种和由种子繁殖的 F_1 杂交品种。尽管切花非洲菊、盆栽和园林种植非洲菊品种的育种目标不尽相同，但是培育更多的花色、花型和着色模式，早花或持续开花，高品质，高花朵数或产量，对生物胁迫和非生物胁迫有抗性，以及强壮的植物品性对于所有类型的非洲菊都十分重要。杂交与近交是非洲菊的主要育种途径，并仍将在未来的非洲菊育种工作中承担关键角色。在最近几年，白粉病抗性、灰霉病抗性、多叶茎、卷曲花瓣等新特性或特征已经被发现。遗传学、基因组学和分子学研究在过去 20 年中引导了很多新工具与资源的发展，包括白粉病和灰葡萄孢抗性分子标记、局部和全基因组遗传连锁图谱、花型和抗病性候选基因、花色苷合成和调控基因克隆、基因组和转录组序列及转基因技术。基于已知的和其他新特性在新品种中的整合及新的育种工具的广泛应用，非洲菊育种有望达到一个新的高度。

关键词 育种；重瓣花；品种改良；切花；花型；非洲菊；杂交；遗传

17.1 引 言

栽培非洲菊因其迷人的花朵、多彩而绚烂的颜色、丰富的着色模式和样式而备受推崇。栽培非洲菊最常用作切花，也常用于盆栽，作为露台、花园和景观植物种植。非洲菊为二倍体，其基因组较大，有 50 条染色体（$2n = 2x = 50$）。栽培非洲菊由野生非洲品种 *Gerbera jamesonii* 和 *G. viridifolia* 杂交而得，被暂且命名为 *Gerbera hybrida*（Bremer 1994；Hansen 1999）。

林奇（Lynch）是非洲菊育种的先锋，1890 年左右，他在英格兰使用南非的非洲菊开创了育种先河。如今，绝大多数非洲菊品种是由荷兰的一些私有育种公司培育和发布的。在美国、加拿大、以色列和日本，较小规模的私人育种工作也很活跃。而政府支持的非洲菊育种工作在美国、韩国、巴西、中国和印度等有报道。据估计，全球范围内，成千上万的非洲菊品种在过去的 125 年中被培育，超过 1000 个非洲菊品种被投入商业化生产，其中约 1/4 具有比较大的商业化生产规模。消费者对于非洲菊，尤其是切花非

Z. Deng (✉), K. Bhattarai
University of Florida, IFAS, Department of Environmental Horticulture, Gulf Coast Research and Education Center, Wimauma, FL, USA
e-mail: zdeng@ufl.edu; krishnabhattarai@ufl.edu

洲菊有很大需求。私有育种公司在近几年推出了许多非洲菊新品种。仅在 2016 年，就有超过 100 种新非洲菊品种在荷兰花卉拍卖会上被推介。

新改良品种的推出极大地丰富了切花非洲菊的花色，提高了花朵质量，增加了花产量，并且延长了瓶插期。经选育，包括重瓣、完全重瓣、拉丝花型和深色花心等在内的大量新特征被发现，并应用于新品种培育。非洲菊种子繁殖的 F_1 盆栽品种通过育种和选择被改良。近年来，抗白粉病和灰葡萄孢、多叶茎、卷曲花瓣和其他新特征和特性被发现。由于这些特征和其他新特征被整合到新品种中，加之新育种工具的研发和应用，非洲菊育种的未来如其花般一片光明。

17.1.1　商业化生产和经济价值

非洲菊于 20 世纪 30 年代商业化，自此，其地位在全球花卉贸易中日渐上升。据报道，1999 年荷兰花卉拍卖会上售出了不少于 6.26 亿支的切花非洲菊。2010 年，这个数值达到了 9.38 亿，一年产生了 1.4 亿欧元的销售额，基于销售数量和销售额，非洲菊荣居世界五大切花之一（FloraHolland 2014）。

非洲菊在 80 多个国家商业化生产，世界上主要的切花非洲菊生产国包括荷兰、哥伦比亚、美国、加拿大、以色列、厄瓜多尔、墨西哥、中国、印度、韩国和日本等，仅荷兰一年就生产多达 4.2 亿支非洲菊。过去几年，美国切花非洲菊的年销售额维持在 3200 万～3600 万美元（USDA 2010，2016）。加利福尼亚州是美国切花非洲菊的首要生产州，年销售 1 亿～1.2 亿支非洲菊，销售额 3200 万～3500 万美元（USDA 2010，2016）。在美国，非洲菊盆栽在美国的生产规模较小但更为广泛，得克萨斯州、加利福尼亚州、伊利诺伊州、佛罗里达州和密歇根州为前五大生产州（Granke et al. 2012），年销售额为 400 万～800 万美元。

从简单的对生长条件控制较少的塑料拱棚（或大棚）到能出色控制温度、光照、光周期、辐射、肥料、pH 和盐的高度自动化先进温室，许多设施系统已用于商业化生产切花非洲菊。水培系统和无土栽培是切花非洲菊商业化生产的一大趋势，它们可以大大降低土传病害的压力并提高花卉产量。随着这些体系的广泛应用，为了使新品种能适应新种植体系，保证预期的生产效率、花朵质量并进行虫害防治，切花非洲菊的育种目标和选择标准将随之调整。由于越来越多非洲菊盆栽也开始用无土栽培和温室生产，因此相似地调整对于盆栽非洲菊品种的培育亦然。

17.1.2　繁殖

传统非洲菊的繁殖方式包括播种、扦插和分株。在非洲菊发展的早期，种子繁殖是产生非洲菊的唯一方式。由于非洲菊是异花授粉且高度杂合，种子繁殖就存在一个严重的缺点，即后代植株形态多变，叶片和花朵在形态上变化极大。在商业化生产中，直到培育盆栽和园林栽培非洲菊近交系与 F_1 杂交品种，种子繁殖才被广泛使用。种子繁殖降低了这些品种的生产成本，且能在特定市场窗口期大量生产一致的植株。分株和扦插在非洲菊早期商业生产中发挥了重要作用。由于分株和扦插扩繁效率低下且容易传播土传病害，如今，除了小规模亲本植株或育种株系繁殖和进行遗传学研究，这些方法鲜有被使用。

自20世纪70年代以来，组织培养已成为切花非洲菊和非种子繁育的盆栽、园林非洲菊的主要繁殖方式。体外培养的启动是一个非常缓慢且昂贵的过程，然而一旦建立了体外培养体系和操作流程，就能年生产数以万计的苗木。非洲菊组织培养生产主要在中国和印度，年生产千万株种苗。茎尖和未成熟的头状花序是最常用的外植体。高浓度细胞分裂素导致表观遗传变异，但是通常来说体细胞无性系变种极为少见，可以保证非洲菊的遗传保真度（Cardoso and Da Silva 2013）。不同非洲菊品种对体外培养条件的反应不同，因此，培养基成分和配方需要经常调整，否则新品种会因对培养条件低响应或响应不稳定而无法再生。

17.1.3 相关种和潜在种质

1737年，Gronovius 建立大丁草属以纪念德国自然学家和医师 Traugott Gerber。大丁草属包含的种的数量一直没有明确。Hansen（1985）鉴定了 21 个和 6 个分别来自非洲和亚洲的种，共计 27 个种。而 Katinas（2004）鉴定出了 29 个种，新增了原属于美洲的 *Chapitalia* 属和 *Tricholine* 属的各一个种。目前，植物名录数据库收录了 40 个被确认为大丁草属的种（http://www.theplantlist.org/）。

大丁草属被分为 7 个部分，其中非洲菊（*Gerbera jamesonii*）、绿叶毛足菊（*G. viridifolia*）与其他三个东非热带的种即毛足菊（*G. ambigua*）、橙黄毛足菊（*G. aurantiaca*）和长叶毛足菊（*G. galpinii*）属于毛足菊组（Hansen 1985）。该组中的种非常相近，仅以一些形态学性状相区分（Hansen 1985）。橙黄毛足菊序大，舌状花长，为深红色，花心黑色，如今在南非是濒危品种（Johnson et al. 2004）。长叶毛足菊原产栖息地潮湿，全叶光滑。这些种可能是提高非洲菊观赏价值、抗逆性、环境适应能力非常有用的种质资源。*Isanthus* 组中的 6 个种均原产亚洲，起源于寒冷高纬度环境。这些亚洲种可能具备抗寒性。其中，原属 *Chapitalia* 的 *G. gossypina* 仅含 9 条染色体（Bala and Gupta 2013）。由此可推测，与非洲菊（$2n = 50$，5.12pg/2C）和绿叶毛足菊（7.68pg/2C）相比，该种可能有更小的基因组，它会是非洲菊基因组测序的优秀备选（Marie and Brown 1993）。

非洲菊和绿叶毛足菊是早期非洲菊育种的亲本，但它们在现代非洲菊品种中的占比有待了解。现代非洲菊品种中是否含有其他种的基因是个有趣的问题。一些花朵特性，如深色花心，是现代非洲菊品种常见的，但非洲菊和绿叶毛足菊植株并无此性状。这是否是其他种，如橙黄毛足菊的基因导致了该特性，以及野生非洲菊种在非洲菊品种改良中发挥了什么作用，未来的基因组测序可以回答这些问题。

大多数大丁草属中的种自然存在于非洲，大丁草属被认为起源于南冈瓦纳大陆，非洲南部是其多样性中心（Hansen 1985）。近来，对大丁草属复合体的遗传发生分析和生物地理学研究为大丁草属及大丁草属复合体的广泛传播提供了新见解（Pasini et al. 2016）。该复合体包括大丁草属和菊科帚菊木族中其他 7 个属：*Amblysperma*、*Chaptalia*、*Leibnitzia*、*Trichocline*、*Uechtritzia* 等。生物地理学研究表明，该复合体的祖先分布区在美洲西南部，后传播至亚洲南部和西南部，再传播至非洲和澳大利亚。*Amblysperma* + *Gerbera* + *Leibnitzia* + *Uechtritzia* 的复合体主要分布范围在亚洲南部和东南部，广义大丁草属（除亚洲种外）随后传播至东非并在此形成祖先分布区。属于非洲广义大丁草属的种大约在 849 万年前的中新世晚期从澳大利亚 *Amblysperma* 分离而来，此推断的自然传播途径与近几世纪非洲菊人工传播途径（从非洲传播至欧洲，后传播到美洲、澳大利亚和亚洲）完全不同。

17.1.4　非洲菊作为模式系统

非洲菊已被作为花朵和花序发育及次生代谢基础研究的模式系统（Teeri et al. 2006）。这些领域的基础研究产生了大量的论文、先进工具和遗传（基因组）资源，对于非洲菊的遗传改良和育种价值颇大。

17.2　主要性状和育种目标

17.2.1　栽培类型

非洲菊育种聚焦于改良三个类型的栽培种：切花、盆栽和园林用花。尽管主要育种目标因栽培类型不同而有所差异，但是很多特征特性对于任何栽培种都是十分重要的。花茎长（60cm 或更长）、颜色鲜艳、花朵优质、产量高和瓶插期长是最重要的切花非洲菊特性。生长紧凑、矮茎、花朵多而鲜艳、持续开花和能种子繁殖是盆栽类型品种最重要的特性。此外，植株强健、对不同土壤条件有良好适应性、抗生物与非生物胁迫及持续开花是园林花床、门廊、露台种植非洲菊所追求的特征和特性。

17.2.2　花径

花径是所有非洲菊栽培种非常重要的性状，尤其是对于切花品种。不同品种切花非洲菊的花径通常可以分为标准花径或小花径两类。标准品种花径 10～13cm，一些品种花径达 15cm，如 Alain Ducasse®、Amico®、Ankur®、Apollo®、Balance®、Big Deal®、Commander®、Debut®、Inferno®、Snowball® 和 Wannabe®。

迷你品种的培育是非洲菊育种的一大突破，此类品种可生产大量直径在 6～9cm 的花朵。最近几年，迷你非洲菊在荷兰花卉拍卖中的占比已经从 1993 年的 33%提高到了大约 60%。迷你非洲菊的广泛流行为切花非洲菊在全球花卉交易中占比提高贡献良多。

17.2.3　花型

每一朵非洲菊的花都是复合花序（头状花序），由舌状花、舌管状花和管状花组成。在植物学上，每一朵小花是一个单独的花朵。这三种类型的小花指代三种不同类型的花朵。为了避免混淆，在这一章节中，小花将用于指代单花，而花朵或花头被用于指代头状花序。这三类小花在形态、性别和着色等多方面存在差异（Teeri et al. 2006）。外围舌状花较大（长 7～8cm）并且有显眼的、由三片花瓣融合而成的舌片。管状花生长于头状花序的中心，不显眼。舌状花为雌花，有功能性雌蕊，雄蕊退化。管状花为两性花，但在功能上是雄性的，因为它的花药融合在一起并包裹着雌蕊。舌管状花位于舌状花和管状花之间，功能与舌状花相似，并有较短的不显眼的舌片。

非洲菊头状花序由一到几轮舌状花和几轮舌管状花与管状花组成。不同非洲菊品种的舌管状花长度存在可见差异，甚至一朵花内的外围和内围舌管状花长度也不尽相同。基于舌管状花和舌状花的相对长度，非洲菊花序可以分成单瓣、半重瓣和重瓣类型（图 17.1a，b）。重瓣花型的其他类型——大重瓣花型和完全重瓣花型已被发现（Kloos et al. 2004）。这两种花型都有多轮增大的舌管状花，其中完全重瓣花型的最外圈舌管状花长于舌状花的 1/2，大重瓣花型的最外圈舌管状花小于或等于舌状花的 1/2。许多大重瓣花型和完全重瓣花型的非洲菊品种已被推广，包括 Gerrondo®和 Pomponi®系列，这些品系的花朵可含高达 500 个花瓣，加上增大的舌管状花和舌状花，其单花重量明显增加（Kloos et al. 2004）。因此，这些类型的花需要更强的花茎支撑，或减小花径以降低重量。

图 17.1 （a）深色花心的单瓣花。（b）浅色花心的重瓣花。（c）抗白粉病植株（左）和白粉病感病植株（右）。（d）26 个非洲菊基因型简单重复序列标记 PCR 条带，每一列表示一个非洲菊基因型

拉丝花型有锯齿状（分离或裂口）外唇，形成丝带或流苏状花头（Kloos et al. 2004）。不同品种的裂片数量和裂深迥异，甚至同一朵花的不同小花也会存在差异，在一些拉丝花型中，舌状花花瓣分裂成 8 片。裂深从花冠尖浅凹刻到深裂至花冠底部不等，若小花深裂，则花头褶皱蓬松，而浅裂舌状花形成的花朵则为流苏状（Kloos et al. 2004）。裂片和裂深高度相关。

17.2.4 花色

花色是非洲菊育种和花卉贸易追求的最重要的性状之一。非洲菊闻名于舌状花和舌管状花丰富多彩的颜色，包括渐变白、奶油色、黄色、橙色、红色、粉色、浅橙色、猩红色、淡紫色、紫色、紫罗兰色等。非洲菊花色唯一缺少的似乎是蓝色。非洲菊花心（花眼或中心管状花）颜色有浅色（绿-黄、黄-绿或浅黄）或深色（黑-棕、棕-黑或黑-紫）。深色花心非洲菊的舌状花和舌管状花颜色形成鲜明对比，许多具有深色花心的非洲菊品种已经被培育。深色花心是由管状花基部直立冠毛中花青素的积累导致的（Helariutta et al. 1995）。

最近，双色花成为非洲菊栽培品种新特性。这个着色模式是由不同色素在同一朵花的不同类型小花、小花不同部位或花瓣不同区域积累导致的。例如，明黄色花朵品种'Nero'的花冠中含类胡萝卜素，而冠毛中含花青素；乳黄色花朵品种'Mix'在所有类型小花的花冠底部积累色素（the Gerbera Laboratory；http://www.helsinki.fi/gerberalab/）。商业非洲菊品种的双色模式广义上可以分为三类：舌状花双色、舌状花和舌管状花花瓣各呈一种颜色、舌状花和舌管状花均为双色。

在非洲菊育种中，皇家园艺协会（RHS）比色卡或其他比色卡通常用于描述花朵的颜色类别和差异，以进行DUS（特异性、一致性和稳定性）测试，或者申请植物专利或植物育种者权利。比色分析已用于对细微差别的颜色进行更准确的描述。利用该技术，花色通过比色计测量，结果将被转换成国际照明委员会（CIE）*L*（光亮度）* *a*（空间的红-绿色轴）* *b*（空间的蓝-黄色轴）坐标。这样的坐标证明足以申获植物品种权（PBR）。花色也可以用反射分光光度计测量并转换为CIE *L* * *C*（色度）* *h*（色调）色彩空间坐标。Tourjee等（1993）的研究表明在整个生长季，*L**、*C* *和*h*值的变异系数小且一致性高。在非洲菊育种种群中，*L* *、*C* *和*h*值不断变化。*h*值范围在0～100，并显示双峰分布，与非洲菊的色素系统相对应。明亮、强烈而又深的颜色具有较高的*C* *值，而较浅、较弱和暗淡的颜色具有较低的*C* *值。

17.2.5 花产量

对于切花种植者来说，产花量是最重要的性状。不同非洲菊品种的产花量差异很大。Prajapati等（2017）研究表明，当生长在印度的温室中时，'Stanza'每年每株能生产42支花[251支/(m^2·年)]，而'Cherany'每年每株生产20支花[122支/(m^2·年)]。生长条件也显著影响非洲菊的产花量（品质），在日温22～25℃、夜温12～25℃、大于12h的光照的条件下，光照强度250～450μmol/(m^2·s)是非洲菊生长的最佳条件（Hopper et al. 1997）。

非洲菊育种已朝着提高产花量迈出了一大步。当前许多标准的切花非洲菊栽培品种在土壤栽培时可生产250～270支/(m^2·年)，无土栽培时可生产310～330支/(m^2·年)。迷你切花非洲菊品种在土壤栽培时可生产460～500支/(m^2·年)，无土栽培时可生产600～650支/(m^2·年)。高产迷你品种包括Babble®、Ballerina®、Banana®、Bing®、Boost®、Choiz®、

Del Monte®、Garfield®、Genius®、Hot Sundae®、Kitty®、Lemon Ice®、Lobby®、Loveland®、Okidaki®、Rapido®、Saga®和 White Star®。

17.2.6 瓶插期

瓶插期是消费者非常重视的切花非洲菊性状。长瓶插期是切花非洲菊最重要的育种目标之一，而在此目标上，非洲菊育种取得了巨大成果。早期非洲菊品种只有 3～5 天的瓶插期，如今一些品种的瓶插期已延长至 24 天。大多数切花非洲菊品种的瓶插期为 2 周或更长时间。

不同非洲菊品种的瓶插期差异很大（Reid 2004）。花茎弯曲（花葶弯曲或花茎塌陷）是瓶插期缩短的一个主要因素。与底部花茎相比，非洲菊花头下 10～15cm 处的花茎木质化程度和硬度都低得多。当此部分花茎由于底部花茎呼吸作用和细菌阻塞而失水时，细胞将无法维持所需的膨大，花茎将在花头的重力作用下弯曲。非洲菊花茎弯曲趋势在不同品种中差异也很大，'Amber'、'Mickey'、'Nikita'、'Ruby Red'、'Simonetta' 和 'Terra Fame' 的花茎几乎不弯曲，而 'Cora'、'Liesbeth' 和 'Tamara' 的花茎则容易在瓶插时弯曲（de Witte et al. 2014；Perik et al. 2012；van Doorn and De Witte 1994）。Perik 等（2012）通过测试 'Tamara' 花茎弯曲花朵和不弯曲花朵的花茎组织，发现花茎弯曲可能是由缺乏厚实的厚壁组织导致的茎上部机械强度不足造成的。非洲菊的花茎有时在花头下 0～5cm 处弯曲，导致所谓的花颈弯曲。'Tamara' 时常发生花茎弯曲，但花颈弯曲并不常见，发生不规律（Perik et al. 2012）。花瓣枯萎和真菌病害也会影响非洲菊的瓶插期。非洲菊花朵属于非呼吸跃变型，乙烯并不能诱导花瓣枯萎（Reid 2004）。非洲菊花朵最常见的病害是灰霉病，其可大大缩短非洲菊的瓶插期。

17.2.7 抗病性

非洲菊的花、叶、根和冠能被许多病原体感染，会出现严重病害症状，抑制非洲菊植株发育，降低花朵品质，减少产花量，并影响植株性能（Reddy 2016）。使用杀菌剂、改进栽培技术和生长条件是病害防治的主要方法。介于多种因素，如病原体对杀菌剂会逐渐产生耐药性、社会和工人对杀菌剂潜在的健康危害越来越关注、强制执行的环保政策逐渐严格及消费者对于高品质花卉的需求渐渐提高，抗病品种的应用在盆栽、切花等不同类型的非洲菊生产中的关注度日益提升。近年来，鉴定抗病资源并将其整合到新的品种中已成为非洲菊育种和遗传研究的主攻目标。

17.2.7.1 白粉病抗性

白粉病是非洲菊最常见的病害之一，主要由单囊壳白粉菌引起。真菌在植物表面形成菌丝网络，最终叶片和花朵被白色滑石样菌落覆盖，使得植物或切花失去销售价值。随着病害发展，叶片坏死且植株渐弱。提高非洲菊的白粉病抗性已成为几个非洲菊育种计划的主要育种目标（Deng and Harbaugh 2010，2013；Kloos et al. 2005；Song and Deng 2013）。在商业化非洲菊品种或育种系中已经鉴定了一些白粉病的抗性资源。非洲菊品种‘Terra Fame’（切花）和‘Festival Semi-Double Orange’（盆花）具有中等白粉病抗性（Hausbeck et al. 2002；Hausbeck 2004）。一些非洲菊育种系（176 和 214、UFGE 31-19、UFGE 5-23）具有较高的白粉病抗性（Deng and Harbaugh 2010；Kloos et al. 2005）。UFGE 31-19 的白粉病抗性已经被整合到许多新的非洲菊育种系（包括 UFGE 4033）（图 17.1c）和栽培品种中（Deng and Harbaugh 2010，2013）。Song 和 Deng（2013）指出，与感病非洲菊品种‘Sunburst Snow White’相比，抗性株系 UFGE 4033 及其后代植株中黄腐菌的菌丝分枝少而短，因而在叶片表面产生较少的分生孢子（减少 77%～94%）。

17.2.7.2　灰霉病抗性

非洲菊灰霉病是由腐生真菌灰葡萄孢引起的，该病原体寄主广泛，危害深远。灰霉病主要发生在生产期及采后的非洲菊花朵上。灰霉病导致非洲菊舌状花和舌管状花上形成坏死斑点或管状花发生腐烂（Fu et al. 2017）。在冬季温室生产和低温贮藏或运输过程中，灰霉病的发生率升高，后果也更为严重。由于拍卖市场上有大量非洲菊出售，一小部分非洲菊的感染就将导致严重的经济损失，荷兰花卉拍卖对携带灰葡萄孢的非洲菊零容忍，所有感染灰葡萄孢的非洲菊鲜花均会被拒绝并销毁（Fu et al. 2017）。因此提高非洲菊对灰葡萄孢的抗性，对降低由灰葡萄孢导致的切花非洲菊经济损失是急需的。

在自然灰葡萄孢的胁迫下，某些非洲菊品种（如‘Fiction’）和育种系的感病程度不同。通过将灰葡萄孢接种到非洲菊全花、舌状花花瓣和管状花底部，Fu 等（2017）阐述了评估非洲菊灰葡萄孢抗性的方案。接种花朵或小花的灰霉病害程度被分为等级 0（无明显症状）～5（完全坏死）。在两个分离种群中，感病指数从整朵花到管状花底部有中到高的相关性（$r = 0.67$～0.83），但是全花和舌状花的感病指数没有显著相关性（Fu et al. 2017），这表明不同非洲菊小花的抗病机制并不相同。

17.2.7.3　疫霉病抗性

隐地疫霉感染非洲菊的根，导致根腐和冠腐。伴随冠腐的发展，叶片变成棕褐色并枯萎，最终整个植株倒塌。多年来，隐地疫霉是引发非洲菊分株和扦插繁殖时疫病的元凶。组织培养繁殖和无土基质的广泛使用限制了疫霉病的传播，减少了其发生率，但是在土壤栽培体系中，它仍然会引起植株枯萎和植株损失。隐地疫霉还是引发园林栽培非洲菊植株损失的主要病原体。Sparnaaij 等（1975）描述了通过一个接种程序和两种测试类型（克隆和后代）来确定非洲菊对隐地疫霉的抗性。在克隆测试中，每一个非洲菊品种有 30～40 个由克隆繁殖所得的幼株被接种。后代测试中，标准品种被用作母本，与待测品种进行杂交，后代植株将被接种以评估品种的抗病性。该标准品种应该为自交不亲和但杂交亲和，且对疫霉病不应太具抗病性或感病性。与克隆测试相比，后代测试似乎更简单且更灵活。Li 等（2008）筛选了 13 个非洲菊品种并对一些本地开发的品种进行了抗病性鉴定。

17.2.7.4 叶斑病抗性

非洲菊上的叶斑是由腐生真菌链格孢菌引起的。植株感染后在下部较老叶片上产生水渍状小斑点，后发展成具有同心轮纹、黄色晕环的凹陷棕色病变。严重侵染会导致非洲菊叶片萎黄和脱落，导致花朵数量减少、扭曲且无法销售（Farhood and Hadian 2012）。在印度，病害筛查研究鉴定了两个抗病品种‘Mammoth’和‘Tiramisu’（Kulkarni et al. 2009；Nagrale et al. 2013）。

17.2.7.5 菊花枯萎病抗性

菊花枯萎病由尖孢镰刀菌菊花专化型引起。该病害症状始于非洲菊植株基部和冠部，向叶和花发展，导致植株矮化、叶片萎黄。严重受害植物会枯萎并最终死亡（Pataky 1988）。非洲菊品种的筛选表明有 47%～75%的受测品种在 2004～2006 年的试验中具备抗病性（Minuto et al. 2007）。

17.2.8 西花蓟马抗性

西花蓟马（*Frankliniella occidentalis*）以非洲菊的花朵为食，在小花上造成难看的斑点和微粒，并破坏小花和花朵的结构。这些蓟马也是凤仙花坏死斑病毒和番茄斑萎病毒的载体。它们主要生活在叶片和花朵的缝隙中，或在小花之间，因此喷施杀虫剂也很难控制。在佛罗里达存在天然且高密度西花蓟马的温室中，一些盆栽非洲菊品种所受虫害较轻，这可能表明商业化非洲菊品种对西花蓟马存在部分抗性（Z. Deng，未发表）。

17.3 遗传防治和遗传模式

17.3.1 花型

17.3.1.1 重瓣花型

以超过 500 种不同花型的非洲菊品系为材料，通过经典遗传学研究，Kloos 等（2004）揭示了非洲菊的花型遗传模式。在自交或异花授粉时，单瓣非洲菊只产生单瓣后代，而重瓣非洲菊后代的花型则出现性状分离。单瓣花和重瓣花的杂交测试中，后代或全为重瓣花，或重瓣花与单瓣花 1∶1 分离。在 F_2 代中，重瓣花与单瓣花 3∶1 分离。单瓣花是隐性性状，而重瓣花由单显性等位基因 *Cr* 控制（Kloos et al. 2004）。

完全重瓣花型的非洲菊后代为重瓣花型，且分离出两种重瓣类型：雄性不育的大重瓣花型和雄性可育的半重瓣花型，两者比例为 1∶1。这是由位于 *Cr* 位点的另一个显性等位基因 Cr^d 控制的，完全重瓣的非洲菊基因型为杂合的 Cr^dCr（Kloos et al. 2004）。三个等位基因组成复等位基因系，Cr^d 控制舌管状花和管状花增大的雄性不育重瓣花型，*Cr* 控制舌管状花变大的重瓣花型，而 *cr* 控制舌管状花和管状花不显著的单瓣花型。它

们的等位性为 $Cr^d > Cr > cr$。*Cr* 对 *cr* 不完全显性，纯合体头状花序的舌管状花比杂合体头状花序的更长。绝大多数的商业化非洲菊品种是单瓣花型或半重瓣花型。因此，*Cr* 和 *cr* 被认为是控制非洲菊花型的最常见等位基因。携带 Cr^d 等位基因的成熟非洲菊花朵会更重瓣，因此存在很高的花茎弯曲或破损的风险。不能排除有相关修饰基因影响 *Cr* 的表达，进而改变完全重瓣花型的重瓣化和雄性不育程度（Kloos et al. 2004）。

Cr 基因座尚未被分子标记或遗传/物理标记定位，也没有任何等位基因被克隆。但是，过去 15 年的分子生物学研究鉴别了该基因座的几个候选基因。非洲菊舌状花和管状花基因表达微阵列分析鉴定出许多 MADS-box 转录因子基因（*GGLO1*、*DGEF2*、*GAGA1*、*GAGA2* 和 *GRCD1*），它们在两种类型小花之间的表达不同（Laitinen et al. 2006），该组基因在模式植物中调控花器官发育。被鉴定的 MADS-box 基因在非洲菊头状花序中沿半径呈梯度表达，极有可能参与非洲菊小花类型发育调控。

CYCLOIDEA/TEOSINTE BRANCHED1 (*CYC/TB1*)-like TCP 域转录因子基因家族在非洲菊中至少有 10 个基因成员，其中包括 *GhCYC2*、*GhCYC3* 和 *GhCYC4*（Tähtiharju et al. 2011）。*CYC1/TB1*-like 基因参与调控叶腋分生组织的发育，决定叶腋分生组织发育成枝或花。*GhCYC2* 在非洲菊头状花序中辐射表达，主要表达于边缘发育的微小舌状花中，而在中心管状花中无表达（Broholm et al. 2008）。由强启动子 35S 异位过表达 *GhCYC2* 的转基因非洲菊，其管状花转变成了花瓣变大、雄蕊发育不良的类舌状花（Broholm et al. 2008）。抑制 *GhCYC2* 的表达导致舌管状花的花瓣变短。过表达 *GhCYC2* 的表型与重瓣性状类似，而抑制 *GhCYC2* 表达导致舌管状花不显著，这与隐形重瓣等位基因 *crcr* 控制的单瓣花型类似。然而，重瓣非洲菊中的 *GhCYC2* 表达与半重瓣非洲菊和单瓣非洲菊中的并无差异（Broholm et al. 2008）。*GhCYC2* 更像是修饰基因，并在非洲菊小花类型的遗传控制中起重要作用。

在非洲菊中，*GhCYC4* 与 *GhCYC2* 具有相似的表达模式但其表达量更高。在非洲菊中异位过表达 *GhCYC4*，管状花转变为类舌状花且舌管状花增长，形成完全重瓣或大重瓣花型（Juntheikki-Palovaar et al. 2014）。*GhCYC3* 在非洲菊花发育阶段和不同类型小花间表达模式显著不同，它在舌状花中的表达量是 *GhCYC2* 和 *GhCYC4* 的很多倍，并且在早期舌状花发育阶段表达量最高（阶段 1），然后逐渐减少至花朵完全开放（阶段 9）。异位激活 *GhCYC3* 引起与 *GhCYC4* 相似的表型变化。*GhCYC3* 是唯一一个在完全重瓣花型非洲菊栽培种的中央花原基中上调表达的基因。*GhCYC4*，尤其是 *GhCYC3*，在非洲菊小花类型的基因决定中起着关键作用。

17.3.1.2　拉丝花型

Kloos 等（2004）研究发现，拉丝花型是由单显性基因 *Sp* 控制的。非拉丝花型的非洲菊互相杂交时仅产生非拉丝类型后代。拉丝花型非洲菊互相杂交的后代要么仅为拉丝花型，要么拉丝花型与非拉丝花型以 3∶1 分离。拉丝花型和非拉丝花型杂交后代中呈现拉丝花型与非拉丝花型 1∶0（即后代全为拉丝花型）或 1∶1 的分离比。在 F_2 代中，拉丝花型与非拉丝花型分离比为 3∶1。*Sp* 和 *Cr* 位点独立分离并且位于不同的连锁组或染色体上（Kloos et al. 2004）。

拉丝花型的发育基础尚待确定。舌状花和舌管状花的舌片由三个花瓣裂片融合而成（Breme 1994; Kotilainen et al. 1999）。拉丝花型可能是由花瓣裂片融合失败导致的（Kloos et al. 2004）。在 *GhCYC2* 表达受抑制的转基因非洲菊中，舌管状花的舌片偶尔分裂成 5～8 个分离花瓣（Broholm et al. 2008）。*GhCYC* 家族中的某些基因可能参与非洲菊拉丝花型的形成。

17.3.2 花色

17.3.2.1 花心颜色

非洲菊花心大致可分为浅色和深色，由管状花冠毛中的色素沉着决定。浅色花心非洲菊品种‘Regina’和深色花心品种‘Nero’杂交，后代浅色花心和深色花心的分离比为 1∶1（Elomaa et al. 1998）。Kloos 等（2005）研究了近 550 个育种系的 F_1、F_2 代和测试杂交的花心颜色分离，结果表明深色花心由单显性基因 *Dc* 控制。多数非洲菊是杂合基因型（*Dcdc*），一些是纯合基因型（*DcDc*）。苞片边缘色素沉着与深色花心表型息息相关，Kloos 等（2005）发现在早期花芽发育中，超过 98%具有深色花心的非洲菊苞片边缘有略带紫色的色素沉着。这种色素沉着在浅色花心花朵的苞片和不到 2%的杂交产生的深色花心后代的花芽的苞片中是缺失的。

17.3.2.2 舌状花和舌管状花花色

非洲菊的花色来源于舌状花和舌管状花中类胡萝卜素和/或花色苷的积累。类胡萝卜素是黄色和橙色非洲菊中的主要色素，而花色苷是红色和粉红色非洲菊的主要色素。非洲菊有天竺葵素（Pg）和矢车菊素（Cy）两类花色苷。基于花朵中色素类型，非洲菊品种被分为 4 类：无花色苷型、天竺葵素型、矢车菊素型、天竺葵素和矢车菊素混合型。白色和黄色的非洲菊不含花色苷，属无花色苷型。其他三类含天竺葵素和/或矢车菊素的为花色苷型。花色苷的不同种类和两种相关的类黄酮（黄酮和黄酮醇）在一些非洲菊栽培种与实验系中可以被找到（表 17.1）。无花色苷的非洲菊有多种着色模式（表 17.1）（Bashandy et al. 2015；Tyrach and Horn 1997）。色彩相近的非洲菊栽培种可能含有不同的类黄酮成分（Asen 1984）。非洲菊花瓣中黄酮的存在降低了花色苷的含量，使非洲菊花色呈橙色而非明显的红色（Martens and Forkmann 2000）。

表 17.1 一些非洲菊品种和实验系的花色及花瓣中包含的花色苷、黄酮、类黄酮种类（通过花色排列）

品种/种系	花色	花色苷	黄酮	类黄酮	参考文献
Delphi	白色	—	（Km）	Ap	4、5
Ivory（sport of Estelle）	白色	—	Km	Ap	2
T1	奶油色	—	Km	—	4、5
S-14	黄色	—	Km	—	1
S-103	黄色	—	Km，Qu	—	1
S-133	黄色	—	Km，Qu	（Lu）	1
S-196	黄色	—	Km，Qu	Ap	1

续表

品种/种系	花色	花色苷	黄酮	类黄酮	参考文献
SC-202	黄色	—	Km，Qu	Ap	1
Simm	黄色	—	（Km）	Ap	4、5
T3	黄色	—	Km	—	4、5
Ceres-2000	橙红色	Cy（Pg）	Qu，Km	—	1
S-4	橙红色	Pg（Cy）	Km，Qu	—	1
S-39	橙红色	Pg	Km，Qu	—	1
S-92	橙红色	Pg	Km，Qu	（Lu）	1
Th 58	橙色	Pg（Cy）	Km（Qu）	Ap（Lu）	4
D1	红色	Pg，Cy	Km，Qu	Ap	4、5
Murphy	浅红色	Pg	?	?	2
Passion	深红色	Pg，Cy	?	?	2
Terra Regina	红色	Pg	?	?	2、3
Clivia	粉色	Cy	Qu	?	4
Estelle	粉色	Pg	Km	Ap	2
Parade	粉紫色	Cy	?	?	2
President	粉紫色	Cy	?	?	2、3

参考文献：1. Asen（1984），2. Bashandy 等（2015），3. Deng 等（2014），4. Martens 和 Forkmann（2000），5. Tyrach 和 Horn（1997）

“—”和“？”分别表示较低含量、缺失或无数据

Pg. 天竺葵素苷，Cy. 矢车菊素苷，Ap. 芹菜苷元，Lu. 木犀草素，Km. 山奈酚，Qu. 槲黄素

20 世纪 60 年代的早期花色分类表型分离研究表明非洲菊的花色由 2～3 个主要基因控制。随后在 70 年代，基于对非洲菊花朵的色素分析和花色苷合成知识的了解，4 个主要基因和一些多基因控制非洲菊花色的假设被提出。Tyrach 和 Horn（1997）对非洲菊自交和杂交后代的色系与花色苷分离进行了广泛的调查（表 17.2）。

表 17.2　花色系分离和杂交、自交非洲菊后代中的花色苷（数据来源于 Tyrach and Horn 1997）

杂交类型	后代分离
无花色苷型自交	仅无花色苷型
无花色苷型×花色苷型	仅无花色苷型，仅花色苷型，或花色苷型：无花色苷型 1：3
花色苷型×无花色苷型	仅花色苷型，或花色苷型：无花色苷型 1：1
花色苷型自交	仅花色苷型，或花色苷型：无花色苷型 3：1
花色苷型×花色苷型	仅花色苷型，或花色苷型：无花色苷型 3：1
天竺葵素型自交	仅天竺葵素型，或天竺葵素型：矢车菊素型 3：1
矢车菊素型自交	仅矢车菊素型，或天竺葵素型：矢车菊素型 1：3
天竺葵素型×矢车菊素型	仅天竺葵素型，或天竺葵素型：矢车菊素型 1：1

多数情况下，无花色苷的非洲菊对有花色苷的非洲菊是隐性的，但在少数情况下，无花色苷型对花色苷型表现出隐性上位（Tyrach and Horn 1997）。这些隐性上位的非洲菊杂交后会产生花色苷型的后代。从生物学的角度来看，花色苷合成通路中的一步或多

步被阻断是无色苷型非洲菊中花色苷缺失的原因（图 17.2）。Martens 和 Forkmann（2000）观察到：①黄烷酮-3-羟化酶（F3H）活性阻断；②二氢黄酮醇-4′-还原酶（DFR）或花青素合成酶（ANS）活性阻断；③步骤①和②（双突变体）同时阻断。

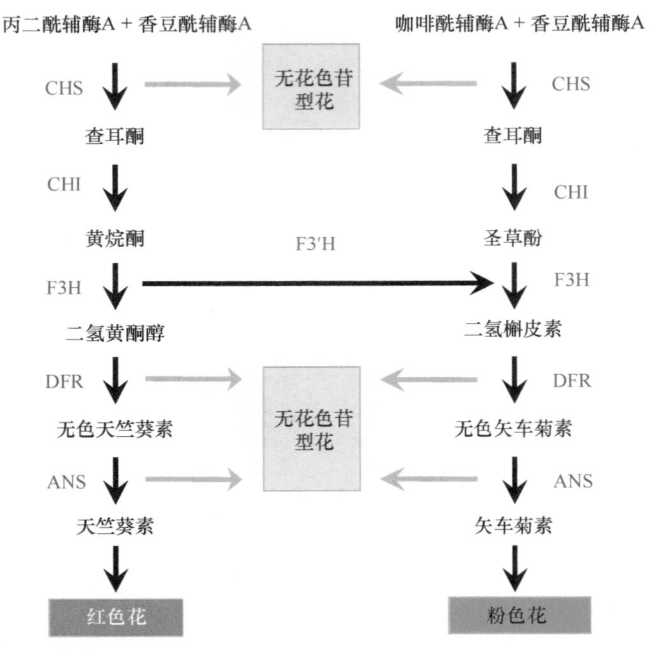

图 17.2 非洲菊花色苷生物合成途径示意图。ANS. 花青素合成酶，CHI. 查耳酮异构酶，CHS. 查耳酮合酶，DFR. 二氢黄酮醇-4′-还原酶，F3′H. 类黄酮-3′-羟化酶，F3H. 黄烷酮-3-羟化酶。查耳酮异构酶、二氢黄酮醇还原酶和/或花青素合成酶活性的缺失或阻断导致花色苷合成的缺失，形成无花色苷型非洲菊花朵（白色、奶油色或黄色）（Martens and Forkmann 2000；Tyrach and Horn 1997）

参与花色苷生物合成途径的酶包括查耳酮合酶（CHS）、查耳酮异构酶（CHI）、黄烷酮-3-羟化酶（F3H）、类黄酮-3′-羟化酶（F3′H）、类黄酮-3′5′-羟化酶（F3′5′H）、二氢黄酮醇-4′-还原酶（DFR）和花青素合成酶（ANS）（图 17.2）。非洲菊中的查耳酮合酶由 *GCHS1*、*GCHS3* 和 *GCHS4* 编码。*GCHS1* 控制非洲菊花瓣中所有类黄酮类物质的生物合成。*GCHS4* 在非洲菊花瓣中表达较强，但与花瓣色素沉着无关（Deng et al. 2014）。4 个 *DFR* 等位基因（*GDFR1-1*、*GDFR1-2*、*GDFR1-2 m* 和 *GDFR1-3*）已在品种‘Terra Regina’、‘Estelle’和‘Ivory’中被鉴定（Bashandy et al. 2015）。*GDFR1-2 m* 是 *GDFR1-2* 的突变形式，其在 DFR 酶编码基因高度保守区域存在单碱基突变，使 DFR 酶失活。DFR 酶在非洲菊体内有很强的底物偏好：*GDFR1-2* 特异性参与天竺葵素生物合成，*GDFR1-3* 特异性参与矢车菊素生物合成（Bashandy et al. 2015）。

花色苷生物合成的结构基因由 MYB、bHLH 和 WD40 蛋白三种转录因子调控。在非洲菊中，已经有两个调控基因被鉴定（Elomaa et al. 1998，2003）。*GMYC1* 和 *GMYB10* 分别编码一个 bHLH 和一个 R2R3 型 MYB 转录因子。这两个基因在决定非洲菊色素沉着强度和着色模式中都发挥关键作用。

17.3.3　切花产量

非洲菊切花产量是数量性状，受很多次要基因调控且很大程度上受环境影响（Harding et al. 1996）。非洲菊“戴维斯”种群超过 16 代的切花产量数据为计算每一代的狭义遗传力和评估代际间遗传力变化提供了数据（Harding et al. 1996）。在早期二、三代中，狭义遗传率非常高（>0.80），在第三代之后，它显著下降，但在第 8～16 代仍然保持在 0.30 左右。遗传力波动受包括重组、自然或人工选择、环境影响变化、近亲繁殖和采样在内的多种因素影响。基于最小二乘法，“戴维斯”种群 16 代切花产量总体估计遗传力为 0.33，基于限制极大似然法则为 0.31（Harding et al. 1996）。估算表明大约 1/3 的切花产量变异是加性的，可以在代际间遗传。

17.3.4　瓶插期

非洲菊实验群体的瓶插期为 8.5～15.5 天，相互差异并不显著，一般配合力差异显著，特殊配合力差异不显著。在不同研究中，非洲菊瓶插期的狭义遗传力为 0.15～0.38，平均为 0.22（Harding and Bryne 1981；Wernett et al. 1996）。这表明非洲菊瓶插期的遗传方差在表型方差中占比较低，且与样本非洲菊的种群遗传多样性、育种途径和环境无关。在大多数研究中，狭义遗传力约等于广义遗传力，这表明非洲菊瓶插期的遗传变异主要受基因加性效应控制。另外，非洲菊瓶插期的评估实验可重复性中等偏高（$r = 0.57$），每一后代中两到三朵花足以决定每株植物的平均瓶插期。

17.3.5　白粉病抗性

在数个非洲菊育种系中，三个白粉病抗性基因座被报道。Kloos 等（2005）发现了控制非洲菊育种系 176 号植株和 214 号植株白粉病抗性的显性基因 *Pmr1*（powdery mildew-resistant）。该基因在优势度和表现度上存在显著变异，尤其在亲本为白粉病高感背景的后代中（Kloos et al. 2005）。Kloos 等（2005）认为这些非洲菊育种系中存在其他能修饰植株及其后代白粉病抗性的基因。非洲菊育种系 UFGE 31-19（及 UFGE 4033）的白粉病抗性是数量性状（Song and Deng 2013）。UFGE 4033 和感病品种‘Sunburst Snow White’杂交后代的白粉病感病程度分布呈现两个峰值。利用分子标记和 QTL 分析鉴定出一个连锁群中的两个区段，即 *Rpx1* 和 *Rpx2*（抗白粉病），它们能解释分离群体中大约 71%的表型方差。

17.3.6　灰霉病抗性

非洲菊对灰霉菌的抗性是受多基因控制的数量性状（Fu et al. 2017）。Fu 等（2017）在灰葡萄孢抗性分离的两个 F_1 种群中鉴定出 20 个引起变异的 QTL 位点，它们分散在 12 个连锁群（LG）中。三个在不同亲本图谱上检测到并共定位在 LG23 区域的 20cM 的 QTL（*RBQB4*、*RBQI4* 和 *RBQWI6*）有可能是单数量性状位点。当大多数的基因座只能解释不到10%（5.7%～8.9%）的表型方差时，有3个位点（*RBQB4*、*RBQWI4* 和 *RBQWI6*）作用更大，每个都解释超过10%（10.3%～11.4%）的表型方差。大量QTL的存在反映出非洲菊对灰葡萄孢的部分抗性的防御机制极为复杂。同一朵花的舌状花和舌管状花对灰葡萄孢的抗性存在一定程度的差异，表明这两类小花调控灰霉病抗性的基因并不相同。

2-PS（2-吡咯酮合成酶）基因编码合成三乙酰内酯的聚酮合成酶。三乙酰内酯是非洲菊中两种丰富的植保素gerberin和parasorboside的潜在前体。敲除 *2-PS* 基因的转基因非洲菊花朵更易感灰葡萄孢（Koskela et al. 2011）。非洲菊管状花的灰葡萄孢抗性与 *2-PS* 基因的序列多态性息息相关。Fu（2017）的候选基因研究表明，除上文提到的20个QTL，其他19个基因的序列多态性与非洲菊整花、舌状花、舌管状花的葡萄孢菌抗性也息息相关。

17.3.7　隐地疫霉抗性

Sparnaaij等（1975）对6个非洲菊株系进行双列杂交以研究隐地疫霉抗性这一数量性状。杂交后代由人工接种疫霉，在接种后2～4个月评估植物的存活率。杂交后代的疫霉抗性方差主要受一般配合力（gca）的影响。狭义遗传率估计在0.61～0.82，属于中等偏高。因此，基因加性效应为非洲菊对隐地疫霉抗性的遗传变异的主要因素。

17.4　育种途径和技术

17.4.1　人工杂交

人工杂交是所有切花非洲菊品种、大多数园林品种、部分盆栽品种改良都会用到的育种途径，一般通过组织培养培育得到无性繁殖品种。杂交育种丰富了非洲菊的花型和花色，产生了具备新的整合质量性状的新品种，改善了诸如切花品质、花产量和瓶插期等对于非洲菊切花十分重要的数量性状。

17.4.1.1　育种亲本和杂交的选择

亲本和杂交组合的选择是杂交育种的重要决策之一。表型与期望表型相近或与目标性状水平接近的非洲菊通常会被选作杂交亲本。杂交亲本通常有互补的性状，如此配对将有更大的潜力产生具有期望综合性状的后代。到目前为止，在非洲菊育种中，亲本的选择和配对主要基于表型信息与试错法。在未来，对目标性状的选择将更多地基于遗传控制、遗传方式、连锁关系、基因型和单倍型、结合力、生物合成途径、基因网络、分子标记，甚至是全基因组预测。此外，包括亲本育性水平在内的其他因子在育种时也应该考虑。例如，大重瓣、完全重瓣非洲菊植株雄性不育，仅能产生少量花粉，因此在育种时只能用作母本。

17.4.1.2　授粉

非洲菊的每一朵花包含三种类型的小花。舌状花、舌管状花都具有功能性雌蕊，而管状花（除了重瓣花型中的管状花）具有功能性雄蕊。头状花序上最外轮的舌状花最先成熟，接着内轮的舌状花成熟。当花药外露、花粉开始脱落时，我们可以从管状花上收集花粉并置于纸或培养皿上。用拇指和食指挤压总苞苞片可以协助提取花粉。当柱头分裂成“Y”形时，舌状花和舌管状花中的雌蕊可接受花粉。在舌状花和舌管状花的柱头外露 1～2 天后，我们可以用铅笔上的橡皮头对其进行手动授粉。在授粉时，可以通过重复操作一两次或更多次来提高种子数量，这对于重瓣花来说尤其重要。非洲菊花朵经授粉后可以产生几到 100 粒种子，具体数量取决于亲本育性、基因型、生长季节和生长条件。授粉 30～40 天后可以收获成熟种子，种子需要在 37℃干燥 2～4 天并去渣，然后储存在低温（4～10℃）干燥环境。

17.4.1.3　种子萌发

播种到适合的发芽培养基表面后（Deng and Harbaugh 2010），非洲菊种子可以在 20～24℃（日）和 18～20℃下发芽（夜）。在 50～60μmol/(m^2·s)冷白光灯下人工光照 12h 可促进非洲菊种子发芽。在这些条件下，种子 7 天内可萌发，当幼苗长到大约一个月大时，可以移植到装有无土基质的容器中在温室中生长。

17.4.1.4　后代筛选

有性杂交后代的筛选将经过以下几个程序。首先是单株筛选，然后是无重复无性系试验和重复无性系试验。单株筛选适用于诸如花型、花色等质量性状和遗传能力强、受环境因子影响小的特征。对于中到低遗传能力的数量性状，无重复和重复无性系试验是必不可少的。在非洲菊新品种发布并投入商业生产之前，必须进行多地点和/或多季节试验以确保新品种表现良好，且典型特征表现独特、持续且稳定。

生长条件能引起许多控制数量性状的基因显著差异表达，如调控切花产量和瓶插期这两个对切花非洲菊品种非常重要的性状。因此，在通常使用的商业化生产系统和生长条件下筛选育种群体就十分重要。

轮回选择是改良非洲菊数量性状的有效途径。Harding 等（1996）进行了 16 代选择实验以提高非洲菊“戴维斯”种群的切花产量。在每一代中，研究者选择了大约 40 个非洲菊株系，并将它们进行杂交产生下一代。切花产量（每一株）由第 1 代的 15.3 朵提高到了第 16 代的 28.3 朵。经过 16 代以上的轮回选择，育种群体的遗传力保持在 0.27。尽管增长幅度有所减弱，但切花产量的提高仍有望持续响应选择（Harding et al. 1996）。

组织培养是非洲菊新品种繁育的主要方法，因此，测试新品种的离体繁殖能力和在体外繁殖时的遗传稳定性和保真性尤为重要。

越来越多的非洲菊育种项目将病害筛查纳入了后代定期筛选的过程中。这种筛选可以在自然病害压力下进行，也可以通过人工接种病原体孢子进行。人工接种需要多种能够代表流行性病害的病原体、剧毒真菌分离物或菌株。有一些为活体营养型的病原体，如引起白粉病的病菌，需要在非洲菊活体组织上进行培养。还有些病原体容易发生变异而产生新的特异性。与植物病理学家密切合作将是非常有益的。非洲菊对主要病害的抗性通常是易遭受环境影响的数量遗传性状。抗性育种的筛选应对测试品种和株系重复进行多次试验，以保证品种的稳定性抗病数据的可靠性。

17.4.2 近交系和 F_1 杂交品种的培育

近交系和 F_1 杂交品种被广泛应用于盆栽品种和部分园林品种的非洲菊选育，由此产生的非洲菊品种可以通过种子进行繁殖，如此能在消费需求较高的特定市场窗口期生产大量的开花统一的盆栽非洲菊植株。与组织培养相比，由种子繁育的非洲菊在单位基础上更具成本效益。此外，育种者和育种公司通过控制自交亲本，可以更好地控制这些品种的生产和商业化。

非洲菊近交培育的交配体系包括自花授粉、全同胞交配、半同胞交配和回交，不同交配体系得到的纯合率不尽相同，自花授粉和回交能达到最高水平的近交率与纯合性。但许多非洲菊自交不亲和或几乎不亲和，在自花授粉后只能产生少量种子，甚至不产生种子。因此，同胞交配常用于非洲菊近交系培育。

在许多非洲菊近交系中，近交衰退的情况普遍且严重，甚至在交配一到两代后就会发生。为了保持非洲菊近交系有足够的植物活力和育性，近亲交配必须限制在 4 个周期之内，这种限制导致在生产杂交品种时，只能用不完全近交系。另外，被用作杂交亲本的近交系在诸如植株大小、叶片大小、开花时间、花色、花型和茎长等可视性特征上必须达到相当程度的纯合率，这对于非洲菊盆栽杂交品种是否能达到视觉一致的开花至关重要。由于近交系只是部分纯合，它们仍会在许多数量性状上产生变异。因此，精选的近交非洲菊株系通常会通过组织培养繁育以避免近交衰退。

双等位基因或顶交法通常被用于评价非洲菊近交系是否适合用作杂交亲本。这些杂交后代将被用于进行非重复和重复温室试验以测试其主要形态特征与园艺学特性。在这些实验中，后代植株表现和视觉一致性将被严密测试。能产生符合育种目标的后代的杂交将被选择并收获更多的种子。而这些符合条件的后代在被选作最终杂交品种前将进行多地域和多季节试验。

17.4.3 诱导突变

突变诱导或突变发生常被用来修饰和改良具有一个或几个弱势特征的精选品种（Broertjes and Van Harten 1978）。Mba 等（2005）在众多观赏植物中列出了 552 个通过诱变培育的品种。非洲菊是高度杂合但可通过组织培养繁育的园艺作物，因此十分适用于诱导和发现突变。然而，除了有限的关于非洲菊的突变诱导信息外，几乎没有非洲菊突变品种被报道。Ghani 等（2013）将体外培养的幼小非洲菊叶柄和茎外植体用 γ 射线照射或甲基磺酸乙酯（EMS）处理，观察到叶片（数量、大小和形态）和花特性（直径和颜色）、花期和茎长的变异。部分经处理的外植体再生得到的非洲菊植株叶片更大、开花提前、花朵增大、茎也增长。一些诱导的变种或许能提高非洲菊的观赏价值。Laneri 等（1990）发现经 γ 射线诱导产生的非洲菊突变体有 97% 呈现不同程度的嵌合。

17.4.4　倍性操作

17.4.4.1　单倍体和双单倍体

单倍体可转化为双单倍体，可用来代替近交系作为杂交亲本选育高度纯合的 F_1 杂交品种。单倍体和双单倍体也是遗传研究与基因组测序、序列组装、单体分析等基因组学研究的优良材料。尽管我们并未发现自然存在的非洲菊单倍体，但是部分非洲菊品种和育种系的单倍体已经被人工诱导。孤雄生殖如体外花药培养对于非洲菊来说十分困难且效率低下，但体外培养未受精胚珠这一孤雌生殖方式是可行的。非洲菊胚珠包括一个典型的蓼型胚囊，在成熟时胚囊中包含 8 个细胞核（3 个反足细胞、2 个极性细胞、1 个卵细胞和 2 个助细胞）。Sitbon（1981）发现将胚囊未完全成熟的非洲菊胚珠分离培养可得到单倍愈伤组织和不定芽，最终得到完整的植株。在 20 世纪 80 年代和 90 年代，大量旨在改善非洲菊孤雌生殖能力的研究被报道（Ahmin and Vieth 1986；Cappadocia and Vieth 1990；Cappadocia et al. 1988；Honkanen et al. 1991；Meynet and Sibi 1984；Miyoshi and Asakura 1996；Sitbon 1981；Tosca et al. 1990，1995，1999）。这些研究表明，50%～80%的非洲菊基因型可在体外诱导产生单倍体愈伤组织，30%的非洲菊基因型可产生单倍体芽。不同基因型植株通过未受精胚株诱导单倍体的效率存在显著差异，在 0～4%。影响孤雌再生单倍体的主要因素包括供体年龄、花朵和胚珠发育时期、生长季节（春、夏、秋）及诱导愈伤组织和不定芽再生的培养基基质与激素组成（Cappadocia and Vieth 1990）。

大多数（60%～100%）由非洲菊胚珠再生获得的植株是单倍体。用抑制有丝分裂的药剂处理单倍体植株将使染色体数量加倍，形成双单倍体植株。通常，将低浓度秋水仙碱或氨磺乐灵添加于培养单倍体植株的液态或固态培养基中，可以使 20%～60%的单倍体植株转变成双单倍体植株（Miyoshi and Asakura 1996；Tosca et al. 1995）。

与二倍体供体或亲本相比，单倍体非洲菊植株要小得多，叶片数量更少，花朵更小（Honkanen et al. 1991；Miyoshi and Asakura 1996），因此需要额外的管护以确保它们能生存、生长和发育。双单倍体植株的生长和植株活力得以提高，但与二倍体亲本比仍要小许多。在大多数双单倍体中，花粉和雄蕊的发育被破坏，植株雄性不育。双单倍体的大多数胚珠很小且在花期皱缩，仅有小部分非洲菊双单倍体是可育的（Honkanen et al. 1991）。非洲菊单倍体和双单倍体在花色与其他形态特征上常出现分离。基于孤雌生殖的非洲菊育种似乎是可行的（Miyoshi and Asakura 1996；Tosca et al. 1990）。

17.4.4.2　多倍体

通过秋水仙碱或氨磺乐灵处理外植体、萌发中的种子或幼苗可得到四倍体非洲菊。这两种药剂都可抑制有丝分裂，使得染色体不能分离并平均分配到两个子细胞中。常用的秋水仙碱浓度是 0.05%～0.2%。与二倍体相比，同源四倍体非洲菊植物在几个特征特性上表现出显著差异，包括更宽更厚的叶片、增厚的花茎（大约 40%）和更大的花心（大约 35%）（Bhattarai et al.，未发表；Gantait et al. 2011；Li et al. 2009），其叶片具有更少但更大的气孔。虽然没有实验数据，但这些形态变化似乎影响了其植物生理、生物和非生物胁迫抗性、花朵产量和瓶插期等。同源四倍体非洲菊的价值仍需进一步阐述。

作为四倍体诱导的副产物，非洲菊八倍体在文献中也被报道。它们的叶片更厚且常常矮化。三倍体非洲菊还未被报道或发现，但由于三倍体高度败育，它们在产生无花粉花朵上具有潜在价值。同源四倍体非洲菊的成功诱导为三倍体的生产和评估提供可能，从而实现其在非洲菊育种中应用的价值。

17.5 生物技术工具的开发与应用

17.5.1 分子标记

随机扩增多态性 DNA（RAPD）、序列特异性扩增区域（SCAR）、酶切扩增多态序列（CAPS）、简单序列重复（SSR）、简单重复序列区间（ISSR）和单核苷酸多态性（SNP）等分子标记方法已被用于非洲菊研究（de Pinho Benemann et al. 2012；Fu et al. 2017；Gong and Deng 2010，2012；Seo et al. 2012）。非洲菊拥有丰富且多态的简单重复序列，每 5.2kb 的表达序列标签（EST）就可以找到一个简单重复序列（Gong and Deng 2010）。在多态的简单重复序列位点，2～8 个等位基因被观察到（de Pinho Benemann et al. 2012；Gong and Deng 2010）（图 17.1d）。SNP 的出现更加频繁，在每 1kb 的表达序列标签中，平均有 7.8 个 SNP 被发现（Fu et al. 2016）。非洲菊基因组和转录组序列中有成百上千的 SNP 可被鉴定。

在评估遗传多样性，理解不同种质资源的遗传关系，评价组织培养繁育非洲菊植株、不同品种的遗传稳定性，绘制遗传连锁图谱和抗病性状 QTL 标签等多个方面，分子标记都发挥了巨大作用（Da Mata et al. 2009；de Pinho Benemann et al. 2012；Fu et al. 2017；Gong and Deng 2010，2012；Seo et al. 2012；Song and Deng 2013）。Bhatia 等（2009）使用 ISSR 标记，观察到三种类型非洲菊外植体在组织培养过程中产生体细胞无性系变异的差异。由非洲菊头状花序和茎尖外植体再生的苗木变异条带少，而由叶片组织诱导再生的苗木产生更多的体细胞无性系变异。Gong 和 Deng（2012）用 SSR 标记分析了 40 个非洲菊品种和育种系的 DNA 指纹，发现这些品种和育种系形成了两个主要类群，分类与育种方式和繁殖方式息息相关，与育种公司相关性不高。使用来源于抗性基因（*R* 基因）模拟序列的分子标记，Song 和 Deng（2013）绘制了第一个非洲菊白粉病抗性两个主要数量性状基因座的局部连锁图谱。QTL 被若干特殊标记追踪。Fu 等（2017）开发了数百个 SNP 标记，绘制了第一个非洲菊全基因组遗传连锁图谱。该图谱包含 24 个连锁群，覆盖 1601cM。利用这些遗传图谱（和表型数据）辅助鉴定了 20 个灰葡萄孢抗性 QTL。

17.5.2 遗传转化和转基因

非洲菊农杆菌转化最早报道于 20 世纪 90 年代早期。自此，遗传转化成为研究非洲菊基因功能和调控不可或缺的工具。非洲菊品种‘Terra Regina’的遗传转化体系已被建立，遗传转化效率为 0.1%～2%（Elomaa and Teeri 2001）。但其他非洲菊品种或品系的遗传转化体系难以建立（Z. Deng，未出版）。目前仍没有转基因非洲菊品种，但是转基因非洲菊的特征描述为非洲菊遗传改良提供了十分有价值的信息。

遗传转化实现了对非洲菊抗性基因表达模式和表达水平的编辑，还能将外源基因引入非洲菊中，从而获得表型改良或具新表型的转基因植株。通过引入反义基因载体，Elomaa 等（1996）降低了编码类黄酮生物合成酶 DFR 的 *dfr* 基因的表达量，使‘Terra Regina’的花色从红色变成了粉红色。通过相同的手段改变 *CHS* 基因的表达量，Elomaa 等（1996）获得了奶油色或粉红色的转基因‘Terra Regina’。Laitinen 等（2008）将非洲菊转录因子（*GMYB10*）过表达，在‘Terra Regina’中观察到了矢车菊素的合成，而野生型的植株只能合成天竺葵素。通过提高或抑制非洲菊转录因子 *GhCYC2*、*GhCYC3*、*GhCYC4* 的表达，Broholm 等（2008）和 Juntheikki-Palovaara 等（2014）使非洲菊的花型从半重瓣变成了重瓣或单瓣。Deng 等（2015）将拟南芥基因（*NPR1*）在非洲菊中过表达，以提高非洲菊对白粉病的抗性。烟草响尾蛇病毒（TRV）、黄瓜花叶病毒（CMV）、凤仙花坏死斑病毒（INSV）和番茄斑点病毒（TSWV）等一些病毒会感染非洲菊植物并对其造成损害。例如，TSWV 感染导致坏叶上出现斑点和花的变形。目前，在非洲菊中并没有发现对这些病毒有抗性的资源，Korbin（2006）将 TSWV 的核蛋白基因转化到 4 个非洲菊品种中后发现转基因植株对 TSWV 的抗性增强。

17.5.3　基因组和转录组测序

基因组和转录组序列是非洲菊遗传研究和改良的宝贵资源。Laitinen 等（2005）用 Sanger DNA 测序仪测定了非洲菊 cDNA 序列，建立了第一个含将近 17 000 个序列的非洲菊表达序列标签数据库。该数据库使非洲菊花发育相关基因的挖掘和基因芯片的构建成为可能。通过微阵数据，Laitinen 等（2005）识别了在非洲菊花茎、冠毛、雄蕊和花瓣等特定部位大量表达的特异基因与标记基因。还有人发掘了包含简单重复序列的表达序列标签，建立了可用于非洲菊的高度特异和多样化的简单重复序列标记（de Pinho Benemann et al. 2012；Gong and Deng 2010）。近年来，第一代 DNA 测序平台之一的 Illumina HiSeq 已经投入非洲菊转录组测序。Kuang 等（2013）利用该技术构建了非洲菊舌状花的转录组，并鉴定出了与非洲菊花瓣中赤霉素代谢和信号转导途径相关的基因。Fu 等（2016）对 4 个非洲菊株系的叶片和花芽的转录组进行了测序，获得了近 37 000 个一致的重叠群。通过绘制这些株系的 cDNA 序列图谱，Fu 等（2016）鉴定出了与茉莉酸生物合成和信号网络相关的 SNP 与基因，它们或许在非洲菊对灰葡萄孢和其他病原体的抗性中发挥重要作用。这些转录组序列使通过候选基因开发非洲菊灰葡萄孢抗性分子标记成为可能（Fu 2017）。

由于非洲菊的基因组非常大（大约 5Gb）并且高度杂合与重复，其测序和组装十分困难。目前，非洲菊基因组测序工作侧重于编码基因、基因空间的测序进行中（Still；http://compgenomics.ucdavis.edu/compositae_data.php?name=Gerbera+hybrida）。此前，Seo 等（2012）将 454 GS-FLX 钛焦磷酸测序方法应用于非洲菊基因组，获得了 22Mb 的非冗余基因组 DNA 序列。这些序列中的鸟嘌呤和胞嘧啶含量大约为 39%，约 3.7%的组装序列包含简单重复序列。这些序列为开发 SSR 标记提供了模板。

17.6 展　　望

基于现有基因型的经典育种已经取得了巨大成就，培育出了具备改良或新颖园艺学特性和观赏品质的非洲菊新品种。为了延续非洲菊育种的成就、满足花农和消费者的需求，新的性状需要被发现或创造，并引入非洲菊中，新型育种工具急需被研发和采用。对主要病虫害，如白粉病、灰霉病、根腐病和冠腐病及西花蓟马的抗性是花农和消费者追求的最重要的性状之一。这些病虫害不仅给花农造成了重大的经济损失，导致管理困难，也使消费者的花和植株遭受干扰。分子标记将是非洲菊育种中用于提高病虫害抗性的有力工具，它将帮助育种者鉴定、定位、标记、组合和堆积已有的抗性基因。基因组、转录组测序和分析将挖掘大量序列多态性，有助于分子标记发展、全基因组遗传图谱构建，并提供基因模板序列。目前尚无非洲菊基因编辑报道，但它仍被期望在先进非洲菊育种工作中发挥作用。改良的非洲菊转基因技术会使将核酸酶（如 Cas9）基因和引导分子（如 gRNA）有效转导进非洲菊细胞核成为可能，最终应用于基因编辑。随着这些技术的完善、成熟和经济适用，它们将被非洲菊育种者和育种公司采用。新性状的整合和新生物技术的应用被寄望于促进非洲菊育种达到新的高度。

本章译者：

金春莲，李　帆，李绅崇

云南省农业科学院花卉研究所，国家观赏园艺工程技术研究中心，云南省花卉育种重点实验室，昆明 650200

参 考 文 献

Ahmin M, Vieth J (1986) Production de plantes haploides de *Gerbera jamesonii* par culture *in vitro* d'ovules. Can J Bot 64:2355–2357

Asen S (1984) High pressure liquid chromatographic analysis of flavonoid chemical markers in petals from *Gerbera* flowers as an adjunct for cultivar and germplasm identification. Phytochemistry 23(11):2523–2526. https://doi.org/10.1016/S0031-9422(00)84090-X

Bala S, Gupta RC (2013) Male meiosis and chromosome number in Asteraceae family from district Kangra of H.P. (Western Himalayas). Int J Bot Res 3(1):43–58

Bashandy H, Pietiäinen M, Carvalho E, Lim KJ, Elomaa P, Martens S, Teeri TH (2015) Anthocyanin biosynthesis in gerbera cultivar 'Estelle' and its acyanic sport ''Ivory'. Planta 242(3):601–611. https://doi.org/10.1007/s00425-015-2349-6

Bhatia R, Singh KP, Jhang T, Sharma TR (2009) Assessment of clonal fidelity of micropropagated gerbera plants by ISSR markers. Sci Hortic 119:208–211

Bremer K (1994) Asteraceae, cladistics and classification. Timber Press, Portland

Broertjes C, Van Harten AM (1978) Application of mutation breeding methods in the improvement of vegetatively propagated crops: an interpretive literature review. Elsevier Scientific Pub. Co., Amsterdam/New York

Broholm SK, Tähtiharju S, Laitinen RA, Albert VA, Teeri TH, Elomaa P (2008) A TCP domain transcription factor controls flower type specification along the radial axis of the *Gerbera* (Asteraceae) inflorescence. Proc Natl Acad Sci 105:9117–9122

Cappadocia M, Vieth J (1990) *Gerbera jamesonii* H. Bolus ex Hook: In vitro production of haploids. In: Bajaj YPS (ed) Biotechology in Agriculture and Forestry, Vol. 12 – Haploids in Crop

Improvement. Springer-Verlag, Berlin/Heidelberg, pp 417–427
Cappadocia M, Chretien L, Laublin G (1988) Production of haploids in *Gerbera jamesonii* via ovule culture: influence of fall versus spring sampling on callus formation and shoot regeneration. Can J Bot 66:1107–1110
Cardoso JC, Da Silva JAT (2013) Gerbera micropropagation. Biotech Adv 31(8):1344–1357
Da Mata TL, Segeren MI, Fonseca AS, Colombo CA (2009) Genetic divergence among gerbera accessions evaluated by RAPD. Sci Hortic 121:92–96
de Pinho Benemann D, Machado LN, Arge LWP, Bianchi VJ, de Oliveira AC, da Maia C, Peters JA (2012) Identification, characterization and validation of SSR markers from the gerbera EST database. Plant Omics J 15(2):159–166
De Witte Y, Harkema H, van Doorn WG (2014) Effect of antimicrobial compounds on cut Gerbera flowers: poor relation between stem bending and numbers of bacteria in the vase water. Postharvest Bio Technol 91:78–83
Deng X, Bashandy H, Ainasoja M, Kontturi J, Pietiäinen M, Laitinen RAE, Albert VA, Valkonen JPT, Elomaa P, Teeri TH (2014) Functional diversification of duplicated chalcone synthase genes in anthocyanin biosynthesis of *Gerbera hybrida*. New Phytol 201:1469–1483
Deng Z, Harbaugh BK (2010) UFGE 4141, UFGE 7014, UFGE 7015, UFGE 7023, UFGE 7032, and UFGE 7034: six new gerbera cultivars for marketing flowering plants in large containers. Hortscience 45:971–974
Deng Z, Harbaugh BK (2013) UFGE 7031 and UFGE 7080 Gerbera cultivars. Hortscience 48:659–663
Deng Z, Zhonglin M, Peres NA (2015) Special research report #309: powdery mildew resistance in transgenic gerbera plants. American Floral Endowment
Elomaa P, Mehto M, Kotilainen M, Helariutta Y, Nevalainen L, Teeri TH (1998) A bHLH transcription factor mediates organ, region and flower type specific signals on dihydroflavonol-4-reductase (*dfr*) gene expression in the inflorescence of *Gerbera hybrida* (Asteraceae). Plant J 16(1):93–99
Elomaa P, Helariutta Y, Kotilainen M, Teeri TH (1996) Transformation of antisense constructs of the chalcone synthase gene superfamily into *Gerbera hybrida*: differential effect on the expression of family members. Mol Breed 2(1):41–50
Elomaa P, Teeri TH (2001) Transgenic gerbera. In: Bajaj YPS (ed) Transgenic crops III. Biotechnology in agriculture and forestry, vol 48. Springer, Berlin/Heidelberg, pp 139–154
Elomaa P, Uimari A, Mehto M, Albert VA, Laitinen RA, Teeri TH (2003) Activation of anthocyanin biosynthesis in *Gerbera hybrida* (Asteraceae) suggests conserved protein-protein and protein-promoter interactions between the anciently diverged monocots and eudicots. Plant Physiol 133:1831–1842
Farhood S, Hadian S (2012) First report of *Alternaria* leaf spot on gerbera (*Gerbera jamesonii* L.) in North of Iran. Adv Environ Bio 6:621–625
FloraHolland (2014) Facts and figures 2014. https://www.royalfloraholland.com/media/3949227/Kengetallen-2014-Engels.pdf
Fu Y (2017) Unraveling the genetics of *Botrytis cinerea* resistance in *Gerbera hybrida*. Thesis, Wageningen University
Fu Y, Esselink GD, Visser RG, van Tuyl JM, Arens P (2016) Transcriptome analysis of *Gerbera hybrida* including in silico confirmation of defense genes found. Front Plant Sci 7:247
Fu Y, Van Silfhout A, Shahin A, Egberts R, Beers M, Van der Velde A, Van Houten A, Van Tuyl JM, Visser RG, Arens P (2017) Genetic mapping and QTL analysis of Botrytis resistance in *Gerbera hybrida*. Mol Breed 37:13. https://doi.org/10.1007/s11032-016-0617-1
Gantait S, Mandal N, Bhattacharyya S, Das PK (2011) Induction and identification of tetraploids using in vitro colchicine treatment of *Gerbera jamesonii* bolus cv. Sciella. Plant Cell Tissue Organ Cult 106:485. https://doi.org/10.1007/s11240-011-9947-1
Ghani M, Kumar S, Thakur M (2013) Induction of novel variants through physical and chemical mutagenesis in Barbeton daisy (*Gerbera jamesonii* Hook.). J Hortic Sci Biotechnol 88:585–590
Gong L, Deng Z (2010) EST-SSR markers for gerbera (*Gerbera hybrida*). Mol Breed 26:125–132
Gong L, Deng Z (2012) Selection and application of SSR markers for variety discrimination, genetic similarity and relation analysis in gerbera (*Gerbera hybrida*). Sci Hortic 138:120–127
Granke LL, Crawford LE, Hausbeck MK (2012) Factors affecting airborne concentrations of *Podosphaera xanthii* conidia and severity of gerbera powdery mildew. Hortscience 47(8):1068–1072
Hansen HV (1985) Taxonomic revision of the genus *Gerbera (*Compositae, Mutisieae*)* sections *Gerbera, Parva, Piloselloides* (in Africa), and *Lasiopus*. Council Nordic Pub Bot

Hansen HV (1999) A story of the cultivated gerbera. New Plantsman (Royal Hort Soc) 6:85–95
Harding J, Byrne T (1981) Heritability of cut-flower vase longevity in *Gerbera*. Euphytica 30:653–657
Harding J, Huang H, Byrne T (1996) Estimation of genetic variance components and heritabilities for cut-flower yield in gerbera using least squares and maximum likelihood methods. Euphytica 88(1):55–60
Hausbeck MK (2004) Take a long-range approach to powdery mildew resistance. GMPro 24:68–69
Hausbeck MK, Quackenbush WR, Linderman SD (2002) Evaluation of cultivars of African daisy for resistance to powdery mildew. B&C Tests 18:O0004
Helariutta Y, Elomaa P, Kotilainen M, Griesbach RJ, Schröder J, Teeri TH (1995) Chalcone synthase-like genes active during corolla development are differentially expressed and encode enzymes with different catalytic properties in *Gerbera hybrida* (Asteraceae). Plant Mol Biol 28:47–60
Honkanen J, Aapola A, Seppanen P, Tormala T, de Wit JC, Esendam HF, Stravers LJM (1991) Production of doubled haploid gerbera clones. Acta Hortic 300:341–346
Hopper DA, Stutte GW, McCormack A, Barta DJ, Heins RD, Erwin JE, Tibbitts TW (1997) Crop growth requirements. Plant growth chamber handbook. North Cent Reg Res Pub 340:217–225
Johnson S, Collin CL, Wissman HL, Halvarsson E (2004) Factors contributing to variation in seed production among remnant populations of the endangered daisy *Gerbera aurantiaca*. Biotropica 36(2):148–155
Juntheikki-Palovaara I, Tähtiharju S, Lan T, Broholm SK, Rijpkema AS, Ruonala R, Kale L, Albert VA, Teeri TH, Elomaa P (2014) Functional diversification of duplicated CYC2 clade genes in regulation of inflorescence development in *Gerbera hybrida* (Asteraceae). Plant J 79:783–796
Katinas L (2004) The *Gerbera* complex (Asteraceae: Mutisieae): to split or not to split. SIDA Contrib Bot 21:935–940
Kloos WE, George CG, Sorge LK (2004) Inheritance of the flower types of *Gerbera hybrida*. J Am Soc Hortic Sci 129(6):803–810
Kloos WE, George CG, Sorge LK (2005) Inheritance of powdery mildew resistance and leaf macrohair density in *Gerbera hybrida*. Hortscience 40:1246–1251
Korbin M (2006) Assessment of gerbera plants genetically modified with TSWV nucleocapsid gene. J Fruit Ornament Plant Res 14:1
Koskela S, Söderholm PP, Ainasoja M, Wennberg T, Klika KD, Ovcharenko VV, Kylänlahti I, Auerma T, Yli-Kauhaluoma J, Pihlaja K, Vuorela PM (2011) Polyketide derivatives active against *Botrytis cinerea* in *Gerbera hybrida*. Planta 233:37–48
Kotilainen M, Helariutta Y, Mehto M, Pollanen E, Albert VA, Elomaa P, Teeri TH (1999) GEG participates in the regulation of cell and organ shape during corolla and carpel development in *Gerbera hybrida*. Plant Cell 11(6):1093–1104
Kuang Q, Li L, Peng J, Sun S, Wang X (2013) Transcriptome analysis of *Gerbera hybrida* ray florets: putative genes associated with gibberellin metabolism and signal transduction. PLoS One 8:e57715
Kulkarni BS, Thammaiah N, Reddy BS, Kulkarni MS (2009) Evaluation of gerbera genotypes against *Alternaria* leaf spot under protected structure. Haryana J Hortic Sci 38:1–2
Laitinen RAE, Immanen J, Auvinen P, Rudd S, Alatalo E, Paulin L, Ainasoja M, Kotilainen M, Koskela S, Teeri TH, Elomaa P (2005) Analysis of the floral transcriptome uncovers new regulators of organ determination and gene families related to flower organ differentiation in *Gerbera hybrida* (Asteraceae). Genome Res 15:475–486
Laitinen RAE, Broholm S, Albert VA, Teeri TH, Elomaa P (2006) Patterns of MADS-box gene expression mark flower-type development in *Gerbera hybrida* (Asteraceae). BMC Plant Bio 6:11. https://doi.org/10.1186/1471-2229-6-11
Laitinen RAE, Ainasoja M, Broholm SK, Teeri TH, Elomaa P (2008) Identification of target genes for a MYB-type anthocyanin regulator in *Gerbera hybrida*. J Exp Bot 59:3691–3703
Laneri U, Franconi R, Altvista P (1990) Somatic mutagenesis of *Gerbera jamesonii* hybr.: irradiation and in vitro culture. I Int Symp In Vitro Culture Hortic Breed 280:395–402
Li H, Yan B, Zhang T, Jiang Y, Zhang H, Yu L, Li S (2009) Preliminary studies on polyploidy mutation of cut flower *Gerbera jamesonii* bolus. Acta Hortic Sinica 36:605–610
Li Y, Liu Y, Li F, Tang X, Chen J, Chen H (2008) Identification of the disease resistance of *Gerbera jamesonii* cultivars to root rot and pathogenicity differentiation of *Phytophthora cryptogea*. J Yunnan Agric Univ 23(1):33

Marie D, Brown SC (1993) A cytometric exercise in plant DNA histograms, with 2C values for 70 species. Biol Cell 78:41–51

Martens S, Forkmann G (2000) Flavonoid biosynthesis in gerbera hybrids: genetics and enzymology of flavones. Acta Hortic 521:67–72

Mba C, Afza R, Lagoda PJL, Darwig J (2005) Strategies of the joint FAO/IAEA programme for the use of induced mutations for achieving sustainable crop production in member states. In: Proceedings of the second international seminar on Production, Commercialisation and Industrialization of Plantain

Meynet J, Sibi M (1984) Haploid plants from *in vitro* culture of unfertilized ovules in *Gerbera jamesonii*. Z Pflanzenzuchtg 93:78–85

Minuto A, Gullino LM, Garibaldi A (2007) Susceptibility of gerbera and chrysanthemum varieties (*Gerbera jamesonii* and *Chrysanthemum morifolium*) to *Fusarium oxysporum f.sp. chrysanthemi*. Comm Agric Appl Biol Sci 72:715–721

Miyoshi K, Asakura N (1996) Callus induction, regeneration of haploid plants and chromosome doubling in ovule cultures of pot gerbera (*Gerbera jamesonii*). Plant Cell Rep 16:1–5

Nagrale DK, Gaikwad AP, Sharma L (2013) Morphological and cultural characterization of *Alternaria alternata* (Fr.) Keissler blight of gerbera (*Gerbera jamesonii* H. Bolus ex J.D. Hook). J Appl Nat Sci 5:171–178

Pasini E, Funk VA, de Souza-Chies TT, Miotto STS (2016) New insights into the phylogeny and biogeography of the *Gerbera*-complex (Asteraceae: Mutisieae). Taxon 65(3):547–562

Perik RRJ, Rae D, Harkema H, Zhong Y, van Doorn WG (2012) Bending in cut *Gerbera jamesonii* flowers relates to adverse water relations and lack of stem sclerenchyma development, not to expansion of the stem central cavity or stem elongation. Postharvest Biol Technol 74:11–18

Pataky NR (1988) Fusarium wilt of herbaceous ornamentals. University of Illinois Extension RPD No. 650. https://ipm.illinois.edu/diseases/rpds/650.pdf

Prajapati P, Singh A, Jadhav PB (2017) Studies on growth, flowering and yield parameters of different genotypes of gerbera (*Gerbera jamesonii* bolus). Int J Curr Microbiol App Sci 6(4):1770–1777

Reddy PP (2016) Gerbera. In: Sustainable crop protection under protected cultivation. Springer, Singapore, pp 355–362

Reid MS (2004) Gerbera, Transvaal daisy: recommendations for maintaining postharvest quality. http://postharvest.ucdavis.edu/Commodity_Resources/Fact_Sheets/Datastores/Ornamentals_English/?uid=17&ds=801. Accessed 18 Apr 2017

Seo KI, Lee GA, Park SK, Yoon MS, Ma KH, Lee JR, Choi YM, Jung YJ, Lee MC (2012) Genome shotgun sequencing and development of microsatellite markers for gerbera (*Gerbera hybrida* H.) by 454 GS-FLX. Afric J Biotechnol 11(29):7388–7396

Sitbon M (1981) Production of haploid *Gerbera jamesonii* plants by in vitro culture of unfertilized ovules. Agronomie 1:807–812

Song X, Deng Z (2013) Powdery mildew resistance in gerbera: mode of inheritance, quantitative trait locus identification, and resistance responses. J Am Soc Hortic Sci 138:470–478

Sparnaaij LD, Garretsen F, Bekker W (1975) Additive inheritance of resistance to *Phytophthora cryptogea* Pethybridge & Lafferty in *Gerbera jamesonii* bolus. Euphytica 24:551–556

Tähtiharju S, Rijpkema AS, Vetterli A, Albert VA, Teeri TH, Elomaa P (2011) Evolution and diversification of the CYC/TB1 gene family in Asteraceae—a comparative study in *Gerbera* (Mutisieae) and sunflower (Heliantheae). Mol Biol Evol 29:1155–1166

Teeri TH, Elomaa P, Kotilainen M, Albert VA (2006) Mining plant diversity: Gerbera as a model system for plant developmental and biosynthetic research. BioEssays 28:756–767

Tosca A, Arcara L, Frangi P (1999) Effect of genotype and season on gynogenesis efficiency in *Gerbera*. Plant Cell Tissue Organ Cult 59:77–80

Tosca A, Lombardi M, Conti L, Frangi P (1990) Genotype response to in vitro gynogenesis technique in *Gerbera jamesonii*. Acta Hortic 280:337–340

Tosca A, Pandolfi R, Citterio SA, Fasoli A, Sgorbati S (1995) Determination by flow cytometry of the chromosome doubling capacity of colchicine and oryzalin in gynogenetic haploids of Gerbera. Plant Cell Rep 14:455–458

Tourjee KR, Harding J, Byrne TG (1993) Calorimetric analysis of gerbera flowers. HortSci 28(7):735–737

Tyrach A, Horn W (1997) Inheritance of flower colour and flavonoid pigments in Gerbera. Plant Breed 116:377–381

United States Department of Agriculture (2010) Floriculture Crops 2009 Summary National Agricultural Statistics Service Sp Cr 6-1 (10)
United States Department of Agriculture (2016) Floriculture Crops 2009 Summary National Agricultural Statistics Service ISSN: 1949-091
van Doorn WG, de Witte Y (1994) Effect of bacteria on scape bending in cut *Gerbera jamesonii* flowers. J Am Soc Hortic Sci 119(3):568–571
van Eck JW, Franken AAJM (1995) Colours of florets of several gerbera (*Gerbera jamesonii* Bolis ex Adlam) cultivars measured with a colorimeter. Euphytica 84(1):49–55
Wernett HC, Sheehan TJ, Wilfret GJ, Marousky FJ, Lyrene PM, PM KDA (1996) Postharvest longevity of cut-flower *Gerbera*. I. Response to selection for vase life components. J Am Soc Hortic Sci 121(2):216–221

第 18 章　铁　筷　子

Emmy Dhooghe, Julia Sparke, Peter Oenings, Thierry Van Paemel,
Marie-Christine Van Labeke, and Traud Winkelmann

摘要　铁筷子属（*Helleborus*）植物，特别是黑嚏根草（*Helleborus niger*）、铁筷子（*H.* × *hybridus*）和一些种间杂交品种，是经济价值日益增加的观赏植物，可用作园林植物、室内盆栽植物和切花。其他几种观赏价值较低的铁筷子在花朵大小、花色、叶片、香味和抗病能力等方面展示出了各种有趣的性状。因此，在该属中利用先进的育种技术将这些性状融合起来可以满足人们对铁筷子产品日益增长的需求。新的育种产品在投放市场前必须满足许多生产和产品质量标准。例如，10 年前，*H.* × *hybridus* 可以作为三年生的开花植物来销售，但现在这些植物在种植 1 年或 2 年后必须开花。同样，对于 *H. niger* 来说，最好是圣诞节前开花。本章结合最新的育种成果综述了铁筷子种类、亲缘关系和现有育种产品、育种目标及实现这些目标的现代技术手段。

关键词　铁筷子；种间杂交；育种；染色体加倍

18.1　铁筷子属的介绍

铁筷子是毛茛科多年生植物，冬季或早春开花。铁筷子作为园林植物、室内盆栽植物和切花在园艺上变得越来越重要（Zonneveld 2001；Susek 2016）。铁筷子遍布整个欧洲。该属大约有 22 种（Meiners et al. 2011），其中一种原产东亚[铁筷子（*H. thibetanus*）]（Tamura 1993）。其属名 *Helleborus* 源自希腊语“elein”（伤害）和“bora”（食物），指的是这种植物的有毒特性（Tamura 1995）。铁筷子属分为 6 个组：*Chenopus* 组（*H. argutifolius* 和 *H. lividus*）、*Griphopus* 组（*H. foetidus*）、*Helleborus* 组（*H. niger*）、*Helleborastrum* 组（包括

E. Dhooghe (✉)
Flanders Research Institute for Agriculture, Fisheries and Food (ILVO), Plant Sciences Unit, Applied Genetics and Breeding, Melle, Belgium
e-mail: emmy.dhooghe@ilvo.vlaanderen.be
J. Sparke
Boehringer Ingelheim Pharma GmbH & Co. KG, Ingelheim am Rhein, Germany
P. Oenings
Heuger, Glandorf, Germany
T. Van Paemel
Het Wilgenbroek, Oostkamp, Belgium
M.-C. Van Labeke
Ghent University, Department Plants and Crops, Ghent, Belgium
T. Winkelmann
Institute of Horticultural Production Systems, Woody Plant and Propagation Physiology, Leibniz Universität Hannover, Hannover, Germany

H. orientalis）、*Dicarpon* 组（*H. thibetanus*）和 *Syncarpus* 组（*H. vesicarius*）（图 18.1）（Zonneveld 2001）。最受欢迎的种是 *Helleborus niger* L.（圣诞玫瑰，占 70%的市场份额）和 *Helleborus orientalis* Lam.（杂交种）（四旬期玫瑰，占 5%的市场份额）（图 18.2）（Rice and Strangman 1993）。此外，一些种间杂交品种（占 25%的市场份额）正日益商业化（表 18.1，图 18.3）。*H. orientalis* 是一种在 2～4 月开花的园林植物。*H. niger* 开花更早——11 月到次年 4 月（Salopek-Sondi et al. 2002；Susek and Ivancic 2006），现在越来越多地被当作室内盆栽植物出售。*Helleborastrum* 组的 *H. orientalis* 被广泛用于种间杂交，从而形成了被称为“东方杂种系”（orientalis hybrids）的群体。实际上它们的植物学名称是 *H.* × *hybridus*（表 18.1），而 *H. orientalis* 是指在自然栖息地分布的原生种 *H. orientalis*。

图 18.1 铁筷子属植物。（a）*H. thibetanus*（*Dicarpon* 组）；（b）*H. atrorubens*（*Helleborastrum* 组）；（c）*H. argutifolius*（*Chenopus* 组）；（d）*H. foetidus*（*Griphopus* 组）；（e）*H. dumetorum*（*Helleborastrum* 组）；（f）*H. vesicarius*（*Syncarpus* 组）

图 18.2 *Helleborus* × *hybridus* 或‘东方杂种系’（orientalis hybrids）或‘庭院杂种系’（garden hybrids）。这些品种都是由 *H. orientalis* 所在的 *Helleborastrum* 组内种间杂交而成的。（a）‘Anemone Magic Picotee’；（b）‘WB Pink Selections’；（c）‘Magic White Spotted’；（d）‘SP Conny’；（e）‘WB Picotee Selections’；（f）‘SP Lily’

表 18.1 铁筷子杂种、假定杂种报道（Mathew 1989；Rice 2009；Rottensteiner 2016）及笔者工作

	名称	杂交组合	特征组合
确定的杂种	东方杂种系、庭院杂种系	包括 *H. orientalis* 在内的 *Helleborastrum* 组内种间杂交	萼片形状和颜色变异，重瓣花，半重瓣花
	H. × *sternii*	*H. argutifolius* × *H. lividus*，也可能是反交：*H. lividus* × *H. argutifolius*	生长类型和花朵性状介于亲本之间的可育的中间材料
	H. × *ballardiae*	*H. niger* × *H. lividus*	
	H. × *nigercors*	*H. niger* × *H. argutifolius*	
	H. × *ericsmithii*	*H. niger* × *H.* × *sternii*	
不确定的假定杂种	*H.* × *jourdanii*	*H. foetidus* × *H. viridis*	生长类型和花朵形态不同
	—	*H. foetidus* × *H. argutifolius*	后代不育，生长类型和花朵形态不同
	H. × *sahinii*	*H. foetidus* × *H. niger*	叶片和花朵形态不同
	—	*H. niger* × *H. orientalis*	生长类型、叶片和花朵形态不同，萼片的颜色和形状也不同
	H. × *lemonnierae*	*H. niger*（var. *macranthus*）× *H.* × *hybridus*（subsp. *abchasicus*）	
	‘Walberton’s Rosemary’	*H. niger* × *H.* × *hybridus*	
	—	*H. niger* × *H. viridis*	
	—	*H. niger* × *H. purpurascens*	
	—	*H. niger* × *H. atrorubens*	
	—	*H. atrorubens* × *H. odorus*	生长类型、叶片和花朵形态不同
	H. × *viridescens*	*H. atrorubens* × *H. viridis*	
	H. × *mucheri*	*H. dumetorum* × *H. multifidus*	
	H. × *tergestinus*	*H. dumetorum* × *H. odorus*	
	—	*H. dumetorum* × *H. viridis*	
	—	*H. odorus* × *H. viridis*	
	—	*H. multifidus* × *H. viridis*	
	—	*H. multifidus* × *H. odorus*	
	H. × *glandorfensis*	*H.* × *ericsmithii* × *H.* × *hybridus*	生长类型、叶片和花朵形态不同，萼片的颜色和形状也不同

图 18.3 铁筷子种间杂种。(a)*H.* × *sternii*'Flame';(b)*H.* × *ballardiae*'HGC Maestro';(c)*H.* × *nigercors*'Snow Love';(d)*H.* × *ericsmithii* 'Candy Love';(e)*H.* × *lemonnierae*'HGC Madame Lemonnier';(f) *H. niger* × *H. atrorubens*

除了将它们作为观赏植物商业化应用外，由于其具有一定的细胞毒性成分且显示出可用于癌症治疗的应用潜力，因此有研究还介绍了某些种的药用价值（如 Maior and Dobrota 2013；Cakar et al. 2014；Tan et al. 2014；Schink et al. 2015）。

18.2 花部形态

第一批花蕾可能出现在 10～11 月，依种类而定。尽管颜色可能有所不同，但该属所有种的主要花部形态都是相同的，某些种的花是花序而不是单生花。商业价值最高的两个种 *H. niger* 和 *H.* × *hybridus* 在花朵结构上有一些重要的差异。对于 *H. niger*，一个花序可以包含 1～4 朵花，但它们大多是顶生、两性和辐射对称的单生花。*H. niger* 的每朵花都有 5～10 个发达的钝（钝尖）萼片，这些萼片是宿存的，可能被错认为是花瓣（图 18.4）。萼片最初是白色的，到花期结束时是绿色的。*H.* × *hybridus* 的花序由 1～7 朵花组成（Rottensteiner 2016）。'东方杂种系'是花朵性状变异最多的类群。花朵颜色可以从白—绿—黄到粉—紫—黑转变，并且它们的花朵形态变异最大（重瓣花、镶边等）。与 *H. niger* 相反，*H.* × *hybridus* 的花大多是下垂的（图 18.2）。

图 18.4 *Helleborus niger* 花形态。(a) *Helleborus niger* ‘HGC Wintergold’，每朵花都有小漏斗状的器官或蜜腺及萼片，最初呈白色至花期结束时变为绿色；(b) *Helleborus niger* ‘HGC Wintergold’，雄蕊开始向心成熟

除萼片外，还观察到小的漏斗状器官（图 18.4）。1887 年在 Prantl 的一项研究中，这些花器官被称为蜜叶。后来（1949 年）Janchen 称它们为蜜腺叶，因为它们通常会分泌花蜜（Erbar et al. 1998）。在柱头接受花粉之后雄蕊才开始向心成熟（雌蕊先熟花）。心皮多胚珠，发育成具有种子的卵泡（图 18.5）。萼片通常在花期结束时重新变绿。这些绿色萼片甚至在花授粉时会形成出一个具有功能的光合作用系统，这种系统会一直存在直至结实（Salopek-Sondi et al. 2000；Shahri et al. 2011；Brcko et al. 2012；Schmitzer et al. 2013）。结实后，整个（绿色）花从植株上脱落。尽管开花持续时间在很大程度上取决于环境气温（Susek and Ivancic 2006），但是单朵花在完全开放后的平均寿命约为 6 天（Shahri et al. 2011）。具干柱头的心皮在基部稍合生。

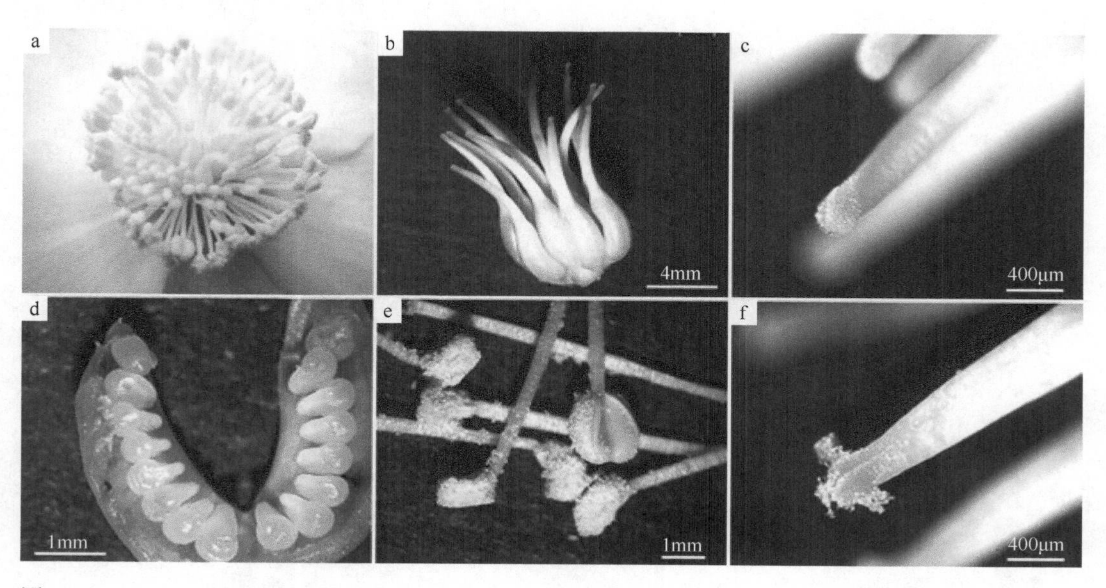

图 18.5 *H. niger* 花器官特写视图。(a) 心皮、雄蕊、位于雄蕊下面的绿蜜腺及白色萼片的全视图；(b) 离生心皮；(c) 具乳头的柱头；(d) 具有单个胚珠的子房的内视图；(e) 刚开裂花药的雄蕊；(f) 授过粉的柱头

Niimi 等（2006）研究了胚胎的形成，他们得出的结论是 *H. niger* 传播的种子中含有处于心形胚期发育不完全的原始胚。据我们所知，所有铁筷子属物种都是自然二倍体（或古四倍体），染色体数目为 32（Zonneveld 2001）。然而，这些种的 DNA 含量差异很大：所有物种的 DNA 含量都很高，但是变化很大，从 *H. argutifolius* 的 18.3pg/2C 到 *H. thibetanus* 的 33.2pg/2C 不等（Meiners et al. 2011）。

18.3 繁殖与栽培

铁筷子通过分株、种子或植物组织培养进行繁殖。分株繁殖产生的后代有限。而组织培养由于其离体培养启动困难（污染率高）、增殖率低、生根和驯化困难等，往往耗时较长（Caesar and Adelberg 2015；Matysiak and Gabryszewska 2016；Gabryszewska 2017）。*H. niger* 通过种子（约 60%）和植物组织培养（约 40%）进行繁殖。‘东方杂种系’主要通过种子繁殖，但市场上也可以找到无性繁殖品种（如 Heuger 的 Spring Promise 系列）。种间杂种只能通过微体快繁进行繁殖。

根据环境条件的不同，受精后的种子成熟需要 10～12 周。在商业生产中，由于种子休眠，从播种到萌发的时间为 26～34 周。要想解除这种休眠，夏季需要高温，秋冬季节需要低温。因此，在自然界中，种子在第二年的秋季或冬季萌发。发芽后，铁筷子幼苗大约在 1 月移植。从种子萌发到出现第一朵花需要 1～2 年。

18.4 育 种 目 标

铁筷子已经从一种园林植物演变成一种具有多种用途的花卉产品，使得近年来铁筷子的销量有所增加。根据植物的最终用途（室内、室外等），必须考虑特定的选择标准。

对于园林用途，育种者主要是对铁筷子旺盛的生长势感兴趣。其他标准是抗霜冻性和对直射阳光的耐受性，因为铁筷子是一种半遮阴植物。适于在露台、阳台、庭院等地种植的种间杂交品种（如 *H.* × *ericsmithii* 或者 *H.* × *nigercors* 类型）的育种也采用了类似的标准。相比之下，室内盆栽 *H. niger* 必须在圣诞节前开花，且必须具有紧凑的株型。当用作切花时，铁筷子种和种间杂交种必须具有长而强壮的花梗与良好的采后品质（如瓶插寿命）。

铁筷子育种的主要目标是花的新颖性：新的花色、颜色图案及重瓣或半重瓣。另外的目标是开花期长（延长至春季）和花朵寿命长。对于 *H. niger* 来说，过去几乎所有的 *H. niger* 植物从 1 月开始开花，但现在主流的选择标准是在圣诞节之前开花，甚至有一种趋势是在 11 月初就开花。同样，过去对 *H.* × *hybridus* 的要求是两年开花，而现在育种者试图选育一年就开花的植株。花头下垂不受欢迎，因此育种者专注于选育花梗较短或营养繁殖期较短/花期更早的品种，这使得花序更直立、更易于观赏。

其他与植物对恶劣的室内/室外环境条件（非生物胁迫）的耐受性和对病虫害（生物胁迫）的耐受性有关的选择标准，如蚜虫、红蜘蛛和蓟马及真菌（腐霉菌、疫霉菌、丝核菌、盾壳霉、枝孢菌、灰霉病），以及最近出现的铁筷子网状坏死病毒，也变得很重要。

目前的市场趋势是引入通过倍性和种间杂交获得的杂交种。这些不同倍性杂交大多是在人工诱导或偶然获得的四倍体和二倍体植株之间进行，从而产生了新的三倍体。这些不同倍性间的杂交可将颜色导入种间杂种中。铁筷子中许多可能的种间组合（表 18.1）还有待探索。

所有这些育种目标目前都是通过后代之间的杂交和选择来实现的，包括种间杂交（见下文）。突变育种，主要涉及多倍化，是一种有趣但很少使用的育种工具。

18.5　铁筷子的系统发育关系

育种者关注种间和种内关系以便选择亲本基因型。系统发育关系已通过 ITS 序列和叶绿体序列分析（Sun et al. 2001）、形态学描述（Susek 2016）、全基因组 RAPD（随机扩增多态性 DNA）（Fico et al. 2005）和 AFLP（扩增片段长度多态性）标记（Meiners et al. 2011）分析得以解析。Sun 等（2001）和 Meiners 等（2011）的分子水平研究都支持将铁筷子属划分为 6 个组（见上文），但在解析 *Helleborastum* 组中的物种方面遇到一些困难。基于 AFLP 标记数据，计算两两 Nei 和 Li（1979）遗传距离，揭示了 *Helleborastrum* 组内的遗传距离高达 0.195，而 *Chenopus* 组和 *Syncarpus* 组之间的遗传距离高达 0.319（Meiners et al. 2011）。

18.6　种间杂交障碍的鉴定与克服

种间杂交是将人们感兴趣的性状，如花色、香味和抗病性，融入当前品种基因库的宝贵工具。因此，无论是过去还是现在，种间杂交都是铁筷子的主要育种工具。为了确保获得可存活的杂种，杂交亲本（母本和父本）之间要互补且具有一定的亲和力。可能的杂交障碍分为受精卵形成前的障碍（受精前的杂交障碍）和受精后的障碍。对障碍类型和原因的了解决定了克服障碍的策略。首先，必须了解两种配子的生存能力。通过噻唑蓝（MTT）染色和原位萌发试验，已经证明许多不同种和基因型的花粉活力都很高，但离体萌发试验没有显示出可靠的结果（Winkelmann et al. 2015）。铁筷子花粉可以在冷冻温度（−18℃）下保存数周，通过冷处理或赤霉素处理，不同物种的开花时间可以同步（Christiaens et al. 2012）。授粉后，雄配子体和母体组织之间发生了复杂的分子与生物化学反应，从花粉黏附到受体柱头上，花粉水合、萌发，到花粉管生长至胚珠（Dresselhaus and Franklin-Tong 2013）。对 44 个不同的铁筷子种间杂交组合的花粉生长进行原位观察，表明尽管花粉管的生长比亲和的种内组合慢且数量少，但绝大多数（39 个组合）的花粉管到达了珠孔（Meiners and Winkelmann 2012）。在这些情况下，显然存在合子前障碍，但在大多数情况下，可以认为发生了受精。然而，用苯胺蓝染色观察花粉管生长并不能检测出精子的传递和配子的融合。在这里，微分干涉显微镜可以弥补这一知识空白。它不仅可以监测早期胚胎发生（Braun and Winkelmann 2016），而且可用于铁筷子物种杂交后的观察。

受精后，胚乳败育或退化可能是由基因组大小存在差异（Haig and Westoby 1991）、表观遗传变化（Michalak 2009）或 Bateson-Dobzhansky-Muller（BDM）不亲和性（Rieseberg and Carney 1998）引起的。这可能会导致胚败育。此外，后期起作用的合子后障碍可能导致杂种活力降低、白化病、杂种衰败或杂种不育（Bomblies and Weigel 2007）。在铁筷子中，使用胚挽救技术可以克服几个物种组合的合子后杂交障碍（Meiners and Winkelmann 2012）。图 18.6 展示了胚珠培养所采用的胚挽救技术的工作步骤。影响成败的重要因素有采样时间（授粉后 6～7 周效果最佳）、杂交方向、亲本的遗传距离和培养期间的温度状况（即打破休眠的低温时期）（Meiners and Winkelmann 2012）。根据亲本基因型的遗传距离对获得的种间杂种的综述表明，即使遗传距离为 0.264 也能产生杂种后代（表 18.2）。

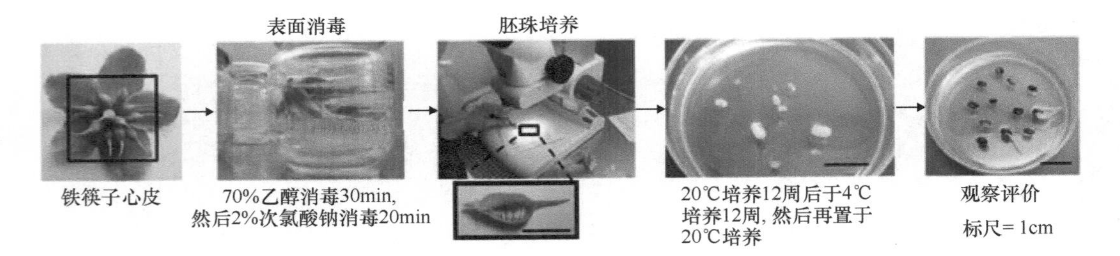

图 18.6 从人为控制杂交到胚珠培养评价的铁筷子种间杂种产生示意图

表 18.2 从胚挽救试验中获得的种间杂种后代，按亲本之间的遗传距离由低到高排列

遗传距离	杂交组合	数量	
		胚珠培养	杂交后代
0.069	*H. argutifolius* × *H. lividus*	114	37
0.069	*H. lividus* × *H. argutifolius*	258	60
0.081	*H.* × *hybridus* × *H. torquatus*	700	6
0.082	*H.* × *hybridus* × *H. cyclophyllus*	403	35
0.093	*H.* × *hybridus* × *H. atrorubens*	431	20
0.097	*H. croaticus* × *H. multifidus*	112	1
0.105	*H.* × *hybridus* × *H. odorus*	743	6
0.106	*H. torquatus* × *H. croaticus*	44	2
0.112	*H.* × *hybridus* × *H. croaticus*	486	14

续表

遗传距离	杂交组合	数量	
		胚珠培养	杂交后代
0.114	*H.* × *hybridus* × *H. multifidus*	559	8
0.115	*H. croaticus* × *H. odorus*	61	2
0.115	*H. odorus* × *H. croaticus*	56	2
0.117	*H.* × *hybridus* × *H. dumetorum*	509	1
0.130	*H.* × *hybridus* × *H. purpurascens*	539	8
0.141	*H. purpurascens* × *H. croaticus*	8	1
0.241	*H. foetidus* × *H. argutifolius*	307	2
0.255	*H.* × *hybridus* × *H. argutifolius*	366	1
0.264	*H.* × *hybridus* × *H. niger*	4640	13

最后是杂种验证。这可以通过形态学标记或流式细胞术（如果亲本的 DNA 含量不同）或分子标记（RAPD 或 AFLP）来验证（Meiners and Winkelmann 2012）。共显性简单重复序列（SSR）标记在铁筷子育种的各个方面也具有重要价值（Heinrich et al. 2012），但目前尚未使用。

18.7 倍 性 改 造

目前的育种工作侧重于增加具有重要商业价值的重要物种的遗传变异。多倍化是实现这一目标的重要方法之一（Mears 1980；Tal 1980；Soltis and Soltis 2000）。与它们的二倍体祖先相比，多倍体通常表现出相当大的生态、形态和遗传差异（Stebbins 1971；Mears 1980；Tal 1980）。如上所述，铁筷子的特征是所有自然物种的染色体数目都是 32。Dhooghe 等（2008）报道在固体培养基中添加氨磺乐灵（oryzalin）（3μmol/L）和氟乐灵（trifluralin）（3μmol/L 和 10μmol/L）培养 12 周可导致 *H. niger* 产生四倍体植株。用 10μmol/L 氟乐灵进行类似处理，获得 1 株四倍体 *H.* × *nigercors* 植株。用 100μmol/L 秋水仙碱溶液处理 16h 或 24h 来诱导不同种类铁筷子的多倍化无效。对于 *H.* × *hybridus*，所有处理均未成功诱导出四倍体。利用流式细胞仪分析秋水仙碱、氨磺乐灵和氟乐灵的处理效果，通过分析细胞核 DNA 含量，可以快速、准确地测定细胞倍性。然而，植物确切的染色体数目通常是通过染色体计数来确定的（Dart et al. 2004）。为了验证获得的流式细胞仪数据，对诱导获得的铁筷子四倍体植株的根尖细胞进行染色体计数，与未处理的对照植株的根尖细胞进行比较。这些分析证实了多倍化。根系组织起源于 LIII 层，双子叶植物的叶组织主要由 LI 层和 LII 层组成。这些结果表明，三个胚层（LI、LII 和 LIII）均为四倍体。

18.8 *H. niger* × *H.* × *hybridus*：胜过一切?

尽管人们进行了许多尝试以期获得将 *H. niger* 的株型结构、花期和 *H.* × *hybridus* 的花朵性状（颜色、形态）结合起来的种间杂交品种，但是商业化的杂交品种仍然屈指可数。在许多情况下，这些杂交会发生花粉萌发和受精，但是杂种发生了受精后障碍。有少数成功的例子：铁筷子‘Snow White’是日本横滨苗圃成功培育的第一个不育杂种。这是以 *H.* × *hybridus* 为母本、*H. niger* 为父本杂交的产物，它的特征是纯白色的花开放后会变成绿色再到淡淡的粉红色（Rice 2009）。2000 年法国培育了由 *H. niger* var. *macrantus* 和 *H.* × *hybridus* subsp. *abechicus* 杂交而成的品种 *H.* × *lemonnierae*（Rice 2009）。*H. niger* 和 *H.* × *hybridus* 的另一个杂交品种是英国苏塞克斯沃尔伯顿苗圃培育的铁筷子‘Walberton's Rosemary’（‘Walhero’）。这些例子，尽管并非全部被科学证实为杂种，但表明 *H. niger* 和 *H.* × *hybridus* 之间的杂交是可行的。然而，观赏性并不总是如预期的那样美好。其他组合也获得了成功的产品：*H.* × *ericsmithii* 和 *H.* × *hybridus*（*H.* × *glandorfensis*）（表 18.1）之间的杂交，目前在市场上被称为“‘HGC Ice N’玫瑰”系列。它们的特点是即使在阳光充足的地方也表现良好，对上述大多数疾病和非生物胁迫都具有较强的耐受性。而且它们连续开花，在同一个花茎上的老花上方产生新花芽，花朵开放后不会褪色。另一个有意思的种间杂交组合产生了品种‘Anna's Red’和‘Penny's Pink’。这些品种是由 Rodney Davey 经过多年培育获得的，其特征是综合了形状良好的红色或粉红色萼片和有斑叶片。但有时，更常见的种间杂交也会取得成功：*H.* × *ericsmithii* 栽培种‘Candy Love’因其粉红色茎上的花高过叶子，在 2010 年冬奥会被选来装饰温哥华的公园。

18.9 结　论

这些例子说明了铁筷子属物种具有巨大的繁育潜力。为了开发高效和强大的分子标记，如简单序列重复（SSR = 微卫星），并将它们与那些难以获得的表型性状联系起来还需要进一步的研究。如果分子序列数据可用，通过基因组编辑进行定向突变将成为一种有意思的育种工具。在这成为现实之前，必须先克服由单个或少数细胞真正再生的难题。最后，试着建立单倍体技术以便利用双单倍体，这将为以后的育种工作奠定基础。

本章译者：

段　青

云南省农业科学院花卉研究所，昆明 650200

参考文献

Bomblies K, Weigel D (2007) Hybrid necrosis: autoimmunity as a potential gene-flow barrier in plant species. Nat Rev Genet 8:382–389

Braun P, Winkelmann T (2016) Localization and overcoming of hybridization barriers in *Delosperma* and *Lampranthus* (Aizoaceae). Euphytica 211:255–275

Brcko A, Pencik A, Magnus V, Prebeg T, Mlinaric S, Antunovic J, Lepedus H, Cesar V, Strnad M, Rolcik J, Salopek-Sondi B (2012) Endogenous auxin profile in the Christmas rose (*Helleborus niger* L.) flower and fruit: free and amide conjugated IAA. J Plant Growth Regul 31:63–78

Caesar L, Adelberg J (2015) Using a multifactor approach for improving stage II responses of *Helleborus* hybrids in micropropagation. Propag Ornamental Plants 15(4):125–135

Cakar J, Haveric A, Haveric S, Maksimovic M, Paric A (2014) Cytotoxic and genotoxic activity of some *Helleborus* species. Nat Prod Res 28:883–887

Christiaens A, Dhooghe E, Pinxteren D, van Labeke MC (2012) Flower development and effects of a cold treatment and a supplemental gibberellic acid application on flowering of *Helleborus niger* and *Helleborus* × *ericsmithii*. Sci Hortic 136:145–151

Dart S, Kron P, Mable BK (2004) Characterizing polyploidy in *Arabidopsis lyrata* using chromosome counts and flow cytometry. Can J Bot 82:185–197

Dhooghe E, Grunewald W, Leus L, Van Labeke M-C (2008) In vitro polyploidisation of *Helleborus* species. Euphytica 165:89–95

Dresselhaus T, Franklin-Tong N (2013) Male-female crosstalk during pollen germination, tube growth and guidance, and double fertilization. Mol Plant 6:1018–1036

Erbar C, Kusma S, Leins P (1998) Development and interpretation of nectary organs in Ranunculaceae. Flora 194:317–332

Fico G, Servettaz O, Caporali E, Tomè F, Agradi E (2005) Investigation of *Helleborus* genus (Ranunculaceae) using RAPD markers as an aid to taxonomic discrimination. Acta Hortic 675:205–209

Gabryszewska E (2017) Propagation in vitro of hellebores (*Helleborus* L.) review. Acta Sci Pol Hortorum Cultus 16(1):61–72

Haig D, Westoby M (1991) Genomic imprinting in endosperm: its effect on seed development in crosses between species, and between different ploidies of the same species, and its implications for the evolution of apomixes. Philos Trans R Soc Lond B 333:1–13

Heinrich R, Klein F, Hohe A (2012) Use of AFLP-markers for estimation of the inbreeding level in *Helleborus orientalis*. Acta Hortic 961:205–210

Maior MC, Dobrota C (2013) Natural compounds with important medical potential found in *Helleborus* sp. Cent Eur J Biol 8:272–285

Mathew B (1989) Hellebores. Alpine Garden Society Publications, Woking

Matysiak B, Gabryszewska E (2016) The effect of in vitro culture conditions on the pattern of maximum photochemical efficiency of photosystem II during acclimatisation of *Helleborus niger* plantlets to ex vitro conditions. Plant Cell Tissue Organ Cult 125(3):585–593

Mears JA (1980) Chemistry of polyploids. In: Lewis WH (ed) Polyploidy: biological relevance, vol 13. Plenum Press, New York, pp 77–102

Meiners J, Winkelmann T (2012) Evaluation of reproductive barriers and realisation of interspecific hybridisations depending on the genetic distances between species in the genus *Helleborus*. Plant Biol 14:576–585

Meiners J, Debener T, Schweizer G, Winkelmann T (2011) Analysis of the taxonomic subdivision within the genus *Helleborus* by nuclear DNA content and genome-wide DNA markers. Sci Hortic 128(1):38–47

Michalak P (2009) Epigenetic, transposon and small RNA determinants of hybrid dysfunctions. Heredity 102:45–50

Nei M, Li WH (1979) Mathematical model for studying genetic variation in terms of restriction endonucleases. Proc Natl Acad Sci U S A 76:5269–5273

Niimi Y, Han DS, Abe S (2006) Temperatures affecting embryo development and seed germination of Christmas rose (*Helleborus niger*) after sowing. Sci Hortic 107:292–296

Rice G (2009) Hybridising *Helleborus niger*. Plantsman 2009:212–215

Rice G, Strangman E (1993) The gardener's guide to growing hellebores. David & Charles, Devon. 160 p

Rieseberg LH, Carney SE (1998) Tansley review no. 102. Plant hybridization. New Phytol 140:599–624

Rottensteiner WK (2016) Attempt of a morphological differentiation of *Helleborus* species in the Northwestern Balkans. Mod Phytomorphol S9:17–33

Stebbins GL (1971) Chromosome evolution in higher plants. Edward Arnold, London

Schink M, Garcia-Kaufer M, Bertrams J, Duckstein SM, Müller MB, Huber R, Stintzing FC, Grundemann C (2015) Differential cytotoxic properties of *Helleborus niger* L. on tumour and immunocompetent cells. J Ethnopharmacol 159:129–136

Schmitzer V, Mikulic-Petkovsek M, Stampar F (2013) Sepal phenolic profile during *Helleborus niger* flower development. J Plant Physiol 170:1407–1415

Salopek-Sondi B, Kovac M, Ljubesic N, Magnus V (2000) Fruit initiation in *Helleborus niger* L. triggers chloroplast formation and photosynthesis in the perianth. J Plant Physiol 157:357–364

Salopek-Sondi B, Kovac M, Prebeg T, Magnus V (2002) Developing fruit direct post-floral morphogenesis in *Helleborus niger* L. J Exp Bot 53:1949–1957

Shahri W, Tahir I, Islam ST, Bhat MA (2011) Physiological and biochemical changes associated with flower development and senescence in so far unexplored *Helleborus orientalis* Lam. cv. Olympicus. Physiol Mol Biol Plants 17(1):33–39

Soltis PS, Soltis DE (2000) The role of genetic and genomic attributes in the success of polyploids. Proc Natl Acad Sci U S A 97:7051–7057

Sun H, McLewin W, Fay F (2001) Molecular phylogeny of *Helleborus* (Ranunculaceae), with an emphasis on the East Asian-Mediterranean disjunction. Taxon 50:1001–1018

Susek A (2016) Perspectives of Christmas rose (*Helleborus niger* L.) genetic improvement. Agricultura 13(1–2):11–19. https://doi.org/10.1515/agricultura-2017-0003

Susek A, Ivancic A (2006) Pollinators of *Helleborus niger* in Slovenian naturally occurring populations. Acta Agric Slov 87(2):205–211

Tal M (1980) Physiology of polyploids. In: Lewis WH (ed) Polyploidy: biological relevance, vol 13. Plenum Press, New York, pp 61–76

Tamura M (1993) Ranunculaceae. In: Kubitski K, Rohwer JG, Bittrich V (eds) The families and genera of vascular plants: flowering plants - Dicotyledons, vol II. Springer-Verlag, Berlin, pp 563–583

Tamura M (1995) Systematic part. In: Engler A, Prantl K (eds) Die Natürlichen Pflanzenfamilien. Bd. 17 a IV Angiospermae. Ordnung Ranunculales. Fam. Ranunculaceae, 2nd edn. Duncker & Humblot, Berlin, pp 220–519

Tan CW, Tian YF, Gong HY, Chen XW, Jiang KJ, Wang R (2014) Two new bufadienolides from the rhizomes of *Helleborus thibetanus* with inhibitory activities against prostate cancer cells. Nat Prod Res 28:901–908

Winkelmann T, Hartwig N, Sparke J (2015) Interspecific hybridisation in the genus *Helleborus*. Acta Hortic 1087:301–308

Zonneveld BJM (2001) Nuclear DNA contents of all species of *Helleborus* (Ranunculaceae) discriminate between species and sectional divisions. Plant Syst Evol 229:125–130

第19章 伽 蓝 菜

Kathryn Kuligowska Mackenzie, Henrik Lütken, Lívia Lopes Coelho,
Maja Dibbern Kaaber, Josefine Nymark Hegelund, and Renate Müller

摘要 伽蓝菜属植物主要由原产于马达加斯加及非洲东部和南部的多肉植物组成。该属中最重要的种是圣诞伽蓝菜，目前已培育出大量商业品种。伽蓝菜种类因具有丰富的花量、单个花朵寿命长、采后性能佳及低运输要求被开发为盆栽植物。该属植物也越来越多地被当作室外花卉和切花使用。如今，伽蓝菜已成为欧洲第二流行的盆栽植物，仅在2016年年营业额就为6700万欧元。该属的育种工作开始于20世纪30年代，主要集中在圣诞伽蓝菜上，目的是培育花色多样的紧凑型栽培品种。目前，伽蓝菜品种的改良是通过种内杂交、种间杂交和基因工程实现的。其育种的目标包括一系列性状的改变，包括表型性状及采后寿命，同时最大程度地减少对环境的影响。

关键词 紧凑型盆栽植物；景天科酸代谢；乙烯耐受性；基因工程；赤霉素信号传递；种间杂交；*rol* 基因；短日照植物；野生种

19.1 引 言

圣诞伽蓝菜由 Perrier de la Bâthie 于1924年在马达加斯加中北部的一座山坡上发现（van Voorst and Arends 1982）。他将该物种命名为 *Kalanchoe globulifera* var. *coccinea*，并把它送到了巴黎的植物园，该植物从那里再传到欧洲和美国的植物园。1934年，德国植物学家 von Poellnitz 将其归类为一个单独的种并定名为圣诞伽蓝菜，并由保育员 Blossfeld 首次将其种植成盆栽植物（van Voorst and Arends 1982）。圣诞伽蓝菜品种选育开始于20世纪30年代。由于最初的育种材料是从单个植物的后代中选择产生的，因此遗传背景很窄（van Voorst and Arends 1982）。早期品种多为通过种子繁殖的二倍体，它们的生长期长、成品高大且异质（Alton and Pertuit 1992）。因此，首要育种目标是选择紧凑型植株。后来，从最初的红色品种中筛选出橙色和黄色突变株并进行种植。第一个种间杂种是在1939年，由圣诞伽蓝菜和矮生伽蓝菜杂交获得（van Voorst and Arends 1982）。

如今，伽蓝菜已成为欧洲第二大盆栽植物，年营业额达6700万欧元，年市场增长

K. K. Mackenzie (✉), H. Lütken, L. L. Coelho, M. D. Kaaber, J. N. Hegelund, R. Müller
Department of Plant and Environmental Sciences, University of Copenhagen, Tåstrup, Denmark
e-mail: kku@plen.ku.dk; hlm@plen.ku.dk; livia@plen.ku.dk; kaaber@plen.ku.dk; jnh@plen.ku.dk; ren@plen.ku.dk

率为 8.1%（FloraHolland 2016）。根据欧盟植物新品种保护办公室（CPVO）的数据，全球已注册的伽蓝菜品种共有 706 个。但是，它们仅包括除圣诞伽蓝菜以外的其他几个种类，如紫武藏（*K. humilis*）、伽蓝菜（*K. laciniata*）、红提灯（*K. manginii*）、江户紫（*K. marmorata*）、唐印（*K. thyrsiflora*）和独叶草（*K. uniflora*）（CPVO 2017）。

由于丰富的开花量及对水和养分的要求低，圣诞伽蓝菜品种作为盆栽植物很受欢迎。营养繁殖的杂交品种因其植株的整齐度和花朵颜色而占市场主导地位（Alton and Pertuit 1992）。除了被用作室内盆栽植物外，伽蓝菜被用作室外植物和作为切花使用时也广受欢迎（Kalanchoe Growers Holland 2017，http://www.kalanchoe.nl；Queen®，2017，http://www.queen.dk）。此外，在马达加斯加和印度，少量种类还被作药用（Descoings 2003）。

一般而言，伽蓝菜品种的育种目的是由种植者、消费者和市场需求共同决定的。在过去的几十年中，多项研究调查了影响伽蓝菜花期的生产和采后条件。乙烯敏感度是影响圣诞伽蓝菜花期的重要因素之一（Marousky and Harbaugh 1979；Høyer and Nell 1995；Willumsen and Fjeld 1995；Serek and Reid 2000；Sanikhani et al. 2008）。在运输过程中适当的温度和光照及有效的通风等因素可以提高圣诞伽蓝菜的花朵寿命（Leonard and Nell 1998）。改善伽蓝菜品种的其他育种目标还包括茂密的枝条和芽，快速的营养繁殖和较短的生产时间，节间短和分枝多的紧凑型株型，新的花型，如多瓣花、花色、花期长、易于开花、叶色和叶形及抗病虫害（Nielsen et al. 2005；Jepsen and Christensen 2006；Christensen et al. 2008；Lütken et al. 2010；Currey and Erwin 2011；Coelho et al. 2015；Kuligowska et al. 2015a；Madriz-Ordeñana et al. 2016；Huang and Chu 2017；Wick 2017）。

19.2 花诱导的起源、生态和环境因素

伽蓝菜属约有 140 个种，主要分布在马达加斯加、非洲东部和南部，并延伸到热带非洲、阿拉伯、热带和东南亚。如今，某些种类在整个热带地区都以新生物种的形式存在（Descoings 2003）。马达加斯加是该属物种数量最多的国家，也是该属物种形态多样性最丰富的国家，表明这里是该物种的形成中心或起源中心（Boiteau and Allorge-Boiteau 1995）。伽蓝菜属植物是多年生灌木，很少长成小树。它们多数是两年至多年生植物，少数是一年生草本植物。该属的特征是具四瓣二齿花（Descoings 2003）。

19.2.1 系统学和系统发育学研究

伽蓝菜属的系统学研究尚不清楚。在对形态学、解剖学、胚胎学、核型、植物地理和分子数据分析的基础上，存在两种截然不同的观点：一种观点是建立单独的属，即落地生根属（*Bryophyllum*）、伽蓝菜属（*Kalanchoe*）和金钟伽蓝属（*Kitchingia*）；另一种观点是将所有物种合并成一个伽蓝菜属，然后再分为亚属（落地生根亚属、*Calophygia*、伽蓝菜亚属）或组（落地生根组、伽蓝菜组和金钟伽蓝组）（Chernetskyy 2011）。对来自

叶绿体 DNA 的分子数据分析表明，落地生根属和金钟伽蓝属属于伽蓝菜属（Van Ham and Hart 1998；Mort et al. 2001）。通过对伽蓝菜属基因的 RAPD 标记和 ITS 序列分析，也可得出类似结论（Gehrig et al. 1997，2001）。Descoings（2003）在 2003 年提出了对伽蓝菜属的简易划分，分为两个部分，即伽蓝菜和落地生根（包括金钟伽蓝）。他认为所有分类都过于多样和人为化，不能很好地解释本属特征。花的特性和胎生植物的繁殖能力，即附着于亲本植物上生长的后代小苗，是这种分类法的基础。部分伽蓝菜的特征是花直立，在花冠筒的中间或上方插入花丝，花萼管通常比花萼裂片短，并且萼片通常是游离的或几乎没有游离的（图 19.1a 上图）。存在于马达加斯加和非洲其他地区的部分伽蓝菜种类不会产生胎生植物。亚洲也有 14 个特有种（Descoings 2003）。属于落地生根属的种有垂悬的花，在花冠中间下方花丝的花萼管通常长于花萼裂片（图 19.1a 下图）。落地生根植物的特征是能够在叶缘或花序形成胎生植物（图 19.1b）。落地生根种是马达加斯加特有的（Descoings 2003；Izumikawa et al. 2007）。在伽蓝菜属中也有一些物种表现出这两个部分的典型特征，这些种类被归类为同一分类群（Descoings 2003）。

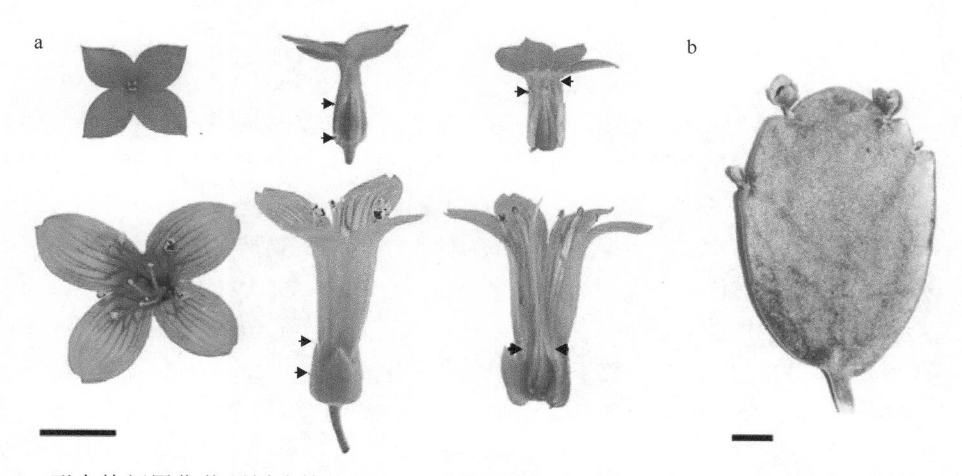

图 19.1 形态特征用作截面划分的基础。（a）伽蓝菜花的形态。上图：伽蓝菜属的圣诞伽蓝菜（*K. blossfeldiana*），下图：落地生根亚属的毛落地生根（*K. pubescens*）；俯视图和侧视图显示了花萼裂片的长度和花被的切面及花丝附着的位置（标尺 = 1cm）。（b）白姬之舞（*K. marnieriana* H. Jacobsen）叶片边缘形成的胎生植物（标尺 = 1cm）

在以后的工作中，不同的研究者建议将伽蓝菜分为三个亚属或三个部分是最合适的分类方法（Chernetskyy 2011）。在现在关于伽蓝菜的研究中，有许多方法可以对其进行分类。在本章中，我们遵循的分类法是伽蓝菜属分为两个部分，因为这种分类法是解释该属植物之间关系最清晰的方法。

19.2.2 景天科酸代谢

伽蓝菜与伽蓝菜属隶属的景天科的其他种类一样，对光合作用进行一种与生态相关的调节，称为景天科酸代谢（CAM）（Gehrig et al. 2001）。CAM 植物通过夜间气孔开放吸收 CO_2，并以有机酸的形式将 CO_2 储存在液泡中，白天则关闭气孔以防止蒸发，并且通过光合作用利用在液泡中重新活化的 CO_2（Lüttge 2004）。伽蓝菜属植物对干旱地区环境的适应性归因于 CAM 代谢的演变（Boiteau and Allorge-Boiteau 1995；Gehrig et al. 1997，2001）。所有的伽蓝菜属植物都可进行 CAM，并且该属中的 CAM 模式非常灵活，在物种内部、物种之间甚至根据叶片年龄的变化而变化。此外，根据其栖息地的不同，已经确定了三种 CAM 类型（Gehrig et al. 1997）。伽蓝菜属植物主要分布在马达加斯加的南部干旱地区和非洲东部干旱地区，它们表现出强大的叶肉性习性（主要指伽蓝菜部分），此类植物的 CAM 仅在夜间进行不依赖于水的 CO_2 固定。落地生根属植物（狭义的落地生根）主要分布在干旱或局部干旱的地区，并且随着有规律的干湿季节变化，它们具有专一但灵活的 CAM。此类植物 CO_2 的固定主要发生在夜间，但如果有足够的水，它们在白天也可能会吸收大量的 CO_2。最后一类包含马达加斯加特有的薄叶植物，主要分布在潮湿的地区（大部分隶属于伽蓝菜属的金钟伽蓝亚属部分）。尽管此类植物有能力进行 CAM，但它们在生命中的大部分时间内都进行 C3 途径的光合作用（Gehrig et al. 1997，2001）。

19.2.3 开花诱导

光周期是诱导景天科植物开花的一种常见机制。在伽蓝菜属中，已经鉴定出两种光周期诱导类型。它们包括短日照（SD）植物，即暴露于短日照下开花的植物，以及长短日照（LSD）植物，即需要特定的双光周期照射才能开花的植物（Zeevaart 1985；Currey and Erwin 2011）。在伽蓝菜属植物内，最重要的种类圣诞伽蓝菜被归类为 SD 植物（Currey and Erwin 2011），而 LSD 植物包括落地生根属植物（Zeevaart 1985）。

通常，伽蓝菜属中的许多种类在温室条件下既难以诱导开花又难以控制开花时间。因此，在育种中很难使用这些种类（Currey and Erwin 2011；Coelho et al. 2015）。与光周期有关的不同因素均会影响伽蓝菜的花诱导。这些因素包括特定的日照长度（允许开花的光周期）和光诱导 SD 周期数（Currey and Erwin 2010）。结果表明，不同伽蓝菜属植物的光周期敏感性不同（Zeevaart 1976）。Currey 和 Erwin（2010）观察到不同种类之间的开花数量在光周期诱导条件下有很大的差异。此外，Huang 和 Chu（2012）还报道了开花时间可能与品种有关。

即使伽蓝菜的开花诱导中光周期起主要作用，但其他因素也会改变开花响应。夜间的低温有效地促进了玉吊钟（*K. fedtschenkoi*）的开花时间，并增强了圣诞伽蓝菜（*K. blossfeldiana*）、棒叶落地生根（*K. verticaillata*）、大叶落地生根（*K. daigremontiana*）、落地生根（*K. pinnata*）和伽蓝菜（*K. laciniata*）的开花数量（Sharma 1970）。此外，将 *K. velutina* 置于较低的夜间温度时会开花，而较高的夜间温度则出现营养生长（Sharma 1973）。有趣的是，早期研究结果表明夜间温度影响圣诞伽蓝菜的碳吸收，在高温环境中该种类的开花数量急剧减少（Spear 1959）。然后，有研究报道了圣诞伽蓝菜花的形成和 CAM 代谢活性相关，包括这两个过程都被短日的光周期所促进，被暗周期的中断所抑制。此外，拟制开花的同时也会拟制 CAM 活性。然而，它们的生理相关性尚不清楚（Alton and Pertuit 1992）。

光照强度可改变伽蓝菜开花诱导。研究表明，与种植在弱光下相比，圣诞伽蓝菜种植在强光下的花枝数量更多，花序更大（Eveleens-Clark et al. 2004），并且开花期提前（Mortensen 2014）。因此，适当增加光照强度可以提高圣诞伽蓝菜的品质（Eveleens-Clark et al. 2004；Carvalho et al. 2006；Mortensen 2014）。

在伽蓝菜属的开花诱导研究中，最大的挑战之一是缺乏对植物幼年生长期时间的了解。Khoury 和 White（1980）观察到伽蓝菜属植物只有在开花诱导之前或期间植物成熟后才会开花。此外，伽蓝菜属幼年期时间长短具有品种特异性。不同品种的圣诞伽蓝菜将在扦插后的 10～14 周成熟（Khoury and White 1980）。Rünger（1966）报道，不同品种的圣诞伽蓝菜成熟度取决于展开的叶对的数量，根据 Schwabe（1969）的报道，该种需要 8 对叶才能开花。江户紫（*K. marmorata*）和落地生根（*K. pinnata*）分别需要 10 对和 37 对叶才能达到成熟（Wadhi and Ram 1967；Zimmer 1996）。

19.3　伽蓝菜属的繁殖

改良伽蓝菜品种的育种方法包括从传统的杂交育种到分子育种的多种技术。下面我们概述最重要的方法。

19.3.1　异花授粉和种间杂交

19.3.1.1　有性生殖

关于伽蓝菜属生殖生物学和授粉生态学研究的报道较少。但有人认为鸟类和昆虫可能是伽蓝菜属的授粉媒介（Hickey and King 1988）。圣诞伽蓝菜的花为雄蕊先熟花（雄性生殖器官比雌性先成熟），即花粉的成熟发生在柱头成熟之前（Hickey and King 1988；Traoré et al. 2014）。对伽蓝菜属植物柱头形态和生理变化进行观察可确定异花授粉与种间杂交的时间（Traoré et al. 2014；Mackenzie et al.，未发表）。

一般来说，伽蓝菜属植物具有自交亲和性或部分自交亲和性，并能够自主授粉（Herrera and Nassar 2009；Kuligowska et al. 2015b；González de León et al. 2016）。伽蓝菜植物有两轮雄蕊：如果不发生异花授粉，则雄蕊下部先开裂，柱头在花成熟过程中逐渐伸长，雄蕊上部完成自花授粉（Hickey and King 1988）。某些种类自花授粉后的种子活力较低，这可能表明异花授粉是主要的育种方式（Kuligowska et al. 2015b；González de León et al. 2016）。

19.3.1.2　种间杂交

自伽蓝菜属植物开始育种至今，圣诞伽蓝菜一直与远缘伽蓝菜通过种间杂交进行遗传改良。表 19.1 列出了伽蓝菜属中产生的种间杂种。从 20 世纪初起，开始在其他伽蓝菜种类之间进行杂交，所获得的种间杂种商业价值不大（Baldwin Jr 1949；Descoings 2003）。

表 19.1　伽蓝菜属的种间杂种

母本	父本	物种名称[a]	参考文献
组内杂交			
极乐鸟伽蓝菜（*K. beauverdii* Hamet）	圣诞伽蓝菜（*K. blossfeldiana* von Poellnitz）		Vlielander 2007
K. blossfeldiana von Poellnitz	*K. citrina* Schweinfurth		Izumikawa et al. 2008
	圆叶景天（*K. farinacea* Balfour）		Izumikawa et al. 2008
	矮生伽蓝菜（*K. glaucescens* Britten）		van Voorst and Arends 1982
	盾叶伽蓝菜（*K. grandiflora* Wight and Arnott）	*K.* × *vadensis*	Descoings 2003
	伽蓝菜（*K. laciniata* De Candolle）		Jepsen and Christensen 2006
	天之羽袖（*K. nyikae* Engler）		Izumikawa et al. 2008
	白粉叶伽蓝菜（*K. pumila* Baker）		Izumikawa et al. 2008
	圆叶伽蓝菜（*K. rotundifolia* Haworth）		Jepsen and Christensen 2006
	匙叶伽蓝菜（*K. spathulata* De Candolle）		Izumikawa et al. 2008
大叶落地生根[b]（*K. daigremontiana* Hamet and H. Perrier[b]）	***K. delagoensis* Ecklon and Zeyher**		Baldwin Jr 1949
	变叶落地生（*K. rosei* Hamet and H. Perrier）		Descoings 2003
台南伽蓝菜（*K. garambiensis* Kudo）	*K. laciniata* De Candolle		Huang and Chu 2017
	K. spathulata De Candolle		Huang and Chu 2017
K. glaucescens Britten	*K. bentii* Wight ex Hooker	*K.* × *kewensis*	Descoings 2003
	K. grandiflora Wight and Arnott	*K.* × *ena*	van Voorst and Arends 1982
	K. lateritia Engler	*K.* × *felthamensis*	Descoings 2003
	K. nyikae Engler		Descoings 2003
青蔓花（*K. gracilipes* Baker）	**红提灯（*K. manginii* Hamet and H. Perrier）**		Cullen et al. 2011
K. grandiflora Wight and Arnott	*K. glaucescens* Britten		Descoings 2003
K. laciniata De Candolle	*K. garambiensis* Kudo		Huang and Chu 2017
	K. lanceolata Persoon		Descoings 2003
***K. laxiflora* Baker**	***K. daigremontiana* Hamet and H. Perrier**		Izumikawa et al. 2007
朱红宫灯长寿花（*K. miniata* Hils and Bojer）	**大宫灯长寿花（*K. porphyrocalyx* Baker）**		Cullen et al. 2011
K. nyikae Engler	*K. blossfeldiana* von Poellnitz		Kuligowska et al. 2015b
白粉叶伽蓝菜（*K. pumila* Baker）	*K. blossfeldiana* von Poellnitz		Izumikawa et al. 2008
K. spathulata De Candolle	*K. blossfeldiana* von Poellnitz		Izumikawa et al. 2008
	K. garambiensis Kudo		Huang and Chu 2017

续表

母本	父本	物种名称 [a]	参考文献
组间杂交			
K. blossfeldiana von Poellnitz	***K. aromatica* H. Perrier**		Jepsen and Christensen 2006
	***K. daigremontiana* Hamet and H. Perrier**		Izumikawa et al. 2008
	***K. laxiflora* Baker**		Izumikawa et al. 2008
	白姬之舞（*K. marnieriana* H. Jacobsen）		Kuligowska et al. 2015b
	毛落地生根（*K. pubescens* Baker）		Izumikawa et al. 2008
K. garambiensis Kudo	***K. manginii* Hamet and H. Perrier**		Huang and Chu 2017
K. spathulata De Candolle	***K. laxiflora* Baker**		Izumikawa et al. 2007

a. 如果列出了物种名

b. 以粗体显示的物种属于落地生根属

尽管种间杂交是新的伽蓝菜品种的主要来源，但目前只有少数工作研究了伽蓝菜属种内和种间的生殖隔离。我们先前对伽蓝菜种间杂交发生的杂交障碍研究表明，在人工授粉过程中，花粉质量是影响杂交成功的重要因素。此外，我们还观察到与抑制花粉萌发和雌蕊花粉管异常生长有关的受精前障碍（Kuligowska et al. 2015b）。在伽蓝菜与落地生根植物之间进行杂交时观察到单侧杂交亲和性（Izumikawa et al. 2007，2008）。在对正反交的花粉管发育进行研究后，我们得出结论，单侧杂交亲和性出现的原因是雌蕊长度存在差异，因为伽蓝菜植物具有明显较短的花柱，不能产生能到达落地生根植物子房的花粉管（Kuligowska et al. 2015b）。

伽蓝菜的受精后障碍可能与种子的胚乳退化有关。关于异常皱纹种子产生（Kuligowska et al. 2015b）及在胚培养后才能获得杂交种的研究（Izumikawa et al. 2008）证实了这一猜想。有趣的是，倍性或染色体数目存在差异不会影响种间杂交的成功率（Izumikawa et al. 2008；Kuligowska et al. 2015b）。对伽蓝菜植物的细胞学研究表明其染色体组数是 17（为主）、18 和 20（为辅）（Baldwin 1938；Uhl 1948）。圣诞伽蓝菜最初是二倍体植物，$2n = 34$。随着杂交选育，出现许多多倍体品种，通常品种是四倍体（$2n = 68$），但也存在 $2n = 72$、75、84、85 和 96 的品种（van Voorst and Arends 1982）。我们对圣诞伽蓝菜品种和伽蓝菜属其他种类的细胞学分析表明，物种内部同时存在二倍体和四倍体植物，其染色体数分别为 $2n = 34$ 和 68。这些杂种是由两个二倍体杂交获得的（Kuligowska et al. 2015b）。

我们研究的伽蓝菜物种间生殖隔离（Kuligowska et al. 2015b）与其他杂交研究结果一致，这表明杂交障碍在近缘物种杂交时发生的频率较低（Kuligowska et al. 2016）。然而，在种间杂交中，特定基因型的遗传背景对杂交有很大影响。因此，遗传距离与生殖隔离仅表现出粗略的相关性（Kuligowska et al. 2015b）。

19.3.1.3　种间杂种的特征

通过对经不同杂交方式获得的种间杂种分析可揭示后代的主要中间特征（Izumikawa

et al. 2007，2008；Kuligowska et al. 2015a）。杂种还表现出两个亲本的中间核 DNA 含量和染色体数目（Izumikawa et al. 2007，2008）。

在我们的研究中，可根据形态特征确定 F_1 代杂种特性。可在发育早期阶段用杂种特有的形态特征来识别。圣诞伽蓝菜（*K. blossfeldiana*）品种和毛落地生根（*K. pubescens*）的杂交后代表现出叶片表面具有绒毛，该特征显然是遗传自父本。叶缘基部的紫色斑点是由在杂交中作为父本的白姬之舞（*K. marnieriana*）遗传而来，而圣诞伽蓝菜则作为母本。在早期很难评估圣诞伽蓝菜和天之羽袖（*K. nyikae*）之间的杂交。但随着植株的发育，植物的结构、叶片形态及花朵颜色是评估植物杂交的显著证据（Kuligowska et al. 2015a，2016）。所选亲本和杂种后代花的特征见图 19.2。

在伽蓝菜杂种中，还可观察到与株高或花径相关的越冬性状的表达。与亲本相比，杂种表现出旺盛的生长势，如几个杂种系都提前开花。此外，杂种的花期也更接近于花期较长的圣诞伽蓝菜（*K. blossfeldiana*）品种。把红花的圣诞伽蓝菜和黄花的天之羽袖（*K. nyikae*）杂交，杂交后代产生了一个带有粉红色花朵的杂种，而其余后代花朵颜色则呈现介于亲本植物之间的红橙色系（Kuligowska et al. 2015a，2016）。

圣诞伽蓝菜（*K. blossfeldiana*）的不同野生种杂交后，获得的杂交种的花粉表现为育性较低或败育（Izumikawa et al. 2008；Kuligowska et al. 2015a）。从匙叶伽蓝菜（*K. spathulata*）和 *K. laxiflora* 之间的杂交后代中也得到了类似的观察结果（Izumikawa et al. 2007）。有趣的是，在我们的研究中发现相同倍性的种杂交产生的后代是不育的，而不同倍性间杂交产生的后代既有可育的又有不育的。因对其杂种没有进行细胞遗传学观察，通过染色体加倍自发恢复育性和未减数配子形成都是对观察到的现象的合理解释（Kuligowska et al. 2015a）。当计划将 F_1 代用于进一步育种时，杂种不育可能成为主要障碍。染色体加倍可以克服与染色体重排相关的杂种不育。有趣的是，事实证明，在不使用抗有丝分裂剂的情况下，由伽蓝菜的叶片再生成植株是一种有效的获得多倍体的方法（Aida and Shibata 2002）。

19.3.1.4 野生种及其特性

伽蓝菜属的野生种具有许多有趣的特征，它们可作为新型观赏性盆栽、园林植物或者切花使用。还可以作为特定性状的供体，通过种间杂交或体细胞融合渗入现有品种。表 19.2 列出了伽蓝菜新品种选育中可选择的潜在性状。

伽蓝菜属中有一些具有香气的种类，这为将香气性状引入圣诞伽蓝菜品种提供了巨大的可能性。但在目前的情形下，由于其生长周期长、花期短和采后性能差，还不适合开发成商业品种（Descoings 2003；Mackenzie et al.，未发表）。花带香味的种类包括 *K.* × *ena*、盾叶伽蓝菜（*K. grandiflora*）、矮生伽蓝菜（*K. glaucescens*）、朱莲（*K. petitiana*）和唐印（*K. thyrsiflora*），而 *K. aromatica* 叶片带有香味。已作为商业品种在销售的带香味的伽蓝菜品种的香味被形容为散发出令人愉悦的柠檬花香（LPlants 1975；Irwin 1976；Pittman 1977；Dilworth 1982）。这些品种是名为‘金尘’的伽蓝菜品种和未命名的香味品种经异花授粉选育的杂交一代或杂交二代。尚不清楚哪种野生物种（如果有的话）促进了这些品种选育（LPlants 1975；Irwin 1976；Pittman 1977；Dilworth 1982）。

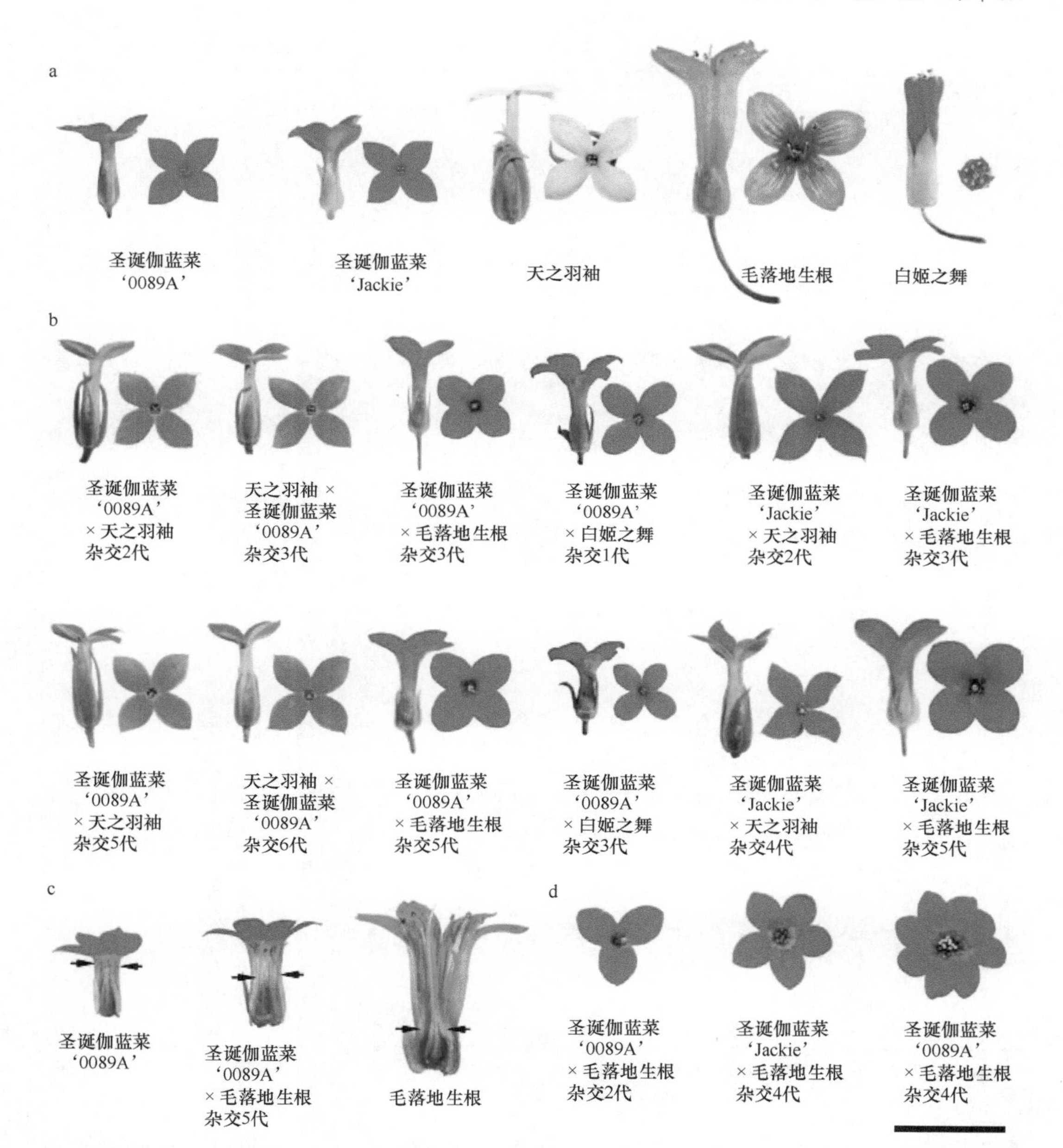

图 19.2　伽蓝菜杂种及其亲本的花朵特征（摘自 Kuligowska et al. 2015a）。(a) 亲本植物的花朵的侧视图和俯视图。(b) 所选杂种花朵的侧视图和俯视图。(c) 花的纵向内视图：圣诞伽蓝菜品种 '0089A'（左）、圣诞伽蓝菜 '0089A' × 毛落地生根杂交第 5 代（中）、毛落地生根（右），箭头指示花丝附着在花冠管上的位置。(d) 圣诞伽蓝菜和毛落地生根杂交后代的花瓣数量变化（标尺 = 2cm）

表 19.2　具有观赏价值的特征和种类

目标性状	种类	应用
钟型花	大宫灯长寿花（*K. porphyrocalyx*）、*K. jongmansii*、落地生根（*K. pinnata*）、青蔓花（*K. gracilipes*）	盆栽和切花
具有攀岩性	极乐鸟伽蓝菜（*K. beauverdii*）、*K. shizophylla*	盆栽

续表

目标性状	种类	应用
耐冷	唐印（*K. thyrsiflora*）	室外盆栽
株型紧凑	千兔耳（*K. millotii*）、苍白旗（*K. integeifolia*）、*K. porphyrocalyx*、白兔耳（*K. eriophylla*）	盆栽
双色花	*K. aromatica*、*K. campanulata*、毛落地生根（*K. pubescens*）	盆栽和切花
花期容易调控	白姬之舞（*K. marnieriana*）、矮生伽蓝菜（*K. glaucescens*）、伽蓝菜（*K. laciniata*）、*K. manginii*	盆栽和切花
高花枝	天之羽袖（*K. nyikae*）、*K. campanulata*、*K. laciniata*、*K. pritwitzii*	切花
附生生长	*K. gracilipes*、*K. porphyrocalyx*、朱红宫灯长寿花（*K. miniata*）、独叶草（*K. uniflora*）	悬挂盆栽
多分枝	*K. marnieriana*、*K. eriophylla*、*K. aromatica*、*K. citrina*	盆栽
速生型	*K. aromatica*、*K. pritwitzii*	盆栽和切花
有香味	*K.* × *ena*、盾叶伽蓝菜（*K. grandiflora*）、*K. glaucescens*、朱莲（*K. petitiana*）、*K. thyrsiflora*	盆栽和切花（生产阶段）
萼片融合	*K. pinnata*、*K. gastonis-bonnieri*、*K. laxiflora*	盆栽和切花
叶色	紫武藏（*K. humilis*）、江户紫（*K. marmorata*）、玉吊钟 'Variegata'（*K. fedtschenkoi* 'Variegata'）、*K. pinnata*	盆栽
叶形	*K. miniata*、*K. laciniata*、*K. daigremontiana*、*K. gracilipes*	盆栽
花瓣	*K. aromatica*、*K. pumila*	盆栽和切花
具小叶	*K. pumila*、*K. jongmansii*、*K. marnieriana*、*K. gracilipes*	盆栽
匍匐茎	趣蝶莲（*K. synsepala*）	盆栽
叶与茎紧密结合	*K. marmorata*、*K. porphyrocalyx*、*K. laciniata*、*K. citrina*	盆栽和切花
直立性强	*K. marmorata*、*K. pinnata*、*K. daigremontiana*	盆栽和切花
叶片具绒毛	*K. aromatica*、*K. pubescens*、仙女之舞（*K. beharensis*）、褐斑伽蓝菜（*K. tomentosa*）	盆栽和室外盆栽（抗虫型）
易生根	*K. aromatica*、*K. grandifiorn*、金丝长寿花（*K. jongmansii*）、*K. citrina*	盆栽和切花（生产阶段）
胎生	落地生根属植物	盆栽
花左右对称	*K. elizae*、*K. robusta*	盆栽

资料来源：Descoings 2003；Jepsen and Christensen 2006；Akulova-Barlow 2009；Currey and Erwin 2011；Mackenzie et al.，未发表

虽然伽蓝菜属主要分布于热带和亚热带，但也有一些种类能够承受较低温度。据报道，雀扇（*K. rhombopilosa*）、江户紫（*K. marmorata*）和趣蝶莲（*K. synsepala*）可以耐受 10℃左右的温度，而最耐寒种类唐印（*K. thyrsiflora*）可以在−3.5℃的温度下生存（Akulova-Barlow 2009）。在育种中使用这些耐低温种类可以研发新的户外品种，以期在世界上较冷的地区使用。此外，还可以在较低温度下生产新品种，从而降低其在北欧国家的生产成本。

伽蓝菜品种传统意义上的育种着眼于花的特性。因此，与叶片特征相关的育种开展很少。有性和无性杂交为选育具有叶片新特征如叶片边缘变色和叶片被绒毛的新品种提供了可能（Kuligowska et al. 2015a，2015b）。一项研究报道了在台南伽蓝菜（*K.*

garambiensis)、伽蓝菜（*K. laciniata*）（报道为 *K. gracilis*）、红提灯（*K. manginii*）和匙叶伽蓝菜（*K. spathulata*）的杂交后代中叶片性状如叶片颜色和叶片形状的遗传模式。文中指出绿叶和深裂叶是由等位基因控制的显性遗传（Huang and Chu 2017）。这为将来育种中品种选择提供了信息。在伽蓝菜属中，还有其他具有潜在性状的种类，如紫武藏（*K. humilis*）、江户紫（*K. marmorata*）、玉吊钟 'Variegata'（*K. fedtschenkoi* 'Variegata'）、落地生根（*K. pinnata*）的叶片带斑纹或 *K. aromatica*、毛落地生根（*K. pubescens*）、仙女之舞（*K. beharensis*）和褐斑伽蓝菜（*K. tomentosa*）的叶片被绒毛（Descoings 2003；Mackenzie et al.，未发表）。具有特征性状的种类如图 19.3 所示。

19.3.2　伽蓝菜属的基因工程

19.3.2.1　紧凑型植株的生产

株型紧凑对于众多具有重要经济价值的观赏盆栽植物如伽蓝菜至关重要。研究表明，与大型盆栽品种相比，消费者通常更喜欢紧凑型的盆栽品种。与大型盆栽植物相比，紧凑型盆栽的优势在于更易于机械操作和耐储运。此外，在昂贵的生产设施中生产紧凑型盆栽需要的空间更小，易于处理，运输成本降低，并且对零售商更有利（Christensen et al. 2008；Lütken et al. 2010，2011，2012a，2012b，2012c）。然而许多观赏植物（如伽蓝菜）都表现出大株型的生长习性，因此必须按照消费者的要求将其控制以生产紧凑型盆栽植物。目前，紧凑的株型通常是通过施用化学生长抑制剂，如矮肽素、丁酰肼或多效唑来实现的。因为人们长期以来一直担心其可能存在毒性（Mullins 1989）及潜在的致癌风险（Yamada et al. 2001）和对人体健康有负面影响，同时对环境不利（Lütken et al. 2012a），这些化学试剂在欧洲等地已不再允许使用（Andersen et al. 2002；Bhattacharya et al. 2010）。尽管常规育种产生了紧凑的伽蓝菜品种，但是仍然需要其他策略，如下述的基因工程。

用 *rol* 基因介导的遗传转化生产紧凑型伽蓝菜品种

发根农杆菌（*Agrobacterium rhizogenes*）是一种植物病原菌，在侵染部位会引起“发根”的产生（Riker et al. 1930）。在植物感染的过程中，作为对根瘤菌的反应，大量的小根作为细“毛”直接从感染部位伸出，这种现象称为“发根”（Tepfer 1990）。细菌感染后，将位于根诱导（Ri）质粒上的转移 DNA（T-DNA）插入并整合到植物宿主基因组 DNA 中（Chilton et al. 1982）。 T-DNA 片段上存在许多基因，并且已知至少有 18 个可读框（ORF）。最广为人知的基因是命名为 *rol*A、*rol*B、*rol*C 和 *rol*D 的根瘤基因位点（*rol*）基因，分别对应于 ORF10、11、12 和 15（White et al. 1985；Slightom et al. 1986）。*rol*A、*rol*B 和 *rol*C 基因联合作用会诱导毛状根发生（Schmulling et al. 1988；Mariotti et al. 1989）。相比之下，其他 ORF 主要影响植物的形态及激素敏感性（Lemcke and Schmulling 1998）。发根农杆菌根据其感染宿主产生的冠瘿碱（作为细菌的营养源）类型不同而进行分类，如黄瓜碱型和甘露碱型的 Ri 质粒含有一个 T-DNA 区（Hansen et al. 1991；Moriguchi et al. 2000）。相反，农杆碱型的 Ri 质粒还存在第二个 T-DNA（图 19.4）。同时存在左（T_L）和右（T_R）T-DNA 的类型被称为分裂 T-DNA 类型。两个 T-DNA 通常在 15～20kb，并被一个大约 16kb 的 DNA 片段隔开，该片段未整合到植物基因组中（White et al. 1985）。T_L-DNA 包含 *rol* 基因，而 T_R-DNA 上存在两个生长素基因 *aux1* 和 *aux2*（Jouanin et al. 1987）（图 19.4）。另外，有文献表明这两个 T-DNA 是独立转移并整合到宿主植物基因组中的（Jouanin et al. 1987；Christensen et al. 2008）。

图 19.3 伽蓝菜属的野生种。(a) 极乐鸟伽蓝菜（*K. beauverdii*）；(b) 白粉叶伽蓝菜（*K. pumila*）；(c) 趣蝶莲（*K. synsepala*）产生匍匐茎；(d) 青蔓花（*K. gracilipes*）的叶片形态；(e) 毛落地生根（*K. pubescens*）的叶片形态及毛状体特写；(f) 褐斑伽蓝菜（*K. tomentosa*）的叶片形态；(g) 极乐鸟伽蓝菜（*K. beauverdii*）的胎生叶片；(h) 玉吊钟（*K. fedtschenkoi*）的花序；(i) 天之羽袖（*K. nyikae*）的花序；(j) 毛落地生根（*K. pubescens*）的花序（标尺 = 1cm）

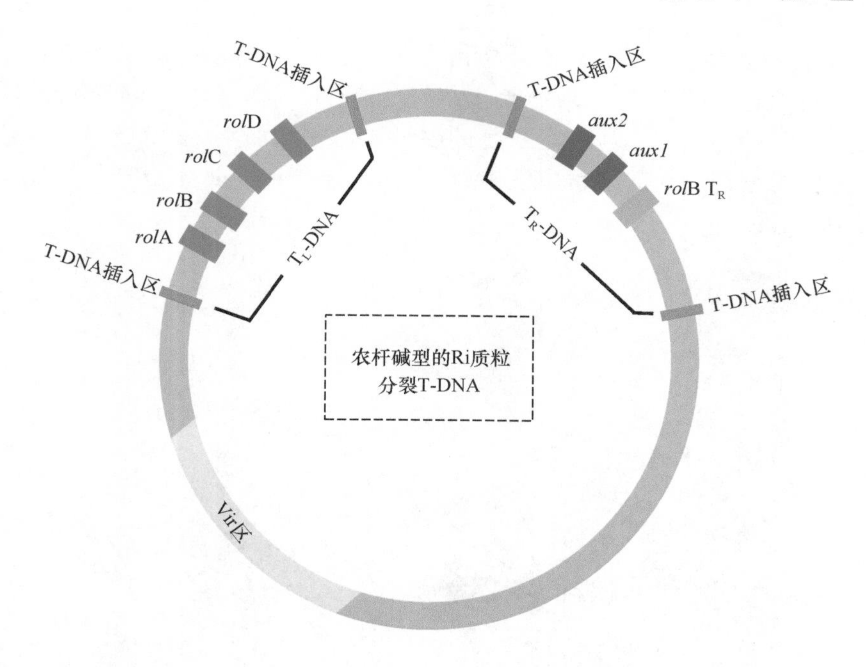

图 19.4　带有分裂的 T-DNA 的农杆碱型的 Ri 质粒图。*rol*A～D 和 *aux*、*rol*B T_R 基因分别在 T_L-DNA 和 T_R-DNA 上的相对位置

通常使用 *rol* 基因对盆栽植物进行遗传转化会导致植物的株型紧凑、节间距缩短、顶端优势减少及侧枝的增加。而且叶片变小、变厚和起皱。*rol* 基因在伽蓝菜属中的作用得到了广泛研究。最初，由组成型 35S 启动子驱动的单个 *rol* 基因载体被应用于遗传转化中以生产紧凑型伽蓝菜品种（Spena et al. 1987）。通过使用农杆碱型发根农杆菌转化的圣诞伽蓝菜‘Molly’品种，由此获得的 T1 品系表现出较高的生长势、节间短、分枝增加和根系增加的生长特性（Christensen et al. 2008）（图 19.5）。但是也出现了负面的影响，如花径变小和非典型（皱纹）叶片的出现（图 19.5）（Christensen et al. 2008），从而降低了其原来的盆栽品质。对 T1 品系的进一步研究表明，因为乙烯耐受性增加而采后性能提高（Christensen and Müller 2009）。随后，我们在商业育种研究中对上述品系进行了进一步的研究。开红花的圣诞伽蓝菜‘Molly’品种（图 19.5）与开橙花的圣诞伽蓝菜‘Sarah’品种杂交所产生的后代 T1 品系中，由发根农杆菌对其进行遗传转化而获得品种 331。PCR 研究表明，*rol* 基因在 F_1 子代中得到遗传，并且在所有表现出紧凑型株型的 F_1 代中都证实了 *rol* 基因的存在（Lütken et al. 2012c）。之后，通过 F_1 代自交而产生 F_2 代，从而获得 F_2 代种群。同样，在表现出紧凑型株型的 F_2 代植株中 *rol* 基因的存在得到了证实（Lütken et al. 2012b，2012c）。与 T1 品系 331、‘Molly’和‘Sarah’相比，除了植株高度降低外，一些含有 *rol* 基因的 F_1 代和 F_2 代还表现出植物冠幅、枝条数、花径、首次开花时间和花期的变化。此外，与‘Sarah’相比，一些含有 *rol* 基因的植株，对乙烯的耐受性也增强（Christensen et al. 2009；Lütken et al. 2012c）。还有研究报道指出由 Ri 介导产生的观赏植物的育性降低甚至出现败育。作为育种策略的延伸，*rol* 基因的功能在 F_1 代与其他商业品种的杂交后代得到了验证。后

代中同样包含 *rol* 基因，可以据此选择所需中间表型的多个品系（Lütken et al. 2012b）。由于伽蓝菜可通过营养繁殖，育性低似乎不是一个大问题（Christensen et al. 2008）。尽管如此，在育种过程中需要一定程度的生育能力来增加用于培育新品种的基因库，上述 Ri 植物的可育后代的花粉表现出育性（Christensen et al. 2010；Lütken et al. 2012b，2012c）。

图 19.5 *rol* 基因在圣诞伽蓝菜中的作用。(a) 圣诞伽蓝菜 'Molly'；(b) 由发根农杆菌介导产生的圣诞伽蓝菜 'Molly' 转化 T1 品系 331

赤霉素信号转导的修饰

由于一些生长延缓剂抑制了赤霉素（GA）激素的合成（Rademacher 2000），因此对该途径的改变是发展紧凑型盆栽植物的一个手段。其可以通过改变 GA 响应来实现。Fridborg 等（1999）在拟南芥中发现了参与 GA 信号途径的 SHI 转录因子。SHI 转录因子隶属于 SHI 家族，是 DNA 转录激活因子，其功能主要是维持植物生长素含量正常（Sohlberg et al. 2006; Staldal et al. 2008）。*SHI* 基因过表达的表型首先在拟南芥中被报道，即促进植物生长、GA 水平升高、顶端优势减弱和花的数量增加。有趣的是，人们发现在植物保持矮化的情况下，通过添加 GA 可以恢复其正常的开花时间（Fridborg et al. 1999）。因此，在 *shi* 突变体中开花时间和植株伸长机制是分开的，这也反映了实现生产紧凑型盆栽植物育种目标的可能性。我们使用组成型 35S 启动子在一些伽蓝菜品种中异位表达 *AtSHI*，这导致了产生花序节间缩短的紧凑型伽蓝菜品种（图 19.6）（Lütken et al. 2010）。有趣的是，将转基因伽蓝菜与对照的内源性基因表达进行比较时发现，一些基因数有所不同。幸运的是，没有观察到叶片形态发生变化。尽管如此，一些转基因伽蓝菜品种显示出较少的分枝（图 19.6）。最重要的是，与拟南芥相似，转基因伽蓝菜株系的开花没有延迟，并且低浓度外源 GA_3 的施用可使开花完全恢复。此外，当拟南芥自身的内源 *SHI* 启动子启动时，开花期延长。最后，我们在伽蓝菜中鉴定了两个类似 *AtSHI* 基因，表明了该转录因子广泛存在（Lütken et al. 2010）。

图 19.6　圣诞伽蓝菜中 *AtSHI* 基因的过表达。（a）圣诞伽蓝菜‘1998-469’；（b）圣诞伽蓝菜‘Sarah’

有学者研究了参与 GA 前体失活的 GA_2 氧化酶（GA_{2OX}）的作用。Gargul 等（2013）开展在圣诞伽蓝菜‘1998-469’中持续高表达烟草 GA_{2OX} 的研究。表型分析表明，转基因株系的平均株高是野生对照株系的 1/3，而平均节数却相似（Gargul et al. 2013）。

另一方法集中在 GA 生物合成的最后阶段，其中 GA 的活性由 GA_{20} 氧化酶（GA_{20OX}）和 GA_3-β 羟化酶决定（Rademacher 2000）。在这种方法中，GA_{20OX} 基因的下调被用来抑制圣诞伽蓝菜‘Molly’的伸长生长。该策略涉及一个乙醇诱导型启动子系统，该系统用于沉默 GA_{20OX} 生物合成途径的基因。在乙醇诱导型启动子系统被激活之前，转化植物在表型上与对照‘Molly’没有区别。通过使用低浓度乙醇使启动子沉默后，植物的株高降低、开花延迟（Topp et al. 2008）。结构同源基因的修饰同样代表了一种改变植物表型的方法。14 类植物的同源异型基因被描述（Mukherjee et al. 2009），如 *KNOX* 基因在器官分化和分生组织建立等发育过程中起着核心作用。激活 *KNOX* 基因通常会导致细胞分裂素生物合成的增加，并通过抑制 GA_{20OX} 来调节 GA 的生物合成（Leibfried et al. 2005；Gordon et al. 2009）。*KNOX* 基因首次在拟南芥和其他农作物中被鉴定，并已成功应用于伽蓝菜植物中来开发紧凑型盆栽品种。从箭叶落地生根的后代中分离出 *KxHKN5* 基因，其过表达时将引起株型紧凑度及叶片形态的改变（Laura et al. 2009）。同样，在圣诞伽蓝菜‘Molly’中，当 *KxHKN5* 基因过表达时，株型紧凑度及叶片形态也发生改变（Lütken et al. 2011）。另外，沉默该基因则导致植物的致密紧凑且叶片形态不变。与对照相比，在圣诞伽蓝菜‘Molly’中过表达 *KxHKN4* 基因导致紧凑型植株的相对花序分枝数显著提高，且因叶绿素含量升高而形成深绿色叶子（Lütken et al. 2011）。

生产紧凑型伽蓝菜的其他分子育种策略

另一种获得紧凑型伽蓝菜的方法是通过根癌农杆菌介导的转化，在伽蓝菜‘1998-469’中表达拟南芥的促分裂原活化蛋白激酶（MAP）的 4 核底物 1（MKS1）。MKS1 表达将导致植株矮化和开花延迟（Gargul et al. 2015）。在节间数相同的情况下，伽蓝菜的转基因植株的花表现出茎、节间和花序长度减少。

总的来说，紧凑的伽蓝菜品种可以通过一些基因工程手段来开发。尽管某些方法有不良反应，如叶子起皱、枝条数量减少及开花延迟，但它们都成功降低了株高，从而证明可通过非化学药品喷洒来获得株型紧凑植株的方法。各种策略实施的成功与否取决于全球范围内各国的转基因法规及其定义。

19.3.2.2 增强乙烯耐受性

乙烯是一种植物激素，可控制植物生长和发育的各个方面。乙烯会对许多观赏植物的开花寿命和采后性能产生不利影响，如导致衰老、芽和花脱落（Lütken et al. 2012a；Olsen et al. 2015）。在伽蓝菜中，乙烯导致花团聚（花瓣闭合），从而使其失去观赏价值。在采后处理过程中添加硫代硫酸银（STS）可有效降低乙烯的影响。但是，该化合物含有银离子，对人体有毒，对环境有害（Nell 1992）。目前已经开发出危害较小的替代品，如 1-甲基环丙烯（1-MCP）；然而，与 STS 相比，在许多情况下 1-MCP 效果不好（Blankenship and Dole 2003；Müller 2011；Lütken et al. 2012a；Olsen et al. 2015）。

分子方面主要集中在对乙烯受体结合位点突变的研究上。在拟南芥中，乙烯响应基因 1（*ETR1*）编码一种组氨酸激酶样乙烯受体，在 *etr1-1* 突变体中不能与乙烯结合（Bleecker et al. 1988；Chang et al. 1993）。乙烯受体基因的分离和对乙烯不敏感显性突变基因的鉴定，使抑制植物中的乙烯反应成为可能。在观赏植物中首次开展与 *etr1-1* 基因相关的组成型启动子的研究，随后在转基因植物中观察到了副作用，如生根差、种子重量小、种子萌发率降低和花粉活力降低（Wilkinson et al. 1997；Clark et al. 1999；Gubrium et al. 2000；Clevenger et al. 2004）。为了避免这些不利的副作用，Bovy 等（1999）有效地使用了从矮牵牛中获得的特定花启动子（fbp1）。后来，该基因构建体作为降低乙烯敏感性的工具成功转入圣诞伽蓝菜中（Sanikhani et al. 2008）。转基因伽蓝菜花朵闭合现象显著减少。如上所述，其他方法也产生了乙烯耐受性伽蓝菜，尽管其中一些研究并未将此特性作为育种质量改进的主要目标，如用 *rol* 基因进行遗传转化（Christensen and Müller 2009；Lütken et al. 2012c）。

19.3.3 其他育种方法

19.3.3.1 诱变

尽管自然突变是获得圣诞伽蓝菜新性状的重要途径，但缺乏诱变方面研究的报道。Broertjes 和 Leffring（1972）对两个圣诞伽蓝菜品种进行 X 射线辐射以期获得突变体。该研究报道了杂交后代在花色和大小、开花时间、花序类型、叶形、大小和叶色及植物习性方面都发生了改变。在棒叶不死鸟（*K. tubiflora*）（Johnson 1948）和伽蓝菜（*K. laciniata*）（Nakornthap 1973）中也通过 X 射线辐射成功诱导出突变体。在伽蓝菜杂种（未命名）中使用了 4 种不同的化学诱变剂进行诱变。这些诱变剂导致花型（包括花托、形状和花瓣的数量）和花色的变化，并且对株高和花的大小有影响。还有学者报道后代中新表型出现的频率取决于所使用的诱变剂种类及其诱变浓度（Krupa-Malkiewicz 2010）。尽管突变诱导是一种无目的的育种方法，但它对选育具有表型变异的伽蓝菜新品种更具有吸引力。

19.3.3.2　离体培养

离体培养可用于获得大规模的无病克隆及基因库的保存。它们还可以促进分子遗传修饰。在伽蓝菜属中，这对培育新品种很有意义（Varga et al. 1988；Sanikhani et al. 2006）。圣诞伽蓝菜的离体培养研究表明，培养基中同时添加生长素（NAA 或 IAA）和细胞分裂素（BAP 或 Kt）或仅添加细胞分裂素（TDZ）促进了以叶和叶柄或叶和节间为外植体的芽的再生（Sanikhani et al. 2006；Lütken et al. 2011）。有学者还报道通过对生长素和细胞分裂素的不同组合研究，确定了从叶片经愈伤组织到芽再生的 IAA 和玉米素的最佳使用浓度（Varga et al. 1988）。在其他伽蓝菜种类中也有体外芽再生的报道。在对箭叶落地生根（*K.* × *houghtonii*）的叶片进行培养时，添加 IAA 和 BAP 可获得芽（Laura et al. 2013）。在对褐斑伽蓝菜（*K. tomentosa*）的茎尖进行培养时，与添加 BAP 和 NAA 或其他激素的培养基相比，在无激素的 MS 培养基上可获得更多的芽（Khan et al. 2006）。

体细胞杂交是获得种间杂种的另一种方法，在体外培养的帮助下可融合野生种和商业品种的优势性状（Kuligowska et al. 2016）。此外，通过开展圣诞伽蓝菜原生质体融合等基因工程，可有助于新性状的引入（Castelblanque et al. 2010）。迄今尚无关于伽蓝菜体细胞融合后成功再生植株的报道。然而，有研究报道圣诞伽蓝菜原生质体可再生植物（Castelblanque et al. 2010）。由原生质体培养再生植株被认为是体细胞杂交的关键（Eeckhaut et al. 2013），因此在将来可开展伽蓝菜属体细胞杂交的研究。

如前所述，伽蓝菜的多倍体可从由叶片节间组织培养再生出的植株中获得（Aida and Shibata 2002）。多倍性是指同一器官中存在不同倍性细胞，是在各种伽蓝菜种类中经常观察到的现象（Castro et al. 2007；Izumikawa et al. 2007；Kuligowska et al. 2015b）。多倍性是由核内染色体复制形成的，即 DNA 重复循环合成而没有细胞分裂。据报道，这种过程主要在基因组小的植物中发生，包括各种肉质植物（Castro et al. 2007）。根据我们对伽蓝菜的研究，流式细胞仪结果显示 6 个不同种的 7 个基因型的叶片组织中都有多倍性现象（Mackenzie et al.，未发表）。图 19.7 中显示的是青蔓花（*K. gracilipes*）多倍性现象。

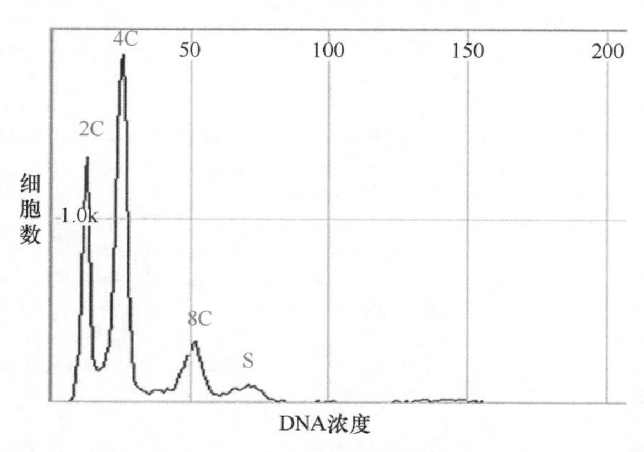

图 19.7　通过流式细胞仪，在青蔓花（*K. gracilipes*）中观察到伽蓝菜细胞的多倍体倍性水平表示为 2C（二倍体）、4C（四倍体）和 8C（八倍体）；S 表示标准玉米（2C = 5.45pg）

因此，染色体自发加倍可能是在植物细胞中进行核内复制的结果。另外，在离体再生体系中可能发生染色体加倍。

19.4 结　论

整体来说，在伽蓝菜的商业生产中，传统育种技术仍然被大多数育种者所选择。但是，从长远来看，基因组学和生物技术将极大地改变育种方法。因此，可以通过新的分子技术如基因编辑技术等的应用来进一步提高伽蓝菜的品质。伽蓝菜有着很高的观赏价值，且易于栽培，在杂交和基因工程方面具有巨大研究潜力。综上所述，这为伽蓝菜成为（除矮牵牛外）未来观赏植物的新的模式植物奠定了基础。

本章译者：
曹　桦
云南省农业科学院花卉研究所，国家观赏园艺工程技术研究中心，云南省花卉育种重点实验室，昆明 650200

参考文献

Aida R, Shibata M (2002) High frequency of polyploidization in regenerated plants of *Kalanchoe blossfeldiana* cultivar 'Tetra Vulcan'. Plant Biotechnol 19:329–335

Akulova-Barlow Z (2009) *Kalanchoe*: Beginner's delight, collector's dream. Cact Succ J 81:268–276

Alton J, Pertuit J (1992) Kalanchoe. In: Larson R (ed) Introduction to floriculture. Academic press, New York, pp 429–450

Andersen HR, Vinggaard AM, Rasmussen TH, Gjermandsen IM, Bonefeld-Jørgensen EC (2002) Effects of currently used pesticides in assays for estrogenicity, androgenicity, and aromatase activity in vitro. Toxicol Appl Pharmacol 179:1–12

Baldwin JT (1938) *Kalanchoe*: the genus and its chromosomes. Am J Bot 25:572–580

Baldwin J Jr (1949) Hybrid of *Kalanchoe daigremontiana* and *K. verticillata*. Bull Torrey Bot Club 1:343–345

Bleecker AB, Estelle MA, Somerville C, Kende H (1988) Insensitivity to ethylene conferred by a dominant mutation in *Arabidopsis thaliana*. Science 241:1086–1089. https://doi.org/10.1126/science.241.4869.1086

Bhattacharya A, Kourmpetli S, Davey MR (2010) Practical applications of manipulating plant architecture by regulating gibberellin metabolism. J Plant Growth Regul 29:249–256. https://doi.org/10.1007/s00344-009-9126-3

Blankenship S, Dole J (2003) 1-Methylcyclopropene: a review. Postharvest Biol Technol 28:1–25

Boiteau P, Allorge-Boiteau L (1995) *Kalanchoe* (Crassulacées) de Madagascar: systématique, écophysiologie et phytochimie. KARTHALA Editions

Bovy AG, Angenent GC, Dons HJ, van Altvorst A-C (1999) Heterologous expression of the *Arabidopsis etr1-1* allele inhibits the senescence of carnation flowers. Mol Breed 5:301–308

Broertjes C, Leffring L (1972) Mutation breeding of *Kalanchoë*. Euphytica 21:415–423

Carvalho SM, Wuillai SE, Heuvelink E (2006) Combined effects of light and temperature on product quality of *Kalanchoe blossfeldiana*. Acta Hortic 711:121–126

Castelblanque L, García-Sogo B, Pineda B, Moreno V (2010) Efficient plant regeneration from protoplasts of *Kalanchoe blossfeldiana* via organogenesis. Plant Cell Tissue Organ Cult 100:107–112. https://doi.org/10.1007/s11240-009-9617-8

Castro S, Loureiro J, Rodriguez E, Silveira P, Navarro L, Santos C (2007) Evaluation of polysomaty and estimation of genome size in *Polygala vayredae* and *P. calcarea* using flow cytometry. Plant Sci 172:1131–1137

Chang C, Kwok SF, Bleecker AB, Meyerowitz EM (1993) *Arabidopsis* ethylene-response gene *ETR1*: similarity of product to two-component regulators. Science 262:539–544

Chernetskyy M (2011) Problems in nomenclature and systematics in the subfamily kalanchoideae (*Crassulaceae*) over the years. Acta Agrobot 64:67–74

Chilton MD, Tepfer DA, Petit A, David C, Casse-Delbart F, Tempe J (1982) *Agrobacterium rhizogenes* inserts T-DNA into the genome of the host plant root cells. Nature 295:432–434

Christensen B, Müller R (2009) *Kalanchoe blossfeldiana* Transformed with *rol* genes exhibits improved postharvest performance and increased ethylene tolerance. Postharvest Biol Technol 51:399–406

Christensen B, Sriskandarajah S, Serek M, Müller R (2008) Transformation of *Kalanchoe blossfeldiana* with rol-genes is useful in molecular breeding towards compact growth. Plant Cell Rep 27:1485–1495. https://doi.org/10.1007/s00299-008-0575-0

Christensen B, Sriskandarajah S, Müller R (2009) Biomass distribution in *Kalanchoe blossfeldiana* transformed with *rol*-genes of *Agrobacterium rhizogenes*. Hortscience 44:1233–1237

Christensen B, Sriskandarajah S, Jensen E, Lütken H, Müller R (2010) Transformation with *rol* genes of *Agrobacterium rhizogenes* as a strategy to breed compact ornamental plants with improved postharvest quality. Acta Hortic 855:69–75

Coelho L, Kuligowska K, Lütken H, Müller R (2015) Photoperiod and cold night temperature in control of flowering in *Kalanchoë*. Acta Hortic 1087:129–134. https://doi.org/10.17660/ActaHortic.2015.1087.14

CPVO varieties database (2017) http://cpvo.europa.eu. Accessed 12.04.2017

Clark DG, Gubrium EK, Barrett JE, Nell TA, Klee HJ (1999) Root formation in ethylene-insensitive plants. Plant Physiol 121:53–60

Cullen J, Knees SG, Cubey HS (2011) Crassulaceae. In: Cullen J, Knees SG, Cubey HS (eds) The European garden flora flowering plants: a manual for the identification of plants cultivated in Europe, both out-of-doors and under glass, vol 3. Cambridge University Press, New York, pp 19–94

Currey C, Erwin J (2010) Variation among *Kalanchoe* species in their flowering responses to photoperiod and short-day cycle number. J Hortic Sci Biotechnol 85:350–355

Currey C, Erwin J (2011) Photoperiodic flower induction of several *Kalanchoe* species and ornamental characteristics of the flowering species. Hortic Sci 46:35–40

Clevenger D, Barrett J, Klee H, Clark D (2004) Factors affecting seed production in transgenic ethylene-insensitive petunia. J Am Soc Hortic Sci 129:401–406

Descoings B (2003) Kalanchoe. In: Eggli U, Hartmann HEK (eds) Illustrated handbook of succulent plants. Crassulaceae. Springer Verlag, New York, pp 143–181

Dilworth WL (1982) Yellow kalanchoe plant. US Patent 4825 P

Eeckhaut T, Lakshmanan PS, Deryckere D, Van Bockstaele E, Van Huylenbroeck J (2013) Progress in plant protoplast research. Planta 238:991–1003

Eveleens-Clark B, Carvalho S, Heuvelink E (2004) A conceptual dynamic model for external quality in kalanchoe. Acta Hortic 654:263–270

FloraHolland (2016) FloraHolland. Facts and figures 2016. http://annualreport.royalfloraholland.com. Accessed 14.07.2017

Fridborg I, Kuusk S, Moritz T, Sundberg E (1999) The *Arabidopsis* dwarf mutant *shi* exhibits reduced gibberellin responses conferred by overexpression of a new putative zinc finger protein. Plant Cell 11:1019–1032

Gargul J, Mibus H, Serek M (2013) Constitutive overexpression of *Nicotiana* GA2ox leads to compact phenotypes and delayed flowering in *Kalanchoë blossfeldiana* and *Petunia hybrida*. Plant Cell Tissue Organ Cult 115:407–418

Gargul JM, Mibus H, Serek M (2015) Manipulation of *MKS1* gene expression affects *Kalanchoë blossfeldiana* and *Petunia hybrida* phenotypes. Plant Biotechnol J 13:51–61

Gehrig H, Rosicke H, Kluge M (1997) Detection of DNA polymorphisms in the genus *Kalanchoe* by RAPD-PCR fingerprint and its relationships to infrageneric taxonomic position and ecophysiological photosynthetic behaviour of the species. Plant Sci 125:41–52

Gehrig H, Gaußmann O, Marx H, Schwarzott D, Kluge M (2001) Molecular phylogeny of the genus *Kalanchoe* (Crassulaceae) inferred from nucleotide sequences of the ITS-1 and ITS-2 regions. Plant Sci 160:827–835

González de León S, Herrera I, Guevara R (2016) Mating system, population growth, and management scenario for Kalanchoe pinnata in an invaded seasonally dry tropical forest. Ecol Evol 6:4541–4550

Gordon SP, Chickarmane VS, Ohno C, Meyerowitz EM (2009) Multiple feedback loops through cytokinin signaling control stem cell number within the *Arabidopsis* shoot meristem. Proc Natl Acad Sci U S A 106:16529–16534. https://doi.org/10.1073/pnas.0908122106

Gubrium E, Clevenger D, Clark D, Barrett J, Nell T (2000) Reproduction and horticultural performance of transgenic ethylene- insensitive petunias. J Amer Soc Hort Sci 125:277–281

Hansen G, Larribe M, Vaubert D, Tempe J, Biermann BJ, Montoya AL, Chilton MD, Brevet J (1991) *Agrobacterium rhizogenes* pRi8196 T-DNA: mapping and DNA sequence of functions involved in mannopine synthesis and hairy root differentiation. Proc Natl Acad Sci U S A 88:7763–7767

Herrera I, Nassar J (2009) Reproductive and recruitment traits as indicators of the invasive potential of *Kalanchoe daigremontiana* (*Crassulaceae*) and *Stapelia gigantea* (*Apocynaceae*) in a Neotropical arid zone. J Arid Environ 73:978–986

Hickey M, King C (1988) *Kalanchoë blossfeldiana* Poelln. In: 100 families of flowering plants, 2nd edn. Cambridge University Press, New York, p 174

Høyer L, Nell TA (1995) Plants respond differently to either dynamic or stationary ethylene exposure. VI Int Symp Postharvest Phys Ornamental Plants 405(1995):277–283

Huang C-H, Chu C-Y (2012) The flower development and photoperiodism of native *Kalanchoe* spp. in Taiwan. Sci Hortic 146:59–64

Huang C-H, Chu C-Y (2017) Inheritance of leaf and flower morphologies in *Kalanchoe* spp. Euphytica 213:4

Irwin LT (1976) Kalanchoe plant US Patent 3992 P

Izumikawa Y, Takei S, Nakamura I, Mii M (2007) Production and characterization of intersectional hybrids between *Kalanchoe spathulata* and *K. laxiflora* (= *Bryophyllum crenatum*). Euphytica 163:123–130. https://doi.org/10.1007/s10681-007-9619-8

Izumikawa Y, Nakamura I, Mii M (2008) Interspecific hybridization between *Kalanchoe blossfeldiana* and several wild *Kalanchoe* species with ornamental value. Acta Hortic 743:59–66

Jepsen K, Christensen E (2006) Double-type kalanchoe interspecific hybrids. US 7453032 B2

Johnson EL (1948) Response of *Kalanchoe tubiflora* to X-radiation. Plant Physiol 23:544

Jouanin L, Guerche P, Pamboukdjian N, Tourneur C, Delbart F, Tourneur J (1987) Structure of T-DNA in plants regenerated from roots transformed by *Agrobacterium rhizogenes* strain A4. Mol Gen Genet 206:387–392

Kalanchoë Growers Holland (2017) http://www.kalanchoe.nl. Accessed 14.07.2017

Khan S, Naz S, Ali K, Zaidi S (2006) Direct organogenesis of *Kalanchoe tomentosa* (Crassulaceae) from shoot-tips. Pak J Bot 38:977

Khoury N, White J (1980) Juvenility and response time of kalanchoe cultivars. J Am Soc Hortic Sci 105:724–726

Krupa-Malkiewicz M (2010) Influence of chemical mutagens on morphological traits in kalanchoe (*Kalanchoe hybrida*). Folia Pomeranae Univ Technol Stetin Agric Aliment Piscaria Zootech 279:11–18

Kuligowska K, Lütken H, Christensen B, Müller R (2015a) Quantitative and qualitative characterization of novel features of *Kalanchoë* interspecific hybrids. Euphytica 205:927–940. https://doi.org/10.1007/s10681-015-1441-0

Kuligowska K, Lütken H, Christensen B, Skovgaard I, Linde M, Winkelmann T, Müller R (2015b) Evaluation of reproductive barriers contributes to the development of novel interspecific hybrids in the *Kalanchoë* genus. BMC Plant Biol 15:15. https://doi.org/10.1186/s12870-014-0394-0

Kuligowska K, Lütken H, Müller R (2016) Towards development of new ornamental plants: status and progress in wide hybridization. Planta 244:1–17

Laura M, Borghi C, Regis C, Casetti A, Allavena A (2009) Overexpression and silencing of *KxhKN5* gene in *K x houghtonii*. Acta Hortic 836:265–269

Laura M, Borghi C, Regis C, Cassetti A, Allavena A (2013) Ectopic expression of *Kxhkn5* in the viviparous species *Kalanchoe* × *Houghtonii* induces a novel pattern of epiphyll development. Transgenic Res 22:59–74

Leibfried A, To JPC, Busch W, Stehling S, Kehle A, Demar M, Kieber JJ, Lohmann JU (2005) WUSCHEL controls meristem function by direct regulation of cytokinin-inducible response regulators. Nature 438:1172–1175. https://doi.org/10.1038/nature04270

Lemcke K, Schmulling T (1998) Gain of function assays identify non-*rol* genes from *Agrobacterium rhizogenes* TL-DNA that alter plant morphogenesis or hormone sensitivity. Plant J 15:423–433

Leonard R, Nell T (1998) Effects of production and postproduction factors on longevity and quality of *Kalanchoe*. Acta Hortic 518:121–124

LPlants (1975) Kalanchoe plant US Patent 3821 P
Lütken H, Jensen LS, Topp SH, Mibus H, Müller R, Rasmussen SK (2010) Production of compact plants by overexpression of *AtSHI* in the ornamental *Kalanchoë*. Plant Biotechnol J 8:211–222
Lütken H, Laura M, Borghi C, Orgaard M, Allavena A, Rasmussen SK (2011) Expression *of KxhKN4* and *KxhKN5* genes in *Kalanchoe blossfeldiana* 'Molly' results in novel compact plant phenotypes: towards a cisgenesis alternative to growth retardants. Plant Cell Rep 30:2267–2279. https://doi.org/10.1007/s00299-011-1132-9
Lütken H, Clarke JL, Muller R (2012a) Genetic engineering and sustainable production of ornamentals: current status and future directions. Plant Cell Rep 31:1141–1157. https://doi.org/10.1007/s00299-012-1265-5
Lütken H, Jensen EB, Wallstrom S, Müller R, Christensen B (2012b) Development and evaluation of a non-GMO breeding technique exemplified by *Kalanchoë*. Acta Hortic 961:51–58
Lütken H, Wallström SV, Jensen EB, Christensen B, Müller R (2012c) Inheritance of rol-genes from *Agrobacterium rhizogenes* through two generations in *Kalanchoë*. Euphytica 188:397–407
Lüttge U (2004) Ecophysiology of crassulacean acid metabolism (CAM). Ann Bot 93:629–652
Madriz-Ordeñana K, Jørgensen H, Nielsen K, Thordal-Christensen H (2016) First report of *Kalanchoe* leaf and stem spot caused by *Corynespora cassiicola* in Denmark. Plant Dis 101:505–505
Mariotti D, Fontana GS, Santini L, Constantino P (1989) Evaluation under field conditions of the morphological alterations (`hairy root phenotype') induced on *Nicotiana tabacum* by different Ri plasmid T-DNA genes. J Genet Breed 43:157–164
Marousky F, Harbaugh B (1979) Ethylene-induced floret sleepiness in *Kalanchoe blossfeldiana* Poelln. Physiological disorders. HortSci 14:505–507
Moriguchi K, Maeda Y, Satou M, Kataoka M, Tanaka N, Yoshida K (2000) Analysis of unique variable region of a plant root inducing plasmid, pRi1724, by the construction of its physical map and library. DNA Res 7:157–163
Mortensen LM (2014) The effect of wide-range photosynthetic active radiations on photosynthesis, growth and flowering of *Rosa* sp. and *Kalanchoe blossfeldiana*. Am J Plant Sci 5:1489–1498
Mort M, Douglas E, Soltis E, Soltis P, Francisco-Ortega J, Santos-Guerra A (2001) Phylogenetic relationships and evolution of *Crassulaceae* inferred from *matK* sequence data. Am J Bot 88:76–91
Mukherjee K, Brocchieri L, Burglin TR (2009) A comprehensive classification and evolutionary analysis of plant homeobox genes. Mol Biol Evol 26:2775–2794. https://doi.org/10.1093/molbev/msp201
Mullins MG (1989) Growth regulators in the propagation and genetic improvement of fruit crops. Acta Hortic 239:101–108
Müller R (2011) Physiology and genetics of plant quality improvement. Doctoral dissertation, University of Copenhagen
Nakornthap A (1973) Radiation-induced somatic mutations in *Kalanchoe* (*Kalanchoe laciniata*). Kasetsart 7:13–18
Nielsen AH, Olsen CE, Møller BL (2005) Flavonoids in flowers of 16 *Kalanchoe blossfeldiana* varieties. Phytochemistry 66:2829–2835
Nell T (1992) Taking silver safely out of the longevity picture. Grower Talks June 35:41–42
Olsen A, Lütken H, Hegelund JN, Müller R (2015) Ethylene resistance in flowering ornamental plants-improvements and future perspectives. Hortic Res 2:15038
Pittman RN (1977) Kalanchoe plant US Patent 4062 P
Queen® (2017) http://www.queen.dk. Accessed 14.07.2017
Rademacher W (2000) Growth retardants: effects on gibberellin biosynthesis and other metabolic pathways. Annu Rev Plant Physiol Plant Mol Biol 51:501–531. https://doi.org/10.1146/annurev.arplant.51.1.501
Riker AJ, Banfield WM, Wright WH, Keitt GW, Sagen HE (1930) Studies on infectious hairy-root of nursery apple trees. J Agric Res 41:507–540
Rünger W (1966) Über die Wirkung von Lang-und Kurztagen auf das Wachstum noch nicht blühfähiger *Kalanchoë*. Gartenbauwissenschaft 1:429–436
Sanikhani M, Frello S, Serek M (2006) TDZ induces shoot regeneration in various *Kalanchoe blossfeldiana* Poelln. Cultivars in the absence of auxin. Plant Cell Tissue Organ Cult 85:75–82
Sanikhani M, Mibus H, Stummann BM, Serek M (2008) *Kalanchoe blossfeldiana* plants expressing the *Arabidopsis etr1-1* allele show reduced ethylene sensitivity. Plant Cell Rep 27:729–737. https://doi.org/10.1007/s00299-007-0493-6
Schmulling T, Schell J, Spena A (1988) Single genes from *Agrobacterium rhizogenes* influence

plant development. EMBO J 7:2621–2629
Schwabe WW (1969) *Kalanchoe blossfeldiana* Poellniz. In: Evans LT (ed) The induction of flowering. Macmillan of Australia, Melbourne, pp 227–246
Serek M, Reid MS (2000) Ethylene and postharvest performance of potted kalanchoe. Postharvest Biol Technol 18:43–48
Sharma GK (1970) Effects of cool nights on flowering of *Kalanchoe fedschenkoi*. Trans Missouri Acad Sci 3:22–28
Sharma GJ (1973) Flower formation in *Kalanchoe velutina* induced by low night temperature. Southwest Nat 18:331–334
Slightom JL, Durand-Tardif M, Jouanin L, Tepfer D (1986) Nucleotide sequence analysis of TL-DNA of *Agrobacterium rhizogenes* agropine type plasmid. Identification of open reading frames. J Biol Chem 261:108–121
Sohlberg JJ, Myrenas M, Kuusk S, Lagercrantz U, Kowalczyk M, Sandberg G, Sundberg E (2006) *STY1* regulates auxin homeostasis and affects apical-basal patterning of the *Arabidopsis* gynoecium. Plant J 47:112–123. https://doi.org/10.1111/j.1365-313X.2006.02775.x
Spear I (1959) Metabolic aspects of photoperiodism. In: Withrow RB (ed) Photoperiodism. Amer Assoc Adv Sci, Washington, DC, pp 289–300
Spena A, Schmulling T, Koncz C, Schell JS (1987) Independent and synergistic activity of *rol A*, *B* and *C* loci in stimulating abnormal growth in plants. EMBO J 6:3891–3899
Staldal V, Sohlberg JJ, Eklund DM, Ljung K, Sundberg E (2008) Auxin can act independently of *CRC, LUG, SEU, SPT* and *STY1* in style development but not apical-basal patterning of the *Arabidopsis* gynoecium. New Phytol 180:798–808. https://doi.org/10.1111/j.1469-8137.2008.02625.x
Tepfer D (1990) Genetic transformation using *Agrobacterium rhizogenes*. Physiol Plant 79:140–146
Topp SH, Rasmussen SK, Sander L (2008) Alcohol induced silencing of gibberellin 20-oxidases in *Kalanchoe blossfeldiana*. Plant Cell Tissue Organ Cult 93:241–248
Traoré L, Kuligowska K, Lütken H, Müller R (2014) Stigma development and receptivity of two *Kalanchoë blossfeldiana* cultivars. Acta Physiol Plant 36:1763–1769. https://doi.org/10.1007/s11738-014-1550-8
Uhl CH (1948) Cytotaxonomic studies in the subfamilies Crassuloideae, Kalanchoideae, and Cotyledonoideae of the Crassulaceae. Am J Bot 35:695–706
Van Ham R, Hart H (1998) Phylogenetic relationships in the Crassulaceae inferred from chloroplast DNA restriction-site variation. Am J Bot 85:123–134
Varga A, Thoma L, Bruinsma J (1988) Effects of auxins and cytokinins on epigenetic instability of callus-propagated *Kalanchoe blossfeldiana* Poelln. Plant Cell Tissue Organ Cult 15:223–231
Vlielander I (2007) *Kalanchoe* plant named 'Fiveranda Orange'. US Patent 17917 P2
van Voorst A, Arends JC (1982) The origin and chromosome numbers of cultivars of *Kalanchoe blossfeldiana* Von Poelln.: their history and evolution. Euphytica 31:573–584. https://doi.org/10.1007/BF00039195
Wadhi M, Ram HM (1967) Shortening the juvenile phase for flowering in *Kalanchoe pinnata*. Pers. Planta 73:28–36
White FF, Taylor BH, Huffman GA, Gordon MP, Nester EW (1985) Molecular and genetic analysis of the transferred DNA regions of the root-inducing plasmid of *Agrobacterium rhizogenes*. J Bacteriol 164:33–44
Wick RL (2017) Diseases of *Kalanchoe*. In: McGovern RJ, Elmer WH (eds) Handbook of Florists' crops diseases. Springer International Publishing, Cham, pp 1–13. https://doi.org/10.1007/978-3-319-32374-9_37-1
Wilkinson JQ, Lanahan MB, Clark DG, Bleecker AB, Chang C, Meyerowitz EM, Klee HJ (1997) A dominant mutant receptor from *Arabidopsis* confers ethylene insensitivity in heterologous plants. Nat Biotechnol 15:444–447. https://doi.org/10.1038/nbt0597-444
Willumsen K, Fjeld T (1995) The sensitivity of some flowering potted plants to exogenous ethylene. Acta Hortic 405:362–371
Yamada K, Honma Y, Asahi KI, Sassa T, Hino KI, Tomoyasu S (2001) Differentiation of human acute myeloid leukaemia cells in primary culture in response to cotylenin A, a plant growth regulator. Br J Haematol 114:814–821
Zeevaart JA (1976) Physiology of flower formation. Annu Rev Plant Physiol 28:321–348
Zeevaart JAD (1985). Bryophyllum. In: (ed.) Haley, A. H. CRC handbook of flowering, vol. 2. CRC Press, Boca Raton, Florida. In: CRC handbook of flowering, vol 5. pp 89–100
Zimmer K (1996) Untersuchungen zur Blühinduktion bei *Kalanchoë marmorata* Baker. Kakteen und andere Sukkulenten 47:188–191

第20章 百　　合

Jaap M. Van Tuyl, Paul Arens, Arwa Shahin, Agnieszka Marasek-Ciołakowska,
Rodrigo Barba-Gonzalez, Hyoung Tae Kim, and Ki-Byung Lim

摘要　百合属主要由分布在北半球温带、寒温带部分地区的大约100种植物组成，具体可分为7个组：轮叶组（*Martagon*）、根茎组（*Pseudolirium*）、百合组（*Lilium*）、具叶柄组（*Archelirion*）、卷瓣组（*Sinomartagon*）、喇叭花组（*Leucolirion*）和斑瓣组（*Oxypetalum*）。各组内的主要种类具有各自的描述特征。这些物种在百合育种中的作用以更新的多边形杂交呈现。百合是用于种间杂交研究的模式作物，用其作为研究材料产生了一系列技术，如克服花粉管生长抑制的授粉方法、防止胚败育的胚挽救方法和克服 F_1 不育性的染色体多倍化方法。利用减数分裂和有丝分裂多倍体化获得多倍体及基因组原位杂交（GISH）的应用证明了基因组重组对于性状渗入的重要性。虽然分子育种尚未在实际育种中实施，但在百合育种中分子标记的开发方面取得了进展，特别是大规模平行测序方法（NGS）的发展及高通量测序的出现，使百合新遗传图谱中的标记能够与抗性基因[百合斑驳病毒（LMoV）和镰刀菌]紧密连锁。在过去的50年中，百合已成为全球最重要的球根花卉和切花之一。百合的培育主要在荷兰开展，其百合种球的生产面积已超过5000hm^2。百合品种包括成千上万个多倍体品种，可分为不同的已育成的杂交系（包括亚洲杂种系、LA杂种系、OT杂种系、OA杂种系、LO杂种系、铁炮杂种系和东方杂种系）。最近几年，重瓣百合是新的发展方向。

关键词　种间杂交；百合；授粉方法；胚挽救；多边形杂交；基因组原位杂交；基因组分化；未减数配子；有性多倍体；三倍体杂种群；遗传图谱；SNP 标记

20.1 引　　言

近50年来，百合已成为世界上最重要的观赏花卉之一。百合主要在荷兰进行培育，

J. M. Van Tuyl (✉), P. Arens, A. Shahin
Wageningen University & Research, Wageningen, The Netherlands
e-mail: jaap.vantuyl@wur.nl
A. Marasek-Ciołakowska
Research Institute of Horticulture, Skierniewice, Poland
R. Barba-Gonzalez
Centro de Investigación y Asistencia en Tecnología y Diseño del Estado de Jalisco A.C, Guadalajara, Mexico
H. T. Kim, K.-B. Lim
Department of Horticultural Science, Kyungpook National University, Daegu, South Korea

其种球种植面积已超过 5000hm^2。百合种球主要用于切花生产，尤其是在荷兰、美国、日本和中国，而作为盆栽或园林花卉应用少一些。百合的数千个栽培品种可被分为 9 个类群。英国皇家园艺协会（RHS）是百合新品种的注册机构，由其公布了《2007 年国际百合登记和检查表》及后续的 5 份补充材料（https://www.rhs.org.uk/plants/ plantsmanship/plant-registration/Lily-cultivar-registration/Lily）。截至 2017 年，由位于 Roelofarendsveen 的荷兰园艺作物检查服务中心对 3499 份荷兰植物育种家的权利申请进行了调查，并有超过 50%的申请获得国家植物品种权登记（https://nederlandsrassenregister.nl/）。此外，除了申请欧洲国家品种权，也可申请植物育种者权（PBR）（http://www.cpvo.europa.eu/main/en/）。

20.2 百合育种史

现今百合栽培种中的亚洲百合、东方百合、铁炮百合和喇叭百合及它们之间的组间百合杂种 LA 百合（Longiflorum × Asiatic）、LO 百合（Longiflorum × Oriental）、OA 百合（Oriental × Asiatic）和 OT 百合（Oriental × Trumpet）大部分都是由亚洲起源的祖先亲本育成。Okubo（2014）在他的《亚洲的百合属历史》中报道了大约 2000 年前中国对百合的首次描述。但是，百合的育种史只有不到 100 年的时间。百合的组内种间杂交始于在日本和美国以卷瓣组百合开展的杂交，以（岩百合 *L. maculatum*、川百合 *L. davidii*、毛百合 *L. dauricum*、卷丹 *L. tigrinum* 和珠芽百合 *L. bulbiferum*）杂交获得了所谓的亚洲百合杂种。同时，有更多的杂种系被育成，如普雷斯顿杂种系（Preston hybrids）、帕特森杂种系（Patterson hybrids）和中世纪杂种系（Mid-century hybrids）。20 世纪 30 年代至 40 年代间，在俄勒冈州的种球农场由 Jan de Graaff（McRae 1998）培育出中世纪百合杂种系。于 1944 年育成的百合品种‘Enchantment’是百合育种上的突破，该品种也是 1970～1996 年荷兰百合花生产中最重要的品种。大约在 50 年前，荷兰公司开始涉足百合育种，育种的重点首先放在了亚洲百合杂种系上，继而是东方百合杂种系。东方百合杂种系主要来源于具叶柄组（*Archelirion* section）内天香百合（*L. auratum*）与鹿子百合（*L. speciosum*）之间的杂交。由 Leslie Woodriff 培育的‘Stargazer’是近 25 年来东方百合杂种系内最重要的、第一个花型向上的品种。同时，通过铁炮百合与亚洲百合之间的组间杂交，LA 百合杂种系首次被育成，Peter Schenk 是这个杂种系商业育种的先驱（Schenk 1990）。LA 百合品种主要是来自二倍体 LA 百合的 F_1 代与亚洲百合回交产生的三倍体。2015 年这个杂交系的种球产量在荷兰已超过 1000hm^2，在很大程度上取代了亚洲百合杂种系。在 LA 百合的培育和栽培取得较大进展时，其他组间三倍体杂种系即 LO 百合（Longiflorum × Oriental）、OT 百合（Longiflorum × Trumpet）和 OA 百合（Oriental × Asiatic）也得到快速发展。LA 百合主要取代亚洲百合，LO 百合取代铁炮百合，而东方百合则被 OT 百合部分取代。过去 20 年中，百合品种的分类发生了较大变化（Van Tuyl and Arens 2011），从当前发展看来，百合种间杂交上取得的重大进展已成为过去时，现在的育种重点应该更直接地放在一些特殊性状上，为此则需要更多先进的检测技术。有关现代品种分类的发展历程，请参见第 20.6 节。

20.3 百 合 属

百合属作为一个最重要的观赏植物类群，是百合科的模式属。该属由大约100种分布在北半球寒温带的植物组成（Haw and Liang 1986；McRae 1998）。这些野生种经过筛选和杂交成为现今的园林百合（McRae 1998）。因此，了解野生百合对于百合的育种具有重要意义。在这一章中，我们将回顾百合属植物的系统发育简史，并简要描述百合属植物的形态特征，以帮助读者了解百合属植物是如何产生的，以及哪些特征长期以来吸引了园艺家。

20.3.1 百合的形态特征

典型的百合植物可分为花序、茎和叶及地下三部分。

花序：百合花序为总状花序或花单生，很少有伞形花序（McRae 1998）。花的形状非常多样，但可以分为土耳其帽状花和喇叭形花，前者反折的花瓣向后弯曲向茎，喇叭形花则向花瓣顶端绽放（Comber 1949）。花的朝向是区别不同种百合的另一个特征。例如，在轮叶组百合中，青岛百合的花朵是直立的，而东北百合和欧洲百合的花型则分别朝外和向下。百合花颜色变化多样，不同百合花也存在斑点有无、条纹和斑块等不同变化。在喇叭花组中，麝香百合和台湾百合是近缘种，在外观上较为相似（Hiramatsu et al. 2001），但前者的花瓣白色，没有斑点，后者的白色花瓣有淡黄色的喉部。

茎和叶：有些茎会直接从种球中长出来，而另一些在萌发之前就在土面下长出。百合的叶序可分为散生和轮生两种类型，典型的轮生叶出现在轮叶组和根茎组两组的植物中（Comber 1949）。一些种类如卷丹和通江百合在叶腋处会长出珠芽。

地下部分：百合会长出籽球（一个小的种球结构）、茎生根、种球茎（地下芽具加厚的肉质鳞片）和基生根（McRae 1998）。

20.3.2 百合的繁殖

百合有多种繁殖方式，对于大多数植物来说种子繁殖是一种典型的繁殖方式，具有防止病毒感染和更经济有效的优点。另外，与其他繁殖方法相比，它是一种保持或增加百合遗传多样性的方法。鳞片繁殖是最经济、最快速的克隆单棵植株的方法，可应用于植物新品种权的申请，但同时增加了植株感染病毒病的问题。组织培养是一种生产无病毒植株的好方法，但它比鳞片繁殖的成本更高，当然人们也可以利用茎生小鳞茎、珠芽鳞茎和分球繁殖来繁育百合种球。

20.3.3 基于系统发育研究的百合分类

百合科由6族15属大约900种植物组成（Peruzzi 2016）。百合属及假百合属、大百合属和贝母属植物同属于百合族（Peruzzi 2016）。尽管豹子花属的某些形态特征与百合属内的一些物种有所区别（Sealy 1983），但最近的系统发育研究支持将豹子花属的物种嵌套在百合属内（Nishikawa et al. 1999；Gao et al. 2012；Du et al. 2014），因此它们也被视为属于百合属（Peruzzi 2016）。

当 Endlicher（1836）首次将百合属分类时，该属被分为 5 个组：金百合组、大百合组、麝香百合组、轮叶组和根茎组。但是后来前两个组分别被转移到了贝母属（Rix et al. 2001）和大百合属（Liang 1980）。剩下的三个组根据形态特征和地理分布特点被重新分成了 7 个组（Comber 1949）。在那之后，De Jong（1974）在前人的研究基础上修订了 Comber（1949）的分类。尽管最近的系统发育分析需要对百合属进行新的界定，因为大多数组并不是单系类群（Nishikawa et al. 1999; Hayashi and Kawano 2000; Lee et al. 2011; Du et al. 2014），但识别新组的形态学特征尚未被提出，因此，我们将遵循 De Jong（1974）的分类法来描述百合属植物的分类及分类中的一些问题。

20.3.3.1 百合组

百合组包括所有欧洲、土耳其和高加索物种，但不包括属于轮叶组的欧洲百合（İkinci et al. 2006）。该组植物叶散生，整个种球由鳞片组成，具土耳其帽状花（除白花百合 *L. candidum*）、子叶出土（除多叶百合 *L. polyphyllum* 和高加索百合 *L. monadelphum* 外），鳞片众多且延迟发芽（白花百合 *L. candidum*）（Comber 1949）。白花百合虽是本组的典型物种，但其喇叭形花和延迟发芽的特征与该组中的其他种类稍有区别（Comber 1949），然而由于许多系统发育分析支持该种作为单系物种，因此白花百合那些不同于其他物种的特征似乎具有自分离特性（Nishikawa et al. 2001；İkinci et al. 2006；Gao et al. 2013；Du et al. 2014）。

20.3.3.2 轮叶组

轮叶组有 5 个种，大部分分布在韩国、日本和中国，但欧洲百合的分布最广，遍布整个欧洲（İkinci et al. 2006）。该组物种通常有 1～3 片轮生叶和一些散生叶。就地理分布和表型而言，仅在韩国郁陵岛分布的汉森百合（*L. hansonii*）变异程度小于欧洲百合（Pelkonen and Pirttila 2012）。然而，最近的调查显示，欧洲百合、青岛百合、东北百合和浙江百合具有 8 个 45S rDNA 位点，而汉森百合则具有 17 个 45S rDNA 位点（Ahn et al. 2017）。有报道认为汉森百合在第四纪时从东北百合中分离出来（Gao et al. 2013），因此汉森百合中的 45S rDNA 位点的增加被看作是近期发生的独立事件。另外，汉森百合的鳞片为完整鳞片，而其他轮叶组百合的鳞片上有节。因此，汉森百合如何在一个小岛上而不是在大陆上产生遗传和形态变异值得进一步深入研究。

20.3.3.3 根茎组

根茎组由大约 20 种分布在北美洲的物种组成。该组植物有根状鳞茎和轮生叶，除其中 4 个种外其他种类均为子叶留土型（Comber 1949）。Comber（1949）认为该组与轮叶组在种球习性差异方面有着密切的关系。然而，最近利用分子标记进行的系统发育分析表明，根茎组与轮叶组的亲缘关系较远（Hayashi and Kawano 2000；Lee et al. 2011；

Gao et al. 2013；Du et al. 2014）。此外，具有轮叶性状的墨脱百合并未被包含在轮叶组和根茎组中（Liang 1985），而且一些贝母属的植物也与百合属植物关系紧密（Turrill 1950；Hayashi and Kawano 2000；Rønsted et al. 2005）。因此，至少百合属的轮叶性状似乎具有趋同性或祖征性而不是共衍性状。尽管一些分子系统发育研究表明根茎组是单系群（Nishikawa et al. 1999，2001），但其他的研究又将费城百合亚组（subsect. 2d Comber）从根茎组剩余的种类中分离出来（Hayashi and Kawano 2000；Du et al. 2014）。有趣的是，该亚组中的两个种——松树百合和费城百合以花朵直立和子叶出土型萌发区别于根茎组中的其他物种（Pelkonen and Pirttila 2012），并且松树百合是根茎组中唯一具有叶片交替特性的物种。由于费城百合标本是该组的一个模式选型，因此，根茎组不能仅基于地理特征而应采用更精确的分子标记来重新划定。

20.3.3.4　具叶柄组

在 Comber 的分类中，具叶柄组包括 7 个种：奄美百合（*L. alexandrae*）、天香百合（*L. auratum*）、野百合（*L. brownii*）、日本百合（*L. japonicum*）、香华丽百合（*L. nobilissimum*）、乙女百合（*L. rubellum*）和鹿子百合（*L. speciosum*）（Comber 1949）。后来根据 De Jong（1974）的意见，乔木百合（*L. arboricola*）、滇百合（*L. bakerianum*）、湖北百合（*L. henryi*）、紫斑百合（*L. nepalense*）、报春百合（*L. primulinum*）和 *L. tenii* 也被归入这个组中。然而，迄今任何系统分类学的研究都不支持 De Jong 的建议，并且野百合的分类地位也远不同于其他 6 个种（Nishikawa et al. 2001；Lee et al. 2011；Du et al. 2014）。因此，应该重新划定在日本和中国分布的 6 种具叶柄组百合（奄美百合 *L. alexandrae*、天香百合 *L. auratum*、日本百合 *L. japonicum*、香华丽百合 *L. nobilissimum*、乙女百合 *L. rubellum* 和鹿子百合 *L. speciosum*），它们的叶普遍散生，叶柄明显，鳞片完整，花喇叭形且具子叶留土特征（Comber 1949）。本节中另一个系统分类地位值得关注的种是天香百合，该种是日本的特有种，也是东方百合杂种系中的重要遗传资源（Okazaki et al. 1994；Lim et al. 2008），它有两个变种，天香百合原变种（*L. auratum* var. *auratum*）和山百合（*L. auratum* var. *platyphyllum*）在形态上具有许多共同特征，二者的区别在于叶片形状和花朵上斑点的数量等（Nishikawa et al. 2002）。但是，这两个变种之间的形态相似性很可能是由趋同进化产生的，因为这两个材料分别与乙女百合及具叶柄组的其他百合种类有紧密关系（Nishikawa et al. 2002）。由此，Makino（1914）的分类中认为如果将作百合看成一个独立的种还需要更深入的研究。

20.3.3.5　卷瓣组

卷瓣组由 30 种原产中国的百合组成，可分为 3 个亚组（Comber 1949；Nishikawa et al. 2001）。该组的特征是叶散生、鳞片完整、种子轻（除朝鲜百合 *L. amabile* 外）且不休眠，子叶出土，花形为土耳其帽形（Comber 1949）。依据先前的百合分类系统（Comber 1949；De Jong 1974），卷瓣组不同的种类在系统发育关系上非常复杂，可以分成 4 个大的亚组（Nishikawa et al. 2001；Du et al. 2014）。大部分卷瓣组（Van Tuyl and Arens 2011）进化聚类支第二支（Du et al. 2014）内的物种（珠芽百合 *L. bulbiferum*、垂花百合 *L.*

cernuum、渥丹 *L. concolor*、毛百合 *L. dauricum*、川百合 *L. davidii*、大花卷丹 *L. leichtlinii*、山丹 *L. pumilum* 和卷丹 *L. tigrinum*）可用于杂交育种。相较于卷瓣组内的其他进化支，第二支的物种可能适合于更广泛的杂交。

20.3.3.6 喇叭花组

喇叭花组植物叶散生，鳞片完整，花喇叭形，种子不休眠，幼苗为子叶出土型（Comber 1949）。Comber（1949）将这个组分为两个亚组：通江-王百合型具有深紫色或棕色种球，麝香-菲律宾型百合具有白色种球。尽管它们之间有密切的联系，但两个进化支都得到了分子系统发育数据的大力支持（Nishikawa et al. 1999，2001；Du et al. 2014）。在杂种育种方面，喇叭花组的物种非常重要，因为麝香百合是百合组间杂交及铁炮杂交系的主要育种资源（Van Tuyl and Arens 2011）。与形态特征相反，许多其他研究认为湖北百合属于喇叭花组，首先，湖北百合的 C 带特征与喇叭花组的 C 带模式匹配（Smyth et al. 1989）。其次，分子系统发育分析表明，湖北百合归类于喇叭花组（Nishikawa et al. 1999；Du et al. 2014）。湖北百合与喇叭花组物种间的种间杂交也支持这一观点（Van Tuyl and Arens 2011）。

20.3.3.7 斑瓣组

与 Comber（1949）的分类相反，De Jong（1974）将该组与卷瓣组分开。Wang 和 Tang（1980）将斑瓣组视为钟花组（包括具有钟状花形的百合种类）。之后，Liang（1980）及 Haw 和 Liang（1986）分析认为现今百合属内钟花组中花头下垂，钟状花型的所有物种都与豹子花属密切相关（Peruzzi 2016）。最新的系统发育研究表明，本组可分为三个亚组，其中两个亚组与卷瓣组密切相关（Du et al. 2014）。

与依赖形态特征的传统分类方法不同，分子数据重建的百合系统分类表明，百合许多引人关注的形态特征并非来自共同的祖先。但是，先前的分子系统发育中分子标记数量较少，导致物种水平的分辨率降低。因此，百合之间的某些关系仍然存在争议。尽管育种方法的发展产生了一系列的百合种间杂交种，但组内杂交种仍然是市场上最重要的百合类群及当前百合育种的基础（Van Tuyl and Arens 2011）。因此，提高百合系统发育信息的准确性对于创造百合新组合、提供育种资源非常重要。

20.4 种间杂交

种间杂交是植物育种中最重要的手段之一，它可以创建全新的杂种并改良大多数农作物的重要农学性状。通过常规的授粉可以相对容易地实现分类地位上不同的种间杂交（包括组间或种间杂交）。利用这些杂交原理，来自三个组的百合通过种间杂交促进了现代品种的分化。

1）卷瓣组，亚洲百合杂种系。最常见的百合杂种系之一，已注册的品种有数百个。该杂种系花色变化较大，具有从白色到红色包括黄色的颜色，花径比其他杂种系的百合小。这个杂种系起源于 11 种野生百合的杂交，包括毛百合、渥丹、山丹、垂花百合、朝鲜百合、大花卷丹、卷丹、匍茎百合、宝兴百合、珠芽百合与川百合（Van Tuyl et al. 2011）。

2）具叶柄组，东方百合杂种系。在过去的几十年中，这个杂种系变得越来越重要，其产量持续增加且部分替代了亚洲百合杂种系。东方百合杂种系花朵大而芬芳，花色为白色或粉红色。东方百合杂种系衍生自至少 6 种百合的杂交：奄美百合、天香百合、日本百合、香华丽百合、乙女百合和鹿子百合（Van Tuyl et al. 2011）。

3）喇叭花组，铁炮-喇叭杂种系。铁炮百合杂种系具喇叭形花，香味独特，包括“新铁炮杂种系”，一个由台湾百合与麝香百合（Wada 1951）杂交成的种间杂种。喇叭百合杂种系由王百合、通江百合和淡黄花百合经过种间杂交形成（Van Tuyl et al. 2011）。

20.4.1 广义杂交

尽管我们已经在创造新的百合杂种系方面取得了成功，已育成数千个亚洲百合、东方百合、铁炮百合和喇叭百合品种并进行了登记注册，但一个新品种需要集中各种优良的园艺性状、花色、花形和抗病性等。例如，具有亚洲百合颜色和东方百合的花径及同时具有亚洲百合花色、东方百合花径、喇叭百合和铁炮百合花形的杂种株系。此外，一些卷瓣组百合对尖孢镰刀菌和某些病毒具有抗性（Straathof and Van Tuyl 1994），而一些具叶柄组百合对椭圆葡萄孢霉具有抗性。这些性状对于百合新品种的培育具有重要意义（Lim et al. 2001a）。

来自不同分类组的杂种或种的种间杂交（组间的种间杂交）很困难，需要利用特殊的生物技术手段。难以获得组间的种间杂种的原因是杂交过程存在受精前和受精后障碍。最初的杂交障碍由柱头不亲和引起的花粉管生长不良所导致（Asano and Myodo 1977a；Asano 1980c），花粉粒在不亲和物种的柱头上无法被某些物质识别代谢，耗尽其自身物质能量储备后就会导致生长不良（Ascher 1966）。为了克服受精前障碍，可采用不同的技术：切割花柱授粉技术（Myodo 1962）、花柱内授粉（Asano and Myodo 1977b）、蒙导授粉（Van Tuyl et al. 1982）及离体授粉（Van Tuyl et al. 1991）。利用这些技术，可以实现百合组间杂交。尽管使用这些技术可以进行百合不同组间的种间杂交并获得组间杂种，但杂交有时仅能在某些组合上成功。例如，麝香百合不能作为父本为亚洲百合或东方百合授粉，但可以作为母本进行杂交。同样的是使用东方百合为父本对亚洲百合进行授粉时，杂交可能不会成功，但反过来使用东方百合作母本时杂交可以实现。在以东方百合和喇叭百合作亲本时，如果将东方百合用作母本，杂交可能会成功。此外，发现了百合杂交能否成功依赖于基因型组合。

受精后障碍是由于种子缺乏胚乳或胚在早期发育阶段败育（Myodo 1975；Asano and Myodo 1977b）。为了克服这些障碍，子房培养、胚珠培养（Van Tuyl et al. 1982，1988，1991；Van Creij et al.1992）和胚挽救（Myodo 1975；Asano and Myodo 1977b，1980；Myodo and Asano 1977；Asano 1978，1980a，1980b）技术被研发出来，并由此获得了数百个百合新品种。

随着新的百合组间杂种的诞生，关于这些新百合杂种的分类也被引入。

LA 百合杂种系。这个杂种系的品种是由铁炮百合与亚洲百合杂种系的品种杂交后获得（如亚洲百合品种‘Whilito’）（Van Tuyl et al. 1991；Van Creij et al. 1992）。该杂种系结合了铁炮百合的花形和亚洲百合的花色。如今，私人公司已经培育出数百个 LA 百合品种，这个杂种系已成为最常见的广义种间杂种之一。此外，还报道了铁炮百合与分类学为卷瓣组百合的杂交（表 20.1），其本质上与 LA 杂交品种不同，如铁炮百合与滇百合（Lim et al. 2008）、珠芽百合（Van Tuyl et al. 2000）、垂花百合（Myodo and Asano 1977）、渥丹（Van Tuyl et al. 2002；Arzate-Fernandez et al. 2006）、毛百合（Asano and Myodo 1980；Van Tuyl et al. 2000）、宝兴百合（Van Tuyl et al. 2011）、紫斑百合（Van Tuyl et al. 2011）和山丹（Asano and Myodo 1980）的杂交。

表 20.1　百合属中涉及分类种的广泛种间杂交实例

基因组型	杂交组合
HeLi	*L. henryi* × *L. candidum*
AT	*L. tigrinum* × *L. regale*
LA	*L. longiflorum* × *L. bakerianum*
LA	*L. longiflorum* × *L. bulbiferum*
LA	*L. longiflorum* × *L. cernuum*
LA	*L. longiflorum* × *L. concolor*
LA	*L. longiflorum* × *L. dauricum*
LA	*L. longiflorum* × *L. lankongense*
LA	*L. longiflorum* × *L. nepalense*
LA	*L. longiflorum* × *L. pumilum*
LHe	*L. longiflorum* × *L. henryi*
LCa	*L. longiflorum* × *L. candidum*
LMa	*L. longiflorum* × *L. martagon*
LMa	*L. longiflorum* × *L. hansonii*
LO	*L. longiflorum* × *L. brownii*
LRu	*L. longiflorum* × *L. rubellum*
LO	*L. formosanum* × *L. speciosum*
LOx	*L. longiflorum* × *L. sempervivoideum*
LOx	*L. longiflorum* × *L. lophophorum*
LPs	*L. longiflorum* × *L. canadense*
LPs	*L. longiflorum* × *L. kelloggii*
LPs	*L. longiflorum* × *L. pardalinum*
MaA	*L. hansonii* × *L. cernuum*
MaA	*L. martagon* × Asiatic
MaPs	*L. martagon* × *L. hansonii*
ONep	Oriental × *L. nepalense*
Ops	Oriental × *L. pardalinum*
TA	*L. regale* × *L. leichtlinii*

A. 亚洲系（卷瓣组）；Ca. 白花百合（百合组）；He. 湖北百合；Li. 百合组；L. 铁炮百合；Ma. 轮叶组；Nep. 紫斑百合（卷瓣组）；O. 东方系（具叶柄组）；Ox. 斑瓣组；Ps. 根茎组；Ru. 乙女百合（具叶柄组）；T. 喇叭花组

LO 百合杂种系。LO 杂种是铁炮百合与东方百合杂交（Van Tuyl et al. 2000）及新铁炮百合与东方百合杂交（Rhee and Kim 2008）形成的杂种系。这个杂种系的品种结合了东方百合的花色和铁炮百合的花径。其中一些杂交种保留了铁炮百合的喇叭花形，而另一些则显示出东方百合的花形，无论哪种花形，花径都比其他杂交系大。还有其他的杂交种涉及铁炮百合与具叶柄组的种质杂交（表 20.1），如与野百合（Van Tuyl 1980）、湖北百合（McRae 1991）和乙女百合（Van Tuyl et al. 2000）的杂交。另一个例子是台湾百合与鹿子百合的杂交（Myodo and Asano 1977）。

OA 百合杂种系。OA 百合是东方百合品种与亚洲百合品种之间进行组间杂交形成的杂种系（Van Tuyl et al. 1991，2000；Barba-Gonzalez et al. 2004）。OA 杂种结合了东方百合的大花径和亚洲百合的花色。

除了栽培品种杂交形成的 OA 杂种之外，还有一些组间野生种杂交形成的种间杂种，涉及分类学上的其他组物种（表 20.1、图 20.1）。一些例子包括汉森百合 × 垂花百合，欧洲百合 × 汉森百合（Van Tuyl et al. 2011），欧洲百合 × 亚洲百合品种，东方百合品种 × 豹纹百合（Van Tuyl et al. 2000，2011），东方百合品种 × 紫斑百合（Hyde 2009），汉森百合、湖北百合 × 白花百合（Van Tuyl et al. 2000），卷丹 × 岷江百合（McRae 1991），以及岷江百合 × 大花卷丹（Matsumoto 1992）。许多这种广泛的种间杂交涉及铁炮百合，它已被用于与湖北百合杂交（Myodo and Asano 1977），与白花百合杂交（Van Tuyl et al. 1991），与欧洲百合、汉森百合杂交（Van Tuyl et al. 2000），与高加索百合杂交（Van Tuyl and Van Holsteijn 1996），与蒜头百合杂交（Van Tuyl et al. 2011），与尖被百合杂交（Wang et al. 2009），与加拿大百合（Van Tuyl et al. 2000）和与凯洛基百合杂交（Fox 1974）。

20.4.2 杂交不育及多倍化

百合经种间、组间和组内杂交育成无数品种，然而，种间杂交存在的最主要缺点就是 F_1 代杂种不育。这种不育由减数分裂过程中染色体不能完全配对所引起（Ohri and Khoshoo 1983；Ishizaka 1994；Yabuya 1991）。此外，还有许多其他染色体异常，如染色体畸变、遗传不一致、染色体分类不平衡、染色体桥、减数分裂过程中染色体滞后及染色体移动的时间差异（Asano 1982a）。所有这些改变都是致命的，并导致不育，阻碍了下一步的育种工作（Asano 1982a；Hermsen 1984；Xie 2012）。

为了克服种间杂种的不育性，传统方法是利用有丝分裂多倍化，将体细胞染色体加倍来恢复育性。“纺锤体毒剂”被用于染色体加倍，这些毒剂通过抑制减数分裂过程中纺锤体的形成而起作用（Barba-Gonzalez et al. 2008）。利用秋水仙碱（Asano 1982b）和氨磺乐灵可恢复百合 F_1 代的育性（Van Tuyl 1990），但即使恢复育性，这种方法也被定义为实现进一步育种计划的“死胡同”，因为在亲本基因组之间没有由于自动同步配对而产生的重组。因此，染色体加倍形成的杂种被称为“永久杂种”，几乎不可能产生遗传变异（Lim et al. 2000；Wendel 2000；Van Tuyl et al. 2002；Ramanna and Jacobsen 2003）。

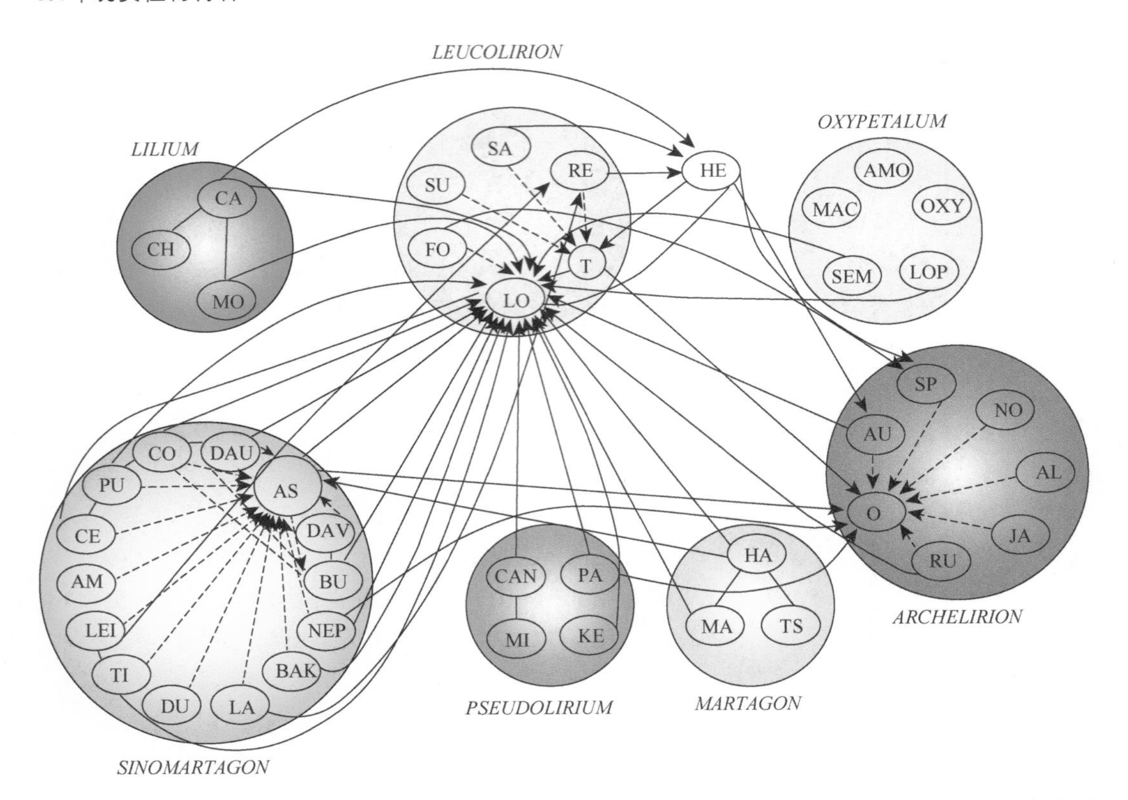

图 20.1　百合属不同种间杂交。不同颜色表示不同组根据 Comber（1949）对百合的分类。AL. 奄美百合；AM. 朝鲜百合；AMO. 玫红百合；AU. 天香百合；BAK. 滇百合；BU. 珠芽百合；CAL. 条叶百合；CA. 白花百合；CAN. 加拿大百合；CE. 垂花百合；CH. 加尔亚顿百合；CO. 渥丹；DAU. 毛百合；DAV. 川百合；DU. 宝兴百合；FO. 台湾百合；HA. 汉森百合；HEN. 湖北百合；JA. 日本百合；KE. 凯洛基百合；LA. 匍茎百合；LEI. 大花卷丹；LOP. 尖被百合；LO. 麝香百合；MAC. 印度百合；MAR. 欧洲百合；MI. 密歇根百合；MO. 高加索百合；NEP. 紫斑百合；NO. 香华丽百合；OXY. 斑瓣百合；PA. 豹纹百合；PU. 山丹；RE. 王百合；RU. 乙女百合；SA. 通江百合；SEM. 蒜头百合；SP. 鹿子百合；TA. 大理百合；T. 喇叭百合；TI. 卷丹；TS. 青岛百合

有丝分裂多倍化的另一种选择是利用未减数的 2*n* 配子（减数分裂多倍化）。2*n* 配子由植物减数分裂发生偏差而产生，导致 2*n* 配子形成的过程称为减数分裂核重组，核重组发生在小孢子或大孢子发生的过程中（Ramanna and Jacobsen 2003）。2*n* 配子可通过两套染色体组传递给子代，大多数时候产生的是三倍体或四倍体后代，取决于“*n*”或“2*n*”配子是由一个或两个亲本贡献而来。在许多 LA 和 OA 百合杂种中，报道了 2*n* 配子的自发发生及其在后代产生中的作用（Lim et al. 2001a，2001b；Barba-Gonzalez et al. 2004；Zhou et al. 2008a）（参见第 20.5 节）。

减数分裂多倍化在育种中应用具有极大的可行性，因为利用该技术产生的后代杂合性不固定并且有性多倍体中存在亲本基因组之间的重组（Bretagnolle and Thompson 1995；Soltis and Soltis 2000）。2*n* 配子的主要属性是实现了亲本染色体片段的渐渗，即使进行回交，渐渗的染色体片段仍保留在子代中（Lim et al. 2001a，2001b；Barba-Gonzalez et al. 2005b；Khan et al. 2012；Luo et al. 2012）。尽管未减数配子出现的概率较低，但最

近随着N2O的应用，在不育的OA和OT百合杂交种中诱导$2n$配子形成已经成为可能（Barba- Gonzalez et al. 2006a；Luo et al. 2016）。

20.5　百合的细胞遗传学

20.5.1　百合野生种的基因组、染色体和倍性

百合属植物拥有植物中最大的基因组，其单倍体DNA含量（1C）范围从比利牛斯百合（Bennett 1972）的32.75pg至加拿大百合的46.92pg（Zonneveld et al. 2005）。百合的基本染色体数$x = 12$，除了自然界中存在的三倍体卷丹和珠芽百合（$2n = 2x = 36$）外，百合属的所有种均为二倍体（$2n = 2x = 24$）（Noda 1978；Noda and Schmitzer 1992；Kim et al. 2006）。在一些物种中发现了带有额外染色体或染色体片段的非整倍体形式（Stewart 1943）。至少在17种百合属植物中报道了额外的B染色体（Brandram 1967），如在条叶百合中（Kayano 1962）。这些所谓的辅助染色体的大小变化范围很大，可从非常小到与正常的A染色体一样大，它们的数量可从1个，如兰州百合，到多达8个，如在大花卷丹与朝鲜百合的杂交后代中（Xie et al. 2014a）。百合因其染色体大而成为细胞学研究的理想模式植物。它们被用于研究减数分裂的各个方面：减数分裂持续时间、染色体配对、交叉形成和互换等（Bennett and Stern 1975；Stern and Hotta 1977）。许多早期的研究集中在百合染色体形态、染色体鉴定和核型分析上（Stewart 1947；Noda 1991）。百合染色体形态包括长度和着丝粒位置，在物种内和物种间都高度保守。因此，基于上述特征只有少数染色体可识别（Lim et al. 2001c；Marasek and Orlikowska 2003）。所有物种都有2对携带近中部着丝粒的长染色体和10对携带近端部着丝粒的染色体。一般来说，物种之间的差异是指次缢痕的数量和位置。Lim（2000）总结概括了44个百合属植物的次缢痕数目，其数目在每一个单倍体基因组上为1～8，因此次缢痕被用作百合种质的染色体鉴定和杂种身份的验证标记（Fernandez et al. 1996；Obata et al. 2000；Marasek and Orlikowska 2003；Marasek et al. 2004）。

20.5.2　带型技术与荧光原位杂交鉴定个体染色体

利用荧光染色处理（Q带、CMA3带、DAPI带）和异染色质切片Giemsa染色（C带）等染色体纵向分化技术可改进染色体鉴定。对于许多百合品种和种，已经根据染色体臂长和染色体分带构建了详细的核型（Marasek et al. 2005）。C-Giemsa染色带技术可用于物种分类研究（Holm 1976；Von Kalm and Smyth 1984；Smyth et al. 1989；Smyth 1999）和杂种鉴定（Smyth and Kongusuwan 1980）。Smyth等（1989）描述了代表6个组20种百合属植物的C带带型，表明染色体带的种间差异较大。一般来说，细的C带散布在染色体的长臂上，而很少有染色体在端粒和着丝粒上有条带。Marasek等（2005）基于Ag-NOR、CMA3/DA/ DAPI（CMA3，色霉素A3；DA，偏端霉素A；DAPI，4′-6-二氨基-2-苯基吲哚）和C-Giemsa染色建立了4种百合（白花百合、新铁炮百合、湖北百合和山丹）的染色体分带标记，根据百合的基因型和各种技术的特点开发出用于识别

24 条染色体的 4～17 条染色体的标记（图 20.2）。

图 20.2 45S rDNA 探针（橙色或红色）和端粒拟南芥序列 TTTAGGG（绿色）与 022538-3 AOA 三倍体杂种的体细胞中期染色体进行荧光原位杂交（照片由 Rodrigo Barba-Gonzalez 博士提供）

近 20 年来，随着 DNA 原位杂交技术的发展，百合细胞遗传学的研究取得了长足的进展，揭示了一些染色体结构的细节（Lim et al. 2000）。荧光原位杂交（FISH）一般用于定位 rRNA 基因位点和基因组中的特异 DNA 序列，而 GISH 可以提供更多关于基因组起源和多样性的信息。Lim 等（2001c）根据染色体臂长、C 带、$AgNO_3$ 染色和 PI-DAPI 带，结合以 5S 和 45S rDNA 为探针的 FISH 技术，构建出铁炮百合和乙女百合的详细核型。C-Giemsa 带和以 45S rDNA 为探针的 FISH 技术被用于分析岷江百合的染色体（Guangxin et al. 2008）。基于染色体臂长和以 45S 和 5S rDNA 基因为探针的荧光原位杂交，宝兴百合、龙牙百合、紫脊百合和岷江百合的详细核型被绘制出来（Wang et al. 2012）。5S 和 45S rRNA 基因位点定位的物理图谱被用于构建卷丹的二倍体（$2n = 2x = 24$）和三倍体（$2n = 3x = 36$）核型（Hwang et al. 2011）。以 45S rDNA 为探针进行定位分析揭示了亚洲百合四倍体品种‘Tresor’和‘Val di Sole’及其后代之间在 FISH 染色体核型上的相似性（Zhou et al. 2015）。Truong 等（2015）对韩国 30 个大花卷丹居群染色体进行包括染色体长度、臂长比、次缢痕数量、DAPI 条带和 FISH 标记（45S 和 5S rRNA 为探针）方面的详细核型分析，发现该种的二倍体和三倍体核型相同。

荧光原位杂交还被成功地用于评估杂交真实性。Marasek 等（2004）用带有 25S rDNA 探针的 FISH 验证了东方百合品种‘Marco Polo’和‘Expression’与湖北百合杂交获得的 F_1 代植株的杂种身份，携带有亲本信号位点的染色体可证实杂种的真实性。同样，有研究利用以 45S rDNA 作为探针的 FISH 荧光原位杂交分析了 7 个亚洲百合品种的核型多样性，并鉴定了 21 个不同倍性水平的亚洲百合杂交后代（Wang et al. 2015）。

20.5.3 百合染色体核型的形态分析

近年来，人们对百合属植物进行了不同研究。例如，Siljak-Yakovlev 等（2003）利用银染、荧光显带和 FISH 技术研究了百合属 3 个在分类学上相近的二倍体种（比利牛斯百合、绒球百合和金苹果百合）的染色体分化。这些技术的结合能够在染色体水平上区分这 3 个物种，将两个核糖体基因家族 18S-5.8S-26S（18S）和 5S rRNA 基因分离到不同的染色体对上，这些发现在染色体水平上揭示了一个清晰的种间分化关系。此外，利用 rRNA 基因位点的物理定位，也阐明了韩国本地百合属植物的细胞遗传学关系（Sultana et al. 2010）。

Muratović 等（2005）揭示了两个密切相关的波斯尼亚和黑塞哥维那特有种（金苹果百合和金百合）之间明显的种间分化，分化涉及 18S-5.8S-26S rRNA 基因位点的数量和位置。采用 FISH 技术以 5S 和 45S rDNA 为探针，对 8 种百合属物种的染色体多样性进行研究，系统发育分析结果显示渥丹、朝鲜百合和条叶百合亲缘关系密切，而卷丹的二倍体和三倍体亲缘关系较远（Lee et al. 2014）。为了解百合科植物的染色体进化模式，有研究者对百合科植物进行了大规模核型分析（Peruzzi et al. 2009）。

20.5.4　基于 GISH 分析的基因组分化

基因组原位杂交（GISH）是一种以总基因组 DNA 为探针的技术，可以对与所用探针同源的基因组进行原位检测（Schwarzacher et al. 1989）。在百合属植物中，GISH 已成功用于种间和种间杂种的基因组研究，从而能够鉴定亲本基因组（图 20.3）（Lim et al. 2000；Marasek et al. 2004），并能鉴定多倍体品种的基因组组成（Zhou et al. 2008b），许多组间杂种被鉴定，如 LA 杂种（Khan et al. 2009a，2010；Zhou et al. 2008b）、OA 杂种（Barba-Gonzalez et al. 2004，2005b）和 OT 杂种（Luo et al. 2012）。

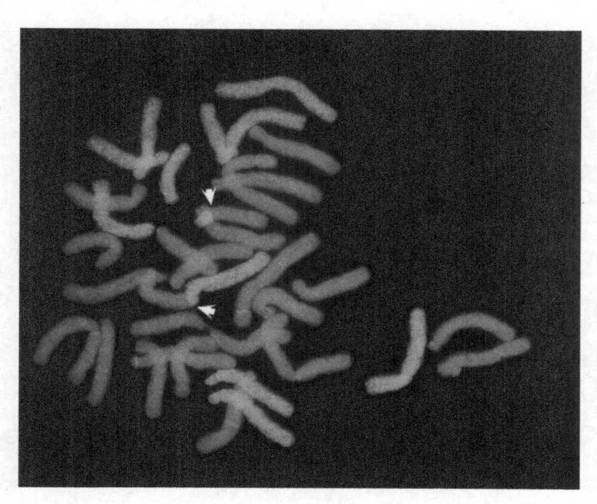

图 20.3　AOA 三倍体杂种‘012168-1’体细胞分裂中期染色体的单靶基因组原位杂交。东方百合基因组用 Cy3 链霉亲和素系统（粉红色染色体）检测，亚洲百合基因组用 DAPI（蓝色染色体）检测；箭头表示重组染色体（照片由 Rodrigo Barba-Gonzalez 博士提供）

Lim 和 Van Tuyl（2002）采用多色原位杂交（GISH/FISH）方法，对由铁炮百合（L）、亚洲百合（A）和东方百合（O）基因组组成的种间杂交个体的染色体与杂种性质进行了鉴别。GISH 在百合中也被广泛用于追踪 LA 和 OA 杂种的 BC1 与 BC2 后代中的基因组重组位点（Karlov et al. 1999；Lim et al. 2001b，2003；Barba-Gonzalez et al. 2005b；Zhou et al. 2008b）。

20.5.5 种间杂种减数分裂

百合大多数种间杂种表现为不育，会对进一步杂交造成阻碍。百合远缘杂种的育性与其减数分裂过程中染色体能否成功配对结合相关。Luo 等（2013）利用 GISH 和常规细胞学方法，观察了东方百合 × 喇叭百合（OT）不育杂种的小孢子发生过程中的染色体异常减数分裂行为。在中期 I 阶段，除二价体外，还形成了异常的染色体组合，包括单价体、三价体、四价体和环八价体。在后期至末期，观察到大量的落后染色体和不同类型的染色体桥结构。Xie（2012）在百合种间杂种减数分裂过程中也发现了一些异常现象，包括染色体断裂和融合，导致异染色体产生。Zhou 等（2008a）分析了铁炮百合 × 亚洲百合（LA）杂种 F_1 的同源染色体行为和杂交，观察到不同基因型之间染色体关联的差异。对 LA 杂种减数分裂进行 GISH 分析显示了不同类型的交换，包括单交换、两链双交换、三链双交换、四链双交换及非姐妹染色单体之间的四链多交换（Zhou et al. 2008a；Xie et al. 2014b）。

20.5.6 未退化配子的起源及其在百合育种中的应用

F_1 杂种的不育性是小孢子和大孢子发生过程中异常减数分裂的结果，减数分裂自发产生具有体细胞染色体数目的未减数配子（$2n$ 配子）。在百合中，已经发现了一些组间的种间杂种可产生这种配子。例如，Barba-Gonzalez 等（2005a）对 OA 百合杂种进行筛选，发现 3%的基因型产生 $2n$ 配子。产生 $2n$ 配子的植株已被成功地用于获得 BC1 后代，以不同的回交组合（A × OA、O × OA、OA × A、LA × A、O × LA、A × LA、O × AuH、OT × O 和 MA × A）获得二倍体品种，以 OAOA × OA 组合获得四倍体品种（Lim et al. 2003；Barba-Gonzalez et al. 2006b；Khan et al. 2009a；Luo et al. 2012；Chung et al. 2013）。

功能性 $2n$ 配子在多倍体品种形成中起重要作用。产生 $2n$ 配子的 LA 百合杂种（Lim et al. 2000，2001b）和 OA 百合杂种（Barba-Gonzalez et al. 2004，2005b）通过与亲本的成功回交获得了数百个三倍体后代植株。异源三倍体百合可产生非整倍体或具整倍体功能的雌配子，可作为百合渐渗育种的母本。在 OTO × OO 组合（Zhou et al. 2015）及 LLO × OTOT 和 LLO × TTTT 组合（Xie et al. 2014a）的杂交后代中均发现了非整倍体。

Khan 等（2009a）以两个不同的 F_1 LA 杂种作为亲本，一个亲本提供 $2n$ 配子卵细胞，另一个亲本提供 $2n$ 配子花粉，最终产生了四倍体后代。一些三倍体 ALA 基因型百合已成功地作为亲本与二倍体和四倍体亲本杂交，产生了大量的近二倍体和近五倍体后代（Lim et al. 2003）。

GISH 分析表明，将产生功能性 $2n$ 配子的 LA 杂交种（Longiflorum × Asiatic）与二倍体亚洲百合（A）品种回交后，大多数后代是三倍体，并且发生了大量的基因组重组（Lim et al. 2003）。同样，由天香百合 × 湖北百合杂交产生的 F_1 杂种及其与东方杂种亲本产生的三倍体 BC1 后代的基因组组成也可用 GISH 进行评估（Chung et al. 2013）。AuH 杂种的 F_1 包含两个亲本的各一套染色体，而 BC1 三倍体 O × AuH 杂种显示有 12 条来自母本的染色体（二倍体东方杂种）和 24 条来自父本的染色体（二倍体 F_1 AuH 杂种），大多数 BC1 杂种具有重组染色体。

20.5.7　利用 GISH 揭示百合 2*n* 配子形成机制

如前所述，2*n* 配子是由减数分裂异常所致，如第一次或第二次减数分裂的缺失、第二次减数分裂的纺锤体形态异常及胞质分裂紊乱（Ramanna and Jacobsen 2003）。采用 GISH 可以描述百合 2*n* 配子形成的三种不同机制，每种机制都有不同的遗传后果（Lim et al. 2001a；Barba-Gonzalez et al. 2005a）。在第一次减数分裂重组（FDR）中，同源染色体在减数分裂 I 时无法配对或分离。在 FDR 中，杂合度得以保持，并且双亲基因组的两套染色体都被传递给后代。在第二次减数分裂重组（SDR）中，第一次减数分裂表现正常，即在减数分裂 I 期间同源染色体的配对和分离正常发生。尽管 SDR 是配子起源，但其可能高度异质。如果二价体在后期 I 分离并且单价体同时分裂，则这种减数分裂会导致不确定减数分裂重组（IMR），并产生两个 2*n* 的 IMR 配子（Lim et al. 2001a）。IMR 配子具有最高程度的遗传变异。

研究发现，在经单性有性多倍体化后获得的 LA 和 OA 杂种群体中，大多数不同的后代是通过 FDR 机制利用功能性 2*n* 配子产生的，无论有无交叉发生（Lim et al. 2001a；Barba-Gonzalez et al. 2005a，2005b；Zhou et al. 2008b）。同样，GISH 评估表明 AuH-F_1 杂种通过 FDR 机制产生了 2*n* 配子（Chung et al. 2013）。

与有丝分裂加倍的杂种相比，未减数配子的优势在于减数分裂过程中亲本染色体之间存在重组，导致后代中存在重组染色体。Khan 等（2009a，2009b）揭示了在 LA 杂种二倍体和三倍体回交后代的染色体中存在广泛的基因间重组。GISH 分析也证实了 LA 和 OA 杂种有性多倍体后代中存在重组染色体（Xie et al. 2010）。在 LA 杂种中，重组染色体的数量和每条染色体的重组位点数量高于其他杂种（Khan et al. 2009b，2010）。

20.5.8　二倍体回交后代的产生及其相关性

大多数种间（组间）品种均是利用杂种 F_1 的 2*n* 配子再次杂交而得到。然而，有研究表明，一些 F_1 杂种可以产生可育的单倍体（*n*）配子，从而在随后的世代中产生二倍体。在 F_1 种间杂种——LA 杂种中可观察到正常染色体配对的发生和单倍体配子与 2*n* 配子的形成（Zhou 2007；Zhou et al. 2008a）。在 Khan 等（2009b）的研究中，由 LA 杂种产生的 104 个回交一代中有近 30%是二倍体或近二倍体。这种正常的减数分裂导致具有同源染色体重组的二倍体 BC1 后代形成，对实现基因渗入可能会有价值（Khan et al. 2009b）。同样，在其他观赏植物中也报道过 *n* 配子和 2*n* 配子的出现，如 *Alstroemeria aurea* × *A. inodora*（Kamstra et al. 1999）。

20.6 现代品种分类

如第 20.2 节所述，百合品种主要是在过去 50 年中通过不同种的杂交培育而成。根据荷兰花卉种球检验机构（BKD）的数据，百合品种可分为 7 个主要类群（表 20.2）。荷兰育种家是这一领域的主要参与者。现今，最主要的荷兰育种公司是 Vletter & Den Haan Brothers、World Breeding BV、Mak Breeding BV、De Jong Lelies BV、Van Zanten Flowerbulbs BV 和 De Looff Innovation BV。De Looff Innovation BV 专门研发重瓣东方百合，这类百合正成为一个新兴的百合类群（图 20.4）。表 20.2 列出了荷兰最重要的百合种球种植面积。在表 20.3 中，18 个最重要的百合品种按其杂种起源类别、倍性水平及 2010 年和 2017 年种植面积进行排序。亚洲百合（A）的重要性正在下降，现在栽培的只包括四倍体品种‘Brunello’（图 20.5f）、‘Tresor’和三倍体品种‘Navona’。近 20 年来，亚洲百合品种在很大程度上被 LA 百合杂种所取代。为了培育 LA 百合，育种家使用自己培育的二倍体亚洲系列杂种，育成许多颜色不同的三倍体LA 品种（黄色，‘Pavia’、‘Nashville’；白色，‘Litouwen’（图 20.5e）；橙色，‘Honesty’；粉色，‘Indian Summerset’、‘Brindisi’）。LA 三倍体品种由两套亚洲百合和一套铁炮百合的基因组组成。由两套铁炮百合和一套东方百合基因组组成的LO 杂种也是三倍体，代表品种是‘White Triumph’。最新的种间杂种类群 OT 百合通常也是三倍体，包含两套东方百合基因组和一套喇叭百合基因组。过去 10 年，OT 百合的发展非常迅速（表 20.2），该杂种群在一定程度上将会取代东方百合。OT 百合中最重要的品种是‘Conca dór’、‘Robina’（图 20.5c）和‘Zambesi’（图 20.5d）。迄今，东方百合（O）仍然是种植面积最大的百合种类（表 20.2），白色的‘Siberia’（图 20.5b）和粉红色的‘Sorbonne’（图 20.5a）仍是东方百合中最流行的品种。东方百合最新的培育方向是重瓣花型。现在，玫瑰百合系列（图 20.4）已经种植了 $30hm^2$ 以上，预计几年内将增长到 $100hm^2$ 或更多。

表 20.2　1972 年、1994 年、2010 年和 2017 年主要百合品种的种球栽培面积（BKD）

Group	1972 年	1994 年	2010 年	2017 年
A	204	1627	436	247
O		1155	1799	1661
L	1	137	46	37
LA		21	886	1271
LO			53	41
OT			349	1203
OA			2	3
其他	59	80	53	33
合计	264	3020	3624	4494

A. 亚洲百合；L. 铁炮百合；O. 东方百合；T. 喇叭百合

图 20.4 新型的重瓣东方百合

表 20.3 2010 年和 2017 年种植的 18 个最重要的百合品种及其杂种系、倍性水平和种球面积（hm^2）

品种	杂种系	倍性	2010 年	2017 年
Sorbonne	O	2	200	170
Siberia	O	2	193	187
Robina	OT	3	84	117
Tiber	O	2	71	57
Conca dór	OT	3	67	129
Brindisi	LA	3	67	52
Santander	O	2	61	68
Litouwen	LA	3	59	153
Tresor	A	4	56	34
Tabledance	OT	3	2	74
Pavia	LA	3	50	65
Yelloween	OT	2	50	52
Zambesi	OT	3	0	116
Indian Summerset	LA	3	14	63
White Heaven	L	2	38	20
Star Gazer	O	2	37	21
White Triumph	LO	3	30	28
Nashville	LA	3	23	75

图 20.5 （a）‘Sorbonne’（O）；（b）‘Siberia’（O）；（c）‘Robina’（OT）；（d）‘Zambesi’（OT）；（e）‘Litouwen’（LA）；（f）‘Brunello’（A）

20.7 分 子 育 种

尽管百合是种间杂交研究的模式作物，但分子育种在实际育种中尚未得到应用（Arens et al. 2014）。目前已发表的百合遗传图谱很少，且大多是密度相对低的标记图谱。Abe 等（2002）利用随机扩增多态性 DNA（RAPD）和简单重复序列区间（ISSR）标记，在两个亚洲品种‘Montreux’和‘Connecticut King’之间构建亲本连锁图谱，以阐明花色苷着色的遗传学机制。Van Heusden 等（2002）用扩增片段长度多态性（AFLP）标记在亚洲回交群体（AA 群体，‘Connecticut King’×‘Orlito’）中绘制了两种重要病原体：尖孢镰刀菌（*Fusarium oxysporum*）和百合斑驳病毒（LMoV）抗性基因的遗传图谱。Van Heusden 等（2002）后续的研究工作由 Shahin 等（2010）完成。在相同的亚洲回交群体中，通过补充先前使用的 AFLP 标记并进行核苷酸结合位点（NBS）分析（Van der Linden et al. 2004）与多样性阵列技术（DArT）标记（Mace et al. 2008），可以实现图谱标记密度的显著提高。同样，Shahin 等（2010）还绘制了用前述的亚洲百合品种‘Connecticut King’与铁炮品种‘White Fox’进行种间杂交获得的 LA 百合种间杂种的 NBS 图谱和 DArT 标记（Khan 2009）（LA 杂种群：以铁炮品种‘White Fox’与亚洲品种‘Connecticut King’杂交获得的种间杂种群）。

第二代测序技术（NGS）的发展极大地改变了生命科学研究的可能性，也显示出对于百合的重要性。例如，Shahin 等（2010）研究表明两个后代种群的标记密度都相对较低，因此采用了一项关于将 NGS 用于单核苷酸多态性（SNP）标记基因分型的技术。为此，使用 454 GS-FLX 钛焦磷酸测序（Roche）对育种中使用的每种百合杂种类型的代表（亚洲品种‘Connecticut King’、铁炮品种‘White Fox’、东方品种‘Star Gazer’和喇叭百合品种）进行 cDNA 文库测序。每种基因型的 454 平台焦磷酸序列已被组装，并已鉴定出多个 SNP（Shahin et al. 2012a，2012b）。在这 4 种基因型百合杂种中，已鉴定的 SNP 数量范围为 4000～11 000 个及以上。已鉴定出大量的 SNP 表明基因型内的多态性水平很高，即使如此，并非所有的单核苷酸多态性位点都能用于 SNP 标记的开发。

SNP 标记基因分型的一个主要挑战是目标 SNP 的侧翼序列中的 SNP，这可能对 SNP 基因分型性能产生重大影响。前人研究发现 4 个百合品种间的多态性率非常高（每 26bp 1 个 SNP），这会使 SNP 基因分型复杂化（更多的 SNP 标记将丢失），并且更可能出现无效等位基因（Shahin 2012）。因此，不应选择具有侧翼的 SNP 用于标记基因分型。在目标 SNP 的 50bp 侧翼区域中，只有 400～700 个 SNP 没有二级 SNP，1000～2000 个 SNP 最多具有一个二级 SNP（Shahin et al. 2012b）。随着当今测序技术能力的增强，包括如今改进的生物信息学和基因分型平台，可能会获得更高的标记数。

在每个已测序基因型中鉴定出的 SNP 标记大多数都是特定基因组所独有（如‘Connecticut King’中的 SNP 在铁炮百合中不具多态性，反之亦然）。为了验证从 NGS 测序中鉴定出的 SNP 是否为有效的 SNP 标记进行作图研究，采用 KASP（KBiosciences competitive allele-specific PCR）技术，以包含 225 个 SNP 标记的测试集对 AA 和 LA 群体进行基因分型，最后分别在 LA 和 AA 群体的遗传图谱上绘制了 94 个和 85 个 SNP 标记（Shahin 2012）。然而，考虑到成功的 SNP 的数量和鉴定出的其他 SNP 的总数，可以通过这种方式生成足够多的标记来开发高密度遗传图谱。

最新的 LA 百合遗传图谱有 565 个标记（NBS、DArT 和 SNP），覆盖 2438cM，标记密度为每 4.3cM 1 个标记。AA 遗传图谱有 409 个标记（AFLP、NBS、DArT 和 SNP），覆盖 2035cM，标记密度为每 5.2cM 1 个标记。据估计，百合的基因组大小为 2740cM；因此，目前的 LA 和 AA 种群图谱分别覆盖了百合基因组的 89%和 74%（Shahin 2012），常见的 DArT、NBS 和 SNP 标记被用于比较 LA 与 AA 两个群体的遗传图谱（‘Connecticut King’是两个群体的一个共同亲本）（Shahin 2012）（图 20.6）。

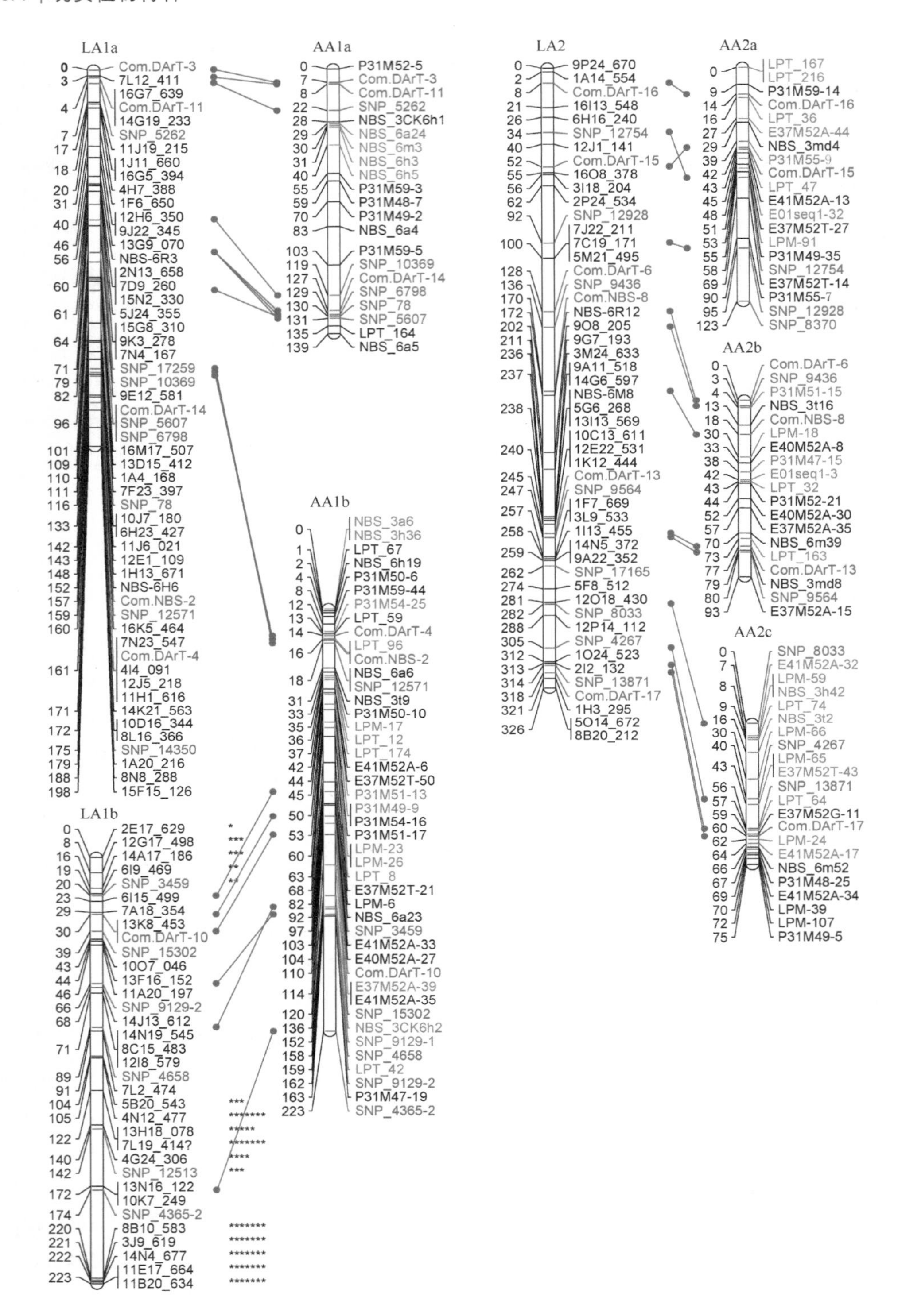

图 20.6 LA 和 AA 群体的连锁群 1 与 2 根据常见的 DArT、NBS 和 SNP 标记排列（Shahin 2012）。所有 SNP 标记及常见的 DArT 和 NBS 标记均为绿色；<AB × AB>标记为红色；偏斜的区域以星号表示

对百合的遗传图谱进行改进和对齐，可使一些重要性状的遗传图谱得到更好的定位。百合斑驳病毒(LMoV)抗性在AA群体遗传图谱(LG AA10)(Van Heusden et al. 2002; Shahin et al. 2010）上被定位为单基因座。SNP标记的加入导致在定位抗性位点（LG AA10）3cM内出现了SNP标记。然而，具有此类gab（3cM）限制了该标记在育种中的实用性，因为在3cM内仍可能发生重组，导致假阳性。在这种情况下，抗性位点两边（紧密连锁）的标记是首选的，有助于识别重组子。标记该图谱区域需要更多的SNP标记。

尖孢镰刀菌的抗性被定位在AA群体的遗传图谱上，Van Heusden等（2002）鉴定了4个推定的QTL，之后当标记密度提高时，Shahin等（2010）在同一群体中鉴定了6个推定的QTL。QTL1是一个很强的QTL，在多年的抗病性鉴定中显示与镰刀菌抗性有关。此外，对LA群体的镰刀菌抗性进行定位作图，确认了在AA群体中检测到的主要QTL（Shahin 2012）。在两个群体中都鉴定出的QTL1为开发能够在分子标记辅助育种（MAB）中应用的镰刀菌抗性分子标记提供了一个良好的开端。但是，由于此QTL的分辨率仍然很低，需要更多标记才能提高定位分辨率。目前所定位的少数SNP标记均未与该QTL连锁。

继对镰刀菌和百合斑驳病毒（LMoV）抗性进行图谱定位后，研究者还对一些观赏性状进行了遗传图谱绘制。在由Shahin等（2010）构建的AA和LA群体及由Abe等（2002）构建的群体中绘制的观赏性状图谱是关于花青素的性状、花被片类胡萝卜素的色素沉着性状（Shahin et al. 2010；Yamagishi 2013)、无斑性状（Shahin et al. 2010；Abe et al. 2002)、茎秆颜色性状（Shahin et al. 2010)、无花药（Shahin et al. 2010）和花朵方向（Shahin et al. 2010）的。

百合有很长的种球培育期（2～3年)，为了将不同亲本的理想性状组合到一个品种中，选择百合往往需要多个育种周期。因此，开发性状标记有助于育种者开展百合育种，其不仅可以加快后代的选择，而且可以为育种计划选择亲本。当性状由隐性等位基因控制时，非常有利于通过标记选择具有正确基因/等位基因的亲本，因为有些性状只有当隐性等位基因以纯合状态（aa）存在时才会表达，如花斑和花朵方向（Shahin et al. 2014)。由此，拥有能够使这些性状在AA和Aa之间进行区分的标记(具有相同的表型）变得非常重要（Shahin et al. 2014)。总体而言，性状标记的有效性对于加快育种进程非常有价值，而对于隐性性状更是如此，因为这大大提高了育种程序的效率。

20.8 结 论

与其他观赏作物相比，百合的育种历史很短。但是由于育种技术在克服杂交障碍方面的发展，在过去的50年中百合的类群发生了戏剧性变化，育成了主要的三倍体杂交系列（LA、LO、OA、OT)。目前，重瓣百合类型品种前景广阔，并已在市场上崭露头角。分子育种尚未在实际育种中得到应用。今后的研究重点为利用针对百合斑驳病毒（LMoV）和镰刀菌抗性的分子标记加速抗病育种。

本章译者：
崔光芬
云南省农业科学院花卉研究所，国家观赏园艺工程技术研究中心，云南省花卉育种重点实验室，昆明 650200

参考文献

Abe H, Nakano M, Nakatsuka A, Nakayama M, Koshioka M, Yamagishi M (2002) Genetic analysis of floral anthocyanin pigmentation traits in Asiatic hybrid lily using molecular linkage maps. Theor Appl Genet 105:1175–1182

Ahn TJ, Hwang HJ, Younis A, Sung MS, Ramzan F, Kwon YI, Kim CK, Lim KB (2017) Investigation of karyotypic composition and evolution in *Lilium* species belonging to the section *Martagon*. Plant Biotechnol Rep 11:407–416

Arens A, Shahin A, Van Tuyl JM (2014) (Molecular) breeding of *Lilium*. Acta Hortic 1027:113–128

Arzate-Fernandez AM, Nakazaki T, Tanisaka T (2006) Production of diploid and triploid interspecific hybrids between *Lilium concolor* and *L. longiflorum* by in vitro ovary slice culture. Plant Breed 117:479–484

Asano Y (1978) Studies on crosses between distantly related species of lilies. III. New hybrids obtained through embryo culture. J Jpn Soc Hortic Sci 47:401–414

Asano Y (1980a) Studies on crosses between distantly related species of lilies. IV. The culture of immature hybrid embryos 0.3~0.4 mm long. J Jpn Soc Hortic Sci 49:114–118

Asano Y (1980b) Studies on crosses between distantly related species of lilies. V. Characteristics of newly obtained hybrids through embryo culture. J Jpn Soc Hortic Sci 49:241–250

Asano Y (1980c) Studies on crosses between distantly related species of Lilies. VI. Pollen-tube growth in interspecific crosses on Lilium longiflorum (L). J Jpn Soc Hortic Sci 49:392–396

Asano Y (1982a) Chromosome association and pollen fertility in some interspecific hybrids of *Lilium*. Euphytica 31:121–128

Asano Y (1982b) Overcoming interspecific hybrid sterility in *Lilium*. J Jpn Soc Hortic Sci 51:75–81

Asano Y, Myodo H (1977a) Studies on crosses between distantly related species of Lilies I. For the interstylar pollination technique. J Jpn Soc Hortic Sci 46:59–65

Asano Y, Myodo H (1977b) Studies on crosses between distantly related species of Lilies II. The culture of immature hybrid embryos. J Jpn Soc Hortic Sci 46:267–273

Asano Y, Myodo H (1980) Lily hybrids newly obtained by the technique combining cut-style pollination with embryo culture (II). Lily Yearb North Am Lily Soc 33:7–13

Ascher PD (1966) A gene action model to explain gametophytic self- incompatibility. Euphytica 15:179–183

Barba-Gonzalez R, Lokker BH, Lim K-B, Ramanna MS, Van Tuyl JM (2004) Use of 2n gametes for the production of sexual polyploids from sterile Oriental × Asiatic hybrids of lilies (*Lilium*). Theor Appl Genet 109:1125–1132

Barba-Gonzalez R, Lim K-B, Ramanna MS, Visser RGF, Van Tuyl JM (2005a) Occurrence of 2n gametes in the F1 hybrids of Oriental × Asiatic lilies (*Lilium*): relevance to intergenomic recombination and backcrossing. Euphytica 143:67–73

Barba-Gonzalez R, Ramanna MS, Visser RGF, Van Tuyl JM (2005b) Intergenomic recombination in F1 lily hybrids (*Lilium*) and its significance for genetic variation in the BC1 progenies as revealed by GISH and FISH. Genome 48:884–894

Barba-Gonzalez R, Miller CT, Ramanna MS, Van Tuyl JM (2006a) Nitrous oxide (N2O) induces 2n gametes in sterile F1 hybrids between Oriental × Asiatic lily (*Lilium*) hybrids and leads to intergenomic recombination. Euphytica 148:303–309

Barba-Gonzalez R, Miller CT, Ramanna MS, Van Tuyl JM (2006b) Induction of 2n gametes for overcoming F1-sterility in lily and tulip. Acta Hortic 714:99–106

Barba-Gonzalez R, Lim K-B, Zhou S, Ramanna MS, Van Tuyl JM (2008) Interspecific hybridization in lily: the use of 2n gametes in interspecific lily hybrids. In: Teixeira da Silva JA (ed) Floriculture, ornamental and plant biotechnology. Advances and topical issues, vol V. Global Science Books Ltd., Middlesex, UK

Bennett MD (1972) Nuclear DNA content and minimum generation time in herbaceous plants. Proc R Soc Lond B Biol Sci 181:109–135
Bennett MD, Stern H (1975) The time and duration of female meiosis in *Lilium*. Proc R Soc Lond B Biol Sci 188:459–475
Brandram SN (1967) Cytogenetic studies of the Genus *Lilium*. MSc thesis, The University of London, pp 51–57
Bretagnolle F, Thompson JD (1995) Tansley review no. 78. Gametes with somatic chromosome number: mechanisms of their formation and role in the evolution of autopolyploid plants. New Phytol 129:1–22
Chung MY, Jae-Dong Chung DJ, Ramanna M, van Tuyl JM, Lim KB (2013) Production of polyploids and unreduced gametes in *Lilium auratum* × *L. henryi* hybrid. Int J Biol Sci 9:693–701
Comber HF (1949) A new classification of the *Lilium*. Lily Yearb R Hortic Soc 13:86–105
De Jong PC (1974) Some notes on the evolution of lilies. Lily Yearb North Am Lily Soc 27:23–28
Du YP, He HB, Wang ZX, Li S, Wei C, Yuan XN, Cui Q, Jia GX (2014) Molecular phylogeny and genetic variation in the genus *Lilium* native to China based on the internal transcribed spacer sequences of nuclear ribosomal DNA. J Plant Res 127:249–263
Endlicher S (1836) Genera plantarum. Beck, Vienna
Fernandez AM, Nakazaki T, Tanisaka T (1996) Development of diploid and triploid interspecific hybrids between *Lilium longiflorum* and *L. concolor* by ovary slice culture. Plant Breed 115:167–171
Fox D (1974) The hybrids of West American lilies. Lily Yearb North Am Lily Soc 27:7–12
Gao YD, Zhou SD, He XJ, Wan J (2012) Chromosome diversity and evolution in tribe Lilieae (Liliaceae) with emphasis on Chinese species. J Plant Res 125:55–69
Gao YD, Harris AJ, Zhou SD, He XJ (2013) Evolutionary events in *Lilium* (including *Nomocharis, Liliaceae*) are temporally correlated with orogenies of the Q-T plateau and the Hengduan Mountains. Mol Phylogenet Evol 68:443–460
Guangxin L, Fengrong H, Mengli X, Zhuhua W, Jian C, Shi J (2008) Giemsa C-banding and FISH analysis of Minjiang lily chromosome by means of root tips. Mol Plant Breed 6:95–99
Haw SG, Liang S-Y (1986) The lilies of China: the genera *Lilium, Cardiocrinum, Nomocharis* and *Notholirion*. Timber Press, Portland, OR
Hayashi K, Kawano S (2000) Molecular systematics of *Lilium* and allied genera (*Liliaceae*): phylogenetic relationships among *Lilium* and related genera based on the rbcL and matK gene sequence data. Plant Species Biol 15:73–93
Hermsen JGT (1984) The potential of meiotic polyploidization in breeding allogamous crops. Iowa State J Res 58:435–448
Hiramatsu M, Ii K, Okubo H, Huang KL, Huang CW (2001) Biogeography and origin of *Lilium longiflorum* and *L. formosanum* (*Liliaceae*) endemic to the Ryukyu Archipelago and Taiwan as determined by allozyme diversity. Am J Bot 88:1230–1239
Holm PB (1976) The C and Q banding patterns of the chromosomes of *Lilium longiflorum* (Thunb.). Carlsb Res Commun 41:217–224
Hwang YJ, Kim HH, Kim JB, Lim KB (2011) Karyotype analysis of *Lilium tigrinum* by FISH. Hortic Environ Biotechnol 52(3):292–297
Hyde R (2009) 'Kushi Maya' the story so far. North Am Lily Soc Q Bull 63(4):10–11
İkinci N, Oberprieler C, Güner A (2006) On the origin of European lilies: phylogenetic analysis of *Lilium* section *Liriotypus* (*Liliaceae*) using sequences of the nuclear ribosomal transcribed spacers. Willdenowia 36:647–656
Ishizaka H (1994) Chromosome association and fertility in the hybrid of *Cyclamen persicum* Mill. *C. hederifolium* Aiton and its amphidiploid. Breed Sci 44:367–371
Kamstra SA, Ramanna MS, De Jeu MJ, Kuipers AGJ, Jacobsen E (1999) Homoeologous chromosome pairing in the distant hybrid *Alstroemeria aurea* × *A. inodora* and the genome composition of its backcross derivatives determined by fluorescence in situ hybridization with species-specific probes. Heredity 82:69–78
Karlov GI, Khrustaleva LI, Lim KB, Van Tuyl JM (1999) Homoeologous recombination in 2n-gamete producing interspecific hybrids of *Lilium* (*Liliaceae*) studied by genomic in situ hybridization (GISH). Genome 42:681–686
Kayano H (1962) Cytogenetic studies in *Lilium callosum*. V. Supernumerary B chromosome in wild populations. Evolution XVI:246–253
Khan N (2009) A molecular cytogenetic study of intergenomic recombination and introgression of chromosomal segments in lilies (*Lilium*). PhD thesis, Wageningen University & Research
Khan N, Barba-Gonzalez R, Ramanna MS, Visser RGF, Van Tuyl JM (2009a) Construction of

chromosomal recombination maps of three genomes of lilies (*Lilium*) based on GISH analysis. Genome 52:238–251

Khan N, Zhou S, Ramanna MS, Arens P, Herrera J, Visser RGF, Van Tuyl JM (2009b) Potential for analytic breeding in allopolyploids: an illustration from Longiflorum × Asiatic hybrid lilies (*Lilium*). Euphytica 166:399–409

Khan N, Barba-Gonzalez R, Ramanna MS, Arens P, Visser RGF, Van Tuyl JM (2010) Relevance of unilateral and bilateral sexual polyploidization in relation to intergenomic recombination and introgression in *Lilium* species hybrids. Euphytica 171:157–173

Khan N, Marasek-Ciolakowska A, Xie S, Ramanna MS, Arens P, Van Tuyl JM (2012) A molecular cytogenetic analysis of introgression in backcross progenies in intersectional *Lilium* hybrids. Floricult Ornamental Biotechnol 6(2):13–20

Kim JH, Kyung HY, Choi YS, Lee JK, Hiramatsu M, Okubo H (2006) Geographic distribution and habitat differentiation in diploid and triploid *Lilium lancifolium* of South Korea. J Fac Agric Kyushu Univ 51:239–243

Lee CS, Kim SC, Yeau SH, Lee NS (2011) Major lineages of the genus *Lilium* (*Liliaceae*) based on nrDNA ITS sequences, with special emphasis on the Korean species. J Plant Biol 54:159–171

Lee HI, Younis A, Hwang YJ, Lim KB (2014) Molecular cytogenetic analysis and phylogenetic relationship of 5S and 45S ribosomal DNA in *Sinomartagon Lilium* species by fluorescence in situ hybridization (FISH). Hortic Environ Biotechnol 55:514–523

Liang SY (1980) Flora Reipublicae Popularis Sinicae, vol 14, Anagiospermae, Monocotyledoneae Liliaceae (I). Science Press, Beijing

Liang SY (1985) A new species *Lilium* from Xizang. Acta Phytotax Sin 5:392–393

Lim KB (2000) Introgression breeding through interspecific polyploidisation in lily: a molecular cytogenetic study. PhD thesis, Wageningen University & Research, 120 pp

Lim KB, Van Tuyl JM (2002) Identification of parental chromosomes and detection of ribosomal DNA sequences in interspecific hybrids of *Lilium* revealed by multicolor in situ hybridization. Acta Hortic 570:403–408

Lim KB, Chung J-D, Van Kronenburg BCE, Ramanna MS, de Jong JH, Van Tuyl JM (2000) Introgression of *Lilium rubellum* Baker chromosomes into *L. longiflorum* Thunb.: a genome painting study of the F1 hybrid, BC1 and BC2 progenies. Chromosom Res 8:119–125

Lim KB, Ramanna MS, de Jong JH, Jacobsen E, Van Tuyl JM (2001a) Indeterminate meiotic restitution (IMR): a novel type of meiotic nuclear restitution mechanism detected in interspecific lily hybrids by GISH. Theor Appl Genet 103:219–230

Lim KB, Ramanna MS, Van Tuyl JM (2001b) Comparison of homoeologous recombination frequency between mitotic and meiotic polyploidization in BC1 progeny of interspecific lily hybrids. Acta Hortic 552:65–72

Lim KB, Wennekes J, De Jong JH, Jacobsen E, Van Tuyl JM (2001c) Karyotype analysis of *Lilium longiflorum* and *Lilium rubellum* by chromosome banding and fluorescence in situ hybridization. Genome 44:911–918

Lim KB, Ramanna MS, Jacobsen E, van Tuyl JM (2003) Evaluation of BC2 progenies derived from 3x-2x and 3x-4x crosses of *Lilium* hybrids: a GISH analysis. Theor Appl Genet 106:568–574

Lim KB, Barba-Gonzalez R, Zhou S, Ramanna MS, van Tuyl JM (2008) Interspecific hybridization in Lily (*Lilium*): taxonomic and commercial aspects of using species hybrids in breeding. In: Teixeira da Silva JA (ed) Floriculture, ornamental and plant biotechnology. Advances and topical issues, vol V. Global Science Books, Isleworth, UK, pp 146–155

Luo JR, Ramanna MS, Arens P, Niu LX, Van Tuyl JM (2012) GISH analyses of backcross progenies of two *Lilium* species hybrids and their relevance to breeding. J Hortic Sci Biotechnol 87:654–660

Luo JR, Van Tuyl JM, Arens P, Niu LX (2013) Cytogenetic studies on meiotic chromosome behaviors in sterile Oriental x Trumpet lily. Genet Mol Res 12(4):6673–6684

Luo JR, Arens P, Niu LX, Van Tuyl JM (2016) Induction of viable 2n pollen in sterile Oriental × Trumpet *Lilium* hybrids. J Hortic Sci Biotechnol 91(3):258–263

Mace E, Xia L, Jordan D, Halloran K, Parh D, Huttner E, Wenzl P, Kilian A (2008) DArT markers: diversity analyses and mapping in *Sorghum bicolor*. BMC Genomics 9:26

Makino T (1914) Observations on the Flora of Japan. Bot Mag Tokyo 28:20–30

Marasek A, Orlikowska T (2003) Morphology of chromosomes of nine lily genotypes and usefulness of morphological markers for hybrid verification. Acta Biol Cracov Ser Bot 45:159–168

Marasek A, Orlikowska T, Hasterok R (2004) The use of chromosomal markers linked with nucleoli organisers for F1 hybrid verification in *Lilium*. Acta Hortic 61:77–82

Marasek A, Sliwinska E, Orlikowska T (2005) Cytogenetic analysis of eight lily genotypes. Caryologia 59:359–366

Matsumoto M (1992) The present situation of commercial cultivation and lily breeding in Japan. Lily Yearb North Am Lily Soc 45:7–12

McRae EA (1991) Back to a true line. Lily Yearb North Am Lily Soc 44:85–89

McRae EA (1998) Lilies: a guide for growers and collectors. Timber Press, Portland, OR

Muratović E, Bogunic´ F, Soljan D, Siljak-Yakovlev S (2005) Does *Lilium bosniacum* merit species rank? A classical and molecular-cytogenetic analysis. Plant Syst Evol 252:97–109

Myodo H (1962) Experimental studies on the sterility of some *Lilium* species. J Fac Agric Hokkaido Univ 52:71–122

Myodo H (1975) Successful setting and culture of hybrid embryos between remote specie of the genus Lilium. Lily Yearb North Am Lily Soc 30:7–17

Myodo H, Asano Y (1977) Lily hybrids newly obtained by the technique combining cut-style pollination with embryo culture. Lily Yearb North Am Lily Soc 30:7–17

Nishikawa T, Okazaki K, Uchino T, Arakawa K, Nagamine T (1999) A molecular phylogeny of *Lilium* in the internal transcribed spacer region of nuclear ribosomal DNA. J Mol Evol 49:238–249

Nishikawa T, Okazaki K, Arakawa K, Nagamine T (2001) Phylogenetic analysis of section *Sinomartagon* in genus *Lilium* using sequences of the internal transcribed spacer region in nuclear ribosomal DNA. Breed Sci 51:39–46

Nishikawa T, Okazaki K, Nagamine T (2002) Phylogenetic relationships among *Lilium auratum* Lindley, *L. auratum* var. *platyphyllum* Baker and *L. rubellum* Baker based on three spacer regions in chloroplast DNA. Breed Sci 52:207–213

Noda S (1978) Chromosomes of diploid and triploid forms found in the natural population of tiger lily in Tsushima. Bot Mag Tokyo 91:279–283

Noda S (1991) Chromosomal variation and evolution in the genus Lilium. In: Tsuchiya T, Gupta PK (eds) Chromosome engineering in plants: genetics, breeding, evolution. Part B. Elsevier, Amsterdam, pp p507–p524

Noda S, Schmitzer E (1992) Natural occurrence of triploid *Lilium bulbiferum* native to Europe. Lily Yearb North Am Lily Soc 43:78–81

Obata Y, Niimi Y, Nakano M, Okazaki K, Miyajima I (2000) Interspecific hybrids between Lilium nobilissimum and L. regale produced via ovules-with-placenta-tissue culture. Sci Hortic 84:191–204

Ohri D, Khoshoo TN (1983) Cytogenetics of garden Gladiolus. Origin and evolution of ornamental taxa. IV. Proc Indian Natl Sci Acad 3:279–294

Okazaki K, Asano Y, Oosawa K (1994) Interspecific hybrids between *Lilium*-Oriental hybrid and L-Asiatic hybrid produced by embryo culture with revised media. Breed Sci 44:59–64

Okubo H (2014) History of *Lilium* species in Asia. Acta Hortic 1027:11–26

Peruzzi L (2016) A new infrafamilial taxonomic setting for Liliaceae, with a key to genera and tribes. Plant Biosyst 150:1341–1347

Peruzzi L, Leitch IJ, Caparelli KF (2009) Chromosome diversity and evolution in Liliaceae. Ann Bot 103:459–475

Pelkonen V, Pirttila A (2012) Taxonomy and phylogeny of the genus *Lilium*. Floricult Ornamental Biotechnol 6:1–8

Ramanna MS, Jacobsen E (2003) Relevance of sexual polyploidization for crop improvement – a review. Euphytica 133:3–18

Rhee H-K, Kim K-S (2008) Interspecific hybridization and polyploidization in lily breeding. Acta Hortic 766:441–445

Rix M, Frank E, Webster G (2001) Fritillaria: a revised classification: together with an updated list of species. Fritillaria Group of the Alpine Garden Society, Edinburgh, UK

Rønsted N, Law S, Thornton H, Fay MF, Chase MW (2005) Molecular phylogenetic evidence for the monophyly of *Fritillaria* and *Lilium* (*Liliaceae; Liliales*) and the infrageneric classification of *Fritillaria*. Mol Phylogenet Evol 35:509–527

Schenk PC (1990) Modern trends in lily breeding. In: Hayward A-F (ed) Lilies and related plants supplement: the proceedings of the 5th international Lily conference, 18–20 July 1989. Modern Trends in Lily Breeding, London, pp 41–49

Schwarzacher T, Leitch AR, Bennett MD, Heslop-Harrison JS (1989) In situ localization of parental genomes in a wide hybrid. Ann Bot 64:315

Sealy JR (1983) A revision of the genus *Nomocharis* Franchet. Bot J Linn Soc 87:285–323

Shahin A (2012) Development of genomic resources for ornamental lilies (Lilium L.). PhD thesis, Wageningen University & Research

Shahin A, Arens P, Van Heusden AW, Van der Linden G, Van Kaauwen M, Khan N, Schouten H, Van de Weg WE, Visser RGF, Van Tuyl JM (2010) Genetic mapping in *Lilium*: mapping of major genes and QTL for several ornamental traits and disease resistances. Plant Breed 130(3):372–382

Shahin A, Van Gurp T, Peters SA, Visser RGF, Van Tuyl JM, Arens P (2012a) SNP markers retrieval for a non-model species: a practical approach. BMC Res Notes 5:79

Shahin A, van Kaauwen M, Esselink D, Bargsten J, van Tuyl J, Visser R, Arens P (2012b) Generation and analysis of expressed sequence tags in the extreme large genomes Lilium and Tulipa. BMC Genomics 13:640

Shahin A, Arens P, Van de Weg WE, Van Tuyl JM (2014) Molecular markers as a tool for parental selection for breeding in Lilium. NALS Yearb 63:106–112

Siljak-Yakovlev S, Peccenini S, Muratović E, Zoldoš V, Robin O, Vallès J (2003) Chromosomal differentiation and genome size in three European mountain *Lilium* species. Plant Syst Evol 236:165–173

Smyth DR (1999) *Lilium* chromosomes. Lily Yearb North Am Lily Soc 52:66–76

Smyth DR, Kongusuwan K (1980) C-banding in Lily chromosomes, and their use in identification of hybrids. Lily Yearb North Am Lily Soc 33:83–86

Smyth DR, Kongusuwan K, Wisudharomn S (1989) A survey of C-band patterns in chromosomes of *Lilium* (*Liliaceae*). Plant Syst Evol 163:53–69

Soltis PS, Soltis DE (2000) The role of genetic and genomic attributes in the success of polyploids. Proc Natl Acad Sci USA 97:7051–7057

Stern H, Hotta Y (1977) Biochemistry of meiosis. In: Darlington CD (ed) Meiosis in perspective, vol 277. Philosophical Transactions Royal Society, London, pp 277–294

Straathof TP, Van Tuyl JM (1994) Genetic variation in resistance to *Fusarium oxysporum* f. sp. *lilii* in the genus *Lilium*. Ann Appl Biol 125:61–72

Sultana S, Lee SH, Bang JW, Choi HW (2010) Physical mapping of rRNA gene loci and inter-specific relationships in wild *Lilium* distributed in Korea. J Plant Biol 53:433–443

Stewart RN (1943) Occurrence of aneuploids in *Lilium*. Bot Gaz 105:620–626

Stewart RN (1947) The morphology of somatic chromosomes in *Lilium*. Am J Bot 34:9–26

Truong NX, Kim JY, Rai R, Kim JH, Kim NS, Wakana A (2015) Karyotype analysis of Korean *Lilium maximowiczii* Regal populations. J Fac Agric Kyushu Univ 60(2):315–322

Turrill W (1950) Character combinations and distribution in the genus Fritillaria and allied genera. Evolution 4:1–6

Van Creij MGM, Van Raamsdonk LWD, Van Tuyl JM (1992) Wide interspecific hybridization of *Lilium*: preliminary results of the application of pollination and embryo-rescue methods. Lily Yearb North Am Lily Soc 43:28–37

Van Heusden AW, Jongerius MC, Van Tuyl JM, Straathof TP, Mes JJ (2002) Molecular assisted breeding for disease resistance in lily. Acta Hortic 572:131–138

Van der Linden CG, Wouters DCAE, Mihalka V, Kochieva EZ, Smulders MJM, Vosman B (2004) Efficient targeting of plant disease resistance loci using NBS profiling. Theor Appl Genet 109:384–393

Van Tuyl JM (1980) Lily breeding research at IVT in Wageningen. Lily Yearb North Am Lily Soc 33:75–92

Van Tuyl JM (1990) Research on mitotic and meiotic polyploidization in lily breeding. Herbertia 45:97–103

Van Tuyl JM, Arens P (2011) Lilium: breeding history of the modern cultivar assortment. Acta Hortic 900:223–230

Van Tuyl JM, Van Holsteijn HCM (1996) Lily breeding research in the Netherlands. Acta Hortic 414:35–45

Van Tuyl JM, Marcucci MC, Visser T (1982) Pollen and pollination experiments. VII. The effect of pollen treatment and application method on incompatibility and incongruity in *Lilium*. Euphytica 31:613–619

Van Tuyl JM, Keijzer CJ, Wilms HJ, Kwakkenbos AAM (1988) Interspecific hybridization between *Lilium longiflorum* and the white Asiatic hybrid ‘Mont Blanc’. Lily Yearb North Am Lily Soc 41:103–111

Van Tuyl JM, Van Diën MGM, Van Creij TCM, Van Kleinwee JF, Bino RJ (1991) Application of in vitro pollination, ovary culture, ovule culture and embryo rescue for overcoming incongruity barriers in interspecific *Lilium* crosses. Plant Sci 74:115–126

Van Tuyl JM, van Dijken HS, Chi HS, Lim K-B (2000) Breakthroughs in interspecific hybridization of lily. Acta Hortic 508:83–88

Van Tuyl JM, Chung M-Y, Chung J-D, Lim K-B (2002) Introgression studies using GISH in interspecific *Lilium* hybrids of *L. longiflorum* × Asiatic, *L. longiflorum* × *L. rubellum* and *L. auratum* × *Lilium henryi*. Lily Yearb North Am Lily Soc 55:17–22

Van Tuyl JM, Khan N, Xie S, Marasek-Ciolakowska A, Lim K-B, Barba-Gonzalez R (2011) *Lilium*. In: Kole C (ed) Wild Crop Relatives: Genomic and Breeding Resources. Plantation and Ornamental Crops. Springer, Heidelberg, Dordrecht, London, New York, pp 161–183

Von Kalm L, Smyth DR (1984) Ribosomal RNA genes and the substructure of nucleolar organizing regions in *Lilium*. Genome 26:158–166

Wada K (1951) *Lilium formolongi*. Lily Yearb North Am Lily Soc 4:73–76

Wang F, Tang J (1980) *Lilium* L. In: Liang SY (ed) Flora Republicae Popularis Sinicae, vol 14. Science Press, Beijing, pp 116–157

Wang J, Huang L, Bao M-Z, Liu G-F (2009) Production of interspecific hybrids between *Lilium longiflorum* and *L. lophophorum* var. *Linearifolium* via ovule culture at early stage. Euphytica 167:45–55

Wang X, Xie S, Zhang Y, Niu L (2012) Chromosomal analysis and mapping of ribosomal genes by Fluorescence in situ hybridization (FISH) in four endemic lily species (*Lilium*) in Qinling Mountains. China Pak J Bot 44(4):1319–1323

Wang Q, Wang J, Zhang Y, Zhang Y, Xu S, Lu Y (2015) The application of fluorescence in situ hybridization in different ploidy levels cross-breeding of lily. PLoS One 10(5):e0126899. https://doi.org/10.1371/journal.pone.0126899

Wendel JF (2000) Genome evolution in polyploids. Plant Mol Biol 42:225–249

Xie S (2012) A molecular cytogenetic analysis of chromosome behavior in *Lilium* hybrids. PhD thesis, Wageningen University & Research, 115 pp

Xie S, Khan N, Ramanna MS, Niu L, Marasek-Ciolakowska A, Arens P, Van Tuyl JM (2010) An assessment of chromosomal rearrangements in neopolyploids of *Lilium* hybrids. Genome 53:439–446

Xie S, Marasek-Ciolakowska A, Ramanna MS, Arens P, Visser RGF, Van Tuyl JM (2014a) Characterization of B chromosomes in *Lilium* hybrids through GISH and FISH. Plant Syst Evol 300(8):1771–1777

Xie S, Ramanna MS, Arens P, Van Tuyl JM (2014b) GISH investigation of crossover events during meiosis of interspecific hybrids of lily. Acta Hortic 1027:143–148

Yabuya T (1991) Chromosome associations and crossability with *Iris ensata* Thunb. In induced amphidiploids of *I. laevigata* Fisch. × *I. ensata*. Euphytica 55:85–90

Yamagishi M (2013) How genes paint lily flowers: regulation of colouration and pigmentation patterning. Sci Hortic 163:27–36

Zhou S (2007) Intergenomic recombination and introgression breeding in Longiflorum × Asiatic lilies. PhD thesis, University Wageningen, 111 pp

Zhou S, Ramanna MS, Visser RGF, Van Tuyl JM (2008a) Analysis of the meiosis in the F1 hybrids of Longiflorum × Asiatic (LA) of lilies (*Lilium*) using genomic in situ hybridization. J Genet Genomics 35(11):687–695

Zhou S, Ramanna MS, Visser RGF, Van Tuyl JM (2008b) Genome composition of triploid lily cultivars derived from sexual polyploidization of Longiflorum × Asiatic hybrids (*Lilium*). Euphytica 160:207–215

Zhou S, Zhong L, Zhang L, Xu Z, Liu X, Li K, Zhou G (2015) Study on the homology of the genomes of tetraploid Asiatic lilies (*Lilium*) using FISH (fluorescence in situ hybridization). Genome 58(11):453–461

Zonneveld BJM, Leitch IJ, Bennett MD (2005) First nuclear DNA amounts in more than 300 angiosperms. Ann Bot 6:229–244

第21章 补 血 草

Ed Morgan and Keith Funnell

摘要 补血草属（*Limonium*）少量物种产生的杂种和品种是广为人知的鲜切花与干花植物。在不断追求新颖性的花卉行业中，人们不断寻找创造新品种的机会。尽管补血草属植物有 150 个种以上，但在商业中应用的仅有 15～20 个种或它们的杂种。由于对补血草属植物的分类研究较少，对其物种数量的估计存在很大误差。补血草属中许多物种存在自然杂交现象，这也是育种家数十年来成功利用的育种方式。但是，在补血草的创新育种中还有其他方法可以应用，并可为育种家、种植者和消费者带来益处。拓宽补血草常规育种方案的多项技术被报道，但公开发表的论文相对较少。在本章中，我们概述了如何利用补血草属（*Limonium*）公认的多样性为国际切花市场引入新颖花卉的方法，还介绍了补血草属的多样性，并就其在育种体系中的应用进行了简要讨论。同时，本章还综述了体外培养技术在补血草属中的应用，包括杂交、胚拯救、倍性操纵、诱变和分子育种技术。

关键词 补血草属；星辰花；育种；胚培养；原生质体；植物遗传转化；诱变；种间杂交

21.1 引 言

21.1.1 概述

补血草属（*Limonium*，又名星辰花或海洋薰衣草）是世界观赏植物产业中广为人知的花卉，其主要作为切花植物，而且在盆栽和景观市场也占有重要地位。补血草以观花为主，特别是其真花凋谢后仍然在植物上保持开放和着色状态的花萼部分。因此，补血草也常被用于制作干花。

目前被商业化利用的杂种和品种的亲本来源于少量补血草属的种。然而结合鲜为人知或者新鉴定的种的特征能增加创造出更多新颖杂交组合的机会。在栽培上常用的有15～20 个种，其中最常用的种包括深波叶补血草（*L. sinuatum*，最常见）、阔叶补血草（*L. latifolium*）和 *L. bellidifolium* 的杂种，以及蓝烟小星辰花（*L. perezii*）和中华补血草（*L. sinense*）的杂种。阿尔泰补血草（*L. altaica*）、*L. bonduelli*、*L. ferulaceum*、大叶补血草（*L. gmelinii*）和耳叶补血草（*L. otolepis*）等其他种也可以直接使用或纳入育种计划。

E. Morgan (🖂), K. Funnell
The New Zealand Institute for Plant & Food Research Ltd, Palmerston North, New Zealand
e-mail: ed.morgan@plantandfood.co.nz

有时也利用“栽培植物分类法”来命名补血草属植物。例如，“*L. altaica*”不是被认同的植物学名称或同义名（如 http://www.theplantlist.org/tpl1.1/record/kew-2599634）。通常育种公司是根据补血草植物的形态特征，将其切花品种主要分为 4 种类型：星辰花（statice）、阿尔泰（altaica）、中华（sinensis）和杂种（基于杂种补血草）（Burchi et al. 2006）。然而事实上，绝大部分应用于育种工作的特定遗传来源和亲本并没有被公开。

因补血草在市场上备受关注，可以找到相当多的描述其繁殖、栽培及采后生理的知识。快速繁殖技术已经公开发表，用于进行其植物材料的扩繁（Hosni et al. 2000）或者保存（Casazza et al. 2002）。关于补血草栽培的各个方面有着大量的报道。Biruk 等（2013）阐述了品种和种植密度对产量与质量的影响。Chen 等（2010）研究了光照和温度对开花的影响。Doi 和 Reid（1995）、Shimamura 等（1997）、Burge 等（1998）、Ichimura（1998）和 Philosoph-Hadas 等（2005）都发表了对补血草切花的花枝采后生理研究的文章。此外，还有许多关于描述补血草生长的综述类文章，其中包括 Dole 和 Wilkins（2005），以及育种公司为种植者提供的信息。

21.1.2 补血草国际市场

补血草是全球销量最高的 20 种切花之一，但是尚无法获得关于植株销售、花枝销售等一致的年同比数据。数据收集和分类的方法似乎在经常变化，尽管可以通过互联网获取到大量的数据资源，但此处呈现的数据不应视为对销售数据的全面统计值而应视为全球价值的一个指标。因为无法查找到可以进行直接比较的年同比数据或者市场同比数据，所以很难构建出一个全面的、长期的销售数据图。

在花束中，补血草通常被作为低成本的“填充花材”，与满天星（*Gypsophila*）等植物共同竞争市场份额。FloraHolland（荷兰）的营业额（时钟销售）数据提供了一种便利的方法来计算一个市场在 4 年内的花枝销售量和销售额。花枝销售量从 2010 年至 2014 年，由 5000 万枝增长至 7200 万枝，分别代表了销售额 1080 万欧元和 1500 万欧元。在 2013 年、2014 年和 2015 年，补血草在荷兰进口产品中排名前十，其进口额在 2013 年和 2014 年为 1000 万欧元，到 2015 年增加至 1100 万欧元。虽然没有得到全部生产中心的综合数据，但在某些国家（如墨西哥和日本），其生产面积呈显著上升的趋势。2006～2008 年，日本的补血草花枝生产量在 1.18 亿～1.27 亿枝（Hanks 2015）。

就育种角度而言，衡量市场活力更好的方法或许是考虑市场上能使用的品种数量。表 21.1 中所示不同杂交类型的补血草切花品种数量只作说明之用，并不是对市场的全面分析，因为还有其他育种公司提供的品种超出所列范围。

由于品种保护申请和品种权授予涉及全球 10 个地区/辖区，因此补血草是一个国际性的商品植物（表 21.2）。每年的授权记录清晰地反映了新品种引进工作已经开展了很长时间并且还在不断地进行中（图 21.1）。从 1951 年到 2016 年中期，约 680 个品种获得品种权保护，截至 2016 年底，又有 209 项申请品种权保护，在撰写这篇文章前，这些品种没有或尚未获得品种权保护（http://www.upov.int/portal/index.html.en）。如果随意选择对 2010～2016 年（含）这一相对较近的时间段做进一步分析，则有 170 项品种权保护申请，61 项被授权（表 21.2）。在此期间，日本提交了 120 项品种权保护申请，欧盟提交了 29 项。这标志着新品种正被持续地引进，2010～2016 年，平均每年有 20 个新品种获得授权（图 21.1）。

表 21.1 一系列育种公司销售的补血草品种数量。这些数据是在 2018 年 1 月从相关公司的网站上收集的

公司名称	品种数量
Ball Colombia	16
Danziger	18
Hilverda	34
Miyoshi	24
Royal Van Zanten	29

表 21.2 从 2010 年初到 2016 年底补血草品种保护的申请和授权。在这个阶段，只有 6 个品种在不同地区获得了两个或者更多的品种权申请。这些数据从 PLUTO 数据库和 UPOV 网站获得（http://www.upov.int/portal/index.html.en）。未获得关于是否某个基因型在其他地区以其他名称进行保护或销售的信息

管辖区	申请项目数	授权项目数
澳大利亚	1	
巴西	2	
厄瓜多尔	1	
以色列	3	2
日本	120	38
肯尼亚	6	
韩国	2	2
墨西哥	4	3
欧盟	29	14
美国	2	2
总计	170	61

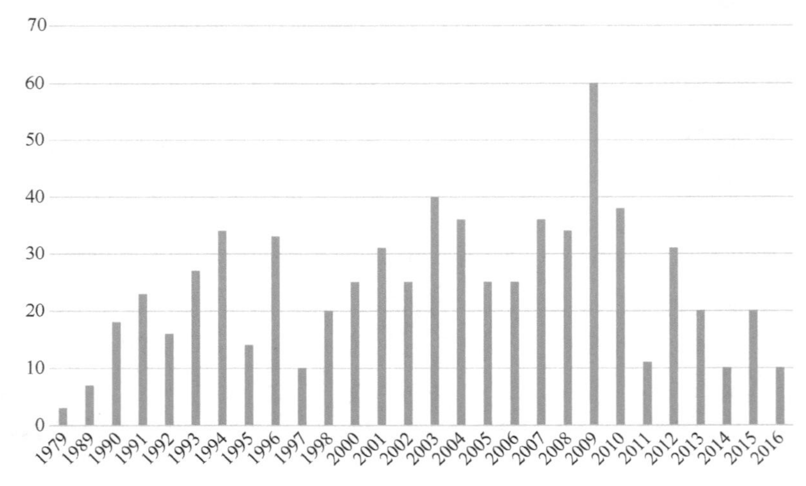

图 21.1 根据 UPOV 记录的每年被授予品种权的补血草栽培品种数量。1979 年的数据是直到 1979 年注册的品种数量；1989 年的数据是 1980～1989 年累计注册的品种数量。这些被授权的品种数据来自 PLUTO 数据库（http://www.upov.int/portal/index.html.en）覆盖的任何地区，且任何品种名称在此图中仅出现一次

日本大量的申请数量似乎有些异常（表 21.2），尤其是对比其他多个地区获得授权的基因型数量很少的情况。据推测，日本大量超出经济学范畴的品种权保护申请和授予是有原因的。Hanks（2015）提到日本补血草生产时指出，尽管日本总的生产面积较小（17 000hm^2），但是与其他国家补血草的生产面积相比较，日本补血草 200hm^2 的生产面积已经很可观了。除了育种家权益（PBR）或植物品种权（PVR）或专利以外，还有其他机制来保护新品种，可能许多品种在公司特定的战略区域受到保护，但是在其他地区经许可可以进行生产和销售。育种者很有可能在市场上销售了比这里认定的更多的基因型，是因为在不同地区使用了其他的名称。这或许解释了在多个管辖区只有相对少数几个品种注册在同一个名称下的现象。在这方面，著名的‘蓝雾’（‘Misty Blue’）是唯一在 7 个地区受到保护的品种。在 570 个品种中大约有 80 个品种在一个以上的地点受到保护。

21.2 补 血 草 属

21.2.1 概述

补血草属（*Limonium* Mill.）为白花丹科（Plumbaginaceae）植物，自然分布于欧洲、美洲、亚洲、非洲及澳大利亚（Baker 1948，1953b）。因此该物种常见于热带和温带地区。补血草通用名称之一为海洋薰衣草（sea lavender），反映了它常生长于沿海环境。该属的许多物种被发现于海岸附近的盐沼中或者远离海岸的盐土、石膏土或碱性土壤中。补血草物种的生长习性和生长周期各不相同，包括一年生草本（如 *L. lobatum*、*L. sinuatum* ssp. *beaumierianum*）、多年生草本[如深波叶补血草（*L. sinuatum* ssp. *sinuatum*）]，以及高达 2m 的灌木（如 *L. arborescens*）（Karis 2004）。

叶基生，呈莲座状，叶片单生，全缘或浅裂（Karis 2004）。小花（长 4～10mm）形成圆锥花序或伞状花序。花茎只含鳞片叶（苞片），花冠具有持久不落的 5 裂花萼、5 个花瓣、雄蕊及雌蕊。花冠颜色通常是从粉红色或紫罗兰色到紫色，尽管在深波叶补血草中颜色范围延伸至红色和蓝色，在黄花补血草（*L. aureum*）中引入了黄色。在许多物种当中，短时间存留的花瓣颜色仅有白色或黄色，然而在如 *L. peregrinum* 和 *L. purpuratum* 等物种中，花瓣颜色呈现粉红色。授粉成功后可结出含有单粒种子的果实。

补血草属的分类很复杂，物种数量估计在 150 多种（Baker 1953b）到 400 种（Khan et al. 2012）。被报道的物种数量被描述为“猜测的”（Palacios et al. 2000）。频繁的种间杂交，有时还伴随着多倍体、非整倍体、孤雌生殖，使得该属的分类困难重重（Lledo et al. 2005；Cowan et al. 1998）。对新物种的频繁描述接连出现，如 *L. maritimum*（Cortinhas et al. 2015），或许最好将这些物种认为是“小种”（“microspecies”）（Cowan et al. 1998），但该术语其实是用于描述分布和种群数量都非常有限的物种。Cowan 等（1998）认为大多数的这些小种很可能是进行无融合生殖的。例如，*L. perplexum* 是一种三倍体无融合生殖植物，仅出现在沿海悬崖的一个小露头上（40m^2），在保护工作开始前，其报道的数量在 19～383 个（Khan et al. 2012）。

经常使用的异名也使得补血草分类工作更为复杂。例如，阿尔泰补血草（*L. altaica*）和 *L. fortunei* 均不是被认可的种名。反复搜索 *L. altaica* 或 *Statice altaica* G. Don 表明这既不是被接受的名称也不是异名（见 http://www.theplantlist.org/tpl1.1/record/ kew-2599634）。

利用新的分子技术对整个属的亲缘关系进行鉴别和修正很有必要。预计此类工作应该能够引起关于所有的植物属是如何构成的讨论，尤其像一些"种"（"小种"）的分布非常有限，而且似乎只能通过无融合生殖得以延续。除了在这里讨论的之外，有性生殖的物种与完全或部分无融合生殖的物种是否完全等价又是一个单独的问题。

21.2.2 生殖生物学

许多学者都曾研究过补血草的生殖生物学（Baker 1948，1953a，1953b；Dulberger 1975）。有性生殖和无融合生殖物种均被报道（Baker 1953a；Cowan et al. 1998），其中无融合生殖仅存在于三个补血草亚组中（Baker 1953a）。

大多数有性生殖物种表现为异型孢子体自交不亲和（Baker 1966）。二态性柱头（玉米状/乳头状）的物种呈现出花粉二态性（A/B 花粉型）。在二态性物种中，根据柱头的外观给不同的柱头类型进行命名：具有玉米状柱头的植物产生 A 型花粉，具有乳头状柱头的植物产生 B 型花粉（Baker 1948，1953a）。A 型花粉与乳头状的柱头杂交亲和，B 型花粉与玉米状的柱头杂交亲和。A 型花粉与 B 型花粉在花粉外壁结构上有所不同（Baker 1948）。

许多有性生殖物种如 *L. peregrinum* 是单态性的。单态性的物种具有头状柱头（Baker 1953a），尽管可以通过自花授粉进行繁殖，但是由这种受精方式产生的种子大多数（70%）不能正常成熟（Burge and Morgan 1993）。

如果使用合适的花粉、柱头组合，种间杂交授粉是可能的（Baker 1966）。有学者对多个物种之间的异花授粉也进行了研究（Zhang 1995）。在组间杂交中，花粉管可进入子房，如蓝烟小星辰花 × 深波叶补血草（*Pteroclados* 组）和中华补血草 × 黄花补血草[宽檐组（*Playthymenium*），都属于 Chrysantheae 亚组]。在某些组间杂交组合中，可以看到花粉管进入卵巢，但频率较低；更常见的是，花粉管的生长在花柱中就被阻止了。当用作雄性亲本时，*L. peregrinum*（单态性）花粉管偶尔会穿过二态性物种（如蓝烟小星辰花和中华补血草）的子房。在许多物种中观察到的花和花粉二态性意味着无须在人工授粉前对小而寿命短的花做去雄处理。

21.2.3 染色体数目

补血草属基本染色体数目为 x = 6、7、8 和 9（Darlington and Wylie 1956）。多倍体广泛地存在于该属中，可能是由未减数配子形成导致的。这种现象在蓝烟小星辰花和深波叶补血草的杂种中可以看到，其中从三倍体幼苗群体中鉴定出具有六倍体核 DNA 含量的幼苗（数据未发表）。如上所述，有许多非整倍体或多倍体物种被报道是通过无融合生殖种子得以延续的（Cortinhas et al. 2015；Lledo et al. 2005；Laguna et al. 2016）。

21.3　补血草的体外繁殖技术

为了引入新颖性，多种体外技术已在育种中应用，其中包括胚培养、染色体加倍、诱变及遗传转化。

21.3.1　杂交育种

杂交通常与染色体加倍结合使用，是一种将新颖性引入观赏植物的广为人知的方法，并且广泛应用于补血草育种。目前有许多关于利用杂交手段培育新品种的报道。种间杂交面临的挑战可能是后代不育和不“适应”，表现出一种或多种不育或适应性不良的症状，如种子败育、生长不良、植株白化或叶片失色（Morgan et al. 2011）。随着物种之间系统发育距离的增加，产生杂种的机会减少，产生不适应后代的可能性也增加。除了通常意料之中的受精前和受精后杂交障碍之外，在预期种子亲本中的无融合生殖进一步阻碍了杂交的进行（未公开数据）。

育种者长期对补血草属相关种进行杂交，培育出了许多商业化品种。例如，Harada 提供了一些种间杂种及其育种背景的列表（Harada 1992）。鉴于目前在商业中使用的近缘种的范围相对较小，所提供的大多数品种在园艺学分类中分为：阿尔泰、杂种、中华及星辰花（或深波叶补血草）。这些类别包括以下种和杂种（Burchi et al. 2006）。

- ‘阿尔泰’包括大叶补血草、阿尔泰补血草、杂种补血草、*L. serotinum* 种间和种内杂种，如‘艾米尔’（‘Emille’）品种。
- ‘杂种’包括杂种补血草和耳叶补血草的杂交种，或者杂种补血草和 *L. caspia* 的杂交种，如‘蓝雾’和‘Beltlaard’。
- ‘星辰花’是深波叶补血草的种和杂种。
- ‘中华’包括中华补血草、*L. fortunei*、*L. tetragonum*、黄花补血草之间的杂种。

尽管许多种之间能自然杂交，但是使用胚培养获得种间杂种为育种者创造了新的拓宽遗传基础的机会，值得注意的是，历史上的许多商业品种均来源于近缘种。Morgan 和同事成功地利用胚珠培养技术创造了 *L. peregrinum* × *L. purpuratum* 的杂种（Morgan et al. 1995）。利用胚培养技术获得了蓝烟小星辰花 × 深波叶补血草的杂交种（Morgan et al. 1998）。使用胚培养的优点是它是一种相对直接的技术，它增加了获得杂种植物的可能性，特别是获得较远缘的杂种。该技术的另一个间接优势是，它提供了健康度较高的植物，如果打算将植株进行国际商业化，健康度可能是一个重要的考虑因素，因商品必然需要满足植物检疫要求。

通过胚珠培养技术，*L. peregrinum* × *L. purpuratum* 杂交产生了许多杂种（Morgan et al. 1995），其中之一‘Chorus Magenta’获得新品种权。这些植株生长良好，至少部分（雌性）可育，并产生了许多回交后代。‘Chorus Magenta’生长在除新西兰以外的几个国家。不幸的是，在投放市场的大约 2 年内，新西兰种植者开始报道高水平的死亡率，植物死于以前未报道的拟茎点霉属（*Phomopsis*）物种（Harvey et al. 2000）。尽管制定了控制措施（Long et al. 2001）和母株管理协议（Funnell et al. 2003），但是在这些物种之间没有进一步开展杂交工作。

蓝烟小星辰花和深波叶补血草杂交产生了许多不育的杂种植株（Morgan et al. 1998）。只有约 10%人工授粉的花产生增大的胚珠。授粉后约 14 天，胚珠内的胚被转移到体外培养。此时，只有小部分足够大的胚珠在胚囊中可见（在显微镜下）。转移到体外培养并从胚囊中取出后，胚迅速生长了约 2 周，但达到约 10mm 的长度后频繁死亡。解决这种停止生长和继而死亡问题的方法是首先将胚转移至体外培养数周，然后先放在含有 TDZ 的琼脂培养基上刺激 24h（Seelye et al. 1994），再放回原来的胚培养基上。这样能诱导愈伤组织的形成，并可以从中生长出芽，随后产生新的植株。一旦转移到温室中，F_1 代杂交种似乎没有什么价值，因为它们呈浅绿色，生长缓慢且不育。但是，它们确实可以呈现出不同于深波叶补血草亲本的花色范围。在体外利用体细胞进行染色体加倍后（见第 21.3.2 节），在四倍体杂种中至少恢复了部分育性。随后杂交工作在蓝烟小星辰花和四倍体杂种之间开展。利用已建立的实验方案（Morgan et al. 1998），胚珠内的胚被转移到体外进行培养，所得的回交后代表现出一定的繁殖力。随后就产生了一系列补血草新品种，被命名为‘Sinzii’在市场上进行销售。这些新的杂种将诸如蓝烟小星辰花的花茎分枝（图 21.2）和茎长与深波叶补血草的花色等特征结合了起来。

图 21.2 蓝烟小星辰花（*Limonium perezii*）(a)、深波叶补血草（*L. sinuatum*）(b) 及蓝烟小星辰花的回交（BC1）杂种（c）的茎和叶，显示三种植物的不同分枝习性和叶片形状

关于原生质体再生成功的报道（Kunitake and Mii 1990）意味着体细胞杂交可能是未来育种的一种选择，特别是想尝试结合不育或无融合生殖植物时。然而，鉴于通过有性杂交可能获得丰富的多样性，尤其是如果利用了胚培养和染色体加倍，育种者或许不需要采取体细胞杂交这种方法。

21.3.2 多倍体育种

在许多补血草属物种中观察到的多倍体现象可能源于未减数配子的形成，从而使染色体加倍。蓝烟小星辰花和深波叶补血草的一个杂交植株的多倍体化就产生于未减数配子（未发表数据）。检测这个杂交植株是否为多倍体是因为它具有区别于其他植株的较厚、颜色较深的叶子。以下两个是关于诱导补血草染色体加倍的报道：一个是利用体外培养进行体细胞加倍，另一个是用秋水仙碱处理种子。

补血草属种间杂交的不育种被放在添加了 15.5mg/L 氨磺乐灵的改良 MS 培养基上生长 14 天，然后被转移至缺少氨磺乐灵的增殖培养基中进行进一步生长（Morgan et al. 2001）。生长 6 周后，新芽被转移到根诱导培养基中生长 6 周，然后从瓶中取出来。使用先前建立的实验方案，采用流式细胞仪对植物的多倍体进行筛选（Morgan et al. 1998）。二倍体杂种的平均核 DNA 含量通常约为 7.6pg DNA（Morgan et al. 1998），四倍体杂种的核 DNA 含量为 16.3pg DNA（Morgan et al. 2001）。在体外利用氨磺乐灵处理产生的四倍体植株生长至开花。花粉活力利用亚历山大染色法（Alexander 1969）进行评估，花粉被转移到相配的蓝烟小星辰花植株柱头后，利用荧光显微镜进行检查花粉萌发率（Kho and Baer 1968）。无菌 F_1 杂种含有 15 条染色体，花粉染色率小于 1%，核 DNA 含量为 7.6pg DNA。一个四倍体杂种含有 30 条染色体，花粉染色率超过 80%，核 DNA 含量为 16.3pg DNA。四倍体杂种已成功与蓝烟小星辰花和深波叶补血草进行双向回交，产生了新的杂种（Morgan et al. 2009）。

秋水仙碱处理种子是众所周知的染色体加倍方案。利用秋水仙碱处理 *Limonium bellidifolium* 的种子使其染色体加倍（Mori et al. 2016），多倍体植株利用流式细胞仪进行鉴定。在处理后恢复生长的植株中，有 2.5%～5%的染色体加倍，为四倍体或混倍体（$2x + 4x$），使用 0.05%秋水仙碱处理 72h 后能得到最高的染色体加倍率（5%）。与二倍体植株相比，四倍体植株表现出四倍体的典型特征，如叶片更宽更厚、花朵更大、气孔明显比二倍体植株大 1.1～1.5 倍（Mori et al. 2016）。

21.3.3 诱变育种

诱导或自发的突变偶尔也能为育种者提供机会。通过自发或诱导的突变可产生大量的观赏植物品种，在补血草中也不例外。‘高艾米尔’（‘Tall Emille’）和‘粉红艾米尔’（‘Pink Emille’）是‘艾米尔’（‘Emille’）自发突变后分别根据植物的高度或花朵的颜色选育得到的。‘海洋白’（‘Oceanic White’）是由 X 射线诱导的‘海洋蓝’（‘Oceanic Blue’）的白花突变品种，‘海洋蓝’是杂种补血草和 *L. bellidifolium* 的种间杂种。

补血草通常由组织培养技术进行繁殖。体细胞无性系变异，即在组织培养过程中产生的遗传变异，可以创造出更多的突变。尽管使用组织快繁技术产生了大量的补血草，然而使用该技术产生体细胞无性系变异的报道很少。原生质体再生有望为由单细胞获得的植株体细胞无性系变异的产生提供特殊的机会。然而，在蓝烟小星辰花的原生质体再生中几乎观察不到变化（Kunitake and Mii 1990），随后有学者鉴定了 56 种原生质体衍生植物的形态和细胞学特征（Kunitake et al. 1995）。与对照植物相比，一些植物的叶长/叶宽的值降低，茎长缩短。其中一株植物的花瓣异常且花粉育性低，但花萼或花瓣颜色没有差异。

不定芽的产生通常被认为是体细胞无性系变异的来源，特别是如果芽是由先前的愈伤组织阶段发育而来。不定芽可从体外培养的 *L. altaica*‘Emille’叶段中再生（Jeong et al. 2001）。Jeong 等（2001）发现芽的再生是直接的，即没有经过愈伤组织阶段。他们报道再生植物与亲本植物之间没有差异，并将其部分归因于新芽的直接再生。

利用 γ 射线诱变处理深波叶补血草（Cardarelli et al. 2002），可影响其叶片数量和茎长，但不影响茎产量或花色，而对于 *L. bellidifolium* 的种子，我们注意到第一代内没有表型变异（未公开数据）。化学诱变（叠氮钠）导致了植物大小和茎长的变化，虽然对花的产量没有影响，也没有提到花色的变化（Vitti et al. 2002）。

在这些报道中，突变表型的鉴定是基于极少量的植物的。在 20～30 年前，诱变育种显然已经被用在补血草中并产生了新颖的植株，如上文提到的‘海洋白’，但目前尚不清楚是否有任何育种者正在大种群中使用诱变来繁殖和寻求变异植株。我们也许遗漏了这部分育种者的信息。

21.3.4 遗传转化

随着新技术的迅速出现，如用于基因编辑的 CRISPR/Cas9 和最近宣布的用于简化遗传转化方法的磁转染，遗传转化为创造新颖植株提供了巨大的机会。直到最近，监管要求对遗传转化技术的应用提出了严重的限制；然而，似乎基因编辑可能受到较少的限制。鉴于许多补血草可以很轻松地进行体外培养操作，如下所述，有许多成功转化的报道也就不足为奇了，尽管并非全部针对出于观赏目的的物种。据推测，目前运用该方法培育补血草会在世界范围内被控制转基因商业化释放的法规所限制。

补血草的根癌农杆菌介导转化已经在许多物种和基因型中被报道。GUS 报告基因的稳定表达已在深波叶补血草的愈伤组织中得到证实（Kimizu et al. 2001）。农杆菌介导的转化技术还被用于在耳叶补血草和杂种补血草之间的无菌种间杂种中产生转基因芽（Mercuri et al. 2001）。T-DNA 片段带有发根农杆菌的 *rolA*、*rolB* 和 *rolC* 基因。转基因植物具有适合切花行业的早开花特性，但是植株矮化了。

Mercuri 等（2003）报道了大叶补血草的农杆菌介导转化，即转基因植物携带 *rolA*、*rolB* 和 *rolC*。其根据表型选择了三种基因型进行详细分析，“超级紧凑”、“紧凑”和“半紧凑”：分析显示每个基因型携带两个拷贝的 *rol* 基因。转基因植物表现出理想的观赏性状，如早花，但是像上面报道的耳叶补血草和杂种补血草的杂种一样，由于植株矮化，它们可能更适合盆栽市场。

尚未有将 CRISPR/Cas9 应用于补血草的报道。如果这些技术的使用确实在转基因生物的监管范围之外，至少在某些管辖权外，育种公司应该在不久的将来可看到应用这些新技术带来的益处。

21.4 结论性意见

补血草属植物的多样性体现在包括花的颜色、植株形态、高度及寿命中，为育种者提供了许多可以纳入杂交育种目标的特征。杂交育种一直是开发补血草新品种的重要方式，而且很可能会保持下去。尽管胚培养可用于从较远缘的杂交中获得杂种植株，但是利用这种方法成功引入感兴趣的性状的报道很少。在远缘杂种中经常观察到的一个特征是不育，当在一组补血草杂种中遇到时，可以使用经典的染色体加倍方法加以克服。回交杂种部分可育，可用于产生更多杂种。在无性繁殖植物（如补血草）中，不育实际上可以成为重要的特征，会减缓新品种进入竞争性育种计划的速度。

在补血草育种中可能未得到充分利用的技术是诱变育种。自发或诱导的突变在之前已经创造了新品种。利用体外技术从原生质体或愈伤组织中再生植物，使诱变育种成为可能。此外，补血草开花期短意味着根据表型筛选大量种群可能是一种经济有效的育种方法。鉴于许多补血草品种通过组织培养进行营养繁殖，利用不育基因型进行诱变育种可能会提供一种相对简单的产生新品种的方法。

尚未发现基于基因组学的育种报道。基因组学育种相关技术正迅速变得越来越便宜，但即便如此，对于补血草这种小作物，此类技术的成本可能仍然不划算。例如，研究人员要解决属的分类，才可能会开发一些育种公司随后便可以利用的基本工具。据报道，虽然植物遗传转化已在多个物种中开展，但是到目前为止，由于监管方面的障碍，很难为商业育种者证明其遗传转化的合法性。“基因编辑”的出现及与该技术相关的监管要求的变化（降低）可能意味着该方法对于育种者具有商业可行性。

对于补血草育种者来说有许多的育种工具，这些可用的技术将使努力提供更好品种的育种者和获得这些新品种的种植者受益。尽管为了满足回报利润的需要，育种者们将务实地选择可使用的育种工具组合，但是育种者可利用的种类繁多，育种工具广泛，这意味着生产者可以在未来的几年里获得令人惊喜的补血草新品种。

本章译者：

欧阳琳[1]，李　帆[2]

1. 中国农业科学院都市农业研究所，成都 610015

2. 云南省农业科学院花卉研究所，国家观赏园艺工程技术研究中心，云南省花卉育种重点实验室，昆明 650200

参 考 文 献

Alexander M (1969) Differential staining of aborted and nonaborted pollen. Stain Technol 44(3):117–122

Baker HG (1948) Dimorphism and monomorphism in the Plumbaginaceae .1. A survey of the family. Ann Bot 12(47):207–219

Baker HG (1953a) Dimorphism and monomorphism in the Plumbaginaceae. 2. Pollen and stigmata in the genus Limonium. Ann Bot 17(67):433–445

Baker HG (1953b) Dimorphism and monomorphism in the Plumbaginaceae .3. Correlation of geographical distribution patterns with dimorphism and monomorphism in Limonium. Ann Bot 17(68):615–627
Baker HG (1966) Evolution functioning and breakdown of heteromorphic incompatibility systems .I. Plumbaginaceae. Evolution 20(3):349
Biruk M, Nigussie K, Ali M (2013) Yield and quality of statice Limonium sinuatum (L.) Mill as affected by cultivars and planting densities. African J Plant Sci 7(11):528–537
Burchi G, Mercatelli E, Maletta M, Mercuri A, Bianchini C, Schiva T (2006) Results of a breeding activity on Limonium spp. In: Schiva T (ed) Proceedings of the 22nd international Eucarpia symposium section ornamentals: breeding for beauty. Acta horticulturae, vol 714. pp 43–49
Burge GK, Morgan ER (1993) Postpollination floral biology of Limonium-Perigrinum (Bergius). N Z J Crop Hortic Sci 21(4):337–341
Burge GK, Morgan ER, Konczak I, Seelye JF (1998) Postharvest characteristics of Limonium 'Chorus magenta' inflorescences. N Z J Crop Hortic Sci 26(2):135–142
Cardarelli M, Temperini M, Vitti D, Saccardo F (2002) Gamma rays induced mutagenesis in Limonium sinuatum. http://agris.fao.org/agris-search/search.do?recordID=IT200306280
Casazza G, Savona M, Carli S, Minuto L, Profumo P (2002) Micropropagation of Limonium cordatum (L.) Mill for conservation purposes. J Hortic Sci Biotechnol 77(5):541–545
Chen JY, Funnell KA, Morgan ER (2010) A model for scheduling flowering of a Limonium sinuatum × Limonium perezii hybrid. Hortscience 45(10):1441–1446
Cortinhas A, Erben M, Paes AP, Santo DE, Guara-Requena M, Caperta AD (2015) Taxonomic complexity in the halophyte Limonium vulgare and related taxa (Plumbaginaceae): insights from analysis of morphological, reproductive and karyological data. Ann Bot 115(3):369–383. https://doi.org/10.1093/aob/mcu186
Cowan R, Ingrouille MJ, Lledo MD (1998) The taxonomic treatment of agamosperms in the genus Limonium Mill. (Plumbaginaceae). Folia Geobot 33(3):353–366
Darlington CD, Wylie AP (1956) Chromosome atlas of flowering plants, 2nd edn. McMillan, New York
Doi M, Reid MS (1995) Sucrose improves the postharvest life of cut flowers of a hybrid limonium. Hortscience 30(5):1058–1060
Dole JM, Wilkins HF (2005) Floriculture: principles and species. Prentice-Hall Inc., Upper Saddle River
Dulberger R (1975) Intermorph structural differences between stigmatic papillae and pollen grains in relation to incompatibility in Plumbaginaceae. Proc R Soc Lond B 188(1092):257–274. https://doi.org/10.1098/rspb.1975.0018
Funnell KA, Bendall M, Fountain WF, Morgan ER (2003) Maturity and type of cutting influences flower yield, flowering time, and quality in Limonium 'Chorus Magenta'. N Z J Crop Hortic Sci 31(2):139–146
Hanks G (2015) A review of production statistics for the cut flower and foliage sector 2015 (part of AHDB Horticulture funded project PO BOF 002a). The National Cut Flower Centre, AHDB Horticulture:102
Harada D (1992) How to grow Limonium. FloraCulture International Nov–Dec 1992, pp 22–25
Harvey IC, Morgan ER, Burge GK (2000) A canker of Limonium sp. caused by Phomopsis limonii sp nov. N Z J Crop Hortic Sci 28(1):73–77
Hosni AM, Hosni YA, Ebrahim MA (2000) In vitro micropropagation of Limonium sinnuatum 'Citron Mountain', a hybrid statice newly introduced in Egypt. Ann Agric Sci (Cairo) 45(1):327–339
Ichimura K (1998) Improvement of postharvest life in several cut flowers by the addition of sucrose. Jarq-Japan Agric Res Q 32(4):275–280
Jeong JH, Murthy HN, Paek KY (2001) High frequency adventitious shoot induction and plant regeneration from leaves of statice. Plant Cell Tissue Org Cult 65(2):123–128
Karis PO (2004) Taxonomy, phylogeny and biogeography of Limonium sect. Pteroclados (Plumbaginaceae), based on morphological data. Bot J Linn Soc 144(4):461–482
Khan Z, Santpere G, Traveset A (2012) Breeding system and ecological traits of the critically endangered endemic plant Limonium barceloi (Gil and Llorens) (Plumbaginaceae). Plant Syst Evol 298(6):1101–1110. https://doi.org/10.1007/s00606-012-0619-3
Kho YO, Baer J (1968) Observing pollen tubes by means of fluorescence. Euphytica 17(2):298–302
Kimizu M, Yamamoto Y, Iwasaki T, Ohki S (2001) Agrobacterium-mediated transient expression of beta-glucuronidase and luciferase genes in Limonium sinuatum. In: Sorvari S, Karhu S, Kanervo

E, Pihakaski S (eds). Proceedings of the Fourth International Symposium on In Vitro Culture and Horticultural Breeding, Tampere, Finland July 2000. Acta Horticulturae 560:181–184.
Kunitake H, Mii M (1990) Plant regeneration from cell culture-derived protoplasts of statice (Limonium perezii Hubbard). Plant Sci (Limerick) 70(1):115–119. https://doi.org/10.1016/0168-9452(90)90039-q
Kunitake H, Koreeda K, Mii M (1995) Morphological and cytological characteristics of protoplast-derived plants of statice (Limonium perezii Hubbard). Sci Hortic 60(3–4):305–312. https://doi.org/10.1016/0304-4238(94)00713-P
Laguna E, Navarro A, Perez-Rovira P, Ferrando I, Pablo Ferrer-Gallego P (2016) Translocation of Limonium perplexum (Plumbaginaceae), a threatened coastal endemic. Plant Ecol 217(10):1183–1194
Lledo MD, Crespo MB, Fay MF, Chase MW (2005) Molecular phylogenetics of Limonium and related genera (Plumbaginaceae): biogeographical and systematic implications. Am J Bot 92(7):1189–1198. https://doi.org/10.3732/ajb.92.7.1189
Long PG, Funnell KA, Fountain WF, Bendall M, Morgan ER (2001) Control of a stem canker caused by Phomopsis limonii on Limonium 'Chorus Magenta'. N Z J Crop Hortic Sci 29(4):247–253
Mercuri A, Bruna S, Ld B, Burchi G, Schiva T (2001) Modification of plant architecture in Limonium spp. induced by rol genes. Plant Cell Tissue Organ Cult 65(3):247–253. https://doi.org/10.1023/a:1010623309432
Mercuri A, Anfosso L, Burchi G, Bruna S, De Benedetti L, Schiva T (2003) Rol genes and new genotypes of *Limonium gmelinii* through *Agrobacterium*-mediated transformation. In: Blom T, Criley R (eds) Proceedings of 26th International Horticultural Congress. Elegant Science in Floriculture. Acta Horticulturae 624: 455-462.
Morgan ER, Burge GK, Seelye JF, Grant JE, Hopping ME (1995) Interspecific hybridization between Limonium perigrinum Bergius and Limonium purpuratum L. Euphytica 83(3):215–224
Morgan ER, Burge GK, Seelye JF, Hopping ME, Grant JE (1998) Production of inter-specific hybrids between Limonium perezii (Stapf) Hubb. and Limonium sinuatum (L.) Mill. Euphytica 102(1):109–115
Morgan ER, Burge GK, Seelye JF (2001) Limonium breeding: new options for a well known genus. In: Proceedings of the twentieth international Eucarpia symposium, section ornamentals: strategies for new ornamentals. Acta horticulturae, pp 39–42
Morgan ER, Burge GK, Timmerman-Vaughan G, Grant JE (2009) Generating and delivering novelty in ornamental crops through interspecific hybridisation: some examples. In: VanTuyl JM, DeVries DP (eds) Xxiii international Eucarpia symposium, section ornamentals: colourful breeding and genetics, vol 836. Acta horticulturae, pp 97–103
Morgan E, Timmerman-Vaughan G, Conner A, Griffin W, Pickering R (2011) Plant interspecific hybridization: outcomes and issues at the intersection of species. Plant Breed Rev 34:161–220
Mori S, Yamane T, Yahata M, Shinoda K, Murata N (2016) Chromosome doubling in Limonium bellidifolium (Gouan) Dumort. by colchicine treatment of seeds. Hortic J 85(4):366–371. https://doi.org/10.2503/hortj.MI-117
Palacios C, Rossello JA, Gonzalez-Candelas F (2000) Study of the evolutionary relationships among Limonium species (Plumbaginaceae) using nuclear and cytoplasmic molecular markers. Mol Phylogenet Evol 14(2):232–249. https://doi.org/10.1006/mpev.1999.0690
Philosoph-Hadas S, Golan O, Rosenberger I, Salim S, Kochanek B, Meir S (2005) Efficiency of 1-MCP in neutralizing ethylene effects in cut flowers and potted plants following simultaneous or sequential application. In: Marissen N, VanDoorn WG, VanMeeteren U (eds) Proceedings of the Viiith international symposium on postharvest physiology of ornamental plants. Acta horticulturae, vol 669, pp 321–328
Seelye JF, Maddocks DJ, Burge GK, Morgan ER (1994) Shoot regeneration from leaf-disks of Limonium-perigrinum using thidiazuron. N Z J Crop Hortic Sci 22(1):23–29
Shimamura M, Ito A, Suto K, Okabayashi H, Ichimura K (1997) Effects of alpha-aminoisobutyric acid and sucrose on the vase life of hybrid Limonium. Postharvest Biol Technol 12(3):247–253. https://doi.org/10.1016/s0925-5214(97)00062-8
Vitti D, Fiocchetti F, Tucci M (2002) Chemical mutagenesis of Limonium sinuatum. http://agris.fao.org/agris-search/search.do?recordID=IT2003062030
Zhang C (1995) Exploration of Limonium interspecific breeding possibility: a thesis presented in partial fulfilment of the requirement for the degree of Master of Science in Plant Science at Massey University, Palmerston North, New Zealand. Massey University

第22章 观赏辣椒

Elizanilda Ramalho do Rêgo and Mailson Monteiro do Rêgo

摘要 观赏辣椒的销售是农业人口的重要收入来源之一。它们用于装饰和消费可为该产品增值，从而增加了对生产者的经济回报。辣椒的果实被认为是各种营养化合物的良好来源，包括类胡萝卜素、类黄酮和必需的矿物质。辣椒植物还可以在香料花园、药用植物花园和芳香花园中用作替代品。开发具有高品质特性的新品种是任何育种计划的主要目标之一，而遗传多样性的存在是育种计划选择和成功的主要标准。育种方法的选择基于多种因素，如生殖体系、目标性状的遗传力、育种目标和基本种群变异性。此外，辣椒的体外培养、基因组学、蛋白质组学、代谢组学、诱变和嫁接等方面的最新研究进展也为育种者的工作提供了帮助。本章将探讨雄性单倍体胚诱导、胚挽救及它们在缩短观赏辣椒育种程序中的应用和贡献。在这个简短的概述中，我们汇总了不同辣椒研究小组获得的研究结果和信息。

关键词 育种；辣椒；多样性；遗传效应；组织培养；基因组学；蛋白质组学；代谢组学

22.1 引 言

辣椒属（Capsicum）属于茄科（Solanaceae），有 31 个野生种和 5 个驯化种即牛角椒、中国辣椒、小米椒、风铃辣椒和毛辣椒（Moscone et al. 2007）。辣椒的野生物种主要集中在巴西南部的山区（Pickersgill 1997）。

近年来，辣椒在烹饪的多功能性、工业和药用特性及观赏性上获得了发展（Stommel and Bosland 2006；Rêgo et al. 2012a，2015a；Rêgo and Rêgo 2016）。

辣椒植物具有观赏价值，主要是由于其美学特征，如杂色的叶片和与叶片形成鲜明对比的深色果色，以及种子易于繁殖、生长期短、耐热和耐旱性及较高的后期生产质量（Stommel and Bosland 2006；Rêgo et al. 2009b，2011a，2011b，2011c，2012b，2012c，2012d；Segatto et al. 2013；Rêgo and Rêgo 2016）。

辣椒品种的巨大变异性使得观赏性状的研究成为可能，如大小、生长习性、早熟性、果实和后期生产质量（Rêgo et al. 2009b，2011c，2012c，2015b；Segatto et al. 2013；Rêgo and Rêgo 2016）。

E. R. do Rêgo (✉), M. M. do Rêgo
Research Productivity, Centro de Ciências Agrárias, Universidade Federal da Paraíba-CCA-UFPB, Areia, Brazil
e-mail: elizanilda@cca.ufpb.br; mailson@cca.ufpb.br

观赏辣椒在消费者市场上具有突出的地位和良好的接受度；它们在某些地区和国家很受欢迎，如欧洲和美国（Bosland 1999）。在巴西，观赏辣椒的销售仍然仅限于街头市场和一些超市，但这种情况正在发生变化，具有较高购买力的消费者已经在花店购买辣椒（Rêgo et al. 2011c，2015a，2016）。

辣椒种植这项业务是农业人口的重要收入来源（Stommel and Bosland 2006）。在巴西的一些州，家庭农业主要任务是经营种植面积不断扩大的辣椒（Rêgo et al. 2009a，2009b，2015a；Finger et al. 2012；Rêgo and Rêgo 2016）。

使用装饰用辣椒粉进行装饰和消费可增加该产品的价值，增加生产者的经济回报（Finger et al. 2012）。据 Reyes-Escogido 等（2011）、Rodriguez-Burruezo 等（2009）和 Rêgo 等（2012a）报道，辣椒的果实也可以用作功能性食品或保健食品。它们被认为是各种营养化合物的良好来源，包括类胡萝卜素、类黄酮和必需的矿物质（Rêgo et al. 2012a）。辣椒还可以在香料花园、药用植物花园和芳香花园中用作替代品。

观赏盆栽植物的销售已变得比切花更广泛。在观赏辣椒行业中，新品种的多样性可打开新的市场（Casali and Couto 1984；Rêgo et al. 2009b，2011c，2015a，2015b，2016）。开发具有高品质特性的新品种是任何育种计划的主要目标之一，而遗传多样性的存在对于育种计划的选择和成功至关重要。

根据 Heiser（1979）的研究，美洲土著是辣椒属植物的第一批育种者，他们驯化了一些种并发展了当今现存的多种果型。辣椒的育种还可通过群体选择和杂交来进行（Patil and Salimath 2008；Rêgo et al. 2009b，2012b，2012c，2015a；Nascimento et al. 2014；Ferreira et al. 2015；Fortunato et al. 2015）。在遗传改良计划中可以采用几种方法来开发新品种。育种方法的选择基于多种因素，如生殖体系、目标性状的遗传力、育种目标和基本种群变异性（Allard 1971；Fehr 1987）。

在观赏辣椒中发展近交系品种使用最广泛的方法是群体选择、系谱、回交、重复选择和单种子后代（SSD）（Bosland 1996；Stommel and Bosland 2006；Rêgo et al. 2011c，2015a；Rêgo and Rêgo 2016）。

高通量方法的发展为辣椒的传统育种提供了多种选择。其中，组织培养和分子生物学显著促进了辣椒的育种（Kothari et al. 2010）。

在这里，我们将尝试展示辣椒体外（组织）培养中的一些困难和最新进展，并报道我们在诱导雄性单倍体和胚挽救研究中的经验与贡献，以及这些技术在巴西观赏辣椒育种计划中的应用。

组织培养技术与分子生物学的结合，可以通过基因工程整合基因及对应的新性状。我们将探讨基因组学、转录组学、蛋白质组学和代谢组学及诱变育种与嫁接在辣椒育种中的应用、最新进展和前景。

22.2 个体表型

辣椒大小、形状、荚果类型和植物习性的多样性使它们能够用作盆栽、花坛和园林及切茎植物（Stommel and Bosland 2006；Reyes-Escogido et al. 2011；Rêgo et al. 2011c，

2012c，2015a；Rêgo and Rêgo 2016）。每一个性状都有许多在商业上可以接受的特征（Stommel and Bosland 2006；Barroso et al. 2012；Santos et al. 2013a，2013b；Silva Neto et al. 2014；Ferreira et al. 2015；Rêgo et al. 2015a，2016）。育种人员必须根据表型将研究重点放在重要性状上（表 22.1）。

表 22.1 各种类型辣椒的观赏性状

性状	类型	引用
株高	盆栽、地被和园林	Stommel and Bosland 2006；Barroso et al. 2012；Rêgo et al. 2012b；Silva Neto et al. 2014；Fortunato et al. 2015；Silva et al. 2015，2017；Rêgo and Rêgo 2016
株辐	盆栽、地被和园林	Barroso et al. 2012；Rêgo et al. 2012b；Silva Neto et al. 2014；Fortunato et al. 2015；Rêgo and Rêgo 2016；Silva et al. 2017
茎秆粗度	盆栽、地被和园林	Barroso et al. 2012；Nascimento et al. 2012；Silva Neto et al. 2014；Fortunato et al. 2015
茎秆长度	盆栽和切枝	Rêgo et al. 2012b；Rêgo and Rêgo 2016
生长习性（多元长势）	盆栽	Stommel and Bosland 2006
株高（4～6 英尺[a]）/株辐（10.16～15.24cm），小/矮化植株；紧凑株型；彩色果	盆栽	Stommel and Bosland 2006
株高（9.4～22.3cm）/株辐（18.5～31.4cm）	盆栽、地被和园林	Bărcanu et al. 2017
株高（22.5～26.5cm）/株辐（24～27cm）	盆栽	Barroso et al. 2012；Silva Neto et al. 2014
株高（14.5～25cm）/株辐（17.2～32.2cm）	盆栽	Neitzke et al. 2016
矮化植株，彩色果	盆栽	Rêgo et al. 2012d；Bărcanu et al. 2017
抗热，耐旱	盆栽、地被和园林	Stommel and Bosland 2006
彩色果（果实成熟后 4 种以上颜色）	盆栽、地被和园林	Stommel and Bosland 2006；Rêgo and Rêgo 2016
有对比色的枝叶	盆栽、地被和园林	Stommel and Bosland 2006；Barroso et al. 2012；Rêgo and Rêgo 2016
果实大小（长、宽和果重）	盆栽、地被和园林	Silva et al. 2015，2017；Bărcanu et al. 2017
单株果实产量	盆栽、地被和园林	Silva et al. 2015；Rêgo and Rêgo 2016
果实辣味	盆栽、地被和园林	Rêgo and Rêgo 2016
叶片和果实挂在植株上的持久性（定植后）	盆栽	Segatto et al. 2013；Rêgo et al. 2015a；Rêgo and Rêgo 2016
对乙烯的耐受性（花、果和叶）	盆栽	Segatto et al. 2013；Rêgo et al. 2015a，2015b；Rêgo and Rêgo 2016
叶片长度和宽度	盆栽	Barroso et al. 2012；Rêgo et al. 2012b；Ferreira et al. 2015
花的长度和宽度	盆栽	Nascimento et al. 2012；Santos et al. 2013a；Silva Neto et al. 2014；Ferreira et al. 2015；Fortunato et al. 2015；Rêgo et al. 2015a，2015b；Silva et al. 2017
花构造（花瓣数、花冠长度、花瓣宽度、花药长度、雌蕊数、花丝长度、花柱长度）	盆栽、地被和园林	Nascimento et al. 2012；Santos et al. 2014；Silva Neto et al. 2014；Ferreira et al. 2015；Fortunato et al. 2015
开花天数	盆栽	Rêgo et al. 2012b，2012c；Silva et al. 2015
结果天数	盆栽	Nascimento et al. 2014；Silva et al. 2015，2017

续表

性状	类型	引用
果实质量性状（维生素 C、可滴定酸和可溶性固形物）	盆栽、地被和园林	Nascimento et al. 2014
室温内植株寿命	盆栽	Rêgo et al. 2009a，2009b，2010，2015a；Rêgo and Rêgo 2016
对乙烯的耐受性	盆栽	Segatto et al. 2013；Rêgo et al. 2015a，2015b；Rêgo and Rêgo 2016
矮化植株，果实直立（隔离用）	地被和园林	Stommel and Bosland 2006
圆果或标准色果	地被和园林	Stommel and Bosland 2006
花、叶和果的颜色、形状与大小	地被和园林	Rêgo and Rêgo 2016
直立彩色果	切枝	Stommel and Bosland 2006
长果柄上簇生果	切枝	Stommel and Bosland 2006
从生茎秆，每茎秆 6 个以上果	切枝	Stommel and Bosland 2006
成熟果实的颜色、形状与大小	切枝	Rêgo and Rêgo 2016
果实硬度	切枝	Rêgo and Rêgo 2016
果实寿命	切枝	Rêgo and Rêgo 2016
叶片脱落（乙烯敏感）	切枝	Rêgo and Rêgo 2016

1 英尺 = 0.3048m

原则上，任何一种辣椒都可以用作观赏植物，但并非每种辣椒品种都适合在盆中种植（图 22.1）（Rêgo and Rêgo 2016）。只有那些在花盆中显示出较小的株型和协调性的品种才能作为盆栽植物进行种植与销售（图 22.1）。考虑的性状包括植物高度、总高度（植物高度和盆高）、冠层宽度、颜色和叶片位置。果实和花朵也是消费者购买时考虑的性状（表 22.1）（Barroso et al. 2012；Nascimento et al. 2013；Rêgo et al. 2015a；Rêgo and Rêgo 2016）。

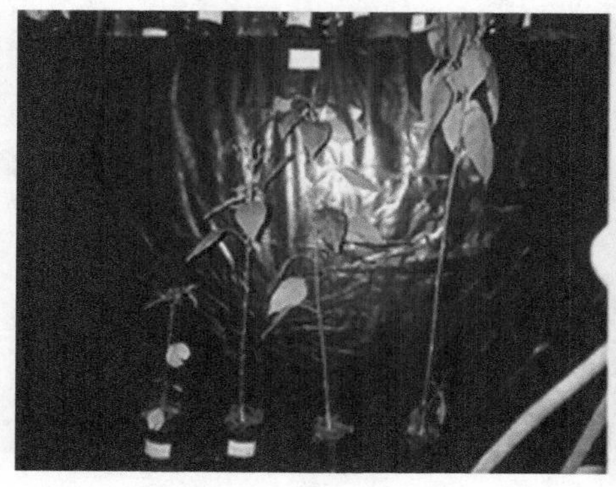

图 22.1　不同的辣椒品种。左侧品种不适于盆栽，右侧品种适于盆栽

为了获得良好的盆栽效果建议将株高与冠层宽度分别是盆高与盆宽的 1.5～2 倍（Stommel and Bosland 2006；Barroso et al. 2012；Silva Neto et al. 2014；Neitzke et al. 2016；Bărcanu et al. 2017）。容量为 900mL 的花盆通常成功地用于生产观赏辣椒（图 22.1）。进

一步研究最佳容器及其尺寸非常重要，这样可减少不必要的支出从而降低盆栽辣椒的最终生产成本。

适合作为园林植物种植的辣椒品种应具有一些性状（表 22.1）。Stommel 和 Bosland（2006）强调了直立花朵和果实对于园林种植植物的重要性。根据这些学者的说法，辣椒可用作边界植物或标本植物或者用于大规模种植。带有垂坠果实的品种也可以用作园林植物（Rêgo et al. 2012d；Cavalcante 2015）（图 22.2）。

图 22.2 观赏辣椒‘pimenta biquinho’变异株（UFPB M2 和 UFPB M3），锥形垂坠的果实数减少

一些性状比其他性状更难以改良，就像数量遗传性状一样。因此，了解这些性状表达所涉及的遗传效应对于辣椒改良计划的成功至关重要。

22.3 数量性状的遗传及其对观赏辣椒改良的应用

一些研究者估算了观赏辣椒多种性状的广义遗传力（Nascimento et al. 2012；Silva Neto et al. 2014；Pessoa et al. 2015；Da Silva et al. 2013；Silva et al. 2016；Naegele et al. 2016；Devi et al. 2017）。近年来，几项研究指出了数量性状表达涉及的遗传效应和狭义遗传力（h_n^2）（表 22.2）。

表 22.2 观赏辣椒的作用效应（AE）、加性（A）、显性（D）和显性遗传效应（PGE）

性状	AE	PGE	引用
发芽	A/D	A	Medeiros et al. 2015
		D	Barroso et al. 2017
胚轴宽度	A/D	A	Ferreira et al. 2015；Barroso et al. 2017
		D	Medeiros et al. 2015
胚轴长度	A/D	A	Medeiros et al. 2015
		D	Barroso et al. 2017
幼苗茎秆长度	A/D	D	Rêgo et al. 2012b

续表

性状	AE	PGE	引用
幼苗茎秆粗度	A/D	A	Rêgo et al. 2012b
畸形苗	A/D	A	Medeiros et al. 2015
		D	Barroso et al. 2017
幼苗株高	A/D	D	Ferreira et al. 2015
胚根散射度	A/D	A	Medeiros et al. 2015；Barroso et al. 2017
胚根长度	A/D	A	Medeiros et al. 2015；Barroso et al. 2017
胚根数量	A/D	A	Medeiros et al. 2015；Barroso et al. 2017
子叶长度	A/D	A	Ferreira et al. 2015
子叶宽度	D	D	Ferreira et al. 2015
株高	A/D	D	Ferreira et al. 2015
	A/D	A	Rêgo et al. 2012b；Santos et al. 2014；Fortunato et al. 2015；Silva et al. 2017
第一分叉高度	A/D	A	Ferreira et al. 2015
茎秆长度	A/D	D	Ferreira et al. 2015
		A	Rêgo et al. 2012b；Santos et al. 2014；Fortunato et al. 2015
茎秆宽度	A/D	D	Ferreira et al. 2015
		A	Santos et al. 2014；Fortunato et al. 2015
冠幅	A/D	D	Ferreira et al. 2015；Silva et al. 2017
		A	Santos et al. 2014；Rêgo et al. 2012b
雄蕊数量	A	A	Ferreira et al. 2015；Fortunato et al. 2015
花冠长度	A/D	A	Rêgo et al. 2012b；Ferreira et al. 2015；Fortunato et al. 2015
		D	Santos et al. 2014
花瓣数量	A/D	D	Ferreira et al. 2015
花瓣宽度	A/D	D	Ferreira et al. 2015
		A	Santos et al. 2014；Fortunato et al. 2015
花药长度	A/D	A	Rêgo et al. 2012b；Ferreira et al. 2015；Fortunato et al. 2015
花柱长度	A/D	A	Fortunato et al. 2015
花丝长度	A/D	D	Rêgo et al. 2012b
叶长	A/D	D	Rêgo et al. 2012b；Santos et al. 2014；Ferreira et al. 2015；Fortunato et al. 2015；Medeiros et al. 2015；Barroso et al. 2017
叶宽	A/D	D	Rêgo et al. 2012b；Ferreira et al. 2015；Barroso et al. 2017
		A	Santos et al. 2014；Fortunato et al. 2015；Medeiros et al. 2015
叶片密度/叶片数	A/D	D	Medeiros et al. 2015
		A	Rêgo et al. 2012b
开花天数	A/D	D	Rêgo et al. 2012c
		A	Silva et al. 2017
挂果天数	A/D	D	Nascimento et al. 2014；Silva et al. 2017
花梗长度	A/D	D	Nascimento et al. 2014；Ferreira et al. 2015
		A	Santos et al. 2014
果实长度	A/D	A	Nascimento et al. 2014；Santos et al. 2014；Silva et al. 2017
果实宽度	A/D	A	Nascimento et al. 2014；Santos et al. 2014；Silva et al. 2017
鲜果重	A	A	Nascimento et al. 2014；Santos et al. 2014

续表

性状	AE	PGE	引用
果实干物质	A/D	A	Nascimento et al. 2014
果实干物质含量	A/D	D	Nascimento et al. 2014
总可溶性固形物含量	A/D	D	Nascimento et al. 2014
果皮厚度/种皮厚度	A/D	D	Nascimento et al. 2014
		A	Santos et al. 2014
胎盘长度	A/D	A	Nascimento et al. 2014
维生素 C 含量	A/D	A	Nascimento et al. 2014
可滴定酸含量	A/D	A	Nascimento et al. 2014
单株结果数	A/D	A	Silva et al. 2017
		D	Nascimento et al. 2014
产量	A/D	A	Nascimento et al. 2014
单果种子产量	A/D	A	Nascimento et al. 2014；Santos et al. 2014

具有非加性效应的基因起主导作用使得目标基因的汇集非常困难，因为这些基因在群体中并不固定（Reddy et al. 2008；Rêgo et al. 2009b）。

对于具有加性遗传效应优势的性状，如苗茎粗度、胚根散射度、胚根长度、胚根数量、第一分叉高度、雄蕊数量、花药长度、花柱长度、果实鲜重、果实宽度、果实长度、果实干物质含量、胎盘长度、维生素 C 含量、可滴定酸含量、产量和单果种子量（表 22.2），建议使用回交或基于选择的方法汇集目标性状控制基因。对于这些性状的另一种育种策略是在早期进行选择。然而，对于具有非加性遗传效应优势的性状，如植株茎长、植株高度、子叶长度和宽度、花瓣数量、花丝长度、结果天数、果实干物质含量和总可溶性固形物含量（表 22.2），探索特定的杂交生产计划可能是一个好策略（Santos et al. 2014；Nascimento et al. 2014；Ferreira et al. 2015；Fortunato et al. 2015；Rêgo et al. 2015a；Silva et al. 2015，2017；Rêgo and Rêgo 2016）。

相互效应可能会影响果实、花朵和植物的性状。一些研究者详述了影响以下性状的相互效应的存在，植物茎长、叶片密度、花丝长度、花冠长度和开花天数（Rêgo et al. 2012b）；株高、叶长、叶宽和下胚轴长度（Rêgo et al. 2012b；Ferreira et al. 2015）；植株高度、下胚轴直径、子叶长度和宽度、茎宽、第一次分叉高度和花瓣数量（Ferreira et al. 2015）；果实宽度、果实长度、果实新鲜物质含量、果实干物质含量、结果天数、产量、维生素 C 含量、可滴定酸度含量、总可溶性固形物含量和单株果实产量（Nascimento et al. 2014）；花梗长度（Nascimento et al. 2014；Ferreira et al. 2015）。

这些结果表明了除了采用人工杂交外，在双列杂交方案中使用正反交的重要性。

要强调的另一个方面是间接选择在育种中的使用。性状之间关联的知识在育种工作中非常重要，尤其是当其中的一个性状由于遗传力低或测量和鉴定问题而难以选择时。

根据 Rêgo 和 Rêgo（2016）的观点，简单的相关系数可能无法完全说明两个变量之间的关系，因为其他变量引起的影响可能会使这些值混淆。偏相关系数消除了其他性状对所研究关联的影响，而利用相关系数来直接和间接影响基本变量的通径分析是关联研究的辅助手段。

在果实性状中利用路径分析，Da Silva 等（2013）确定水果干物质与花梗和果实长度、果实直径、果皮厚度和平均果实重量呈负相关。他们还使用路径分析研究了果实性状（花梗长度、果实长度、果实纵径、果实重量和果皮厚度）对观赏辣椒果实干物质含量（FDMC）的直接和间接影响。他们展示了通过果实纵径间接选择 FDMC 的可能性及结果。FDMC 由显性效应决定，果实纵径由加性遗传效应决定，这使得在育种的早期通过果实纵径来选择 FDMC 有效。如果目的是选择果实中干物质含量较高的植物，则应选择果实纵径较小的植物。

Rêgo 等（2015b）还报道了果实性状（果实宽度、果实重量、果皮厚度、子房室数和单果种子产量）与叶片长度和叶片宽度之间呈正相关。这些性状与总可溶性固形物含量、可滴定酸度含量、干物质含量及单株果实产量和结果天数呈负相关。他们认为，要获得高产量和果实质量，育种者应该选择坐果率最高、果实纵径最小的植物。

果实性状与乙烯引起的叶片脱落之间存在高度正相关关系。通过选择果实较小、果皮较薄且干物质含量较低的植物，可以选择对乙烯具有更高抗性的植物。暴露于乙烯后，果实掉落与叶片衰老之间没有相关性（Nascimento et al. 2015）。

在这种情况下，关联性状在筛选中的使用是改良性状的有效策略。

多元分析的新手段在以下研究中已经得到应用。

Silva 等（2016）试图建立与辣椒种苗、植株、花序和果实、叶片尺寸相关的 28 个形态和农艺性状之间的关联结构和模式。表型和基因型信息均显示出总体相同的关联结构。他们强调了受测基因型对农艺和观赏性状育种的重要性。

Da Silva 等（2015）使用多元技术来确定辣椒果实形态特征的样本数量。该技术考虑了响应变量之间的相关性，比 Silva 等（2011）的用单变量形式来确定辣椒果实有效样本数量更有效。

Devi 等（2017）提出基于分类的聚类分析和主成分分析是对辣椒品种进行分类的好方法。De Mesquita 等（2016）研究表明基于典型判别变量比基于马氏距离的沃德算法的聚类分析更有效。

Van Zonneveld 等（2015）使用主成分分析发现辣椒不同优株间，辣椒素水平与果实形状和重量及抗氧化能力和辣椒中多酚浓度存在很大差异。

为了确定变异系数的分类，需要进行更多的研究。Silva 等（2011）发表了根据实验误差对变异系数进行分类的最精确形式。他们的研究还表明实验的精度取决于变量。

尽管这些研究人员做了许多努力，但仍需要进行更多的研究来了解确定数量性状的遗传学。

22.4 可利用种质资源

遗传变异性对于开发具有优异遗传潜能的新品种至关重要，便于开发具有抗生物和非生物逆境的、高产量和高品质的品种。育种计划中要考虑的另一个重要因素是可利用的种质资源。

国际植物遗传资源委员会（IBPGR 1983）曾经组织在全球范围内对辣椒遗传资源进

行勘探和收集。

根据 Bosland（1996）的报道，热带森林砍伐是辣椒种质资源保护面临的最严重、最紧迫的环境问题之一。

半驯化和野生型的几种辣椒种类仍未知且尚未开发（Rêgo et al. 2012d）。*Capsicum flexuosum*、*C. campylopodium*、*C. parvifolium*、*C. schottianum*、*C. recurvatum*、*C. villosum*、*C. cornutum*、*C. buforum*、*C. pereirae*、*C. friburgense* 和 *C. hunzikerianum* 是较少使用的野生物种。后三种是巴西东部沿海地区的辣椒品种（Barboza and Bianchetti 2005；http://www.wildchilli.eu/index.php/wildchilli，2011）。它们用作栽培种新基因来源的潜力尚不清楚。

正如巴西的一些州，世界各地农业的扩张可能导致几种辣椒特有物种的地方品种灭绝。此外，甘蔗、咖啡和大豆种植园的乱砍滥伐导致巴西南部和西南部的大西洋森林减少，在研究人员甚至没有机会对其进行评估之前，非驯化辣椒品种的遗传资源库已经减少（Rêgo et al. 2012d）。

巴西研究小组正在努力评价其收集的辣椒种质资源植株群体的各性状变异性（Rêgo et al.2012a）。

一些以提高观赏性为目标的育种计划已探索驯化物种之间和内部的遗传多样性。研究结果大多表明，可以由种质库中保存的原始地方品种开发出盆栽和花园专用辣椒品种（Rêgo et al. 2009a，2010，2011c，2012d，2015a，2015b；Nascimento et al. 2012，2013；Santos et al. 2013a，2013b；Silva Neto et al. 2014；Pessoa et al. 2015；Rêgo and Rêgo 2016）。

Bosland（1996）强调了利用外来种质作为遗传多样性的来源，以改良新墨西哥州立大学的辣椒育种和遗传学计划开发的商业辣椒。根据他的说法，利用外来种质改良商业辣椒的潜在价值正在开发中。

USDA 收集的材料是全世界育种和研究计划的种质资源库（Bosland 1996）。

根据 Van Zonneveld 等（2015）报道，秘鲁国家创新研究所（INIA）收集保存了 106 份种质材料，玻利维亚的 Centro de Investigaciones Fitoecogenéticas de Pairumani 收集保存了 396 份种质材料。他们研究评价了在秘鲁和玻利维亚收集保存的资源，并鉴定出有利用前景的辣椒品种资源。

Lee 等（2016）筛选评价了来自 97 个国家的 4652 份辣椒种质资源。种质资源的地理来源和基本数据从韩国农村振兴厅（3599 份种质）（RDA，韩国全州市）和首尔国立大学（1053 份种质）（SNU，韩国首尔市）获得。他们建立了一个辣椒核心（CC240）种质库，由来自 44 个地理位置的 6 个植物学种组成，可代表整个种质库的资源多样性。

巴西研究机构如 Embrapa Cenargen、IAC 及巴西的几所大学收集了大量辣椒种质资源，其中包含巴西大部分栽培和野生物种（Rêgo and Rêgo 2016）。

在罗马尼亚，Bărcanu 等（2017）以盆栽应用为目的对 Buzău 研究与开发站的 214 份种质资源进行了评价。

尽管研究人员努力收集评价辣椒属种群内部和种群之间的变异性，但仍有许多种质未知。

22.5　种子生产

世界各地对辣椒新品种的需求不断增长，如具有色彩鲜艳、从叶子中脱颖而出引人注目的果实和花、较好的采后品质。

利用杂交可将目标性状的基因插入栽培品种中。然而，特定品系和品种内自然杂交不会产生有意义的变异。

辣椒品种中的异花授粉并非大家都熟知。在实践中，经过几代开放授粉，我们常发现甜椒地里有被辛辣品种杂交感染的植株。防止异花授粉的一种方法是单独覆盖植物，用一个布料的笼子覆盖一棵植物（图22.3）（Rêgo and Rêgo 2016）、用一个布料的笼子覆盖多棵植物（Bosland 1993）或在花朵开放前用胶水封闭花朵（图22.4）（Rêgo et al. 2012e）。

图22.3　辣椒隔离自交授粉（Rêgo and Rêgo 2016）

图22.4　自花授粉阶段：（a、b）适龄花蕾鉴定标记；（c、d）胶水封闭花蕾；（e）已被胶水封闭的花蕾；（f）自花授粉3天后的花朵；（g）自花授粉5天后的花朵；（h）处于中间成熟阶段的果实（Rêgo et al. 2012e；Rêgo and Rêgo 2016）

Rêgo 等（2012f）和 Barroso 等（2015a）的研究表明种子质量对种苗的培育至关重要。他们的研究显示发芽率（播种 14 天后）这个性状遗传力较低，具有显性/上位效应。然而，Medeiros 等（2015）发现在组培条件下，与发芽有关的性状的遗传力很高，具有加性效应。

涉及辣椒不同生物学种、类型或品种间的杂交尚未被广泛研究（Rêgo et al. 2009b）。根据 Rêgo 等（2012d）的研究，在辣椒育种过程中限制广泛使用杂交的原因是杂交的困难和单果结籽数很低。根据 Crispim 等（2017）的研究，观赏辣椒自花蕾期开始柱头具有一定的花粉接受能力，但开花后才达到最高的接受水平。他们还报道，选择开花前的花蕾开展杂交授粉是最重要的，因为它显示出高的柱头接受能力并且更易于处理。

尽管某些种内杂交显示出较低的坐果率，大约 20%，但同一物种的品种间杂交通常会产生足够数量的种子（Nascimento et al. 2015）。由于种间不亲和/不协调，很难获得源自种间杂交的种子（Bosland and Votava 2003；Nascimento et al. 2012，2015；Rêgo et al. 2012d）。

雄性不育是辣椒育种中的一个重要特征，因为母株上没有成活的花粉，很容易获得杂交种子（Shifriss and Frankel 1969；Corrêa et al. 2007；Monteiro et al. 2011）。Cavalcante（2015）通过 γ 射线辐射诱导‘pimenta biquinho’（*C. chinense*）植物获得雄性不育材料。

通过与一个或两个父母的回交可以很容易地恢复雄性不育植株的生育能力。Shifriss（1997）在其综述中详细介绍了如何保持雄性不育和恢复生育能力的方法。

Barchenger 等（2016）提出了一种利用乙烯利有效促进辣椒种子生产的方法。其研究表明，该方法能有效减少（但不是完全消除）人工蔬花和蔬果的工作量。用 1000ppm 和 2000ppm 的乙烯利处理会有效减少花的数量，而使用 2000ppm 的乙烯利处理会有效减少果实的数量。他们还发现，对于乙烯利的成功应用还需进行多地点和多基因型评估。

特定的传粉媒介可以提高辣椒品种的果实和种子产量。在一年生辣椒（*C. annuum*）中使用膜翅目昆虫能增强传粉效率（Crispim et al. 2017）。

22.6 产 后

开发观赏辣椒品种的任何育种计划都应包括选择快速生长的基因型，具有抗衰老性和改善的采后货架期（Rêgo et al. 2015a）。

关于影响观赏辣椒生产因素的研究报道较少，如尺寸、早熟性、盆中的老化能力及生产后因素，如对乙烯的敏感性，在低和高光照条件下维持光合作用的能力及使用抑制剂来提高盆栽寿命（Rêgo et al. 2010；Segatto et al. 2013；Santos et al. 2013a）。

Rêgo 等（2010）研究表明，在经过模拟运输 48h 后，不同品种的盆栽观赏辣椒的寿命在 13～72 天。

Segatto 等（2013）研究了观赏辣椒对乙烯刺激的敏感性。根据其研究，10μL/L 的乙烯处理 48h，导致某些辣椒品种的叶片叶绿素含量降低。商业品种‘Calypso’是最敏感的，在乙烯存在下叶片 100%掉落。另外，株系‘BGH 1039’受到的影响较小，暴露于乙烯后叶片仅 25%掉落。

Santos 等（2013a）在观赏辣椒的 F_2 种群中发现，不同群体及不同组织、叶片和果实对乙烯的敏感性存在显著差异（图 22.5）。

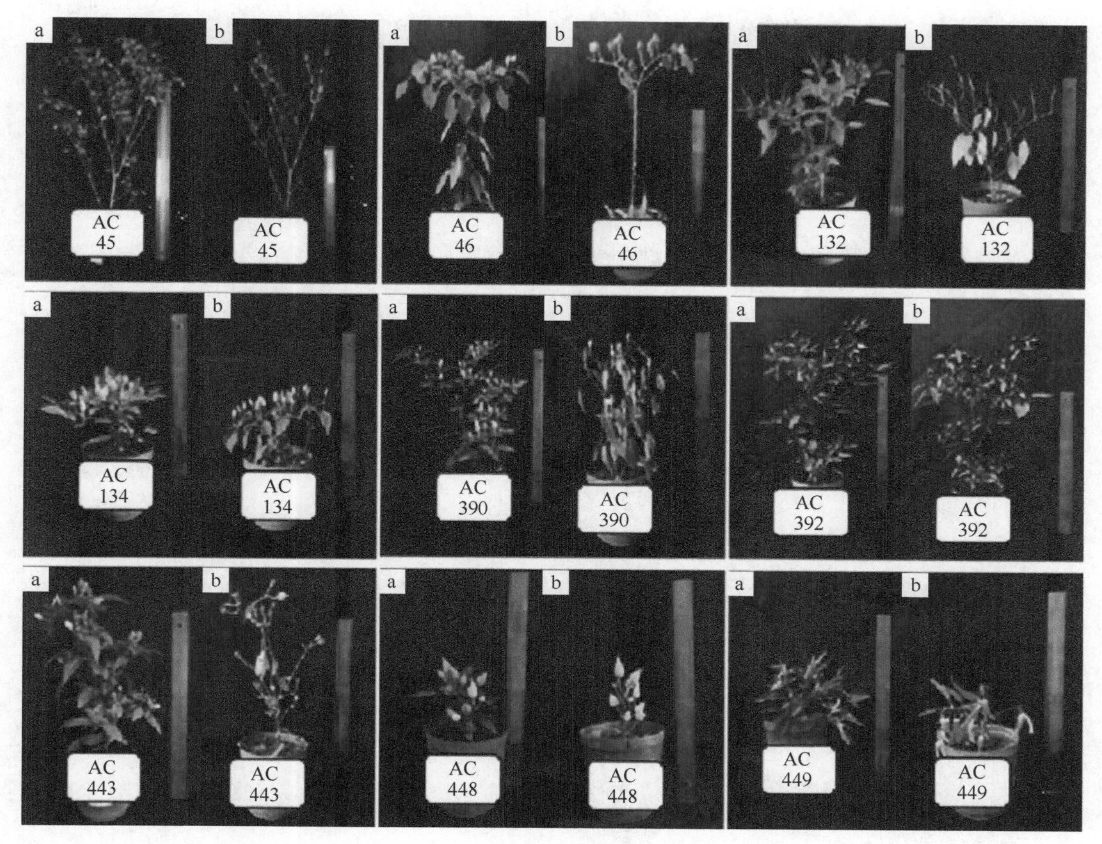

图 22.5　F_2 代观赏一年生辣椒（*Capsicum annuum*）对乙烯（10μL/L）的敏感性。a. 乙烯处理前；b. 乙烯处理 144h 后。AC 443、448、449、46 和 45 表现为敏感；AC 390 和 132 表现为中等敏感；AC 134 和 392 表现为抗性

乙烯引起的植株叶片脱落是一种遗传力较低的性状，由两个显性基因互作控制。根据 Rêgo 和 Rêgo（2016）的研究，乙烯引起植株果实脱落的性状具有很高的遗传力。

插入水中的辣椒品种‘Rio Light Orange’切茎可以冷藏保存长达 1 周，而品种‘Cappa Round Red’的切茎可以冷藏保存长达 2 周（De França et al. 2017）。这些研究者认为，这些品种似乎对乙烯不敏感，抗乙烯剂对切茎的采收后特性影响很小。他们的研究显示保持液对观赏辣椒品种‘Black Pearl’、‘Rooster’和‘Stromboli’瓶插寿命有积极的影响。对于 9 个实验品种，使用商业水化器后，瓶插寿命减少的有 1 个品种，没有影响的有 8 个品种。

22.7　生物技术在辣椒育种中的应用及其研究进展

22.7.1　辣椒的组织培养

Rêgo 等（2016）综述了辣椒属植物组织培养的最新进展和应用。他们讨论了辣椒

体外培养的困难和进展，在诱导雄性单倍体胚胎方面的经验和贡献，辣椒对乙烯的敏感性和抗性，组培开花、结果和种子生产，以及胚挽救及其在巴西观赏辣椒育种计划中的应用。这里讨论了我们团队实验室取得的研究进展，如雄性单倍体胚胎培养、遗传转化、^{60}Co γ 射线照射诱导突变，嫁接诱导的遗传多样性及使用不成熟合子胚（IZE）系统缩短选择周期。

22.7.1.1 花药培养

影响花药培养诱导单倍体的关键因素是基因型、单核小孢子的成熟阶段、培养基、培养条件和植物生长调节剂。

花药培养诱导单倍体的开创性工作是由 Guha 和 Maheshwari（1964）在毛曼陀罗中首先完成的。在辣椒属的物种中有自发形成单倍体的报道（Pochard and Dumas de Vaulx 1979）。

然而，有关辣椒花药培养和单倍体植株再生的第一个报道由 Wang 等（1973）完成。将具有单核小孢子的花药置于 MS 培养基中生长，其中对某些微量营养素和维生素进行了修饰，并补充了激动素、ANA 或 2,4-D。生长 33 天后，花药囊中开始出现绿色小植株。具有单核小孢子的花药已被优选为进行体外培养的材料（Lantos et al. 2009；Irikova et al. 2012）。

有研究进行了花蕾大小与小孢子发育阶段的关联分析（Barroso et al. 2015b）。同时对小孢子发育阶段、花蕾长度（FBL）和花药长度（ANL）与大多数单核小孢子的发育阶段也做了关联分析（Barroso et al. 2015b）。通过相关分析，分析了芽的形态与小孢子发育阶段的关系。均值测试证实单核小孢子的最大数量出现在第 2 阶段（花蕾的花瓣和萼片长度大致相同）。

花药培养规程的关键步骤是选择在最后一个单细胞阶段具有超过 50%的小孢子的花蕾，并在 4℃下将花蕾预处理 1 天，然后在 9℃下于双层系统进行花药培养 1 周，随后在 28℃下连续暗培养。

关于小孢子或花药培养的更多研究可见 Supena 和 Custers（2011）、Cheng 等（2013）、Olszewska 等（2014）及我们实验室（Barroso 2016）。

我们已经应用 Supena 等（2006）及 Supena 和 Custers（2011）开发的培养规程，筛选来自 Paraiba 联邦大学农业科学中心种质库的几份种质，我们获得了高频率的花药培养胚（未发表）。观察到雄性单倍体培养对基因型依赖性很大，以及观赏辣椒品种间雄性单倍体培养成效存在明显差异。

尽管具有不同果实性状的株系属于同一类型，但其基因型对不同培养规程的反应不同，如株系‘UFPB-132’对 Dumas de Vaulx 等（1981）描述的方法没有反应；但对 Supena 等（2006）描述的方法有积极的反应。相反，株系‘UFPB-001’对第二种培养方法无响应，但对第一种培养方法有积极的响应。

因此，在开始基于双单倍体（DH）培养的育种计划之前，建议评估每个品种对不同培养规程的反应。

22.7.1.2　辣椒的遗传转化

自 Liu 等（1990）首次报道辣椒转基因研究，各种辣椒转基因实验方法已出现但没有得到普遍使用（见 Rêgo and Rêgo 2016；Min et al. 2015；Heidmann and Boutilier 2015）。Min 等（2015）指出，辣椒的遗传转化存在一些障碍，包括极低的转化率、正确基因型的选择及由于直接芽形成而导致的高频率假阳性。Heidmann 和 Boutilier（2015）认为，辣椒遗传转化的主要瓶颈不仅是容易受到农杆菌感染，而且得到具有再生能力细胞的频率非常低。

尽管有上述困难，Min 等（2015）开发了一种有用的培养实验方法及合适的选择方法。辣椒遗传转化方案最重要的方面是对由愈伤组织再生的芽进行选择，这称为愈伤组织介导的芽形成。据 Min 等（2015），他们用于辣椒的遗传转化方法具有可复制性和可靠性。

Heidmann 和 Boutilier（2015）提出了另一个有望实现高效且可重复操作的方法，该方法利用了可诱导 BBM 的组成型表达。该方法用于转化甜椒 F_1 杂交品种 ‘Fiesta’ 和 ‘Ferrari’ 的子叶外植体，以及非商业上可获得的亲本。转化效率为 0.6%～4%，从种子到种子整个过程大约需要 9 个月。Heidmann 和 Boutilier（2015）使用这种方法，已经再生了 100 多种来自不同基因型的可育转基因植物，并且这些植株可将转入的基因传递给了下一代。

22.7.1.3　通过胚挽救缩短观赏辣椒选育周期的策略

缩短选育周期的策略非常有用的两个原因：①加快育种程序；②间接地降低植物材料的养护成本。该方法包括切除和体外培养未成熟果实的合子胚。该方法可能包括针对特定性状的分子标记辅助选择，可使育种者避免费时的评价筛选，并使用早熟世代的第一批果实。

对未成熟合子胚进行分离和体外（组培）萌发可能有助于缩短育种周期并加快育种程序。这种方法在辣椒属中首次应用是由 Manzur 等（2014）完成的，其评估了不同生长季节辣椒未成熟胚分离和培养的效率。与常规育种系统相比，胚培养的方法可缩短育种周期 70 天。它使辣椒的育种者每年可繁殖三代，甚至四代（对于卡宴辣椒），而传统系统只能繁殖两代。

Barroso（2016）评估了 Manzur 等（2014）描述的方法，研究材料是三种不同基因型的观赏辣椒（UFPB 001、UFPB 004 和 UFPB 099）（图 22.6）。结果表明，采用 Manzur 等（2014）建议的分子标记辅助选择，每年最多可以繁殖四代，不采用分子标记辅助选择育种可以繁殖三代。两种系统之间的主要区别在于：第一阶段，在未成熟合子（IZE）系统中，从自花授粉到未成熟的果实即未成熟合子胚需要 30 天；而在常规系统中，这一阶段一直持续到果实和种子成熟，需要 69 天，增加了 39 天。第二阶段（图 22.6），IZE 系统中未成熟合子胚的体外（组培）萌发和小植株的发育需要 28 天；而在常规系统中，该阶段从在聚乙烯托盘中播种直到可移植时（三片真叶）为止，需要 52 天，增加了 21 天。第三阶段是炼苗，IZE 系统比常规系统增加 1 天。最后一个阶段即第四阶段在 Manzur 的方法中没有介绍，但由于特征评价的需要，我们在此进行介绍。IZE 系统（45

天）和常规改进（48 天）这两个系统之间的差异很小。与常规系统相比，IZE 系统一个育种周期可节省 63 天的时间，每年可完成三个育种周期。

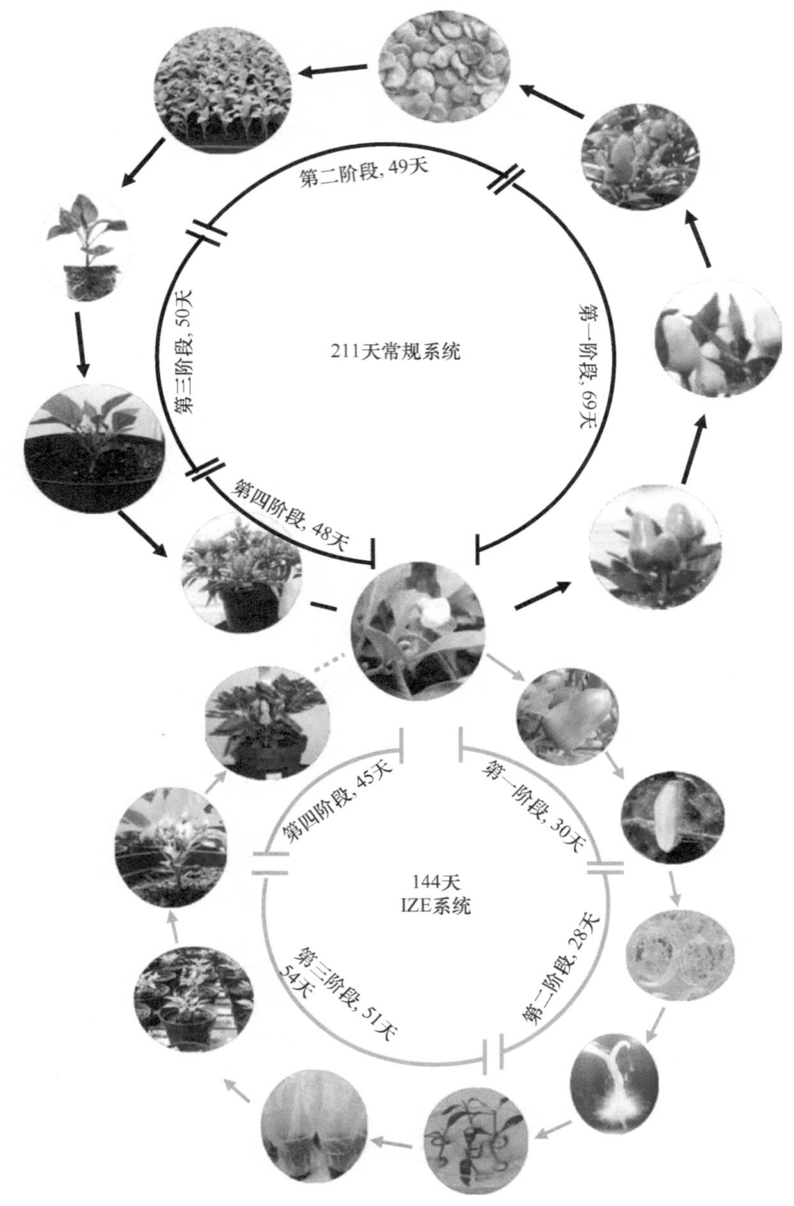

图 22.6　观赏辣椒两个育种系统育种周期的比较。常规系统（217 天）（黑色箭头）和未成熟合子（IZE）系统（154 天）（橙色箭头）；IZE 系统每个周期缩短 63 天（改编自 Manzur et al. 2014）

使用 IZE 系统，所用的培养基种类和未成熟合子胚的年龄也决定了开花的早晚（图 22.7）。

与对照植株相比，正如 Gopalkrishnan 等（1993）报道，在体外（组培）再生的幼苗显示出相同的育性，但导致植株矮化和早开花（图 22.7）。

图 22.7 由未成熟合子胚（IZE）系统经体外（组培）培养获得的观赏辣椒（*Capsicum annuum* L.）植株。(a) 花期后（DAA）30 天（左）和 25 天（右）未成熟合子胚（IZE）产生的植株在植株高度和果实发育方面的差异，右侧植株表现出类似侏儒的表型。(b) MS（M1 和 M2）培养基对 UFPB 099 基因型植株高度和果实发育的影响。与 M1 培养基相比，M2 培养基培养出的植株表现出侏儒症。取开花 25 天未成熟合子胚培养和使用 1/2 MS 培养基均可减小观赏辣椒的株型

未成熟合子胚挽救技术已被用于扩展辣椒属中的种间杂交（Debbarama et al. 2013）。他们实现了在 *C. chinense*、*C. annuum* 和 *C. frutescens* 之间相互杂交，并培养挽救了杂交授粉后 27 天的未成熟合子胚。未成熟胚的成活率达到 84%，获得了杂交植物，并使用形态学和分子标记证实了杂交性。

22.7.2 辣椒的基因组学

22.7.2.1 辣椒中的分子标记和基因定位

Rodrigues 等（2016）最新修改了辣椒育种中应用的分子标记。他们正在使用包括 RAPD、RFLP、AFLP、SSR 和 ISSR 在内的分子标记来辅助辣椒育种。这些分子标记已被用于种质的遗传多样性估计、品种的 DNA 指纹图谱、种子的遗传纯度评估、育种程序中的选择、基因分离及具有经济价值的几个性状的遗传作图。这些分子标记之间的主要区别在于检测多态性的能力、应用成本、简便性和结果稳定性。

自从辣椒的第一个遗传图谱（Tanksley et al. 1988）构建以来，辣椒属的分子育种有了很大进展（见 Rodrigues et al. 2016；Paran et al. 2007）。以定位主要经济性状基因为目标的第一代（Tanksley et al. 1988）和第二代（Paran et al. 2004）遗传图谱已见不少报道，如抗病性（Kyle and Palloix 1997）、对根结线虫的抗性（Djian-Caporilano et al. 2007）、对细菌斑点的抗性（Jones et al. 2002），果实性状如辛辣味（Lefebvre et al. 1995；Ben-Chaim et al. 2001）、果实颜色（Hurtado-Hernandez and Smith 1985）、果实质地（Rao and Paran 2003）和植物结构（Paran 2003）。对 QTL（数量性状基因座）的研究也见不少报道，主要发现与性状关联的是 QTL 位点，如疫霉病（Quirin et al. 2005）、炭疽病（Voorrips et al. 2004）、果实大小（Ben-Chaim et al. 2001；Rao et al. 2003）、果实形状（Ben Chaim et al. 2003）和辛辣味（Ben Chaim et al. 2006）。

单核苷酸多态性（SNP）标记也被用于观赏辣椒的育种程序中（Garcés-Claver et al. 2007）。该研究报道了一种与辣椒果实的胎盘辛辣有关的 SNP，该 SNP 在 307pb 的表达

序列标签中被检测到。

Jeong 等（2010）由核和细胞质 DNA 序列开发了 SNP，以鉴定辣椒的不同种类。使用高分辨率融和分析（HRM），他们检测了等位基因变异。使用该方法分析了属于 6 个物种的总共 31 个辣椒品种，结果表明所开发的标记可用于快速鉴定辣椒新种质。

22.7.2.2 辣椒的基因组序列

在完成第一个植物基因组（拟南芥基因组计划）的测序后仅 14 年（The *Arabidopsis* Genome Initiative 2000），Kim 等（2014）完成了辣椒‘CM334’的全基因组测序。

‘CM334’是从墨西哥的莫雷洛斯州收集的一个地方品种，对各种病原体、病毒和根结线虫都表现出很高的抗性。表 22.3 给出了辣椒基因组数据的摘要。首先，Kim 等（2014）估计其基因组的大小为 3.06Gb，包括 37 989 个长序列片段（N50 = 2.47Gb）（表 22.3）。他们还对两个辣椒品种（‘Perennial’和‘Dempsey’）进行了重测序，并对野生种 *C. chinense* 进行了测序。比较品种‘CM334’和其他三个辣椒品种的基因组后，他们证实‘Perennial’、‘Dempsey’和 *C. chinense* 的基因组分别有 0.35%、0.39%和 1.85%（1090 万、1190 万和 5660 万个 SNP）存在差异。基因组之间的大部分遗传差异（超过 70%）归因于转座子，尤其是长末端重复序列（LTR）类别（Kim et al. 2014）。

表 22.3 辣椒基因组和基因注释信息的统计摘要

长序列片段数	37 989
片段总长	3.06Gb
定位到的长序列片段	2.63Gb（86.0%）
Contig 数	337 328
GC 含量	35.03%
基因数	34 903
编码序列平均/总长度	1 009.9/35.2Mb
平均外显子/内含子长度	286.5bp/541.6pb
转座元件总长度	2.34Gb（76.4%）

修改自 Kim 等（2014）

在 PGA 流程中，总共预测了 34 903 个蛋白质编码基因（Pepper Genome Annotation v 1.5）。基因和数量与番茄与马铃薯的大致相同，表明茄科中的基因数量相似。但是，辣椒基因组是番茄基因组的 4 倍，这是由于 LTR 逆转座子在异染色质和常染色质区域中大量存在。

正如 Wang 等（2006）先前的报道，Kim 等（2014）还确定了辣椒和番茄的物种形成时间是 1910 万年前。研究结果还表明辣椒的辛辣味是通过新基因的进化和物种形成后基因表达变化而产生的，新基因的进化是由原有基因的不均等复制引起的。

Kim 的研究小组认为，获得辣椒基因组将有利于育种手段的创新，如针对观赏辣椒重要性状（果实大小、产量、辣味、对生物和非生物胁迫的耐受性、营养成分含量和对病害的抗性）探索基因组选择和全基因组关联性分析。

22.7.2.3　植物基因组编辑

最近，用于研究基因功能和分子育种的强大工具已被开发出来，称为植物基因组编辑（Zhang et al. 2017）。这些技术允许使用人工核酸内切酶在目标基因座上产生DNA双链断裂（DBS）。DSB通过NHEJ（非同源末端连接）和HDR（同源性定向修复）等细胞中的途径得到修复，分别导致突变和序列置换（Zhang et al. 2017）。实际上，最常用的基因编辑系统是CRISPR/Cas9（Peng et al. 2017）。在此系统中，核酸内切酶Cas9通过CRISPR小RNA靶向目标DNA序列。ZFN（Townsend et al. 2009）、TALEN（Demorest et al. 2016）和CRISP/Cpf1（Zhang et al. 2017）是已经报道的其他三个基因组编辑工具，它们显示出了在功能基因组学和精确分子改良中应用的潜力；但到目前为止，尚未在辣椒属中使用这些基因组编辑工具。

22.7.3　辣椒的转录组学

转录组定义为在给定的一组环境条件下，在特定时间点，在器官或组织中转录的所有RNA分子的集合（Martínez-López et al. 2014）。

在辣椒属中，关于转录组的研究很少并且最近才有。通常，它随以下因素变化，植株发育和果实成熟度（Martínez-López et al. 2014）、非生物胁迫、热（Li et al. 2015）和低温胁迫等（Li et al. 2016）。

Martínez-López等（2014）评估了开花后（DAA）10天、20天、40天和60天*Capsicum annuum* L.'Tampiqueño 74'不同发育阶段果实的转录组。他们使用RNA-Seq方法，并报道了与辣椒素和抗坏血酸生物合成相关的基因表达发生在开花后20天，而类胡萝卜素生物合成的相关基因表达发生在开花后60天。

为了分析转录组，将基因按生物学过程（BP）和代谢途径（MP）进行分组（Martínez-López et al. 2014）。参照模式植物拟南芥同源基因，在第一类中，与相应拟南芥直系同基因的分类相比，总共将8628个辣椒基因分为875个生物学过程，而在第二类（MP）中，将1794个基因分为152种代谢途径。最后，研究报道了辣椒果实转录组在开花后10～20天和40～60天发生了深刻的变化。最后一个时间点时显著差异表达的基因数占总显著差异表达基因数的49%，这些基因主要是表达量减少的与果实成熟相关的基因和表达量增加的与果实衰老相关的基因。开花后60天时的果实转录组是所有分析阶段中最功能化但最不多样化的。

Li等（2015）研究了在热胁迫下热敏辣椒品种'S590'（CaS）和耐热品种'R597'（CaR）幼苗转录组。他们的结果证实35个基因参与热应激反应，并表明大多数热激蛋白在CaS中出现高表达，而不在CaR中高表达；而转录因子和激素信号基因在CaR中的表达水平高于CaS。这些发现代表了对辣椒耐热分子机制研究和了解的第一步。

最近的研究还分析了涉及寒冷胁迫和24-表油菜素内酯（EBR）抗冷作用的辣椒转录组（Li et al. 2016）。研究者使用RNA-Seq分析，得到油菜素内酯（BR）诱导的耐冷植株基因表达谱，并与对照植株对比。结果表明，EBR诱导了656个基因差异表达，其中335个基因上调，而321个基因下调。此外，研究发现在寒冷胁迫下，EBR积极调节与光合作用相关的基因，从而促进净光合速率（P_n）、F_v/F_m和叶绿素含量的显著增加。同样，它也诱导了纤维素合酶类似蛋白、UDP-糖基转移酶和细胞氧化还原稳态相关基因的合成。在EBR存在下，水杨酸（SA）和茉莉酸（JA）的内源性水平提高，而乙烯（ETH）的生物合成途径受到抑制，这表明油菜素内酯通过与水杨酸、茉莉酸和乙烯信号途径的相互作用响应寒冷。总的来说，这项研究为表明EBR在转录水平及在应对辣椒低温胁迫的反应中具重要作用提供了第一证据，涉及转录、信号转导和代谢稳态的调节。

22.7.4 辣椒的蛋白质组学

在过去的10年中，与分子标记结合，用于基因测序、微阵列实验及在生物体细胞内表达基因和蛋白质的新工具在鉴定染色体上特定性状相关位置时非常有用，有利于准确性更高的植物遗传改良筛选（Nimisha et al. 2013）。

关于辣椒和甜椒中蛋白质组学的报道很少，而且非常新，都是基于比较蛋白质组学分析（Siddique et al. 2006；Wongpia and Lonthaisong 2010；Wu et al. 2013；Xie et al. 2017）。

Siddique等（2006）提供了第一个全面的灯笼椒色素母细胞蛋白质组学分析。该研究小组鉴定出151种具有高置信度的蛋白质。大多数鉴定出的蛋白质在质体碳水化合物和氨基酸代谢中具有活性。最丰富的蛋白质是辣椒红素/辣椒红素合酶和原纤维蛋白，它们参与类胡萝卜素的合成和储存，这些类胡萝卜素在色母细胞中积累水平高。色素母细胞蛋白质组的分析支持这样的观点，即植物中存在质体特异性代谢网络。

尖孢镰刀菌引起的枯萎病是全世界辣椒和甜椒生产的主要问题。Wongpia和Lonthaisong（2010）报道了两个辣椒品种的蛋白质谱，一个抗枯萎病品种（‘Mae Ping 80’）和一个易感枯萎病品种（‘Long Chili 445’）。他们在抗性品种中发现了9个斑点差异表达，而在易感品种中有37个斑点差异表达。这些蛋白质被鉴定为涉及植物防御机制的NADPH HC毒素还原酶、丝氨酸/苏氨酸蛋白激酶和1-氨基环丙烷-1-羧酸合酶。最后，辣椒的抗性与和ROS解毒有关的蛋白质的较高表达相关，而辣椒植物抗枯萎病的能力与非诱导性免疫-1-蛋白的表达有关。

在辣椒中，细胞质雄性不育（CMS）是由核和线粒体基因之间相互作用引起的（Martin and Grawford 1951；Peterson 1958）。Wu等（2013）在辣椒（*Capsicum annuum* L.）的雄性不育系（指定为NA3）和其保持系（指定为NB3）之间进行了比较蛋白质组学分析和蛋白质谱分析。

基于质谱分析，有研究者确定了代表23种不同蛋白质的27个斑点。雄性不育系花药/芽中有超过14种蛋白质被下调；另外，多酚氧化酶、ATP合酶亚基β和肌动蛋白被上调。NA3的雄性不育可能与能量代谢紊乱、过量的乙烯合成和淀粉合成受抑制有关。

Xie等（2017）使用相同的技术探索并鉴定了太空飞行（诱变）后辣椒的差异表达蛋白。将空间诱变后获得了新的性状的突变体（Y1、Y2和Y3）与它们的野生型

（对照）（W1 和 W2）进行比较。在成功测序的 39 种差异表达蛋白质中，有 31 种与已知蛋白质同源，其余被推测为 6 个不同的组，包括蛋白质代谢、能量代谢和光合作用。

这些研究为进一步研究与突变体、质体和细胞质雄性不育有关的基因功能奠定了基础，并探索了它们的分子基础。

22.7.5 辣椒的代谢组学

代谢组学被定义为对内源性代谢物的全面分析，并试图系统地鉴定和量化生物样品中的代谢物（Zhang et al. 2012）。代谢物代表了各种各样的低分子量物质，如脂质、氨基酸、蛋白质、核酸、维生素、硫醇和碳水化合物，这使得整体的分析成为一项艰巨的挑战。

尽管代谢组学领域在技术上有了进步，但在辣椒属上的研究进展甚微，特别是在观赏辣椒方面，正如我们将在下面讨论的那样。

Wahyuni 等（2013）评估了 32 个辣椒株系成熟果实的代谢多样性，这些辣椒属 *C. annuum*、*C. chinense*、*C. frutescens* 和 *C. baccatum* 不同的种。并使用非目标液相色谱-质谱（LC-MS）（针对半极性化合物）和顶空气相色谱-质谱（GC-MS）平台（针对挥发性化合物）测定了果皮中的代谢组谱。

半极性代谢物的组分与物种组相关性很好，表明物种之间的遗传差异反映在代谢差异中（Wahyuni et al. 2013），因此可以将 *C. annuum* 与 *C. chinense*、*C. frutescens* 和 *C. baccatum* 区分开来。这种有效的区分可能是由物种驯化和选择历史所致。

对于挥发性化合物，研究的种质以是否具有辛辣味的性状能区分开来，而不是以物种区分（Wahyuni et al. 2013）。辛辣味性状受编码辣椒素合酶的 *Pun1* 基因控制（Mazourek et al. 2009；Stewart et al. 2005）。因此，辣椒素类成分的缺失和辛辣味的缺失是由该基因（*Pun1*）功能的丧失导致的。

Jang 等（2015）对辣椒（*C. annuum*）果实发育过程中的代谢组特性进行了有趣的研究。他们在开花后（DPA）16 天、25 天、36 天、38 天、43 天和 48 天的 6 个发育阶段进行了非目标代谢组分析，以评估代谢物的变化。不同发育阶段果皮颜色、代谢物、基因表达和抗氧化活性的变化是不同的。在 16DPA 和 25DPA 时，大多数有机酸（如草酸、苹果酸和琥珀酸）及腐胺和麦芽糖的含量较高，并逐渐降低。有机酸有助于提高水果的味道、风味和整体品质（Shin et al. 2015），而苹果酸和柠檬酸会受到水果成熟度的强烈影响（Osorio et al. 2012）。腐胺与生物和非生物胁迫程度（Aman and Zora 2004）、乙烯产量（Aman et al. 2005）和植物生长（Yahia et al. 2001）、开花和果实发育（Aman et al. 2005）有关。

在 36DPA 和 38DPA 时观察到氨基酸和柠檬酸水平的变化。在此阶段，水果的颜色为橙色。氨基酸在保持水果品质和营养价值方面起着重要作用（Glew et al. 2003）。L-缬氨酸是辣椒素链结构一端的氨基酸前体，缬氨酸途径是辣椒素生物合成中的重要组分（Prasad et al. 2006）。

在 43DPA 和 48DPA 阶段，疏水氨基酸，如 L-亮氨酸、L-异亮氨酸和 L-苯丙氨酸，以及 L-脯氨酸、L-天冬氨酸、山奈酚苷、辣椒素和二氢辣椒素的水平明显高于早期阶段。重要的是观察到抗氧化活性的变化高度反映了代谢物的变化。总之，这项研究确实提供了有关辣椒发育各个阶段果实营养成分的有用信息。非靶向代谢组学和基因调控网络分析是解释辣椒（包括观赏辣椒）代谢过程中生化物质变化的有趣工具（Jang et al. 2015）。

22.8 观赏辣椒的变异

自使用 X 射线诱导辣椒突变的第一个报道以来（Raghavan and Venkatasubban 1940），几项有关的研究已见报道（有关综述请参见 Daskalov 1986）。Daskalov（1986）将干辣椒种子暴露于 X 射线进行诱变，并成功地鉴定出几种植物大小、分枝、叶绿素含量及叶片和果实大小突变体。

通过将辣椒植物的种子暴露于 γ 射线、X 射线和 EMS 中获得了几种有用的突变体（Daskalov 1974，1986），包括雄性不育、无花色苷、果实形态和颜色改变及矮化突变体，可直接用于杂交育种。

我们实验室使用辣椒 *Capsicum chinense* Jacq 两个种（G1，红色果实；G2，黄色果实）的干种子诱导突变体。使用 ^{60}Co γ 射线辐照器 Gamma Cell 220（Cavalcante 2015）对种子进行 0Gy、25Gy、50Gy、100Gy、200Gy、400Gy 和 800Gy 不同剂量的 γ 射线辐照处理。结果表明，播种前对辣椒种子进行 γ 射线辐照会导致幼苗和植物的营养、生殖和生产性状发生变化。突变影响了几个特征，如植物的大小、分枝、叶绿素含量、无花色苷、叶子和果实的大小、果实形态和颜色的改变及矮化植物，可直接用于杂交育种。此外，形态突变的频率与诱变剂的剂量呈正比（Terzyan et al. 1974）。

γ 射线剂量的选择取决于基因型。通常，应根据观赏育种计划的目标使用 100Gy、200Gy 和 400Gy 的剂量。100Gy 和 200Gy 的剂量分别诱导了最矮的植物和更大的花朵，这是观赏植物两个重要性状。200Gy 的剂量还诱导了 G2（黄色果实）的雄性不育（图 22.8b），与对照（图 22.8e）相比，花药闭合、花粉少或死亡（图 22.8d）。Daskalov（1968，1973）也报道了类似的研究结果。

剂量 400Gy 导致 G2 的果实产量较低、花朵变小（图 22.8a）、种子消融和单性结实果实的产生（图 22.8f、g），以及株高降低和果实变小（图 22.8c），这些都是观赏植物的重要特征。Daskalov（1986）建议选用 50～100Gy 诱导辣椒突变。最近，García-Gaytán 等（2017）将智利辣椒（*Capsicum annuum* L.）的种子暴露于不同的 γ 射线（GR）剂量下（0Gy、10Gy、80Gy 和 120Gy）。他们的结果提供了证据，证明 γ 射线辐射剂量及用于植物生长的营养液的渗透势会影响辣椒果实中的大量营养素浓度和质量指标，如 pH 和电导率。

800Gy 剂量可促使 *C. chinense* Jacq 幼苗的过早死亡。可能是由于 γ 射线辐射（^{60}Co）对种子细胞的 DNA 双螺旋结构直接产生不利影响（García-Gaytán et al. 2017）。

图 22.8　不同剂量的 γ 射线（^{60}Co）诱变后，两种辣椒基因型的花、花粉粒和果实的发育。(a) 不同剂量（b）雄性不育（花药封闭）；(c) 可育（花药开放）花；(d、e) 不能存活和能存活的花粉；(f、g) 大和小的果实；(h) 无种子果实（红色和黄色水果）；(i) 退化的种子（Cavalcante 2015）

22.9　辣椒嫁接引起的遗传多样性

自第一篇科学论文表明将西瓜（*Citrullus lanatus*）嫁接到南瓜（*Cucurbita moschata* Duch.）上可增加对病原体（镰刀菌和叶甲虫幼虫）的抗性和果实产量以来（Tateishi 1927），嫁接技术已迅速应用于高价值茄科和葫芦科作物来提高产量与增加抗性（King et al. 2010）。

嫁接提供了通过探索砧木根系性状自然遗传变异来影响商业化接穗表型的机会（Albacete et al. 2009）。

有关嫁接的综合评论已汇编了有关草本植物和观赏植物的嫁接历史与当前应用信息，也包括实践和农艺方面，如操作、砧木、种类和农作物性能（Lee and Oda 2010；

Albacete et al. 2015）。在辣椒中，嫁接被用来控制辣椒疫霉（Morra 2004）和线虫(Kokalis-Burelle et al. 2009)，而且独特的是，该作物是我们唯一可以找到的研究砧木/接穗相容性的材料。根据 Kormos 和 Kormos（1955）的研究，单个显性基因控制着嫁接在其他辣椒、番茄、曼陀罗或蒜芥茄上的辣椒接穗的亲和性。

关于嫁接诱导的遗传变化，Taller 等（1998）声称他们成功地诱导了辣椒 *C. annuum* 果实性状及其遗传的一些变化。

我们实验室评估了嫁接诱导观赏辣椒果实性状遗传变异的效率。属于两个物种的 4 个种质，*C. chinense* Jacq.（UFPB-04、UFPB-02 和 UFPB-05）和 *C. frutescens* L.（UFPB-03），以所有可能的组合（砧木或接穗）相互嫁接，采用孔嫁接（Lee and Oda 2010）。在温室中总共进行了 80 次嫁接（4 个砧木 × 4 个接穗 × 5 个重复），成功率为 63.75%。当分别将 UFPB-02 和 UFPB-04 作为砧木、UFPB-03 和 UFPB-05 作为接穗时，我们诱导了果实性状的遗传变异（表 22.4、图 22.9）。

表 22.4　4 种观赏辣椒材料的嫁接成功率和失败率

砧木＼接穗	UFPB-02	UFPB-03	UFPB-04	UFPB-05
UFPB-02	+++––	+++––	+++––	++++–
UFPB-03	++++–	+++++	++–––	+––––
UFPB-04	+++––	+++––	+++––	+++––
UFPB-05	+––––	+++++	+––––	+++++

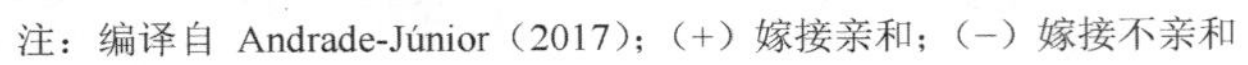
注：编译自 Andrade-Júnior（2017）；（+）嫁接亲和；（–）嫁接不亲和

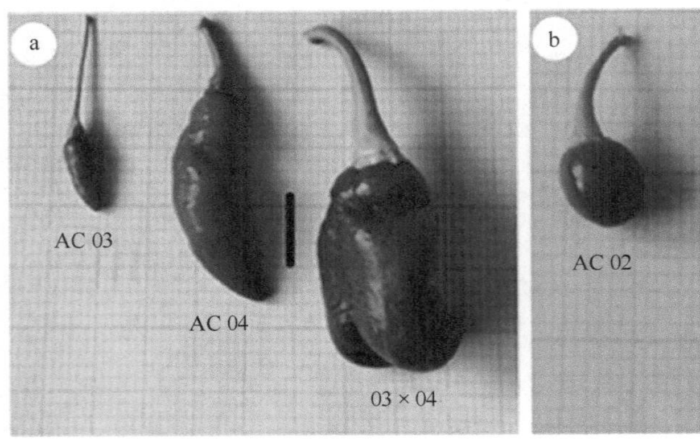

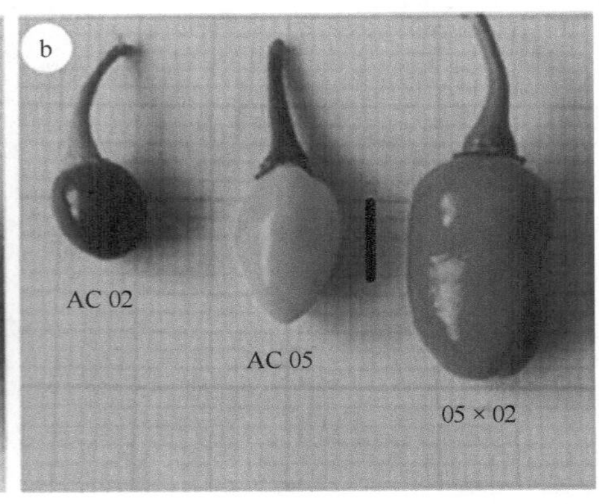

图 22.9　观赏辣椒嫁接诱导的果实形状的遗传变异。(a) 左侧：株系 UFPB-03（AC 03）(*C. frutescens* L.) 为砧木，小果；中间：株系 UFPB-04（AC 04）(*C. chinense* Jacq.) 为接穗，大果；右侧：来自 03 × 04 嫁接组合的最大果，比砧木和接穗果实大。(b) 左侧：株系 UFPB-02（AC 02）(*C. chinense* Jacq.) 为砧木，红色小球形果；中部：株系 UFPB-05（AC 05）(*C. chinense* Jacq.) 为砧木，黄色细长果实；右侧：来自 05 × 02 嫁接组合的最大果（标尺 = 1.0cm）(编译自 Andrade-Júnior 2017)

据 Albacete 等（2015），采用砧木可以有助于产生新的有用的基因型变异（通过表观遗传学）和创造质量更高的新产品。嫁接的潜力与遗传变异一样强大，能够跨越砧木和接穗之间的潜在不相容性障碍。

此外，可耕作物的表型变异通常反映了基因型 × 环境（G × E）相互作用；而在嫁接植物中，由于结合了两种不同的基因型，表型更加复杂，砧木（R）和接穗（S）之间的交流驱使 R × S × E 相互作用（Albacete et al. 2015）。

22.10 结 论

尽管世界各地的研究人员做了种种努力，但仍需要进行更多的研究来解析决定数量性状和辣椒属种群内和种群间变异性的遗传规律，其中许多仍然未知。同样对于辣椒属的基因组学、转录组学、蛋白质组学和代谢组学，从消费市场对更新、更优品种的需求来判断，我们知道了生物技术在观赏辣椒遗传改良中的应用潜力。我们的研究小组利用生物技术开发了新产品，使用不成熟合子胚技术开发了纯合品系和新品种，并缩短了繁殖周期，但仍有许多工作要做，特别是在测序技术普遍应用的大背景下。

本章译者：
阮继伟
云南省农业科学院花卉研究所，国家观赏园艺工程技术研究中心，云南省花卉育种重点实验室，昆明 650200

参考文献

Albacete A, Martínez-Andújar C, Ghanem ME, Acosta M, Sánchez- Bravo J, Asins MJ, Cuartero J, Lutts S, Dodd IC, Pérez-Alfocea F (2009) Rootstock-mediated changes in xylem ionic and hormonal status are correlated with delayed leaf senescence, and increased leaf area and crop productivity in salinized tomato. Plant Cell Environ 32:928–938

Albacete A, Martínez-Andújar C, Martínez-Pérez A, Thompson AJ, Dodd IC, Pérez-Alfocea F (2015) Unravelling rootstock x scion interactions to improve food security. J Exp Bot 66(8):2211–2226

Allard RW (1971) Princípios do melhoramento genético das plantas. Edgard Blucher, São Paulo. 381p

Aman UM, Zora S (2004) Endogenous free polyamines of mangos in relation to development and ripening. J Am Soc Hortic Sci 129:280–286

Aman UM, Zora S, Ahmad SK (2005) Role of polyamines in fruit development, ripening, chilling injury, storage and quality of mango and other fruits: a review. In: Proceedings of the international conference on mango and date palm: culture and export, University of Agriculture, Faisalabad, Pakistan, 20–23 June 2005, pp 182–187

Barboza GE, Bianchetti LB (2005) Three new species of *Capsicum* (Solanaceae) and a key to the wild species from Brazil. Syst Bot 30(4):863–871

Bărcanu E, Vînătoru C, Zamfir B, Bratu C, Drăghici E (2017) Characterization of new ornamental chilli genotypes created at VRDS Buzău. Scientific papers. Series B, Horticulture, vol LXI. Print ISSN 2285-5653, CD-ROM ISSN 2285-5661, Online ISSN 2286-1580, ISSN-L2285-5653

Barchenger DW, Coon DL, Bosland PW (2016) Efficient breeder seed production utilizing ethephon to promote floral and fruit abscission in ornamental chile peppers. HortTechnology 26:30–35

Barroso, PA (2016) Cultura de anteras e de embriões zigóticos imaturos no melhoramento de pimenteiras ornamentais (Capsicum annuum L.). PhD thesis, UFPB, Areia, PB, 75p

Barroso PA, Rêgo ER, Rêgo MM, Nascimento KS, Nascimento NFF, Nascimento MF, Soares WS, Ferreira KTC, Otoni WC (2012) Analysis of segregating generation for components of seedling and plant height of pepper (*Capsicum annuum L.*) for medicinal and ornamental purposes. Acta Hortic 953:269–275

Barroso PA, Dos Santos-Pessoa AM, Medeiros GDA, Da Silva-Neto JJ, Rêgo ER, Rêgo MM (2015a) Genetic control of seed germination and physiological quality in ornamental pepper. Acta Hortic 1087:409–413

Barroso PA, Rego MM, Rego ER, Soares WS (2015b) Embryogenesis in anthers from ornamental pepper plants (*Capsicum annum* L.). Genet Mol Res 14(4):13349–13363

Barroso PA, Rêgo MM, Rêgo ER, Ferreira KTC (2017) Genetic effects of in vitro germination and plantlet development in chilli pepper. Genet Mol Res 16(3). https://doi.org/10.4238/gmr16038869

Ben-Chaim A, Grube R, Lapidot M, Jahn M, Paran I (2001) Identification of quantitative trait loci associated with resistance to cucumber mosaic virus in Capsicum annuum. Theor Appl Genet 102:1213–1220

Ben-Chaim A, Borovsky E, Rao GU, Tanyolac B, Paran I (2003) *fs*3.1: a major fruit shape QTL conserved in Capsicum. Genome 46:1–9

Ben-Chaim A, Borovsky Y, Falise M, Mazourek M, Kang B-C, Paran I et al (2006) QTL analysis for capsaicinoid content in Capsicum. Theor Appl Genet 113:1481–1490

Bosland PW (1993) An effective plant field-cage to increase the production of genetically pure chile (Capsicum spp.) seed. HortScience 28:1053

Bosland PW (1996) Capsicums: innovative uses of an ancient crop. In: Janick J (ed) Progress in new crops. ASHS Press, Arlington, VA, pp 479–487

Bosland PW (1999) Encyclopedia of chiles. In: Hanson B (ed) Chile peppers, Brooklyn Botanical Garden, handbook series. Brooklyn Botanical Garden, Brooklyn, NY, pp 17–21

Bosland PW, Votava EJ (2003) Peppers: vegetable and spice capsicums. CABI Publishing, Wallingford, UK. 204p

Casali VWD, Couto FAA (1984) Origem e botânica de *Capsicum*. Inf Agrop 10(113):8–10

Cavalcante LC (2015) Caracterização morfoagronômica de mutantes de pimenta biquinho (Capsicum chinence Jacy) submetidas a radiação gama. Dissertação de mestrado, Programa de Pós-Graduação em Agronomia-Centro de Ciências Agrárias-Universidade Federal da Paraíba

Cheng Y, Ma R, Jiao Y, Qiao N, Li T (2013) Impact of genotype, plant growth regulators and activated charcoal on embryogenesis induction in microspore culture of pepper (*Capsicum annuum* L.). S Afr J Bot 88:306–309

Corrêa LB, Barbieri RL, Silva JB (2007) Caracterização da viabilidade polínica em acessos de Capsicum (Solanaceae). Rev Bras Biosci Porto Alegre 5(supl. 1):660–662

Crispim JG, Rêgo ER, Rêgo MM, Nascimento NFF, Barroso PA (2017) Stigma receptivity and anther dehiscence in ornamental pepper. Hortic Bras 35(4):609–612

Da Silva AR, Nascimento M, Cecon PR, Sapucay MJ, Do Rêgo ER, Barbosa LA (2013) Path analysis in multicollinearity for fruit traits of pepper. IDESIA (Chile) 31(2):55–60

Daskalov S (1968) A male sterile pepper (*C. annuum* L.) mutant. Theor Appl Genet 3(8):370–372

Daskalov S (1973) Investigation of induced mutants in Capsicum annuum L. III. Mutants in the variety Zlaten medal. Genet Plant Breed 6:419–429. [in Bulgarian]

Daskalov S (1974) Investigations on induced mutants in sweet pepper (*Capsicum annuum* U). In: EUCARPIA (ed) Genetics an breeding of *Capsicum*. Horticultural Research Institute, Budapest, pp 81–90

Daskalov S (1986) Mutation breeding in pepper. IAEA, Vienna, pp 1–27

De França CFM, Dole JM, Carlson AS, Finger FL (2017) Effect of postharvest handling procedures on cut Capsicum stems. Sci Hortic 220:310–316

De Mesquita JCP, Rêgo ER, Silva AR, Silva Neto JJ, Cavalcante LC, Rêgo MM (2016) Multivariate analysis of the genetic divergence among populations of ornamental pepper (*Capsicum annuum* L.). Afr J Agric Res 11(42):4189–4194

Debbarama C, Khanna VK, Tyagi W et al (2013) Wide hybridization and embryo-rescue for crop improvement in *Capsicum*. Agrotechnology S11:003. https://doi.org/10.4172/2168-9881S11–003

Demorest ZL, Coffman A, Baltes NJ, Stoddard TJ, Clasen BM, Luo S, Retterath A, Yabandith A, Gamo ME, Bissen J, Mathis L, Voytas DF, Zhang F (2016) Direct stacking of sequence-specific nuclease-induced mutations to produce high oleic and low linolenic soybean oil. BMC Plant Biol 16:225

Devi AA, Singh NB, Singh MD (2017) Classification and characterization of chilli (Capsicum annuum L.) found in Manipur using multivariate analysis. Electron J Plant Breed 8(1):324–330

Djian-Caporilano C, Lefebvre V, Sage-Daubeze AM, Palloix A (2007) Capsicum. In: Singh RJ (ed) Genetic resources, chromosome engineering, and crop improvement: vegetable crops 3. CRC Press, Boca Raton, pp 185–243

Dumas de Vaulx R, Chambonnet D, Pochard E (1981) Culture *in vitro* d'antheres de piment (Capsicum annuum): amelioration des taux d'obtention de plantes chez different genotypes par traitments a+35 °C. Agronomie 1:859–864

Fehr WR (1987) Principles of cultivar development: theory and technique, vol 1. Macmillian Publication, New York. 736p

Ferreira KTC, Rêgo ER, Rêgo MM, Fortunato FLG, Nascimento NFF, De Lima JAM (2015) Combining ability for morpho-agronomic traits in ornamental pepper. Acta Hortic 1087:187–194

Finger FL, Rêgo ER, Segatto FB, Nascimento NFF, Rêgo MM (2012) Produção e potencial de mercado para pimenta ornamental. Inf Agrop 33(267):14–20

Fortunato FLG, Rêgo ER, Rêgo MM, Pereira Dos Santos CA, Gonçalves De Carvalho M (2015) Heritability and genetic parameters for size-related traits in ornamental pepper (*Capsicum annuum* L.). Acta Hortic 1087:201–206

Garcés-Claver A, Fellman SM, Gil-Ortega R, Jahn M, Arnedo-Andrés MS (2007) Identification, validation and survey of a single nucleotide polymorphism (SNP) associated with pungency in *Capsicum* spp. Theor Appl Genet 115:907–916

García-Gaytán V, Trejo-Téllez LI, Gómez-Merino FC, García-Morales S, Tejeda-Sartorius O, Ramírez-Martínez M, Delgadillo-Martínez J (2017) Gamma radiation and osmotic potential of the nutrient solution differentially affect macronutrient concentrations, pH and EC in chilhuacle pepper fruits. J Radioanal Nucl Chem. https://doi.org/10.1007/s10967-017-5655-6 315:145–156

Glew RH, Ayaz FA, Sanz C, VanderJagt DJ, Huang HS, Chuang LT, Strnad M (2003) Changes in sugars, organic acids and amino acids in medlar (Mespilus germanica L.) during fruit development and maturation. Food Chem 83:363–369

Gopalkrishnan KM, Naidu R, Sreedhar D (1993) Shortening breeding cycle through immature embryo culture in sunflower (*Helianthus annuus* L.). Helia 16:61–68

Guha S, Maheshwari SC (1964) *In vitro* production of embryos from anthers of *Datura*. Nature 204:497

Heidmann I, Boutilier K (2015) Pepper, sweet (*Capsicum annuum*). In: Wang K (ed) *Agrobacterium protocols: volume 1*, methods in molecular biology, vol 1223. Springer Science Business Media, New York, pp 321–334. https://doi.org/10.1007/978-1-4939-1695-5_25

Heiser CB Jr (1979) Peppers – *Capsicum* (Solanaceae). In: Simmonds NW (ed) Evolution of crop plants. Longman, Harlow, pp 265–273

Hurtado-Hernandez H, Smith PG (1985) Inheritance of mature fruit color in *Capsicum annuum* L. J Hered 76:211–213

International Board for Plant Genetic Resources (IBPGR) (1983) Genetics resources of *Capsicum*, a global plan and action. IBPGR, Rome. 49p

Irikova T, Grozeva S, Denev I (2012) Identification of BABY BOOM and LEAFY COTYLEDON genes in sweet pepper (*Capsicum annuum* L.) genome by their partial gene sequences. Plant Growth Regul 67:191–198

Jang YK, Jung ES, Hun-Ah L, Choi D, Lee CH (2015) Metabolomic characterization of hot pepper (*Capsicum annuum* "CM334") during fruit development. J Agric Food Chem 63(43):9452–9460

Jeong H-L, Jo YD, Park S-W, Kang B-C (2010) Identification of *Capsicum* species using SNP markers based on high resolution melting analysis. Genome 53:1029–1040

Jones JB, Minsavage GV, Roberts PD, Johnson RR, Kousik CS, Subramanian S, Stall RE (2002) A nonhypersensitive resistance in pepper to the bacterial spot pathogen is associated with two recessive genes. Phytopathology 92:273–277

Kim S et al (2014) Genome sequence of the hot pepper provides insights into the evolution of pungency in *Capsicum* species. Nat Genet 46(3):270–279

King SR, Davis AR, Zhang X, Crosby K (2010) Genetics, breeding and selection of rootstocks for Solanaceae and Cucurbitaceae. Sci Hortic 127:106–111

Kokalis-Burelle N, Bausher MG, Rosskopf EN (2009) Greenhouse evaluation of *Capsicum* rootstocks for management of *Meloidogyne incognita* on grafted bell pepper. Nematropica 39(1):121–132

Kormos J, Kormos J (1955) Affinity in grafting. Ann Inst Biol Tihany 23:161–175

Kothari SL, Joshi A, Kachhwaha S, Ochoa-Alejo N (2010) Chilli peppers – a review on tissue culture and transgenesis. Biotechnol Adv 28:35–48

Kyle MM, Palloix A (1997) Proposed revision of nomenclature for potyvirus resistance genes in Capsicum. Euphytica 97:183–188

Lantos C, Juhász AG, Somogy G, Ötvös K, Vági P, Mihály R et al (2009) Improvement of isolated microspore culture of pepper (*Capsicum annuum* L.) via co-culture with ovary tissues of pepper or wheat. Plant Cell Tissue Organ Cult 97:285–293

Lee JM, Oda M (2010) Grafting of herbaceous vegetables and ornamental crops. Hortic Rev 28:62–124

Lee H-Y, Ro N-Y, Jeong H-J et al (2016) Genetic diversity and population structure analysis to construct a core collection from a large *Capsicum* germplasm. BMC Genet 17:142. https://doi.org/10.1186/s12863-016-0452-8

Lefebvre V, Palloix A, Caranta C, Pochard E (1995) Construction of an intraspecific integrated linkage map of pepper using molecular markers and double haploid progenies. Genome 38:112–121

Li T, Xu X, Li Y, Wang H, Li Z, Li Z (2015) Comparative transcriptome analysis reveals differential transcription in heat-susceptible and heat-tolerant pepper (*Capsicum annum* L.) cultivars under heat stress. J Plant Biol 58:411–424

Li J, Yang P, Kang J, Gan Y, Yu J, Calderón-Urrea A, Lyu J, Zhang Z, Feng Z, Xie J (2016) Transcriptome analysis of pepper (*Capsicum annuum*) revealed a role of 24-epibrassinolide in response to chilling. Front Plant Sci 7:1–17

Liu W, Parrott WA, Hildebrand DF, Collins GB, Williams EG (1990) *Agrobacterium* induced gall formation in bell pepper (*Capsicum annuum* L.) and formation of shoot-like structures expressing introduced genes. Plant Cell Rep 9:360–364

Manzur JP, Oliva-Alarcón M, Rodríguez-Burruezo A (2014) In vitro germination of immature embryos for accelerating generation advancement in peppers (*Capsicum annuum* L.). Sci Hortic 170:203–210

Martin J, Grawford JH (1951) Several types of sterility in *Capsicum frutescens*. Proc Am Soc Hortic Sci 57:335–338

Martínez-López LA, Ochoa-Alejo N, Martínez O (2014) Dynamics of the chili pepper transcriptome during fruit development. BMC Genomics 15:143–161

Mazourek M, Pujar A, Borovsky Y, Paran I, Mueller L, Jahn MM (2009) A dynamic interface for capsaicinoid systems biology. Plant Physiol 150(4):1806–1821

Medeiros GDA, Rêgo ER, Barroso PA, Ferreira KTC, Pessoa AMDS, Rêgo MM, Crispim JG (2015) Heritability of traits related to germination and morphogenesis in vitro in ornamental peppers. Acta Hortic 1087:403–408

Min J, Shin SH, Hyun JY, Harn CH (2015) Pepper, chili (*Capsicum annuum*). In: Wang K (ed) *Agrobacterium protocols: volume 1*, methods in molecular biology, vol 1223. Springer, New York, pp 311–320. https://doi.org/10.1007/978-1-4939-1695-5_24

Monteiro CES, Pereira TNS, Campos KP (2011) Reproductive characterization of interspecific hybrids among Capsicum species. Crop Breed Appl Biotechnol 11(3):241–249

Morra L (2004) Grafting in vegetable crops. In: Tognoni F, Pardossi A, Mensuali-Sodi A, Dimauro B (eds) The production in the greenhouse after the era of the methyl bromide, Comiso, pp 147–154 Symposium in Comiso City, Italy.

Moscone EA, Escaldaferro MA, Gabrielle M, Cecchini NM, García YS, Jarret R, Daviña JR, Ducasse DA, Barboza GE, Ehrendorfer F (2007) The evolution of the chili pepper (*Capsicum* – Solanaceae): a cytogenetic perspective. Acta Hortic 745:137–169

Naegele RP, Mitchell J, Hausbeck MK (2016) Genetic diversity, population structure, and heritability of fruit traits in Capsicum annuum. PLoS One 11(7):e0156969

Nascimento MF, Finger FL, Bruckner CH, Silva-Neto JJ, Rêgo MM, Nascimento NFF, Rêgo ER (2012) Heritability and variability of morphological traits in a segregating generation of ornamental pepper. Acta Hortic 953:299–304

Nascimento NFF, Nascimento MF, Santos RMC, Bruckner CH, Finger FL, Rego ER, Rego MM (2013) Flower color variability in double and three-way hybrids of ornamental peppers. Acta Hortic 1000:457–464

Nascimento NFF, Rêgo ER, Nascimento MF, Bruckner CH, Finger FL, Rêgo MM (2014) Combining ability for yield and fruit quality in the pepper *Capsicum annuum*. Genet Mol Res 13:3237–3249

Nascimento MF, Rêgo ER, Nascimento NF, Santos R, Bruckner CH, Finger FL, Rêgo MM (2015) Correlation between morphoagronomic traits and resistance to ethylene action in ornamental peppers. Hortic Bras 33(2):151–154

Neitzke RS, Fischer SZ, Barbieri R, Treptow R (2016) Pimentas ornamentais: aceitação e preferências do público consumidor. Hortic Bras 34:102–109

Nimisha S, Kherwar D, Ajay KM et al (2013) Molecular breeding to improve guava (*Psidium guajava* L.): current status and future prospective. Sci Hortic 164:578–588

Olszewska D, Kisiała A, Niklas-Nowak A, Nowaczyk P (2014) Study of in vitro anther culture in selected genotypes of genus *Capsicum*. Turk J Biol 38:118–124

Osorio S, Alba R, Nikoloski Z, Kochevenko A, Fernie AR, Giovannoni JJ (2012) Integrative comparative analyses of transcript and metabolite profiles from pepper and tomato ripening and development stages uncovers species-specific patterns of network regulatory behavior. Plant Physiol 159:1713–1729

Paran I (2003) Marker assisted utilization of exotic germplasm. In: Nguyen HT, Blum A (eds) Physiology and biotechnology integration for plant breeding. Marcel Dekker, New York

Paran I, Rouppe van der Voort J, Lefebvre V, Jahn M, Landry L, van Schriek M, Tanyolac B, Caranta C, Ben Chaim A, Livingstone K, Palloix A, Peleman J (2004) An integrated genetic linkage map of pepper (*Capsicum* spp.). Mol Breed 13:251–261

Paran I, Ben-Chaim A, Kang B-C, Jahn M (2007) Capsicums. In: Kole C (ed) Genome mapping and molecular breeding in plants, vol 5: Vegetables, pp 209–226

Patil SSA, Salimath PM (2008) Estimation of gene effects for fruit yield and its components in chili (*Capsicum annuum L.*). J Agric Sci 21(2):181–183

Peng A, Chen S, Lei T, Xu L, He Y, Wu L, Yao L, Zou X (2017) Engineering canker-resistant plants through CRISPR/Cas9-targeted editing of the susceptibility gene CsLOB1 promoter in citrus. Plant Biotechnol J 15:1509. https://doi.org/10.1111/pbi.12733

Pessoa AM, Rêgo ER, Barroso PA, Rêgo MM (2015) Genetic diversity and importance of morpho-agronomic traits in a segregating F_2 population of ornamental pepper. Acta Hortic 1087:195–200

Peterson PA (1958) Cytoplasmically inherited male sterility in *Capsicum*. Am Nat 92:111–119

Pickersgill B (1997) Genetic resources and breeding of *Capsicum* spp. Euphytica 96:129–133

Pochard E, Dumas de Vaulx R (1979) Haploid parthenogenesis in *Capsicum annuum* L. In: Hawkes JG, Lester RN, Skelding AD (eds) The biology and taxonomy of the Solanaceae. Academy Press, London

Prasad BCN, Gururaj HB, Kumar V, Giridhar P, Ravishankar GA (2006) Valine pathway is more crucial than phenyl propanoid pathway in regulating capsaicin biosynthesis in Capsicum frutescens mill. J Agric Food Chem 54:6660–6666

Quirin EA, Ogundiwin EA, Prince JP, Mazourek M, Briggs MO, Chlanda TS, Kim KT, Falise M, Kang BC, Jahn MM (2005) Development of sequence characterized amplified region (SCAR) primers for the detection of Phyto.5.2, a major QTL for resistance to Phytophthora capsici Leon. in pepper. Theor Appl Genet 110:605–612

Raghavan T, Venkatasubban K (1940) Studies in the South Indian Chillies. I. A description of the varieties, chromosome numbers and the cytology of some X-ray derivatives in *Capsicum annuum* L. Proc Indian Acad Sci B 12:29–46

Rao GU, Ben Chaim A, Borovsky E, Paran I (2003) Mapping of yield related QTLs in pepper in an inter-specific cross of Capsicum annuum and C. frutescens. Theor Appl Genet 106:1457–1466

Rao GU, Paran I (2003) Polygalacturonase: a candidate gene for the soft flesh and deciduous fruit mutation in Capsicum. Plant Mol Biol 51:135–141

Reddy MG, Kumar HDM, Salimath PM (2008) Combining ability analysis in chilli (*Capsicum annuum* L.). Karnataka J Agric 21:494–497

Rêgo ER, Rêgo MM (2016) Genetics and breeding of chili pepper *Capsicum* spp. In: Rêgo ER, Rêgo MM, Finger FL (eds) Production and breeding of chilli peppers (*Capsicum* spp.). Springer International Publishing, Cham, pp 1–129

Rêgo ER, Rêgo MM, Silva DF, Cortez RM, Sapucay MJLC, Silva DR, Silva Junior SJ (2009a) Selection for leaf and plant size and longevity of ornamental peppers (Capsicum spp.) grown in greenhouse condition. Acta Hortic 829:371–375

Rêgo ER, Rego MM, Finger FL, Cruz CD, Casali VWD (2009b) A diallel study of yield components and fruit quality in chilli pepper (Capsicum *baccatum*). Euphytica 168:275–287

Rêgo ER, Silva DF, Rêgo MM, Santos RMC, Sapucay MJLC, Silva DR, Silva Júnior SJ (2010) Diversidade entre linhagens e importância de caracteres relacionados à longevidade em vaso de linhagens de pimenteiras ornamentais. Rev Bras Hortic Ornam 16:165–168

Rêgo ER, Rêgo MM, Matos IWF, Barbosa LA (2011a) Morphological and chemical characterization of fruits of *Capsicum* spp. accessions. Hortic Bras 29:364–371

Rêgo ER, Rêgo MM, Cruz CD, Finger FL, Casali VWD (2011b) Phenotypic diversity, correlation and importance of variables for fruit quality and yield traits in Brazilian peppers (*Capsicum baccatum*). Genet Resour Crop Evol 58:909–918

Rêgo ER, Finger FL, Nascimento MF, Barbosa LAB, Santos RMC (2011c) Pimenteiras ornamentais. In: Rêgo ER, Finger FL, Rêgo MM (eds) Produção, Genética e Melhoramento de Pimentas (Capsicum spp.), vol 1. Imprima, Recife, pp 205–223
Rêgo ER, Finger FL, Rêgo MM (2012a) Consumption of pepper in Brazil and its implications on nutrition and health of humans and animals. In: Salazar MA, Ortega JM (eds) Pepper: nutrition, consumption and health, vol 1. Nova Science Publishers, Inc, New York, pp 159–170
Rêgo ER, Rêgo MM, Costa FR, Nascimento NFF, Nascimento MF, Barbosa LA, Fortunato FLG, Santos RMC (2012b) Analysis of diallel cross for some vegetative traits in chili pepper. Acta Hortic 937:297–304
Rêgo ER, Fortunato FLG, Nascimento MF, Nascimento NFF, Rêgo MM, Finger FL (2012c) Inheritance of earliness in ornamental pepper (*Capsicum annuum*). Acta Hortic 961:405–410
Rêgo ER, Finger FL, Rêgo MM (2012d) Types, uses and fruit quality of Brazilian chili peppers. In: Johnathan F (ed) Spices: types, uses and health benefits, vol 1. Nova Science Publishers, Inc, New York, pp 1–70
Rêgo ER, Nascimento MF, Nascimento NFF, Santos RMC, Fortunato FLG, Rêgo MM (2012e) Testing methods for producing self-pollinated fruits in ornamental peppers. Hortic Bras 30:708–711
Rêgo MM, Barroso PA, Rêgo ER, Santos WS, Nascimento KS, Otoni WC (2012f) Diallelic analysis during in vitro seed germination in ornamental chili pepper. Acta Hortic 1099:765–769
Rêgo ER, Rêgo MM, Finger FL (2015a) Methodological basis and advances for ornamental pepper breeding program in Brazil. Acta Hortic 1087:309–314
Rêgo MM, Sapucay MJLC, Rêgo ER, Araújo ER (2015b) Analysis of divergence and correlation of quantitative traits in ornamental pepper (*Capsicum* SPP.). Acta Hortic 1087:389–394
Rêgo ER, Rêgo MM, Barroso PA (2016) Tissue culture of *Capsicum* spp. In: Rêgo ER, Rêgo MM, Finger FL (eds) Production and breeding of chilli peppers (*Capsicum* spp.). Springer International Publishing, Cham, pp 97–127
Reyes-Escogido MDL, Gonzalez-Mondragon EG, Vazquez-Tzompantzi E (2011) Chemical and pharmacological aspects of capsaicin. Molecules 16(2):1253–1270
Rodrigues R, Batista FRC, Moulin MM (2016) Molecular markers in Capsicum spp. breeding. In: Rêgo ER, Rêgo MM, Finger FL (eds) Production and breeding of chilli peppers (*Capsicum* spp.). Springer Verlag, Heidelberg, pp 121–142
Rodriguez-Burruezo A, Prohens J, Raigón MD, Nuez F (2009) Variation for bioactive compounds in ají (*Capsicum baccatum* L) and rocoto (*C. pubescens* R and P) and implications for breeding. Euphytica 170:169–181
Santos RMC, Rêgo ER, Nascimento MF, Nascimento NFF, Rêgo MM, Borém A, Finger FL, Costa DS (2013a) Ethylene resistance in a F_2 population of ornamental chili pepper (*Capsicum annuum*). Acta Hortic 1000:433–438
Santos RMC, Rêgo ER, Borém A, Nascimento NFF, Nascimento MF, Finger FL, Carvalho GC, Lemos RC, Rêgo MM (2013b) Ornamental pepper breeding: could a chili be a flower ornamental plant. Acta Hortic 1000:451–456
Santos RMC, Rêgo ER, Borém A, Nascimento MF, Nascimento NFF, Finger FL, Rêgo MM (2014) Epistasis and inheritance of plant habit and fruit quality traits in ornamental pepper (*Capsicum annuum* L.). Genet Mol Res 13(4):8876–8887
Segatto FB, Finger FL, Rêgo ER, Pinto CMF (2013) Effects of ethylene on the post-production of potted ornamentals peppers (*Capsicum annuum*). Acta Hortic 1000:217–222
Shifriss C (1997) Male sterility in pepper (*Capsicum annuum* L.). Euphytica 93:83–88
Shifriss C, Frankel R (1969) A new male sterility gene in *Capsicum annuum* L. J Am Soc Hortic Sci 94:385–387
Shin GR, Lee S, Lee S, Do SG, Shin E, Lee CH (2015) Maturity stage-specific metabolite profiling of *Cudrania tricuspidata* and its correlation with antioxidant activity. Ind Crop Prod 70:322–331
Siddique MA, Grossmann J, Gruissem W, Baginsky S (2006) Proteome analysis of bell pepper (*Capsicum annuum* L.) chromoplasts. Plant Cell Physiol 47(12):1663–1673
Silva Neto JJ, Rêgo ER, Nascimento MF, Silva Filho VAL, Almeida Neto JX, Rêgo MM (2014) Variabilidade em população base de pimenteiras ornamentais (*Capsicum annuum* L.). Rev Ceres 61(1):84–89
Silva AR, Rêgo ER, Cecon PR (2011) Sample size for morphological characterization of pepper fruits. Hortic Bras 29:125–129

Silva CQ, Jasmim JM, Santos JO, Bento CS, Sudré CP, Rodrigues R (2015) Phenotyping and selecting parents for ornamental purposes in chili pepper accessions. Hortic Bras 33(1):66–73

Silva ARD, Rêgo ER, Pessoa AMDS, Rêgo MM (2016) Correlation network analysis between phenotypic and genotypic traits of chili pepper. Pesq Agrop Bras 51(4):372–377

Silva CQ, Rodrigues R, Bento CS, Pimenta S (2017) Heterosis and combining ability for ornamental chili pepper. Hortic Bras 35(3):349–357

Stewart C Jr, Kang BC, Liu K, Mazourek M, Moore SL, Yoo EY, Kim BD, Paran I, Jahn MM (2005) The Pun1 gene for pungency in pepper encodes a putative acyltransferase. Plant J 42(5):675–688

Stommel JR, Bosland PW (2006) Ornamental pepper, *Capsicum annuum*. In: Anderson N (ed) Flower breeding and genetics: issues, challenges and opportunities for the 21st century. Springer, Dordrecht, pp 561–599

Supena EDJ, Custers JBM (2011) Refinement of shed-microspore culture protocol to increase normal embryos production in hot pepper (*Capsicum annuum* L.). Sci Hortic 130(4):769–774

Supena EDJ, Suharsono S, Jacobsen E, Custers JBM (2006) Successful development of a shed microspore culture protocol for doubled haploid production in Indonesian hot pepper (*Capsicum annuum* L.). Plant Cell Rep 25:1–10

Taller J, Hirata H, Yagishita N, Kita M, Ogata S (1998) Graft-induced genetic changes and the inheritance of several characteristics in pepper (*Capsicum annuum* L.). Theor Appl Genet 97(5):705–713

Tanksley SD, Bernatzky R, Lapitan NL, Prince JP (1988) Con- servation of gene repertoire but not gene order in pepper and tomato. Proc Natl Acad Sci U S A 85:6419–6423

Tateishi K (1927) Grafting watermelon onto pumpkin. J Jpn Hortic 39:5–8

Terzyan P, Batikyan H, Sahakyan T (1974) The effect of X-rays on mitotic activity and frequency of structural chromosome rearrangements in root cells of the species *C. annuum* L. (*in Armenian*). Biologicheskij Zhurnal Armenii 27:35–39

Townsend JA, Wright DA, Winfrey RJ, Fu F, Maeder ML, Joung JK, Voytas DF (2009) High-frequency modification of plant genes using engineered zinc-finger nucleases. Nature 459:442–445

Van Zonneveld M, Ramirez M, Williams DE, Petz M, Meckelmann S, Avila T, Bejarano C, Rios L, Peña L, Jäger M, Libreros D, Amaya K, Scheldeman X (2015) Screening genetic resources of *Capsicum* peppers in their primary center of diversity in Bolivia and Peru. PLoS One 10(9):e0134663

Voorrips RE, Finkers R, Sanjaya L, Groenwold R (2004) QTL mapping of anthracnose (Colletotrichum spp.) resistance in a cross between *Capsicum annuum* and *C. chinense*. Theor Appl Genet 109:1275–1282

Wahyuni Y, Ballester AR, Tikunov Y, de Vos RCH, Pelgrom KTB, Maharijaya A, Sudarmonowati E, Bino RJ, Bovy AG (2013) Metabolomics and molecular marker analysis to explore pepper (Capsicum sp.) biodiversity. Metabolomics 9:130–144

Wang YY, Sun CS, Wang CC, Chien NJ (1973) The induction of pollen plantlets of *Triticale* and *Capsicum annuum* anther culture. Sci Sinica 16:147–151

Wang Y, Tang X, Cheng Z, Mueller L, Giovannoni J, Tanksley SD (2006) Euchromatin and pericentromeric heterochromatin: comparative composition in the tomato genome. Genetics 172(4):2529–2540

Wongpia A, Lomthaisong K (2010) Changes in the 2DE protein profiles of chilli pepper (*Capsicum annuum*) leaves in response to *Fusarium oxysporum* infection. ScienceAsia 36:259–270

Wu Z, Cheng J, Qin C, Hu Z, Yin C, Hu K (2013) Differential proteomic analysis of anthers between cytoplasmic male sterile and maintainer lines in *Capsicum annuum* L. Int J Mol Sci 14:22982–22996. https://doi.org/10.3390/ijms141122982

Xie LB, Wang X, Peng M, Zhou Y, Chen LX, Liu LX, Gao YL, Guo YH (2017) Comparative proteome analysis in hot pepper (*Capsicum annuum* L.) after space flight. ΦYTON 86:236–245

Yahia EM, Contreras-Padilla M, Gonzalez-Aguilar G (2001) Ascorbic acid content in relation to ascorbic acid oxidase activity and polyamine content in tomato and bell pepper fruits during development, maturation and senescence. LWT Food Sci Technol 34:452–457

Zhang A, Sun H, Wang P, Han Y, Wang X (2012) Modern analytical techniques in metabolomics analysis. Analyst 137(2):293–300

Zhang H, Zhang J, Lang Z, Botella JR, Zhu J-K (2017) Genome editing – principles and applications for functional genomics research and crop improvement. Crit Rev Plant Sci 36(4):291–309

第23章 蝴 蝶 兰

Chia-Chi Hsu, Hong-Hwa Chen, and Wen-Huei Chen

摘要 蝴蝶兰是世界上最受欢迎的栽培兰花之一。迄今为止，英国皇家园艺协会（RHS）已注册了蝴蝶兰（*Phalaenopsis*）的92个原生种和34 112个杂交种，但仅有18个原生种被频繁用于蝴蝶兰育种。在蝴蝶兰市场上，颜色丰富的大花型最受欢迎。南洋白花蝴蝶兰（*Phal. amabilis*）和菲律宾白花蝴蝶兰（*Phal. aphrodite*）是用于繁殖白色大花型蝴蝶兰的主要品种。杂交蝴蝶兰颜色丰富，有粉红色、紫红色、绿色、橙黄色和黑色，而且带红色唇瓣的花瓣也有多种颜色。西蕾丽蝴蝶兰（*Phal. schilleriana*）和桑德蝴蝶兰（*Phal. sanderiana*）是大红花的主要亲本，而桃红蝴蝶兰（*Phal. equestris*）和五唇兰（*Phal. pulcherrima*）对中小型红花蝴蝶兰的发展很重要。裂唇蝴蝶兰亚属（*Polychilos*）的品种是橙黄色花蝴蝶兰最重要的祖先。最近，人们对带有红色条纹、红色斑点的白花、粉花或黄花的兴趣日益增加。这些特征是由林登蝴蝶兰（*Phal. lindenii*）、史塔基蝴蝶兰（*Phal. stuartiana*）和安汶蝴蝶兰（*Phal. amboinensis*）导入的，并且与畸变花和大脚花一样，具有小丑斑点和复杂颜色图案的杂色花已投放到市场。除颜色外，香味和抗性也逐渐成为蝴蝶兰育种的重要目标。

关键词 大脚花；育种；杂色花；兰花；畸变花；蝴蝶兰

23.1 引 言

蝴蝶兰因其外形而被命名为“phalaina”和“opsis”，通常称为“蝴蝶兰”。蝴蝶兰的花形独特而美丽，外侧具三片花萼，轮生，内侧具两片花瓣和一片较为特殊的唇瓣。雄蕊和心皮融合成的圆柱位于花朵中心。萼片和花瓣的形态与颜色非常相似，统称为花被。相反，唇瓣虽由特殊的花瓣衍生而来，但其形状和颜色与花瓣截然不同。特殊的唇瓣使花朵呈对称形，整朵花好似一只飞舞的蝴蝶。蝴蝶兰为总状花序，开花数量多，花期长。这一系列的特征使得蝴蝶兰成为目前兰花市场上最成功和最受欢迎的种属。

C.-C. Hsu
Department of Life Sciences, National Cheng Kung University, Tainan, Taiwan
H.-H. Chen
Department of Life Sciences, National Cheng Kung University, Tainan, Taiwan
Orchid Research and Development Center, National Cheng Kung University, Tainan, Taiwan
e-mail: hhchen@mail.ncku.edu.tw
W.-H. Chen (✉)
Orchid Research and Development Center, National Cheng Kung University, Tainan, Taiwan

根据英国皇家园艺协会的记录，并查询了兰花数据库软件可知，全世界共有 92 个蝴蝶兰原生种，包括合并基因的五唇兰（*Doritis*）、湿唇兰（*Hygrochilus*）、尖囊兰（*Kingidium*）、袋距兰（*Lesliea*）、象鼻兰（*Nothodoritis*）、羽唇兰（*Ornithochilus*）和萼脊兰（*Sedirea*）等品种（OrchidWiz 2017）。根据 Christenson（2001）的描述，这些原生种可分为 5 个亚属，即落叶蝴蝶兰亚属（*Aphyllae*）、柏氏蝴蝶兰亚属（*Parishianae*）、蝴蝶兰亚属（*Phalaenopsis*）、裂唇蝴蝶兰亚属（*Polychilos*）和长吻蝴蝶兰亚属（*Proboscidioides*）。只有 18 个种属于蝴蝶兰亚属和裂唇蝴蝶兰亚属这两个亚属，但这些品种是众多品种最常用的育种亲本。目前，已经繁殖出 34 112 个杂交种并正式注册于英国皇家园艺协会数据库中（OrchidWiz 2017）。

23.2　世界蝴蝶兰市场

蝴蝶兰是世界兰花市场上排名第一的盆栽观赏植物（Chen 2017）。根据 2014 年中国台湾蝴蝶兰种植者协会的调查数据，荷兰和中国台湾的蝴蝶兰种植者将大规模种植的蝴蝶兰几乎销售到全世界。

23.2.1　欧盟蝴蝶兰市场

欧洲市场是全球蝴蝶兰销售量最大的市场，达 1.6 亿盆/年。在欧洲，荷兰是最大的蝴蝶兰生产国。不同国家有不同的需求，在荷兰，具 2～3 个花箭的不同花型和颜色蝴蝶兰更受欢迎，而英国的消费者偏爱单花箭、大花型（花径＞10cm）品种，北欧国家的人们喜欢具 2 个花箭、中型花朵（花径 7～10cm）品种，意大利北部偏爱大花型品种，意大利南部的人们更喜爱小花型（花径 2～7cm）品种。

纵观欧洲蝴蝶兰市场，蝴蝶兰花型的主要发展趋势为多花、具 2～3 个花箭的中型花。大约 65%的蝴蝶兰产品是由 12cm 盆种植的 6.5～10cm 的花，另各有 15%的蝴蝶兰产品为由 9cm 盆种植的 6.5～10cm 的花或者由 6cm 盆种植的 4～5.5cm 的花，大花型（10～12cm）蝴蝶兰占比约 10%，种植于 15cm 的花盆中。蝴蝶兰价格因规格不同而异，一个、二个、三个花箭的售价分别为 1.5～2.5 欧元、3.5 欧元和 4～5 欧元。因此，易于产生多花箭和花朵的杂交种更受欢迎。

尽管高饱和度的深色花品种是市场首选，但其实各种颜色的蝴蝶兰在欧洲都很流行。其他一些重要选择要素则是易种植、易开花、抗病性、多花箭和多花等。

23.2.2　美国蝴蝶兰市场

在美国花卉市场，蝴蝶兰每年售出约 2400 万盆。美国兰花学会（AOS）是最大的蝴蝶兰育种协会。总体而言，美国人更喜欢大植株、大花型（花径 8～13cm）的品种（表 23.1）。美洲东部的人们喜欢大花型、白色花朵的品种，而美国西部的人们则喜欢各种颜色的大花型品种。

表 23.1　美国市场主流蝴蝶兰花型及其占比

	花盆型号	市场占比（%）	花径（cm）	花箭高度（cm）
东部	2″（6cm）	5	4.5～5	25～28
	3″（9cm）	20	6～7.5	42～45
	4″（12cm）	60	8～10	55～60
	5″（15cm）	15	10～13	60～70
西部	2″（6cm）	5	4.5～5	25～30
	3″（9cm）	30	5.8～7	35～42
	4″（12cm）	50	8.5～10	55～60
	5″（15cm）	15	10～13	60～70

对于大花型品种来说，具有 2 个花箭的植株比较畅销。花序需包含 9 朵以上的花，且花径应在 11cm 以上。白色花占据了 50%～70%的东方市场，其中 5%为具红色唇瓣的白花，10%～20%为粉色花和红色花。近年来，带红色唇瓣的黄色花朵和黑色花朵的比例有所增加。

中花型蝴蝶兰（花径 5～9cm）一般具有 2 个花箭，每箭上有 7 朵花，并有各种花色，包括粉红色、红色、紫红色、黄色和橙色、白花红唇或斑点、黄花红唇具条纹或斑点。小花型蝴蝶兰在美国并不流行。

23.2.3　亚洲蝴蝶兰市场

总体而言，亚洲蝴蝶兰市场每年估计有 8400 万盆销售量，其中中国占有 6000 万盆，日本占 1400 万盆，东南亚占 1000 万盆。中国超过 85%的蝴蝶兰是红色的大花型品种，而日本人则喜欢白色大花型品种（占市场份额的 80%）。中国台湾是亚洲地区的主要出口地，也是最新的蝴蝶兰杂交育种基地（表 23.2）。

表 23.2　2014 年中国台湾蝴蝶兰出口前十名的国家或地区

序号	国家或地区	重量（t）	金额（千美元）	市场占比（%）
1	美国	4 890	53 943	41.90
2	日本	2 750	34 284	26.60
3	越南	1 133	6 814	5.30
4	澳大利亚	525	6 762	5.30
5	荷兰	355	5 867	4.60
6	加拿大	482	3 734	2.90
7	英国	124	2 919	2.30
8	韩国	152	1 913	1.50
9	法属留尼汪	175	1 678	1.30
10	中国香港	243	1 629	1.30
11	其他	897	9 177	7.10

23.3　蝴蝶兰的育种研究

23.3.1　蝴蝶兰的主要观赏性状

花色是蝴蝶兰最重要的观赏性状。尽管白花蝴蝶兰仍在市场上占主导地位，但其他各种花色（包括粉色、红色、紫红色、黄色、橙色、绿色和黑色）的蝴蝶兰同样在不断培育中。人们对复杂的花色越来越有兴趣，带有不同斑点或图案的花朵是两种或多种颜色在同一朵花上表达的结果，如带有红色唇瓣的它色花被，或带有红色条纹或斑点的白花或黄花。

根据杂交育种得到花朵大小的不同，可将它们分为三组——大花型、中花型和小花型，其花径分别为＞10cm、7～10cm 和＜7cm。有时会发现形态杂乱的花朵，但只保留了性状稳定和漂亮的花型品种，其他的则被当作突变体丢弃了。畸变花（peloric flower）和大脚花是经常使用的两种花型，畸变花拥有两片转化成唇瓣的花瓣，而大脚花（bigfoot flower）则拥有一片像花瓣的唇瓣。

23.3.2　蝴蝶兰原生种的育种

在 92 个原生种中，只有 18 个种经常被用作育种亲本。它们属于蝴蝶兰亚属和裂唇蝴蝶兰亚属。蝴蝶兰亚属包括南洋白花蝴蝶兰、菲律宾白花蝴蝶兰、桃红蝴蝶兰、林登蝴蝶兰、五唇兰、桑德蝴蝶兰、西蕾丽蝴蝶兰和史塔基蝴蝶兰。裂唇蝴蝶兰亚属包括安汶蝴蝶兰、贝丽娜蝴蝶兰（*Phal. bellina*）、鹿角蝴蝶兰（*Phal. cornu-cervi*）、横纹蝴蝶兰（*Phal. fasciata*）、象耳蝴蝶兰（*Phal. gigantea*）、露德蝴蝶兰（*Phal. lueddemanniana*）、曼式蝴蝶兰（*Phal. mannii*）、苏门答腊蝴蝶兰（*Phal. sumatrana*）、红脉蝴蝶兰（*Phal. venosa*）和紫纹蝴蝶兰（*Phal. violacea*）（表 23.3）。这 18 个原生种的植株大小、花朵形态、花色和香味都各不相同。

表 23.3　用于亲本繁殖的蝴蝶兰原生种

种名	一代杂交种数量	杂交衍生种总数	获奖次数	平均每箭花朵数	平均花径（cm）
Phal. amabilis	475	31 573	94	13.4	7.8
Phal. amboinensis	532	14 951	102	3.6	5.5
Phal. aphrodite	58	31 460	39	16.8	7.6
Phal. appendiculata	6	6	7	2.7	1.1
Phal. bastianii	18	18	13	7.7	3.9
Phal. bellina	104	245	130	2.5	5.1
Phal. braceana	5	5	6	8.5	3.3
Phal. buyssoniana	35	490	15	20.8	4.1
Phal. celebensis	45	65	24	54.8	3.1
Phal. chibae	18	20	11	22.5	1.0
Phal. cochlearis	27	75	3	5.4	4.0
Phal. corningiana	63	215	17	3.4	5.8

续表

种名	一代杂交种数量	杂交衍生种总数	获奖次数	平均每箭花朵数	平均花径（cm）
Phal. cornu-cervi	99	372	116	4.5	3.0
Phal. deliciosa	31	49	13	18.5	1.6
Phal. doweryensis	5	5	4	5.2	5.2
Phal. equestris	550	21 805	127	28.0	2.9
Phal. fasciata	124	9 729	7	3.8	5.0
Phal. fimbriata	56	169	6	4.7	3.8
Phal. finleyi	25	28	5	3.5	1.6
Phal. floresensis	36	77	6	3.3	4.0
Phal. fuscata	103	273	4	13.1	3.8
Phal. gibbosa	4	5	4	3.9	1.1
Phal. gigantea	206	5 270	84	22.8	5.4
Phal. hieroglyphica	20	8 514	71	4.1	5.6
Phal. honghenensis	22	23	6	12.1	2.9
Phal. hygrochila	43	66	5	5.3	3.5
Phal. inscriptiosinensis	20	21	2	3.1	3.8
Phal. intermedia	6	8	2	10.0	4.0
Phal. javanica	99	187	9	3.4	2.7
Phal. kunstleri	8	9	3	5.9	3.9
Phal. lindenii	115	609	40	22.6	3.6
Phal. lobbii	55	104	64	4.9	2.0
Phal. lowii	10	16	8	7.8	3.9
Phal. lueddemanniana	325	20 024	56	3.3	5.5
Phal. maculata	34	79	8	4.3	3.2
Phal. malipoensis	3	3	2	6.0	1.6
Phal. mannii	157	1 212	46	10.7	3.4
Phal. mariae	107	1 224	20	10.2	3.7
Phal. marriottiana	7	7	3	6.8	4.8
Phal. mentawaiensis	5	16	1	3.0	4.4
Phal. micholitzii	50	1 211	5	1.6	4.9
Phal. modesta	40	48	7	3.8	3.1
Phal. pallens	25	497	7	2.7	5.1
Phal. pantherina	16	29	3	2.2	4.4
Phal. parishii	31	73	24	5.3	1.6
Phal. philippinensis	106	162	17	26.5	8.1
Phal. pulcherrima	278	9 466	143	22.1	3.2
Phal. pulchra	37	47	5	12.7	5.0
Phal. reichenbachiana	5	5	2	1.8	4.7
Phal. sanderiana	114	25 426	6	16.3	7.9
Phal. schilleriana	246	27 991	71	34.7	7.0
Phal. speciosa	26	112	10	2.5	4.8

续表

种名	一代杂交种数量	杂交衍生种总数	获奖次数	平均每箭花朵数	平均花径（cm）
Phal. stobartiana	10	10	7	31.6	3.1
Phal. stuartiana	392	26 377	42	34.7	6.3
Phal. sumatrana	166	4 485	28	4.1	5.8
Phal. taenialis	10	13	6	13.3	3.0
Phal. tetraspis	71	129	31	3.5	4.5
Phal. thailandica	10	12	1	6.3	1.3
Phal. venosa	333	3 161	29	3.2	4.2
Phal. violacea	472	5 040	197	2.7	5.0
Phal. viridis	14	16	4	4.9	4.1
Phal. wilsonii	33	33	8	8.8	3.0

数据来源于 OrchidWiz X3.3，2017

在蝴蝶兰亚属中，南洋白花蝴蝶兰和菲律宾白花蝴蝶兰是白花、中花型（花径 7.5～7.8cm）品种，与花径超过 10cm 的大花型品种一起用于繁殖大花型白花品种（表 23.3）。桑德蝴蝶兰、西蕾丽蝴蝶兰和史塔基蝴蝶兰为粉色中花型（花径 6.3～7.9cm）品种，用于繁殖红花和大花型杂交种。另外，史塔基蝴蝶兰也有带红色斑点的白花，是红色斑点特征的主要繁殖亲本。桃红蝴蝶兰、林登蝴蝶兰和五唇兰为小花型（2.9～3.6cm）、多花色品种，这些品种主要用作中小花型的育种亲本。桃红蝴蝶兰提供多花的特征，林登蝴蝶兰提供红色条纹的特征，五唇兰提供深紫红色的特征。

裂唇蝴蝶兰亚属的品种通常具有强烈的气味，如安汶蝴蝶兰、贝丽娜蝴蝶兰、露德蝴蝶兰、红脉蝴蝶兰和紫纹蝴蝶兰。因此，它们是培育带香味杂交种的主要亲本。另外，安汶蝴蝶兰、横纹蝴蝶兰、象耳蝴蝶兰、露德蝴蝶兰、曼式蝴蝶兰、苏门答腊蝴蝶兰和红脉蝴蝶兰有黄色花，它们被用作培育黄花杂交种的亲本。从贝丽娜蝴蝶兰和紫纹蝴蝶兰的两个变种里已经发现了蓝具紫色花，因此这两个种被用来培育梦想中的蓝花杂交种。

这两个亚属的品种除了花形上的差异外，染色体的大小也不同。蝴蝶兰亚属的染色体较小，而裂唇蝴蝶兰亚属的染色体中等偏大（见第 23.7 节）。由于染色体尺寸不一致，减数分裂过程中其不能配对，因此这两个亚属之间的不同种不能成功杂交。

23.3.3　蝴蝶兰的育种方向

蝴蝶兰的育种方向可以大致分为三个方面：花型、花色和气味。另外，在培育杂交种的过程中，植株的抗病性和对极端气候抗性的增强也是一大选择趋势。但是到目前为止，这并不是大多数育种计划的主要目标。

在花型上，大花型是主要的育种方向，同时小花型品种的选育开始成为新的发展趋势。近年来开始培育具有变化形态的花，如花瓣像唇瓣的畸变花和唇瓣像花瓣的大脚花。

在花色育种中，白色花、粉色花和红色花的育种已趋于稳定，现在蝴蝶兰育种研究者开始寻找具有多种颜色和图案的花朵，如分布在不同区域的红色斑点和条纹图案。不常见的和新的颜色也很受欢迎，如黄色、橙色、绿色和黑色，以及梦想中的纯蓝色，注意不是蓝紫色。到目前为止，在蝴蝶兰原生种和杂交种中都未曾发现纯蓝色花。

蝴蝶兰不仅有各种各样的花色，也有与众不同的花型，气味可以作为育种的附加特征。然而，香味特征不能稳定遗传给后代，因此大多数杂交种并没有香味。可以将香味基因位点标记下来的分子育种也许能提高香味杂交种的育种成功率。

23.4 蝴蝶兰各种花色的育种

23.4.1 白花

白色是蝴蝶兰最重要的颜色。在育种工作中，尽管中小花型和带红色唇瓣的白花越来越流行，但主要焦点仍然集中在大花型（花径＞10cm）上（图 23.1）。白色大花的主要特征是纯白色、花朵大、花型丰满、花序整齐。

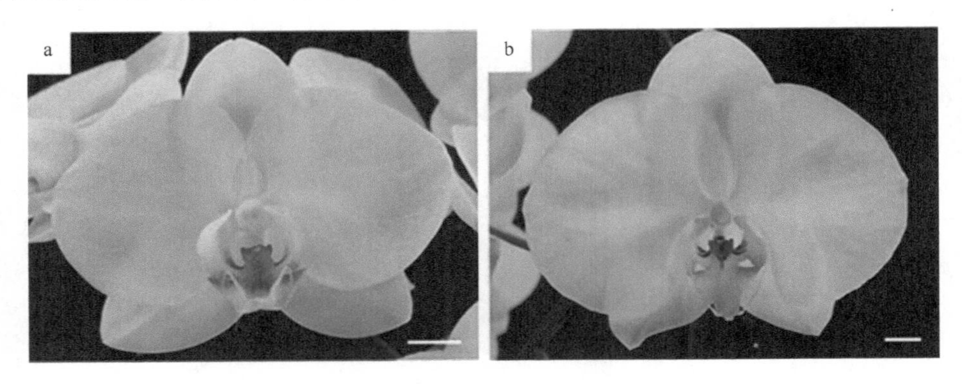

图 23.1 白色大花型蝴蝶兰杂交种的各种表现。（a）蝴蝶兰‘巴丹’（*Phal.*‘Bataan’），（b）蝴蝶兰‘团结精神’（*Phal.*‘Join Spirit’），鼻宽 = 1cm

另外，白色大花型品种是大多数大花型杂交种的主要繁殖亲本，即使一些五颜六色的品种也是如此。所有的大花型蝴蝶兰杂交种都含有 50%以上白色大花的血统。

23.4.1.1 原生种对白色大花型杂交种的贡献

蝴蝶兰杂交种白色大花的育种是蝴蝶兰育种历史的先驱，它已经持续了 100 多年。对白色大花有贡献的主要原生种有南洋白花蝴蝶兰和菲律宾白花蝴蝶兰，占 90%以上的血统。南洋白花蝴蝶兰和菲律宾白花蝴蝶兰比较相似，后者在早期是被归类到南洋白花蝴蝶兰种下的菲律宾白花蝴蝶兰亚种的。它们都具有白色的花，花径 7～8cm，黄色唇瓣上有红色斑点，单花序中约含有 30 朵花。南洋白花蝴蝶兰和菲律宾白花蝴蝶兰之间主要的区别是花朵中心裂片与唇瓣的连接，菲律宾白花蝴蝶兰的唇瓣呈指状三角形，南洋白花蝴蝶兰的唇瓣呈椭圆盾形。

23.4.1.2 重要的白色大花型杂交种

在白色大花蝴蝶兰的繁育历史中，有几个杂交种扮演了至关重要的角色。1940 年注册的蝴蝶兰‘多丽丝’（*Phal.*‘Doris’）[蝴蝶兰‘伊丽莎白’（*Phal.*‘Elisabethae’）× 蝴蝶兰‘凯瑟琳 • 西格沃特’（*Phal.*‘Katherine Siegwart’）]是其中非常重要的杂交种之一。根据 OrchidWiz（2017）的数据，在所有后代中，该品种促进了 262 个子代（G1）杂交种和 30 266 个后代杂交种的发展（表 23.4）。通过‘多丽丝’四倍体连续 15 代的反向杂交来培育大多数拥有纯净基因的大白花杂交种。

表 23.4 ‘大白花’育种中，多丽丝及其后代衍生而来的重要杂交种

杂交种名	杂交代数[a]	英国皇家园艺协会登记年份	亲本	子代杂交种数	杂交后代总数	获奖数	平均花径（cm）	花色[b]
Phal. Doris	G0	1940	(Elisabethae × Katherine Siegwart)	262	30 266	34	9.5	WWy
Phal. Grace Palm	G1	1950	(Doris × Winged Victory)	138	24 647	29	12.3	WWy
Phal. Juanita	G2	1957	(Chief Tucker × Grace Palm)	87	20 470	8	12.3	WWy
Phal. Dos Pueblos	G1	1956	(Doris × Grace Palm)	113	11 826	5	—[c]	WWy
Phal. Yukimai	G2	1983	(Musashino × Grace Palm)	50	661	3	12.2	WWy
Phal. Sogo Yukidian	G3	1998	(Yukimai × Taisuco Kochdian)	143	325	31	12.5	WWy
Phal. Cast Iron Monarch	G1	1957	(Louise Georgianna × Doris)	69	18 971	3	11.5	WWy
Phal. Palm Beach	G2	1958	(Doris × Cast Iron Monarch)	44	17 394	5	—	WWy
Phal. Susan Merkel	G3	1960	(Chieftain × Palm Beach)	39	13 214	2	—	WWy
Phal. Elinor Shaffer	G1	1960	(Juanita × Doris)	74	8 670	20	12.4	WWy
Phal. Mount Kaala	G2	1966	(Doreen × Elinor Shaffer)	98	4 938	2	11.7	WWy
Phal. Joseph Hampton	G1	1966	(Monarch Gem × Doris)	178	1 712	9	12.1	WWy

数据来源于 OrchidWiz X3.3，2017

a. 世代相传。G0 为蝴蝶兰‘多丽丝’。G1～G3 表示‘多丽丝’衍生出的不同后代

b. 栽培品种的花色。第一个词为花被的颜色，第二个词为唇瓣的颜色。W 代表白色，Wy 代表白色带淡黄色

c. OrchidWiz X3.3，2017 中未查找到

‘多丽丝’最重要的子代杂交种是‘优雅棕榈’（*Phal.*‘Grace Palm’）[‘多丽丝’ × ‘胜利女神’（*Phal.*‘Winged Victory’）]及其回交后代 *Phal.*‘Dos Pueblos’（‘多丽丝’ × ‘胜利女神’）。经统计，这 2 个杂交种分别培育出 138 个和 113 个子代杂交种，以及为 24 647 个和 11 826 个杂交后代做出了贡献（表 23.4）。它们的后代‘胡安妮塔’（*Phal.*‘Juanita’）[‘希尔斯酋长’（*Phal.*‘Chief Tucker’）× ‘胜利女神’]、‘新时代’（*Phal.*‘New Era’）（‘胜利女神’ × *Phal.*‘Sally Lowrey’）、‘安 • 哈特’（*Phal.*‘Ann Hatter’）（‘胡安妮塔’ × ‘新时代’）和‘普韦布宝石’（*Phal.*‘Pueblo Jewel’）[*Phal.*‘Dos Pueblos’ × ‘粉宝石’（*Phal.*‘Pink Jewel’）]由‘多丽丝’衍生出的杂交种数量最多，但只有‘胡安妮塔’开白花，而其他则是带红色唇瓣的白花，这意味着它们的后代大多数是具白色花被或红色唇瓣的杂交种（见第 23.4.2.3 节）。

‘多丽丝’的其他子一代杂交种‘铸铁君主’（*Phal.*‘Cast Iron Monarch’）[‘路易

斯·乔治亚娜’（*Phal.* ‘Louise Georgianna’）×‘多丽丝’]、‘埃莉诺·谢弗’（*Phal.* ‘Elinor Shaffer’）（‘胡安妮塔’×‘多丽丝’）和‘约瑟夫·汉普顿’（*Phal.* ‘Joseph Hampton’）[‘帝王宝石’（*Phal.* ‘Monarch Gem’）×‘多丽丝’]，分别有 69 个、74 个和 178 个子一代杂交种以及总数为 18 971 个、8670 个和 1712 个杂交种，其中包含了具白色唇瓣的白花（表 23.4）。它们的后代‘棕榈滩’（*Phal.* ‘Palm Beach’）（‘多丽丝’×‘铸铁君主’）、‘苏珊·默克尔’（*Phal.* ‘Susan Merkel’）[‘酋长’（*Phal.* ‘Chieftain’）×‘棕榈滩’]和‘芒特·卡拉’（*Phal.* ‘Mount Kaala’）[‘多琳’（*Phal.* ‘Doreen’）×‘埃莉诺·谢弗’]同样拥有白花，并且在蝴蝶兰的进一步发展中起着重要作用。

此外，“大白花”中最著名的杂交种是 *Phal.* ‘Sogo Yukidian’“V3”（图 23.2），该品种是由世界兰业有限公司在中国台湾用 *Phal.* ‘Yukimai’（母本）和 *Phal.* ‘Sogo Kochdian’（父本）杂交而来的后代品种（表 23.4）。*Phal.* ‘Sogo Kochdian’“V3”拥有完美丰满的花型、稳健弯曲的花箭以及整齐的花序。作为一株完美的杂交种，*Phal.* ‘Sogo Yukidian’“V3”的后代性状都不如其自身性状好，所以它只有 143 个子一代和总共 308 个杂交后代。因此，“V3”几乎垄断了蝴蝶兰市场上的“大白花”。而且，*Phal.* ‘Sogo Yukidian’“V3”和‘世界一流’“大脚花”[*Phal.* World Class “Bigfoot”]的杂交后代‘育品复活岛’（*Phal.* ‘Yu Pin Easter Island’）是大脚花杂交种的主要育种亲本（见第 23.5.3 节）。

图 23.2 “大白花”杂交种 *Phal.* ‘Sogo Yukidian’ 的花朵，（a）全株，（b 和 c）花朵，鼻宽 = 1cm

家谱分析表明，*Phal.* ‘Sogo Yukidian’“V3”含有 41.44%爪哇白花蝴蝶兰（*Phal. rimestadiana*）（现分类为南洋白花蝴蝶兰的变种）的血统、40.14%南洋白花蝴蝶兰的血统、15.30%菲律宾白花蝴蝶兰的血统、2.15%史塔基蝴蝶兰的血统、0.59%西蕾丽蝴蝶兰的血统和0.39%桑德蝴蝶兰的血统。由于 *Phal.*‘Sogo Yukidian’拥有96.88%的白色原生种血统，因此才拥有硕大而丰满的花朵，以及整齐有序的花序。少量的史塔基蝴蝶兰、西蕾丽蝴蝶兰和桑德蝴蝶兰血统贡献了其纯净的白色。

23.4.1.3 白色大花型育种

在分子遗传学中，“白色”表示“无色”，这意味着与颜色相关的基因不起作用。唇瓣上带有红点点的黄色花朵表明白色花朵含有与颜色相关的基因，但这些基因在白色花朵中不表达（Hsu et al. 2015a）。因此，将白花与任何其他颜色的杂交种进行杂交可以导入功能性的颜色基因，并产生带颜色的杂交种。繁殖“大白花”的要点是只能选择白花品种进行杂交，不能使用有色花品种。

白色大花型育种的其他重要特征是花箭具有良好的弯曲度、整齐的花序、有序的花序和许多饱满的花朵。这些特征已经存在于南洋白花蝴蝶兰、菲律宾白花蝴蝶兰和大多数“大白花”杂交品种的血统中。然而，培育出比 *Phal.*‘Sogo Yukidian’“V3”更优秀的杂交蝴蝶兰品种仍然是一项具有挑战性的任务。当前提高“大白花”品质的方向主要集中在克服远距离运输、提高由生物和非生物胁迫所引起的抵抗力以及培育多花箭品种上。

23.4.2 红色花

蝴蝶兰红花杂交种是春末繁殖数量最多的蝴蝶兰品种。它们的范围从粉红色、红色、淡紫色到紫红色，以及各种色素沉着模式，包括带有红唇的花被和各种带有红色斑点或条纹的白花(称为色素沉着图案)。蝴蝶兰红色花朵的育种可以分为红色大花型(表23.5，图23.3)、红色中小花型（见第23.5.1节）、具有红色唇瓣的白色花（表23.6），以及带有红色斑点或条纹的白花或粉色花（表23.7）。为研发大红色花朵，‘多丽丝’及其子一代杂交品种 *Phal.*‘Zada’是优良的育种亲本，它们在多花、花型饱满和花序整齐方面表现出色（表23.5）。如今，红花蝴蝶兰杂交种的两种主要育种策略为：大花型红花和包含各种色素沉着模式的花朵。

表23.5 红色大花型育种的重要杂交种

杂交种名	杂交代数[a]	英国皇家园艺协会登记年份	亲本	子一代杂交种数	杂交后代总数	获奖数	平均花径（cm）	花色[b]
Phal. Doris	G0	1940	（Elisabethae × Katherine Siegwart）	262	30 266	34	9.5	WW
Phal. Zada	G1	1958	（San Songer × Doris）	206	15 010	9	8.4	RR
Phal. Lipperose	G2	1968	（Ruby Wells × Zada）	139	9 124	0	—[c]	RR
Phal. Lippezauber	G2	1969	（Doris Wells × Zada）	20	6 446	0	—	—[d]
Phal. Lippstadt	G3	1971	（Doris Wells × Lipperose）	52	5 749	1	11.1	PR
Phal. Abendrot	G3	1974	（Lippezauber × Lippstadt）	229	3 723	17	10.8	PR
Phal. Flor de Mato	G2	1972	（Zada × Satin Rouge）	91	2 927	5	9.4	PR
Phal. Herbert Hager	G3	1977	（Dear Heart × Flor de Mato）	50	2 265	14	9.2	PR
Phal. Pinlong Cinderella	G4	1983	（Morgenrot × Herbert Hager）	22	1 734	0	—	—

续表

杂交种名	杂交代数[a]	英国皇家园艺协会登记年份	亲本	子一代杂交种数	杂交后代总数	获奖数	平均花径（cm）	花色[b]
Phal. New Cinderella	G4[e]	1997	（Pinlong Cinderella × New Eagle）	47	1 452	2	11.7	PstR
Phal. Barbara Beard	G2	1962	（Virginia × Zada）	48	8 382	2	9.8	RR
Phal. Lois Jansen	G2[f]	1969	（Barbara Beard × Ruby Lips）	14	5 469	2	8.9	WstR
Phal. Carter Shenk	G3	1972	（Lois Jansen × Suemid）	16	4 239	0	—	—
Phal. Irene Van Alstyne	G3	1967	（Carol Brandt × Barbara Beard）	10	5 615	0	—	—
Phal. Terry-Beth Ballard	G3[g]	1972	（Irene Van Alstyne × Ruby Zada）	40	3 981	1	8.5	WstR
Phal. Raycraft	G1	1962	（Aalsmeer Rose × Doris）	23	4 456	5	8.7	PstR
Phal. Otohime	G2	1973	（Grace Palm × Raycraft）	45	4 169	0	—	—
Phal. Happy Valentine	G3	1983	（Otohime × Odoriko）	199	3 278	17	10.8	PstR
Phal. Chia Lin	G4[h]	1982	（James Hall × Johanna）	39	2 919	1	6.5	WstR
Phal. Ta Bei Chou	G4[i]	1987	（Abendrot × Chia Lin）	39	2 654	0	—	—
Phal. King Shiang's Coral	G3[j]	1991	（Ta Bei Chou × Otohime）	19	1 690	0	—	—
Phal. King Shiang's Kide	G5	1991	（Ta Bei Chou × Paifang's Sardonyx）	3	1 261	0	—	—
Phal. King Shiang's Rose	G4[k]	1992	（King Shiang's Kide × King Shiang's Coral）	120	1 258	2	8.7	PstR

数据来源于 OrchidWiz X3.3，2017

a. 世代相传。G0 为蝴蝶兰‘多丽丝’。G1～G5 表示‘多丽丝’衍生出的不同后代

b. 栽培品种的花色。第一个词为花被的颜色，第二个词为唇瓣的颜色。W 代表白色，Wst 代表白花带红条纹，P 代表粉色，Pst 代表粉花带红条纹，R 代表红色

c. OrchidWiz X3.3，2017 中未查找到

d. OrchidWiz X3.3，2017 中未查找到

e. *Phal.*‘Pinlong Cinderella’和 *Phal.*‘New Eagle’分别是‘多丽丝’的第四与第三代杂交后代，因此它们的杂交后代新灰姑娘（*Phal.*‘New Cinderella’）是‘多丽丝’的第四代杂交后代

f. 芭芭拉胡子（*Phal.*‘Barbara Beard’）和 *Phal.*‘Ruby Lips’分别是‘多丽丝’的第二代与第一代杂交后代，因此它们的杂交后代 *Phal.*‘Lois Jansen’是‘多丽丝’的第二代杂交后代

g. *Phal.*‘Irene Van Alstyne’和 *Phal.*‘Ruby Zada’分别是多丽丝的第三与第二代杂交后代，因此它们的杂交后代 *Phal.*‘Terry-Beth Ballard’是‘多丽丝’的第三代杂交后代

h. *Phal.*‘Chia Lin’是由‘多丽丝’的第五代杂交后代 *Phal.*‘James Hall’（*Phal.*‘Red Lip’× *Phal.*‘Barbara Moler’）与第三代杂交后代 *Phal.*‘Johanna’（*Phal.*‘Ella Freed’× *Phal.*‘Jiminy Cricket’）杂交而来的第四代杂交后代

i. *Phal.*‘Abendrot’和 *Phal.*‘Chia Lin’分别是‘多丽丝’的第三与第四代杂交后代，因此它们的杂交后代‘周大北’（*Phal.*‘Ta Bei Chou’）是‘多丽丝’的第四代杂交后代

j.‘周大北’和 *Phal.*‘Otohime’分别是‘多丽丝’的第四与第二代杂交后代，因此它们的杂交后代 *Phal.* King Shiang's Coral 是‘多丽丝’的第三代杂交后代

k. *Phal.*‘King Shiang's Coral’和 *Phal.*‘King Shiang's Kide’分别是‘多丽丝’的第三和第五代杂交后代，因此它们的杂交后代 *Phal.*‘King Shiang's Rose’是‘多丽丝’的第四代杂交后代

23.4.2.1 有助于红色花繁育的原生种

对于红色大花型花朵，西蕾丽蝴蝶兰和桑德蝴蝶兰是用作育种亲本的主要原生种。西蕾丽蝴蝶兰的花朵带香味，花径 5～8cm，花色粉红色至大红色，多花花序以及花箭有分枝（图 23.4）。桑德蝴蝶兰花径 5～8cm，花色粉红色至紫红色，多花花序。由于导入了‘多丽丝’的基因，南洋白花蝴蝶兰和菲律宾白花蝴蝶兰的血统对红色大花型也很重要。

图 23.3 具有大红色花朵的蝴蝶兰杂交种的各种表型。(a) *Phal.* 'Taisuco Peace',(b) *Phal.* 'Nobby's Spring Alice',(c) *Phal.* 'Taida New Luchia',(d) *Phal.* 'Hong Lin Jewelry',(e) *Phal.* 'OX Pink Yukidian',(f) *Phal.* 'Dragon Tree Firerose',鼻宽 = 1cm

表 23.6 育种目标为白瓣红唇的重要杂交后代

杂交种名	杂交代数[a]	英国皇家园艺协会登记年份	亲本	子一代杂交种数	杂交后代总数	获奖数	平均花径(cm)	花色[b]
Phal. Doris	G0	1940	(Elisabethae × Katherine Siegwart)	262	30 266	34	9.5	WW
Phal. Sally Lowrey	—[c]	1954	(Pua Kea × *equestris*)	32	17 085	2	—[d]	—[e]
Phal. Judy Karleen	G2	1957	(Chieftain × Sally Lowrey)	17	12 785	3	9.5	WR
Phal. New Era	G2	1958	(Grace Palm × Sally Lowrey)	9	14 378	2	—	WR
Phal. Ann Hatter	G3	1962	(Juanita × New Era)	21	13 531	4	9.8	WR
Phal. Cover Girl	G1	1958	(Sally Lowrey × Doris)	4	136	1	8.0	WO
Phal. Pueblo Jewel	G2	1968	(Dos Pueblos × Pink Jewel)	22	6 588	1	8.9	WR
Phal. Luchia Lady	G4	1991	(Pamela Lady × Pinlong Cardinal)	7	222	1	11.0	WR

续表

杂交种名	杂交代数[a]	英国皇家园艺协会登记年份	亲本	子一代杂交种数	杂交后代总数	获奖数	平均花径（cm）	花色[b]
Phal. Hsinying Lip	G4	1993	（Musashino × Su's Red Lip）	8	135	1	11.9	WR
Phal. Luchia Lip	G5	1998	（Luchia Lady × Hsinying Lip）	32	106	3	10.0	WR
Phal. Mount Beauty	G3	1993	（Mount Kaala × Hamakita Beauty）	26	766	0	—	—
Phal. Su's Red Lip	—	1986	（South Cha-Li × Lucky Lady）	70	950	1	9.5	WR
Phal. Hsin Red Lip	G4	1993	（Mount Beauty × Su's Red Lip）	7	10	0	—	WR
Phal. Mount Lip	G3	1997	（South Cha-Li × Mount Beauty）	88	348	8	10.4	WR
Phal. Hsinying Mount	G4	2000	（Mount Lip × Tinny Ace）	18	64	4	10.6	WR

数据来源于 OrchidWiz X3.3，2017

a. 世代相传。G0 为蝴蝶兰多丽丝。G1～G5 表示多丽丝衍生出的不同后代

b. 栽培品种的花色。第一个词为花被的颜色，第二个词为唇瓣的颜色。W 代表白色，O 代表橙色，R 代表红色

c. 不是 G0 代多丽丝的繁殖后代

d. OrchidWiz X3.3，2017 中未查找到

e. OrchidWiz X3.3，2017 中未查找到

表 23.7　繁育红色条纹（脉）或红色斑点的重要杂交种

杂交种名	杂交代数[a]	英国皇家园艺协会登记年份	亲本	子一代杂交种数	杂交后代总数	获奖数	平均花径（cm）	花色[b]
Phal. Robert W. Miller	—[c]	1960	（*lindenii* × *sanderiana*）	2	3	0	—[d]	—[e]
Phal. Peppermint	—	1964	（*lindenii* × Pink Profusion）	40	81	6	4.7	PstR
Phal. Baguio	—	1966	（*schilleriana* × *lindenii*）	15	118	2	—	WstP
Phal. Doris	G0	1940	（Elisabethae × Katherine Siegwart）	262	30 266	34	9.5	WW
Phal. Star of Rio	—	1956	（Bataan × *lueddemanniana*）	13	10 635	4	10.2	WspR
Phal. Samba	—	1963	（Star of Rio × *amboinensis*）	87	7 735	20	7.3	WspR
Phal. Ella Freed	G2	1970	（Show Girl × Samba）	129	7 268	8	9.8	WstR
Phal. Freed's Danseuse	G3	1975	（Ella Freed × Career Girl）	54	4 180	5	9.0	WstR
Phal. Modern Stripes	G4	1987	（Freed's Danseuse × Chiali Stripe）	21	3 513	0	—	—
Phal. Okay Seven	G5	1992	（Modern Stripes × Houpi Beauty）	24	2 170	1	8.2	WstR
Phal. Taisuco Stripe	G6	1994	（Okay Seven × Taisuco Gaster）	13	318	2	6.7	WstR
Phal. Little Gem Stripes	G7	1997	（Taisuco Stripe × Taisuco Gem）	144	292	16	5.4	WstR
Phal. Lucky Shenk	G4	1979	（Lucky Lady × Carter Shenk）	4	4 183	0	—	—
Phal. Chiali Stripe	G4[f]	1983	（Cindy Tsai × Lucky Shenk）	21	4 175	0	—	—
Phal. Houpi Beauty	G5	1987	（Tsuei You Queen × Chiali Stripe）	5	3 441	0	—	—
Phal. Sun Prince	G5[g]	1995	（Cypress Pink × Houpi Beauty）	39	2 148	0	—	—
Phal. Leopard Prince	G6	1997	（Sun Prince × Ho's French Fantasia）	221	567	32	9.3	WspR

数据来源于 OrchidWiz X3.3，2017

a. 世代相传。G0 为蝴蝶兰‘多丽丝’。G1～G7 表示‘多丽丝’衍生出的不同后代

b. 栽培品种的花色。第一个词为花被的颜色，第二个词为唇瓣的颜色。Wst 代表白花带红色条纹，Wsp 代表白花带红色斑点，P 代表粉色，R 代表红色

c. 不是 G0 代‘多丽丝’的繁殖后代

d. OrchidWiz X3.3，2017 中未查找到

e. OrchidWiz X3.3，2017 中未查找到

f. *Phal.*‘Cindy Tsai’和 *Phal.*‘Lucky Shenk’分别是‘多丽丝’的第三代与第四代杂交后代，因此它们的杂交后代 *Phal.*‘Chiali Stripe’是‘多丽丝’的第四代杂交后代

g. *Phal.*‘Cypress Pink’和 *Phal.*‘Houpi Beauty’分别是‘多丽丝’的第四与第五代杂交后代，因此它们的杂交后代 *Phal.*‘Sun Prince’是‘多丽丝’的第五代杂交后代

图 23.4 红花蝴蝶兰品种，西蕾丽蝴蝶兰，鼻宽 = 1cm

另外，桃红蝴蝶兰和五唇兰是繁殖小型与中型红色花杂交种的主要野生种。桃红蝴蝶兰具有 2～4cm 的花朵，颜色多样，包括白色、粉红色、红色、紫红色和蓝紫色。桃红蝴蝶兰也具多花箭、花箭分枝和多花花序的特征。五唇兰花径 2～3cm，多花，花色有白色、粉红色、紫红色、深红色和蓝紫色。

此外，其他原生种可提供不同的颜色特征，如各种色素。史塔基蝴蝶兰的花朵为带红色斑点的白花，是培育红色斑点性状的主要亲本。林登蝴蝶兰导入了红色条纹的特征。桃红蝴蝶兰有红黑色和紫色的花朵。

23.4.2.2 重要红色大花型杂交种

白花品种‘多丽丝’（父本）和红花品种 *Phal.*‘San Songer’（母本）于 1958 年杂交得到的 *Phal.*‘Zada’已成为红色大花型的主要育种亲本，分别得到了 206 个子一代杂交种和 15 010 个杂交种品种（表 23.5）。它的子一代杂交种 *Phal.*‘Lipperose’（*Phal.*‘Ruby Wells’× *Phal.*‘Zada’）、‘弗洛尔德马托’*Phal.*‘Flor de Mato’（*Phal.*‘Zada’× *Phal.*‘Satin Rouge’）和‘芭芭拉胡子’（*Phal.*‘Virginia’× *Phal.*‘Zada’）是最常用的育种亲本（表 23.5）。

Phal.‘Lipperose’属于蝴蝶兰亚属，开红色大花，是 Hark-Orchideen 于 20 世纪 70 年代在德国注册的系列品种。该系列中的其他杂交种有 *Phal.*‘Lippezauber’（*Phal.*‘Doris Wells’× *Phal.*‘Zada’）、*Phal.*‘Lippstadt’（*Phal.*‘Doris Wells’× *Phal.*‘Lipperose’）和 *Phal.*‘Abendrot’（*Phal.*‘Lippezauber’× *Phal.*‘Lippstadt’），它们都常被用作育种亲本。所有这些品种的花径都在 10.8～11.1cm，具带红色唇瓣的粉色到红色花被。其中，*Phal.*‘Abendrot’是最常用的杂交亲本（表 23.5）。

‘弗洛尔德马托’、‘赫伯特黑格’（*Phal.*‘Herbert Hager’）[‘甜心’（*Phal.*‘Dear Heart’）×‘弗洛尔德马托’]、*Phal.*‘Pinlong Cinderella’（*Phal.*‘Morgenrot’×‘赫伯特黑格’）和‘新灰姑娘’（*Phal.*‘Pinlong Cinderella’× *Phal.*‘New Eagle’）产生了一系列的杂交后代，分别有 5022 个子代和 47 个子一代杂交种，以及 2265 个、1734 个和 1452 个杂交后代。‘赫伯特黑格’的花朵为花径 9.2cm 的大花，唇瓣红色，花被粉

色，而‘新灰姑娘’花径为 11.7cm，花朵具有带红色条纹和红色唇瓣的粉色花被。

另一个重要的育种系起源于‘芭芭拉胡子’、‘洛伊丝扬森’（*Phal.*‘Lois Jansen’）[‘芭芭拉胡子’ × ‘宝石红唇’（*Phal.*‘Ruby Lips’）]和‘艾琳 • 范 • 阿尔斯汀’（*Phal.*‘Irene Van Alstyne’）[‘卡萝尔 • 布兰特’（*Phal.*‘Carol Brandt’）× ‘芭芭拉胡子’]，以及它们的子一代杂交种‘卡特申克’（*Phal.*‘Carter Shenk’）（‘洛伊丝扬森’ × *Phal.*‘Suemid’）和 *Phal.*‘Terry-Beth Ballard’（‘艾琳 • 范 • 阿尔斯汀’ × *Phal.*‘Ruby Zada’）（表 23.3）。所有这些杂交种的花朵大小相似，花径为 8.5～9.8cm，但‘芭芭拉胡子’开红色花，而‘洛伊丝扬森’和 *Phal.*‘Terry-Beth Ballard’的花朵为带红色条纹的白色或粉色花。

除了 *Phal.*‘Zada’、*Phal.*‘Raycraft’（*Phal.*‘Aalsmeer Rose’ × ‘多丽丝’）及其子代的血统，*Phal.*‘Otohime’（‘优雅棕榈’ × *Phal.*‘Raycraft’）和‘快乐情人节’（*Phal.*‘Happy Valentine’）（*Phal.*‘Otohime’ × *Phal.*‘Odoriko’）为红色花朵品种的开发做出了重要贡献。这些杂交种的花径为 8.7～10.8cm，粉红色花被带红色条纹和红色唇瓣。

用导入桃红蝴蝶兰基因的方法获得红花蝴蝶兰始于‘红珊瑚’（*Phal.*‘Red Coral’）[四倍体五唇兰（*Phal. pulcherrima* var. *buyssoniana*）×‘多丽丝’]和 *Phal.*‘Memoria Clarence Schubert’（四倍体五唇兰 × *Phal.*‘Zada’），分别用作 81 个和 62 个子一代杂交种与 289 个及 221 个杂交种的育种亲本。这两个杂交种花朵尺寸相似，花径为 6.5～7.6cm，花色为红色。

Phal.‘Chia Lin’（*Phal.*‘James Hall’ × *Phal.*‘Johanna’）是‘多丽丝’的第四代杂交种，拥有两种不同颜色的花朵，紫红色花或带红色唇瓣的黄色花（表 23.5）。因此 *Phal.*‘Chia Lin’是繁育紫红色花和红唇黄花的主要亲本（见第 23.4.3.3 节）。*Phal.*‘Chia Lin’有 39 个子一代杂交种和 2919 个杂交后代，但是除了‘周大北’（*Phal.*‘Abendrot’ × *Phal.*‘Chia Lin’）、‘卡迪纳尔兄弟’（*Phal.*‘Brother Cardinal’）（*Phal.*‘Chia Lin’ × *Phal.*‘Pinlong Major’）和‘泰达戴维’（*Phal.*‘Taida David’）（*Phal.*‘Chia Lin’ × *Phal.*‘Sogo David’）以外，大多数子一代杂交种都具有黄色、中型花朵。其中，‘周大北’的后代最多，其后代 *Phal.*‘King Shiang's Coral’（‘周大北’ × *Phal.*‘Otohime’）、*Phal.*‘King Shiang's Kide’（‘周大北’ × *Phal.*‘Paifang's Sardonyx’）和 *Phal.*‘King Shiang's Rose’（*Phal.*‘King Shiang's Kide’ × *Phal.*‘King Shiang's Coral’）是红花蝴蝶兰的重要亲本。*Phal.*‘King Shiang's Rose’的花径为 8.7cm，花朵为粉色，带红色条纹和红色唇瓣。

23.4.2.3 红唇白花蝴蝶兰的重要杂交种

大多数具有红唇白花的蝴蝶兰杂交种来自“小红花”品种桃红蝴蝶兰和黄花或红唇黄花品种露德蝴蝶兰。例如，*Phal.*‘Sally Lowrey’[‘普卡’（*Phal.*‘Pua Kea’）× 桃红蝴蝶兰]开粉色小花，可与大白花杂交以增加花朵大小（表 23.6）。因此，拥有红唇白花的 *Phal.*‘Judy Karleen’（‘酋长’ × *Phal.*‘Sally Lowrey’）、‘新时代’（‘胜利女神’ × *Phal.*‘Sally Lowrey’）和‘封面女郎’（*Phal.*‘Cover Girl’）（*Phal.*‘Sally Lowrey’ × ‘多丽丝’）被进一步用作育种亲本，分别得到 17 个、9 个和 4 个子一代杂交种以及

12 785 个、14 378 个和 136 个杂交后代。

在蝴蝶兰市场上，最有名的红唇白花品种是‘新红唇’（*Phal.*‘Hsin Red Lip’）[‘美景山’（*Phal.*‘Mount Beauty’）×‘苏的红唇’（*Phal.*‘Su's Red Lip’）]、‘露西亚唇’（*Phal.*‘Luchia Lip’）[‘露西亚女士’（*Phal.*‘Luchia Lady’）× *Phal.*‘Hsinying Lip’]、*Phal.*‘Mount Lip’（*Phal.*‘South Cha-Li’×‘美景山’）和 *Phal.*‘Hsinying Mount’（*Phal.*‘Mount Lip’× *Phal.*‘Tinny Ace’）（表 23.6）。‘露西亚唇’是从露德蝴蝶兰与白色大花型品种杂交数代的红唇白花后代中选出的，‘露西亚唇’花径约 10cm，拥有 32 个子一代和 106 个杂交后代。‘新红唇’是由‘苏的红唇’（父本）与‘美景山’（母本）的白花或红唇白花品种历经数代杂交得来的。

Phal.‘Mount Lip’（图 23.5）和 *Phal.*‘Hsinying Mount’以其白色花朵带红色唇瓣且花瓣基部呈粉色形成白花红心的花色而闻名。*Phal.*‘Mount Lip’和 *Phal.*‘Hsinying Mount’都被用于进一步的育种工作中（表 23.6）。家谱分析表明，*Phal.*‘Mount Lip’含有 37.0%的爪哇白花蝴蝶兰基因、35.4%的南洋白花蝴蝶兰基因、17.2%的菲律宾白花蝴蝶兰基因、3.5%的桃红蝴蝶兰基因、2.0%的西蕾丽蝴蝶兰基因、1.9%的桑德蝴蝶兰基因、1.6%的露德蝴蝶兰基因以及其他基因。

图 23.5　红唇白花蝴蝶兰。（a）*Phal.*‘Mount Lip’，（b）*Phal.*‘Fuller's Pink Rose’，鼻宽 = 1cm

23.4.2.4　花朵带红色条纹的重要杂交种

带红色条纹的花朵性状来自原生种林登蝴蝶兰，其花朵为带红色条纹的白花，并带有潜脉，这就是所谓的叶脉模式。林登蝴蝶兰已被用作红色条纹性状的育种亲本，如 *Phal.*‘Robert W. Miller’（林登蝴蝶兰 × 桑德蝴蝶兰）、‘薄荷’（*Phal.*‘Peppermint’）（林登蝴蝶兰 × *Phal.*‘Pink Profusion’）和‘碧瑶’（*Phal.*‘Baguio’）（西蕾丽蝴蝶兰 × 林登蝴蝶兰）（表 23.7）。

大多数带有红色条纹的杂交种都源自具带红色斑点花朵的品种，红色斑点沿脉络连成线排列，看起来就像条纹。例如，‘里约之星’（*Phal.*‘Star of Rio’）[‘巴丹’（*Phal.*‘Bataan’）× 露德蝴蝶兰]就是白花或带红色条纹的露德蝴蝶兰和白花蝴蝶兰‘巴丹’（南洋白花蝴蝶兰 × 菲律宾白花蝴蝶兰）的杂交种，其红色白点从花瓣中心呈辐射状分布。它的子一代杂交种‘桑巴’（*Phal.*‘Samba’）（‘里约之星’× 安汶蝴蝶兰）也具带红色斑点的白花。‘里约之星’和‘桑巴’被进一步用于繁育新的杂交种。然而，它们的后代‘埃拉•弗里德’（*Phal*.‘Ella Freed’）（*Phal*.‘Show Girl’×‘桑巴’）、*Phal*.‘Freed's Danseuse’［‘埃拉•弗里德’×‘职业女孩’（*Phal*.‘Career Girl’）］、‘摩登条纹’（*Phal*.‘Modern Stripes’）（*Phal*.‘Freed's Danseuse’×*Phal*.‘Chiali Stripe’）、*Phal*.‘Okay Seven’（‘摩登条纹’×*Phal*.‘Houpi Beauty’）、*Phal*.‘Taisuco Stripe’（*Phal*.‘Okay Seven’×*Phal*.‘Taisuco Gaster’）和 *Phal*.‘Little Gem Stripes’（*Phal*.‘Taisuco Stripe’×*Phal*.‘Taisuco Gem’）都具粉色至红色的条纹、白色花瓣、红色唇瓣（表 23.7）。其中，*Phal*.‘Little Gem Stripes’（图 23.6）仍被经常用作繁育带红色条纹的白花和黄花亲本。

图 23.6 白底红色条纹蝴蝶兰杂交种的各种表型。（a）*Phal.*‘Little Gem Stripes’，（b）*Phal.*‘Taida Salu’，（c）*Phal.*‘Taisuco Glitter’，（d）*Phal.*‘Fuller's Pink Gem’，（e）*Phal.*‘OX Leo Prince’，（f）*Phal.*‘I-Hsin The Big Bang’，鼻宽 =1cm

为了导入大花性状，使用“大红花”杂交种‘卡特申克’与‘幸运女士’（*Phal.*‘Lucky Lady’）杂交得到了‘幸运申克’（*Phal.*‘Lucky Shenk’）。随后，该杂交种被

进一步杂交得到了 *Phal.*‘Chiali Stripe’（*Phal.*‘Cindy Tsai’×‘幸运申克’）、*Phal.*‘Houpi Beauty’（*Phal.*‘Tsuei You Queen’ × Phal.‘Chiali Stripe’）、‘太阳王子’（*Phal.*‘Sun Prince’）（*Phal.*‘Cypress Pink’× *Phal.*‘Houpi Beauty’）和‘花豹王子’（*Phal.*‘Leopard Prince’）（‘太阳王子’×*Phal.*‘Ho’s French Fantasia’）。另外，‘花豹王子’拥有带红色条纹的白色花朵，荣获了 32 个奖项。‘花豹王子’已经繁育出了 221 个子一代杂交种和 567 个杂交后代（表 23.7）。

家谱分析表明，*Phal.*‘Little Gem Stripes’含有 29.3%的桃红蝴蝶兰基因、21.7%的南洋白花蝴蝶兰基因、15.9%的爪哇白花蝴蝶兰基因、9.6%的菲律宾白花蝴蝶兰基因、7.9%的西蕾丽蝴蝶兰基因、5.3%的桑德蝴蝶兰基因、4.4%的露德蝴蝶兰基因、4.3%的安汶蝴蝶兰基因、1.0%的史塔基蝴蝶兰基因以及其他基因。

23.4.2.5　繁殖红花蝴蝶兰的分子遗传学

花青素是几乎所有植物都存在的水溶性色素，是花朵呈现橙色、红色、紫色和蓝色的主要原因。花青素积累的生物合成途径是植物次生代谢中研究得最广泛的途径之一，包括查耳酮合酶（CHS）、查耳酮异构酶（CHI）、黄烷酮 3-羟化酶（F3H）、类黄酮 3′-羟化酶（F3′H）、类黄酮 3′5′-羟化酶（F3′5′H）、二氢黄酮醇 4-还原酶（DFR）、花青素合酶（ANS）和 UDP-葡萄糖：类黄酮 3-*O*-葡萄糖基转移酶（UFGT）（Grotewold 2006）。另外，许多涉及花青素生物合成的调控基因也已从多种植物中克隆并鉴定。R2R3-MYB 和碱性螺旋-环-螺旋（bHLH）转录因子及 WD40 重复序列（WDR）蛋白是花青素生物合成的三个主要调控蛋白家族（Winkel-Shirley 2001；Koes et al. 2005；Feller et al. 2011；Hichri et al. 2011；Petroni and Tonelli 2011）。其中，这些 R2R3-MYB 转录因子为下游基因提供特异性并导致组织特异性积累花青素的关键成分（Borevitz et al. 2000；Zhang et al. 2003）。

在西蕾丽蝴蝶兰中，*PsDFR* 和 *PsMYB* 在‘常春仙子’（*Phal.*‘Ever-spring Fairy’）的紫色花朵和带黑点花瓣中表达（Ma et al. 2009）。*PeUFGT3*（桃红蝴蝶兰 UFGT3）在‘露西亚女士’和其他红花杂交种的红色唇瓣中高表达，但在白色花被中不表达（Chen et al. 2011）。

已对参与花青素生物合成途径的基因进行了全基因组筛选，结果发现三个结构基因 *PeF3H5*、*PeDFR1* 和 *PeANS3* 以及三个调控基因 *PeMYB2*、*PeMYB11* 与 *PeMYB12* 调控蝴蝶兰中花青素的积累（Hsu et al. 2015a）。然而，这三个 *PeMYB* 基因在同一朵花中调节着不同的色素沉着模式，而且在花被和唇上，*PeMYB* 基因对色素沉着模式的调节也大不相同。在被片中，*PeMYB2*、*PeMYB11* 和 *PeMYB12* 分别控制全红色色素沉着、红色斑点和静脉纹。在唇瓣中，*PeMYB11* 调节愈伤组织中的红斑，*PeMYB12* 是中央叶色素沉着的主要调节因子。这三个 *PeMYB* 基因的调节作用结合导致了蝴蝶兰中非常复杂的花卉色素沉着模式（Hsu et al. 2015a）。

23.4.3　黄花蝴蝶兰

黄色的杂交蝴蝶兰只占蝴蝶兰市场的一小部分，然而黄色的明亮感和杂交种数量的稀少使得黄花杂交种的育种也很重要。黄色杂交种可细分为纯黄色的花朵（表 23.8，图 23.7）、带红色唇瓣的黄色花被和带红色条纹的黄色花朵（表 23.8）以及橙色和绿色花朵（表 23.9）。

表 23.8 繁育黄花蝴蝶兰的重要杂交种‘金沙’（*Phal.*‘Golden Sands’）、‘台北金’（*Phal.*‘Taipei Gold’）及其后代

杂交种名	杂交代数[a]	英国皇家园艺协会登记年份	亲本	子一代杂交种数	杂交后代总数	获奖数	平均花径（cm）	花色[b]
黄色花								
Phal. Golden Sands	G0	1964	（Fenton Davis Avant × *lueddemanniana*）	137	6572	18	8.4	YspO
Phal. Golden Amboin	G1	1976	（Golden Sands × *amboinensis*）	113	2845	11	7.0	YspO
Phal. Salu Spot	G2	1992	（Paifang's Auckland × Golden Amboin）	44	1467	3	7.7	YspO
Phal. Golden Sun	G3	1995	（Salu Spot × Sentra）	102	1099	5	8.0	YspO
Phal. Solar Flare	G1	1979	（Golden Sands × Golden Pride）	19	71	29	6.8	YspO
Phal. Liu Tuen- Shen	G1	1979	（*gigantea* × Golden Sands）	80	4528	6	8.0	YspO
Phal. Fortune Saltzman	G2	1983	（Liu Tuen-Shen × Barbara Freed Saltzman）	16	44	3	7.6	GR
Phal. Golden Peoker	G2	1983	（Misty Green × Liu Tuen-Shen）	208	3227	9	7.1	WspO
Phal. Taipei Gold	G0	1984	（Gladys Read × *venosa*）	152	1776	14	8.0	YspR
Phal. Brother Lawrence	G1	1995	（Taipei Gold × Deventeriana）	108	762	27	7.7	YspR
Phal. Brother Nugget	G1	1995	（Taipei Gold × Brother Imp）	8	192	0	—[c]	YspR
Phal. I-Hsin Sunflower	G1	2001	（Taipei Gold × Brother Nugget）	77	182	2	7.0	YspR
Phal. Sogo Manager	G2	1997	（Brother Lawrence × Autumn Sun）	92	455	12	6.8	YspR
Phal. Sogo Pride	G2	2000	（Brother Lawrence × Sogo Manager）	87	172	5	6.4	YspR
红唇黄花								
Phal. Chia Lin	G4	1982	（James Hall × Johanna）	39	2919	1	6.5	WstR
Phal. Sunrise Star	G5	2003	（Chia Lin × Tinny Honey）	58	102	0	—	YR
Phal. Chian Xen Queen	G5	2004	（Chia Lin × Mount Beauty）	44	170	0	—	YstR
Phal. Fuller's Sunset	G6	2004	（Taisuco Date × Chian Xen Queen）	24	26	17	8.9	YR
具红色条纹的黄花								
Phal. Taida Salu	—[d]	1997	（Salu Spot × Happy Beauty）	29	38	8	7.8	YstR
Phal. Little Gem Stripes	—	1997	（Taisuco Stripe × Taisuco Gem）	144	292	16	5.4	WstR
Phal. Taida Golden Gem	—	2004	（Little Gem Stripe × Salu Peoker）	0	0	0	—	—[e]
Phal. Fangmei Sweet	—	2009	（Little Gem Stripe × Dou-dii Golden Princess）	5	5	15	6.1	YstR

数据来源于 OrchidWiz X3.3，2017

a. 世代相传。G0 为蝴蝶兰‘金沙’或蝴蝶兰‘台北金’。G1～G6 表示‘金沙’或‘台北金’衍生出的不同后代

b. 栽培品种的花色。第一个词为花被的颜色，第二个词为唇瓣的颜色。Wst 代表白花带红色条纹，Wsp 代表白花带红色斑点，G 代表黄绿色，O 代表橙色，R 代表红色，Y 代表黄色，Ysp 代表黄花带红色斑点，Yst 代表黄花带红色条纹

c. OrchidWiz X3.3，2017 中未查找到

d. 不是 G0 代‘金沙’或‘台北金’的繁殖后代

e. OrchidWiz X3.3，2017 中未查找到

图 23.7 黄花蝴蝶兰杂交种的各种表型。（a）*Phal.* ‘Chian Xen Gold’，（b）*Phal.* ‘Golden Apollon’，（c）*Phal.* ‘I-Hsin Bright Star’，（d）*Phal.* ‘Taisuco Date’，鼻宽 = 1cm

表 23.9 橙色和绿色花朵育种的重要杂交种

杂交种名	杂交代数[a]	英国皇家园艺协会登记年份	亲本	子一代杂交种数	杂交后代总数	获奖数	平均花径(cm)	花色[b]
橙色花								
Phal. Princess Kaiulani	—	1961	（*violacea* × *amboinensis*）	225	2545	29	—[c]	RR
Phal. Luedde- violacea	—	1895	（*lueddemanniana* × *violacea*）	82	2197	31	5.3	RR
Phal. George Vasquez	—	1974	（*violacea* × Luedde-violacea）	128	1692	11	5.0	RR
Phal. Brother Sara Gold	—	1997	（Sara Lee × Taipei Gold）	91	351	13	6.0	OspR
Phal. Salu Peoker	—	2000	（Golden Sun × Yellow Peoker）	86	289	4	7.2	OspR
Phal. KV Golden Star	—	2006	（Salu Peoker × Brother Sara Gold）	6	7	4	6.3	OspR
Phal. Sogo Lawrence	—	2002	（Brother Peter Star × Brother Sara Gold）	66	161	7	5.6	OspR
Phal. Surf Song	—	2004	（Taipei Gold × King Shiang’s Rose）	14	14	3	7.9	OR
绿色花								
Phal. Gelblieber	—	1984	（amboinensis × micholitzii）	36	1064	1	6.1	GspW
Phal. Yungho Gelb Canary	—	1995	（Gelblieber × Princess Kaiulani）	86	296	4	5.1	YY
Phal. Su-An Cricket	—	1986	（Gelblieber × mannii）	45	270	4	4.2	GW
Phal. Hannover Passion	—	1994	（Gelblieber × mariae）	41	140	2	5.8	GspW
Phal. Nobby’s Green Eagle	—	2005	（Gelblieber × Nobby’s Fox）	10	13	17	5.5	GW
Phal. Fortune Saltzman	—	1983	（Liu Tuen-Shen × Barbara Freed Saltzman）	16	44	3	7.6	GR

数据来源于 OrchidWiz X3.3，2017

a. 非常规亲本

b. 栽培品种的花色。第一个词为花被的颜色，第二个词为唇瓣的颜色。W 代表白色，G 代表黄绿色，Gsp 代表绿花带斑点，O 代表橙色，Osp 代表橙花带斑点，R 代表红色，Y 代表黄色

c. OrchidWiz X3.3，2017 中未查找到

繁育黄色花朵的主要难题是花朵开放后颜色变淡，甚至接近白色。因此，对于一个好的杂交种来说，黄色必须明亮或浓密，并能持续所有开花阶段。此外，由于黄花源于野生种，大多数黄花杂交种只有一个花箭、花少、花呈星形。因此，黄花杂交种需要与白花杂交种杂交，以改变花的形态特征并获得完整的花朵。

23.4.3.1　原生种对黄花杂交种选育的贡献

在 92 个原生种中，只有约 20 个物种具有黄色的花色，都属于裂唇蝴蝶兰亚属。其中，安汶蝴蝶兰、鹿角蝴蝶兰、横纹蝴蝶兰、象耳蝴蝶兰、露德蝴蝶兰、曼氏蝴蝶兰、苏门答腊蝴蝶兰和红脉蝴蝶兰最常用作亲本（图 23.8）。这些开黄花的原生种大多只开很少的星形花，因此需要与白花杂交以增加花的数量并创造更好的花形态。然而，由于开黄花的裂唇蝴蝶兰亚属和白色大花蝴蝶兰亚属在染色体大小上存在差异，这两个亚属之间的杂交很难获得成功，而且大多是不育的杂交种（见第 23.7 节）。

图 23.8　用于繁殖黄花蝴蝶兰杂交种的原生种。（a）露德蝴蝶兰，（b）安汶蝴蝶兰，（c）象耳蝴蝶兰，（d）曼氏蝴蝶兰，鼻宽 = 1cm

23.4.3.2　黄花蝴蝶兰的主要杂交种

第一个著名的黄花杂交种是‘金色大帝’（*Phal.*‘Golden Emperor’）[*Phal.*‘Snow Daffodil’× 曼波舞曲（*Phal.*‘Mambo’）]，其具有圆润的花朵形状和花呈纯黄色。它于 1982 年登录。然而，‘金色大帝’的染色体数目是 $2n = 3x$，所以‘金色大帝’的后代罕见，只获得了 7 个杂交后代。

‘金沙’（*Phal.*‘Fenton Davis Avant’× 露德蝴蝶兰）具有露德蝴蝶兰黄花的血统，

是137个子一代杂交种和6572个杂交后代的主要育种亲本（表23.8）。‘金沙’及其大部分后代都具有带有红色斑点和橙色唇瓣的黄色花朵。它的后代*Phal.*‘Golden Amboin’（‘金沙’× 安汶蝴蝶兰）、*Phal.*‘Salu Spot’（*Phal.*‘Paifang's Auckland’× *Phal.*‘Golden Amboin’）和‘金色太阳’（*Phal.*‘Golden Sun’）（*Phal.*‘Salu Spot’× *Phal.*‘Sentra’）是重要的育种亲本。在‘金沙’的所有杂交后代中，子一代杂交种‘日辉’（*Phal.*‘Solar Flare’）（‘金沙’× *Phal.*‘Golden Pride’）和子四代杂交种*Phal.*‘Ambo Buddha’（*Phal.*‘Brother Buddha’× 安汶蝴蝶兰）获得奖项最多。

家谱分析表明，*Phal.*‘Ambo Buddha’含有56.3%的安汶蝴蝶兰基因、7.3%的爪哇白花蝴蝶兰基因、6.3%的横纹蝴蝶兰基因、6.3%的象耳蝴蝶兰基因、6.3%的露德蝴蝶兰基因、4.0%的南洋白花蝴蝶兰基因、3.2%的菲律宾白花蝴蝶兰基因、2.3%的桃红蝴蝶兰基因和1.2%的桑德蝴蝶兰基因。

‘金沙’(母本)和具蜡质花朵的象耳蝴蝶兰(父本)杂交获得了*Phal.*‘Liu Tuen-Shen’，其花朵具有更好的质感，并被进一步用作育种亲本，获得了80个子一代杂交种和4528个杂交后代（表23.8）。从其后代的体细胞“意外”突变体中选出了‘金色小丑’（*Phal.*‘Golden Peoker’）（*Phal.*‘Misty Green’× *Phal.*‘Liu Tuen-Shen’），从而开启了丑角花的育种（见第23.4.4节）。另一个后代*Phal.*‘Fortune Saltzman’（*Phal.*‘Liu Tuen-Shen’× *Phal.*‘Barbara Freed Saltzman’）具绿色花被和红色唇瓣（见第23.4.3.6节）。

红脉蝴蝶兰参与黄花蝴蝶兰的育种是1984年从‘台北金’（*Phal.*‘Gladys Read’× 红脉蝴蝶兰）开始的，获得了152个子一代杂交种和1776个杂交后代。其中，‘劳伦斯兄弟’（*Phal.*‘Brother Lawrence’）（‘台北金’× *Phal.*‘Deventeriana’）、*Phal.*‘Brother Nugget’[‘台北金’×‘淘气兄弟’（*Phal.*‘Brother Imp’）]、*Phal.*‘I-Hsin Sunflower’（‘台北金’× *Phal.*‘Brother Nugget’）、*Phal.*‘Sogo Manager’[‘劳伦斯兄弟’×‘秋日’（*Phal.*‘Autumn Sun’）]和*Phal.*‘Sogo Pride’（‘劳伦斯兄弟’× *Phal.*‘Sogo Manager’）是主要的育种亲本。

23.4.3.3 红唇黄花蝴蝶兰的重要杂交种

Phal.‘Chia Lin’（*Phal.*‘James Hall’× *Phal.*‘Johanna’）是繁殖具紫红色花品种和红唇黄花蝴蝶兰的主要亲本（表23.8）。它的大多数子一代杂交种都具红唇黄花，少数品种开紫红色花。其中，‘黎明之星’（*Phal.*‘Sunrise Star’）[*Phal.*‘Chia Lin’×‘小宝贝’（*Phal.*‘Tinny Honey’）]和*Phal.*‘Chian Xen Queen’（*Phal.*‘Chia Lin’×‘美景山’）是著名的红唇黄花杂交种。*Phal.*‘Chian Xen Queen’拥有非常著名的子一代杂交种*Phal.*‘Chian Xen Queen’，该品种有亮黄色花瓣和红色唇瓣，拥有24个子一代杂交种和26个杂交后代。*Phal.*‘Fusheng's Golden Age’（*Phal.*‘Fong-Tien's Yellow Butterfly’× *Phal.*‘Chian Xen Queen’）也是*Phal.*‘Chian Xen Queen’的子一代杂交种，也拥有亮黄色花瓣和红色唇瓣（图23.9）。家谱分析表明，*Phal.*‘Fuller's Sunset’含有28.0%的南洋白花蝴蝶兰基因、19.8%的爪哇白花蝴蝶兰基因、14.1%的安汶蝴蝶兰基因、9.7%的菲律宾白花蝴蝶兰基因、9.4%的红脉蝴蝶兰基因、5.5%的露德蝴蝶兰基因以及其他基因。黄色可能是来自红脉蝴蝶兰的血统，红色唇瓣可能是来自南洋白花蝴蝶兰的血统。

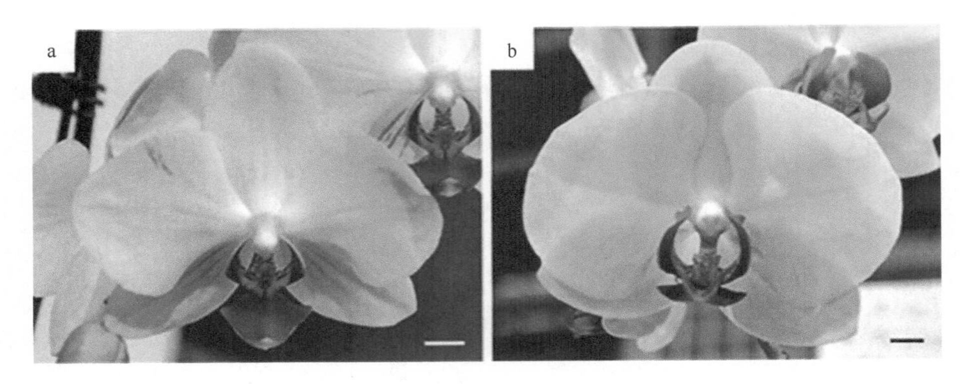

图 23.9　具黄色花被和红色唇瓣的蝴蝶兰杂交种表型。
（a）*Phal.*‘Taisuco Gloria’，（b）*Phal.*‘Fusheng's Golden Age’，鼻宽 = 1cm

23.4.3.4　具红色条纹的黄花蝴蝶兰的重要杂交种

大多数带有红色条纹的黄色花朵由带红色斑点的黄色花朵与带红色条纹的白色花朵杂交而来。*Phal.*‘Taida Salu’是由红斑黄花品种 *Phal.*‘Salu Spot’（母本）与红纹白花品种‘幸福美人’（*Phal.*‘Happy Beauty’）（父本）杂交而来的，它被用作育种亲本获得了 29 个子一代杂交种（表 23.8）。红纹白花品种 *Phal.*‘Little Gem Stripe’被用作杂交亲本，获得了 *Phal.*‘Taida Golden Gem’（*Phal.*‘Little Gem Stripe’× *Phal.*‘Salu Peoker’）和 *Phal.*‘Fangmei Sweet’（*Phal.*‘Little Gem Stripe’× *Phal.*‘Dou-dii Golden Princess’）（图 23.10），它一共有 5 个子一代杂交种，并获得了 15 个奖项。

图 23.10　具红色条纹的黄花蝴蝶兰杂交种的各种表型。（a）*Phal.*‘Fangmei Sweet’，
（b）*Phal.*‘Fuller's Gold Stripes’，鼻宽 = 1cm

23.4.3.5　橙花的重要杂交种

橙花蝴蝶兰由黄花蝴蝶兰和红花蝴蝶兰杂交而来。要获得橙花，必须实现以下三个条件：首先，黄色稳定显色；其次，红色均匀地分布在花朵上；最后，黄色和红色能相互作用显色。总体而言，蝴蝶兰的黄色源自红脉蝴蝶兰和安汶蝴蝶兰，红色来自紫纹蝴蝶兰和露德蝴蝶兰。

1961 年，紫纹蝴蝶兰和红脉蝴蝶兰的杂交种‘凯卢拉尼公主’（*Phal.*‘Princess Kaiulani’）登记注册（表 23.9）。1974 年，紫纹蝴蝶兰和 *Phal.*‘Luedde-violacea’杂交

获得了‘乔治·瓦斯奎兹’（*Phal*.‘George Vasquez’）。最近，*Phal*.‘Brother Sara Gold’（*Phal*.‘Sara Lee’×‘台北金’）、*Phal*.‘Salu Peoker’[‘金色太阳’×‘黄色小丑’（*Phal*.‘Yellow Peoker’）]和 *Phal*.‘Sogo Lawrence’[‘彼得之星兄弟’（*Phal*.‘Brother Peter Star’）×*Phal*.‘Brother Sara Gold’]成为重要的育种亲本，分别繁育出了 91 个、86 个和 66 个子一代杂交种以及总共 351 个、289 个和 161 个杂交后代。然而，这些品种的花大多数为黄色的花瓣上均匀分布了红色斑点，而不是真正的橙色。

此外，*Phal*.‘Surf Song’（‘台北金’×*Phal*.‘King Shiang's Rose’）具带红色条纹的粉色花朵，*Phal*.‘OX Gold Orange’为牛兰花农场的一个无性繁殖品种，其被选出用于橙花蝴蝶兰的培育（图 23.11）。*Phal*.‘Surf Song’有 14 个子一代杂交种，并获得了

图 23.11　橙花蝴蝶兰杂交种的各种表型。(a) *Phal*.‘Brother Sara Gold’，(b) *Phal*.‘Kdares Orange Lover’，(c) *Phal*.‘Surf Song’，(d) *Phal*.‘KS Orange’，(e) *Phal*.‘Charming Fortune’，(f) *Phal*.‘Fangmei A Hot’，鼻宽 =1cm

3 个奖项。家谱分析表明，*Phal.*‘Surf Song’含有 25.0%的红脉蝴蝶兰基因、23.3%的南洋白花蝴蝶兰基因、22.8%的爪哇白花蝴蝶兰基因、10.1%的菲律宾白花蝴蝶兰基因、9.1%的西蕾丽蝴蝶兰、4.2%的桑德蝴蝶兰基因等。黄色是来自红脉蝴蝶兰的血统，红色是来自西蕾丽蝴蝶兰和桑德蝴蝶兰的血统。

23.4.3.6 绿花蝴蝶兰的重要杂种

当花瓣中含有不会降解的叶绿体时，就会形成绿色花朵。培育绿花的主要原生种为 *Phal. cornu-cervi* var. *alba*、*Phal. mannii* var. *alba*、紫纹蝴蝶兰和象耳蝴蝶兰。1984 年，*Phal.*‘Gelblieber’[安汶蝴蝶兰 × 米库氏蝴蝶兰（*Phal. micholitzii*）]被用作培育绿花蝴蝶兰的重要亲本（表 23.9）。它的子一代杂交种 *Phal.*‘Yungho Gelb Canary’（*Phal.*‘Gelblieber’×‘凯卢拉尼公主’）、*Phal.*‘Su-An Cricket’（*Phal.*‘Gelblieber’× 曼氏蝴蝶兰）、*Phal.*‘Hannover Passion’[*Phal.*‘Gelblieber’×‘玛莉亚蝴蝶兰’（*Phal. mariae*）]和 *Phal.*‘Buena Jewel’（紫纹蝴蝶兰 × *Phal.*‘Gelblieber’）都有绿色的花瓣和白色的唇瓣，是重要的育种亲本。

从‘金沙’和具蜡质花朵的象耳蝴蝶兰的杂交种中选择了 *Phal.*‘Liu Tuen-Shen’，并获得了花朵质地更好的杂交后代‘幸运的萨尔茨曼’（*Phal.*‘Fortune Saltzman’）（*Phal.*‘Liu Tuen- Shen’× *Phal.*‘Barbara FreedSaltzman’），其花色为纯浅绿色，在市场上是比较受欢迎的品种（图 23.12）。家谱分析表明，‘幸运的萨尔茨曼’中含有 20.5%的象耳蝴蝶兰基因、23.3%的南洋白花蝴蝶兰基因、20.8%的爪哇白花蝴蝶兰基因、17.2%的露德蝴蝶兰基因、10.1%的菲律宾白花蝴蝶兰基因、3.1%的桃红蝴蝶兰基因、0.2%的西蕾丽蝴蝶兰基因和 0.2%的史塔基蝴蝶兰基因。花朵的绿色应该是来自象耳蝴蝶兰的血统。

23.4.4 丑角花（黑色）

之所以称为“丑角花”是因为花朵的斑纹与小丑脸上涂抹的黑色斑点相似。丑角花具有真正的红黑色或紫黑色，通常会形成非常复杂的色素沉着模式（图 23.13）。在市场上，丑角花是蝴蝶兰的新花色，但在育种早期，这种花色并不太受欢迎。造成这种现象的两个原因是，丑角花的颜色看起来很脏，并且与其他花朵相比，丑角花在外植体繁殖中的突变率相对较高。

丑角花的起源可以追溯到被兄弟国际兰园有限公司注册的突变体品种‘金色小丑’“兄弟”（*Phal.*‘Golden Peoker’“Brother”）（*Phal.*‘Misty Green’× *Phal.*‘Liu Tuen- Shen’）。它的亲本 *Phal.*‘Misty Green’花色为红点绿花，*Phal.*‘Liu Tuen-Shen’花色为红点黄花。它们的杂交种‘金色小丑’花色为红点白花，“兄弟”是该杂交种中表现最佳的。在蝴蝶兰‘金色小丑’“兄弟”的外植体繁殖过程中，Ever Spring Orchids Nursery 发现了一个无性突变体，并将其命名为‘金色小丑’“常春”[*Phal.*‘Golden Peoker’“Ever-spring（ES）”]，其花瓣中有紫色斑点，而且几乎没有斑点融合在一起。在‘金色小丑’“常春”的外植体繁殖中，可以获得三种不同的表型。大约 40%的外植体繁殖植株花朵与‘金色小丑’“兄弟”的相似，约 30% 的植株保持了‘金色小丑’“ES”的性状，另外 30%的植株表现出了成片紫黑色斑点的新性状，这一新品种被命名为‘金色小丑’“BL”。‘金色小丑’“ES”和‘金色小丑’“BL”掀开了丑角花蝴蝶兰育种的新篇章。

图 23.12　绿花蝴蝶兰杂交种的各种表型。(a) *Phal.* ‘Lioulin Green Eagle’，(b) *Phal.* ‘Nobby’s Green Eagle’，(c) *Phal.* ‘Heliodor’，(d) *Phal.* ‘Fortune Saltzman’，(e) *Phal.* ‘Unimax Moonlight’，(f) *Phal.* ‘ARK’s Green Angel’，鼻宽 = 1cm

没有蝴蝶兰原生种与丑角花有直接关联。蝴蝶兰‘金色小丑’的基因构成为 25.0%的象耳蝴蝶兰基因、15.0%的 *Phal. Rimestadiana* 基因、12.5%的安汶蝴蝶兰基因、12.5%的露德蝴蝶兰基因、10.2%的南洋白花蝴蝶兰基因、7.4%的菲律宾白花蝴蝶兰基因及其他相关基因。

23.4.4.1　丑角花的重要杂交种

虽然繁育出了‘金色小丑’“ES”、‘常春之王’(*Phal.* ‘Ever-spring King’)(*Phal.*

‘Chih Shang’s Stripes’ × ‘金色小丑’）和‘常春之光’（*Phal.* ‘Ever-spring Light’）[‘常春之星’（*Phal.* ‘Ever-spring Star’）× ‘金色小丑’] 等品种，但只有 1/300 个后代保持了丑角花的性状(表 23.10)。1999 年,‘常春王子’（‘金色小丑’× *Phal.*‘Taisuco Beauty’）注册。通过用‘金色小丑’“BL”代替‘金色小丑’“ES”，丑角花的百分比从 1/300（使用“ES”）增加到 1/30（使用“BL”)。因此，相对稳定的丑角花表型促进了黑色花朵蝴蝶兰的持续育种。

图 23.13 蝴蝶兰杂交种“丑角花”。(a)‘常春王子’(*Phal.* ‘Ever-spring Prince’),(b) *Phal.* ‘OX Black Tea’,(c) *Phal.* ‘Chian Xen Diamond’,(d) *Phal.* ‘Chian Xen Piano’,(e) *Phal.* ‘Lianher Focus’,(f) *Phal.* ‘Ebony Sweet Gem’，鼻宽 = 1cm

蝴蝶兰‘金色小丑’“BL”被用于丑角花的进一步育种。*Phal.* ‘I-Hsin Black Jack’[‘金色小丑’×‘花豹王子’(*Phal.* ‘Leopard Prince’)]、*Phal.* ‘Haur Jin Diamond’（‘金色小丑’× *Phal.* ‘Ching-Her Buddha’）和‘黎明红色小丑’（*Phal.* ‘Sunrise Red Peoker’）

（‘金色小丑’ × *Phal.* ‘Kuntrarti Rarashati’）都是重要的育种亲本，它们分别有 26 个、77 个和 43 个子一代杂交种以及 105 个、366 个和 100 个杂交后代。

表 23.10　蝴蝶兰‘金色小丑’及其后代中用于培育丑角花的重要杂交种

杂交种名	杂交代数[a]	英国皇家园艺协会登记年份	亲本	子一代杂交种数	杂交后代总数	获奖数	平均花径（cm）	花色[b]
Phal. Golden Peoker	G0	1983	（Misty Green × Liu Tuen-Shen）	208	3227	9	—[c]	WspW
Phal. Ever-spring King	G1	1992	（Chih Shang's Stripes × Golden Peoker）	74	550	9	8.0	WspR
Phal. Happy Ufo	G2	2000	（King Shiang's Beauty × Ever-spring King）	25	157	3	6.8	—[d]
Phal. Brother Love Rosa	G2	2002	（Ever-spring King × Sinica Knight）	16	21	0	—	—
Phal. Yu Pin Pearl	G2	2001	（Ever-spring King × Musashino）	14	27	1	7.7	WspW
Phal. Ever-spring Light	G1	1992	（Ever-spring Star × Golden Peoker）	26	899	5	7.8	WspR
Phal. Ching Hua Spring	G2	1998	（Minho Princess × Ever-spring Light）	6	841	0	—	WspR
Phal. Chian Xen Pearl	G3	2002	（Ching Hua Spring × Nobby's Pink Lady）	164	724	24	9.2	WspR
Phal. Fusheng Pink Pearl	G4	2005	（Chian Xen Pearl × Fusheng's Purple Gem）	40	110	4	10.9	PspR
Phal. Chian Xen Mammon	G4	2004	（Chian Xen Pearl × Mount Beauty）	28	66	8	10.4	WspR
Phal. Chian Xen Magpie	G4	2004	（Chian Xen Pearl × Tinny White）	21	28	3	11.6	WspR
Phal. Yu Pin Lady	G4	2003	（Yu Pin Lover × Chian Xen Pearl）	20	41	5	10.0	PspR
Phal. Fureshing Speckle	G4	2004	（Chian Xen Pearl × Chian Xen Sweet Valentine）	17	55	1	10.0	PspR
Phal. Taida Pearl	G3	2001	（Ching Hua Spring × Sogo Davis）	54	104	6	8.1	WspR
Phal. KV Charmer	G4	2006	（Salu Peoker × Taida Pearl）	2	2	9	8.8	YspR
Phal. Ever-spring Prince	G1	1999	（Golden Peoker × Taisuco Beauty）	46	100	13	8.1	WspR
Phal. Ever-spring Fairy	G1	1997	（Taisuco Kochdian × Golden Peoker）	9	9	2	10.2	WspR
Phal. I-Hsin Black Jack	G1	1999	（Golden Peoker × Leopard Prince）	26	105	13	7.5	WspP
Phal. Haur Jin Diamond	G1	1999	（Golden Peoker × Ching-Her Buddha）	77	366	4	7.5	YspR
Phal. Tai-I Yellow Bird	G2	2002	（Salu Peoker × Haur Jin Diamond）	40	90	5	5.7	YspR
Phal. Haur Jin Princess	G2	2002	（Haur Jin Diamond × Dou-dii Golden Princess）	39	91	12	7.1	YspR
Phal. Diamond Beauty	G2	2003	（Chingruey's Beauty × Haur Jin Diamond）	21	29	1	—	—
Phal. Shin Yi Diamond	G2	2002	（Haur Jin Diamond × Ching-Her Buddha）	12	25	4	8.2	YspR
Phal. Yu Pin Natsume	G2	2003	（Autumn Sun × Haur Jin Diamond）	13	39	4	5.5	YspR
Phal. Chian Xen Diamond	G1	2002	（Golden Peoker × Judy Valentine）	5	6	1	8.6	PspR
Phal. Sunrise Red Peoker	G1	2003	（Golden Peoker × Kuntrartic Rarashati）	43	100	1	4.8	YspP
Phal. Brother Purple	G1	1995	（Golden Peoker × Brother Glamour）	97	329	0	—	WspR
Phal. Brother Peacock	G1	1992	（Paifang's Queen × Golden Peoker）	16	439	0	—	WspR
Phal. Brother Precious Stones	G2	1998	（Brother Fancy × Brother Purple）	45	61	24	7.5	WspR
Phal. Brother Pirate King	G2	1998	（Fortune Buddha × Brother Purple）	39	74	19	7.5	YspR
Phal. Brother Supersonic	G2	1997	（Sara Lee × Brother Purple）	23	30	19	6.6	WspR

数据来源于OrchidWiz X3.3，2017

a. 世代相传。G0 为蝴蝶兰‘金色小丑’。G1～G4 表示‘金色小丑’衍生出的不同后代

b. 栽培品种的花色。第一个词为花被的颜色，第二个词为唇瓣的颜色。W 代表白色，Wsp 代表白花带斑点，P 代表粉色，Psp 代表粉花带斑点，R 代表红色，Y 代表黄色，Ysp 代表黄花带斑点

c. OrchidWiz X3.3，2017 中未查找到

d. OrchidWiz X3.3，2017 中未查找到

‘常春之光’拥有最多的后代，包括‘金色小丑’、26 个子一代杂交种和 899 个杂交后代。它的子一代 *Phal.*‘Ching Hua Spring’（*Phal.*‘Minho Princess’×‘常春之光’）对于 *Phal.*‘Chian Xen Pearl’[*Phal.*‘Ching Hua Spring’×‘时髦粉佳人’（*Phal.*‘Nobby's Pink Lady’）]的繁殖非常重要，是非常重要的育种亲本。*Phal.*‘Chian Xen Pearl’已经在‘金色小丑’的后代中赢得了 24 个奖项。

在花色研究中做出重大贡献的其他突破性选择是‘常春之王’及其后代‘快乐飞碟’（*Phal.*‘Happy Ufo’）（*Phal.*‘King Shiang's Beauty’×‘常春之王’）、*Phal.*‘Haur Jin Diamond’及其杂交种 *Phal.*‘Tai-I Yellow Bird’（*Phal.*‘Salu Peoker’×*Phal.*‘Haur Jin Diamond’）、*Phal.*‘Haur Jin Princess’（*Phal.*‘Haur Jin Diamond’×*Phal.*‘Dou-dii Golden Princess’）（表 23.10）。

‘金色小丑’这一突变体出现之后，兄弟国际兰园有限公司（Brother Orchid Nursery Co.，Ltd.）用它作为亲本繁殖和注册了 81 个子一代杂交种，其中包括了‘紫色兄弟’（*Phal.*‘Brother Purple’）和‘孔雀兄弟’（*Phal.*‘Brother Peacock’）。‘紫色兄弟’的子一代杂交种‘宝石兄弟’（*Phal.*‘Brother Precious Stones’）[‘幻想兄弟’（*Phal.*‘Brother Fancy’）×‘紫色兄弟’]、‘海贼王兄弟’（*Phal.*‘Brother Pirate King’）[‘幸运之佛’（*Phal.*‘Fortune Buddha’）×‘紫色兄弟’]和‘超音速兄弟’（*Phal.*‘Brother Supersonic’）（*Phal.*‘Sara Lee’×‘紫色兄弟’）后来被用于蝴蝶兰杂交种的进一步研发。

23.4.4.2 丑角花育种的分子遗传学

丑角或黑色的花有大量的花青素积累（Kuo and Wu 2011），是由花青素相关基因（如酶和转录因子）高表达所致。在血橙（*Citrus sinensis*）中，类 Copia 的逆转座子控制附近 MYB 转录因子 Ruby 的高表达，导致果实中大量花青素的产生（Butelli et al. 2012）。在花菜中也发现了类似的情况，其中 *Harbinger* DNA 转座子插入了紫色（Pr）基因的调控区域，编码 R2R3-MYB 转录因子，导致 curds 中 Pr 上调形成深紫色（Chiu et al. 2010）。蝴蝶兰中的丑角花形成可能是由转座子插入 R2R3-MYB 转录因子的调控区引起的。

23.4.5 蓝紫色的花朵

尽管在 RHS 中登录了蝴蝶兰杂交种 34 112 个，但是带有蓝色或蓝色至紫色花朵的杂交种非常少见。另外，蝴蝶兰中的“蓝色”花是靛蓝色、蓝紫色或薰衣草色，并不是天蓝色或所谓的真蓝色。在自然界中，大多数蝴蝶兰都没有蓝色的花朵。但是，已经发现了一些蓝紫色的变种，所以梦中的蓝色花朵可能会变成现实的。

23.4.5.1 原生种对蓝紫色杂交种的影响

已经发现了几种带有蓝紫色花朵的蝴蝶兰原生种，如五唇兰蓝化型（*Phal. pulcherrima* var. *coerulea*）、桃红蝴蝶兰蓝化型（*Phal. equestris* var. *coerulea*）、贝丽娜蝴蝶兰和 *Phal. violacea*“Indigo”（图 23.14）。五唇兰有很多花色，从白色、深红色到蓝紫色，五唇兰蓝化型具蓝紫色的花被和深紫色的唇瓣。桃红蝴蝶兰也有多种花色，白色、粉红色和红色，桃红蝴蝶兰蓝化型具白色的花被和蓝紫色的唇瓣。贝丽娜蝴蝶兰具黄色和紫色的花朵，*Phal. bellina*“Indigo”有蓝紫色的花朵，尽管它的后代又恢复了紫色的花朵。紫纹蝴蝶兰具深紫色的花朵，*Phal. violacea*“Indigo”具蓝紫色的花朵。

图 23.14 蓝紫色蝴蝶兰杂交种的花。（a）桃红蝴蝶兰，（b）紫纹蝴蝶兰，（c）贝丽娜蝴蝶兰，（d）*Phal.*‘Kenneth Schubert’，（e）‘紫色马丁’，鼻宽 = 1cm

兰科万代兰属的花色比蝴蝶兰具有更接近蓝紫色的颜色，因此，将万代兰的蓝色基因引入蝴蝶兰中将是培育蓝紫色蝴蝶兰的另一种途径。

23.4.5.2 主要的蓝紫色花蝴蝶兰杂交种

Phal.‘Kenneth Schubert’（五唇兰 × 紫纹蝴蝶兰）是第一个登录的具蓝紫色花朵的杂交种，并被用作主要的育种亲本（图 23.14）。它的后代‘紫色马丁’（*Phal.*‘Kenneth Schubert’× 紫纹蝴蝶兰）和‘小蓝鸟’（*Phal.*‘Little Blue Bird’）（*Phal.*‘Kenneth Schubert’× 五唇兰）来自 *Phal.*‘Kenneth Schubert’的逆代回交。到目前为止，‘紫色马丁’是蝴蝶兰中花朵颜色最接近蓝色的品种。近年来，随着原生种新品种的发现，蓝紫色和蓝色花朵的育种有了新的动力。

23.4.5.3 培养蓝紫色的分子遗传学

形成花色的主要色素化合物是花青素和类胡萝卜素。花青素产生橙色、红色和蓝色，而类胡萝卜素表现为黄色。植物中存在三种花青素，它们与不同的颜色有关，分别为花葵素、花青素和花翠素，分别代表橙色、红色和蓝色。从天竺葵素（pelargonidin）到花青素（cyanidin）或从花青素（cyanidin）到飞燕草素（delphinidin）的转变分别需要类黄酮3'-羟化酶（F3'H）和类黄酮3'5'-羟化酶（F3'5'H）的酶活性（Sasaki and Nakayama 2015）。例如，鼠尾草不含F3'H和F3'5'H，产生天竺葵苷类衍生物，呈橙红色（Tanaka and Ohmiya 2008）。玫瑰有F3'H而没有F3'5'H，所以它产生花青素（cyanidin），呈红色（Tanaka and Ohmiya 2008）。薰衣草同时含有F3'H和F3'5'H，可产生飞燕草素（delphinidin）型花青素，呈蓝色（Tanaka and Ohmiya 2008）。因此，如果想要培养出蓝色花蝴蝶兰，需要F3'5'H在花朵中表达。已有研究表明，从黑色花和红色花蝴蝶兰杂交种中克隆了编码F3'5'H酶的基因，并通过瞬时表达测定法研究了其功能，尽管转化后的花朵没有呈现蓝色（Su and Hsu 2003；Wang et al. 2006）。

另外，据报道花青素所在的液泡中的pH会影响蓝紫色，这意味着当液泡中的pH升高时，含花青素的花看起来更蓝，而pH降低则变红（Griesbach 2005）。

23.5 不同形态蝴蝶兰的育种

除了花色，花的形态也是蝴蝶兰育种最重要的目标之一。在过去的几十年里，培育更小的花和不同形状的花变得越来越重要。

23.5.1 中小型花

多数蝴蝶兰育种都注重各种颜色的大花（花径＞10cm），见第23.4.1节和第23.4.2节。随着大型花卉育种的饱和，近年来出现了一种新的育种方向，即中小型花，中小型花植物一般高约20cm，花径3～7cm。中小型蝴蝶兰单位面积可以种植更多，且易于运输，所以经济效益更高。此外，小型花和中型花的杂交种适合作为礼物或用来装饰办公室的桌子或写字台，这意味着这些杂交种与大花的杂交种有着不同的市场。

23.5.1.1 对中小型花有贡献的原生种

小花型原生种主要有桃红蝴蝶兰、五唇兰和林登蝴蝶兰。桃红蝴蝶兰一朵花序上开许多花，花径2.5cm，颜色各异，有白色、粉红色、红色及紫红色（图23.15）。用桃红蝴蝶兰作为亲本的育种目标是植株紧凑、花小、花箭多、花多、花色多样（Hsu and Chen 2015）。五唇兰花径2～3cm，颜色各异，主要提供深红色和深紫色的表型特征。林登蝴蝶兰花径约为3cm，该品种是引入红色条纹特征的主要资源。

23.5.1.2 中小型蝴蝶兰的重要杂交种

中小型花蝴蝶兰杂交种的主要育种亲本是 *Phal.* ‘Cassandra’（桃红蝴蝶兰 × 史塔

基蝴蝶兰）。该杂交种于 1896 年登录。它的花径 3～5cm，粉红色，有红色斑点。截至目前，该杂交种已贡献了总共 222 个子一代和 3305 个子代（表 23.11，图 23.16）。它的后代中，*Phal.*‘Timothy Christopher’（*Phal.*‘Cassandra’× 菲律宾白花蝴蝶兰）、*Phal.*‘Carmela's Pixie’（*Phal.*‘Terilyn Fujitake’× *Phal.*‘Cassandra’）和 *Phal.*‘Be Glad’（*Phal.*‘Swiss Miss’× *Phal.*‘Cassandra’）也都有最新一代的中小型花蝴蝶兰后代。

图 23.15　小花型蝴蝶兰品种。（a）白色萼片、白色花瓣和白色唇瓣，（b）白色萼片、白色花瓣和黄色唇瓣，（c）红色萼片、红色花瓣和红色唇瓣，（d）红色萼片、红色花瓣和橙色唇瓣，鼻宽 = 1cm

表 23.11　用于培育中小型花蝴蝶兰的重要杂交种 *Phal.*‘Cassandra’及其后代

杂交种名	杂交代数[a]	英国皇家园艺协会登记年份	亲本	子一代杂交种数	杂交后代总数	获奖数	平均花径（cm）	花色[b]
Phal. Cassandra	G0	1896	（*equestris* × *stuartiana*）	222	3305	25	4.5	PR
Phal. Timothy Christopher	G1	1982	（Cassandra × *aphrodite*）	197	615	8	5.5	WW
Phal. Liu's Twilight Rainbow	G2	2004	（Rainbow Chip × Timothy Christopher）	46	71	3	3.6	WstR
Phal. Liu's Bright Ruby	G2	2000	（*lindenii* × Timothy Christopher）	30	76	1	4.0	WstR
Phal. Rong Guan Amah	G2	1996	（Timothy Christopher × *amabilis*）	29	86	11	5.2	WY
Phal. Tying Shin Fantastic World	G2	2008	（Chian Xen Pearl × Timothy Christopher）	3	3	17	7.1	WspR
Phal. Carmela's Pixie	G1	1990	（Terilyn Fujitake × Cassandra）	159	1221	22	5.5	PR
Phal. Zuma's Pixie	G2	1992	（Carmela's Pixie × *equestris*）	131	710	25	3.8	RR
Phal. Ho's Little Caroline	G2	1993	（Be Glad × Carmela's Pixie）	42	89	1	4.0	WstR
Phal. Taisuco Pixie	G2	1994	（Sun-Chen Beauty × Carmela's Pixie）	41	165	6	6.1	RR
Phal. Sogo Mini Dog	G2	1996	（Happy Lip × Carmela's Pixie）	29	46	1	5.6	WR

续表

杂交种名	杂交代数[a]	英国皇家园艺协会登记年份	亲本	子一代杂交种数	杂交后代总数	获奖数	平均花径（cm）	花色[b]
Phal. Super Pixie	G2	1996	（Carmela's Pixie × Zuma's Pixie）	18	28	0	—[c]	—[d]
Phal. Be Glad	G1	1978	（Swiss Miss × Cassandra）	131	1351	18	4.9	WR
Phal. Be Tris	G2	1989	（Be Glad × *equestris*）	75	351	19	4.0	PR
Phal. Nobby's Amy	G2	1998	（Be Glad × Rothschildiana）	51	81	14	5.8	WspW
Phal. Ho's Little Caroline	G2	1993	（Be Glad × Carmela's Pixie）	42	89	1	4.0	WstR
Phal. Ho's Amaglad	G2	1990	（Be Glad × *amabilis*）	40	145	13	6.2	WspR
Phal. Liu's Rainbow	G2	1995	（Be Glad × *lindenii*）	38	67	0	—	WspY
Phal. Glad Melinda	G2	1985	（Be Glad × Melinda Nan）	34	193	2	8.1	WR
Phal. Pixie Star	G5	1997	（*pulcherrima* × Joyful）	5	5	21	3.8	RR
Phal. Nankung's 4.55 PM	G1	2008	（Cassandra × Nankung's Beauty Girl）	38	42	13	6.4	WspW
Phal. Caribbean Sunset	G1	1970	（Cassandra × Mambo）	37	96	0		PP
Phal. Purple Gem	—[e]	1963	（*pulcherrima* × *equestris*）	29	41	32	3.2	RR
Phal. Eduardo Quisumbing	—	1956	（*amabilis* × *pulcherrima*）	5	6	1	—	WW

数据来源于 OrchidWiz X3.3，2017

a. 世代相传。G0 为 *Phal.*'Cassandra'。 G1～G5 表示 *Phal.*'Cassandra' 衍生出的不同后代

b. 栽培品种的花色。第一个词为花被的颜色，第二个词为唇瓣的颜色。W 代表白色，Wsp 代表白花带斑点，Wst 代表白花带条纹，P 代表粉色，R 代表红色，Y 代表黄色

c. OrchidWiz X3.3，2017 中未查找到

d. OrchidWiz X3.3，2017 中未查找到

e. 不是 G0 代 *Phal.*'Cassandra' 的繁殖后代

图 23.16 小花型蝴蝶兰杂交种的各种表型。（a）*Phal.*'Cassandra'，（b）*Phal.*'Liu's Hua Lien Red-Carpet'，（c）*Phal.*'Taida Little Vivien'，（d）*Phal.*'Brother Spring Dancer'，鼻宽 = 1cm

Phal.‘Timothy Christopher’花径 4.0～6.2cm，白色，它的子二代（G2）杂交种 *Phal.*‘Liu’s Twilight Rainbow’（*Phal.*‘Rainbow Chip’× *Phal.*‘Timothy Christopher’）是所有已注册蝴蝶兰杂交种中花最多的，一株植株总共有 248 朵花，比较所有由 *Phal.*‘Timothy Christopher’产生的后代，*Phal.*‘Tying Shin Fantastic World’（*Phal.*‘Chian Xen Pearl’× *Phal.*‘Timothy Christopher’）获得了 17 个奖项。

Phal.‘Carmela’s Pixie’花径 5cm，颜色从粉红色、紫红色到深红色。它的继代杂交种 *Phal.*‘Zuma’s Pixie’（*Phal.*‘Carmela’s Pixie’× 桃红蝴蝶兰）的花朵仍然较小（3.2～4.5cm），颜色从紫红色到深紫色，带有深红色的唇瓣。在 *Phal.*‘Carmela’s Pixie’所有后代中，*Phal.*‘Zuma’s Pixie’是最成功的。*Phal.*‘Be Glad’花径为 3.8～5.3cm，白色，基部为淡粉色，唇瓣为黑玫瑰色。它的后代 *Phal.*‘Be Tris’（*Phal.*‘Be Glad’× 桃红蝴蝶兰）花径 3.0～4.6cm，常用作育种亲本。在 *Phal.*‘Be Glad’的后代中，*Phal.*‘Pixie Star’（五唇兰 × *Phal.*‘Joyful’）和 *Phal.*‘Be Tris’获得了最多奖项（表 23.11）。来自两个原生种之间杂交的其他主要杂交种，如‘紫色宝石’（*Phal.*‘Purple Gem’）（五唇兰 × 桃红蝴蝶兰）和 *Phal.*‘Eduardo Quisumbing’（南洋白花蝴蝶兰 × 五唇兰），也用于中小型花的育种。‘紫色宝石’花径 2.5～3.9cm，颜色从深粉红色到深紫色不等，而 *Phal.*‘Eduardo Quisumbing’的特征是白花、直花箭、无分枝。

为研究桃红蝴蝶兰对中小型花蝴蝶兰杂交种发展的影响，对不同的桃红蝴蝶兰后代花的大小进行了分析。在英国皇家园艺协会（OrchidWiz 2017）中登录的 21 805 个后代中，共有 2774 个后代进行了桃红蝴蝶兰血缘关系分析（图 23.17）。数据显示，第 1～3 代的后代花径在 4～6cm，第 5～10 代的后代花径在 6～11cm（图 23.17）。这些结果表明，

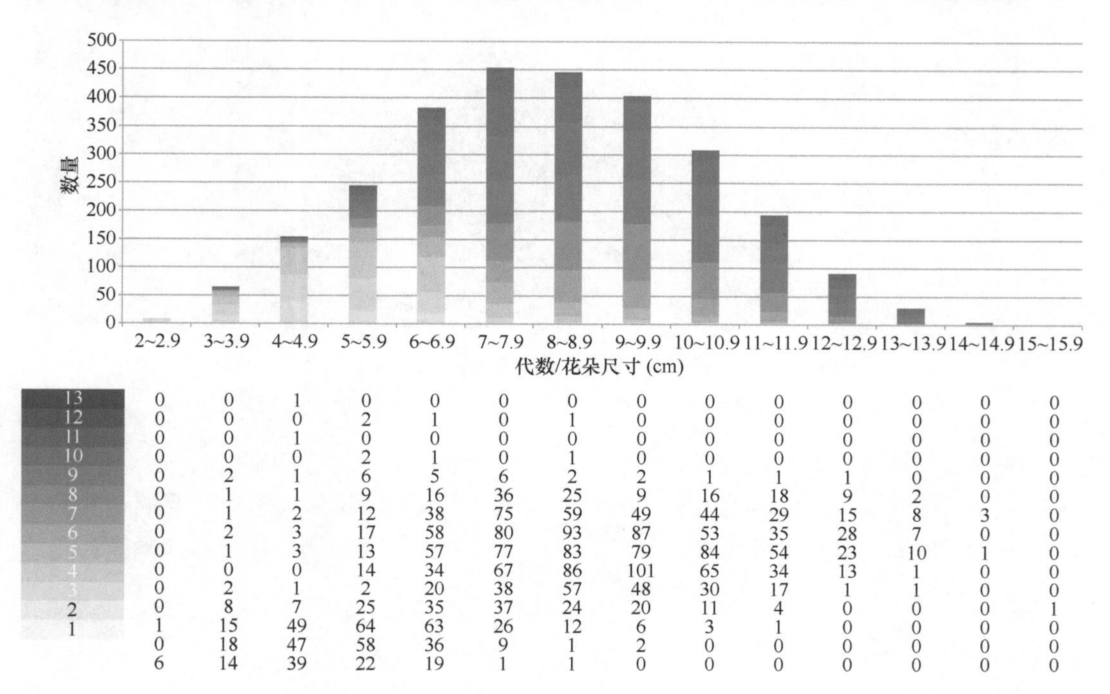

	2~2.9	3~3.9	4~4.9	5~5.9	6~6.9	7~7.9	8~8.9	9~9.9	10~10.9	11~11.9	12~12.9	13~13.9	14~14.9	15~15.9
13	0	0	1	0	0	0	0	0	0	0	0	0	0	0
12	0	0	0	2	1	0	1	0	0	0	0	0	0	0
11	0	0	1	0	0	0	0	0	0	0	0	0	0	0
10	0	0	0	2	1	0	1	0	0	0	0	0	0	0
9	0	2	1	6	5	6	2	2	1	1	1	0	0	0
8	0	1	1	9	16	36	25	9	16	18	9	2	0	0
7	0	1	2	12	38	75	59	49	44	29	15	8	3	0
6	0	2	3	17	58	80	93	87	53	35	28	7	0	0
5	0	1	3	13	57	77	83	79	84	54	23	10	1	0
4	0	0	0	14	34	67	86	101	65	34	13	1	0	0
3	0	2	1	2	20	38	57	48	30	17	1	1	0	0
2	0	8	7	25	35	37	24	20	11	4	0	0	0	1
1	1	15	49	64	63	26	12	6	3	1	0	0	0	0
	0	18	47	58	36	9	1	2	0	0	0	0	0	0
	6	14	39	22	19	1	1	0	0	0	0	0	0	0

图 23.17　桃红蝴蝶兰不同后代花的大小关系。数据来自 OrchidWiz X3.3，2017

与桃红蝴蝶兰杂交后的第一代衍生了中小型花蝴蝶兰杂交种，其中桃红蝴蝶兰血统的百分比越高，小花型特征越显著。

23.5.2 畸变花

畸变花有两瓣变成唇状形态，似乎包含三个萼片、三个唇瓣和一个柱头（图 23.18）。“畸变花”一词可以追溯到 Charles Darwin 的 *The Variation of Animals and Plants under Domestication* for snapdragon（*Antirrhinum*）（Darwin 1868），指的是花朵从正常的两侧对称型向辐射对称型转变的畸变现象。大多数花发生突变后被兰花栽培者丢弃，很少有唇状花瓣轻微改变的畸形花被选为新的杂交种。

图 23.18 畸变花蝴蝶兰杂交种的各种表型。（a）常规形态，（b 和 c）桃红蝴蝶兰的畸变花，（d）*Phal.* ‘Liu's Pale Micholitz’，（e）*Phal.* ‘Fuller's Miss’，鼻宽 = 1cm

畸变花有两片花瓣变成唇状，唇瓣发育的分子机制是兰科植物研究的一个重要而有趣的问题。据报道，一种植物器官识别基因 B 类 *PeMADS4* 可作为唇瓣发育的调控因子（Tsai et al. 2004）。此外，另一组 *AGL6* 基因与 B 类基因相互作用，共同调控蝴蝶兰花的唇瓣或唇瓣发育（Hsu et al. 2015b）。

23.5.3 大脚花

与畸变花相反，大脚花的唇瓣呈花瓣状，因此这些花似乎具有三个萼片、三个花瓣和一个柱头，使花朵接近圆形（图 23.19）。实际上，大脚花是一种畸变花，但是以第一个育种亲本‘世界一流’“大脚花”（*Phal.*‘Mae Hitch’× *Phal.*‘Kathy Sagaert’）命名的。该杂交种是由 Carmela Orchids（HI，USA）在 1991 年筛选出的，其花箭上的 11 朵花都有大脚花表型。‘世界一流’“大脚花”是 *Phal.*‘Mae Hitch’与 *Phal.*‘Kathy Sagaert’的子一代种群中唯一具有大脚花表型的个体，其父母本都没有大脚花的表型，所以大脚花性状可能是由基因突变引起的，而不是由遗传所致。

图 23.19　大脚花杂交种的花朵。(a) *Phal.*‘United White Bear’，(b) *Phal.*‘Fangmei Dream Wedding’，(c) *Phal.*‘Fuller's 3580’，(d) *Phal.*‘Taisuco Sunstone’，(e) *Phal.*‘Lioulin Lovely Lip’，(f) *Phal.*‘Fuller's O-Plus’，鼻宽 = 1cm

大脚花蝴蝶兰的育种是最近才开始的，并且它们在蝴蝶兰市场的销售始于最近 10 年。大脚花表型随唇状花瓣的变化而发生巨大变化，花色有白色、粉红色、红色，并带有红色斑点和条纹。

蝴蝶兰原生种中没有显著的大脚花性状。系谱分析表明，‘世界一流’“大脚花”中包含 26.40%爪哇白花蝴蝶兰的基因、25.30%南洋白花蝴蝶兰的基因、13.60%桃红蝴蝶兰的基因、12.3%西蕾丽蝴蝶兰的基因、11.40%菲律宾白花蝴蝶兰的基因、5.80%桑德蝴蝶兰的基因、4.00%史塔基蝴蝶兰的基因和 1.20%露德蝴蝶兰的基因。

23.5.3.1 大脚花蝴蝶兰的重要杂交种

所谓的“大脚花”始于‘世界一流’“大脚花”，它是 44 个新一代杂交后代和 355 个杂交后代的亲本（表 23.12）。由于很少有育种者拥有‘世界一流’“大脚花”或它的后代，因此大脚花的育种并没有广泛地开展。直到 2011 年，大多数杂交种因 2009 年的‘育品复活岛’（*Phal.*‘Sogo Yukidian’×‘世界一流’）与 2010 年的‘育品烟花’（*Phal.*‘Yu Pin Fireworks’）（*Phal.*‘Sogo Dove’×‘世界一流’）发现和传播而被登录。2012～2016 年，大脚花育种广泛开展，每年登录杂交品种 38～63 个（图 23.20）。

表 23.12 用于培育‘世界一流’大脚花的重要杂交种及其后代

杂交种名	杂交代数[a]	英国皇家园艺协会登记年份	亲本	子一代杂交种数	杂交后代总数	获奖数	平均花径（cm）	花色[b]
Phal. World Class	G0	1990	（Mae Hitch × Kathy Sagaert）	44	355	3	9.1	WstW
Phal. Yu Pin Easter Island	G1	2009	（Sogo Yukidian × World Class）	64	112	11	10.6	WstP
Phal. Fuller's AD-Plus	G2	2000	（*amabilis* × Yu Pin Easter Island）	7	76	1	7.9	WW
Phal. Fuller's D-Plus	G2	2012	（Yu Pin Easter Island × Fuller's Purple Queen）	5	5	0	—[c]	—[d]
Phal. Lioulin Pure Lip	G2	2012	（Yu Pin Easter Island × Sogo Yukidian）	5	5	0	—	—
Phal. Fuller's F-Plus	G2	2012	（Yu Pin Easter Island × Fuller's Cow）	4	4	0	—	—
Phal. Yushan Mongo	G2	2011	（Leopard Prince × Yu Pin Easter Island）	4	4	0	—	—
Phal. Yu Pin Fireworks	G1	2010	（Sogo Dove × World Class）	79	161	8	9.7	WstP
Phal. Tying Shin World Class	G2	2011	（Sogo Vivien × Yu Pin Fireworks）	15	16	3	6.0	WstW
Phal. Lioulin Pretty Lip	G2	2012	（Yu Pin Fireworks × Chian Xen Mammon）	10	10	0	—	—
Phal. Fuller's E-Plus	G2	2012	（Yu Pin Fireworks × Fuller's Milk）	9	9	0	—	—
Phal. Tying Shin Red Emperor	G2	2011	（Champion Lightning × Yu Pin Fireworks）	7	7	0	—	—
Phal. Lioulin Lovely Lip	G2	2012	（Yu Pin Fireworks × Chian Xen Piano	6	7	1	7.8	WstW
Phal. Sasquatch	G1	2005	（World Class × *amabilis*）	7	8	2	7.5	WP
Phal. Tropical Stripes	G1	1994	（Paul Tatar × World Class）	6	9	0	—	—
Phal. Jordan's Pixie	G1	1996	（Carmela's Pixie × World Class）	4	5	1	6.4	PstR
Phal. Krull's Butterfly	G1	1992	（Music × World Class）	0	0	12	10.2	PstP
Phal. Dendi's Yeti	G1	2008	（Taisuco Happybeauty × World Class）	0	0	7	8.2	PstP

数据来源于 OrchidWiz X3.3，2017

a. 世代相传。G0 为蝴蝶兰‘世界一流’。G1～G2 表示‘世界一流’衍生出的不同后代

b. 栽培品种的花色。第一个词为花被的颜色，第二个词为唇瓣的颜色。W 代表白色，Wst 代表白花带条纹，P 代表粉色

c. OrchidWiz X3.3，2017 中未查找到

d. OrchidWiz X3.3，2017 中未查找到

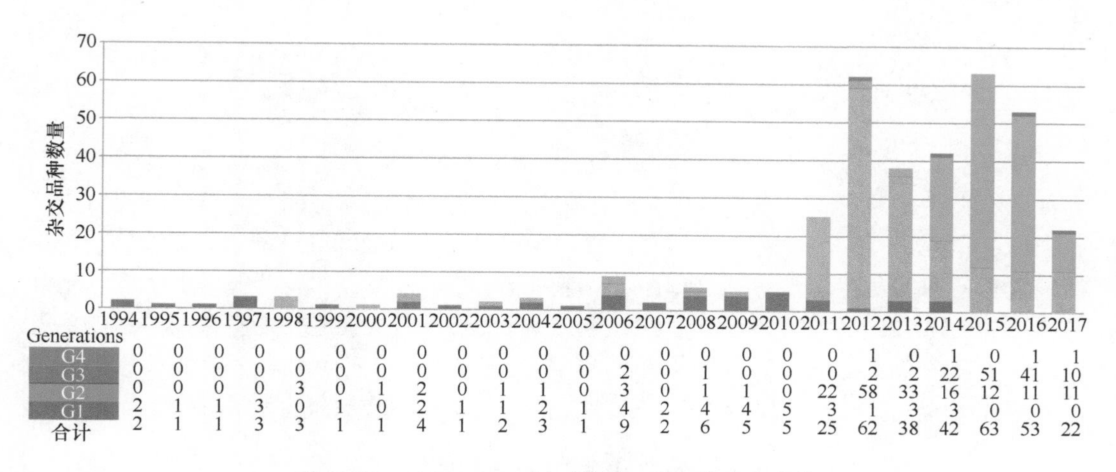

Generations	1994	1995	1996	1997	1998	1999	2000	2001	2002	2003	2004	2005
G4	0	0	0	0	0	0	0	0	0	0	0	0
G3	0	0	0	0	0	0	0	0	0	0	0	0
G2	0	0	0	0	3	0	1	2	0	1	1	0
G1	2	1	1	3	0	1	0	2	1	1	2	1
合计	2	1	1	3	3	1	1	4	1	2	3	1

Generations	2006	2007	2008	2009	2010	2011	2012	2013	2014	2015	2016	2017
G4	0	0	0	0	0	0	1	0	1	0	1	1
G3	2	0	1	0	0	0	2	2	22	51	41	10
G2	3	0	1	1	0	22	58	33	16	12	11	11
G1	4	2	4	4	5	3	1	3	3	0	0	0
合计	9	2	6	5	5	25	62	38	42	63	53	22

图 23.20　1994～2017 年‘世界一流’的杂交种数

‘育品复活岛’和‘育品烟花’是由 Neng-I Chang（育品生物科技股份有限公司，中国台湾）在英国皇家园艺协会中登录的，并作为普通蝴蝶兰出售，这一事实加快了大脚花育种的传播。作为‘世界一流’“大脚花”的子一代，‘育品复活岛’和‘育品烟花’分别拥有 64 个和 79 个继代杂交种以及 112 个和 161 个全代杂交种（表 23.12）。‘育品复活岛’的花径 7～14cm，白色，具深粉红色条纹，‘育品烟花’的花径 8～13cm，白色，具粉红色条纹。

用‘育品复活岛’和‘育品烟花’作为育种亲本，Fuller's Orchid Nursery 分别登录了 13 个和 25 个继代杂交新品种；Tying Shin Orchid 分别登录了 1 个和 13 个杂交新品种；Huang Gao Ming 分别登录了 9 个和 5 个杂交新品种。其中，*Phal.*‘Fuller's AD-Plus’（南洋白花蝴蝶兰 ×‘育品复活岛’）来自 Fuller's Orch.，*Phal.*‘Tying Shin World Class’（*Phal.*‘Sogo Vivien’×‘育品烟花’）来自 Tying Shin Orch.，*Phal.*‘Lioulin Pretty Lip’（‘育品烟花’× *Phal.*‘Chian Xen Mammon’）来自 Mr. Huang Gao Min，Fuller's Orch. 培育出的 *Phal.*‘Fuller's E-Plus’（‘育品烟花’× *Phal.*‘Fuller's Milk’）在育种中最常用（表 23.12）。

23.6　香型蝴蝶兰的育种

香味是植物特别是观赏植物的主要特征。香味是由挥发性芳香化合物组成的复合物产生的，可用来吸引授粉媒介，帮助自然界中的植物授粉。

蝴蝶兰的育种主要集中在选择大花朵和各种颜色上，很少关注培育香型杂交种。此外，大多数香型杂交品种在继代后代身上就失去了香味。主要的瓶颈是气味不是一个显性或稳定的特征，不能稳定遗传给下一代，这意味着大多数后代的气味特征会退化或消失。此外，大多数具有强烈香味的原生种开的花都很小、呈星形、花的数量有限，这些香型原生种的形态特征与市场上大多数大圆花杂交品种的特征相差甚远。为了繁殖香型大花蝴蝶兰杂交种，香型原生种应与大花杂交数代，但在这些世代中，气味可能会丢失。

23.6.1 原生种对香型杂交种的贡献

蝴蝶兰有部分原生种具有带气味的特征，但只有极少数具有强烈香气的品种被用作育种亲本。具有强烈气味的主要原生种有安汶蝴蝶兰、贝丽娜蝴蝶兰、露德蝴蝶兰、红脉蝴蝶兰和紫纹蝴蝶兰（表 23.13，图 23.21）（Hsiao et al. 2008；Yeh et al. 2012）。这些品种含有各种挥发性化合物，如贝丽娜蝴蝶兰以单萜、芳樟醇和香叶醇为主要香气化合物（Hsiao et al. 2006，2008）。但是，大多数具有强烈气味的原生种的花朵是星形的，需要与大圆形花杂交种进行杂交改良。表 23.13 列出了其他一些气味从弱到强的原生种，但大部分气味太弱或香味化合物差异太大的品种，很难用作香气育种的亲本。

表 23.13 香型原生种的气味特征

品种	气味特征
Phal. amabilis	无香或淡香
Phal. amboinensis	浓香
Phal. bellina	浓香
Phal. corningiana	浓香
Phal. cornu-cervi	中香
Phal. equestris	淡香至中香
Phal. fasciata	中香
Phal. fimbriata	淡香
Phal. fuscata	中香
Phal. gigantea	淡香至中香
Phal. hieroglyphica	淡香
Phal. javanica	中香
Phal. kunstleri	淡香
Phal. lobbii	中香至浓香
Phal. lueddemanniana	中香至浓香
Phal. mannii	淡香
Phal. mariae	淡香
Phal. modesta	浓香
Phal. parishii	中香
Phal. pulchra	中香
Phal. reichenbachiana	中香
Phal. schilleriana	中香至浓香
Phal. stuartiana	淡香至中香
Phal. sumatrana	中香
Phal. tetraspis	中香
Phal. venosa	浓香
Phal. violacea	浓香
Phal. viridis	淡香
Phal. wilsonii	中香

数据来源于 Hsiao 等（2008）和 Yeh 等（2012）

图 23.21 香型蝴蝶兰。(a)贝丽娜蝴蝶兰,(b)紫纹蝴蝶兰,鼻宽 = 1cm

23.6.2 重要的香型杂交种

Yeh 等(2012)分析了蝴蝶兰杂交种的香型后代百分比(表 23.14)。香型西蕾丽蝴蝶兰与无香型的 *Phal.*‘Ho's Little Caroline’杂交,有 52.7%的子一代具有香味;但香型西蕾丽蝴蝶兰与无香型的 *Phal.*‘Sogo Berry’、*Phal.*‘Be Glad’和 *Phal.*‘Jiaho Cherry’进行杂交,只有 10.2%~20.0%的子一代具有香味。此外,香型西蕾丽蝴蝶兰与无香型的 *Phal.*‘Rong Guan Mary’进行杂交,子一代杂交种都没有气味特征。在分析这些父本的系谱时发现,*Phal.*‘Ho's Little Caroline’、*Phal.*‘Sogo Berry’、*Phal.*‘Be Glad’、*Phal.*‘Jiaho Cherry’和 *Phal.*‘Rong Guan Mary’分别具有西蕾丽蝴蝶兰 1.6%、6.1%、0、1.3%和 17.4%的血统。所有这些结果表明,香味特征不依赖于母本,也不依赖于西蕾丽蝴蝶兰血统所占的百分比。

表 23.14 蝴蝶兰杂交种的香型后代百分比

亲本	带香味	无香	香味百分比(%)
Phal. schilleriana “HDAIS#2” × *Phal.* Ho's Little Caroline “Ho's Little Zebra”	29	26	52.7
Phal. schilleriana “HDAIS#2” × *Phal.* Sogo Berry	13	52	20.0
Phal. schilleriana “HDAIS#3” × *Phal.* Be Glad “La Flora”	38	333	10.2
Phal. schilleriana “HDAIS#3” × *Phal.* Jiaho Cherry “KF#1”	8	44	15.4
Phal. schilleriana “HDAIS#5” × *Phal.* Rong Guan Mary	0	7	0.0
Phal. Kung's Roth-Fairy × *Phal.* Be Glad “La Flora”	0	49	0.0
Phal. Kung's Roth-Fairy × *Phal.* Lung-An Mist Pink	4	3	57.1
Phal. Kung's Roth-Fairy × *Phal.* Timothy Christopher “KF#18”	3	4	42.9
Phal. Kung's Roth-Fairy × *Phal.* Yu Pin Summer	0	65	0.0

该表根据 Yeh 等(2012)的方法修改

香型 *Phal.*‘Kung's Roth-Fairy’与香型 *Phal.*‘Lung-An Mist Pink’进行杂交,有 57.1%的子一代具有香味;香型 *Phal.*‘Kung's Roth-Fairy’与无香型的 *Phal.*‘Timothy Christopher’的杂交后代具有香味的比例为 42.9%。然而,香型 *Phal.*‘Kung's Roth-Fairy’与无香型的 *Phal.*‘Be Glad’或‘育品夏天’(*Phal.*‘Yu Pin Summer’)进行杂交,杂交后代都不带香味。这些结果表明,两个有香味的父母本之间杂交有气味的子一代杂交种

超过 50%，而在有香味的母本和没有香味的母本之间杂交结果差异很大。

23.6.3 香型蝴蝶兰育种的分子遗传学

在贝丽娜蝴蝶兰中已经研究了香型花的分子机制（Hsiao et al. 2008）。贝丽娜蝴蝶兰的主要香气成分是芳樟醇和香叶醇，均属于单萜。根据转录组分析，已分离出编码香叶基二磷酸合酶（GDPS）的基因，并将其命名为 *PbGDPS*（Hsiao et al. 2008）。*PbGDPS* 的表达与花后第 5 天单萜的最大释放量有关，并且证明了其酶促活性可以催化生成香叶基二磷酸香叶酯（GDP），这是生产芳樟醇和香叶醇的前体（Hsiao et al. 2008）。

尽管 *PbGDPS* 对于贝丽娜蝴蝶兰的香味产生很重要，但对于香气遗传还有许多问题需要解决。例如，是否可以通过遗传转化 *PbGDPS* 使无香型的白花 *Phal.* ‘Sogo Yukidian’“V3”产生香气。其他具有强烈气味的原生种，如红脉蝴蝶兰和安汶蝴蝶兰含有各种芳香化合物，但不含单萜。因此，各种香味花卉的生物合成和调控机制有待进一步研究。

23.7 影响蝴蝶兰育种的物种基因组大小变异

23.7.1 蝴蝶兰属植物的细胞遗传学研究

在蝴蝶兰新品种的选育中，通常将原生种用作引入特定性状的亲本之一，如贝丽娜蝴蝶兰的香味及安汶蝴蝶兰的黄花性状。然而，在种间杂交过程中，无论亲本杂交还是后代的后续杂交，育种者都会遇到不育问题（Chuang et al. 2008）。可以确定两个主要的不育原因：原生种和变种之间的染色体大小存在差异以及倍性水平存在差异（Arends 1970；Griesbach 1985；Chuang et al. 2008）。

尽管所有二倍体蝴蝶兰都有相同数量的 38 条染色体（$2n = 2x = 38$）（Christenson 2001），但有 9 种核型表明它们的染色体大小和着丝粒位置不同。其中，菲律宾白花蝴蝶兰的染色体很小，桃红蝴蝶兰和史塔基蝴蝶兰（＜2μm）、鹿角蝴蝶兰和露德蝴蝶兰（＜2μm 和 2.0～2.5μm）中小，红脉蝴蝶兰、安汶蝴蝶兰、曼氏蝴蝶兰和紫纹蝴蝶兰（＜2μm、2.0～2.5μm 和＞2.5μm）中大小不一（表 23.15）（Shindo and Kamemoto 1963；Kao et al. 2001）。

商品种/杂交种通常是四倍体，而原生种是二倍体（Griesbach 1985；Chuang et al. 2008）。为了实施一个有效的育种计划，有必要对蝴蝶兰原生种和亲本种群的基因组大小与倍性水平信息进行了解。这可以通过使用碘化丙啶（PI）和 4′,6-二氨基-2-苯茚二酮（DAPI）染色的流式细胞术来实现（Lin et al. 2001；Chen et al. 2013b，2014）。通过对蝴蝶兰物种的研究，确定了 50 个物种 2C 值的变化，其中菲律宾白花蝴蝶兰的值为 2.80，紫纹蝴蝶兰的值为 15.03pg。通过对 9 种蝴蝶兰体细胞中期染色体的核型分析，发现基因组大小与染色体大小显著相关（表 23.15）。小染色体组的菲律宾白花蝴蝶兰、桃红蝴蝶兰和史塔基蝴蝶兰的平均 2C 值为 2.8～3.37pg，而中小染色体组的鹿角蝴蝶兰和露德蝴蝶兰的平均 2C 值在 6.44～6.49pg。此外，小中大染色体组的红脉蝴蝶兰、安汶蝴蝶兰、曼氏蝴蝶兰和紫纹蝴蝶兰的 2C 值为 9.52～14.36pg。

表 23.15 9 种蝴蝶兰体细胞中期染色体的核型分析

品种	2*n*	染色体大小（μm）			细胞核DNA含量（pg/2C）
		小 ＜2.0	中 2.0～2.5	大 ＞2.5	
Phal. aphrodite	38	38	0	0	2.80 ± 0.06
Phal. equestris	38	38	0	0	3.37 ± 0.05
Phal. stuartiana	38	38	0	0	3.13 ± 0.07
Phal. cornu-cervi	38	34	4	0	6.44 ± 0.16
Phal. lueddemanniana	38	28	10	0	6.49 ± 0.22
Phal. venosa	38	12	4	22	9.52 ± 0.27
Phal. amboinensis	38	8	4	26	14.36 ± 0.19
Phal. violacea	38	6	6	26	15.03 ± 0.21
Phal. mannii	38	0	6	32	13.50 ± 0.12

根据 Kao 等（2001）和 Lin 等（2001）的研究结果整理而来

总体而言，原生种的染色体大小和基因组含量差异较大，导致了蝴蝶兰育种的不育问题。通过细胞遗传学研究，可以很容易选择更合适的原生种或杂交种作为育种亲本，以提高蝴蝶兰育种的效率。蝴蝶兰物种中多倍体的诱导提供了一个很好的策略，可以与四倍体杂交种进行有效杂交（请参阅下一节）。

23.7.2 蝴蝶兰多倍体诱导

多倍体在许多农作物和园艺植物的育种中都起着重要作用（Sattler et al. 2016），对于改良蝴蝶兰也很有帮助。蝴蝶兰的大多数商业品种是四倍体，而大多数原生种是二倍体，商业种和原生种之间的杂交障碍限制了正常杂交手段的进行。因此，需要一种简单的技术来增加二倍体物种的倍性水平。由于蝴蝶兰的类原球茎和类原球茎样体（PLB）组织中存在多倍体细胞，因此有机会由这些细胞再生多倍体植物。

为了诱导多倍体，将一组类原球茎或 PLB 切成一半长度，然后将具有生长点的部分在 T2 培养基（Hyponex No. 13.5g/L、胰蛋白 11g/L、柠檬酸 0.1g/L、活性炭 1g/L、蔗糖 20g/L、马铃薯 20g/L、香蕉 25g/L 和琼脂 7.5g/L）中持续继代培养 2～3 个月（Chen et al. 2013a）。重复此过程数次，直到鉴定出加倍的下一代 PLB。然后选择这些 PLB，并在 5 个月的时间内再生为幼苗（Chen et al. 2013a）。这是一种简单、有效、可靠的生产蝴蝶兰多倍体植株的方法。这可能对未来兰花新品种的发展产生重大影响。

23.8 结论与展望

蝴蝶兰是最受欢迎的栽培兰花，具有独特而美丽的花朵形态。截至 2017 年 5 月，在英国皇家园艺协会中登录的蝴蝶兰杂交种有 34 112 个。在蝴蝶兰市场上白色和各种颜色的大花最受欢迎。蝴蝶兰花色包括粉红色到红色和紫色，黄色到绿色和橙色，黑色以及各种带有红色唇瓣的花朵颜色。另外，具有各种色素沉着图案（如红色条纹和斑点）的白花，粉红色或黄色花朵是最近的主要育种方向之一，很少发现形态改变的花。而畸形花和大脚花则成为蝴蝶兰育种的新趋势。另外，培育带有香味的且有许多大圆花的杂交种对于育种者来说是一个挑战，可以通过分子生物学和遗传转化手段来实现。

本章译者：

李　涵
云南省农业科学院花卉研究所，国家观赏园艺工程技术研究中心，云南省花卉育种重点实验室，昆明 650200

参 考 文 献

Arends JC (1970) Cytological observations on genome homology in eight interspecific hybrids of *Phalaenopsis*. Gentica 41:88–100
Borevitz JO, Xia Y, Blount J, Dixon RA, Lamb C (2000) Activation tagging identifies a conserved MYB regulator of phenylpropanoid biosynthesis. Plant Cell 12:2383–2393
Butelli E, Licciardello C, Zhang Y, Liu J, Mackay S, Bailey P, Reforgiato-Recupero G, Martin C (2012) Retrotransposons control fruit-specific, cold-dependent accumulation of anthocyanins in blood oranges. Plant Cell 24(3):1242–1255
Chen WH (2017) The development of *Phalaenopsis* orchid industry in Taiwan. Orchid Digest 81-1:22–27
Chen WH, Hsu CY, Cheng HY, Chang H, Chen HH, Ger MJ (2011) Downregulation of putative UDP-glucose: flavonoid 3-O-glucosyltransferase gene alters flower coloring in *Phalaenopsis*. Plant Cell Rep 30:1007–1017
Chen WH, Kao YK, Tang CY (2014) Variation of the Genome Size among *Phalaenospsis* species using DAPI. J Taiwan Soc Hortic Sci 60:115–123
Chen WH, Kao YL, Tang CH (2013a) Method for producing polyploid plants of orchids. US Patent 8,383,881
Chen WH, Kao YK, Tang CY, Tsai CC, Lin TY (2013b) Estimating nuclear DNA content within 50 species of the genus *Phalaenopsis Blume* (Orchidaceae). Sci Hortic (Amsterdam) 161:70–75
Chiu LW, Zhou X, Burke S, Wu X, Prior RL, Li L (2010) The purple cauliflower arises from activation of a MYB transcription factor. Plant Physiol 154(3):1470–1480
Christenson EA (2001) Phalaenopsis. In: Phalaenopsis: a monograph. Timber Press, Portland
Chuang HT, Hsu ST, Shen TM (2008) Breeding barriers in yellow *Phalaenopsis* orchids. J Taiwan Soc Hortic Sci 54:59–66
Darwin C (1868) The variation of animals and plants under domestication, 1st edn. John Murray, London
Feller A, Machemer K, Braun EL, Grotewold E (2011) Evolutionary and comparative analysis of MYB and bHLH plant transcription factors. Plant J 66:94–116
Griesbach RJ (1985) Polyploidy in Phalaenopsis orchid improvement. J Hered 76:74–75
Griesbach RJ (2005) A scientific approach to breeding blue orchid-exploring new frontiers in search of elusive flower colors. Orchids 74:376–379
Grotewold E (2006) The genetics and biochemistry of floral pigments. Annu Rev Plant Biol 57:761–780
Hichri I, Barrieu F, Bogs J, Kappel C, Delrot S, Lauvergeat V (2011) Recent advances in the transcriptional regulation of the flavonoid biosynthetic pathway. J Exp Bot 62:2465–2483
Hsiao YY, Jeng MF, Tsai WC, Chuang YC, Li CY, Wu TS, Kuoh CS, Chen WH, Chen HH (2008) A novel homodimeric geranyl diphosphate synthase from the orchid *Phalaenopsis bellina* lacking a DD(X)2-4D motif. Plant J 55:719–733

Hsiao YY, Tsai WC, Kuoh CS, Huang TH, Wang HC, Wu TS, Leu YL, Chen WH, Chen HH (2006) Comparison of transcripts in *Phalaenopsis bellina* and *Phalaenopsis equestris* (Orchidaceae) flowers to deduce monoterpene biosynthesis pathway. BMC Plant Biol 6:14

Hsu CC, Chen WH (2015) The breeding achievements from *Phalaenopsis equestris*. Malayan Orchid Rev 49:41–47

Hsu CC, Chen YY, Tsai WC, Chen WH, Chen HH (2015a) Three R2R3-MYB transcription factors regulate distinct floral pigmentation patterning in *Phalaenopsis* spp. Plant Physiol 168:175–191

Hsu HF, Hsu WH, Lee YI, Mao WT, Yang JY, Li JY, Yang CH (2015b) Model for perianth formation in orchids. Nature Plants 1:15046

Kao YY, Chiang SB, Lin TY, Hsieh CH, Chen YH, Chen WH, Chen CC (2001) Differential accumulation of heterochromatin as a cause for karyotype variation in Phalaenopsis orchids. Ann Bot 87:387–395

Koes R, Verweij W, Quattrocchio F (2005) Flavonoids: a colourful model for the regulation and evolution of biochemical pathways. Trends Plant Sci 10:236–242

Kuo PC, Wu TS (2011) Biosynthetic pathway of pigments in *Phalaenopsis* species. Orchid Biotechnol II:129–144

Lin S, Lee H, Chen W, Chen C, Kao Y, Fu Y, Chen Y, Lin T (2001) Nuclear DNA contents of *Phalaenopsis* sp. and *Doritis pulcherrima*. J Am Soc Hortic Sci 126:195–199

Ma H, Pooler M, Griesbach R (2009) Anthocyanin regulatory/structural gene expression in *Phalaenopsis*. J Am Soc Hortic Sci 134:88–96

OrchidWiz (2017) OrchidWiz Encyclopedia Version X3.3 May 2017 Database. OrchidWiz Database LLC, Ames

Petroni K, Tonelli C (2011) Recent advances on the regulation of anthocyanin synthesis in reproductive organs. Plant Sci 181:219–229

Sasaki N, Nakayama T (2015) Achievements and perspectives in biochemistry concerning anthocyanin modification for blue flower coloration. Plant Cell Physiol 56:28–40

Sattler MC, Carvalho CR, Clarinde WR (2016) The polyploidy and its role in plant breeding. Planta 243:281–296

Shindo S, Kamemoto H (1963) Karyotype analysis of some species of *Phalaenopsis*. Cytologia 28:390–398

Su V, Hsu BD (2003) Cloning and expression of a putative cytochrome P450 gene that influences the colour of *Phalaenopsis* flowers. Biotechnol Lett 25:1933–1939

Tanaka Y, Ohmiya A (2008) Seeing is believing: engineering anthocyanin and carotenoid biosynthetic pathways. Curr Opin Biotechnol 19:190–197

Tsai WC, Kuoh CS, Chuang MH, Chen WH, Chen HH (2004) Four *DEF*-like MADS box genes displayed distinct floral morphogenetic roles in *Phalaenopsis* orchid. Plant Cell Physiol 45:831–844

Wang J, Ming F, Han Y, Shen D (2006) Flavonoid-3′,5′-hydroxylase from *Phalaenopsis*: a novel member of cytochrome P450s, its cDNA cloning, endogenous expression and molecular modeling. Biotechnol Lett 28:327–334

Winkel-Shirley B (2001) Flavonoid biosynthesis. A colorful model for genetics, biochemistry, cell biology, and biotechnology. Plant Physiol 126:485–493

Yeh YC, Tsai YS, Shih CT, Huang P (2012) The breeding of scented flowers in *Phalaenopsis*. In: Orchid breeding in Taiwan. Published by Taiwan Orchid Breeders Society. pp 86–94

Zhang G, Lu S, Chen TA, Funk CR, Meyer WA (2003) Transformation of triploid bermuda grass (*Cynodon dactylon* × *C. transvaalensis* cv. TifEagle) by means of biolistic bombardment. Plant Cell Rep 21:860–864

第 24 章　报　春　花

Juntaro Kato, Mayuko Inari-Ikeda, Mai Hayashi, Junji Amano,
Hiroaki Ohashi, and Masahiro Mii

摘要　报春花科报春花属大约包含 500 种植物，这些物种大多数位于北半球的凉爽、高海拔山区或高纬度地区。种间杂交发生在报春花属的几个物种之间，用于扩大变异，整合两个亲本的独特性状，并将一个种的有用性状引入另一个种。然而，杂交胚胎败育是种间杂交的严重障碍之一。通过使用我们研究开发的适宜培养基，在多个种间杂交组合中成功地产生了种间杂种。尽管在这些杂交中获得的所有二倍体杂种均显示不育，但三倍体杂种在一定程度上出现了雌性育性。这些三倍体的自然未减数配子和单倍体配子之间受精形成了一些种间杂交组合。一些杂种或 BC1 表现出吸引人的独特性状，可直接用作新品种或作为进一步育种的有用材料。

关键词　胚挽救；杂种不育；种间杂交；报春花；未减数配子；未减数 $3x$ 配子；多倍体

24.1　引　　言

在观赏植物育种中，采用种间杂交有助于培育新品种。兰花、百合、月季、矮牵牛等重要花卉作物的品种都是通过种间杂交培育出来的。在种间杂交中，胚乳发育不良或自身生长能力差会导致杂交胚胎败育（Woodel 1960）。为了克服这些困难获得杂交种，胚挽救技术已成为败育前抢救杂交种胚胎的有效手段。仙客来（Ishizaka 2008）、长寿花（Izumikawa et al. 2008）等的种间杂种都是借助胚挽救技术产生的。

报春花分布于北半球温带地区，包括东亚和欧洲到亚北极带，约 500 种报春花，按形态特征划分为 37 个组（Richards 1993）。由于大多数报春花物种具有美丽迷人的花朵，因此某些具有较高观赏价值以及易于栽培和繁殖的种已被作为重要的观赏植物。例如，

J. Kato
Department of Biology, Aichi University of Education, Kariya, Aichi, Japan
M. Inari-Ikeda
Department of Nutrition, School of Health and Nutrition, Tokai Gakuen University, Nagoya, Aichi, Japan
M. Hayashi • J. Amano
Laboratory of Plant Cell Technology, Graduate School of Horticulture, Chiba University, Matsudo, Chiba, Japan
H. Ohashi
Faculty of Agriculture, Ehime University, Matsuyama, Ehime, Japan
M. Mii (✉)
Laboratory of Plant Cell Technology, Graduate School of Horticulture, Chiba University, Matsudo, Chiba, Japan
Center for Environment, Health and Field Sciences, Chiba University, Kashiwa, Chiba, Japan
e-mail: miim@faculty.chiba-u.jp

P. obconica、*P. malacoides*、*P.* × *polyantha* 被列为盆栽和花园的主要栽培花卉作物。*P. obconica*、*P. malacoides* 的花特征具有较大变异的品种只能通过种内杂交获得，而 *P.* × *polyantha* 则是通过欧洲本土的三种植物，即 *P. vulgaris*、*P. veris* 和 *P. elatior* 经种间杂交获得。大约在 40 年前的日本，已成功地通过 *P.* × *polyantha* 与 *P. juliae* 杂交产生了独特的矮化品种，称为 Julian Hybrid。通过 *P. auricula* 和 *P. hirsta* 之间的种间杂交，培育出了许多称为黑木耳的品种（Richards 1993），并且与欧洲种间杂种品种进行重复杂交。通过种间杂交和品种间的重复杂交，培育出了 *P. marginata* 和 *P. allionii* 的品种（Richards 1993）。

在东亚本地的报春花中，日本本土的四倍体报春花（*P. japonica*）被引入欧洲。通过与属于同一地区并起源于中国的几个种进行种间杂交，产生了许多称为 'japonica' 杂种的花园品种（Ernst 1950）。虽然 *P. sinensis* 也是 19 世纪传入欧洲的亚洲报春花属植物，但只有通过与 *P. obconica* 和 *P. malacoides* 进行种内杂交，才能培育出具有独特花型的品种。在日本，自江户时代（300 年前）开始，使用 *P. sieboldii* 进行种内杂交产生了 400 多个具有各种花色和独特花形的品种。上述大多数重要的报春花主要采用种子繁殖，而 *P. sieboldii* 则通过营养繁殖，即通过根茎分苗进行繁殖。

为了拓宽 *P. sieboldii* 的遗传变异，并利用其他未开发的物种作为新型观赏植物，我们一直在对亚洲本土的几种报春花进行种间杂交与胚挽救研究。在这篇综述中，我们总结了这些种间杂交的研究。

24.2　报春花的异型花柱自交不亲和

在报春花属中进行育种，非常重要且必须记住的是该分类单元的所有物种的二型花都具有异型花柱自交不亲和系统。报春花的花由花冠裂片和花冠筒组成，花冠筒和花冠裂片之间的平衡决定了可见的花形态。大多数报春花属植物具有长的花冠筒，雄蕊花丝的花药附着在花冠筒上。在报春花的二型花中，通常会识别出两种类型的花形态，即针式和线式。针式和线式是以花冠筒中柱头和花药的位置来区分的。在针式花中，柱头位于花冠筒的顶部，花药与花丝基部附着在花冠筒的中部（图 24.1）。在线式花中，柱头

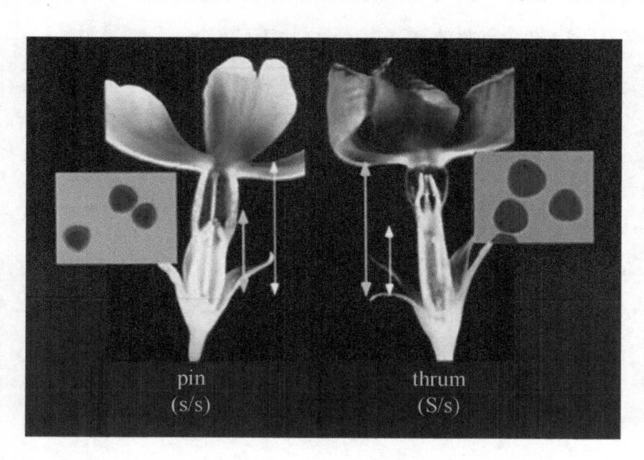

图 24.1　*P. sieboldii* 的针式（pin）和线式（thrum）花结构

位于花冠筒的中部，花丝基部附着在花冠筒的顶部附近（图 24.1）。然而，在同一个种中，针式和线式的整朵花形态没有明显的差异。在报春花中，通常通过不同类型花之间的杂交来成功实现受精，从而产生种子，即针式 × 线式和线式 × 针式，称为合法杂交。同一类型花间的杂交，即针式 × 针式和线式 × 线式，称为不合法杂交。由于这种异型花柱自交不亲和系统存在于报春花中，就如种内杂交一样，应用合法杂交进行种间杂交很重要（Kato and Mii 2006）。

24.3 *Primula sieboldii* 和 *P. kisoana* 之间的种间杂交

P. sieboldii 被归类为 *Cortusoides* 组 *Cortusoides* 亚组。尽管 *P. sieboldii* 原产日本、朝鲜半岛和中国东北部，但仅在日本的江户时代（约 300 年前）（Torii 1985）被驯化。从那以后，通过利用这些种质进行幼苗和种内杂交筛选出的新优良个体，培育出 300 多个品种。通过观察体细胞染色体，在 *P. sieboldii* 形成的品种中，确认有 63 个二倍体、8 个三倍体和 1 个四倍体（Yamaguchi 1973）。此外，我们之前对 200 个品种进行流式细胞仪分析还显示，目前 *P. sieboldii* 形成的品种包括二倍体、三倍体和四倍体。三倍体和四倍体通常都具有比二倍体硬的叶片与较厚的花茎。由于没有关于通过染色体倍增剂如秋水仙碱处理人工产生四倍体的报道，认为四倍体可能是在胚胎发生过程中通过染色体自发倍增产生的。同时，二倍体和四倍体品种间的杂交也没有报道，因此，推测三倍体是由正常配子和未减数配子通过受精作用产生的。

P. kisoana 是日本的另一个本土物种，和 *P. sieboldii* 一样被归为 *Cortusoides* 组。从日本的九州到北海道，*P. sieboldii*（*Cortusoides* 亚组）分布广泛，但 *P. kisoana*（*Geranioides* 亚组）仅限于 Shikoku 和 Kanto 的狭窄山区。尽管 *P. kisoana* 中存在一些品种，但与 *P. sieboldii* 相比，其花的性状变异较小。当这两个物种杂交时，获得了成熟的种子。然而，即使经过 100mg/L 的赤霉素（GA_3）处理，它们中的大多数仍无法发芽。而且，少数发芽的种子死亡时并没有出现真叶。因此，我们使用含 0.3%结冷胶的 1/2 MS 培养基（Murashige and Skoog 1962）（1/2 MS），补充了 5%（*w/V*）蔗糖和 50mg/L GA_3（Kato and Mii 2000），对种间杂交后 4 周的未成熟种子进行了体外培养。在利用 14 个 *P. sieboldii* 和 3 个 *P. kisoana* 总共得到的 34 个杂交组合中，有 10 个交叉组合获得了种间杂交植株。这些种间杂种表现出两种不同的倍性水平，倍性水平取决于用作母本的物种。当以 *P. sieboldii* 作为母本时，只有‘Miyuki’和‘Ooasahi’两个品种产生了杂交后代，且所有的杂交后代都是三倍体。相反，使用 *P. kisoana* 的白花品种‘Shirobana-shikoku’作为反向杂交的母本，则产生正常的二倍体杂种。由于三倍体杂种具有 *P. kisoana* 的一个特定的大卫星染色体和 *P. sieboldii* 的两个特定的小卫星染色体，我们得出结论，三倍体是由未减数分裂的 *P. sieboldii* 雌配子与正常减数分裂的 *P. kisoana* 雄配子受精产生的（图 24.2）。种间杂种表现出亲本的中间性状，但也存在一些变异（图 24.3）。

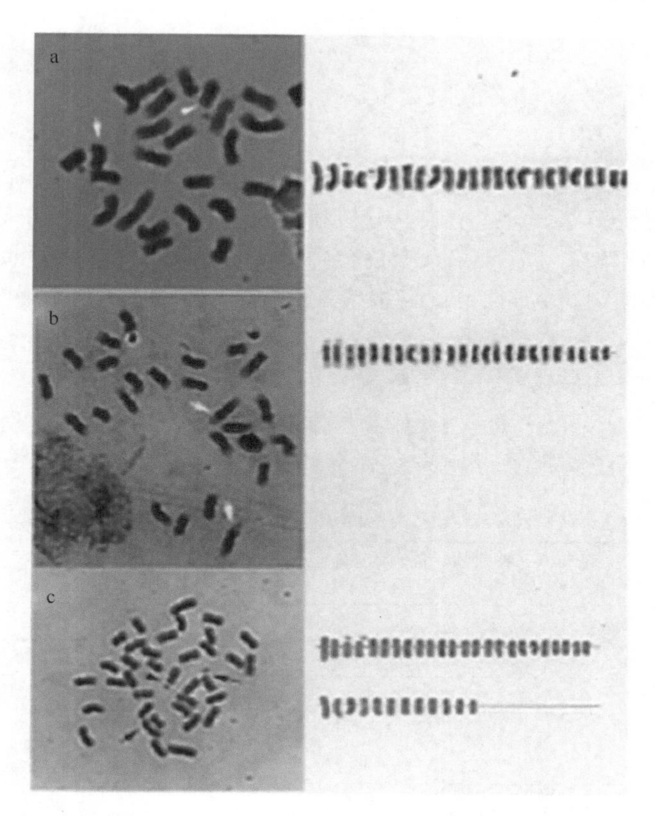

图 24.2　*P. sieboldii*（$2n = 2x = 24$）（a）、*P. kisoana*（$2n = 2x = 24$）（b）和三倍体杂种（$2n = 2x = 36$）（c）的体细胞染色体与核型

图 24.3　*P. sieboldii* 和 *P. kisoana* 之间三倍体种间杂交后代的花型分离。从左到右分别为母本 *P. sieboldii*‘Miyuki’、父本 *P. kisoana*‘Shirobana-shikoku’和 3 个三倍体杂种

24.4　*Primula sieboldii* 和 *P. jesoana* 之间的种间杂交

P. jesoana（JJ 基因组）原产日本，与 *P. sieboldii* 同属于 *Cortusoides* 组，但 *P. jesoana* 分属于 *Geranioides* 亚组。*P. jesoana* 广泛分布于从 Hokuriku 到 Hokkaido 的凉爽地区。据报道，*P. jesoana* 的核型与 *P. sieboldii*（SS 基因组）相似（Yamaguchi 1973），我们期望二者不仅能够成功进行种间杂交，而且可以产生后代。因此，在与 *P. sieboldii* 和 *P.*

kisoana 所用的相同培养基上，我们对 *P. sieboldii* 和 *P. jesoana* 进行种间杂交 4 周后的不成熟种子进行培养。在 63 个杂交组合中，可发芽的种子是从 6 个以 *P. sieboldii* 品种为母本的 15 个杂交组合中获得的。所有的 3 个以 *P. jesoana* 为母本的合法杂交组合也得到了可以萌发的种子（表 24.1）。由这些种间杂交获得的大多数植株在温室里能成功地驯化并正常生长，直至开花。流式细胞仪分析和 ITS 区的 PCR-RFLP 密度分析表明，当使用 *P. sieboldii* 作为母本时，在种间杂交后代中发现了两种倍性水平，即二倍体（SJ 基因组）和三倍体（SSJ 基因组）（图 24.4，图 24.5）。所有由杂交获得的杂种中，使用 *P. sieboldii* 的 4 个品种‘Bijyono-mai’、‘Hatsugoromo’、‘Jintsuriki’和‘Yanagino-yuki’作为母本产生的杂交后代都是二倍体，而以‘Miyuki’和‘Yatsugatake’作为母本杂交产生了三倍体（表 24.2）。然而，当以 *P. jesoana* 作为母本时，除两个四倍体外，所有杂种均为二倍体，四倍体估计是由二倍体合子在有丝分裂时染色体加倍形成的。

表 24.1 *P. sieboldii* 和 *P. jesoana* 种间相互杂交的种子产生与萌发

母本	杂交组合数量	发芽种子的杂交组合数量	产生发芽种子的父本	授粉的花数量	产生种子的花数量	种子总数（*a*）	发芽种子数（*b*）	发芽率（*b*/*a* × 100）
P. sieboldii（pin）/*P. jesoana*（thrum）								
Bijyono-mai	4	1	PUBS04–10	1	1	97	1	0.01
Hatsugotromo	4	1	PUBM04–6	2	2	174	1	0.01
Jintsuriki	7	2	PUBS04–12	2	2	152	5	3.3
			PUBS04–13	1	1	67	1	1.5
Jyanomegasa	6	0	—	—	—	—	—	—
Miyuki	10	9	PUB03–2	3	2	131	2	1.5
			PUBM04–3	2	2	153	7	4.6
			PUBM04–6	3	1	85	3	3.5
			PUBM04–13	2	2	132	2	1.5
			PUBS04–9	2	2	154	1	0.6
			BS04–10	3	3	201	35	17.4
			BS04–11	3	3	257	15	5.8
			BS04–12	2	2	137	11	8
			BS04–13	1	1	78	6	7.7
Yanagino-yuki	3	1	BS04–11	2	2	187	2	1.1
Yaseigatashiro	4	0	—	—	—	—	—	—
P. sieboldii（thrum）/*P. jesoana*（pin）								
Hanaguruma	6	0	—	—	—	—	—	—
Shiro-tonbo	7	0	—	—	—	—	—	—
Sumizomegenji	9	0	—	—	—	—	—	—
Tajimabeni	4	0	—	—	—	—	—	—
Yatsugatake	2	1	PUBM04–12	3	2	98	2	2
P. jesoana/*P. sieboldii*								
pbp03（pin）	2	2	Shiro-tonbo（thrum）	6	6	354	137	38.7
			Hanaguruma（thrum）	3	3	119	33	27.7
pbt03（thrum）	1	1	Iwatokagura（pin）	1	0	0	0	0

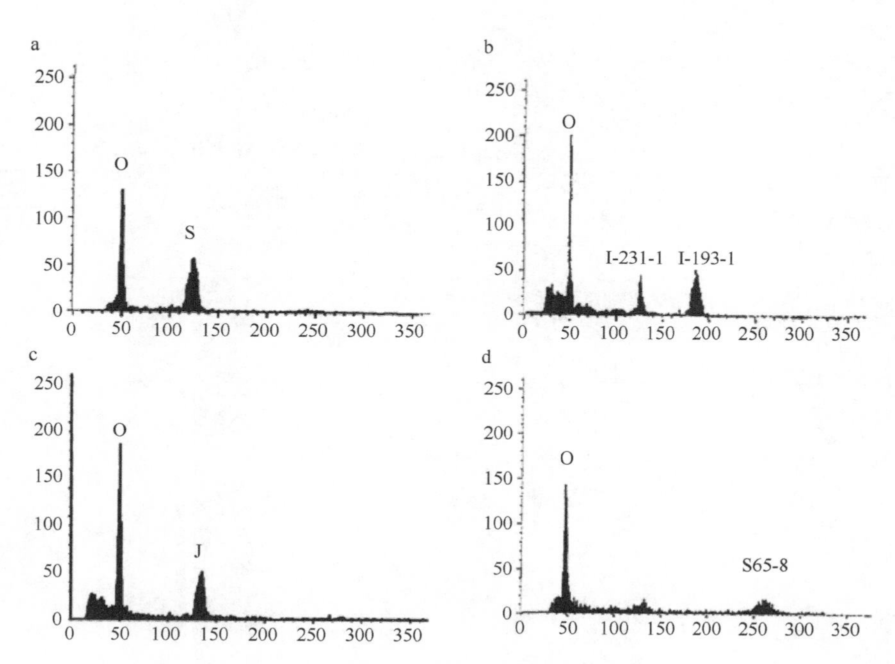

图 24.4 *P. sieboldii* 和 *P. jesoana* 种间杂交后代的流式细胞仪分析。(a)*P. sieboldii*(S 峰);(b)*P. jesoana*(J 峰);(c)二倍体杂交种(I-231-1 峰)和三倍体杂交种(I-192-1 峰);(d)四倍体杂交种(S65-8 峰)。O:*P. obconica* 作为内标

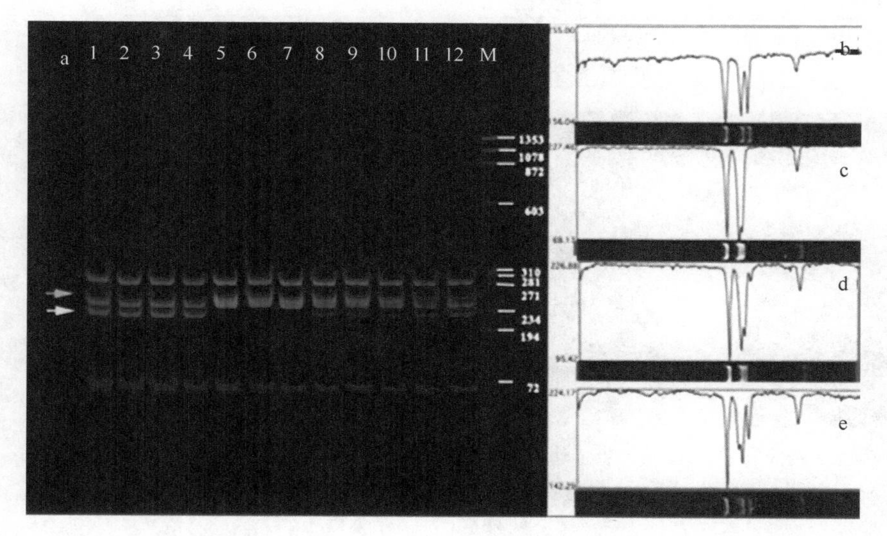

图 24.5 *Hae*Ⅲ消化 ITS 区的 PCR-RFLP 分析。(a)1～4 号道,*P. sieboldii*;5～7 号道,*P. jesoana*;8～10 号道,二倍体杂交种;11～12 号道,三倍体杂交种和 M;箭头指示 ϕx174/*Hae*Ⅲ尺寸标记。(b～d)使用 NIH 图像的带强分析,(b)*P. sieboldii*,(c)*P. jesoana*,(d)二倍体杂种,(e)三倍体杂种

与 *P. sieboldii* 相比,*P. jesoana* 具有明显不同的叶片形状,花也小于 *P. sieboldii*,但具有相似的花型。二倍体种间杂种整个花的形状与两个亲本相似,但花的大小与 *P. sieboldii* 相似。以含锯齿花冠裂片的 *P. sieboldii*‘Shiro tonbo’作为花粉亲本杂交获得的二倍体杂种,分为正常和锯齿两种花瓣类型(图 24.6)。该结果表明 *P. sieboldii* 的锯齿

状花冠可能受一个基因调控，并且该特征在二倍体种间杂种中显著表达。不幸的是，二倍体和三倍体杂种的花粉都不育。

表 24.2 ***P. sieboldii* 与 *P. jesoana* 种间杂交后代倍性的流式细胞仪分析**

交叉组合				后代数量	二倍体杂种	三倍体杂种	四倍体杂种
P. sieboldii（♀）× *P. jesoana*（♂）							
Bijyono-mai	针式	PUBS04-10	线式	1	1		
Hatsugoromo	针式	PUBM04-6	线式	1	1		
Jintsuriki	针式	PUBS04-12	线式	4	4		
		PUBS04-13	线式	1	1		
Miyuki	针式	PUB03-2	线式	2		2	
		PUBM04-3	线式	5		5	
		PUBM04-6	线式	2		2	
		PUBM04-13	线式	2		2	
		PUBS04-9	线式	1		1	
		PUBS04-10	线式	34		34	
		PUBS04-11	线式	11		11	
		PUBS04-12	线式	9		9	
		PUBS04-13	线式	6		6	
Yanagino-yuki	针式	PUBS04-11	线式	2	2		
Yatugatake		PUBM04-12	针式	2		2	
P. jesoana（♀）× *P. sieboldii*（♂）							
pbp03	针式	Hanaguruma	线式	27	27		
		Shiro-tonbo	线式	104	102		2

图 24.6　*P. sieboldii* 与 *P. jesoana* 杂交获得的种间杂种的特性。(a) 正在开花的母本 *P. jesoana*（左）、种间二倍体杂种（中）和父本 *P. sieboldii*‘Shiro tonbo’（右）；(b) 作为母本的 *P. sieboldii*（左）、二倍体杂种（中）和作为父本的 *P.* jesoana（右）的叶形；(c) 作为母本的 *P. sieboldii*（左）、三倍体杂种（中）和作为父本的 *P. jesoana*（右）的叶形。(d) *P. jesoana*（左）和 *P. sieboldii*‘Shiro tonbo’（右）杂交获得的二倍体杂种（中间八朵花）的花形变异

24.5　*P. kisoana* 和 *P. takedana* 以及 *P. cortusoides* 和 *P. takedana* 之间的种间杂交

P. takedana（*Reinii* 组）分布于日本北部岛屿 Hokkaido 的部分地区，喜凉爽气候。

这种植物的茎上有2～3朵白色不完全开放的小花（图24.7a）。*P. takedana*具有独特的掌状叶，但在温暖地区难以生长。因此，开发具有这种叶片特征，同时结合有更丰富多彩的花朵和对温暖气候具有较强抗性的栽培品种是很有意思的。为了达到这样的育种目标，我们在*P. kisoana*（*Cortusoides*组）（图24.7b）和*P. takedana*之间以及*P. cortusoides*（*Cortusoides*组）（图24.7b）和*P. takedana*之间进行了种间杂交。授粉3～4周后，在加入5%蔗糖、0.1mg/L NAA、0.1mg/L 6-BA和50mg/L GA_3、0.25%（*w/V*）结冷胶的1/2 MS培养基中进行胚挽救培养。

图24.7 *P. takedana*（a）、*P. kisoana*（b）、*P. cortusoides*（c）、*P. kisoan*与*P. takedana*（d～f）的二倍体种间杂种，以及*P. cortusoides*和*P. takedana*的二倍体杂种的花形态

总共有1个*P. takedana*（雌性）与*P. kisoana*（雄性）的杂交组合，3个*P. kisoana*（雌性）与*P. takedana*（雄性）的杂交组合，以及2个*P. cortusoides*与*P. takedana*的杂交组合用于产生种间杂种。虽然在*P. takedana*（雌性）与*P. kisoana*（雄性）的杂交组合中没有获得可萌发的种子，但*P. kisoana*（雌性）与*P. takedana*（雄性）之间3个杂交组合中的2个，以及*P. cortusoides*（雌性）与*P. takedana*（雄性）之间的2个杂交组合分别产生了12粒和13粒可萌发的种子（表24.3）。通过ITS区的PCR-RFLP密度和流式细胞仪分析，确定*P. kisoana*（雌性）与*P. takedana*（雄性）之间的杂交后代除1株为二倍体外，其余均被鉴定为双二倍体杂种。一些二倍体杂种被成功地驯化。

从*P. takedana*与*P. kisoana*的杂交中，获得了花冠颜色为红色或粉红色的两种类型的花。这两种花的大小均比*P. takedana*大，接近*P. kisoana*的大小。其中一种类型的开花性状与*P. takedana*相似（图24.7d），而另一种类型的开花性状与*P. kisoana*相似（图24.7e）。*P. cortusoides*和*P. takedana*之间的所有种间杂种均呈粉红色的一致花形。它们的花大小介于亲本之间，花冠筒和花冠裂片的长度大致相同，就如*P. takedana*一样，花冠裂片没有完全打开。这些结果表明，这些杂交组合的白花色是隐性的。*P. takedana*

花冠裂片不完全开放的性状因杂交组合的不同而有不同的遗传规律，在 *P. cortusoides* 和 *P. takedana* 之间的杂种中显性表达，而在 *P. kisoana* 和 *P. takedana* 之间的杂种中隐性表达。利用这些种间杂种，可以阐明报春花中花冠裂片张开角所涉及的基因。

表 24.3　*P. kisoana* 和 *P. takedana* 以及 *P. cortusoides* 和 *P. takedana* 杂交的种子产生与萌发

母本	针式/线式	父本	针式/线式	授粉的花数量	产生种子的花数量	种子总数（*a*）	发芽种子数（*b*）	发芽率（*b*/*a*×100）
P. takedana/P. kisoana								
TAKH03-1P	针式	SSKT03-2T	顺式	1	0	0	0	
P. kisoana/P.takedana								
skt 03-1	顺式	tak p-1	针式	2	1	43	9	14
skt 03-2	顺式	tak p-1	针式	1	0	0	0	
sskt03-1	顺式	TAKH03-1P	针式	1	1	93	3	3.2
P. cortusoides/P. takedana								
sax04-1	针式	tak04-1	顺式	1	1	76	1	1.3
		tak04-2	顺式	2	2	190	12	6.3

24.6　二倍体 *P. rosea* 与四倍体 *P. denticulata* 的种间杂交及未减数 3*x* 雌配子的形成

P. rosea 与 *P. denticulata* 是两种常见的观赏植物。这两个种都原产喜马拉雅山脉，分别属于 *Oreophlomis* 组和 *Denticulata* 组（Richards 1993）。*P. denticulata* 有许多红色、紫色和白色的花聚集在花茎顶部，看起来呈独特的球形排列（图 24.8a），而 *P. rosea* 则有 5～6 朵亮粉色的花在花茎上（图 24.8b）。在 *P. rosea*（二倍体）和四倍体 *P. denticulata* 之间进行种间杂交，并将授粉 4 周后获得未成熟胚珠在前文描述的培养基上培养。从这两个种的反向杂交中获得可萌发的种子。不同杂交组合可萌发种子成功授粉的概率不同。其中以线式 *P. denticulata*（雌性）和针式 *P. rosea*（雄性）的杂交组合的比例最高（72.7%），以针式 *P. rosea*（雌性）和线式 *P. denticulata*（雄性）的杂交组合的比例最低（13.3%）。但是，成功授粉的花产生的可萌发种子数量未发现存在显著差异（Hayashi et al. 2007）。不同杂交组合的幼苗生长存在差异，由 *P. denticulata* 和 *P. rosea* 获得的幼苗显示出旺盛的生长，且大多数后代都被成功地驯化，但是经反向杂交获得的幼苗显示出较差的生长，且大多数幼苗不能驯化。6 个后代产生了花（图 24.8c，d），这些花在大小和颜色（图 24.8e，f）以及中间叶的大小（图 24.8g）上表现出分离。具有较大花瓣的 3 个后代的花较少，而其他后代则有小但数量较多的花。不幸的是，这些杂交种没有可育的花粉或花粉育性很低（Hayashi et al. 2007）。因此，我们用这些杂种作为雌性亲本进行了回交，以确认雌性的可育性。三个子代用于与四倍体 *P. denticulata* 的花粉回交，其他子代与二倍体 *P. rosea* 的花粉回交。随后，从三倍体杂种‘25-5’与四倍体 *P. denticulata* 花粉的回交中获得 11 粒可萌发的 BC1 种子。我们最初预期这些 BC1 杂种表现为超三倍体，其具有 3 个 *P. denticulata* 基因组以及一些 *P. rosea* 的染色体。流式细胞仪分析和体

细胞染色体观察表明，这 11 株 BC1 虽然不是超三倍体，但有 9 株是 DDDDR 基因组的五倍体、亚五倍体或超五倍体，剩下的 2 株植物分别是亚六倍体和八倍体（Hayashi et al. 2009）。这些五倍体和超五倍体或亚五倍体估计是由很少产生的未减数 3*x* 雌配子（在与 DDR 的三倍体杂交中形成）和四倍体 *P. denticulata* 的二倍体配子之间受精所产生。这些五倍体可能是产生超四倍体品种的重要遗传资源，超四倍体品种具有 4 个 *P. denticulata* 基因组，并通过与四倍体 *P. denticulata* 的回交添加一些与 *P. rosea* 相关的性状。

图 24.8　*P. denticulata* 与 *P. rosea* 杂交后代的性状（a）*P. denticulata* 开花（标尺 =2cm）；（b）*P. rosea*（标尺 =2cm）；（c 和 d）三倍体杂种（标尺 =2cm）；（e）*P. denticulata*（左）、两个三倍体杂种（中）、*P. rosea*（右）花瓣形状的正视图（标尺 =1cm）；（f）*P. denticulata*（左）、两个三倍体杂种（中）、*P. rosea*（右）花瓣形状的侧面图（标尺 =1cm）；（g）*P. denticulata*（左）、三倍体杂种（中）、*P. rosea*（右）的叶（标尺 =1cm）（引自 The Journal of Horticultural Science & Biotechnology, 2007, 82(1): 5-10）

24.7 *P. filchnerae* 与 *P. sinensis* 的种间杂交及后续回交

P. sinensis 与 *P. filchnerae* 均为中国特有种。自 19 世纪开始在欧洲栽培 *P. sinensis*（*Auganthus* 组），并繁殖了许多花色各异的品种（图 24.9a）。相反，*P. filchnerae*（*Pinnatae* 组）带有许多花梗和几轮淡粉色的花（图 24.9b）。尽管该物种具有观赏植物的独有特征，但在 20 世纪末引入日本之前并没有广为人知。为了培育出花色丰富的新品种，我们尝试在这两个品种之间培育出种间杂种。尽管这两个物种的现有品种只有针式花，但我们的初步实验表明，每个种的种内不合法杂交（针式/针式）的自花授粉是兼容的。因此，这些物种间的杂交是通过针式/针式杂交组合进行的。在总共的 26 种相互杂交组合中，使用 *P. filchnerae* 作为雌性亲本进行了 17 种杂交组合，使用 *P. sinensis* 作为雌性亲本进行了 9 种杂交组合。在最后这些杂交组合中，三个胚珠在授粉后 40 天于前文描述的胚挽救培养基上不能正常萌发，只有胚根出现，但缺少上部。由于这些不正常的幼苗只表现出胚根伸长或产生愈伤组织，因此将它们转移到含 0.3%结冷胶的 1/2MS 培养基上，并添加 30g/L 蔗糖、0.1mg/L NAA 和 1mg/L 玉米素或噻二唑仑（TDZ）。转移后，从胚根诱导出绿色愈伤组织，并在含玉米素的培养基上用两个胚珠的愈伤组织再生出植株，将这些再生的植物转移到添加 30g/L 蔗糖、含 0.3%结冷胶的不含植物生长调节剂（PGR）的 1/2MS 培养基上，然后这些再生植株成功地被驯化了。

通过流式细胞仪和随机扩增多态性 DNA（RAPD）分析，确定了再生植株的倍性水平和杂交性。以 *P. filchnerae*（FF 基因组）为母本，与具有橙色花瓣的 *P. sinensis* 的花粉（SS 基因组）杂交获得的再生植株中，有一株是具有 FS 基因组的二倍体杂种。通过流式细胞仪分析证实，使用具有红色花瓣的 *P. sinensis* 作为雌性亲本，和二倍体的 *P. filchnerae* 的花粉杂交获得的另一株再生植物是六倍体杂种（SSSSFF 基因组）（Amano et al. 2006）。六倍体杂种有一个花梗，但花不多。然而，具有 FS 基因组的二倍体杂种（后来被命名为 *Primula* ‘Thirty-one’）具有许多花梗和深粉红色的花，其花的大小介于双亲之间（图 24.9c）。

将 *Primula* ‘Thirty-one’进行离体无性繁殖，并作为盆栽观赏植物出售。在这些微繁殖植物中，常常发现四倍体基因型表现出旺盛的生长。这些植物被证实是四倍体（双二倍体），并出现了自发染色体加倍。由于这些四倍体植株的花粉表现出高度可育，因此它们是自花授粉的，后代的花色变化很大，如通过反复自花授粉产生的 F_5 后代所示（图 24.9d）。尽管花色变异，但四倍体植株仍表现出一些不利的性状，如茎秆高、叶片大、花瓣较宽等，使植株整体形态不甚理想。

因此，我们将原二倍体 *Primula* ‘Thirty-one’ 与双亲进行了回交，并通过与 *Primula* ‘Thirty-one’ 所使用的相同胚挽救方法获得了 9 个 BC1 后代。当以 *P. filchnerae* 作母本时，获得了 7 个 BC1，而当以 *Primula*‘Thirty-one’作为母本，并用 *P. filchnerae* 或 *P. sinensis* 的花粉授粉时，每个杂交组合各获得一株 BC1 植物。当以 *P. sinensis* 作为母本时，未获得 BC1。体细胞染色体观察显示，在所有 4 个 BC1 中成功观察到体细胞的 25 条染色体。

由于两个亲本的染色体数目均为 24，因此该结果表明在 *Primula*‘Thirty-one’的减数分裂阶段，两个亲本的同源染色体之间经常发生不分离现象。这种不分离现象可能是 *Primula*‘Thirty-one’花粉育性低的原因之一。回交后代花的大小小于 *P. filchnerae*，并且植物形态不同于 *Primula*‘Thirty-one’和两个祖父母（图 24.9e）。为了进一步改善 *Primula*‘Thirty- one’，目前正在研究二倍体和四倍体水平的后代。

图 24.9　*P. filchnerae* 与 *P. sinensis* 种间杂种后代的变异。（a）种间杂种 *Primula*‘Thirty-one’；（b）母本 *P. filchnerae*；（c）花粉亲本 *P. sinensis*；（d）通过自发染色体加倍形成的四倍体 *Primula*‘Thirty-one’重复自交获得的 F_5 后代的花色分离；（e）*Primula*‘Thirty-one’（雌性，右）与 *P. filchnerae*（雄性，左）回交获得的 BC1 子代（中）

24.8 *P. sieboldii* 与 *P. obconica* 的种间杂交及后续回交

P. obconica 属于 *Obconicolisteri* 组，是作为观赏植物栽培的最重要的报春花属物种之一。我们在 *P. sieboldii* 与 *P. obconica* 之间进行了种间杂交，这是一个交叉的远缘杂交（Kato et al. 2001）。杂交后 3～4 周，在含有 0.25%（*w*/*V*）结冷胶的 1/2MS 培养基中加入 5%蔗糖、20mg/L GA_3、0.1mg/L NAA 和 0.1mg/L 6-BA 后进行胚挽救培养。利用 54 个 *P. sieboldii* 品种和 8 个 *P. obconica* 品种进行的 140 个双向杂交中，我们培养出 15 株植物，是以 *P. sieboldii* 作为母本的 9 个杂交组合产生的（Kato et al. 2001）。由于 *P. sieboldii*（SS 基因组）的 DNA 含量是 *P. obconica*（OO 基因组）的 2.5 倍，我们用流式细胞仪分析了由二者杂交获得的植株的杂交性和基因组组成。发现了两种具有三个基因组组成的倍性水平：10 个二倍体（SO 基因组），4 个三倍体（分别含有两个 *P. sieboldii* 和一个 *P. obconica* 基因组），以及一个具有相反组成的三倍体（SOO 基因组）。

尽管具有 SOO 基因组的三倍体种间杂种是不育的，但是具有 SOO 基因组的三倍体杂种‘280-1’在减数分裂中期 I 每个同源染色体之间都可能形成二价体，因此有望产生一些可育的配子。所以，我们通过在添加 5%蔗糖、0.3%结冷胶和 50mg/L GA_3 的 1/2MS 培养基上来生产‘280-1’和 *P. obconica* 之间的反向杂交 BC1 植株。

虽然‘280-1’表现出相对较高的花粉育性（醋酸洋红染色实验达到约 70%），但在以‘280-1’作为花粉亲本的 12 个杂交组合中并没有产生可萌发的种子。相比之下，在以‘280-1’为母本的 96 个杂交组合中，共获得 133 粒可萌发种子，其中 41 个个体在体外条件下生长正常(表 24.4)。其中的一些植株可以成功地适应环境，且 9 株后代在 3～4 年后开花（图 24.10），这些 BC1 子代的相对 DNA 含量连续地分布在 *P. obconica* 和‘280-1’之间（图 24.11）。

表 24.4　三倍体杂种回交种子的产生和萌发，该三倍体杂种由 *P. sieboldii* 和 *P. obconica* 与 *P. obconic* 杂交而来

母本	针式/顺式	父本	针式/顺式	杂交数量	产生种子的花数量	种子总量	发芽种子数	发芽率（%）
SOO/*P. obconica*								
280-1	针式	*P. obconica*	顺式	66	52	1808	133	9.86
			针式	30	9	21	6	5.61
P. obconica	顺式	280-1	针式	12	2	9	0	0

这 9 个杂种均表现出与 *P. obconica* 之间的不同之处。例如，依花梗长度被分为两种类型，一种类型的长度与‘280-1’或 *P. obconica* 的长度大致相同，而另一种则表现为短花梗。花冠形状在 BC1 后代中也有显著的差异。其中一个 BC1，即‘143-8’与 *P. obconica* 的花冠形状非常相似，但另一些 BC1 在花形上与 *P. obconica* 则有一些差异。‘8-1’（$2n = 26$）的花冠裂片形状，继承了 *P. sieboldii* 如‘280-1’的花冠裂片呈锯齿状。这些结果表明，许多数量性状基因座（QTL）可能位于 *P. sieboldii* 的染色体上，与整个花冠形状有关。

图 24.10　‘280-1’三倍体杂种回交后代的花粉形态变异[SOO 基因组，来自 *P. sieboldii*（SS 基因组）与 *P. obconica*（OO 基因组）杂交，再与 *P. obconica*（OO 基因组）的花粉回交]。(a)‘280-1’；(b)‘8-1’；(c)‘139-2’；(d)‘143-8’；(e)‘141-16’；(f)‘141-10’；(g)‘141-15’；(h)‘139-1’

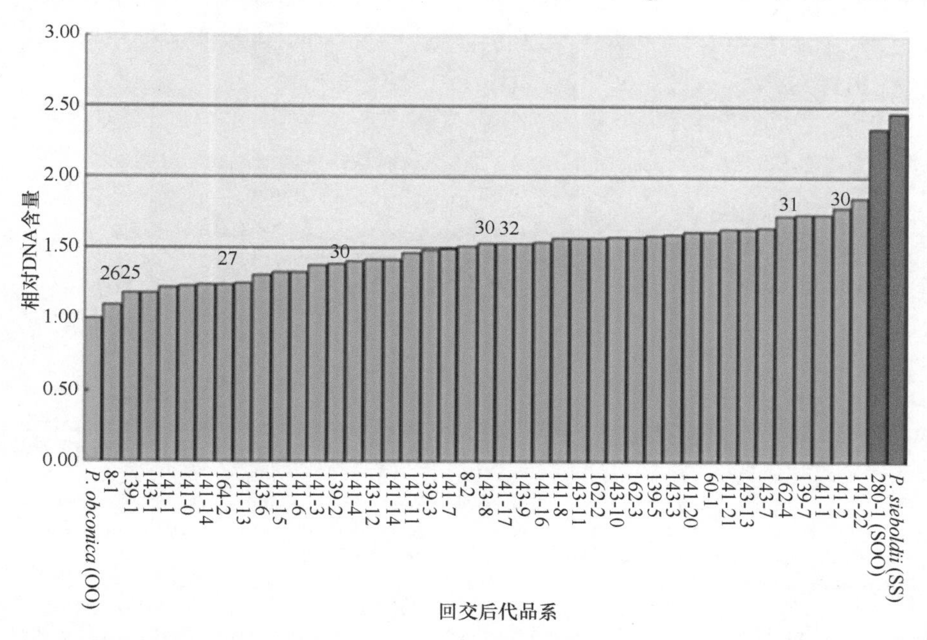

图 24.11　与三倍体杂种（SOO 基因组）‘280-1’之间回交获得的 BC1 后代的相对 DNA 含量分布。‘280-1’是 *P. sieboldii*（SS 基因组）和 *P. obconica*（OO 基因组）之间的种间杂种。一些柱状图上的数字是明确的染色体数目

24.9　结　　论

为了在报春花中培育出种间杂种，胚挽救是一个重要且不可缺少的工具，尤其是胚挽救时间和培养基对杂交种的产生至关重要。一般来说，在由 0.2%结冷胶固化的 1/2MS

培养基中加 5%蔗糖、0.1mg/L NAA、0.1mg/L 6-BA 和 50mg/L GA_3 组成的培养基可有效挽救报春花的未成熟胚。因此，该培养基可用于报春花属的不同种间杂交组合。

系统发育关系对于种间杂种的产生也很重要。虽然近缘种间的杂交组合通常会产生预期的种间杂种，但一些亲缘关系相对较远的杂交组合，如 *P. filchnerae* 和 *P. sinensis*，也会产生杂交植株。在后一种情况中，异常幼苗需要转移到添加 NAA 和玉米素的培养基上，以获得正常的再生芽。

杂种不育是种间杂交中最严重的问题之一。在我们的研究中即使两个亲本的核型相似，获得的大多数二倍体种间杂种显示出不育。这些结果表明，染色体重排已经在进化过程中发生了，并且两个物种的基因组之间丢失了一些同源染色体。尽管产生二倍体是种间杂种恢复育性的有效方法之一，但具有育性的倍半二倍体（3*n*），如 *P. obconica* 的未减数 2*n* 雄配子与 *P. sieboldii* 的单倍体 *n* 配子杂交获得的 SOO 基因组的三倍体杂种，也可能是产生下一代的好工具。在报春花的种间杂交中，各种种间杂交组合可形成三倍体杂种。迄今为止，虽然仅在 *P. sieboldii* 和 *P. obconica* 的种间三倍体后代中证实了可育性，但在其他种间杂交组合中也有可能发现其他具有可育性的三倍体杂种。当三倍体杂种具有雌性可育性时，随后的回交可能只将有用性状从一个亲本引入另一个亲本。

借助于胚挽救培养，利用种间杂交在报春花属的育种中产生各种新品种仍可能具有很大的潜力，就如百合属一样，许多品种是通过种间杂交和多倍体化产生的（van Tuyl and Arens 2011）。

本章译者：

蔡艳飞

云南省农业科学院花卉研究所，国家观赏园艺工程技术研究中心，云南省花卉育种重点实验室，昆明 650200

参 考 文 献

Amano J, Kato J, Nakano M, Mii M (2006) Production of inter-section hybrids between *Primula filchnerae* and *P. sinensis* through ovule culture. Sci Hortic 110:223–227. https://doi.org/10.1016/j.scienta.2006.06.027

Ernst A (1950) Resultate aus Kreuzungen zwischen der tetraploiden, monomorphen *Primula japonica* und diploiden, mono-und dimorphen Arten der Sektion Candelabra. Arch Julius Klaus Stift Vererbungsforsch Sozialanthropol Rassenhyg 25:135–236

Hayashi M, Kato J, Ohashi H, Mii M (2007) Variation of ploidy level in inter-section hybrids obtained by reciprocal crosses between tetraploid *Primula denticulata* (2n=4=x44) and diploid *P. rosea* (2n=2x=22). J Hortic Sci Biotechnol 82:5–10. https://doi.org/10.1080/14620316.2007.11512191

Hayashi M, Kato J, Ohashi H, Mii M (2009) Unreduced 3x gamete formation of allotriploid hybrid derived from the cross of *Primula denticulata* (4x) 3 *P. rosea* (2x) as a causal factor for producing pentaploid hybrids in the backcross with pollen of tetraploid *P. denticulata*. Euphytica 169:123–131. https://doi.org/10.1007/s10681-009-9955-y

Ishizaka H (2008) Interspecific hybridization by embryo rescue in the genus *Cyclamen*. Plant Biotechnol 25:511–519

Izumikawa Y, Takei S, Nakamura I, Mii M (2008) Production and characterization of inter-sectional hybrids between *Kalanchoe spathulata* and *K. laxiflora* (= *Bryophyllum crenatum*). Euphytica 163:123–130. https://doi.org/10.1007/s10681-007-9619-8

Kato J, Mii M (2000) Differences in ploidy levels of inter-specific hybrids obtained by reciprocal crosses between *Primula sieboldii* and *P. kisoana*. Theor Appl Genet 101:690–696. https://doi.org/10.1007/s001220051532

Kato J and Mii M (2006) Production of interspecific hybrid plants in *Primula*. in V M Loyola-Vargas and F Vazquez-Flota (eds) Methods of molecular biology vol. 318., Plant Cell Culture Protocols, 2ndnd edn., Humana Press Inc.., Totowa, , p253–262. https://doi.org/10.1385/1-59259-959-1:253

Kato J, Ishikawa R, Mii M (2001) Different genomic combinations in inter-section hybrids obtained from the crosses between *Primula sieboldii* (section Cortusoides) and *P. obconica* (section Obconicolisteri) by the embryo rescue technique. Theor Appl Genet 102:1129–1135. https://doi.org/10.1007/s001220000516

Murashige T, Skoog F (1962) A revised medium for rapid growth and bioassay with tobacco tissue cultures. Physiol Plant 15:473–497. https://doi.org/10.1111/j.1399-3054.1962.tb08052.x

Richards J (1993) Primula. Timber Press, Portland

Torii T (1985) Sakurasou (in Japanese). Nihon-televi, Akatuki-Insatu press, Tokyo

van Tuyl JM, Arens P (2011) Lilium: breeding history of the modern cultivar assortment. Acta Hortic 900:223–300

Woodel SRJ (1960) Studies in British primulas. VII. Development of normal seed and of hybrid seed from reciprocal crosses between *P. vulgaris* Huds. And *P. veris* L. New Phytol 59:302–313

Yamaguchi S (1973) Cytogenetical studies in *Primula sieboldii* E. Morren. Chromosome numbers of 80 cultivars and related wild species. Jpn J Breed 23:86–92

第 25 章　毛　　茛

Margherita Beruto, Mario Rabaglio, Serena Viglione, Marie-Christine Van Labeke, and Emmy Dhooghe

摘要　花毛茛是唯一具有观赏价值的毛茛属植物。通常它是作为切花种植的，但也可作为盆栽花卉。目前花毛茛在园林和景观设计中的作用越来越重要。在商业生产中，其是通过种子繁殖或在育苗结束时收集其地下贮藏器官（以下称为块根）来实现生产的；后一种方式获得的种苗花期提前且花数量更多，因此在商业生产中块根繁殖是主要的手段。在本章中我们将介绍花毛茛在育种过程中一些重要的细节性形态学特征以及影响试管花粉萌发的一些因素。通常情况下，在异花授粉和种内自花授粉中不存在杂交障碍，但有些品种（如‘Alfa’）表现出自交不亲和性，甚至还有人假设有无融合生殖的发生。同时我们对花毛茛切花和盆花的商业性生产的育种目标与前景进行了探讨。

关键词　育种；切花；种间杂交；种内杂交；组织培养；毛茛

25.1　引　　言

毛茛属有将近 600 个种，分布在全球（Tamura 1995a），花毛茛通常被用于传统药材和观赏。花毛茛属于毛茛科，其野生种分布于地中海东部，在 16 世纪传入西欧。在 18 世纪该种被广泛用于庭院观赏。19 世纪，英国进行了广泛的育种工作，并由此产生了 500 多个品种。在那之后，花毛茛逐渐过时，但是在过去的 30～40 年，它重新流行起来。在地中海自然条件下，荷兰育种家通过种子繁殖，培育出了许多切花和盆栽品种。Meynet（1974）在法国启动了一个育种计划，其主要目标是在这个物种中获得近等基因系，即对近亲繁殖非常敏感。他试图利用亲本的关系建立杂交系，部分是近亲繁殖，他注意到在三到四代之后，后代丧失活力。同时他对经离体花药培养得到的双单倍体进行了初步研究（Meynet and Duclos 1990）。1983 年，Sakata 在美国加利福尼亚州摩根山进行了播

M. Beruto (✉)· S. Viglione
Regional Institute for Floriculture (IRF), Sanremo (IM), Italy
e-mail: beruto@regflor.it
M. Rabaglio
Biancheri Creations, Camporosso (IM), Italy
M.-C. Van Labeke
Ghent University, Department Plants and Crops, Ghent, Belgium
E. Dhooghe
Flanders Research Institute for Agriculture, Fisheries and Food (ILVO), Plant Sciences Unit, Applied Genetics and Breeding, Melle, Belgium

种，并引进了由种子繁殖的 Bloomingdale 系列的 F_1 代。最近，意大利育种家开发出了新型的花毛茛，并在国际市场上获得好评。

现在，花毛茛在不同的地中海国家（如意大利、法国、以色列）、荷兰、南非、美国加利福尼亚和日本栽培。来自荷兰 FloraHoland 的数据显示，2014 年花毛茛占切花和观叶植物总量的 0.4%，据报道，在荷兰花卉拍卖会上其作为夏季顶级的第五大花卉出售大约 6000 万株。从 Royal FloraHoland 收集的数据（2015 年 1 月至 2017 年 5 月）显示花毛茛产品主要来自荷兰、以色列和意大利。一小部分（6%～8%）来自埃塞俄比亚、肯尼亚、突尼斯和土耳其。花毛茛的新兴市场是厄瓜多尔，每年种植 250 万～300 万根块根，日本每年种植面积约 10hm^2，块根每年增长超过 100 万根（Biancheri Creations，个人交流）。在法国 Hyères 市场的 2014～2015 年生产季节（2014 年 9 月至 2015 年 3 月），有不到 100 万根块根被出售，它们属于 35～40 个品种（SICA Marchéaux Fleurs，官方数据）。迄今为止，在意大利花毛茛是继玫瑰、菊花和非洲菊之后的第四大花卉，有 1.32 亿株（来源：ISPF 2014），种植面积 300～350hm^2。Sanremo 花卉市场记录花毛茛作为最受欢迎的切花销售了 1500 万株，栽培面积约 150hm^2（E. Sparago，个人交流）。

花毛茛与纤细毛茛的亲缘关系最为密切。它们是 Tethyan 支系的一部分（Emadzade et al. 2011）。从地理上讲，这个分支位于地中海地区。

在这里我们介绍了毛茛属植物的一些相关特征。由于它是可以营养繁殖的新品种，并且块根采后生命力强，因此这种作物可以在不同的气候条件下种植。我们将介绍目前毛茛属植物科学研究的最新进展，特别是毛茛属植物的育种方法。

25.2　开发周期与商业生产

花毛茛是一种多年生植物，其生命周期表现为在寒冷潮湿的冬季进行营养生长，在炎热干燥的夏季进行休眠。授粉 2～3 个月后，当果实表皮变干、发棕色时，圆柱形的瘦果开裂，种子释放。

在凉爽潮湿的夏季，种子可以在 5 月播种，以便在 11 月初（Meynet 1993）开花；在地中海地区，只有人工将土壤温度降到 18℃时，才可以使用此方法。在夏季高温和干燥条件下，7～8 月前不播种，或播种最好等到初秋，这时胚胎自然地度过了休眠时期并成熟。播种后大约 2 个月，植物开始迅速生长，春天开花。

休眠期间的生存由地下休眠器官（块根）保证，这部分属于营养生长（以下称为“块根”）。块根通常是经过一个生长季后成苗，幼苗发育对块根的形成有很大的影响。Meynet（1974）强调花毛茛幼苗有玫瑰型叶；营养芽位于外生叶处，并产生块根的生长点（图 25.1）。因此，使用块根产生的苗比种子萌发产生的幼苗花产量更高，并且它们受益于“启动”效应，这使得花期提早（Meynet 1993）。

贮藏器官在秋天的再生过程中耗尽。与许多其他的隐芽植物相比，花毛茛的茎尖分生组织一般处在休眠状态，当出现秋季降雨和温度较低时，它能够重新萌发。Kamenetsky 等（2005）研究表明：根中皮层细胞大小增加，并且在细胞壁中有果胶存在。此外，作者发现经过夏季的休眠和随后秋季的补水，块根的淀粉和蛋白质也积累起来，以支持植物后续的生长。块根可以很好适应地下长时间的贮藏，这也是花毛茛被称为“可复活的地质植物”的原因（Beruto et al. 2009；Kamenetsky et al. 2005）。

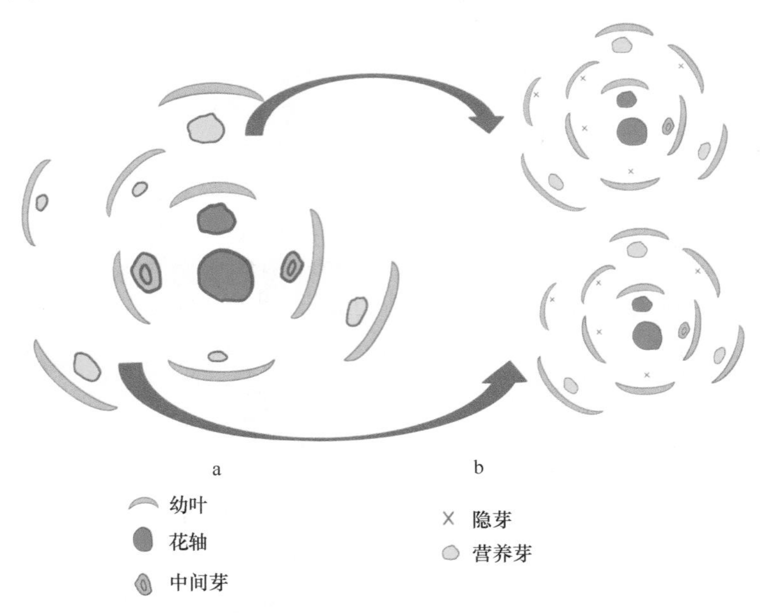

图 25.1　花毛茛幼苗玫瑰型叶的结构（a）与块根中的生长点（b）
[基于 Meynet（1974）的代表图]

为了实现更高的开花率，在商业生产上，往往收获块根，进行干燥处理，并出售给世界各地的销售商。干燥后的块根一般在种植前都要湿润。尽管 Meynet（1993）建议使用干燥的块根进行种植，但在农业种植实践中，还是建议使用湿润的块根，以达到植株的生长更加一致。在文献中有不同的保湿方法。一些作者（De Hertogh 1996；Umiel and Hagiladi 1999）建议在种植前将块根浸泡在缓慢流动的水中 24h；Ohkawa（1986）使用冷水（约 6℃）浸泡 8h；Cerveny 和 Miller（2012）发现采用 15～25℃的水温可以提高田间生长量。本章作者观察到在不同的品种中有不同的反应，结果为幼嫩的块根不易受水温的影响。

8～10℃的低水温预处理块根 5 周后，对不同商业品种的开花有一定的促进作用（Beruto et al. 2009），与未经处理的块根相比，缩短了开花的周数，并且平均开花数更多。这被认为是由于花毛茛需要类似于自然的生长条件，也就是需要低温打破块根的夏季休眠（Kamenetsky et al. 2005）。

此外，发芽受块根年龄和贮藏过程中条件的影响。一般来说，湿度和温度升高对长期贮藏块根的活力不利（Cerveny and Miller 2011）。在我们的实验中，下图显示了温度对块根在室内贮藏 1 年后发芽率的影响（图 25.2）。块根低温（2℃）贮藏 1 年后在开放或改良的气体成分（2% O_2 和 4% CO_2）条件下，虽然有一基因型观察到花期推迟，但保持了较高的萌发能力（Beruto et al. 2009）。通过分析长期贮藏实验块根的成分表明：可溶性碳水化合物中葡萄糖和果糖含量没有显著变化；阿拉伯糖含量的显著性分析表明，该碳水化合物与根系水分获取能力呈显著相关性。从我们的结果来看，阿拉伯糖的存在可以从细胞壁结构以及其在根系适应胁迫过程中的作用两个角度来考虑。由于它的存在，细胞壁的柔韧性可以在脱水/再水化过程中得以保持。它对细胞壁孔隙率的影响可以决定复水生长期细胞与环境互作时细胞的疏水性和营养吸收。

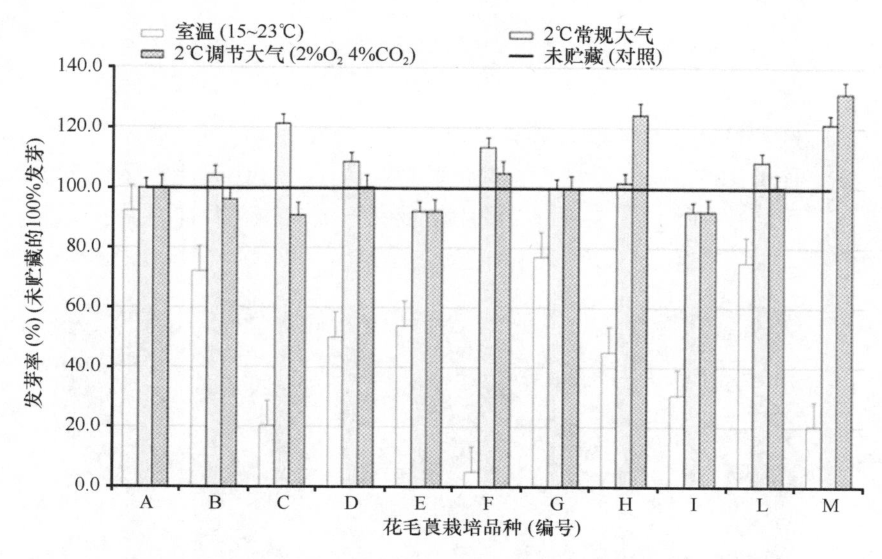

图 25.2　贮藏 1 年对不同花毛茛商业品种块根发芽率的影响（代号 A～M）。未贮藏的块根为对照组，其发芽率为 100% [基于 Beruto 等（2009）的数据]

花毛茛的商品化生产是以分级贮藏为基础的，根据预期的不同开花期，一般贮藏小（2～3cm）和大（5～7cm）的块根。大块根花期从 10 月中旬开始，而小的块根一般在 1～2 月。这是商业生产中在生产季节延迟花期的一个重要方面，可以通过使用不同大小的块根材料来保证。Meynet（1993）发现块根上的芽数影响花的形态，并且近交亲本的后代花期有轻微的延迟。

花毛茛对光热周期非常敏感。不同的研究者发现这种作物同时受温度和光周期的影响（Karlsson 2003；Meynet 1993）。Ohkawa（1986）进行了不同温度和日照长度对花毛茛生长与开花影响的实验。他发现，与自然生长的短日照植物相比，长日照条件可以加速开花，但花的数量较少。在长日照条件下，花毛茛幼苗的叶片生长更为显著，在盆栽条件下也发现了类似的结果。不同的研究表明花毛茛可以被认为是一种长日照植物。

花毛茛可以在不同的条件下（温室或露天）栽培，它可以被认为是一个相当原始的物种。花毛茛对施肥非常敏感（Hassan et al. 1985）。同时对盐度比较敏感（Valdez Aguilar et al. 2009）。

花毛茛最受欢迎的特征之一是其切花的采后表现（Scariot et al. 2009）。花毛茛的衰老表现为花瓣和花药颜色变化，雌蕊和雄蕊间的距离增加。人们发现花的成熟和衰老伴随着花瓣组织中可溶性碳水化合物与酚含量的降低、离子的流失和蛋白质降解的加速（Shahri and Tahir 2011）。切花的收获应该在清晨进行，而且花应在 0～5℃下分级、包装和贮藏。在转移到水中之前，用放线菌酮（0.01μmol/L 和 0.05μmol/L）进行切花预处理，可以提高花毛茛的瓶插寿命和膜完整性。此外，Scariot 等（2009）发现用氨基酸和硫代硫酸银处理可以提高花毛茛寿命。作者总结认为：这种有益的效果应归因于抗菌作用，而不是对乙烯生物合成的抑制作用，因为外源乙烯的应用并没有对10个研究品种中的任何一个产生负面影响。乙烯不是毛茛切花衰老的重要调节因子也被Kenza等（2000）证实。

25.3 繁殖和组织培养的影响因素

花毛茛的繁殖是通过种子和营养繁殖进行的。

对于商业种子生产，人工授粉有不同步骤，要进行花粉处理并收集。去除母本雄蕊，轻轻将父本花粉刷在母本雌性柱头上，并用粗棉布网或袋子盖住雌性柱头。授粉成功后，果实开始生长，最终收集种子（图25.3，图25.4）。

图25.3 花毛茛人工授粉育种的步骤。（a）轻刷雄蕊收集花粉；（b）去雄的花准备授粉；（c）授粉后雌蕊套袋

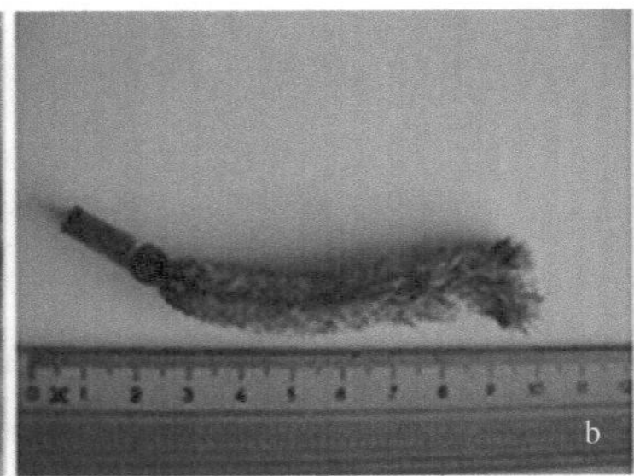

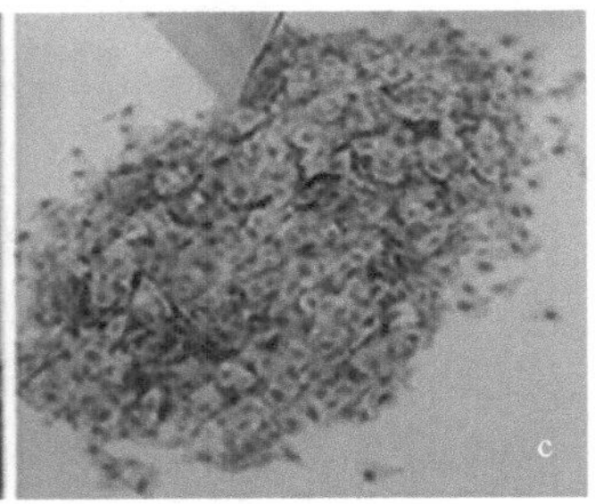

图25.4 花毛茛人工授粉成功后的后续步骤。（a）果实生长；（b）不同瘦果释放的成熟穗；（c）准备商业化的种子

一朵花通常能产生200～700粒种子。商业种子生产有收集、清洁和贮藏种子不同步骤。在收获时，种子处于休眠状态；3～4个月后，种子变得有活性（发芽率为80%～85%），并在一年内保持这种发芽潜力。因此，7月出售的种子来自前一年种植的，而用于秋季播种的种子是同年生产的。种子可在15～16℃和45%相对湿度下贮藏，为了使幼苗生长良好，没有必要对种子进行特殊处理。

为了在一年内能够提供商业种子，世界各地的育种者做了一项重要的工作：9～10月播种，3～4月采花，在培养周期结束时（5～6月）丢弃块根。由于花毛茛主要的病毒和真菌疾病感染不是通过种子传播的，因此种子繁殖体系被用来生产健康的种子（Turia et al. 2006）。然而，在种子繁殖中，冬季生产（比春季生产利润率更高）被阻止。此外，育种计划相当复杂，需要4～5年才能获得一个新的品种，而且杂交种子的一致性并不总是令人满意。同时杂交种子的生产仍是一项高度劳动密集型的劳动。

对于营养繁殖，可以使用分开块根的方法，但是一年生的繁殖率只有200%～500%，病毒和真菌引起的不同疾病是较大的问题（Beruto and Debergh 2004；Meynet 1993）。体外繁殖可以解决这些问题，同时，体外繁殖可提供脱毒的新品种块根和通过种子繁殖观赏性状不稳定的新品种（如一些绿色和奇幻色的花）。意大利Biancheri Creations收集的数据显示，意大利是世界上花毛茛离体培养品系最多的国家（60～70个组织培养品种已经商业化），这表明通过组织培养的植物种苗在销售总额中的占比越来越大（例如，在近5年来，法国的组培花毛茛块根总销售量由23%提高到67%）（Biancheri Creations，个人交流）。Sanremo地区的花卉研究所IRF，与Biancheri Creations公司合作，开发了一种组织培养方法，使花毛茛能够通过无性系繁殖投放市场。通过克隆获得的块根更具活力，与传统方法相比，组织培养能产生更多的花蕾和更大的花，产生具有更好质量和更高品质的均质块根（Beruto et al. 1996）。与种子幼苗相比，组培苗芽数增加，开花早，单株块根产量提高（Beruto et al. 1996）。

如Beruto和Debegh（2004）所述，腋芽诱导已被证明是花毛茛离体繁殖的最佳方法，并从原理上为花毛茛提供了一种真正的外植体材料。这里要考虑继代培养的时间超过12～22个月后，可能导致开花率降低以及一些植株的异常表现，如植物活力的降低，叶和花的变异率增加等（Beruto et al. 1996）。在花色方面，通过无性系繁殖，没有明显变异，但需要注意，与无性繁殖母本相比，后代观赏性状中有条纹或特殊的细微差别（Biancheri Creations，个人交流）。

我们的组织培养系统基于一个集成的生产计划，它保证了健康的无病毒原种可大规模生产（Beruto and Debergh 2004；Beruto 2010）。我们考虑用腋芽接种，芽取自成熟的母株或发芽的块根（Beruto 2010）。根据我们的操作规程，不同基因型和供体母本材料生理状态的差异会有不同的污染比例，为10%～80%。一般来说，当采摘成熟植株中花蕾时最好在开花前1个月采集外植体。从发芽的块根上取芽时，要考虑在什么条件下取，如何形成的芽，并小心取芽。我们于12℃、光强50μmol/(m^2·s)、光照8h条件下在复水块根上取得芽的效果最好（图25.5）。

增殖阶段可以在MS培养基上进行（Murashige and Skoog 1962），其中添加生长调节剂6-BA（6-苄基腺嘌呤）、激动素和NAA（萘乙酸）。在条件成熟的实验室中培养，每4周可产生1.5～4.5个新芽，具体取决于不同的基因型（Beruto and Debergh 2004），但为了达到满意的繁殖效率，应考虑不同的因素。水分条件、营养素配比和琼脂杂质都可能影响组织培养材料（Beruto 1997；Beruto and Curir 2006；Beruto et al. 2001）。改变接种植物的密度和更新培养基的次数会导致组织培养材料质量的不同（Beruto and Portogallo 2000）。培养基中糖的类型是另外一个影响花毛茛组培苗生长的因素，果糖是一种较为理想的糖原（Beruto and Portogallo 2000）。此外，盐配方和蔗糖浓度对离体培养植株的开花具有约束作用（Beruto and Rinino 2009）。

图 25.5 利用腋芽诱导的花毛茛组织培养。(a) 带有芽点的块根准备进行体外接种；(b) 收集健壮植株以获取腋芽（箭头）；(c) 体外繁殖阶段；(d) 组织培养试管苗生根

在培养周期结束时形成了块根。当白天温度较低、日照较短时，光热周期特性使花毛茛还要进行短暂的实验室驯化处理，时间一般在 2～3 个月。因此，在瓶内组培过程中使用生长延缓剂。我们的实验中，在繁殖和生根阶段的 3 个月内，温度升高对花毛茛离体生长有抑制作用，成熟芽和微小芽可以在 2℃下保存（Beruto et al. 2011）。Benelli 等（2012）发现，如果在补充 60g/L 蔗糖、温度范围为 6～10℃条件下，花毛茛体细胞的贮藏可以达到令人满意的水平。

在地中海地区，可以用无纺布覆盖试管苗驯化一周。但是，如果这个阶段是在温暖的秋季进行，则应予以注意（9～10 月），在此期间，建议在塑料大棚中并在遮阴网下进行驯化。7～12 天后，拉开棚膜，并保证一定的湿度（图 25.6）。

图 25.6 (a) 练苗末期；(b) 使用带遮阴网的塑料大棚进行组培苗练苗

Borriello 等（2017）提供了在驯化阶段使用 AMF 的新信息。他在体外证明了花毛茛试管苗在驯化期可被 AMF 定植，但转移到田间时，真菌种类对植物的影响是不同的。

25.4 有性繁殖和开花生物学

花毛茛的茎高达 30～45cm，有大而明亮的多种颜色的花（白色、黄色、粉色、金色、橙色、红色和混合）。这些花序在种植后大约 5 个月出现。花辐射对称，包含卵形萼片和倒卵形花瓣。其雄蕊数目变化很大：有时雄蕊发育良好；有时没有雄蕊。花瓣可以是花药的残余或者是由花药到花瓣的中间花器官演化而来（图 25.7）。显然，在这个种类中，花瓣和雄蕊可以相互转化（Ronse Decraene and Smets 1995；Tamura 1995b；Yuan and Yang 2006）。

图 25.7 花毛茛雄蕊有时和花瓣融为一体的中间花器官（标尺 = 1mm）

球形和多孔花粉粒在果实纵裂后散落（Dhooghe et al. 2012）。许多单胚珠心皮排列在一个发育良好的花托上，在花期时，膨胀形成种子。柱头具单细胞乳突，干燥（图 25.8）。这种果称为有翅瘦果。当成熟的果实脱落时，胚胎尚未完全发育；花毛茛的胚胎处在鱼雷胚时期（图 25.9）。

在育种项目的有性生殖方面，了解花粉质量、柱头质量和花柱的性状是非常重要的。Dhooghe（2009）对花毛茛花粉萌发进行了检测。花毛茛品种‘阿尔法’的花粉在含 100mg/L H_3BO_3、100mg/L $CaCl_2{\cdot}2H_2O$、100mg/L KH_2PO_4 和 150g/L 蔗糖（pH 5.4）的花粉萌发培养基上整体表现良好，发芽率为 36.0%±5.1%（Dhooghe 2009）。花粉密度对其萌发和花粉管延伸长度没有显著影响，但观察到在最高花粉密度 2mg/mL 时呈负趋势。添加水解酪蛋白（CH，1000mg/L）可能有利于‘Alfa’花粉萌发（Dhooghe 2009）。另

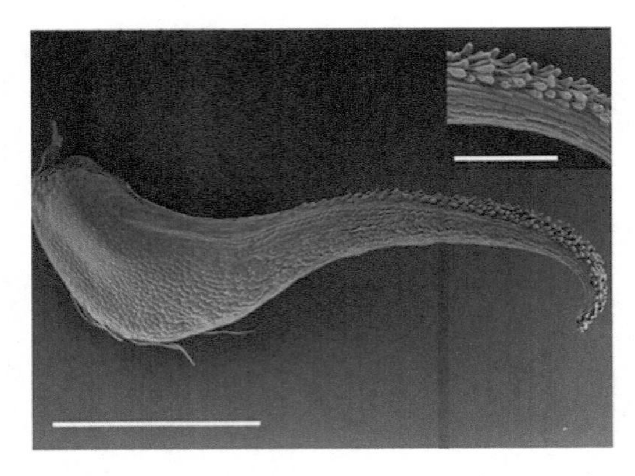

图 25.8　花毛茛品种‘Krisma’心皮（标尺= 1mm），柱头具单细胞乳突（内图，标尺 = 200μm）

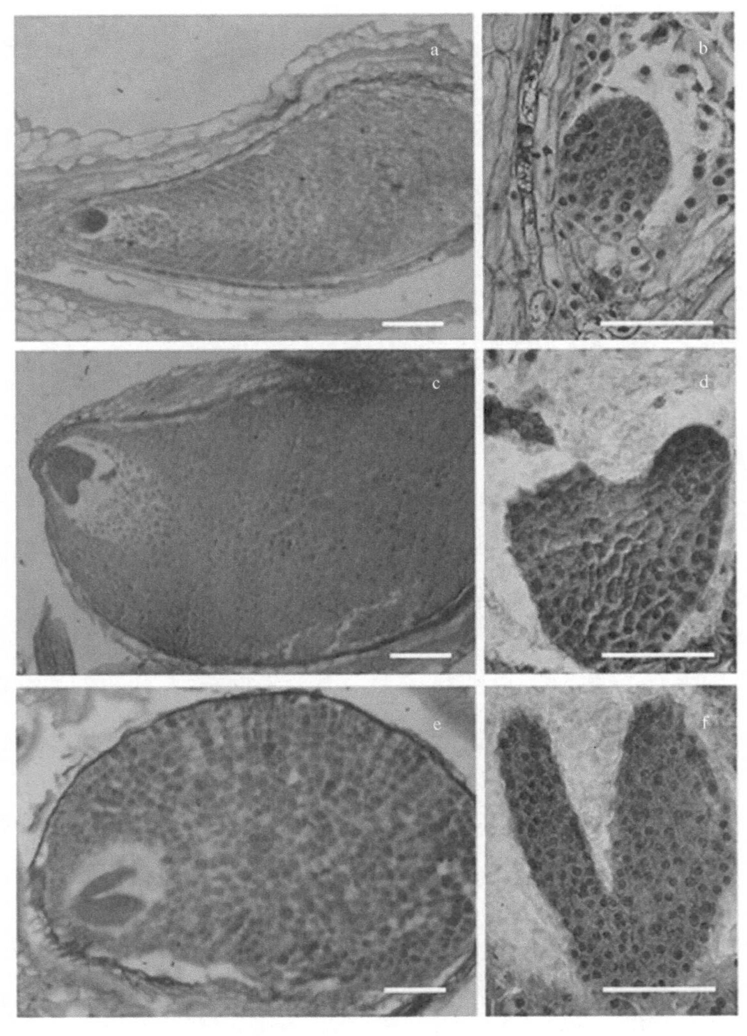

图 25.9　花毛茛杂交品种‘Krisma’×‘Bianco Strié’子代中的胚胎发生（a、c、e 标尺 = 500μm；b、d、f 标尺 = 200μm）。（a 和 b）授粉 4 周后绿色幼果中球形胚；（c 和 d）授粉 5 周后的心形胚；（e 和 f）授粉 8 周后成熟小果中鱼雷胚

一种改善花粉萌发和增加花粉管长度的方法是将钙浓度从100mg/L增加至300mg/L，将糖浓度从150g/L增加至200g/L。花粉萌发的最适pH范围是5.5～6.5。在最低pH（4.5）时，'Alfa'花粉管的萌发受到抑制（Dhooghe 2009）。所有这些实验都优化了培养基，配方为100mg/L H_3BO_3、300mg/L $CaCl_2·2H_2O$、100 mg/L KH_2PO_4、1000mg/L CH和200g/L糖（pH 6.5）。这种培养基可以模拟与花粉在可亲和柱头上萌发相似的条件。

同时对雌花最易授粉的时期进行了研究。我们将花的发育分为6个时期(图25.10)。用苯胺蓝染色证实了在花毛茛花期4时花粉管与花柱总长度的比例最佳，花毛茛中雄蕊在花期5时成熟。雄蕊成熟在心皮（原雌蕊花）最易授粉时期后。花毛茛雄雌蕊成熟时间相差4～14天。雌蕊在开花3～5天可以授粉，花药暴露花粉2～3天。通过苯胺蓝染色结果推断，受精可能发生在授粉后3～4天。花毛茛授粉的时间与温度有关，如果授粉是在3月或4月完成，授粉时气温分别为47（T_{sum} 681℃）和36（T_{sum} 727℃）。

图25.10　花毛茛开花期的研究（成花后的区别）。(a) 花芽可见，但直径不超过0.5cm（花期1）；(b) 花蕾不再被萼片覆盖，但最终花瓣颜色不显现（花期2）；(c) 外部花瓣显现最终颜色（花期3）；(d) 花开放（开花期），但花药仍未成熟（花期4）；(e) 花药释放花粉（花期5）；(f) 花枯萎（花期6）

25.5　受精障碍

为了了解受精障碍，我们使用三个花毛茛品种（'Alfa'、'Bianco Strié'和'Krisma'）作为实验材料，它们是通过自花授粉或异花授粉结实的。这些实验清楚地显示了两个品种（'Bianco Strié'和'Krisma'）通过自花授粉或异花授粉方式都能有效地获得幼果，证明没有任何受精障碍。在异花授粉时，‘Alfa’是一个很好的传粉者或授粉者（图 25.11）。然而，‘Alfa’在自花授粉时并没有形成幼果，或偶尔每次授粉形成 1～5 个幼果。苯胺蓝染色显示‘Alfa’在自花授粉时花粉管的扭曲抑制了其在花柱中的生长，这让我们得出结论：‘Alfa’是自交不亲和的。通过石蜡切片和流式细胞仪对‘Alfa’×‘Alfa’的少量幼果进一步研究发现：每个胚珠都形成一个胚胎，但授粉 5 周后，胚胎发育仍处于球形胚阶段（图 25.12d），而自交亲和的杂交后代的胚胎已经达心脏胚期（图 25.9）。这表明自交不亲和的花毛茛胚胎的生长在这个阶段被抑制。使用‘Alfa’×‘Bianco Strié’异花授粉的幼果作为对照，通过流式细胞仪检测了自花授粉‘Alfa’子代的倍性和‘Alfa’的叶组织。叶组织产生了一个清晰的 2*x* 峰值（图 25.12a），而对照的幼果为 2*x*（胚胎和种皮）和 3*x*（胚乳）峰值（图 25.12b）。自花授粉的‘Alfa’幼果产生了 2*x* 和 4*x* 峰值（图 25.12c）。这些结果让我们怀疑‘Alfa’发生了无融合生殖：2*x* 峰值可以代表没有受精的胚胎和种皮，而 4*x* 峰值可能是二倍体细胞分裂或中心细胞增殖加倍的结果。在花毛茛中观察到配子体发生了无融合生殖现象（Nogler 1984）。在这个物种中，两个精子使两个极核受精，形成正确的亲本胚乳 4m：2p 的基因组比（Koltunow and Grossniklaus 2003）。

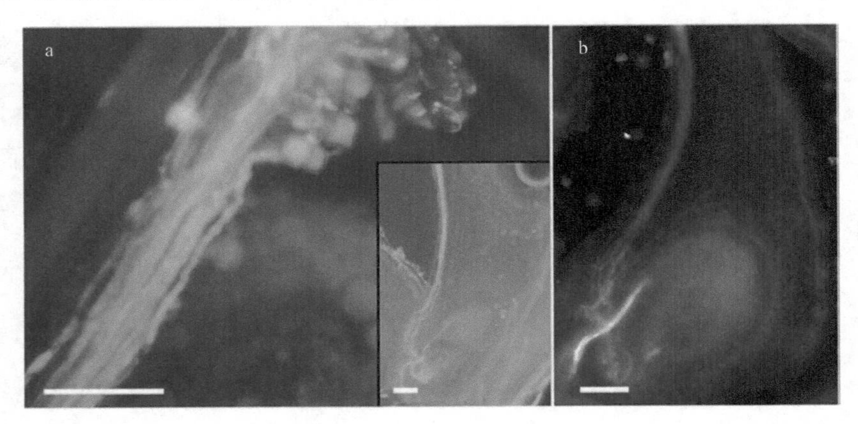

图 25.11 花毛茛授粉 56h 后用苯胺蓝染色，观察花粉管生长（标尺 = 100μm）。(a) 杂交组合‘Krisma’×‘Alfa’花粉管生长情况，花粉管已到达胚珠的珠孔末端（插图）；(b) 杂交组合‘Alfa’×‘Bianco Strié’无抑制花粉管生长情况

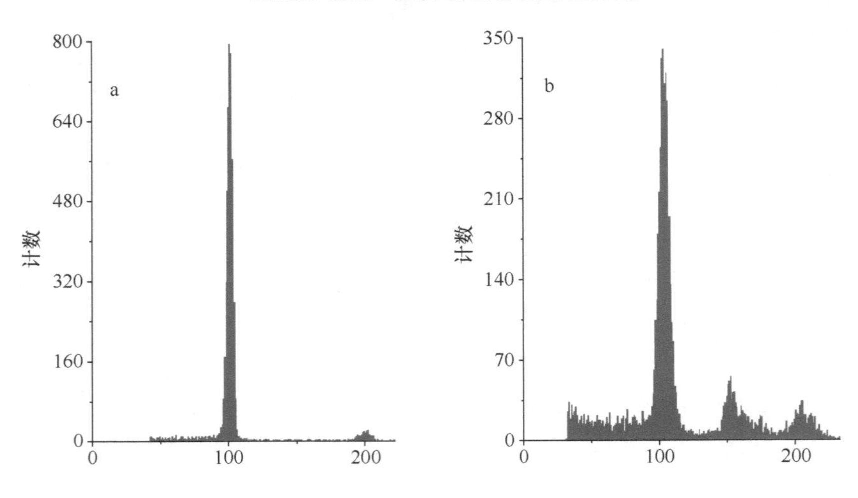

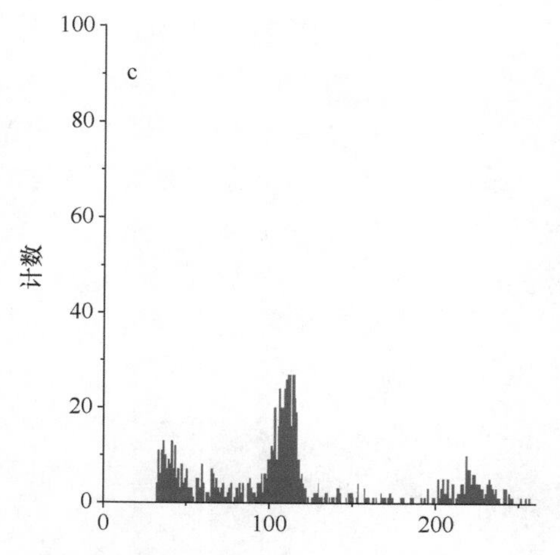

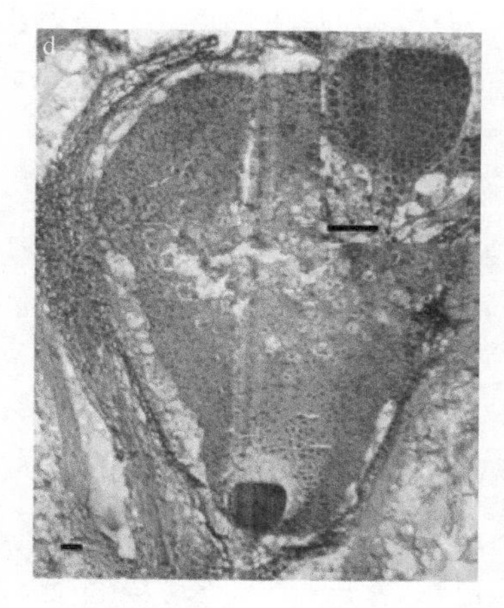

图 25.12　花毛茛的流式细胞仪分析与石蜡切片观察。（a）流式细胞仪测定‘Alfa’叶组织，2*x* 峰定为 100；（b）流式细胞仪分析以‘Alfa’×‘Bianco Strié’杂交组合为参照来测定未成熟幼果，出现 2*x* 峰值（胚和种皮），在胚乳中出现 3*x* 峰值，4*x* 峰值是由二倍体细胞分裂所致；（c）流式细胞仪测定‘Alfa’×‘Alfa’杂交所获得的未成熟幼果，显示 2*x* 峰值（胚胎、种皮）和 4*x* 峰值（二倍体细胞分裂和/或增殖加倍的中央细胞）；（d）授粉 5 周后‘Alfa’×‘Alfa’种子的石蜡切片，显示了一个停止生长的球形胚（内图标尺 = 100μm）

25.6　细胞学和倍性育种

花毛茛的基本染色体数目为 8，在二倍体基因组中产生 16 条 R 型染色体。毛茛属植物的基因组大小变化很大：细胞核 DNA 含量（2C）在 3.9（*R. lateriflorus*）～51.6（*R. lingua*）pg（Smith and Bennett 1975）。Zonneveld 等（2005）检测了一个毛茛属植物的基因组大小（2C）为 16.5pg，属于花毛茛。总染色体长度从 21.5μm ± 2.1μm 至 14.3μm ± 3.0μm，得到了 8 条具中着丝粒、6 条具近中着丝粒和 2 条具近端着丝粒染色体的核型（Dhooghe 2009）。这与以前 Baltisberger 和 Widmer（2005）发表的核型略有不同。

在 F_1 杂交后代中已经成功地验证了可使倍性改变的技术（Ferrero et al. 2006），可通过花粉培养产生纯合二倍体，以及使用秋水仙碱、氨磺乐灵和氟乐灵等抗有丝分裂剂对组织培养材料进行多倍体育种（Dhooghe et al. 2009）。尽管所有抗有丝分裂药物均成功，但在氟乐灵浓度为 2μmol/L，固体培养 8 周后，多倍体化比例是最高的。

25.7　花毛茛商业育种目标和前景展望

花毛茛的商业育种主要在荷兰、法国、意大利和日本。其都参与了培育和繁殖具有特点的、市场潜力大的盆花生产项目。

25.7.1　切花

当花毛茛作为切花栽培时，其花的性状是最重要的。因此，商业育种项目主要集中在花的直径、形态和颜色。最近，特别有趣的是经过长时间的育种和组织培养研发后，由 Biancheri Creations 开发了名为 PONPON 的品系，该品系花瓣双色且呈锯齿状。多样的、奇特的花形和花色使得这一品系在市场上越来越受到重视。Hyères（法国）收集的数据显示，这一品系的花毛茛试管苗约占 30%，而且其影响越来越大（Biancheri Creations，个人交流）。具新颜色和新花形的品系，如有绿色中心线条的花毛茛品系也已为客户提供（图 25.13）。虽然花毛茛的花色已经非常丰富，但还是没有蓝色的。如果能开发出蓝色的花毛茛，很可能会打开新市场并创造新的机遇。种间杂交可以改善一些农艺性状特征，如 *R. asiaticus* × *R. cortusifolius* 这个组合能够形成大的叶片，并使花瓣具有香味（Biancheri Creations，个人交流）。另一个有趣的特性可能是使不同的花枝上开出不同的花，这个是全世界的育种家都在努力的方向。

图 25.13　花毛茛 PONPON 系列（a～c）、绿心系列（d～f）
（源于 Biancheri Creations，Camporosso，IM，Italy）

适合切花生产也意味着必须使产量、花枝长度、成熟期和花的瓶插寿命都达到一定的要求。如前所述，不同品种切花的衰老程度存在显著差异，评为“绿色”品种意味着瓶插寿命最长。

除此之外，花毛茛对白粉病、假单胞菌属细菌性疾病和尖孢镰刀菌尤为敏感。因此，新品种必须具有良好的抗病性。

在许多国家，利润最高的生产是通过块根的种植来实现的，因此，获得易于操作的具有良好贮藏性的块根品种对生产非常有利。

25.7.2　盆花

除了切花市场，花毛茛也作为盆栽植物销售。因此，另外一些育种标准也很重要，如紧凑型植株造型和长的花期。对于盆栽植物来说颜色和开花时间是很重要的。另一种在市场上观察到的趋势是生产无须生长调节剂的紧凑型品系。在监管越来越严的条件下，可以解决由生长调节剂使用限制而导致的生产问题。

这些育种目标是通过花毛茛亲本杂交及后代的选择来完成的。由于可以使用理想的雄性或雌性作为亲本，因此引进体外繁殖技术可以大大增加种子的生产量。在继代培养中植物材料表现出变异性、花粉减少，这些都应该引起重视。

25.8　花毛茛和银莲花的种间杂交

为了扩大遗传基因库，Dhooghe（2009）研究了花毛茛和银莲花的种间杂交。虽然发现花毛茛和银莲花杂交后花粉的捕获、黏附、水化和萌发都正常，但在花柱中花粉管生长有复杂的受精前障碍。这些杂交种的种子败育也暗示着存在受精后障碍。为了防止败育，在授粉后 3～5 周收获种子，利用组织培养进行胚挽救。

从杂交后代的石蜡切片上可以看出，银莲花‘Mistral Wine’× 花毛茛‘Alfa’授粉 20 天后种子的发育仍然正常，杂交后代的发育处于球形胚时期（图 25.14a）。然而，授粉后 34 天，花毛茛‘Krisma’× 银莲花‘Wicabri Blue’的胚胎停止生长（与图 25.9 相比）并停留在球形胚时期（图 25.14b）。

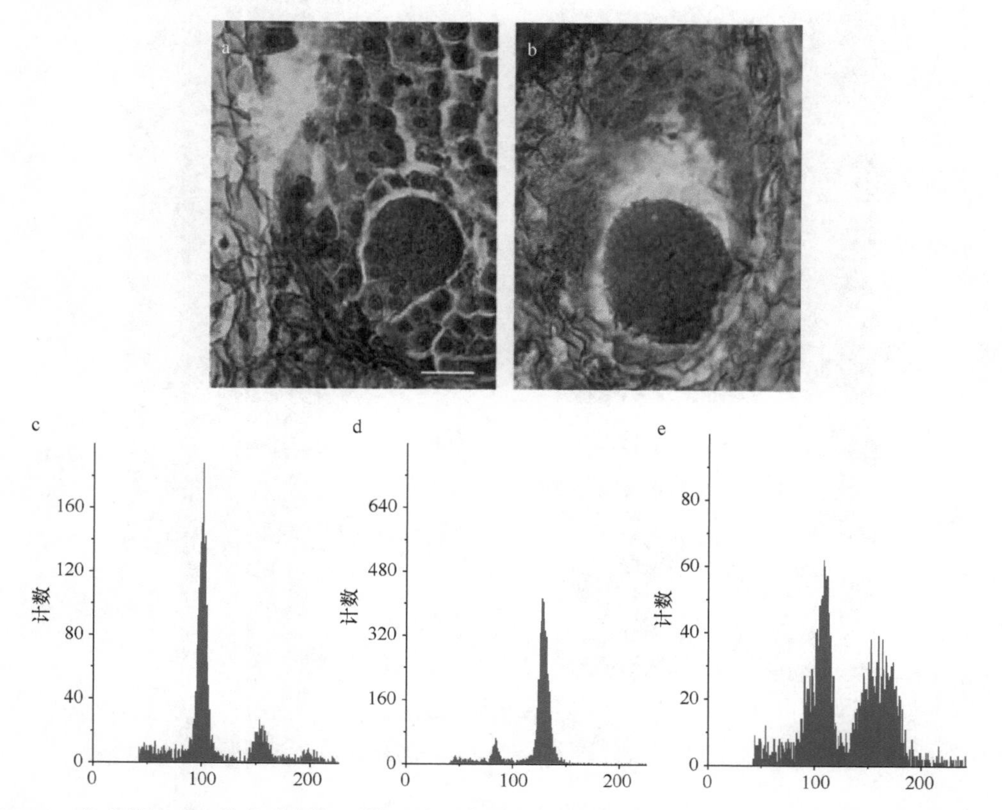

图 25.14　花毛茛流式细胞仪分析与石蜡切片观察（标尺 = 100μm）。(a) 石蜡切片观察，银莲花‘Mistral Wine’× 花毛茛‘Alfa’杂交种子，授粉 20 天后显示正常胚胎发育。(b) 石蜡切片观察，花毛茛‘Krisma’× 银莲花‘Wicabri Blue’杂交种子，授粉 34 天后显示生长停止在球形胚阶段。(c) 流式细胞仪分析，‘Mistral Fucsia’×‘Mistral Fucsia’杂交，显示相对较高的 $2x$ 峰值（胚胎、种子）和较低的 $3x$ 峰值（胚乳）。(d) 流式细胞仪分析，银莲花‘Wicabri blue’×‘Wicabri blue’杂交，显示相对较低的 $2x$ 峰值（胚胎、种皮）和较高的 $3x$ 峰值（胚乳）。(e) 流式细胞仪分析，银莲花‘Mistral Wine’× 花毛茛‘Alfa’杂交，也显示了一个 $2x$ 峰值（胚胎、种皮）和 $3x$ 峰值（胚乳），这表明该组合发育正常

当用流式细胞仪分析这些幼果时，出现了一个可清楚区分的 $2x$ 和 $3x$ 峰值（图 25.14e）。与对照相同时期正常幼果的轮廓相似，但未成熟和成熟幼果之间存在差异（图 25.14c，d）。未成熟的果实有一个显著的 $2x$ 峰值和一个较低的 $3x$ 峰值，而成熟的果实显示出一个相对较高的 $3x$ 峰值和较低的 $2x$ 峰值（图 25.14）。

在花毛茛和银莲花杂交后代中，花毛茛的某些特征转移到银莲花上，可以获得具更多花瓣化萼片的花（图 25.15）。尽管父本的贡献率较低，但采用 AFLP 证实大多数杂交后代含有母本和父本基因（Dhooghe et al. 2012）。通过基因组 GISH 分析和染色体计数，证实了 F_1 代含有不同染色体数目，主要是母本的染色体。

图 25.15　杂交后代及其代表亲本。银莲花‘Wicabri Blue’B19 × 花毛茛‘Alfa’，银莲花‘Mistral Fuchsia’F_5 × 花毛茛‘Alfa’，其 F_1 后代（B19A_231、F5A_450）具有更多花瓣化萼片

25.9　结　　论

花毛茛的育种仍主要通过种内杂交进行。组织培养可以提高育种效率。前人的论文表明：花毛茛倍性育种技术是比较成功的，但据我们所知，在商业运用中还存在一定的问题。目前，传统的育种技术仍然能够提供足够的变异（如 PONPON 系），但是研究银莲花和花毛茛之间的杂交表明：作为种间杂交，通过扩大繁殖群体，使用更多技术及胚

挽救来引入新的基因型是可行的。这些技术需要关注更多的问题，如花粉活力、受精障碍、自交不亲和等。

希望今后在育种上，分子标记辅助选择、抗病性筛选和重要性状遗传等可以发挥更大的作用。

本章译者：

贾文杰

云南省农业科学院花卉研究所，昆明 650200

参 考 文 献

Baltisberger M, Widmer A (2005) Cytological investigations on some *Ranunculus* species from Crete. Candollea 60:335–344

Benelli C, Ozudogru EA, Lambardi M, Dradi G (2012) *In vitro* conservation of ornamental plants by slow growth storage. Acta Hort 961:89–93

Beruto M (1997) Agar and gel characteristics with special reference to micropropagation systems of *Ranunculus asiaticus* L. PhD Thesis, Gent, Faculteit Landbouwkundige en Toegepaste Biologische Wetenschappen, University Gent, Belgium, pp 21–44, 125–184

Beruto M (2010) *In vitro* propagation through axillary shoot culture of *Ranunculus asiaticus* L. In: Jain SM, Ochatt SJ (eds) Protocols for *in vitro* propagation of ornamental plants, methods in molecular biology, vol 589. Humana Press, New York, pp 29–37

Beruto M, Curir P (2006) Effects of agar and gel characteristics on micropropagation: *Ranunculus asiaticus*, a case study. In: Teixeira da Silva JA (ed) Floriculture, ornamental and plant biotechnology advances and topical issues, vol II. Global Science Books, Isleworth, pp 277–284

Beruto M, Debergh P (2004) Micropropagation of *Ranunculus asiaticus*: a review and perspectives. Plant Cell Tiss Org 77:221–230

Beruto M, Portogallo C (2000) Factors affecting the growth of *Ranunculus* in vitro. Acta Hort 520:163–170

Beruto M, Rinino S (2009) Effetto del substrato sulla fioritura in vitro di piante di ranuncolo. Italus Hortus 16(2):32–34. (Abstract in English and figure captions in English)

Beruto M, Cane G, Debergh P (1996) Field performance of tissue-cultured plants of *Ranunculus asiaticus* L. Sci Hortic 66:229–239

Beruto M, La Rosa C, Portogallo C (2001) Effects of agar impurities on *in vitro* propagation of *Ranunculus asiaticus* L. Acta Hort 560:399–402

Beruto M, Fibiani M, Rinino S, Lo Scalzo R, Curir P (2009) Plant development of *Ranunculus asiaticus* L. tuberous roots is affected by different temperature and oxygen conditions during storage period. Israel J Plant Sci 57:377–388

Beruto M, Rinino S, Bisignano A, Fibiani M (2011) Study of slow growth conditions of *Ranunculus in vitro* shoots. Acta Hort 908:391–403

Borriello R, Maccario D, Viglione S, Bianciotto V, Beruto M (2017) Arbuscular mycorrhizal fungi and micropropagation of *Ranunculus asiaticus* L.: a useful alliance? Acta Hort 1155:549–555

Cerveny CB, Miller WB (2011) Storage temperature and moisture content affect respiration and survival of *Ranunculus asiaticus* dry tuberous roots. Hortscience 46(11):1523–1527

Cerveny CB, Miller WB (2012) Soaking temperature of dried tuberous roots influences hydration kinetics and growth of *Ranunculus asiaticus* (L.). Hortscience 47(2):212–216

De Hertogh AA (1996) Holland bulb forcer's guide, 5th edn. Alkemade Printing BV, Lisse

Dhooghe E (2009) Morphological and cytogenetic study of ornamental Ranunculaceae to obtain intergeneric crosses. PhD thesis, Ghent University, p 244

Dhooghe E, Denis S, Eeckhaut T, Reheul D, Van Labeke MC (2009) *In vitro* induction of tetraploids in ornamental *Ranunculus*. Euphytica 168:33–40

Dhooghe E, Reheul D, Van Labeke MC (2012) Cytological and molecular characterization of intertribal hybrids between the geophytes *Anemone coronaria* L. and *Ranunculus asiaticus* L. (*Ranunculaceae*). Floriculture and Ornamental Biotechnology 6:104–107

Emadzade K, Gehrke B, Linder HP, Hörandl E (2011) The biogeographical history of the cosmopolitan genus *Ranunculus* L. (*Ranunculaceae*) in the temperate to meridional zones. Mol

Phylogenet Evol 58:4–21

Ferrero F, Duclos A, Ottenwaelder L, Thiebaut MJ, Jacob Y (2006) Use of homozygosity in the *Ranunculus asiaticus* breeding process. Acta Hort 714:119–128

Hassan HA, Agina EA, Koriesh EM, Mohamad SM (1985) Physiological studies on *Anemone coronaria* L and *Ranunculus asiaticus* L. 3. Effect of foliar nutrition and gibberellic acid. Ann Agr Sci- Moshtohor 22:593–615

Kamenetsky R, Peterson RL, Melville LH, Machado CF, Bewley D (2005) Seasonal adaptations of the tuberous roots of *Ranunculus asiaticus* to desiccation and resurrection by changes in cell structure and protein content. New Phytol 166:193–204

Karlsson M (2003) Producing ravishing *Ranunculus*. Greenhouse Product News, January 2003: 44–48

Kenza M, Umiel N, Borochov A (2000) The involvement of ethylene in the senescence of *Ranunculus* cut flowers. Postharvest Biol Technol 19(3):287–290

Koltunow AM, Grossniklaus U (2003) Apomixis: a developmental perspective. Annu Rev Plant Biol 54:547–574

Meynet J (1974) Research on *Ranunculus* hybrid varieties for growing under protection. Acta Hort 43:191–196

Meynet J (1993) Ranunculus. In: De Hertogh A, Le Nard M (eds) The physiology of flower bulbs. Elsevier Science Publishers, Amsterdam, pp 603–610

Meynet J, Duclos A (1990) Culture *in vitro* de la renoncule des fleuristes (*Ranunculus asiaticus* L.). II. Production de plantes par culture d'anthères *in vitro*. Agronomie 10:213–218

Murashige T, Skoog F (1962) A revised medium for rapid growth and bioassays with tobacco tissue cultures. Physiol Plant 15:473–497

Nogler GA (1984) Genetics of apospory in apomictic *Ranunculus auricomus*. V. Conclusion. Bot Helv 94:411–422

Ohkawa K (1986) Growth and flowering of *Ranunculus asiaticus*. Acta Hort 177:165–177

Ronse Decraene LP, Smets EF (1995) Evolution of the androecium in the Ranunculiflorae. Plant Syst Evol 59:63–70

Scariot V, Larcher F, Caser M, Costa E, Beruto M, Devecchi M (2009) Flower longevity in ten cultivars of cut *Ranunculus asiaticus* L. as affected by ethylene and ethylene inhibitors. Europ J Hort Sci 74(3):137–142

Shahri W, Tahir I (2011) Flower development and senescence in *Ranunculus asiaticus* L. J Fruit Ornamental Plant Res 19(2):123–131

Smith JB, Bennett MD (1975) DNA variation in the genus *Ranunculus*. Heredity 35:231–239

Tamura M (1995a) Systematic part. In: Engler A, Prantl K (eds) Die Natürlichen Pflanzenfamilien. Bd. 17 a IV Angiospermae. Ordnung Ranunculales. Fam. *Ranunculaceae*, 2nd edn. Duncker & Humblot, Berlin, pp 220–519

Tamura M (1995b) Reproductive structures. In: Engler A, Prantl K (eds) Die Natürlichen Pflanzenfamilien. Bd. 17 a IV Angiospermae. Ordnung Ranunculales. Fam. *Ranunculaceae*, 2nd edn. Duncker & Humblot, Berlin, pp 41–76

Tian HQ, Russell SD (1997) Micromanipulation of male and female gametes of *Nicotiana tabacum* .1. Isolation of gametes. Plant Cell Rep 16:555–560

Turina M, Ciuffo M, Lenzi R, Rostagno L, Mela L, Derin E, Palmano S (2006) Characterization of four viral species belonging to the family *Potyviridae* isolated from *Ranunculus asiaticus*. Phytopathology 96:560–566

Umiel N & Hagiladi A (1999) Preparation of *Ranunculus* corms for early flowering in 8 easy steps. Dept Orn Hort Agr Res Org, Bet-Dagan, Israel

Valdez-Aguilar L, Grieve CM, Poss J (2009) Hypersensitivity of *Ranunculus asiaticus* to salinity and alkaline pH in irrigation water in sand cultures. Hortscience 44(1):138–144

Yuan Q, Yang Q (2006) Tribal relationships of *Beesia*, *Eranthis* and seven other genera of *Ranunculaceae*: evidence from cytological characters. Bot J Linn Soc 150:267–289

Zonneveld BJM, Leitch IJ, Bennett MD (2005) First nuclear DNA amounts in more than 300 angiosperms. Ann Bot 96:229–244

第26章　高 山 杜 鹃

Stephen L. Krebs

摘要　杜鹃花属（*Rhododendron*）包括常说的高山杜鹃（rhododendrons）和杜鹃花（azaleas）在内的800余种丰富的观赏物种。在过去的200年中，西方植物育种家利用这些自然物种作为材料，将其观赏性和适应性进行重组，现已培育出25 000余个品种。在此过程中，以爱好者和收集者为主的育种人更多因个人兴趣，而非以市场需求或获得商业报酬为目的，进行着以美学属性为主的杂交育种。本章提请读者注意杜鹃花植物在野生环境中的复杂多样性，能为培育更具适应性，在更具挑战性条件下表现良好的植物品种提供育种材料。提高高山杜鹃和杜鹃花对病原菌与虫害的抗性，并增强其对非生物胁迫（如极端温度和湿度、碱性土壤或高盐度）的耐受性，还将使其更适于景观种植，并适合更广阔的市场，从而使园艺事业受益。

关键词　胁迫耐受性；多倍体；耐寒性；耐热性；耐碱性；盐胁迫；疫病抗性；根腐病抗性；气候变化

26.1　高山杜鹃王国的介绍

高山杜鹃（rhododendrons）被称为灌木之王。有人可能会提出其他明智的说法，但实际上大多数人认为杜鹃花属的观赏多样性超出了现有大多数的其他流行木本园林植物。在一个由50～100种高山杜鹃和其相近种——杜鹃花类组成的温区花园里，能够展示植物的丰富多样性，包括从紧凑、密集的植物灌丛到具高大树形的植物的生长习性，以及具有不同大小、形状、短柔毛和香气的叶子。在丰富的物种收集中，其花色将包括红色、橙色、黄色、紫色和紫罗兰色，加上白色。花序或通常所称的花簇，可小而不规则，或由12～20朵花组合成优美的球形或金字塔形。因物种不同，花冠可能为合瓣或是离瓣，有些花冠内还具有色彩艳丽的彩斑。如果精心设计与搭配，在花园里加入其他物种，还将会给花园增添芬芳，并可确保从早春到夏末的持续花期。杜鹃花属（*Rhododendron*）是一个引人注目的属。尽管绝大多数园丁不会像刚刚描述的那样，去收集大量珍品，但高山杜鹃和杜鹃花确实是很多冷、温带花园里最受喜爱的多年生植物及重要的造园植物。

物种内和物种间的差异只是多样性冰山的一角。两个多世纪以来，西方育种人一

S. L. Krebs (✉)
The Holden Arboretum, Kirtland, OH, USA
e-mail: skrebs@holdenarb.org

直在进行种间异花授粉，对自然多样性进行重组并获得了无数新类型。现在有 25 000 多个高山杜鹃和杜鹃花品种被记录于国际杜鹃花属名册和登记表中（Leslie 2004）。对于园艺产业来说，其品种数量已过多，但是这展现出了人们对这些植物的广泛热爱。高山杜鹃的杂交历史有众多“后院”育种人，这些育种人都是有杜鹃花知识的收集者，他们喜欢收获杂交种子，播种并进行培育，然后对表现最好的植株进行选择并予以命名。

杜鹃花属系属杜鹃花科（Ericaceae），通常被称为荒地植物家族。杜鹃花科在世界范围内拥有 124 个属，包括许多重要的观赏植物和农作物，如 *Vaccinium*、*Gaylussacia*、*Calluna* 和 *Erica*、*Kalmia*，*Pieris*、*Oxydendrum*、*Arctostaphylos* 和 *Enkianthus*。几乎所有的杜鹃花科植物都喜欢长于酸性土壤中，通常被称为嗜酸植物。在花园种植杜鹃花时，通常建议使用 pH 4.5～5.5 的土壤（Leach 1961）。因为酸性条件与关键营养元素（N、P、Ca、Mg）的可利用性降低有关。杜鹃花科植物的根系统与接触其根生长的真菌共生，真菌向寄主植物提供土壤养分，以换取其进行光合作用的碳水化合物。荒地植物还具有许多常绿树种的特征，即通过保留获取的营养来适应贫瘠的土壤（Aerts 1995）。同其他杜鹃花科植物一样，杜鹃花属的根系形成了一个细密的纤维垫，分布于几英寸（1 英寸=2.54cm）的土表上。这一特征决定其需要土壤具备良好的排水性，同时也决定其不具较强的耐旱能力。

杜鹃花属（*Rhododendron*）是个物种具有多样形态的属，其分类学很复杂，经过了几次重大修订。物种数量在 800～1000，具体数量取决于所使用的分类方法。称为“杜鹃花（azaleas）”的植物也包含在该属中，尽管该词常从“高山杜鹃（rhododendrons）”中分离出来，而导致对两者关系的误解。以前，杜鹃花曾被认为是一个单独的属，但现在，根据最新的修订，杜鹃花属被分为 2 个亚属和 4 个组（表 26.1）。

表 26.1　基于 RPB2 分子进化分析的杜鹃花分类法（Goetsch et al. 2005）

分支	亚属	组	种数	描述
A	杜鹃花亚属 *Rhododendron*	杜鹃花组 *Rhododendron*[a]	149	小叶、常绿或有鳞片的杜鹃花（鳞片在叶片下面）
		髯花杜鹃花组 *Pogonatum*	13	
		类越橘杜鹃花组 *Vireya*	300	
	长蕊杜鹃花亚属 *Choniastrum*		11	
B	常绿杜鹃花亚属 *Hymenanthes*	常绿杜鹃花组 *Ponticum*[a]	200	大叶、常绿或无鳞片的杜鹃花（无叶鳞）
		羊踯躅组 *Pentanthera*[a]	23	落叶杜鹃花（无叶鳞）
C	马银花亚属 *Azaleastrum*	映山红组 *Tsutsusi*[a]	81	绝大多数常绿杜鹃花（无叶鳞），部分十花药杜鹃花组中的落叶种类
		十花药杜鹃花组 *Sciadorhodion*	6	
		马银花组 *Azaleastrum*	4	
原始姐妹类群	叶状苞杜鹃花亚属 *Theorhodion*		2	

a. 在温带地区具有园艺价值并应用的主要组类

杜鹃花的 4 个组与园艺应用的 4 个主要组相对应，每个组皆具有如表 26.2 所示的独有特征和适应性。

表 26.2　应用于温带花园中的杜鹃花组的园艺性状描述

<table>
<tr><th>描述</th><th>例子</th></tr>
<tr><td>杜鹃花组 Section Rhododendron（杜鹃花亚属 subgenus Rhododendron）

有鳞（叶有鳞、常绿）杜鹃花　　通常为早开花植物，花色在白-黄-红-紫-紫蓝色范围内。生长习性为紧凑的矮灌丛，特别是当其在高光照条件下生长时。茎和叶通常具芳香气味，这些组织中存在的萜烯产生了如松树般的气味。来自腺叶鳞片的挥发性萜烯与防御诸如黑藤象鼻虫等昆虫有关（Doss 1984）
毛肋杜鹃花（右图）是已知的有鳞杜鹃花物种中唯一一个具有美丽紫蓝色花的物种（照片由 Susan Lightfoot 提供）</td><td>毛肋杜鹃花 R. augustinii
</td></tr>
<tr><td>常绿杜鹃花组 Section Ponticum（常绿杜鹃花亚属 subgenus Hymenanthes）

无鳞（叶无鳞、常绿）杜鹃花　　花期差异很大，从早春到夏季，取决于种类。花色与有鳞杜鹃花的种类重叠，没有紫罗兰色，但有纯黄色和红色。植物生长习性、叶片形状和叶片大小也有很大的不同，类型包括从低矮紧凑的植株到高达 30m 的乔木。杂种植物因其宽大的叶簇、硕大的花序（花簇）和完美外观而备受赞誉。尽管有些无鳞杜鹃花可以在充足的阳光下生长，尤其是在较高的海拔下，但是通常建议在花园中使用时进行局部遮阴以获得最佳效果
酒红杜鹃花（右图）盛开在北卡罗来纳州的罗恩山上。该物种是耐寒育种中最重要的物种之一（照片由 Don Hyatt 提供）</td><td>酒红杜鹃花 R. catawbiense
</td></tr>
<tr><td>羊踯躅组 Section Pentanthera（常绿杜鹃花亚属 subgenus Hymenanthes）

落叶杜鹃花的价值在于其出众的先花后叶特性。开花时间因种而异，可从早春（加拿大杜鹃花 R. canadense）至夏末（桃叶杜鹃花 R. prunifolium）。这个组包含从杜鹃花中发现的最纯正的橙色和黄色，以及白色、粉红色、猩红色和淡紫色。有些种类的花很香。植物生长习性或为紧凑的矮灌丛（某些物种为附生），或为直立且高大的乔木。最好置于强光下，以避免徒长并可确保良好花蕾的形成。
折萼杜鹃花（右图）是原产墨西哥湾沿岸地区的一种芳香物种，但能在美国北部（北美植物耐寒 5 区，最冷区）生长（作者照片）</td><td>折萼杜鹃花 R. austrinum
</td></tr>
<tr><td>映山红组 section Tsutsusi（马银花亚属 subgenus Azaleastrum）

常绿杜鹃花　　这个组里的种类原产中国东南部、日本和韩国。虽然许多生长在温暖气候中，但亦有一些能耐寒。在美国南部通常不能成功种植。常绿杜鹃花是非常艳丽和受人喜爱的春季开花植物。该类群的特征受人喜爱在于紧凑且密集的生长习性和对充足阳光的偏好。常绿杜鹃花的特征由其叶片二态性所致，春季叶片后是夏季叶片的生长，而夏季叶片的生长可以持续至下一个春季前的冬季。花不香，但通常非常大，花色有白色、粉红色、红色或紫色。
九州杜鹃花‘库尔山’（右图）是该物种中的一个双色花品种，原产日本南部，可耐-23℃的低温（作者照片）</td><td>九州杜鹃花‘库尔山’
R. kiusianum‘Komo Kulshan’
</td></tr>
</table>

杜鹃花属的物种遍布北半球，并横跨东南亚附近的赤道地区。在中国西南部和喜马拉雅山脉东部发现了杜鹃花组（sections *Rhododendron*）（有鳞杜鹃花）和常绿杜鹃花组（*Ponticum*）（无鳞杜鹃花）的大量物种。这是一个地形极端起伏的地区，高山山脉紧邻亚热带河谷，由此可能造成地理隔离，限制了基因流动并导致物种形成和局部适应（图 26.1）。尽管该地区被认为是杜鹃花属的起源中心，但有些人认为该地区先由原始物种组成，后则多样化（Irving and Hebda 1993）。整个东南亚出现另一种形态多样的物种聚集区，包括隶属于类越橘杜鹃花组（section *Vireya*）的约 300 种无鳞杜鹃和约 1/3 的杜鹃花属植物。在这个赤道气候带下，它们在较为凉爽的高海拔地区生存，有时也作为附生植物生长。绝大多数的落叶杜鹃花（deciduous azaleas）（羊踯躅组 section *Pentanthera* 和十花药杜鹃花组 section *Sciadorhodion*）原产北美东部，其余的物种分布于亚洲。映山红组中的常绿杜鹃花（evergreen azalea）种类全部原产亚洲。大多数野生杜鹃花生长于气候温和、降雨充沛、空气湿润、养分含量低的酸性土壤中（Cox 1979）。这些生长条件在许多海洋和山区都有发现，但在大陆性气候下的内陆地区没有，而这些内陆地区亦是杜鹃花明显不存在的地区。高山杜鹃和杜鹃花育种的重大挑战之一是培育能够适应寒冷冬季、炎热夏季、较少降雨和干旱、较差土壤的杂交种。幸运的是，物种向中亚、欧洲和北美洲的广泛辐射，加上适应性变化和集中区域的多样性，提供了应对诸多挑战所需的功能多样性。

图 26.1 中国云南大理苍山上的蓝果杜鹃。蓝果杜鹃 [*Rhododendron cyanocarpum* (Franch.) W. W. Smith] 为常绿杜鹃组（*Ponticum*），是云南省大理苍山特有的濒危植物，常见于海拔 3000～4000m 的云杉或冷杉林下、高山杜鹃林中。叶宽倒卵形或近于圆形，花白色或粉红色，花期 4～5 月份

在杜鹃花亚属（subgenus *Rhododendron*）中有个明显例子，有鳞杜鹃花物种分布在从北极圈到澳大利亚北部的广阔范围中。这个类群包括杜鹃花属中的大多数高山物种，

其中大多数是来自东南亚的对霜冻敏感的物种，而在中国-喜马拉雅山脉、南欧和北美洲东部山区分布的其他物种则非常耐寒。

人们对杜鹃花属的园艺兴趣可以追溯到至少几个世纪前，当时从野外选育的常绿皋月杜鹃花（Satsuki azaleas）是日本杜鹃花栽培的全部。在西方，公众对该属及其育种的兴趣与维多利亚时代的植物探索时代以及杜鹃花种类从亚洲引入英国的时间相吻合。到19世纪中叶，高山杜鹃和杜鹃花育种已使用且直至今日仍然持续使用的“熔炉”方法来进行性状重组。杜鹃花育种几乎完全基于种间杂交，利用植物谱系来进行下一代亲本的选择。在A和B物种之间主要进行F_1代杂交，而不是繁育模式化的F_2或回交种群来进行重组，选择的A×B杂交个体通常与C物种杂交以增加其他性状的变异，以此类推。其结果是栽培品种的基因组复杂性达到惊人的程度。Salley和Greer（1986）列出的系谱书揭示了许多具有多代谱系的当代杂种可以追溯到6个或更多的亲本物种。例如，David Leach的强健黄色杂种‘卡皮’（*R.*‘Capistrano’），其遗传背景包含来自北美洲的2个物种和来自亚洲的7个物种，以及一些未记录的祖先物种。

杜鹃花属多样的遗传和生殖特征使该属植物进行异常的杂交配组成为可能。根据经验，同一分类组内的物种间异花授粉可育，并能产生可育的后代。这为繁殖提供了巨大的遗传变异的种子资源库，因为常绿杜鹃花组（section *Ponticum*）（无鳞杜鹃花）内有大约200个物种，杜鹃花组（section *Rhododendron*）（非热带的有鳞杜鹃花）内有近150个物种。尽管它们的基因组复杂，但大多数杂种在遗传上都是稳定的。Sax（1930）对无鳞杜鹃花种间F_1代杂种的细胞遗传学研究发现，减数分裂过程中发生正常二价配对，由此作者得出结论，“亲本染色体具有完全或几乎完全的亲和性，尽管亲本（亚洲和北美高山杜鹃）可能已经分离数百万年”。对北美洲本地落叶杜鹃花物种之间自然杂交的研究也发现类似单价或迟滞染色体这样极少的减数分裂配对异常现象（Li 1957）。对高度种间杂交后分离得到的群体进行同工酶标记分析（Krebs 1996），结果发现，仅有极少的性状发生分离突变，而这皆源于无鳞亲本物种之间的染色体同源性高，以及经过多代育种选择后染色体及遗传的稳定性。

组间交配水平（intersectional level）上则存在较强的生殖障碍，据报道杂交成功的很少（Kehr 1977；Leach 1961）。在野生杜鹃花杂交中，有许多受精前和受精后障碍，如花粉管不亲和与胚败育，都可能阻止或严重减少可育后代的产生（Williams et al. 1990）。然而，在杜鹃花 × 高山杜鹃杂交中，有一些被称为“阿扎利登”（azaleodendron）的杂交成功案例。一个奇特的例子是一个亚属间杂种，由类越橘杜鹃花（*Vireya* rhododendron、*R. retusum*、杜鹃亚属 subgenus *Rhododendron*）和落叶杜鹃花（常绿杜鹃花亚属 subgenus *Hymenanthes*）杂交产生，但花不育（Rouse et al. 1988）。由落叶杜鹃花和无鳞杜鹃花（亚属内的亚组间）产生的阿扎利登相对较成功，尤其是如果亲本之一是多倍体杜鹃花物种，如四倍体火焰杜鹃花（*R. calendulaceum*）和长柱杜鹃花（*R. occidentale*）（Kehr 1977）。这种现象很有趣，因为它暗示了多倍体在克服广泛杂交和育性障碍中的作用，这将在本章中进行详细讨论。此外，常绿杜鹃花组（section *Ponticum*）和羊踯躅组（section *Pentanthera*）之间的杂交能力也证实了两者的一些遗传相似性，而这也支撑了Goetsch等（2005）最近的分类工作，即将落叶杜鹃花从它们先前的亚属中移除，而将其移至常绿杜鹃花亚属（subgenus *Hymenanthes*）下同名的新组。阿扎利登的其他例子还包括 1820～1830 年在西方记录的两个最早的阿扎利登杂交种——‘阿扎丽斯’（*R.*‘Azaleoides’）和‘伯若特尼’（*R.*‘Broughtonii aureum’）。

本章的重点为改善杜鹃花在温带气候下的适应性状。但这并不意味着降低其观赏重要性，因为这对于在自家花园里种植高山杜鹃和杜鹃花的人们来说至关重要。经过 200 年的育种，杜鹃花杂交种的美感得到了极大的提升且实现了种类的多样化，带给育种者的突破观赏性的挑战减少。同时，尽管杜鹃花杂种数量攀升至 25 000 以上，但在美国进行全国性商业销售的高山杜鹃品种数量仍然很少且维持不变，主要由 19 世纪留存至今的“铁甲”（ironclad）系列品种组成。对于普通园丁来说，这个品系的高山杜鹃具有最广泛的适应性和最好的观赏性。随着杜鹃花的社会需求及其环境的变化，这种对精致植物繁育的追求，与公共园艺及其产业对具有新特征的粗放型植物的需求之间的脱节变得越来越严重。高山杜鹃在市场上失去了一些光彩，新一代的园丁似乎不太乐意种植像高山杜鹃这样需要关注的植物，而诸如气候变化和入侵性病虫害等病害将加大高山杜鹃的生存压力，致使其日渐在一个狭窄的适宜区生长。高山杜鹃和杜鹃花许多适应性状都有很大的改善空间，以使其能够应对以上这些挑战，变得更易于消费者种植，并扩大市场。利用杜鹃花自然的物种多样性提供诸多需要的适应性状，通过传统植物育种的手段，重组形成新的杂种。但是，这将需要基于科学的方法来进行育种和评估，通常需要机构资源，如多个田间场所、温室和实验室设施。这些都是本章所述内容涉及的范围。对于那些想深入研究杜鹃花属悠久杂交历史的人，可参考众多相关的优秀著作（Leach 1961；Livingston and West 1978；Nelson 2000）。

26.2 多 倍 性

多倍体植物有多个染色体组，超过二倍体的 $2n = 2x$ 染色体组数，其中 x 是生物染色体基数。多倍体化被证明是一种成功的植物进化策略，根据不同的计算方法，多倍体频率为 35%～70%（Soltis et al. 2009）。基因组大小的增加会导致一系列变化，从而增加植物的适应性和多样性，这些变化包括基因复制、杂合性增加、生理生化变化和生殖隔离（Bennett 2004；Soltis et al. 2014）。

在杜鹃花属的某些自然类群中存在多倍体，其中一些已进行物种选择或繁殖而成为杂种，并进行了商业应用。倍性水平的早期检测依赖于染色体计数，其中 $x = 13$（Ammal 1950；Ammal et al. 1950；Atkinson et al. 2000；Hosoda et al. 1953；Li 1957；Nakamura 1931；Sax 1930）。近年来，流式细胞技术的使用促进了倍性检测并提供了有关杜鹃花物种和杂种基因组大小的新信息（Contreras et al. 2007；de Schepper et al. 2001；Jones et al. 2007；Perkins et al. 2015）。这些研究表明，常绿杜鹃花组（无鳞杜鹃花）内的所有检测物种均为二倍体。奇怪的是，这一组中有很多的杂种却是三倍体或四倍体（Jones et al. 2007；Perkins et al. 2015），这可能是由其复杂的遗传背景，即杂种谱系中存在多个物种（Salley and Greer 1986）所导致的。种间杂种可以形成未减数的 $2n$ 配子（需要通过有性繁殖来增加倍性），即使在亲本中没有观察到该特征（Mason et al. 2011；Ramsey and Schemske

1998)。总体而言，这些多倍体的无鳞杜鹃花具有遗传多样性，对于有兴趣进行多倍体育种，或想尝试对该属进行广度杂交的育种者来说，这是一个充满希望的种质资源。

与无鳞杜鹃花相比，杜鹃花组中的有鳞杜鹃花有众多的多倍体物种，从二倍体到十二倍体(Ammal 1950)。该组的流行杂种中包含有二倍体、三倍体、四倍体和六倍体(Jones et al. 2007)。来自东南亚的27个有鳞杜鹃花物种（类越橘杜鹃花组）被证实都是二倍体(Atkinson et al. 2000)。

在杜鹃花中也有多倍体。尽管所有参试的常绿杜鹃花物种（映山红组）均为二倍体，但根据染色体计数(Ammal 1950；Heursel and De Roo 1981；Hosoda et al. 1953；Pryor and Frazier 1970；Nakamura 1931）或流式细胞仪（Jones et al. 2007）检测出一些种间杂种为三倍体或四倍体。与上述无鳞杜鹃花杂种一样，常绿杜鹃花的种间杂种可能形成了未减数配子，这为今后通过常规植物育种培育多倍体提供了平台。落叶杜鹃花（羊踯躅组）多数原产北美洲，含有相当高的多倍体频率——17个物种中有7个被确证为三倍体或四倍体（Jones et al. 2007），其中包括有名的 *R. colemanii*。由于其中一些物种已广泛用于观赏育种，如橙红色的四倍体火焰杜鹃花，因此有许多的多倍体杂种也就不足为奇了(Jones et al. 2007)。

杜鹃花属的组内杂交最为成功，因为每个组内都有众多的物种，所以杂种的重组就能产生丰富的遗传多样性。但是，某些组具有一些特有特征，希望可将其导入其他组中。例如，在有鳞杜鹃花（杜鹃花组）中发现其具有黑象鼻虫（black vine weevil）抗性，该组的腺鳞片散发着与威慑象鼻虫作用相关的芳香油（Doss 1984），但是无鳞杜鹃花（常绿杜鹃花组）没有这些特性。在这两个组中，使用二倍体物种进行广泛杂交的尝试普遍不成功（Kehr 1977），但是可以使用多倍体亲本，如利用有鳞杜鹃花栽培种‘时代’(*R.* ‘Epoch’)(来自儒小杜鹃花 *R. minus* 的人工四倍体）与一些已确认的四倍体无鳞杜鹃花栽培种（Jones et al. 2007；Perkins et al. 2015）杂交。还应注意，杜鹃花属中最成功的组间杂交品种是落叶杜鹃花和无鳞杜鹃花之间的杂交——所谓的阿扎利登，这里的落叶杜鹃花亲本是四倍体（Kehr 1977）。如果高山杜鹃亲本是二倍体，那么三倍体杂种通常是不育的，由于基因数量的影响，表型通常更接近落叶杜鹃花亲本。出于育种目的，如果父母本双方均为四倍体，则可获得更好的结果，产生的后代更有可育性，并且将与常绿杜鹃花亲本更相似。

对多倍体的园艺兴趣主要集中在受倍性影响的两个性状（生育力和大小）上。例如，一些流行的二倍体观赏植物的易繁殖特性或入侵性特征（如卫矛 *Euonymus alatus* 和浆果金丝桃 *Hypericum androsaemum*）已通过不育三倍体的培育得以消除（Thammina et al. 2011；Trueblood et al. 2010）。尽管杜鹃花在世界许多地方并未发现具有入侵性，但在英国有一个重大例外，英国从伊比利亚半岛（Iberian Peninsula）引种了二倍体的长序杜鹃花（*R. ponticum*），结果导致其在野外大肆繁衍（Milne and Abbott 2000）。$2x \times 4x$ 杂交产生的不育性也可能是杜鹃花属的一个有用的育种目标，特别是如果三倍体的观赏特性也能改善了的话。同时，利用秋水仙碱（colchicine）和氨磺乐灵（oryzalin）可使体细胞组织的染色体数目加倍。该方法被证明有效，如在不育的阿扎利登杂种‘芬芳’(*R.* ‘Fragrgant Affing’）的人工四倍体中将同源染色体进行正常减数分裂配对——使其育性恢复，产生可育植株（Contreras et al. 2007）。

多倍体中基因组大小和 DNA 含量的增加会使细胞变大（称为 gigas），从而产生更理想的组织和器官，如更大的果实、更大的花朵和更厚的叶子。在杜鹃花属中，人们常常提出将增加倍性作为改善观赏性状的一种方法，这是基于偶然的（未经测量的）观察结果，即多倍体拥有较厚的茎和叶，以及较大的花朵（Barlup 2002；Hosoda et al. 1953；Kehr 1976），或者是因为多倍体杜鹃花经常出现在园丁手册中的优质植物名单里（Perkins et al. 2015）。在某些情况下，多倍体植物的花期会晚且长于其二倍体近缘体。这可能是由多倍体的细胞表面/体积比较小导致其较慢的新陈代谢和发育速度（Levin 1983；Otto 2007）。

化学诱导或人工加倍获得的多倍体的遗传特征可能会比其二倍体祖先逊色。如在杜鹃花属中，因为体细胞倍增产生的重复等位基因造成了近交衰退，从而增加了基因纯合性和有害基因在异型杂交中的表达。诱导多倍体的近交抑制可能更强或更弱，这取决于单个二倍体祖先的杂合度，因此希望由具有遗传差异的多个个体而不是单个个体中产生多倍体，并对最有活力的个体进行选择。如果结果仍然很差，则这些人工多倍体或许可以与其他，最好是天然多倍体进行杂交授粉，以提高其杂合度。

通过人工手段增加染色体数目可能会对植物的适应性产生负面影响（Levin 1983）。使用美国杜鹃花协会的在线植物耐寒数据进行分析，得到了北美洲常见的 1775 个已登记杜鹃花物种和栽培品种的带花蕾最高耐寒温度分布频率图（图 26.2），发现化学诱导得来的四倍体的耐寒性比其同源二倍体品种低。通过离体冷冻实验研究，Väinölä 和 Repo（1999）发现两个人工诱导四倍体的冷驯化组织比其二倍体祖先品种的耐寒性低，表现

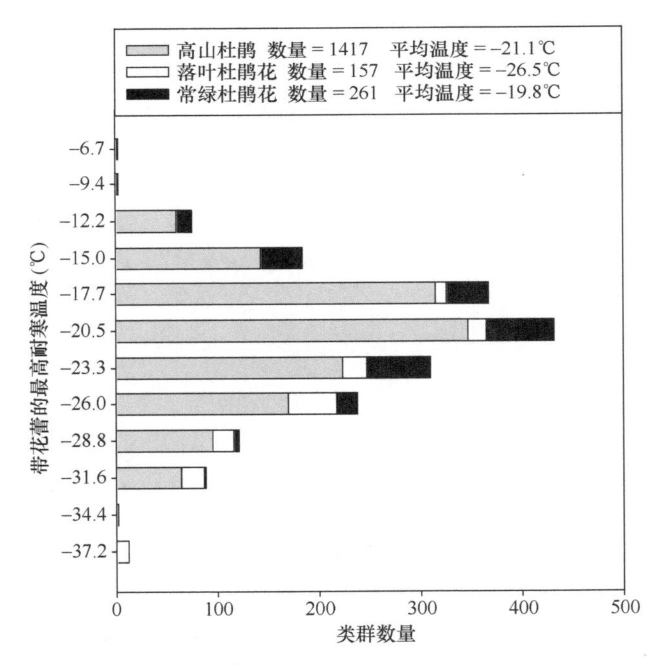

图 26.2 北美常见的 1775 个已登记杜鹃花物种和栽培品种的带花蕾最高耐寒温度分布图。使用美国杜鹃花协会的在线植物耐寒数据来确定不同温度的分布频率

为叶片耐10℃低温和花蕾耐3℃低温。通过比较两个寒冷冬季后的露天春季开花情况，Krebs（2005）认为人工诱导四倍体‘超级诺娃’（*R.*‘Supernova’）的花蕾比其二倍体亲本植物‘诺娃’（*R.*‘Nova Zembla’）的耐寒性低约15℃。这种耐寒性降低的原因尚不可知——这可能是由于近交衰退，或是在叶片或树体适应寒冷的情况下，多倍体中存在较多的木质素，从而加速了冻融诱导的栓塞和冷冻后的伤害（Lipp and Nilsen 1997）。

多倍体仍然是杜鹃花属今后育种和研究中充满希望的领域。来自其他物种的基因组大小的信息（已研究的还不及该属的一半）将增进对该属生物学和进化的了解。需要告知那些从事较大基因组研究的植物育种者，具有多样遗传类型的天然多倍体模式（同源多倍体与异源多倍体）还尚未确定。目前，多倍体高山杜鹃和杜鹃花的园艺优势还未建立，并且目前在商业应用上尚未对多倍体品种产生明确的偏好。随着今后进一步的育种，将会有更多用于评估其观赏性和功能性的多倍体植物，因此，相对于栽培二倍体的基准，对多倍体的性状进行更加客观的定量测量将非常重要。这些数据不仅对于植物选择至关重要，而且对于证明多倍体对消费者和整个产业的益处都是至关重要的。

26.3 耐 寒 性

为扩大应用范围，耐寒育种及其研究一直是高山杜鹃和杜鹃花杂交种培育的核心。例如，在有鳞杜鹃花（常绿杜鹃花组）中，在亚洲发现的绝大多数物种都有很高的观赏性——具有异国情调的叶形与纯正的红色、黄色和紫色的花色，但这些物种仅适度耐寒，带花蕾的耐寒范围为–25～–20℃。尽管它们适宜北部海洋性气候，但它们缺乏适应大陆性气候所需的更高水平的抗冻性。在亚洲和北美洲皆有少数耐寒的有鳞杜鹃花，其带花蕾的耐寒范围为–35～–25℃，但它们的观赏性较低，花白色或粉红色。自西方开始对高山杜鹃和杜鹃花进行育种以来，人们常用这些浅色花的耐寒物种与其他较具观赏性但较不耐寒的物种进行种间杂交。

杜鹃花属物种分布于从赤道到北极圈，从海平面到高于海拔4000m的区域，故其在耐冻性上有较大的差异。Sakai等（1986）对69个物种和32个栽培品种进行了迄今为止最全面的杜鹃花耐寒性调查。采用受控的离体冷冻实验对引种的组织进行比较，研究其花和营养组织的反应。花蕾对冷冻的敏感性较营养性芽、叶或茎（皮层和木质部）更强。在寒冷加剧的情况下，花与营养组织之间的耐寒性差距增加。在无鳞杜鹃花中，常绿杜鹃花组（section *Ponticum*）常绿杜鹃花亚组（subsection *Pontica*）中的种类最耐寒，可带花蕾在低于–25℃或带叶在低于–40℃的霜冻下存活。这些种类包括北美洲的酒红杜鹃花（*R. catawbiense*）和极大杜鹃花（*R. maximum*），以及亚洲的短果杜鹃花（*R. brachycarpum*）和*R. yakushimanum*，这些都可作耐寒育种的基础种质。杜鹃花组中最耐寒的无鳞杜鹃花包括来自北美洲的儒小杜鹃花（*R. minus*），以及分布于中国北方、西伯利亚、韩国和俄罗斯东部的相关种类如兴安杜鹃花（*R. dauricum*）与迎红杜鹃花（*R. mucronulatum*），还有在南极延伸呈环极分布的高山杜鹃花（*R. lapponicum*）。北美洲的落叶杜鹃花（羊踯躅组），如粘杜鹃花（*R. viscosum*）、甜杜鹃花（*R. arborescens*）和加拿大杜鹃花（*R. canadense*）带花蕾的耐寒温度可至–30℃，比一些亚洲寒冷地区的常绿杜鹃花物种（映山红组）都耐寒，包括 *R. yedoense* var. *poukhanense*（韩国）和九州杜鹃花（*R. kiusianum*）（日本高海拔地区）。作者提到，基于纬度和/或海拔的耐寒性变化与气候差异之间存在良好的对应关系，并观察到最耐寒的物种位于中国-喜马拉雅地区（Sino-Himalayan region）多样性中心的地理边缘上，那里气候较温和，物种处于高海拔地区，如朱红大杜鹃花（*R. griersonianum*）、硬刺杜鹃花（*R. barbatum*）和树形杜鹃花（*R. arboreum*）都被归类为冷敏感型，带花蕾的耐寒温度不超过–18℃（Sakai et al. 1986）。

记录于国际杜鹃花属名册和登记表（Leslie 2004）的所有品种都有一个描述为“耐寒温度至 *X*”的带花蕾耐寒性评价，其中的 *X* 是最冷温度，在该温度下于冬季观察到的休眠花蕾损害为零或接近零。由于该观察通常基于命名和注册前某地的一段相对短的时间内（＜10 年），因此应将报道的耐寒温度视为园丁的种植指南，而不是准确的生理指标。除绝对温度外，冬季花蕾的损害还可以归因于许多因素，如秋季植物的冷适应水平，土壤冻结导致的干燥压力或冬季解冻期间植物的冷适应水平降低。理想情况下，植物描述还应包括花蕾的田间受损温度，该温度可用于在没有损坏的最冷温度和有损坏的最暖温度之间区分“实际”的耐寒性。育种者通常需要等待数年甚至数十年才能得到冬季天气的测评结果，从而才可以提供此信息。实际结果是绝大多数高山杜鹃和杜鹃花的注册与引种都是在有限的耐寒性知识基础上进行的，因此需要将其无性系种植于具有不同条件和气候的花园中，最终获得其冬季表现的全面认识。然后，适应性最强的植物就会出现在由美国杜鹃花协会等园艺俱乐部针对不同地区编制的“出色表现”名单中。

为了解这些信息，可以使用美国杜鹃花协会网站（https://www.rhododendron.org/search_intro.asp）查看 1775 个杜鹃花种类从–6.7℃到–37.2℃的花蕾耐寒性评价和分级信息（图 26.2）。以上这些样本的带芽平均加权耐寒温度为–23℃，表明大多数类群适合于北美植物耐寒 6a 区的中冷地区。原产亚洲的常绿杜鹃花是耐寒性最低的群体，带芽的平均耐寒温度为–19.8℃（北美植物耐寒 6b 区）。在美国南部，常绿杜鹃花代替了花园中的高山杜鹃，因为它们不经历霜冻，且更适合于 8 区和 9 区的夏季高温。相反，源自北美和亚洲的落叶杜鹃花是最抗冻的群体，带芽的平均抗冻温度为–26.5℃（北美植物耐寒 5a 区）。在园丁中最受欢迎的常绿有鳞杜鹃花和无鳞杜鹃花种类均具有适应–21.1℃的中等耐寒性，对应为北美植物耐寒 6a 区。在较高的冷冻水平（低于–28℃），物种几乎全部由高山杜鹃和落叶杜鹃花组成，并有少量极度耐寒的杜鹃花（如下所述），其带芽的耐寒温度为–37.2℃，适合北美植物耐寒 3 区和 4 区。在一定程度上，有限的杜鹃花极端耐寒品种也反映了物种中有强耐寒性，却缺乏观赏性的特性。但这也是由缺乏针对严寒条件育种的计划导致。缺乏此类育种计划是没有远见的，因为美国北部（如美国中西部和安大略省南部）可能是高山杜鹃和杜鹃花育种计划中的巨大的目标市场。

值得一提的是，有一些植物育种工作已经选育出了非常耐寒的品种。落叶杜鹃花的‘北极光’（‘Northern Lights’）系列是由明尼苏达大学景观植物园在冬季严酷、夏季炎热的大陆性气候条件下培育而来。它们适应北美植物耐寒 3 区，这使它们跻身于世界上最耐霜冻的杜鹃花杂交品种中。该组中最主要的耐寒性（和香气）来源于 *R. prinophyllum*，它是一种粉红色的北美洲物种，其带芽的耐寒温度可至–35℃（Moe and Pellett 1986; Susko et al. 2016）。为了培育具有更纯正颜色的杂种，将 *R. prinophyllum* 与羊踯躅（mollis azaleas）中精选的 *R.* × *kosteranum* 杂交，而 *R.* × *kosteranum* 是从以前欧洲使用的日本杜鹃花莲华踯躅（*R. japonicum*）中培育出来的。‘北极光’杜鹃花育种计划正在积极引入新物种（Hokanson et al. 2005），扩大遗传基础，使其涵盖其他耐寒的本地物种，如大西洋杜鹃花（*R. atlanticum*）和加拿大杜鹃花，以及其他彩色和大花资源，如埃克斯伯瑞杂种（Exbury azalea hybrids）。

芬兰赫尔辛基大学和植物园的耐寒性育种与研究已培育出了有鳞杜鹃花杂种，尽管它们是常绿植物，但仍适合北美植物耐寒4区（Tigerstedt and Uosukainen 1996）。其耐寒性育种计划的基础种质是 *R. brachycarpum* ssp. *tigerstedtii*，它被认为是常绿杜鹃花组常绿杜鹃花亚组（subsection *Pontica*）中最耐寒的成员。该类群由13个亚洲和北美洲物种组成，可用于耐寒性和其他适应性育种。更进一步，短果杜鹃花被证明是具有良好育种价值的亲本，能够向其后代传递较高的耐寒性（Uosukainen 1992）。第一代杂种中的其他亲本由杜鹃花属物种和栽培品种组成，这些品种（多数）耐寒，但颜色范围有限——*R. catawbiense* var. *album* 白色、*R. yakushimanum* 白色、*R. smirnowii* 玫瑰粉色、紫背杜鹃花（*R. forrestii*）红色和‘德雷休斯博士’（*R.*‘Dr. H.C. Dresselhuys’）粉红色，最初引入的‘哈嘎’（*R.*‘Haaga’）和‘赫尔辛基大学’（*R.*‘Helsinki University’）的带花蕾耐寒温度为−37～−34℃。

如前所述，杜鹃花属的杂种经常在不知其耐寒适应性的条件下进行商业应用。一旦将它们移栽到其他位置，它们有时会达不到期望，有时甚至超出预期。后者一个很好的例子是19世纪中叶安东尼·沃特（Anthony Waterer）在英格兰的耐普山（Knap Hill）苗圃进行的育种工作。那是一个活跃的植物探索时期，高山杜鹃也越来越引起公众的关注，为了提高杂种的耐寒性以适合英国种植，沃特先生使用来自美国东部山区的酒红杜鹃进行了广泛杂交。尽管没有保存任何记录，但沃特（Waterer）使用的其他物种可能还包括常绿杜鹃花亚组，如来自北美洲的极大杜鹃花与中亚物种高加索杜鹃花（*R. caucasicum*）和长序杜鹃花（*R. ponticum*），它们现在是英国的归化入侵植物。在英格兰流行之后，诸如‘赤眉’（*R.*‘Atrosanguineum’）、‘酒会’（*R.*‘Catawbiense Album’）和‘紫水晶’（*R.*‘Roseum Elegans’）等品种在马萨诸塞州阿诺德植物园的较冷环境中进行了实验，证明它们皆具有耐寒性，其芽的耐寒温度到−32℃（Madsen 2000）。即使它们在英格兰沃金的8区气候中进行了评估和选择，但也很难进入4区。这一小类植物通常被称为“铁甲”（ironclads），即使经过了一个多世纪的新杂种培育，它们仍然是美国所有杜鹃花商业栽培中最成功的。尽管‘铁甲’的观赏性不如现代杂种，但其更容易被普通的园丁种植，且适应性更强，部分原因是具有酒红杜鹃的耐寒性和长序杜鹃花的旺盛生长习性。

19世纪的耐寒性育种在不了解孟德尔定律的情况下，通过有性繁殖进行了基因的重组。现在，对植物耐寒性的遗传学和生理学，尤其是木本植物的遗传学和生理学，有了更多的了解。这些知识对于优化计划育种程序、选择优良的亲本材料以及杂交后代评估都非常有用。木本植物田间适应性的现场评估具有挑战性，因为这很耗时，并且不可预测的环境条件会发生变化，而且并非总是有适合植物选择的理想方法。如果没有更好的筛选后代的方法，尤其是在幼年时，木本植物的育种就会变得效率低下，需要额外的资源和时间，以及整个群体开花（生理成熟）所需的田间时间与自然环境提供的关键压力相关。

基于实验室的抗冻性测定为预测田间表现及植物对抗冻胁迫的遗传和生理基础调查提供了有用的工具。Sakai 等（1986）为研究高山杜鹃和杜鹃花的耐寒性，对其不同组织的最低存活温度（LST）进行了离体冷冻检测。该方法通过冷冻机编程模拟自然的冻融循环程序，确定发生最小伤害的最低测试温度，即 LST。通过肉眼测定冻伤，如冻融后组织的坏死（氧化）。或者，在冷冻程序中将热电偶插入测试组织中，检测组织中发生冷冻时释放出的水熔化所产生的热量，这种方法称为差热分析法或 DTA（Sakai et al. 1986）。一种最新的定量测定离体耐寒性的方法是利用电解质渗透比较，测定冻伤、未受伤和煮沸组织的水溶液电导率，并计算在一定温度范围内的相对损伤百分率，由此得到一个可以用于评估 LT_{50}（半致死温度）的“S”形响应曲线，或以杜鹃花属的叶子为例，计算出现最大伤害率时的温度，即 T_{max}（Lim et al. 1998a）。Väinölä 和 Repo（2000）使用阻抗光谱法获得胞外电阻率（r_e）和弛豫时间（τ）这两种冻伤指标，进一步研究了杜鹃花属叶片的抗冻性。他们的研究结果确定了通过阻抗光谱法测得的耐冻性估计值与使用视觉评级和电解质渗透的传统方法之间有较好的相关性，但是光谱分析得出的耐寒性较低。

通过进一步分析 Sakai 等（1986）提供的数据，证明了这些方法作为辅助育种计划和基础研究的重要性。由于 LST 值是针对不同组织计算的，因此可以绘制出花蕾的耐寒性（用于观赏植物的度量标准）与营养组织，如叶片耐寒性之间的关系，图 26.3 所示为常绿杜鹃组类群的相关研究结果。

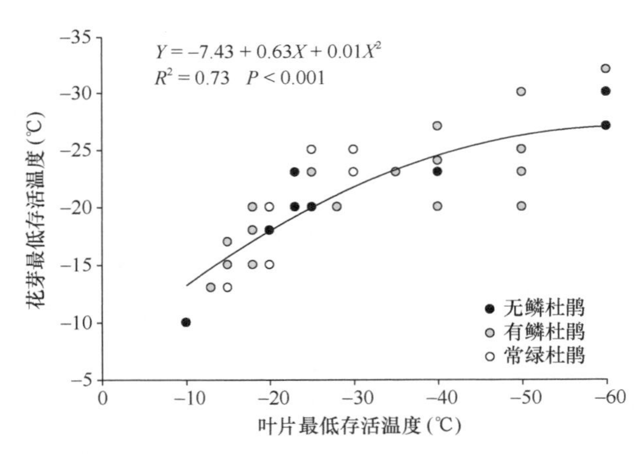

图 26.3 利用 Sakai 等（1986）报道的 53 种杜鹃花属物种数据，作者对叶片 LST 下的花蕾最低存活温度（LST）进行回归处理。某些坐标代表多个数据点

分析表明，花蕾与叶片的耐寒性之间存在显著的非线性关系，在低于−25℃的温度下，花蕾的耐寒性低于叶片，而在−34℃的最低温度附近则有较小的增量，耐寒性亦有增强。高山杜鹃和杜鹃花花蕾的抗冻能力是通过过冷却这种避免霜冻作用的机制来实现的。该机制可防止花原基结冰（George et al. 1974；Ishikawa and Sakai Sakai 1981）。对于大多数耐寒的热带植物，过冷却可以抑制结冰，直到组织中的水温达到−38℃为止，

在此温度下组织会出现均匀的冰核并造成损伤（Wisniewski et al. 2004）。因此，耐寒杜鹃花属物种，如酒红杜鹃花、短果杜鹃花和兴安杜鹃花，其花芽冻结耐受性接近其生理极限。与花蕾相反，杜鹃花属和其他木本植物的营养组织在低于 0℃的温度下就表现出胞外结冰，但不出现胞内结冰（Sakai and Larcher 1987）。在低于开始结冰的缓慢冷却速率下，水从细胞移入质外体导致细胞脱水应激，直到达到破坏细胞膜或胞质溶解的阈值温度为止。这种低温驯化方法比过冷却具更高的耐寒性，某些物种的叶片 LST 为−60℃（图 26.3）。

花蕾与叶片抗冻性之间的非线性关系可以通过所涉及的不同生理机制来解释。但是，根据叶片适应冬季的耐寒性可以对花蕾的耐寒性进行很好的预测。建议通过确定叶片的 LT_{50} 或 T_{max} 来加快耐寒性育种周期。尽管通常使用花蕾存活温度来描述杜鹃花和其他观赏植物的耐寒性，但是从种子至植物开花可能需要 4～5 年时间，因此对营养组织进行早期筛选是理想的选择。利用叶片组织的受控冻融应力对耐寒性进行实验室评价的结果显示，叶片组织的受控冻融应力与包括杜鹃花属（Johnson and Hirsch 1995；Lim et al. 1998a）在内的阔叶常绿植物类之间的露地耐寒性测定结果相对应。Sakai 等（1986）也测量了其他组织，诸如分生组织、皮层和木质部的耐寒性，它们也与花蕾耐寒性相关（数据未显示）。但是，使用叶片可以测量“抗冻性”而不是“规避低温”，因为其他营养组织（如木质部）也出现过冷却（Arora et al. 1992；Die et al. 2017）。冬季杜鹃花属叶片的另一个优点是它们不表现内眠性，但在此期间保持生态休眠状态，当暴露于暖温下时能够进行生理活动（如光合作用）（Harris et al. 2006）。花蕾和木质部等组织中休眠与低温驯化同时转化出现，使之难以将其生理变化归因于某种具体的物候变化（Arora et al. 1997；Die et al. 2017），而冬季叶片的分析研究则避免了以上问题。

在木本植物中群体发育早期的表型筛选可能会因其幼龄与成年（开花）植株之间不同的生理和/或苗龄而有所差异。在杜鹃花属之前的研究中，将 2～3 年生后代分离群体叶片平均 T_{max} 与亲本（成熟植物）值进行比较后，观察到其表型的耐寒性与苗龄相关（Lim et al. 1998b）。例如，在耐寒杂种 *R.*‘Ceylon’（叶片 T_{max} = −43.2℃）的自交 F_2 代群体中，3 年生后代的叶片平均 T_{max} 为−27.6℃，而 42 个后代中只有一个的耐寒性水平相当于 30 年生的亲本植物。回交群体的相似结果表明，耐寒 × 不耐寒杂交的大多数后代比幼嫩的亲本更敏感（或较不耐寒）。这表明幼嫩群体中耐寒性表型分布依其苗龄倾向于冷敏感。随后对极大杜鹃花的野生种群苗龄对其耐寒性的影响进行了研究，结果显示成熟个体（约 30 年生）的叶片 T_{max} 比未成年幼苗的 T_{max} 平均低（更耐寒）9.2℃，证实了苗龄对耐寒性的影响（Lim et al. 2014）。该研究还确定，3～5 年生幼苗，其 T_{max} 在 3.7～6.7℃，耐寒性随苗龄显著增加。与此同时，同一组子代的耐寒性等级适度相关，这表明通过耐寒性的早期选择可以实现 1～2 年后的表型预测。但是，因为这些群体很难保持 5 年以上的时间，所以尚未确定幼苗的耐寒性评估与其成熟植物的冬季表现之间的关系。

杜鹃花群体分离的叶片耐冻性的世代平均值比较结果表明，该性状是多基因控制的（Lim et al. 1998b）。这一发现与其他草本和木本植物上的研究结果相符，该研究发现，耐寒性由多个基因控制，也可由多个微效基因叠加作用或少数几个主效基因调控（Guy 1990；Thomashaw 1990；Väinölä 2000）。使用现代基因组学和蛋白质组学工具进行的性状深入研究证实了这些耐寒性遗传的经典分析。

随着差异表达基因分析工具的出现，证实了在低温环境下会有多个基因和蛋白表达，而在正常条件下这些基因或蛋白不表达或仅以低水平表达。通过比较酒红杜鹃花在低温处理叶片和正常叶片的表达序列标签（EST）文库，从低温处理叶片组织中鉴定出308个EST cDNA转录本，并且在低温处理组织和正常组织中同时出现的转录本非常少（仅见6.3%）（Wei et al. 2005）。这表明酒红杜鹃花的正常叶片和低温处理叶片之间的基因表达模式存在显著差异。值得注意的是，有4个基因或基因家族在冬季组织中大量表达，按表达量进行降序排列，它们分别是ELIPS（early light-inducible proteins）（在减少光氧化应激中起作用的早期光诱导蛋白）、LEA（late embryogenesis abundant）/脱水蛋白（胚胎发生晚期富集基因，包括因脱水和冷冻等干燥胁迫诱导的脱水蛋白）、细胞色素P450（可能参与了冬季光合作用过程中的光合调节）和β淀粉酶（分解淀粉，并参与温度胁迫期间的渗透调节）（Wei et al. 2005）。最近，对酒红杜鹃花的低温处理叶片和正常叶片进行了蛋白质组学分析比较（Die et al. 2017）。该研究鉴定了54种受季节调节的蛋白质，并且与胁迫相关的蛋白质丰度增幅最大，其中包括erd-10-like脱水蛋白、一个冷休克结构域蛋白3（CSDP）和两个可能参与减少叶绿体基质中光反应中心附近的冬季光氧化的脱氢抗坏血酸还原酶（MDAR）。植物的冬季组织中其他上调的蛋白质还包括一些参与能量和碳水化合物代谢、调节/信号转导、次生代谢（可能涉及细胞壁重塑）和细胞膜通透性的蛋白质。与EST分析相比，蛋白质组学无法识别特定的ELIP，但在低温处理组织中发现了两个叶绿素a/b（CAB）蛋白。ELIP是CAB基因家族的瞬时表达成员，其作用可能有助于最大程度地减少常绿杜鹃花冬季叶片吸收多余太阳能所引起的氧化应激（Die et al. 2017；Peng et al. 2008a）。

利用基因差异表达和转录组分析鉴定出来的候选基因或蛋白质，可作为杜鹃花属的耐寒性标记来帮助育种者进行间接选择，而不是仅仅依赖于表型的选择。迄今为止，大多数研究都集中使用特定的杜鹃花属脱水蛋白作为生化和遗传标记。Lim等（1999年）鉴定出一个积聚在冷适应叶片组织中的25kDa脱水蛋白，该蛋白质在强耐寒的北美洲物种酒红杜鹃花中的含量高于较不耐寒的亚洲物种云锦杜鹃花（*R. fortunei*）。在酒红杜鹃花×云锦杜鹃花杂交F_1代中的一个小隔离群体中，子代叶片的T_{max}值与25kDa脱水蛋白的表达相关，这表明它可能在表型变异中起主要作用，并可用作遗传标记。随后的研究表明，25kDa脱水蛋白在常绿杜鹃花组和杜鹃花组内的21个不同杜鹃花属类群中具有独特的保守性，共检测到11种脱水蛋白，范围在25～72kDa（Marian et al. 2004）。25kDa脱水蛋白的变异在近缘属山月桂属（*Kalmia*）（月桂树，mountain laurel）中也有发现，但在类越橘杜鹃花组的*R. brookeanum*这一热带杜鹃花中并未发现。脱水蛋白在耐寒性中作用的更多直接证据来自转基因植物，这些植物过表达其他物种的脱水蛋白基因，包括RcDHN5，该基因编码来自酒红杜鹃花的一个约27.1kDa的脱水蛋白（Peng et al. 2008b）。这些研究表明，组成型脱水蛋白的表达提高了转基因植物的耐寒性（Bao et al. 2017；Houde et al. 2004；Peng et al. 2008b；Qiu et al. 2014；Yin et al. 2006）。

图谱研究的结果在证明脱水蛋白预测耐寒性表型的能力方面更加模棱两可。豇豆的近等基因系和编码35kDa蛋白质的脱水蛋白与幼苗萌发时的耐冷性有关（Ismail et al.

1999）。在大麦（*Hordeum vulgare*）中，脱水蛋白 Dhn1 和 Dhn2 被定位到第 7 染色体的数量性状基因座（QTL）上（Pan et al. 1994）。但是，随后的研究表明，这两个脱水蛋白的表达不是由冷诱导的，而大麦脱水蛋白基因家族的另一个冷诱导成员——Dhn5，没有定位到与耐寒性表型相关的 QTL 区域上（van Zee et al. 1995）。在越橘（*Vaccinium caesariense*）（蓝莓）的一个实验杂交群体中，编码 60kDa 脱水蛋白的 2.0kb cDNA 克隆的序列被用于第 12 个连锁组的基因定位，但该标记并未与花芽耐寒性基因共分离（Panta et al. 2004）。

尽管高山杜鹃和杜鹃花品种的整体耐寒性通过植物育种得以提高，但其仍有进一步改善的空间（图 26.2）。现代植物育种者具备更多杜鹃花属耐寒性的生理学和遗传学知识，更能胜任这项任务。现在可以通过实验室技术，对营养组织和花组织进行有效且可重复的可控冷冻测试来进行表型鉴定，为田间表型选择提供依据与指导。通过幼苗评价可以提高木本植物的育种效率，杜鹃花属和其他属还需要更多的研究来确定其幼苗与成年植株之间对应的耐寒性关系，以便获得适用于幼苗群体的精准筛选。随着耐寒性基因型选择的增多，以及更加实惠和精湛的候选基因鉴定和分子标记作图技术的应用，使得杜鹃花育种工作更加充满希望。

26.4　耐　碱　性

与杜鹃花科的许多物种一样，高山杜鹃喜欢排水良好的酸性土壤，在适宜的土壤中才能达到最佳生长状态。一般认为，大多数高山杜鹃适宜的土壤 pH 是 4.5～5.5（Cox 1993；Leach 1961）。在许多本来适合高山杜鹃栽培的温带地区，由于没有酸性土壤而限制了高山杜鹃的园林应用和市场销售。因此，人们对鉴别具有较高 pH 耐受性的杜鹃花属物种以及培育“耐碱”的高山杜鹃产生了浓厚的兴趣。

在世界的某些地区，杜鹃花属物种自然生长于一些 pH 呈碱性的石灰石或蛇纹石土壤上（表 26.3）。研究人员研究并提出了许多有关耐碱的假设，以期使这种显著的适应性与普遍认为的酸性利于生长的园艺实践相协调，McAleese 和 Rankin（2000）以及 Kaisheva（2006）均对此进行了阐述。现已提出的几种避免高 pH 的策略有：①植物生长在有机层上，该有机层可以缓冲石灰岩基岩的碱性作用，②季风期间大量的降水防止了钙在根部区域积聚，③存在的白云质石灰岩（而不是纯碳酸钙）提供了镁，可置换土壤中的钙。然而，对生长在中国云南省石灰岩上或美国加利福尼亚州蛇纹石土壤上的杜鹃花属物种进行仔细的检查与分析后，结果表明，其根部渗透到的石灰岩土壤，pH 通常＞7.0，有机质含量有限，镁含量低（Leiser 1957；McAleese and Rankin 2000；Kaisheva 2006）。表 26.3 中列出的物种可被认为是真正适应钙质的，在野外它们与适应偏中性土壤的其他植物类群共生，如云南高山上多年生的报春花属（*Primula*）和瑞香属（*Daphne*）物种。

对杜鹃花属碱耐受性和不耐受性的研究已经在其所涉及的生理学方面取得了一些见解。高钙似乎不能抑制杜鹃花的生长（Hanger et al. 1981；Mordhorst et al. 1993），但钙源会影响其组织的正常生长。高浓度硫酸钙（$CaSO_4$）处理的植株长势优于同浓度的碳酸钙

表 26.3 在接近中性或更高 pH 的土壤中生长的杜鹃花属物种

分类学类群	物种	pH 上限	耐受低温[a]
常绿杜鹃花亚属 常绿杜鹃花组 无鳞杜鹃花	亮叶杜鹃花 *R. vernicosum*	6～6.9[b]	5°F/−26℃
	腺房杜鹃花 *R. adenogynum*	6～6.9[b]	−15°F/−26℃
	R. agganiphum	6～6.9[b]	0°F/−18℃
	紫玉盘杜鹃花 *R. uvariifolium*	6～6.9[b]	5°F/−15℃
	栎叶杜鹃花 *R. phaeochrysum*	6.8[c]	−5°F/−21℃
	粉钟杜鹃花 *R. balfourianum*	7.4[c]	−5°F/−21℃
常绿杜鹃花亚属 羊踯躅组 落叶映山红	长柱杜鹃花 *R. occidentale*	＞7.2～8.6[d]	−5 °F/−21℃
	R. prinophyllum	6.9～7.1[e]	−25°F/−32℃
杜鹃花亚属 杜鹃花亚组 有鳞杜鹃花	照山白 *R. micranthum*	＞7.0[f]	−15°F/−26℃
	欧洲高山杜鹃花 *R. hirsutum*	7.6[c, g]	−10°F/−23℃
	欧洲锈色杜鹃花 *R. ferrugineum*	6.8[c]	−10°F/−23℃
	多色杜鹃花 *R. rupicola*	＞7.0[b]	−10°F/−23℃
	毛嘴杜鹃花 *R. trichostomum*	6.8	−5°F/−21℃
	红棕杜鹃花 *R. rubiginosum*	6～7.3[b, c]	0°F/−18℃
	楔叶杜鹃花 *R. cuneatum*	7.0–7.9[b, c]	−10°F/−23℃
	樱草杜鹃花 *R. primuliflorum*	7.0～7.5[b, c]	−5°F/−21℃
	草原杜鹃花 *R. telmateium*	7.0～7.9[b, c]	5°F/−15℃
	北方雪层杜鹃花 *R. nivale* ssp. *boreale*	7.9	−10°F/−23℃
	云南杜鹃花 *R. yunnanense*	6.8～7.3[b, c]	0°F/−18℃
	糙毛杜鹃花 *R. trichocladum*	6.8[c]	−5°F/−21℃
	光亮杜鹃花 *R. nitidulum*	7.7[c]	−10°F/−23℃

a. 美国杜鹃花协会 网址：https://www.rhododendron.org/index.htm
b. McAleese and Rankin 2000
c. Kaisheva 2006
d. Leiser 1957
e. Widrlechner and Larson 1993
f. Chaanin 1998
g. Cox 1985

（$CaCO_3$）处理的植株（Chaanin 1998；Giel and Bojarczuk 2011）。一些研究人员将此现象归因于碳酸钙施用后引起的碳酸氢根（HCO_3^-）毒性（Chaanin and Preil 1994；Chaanin 1998）。碳酸氢根是钙质土壤中的主要阴离子成分，在钙质土壤中可能存在的浓度下，它可以抑制不适应的嫌钙植物的根生长（细胞伸长）并破坏铁的吸收，从而导致典型的由碳酸钙诱导的黄化症状（Lee and Woolhouse 1969）。Chaanin（1998）对 200 种杜鹃花属物种及其杂种进行碳酸钙处理后认为，在低碳酸钙水平（pH 4.2，32mg/L HCO_3^-）下，所有植物均生长良好，但在中等水平时大多数表现出生长迟缓和缺铁性症状（pH 6.4，814mg/L HCO_3^-）。在最高的碳酸钙处理（pH 7.1，1554mg/L HCO_3^-）下，除照山白（*R. micranthum*）、长柱杜鹃花（*R. occidentale*）和大字杜鹃花（*R. schlip-penbachii*）的少量幼苗外，其他所有的植物均死亡。对照山白的进一步研究表明，该物种能在 3000mg/L HCO_3^-下健康生长，该浓度约为其他杜鹃花碳酸氢盐害浓度的 2 倍（Chaanin 1998）。

在对长于云南石灰岩上的杜鹃花属物种进行野外调查的同时，增加了对如表26.3所示的耐碱树种的研究。这项研究使我们对某些分类单元如何适应这些条件有了更好的理解。McAleese和Rankin（2000）将这一地区土壤pH范围内的碳酸氢盐水平定义为“充沛”，随后的研究集中于pH对植物吸收关键元素的影响上。对pH 4.0～8.0的土壤和植物进行分析的结果表明，叶片中钙和镁的含量受到严格调节且不受pH的影响，如它们不会随着pH和微量营养成分含量的增加而增加（Kaisheva 2006；McAleese and Rankin 2000）。有效铁随pH的增加而降低，但是叶片分析的结果表明，在所有不同pH的土壤中，铁的含量均低，而不仅仅是在最碱性的土壤中。相反，叶片中锰富集与黄化和不良生长紧密相关——叶片组织中的Mn含量在pH 4.5～7.0时呈直线下降，即使在pH约为6.8时，Mn含量增加至最高点，此后急速下降（Kaisheva 2006）。像其他一些菊科植物一样，杜鹃花也不调控其所吸收的锰，而使其以高达4000ppm的浓度存储在叶片中，该浓度远超出其他植物公认的有害浓度（Kaisheva 2006）。石灰岩无法提供Mn，其主要来源可能是杜鹃花的落叶（McAleese and Rankin 2000）。

虽然已经鉴定出耐碱的物种，但迄今为止，作为育种资源进行应用的甚少。与大叶的无鳞杜鹃花（常绿杜鹃花亚组）相比，小叶的有鳞杜鹃花（杜鹃花组）可选资源更多，如果将耐寒性也作为目标性状的话，则可选择的资源将大大减少（表 26.3）。照山白和欧洲高山杜鹃花（*R. hirsutum*）（Chaanin 1998）这两个耐碱物种间杂交，培育出一个名为‘盛开的巴克斯’（*R.* Bloombux®）的杂种，据传该杂种可耐受pH 7.5的土壤。因此将耐碱性状渗入缺乏高 pH 耐受性但具有其他理想性状的遗传背景中可能会具有挑战性，因为耐碱性状的遗传力未知。通过在生长条件中添加碳酸钙来进行高pH的筛选是一种“黑匣子”方法，因为选择的标准尚不明确——在高pH条件下，植物是否耐受碳酸氢盐和/或有低养分利用率？利用加钙的生长培养基进行离体筛选的方法被用于高山杜鹃育种群体中的耐碱幼苗鉴定上。这种方法是基于早期杜鹃花属Ca毒害的报道而来，根据相反的最新信息，作者认为这是“错误”的方法，但将其作为对碳酸氢盐的耐受性进行筛查可能会更有效。随后，基于幼苗叶片的黄化程度，在碳酸钙改良的栽培容器中，对耐碱 × 不耐碱 F_1 群体的分离后代进行pH耐受性的表型性状分析，利用这些表型性状将高pH耐受性状定位于两个QTL位点（Dunemann et al. 1999）。虽然研究人员指出，进一步的标记开发可以提高间接选择的准确性和效率，但幼苗检测本身似乎提供了一种相当有效、高通量的筛选大量群体的方法。

Preil和Ebbinghaus（1994）进行的耐碱性筛选最终被用于高山杜鹃砧木的筛选鉴定中。来自德国的‘英卡罗’（INKARHO®）砧木——耐碱的高山杜鹃共同体（interessengemeinschaft kalktolerance *Rhododendron*）可耐受pH为6.5～7.0、碳酸氢盐浓度约为400mg/kg的土壤（Chaanin 1998）。‘英卡罗’（INKARHO®）砧木来源于云锦杜鹃花 × ‘坎宁安之白’（*R.* ‘Cunningham's White’）之间的杂交，‘坎宁安之白’是一个19世纪的中等耐碱品种（Dunemann et al. 1999；Giel and Bojarczuk 2011），在欧洲广泛作为砧木应用。在德国的田间试验表明，在pH为6.6时，‘英卡罗’的接穗亲和性和根系生长均优于‘坎宁安之白’砧木（Preil and Ebbinghaus 1994）。英国皇家园艺协会随后的试验得出以下结论，在pH为5.0和8.0时，两种砧木的健康指标SPAD值、叶绿素含量和叶片绿色程度没有显著差异（Alexander 2008）。在低 pH 下缺乏表型差异可能并不奇怪，因为两种砧木都在或接近大多数杜鹃花所需的最佳条件下生长。在 pH 8.0，远高于‘英卡罗’的推荐临界值 7.0 时，两种砧木可能受到同等压力。RHS 的报告省略了 pH 6.0～7.0 的关键中间值，‘英卡罗’在此范围内的表现可能优于传统砧木。现在，北美市场正在引种‘英卡罗’，且正在进行温室和田间试验（作者未发表数据），以测试‘英卡罗’在 pH 4.5～7.5 土壤条件下对大陆性气候的适用性。

26.5 耐 盐 性

高山杜鹃是公认的对盐敏感的植物，在盐渍地区通常避免将其进行景观应用。盐渍地区也经常干旱且碱性更强，致使高山杜鹃无法成为花园可持续应用的选择。但是，在某些情况下，原本有利于高山杜鹃生长的土壤会暴露于较高盐浓度下，如沿海地区在飓风风暴潮来时得到的海水，或者，冬季撒在道路上的盐在扫雪车的作用下或待雪融化后会渗透到邻近土壤中。在生产过程中，必须注意避免灌溉水和肥料在容器基质中积累盐分。苗圃中普遍通过反渗透或其他去除有害盐分的方法来过滤灌溉水，现在越来越多的绿色产业受到法规的强制性要求，来回收盐分较高的灌溉水。对于高山杜鹃而言，耐盐性可能不是当下所必需的一个性状，但它却是一个可以提高生产者和消费者的效益，并可能扩大其地理市场的性状。

杜鹃花属物种的耐盐多样性尚未探索。仅有的少量报道集中在常绿杜鹃花品种上（Lunin and Stewart 1961；Milbocker 1988）。Milbocker（1988）对具有遗传多样性的 101 种杜鹃花杂种进行了一项调查，确定了不同氯化钠浓度处理土壤后杜鹃花根细胞的质壁分离和“叶片灼烧”程度。在该研究中观察到的耐盐性水平差异高达 9 倍，其中包括 64 个品种（63%）被列为盐敏感型品种（在 NaCl＜20mmol/L 时细胞质壁分离），29 个（29%）盐半耐受型（在 NaCl 为 30～60mmol/L 时细胞质壁分离）品种和 8 个（8%）能够耐受 70～100mmol/L NaCl 且未发生根细胞质壁分离的品种。这些数据证实了映山红组整体的盐敏感性，且同时揭示了有耐中等盐度条件（100mmol/L NaCl）的变异体存在，这可能对将来的育种有用。杜鹃花属中的其他主要分类群没有耐盐性的可比信息。但是，一些物种的沿海分布表明有耐盐类群的存在。

常绿杜鹃花耐盐物种的起源尚不清楚。Milbocker（1988）指出，许多高度盐敏感的类群被归类为久留米杜鹃花（Kurume azaleas），其起源在日本南部九州（Kyushu）山区发现的 3 种杜鹃花之间的杂交。相反，最耐盐的品种则为皋月杜鹃花类群（Indica azalea group）的成员。皋月杜鹃花（Indica azaleas）包括相关的皋月杜鹃花杂种（Satsuki hybrid group），具有与久留米杂种（Kurume hybrids）不同的血统。皋月杜鹃花杂种可能起源于两个物种——皋月杜鹃花（*R. indicum*）和圆叶杜鹃花（*R. eriocarpum*）（同义：*R. tamurae*），两者都是九州（Kyushu）南部屋久岛（Yakushima）上的植物。这两个物种关系紧密——圆叶杜鹃花被一些植物学家认为是皋月杜鹃的一个品种，通常被称为“矮皋月杜鹃”（dwarf Indica azalea）。尽管皋月杜鹃花常常生长于山区，但其分布范围有时会与岛上的圆叶杜鹃花重叠。圆叶杜鹃花种群常分布于岛沿岸海平面那些暴露于海风中的炎热潮湿的恶劣环境中（Creech 1978；Galle 1985）。该地区重叠物种间的自然杂交已有记载（Tagane et al. 2008），并被认为是皋月杜鹃花杂种（Satsuki azaleas）的原始来源，由当地居民首次欣赏并收集，并在几个世纪后被引至西方。巧合的是，皋月杜鹃花杂种可能也通过圆叶杜鹃花，成为皋月杜鹃花耐盐性的一个来源（Kaku 1993）。

重要的是，常绿杜鹃花中的皋月杜鹃花类群除耐盐性外，还以其适应性著称。对 73 个品种由疫霉真菌（*Phytophthora cinnamomi*）引起的根腐病的抗性进行评估，根据根部病症的目测评分结果，有 20 种（27%）有抗性，而其余的为中等抗性和易感类群（Benson 1980）。与盐胁迫实验（Milbocker 1988）一样，久留米杜鹃花的耐受性最低（最易感），而皋月杜鹃花的耐受性最高（抗性）。由于在两个研究中使用的品种有一些重叠，因此病害（生物）和盐（非生物）胁迫耐受性之间的关系得以说明，如图 26.4 所示。

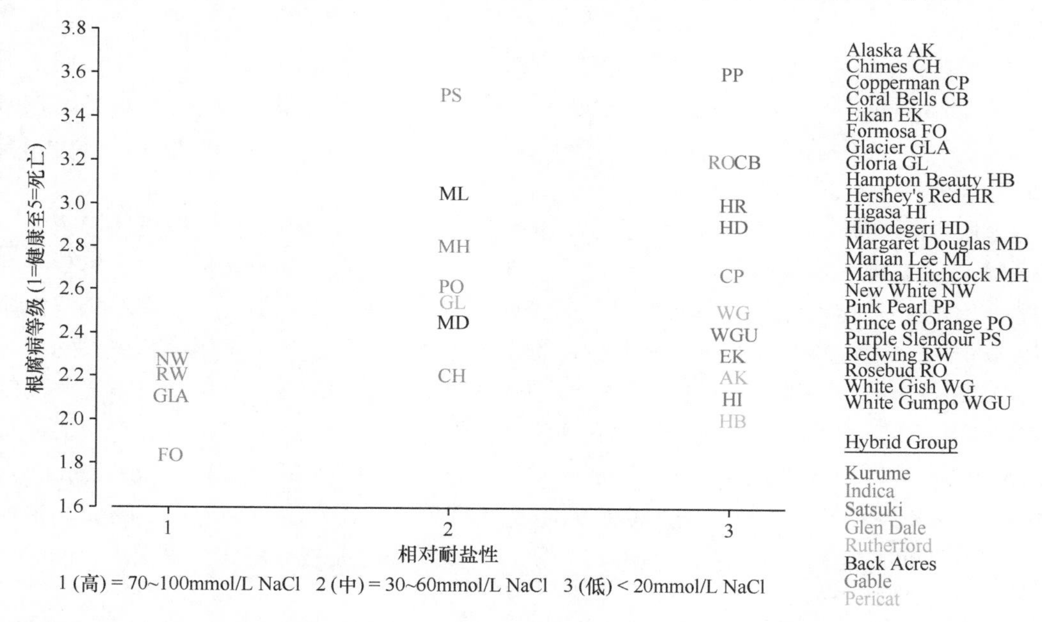

图 26.4　作者指出的常绿杜鹃花品种的耐盐性（Milbocker 1988）和疫霉菌根腐病抗性（Benson 1980）之间的关系。较低数字表示较高的抗病性（*y* 轴）或较高的耐盐性（*x* 轴）；品种用 2～3 个字母表示，颜色代表它们归属的杜鹃花类群

由于盐敏感杜鹃花对疫霉根腐病具有抗性或易感性，因此总体关系不强（$r^2 = 0.12$）。更令人感兴趣的是某些杂种群体性状的差异，如与最耐盐和最抗病的皋月杜鹃杂种群体（左下象限）相比，久留米杜鹃花的一些对盐敏感且对病害敏感的品种（右上象限）在经过耐盐和抗病测试后，有一些具有了中等水平的抗病和耐盐能力。耐盐性和抗病性之间的强相关性暗示了这两个性状之间具功能联系。由于这些实验都是在单独的研究中对这些性状进行独立分析，因此一个重要的问题是，当植物同时受到两种胁迫时是否能保持植株健康。

由于人们认识到植物经常面临环境胁迫和生物胁迫这两种胁迫的共同压力，同时人们还担心气候变化可能会使植物承受超出其适应能力范围内的压力程度，因此对多重胁迫的耐受性是当前植物研究的一个活跃领域。它们可能无法适应的原因之一是非生物胁迫——包括高温、洪水、干旱和高盐浓度，导致植物更容易受到生物胁迫的影响，这种现象称为病害易感性。例如，暴露于亚致死浓度的盐条件下会加剧许多一年生作物和多年生植物（包括杜鹃花）的病害。在接种疫霉真菌（*Phytophthora ramorum*，栎树猝死病菌）之前，用 0.2mol/L NaCl 对荚蒾‘春美人’（*Viburnum tinus*‘Spring beauty’）和杜鹃花‘坎宁安之白’的根进行 24h 处理，结果显示，与接种前未接受盐胁迫的对照组相比，该病的发病率和死亡率更高（Roubtsova and Bostock 2009）。与未接受胁迫的接种植物相比，在接种疫霉真菌（*Phytophthora cryptogea*，一种根腐病病原）之前，采用氯化钠脉冲处理菊花根部的方法来也可加速病害症状的发展（MacDonald 1982）。

近年来，人们对胁迫引发的病害易感性的生理和分子基础的了解有所加深（Bostock et al. 2014）。很显然，这种现象主要归结于寄主的防御信号和基因表达。植物激素脱落酸（ABA）与水杨酸（SA）在植物抵御非生物（ABA）和生物（SA）胁迫中起着重要作用，但是当植物受到两种胁迫时，它们通常呈负相关或“拮抗”作用。ABA 在某些系统中是 SA 的负调节剂。例如，在盐胁迫、干旱胁迫或外源施用后，ABA 水平的升高会抑制病原体防御基因局部和全株表达所需的 SA 的产生（DiLeo et al. 2010；Mohr and Cahill 2007；Yasuda et al. 2008）。

图 26.4 中的杜鹃花列阵提供了可用于确定盐胁迫对木本植物系统抗病性影响的一系列表型。易感实验的结果在很大程度上取决于杜鹃花的杂交起源。例如，如果皋月杜鹃花的耐盐性与抗性相结合是自然进化的条件，那么盐胁迫可能不会使它们易患疫霉属（*Phytophthora*）根腐病，因为这两个性状的联合选择可能早已产生于病原体生长的屋久岛（Yakushima）沿海地区皋月杜鹃花 × 圆叶杜鹃花的自然杂交种群中。日本屋久岛气候温暖湿润，对疫霉属（*Phytophthora*）真菌有利，而且在地理上与中国台湾相邻，与东南亚其他地区一起被认为是 *P. cinnamomi* 的可能起源地（Kho et al. 1978；Zentmyer 1988）。在易感条件下维持对两种胁迫的耐受性将是一个新发现，因为在生物和非生物联合胁迫期间，能够规避防御系统并抑制新陈代谢作用的天然突变体似乎很少，但对于植物育种非常有价值。

或者，常绿杜鹃花可能来自最近的物种杂交，结合了未经共同进化而来的性状。在这种情况下，耐盐性表型可能因盐胁迫而易感病。在其他植物（水稻和玉米）中，盐敏感品种和耐盐品种的比较表明，在盐胁迫处理后，耐性较高的基因型中的 ABA 含量亦较高（Moons et al. 1995；Zörb et al. 2013）。ABA 水平的升高可进一步抑制 SA 介导的寄主防御途径，结果是耐盐品种较盐敏感植物更易受根腐病的侵害。采用联合胁迫的方式做进一步的研究可为这些推测提供答案，而且可以用于常绿杜鹃花和可能的其他杜鹃花类群中的综合耐盐性与抗病性育种。

26.6 樟疫霉（*Phytophthora cinnamomi*）抗性（根腐病）

疫霉属（*Phytophthora*）真菌是世界上最具破坏力的一类植物病原菌，该病源于希腊语，意为“植物破坏者”。尽管疫霉菌常被称为真菌，但它并非真正的真菌，其隶属于包括藻类和硅藻在内的假菌界（Chromista）王国中（Agrios 1997）。樟疫霉（*Phytophthora cinnamomi*）是森林、商业苗圃以及居民区景观中的1000余种木本灌木和树木（McDougall et al. 2005；Thorn and Zentmyer 1952）与草本类群的根腐病、枯死病的直接病原菌。尽管人们认为它起源于东南亚（Kho et al. 1978；Zentmyer 1988），但人类活动已将樟疫霉传播到世界的绝大多数地区，其现已成为一个严重的全球性病害。流行病学家预测，随着植物材料全球性传播的增加和气候变化可能导致的病原体范围的扩大及其活动性的增加，疫霉菌根腐病（*Phytophthora* root rot disease）将在 21 世纪变得更为严重，特别是在温度升高和降雨增多的情况下（Anderson et al. 2004；Bergot et al. 2004；Brasier and Scott 1994；Burgess et al. 2016）。

由疫霉引起的根腐病是包括杜鹃花属在内的许多观赏植物常见且致命的病害（Daugherty and Benson 2001；Hoitink et al. 1986；Leach 1961）。专业种植者已经学会通过使用抑制性堆肥树皮混合物（Hoitink et al. 1977）或杀真菌剂（Benson 1987）来减少容器栽培中的病害发生率，但是对于种植杜鹃花的苗圃、园丁和家庭来说，这种普遍存在的病原体仍然是持续存在的挑战性难题。容器栽培中的根腐病问题可能会随着产业中灌溉水的再循环而再次发生，因为灌溉水池可能是病原体的主要来源（Bush et al. 2003）。

遗传赋予寄主的抗性被证明是对该病害额外且更可持续的防御线。许多研究报道了杜鹃花属类群中的根腐病抗性来源（Benson 1980；Hoitink and Schmitthenner 1974；Krebs and Wilson 2002），表 26.4 列出了具有抗性的物种名单。

表 26.4　抗疫霉的杜鹃花属物种名录[a]

种类[b]	组	分布地点	类群
永宁杜鹃花 *R. glomerulatum* syn. *yungningense*	杜鹃花组	中国云南西北部和四川西南部	有鳞杜鹃花
微笑杜鹃花 *R. hyperythrum*	常绿杜鹃花组	中国台湾	无鳞杜鹃花
高山杜鹃花 *R. lapponicum*	杜鹃花组	极地	有鳞杜鹃花
长柱杜鹃花 *R. occidentale*	羊踯躅	美国西北	落叶杜鹃花
玉山杜鹃花 *R. pseudochrysanthum*	常绿杜鹃花组	中国台湾	无鳞杜鹃花
五叶踯躅花 *R. quinqefolium*	大字杜鹃花组	日本北部	落叶杜鹃花
罗勒杜鹃花 *R. sanctum*	映山红组	日本北部	常绿杜鹃花
映山红 *R. simsii*	映山红组	缅甸，泰国，中国，日本南部	常绿杜鹃花
毛蕊杜鹃花 *R. websterianum*	杜鹃花组	四川西北部	有鳞杜鹃花
碟花杜鹃花 *R. aberconwayi*	常绿杜鹃花组	中国滇中西部	无鳞杜鹃花
雅容杜鹃花 *R. charitopes*	杜鹃花组	中国滇西北，缅甸	有鳞杜鹃花
睫毛杜鹃花 *R. ciliatum*	杜鹃花组	尼泊尔，印度锡金，不丹，中国	有鳞杜鹃花
独龙杜鹃花 *R. nitens* syn. *calostrotum* ssp. *riparium*	杜鹃花组	印度，缅甸东北部，中国西藏东南部和滇西北	有鳞杜鹃花
金毛杜鹃花 *R. oldhamii*	映山红组	中国台湾	常绿杜鹃花
腋花杜鹃花 *R. racemosum*	杜鹃花组	中国云南和四川西南	有鳞杜鹃花
基毛杜鹃花 *R. rigidum*	杜鹃花组	中国云南西和四川南部	有鳞杜鹃花
大字杜鹃花 *R. schlippenbachii*	大字杜鹃花组	韩国，中国，俄罗斯东部	落叶杜鹃花
云南杜鹃花 *R. yunnanense*	杜鹃花组	中国四川西南和云南北部，缅甸北部，西藏东南	有鳞杜鹃花

a. Hoitink and Schmitthenner 1974

b. 美国杜鹃花协会 网址：www.rhododendron.org/

在迄今为止筛选出的 164 个物种中，仅 17%确定为对疫霉有中度或高度抗性，对 350 个高山杜鹃和杜鹃花品种进行的抗性筛选的结果显示，分类群中的抗性率占 6%（Benson 1980；Hoitink and Schmitthenner 1974；Krebs and Wilson 2002）。

总体而言，这些研究结果表明：①杜鹃花属的抗性由多基因控制，因此导致杜鹃花基因型和部分抗性会对病害产生一个连续的抗性级别，而非是完全对病害免疫；②抗性的来源在分类学上具有多样性；③总体来说，完全抗性较为少见。94%的高山杜鹃和杜鹃花品种都易感染根腐病的事实表明，对同时具备观赏性和抗病性的杂交种的需求较为迫切。

在寄主和病原体具有共同进化史的地方，人们有望在野外获得对疫霉有抗性的物种。表 26.4 中除 1 种抗性物种以外，其他所有有抗性的物种都来自亚洲，这与病原体起源于东南亚的假说总体上是一致的。值得注意的是，4 个有抗性且系属不同分类学系统的物种为台湾特有的或台湾为其自然分布地区之一。台湾被认为是疫霉以及相近病原体，如 *P. lateralis* 和栎树疫霉（*P. ramorum*）的起源中心之一，前者可引起雪松枯死，后者则引起栎树猝死（Brasier et al. 2010；Kho et al. 1978；Zentmyer 1988）。中国西南地区（云南和四川）也出现了 1 个具有抗性的物种类群，其在遗传和生理上与在台湾发现的抗性物种截然不同，很可能是根腐病抗性的独立进化谱系。该类群位于中国西南部，多样性较低，主要由有鳞杜鹃花物种组成（杜鹃花组）。尽管这种抗性分类类群的地理集中分布表明其与病原体的远古发生有关系，但迄今为止，在云南和四川尚无发现疫霉的报道。落叶杜鹃花长柱杜鹃花是个特例，因为尽管它具有抗性，但该病原体直到最近才被引入北美洲，而该物种极有可能是在 19 世纪初引进到美国东南部。因为长柱杜鹃花非常适应其他土壤胁迫，特别是高 pH 的蛇纹石土壤，加之其对非生物胁迫的耐受性，所以可能对病害提供了一些交叉保护。

尽管培育具有根腐病抗性的高山杜鹃和杜鹃花会有很高的园艺应用价值，但直到现在还没有正式的育种计划。尽管如此，有证据表明，在过去的杂交过程中已间接进行了抗性育种。例如，亚洲常绿杜鹃花物种——映山红是一个有价值的观赏物种，自其引入西方后就被广泛杂交。它对疫霉具有抗性，很多来源于它的品种，尤其是皋月杜鹃花杂种类群，也具有抗性（Benson 1980），尽管品种选择是根据观赏性状和植株整体活力进行的，而非是针对其病性。

作者于 2000 年在俄亥俄州科特兰（Kirtland，OH）的霍顿植物园（The Holden Arboretum）开始进行有鳞杜鹃花的根腐病抗性育种计划。在随后的几年里，通过有效的温室抗病筛选，有鳞杜鹃花中抗性种质的补充收集（Krebs and Wilson 2002），以及世代抗性提高的评价记录（Krebs 2009），疫霉属（*Phytophthora*）抗性育种机制得以改善。在第一代，以抗性品种为亲本，选择具有观赏性的后代。在第二代（F_2）选择中，抗性提高至 27%～40%。但是，抗性品种的育种值是不可预测的，如使用中等抗性的亲本‘火箭’（*R.*‘Rocket’）（图 26.5a）进行选择获得的抗性后代超过了使用更具抗性的亲本‘卡罗琳’（*R.*‘Caroline’）所获得的（图 26.5b）。

‘火箭’和‘卡罗琳’两个品种都是种间杂种，它们的谱系可以追溯到不同的物种，并且它们之间的抗性在遗传学和生理学上很有可能不同。具有较高抗性水平的亲本获得较低抗性杂交后代的结果表明，‘卡罗琳’的抗性遗传力低于‘火箭’。在许多其他抗性品种中也观察到了较低的育种价值，包括‘卡罗琳’的后代‘迪斯科’（*R.*‘Disca’）。

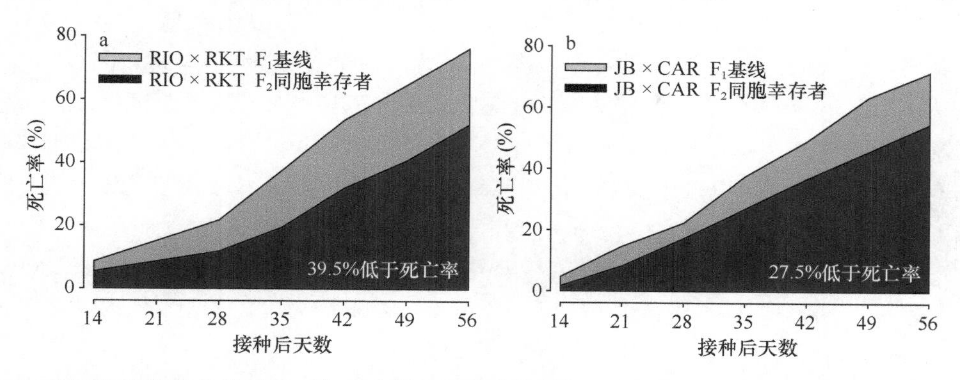

图 26.5　使用抗性栽培品种经过两代后获得具根腐病抗性的结果。进行了两个敏感（S）× 抗性（R）F_1 杂交：（a）‘瑞欧’（‘Rio’）（S，易感）×‘火箭’（‘Rocket’）（RKT，中等抗性）和（b）‘珍妮特’（‘Janet Blair’）（JB，易感）×‘卡罗琳’（‘Caroline’）（CAR，高抗）。将来自 F_1 代的根腐病幸存植株间杂交产生 F_2 代，然后将其与 F_1 基线平行进行筛选。抗病性结果（降低的死亡率）通过病害发展图下方的面积差异进行评价（源自 Krebs 2009）

根据这些研究结果，霍顿植物园的育种计划改为使用杜鹃花属物种来进行抗性育种。特别是使用来自台湾的抗性物种微笑杜鹃花（*R. hyperythrum*）（表 26.4）似乎很有希望，因为它具有一定的耐寒性（带花蕾的耐寒温度约为–20℃），并具有诱人、密集簇生和美丽叶簇的植株表现（图 26.6）。同时它也耐热，是少数能够在北美植物耐寒 9 区（墨西哥湾南部）生长的有鳞杜鹃花物种之一。Thornton（1990）在路易斯安那州（Louisiana）南部进行的育种实验结果表明，微笑杜鹃花是有希望培育出能够适应温暖潮湿的墨西哥湾沿岸气候的杂交种的亲本材料。除抗根腐病外，耐热性也是一个期望的育种目标，因为这将为美国的杜鹃花打开一个新南方市场，目前美国的杜鹃花仅限在北部较凉爽的气候带种植。由于微笑杜鹃花为白花（图 26.6），且其耐寒性不及俄亥俄州

图 26.6　来自台湾的微笑杜鹃花（*R. hyperythrum*），是作者进行疫霉属（*Phytophthora*）根腐病抗性育种计划的基础种质，除了抗性，该种花朵繁茂且叶片光滑、色泽浓绿

北部育种计划所需的耐寒性，故其 F_1 代是通过与那些经过验证的更耐寒且更鲜艳的抗性物种进行杂交后获得的。

微笑杜鹃花的疫霉抗性遗传力相当高（$h^2 = 0.86$，作者未发表数据），温室育种实验结果表明，F_1 代的抗病性得到了合理的提高。例如，在微笑杜鹃花与易感品种‘卡尔萨普’（*R.* ‘Calsap’）杂交的一个群体里面（图 26.7a），在 F_1 代中观察到抗性增强——基于病害发展图下方的面积，死亡率降低了 57%——超过了来自抗性品种的 F_2 群体的增幅（图 26.5）。在相同群体中的根腐病得分分布（图 26.7b）显示为连续的变化，约 25%的后代抗性水平与抗性亲本相当（根系暴露于病原菌中达 4 个月后的根腐病等级为 1～2 级）。值得注意的是，遗传背景中微笑杜鹃花为 25%的回交群体高度易感（图 26.7a），建议在育种计划中，通过 F_1 同系杂交（sib mating）来培育以后的群体，以维持微笑杜鹃花在杂交种中 50%的遗传背景，从而获得最佳的育种进展。

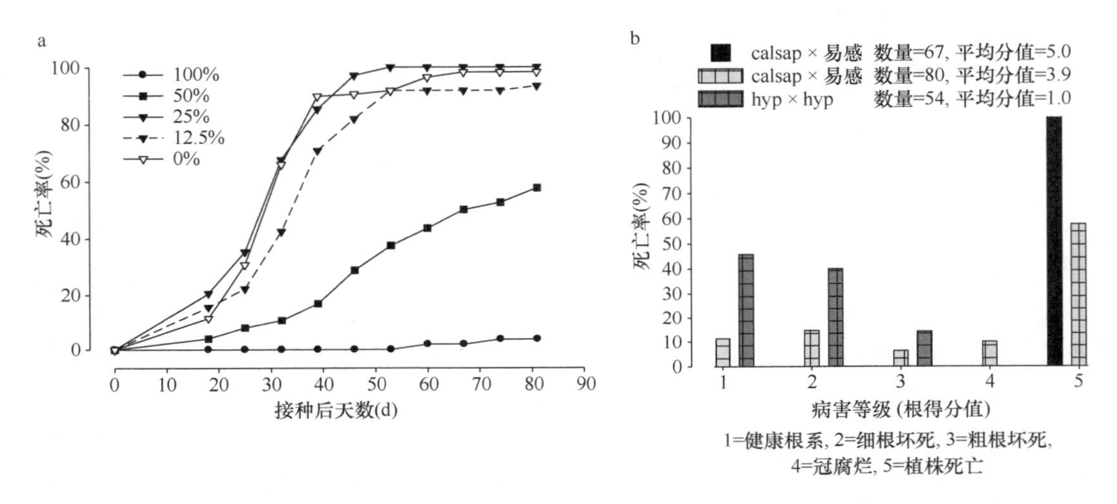

图 26.7 病害发展图。（a）显示了微笑杜鹃花的基因组水平对易感遗传背景的影响——50%是‘卡尔萨普’（*R.* ‘Calsap’）（S）× 微笑杜鹃花（R）杂交 F_1 代，25%是‘卡尔萨普’（*R.* ‘Calsap’）的回交后代。（b）接种后 129 天，‘卡尔萨普’（*R.* ‘Calsap’）× 微笑杜鹃花 F_1 杂交后代的根表型得分情况。所有参试植物均在 4～5 个月大的幼苗阶段进行接种（源自 Krebs 2009）。1 分表示健康根系，2 分表示细根坏死，3 分表示粗根坏死，4 分表示冠腐烂，5 分表示死亡植物

2005 年，在霍顿植物园中，通过将微笑杜鹃花与具紫色、红色、粉红色、黄色和嵌色的白花的强耐寒杜鹃花品种进行杂交，产生了 2500 多个 F_1 后代。此后，在这些植物种于田间后的第一次开花（2009～2010 年）时，根据观赏性对其进行选择。由于 2009 年的冬季霜冻温度为–25℃，因此随同进行了带花蕾的耐寒性选择。选择了大约 170 个 F_1 代杂种，并通过嫁接扦插繁殖（rooting stem cutting），然后运到路易斯安那州（Louisiana）南部，在那进行重复种植。

2012 年，在温暖气候下进行了一个田间试验（北美植物耐寒 9 区）。南部实验是与植物开发服务有限公司（PDSI）（Plant Development Services，Inc.）合作开展的。PDSI 是位于阿拉巴州洛克斯利（Loxley）的花仙子苗圃（flowerwood nursery）的植物引进与营销部门。PDSI 对耐热杜鹃很感兴趣，以将其放入南方市场的植物品种组合中，这其

中也包括备受欢迎的多次开花的安酷杜鹃（Encore^{TM} azaleas）。表 26.5 显示了俄亥俄州（Ohio）和路易斯安那州（Louisiana）测试地点的气候比较。

表 26.5　美国北部和南部杜鹃花测试地点的冬季与夏季气候比较

杜鹃花测试地点	平均日最低温度（1～2 月）（℃）	平均日最高温度（7～8 月）（℃）	平均降雨量（mm）
俄亥俄州麦迪逊	–6.1	26.9	960
路易斯安那州独立村	3.8	33.4	1668

2015 年，在路易斯安那州的田间种植 3 个季节后进行了植物筛选。当时，165 株待选植株中只有 30 株（占 18%）被认为在该地点温暖潮湿的条件下表现良好。其余的待选植株或完全死亡或株系死亡率超过 25%，筛选出的植株作为候选品种。根据症状，根腐病似乎是其死亡原因，从该试验所涉及的其他易感和抗性对照植物与预期相符的试验结果，推断出疫霉病原菌的存在。Kong 等（2003）从坏死的田间试验组织中分离出病原菌并进行了序列分析，在该地点鉴定出了疫霉（作者未发表数据）。这是预料之中的，因为路易斯安那州的测试地点以前曾作为山茶花种植苗圃，而山茶花也是疫霉的寄主植物。这些研究结果表明，路易斯安那州的植物生长特性主要由高病害压力条件决定。利用保存于俄亥俄州的原始植物来繁育株系，再对路易斯安那州存留植株和感病死亡植株进行受控接种并确认病原菌，这是决定我们的杂交种群体存留的主要手段，而在这种南部环境中表现最佳的个体就是最具根腐病抗性的。

下面将对俄亥俄州和路易斯安那州试验中表现最好的高山杜鹃进行介绍，其中许多可以作为培养更优秀的杂交后代的潜在亲本。有些 F_1 代不育，尤其是在以红花品种为亲本的杂种中。与亲本微笑杜鹃花的白色花序相比（图 26.6），F_1 代选择具有更丰富的花色和更显著的球状或金字塔状花序形状这些商业杜鹃花所需具备的重要观赏性状（图 26.8a）。其中选育的一个新品种（图 26.8b）将于 2019 年作为 PDSI 的“南门”（Southgate®）

图 26.8　（a）在俄亥俄州和路易斯安那州的田间试验中表现最好的高山杜鹃实例，所有均为来源台湾的白花、抗根腐病的微笑杜鹃花的第一代杂交后代。（b）作者与‘辉煌’（*R.* Splendor^{TM}），首个霍顿植物园针对南北部市场选育的品种

系列杜鹃花的产品上市，商品名为‘辉煌’（*R.* Splendor^{TM}）。“南门”品牌下的第一批高山杜鹃是由约翰·桑顿（John Thornton）博士培育的，约翰·桑顿博士率先使用微笑杜鹃花来培育耐热杂种（Thornton 1990）。

26.7 疫霉根腐病（*Phytophthora* root rot）抗性与对温暖气候的适应

微笑杜鹃花是一种来自台湾的耐根腐病的物种，也耐热，并且是一种少见的能够在美国南部的墨西哥湾（北美植物耐寒 9 区）沿岸生长的有鳞杜鹃花。育种计划的目标是，抗病性是耐热性的重要组成部分，因为在温暖潮湿的气候中生存是必不可少的，而潮湿的气候为根腐病病原菌提供了有利条件，病害压力很高。将微笑杜鹃花的抗性和耐热性共同转移至其杂种的杂交将是获得理想性状的组合。

性状关联的证据得到了普遍研究结果的支持，即抗病性在那些生长于北美洲较温暖气候下的高山杜鹃和杜鹃花中很常见。在栽培分类学中，亚洲常绿杜鹃花（映山红组）对疫霉的抗性最强（Benson 1980），它们在整个南海湾（the Gulf South）（北美植物耐寒 9 区）遍布生长。相反，大部分易感的有鳞杜鹃花通常被限制在 7 区或较冷的北纬更多地区。除在美国南部表现良好的原产台湾的微笑杜鹃花（Thornton 1990）之外，台湾还有一种，阿里山杜鹃花（*R. pseudochrysanthum*）已被证明对疫霉有抗性（Hoitink and Schmitthenner 1974）。据报道，台湾杜鹃花（*R. formosanum*）和玉山杜鹃花（*R. morii*）是培育耐热植物的优良亲本（Thornton 1990），但它们的抗病性情况未知。这 4 个台湾物种的地理范围相似，在某些情况下是同域的（sympatric）。阿里山杜鹃花、玉山杜鹃花、微笑杜鹃花和 *R. rubropunctatum* 被称为玉山杜鹃花复合体，它们具有许多相似之处，并且似乎密切相关。物种之间的遗传距离接近于零，它们形成了一个单一的系统进化枝，暗示着有一个共同的祖先（Chung et al. 2007）。由于台湾被认为是疫霉的起源中心之一，因此对这种病的抗性可能在这组植物中有了早期进化，并随着种群分化和时间推移而得以维持。

虽然抗病性在适应温暖气候中能发挥关键作用，但其他生理属性也涉及其中。Arisumi 等（1986）在对耐热的有鳞杜鹃花物种和杂交种进行研究时，生长季每月进行 34.4～38.7℃的高温热处理，监测塑料膜盖顶温室中的盆栽苗的存活率和生长活力。台湾的台湾杜鹃花、玉山杜鹃花、阿里山杜鹃花和微笑杜鹃花归于最耐热的物种中。由于植株是在无土盆栽条件下无病生长，因此它们的活力是由根腐病之外的其他因素造成的。与抗热性较低的杜鹃花属物种相比，微笑杜鹃花具有更高的光合最适温度（Ranney et al. 1995），台湾其他密切相关的分类群可能也有该特性。长序杜鹃花（亚洲中部）、云锦杜鹃花、大白杜鹃花（*R. decorum*，来自中国的相关物种）和日本低海拔的东石楠花（*R. degroniuanum*）复合体，特别是冲绳岛（Oki Island）的 *R. metternichii*（以前称为 *R. degronianum* var. *hodoense* f. *brevifolium*，现在分类上被定义为 *R. degronianum* ssp. *Heptamerum* var. *brevifolium*，分布于日本海的海平面地区）都表现出了抗热性（Arisumi et al. 1986）。

微笑杜鹃花突出的根腐病抗性表现于其在环境胁迫条件下仍能得以维持。如上所述，在盐胁迫的背景下，非生物胁迫（如极端水分或温度的胁迫）可能导致寄主发生生理变化。高降雨量和温暖温度的影响特别令人感兴趣，因为这是路易斯安那州测试地点非生物胁迫的两个主要组成部分。以前的研究表明，接种前进行淹水或干旱处理会导致‘卡罗琳’品种抗病性丧失（Blaker and MacDonald 1981）。MacDonald（1991）记录了热诱导对疫霉属（*Phytophthora*）的敏感性，他在接种 *P. cryptogea* 之前先用水浴胁迫盆栽菊花的根，并观察到40℃温度处理下，相对于接种但未胁迫的植物，其根坏死比例增加了4倍。

在俄亥俄州麦迪逊市（Madison）确定有疫霉的田间进行了淹水实验（Krebs 2013）。结果表明，遗传背景中缺乏微笑杜鹃花的抗病性品种在生长季节中反复进行淹水和排水循环处理则容易患病。这些品种包括‘卡罗琳’和‘英格丽’（*R.* ‘Ingrid Melquist’），在温室试验中，其根部环境保持湿润但排水良好时具有抗性，但在田间淹水处理后其根部病害分值在90%以上（更易感）（Krebs and Wilson 2002）。相比之下，淹水的微笑杜鹃花植株的病害分数比它们未淹水的等级平均增加了35%以上，明显低于（更有抗性）抗病性品种的基准。在试验结束时，一组源自微笑杜鹃花的 F_1 杂交种的平均根部病害得分与其抗病性亲本物种没有显著差异。因此，在淹水条件下，与具有不同遗传背景的抗病性基因型相比，微笑杜鹃花更不容易发生根腐病，尽管目前尚不清楚这种应激反应差异的根本原因，但其在 F_1 代中似乎是可遗传的。

类似的，在热激诱发根病害的严重程度上，与另一种抗病性物种 *R. keiskei* 相比，微笑杜鹃花的最小（Krebs 2018）。在该实验中，将幼苗根部进行35℃（2h）/ 40℃（3h）的变温处理接种疫霉，并对一段时间内根部病害的严重程度进行评价。与接种病菌但没有热胁迫的植物相比，高温胁迫使 *R. keiskei* 根部病害分值在接种后的22天之内持续提高，从57%增加到86%。在微笑杜鹃花中观察到相反的趋势，即在大多数的采样时间里，高温胁迫和接种后的根腐病症状极少，等于或低于无胁迫的植株。

淹水和热胁迫实验的综合结果表明，杜鹃花属类群的基因型多样，包括易感和抗性基因型，都容易受到疫霉属（*Phytophthora*）根腐病的胁迫。微笑杜鹃花是一个明显的例外，它不仅具有抗病性，而且对由非生物胁迫（如淹水和高温等）导致的病害具有较低的易感性。这些都是珍贵的性状，特别是在培育可持续和适应能力强的园林植物的情况下。具有抗病性和抗热性等特性将使大叶杜鹃更容易被消费者种植，并在较温暖地区拓展其园艺应用。这些性状还可减少生产苗圃中用于控制疫霉的杀菌剂的使用。更有甚者，路易斯安那州的试验为更多北纬地区提供了气候变化模型，以预测植物在更温暖和更潮湿条件下的表现，并得以全面了解。该试验结果表明，高山杜鹃对更温暖和更潮湿环境的适应能力在很大程度上取决于对根腐病的抵抗力，这一发现引起了更广泛的关注。如流行病学家所预测的，如果气候的变化增加了植物病原菌的地理范围及活性，那么抗病性对于植物健康将变得越来越重要。进一步研究微笑杜鹃花疫霉根腐病抗性的遗传和生理基础及其限制易感环境压力影响的能力，可能会为旨在减轻气候变化影响的其他育种计划提供参考。

26.8 栎树猝死病菌（*Phytophthora ramorum*）抗性（叶枯病）

2002 年，一种新发现的疫霉真菌，栎树疫霉（*Phytophthora*、*P. ramorum*）被发现与两大洲不同植物物种的病害相关（Rizzo et al. 2002；Werres et al. 2001）。在北美洲，该病原菌引起加利福尼亚州沿海栎树物种上的溃疡病，该病被称为栎树猝死病或 SOD。在欧洲（德国和荷兰），最初是在苗圃中观赏植物的树枝和叶上出现干枯症状，该病现在称为叶斑病（ramorum leaf bligh）或枯梢病（ramorum dieback disease）。自发现以来，病原菌的寄主范围已增加到 100 多种物种（APHIS 2013）。高山杜鹃和杜鹃花、荚蒾、山茶花、月桂树、木兰和丁香等许多具有商业价值的园林植物都被列在 APHIS 的监管苗木名录中，因为感病植物的大量运输对自然生态系统构成了重大威胁。

该病原菌已在北加利福尼亚州（Northern California）的森林群落中自然归化到俄勒冈州南部（Southern Oregon），在那里引起几种本地栎树种的树干出现 SOD 病害，且随后死亡。在大多数公认的寄主植物上，该病并不是致命的——症状仅限于叶片斑点和枝条枯死。尽管如此，这些寄主植物在野外和商业环境中都是栎树疫霉的重要传播媒介。在加利福尼亚州（California），当地月桂树（*Umbellularia californica*）的林下被认为是最具栎树疫霉感染力的媒介（Garbelotto and Hayden 2012）。在英国，侵入性杜鹃花物种长序杜鹃花是栎树疫霉和 *P. kernoviae* 两种病原菌的主要载体，其中的 *P. kerorviae* 是一种新发现的疫霉真菌，不仅能引起 SOD，也会对欧洲山毛榉（*Fagus sylvatica*）产生健康危害（Brasier et al. 2005）。受侵染苗木暴发病害已有记录（Lane et al. 2003），高山杜鹃特别受人关注，因为它们是欧洲苗圃植物检疫调查中栎树疫霉的最常见寄主（De Dobbelaere et al. 2010）。

苗圃中病害的控制遵循最佳管理实践指南，旨在避免病害发生或使病害发生最小化。这些措施包括经常进行苗圃检查、灌溉管理、高风险和低风险苗木的混合使用，适当的卫生措施，以控制病害的暴发。对于流行的观赏植物，如作为栎树疫霉主要寄主和潜在传播载体的杜鹃花，对具有叶枯病抗性的商业库存（物种和品种）的鉴定需求及培育新资源以替代高风险基因型的需求均是与日俱增的。

在 Tooley 等（2004）的一项研究中，确定了杜鹃花属和其他杜鹃花科的观赏植物对栎树疫霉的反应差异。De Dobbelaere 等（2010）随后编写了欧洲贸易中常见的杜鹃花属分类筛选手册。综合起来，这些报告包括对 78 个栽培品种和 33 个物种的高山杜鹃与杜鹃花的病害评估，其中一些物种被纳入两次调查，可作交叉参考。

表 26.6 列出了这些研究中最具抗性的分类单元。病害的严重程度受许多变量的影响，包括叶片是否存在伤口、叶龄（幼嫩或完全展开）、接种体（孢子囊）密度、环境条件以及整株植物与分离叶的接种量等（De Dobbelaere et al. 2010；Tooley et al. 2004）。

表 26.6 中，在不带伤口的嫩叶上接种后的第 6 天，叶感染面积≤10%的，确定为抗性植物。相反，＞30%的叶面积有栎树疫霉病灶的为较易感基因型。在杜鹃花属中已报道了来自欧洲（EU A1）和北美洲（NA A1、NA A2）的 3 种栎树疫霉的克隆谱系/交配类型在侵染性上的差异（Elliott et al. 2011）。然而，Tooley 等（2004）并未发现分离株

表 26.6 Tooley 等（2004）和 De Dobbelaere 等（2010）报道的具栎树疫霉抗性的杜鹃花属物种与品种名录

类群（组） N= 参与评价的组所包括的种类数
落叶映山红种类（羊踯躅组） $N=6$ 树形杜鹃花 *R. aborescens* 粘杜鹃花 *R. viscosum*
落叶映山红品种（羊踯躅组） $N=0$
常绿映山红种类（映山红组） $N=3$ 皋月杜鹃花‘马克’*R. indicum*‘Macrantha’ *R. macrosepalum* *R. yedoense* var. *poukhanense*
常绿映山红种类（映山红组） $N=12$ ‘吉拉德的玫瑰’*R.*‘Girard’s Rose’ ‘赫尔穆特’*R.*‘Helmut Vogel’ ‘日野赤红’*R.*‘Hino-Crimson’ ‘克林特夫人’*R.*‘Mrs. Klint’ ‘奥托’*R.*‘Otto’ ‘紫色光彩’*R.*‘Purple Splendor’
有鳞杜鹃花种类（杜鹃花组） $N=12$ 问客杜鹃花 *R. ambiguum* *R. campylogynum* var. *myrtilloides* *R. carolineanum* 朱砂杜鹃花 *R. cinnabarinum* 粉紫杜鹃花 *R. impeditum* *R. keiskei* 儒小杜鹃花 *R. minus* 腋花杜鹃花 *R. racemosum*
有鳞杜鹃花种类（杜鹃花组） $N=9$ ‘PJM’*R.*‘PJM’ ‘三叶草’*R.*‘Shamrock’
无鳞杜鹃花种类（常绿杜鹃花组） $N=14$ *R. arboreum* 不凡杜鹃花 *R. insigne* 极大杜鹃花 *R. maximum* *R. yakushimanum* 圆叶杜鹃花 *R. williamsianum*
无鳞杜鹃花种类（常绿杜鹃花组） $N=56$ ‘阿巴特’*R.*‘Albatross Townhill White’ ‘埃克斯伯瑞杂种’*R.*‘Exbury Hybrid’ ‘奇妙’R.‘Fantastica’ ‘花园丽格’*R.*‘Gartendirektor Rieger’ ‘金火炬’*R.*‘Golden Torch’ ‘哈夫丹’*R.*‘Halfdan Lem’ ‘林姆君主’*R.*‘Lem’s Monarch’ ‘摩根’*R.*‘Morgenrot’ ‘晨云’*R.*‘Morning Cloud’ ‘珀斯’*R.*‘Percy Wiseman’ ‘红杰克’*R.*‘Red Jack’ ‘三叶草’*R.*‘Shamrock’ ‘玳瑁橙’*R.*‘Tortoisechell Orange’

EU A1 和 NA A2 在均衡侵染寄主时导致的病害严重程度间存在显著差异，但是发现了寄主 × 分离株之间的显著互作。此处获得的筛选结果是用 EUA1 分离株接种的（De Dobbelaere et al. 2010）或分别来自 EU A1 和 NA A2 单独接种的平均病害反应（Tooley et al. 2004）。

总体而言，两项研究关于分类单元获得的结果是一致的。在这两个研究中，诸如品种‘诺娃’和‘紫水晶’（*R.*‘Roseum elegans’）均易感病，并且交叉证实了物种 *R. carolineanum* 和 *R. yakushimanum* 对栎树疫霉具有抗性。在所有主要观赏植物中均发现了抗性，但在常绿杜鹃花和无鳞杜鹃花中最为常见（约占所筛选物种和品种的 60%）（表 26.6）。常绿杜鹃花中包括一些总体上抗性最强的类群，感染栎树疫霉的叶面积不到 1%。由于受伤叶片的病害症状较高，因此两项研究均得出结论，抗病性机制是在叶片组织渗透水平上起作用的。抗病性可能来自形态性状，如有鳞杜鹃花上的鳞片和常绿杜鹃花上的叶毛，它们可以提供物理屏障或疏水性表面，从而阻止游动孢子到达叶面。尽管少有无鳞杜鹃（无鳞叶片）对栎树疫霉具有抗性（26%），但是值得注意的是，一些抗病性最强的物种，包括 *R. yakushimanum*、树形杜鹃花和不凡杜鹃花（*R. insigne*），在其叶背面有一层叶毛。

栎树疫霉的抗性育种可以为商业生产提供低风险的品种，同时减少那些威胁野生栎树、山毛榉和其他物种的病害暴发的可能性。目前，对抗病性遗传以及生物安全问题的了解缺乏阻碍了育种进程。由于批准用于田间测试植物的地点有限，通常在受控且封闭的环境中，对离体叶子或整株植物进行病害反应测定，而对田间条件的适用性未知。在苗圃棕栎木（*Notholithocarpus densiflorus*，是原产加利福尼亚的物种，极易感栎树猝死病）评估中，抗性的遗传力很低，但在田间评估中，亲子存活率的对应性更好，表明有用的附加遗传变异可用于育种。即使没有在杜鹃花属上进行正常的杂交试验，但从物种和杂种调查中也可以明显看出，对栎树疫霉的抗性（或敏感性）是可以遗传的。

例如，酒红杜鹃花、长序杜鹃花两个物种均易感病，它们均为易感品种‘诺娃’、‘紫水晶’和‘乔伊’（*R.*‘Chionoides’）提供了部分遗传背景（De Dobbelaere et al. 2010；Tooley et al. 2004）。抗性无鳞杜鹃花 *R. yakushimanum* 为抗性品种‘晨云’（*R.*‘Morning Cloud’）、‘莫格若特’（*R.*‘Morganrot’）、‘珀斯’（*R.*‘Percy Wiseman’）和‘幻想曲’（*R.*‘Fantastica’）贡献了 50%的基因（表 26.6）。其他更多的物种，如 *R. keiskei*（杜鹃花组）、*R. yedoense* var. *poukhanense*（映山红组）和圆叶杜鹃花（常绿杜鹃花组），分别为它们的杂交后代‘三叶草’（*R.*‘Shamrock’）、‘紫色光彩’（*R.*‘Purple Splendor’）和‘离歌’（*R.*‘Gartendirektor Rieger’）提供了抗病性状。虽然映山红未在任何一项研究中进行过测试，但其对疫霉的抗性可从它的 3 个杂交种——‘赫尔穆特’（*R.*‘Helmut Vogel’）、‘奥托’（*R.*‘Otto’）和‘金夫人’（*R.*‘Mrs. Kint’）近乎为零的叶片感染结果中推断得出。因此，抗病性的鉴定和筛选具有多种应用。它提供了一系列低风险的高山杜鹃和杜鹃花品种清单，可立即用于针对病害防控的苗圃计划中，还应确定使用哪个类群进行抗性育种，并通过比较遗传背景和谱系来预测还有哪些未经测试的材料可能具有抗性。

26.9　未来园林中的杜鹃花

高山杜鹃和杜鹃花是需要相对严格的条件才能达到最佳生长状态的植物。在野外，它们分布于气候温和的潮湿山地生态系统下，生长在养分含量低的有机酸性土壤中，从而减少了与其他植物的竞争。杜鹃花属的育种，在适当结合文化习俗发展的同时，已经能够将园艺用途拓展作为次要条件，以使这些美丽的植物可供更广泛的公众使用。然而，如今的气候变化可能给花园景观及野外生长的高山杜鹃和杜鹃花施加了过大的环境压力。非生物胁迫还可伴随生物胁迫，如温度和降雨的增加可能会增加植物病原菌的发生范围与病害压力。此外，全球化带来的异地病虫害的快速传播也构成了严重的健康威胁。这些问题并非杜鹃花属所独有，但该属提供了一个解释，以说明适应性状的自然变异如何减轻气候变化和入侵物种的影响。

当然可能存在局限性。例如，干旱是某些温带地区气候变化的预测结果，杜鹃花属的耐旱性尚未得到科学证明。鉴于其偏爱的生境和浅根系统，耐旱的高山杜鹃和杜鹃花将可能是自然规则的一个意外。但是，有针对性的探索可能会发现野外一些具有一定耐旱性的植物。例如，北美洲的有鳞杜鹃花海滨杜鹃花（*R. macrophyllum*）种群沿着太平洋海岸生长，那里属于地中海气候，夏季非常干燥。把这些种群与那些生长在内陆较高海拔，每年获得更多降雨的同一物种进行比较，将具有指导意义。在缺乏耐旱性的情况下，未来在花园中使用杜鹃花的可持续性可能会大大降低。

降雨的增加，特别是强降雨的增加，也是对气候变化的一种预测，并且已经有文献记载（Min et al. 2011）。在这种情况下，杜鹃花或许可以适应。杜鹃花可以忍受至少 48h 的淹水而不受伤害（Blaker and MacDonald 1981）。但是淹水会加剧根腐病，因为过量的土壤水分促使疫霉病原菌发育至孢子阶段，从而产生易于传播并侵染根茎的游动孢子。上面所述的根腐病抗性育种工作表明，来自微笑杜鹃花的疫霉抗性基因的渗入可以培育出能在有病原菌的温暖湿润环境中存活的杂种。路易斯安那州南部的平均降雨量偏高（测试地点附近为 1668mm）（表 26.5），通常为强降雨，能导致土壤流失，在某些情况下还会发生严重洪灾，如在 2016 年 8 月连续 3 天累积的 762mm 降雨量。除具有抗病性外，微笑杜鹃花还具有其他特性，特别是耐热性，突出表现在其厚实的叶簇和较高的光合最适温度（photosynthetic temperature optimum）上（Ranney et al. 1995）。微笑杜鹃花是野生多样性的一个很好的例子，有兴趣培育更能耐受气候变化相关压力的植物的育种者可以利用它。

尽管有全球温度升高的记录和预测，但在可预见的将来，耐寒性将继续成为温带地区景观的重要特征。在同时考虑平均温度和温度变化增加的气候模型中，寒冷的天气模式不会明显改变，而炎热气候变得愈加普遍。近几十年来，北极出现的迅速变暖现象与北半球异常寒冷的冬季相逢。这些事件是由于极地涡旋向南移动导致中纬度地区出现严寒，这种趋势的起源和持续时间是科学家争论的话题（Overland et al. 2016；Sun et al. 2016）。考虑到未来霜冻的危害，长寿命植物的育种者应采取保守立场，并继续选择可能的最耐寒材料。

杜鹃花未来性状的培育还应考虑替代的育种产品和商业的实践。本章中讨论的许多性状特征都与土壤环境中的胁迫——极端水分（淹水、干旱）、高 pH、盐胁迫和病害有关。鉴于足够的遗传多样性，培育和选择可以承受这些诸多压力的个体植物是有可能的。但是，每次引种都要经过土壤环境的各种胁迫过程，这非常耗时，因为它通常涉及植株繁殖，以及实施多个田间试验并进行数年的试验。对于较早描述的抗根腐病/耐热杜鹃花，从种子起至少需要 10 年才能确定适合商业引种的品种。一种有吸引力且最终更有效的方法是选择个体作为砧木，用于赋予这些植物土壤适应性。

杜鹃花的嫁接在欧洲市场很普遍，而在北美洲则尚未进行。北美洲的品种采用自根繁殖（通过扦插或组织培养），然后在容器中种植直至销售。这种自生根的方法通过迅速将新植物推向市场并保持低成本而使种植者和消费者受益。但是，当将盆栽植物移至家庭景观中“真实的”或不合标准的土壤条件下时，其在无土、抑制病害的树皮混合物中生长的自生根杜鹃的生产益处将不复存在。大多数消费者在种植杜鹃花的过程中并不了解或不愿意正确地栽种和调配土壤，这对许多新品种而言是灾难，比起‘铁甲’被证实的表现，这些品种很少被正确栽种并使得品种特性真实呈现。

使用有活力和适应性强的砧木可以解决杜鹃花所面临的许多与土壤有关的基质问题，并使它们在日渐衰落的北美洲市场上得以恢复。具有诸如高 pH 耐受性、耐旱性和抗病性等特性的砧木除可以提高高山杜鹃与杜鹃花在现有市场中的份额外，还可以为高山杜鹃和杜鹃花开辟新的地理市场。砧木生产方法还可以增加杜鹃花的观赏性，因为可以将新品种嫁接到成熟的砧木上，而不是将相同的自生根植物分配到未经测试的地区。最后，通过嫁接砧木进行繁殖可以显著减少将新型高山杜鹃和杜鹃花推向市场所需的研发时间。虽然砧木的发育（确定适应性和亲和性）本身是一个漫长的过程，但单个终极产品可以广泛用于许多品种的接穗。与评估每个新候选品种在其自生根上的等效适应性相比，这花费的时间要少得多。

本章译者：

解玮佳[1]，张春英[2]

1. 云南省农业科学院花卉研究所，国家观赏园艺工程技术研究中心，云南省花卉育种重点实验室，昆明 650200

2. 上海植物园，上海城市植物资源开发应用工程技术研究中心，上海 200231

参 考 文 献

Aerts R (1995) The advantages of being evergreen. Trends Ecol Evol 10:402–407

Agrios GN (1997) Plant pathology. Academic, New York, pp 266–226

Alexander P (2008) Are INKARHO rhododendrons more lime tolerant than traditional calcifuge rhododendrons? SEESOIL 17:8–17

American Rhododendron Society. https://www.rhododendron.org/search_intro.asp

Ammal EKJ (1950) Polyploidy in the genus *Rhododendron*. Rhododendron Year Book 5:92–98

Ammal EKJ, Enoch IC, Bridgwater M (1950) Chromosome numbers in species of *Rhododendron*. Rhododendron Year Book 5:78–91

Anderson PK, Cunningham AA, Patel NG, Morales FJ, Epstein PR, Daszak P (2004) Emerging infectious diseases of plants: pathogen, pollution, climate change, and agrotechnology drivers.

Trends Ecol Evol 19:535–544
APHIS (2013) www.aphis.usda.gov/plant_health/plant_pest_info/pram/downloads/pdf_files/usdaprlist.pdf
Arisumi K, Matsuo E, Sakata Y, Tottoribe T (1986) Breeding for heat resistant rhododendrons. Part 2: differences in heat resistance among species and hybrids. J Am Rhod Soc 40:215–219
Arora R, Wisniewski ME, Scorza R (1992) Cold acclimation in genetically related (sibling) deciduous and evergreen peach (Prunus persica [L.] Batsch). I. Seasonal changes in cold hardiness and polypeptides of bark and xylem tissues. Plant Physiol 99:1562–1568
Arora R, Rowland LJ, Panta GR (1997) Chill-responsive dehydrins in blueberry: Are they associated with cold hardiness or dormancy transitions? Physiol Plant 101:8–16
Atkinson R, Jong K, Argent G (2000) Chromosome numbers of some tropical rhododendrons (section *Vireya*). Edinburgh J Bot 57:1–7
Bao F, Du AY, Yang W, Wang J, Cheng T et al (2017) Overexpression of *Prunus mume* dehydrin genes in tobacco enhances tolerance to cold and drought. Front Plant Sci 8:151. https://doi.org/10.3389/fpls.2017.00151
Barlup J (2002) Let's talk hybridizing: hybridizing with elepidote polyploid rhododendrons. J Am Rhod Soc 76:75–77
Bennett MD (2004) Perspectives on polyploidy in plants – ancient and neo. Bio J Linn Soc 18:411–423
Benson DM (1980) Resistance of evergreen azalea to root rot caused by *Phytophthora cinnamomi*. Plant Dis 64:214–215
Benson DM (1987) Occurrence of *Phytophthora cinnamomi* on roots of azalea treated with pre-inoculation and post-inoculation applications of metalaxyl. Plant Dis 71:818–820
Bergot M, Cloppet E, Perarnaud V, Deque M, Marcaiss B, Desprez-Loustau ML (2004) Simulation of potential range expansion of oak disease caused by *Phytophthora cinnamomi* under climate change. Glob Chang Biol 10:1539–1552
Blaker NS, MacDonald JD (1981) Predisposing effects of soil moisture extremes on the susceptibility of rhododendron to *Phytophthora* root and crown rot. Phytopathology 71:831–834
Bostock RM, Pye MF, Roubstova TV (2014) Predisposition in plant disease; exploiting the nexus in abiotic and biotic stress perception and response. Annu Rev Phytopathol 52:517–549
Brasier CM, Scott JK (1994) European oak declines and global warming: a theoretical assessment with special reference to the activity of *Phytophthora cinnamomi*. EPPO Bull 25:221–232
Brasier CM, Beales PA, Kirk SA, Denman S, Rose J (2005) *Phytophthora kernoviae* sp. nov., an invasive pathogen causing bleeding stem lesions on forest trees and foliar necrosis of ornamentals in the UK. Mycol Res 109:853–859
Brasier CM, Vettraino AM, Chang TT, Vannini A (2010) *Phytophthora lateralis* discovered in an old growth *Chaemaecyparis* forest in Taiwan. Plant Pathol 59:595–603
Burgess TI, Scott JK, McDougall KL, Stukely MJ, Crane C, Dunstan WA, Brigg F, Andjic V, White D, Rudman T, Arentz F, Ota N, Hardy GE St J (2016) Current and projected global distribution of *Phytophthora cinnamomi*, one of the world's worst plant pathogens. Climate Change Biol 23:1661–1674
Bush EA, Hong C, Stromberg EL (2003) Fluctuations of *Phytophthora* and *Pythium* spp. in components of a recycling irrigation system. Plant Dis 87:1500–1506
Chaanin A (1998) Lime tolerance in rhododendron. Combined Proc Int Plant Prop Soc 48:180–182
Chaanin A, Preil W (1994) Influence of bicarbonate on iron deficiency chlorosis in *Rhododendron*. Acta Hortic 364:71–77
Chung JD, Lin TP, Chen YL, Chen YP, Hwang SY (2007) Phylogeographic study reveals the origin and evolutionary history of a *Rhododendron* species complex in Taiwan. Mol Phylogenet Evol 42:14–24. https://doi.org/10.1016/j.ympev.2006.06.027
Contreras RN, Ranney TG, Tallury SP (2007) Reproductive behavior of diploid and allotetraploid *Rhododendron* L. 'Fragrant Affinity'. Hortscience 42:31–34
Cox PA (1979) The larger species of *Rhododendron*. BT Batsford, London
Cox PA (1985) The smaller rhododendrons. Timber Press, Portland
Cox PA (1993) The cultivation of rhododendrons. BT Batsford, LTD, London
Creech JL (1978) A distribution note on *Rhododendron tamurae*. J Am Rhod Soc 32:100
Daugherty ML, Benson DM (2001) Rhododendron diseases. In: Jones RK, Benson DM (eds) Diseases of woody ornamentals and trees in nurseries. APS Press, St. Paul, pp 334–335
De Dobbelaere I, Vercauteren A, Speybroek N, Berkvens D, Van Bockstaele E, Maes M, Heungens K (2010) Effect of host factors on the susceptibility of *Rhododendron* to *Phytophthora ramorum*. Plant Pathol 59:301–312

De Schepper S, Leus L, Mertens M, Van Bockstaele E, De Loose M, Debergh P, Heursel J (2001) Flow cytometric analysis of ploidy in *Rhododendron* subgenus *Tsustusi*. Hortscience 36:125–127

Die JV, Arora R, Rowland LJ (2017) Proteome dynamics of cold-acclimating *Rhododendron* species contrasting in their freezing tolerance and thermonasty behavior. PLoS One 12(5):e0177389. https://doi.org/10.1371/journal.pone.0177389

DiLeo MV, Pye MF, Roubstova TV, Duniway JM, MacDonald JD, Rizzo DM, Bostock RM (2010) Abscisic acid in salt stress predisposition to *Phytophthora* root rot and crown rot in tomato and chrysanthemum. Phytopathology 100:871–879

Doss RP (1984) Role of glandular scales of lepidote rhododendrons in insect resistance. J Chem Ecol 10:1787–1798

Dunemann F, Kahnau R, Stange I (1999) Analysis of complex leaf and flower characters in *Rhododendron* using a molecular linkage map. Theor Appl Genet 98:1146–1155

Elliott M, Sumampong G, Varga A, Shamoun SF, James D, Masri S, Grünwald NJ (2011) Phenotypic differences among three clonal lineages of *Phytophthora ramorum*. For Pathol 41:7–14. https://doi.org/10.1111/j.1439-0329.2009.00627.x

Galle FC (1985) Azaleas. Timber Press, Portland

Garbelotto M, Hayden KJ (2012) Sudden Oak Death: interactions of the exotic oomycete *Phytophthora ramorum* with naïve North American hosts. Eukaryot Cell 11:1313–1323

George MF, Burke MJ, Weiser CJ (1974) Supercooling in overwintering azalea flower buds. Plant Physiol 54:29–35

Giel P, Bojarczuk K (2011) Effects of high concentrations of calcium salts in the substrate and its pH on the growth of selected rhododendron cultivars. Acta Soc Bot Pol 80:105–111

Goetsch L, Eckert AJ, Hall BD (2005) The molecular systematics of *Rhododendron* (Ericaceae): a phylogeny based upon RPB2 gene sequences. Syst Bot 30:616–626

Guy CL (1990) Cold acclimation and freezing stress tolerance: role of protein metabolism. Ann Rev Plant Physiol 41:187–223

Hanger BC, Bjarnson EN, Osborn RK (1981) The growth of rhododendrons in containers in soil, treated with either $CaCO_3$ or Ca SO_4. Plant Soil 61:479–483

Harris GC, Antoine V, Chan M, Nevidomskyte D, KoÈniger M (2006) Seasonal changes in photosynthesis, protein composition and mineral content in *Rhododendron* leaves. Plant Sci 170:314–325

Hayden KJ, Garbelotto M, Dodd R, Wright JW (2013) Scaling up from greenhouse resistance to fitness in the field for a host of an emerging forest disease. Evol Appl 6:970–982

Heursel J, De Roo R (1981) Polyploidy in evergreen azaleas. Hortscience 16:765–766

Hoitink HAJ, Schmitthenner AF (1974) Resistance of rhododendron species and hybrids to *Phytophthora* root rot. Plant Dis Rep 58:650–653

Hoitink HA, Van Doren JDM Jr, Schmitthenner AF (1977) Suppression of *Phytophthora cinnamomi* in composted hardwood bark potting medium. Phytopathology 67:561–565

Hoitink HAJ, Benson DM, Schmitthenner AF (1986) Diseases caused by fungi: *Phytophthora* root rot. In: Coyier DL, Roane MK (eds) Compendium of rhododendron and azalea diseases. APS Press, St. Paul, pp 4–8

Hokanson SC, McNamara S, Zuzek K, Rose N (2005) *Rhododendron* 'Candy Lights' and 'Lilac Lights'. Hortscience 40:1925–1927

Hosoda T, Moriya A, Sarahima S (1953) Chromosome numbers of satsuki, *Rhododendron lateritium*. Genetica 26:407–409

Houde M, Dallaire S, N'Dong D, Sarhan F (2004) Overexpression of the acidic dehydrin WCOR410 improves freezing tolerance in transgenic strawberry leaves. Plant Biotechnol J 2:381–387

Irving E, Hebda R (1993) Concerning the origin and distribution of *Rhododendron*. J Am Rhod Soc 47:139–162

Ishikawa M, Sakai A (1981) Freezing avoidance mechanisms by supercooling in some *Rhododendron* flower buds with reference to water relations. Plant Cell Physiol 22:953–967

Ismail AM, Hall AE, Close TJ (1999) Allelic variation of a dehydrin gene cosegregates with chilling tolerance during seedling emergence. PNAS 96:1356–13570

Johnson GR, Hirsch AG (1995) Validity of screening for foliage cold hardiness in the laboratory. J Environ Hortic 13:26–30

Jones JR, Ranney TG, Lynch NP, Krebs SL (2007) Ploidy levels and relative genome sizes of diverse species, hybrids, and cultivars of Rhododendron. J Am Rhod Soc 61:220–227

Kaisheva, ME (2006) The effect of metals and soil pH on the growth of *Rhododendron* and other alpine plants in limestone soil. Dissertation. University of Edinburgh www.era.lib.ed.ac.uk/handle/1842/2606

Kaku S (1993) Monitoring stress sensitivity by water proton NMR relaxation times in leaves of azaleas that originated in different ecological habitats. Plant Cell Physiol 34:535–541

Kehr AE (1976) Polyploids in rhododendron breeding. J Am Rhod Soc 50:215–217

Kehr AE (1977) Azaleodendron breeding. J Am Rhod Soc 31:226–232

Kho WH, Chang S, Su CH (1978) Isolates of *Phytophthora cinnamomi* from Taiwan as evidence for an Asian origin of the species. Trans Br Mycol Soc 71:496–499

Kong P, Hong CX, Richardson PA (2003) Rapid detection of *Phytophthora cinnamomi* using PCR with primers derived from the *Lpv* putative storage protein genes. Plant Pathol 52:681–693

Krebs SL (1996) Normal segregation of allozyme markers in complex rhododendron hybrids. J Hered 87:131–135

Krebs SL (2005) Loss of winter hardiness in *R*. 'Supernova', an artificial polyploid. J Am Rhod Soc 59:74–75

Krebs SL (2009) Breeding rhododendrons resistant to *Phytophthora* root rot disease. In: Roy J (ed) The world of Rhododendrons: Proceedings of the 2008 conference at the Royal Botanic Garden Edinburgh. RBGE Yearbook No. 11, Edinburgh, UK, pp 53–58

Krebs SL (2013) Resistance to *Phytophthora* root rot varies among rhododendrons subjected to repeated flooding in the field. Acta Hortic 990:243–252

Krebs, SL (2018). Heat-induced predisposition to *Phytophthora* root rot disease in *Rhododendron*. Acta Horticulturae 1191:59–68

Krebs SL, Wilson MD (2002) Resistance to *Phytophthora* root rot in contemporary rhododendron cultivars. Hortscience 37:790–792

Lane CR, Beales P, Hughes KJD, Webber J (2003) First outbreak *of Phytophthora ramorum* in England, on *Viburnum tinus*. Plant Pathol 52:414. https://doi.org/10.1046/j.1365-3059.2003.00835.x

Leach DG (1961) Rhododendrons of the world and how to grow them. Sribner's Sons, New York

Lee JA, Woolhouse HW (1969) A comparative study of bicarbonate inhibition of root growth in calcicole and calcifuge grasses. New Phytol 68:1–11

Leiser AT (1957) *Rhododendron occidentale* on alkaline soil. Rhododendron Camellia Yearbook 11:47–51

Leslie A (2004) The international *Rhododendron* register and checklist, 2nd edn. Royal Horticultural Society, London

Levin DA (1983) Polyploidy and novelty in flowering plants. Am Nat 122:1–25

Li H (1957) Chromosome studies in the azaleas of eastern North America. Am J Bot 44:8–14

Lim CC, Arora R, Townsend ED (1998a) Comparing *Gompertz* and *Richards* functions to estimate freezing injury in *Rhododendron* using electrolyte leakage. J Am Soc Hortic Sci 123:246–252

Lim CC, Arora R, Krebs SL (1998b) Genetic study of freezing tolerance in Rhododendron populations: implications for cold hardiness breeding. J Am Rhod Soc 52:143–148

Lim CC, Krebs SL, Arora R (1999) A 25-kDa dehydrin associated with genotype- and age-dependent leaf freezing-tolerance in *Rhododendron*: a genetic marker for cold hardiness? Theor Appl Genet 99:912–920. https://doi.org/10.1007/s001220051312

Lim CC, Krebs SL, Arora R (2014) Cold hardiness increases with age in juvenile *Rhododendron* populations. Front Plant Sci 5:542. https://doi.org/10.3389/fpls.2014.00542

Lipp CC, Nilsen ET (1997) The impact of subcanopy light environment on the hydraulic vulnerability of *Rhododendron maximum* to freeze-thaw cycles and drought. Plant Cell Environ 20:1264–1272

Livingston PH, West FH (eds) (1978) Hybrids and Hybridizers; rhododendrons and azaleas for eastern North America. Harrowood Books, Newtown Square

Lunin J, Stewart FB (1961) Soil salinity tolerance of azaleas and camellias. Proc Am Soc Hortic Sci 77:528–532

MacDonald JD (1982) Effect of salinity stress on the development of *Phytophthora* root rot of *Chrysanthemum*. Phytopathology 72:214–219

MacDonald JD (1991) Heat stress enhances *Phytophthora* root rot severity in container-grown chrysanthemums. J Am Soc Hortic Sci 116:36–41

Madsen K (2000) In pursuit of ironclads. Arnoldia 60:29–32

Marian CO, Krebs SL, Arora R (2004) Dehydrin variability among rhododendron species: a 25-kDa dehydrin is conserved and associated with cold acclimation across diverse species. New Phytol 97:773–780. https://doi.org/10.1111/j.1469-8137.2003.01001.x

Mason AS, Nelson MN, Yan G, Cowling WA (2011) Production of viable male unreduced gametes

in *Brassica* interspecific hybrids is genotype specific and stimulated by cold temperatures. BMC Plant Biol 11:103. https://doi.org/10.1186/1471-2229-11-103
McAleese AJ, Rankin DWH (2000) Growing rhododendrons on limestone soils: is it really possible? J Am Rhod Soc 54:126–134
McDougall KL, Hobbs RJ, Hardy GESJ (2005) Distribution of understory species in forest affected by *Phytophthora* in south-western Western Australia. Aus J Bot 55:813–819
Milbocker DC (1988) Salt tolerance of azalea cultivars. J Am Soc Hortic Sci 113:79–84
Milne R, Abbott RJ (2000) Origin and evolution of invasive naturalized material of Rhododendron ponticum L. in the British isles. Mol Ecol 9:541–556
Min SK, Zhang X, Zwiers FW, Hegerl GC (2011) Human contribution to more-intense precipitation extremes. Nature 470:378–381. https://doi.org/10.1038/nature09763
Moe S, Pellett H (1986) Breeding for cold hardy azaleas in the land of the northern lights. J Am Rhod Soc 40:203–205
Mohr PG, Cahill DM (2007) Suppression by ABA of salicylic acid and lignin accumulation and the expression of multiple genes in *Arabidopsis* infected with *Pseudomonas syringae* pv. *tomato*. Funct Integr Genomics 7:181–191
Moons A, Bauw G, Prinsen E, Van Montagu M, Van der Straeten D (1995) Molecular and physiological responses to abscisic acid and salts in roots of salt-sensitive and salt-tolerant Indica rice varieties. Plant Physiol 107:177–186
Mordhorst AP, Kullik C, Preil W (1993) Ca uptake and distribution in *Rhododendron* selected for lime tolerance. Gartenbauwissenschaft 58:111–116
Nakamura M (1931) Cytological studies on the genus *Rhododendron*. J Soc Trop Agric 3:103–109
Nelson S (2000) The Pacific rhododendron story; the hybridizers, collectors, and gardens. Binford and Mort Publishers Hillsboro, Portland
Otto SP (2007) The evolutionary consequences of polyploidy. Cell 131:452–462
Overland JE, Dethloff K, Francis JA, Hall RJ, Hanna E, Kimm SJ, Screen JA, Shepherd TG, Vihma T (2016) Nonlinear response of mid-latitude weather to the changing Arctic. Nat Clim Chang 6:992–999. https://doi.org/10.1038/nclimate3121
Pan A, Hayes PM, Chen F, Blake T, Chen THH, Wright TTS, Karsai I, Bedo Z (1994) Genetic analysis of the components of winter hardiness in barley (*Hordeum vulgare* L.). Theor Appl Genet 89:900–910
Panta GR, Rowland LJ, Arora R, Ogden EL, Lim CC (2004) Inheritance of cold hardiness and dehydrin genes in diploid mapping populations of blueberry. J Crop Improv 10:37–52. https://doi.org/10.1300/J411v10n01_04
Peng Y, Lin W, Wei H, Krebs SL, Arora R (2008a) Phylogenetic analysis and seasonal cold acclimation-associated expression of early light-induced protein genes of *Rhododendron catawbiense*. Physiol Plant 132:44–52
Peng Y, Reyes JL, Wei H, Yang Y, Karlson D, Covarrubias AA, Krebs SL, Fessehaie A, Arora R (2008b) RcDhn5, a cold acclimation-responsive dehydrin from *Rhododendron catawbiense* rescues enzyme activity from dehydration effects in vitro and enhances freezing tolerance in RcDhn5-overexpressing *Arabidopsis* plants. Physiol Plant 134:583–559. https://doi.org/10.1111/j.1399-3054.2008.01164.x
Perkins S, Perkins J, Castro M, De Oliveira JC, Castro S, Loureiro J (2015) More weighings: exploring the ploidy of hybrid elepidote rhododendrons. The Azalean 37:28–42
Preil W, Ebbinghaus R (1994) Breeding of lime tolerant *Rhododendron* rootstocks. Acta Hortic 364:61–70
Pryor RL, Frazier LC (1970) Triploid azaleas of the Belgian-Indian series. Hortscience 5:114–115
Qiu H, Zhang L, Liu C et al (2014) Cloning and characterization of a novel dehydrin gene, *SiDhn2*, from *Saussurea involucrata* Kar. et Kir. Plant Mol Biol 84:707–718. https://doi.org/10.1007/s11103-013-0164-7
Ramsey J, Schemske DW (1998) Pathways, mechanisms, and rates of polyploidy formation in flowering plants. Ann Rev Ecol Syst 29:467–501
Ranney TG, Blazich FA, Warren SL (1995) Heat tolerance of selected species and populations of *Rhododendron*. J Am Soc Hortic Sci 120:423–428
Rizzo DM, Garbelotto M, Davidson JM, Slaughter GW, Koike ST (2002) *Phytophthora ramorum* as the cause of extensive mortality of *Quercus* spp. and *Lithocarpus densiflorus* in California. Plant Dis 86:205–214
Roubtsova TV, Bostock RM (2009) Episodic abiotic stress as a potential contributing factor to onset and severity of disease caused by *Phytophthora ramorum* in *Rhododendron* and *Viburnum*.

Plant Dis 93:912–918
Rouse JL, Williams EG, Knox RB (1988) A vireya azaleodendrdon in flower. J Am Rhod Soc 42:166–167
Sakai A, Fuchigami L, Weiser CJ (1986) Cold hardiness in the genus *Rhododendron*. J Am Soc Hortic Sci 111:273–280
Sakai A, Larcher W (1987) Frost survival of plants: responses and adaptations to freezing stress. Ecol Stud 62. Springer, Berlin
Salley HE, Greer HE (1986) Rhododendron hybrids; a guide to their origins. Timber Press, Portland
Sax K (1930) Chromosome stability in the genus rhododendron. Am J Bot 17:247–251
Shuichiro Tagane, Michikazu Hiramatsu, Hiroshi Okubo, (2008) Hybridization and asymmetric introgression between Rhododendron eriocarpum and R. indicum on Yakushima Island, southwest Japan. Journal of Plant Research 121 (4):387–395
Soltis DE, Albert VA, Leebens-Mack J, Bell CD, Paterson AH, Zheng C, Sankoff D, de Pamphilis CW, Kerr Wall P, Soltis PS (2009) Polyploidy and angiosperm diversification. Am J Bot 96:336–348. https://doi.org/10.3732/ajb.0800079
Soltis PS, Liu X, Marchant DB, Visger CJ, Soltis DE (2014) Polyploidy and novelty: Gottlieb's legacy. Phil Trans R Soc B 369. https://doi.org/10.1098/rstb.2013.0351
Sun L, Perlwitz J, Hoerling M (2016) What caused the recent "Warm Arctic, Cold Continents" trend pattern in winter temperatures? Geophys Res Lett 43:1–8. https://doi.org/10.1002/2016GL069024
Susko AQ, Bradeen JM, Hokanson SC (2016) Towards broader adaptability of North American deciduous azaleas. Arnoldia 74:15–27
Thammina C, He M, Lu L, Cao K, Yu H, Chen Y, Tian L, Chen J, McAvoy R, Ellis D, Wang Y, Zhang X, Li Y (2011) In vitro regeneration of triploid plants of *Euonymus alatus* 'Compactus' (burning bush) from endosperm tissues. Hortscience 46:1141–1147
Thomashaw MF (1990) Molecular genetics of cold acclimation in higher plants. Adv Genet 28:99–131
Thorn WA, Zentmyer GA (1952) Hosts of *Phytophthora cinnamomi* Rands, the causal organism of avocado root rot. Calif Avocado Society Yearbook 37:196–200
Thornton JT (1990) Breeding rhododendrons for the Gulf South. J Am Rhod Soc 44:91–93
Tigerstedt PMA, Uosukainen M (1996) Breeding cold hardy rhododendrons. J Am Rhod Soc 50:185–189
Tooley PW, Kyde KL, Englander L (2004) Susceptibility of selected ericaceous ornamental host species to *Phytophthora ramorum*. Plant Dis 88:993–999
Trueblood CE, Ranney TG, Lynch NP (2010) Evaluating fertility of triploid clones of *Hypericum androsaemum* L. for use as non-invasive landscape plants. Hortscience 45:1026–1028
Uosukainen M (1992) *Rhododendron brachycarpum* sub-sp. *tigerstedtii* Nitz. – a transmitter of extreme frost hardiness. Acta Hortic 320:77–83
Väinölä A (2000) Genetic and physiological aspects of cold hardiness in *Rhododendron*. Dissertation, University of Helsinki https://helda.helsinki.fi/handle/10138/20766
Väinölä A, Repo T (1999) Cold hardiness of diploid and corresponding autotetraploid rhododendrons. J Hortic Sci Biotechnol 74:541–546
Väinölä A, Repo T (2000) Impedance spectroscopy in frost hardiness evaluation of *Rhododendron* leaves. Ann Bot 86:799–805
van Zee K, Chen FQ, Hayes PM, Close TJ, Chen THH (1995) Cold-specific induction of a dehydrin gene family member in barley. Plant Physiol 108:1233–1239
Wei H, Dhanaraj AL, Rowland LJ, Fu Y, Krebs SL, Arora R (2005) Comparative analysis of expressed sequence tags from cold-acclimated and non-acclimated leaves of *Rhododendron catawbiense* Michx. Planta 221:406–416. https://doi.org/10.1007/s00425-004-1440-1
Werres S, Marwitz R, Man In't Veld WA, de Cock AWAM, Bonants PJM, de Weerdt M, Themann K, Ilieva E, Baayen RP (2001) *Phytophthora ramorum* sp. nov., a new pathogen on *Rhododendron* and *Viburnum*. Mycol Res 105:1155–1165
Widrlechner MP, Larson RA (1993) Exploring the deciduous azaleas and elepidote rhododendrons of the Midwestern United States. J Am Rhod Soc 47:153–156
Williams EG, Rouse JL, Palser BF, Knox RB (1990) Reproductive biology of rhododendrons. Hortic Rev 12:1–67
Wisniewski M, Fuller M, Palta J, Carter J, Arora R (2004) Ice nucleation, propagation, and deep supercooling in woody plants. J Crop Improv 10:5–16
Yasuda M, Ishikawa A, Jikumaru Y, Seki M, Umezawa T, Asami T, Maruyama-Nakashita A, Kudo T, Shinozaki K, Yoshida S, Nakashita H (2008) Antagonistic interaction between systemic

acquired resistance and the abscisic acid–mediated abiotic stress response in *Arabidopsis*. Plant Cell 20:1678–1692
Yin Z, Rorat T, Szabala BM, Ziółkowska A, Malepszy S (2006) Expression of a *Solanum sogarandinum* SK3-type dehydrin enhances cold tolerance in transgenic cucumber seedlings. Plant Sci 170:1164–1172. https://doi.org/10.1016/j.plantsci.2006.02.002
Zentmyer GA (1988) Origins and distribution of four species of *Phytophthora*. Trans Br Mycol Soc 91:367–378
Zörb C, Geilfus CM, Mühling KH, Ludwig-Müller J (2013) The influence of salt stress on ABA and auxin concentrations in two maize cultivars differing in salt resistance. J Plant Physiol 170:220–224

第 27 章　月　　季

Leen Leus, Katrijn Van Laere, Jan De Riek, and Johan Van Huylenbroeck

摘要　自古以来，月季作为一种观赏植物装饰着家庭和花园。月季被用于许多不同的观赏用途，如切花、园艺植物和盆栽植物，以及工业（香水）、医药和烹饪。月季在所有的观赏植物中具有最大的经济价值，在选择和育种方面有着悠久与文献记载的传统。月季有 30 000 多个品种，是所有作物中育种产量最大的物种，但新品种需求仍在持续增长。目前，寻找新的观赏性状仍然是主要的育种目标，此外，还寻求切花月季产品的差异化、增产以及对（新）产地的更好适应性特征。在庭院月季中，育种者挑选了对病害适应性较强的品种，包括月季丛矮病（RRD）等新的重要病害。月季育种具有挑战性是因为栽培月季的遗传背景非常狭窄、存在多倍性和/或倍性水平差异、生殖障碍（包括有限的繁殖力和发芽挑战）等。育种者需要掌握的技能知识包括性状背景知识、（新）分子技术和基因组信息。本章概述了月季育种、育种目标、杂交过程以及常规和分子育种工具方面的挑战。我们邀请了几位育种者，从现代月季育种的趋势和演变方面，分享了切花月季和庭院月季育种的实用性观点。

关键词　育种者的观点；切花月季；繁殖能力；庭院月季；月季育种；倍性；细胞遗传学；分子标记

27.1　引　　言

27.1.1　月季的分类

蔷薇属属于蔷薇科。后者按果实类型分为 4 个亚科：绣线菊亚科、李亚科、苹果亚科和包括蔷薇属的蔷薇亚科。大而多样的蔷薇科家族约有 3000 个种，约 100 个属。该植物家族中的 90 多个种具有重要的经济价值，包括水果（如苹果、桃子、草莓、李子）和观赏植物（如月季）（Longhi et al. 2014）。

蔷薇属为半木质的多年生植物，通常是灌木，有 5 个萼片和 5 个花瓣，雄蕊众多，丰富的花柱以黏性柱头为特征。玫瑰染色体（n）的基数为 7，倍性水平通常从二倍体（$2n = 2x = 14$）到四倍体（$2n = 4x = 28$）不等。然而，已经发现了六倍体（$2n = 6x = 42$）、八倍体（$2n = 8x = 56$）甚至十倍体（$2n = 10x = 70$）（Jian et al. 2010a）。在犬蔷薇组的物

L. Leus (✉) · K. Van Laere · J. De Riek · J. Van Huylenbroeck
Flanders Research Institute for Agriculture, Fisheries and Food (ILVO), Plant Sciences Unit,
Applied Genetics and Breeding, Melle, Belgium
e-mail: leen.leus@ilvo.vlaanderen.be

种中发现了不平衡的染色体数目和无融合生殖机制；这些高倍体物种是由五倍体 $2n = 5x = 35$（有时六倍性）演化的（Lim et al. 2005）。

除了一个非洲热带种以外，月季只在北半球温带和亚热带地区的自然植物区系中被发现。月季的分类是由 Crépin（1889）创始的，后来由 Rehder（1940）建立了一个可靠的分类。基于这种分类，Wissemann（2003）进行了补充和修正。对月季的系统发育进行了许多研究，但由于低序列差异的系统发育分辨率较弱，以及种间杂交事件普遍存在，研究结果并不总是一致，而且存在一些问题。此外，月季品种的命名是复杂的，而且采样常常是不完整的或有偏差的，如通过使用栽培品种或对某些群体或地理区域的有限采样（Fougère-Danezan et al. 2015）

蔷薇属由 140～180 个物种组成（根据来源而定），分为 4 个亚属：单叶蔷薇亚属、蔷薇亚属、沙蔷薇亚属和缫丝花亚属。蔷薇亚属共分为 10 组，并且 95%的蔷薇属物种发现于其中（Wissemann and Ritz 2005）。其中有 7 个组在栽培月季系统学中具有重要意义，其中合柱组（二倍体）、法国蔷薇组（四倍体）、月季组（二倍体）和芹叶组（二倍体和四倍体）是最重要的（Smulders et al. 2011）。次要的是桂味组[有时包括北美蔷薇组（二倍体到八倍体）]和犬蔷薇组（四倍体到六倍体）。剩下的三个组，即金樱子组、木香组和硕苞组，仅由一或两个物种组成，在栽培月季史上仅具有一定的重要性。

27.1.2 月季品种的起源

现代月季品种的原始种质只有 8～15 个物种（确切的数量取决于作者的解释以及其在现代月季驯化中所起的重要作用）。Wylie（1954）的综述中提及了来自 4 个不同组的重要物种。月季组具有中国月季（*R. chinensis*）和巨花蔷薇（*R. gigantea*），法国组中有法国蔷薇（*R. gallica*）和大马士革玫瑰（*R. damascena*），合柱组中有野蔷薇（*R.multiflora*）、光叶蔷薇（*R. wichuran*a）和麝香玫瑰（*R. moschata*），芹叶组中仅有异味蔷薇（*R. foetida*）。其他重要性不高的物种分别为来自桂味组的玫瑰（*R. rugosa*）和 *R.cinnamomeae*，芹叶组中的密刺蔷薇（*R. spinosissima*），合柱组中的 *R. phoenicia*、*R. sempervirens* 和 *R. arvensis* 以及犬蔷薇组中的 *R. rubiginosa*。

将现代月季品种祖先的重要性状引入了月季品种基因库中，如中国月季的连续性开花（1800 年）；光叶蔷薇的耐寒性；异味蔷薇的黄花色（1900 年）。在提到的重要祖先中，只有法国蔷薇和异味蔷薇是四倍体，而其余的是二倍体（Wylie 1954）。De Cock 等（2007）综述了野生月季和栽培月季之间的遗传关系。

市场上的月季品种数量不详，现代月季已注册了 30 000 多个品种（Young et al. 2007）。在过去两个世纪中培育了数量众多的月季品种，使得月季成为与任何其他观赏作物相比最为密集的作物（Marriott 2017）。月季品种可作为商业用途的切花、庭院植物和微型盆栽植物。此外，它们还是应用于香水行业的主要花卉之一。同时，它们的重要性还源于其药用价值和烹饪品质（Gudin 1999）。

除植物分类外，栽培月季还经常根据其园艺性状进行分组。园艺分类基于植物习性、花朵以及花朵的形状和颜色。品种分为杂交茶香月季（一朵花）、丰花月季（簇生大花）、多花月季（簇生小花）和微型月季。多数切花月季品种是杂交茶香月季，而庭院月季通常是丰花型的。最初的园艺类别包括波旁月季、波特兰月季、杂交长春月季、诺伊赛特月季、茶香月季等。这些品种具有历史意义，但因集约育种而逐渐衰落，集约育种缩小了现代月季的遗传库（Wylie 1954；Matsumoto et al. 1998；Scariot et al. 2006）。AFLP 标记被用于野生和栽培月季的聚类。该标记将不同的月季品种分为欧洲和东方两个簇，每个品种都与不同的野生物种有关（Koopman et al. 2008）。育种在继续缩小原始基因库，这一现象在切花月季育种中尤为明显。Vukosavljev 等（2016）发现，所有类型的庭院月季之间的遗传分化水平是切花月季之间的 4 倍。园艺群体之间存在多样性重叠，并且育种者认为切花月季实际上是庭院月季的一部分。特别是切花月季，人们质疑现有的遗传变异是否足以满足未来的育种目标。据推测，基于狭窄的遗传背景进行的选择已经导致了严重的遗传侵蚀（de Vries and Dubois 1996）。有些人则反对该理论中关于四倍体栽培品种和无性繁殖的高杂合度（Gudin 2001）。

27.1.3　月季的育种历史

栽培月季的历史证据可追溯到中国、西亚和北非的文明时代，距今已有 5000 年。人们相信，在野外发现的物种的最初定居和繁殖行为导致了自发的种间月季杂交。长期以来，这些月季被视为原始种（Gudin 2017）。一个例子是大马士革玫瑰，它现在被认为是一个结合了三个亲本物种的杂交种：法国蔷薇、麝香蔷薇和腺果蔷薇（*R. fedtschenkoana*）（Iwata et al. 2000）。在 19 世纪和 20 世纪，月季作为园林植物受到广泛欢迎，月季园和月季育种蓬勃发展。在那个时期，真正的红色、黄色（有一些例外）和杏色花是未发现的。花朵通常是单瓣、半重瓣以及经典重瓣。这些古老的一年一次性开花月季仍然可以在许多文献中找到（Marriott 2017）。在月季中定向杂交的最早记录可追溯到 19 世纪初（Gudin 2017）。

Liorzou 等（2016）研究了古代欧洲和亚洲种质的遗传多样性。这些作者重建并证实了月季育种历史文献中所报道的亚洲遗传背景在欧洲栽培种中的渗入。据推测，月季的杂交在东方世界中可能更古老。在植物育种历史中，月季被认为是第一个非食用野生物种。自从开展月季杂交育种以来，月季育种的实践并没有发生革命性的变化（Gudin 2001）。

月季育种研究是由极具竞争力的私人公司进行的。应用的遗传知识是专有的，并且尚未公开（de Vries and Dubois 1996）。为了改善方法，一些公司与研究小组建立了共同的研究计划。然而，这项工作中应用最广泛的部分从未发表过，如关于突变诱导、香味、色素或如无刺等特定性状遗传的研究（Gudin and Mouchotte 1996）。因为盆栽月季、切花月季和庭院月季对各自栽培体系要求不同，不同类型的月季大多由不同的种植者进行栽培，往往育种人员缺乏不同类型月季的专业知识，这导致不同月季类型之间的“异花授粉”。全球共有 25～30 家极具竞争力的月季育种公司，以及更多业余的月季育种公司（Gudin 2017）。

月季育种必须克服高杂合度，从授粉到种子萌发等有性生殖的已知问题以及不同倍性水平的难题（Gudin 1999）。尤其是在切花月季（四倍体）中，遗传的模式是复杂的，且不完全清楚（Koning-Boucoiran et al. 2012）。自 20 世纪 60 年代以来，月季育种已受益于不断增长的知识，主要是关于该物种有性繁殖的知识（Gudin 2001）。如今，针对早期产量预测的程序以及有关抗病性的生物测定法已报道。尽管获得了遗传数据，月季育种仍然非常依赖于育种者的经验，在这些经验中，基于审美特征的主观选择是至关重要的（Gudin 2003）。此外，还采用了与不同的目标相对应的选择程序，如耐低温、抗病性或延长保鲜期等育种目标（Gudin 2001）。

对于切花而言，最重要的育种目标与观赏价值和质量价值相关，如迷人的花色、坚韧的花瓣和重瓣花、茎的质量、花朵大小、瓶插寿命、耐运输性以及同样重要的产量。由于香气与柔软的花瓣和较短的瓶插寿命呈正相关，因此香气几乎在切花月季中完全消失（Chaanin 2003）。切花月季育种中重要的新性状是花朵绽放的形状、花色的演变、叶片的光泽度等（Gudin 2017）。对于盆栽月季，矮化遗传是由一个显性基因控制的。盆栽月季的选择标准是一簇的花朵数量、花朵颜色、花朵大小、花瓣数量和植物习性。其他特性与产量和开花寿命有关（de Vries 2003）。除了重要的审美特征外，抗病性是庭院月季育种的主要目标。对月季芳香的需求通常伴随一些问题，如芳香与对病害的敏感性相关（Gudin 1995，2003）。越来越多的月季被用于园林景观，因此，需要“无忧无虑”的类型，这意味着不需要修剪和作物保护措施。除此之外，理想型的月季还具有四季装饰性的效果，如在冬季具吸引人的蔷薇果（Gudin 2003）。

27.1.4 切花月季的历史

在切花月季的发展过程中，杂交历史以及一系列不同倍性水平物种的应用是非常重要的（Smulders et al. 2011）。杂交茶香月季的四倍体背景可以追溯到杂交长春月季和茶香月季之间的杂交。1840 年至 19 世纪末，这两类月季是主要的庭院月季。茶香月季（$2n = 2x$）和杂交长春月季（$2n = 4x$）之间杂交产生了月季品种‘法兰西’（La France），其由法国育种家 Guillot 于 1867 年首次引入法国。‘法兰西’是第一个现代月季品种（切花月季型），其不仅具有杂交长春月季的习性，而且具有茶香月季的细长花芽特征，还是三倍体（Wylie 1954）。其特征是花枝长且直，花芽形态良好，花朵饱满。这一类杂交茶香月季新类型，仅能通过杂交长春月季和茶香月季原始组合之间的反复杂交获得。

那个时代的育种者抱怨这类新的三倍体杂交茶香月季结实率很低。像其他组群的月季品种一样，最终通过将三倍体与原始四倍体亲本回交克服了倍性差异。 三倍体是中国月季（主要是二倍体）和欧洲古老月季（通常是四倍体）之间的重要桥梁。因此，第一个波旁月季、杂交中国月季和杂交茶香月季均是三倍体（Liorzou et al. 2016）。

第一个四倍体杂交茶香月季，花色有白色、粉红色和红色。黄色的出现，是通过法国育种家 Pernet-Ducher 将异味蔷薇性状引入杂交亲本后获得的。尽管异味蔷薇的繁殖能力低，但经过反复杂交，选育出了第一个黄色杂交茶香月季‘Soleil d’Or’，并于 1900 年进入市场。育种史记载中，没有其他野生种参与该黄色系的演化。

据考证，切花月季的生产始于 1896 年，那年在荷兰建立了第一座专门用于种植切花月季的温室（Marriott 2017）。许多著名的切花月季品种在市场上盛行多年。如今有许许多多不同的品种可供选择，而 40 年前在市场上占据主导地位的仅为少数几个品种（Marriott 2017）。多年以前的一些切花月季品种至今还很有名气，如‘Baccara’（Meilland 1954），一个与红色切花月季紧密联系的名字；‘Super Star’（Tantau 1960），也被用作庭院月季的名字；‘Sonia’（Meilland 1974），被认为是产量最高的切花月季之一；‘Frisco’（Kordes 1986），是一种所谓的黄色甜心月季，具有花小和产量高的特点，以其极长的瓶插寿命而闻名，至今仍在市场上销售；以及广受欢迎的红色月季‘First Red’（NIRP International 1991）。

切花月季根据花朵大小可分为几类。作为奢侈品的月季通常是大花型的；然而，小花型栽培品种的产量很高。小花型中短枝的月季也称为甜心月季。每个花枝上有多个花朵的称多头切花月季，其在切花生产中仅占很小的比例。大花月季通常在花店里出售，而小花月季主要在超市里出售。大花月季生长在传统种植区，如荷兰和南美洲（厄瓜多尔、哥伦比亚）。因为小花月季单枝花的人工成本较高，所以主要集中在非洲地区种植。在欧洲，许多种植者通过温室无土栽培、人工照明（包括 LED 照明）、压枝、加热、添加二氧化碳、气候调节和自动化等技术手段，扩大了小花月季种植面积，提升了种植技术体系。有些欧洲种植者更喜欢降低生产成本，仍然使用土壤栽培，有时在不加热的温室里甚至选择露地栽培。育种者培育的月季新品种需要满足不同类型种植者的需求，这不仅取决于种植者来自不同的国家，而且取决于种植者所使用的种植系统。一些育种公司专门针对特定种植市场选育新品种。例如，Tantau 公司专注于南美洲种植区，而 Kordes 公司正在为非洲种植者开发新基因型。有些育种者专注于月季花形。例如，庭院月季育种公司 David Austin 开发了一些切花月季，其典型风格就是所谓的“古老英国月季”。

27.1.5　月季的经济重要性

关于月季的大多数经济信息都集中在切花月季上。全世界每年销售大约 100 亿枝月季（AIPH 2016）。除此以外，就是盆栽月季（每年 8000 万盆）和庭院月季（每年 2.2 亿株）（Roberts et al. 2003）。月季在切花生产中仍然占有绝对第一的位置。红色月季占市场的 30%（Chaanin 2003）。

出口欧洲的月季最重要的产地是荷兰、肯尼亚和埃塞俄比亚，厄瓜多尔位居第四，但拉丁美洲的月季生产国主要向北美洲国家出口。对于美洲，最重要的进口产品来自厄瓜多尔和哥伦比亚（AIPH 2016）。在拉丁美洲高海拔（2000～3000m）夜晚寒冷地区种植，需要具有良好耐低温能力的品种。运输方面的要求给育种也增加了新的要求，如采后品质和无刺性（Gudin 1999）。研究者还研究了在海运或空运的过程中月季植物生理学的差异。育种公司在这些地区开展品种测试，并选择特定的种植区。

埃塞俄比亚现在是欧盟的第三大月季切花出口国，尽管该国在 2000 年之前尚未开始花卉种植。自 20 世纪 80 年代以来，赤道附近的非洲和美洲热带地区的月季生产面积不断增加（Zieslin 1996；Gudin 1999），非洲地区巩固了它们的市场。全球最大的切花月季生产地区是印度（61 000hm^2）和中国（14 348hm^2）（AIPH 2015），尽管这两个国家有关月季销售的经济数据不能充分统计。因为这两个国家的月季种植主要是为满足国内消费市场的，所以这两个国家生产的月季大多没有反映在出口数据中。全球月季产量的前五大国家还有厄瓜多尔（5472hm^2）、哥伦比亚（2465hm^2）和肯尼亚（2164hm^2）（AIPH 2015）。在荷兰，月季产量正在迅速下降。2004 年，荷兰的月季产量仍然约为 850hm^2，到 2016 年降至 238hm^2（Royal Flora Holland 2016）。在荷兰，切花月季产量的减少与切花总产量的减少是不成比例的。

在经济数据中关于庭院月季生产的讨论较少，庭院月季通常被认为是苗圃植物的标志。针对其他类型的用途，如香水、医药或营养行业的育种很少。与切花月季相比，庭院月季的利润较低，因此有些公司只培育切花月季新品种。

尽管大型生产区主要分布在欧盟以外，但欧盟仍然是培育新切花月季品种的重要据点。自 1948 年以来，植物育种者权利保护制度的实施一直鼓励着月季育种者（Marriott 2017）。除了保护知识产权的某些制度（例如商标）外，还可以在 UPOV（国际植物新品种保护联盟）注册新品种；欧盟内的植物品种权由欧盟植物品种局（CPVO）管理。CPVO 数据库共通过了 61 000 个观赏植物的申请。自 1995 年以来，向 CPVO 一共提出了 4189 份有关月季新品种（所有类型）的申请，这些月季代表了观赏植物育种中最重要的一个属。其中 1742 项申请仍然有效，1780 项曾被授予但现在已经终止，育种者撤回 259 项，69 项被拒绝，还有 339 项正在申请中（CPVO 2018）。

27.2 育 种 目 标

27.2.1 育种计划的建立

育种计划可分为三个主要阶段：①通过杂交创造新的后代；②幼苗的选择；③所选基因型的繁殖和市场推介。在育种公司之间以及月季类型之间，项目的策略、方案和期限是不同的。适用于所有杂交月季的一项基本标准是选择迷人和新的创新性状，如迷人的花色、花形和植物习性（Chaanin 2003）。选育的大多数月季都是切花月季。其他类型的月季，如庭院月季，具有不同的选择标准和不同的市场策略，并且经常采用“风险”杂交方式（通常具有范围更广的不同月季品种和物种）。

确定具体的育种目标后，下一步就是选择相应的亲本植物。育种者知道有关亲本植物特定特性的信息，并将其存储在数据库中。除了表型性状的信息外，还记录了杂交和制种数据。这些为预测可能的杂交后代表型、基因型提供了一个有用的信息来源（Pipino et al. 2011）。

新后代产生的第一阶段包括 4 个方面：授粉、胚胎发育、种子休眠解除和种子萌发，所有这些都在育种过程的第一年进行。在杂交过程中，需要确定最佳开花期来完成受精，确定每个亲本种群进行杂交的次数以及收获/收集瘦果的时期。通过大量幼苗（每个杂交组合至少 200 株）构建杂交大群体，对于观察感兴趣性状的变异和分离是很重要的。具有高结实率性状的这类基因型植株，可通过产生更多的后代来评价其不同性状，从而提高育种计划的效率。月季育种者最感兴趣的性状是整体的观赏性，即花朵颜色和形状、花序结构和植物习性。切花月季的生产能力（每平方米生产的花枝数量）很重要，但这一性状要在后期进行评估。育种者对幼苗进行评估，颜色是第一个特征。花枝的评估包含了花枝的活力、长度和稳定性。切花月季的花朵大小应与花枝长度成正比（Pertwee 1995）。在切花月季中，花朵大小和花枝长短主要划分为 4 种商业类型，大花月季（或长枝月季，如长达 100cm）、中花月季（或中枝月季，如长约 50cm）、甜心月季（或短枝月季）和簇花（多头）月季。

对于切花月季，采后品质是非常重要的。应保持产品在运输期间的质量和瓶插寿命。因此，由于不同基因型品种维持采后品质的能力不同，在选择过程中进行了一些实验。首先，必须确定最佳采收时间。接下来，将花枝浸入 2～4℃的水中保存 1 天，然后干燥保存和运输。再在标准室温（20℃ ± 1℃；相对湿度 60% ± 5%；3W/m^2 的光）下进行瓶插寿命测试。正常情况下，所选品种的瓶插寿命大多为 12～14 天，有些品种的瓶插寿命更长。香味浓郁的品种瓶插寿命较短，即 8～10 天（Chaanin 2003）。

27.2.2　多倍体月季的遗传

所有的商业切花月季都是四倍体；许多庭院月季也是四倍体，但有些是二倍体或三倍体。多倍性导致遗传分离复杂性的增加，从而阻碍了这些作物的遗传研究。在月季遗传研究中，基于性状遗传研究与遗传图谱的发展对月季二倍体研究较为普遍。在多倍体中，遗传背景可以通过基于一个物种（同源多倍体）或基于多个物种（异源多倍体）的背景来定义。此外，染色体的多元二体行为，在同源和异源多倍体中分别发挥了作用。通常，不能严格地将植物归为这两种多倍体中的一种，这也引发了有关其减数分裂行为和遗传的问题。在野生四倍体月季物种中，已经观察到了二体和四体行为（Wissemann and Ritz 2005；Joly et al. 2006）。在栽培月季中，假定了四体遗传（Gar et al. 2011）与优先配对（Koning-Boucoiran et al. 2012）相结合。Bourke 等（2017）研究玫瑰中的优先染色体配对时，发现同源染色体之间以及沿同源染色体的配对行为存在差异。当染色体臂的行为不同时，配对取决于亲本、染色体，甚至取决于单个染色体。这些作者认为，不同来源的不同物种的谱系可能导致了节段性同源多倍体。来自 7～14 个物种的染色体之间的随机配对发生频率异常高。该信息对于根据重组频率绘制遗传图谱非常重要。

在四倍体定位群体中研究了分子标记的分离模式。由于四倍体月季标记-性状关联分析难以建立，在四倍体月季育种中使用分子标记辅助选择可能受到限制。四倍体连锁图谱可用于确定两个形态性状的关联，如花瓣数和茎上的皮刺数以及抗白粉病性之间的关系（Koning-Boucoiran et al. 2012）。对于某些特定性状，已经在二倍体中研究了其遗传模式。Debener（2017）总结了月季的这些遗传特性。表 27.1 给出了一些重要特性的示例。仅有少数性状由某基因控制，如连续开花由 TFL1（Terminal Flower 1）直系同源基因控制。在连续开花的突变体中，基因组中插入了一个转座子（Iwata et al. 2012）。其他性状更为复杂。例如，香气由多个基因控制，因为月季的香气由数百个挥发性分子形成。RhNUDX1 是一种不依赖于萜类合成酶的单萜类生物合成途径中的胞质成分，其产生与月季的香味有关（Magnard et al. 2015）。

表 27.1　月季某些性状的遗传

性状	遗传模式：单基因遗传、数量遗传、隐性遗传和（共）显性遗传	参考文献
粉红色花朵	单基因共显性遗传	Debener 1999
黄色花朵	数量遗传	de Vries and Dubois 1980
重瓣花	单基因显性遗传和数量遗传	Debener 1999 Hibrand-Saint Oyant et al. 2008
连续开花	单基因隐性遗传（TFL1）	Semeniuk 1971a，1971b；Iwata et al. 2012
矮小株型	单基因显性遗传	Dubois and de Vries 1987
有光泽的叶子	单基因显性遗传	Lammerts 1945
白粉病抗性	单基因显性遗传（Rpp1）和数量遗传	Linde and Debener 2003；Hosseini Moghaddam et al. 2012
黑斑病抗性	单基因显性遗传（Rdr1～3）和数量遗传	Von Malek and Debener 1998；Whitaker et al. 2010
耐寒性	数量遗传	Svejda 1979

有关性状遗传的信息是必不可少的，特别是对于通过广泛杂交引进野生物种的基因。二倍体种群将有助于我们在未来获得更多的遗传信息（Debener 2017）。

27.2.3　生物胁迫抗性

病害的流行取决于月季的类型和种植面积。Debener 和 Byrne（2014）提到，每年每公顷的虫害和病害控制成本为 7000～32 000 美元，其中一半用于病害控制。这些数据是基于对中国、南美洲、美国和欧洲种植者调查得出的，成本最高的是欧洲切花月季的生产。

对环境保护的日益关注极大地影响了庭院月季的育种。抗病性成为许多庭院月季育种者的主要课题之一（Gudin 2003）。月季在园林绿化和公共绿地中的使用遵循了这一趋势。Leus（2017）最近对月季的抗病育种进行了概述。Debener 和 Byrne（2014）则主要集中于生物技术的潜力进行了另一番评述。月季上最常见的真菌病是黑斑病（蔷薇双壳菌）、灰霉病（灰葡萄孢）、白粉病（白粉菌）、炭疽病（蔷薇痂圆孢）、叶斑病（*Cercospora* spp.）和铁锈病（*Phragmidium* spp.）以及卵菌霜霉病（*Peronospora sparsa*）。在月季中，发现了 11 种病毒（Horst 1983）。由月季丛矮病毒（RRV）引起的月季丛矮病（RDD）是月季生产日益重要的问题。由根癌农杆菌引起的冠瘿病是最重要的细菌性病害。

有些病害是世界上某些地区特有的。例如，RRV 只发生在加拿大和美国。这种由螨虫传播的病毒对庭院月季有毁灭性的影响。其他的病害只在特定的生长条件下才会快速发展。例如，黑斑病需要叶片湿润才能发生，因此主要发生在庭院月季上。白粉病出现在温室种植的切花月季和庭院月季上。霜霉病也与特定的温度和湿度条件有关。葡萄孢菌与运输有关，尤其是在切花月季中。

对黑斑病（Blechert and Debener 2005）和白粉病（Dewitte et al. 2007）与植物之间的相互作用已进行了显微观察。对病原体特异性的抗性，对于黑斑病（Debener et al. 1998；Leus 2005；Yokoya et al. 2000；Whitaker et al. 2007，2010）和白粉病（Linde and

Debener 2003；Leus et al. 2006），在不同月季基因型之间是不同的。Leus 和 Van Huylenbroeck（2009）对抗病育种（尤其是白粉病）的可能性、生物鉴定方法的使用等进行了描述。

在月季中发现了黑斑病抗性基因 *Rdr1*、*Rdr2* 和 *Rdr3*（Debener and Byrne 2014）。*Rpp1* 赋予了月季对白粉菌生理小种特异性的抗性（Linde and Debener 2003）。

特别是在庭院月季中，抗病性是在田间反复选择的目标。目前没有使用分子标记辅助选择育种，因为尚无与稳定抗病性基因关联的分子标记（Debener and Byrne 2014）。

大多数现代品种都易受黑斑病侵害，但有几个野生种表现出抗性：木香花、加罗林蔷薇、金樱子、野蔷薇、玫瑰、刺梨和光叶蔷薇（Drewes Alvarez 2003）。对于白粉病，有些野生种具有抗性：*R. agrestis*、*R. glutinos* 和峨眉蔷薇被认为具有高抗性（Linde and Shishkoff 2003）。

不同月季品种对线虫的抗性不同。特别是对于砧木来说，抗线虫是很有价值的。在野蔷薇‘K1’、弗吉尼亚蔷薇（*R. virginiana*）、*R. laevigata anemoides* 和一些紫叶蔷薇（*R. glauca*）种质中鉴定出了对穿刺根腐线虫（*Pratylenchus penetrans*）的抗性。这些品种的抗性比流行的 *R. corymbifera*‘Laxa’砧木抗性更高（Peng et al. 2003）。野蔷薇和中国月季对花生根结线虫（*Meloidogyne arenaria*）、南方根结线虫（*Meloidogyne incognita*）和爪哇根结线虫（*M. javanica*）种群具有抗性。所述物种对北方根结线虫（*M. hapla*）的抗性程度较为多变（Wang et al. 2004）。*R. manetii* 似乎具有很高的抗线虫性，而常用的犬蔷薇砧木通常具有良好的抗线虫能力。

自 2011 年以来，月季丛矮病在美国日益严重。2015 年，月季丛矮病毒被确认为 RRD 病原体。众所周知，果叶刺瘿螨是一种传播病毒的载体（Di Bello et al. 2015）。其引起的症状是叶脉红肿、花斑叶病、皮刺过多、斑点病、发育迟缓，最终导致植物死亡。目前正在研究月季原种（Sol et al. 2017）和品种（Di Bello et al. 2018）对瘿螨的易感性差异。在对 20 个月季品种对螨虫和病毒抗性进行评估后，所有品种均显示出对螨虫的易感性，但其中一个品种通过螨虫传播或劈接后对病毒具有抗性（Di Bello et al. 2018）。在抗病育种中，正在研究使用月季抗病原种（如 *R. setigera* 和 *R. palustris*）（Roundey et al. 2017）。

27.2.4　非生物胁迫抗性

对于月季而言，非生物胁迫主要是低温、干旱、高温和盐害。耐寒性对于在寒冷气候下的庭院月季种植至关重要。大家一致认为，原产寒冷气候的野生原种的耐寒性适合引入栽培月季品种中。Zlesak 等（2017）观察到美国不同气候区月季品种的抗寒性存在明显差异。值得注意的是，有些品种在寒冷的冬季会出现枯萎，但能够恢复。这些信息对于种植者和消费者都是有价值的。加拿大的育种者就专门从事耐寒性育种。Svejda（1977，1979）使用 *R. kordesii* 和玫瑰的杂交种，培育耐寒性增强的月季。加拿大探险者和帕克兰月季系列被称为耐寒系列。用叶绿素荧光法评估这些月季的耐寒性，结果表明其与肉眼观察到的组织损伤有很好的相关性。因此，叶绿素荧光量成为评估候选品种耐寒性的合适参数（Hakam et al. 2000）。Ouyang 等（2017）利用水分含量和不同可溶性碳水化合物含量，评估了不同月季基因型在田间条件下的耐寒性和季节性波动。此外，基于茎秆的电解质渗透率，使用了标准化实验室测试，Vukosavljev（2014）研究了两个隔离种群冬季耐寒性的数量性状基因座（QTL）。每个群体都发现了两个 QTL，其中一个是相同的。从研究中可以得出结论，再生能力和对低温损伤的耐受性是独立遗传的。基于转录组分析，研究了月季相关候选基因耐寒性的潜在机制（Zhang et al. 2016）。

在降雨量少和庭院用水受到限制的地区，选育合适的月季品种来适应干旱胁迫是很重要的。一般而言，月季植株具有良好的耐旱性，但当发生干旱胁迫时花的发育受到负面影响（Chimonidou Pavlidou 2004）。Li Marchetti 等（2015）发现了植物基因型、植物结构和与水分亏缺相关生理反应之间的紧密联系。许多参数可用于评价植物的干旱反应。相对含水量和叶水势被认为是评价月季抗旱性的参数（De Dauw et al. 2013）。

在对 4 个月季品种研究时发现，根系发育与干旱胁迫之间存在相关性（Harp et al. 2015）。Liang 等（2017）研究了耐热与非耐热亲本杂交获得的月季种群耐热性的变化，热胁迫导致花直径、花瓣数和花干重下降。然而，最后得出的结论是，短时间（44℃下 1h）暴露在高温下不适合作为耐热性月季的选择条件。

耐盐性变得越来越重要。例如，在人工基质上生产的切花月季，因用水量的减少和循环水源的使用，其耐盐性要求越来越高。在温室耐盐实验中评估得到不同基因型之间的胁迫反应存在明显差异（Cai et al. 2014）。对采用不同砧木的切花月季进行了评价，结果表明，与其他常用砧木基因型相比，‘Manetti’具有更好的耐盐性（Cabrera et al. 2009）。

27.2.5 月季原种的利用

除了抗逆性的利用，野生种还能拓宽栽培月季的遗传基础，同时能增加栽培月季的奇特性状，如 *R. blanda* 的无刺，*R. hugonis*、*R. ecae* 和 *R. primula* 的黄色花，*R. laxa* 的抗寒性，*R. suffulta* 的耐旱性以及 *R. arvensis* 的耐荫性（Spethmann and Feuerhahn 2003）。

在 *R. persica* 中发现了利用月季原种组合新花色的示例。以前，该种被划为一个单独的属，称为 *Hulthemia persica*。其典型性状是双色花瓣，花的中心是较深的颜色，让花具有独特的外观。但其回交是很困难的（Zlesak 2006）。如今，市场有几种来自 *R. persica* 的种间杂交种。

栽培种‘往日情怀’和光叶蔷薇杂交的一个例子如图 27.1 所示。F_1 代杂交种的许多性状出现了分离，如花的颜色和花瓣数量（5 瓣为单瓣，大于或等于 10 瓣为重瓣）。

27.2.6 育种目标取决于月季类型

育种目标取决于月季类型：庭院月季、切花月季、盆栽月季。所有的月季类型都需要某些性状，但是，性状的重要性可能有所不同（表 27.2）。

图 27.1 杂交组合‘往日情怀’（左上）× 光叶蔷薇（右上）。获得的 F_1 杂种在田间评价中观察到的花色和花瓣数的分离情况

表 27.2 不同类型月季育种中的性状评价：1. 次要，2. 中等重要，3. 非常重要

特征	切花月季	庭院月季	盆栽月季
产量	3	3	3
花的类型、颜色、形状等	3	3	3
抗病性	1	3	1
香气	1	2	1
生产区域的适应性	3	1	1
瓶插或保存限期	3	1	3
运输质量	3	2	2
无刺	2	1	1
单枝的花朵数	1	1	3

用于家庭环境的盆栽小型月季，需要紧凑、分枝良好、大而持久的花朵、持续开花、在室内以及整个零售链中具有弱光耐受性和乙烯不敏感性等良好的特性（Zlesak 2006）。从修剪到开花的天数也是很重要的（de Vries 2003）。对于微型月季，植株矮化的表型是源自 *R. chinensis minima* 的单基因显性性状（Dubois and de Vries 1987）。de Vries（2003）

对盆栽月季的选育做了评述。

砧木品种也是可以得到的。最著名的是 *R. canina*、*R. indica*‘Major’、*R. multiflora*、‘Manetti’和‘Dr. Huey’。种子繁殖和插条的使用取决于砧木的类型。对于砧木而言，重要的是易于栽培和发芽、与品种的相容性、无刺、良好的根系发育能力以及对土壤或基质类型的适应性、抗病性、根出条少等（Zlesak 2006）。对于接穗月季，培育与中间砧木或者砧木亲和性高的类型。自生根的月季可能越来越重要，但是这意味着在选择过程中需要评估生根能力。月季的育种者和种植者时常提及砧木需要改良。

27.3 生育力：月季的杂交和种子繁殖

27.3.1 月季花粉

蔷薇科植物的花粉都具有相似的结构。月季的花粉看起来像由许多花粉粒组成的淡黄色灰尘。通过对月季花粉的组织学检查发现其有三条纵沟（triporate）（Reitsma 1966），外壁层有三个孔（triporate）（Caser 2017）。花粉的形状是细长的椭圆形，长宽比为 2（Caser 2017）。

花粉的大小与物种、倍性和环境相关（Zlesak 2009）。Pécrix 等（2011）确定了在二倍体月季中花粉的直径范围为 18～36μm（平均 26.8μm±2.6μm），在三倍体月季中的范围为 24～42μm（平均 35.4μm±3.2μm）。Pipino 等（2011）认为切花月季的生育力与花粉的直径相关。

观察花粉的繁殖力可以提高育种的效率，使用新的花粉供体可以减少花粉不育的风险。Pipino 等（2011）基于育种者的数据库和亲本的生育力数据对花粉形态与体外萌发进行了研究。从这个数据库中选择的基因型，花粉要么是小（平均直径＜30μm）、皱缩和不规则（异常），要么是大（平均直径＞30μm）、椭圆形和有沟壑（正常）。每次杂交产生的种子数与花粉直径（$r = 0.94$）或正常花粉百分比（$r = 0.96$）之间存在高度相关性。该方法对于作为花粉供体的月季基因型具有较好的预测作用。这对于评估浸水花粉直径的增大和其形状从椭圆变成球形非常重要。当使用浸水花粉时，花粉直径大小的差异还是很小的。异常的不育的花粉不会膨胀。因此，将花粉浸入水中可用来作为评估其生育力的非常简单的测试方法。当月季花粉接触到柱头分泌物时膨胀过程就开始了。在花粉生育力测试中另一个众所周知的方法是在人造基质上进行花粉染色和萌发。Caser（2017）检验过用不同的染料评估月季花粉生育力，如醋酸洋红染色、Alexander 染色、苯胺蓝染色。各种信号，特别是在花粉与柱头相互反应中，从花粉粒的生长到形成花粉管诱发花粉粒产生强烈的代谢活动，在受精阶段前期，雌蕊支持花粉萌发和花粉管生长，并定向引导其走向胚珠（Herrero and Hormaza 1996）。

花粉的贮存对于育种者非常重要。为保持花粉的质量最好保存在−80℃。在 4℃或−20℃，花粉质量下降速度更快（Giovannini et al. 2015）。花粉质量的保持也取决于月季的基因型（Macovei et al. 2016）。

未减数配子，特别是未减数的花粉，经常在不同倍性水平的月季中被发现（Crespel

and Gudin 2003；El Mokadem et al. 2002a，2002b；Van Huylenbroeck et al. 2005；Crespel et al. 2006）。来自 *Caninae* 组五倍体月季特殊的不平衡减数分裂被较好地描述（Nybom et al. 2004；Ritz and Wissemann 2011）。未减数配子的形成取决于环境条件，$2n$ 配子形成的模式在遗传学上等同于 FDR（Crespel et al. 2006）。用流式细胞术评估月季花粉的倍性水平（Roberts 2007）。

27.3.2　杂交过程

月季的杂交是通过将父本的花粉授于母本的雌蕊来完成的。杂交最好在清晨进行。为了诱导花粉供体的花药成熟和花粉释放，这些花粉应于前一天收集并在干燥的地方放置 24h。为了避免自花授粉，种子亲本在花粉释放之前做去雄准备（Crespel and Mouchotte 2003）。因为认为 24h 的时间足以诱导产生花粉萌发所必需的柱头分泌物，所以去雄后一天进行授粉（Jacob and Ferrero 2003）。月季杂交通常是利用每朵花进行一次授粉完成。然而，同一朵花重复授粉会导致每穗瘦果更多（de Vries and Dubois 1983）。在北半球，杂交通常是在 5～6 月进行。在其他地区，如厄瓜多尔，授粉全年都可进行。

种植庭院月季的经济重要性较低。大型月季育种公司、专业公司或许多业余育种家有时将庭院月季杂交作为一种辅助活动。与切花月季相比，因为庭院月季使用了大量的不同表型，所以种子繁殖显得并不是那么关键。在切花月季中常进行更多“困难”的杂交，如远缘杂交或不同倍性之间的杂交。它们通常产生具有更高变异性的后代。

在月季中，坐果率和种子的萌发率低于 50%，蔷薇果实里面包含 1～30 粒种子（Gudin 2017）。在切花和庭院月季育种中，成熟的蔷薇果在 9 月底和 10 月初，或授粉后 3～4 个月收集。种子从果实中收集并计数。然后，将种子保持在 2～5℃、中等湿度水平 6～12 周（Zlesak 2006）。根据气候的不同，种子可在发芽基质中进行春化，这样可以直接播种；否则，必须在寒冷的室内保存。杂交种子萌发后获得 12 万株幼苗被认为是现代切花月季育种计划的良好起点。根据可用空间和种子的起始数量，种子通常播种在苗床或工作台上，密度为每平方米 150～400 粒种子。春化被认为在两个月内完成（Pipino 2011）。

27.3.3　月季的胚胎发育

Pipino 等（2013）通过两个四倍体栽培月季品种‘Melglory’和‘Cassandra’来观察胚胎的发育。对于这两个品种，授粉后（DAP）只需 15 天胚胎就发育完全。30DAP 观察到发育完全的种子。在胚胎发育期间测量脱落酸（ABA）的含量，众所周知，脱落酸在诱导和维持种子的休眠以及种子萌发的过程中起着重要的作用。这两个月季基因型，在初始 ABA 浓度非常高（尽管不同），观察到一个四参数对数衰减模式，在 30DAP 时 ABA 基础水平相似。30～60DAP，发育完全的胚胎 ABA 浓度没有显著变化。因此在 15DAP 时，胚胎完全发育，但仍然被非常薄而细腻的种皮所包围。然而，它们的 ABA 含量还没有降到基础水平。另外，在 30DAP 时，胚胎的特征是 ABA 显著降低，但已经被包裹在相当坚硬的果皮中。Bosco 等（2015）对 28～49DAP 时月季种皮和胚中吲哚-3-乙酸（IAA）与 ABA 浓度的组合进行了更详细的研究。种皮中的 IAA 和 ABA 含量均从 28DAP 开始下降。在胚胎中，IAA 和 ABA 的含量分别从 28DAP 和 42DAP 开始增多，但在 49DAP 时开始下降。

胚胎发育信息可用于确定月季早期胚挽救方案中的时间点。这可用于原本会流产的胚胎，即胚乳无法发育并导致胚胎流产的种间或属间杂交中。当成熟胚中 ABA 浓度显著降低，而种子果皮尚未完全硬化时，四倍体杂交月季进行胚挽救的最佳时间点可以在 15～30DAP。从成熟的种子中切除胚胎是精细而费时的。Pipino 等（2013）建议的时间点，Caser 等（2014）在组织培养中进行了测试和验证。在 7DAP 和 14DAP 时，未成熟的胚胎中没有幼苗生长。但是在 21DAP 时，开始了体外发芽。在 28DAP 时，发芽率达 40%以上，在 35DAP 时，发芽率下降。

27.3.4 月季种子休眠

月季的种子是在瘦果中形成的。这是一种干果，仅含有一粒种子，几乎填满了果皮。种子和瘦果在育种与文学中常用作同义词。

Gudin（2017）从生殖繁殖的角度回顾了月季种子的繁殖情况，如砧木繁殖（*Caninae rose*）、香水生产（*R. gallica* 和 *R.* × *damascena*）和园林绿化。根据月季所在地区的冬季气候条件，研究了月季特有的种子休眠适应性。众所周知，有些野生种，如 *R. rugosa*、*R. gallica*、*R. canina*、*R. soulieana*，只有在 1～4℃冷藏后才能打破休眠。有些月季则在发芽之前需要第二个冬天的春化作用（Werlemark et al. 1995）。相比之下，适应远东沙漠地区的 *R. persica* 种子对低温处理反应不足，但能够应对脱水。休眠的程度和类型、克服休眠的时间和休眠温度因月季品种而异（Gudin 2017）。

低温会降低种子中 ABA 的含量。相似的，交替的升高和降低温度也会降低 ABA 的含量。因此，建议在低温层积处理之前，先在 20℃下进行室温层积处理（Werlemark et al. 1995；Alp et al. 2009）。

与野生种的层积处理相比现代月季品种的层积处理通常是一个较短的过程。栽培月季最佳持续层积处理期为 3 个月：23℃下 1 个月，4℃下 2 个月（Gudin 2017）。在月季杂交品种中，在固定和控制温度下进行层积处理比在自然条件下的发芽率要高。

月季品种种子的萌发受到种子硬度（与果皮有关）和初级休眠（在不利条件下防止种子萌发的自然保护）的阻碍；这与月季品种对自然栖息地的适应性有关。种子成熟会打破初级休眠，但是中断过程可能会导致次级休眠。生理休眠涉及 ABA/GA（赤霉素）水平的控制。最后是果实成熟过程中由环境变化（如温度）引起的发芽多态性（de Vries and Dubois 2015）。Gudin 等（1990）证明了月季瘦果的果皮厚度可以决定发芽。果皮由表、中、内果皮层组成。内果皮是瘦果不透水的主要原因。果皮厚度受环境因素控制，尤其是瘦果成熟过程中的温度和遗传因素。

许多研究都尝试了用化学或机械方法去除果皮（Svejda 1968；Densmore and Zasada 1977；Yambe and Takeno 1992；Morpeth and Hall 2000；Bhanuprakash et al. 2004；Zhou et al. 2009；Pipino et al. 2011）。

通过改变温度、湿度和/或激素处理来打破休眠也得到了广泛的研究（Semeniuk and Stewart 1962；Stewart and Semeniuk 1965；Svejda 1968；Densmore and Zasada 1977；Yambe and Takeno 1992；Bo et al. 1995；Bhanuprakash et al. 2004；Zlesak 2005，2008；Anderson and Byrne 2007；Zhou et al. 2009；Pipino et al. 2011）。

这些研究的结果通常具有物种或基因型特异性。尽管基因型与果实采收成熟度、去除种子种皮时间、种子发育所需温度和层积时间等方面有关联，但是月季种子一般受非深层生理休眠的影响（Zlesak 2006）。如前所述，来自 *Systylae* 的月季，如 *R. multiflora* 和 *R. wichurana*，与部分来自 *Caninae* 的月季，如 *R. canina* 的休眠水平相比，表现出相对较低的休眠水平（Ueda 2003）。休眠的原因似乎主要与种皮和果皮有关，而非胚胎。当种皮和果皮被完全切除后，会促进月季离体胚胎的发芽（Barton 1961；Jackson and Blundell 1963；Blundell 1965；Jackson 1968；Yambe et al. 1992；Bo et al. 1995），对 *R. rugosa* 胚胎的研究表明当其种皮的外源 ABA 水平增加时，休眠水平会增加（Jackson 1968）。通过去除果皮会打破休眠，证实了在果皮内表面的 ABA 会抑制种子的萌发（Yambe et al. 1992）。

在月季中，生理休眠与果皮中的 ABA 浓度呈正相关（Jackson 1968；Bo et al. 1995）。通常认为 ABA 与 GA 的含量比可以反映从休眠到准备萌发的程度。当授粉后 10 天或 14 天施用 GA 时，GA 处理增加结实率（de Vries and Dubois 1987；Ogilvie et al. 1991），但在成熟种子播前进行处理没有明确的效果。Jackson（1968）使用 GA 克服了种子休眠的问题，但是 Zlesok（2010）、Zhou 等（2009）使用其处理机械破皮的种子以及低温层积处理的种子，没有显著效果。为了降低瘦果中 ABA 的含量，最常见的处理方法是在20℃进行热层积处理，随后进行 2～4℃的低温层积处理（Densmore and Zasada 1977；Werlemark et al. 1995；Alp et al. 2009）。

最近针对月季种子做了大量实验（Yambe and Takeno 1992；Morpeth and Hall 2000；Zhou et al. 2009），在实验条件下，评估培养基和生长室中种子的发芽情况。然而，育种公司优化发芽条件需要评估温室中的结果。此外，大多数研究是针对月季野生种而不是杂交茶香月季，因此，结果仅对育种者有研究价值。

值得注意的是，月季野生种和杂交月季表现出不同的生理休眠水平。相对于杂交月季的种子，野生种的种子有着较强的休眠特征，一般会持续 5～6 个月甚至更久，育种者为降低发芽障碍会间接地选择种子（Zlesak 2006）。*R. multiflora* 的种子在第一年有90%发芽率，而适应寒冷地区如斯堪纳维亚地区的 *R. canina*，在播种后 2 年的发芽率不超过 30%（Rowley 1956）。杂交茶香月季的发芽率通常低于 50%，且不均匀（Gudin 2003；Ueda 2003；Zlesak 2006；Anderson and Byrne 2007）。根据 Pipino（2011），切花月季和庭院月季种子发芽的不同是不可忽略的。育种者旨在开发每次杂交后能产生有活力胚胎的种子来弥补人工授粉的大量投资。种子萌发是月季新基因型选育过程中的一个瓶颈。

27.4 育 种 技 术

27.4.1 组织培养育种

大多数在其他植物上使用和发展的基于实验室的育种技术已经在月季上进行了实验。然而，由于需要熟练的操作技能和专门的设备，以及构建技术体系需相对较长的时间，这些技术在实际育种工作中的应用通常是有限的（Chaanin 2003）。

在组织培养中，月季的芽、体细胞胚及配子胚的再生是很难控制的（Rout et al. 1999）。在转基因月季的开发利用中，只有体细胞胚的培养被成功应用，从而使体细胞胚的培养成为一项关键技术（Debener and Linde 2009）。体细胞胚发生的成功率高度依赖于品种的基因型（Kim et al. 2003a）。Rout 等（1999）及 Pati 等（2006）阐述了体细胞胚在月季中不同的再生途径。为了建立有效的体外再生体系，许多研究者使用了不同类型的外植体。以叶片为外植体经体细胞胚再生为植株被描述得最多（de Wit et al. 1990；Rout et al. 1991；Arene et al. 1993；Derks et al. 1995；Hsia and Korban 1996；Yokoya et al. 1996；Visessuwan et al. 1997；Kintzios et al. 1999；Kintzios et al. 2000；Li et al. 2002a；Kim et al. 2003a；Lee et al. 2010a；Jang et al. 2016）。其他的月季组织也被使用，如花丝（Noriega and Sondahl 1991）、花瓣（Murali et al. 1996；Estabrooks et al. 2007；Lee et al. 2010a；Vergne et al. 2010；Jang et al. 2016）、茎段（Rout et al. 1991；Arene et al. 1993；Hsia and Korban 1996；Dohm et al. 2002）、合子胚（Kunitake et al. 1993；Kim et al. 2003b，2009）和根（van der Salm et al. 1996；Kamo et al. 2005；Sarasan et al. 2001）。利用节间（Lloyd et al. 1988；Ishioka and Tanimoto 1990；Hsia and Korban 1996）、花丝、叶柄（Burrell et al. 2006）以及未成熟胚（Burger et al. 1990）已经实现了愈伤组织阶段的器官发生。从胚细胞悬浮液中获得的原生质体已经再生为植株（Matthews et al. 1991；Schum et al. 2001）。Matthews 等（1991）通过融合不同月季品种的原生质体获得了再生植株。Squirrell 等（2005）融合了月季、李属和悬钩子属的原生质体。已经由叶片和叶柄直接获得了再生的不定芽（Lloyd et al. 1988；Dubois and de Vries 1995；Ibrahim 1999；Dubois et al. 2000；Ibrahim and Debergh 2001；Pati et al. 2004），有时候愈伤组织的形成早于再生。Pourhosseini 等（2013）比较了不同基因型月季的直接和间接再生途径。尽管它们具有可变性，但是这些再生体系大多数是非常耗时的，而且芽的再生率很低。然而，它们具有高度的品种特异性，不易复制。Nguyen 等（2017）分析了 96 个月季基因型不定芽再生的能力，确定了基因座和候选基因。

27.4.2 转化

有关农杆菌介导的转化方法（Firoozabady et al. 1994；Derks et al. 1995；van der Salm et al. 1997，1998；Dohm et al. 2002；Li et al. 2002b，2003；Condliffe et al. 2003；Kim et al. 2004；Katsumoto et al. 2007；Lee et al. 2010b；Vergne et al. 2010）以及少量通过生

物粒子轰击的转化方法（Marchant et al. 1998b）的文献，均表明转化效率非常低（Debener and Byrne 2014）。转化的方法通常是基于月季叶片和花丝愈伤组织共培养而建立的（Firoozabady et al. 1994；Souq et al. 1996；Li et al. 2002b）。van der Salm 等（1997）利用茎段作为主要外植体。

通过遗传转化所尝试的性状是多样的。Debener 和 Byrne（2014）阐述了其可提高抗病性。利用基因导入技术将编码一种碱性（I 类）几丁质酶的水稻基因导入黑斑病易感月季'Glad Tidings'胚性愈伤组织中（Marchant et al. 1998a）。为了增强对白粉病的抗性，通过农杆菌介导转化将一种抗菌蛋白基因 Ace-AMP1 导入'Carefree Beauty'中（Li et al. 2003）。Pourhosseini 等（2013）通过导入水稻的几丁质基因来提高月季对白粉病的抗性。

基于查耳酮合酶基因（*chs*）对月季进行正义和反义分析，发现暗红色品种的花色发生了变化（Firoozabady et al. 1994；Souq et al. 1996）。正义方法产生的转基因花从红色变为粉色，反义方法产生淡红色到洋红色的花。在这两种情况下均未获得白色花。考虑的其他目标是衰老、乙烯反应（Zakizadeh et al. 2013）、冷应激（Chen et al. 2010）、花青素以及丁香酚含量（Zvi et al. 2012）。月季转化中最著名的例子就是 delphinidin 在花瓣中积累产生了蓝色月季（Katsumoto et al. 2007；Tanaka et al. 2007）。这种月季是市场上罕见的转基因观赏植物之一。蓝色月季 Applause^{TM} 于 2009 年首先在日本上市，随后在美国和加拿大上市。

基于 *Rol* 基因能诱导根的形成，该基因的特异性已被用于提高砧木的生根能力，主要是木本植物；van der Salm 等（1997）已将 *rolB* 和 *rolABC* 基因导入月季砧木'Moneyway'中。

27.4.3　多倍体化和单倍体化

杂交中的问题常由倍性差异引起。通过单倍体降低倍性水平，或通过有丝分裂或减数分裂多倍体化增加倍性这些技术克服二倍体障碍。在月季中，采用抗有丝分裂剂，如秋水仙碱（Fagerlind 1958；Semeniuk and Arisumi 1968；Kermani et al. 2003；Feng et al. 2016）、甲基胺草磷（APM）（Khosravi et al. 2008）、氨磺乐灵（Kermani et al. 2003；Allum et al. 2007；Khosravi et al. 2008）或氟乐灵（Zlesak et al. 2005；Khosravi et al. 2008；Feng et al. 2016），通过有丝分裂体多倍体化获得多倍体。另外一些研究中，利用组织培养技术（Kermani et al. 2003；Allum et al. 2007；Khosravi et al. 2008），而在较早的研究中，采用处理幼苗体内分生组织的方式（Fagerlind 1958；Semeniuk and Arisumi 1968；Meneve 1995；Zlesak et al. 2005；Feng et al. 2016）。通常首选组织培养方法，因为已知原始二倍体的基因型，并且可以重复处理基因型以提高成功率。用不同浓度秋水仙碱、氟乐灵和稻瘟灵处理的体外培养实验表明，对于许多月季基因型，如 Kermani 等（2003）所述 5μmol/L 的稻瘟灵浓度产生的四倍体数量最多。重要的是使用的外植体尺寸应非常小（只有几毫米），以确保抗有丝分裂剂的进入（Allum et al. 2007；Kermani et al. 2003）。

Kermani 等（2003）评估了在植物中，多倍体化对如从二倍体中获得的四倍体以及从三倍体中获得的六倍体的表型和花粉生育力的影响。染色体加倍后，叶厚增加，宽长比增大，叶色呈深绿色。然而，这在所有的六倍体中都没有观察到。四倍体的节间较长，而六倍体的节间较短。在一个四倍体基因型中，花瓣数增加。花粉生育力在四倍体中增加，但其在所有六倍体中没有增加。通过利用未减数的配子（Crespel and Gudin 2003）和原生质体融合进行体细胞多倍体化（Mottley et al. 1996），使月季可能发生减数分裂多倍体化。已经开发和研究了月季中的二倍体植物（El Mokadem et al. 2002a，2002b）。最著名的二倍体是 *R.* × *kordesii*，它是由 *R. rugosa* 和 *R. wichurana* 开发而成的，产生了一组 *R.* × *kordesii* 杂种。特别针对黑斑病抗性，在杂交中使用了 Baye 的二倍体 67-305、86-3 和 86-7（Byrne et al. 1996；Byrne and Crane 2003）。

利用辐射花粉和四倍体月季品种的胚挽救技术，在月季中获得了单倍体植株（Meynet et al. 1994；Crespel et al. 2002；El Mokadem et al. 2002a，2002b）。

27.4.4 二倍体和种间杂交

在月季中种间杂交通常会涉及二倍体杂交。在现代月季的遗传背景中发现了二倍体和种间杂交种。在过去尽管这造成了一些问题，但是许多二倍体和种间杂交种是由庭院月季培育得到的，因为二倍体参与杂交，所以许多三倍体作为新品种而商业化。回顾已有文献记载的月季品种进化史，发现月季品种的倍性变化并不是单向的，而是朝着更高的倍性水平发展。有趣的是，许多二倍体品种的遗传背景中具有四倍体亲本。三倍体可以产生二倍体和四倍体，这取决于另一个亲本的倍性水平。通常三倍体植物被认为是不育的，在育种计划中的应用有限。在月季中，三倍体具有一定的繁殖能力（de Vries and Dubois 1996；Grossi and Jay 2002），并优选作为花粉供体（Van Huylenbroeck et al. 2005）。

一些研究评估了原种和品种之间的杂交（de Vries and Dubois 2001；Van Huylenbroeck et al. 2007；Werlemark et al. 2009；Abdolmohammadi et al. 2014）。Werlemark 等（2009）研究了 *Caninae* 减数分裂对月季品种杂交的影响。以五倍体犬蔷薇为母本，四倍体品种为父本。虽然生育力很低，但还是获得了杂交后代。在形态学和抗病性方面，后代更多地表现出母本的许多特性，如单花。在所获得的后代中，发现了连续开花的后代（Werlemark et al. 2009）。Abdolmohammadi 等（2014）用五倍体的犬蔷薇作为母本，并通过组织培养得到了胚。在使用 *R. canina* 的情况下，没有果或种子的形成，但与 *Caninae* 中的其他原种杂交能获得后代。犬蔷薇的生育力很高，在与其他物种杂交时，它们作为母本时表现为多倍体，作为父本时表现为二倍体（Smulders et al. 2011）。

R. damascena（$2n = 4x$）与四倍体月季品种在四倍体水平上的种间杂交，其组织培养中的胚萌发率较高（Abdolmohammadi et al. 2014）。

在不同欧洲野生种和四倍体品种间的杂交实验中，对育种效率进行了评估。发现实验组合间存在差异。但是，一般而言，不论四倍体作为母本或是父本时，杂交效率与坐果率或发芽率相似。当四倍体品种作为母本时，产生的种子数量更多（Van Huylenbroeck et al. 2007）。de Vries 和 Dubois（2001）对使用二倍体、四倍体、六倍体和八倍体进行种间杂交进行了描述。在二倍体杂交中，通常会发现意外的倍性水平，这可能是由未减数配子的形成引起的（de Vries and Dubois 2001；Van Huylenbroeck et al. 2005）。

MacPhail 和 Kevan（2009）回顾了 94 份月季的繁殖体系。他们提供了关于自花授粉、无性生殖等方面的信息。此外，还根据可用的文献数据给出了坐果率和种子数。在 94 份月季中，其中有 34 份是自交不亲和的；其他 18 份与文献中的数据相矛盾。无丝分裂非常少见，但其的确存在。

27.4.5 芽变和突变育种

在月季中获得新品种的另一个方法就是自然芽变。月季中众所周知的一种株型变化是从直立到藤本，这种变化是由芽变自然产生的。一个著名的例子就是‘Iceberg’（syn. ‘Schneewittchen’；‘Fée des Neiges’），它芽变后的藤本品种称为‘攀登冰山’。Wylie（1954）概述了 1926～1950 年引入的杂种茶香月季的起源。当时，已有 1443 种（2236 种中）引种的栽培种起源记录，其中 411 种是由芽变产生的。其中，140 个芽变品种均表现出藤本这一特征变异，而其他 271 个芽变品种则表现出另一种特征的变化，主要与花有关。芽变仅能引起少量月季品种性状的改变。例如，由‘Ophelia’品种产生的 206 个芽变品种已被注册。据说，切花月季的大量生产是导致芽变品种大量出现的因素之一（Wylie 1954）。

月季的起源并不总是有据可查的，但最近的切花月季也被认为是芽变起源的。据说总共 18 000 个品种中有 10% 来自自然芽变（Haring 1986）。在迪瑞特公司培育的月季品种中可以找到实例，如‘Ravel’和‘Pavarotti’是众所周知的‘Vivaldi’的芽变品种。同样，‘Rossini’可能也是芽变品种，因为其与‘Ravel’的遗传关系非常紧密（Wylie 1954）。

Datta（2012）回顾了观赏植物中月季的诱变育种。自 20 世纪 60 年代以来，就尝试人工诱导月季突变（Chan 1966）。组织培养材料是最常用的诱变育种材料。不同的作者描述了不同类型月季的植物特性、花色和/或其他花朵特征的变化（Walther and Sauer 1986；Arnold et al. 1998；Ibrahim 1999；Senapathi and Rout 2008；Bala and Singh 2015）。基于诱变育种培育了几个商业化的品种。最常见的方法是辐射，使用化学试剂诱导在月季诱变育种中是很少见的。突变育种以前在育种公司很流行。过去，Kordes、Tantau 和 Meilland（Gudin 2000）致力于通过诱导突变来开发更多的变异。最近，Kahrizi 等（2015）评估不同剂量的 γ 射线对组织培养中不同基因型月季的影响。

在 IAEA 突变品种数据库中，注册的具有突变起源的月季品种有 67 个（IAEA 2018）。数据库中的月季数量很有可能不完整。第一个是 1960 年注册的由 John James 培育的‘Pink Hat’品种；最后一个是 2000 年注册的由日本培育的三个突变体。1981～1990 年，注册了 50 个开发的突变品种。多数突变体都是由中国、印度或日本培育的。据报道，通过化学处理获得 4 个品种，通过物理处理获得 57 个品种，其他 6 个品种则没有相关信息的报道。

根据 1991 年 UPOV 公约，突变品种的品种权依然属于原始品种育种人所有（UPOV 2018）。例如，在月季中，一个原始的受欢迎品种‘Frisco’，其已有多达 16 个受保护的不同颜色的突变体（Dubois 2003）。因此，利用分子技术识别现有品种的自然和诱导变异类型是非常重要的（De Riek et al. 2001；Vosman et al. 2004）。

27.4.6 新的育种技术（NBT）

利用核酸酶设计有望在基因组区域引入靶向突变。这些技术包括锌指核酸酶（ZFN）、Talens 和基于 CRISPR/Cas 的系统。最可行的选择是，敲除与病害易感性相关的基因。没有引起基因组 DNA 的变化，可能不会被视为转基因。在多倍体中使用基于核酸酶的技术尤为有趣。但是，仍然需要更多与基因敲除相关的副作用信息。在月季中，敲除 Mildew Locus O（*MLO*）基因，很可能产生抗白粉病的月季品种。在月季中，已对 4 个 *MLO* 同源基因（*RhMLO1*、*RhMLO2*、*RhMLO3* 和 *RhMLO4*）进行了测序和定位（Kaufmann et al. 2012）。*MLO* 基因是所谓的易感基因（S 基因）。其功能丧失导致其他植物物种的广谱抗性。第一种方法使用 RNAi 沉默 4 个 *MLO* 基因的表达（Geike et al. 2015）。但由于这种方法是转基因的，因此预计更好的实际应用是使用核酸酶设计（如 CRISPR/Cas9）敲除 *MLO* 基因。

27.5 分子（细胞）遗传学辅助月季育种

27.5.1 分子和细胞遗传学标记发展

Hubbard 等（1992）、Rajapakse 等（1992）、Torres 等（1993）以及 Ballard 等（1995）通过限制性片段长度多态性（RFLP），在月季中引入了 DNA 指纹识别的概念。自那时以来，已经开发了许多其他基于 DNA 的方法，现在经常使用这些方法来区分品种、进行育种选择和基因库访问，以及评估各种植物的遗传变异性和遗传相关性，包括蔷薇属（Nybom 2017 年修订）。这些技术的成功实际上主要取决于可用的标记数不受限制，以及它们不受环境条件影响的事实。在可用的不同方法中，随机扩增多态 DNA（RAPD）和扩增片段长度多态性（AFLP）是最早应用于月季的方法。它们都产生了多位点和高度多态性的带型，主要起源于核 DNA。进行品种鉴定的早期例子是利用 RAPD 的 Cubero 等（1995）、Reynders-Aloisi 和 Bollereau（1996）、Gallego 和 Martinez（1996）、Matsumoto 和 Fukui（1996），以及使用 AFLP 的 De Riek 等（1997，2001）和 Zhang 等（2000）。后来，出现了更多的位点特异性方法，PCR 扩增重复 DNA 序列，即所谓的微卫星，被经常应用于月季中的核 DNA 和 cpDNA 筛选（Esselink et al. 2003；Yan et al. 2005；Zhang et al. 2006；Hibrand-Saint Oyant et al. 2008）。最近，单核苷酸多态性（SNP）受到更多的关注，并被开发用于同时对多个位点进行大规模筛选（Nybom 2017）。为了开发用于月季的通用大型的 SNP 阵列，Koning-Boucoiran 等（2014）结合了来自不同组的四倍体切花和庭院月季以及二倍体野蔷薇的 EST。对于这个 RNA-Seq 库，使用 Illumina 配对端和 454 测序生成了大约 7 亿个读数。选择四倍体月季中的 SNP 来构建基因分型阵列，该阵列可用于四倍体园林和切花月季的遗传作图与标记-性状关联分析。WagRhSNP Axiom 阵列上总共包含 68 893 个 SNP（Koning-Boucoiran et al. 2014）。

在育种中，植物品种保护是一个热门的话题。虽然用形态特征鉴别不同的月季品种并不难，但是分子标记已广泛应用于月季品种的鉴定。主要的原因是月季品种层出不穷，这使得选择参考对照材料变得复杂化。目前正在努力将大量 SNP 从 68K WagRhSNP 阵列减少为一组可用于实践的 SNP 标记（personal communication of Hedwich Teunissen，NAKTuinbouw，the Netherlands），可用于支持植物品种多样性实验中的差异性评估，以保护植物育种者的权利。

除了分子标记外，有效的细胞遗传学标记也可以帮助我们追溯许多生物学的问题（Jiang and Gill 2006）。荧光原位杂交（FISH）标记已经显示了它们追踪杂交后代个体染色体的有效性。这不仅可以促进目标性状从原种到品种的转移（Szinay et al. 2010），还可以揭示物种和异源多倍体的形成（Badaeva et al. 2016）。但是，由于以下几个原因，月季细胞遗传学的研究具有一定的挑战性：①月季具有较小的基因组大小（二倍体基因组大小在 0.83～1.30pg/2C）（Roberts et al. 2009）和难以区分的小染色体（$x = 7$）；②多倍体在属中常见；③月季根和芽有低的有丝分裂指数，细弱的根很难处理（Ma et al. 1996）。尽管如此，我们仍以 ITR、5S rDNA 和 45S rDNA 为细胞遗传学标记，探索了区分蔷薇属单个染色体的可能性。rDNA（45S，5S）的 FISH 定位已经完成，并利用其对野生种之间的系统发育关系进行了评估（Ma and Chen 1991，1992；Ma et al. 1997a，1997b；Fernandez-Romero et al. 2001；Akasaka et al. 2002，2003；Liu et al. 2008；Jian et al. 2010a，2010b）。从这些研究可以得出结论，几乎所有的二倍体蔷薇属基因组都有一对染色体，在核仁组织区域（NOR）上具有 45S rDNA 信号。但有例外，*R. foliolosa* 有三个染色体对具有 45S rDNA 位点。分别用带有 45S rDNA 的 FISH 标记证明了 *R. chinensis* 和 *R. gallica* 同源三倍体与异源四倍体的特性（Fernandez-Romero et al. 2001）。与 45S rDNA 位点相比，5S rDNA 位点位于蔷薇属的两对染色体上且属于“subset A”（Akasaka et al. 2002）。其中一个 5S rDNA 位点通常与 45S rDNA 基因配置在非携带染色体上（Akasaka et al. 2002；Lim et al. 2005）。Kirov 等（2016）采用 5S rDNA、45S rDNA 和拟南芥型端粒重复进行 FISH，结果表明其能够区分 *Rosa wichurana* 上的 5 条染色体（图 27.2）。通过将 FISH 结果与染色体形态学测量相结合（Kirov et al. 2014a），可以鉴定 *R. wichurana* 的所有 7 条有丝分裂染色体。采用以 45S 和 5S 为探针的双色 FISH 对 *R. canina*（$2n = 5x = 35$）中的染色体配对进行了研究，表明两个染色体组参与了二价染色体形成，其他三个染色体组之间没有发生重组（Lim et al. 2005）。总之，目前月季的细胞遗传学标记被限制在 45S rDNA、5Sr DNA 和保守的 ITS 序列。然而，随着“SteamDrop”协议的发展，利用高效染色体玻片的制备（Kirov et al. 2014b）和生物信息学工具来分离串联重复序列（如 Repeat Explorer）（Novak et al. 2013）、设计更多更好的细胞遗传学标记将变为可能。

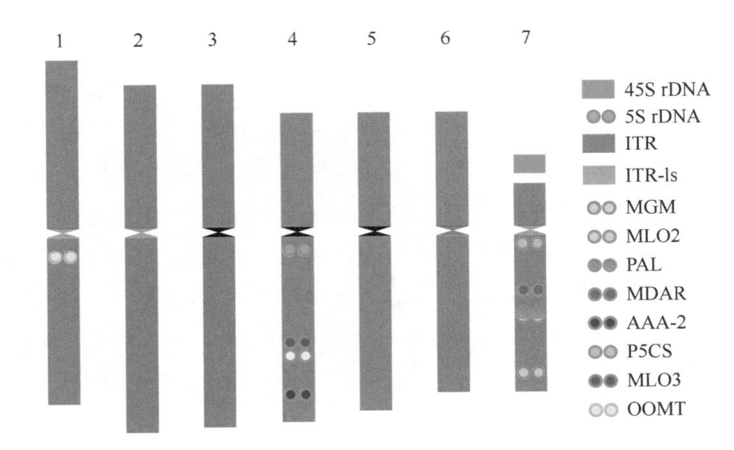

图 27.2 光叶蔷薇 *R. wichurana* 染色体上的重复（5S rDNA，45S rDNA 基因，拟南芥型端粒重复序列）和特异（基因）序列的细胞遗传学图。ITR 和 ITR-1s：分别在 37℃和室温下可见的间质端粒重复信号

27.5.2 遗传连锁图谱的构建及其与物理图谱的整合

有些研究，利用 RAPD 和 AFLP 标记对一个分离群体进行了遗传分析并作图（Debener and Mattiesch 1999）。采用 AFLP、简单序列重复（SSR）、蛋白激酶（PK）、抗性基因类似物（RGA）、RFLP、序列特异性扩增区（SCAR）和形态标记构建了一个完整的图谱（Yan et al. 2005）。基于 4 个二倍体种群和 1000 多个初始标记的信息，构建了第一个完整一致的图谱（ICM）（Spiller et al. 2011）。该整合图谱共包含 597 个标记，其中 206 个基于序列的标记，分布在 530cM 的长度上，位于 7 个连锁群（LG）上。通过使用更大数量的有效种群，使用更高密度的标记，此 ICM 中的标记顺序比单个种群图中的标记顺序更可靠。因此，基于 ICM 建立了月季 LG 的标准命名法，并提出了重要的观赏性状基因的位置，如自交不亲和性、抗黑斑病性（*Rdr1*）、香味和连续开花（Spiller et al. 2011）。总共定位了 10 个表型单一的位点，包括 7 个不同性状的 QTL 和 51 个 EST 或基于基因的分子标记。

在月季育种实践中，二倍体材料的使用是有限的。人们早已认识到，多倍体物种并不总是整齐地属于同源或异源多倍体的类别，因此采用了“节段异源多倍体”一词来描述两者之间的中间物种。这种中间物种的减数分裂行为尚未完全理解，也不能在如何模拟它们的遗传模式方面达成共识（Bourke et al. 2017）。因此，瓦格宁根大学（荷兰）的一组研究人员利用四倍体月季种群开发了一个密集的遗传图谱，为利用 QTL 分析重要性状和分子标记辅助育种的未来实施奠定了基础（Vukosavljev et al. 2016）。在此之前，在四倍体月季中，现有的连锁图谱包括不到 300 个标记，覆盖 28 个连锁群（7 条染色体的 4 个同源组）。用于月季的 68K WagRhSNP Axiom SNP 阵列，在四倍体水平上整合 SNP 数量，由此产生了四倍体庭院月季品种‘Red New Dawn’的基因型后代。不幸的是，后代似乎不是来自一对单一的双亲杂交。意想不到的亲本杂交事件经常发生在月季中。因此，构建了一对双亲和一个以‘Red New Dawn’为共同母本的自交群体的连锁图。自交亲本的图谱是密度最大，在 25 个连锁群上有 1929 个 SNP 标记，覆盖 1765.5cM，平均标记距离为 0.9cM。与近缘草莓基因组有广泛的同源性。总体共线性得到证实，但也揭示了新的基因组重排。假染色体中的一些重排表明，沿着染色体臂，配对亲和力可能变化。这拓宽了我们目前对节段异源多倍体的理解。

利用四倍体切花月季种群，开发了 68K WagRhSNP 阵列，这是一种同源染色体的超高密度连锁图，是基于以前的采用同源四倍体的方法构建的。在这个群体中预测的二价构型使得 Bourke 等（2017）量化了同源染色体之间配对行为的差异，从而重新定位了 7 条月季同源染色体上的所有 25 695 个 SNP 标记，以确定每个亲本中每条染色体的配对行为。通过排斥相连接检测评估值的确定，可以推断出染色体配对行为的差异。

尽管遗传图谱在生物学中很重要，但由于标记重组频率沿染色体的分布不均等，因此它们未显示标记在染色体上的真实位置或基因/标记之间的实际物理距离。遗传图谱上的 1cM 可能相当于几个 kb 以及数百万 bp 的物理距离（Kirov 2017）。此外，越来越多的证据表明，基因的功能与其在染色体上的物理位置密切相关（Mandakova et al. 2015）。此外，可以使用物理图谱更详细地研究基因组之间的同步性和共线性，因为它们揭示了基因组结构和组织（O'Neill and Bancroft 2000）。最后，有关基因组结构和染色体重排的知识已广泛地应用到植物基因克隆实践中，这些事例充分显示了物理图谱的重要性（Himi and Taketa 2015）。识别参与重要经济性状形成途径的植物基因，可以促进植物育种和选择（Dohm et al. 2014）。由于所有这些原因，遗传图谱和物理图谱的整合是不可或缺的。为了将蔷薇属连锁群锚定到物理染色体上，采用 tyramide-FISH 和基于高分辨率熔解曲线（HRM）的分子标记系统相结合的技术，将基因片段（大小介于 1100 ～1700bp）物理映射到 *Rosa wichurana* 有丝分裂的 7 号、4 号和 1 号染色体，并将它们分配给三个连锁群（Kirov et al. 2014a）。此外，使用高分辨率的 tyramide-FISH 和多色 tyramide-FISH，成功地将 8 个基因（1.7～3kb）定位到了 *R. wichurana* 的粗线期 4 号和 7 号染色体上，与有丝分裂染色体相比，分辨率提高了 20 倍（Kirov et al. 2015）（图 27.2）。对 *Rosa wichurana* 和 *Fragaria vesca* 之间的基因组序列进行比对，结果显示 7 号染色体的共线性差，而 4 号染色体的共线性好。

27.5.3　月季基因组

自 Matsumoto 等（1998，2000）以来，除了标记技术外，序列分析已用于月季的系统发育研究。

自 20 世纪 90 年代以来，第一个月季基因组序列在最近几个月发表。在 2017～2018 年，一些月季基因组序列论文已经提交和/或发表。Nakamura 等（2017）发表了野生种 *Rosa multiflora* Thunb. 的基因组序列草案。使用 Illumina MiSeq 和 HiSeq 平台组装序列。组装的序列总长度为 739 637 845bp，由 83 189 个组装序列组成，接近 k-mer 分析估计的 711Mb 长度。组装序列的 N50 平均长度为 90 830bp，其中最长的为 1 133 259bp。组装序列的平均 GC 含量为 38.9%。基因预测发现 67 380 个候选基因与已知基因和结构域具有同源性，包括完整和部分基因结构。这种二倍体植物的大量基因可能反映了来源于 *R. multiflora* 的自交不亲和基因组的异质性，推测涉及花色、花香和开花习性相关基因。

Raymond 等（2018）和 Hibrand Saint-Oyant 等（2018）通过同时对 *R. chinensis* ‘Old Blush’测序，为蔷薇属开发了第一个高质量的参考基因组序列。Raymond 等（2018）使用测序和高密度遗传图（82 个 contigs，N50 为 24Mb，覆盖 503Mb 的锚定组装序列），并通过 Hi-C 染色体接触图谱数据进行验证。鉴定连续开花的新候选基因，并研究花香与颜色的关系。通过对现代月季品种起源中的几个野生种和‘La France’（这是结合欧洲和中国物种背景的第一个杂交种之一）进行重测序，研究了遗传多样性。Hibrand Saint-Oyant 等（2018）将长读和短读测序与锚定结合到高密度遗传图上。利用遗传图谱作为组装序列和染色体之间的中介，让基因组（覆盖基因组的 512Mbp；N50 为 3.4Mb；L75 为 97）在假设的分子标尺上进行排列；这在重组率高的区域是非常准确的。相反，基因组组装在重组率低的区域，如着丝粒和核周区，准确性往往不够。为了填补这些空白，物理地图可能非常有用（Kirov 2017）。因此，Hibrand Saint-Oyant 等（2018）通过注释 TE 家族和使用 FISH 定位中心重复序列来描绘标志性染色体特征，包括核周区。对 8 种蔷薇属植物进行了遗传多样性分析。基因注释得到了广泛的转录数据支持，通过这些数据可以确定关键观赏性状的潜在遗传调节因子，包括刺密度和花瓣数。这些参考序列将成为研究多倍体化、减数分裂和发育过程的重要资源，如花色、花香、花和刺的发育。这些参考序列还将通过与性状相关的分子标记的发展，鉴定潜在相关基因以及利用蔷薇科的共线性来加速育种。

27.6 切花月季的实用育种：育种家的观点

27.6.1 国际背景与产品差异

最大的切花月季育种公司一直位于欧洲（如荷兰、德国、法国），最古老的可以追溯到 100 多年前。今天，一些主要的荷兰育种公司已经被整合到更大的群体中，培育了各种各样的观赏植物。

然而，随着南美洲（厄瓜多尔、哥伦比亚）和非洲（肯尼亚）的重要性日益增加，欧洲的关注焦点开始转移。在过去的 15～20 年里，厄瓜多尔育种者开始了他们自己的计划。2010 年，首批欧洲育种家联合选择将杂交育种新的生产设施迁往肯尼亚。表 27.3 概述了全球切花月季育种者的情况。

切花月季生产区的这种地理差异也导致了针对生产区的特定选择标准。这些选择标准不仅与生产区的不同气候环境直接相关，而且与特定的生产系统直接相关。在欧盟，月季通常种植在具有折枝系统的基质中。经常使用同化光，特别是在北欧国家。生产系统是以土壤栽培或珍珠岩基质为基础的。在厄瓜多尔，发现了不同类型的生长系统。大多数月季仍然种植在土壤中，但还有一些种植者在塑料温室中使用珍珠岩基质。月季生产公司还种植和嫁接砧木。在非洲，月季生产通常种植在塑料大棚的土壤中。

在选择的早期阶段，将在不同的生产系统和环境下测试候选品种。厄瓜多尔育种者在厄瓜多尔以及非洲和哥伦比亚进行测试试验（海拔在这里起着重要的作用）。

表 27.3　全球切花月季育种公司

月季育种者	国家
Brown Breeding	厄瓜多尔
David Austin Roses	英国
De Ruiter	荷兰
Delbard	法国
Delforge Roses	比利时
Dümmen Orange	荷兰
Esmeralda Breeding	厄瓜多尔
Franko	新西兰
Interplant Roses	荷兰
Jan Spek roses	荷兰
Meilland International	法国
NIRP International	意大利
Schreurs	荷兰
Rosen Tantau	德国
United Selections	荷兰/肯尼亚
W. Kordes' Söhne	德国

在人造光的标准生产条件下，月季产量通常为每平方米 200～300 枝。在欧盟，光照条件从传统的补充照明 60～72μmol/(m²·s)（5000～6000lx）已提高到 240μmol/(m²·s)（20 000lx）。非洲或厄瓜多尔不需要人工光照。

在过去的 10 年里，大型月季的产量每平方米增加了 100 枝。产量的增加是由于育种的发展和种植技术的改良，包括施肥、CO_2剂量和照明。特别是在厄瓜多尔，海拔越高，植物就越大。

在 20～15 年前，小花月季（甜心）的受欢迎程度更高，但有一段时间小花月季的市场出现下降。现在，因为各类超市销售小花月季，所以小花月季在非洲的产量再次上升。这些小花月季的年产量可达每平方米 400 支。

27.6.2　月季选育

在荷兰，切花月季育种规模平均为 500～600 个亲本植株。每年都会引进 10～15 个新的亲本。一般来说，90%的亲本是可通过市场获得的品种；所有育种者都倾向于使用相同的基因库。厄瓜多尔的育种者使用较小的基因库，因为他们较少接触到大量的母株。

在月季品种选育的第一年，幼苗基于其特定的生长区域（欧盟、非洲或厄瓜多尔）来播种和繁殖。在荷兰或非洲的条件下，幼苗的繁殖是通过扦插进行的，而在厄瓜多尔，幼苗是在砧木'Natal Briar'上嫁接繁殖的。

第一次选择是在全年对 80 000～100 000 株幼苗进行评价的基础上完成的。一般来说，花朵的特性是苗期的主要选择标准（表 27.4）。幼苗早期的首要选择标准是与饱满的花蕾相关的花瓣数量。平均而言，10%的幼苗已经被这一标准淘汰。第一阶段的选择是在没有人造光的情况下完成的。总共 2%～3%或 3000～4000 株幼苗在第一次选择中保留。选定的幼苗随后被克隆繁殖，以确保每个基因型有 10 株植株，以满足下一步的选择。

经过第二轮筛选和选择，第三轮和第四轮选择分别在 30 株和 300 株上进行。在这些选择周期中，对其他特征进行判断，并收集生产数据（表 27.4）。最后的结果出现在 4～5 年后，有时更快。在那时，有 2～3 年的生产数据，瓶插寿命的表现也有很好的记录。育种的一个重要问题是进一步的选择时机。较长的观察期虽然增加了成本，但对不同基因型各方面的能力有了更好的了解。

表 27.4　基于选育的不同阶段，切花月季的重要评选标准

选择阶段	重要评选标准
幼苗期	花瓣数量 花蕾形态与特征 出色的花朵 感病性
第一次克隆繁殖	花的特征 花枝长度 花枝长与花径之比 皮刺 活力
进一步克隆繁殖	产量 瓶插寿命 运输耐受性 抗病性

最近的研究进展已提高了育种技术的效率。例如，种子萌发率（平均为 35%～40%，在某些基因型中高达 80%）和花粉萌发率在育种者获得的知识基础上有所改善。种子萌发率与播种前种子的层积处理相关。酸处理对发芽有积极的影响。一个较长的层积处理周期导致发芽的时间更整齐。花粉生育实验通常由许多育种者来完成。虽然流式细胞术和分子标记在其他切花中有应用，但在商业上并未用于月季中。流式细胞术不适用于多倍体，因为切花月季通常是四倍体，所以不能通过流式细胞术来估计其倍性。对于四倍体的月季，在分子标记的选择和利用方面也具有一定的障碍。

与欧洲育种家相比，厄瓜多尔有利的气候使育种者全年都可以进行杂交。在厄瓜多尔气候条件下，其还具有一个优势，那就是在第一轮后代中淘汰一些不好的品种。

欧洲育种者在研究过程中利用软件进行数据收集。而在厄瓜多尔，数据收集通常使用纸笔进行。

切花在运输过程中的耐运输能力是切花月季评价的一个重要方面。根据欧盟的运输链，在为期 4 天的测试中模拟水中的运输：首先将切花月季包装好，存放在寒冷的房间中过夜；然后将月季放进水桶中模拟运输。根据非洲情况进行的运输模拟，月季是通过出口公司送往海外的，其中包括花枝不插入水中的时期。

在选择过程中，经常将运输模拟和瓶插寿命测试结合起来评估花朵的变色程度与持久性。月季瓶插寿命测试最少为 12～14 天。质量最好的月季在运输后仍有 14～16 天的瓶插寿命。芳香和花瓣数量较低的月季，花朵往往更快开放。在采后评估期，对于白色月季主要是评估其瓶插的干枯情况以及花色由白变绿的情况。

目前增强的抗病性没有附加的商业价值。因此，没有使用生物测定或其他手段来评估抗病性。在选择和培养过程中，当问题自发发生时，就会进行观察。各生产区的病害压力可能不同。在非洲，主要关注根癌病和黑斑病。与欧洲和非洲区域相比，厄瓜多尔的病原体较少。最重要的病害是灰霉病（主要在运输过程中）、霜霉病和白粉病。生长条件不卫生和环境控制管理不足等增加了灰霉病发生的风险。灰霉病抗性得到了更多的关注，在栽培和运输过程中观察到月季基因型之间的差异。一般情况下，灌丛月季具有更好的抗霉菌性。生物控制被越来越多地用于作物保护，包括在非洲，一些关于病原体的具体专题小型研究项目已经开始。除了抗病性，害虫的抗药性在未来可能会得到更多的关注，但目前还不是一个大问题。月季的虫害问题主要为蓟马、蚜虫和红蜘蛛。

最近，非生物胁迫、盐渍土、热和干旱胁迫，已成为人们评价切花月季适应性时的热点。因为环境立法，所以抗盐性在欧洲将变得更加重要。炎热和干旱是世界上某些地区特有的。

在育种过程中评估月季的其他特征往往涉及小众市场，如双色花、绿心花（在已经开放的花中发生花蕾的增殖）、扭曲的花心（花瓣太多而没有完全开放的花）等。新奇特的尝试，如使用 *Rosa persica*（*Hulthemia* 杂交品种）花，具有一个深色中心的那些品种，对于花瓣极多的切花，此功能不适用，因为其中心区域无法显示。因此，双色花在灌丛月季和盆栽月季中更为受欢迎。

在切花月季中，香味不属于选育特征。偶尔也会考虑这一特性，但种植者和贸易商往往对香味持有偏见，因为他们认为香味对瓶插寿命有负面影响。

27.6.3　月季品种的市场

花的颜色按市场重要性的顺序是红色、白色、粉红色和其他颜色。红色和白色是首选的颜色，特别是在欧盟和俄罗斯；俄罗斯市场更喜欢大花。

目前，欧盟最受欢迎的大型月季品种仍然是 Red Naomi!™（‘Schemocba’），它由 Schreurs 公司于 2006 年推出。在引种时，该品种表现出较好的产量、较好的花蕾大小和较低的病害易感性。以前最受欢迎的是 20 世纪 90 年代末推出的 Grand Prix™（‘Selaurum’）（Select Roses B.V.）。对于白色月季 Avalanche+™（‘Lexhcaep’），由 Lex+now Dümmen Orange 公司培育，并于 2001 年推出，到现在仍然是最受欢迎的。

品种的市场渗透性与地理区域有关。在厄瓜多尔，最受欢迎的红色月季是 Freedom™（‘Tan97544’, Tantau）。其他流行品种是 Explorer™ 和 Brighton™（Interplant）、Mondial™（Kordes）、Tibet™ 和 Scarlata™（Schreurs）、Orange Crush™（De Ruiter）。在非洲，Red Calypso™（‘Panroug’，De Ruiter），一种中等大小的月季，是最受欢迎的。

其他月季在很长一段时间内有一个虽小但稳定的位置。一个例子是粉红色品种

Aqua!™（'Schrenat'，Schreurs）：该品种于 1996 年引进。虽然它从来没有获得很大的人气，但它仍然在市场上占有一席之地。

新引进的红色月季很难发展，因为种植者的保守态度。欧盟的育种者与厄瓜多尔相比，他们培育的品种在市场上仍更具有商业优势。这可能是由于后者最近的育种活动和学习着重于了解亲本植物以及对后代的期望。此外，由于对新的育种者缺乏信心，以及其与种植者的贸易关系，厄瓜多尔的品种更难进入市场。厄瓜多尔原产的最著名品种是 Hot Nina™、Kahala™（Brown Breeding）、Playa Blanca™（Esmeralda）。

新品种的引进取决于地理区域。在欧盟和厄瓜多尔，育种者在展示温室中展示新品种。在墨西哥、巴西和亚洲等新兴生产地区，育种者没有新品种展示室，而是与在不同地点测试几个品种的育种者代理人合作。

一个育种者每年在每个产区大约释放 4 个新品种。平均而言，一个育种者每年引进 15 个品种。在这 15 个品种中，只有一个品种的寿命可以达到数年。品种的寿命受不同因素的影响：市场需求、市场更替和种植者种植品种的更新速度。

在过去，种植者通常每隔 4～5 年更新一次月季种植。然而现在，通过保留月季生产长达 10 年来降低成本。成本节省不仅包括种植公司的成本，也包含品种专利费的成本。在过去，专利费成本是一次性的，但现在变为每年支付。所需的费用取决于品种以及种植者必须承担的种植新品种的风险。一般来说，对于一个新品种，其专利费将低于一个已经在市场上具有一定比重且非常受欢迎的品种的专利费。

种植者现在正在寻求月季品种的一个完整套餐链，包括高产的良好种质，还有完备的销售渠道。种植规模较大的种植者在引进一个新品种时，如果这个新品种具有较低专利费的优势，该品种才能进入市场。

欧洲种植者在几公顷的玻璃温室中种植一个或两个品种。相比之下，南美洲的种植者从不同的育种者那里选择并生产多达 50 个不同的品种。

27.6.4 切花月季育种的未来展望

育种者预测，提高切花月季的产量仍然是可能的，但他们不知道哪些性状将在未来变得更加重要，特别是因为不同的地理文化市场设定了不同的优先事项。根据不同的地理位置，找出月季性状表现不同的原因正变得越来越重要。

数据收集可以帮助育种者评估月季幼苗，并用于更好地选择和组合亲本植物。软件被用来选择特定的性状及为品种性状描述登记注册（如叠加性状）。然而，到目前为止，这些数据只被用于支持杂交组合的选择。记录的性状是花朵颜色、植物类型、刺、结实率、抗病性和发芽能力（作为杂交亲本的发芽情况）。记录的数据仅代表单个植物水平，并不代表种群水平。然而，当一个新的性状特征被引入时，这可能是个有趣的回交程序，如 *Hulthemia* 杂交品种。在未来，更多数据收集和处理的应用将被开发与应用于育种实践。

在一般情况下，切花月季的研究工作是有限的，这主要是由于生产地域的转移。切花生产区越来越多地向南美洲和非洲转移，而主要的育种者仍然位于欧盟。育种区和生产区之间的脱节，降低了国家基金机构资助欧盟月季遗传研究的意愿。

27.7　庭院月季的育种实践

27.7.1　庭院月季育种的挑战

虽然主要的商业庭院月季育种者都在欧盟，但是庭院月季育种活动遍布世界各地。育种爱好者在庭院月季种植者中普遍存在。在商业庭院月季育种中，使用了更广泛的母本，这使得母本的结实率成为关注的焦点。种子供体植株需要具备与大量父本杂交相容的能力。一些育种者从测试杂交开始，以评估母本的结实率，然后再将其用作繁殖计划中的亲本。

对抗病能力的关注取决于栽培实践。黑斑病在庭院月季中无处不在，但随着更多的庭院月季在容器和大棚中生长，白粉病越来越流行。一些育种者在他们的选择方案中引入了黑斑病的人工免疫接种。公司提到，使用具有特征的病理型是不可能的，因为其没有实验室设施。当该公司有实验室设施时，也会进行不同黑斑菌株的第一次测试。一些育种者认为，黑斑病造成的问题已经变得更加严重，而其他问题，如红蜘蛛，已经减少。

特别是在美国，月季丛矮病毒（RRV）是一个主要关注的问题。抗性育种正在进行中，但尚未取得明显成功。一些育种者通过设置实验地点来筛选候选品种。

在美国，新出现的害虫还有月季叶蜂、蔷薇瘿蚊、淤泥寄生虫、蓟马等。这些害虫主要危害月季叶片。在月季品种之间发现了抗虫性的变化，这为在商业品种中引入这些性状的特定育种开辟了前景。

庭院月季中的植物造型正受到人们的关注，尤其是在美国，因为植物造型与植物的退化之间存在相关性。因此，在欧盟和美国种植的月季灌丛具有不同的表型。这不仅仅是一个品味问题，更是植物生长适应性的表现。在欧盟流行的植物类型并不总是适合在美国应用。美国有更广泛的气候类型，如晚春霜冻，这会引起某些类型庭院月季出现问题。

育种家把重点放在抗寒性实验上，他们选择较冷的地区，有时也应用基于电子漏的实验测试来获得抗寒性的信息。

与切花月季相比，庭院月季在商业上的重要性较低，现在更多的研究工作集中在庭院月季的种植。尤其是美国对庭院月季有着浓厚的研究兴趣。

27.8　结　　论

尽管月季种植和育种有着数千年的历史，但育种研究仍然充满挑战。生育力和繁殖力是育种家关注的焦点，也是最近发表的科学研究关注的焦点。月季进行组织培养是困难的，但再生并不是不可能，成功的遗传转化就是证明，可用于创造蓝色月季。突变育种在过去虽然被用于月季，但这种技术不再流行。月季多倍体是个有趣的研究课题，多倍体育种对月季育种进程有很大的影响。虽然切花月季的四倍体状态有碍于标记技术在其选择过程中的发展和应用。然而，有关月季的文献中描述了分子技术，最近发表的月季基因组序列让我们认识到越来越多的月季遗传背景知识。然而，月季育种公司的研究和实际应用之间依然还存在着很大的差距。

在切花月季中，生产正在推进。公司越来越注重差异化，如为特定的生产领域培育月季品种。最大和最古老的育种公司均位于欧洲（荷兰、法国和德国），在过去的 10～15 年新的南美洲月季育种公司进入了市场。到目前为止，一名欧洲育种家已经将其的育种计划从欧洲转移到了非洲。

在庭院月季中，远缘杂交及生殖障碍是研究中最普遍存在的问题。在月季中，为扩大月季狭窄的遗传背景，对原种的利用是非常引人关注的。原种也被应用到特定的育种目标中。正在构建 *R. persica* 杂交组合，以期获得新的月季花色组合。利用原种还可以获得更好的抗病能力。选育出的高抗月季丛矮病毒的月季就是一个例子。

致谢：我们感谢育种者分享他们目前在切花月季育种中的经验和观点。还感谢 ILVO 公司的英文编辑 Miriam Levenson，感谢他对手稿结构、客观问题及英文修改的有用建议。

本章译者：
邱显钦，张　颢，李绅崇
云南省农业科学院花卉研究所，国家观赏园艺工程技术研究中心，云南省花卉育种重点实验室，昆明 650200

参 考 文 献

Abdolmohammadi M, Kermani MJ, Zakizadeh H, Hamidoghli Y (2014) An vitro embryo germination and interploidy hybridization of rose (*Rosa* sp). Euphytica 198:255–264
AIPH (2015) International statistics flowers and plants. Vol 63 International Association of Horticultural Producers
AIPH (2016) International statistics flowers and plants. Vol 64 International Association of Horticultural Producers
Akasaka M, Ueda Y, Koba T (2002) Karyotype analysis of five wild rose species belonging to septet a by fluorescence in situ hybridization. Chromosom Sci 6:17–26
Akasaka M, Ueda Y, Koba T (2003) Karyotype analysis of wild rose species belonging to septet B, C, and D by molecular cytogenetic method. Breed Sci 53:177–182
Allum JF, Bringloe DH, Roberts AV (2007) Chromosome doubling in a Rosa rugosa Thunb. Hybrid by exposure of in vitro nodes to oryzalin: the effects of node length, oryzalin concentration and exposure time. Plant Cell Rep 26:1977–1984
Alp S, Çelik F, Türkoglu N, Karagöz S (2009) The effects of different warm stratification periods on the seed germination of some *Rosa* taxa. Afr J Biotechnol 8:5838–5841
Anderson N, Byrne DH (2007) Methods for *Rosa* germination. Acta Hortic 751:503–507
Arene L, Pellegrino C, Gudin S (1993) A comparison of the somaclonal variation level of *Rosa hybrida* cv. Meirutral plants regenerated from callus or direct induction from different vegetative and embryonic tissues. Euphytica 71:83–92
Arnold NP, Barthakur NN, Tanguay M (1998) Mutagenic effects of acute γ-irradiation on miniature roses. Target theory approach. Hortic Sci 33:127–129
Badaeva ED, Ruban AS, Zoshchuk SA, Surzhikov SA, Knüpffer H, Kilian B (2016) Molecular cytogenetic characterization of *Triticum timopheevii* chromosomes provides new insight on genome evolution of T. zhukovskyi. Plant Syst Evol 302:943–956
Bala M, Singh KP (2015) In vitro mutagenesis in rose (*Rosa hybrida* L.) cv. Raktima for novel traits. Indian J Biotechnol 14:525–531
Ballard RE, Rajapakse S, Abbott AG, Byrne D (1995) DNA markers in rose and their use for cultivar identification and genome mapping. Acta Hortic 424:265–268
Barton LV (1961) Experimental seed physiology at the Boyce Thompson Institute. Proc Int Seed Test Assoc 26:561

Bhanuprakash K, Tejaswini Y, Yogeesha HS, Naik LB (2004) Effect of scarification and gibberellic acid on breaking dormancy of rose seeds. Seed Res 32:105–107

Blechert O, Debener T (2005) Morphological characterisation of the interaction between *Diplocarpon rosae* and various rose species. Plant Pathol 54:82–90

Blundell JB (1965) Studies of flower development, fruit development, and germination in *Rosa*. PhD Dissertation, University of Wales

Bo J, Huiru D, Xiaohan Y (1995) Shortening hybridization breeding cycle of rose - a study on mechanisms controlling achene dormancy. Acta Hortic 404:40–47

Bosco R, Caser M, Ghione GG, Mansuino A, Giovannini A, Scariot V (2015) Dynamics of abscisic acid and indole-3-acetic acid during the early-middle stage of seed development in rosa, hybrida. Plant Growth Regul 75:265–270

Bourke PM, Arens P, Voorrips RE, Esselink GD, Koning-Boucoiran CFS, van't Westende WPC, Santos Leonardo T, Wissink P, Zheng C, van Geest G, Visser RGF, Krens FA, Smulders MJM, Maliepaard C (2017) Partial preferential chromosome pairing is genotype dependent in tetraploid rose. Plant J 90:330–343

Burger DW, Liu L, Zary KW, Lee CI (1990) Organogenesis and plant regeneration from immature embryos of *Rosa hybrida* L. Plant Cell Tissue Organ Cult 21:147152

Burrell AM, Lineberger RD, Rathore KS, Byrne DH (2006) Genetic variation in somatic embryogenesis of rose. HortSci 41:1165–1168

Byrne DH, Crane YM (2003) Biotechnologies for breeding / Amphidiploidy. In: Roberts, Debener, Gudin (eds) Encyclopedia of rose science. Elsevier, Oxford

Byrne DH, Black W, Ma Y, Pemberton HB (1996) The use of amphidiploidy in the development of blackspot resistant rose germplasm. Acta Hortic 424:269–272

Cabrera RI, Solis-Perez AR, Sloan JJ (2009) Greenhouse rose yield and ion accumulation responces to salt stress as modulated by rootstock selection. HortSci 44:2000–2008

Cai X, Sun Y, Starman T, Hall C, Niu G (2014) Response of 18 earth-kind® rose cultivars to salt stress. HortSci 49:544–549

Caser M (2017) Pollen grains and tubes. Reference Module in Life Sciences, Elsevier. https://doi.org/10.1016/B978-0-12-809633-8.05077-9

Caser M, Dente F, Ghione GG, Mansuino A, Giovannini A, Scariot V (2014) Shortening of selection time of *Rosa hybrida* by in vitro culture of isolated embryos and immature seeds. Propag Ornamen Plant 14:139–144

Chaanin A (2003) Breeding / selection strategies for cut roses. In: Roberts, Debener, Gudin (eds) Encyclopedia of rose science. Elsevier, Academic Press, Oxford

Chan AP (1966) Chrysanthemum and rose mutations induced by X-rays. Proc Am Soc Hortic Sci 88:613–620

Chen JR, Lü JJ, Liu R, Xiong XY, Wang TX, Chen SY, Guo LB, Wang HF (2010) DREB1C from *Medicago truncatula* enhances freezing tolerance in transgenic *M. truncatula* and China rose (*Rosa chinensis* Jacq.). Plant Growth Regul 60:199–211

Chimonidou-Pavlidou D (2004) Malfunction of roses due to drought stress. Sci Hortic 99:79–87

Condliffe PC, Davey MR, Power JB, Koehorst-van Putten H, Visser PB (2003) An optimized protocol for rose transformation applicable to different cultivars. Acta Hortic 612:115–120

CPVO (2018) Variety database http://cpvo.europa.eu/en/applications-and-examinations/cpvo-varieties-database Accessed 4 March 2018

Crépin F (1889) Sketch of a new classification of roses. J Roy Hortic Soc Lond 11:217–228

Crespel L, Gudin S (2003) Evidence for the production of unreduced gametes by tetraploid *Rosa hybrida* L. Euphytica 133:65–69

Crespel L, Mouchotte J (2003) Methods of cross-breeding. In: Roberts, Debener, Gudin (eds) Encyclopedia of rose science. Elsevier, Oxford

Crespel L, Gudin S, Meynet J, Zhang D (2002) AFLP-based estimation of 2n gametophytic heterozygosity in two parthenogenetically derived dihaploids of *Rosa hybrida* L. Theor Appl Genet 104:451–456

Crespel L, Ricci S, Gudin S (2006) The production of 2n pollen in rose. Euphytica 151:155–164

Cubero JI, Millan T, Osuna F, Torres AM, Cobos S (1995) Varietal identification in *Rosa* by using isozym and RAPD markers. Acta Hortic 424:261–264

Datta SK (2012) Success story of induced mutagenesis for development of new ornamental varieties. In: Kozgar K (ed) Induced mutagenesis in crop plants: bioremediation, biodiversity and bioavailability. Global Science Books Ltd., UK, Ikenobe, Japan pp 15–26

De Cock K, Scariot V, Leus L, De Riek J, Van Huylenbroeck J (2007) Understanding genetic relationships of wild and cultivated roses and the use of species in breeding. CAB Rev 52:2

De Dauw K, Van Labeke MC, Leus L, Van Huylenbroeck J (2013) Drought tolerance screening of a *Rosa* population. Acta Hortic 990:121–127
De Riek J, Dendauw J, Mertens M, Van Bockstaele E, De Loose M (1997) Use of AFLP for variety protection: some case studies on ornamentals. Medded Fac Landbouwww Univ Gent 62:1459–1466
De Riek J, Dendauw J, Leus L, De Loose M, Van Bockstaele E (2001) Variety protection by use of molecular markers: some case studies on ornamentals. Plant Biosyst 135:107–113
de Vries DP (2003) Breeding / Selection strategies for pot roses. In: Roberts, Debener, Gudin (eds) Encyclopedia of rose science. Elsevier, Academic Press, Oxford
de Vries DP, Dubois LAM (1980) Inheritance of pigments. Am Rose Annu 65:145–148
de Vries DP, Dubois LAM (1983) Pollen and pollination experiments. X. The effect of repeated pollination on fruit- and seed set in crosses between the hybrid tea-rose cvs. Sonia and Ilona. Euphytica 32:685–689
de Vries DP, Dubois LAM (1987) The effect of temperature on fruit-set, seed set and seed-germination in Sonia X Hadley hybrid tea-rose crosses. Euphytica 36:117–120
de Vries DP, Dubois LAM (1996) Rose breeding: past, present, prospects. Acta Hortic 424:241–248
de Vries DP, Dubois LAM (2001) Developments in breeding for horizontal and vertical fungus resistance in roses. Acta Hortic 552:103–112
de Vries DP, Dubois LAM (2015) Factors affection the germination of hybrid rose achenes: a review. Acta Hortic 1064:151–164
de Wit JC, Esendam HF, Honkanen JJ, Tuominen U (1990) Somatic embryogenesis and regeneration of flowering plants in rose. Plant Cell Rep 9:456–458
Debener T (1999) Genetic analysis of horticulturally important morphological and physiological characters in diploid roses. Gartenbauwissenschaft 64:14–20
Debener T (2017) Inheritance of characteristics. Reference Module in Life Sciences, Elsevier. https://doi.org/10.1016/B978-0-12-809633-8.05047-0
Debener T, Byrne DH (2014) Disease resistance breeding in rose: current status and potential of biotechnological tools. Plant Sci 228:107–117
Debener T, Linde M (2009) Exploring complex ornamental genomes: the rose as a model plant. Crit Rev Plant Sci 28:267–280
Debener T, Mattiesch L (1999) Construction of a genetic linkage map for roses using RAPD and AFLP markers. Theor Appl Genet 99:891–899
Debener T, Drewes-Alvarez R, Rockstroh K (1998) Identification of five physiological races of black spot, *Diplocarpon rosae*, wolf on roses. Plant Breed 117:267–270
Densmore RA, Zasada JC (1977) Germination requirements of Alaskan *Rosa acicularis*. Can Field Nat 91:58–62
Derks FHM, van Dijk AJ, Hänisch ten Cate CH, Florack DEA, Dubois LAM, de Vries DP (1995) Prolongation of vase life of cut roses via introduction of genes coding for antibacterial activity. Somatic embryogenesis and *Agrobacterium*-mediated transformation. Acta Hortic 405:205–209
Dewitte A, Leus L, Van Huylenbroeck J, Van Bockstaele E, Höfte M (2007) Characterization of reactions to powdery mildew (*Podosphaera pannosa*) in resistant and susceptible rose genotypes. J Phytopathol 155:264–272
Di Bello PL, Ho T, Tzanetakis IE (2015) The evolution of emaraviruses is becoming more complex: seven segments identified in the causal agent of rose rosette disease. Virus Res 210:241–244
Di Bello PL, Thekke-Veetil T, Druciarek T, Tzanetakis IE (2018) Transmission attributes and resistance to rose rosette virus. Plant Pathol 67:499–504
Dohm A, Ludwig C, Schilling D, Debener T (2002) Transformation of roses with genes for antifungal proteins to reduce their susceptibility to fungal diseases. Acta Hortic 572:105–111
Dohm JC, Minoche AE, Holtgräwe D, Capella-Gutiérrez S, Zakrzewski F, Tafer H, Rupp O, Sörensen TR, Stracke R, Reinhardt R, Goesmann A, Kraft T, Schulz B, Stadler PF, Schmidt T, Gabaldón T, Lehrach H, Weisshaar B, Himmelbauer H (2014) The genome of the recently domesticated crop plant sugar beet (*Beta vulgaris*). Nature 505:546–549
Drewes-Alavarez R (2017) Early embryo rescue. Reference Module in Life Sciences, Elsevier. https://doi.org/10.1016/B978-0-12-809633-8.05002-0
Drewes-Alvarez R (2003) Disease / black spot. In: Roberts, Debener, Gudin (eds) Encyclopedia of rose science. Elsevier, Academic Press, Oxford
Dubois LAM (2003) Intellectual property/plant patents and trademarks. In: Roberts, Debener, Gudin (eds) Encyclopedia of rose science. Elsevier, Oxford
Dubois LAM, de Vries DP (1987) On the inheritance of the dwarf character in polyantha x *Rosa*

chinensis minima (Sims) Voss F1-populations. Euphytica 36:535–539
Dubois LAM, de Vries DP (1995) Preliminary report on direct regeneration of adventitious buds on leaf explants of in vivo grown glass house rose cultivars. Gartenbauwissenschaft 60:249–253
Dubois LAM, de Vries DP, Koot A (2000) Direct shoot regeneration in the rose: genetic variation of cultivars. Gartenbauwissenschaft 65:45–49
El Mokadem H, Crespel L, Meynet J, Gudin S (2002a) The occurrence of 2n-pollen and the origin of sexual polyploids in dihaploid roses (*Rosa hybrida* L.). Euphytica 125:169–177
El Mokadem H, Meynet J, Crespel L (2002b) The occurrence of 2n eggs in the dihaploids derived from *Rosa hybrida* L. Euphytica 124:327–332
Esselink D, Smulders MJM, Vosman B (2003) Identification of cut-rose (*Rosa hybrida*) and rootstock varieties using robust sequence tagged microsatellite markers. Theor Appl Genet 106:277–286
Estabrooks T, Browne R, Dong Z (2007) 2,4,5-Trichlorophenoxyacetic acid promotes somatic embryogenesis in the rose cultivar 'Livin' easy' (*Rosa* sp.). Plant Cell Rep 26:153–160
Fagerlind F (1958) Hip and seed formation in newly formed *Rosa* polyploids. Acta Hortic Berg 17:229–256
Feng H, Wang ML, Cong RC, Dai SL (2016) Colchicine- and trifluralin-mediated polyploidization of *Rosa multiflora* Thunb. Var. inermis and *Rosa roxburghii* f. Normalis. J Hortic Sci Biotechnol 92:279–287
Fernandez-Romero MD, Torres AM, Millan T, Curero JL, Cabrera A (2001) Physical mapping of ribosomal DNA on several species of the subgenus Rosa. Theor Appl Genet 103:835–838
Firoozabady E, Moy Y, Courtneygutterson N, Robinson K (1994) Regeneration of transgenic rose (*Rosa hybrida*) plants from embryogenic tissue. Biotechnology 12:609–613
Fougère-Danezan M, Joly S, Bruneau A, Gao XF, Zhang LB (2015) Phylogeny and biogeography of wild roses with specific attention to polyploids. Ann Bot 115:275–291
Gallego FJ, Martinez I (1996) Molecular typing of rose cultivars using RAPDs. J Hortic Sci 71:901–908
Gar O, Sargent DJ, Tsai CJ, Pleban T, Shalev G, Byrne DH, Zamir D (2011) An autotetraploid linkage map of rose (*Rosa hybrida*) validated using the strawberry (*Fragaria vesca*) genome sequence. PLoS One 6:e20463
Geike J, Kaufmann H, Schürmann F, Debener T (2015) Targeted mutagenesis of *MLO* homologous genes in the rose genome. Acta Hortic 1087:507–513
Giovannini A, Macovei A, Donà M, Valassi A, Caser M, Mansuino A, Ghione GG, Carbonera D, Scariot V, Balestrazzi A (2015) Pollen grain preservation at low temperatures in valuable commercial rose cultivars. Acta Hortic 1064:63–66
Grossi C, Jay M (2002) Chromosome studies of rose cultivars: application into selection process. Acta Bot Gallica 149:405–413
Gudin S (1995) Rose improvement, a breeders experience. Acta Hortic 420:125–128
Gudin S (1999) Improvement of rose varietal creation in the world. Acta Hortic 495:283–291
Gudin S (2000) Rose: genetics and breeding. In: Plant breeding reviews, vol 17, pp 159–189
Gudin S (2001) Rose breeding technologies. Acta Hortic 547:23–33
Gudin S (2003) Breeding / Overview. In: Roberts, Debener, Gudin (eds) Encyclopedia of rose science. Elsevier, Academic Press, Oxford
Gudin S (2017) Seed propagation. Reference Module in Life Sciences, Elsevier. https://doi.org/10.1016/B978-0-12-809633-8.05093-7
Gudin S, Mouchotte J (1996) Integrated research in rose improvement – a breeders experience. Acta Hortic 424:285–291
Gudin S, Arene L, Chavagnat A, Bulart C (1990) Influence of endocarp thickness on rose achene germination: genetic and environmental factors. HortSci 25:786–788
Hakam N, DeEll JR, Khanizadeh S, Richer C (2000) Assessing chilling tolerance in roses using chlorophyll fluorescence. HortSci 35:184–186
Haring PA (1986) Modern Roses 9. American Rose Society, Shreveport
Harp DA, Kay K, Zlesak DC, George S (2015) The effect of rose root size on drought stress tolerance and landscape plant performance. Texas J Agric Nat Resour 28:82–88
Herrero M, Hormaza JI (1996) Pistil strategies controlling pollen tube growth. Sex Plant Reprod 9:343–347
Hibrand Saint-Oyant L, Ruttink T, Hamama L, Kirov I, Lakhwani D, Zhou NN, Bourke P, Daccord N, Leus L, Schulz D, Van de Geest H, Hesselink T, Van Laere K, Debray K, Balzergue S, Thouroude T, Chastellier A, Jeauffre J, Voisine L, Gaillard S, Borm T, Arens P, Voorrips R, Maliepaard C, Neu E, Linde M, Le Paslier MC, Berard A, Bounon R, Clotault J, Choisne N, Quesneville H, Kawamura K, Aubourg S, Sakr S, Smulders R, Schijlen E, Bucher E, Debener

T, De Riek J, Foucher F (2018) A high-quality genome sequence of *Rosa chinensis* to elucidate ornamental traits. Nature Plants. https://doi:10.1038/s41477-018-0166-1
Hibrand-Saint Oyant L, Crespel L, Rajapakse S, Zhang L, Foucher F (2008) Genetic linkage maps of rose constructed with new microsatellite markers and locating QTL controlling flowering traits Tree Genetics & Genomes 4:11
Himi E, Taketa S (2015) Isolation of candidate genes for the barley Ant1 and wheat Rc genes controlling anthocyanin pigmentation in different vegetative tissues. Mol Gen Genomics 290:1287–1298
Horst RK (1983) Compedium of rose diseases. APS Press, St. Paul
Hosseini Moghaddam H, Leus L, De Riek J, Van Huylenbroeck J, Van Bockstaele E (2012) Construction of a genetic linkage map with SSR, AFLP and morphological markers to locate QTLs controlling pathotype-specific powdery mildew resistance in diploid roses. Euphytica 184:413–427
Hsia CN, Korban SS (1996) Organogenesis and somatic embryogenesis in callus cultures of Rosa hybrida and Rosa chinensis minima. Plant Cell Tissue Organ Cult 44:1–6
Hubbard M, Kelly J, Rajapakse S, Abbott AG, Ballard RE (1992) Restriction fragment polymorphism in rose and their use for cultivar identification. HortSci 27:172–173
IAEA (2018) database https://mvd.iaea.org/. Accessed 04 Mar 2018
Ibrahim R (1999) In vitro mutagenesis in rose. PhD Dissertation, Ghent University
Ibrahim R, Debergh P (2001) Factors controlling high efficiency adventitious bud formation and plant regeneration from in vitro leaf explants of roses (*Rosa hybrida* L.). Sci Hortic 88:41–57
Ishioka N, Tanimoto S (1990) Plant regeneration from Bulgarian rose callus. Plant Cell Tissue Organ Cult 22:197–199
Iwata H, Kato T, Ohno S (2000) Triparental origin of damask roses. Gene 259:53–59
Iwata H, Gaston A, Remay A, Thouroude T, Jeauffre J, Kawamura K, Hibrand-Saint Oyant L, Araki T, Denoyes B, Foucher F (2012) The TFL1 homologue KSN is a regulator of continuous flowering in rose and strawberry. Plant J 69:116–125
Jackson GAD (1968) Hormonal control of fruit development, seed development and germination with particular reference to *Rosa*. SCI monogr 31:157–156
Jackson GAD, Blundell JB (1963) Germination in *Rosa*. J Hortic Sci 38:310–320
Jacob Y, Ferrero F (2003) Pollen grains and tubes. In: Roberts, Debener, Gudin (eds) Encyclopedia of rose science. Elsevier, Academic Press, Oxford
Jang HR, Lee HJ, Park BJ, Pee OJ, Paek KY, Park SY (2016) Establishment of embryogenic cultures and determination of their bioactive properties in *Rosa rugosa*. Hortic Environ Biotechnol 57:291–298
Jian HY, Zhang H, Tang KX, Li SF, Wang QG et al (2010a) Decaploidy in *Rosa praelucens* Byhouwer (Rosaceae) endemic to Zhongdian plateau, Yunnan, China. Caryology 63:162–167
Jian HY, Zhang H, Zhang T, Li SF, Wang QG et al (2010b) Karyotype analysis of different varieties of *Rosa odorata* sweet. J Plant Genet Res 11:457–461
Jiang J, Gill BS (2006) Current status and the future of fluorescence in situ hybridization (FISH) in plant genome research. Genome 49:1057–1068
Joly S, Starr JR, Lewis WH, Bruneau A (2006) Polyploid and hybrid evolution in roses east of the Rocky Mountains. Am J Bot 93:412–425
Kahrizi ZA, Kermani MJ, Amiri M, Vedadi S, Hosseini Z (2015) In vitro radio-sensitivity of different genotypes and explants of rose (*Rosa hybrida*). J Hortic Sci Biotechnol 88:47–52
Kamo K, Jones B, Bolar J, Smith F (2005) Regeneration from long-term embryogenic callus of the *Rosa hybrida* cultivar cardinal. In Vitro Cell Dev Biol Plant 41:32–36
Katsumoto Y, Fukuchi-Mizutani M, Fukui Y, Brugliera F, Holton TA, Karan M, Nakamura N, Yonekura-Sakakibara K, Togami J, Pigeaire A, Tao GQ, Nehra NS, Lu CY, Dyson BK, Tsuda S, Ashikari T, Kusumi T, Mason JG, Tanaka Y (2007) Engineering of the rose flavonoid biosynthetic pathway successfully generated blue-hued flowers accumulating delphinidin. Plant Cell Physiol 48:589–1600
Kaufmann H, Qiu X, Wehmeyer J, Debener T (2012) Isolation, molecular characterization, and mapping of four rose *MLO* orthologs. Front Plant Sci 3:244
Kermani MJ, Sarasan V, Roberts AV, Yokoya K, Wentworth J, Sieber VK (2003) Oryzalin-induced chromosome doubling in *Rosa* and its effect on plant morphology and pollen viability. Theor Appl Genet 107:1195–1200
Khosravi P, Kermani MJ, Nematzadeh GA, Bihamta MR, Yokoya K (2008) Role of mitotic inhibitors and genotype on chromosome doubling of *Rosa*. Euphytica 160:267–275

Kim CK, Chung J, Jee S, Oh J (2003a) Somatic embryogenesis from in vitro grown leaf explants of *Rosa hybrida* L. J Plant Biotech 5:169–172

Kim SW, Oh SC, Liu JR (2003b) Control of direct and indirect somatic embryogenesis by exogenous growth regulators in immature zygotic embryo cultures of rose. Plant Cell Tissue Organ Cult 74:61–66

Kim CK, Chung JD, Park SH, Burrell AM, Kamo KK, Byrne DH (2004) *Agrobacterium tumefaciens*-mediated transformation of *Rosa hybrida* using the green fluorescent protein (GFP) gene. Plant Cell Tissue Organ Cult 78:107–111

Kim SW, Oh MJ, Liu JR (2009) Somatic embryogenesis and plant regeneration in zygotic embryo explant cultures of rugosa rose. Plant Biotechnol Rep 3:199–203

Kintzios S, Manos C, Makri O (1999) Somatic embryogenesis from mature leaves of rose (*Rosa* sp.). Plant Cell Rep 18:467–472

Kintzios S, Drossopoulos JB, Lymperopoulos C (2000) Effect of vitamins and inorganic micronutrients on callus growth and somatic embryogenesis from young mature leaves of rose. J Plant Nutr 23:1407–1420

Kirov I (2017) Physical mapping of genes on plant chromosomes. PhD Dissertation, Ghent University

Kirov I, Van Laere K, De Riek J, De Keyser E, Van Roy N, Khrustaleva L (2014a) Anchoring linkage groups of the *Rosa* genetic map to physical chromosomes with tyramide-FISH and EST-SNP markers. PlosONE 9:e95793

Kirov I, Divashuk M, Van Laere K, Soloviev A, Khrustaleva L (2014b) An easy "SteamDrop" method for high quality plant chromosome preparation. Mol Cytogenet 7:1–10

Kirov I, Van Laere K, Khrustaleva L (2015) High resolution physical mapping of single gene fragments on pachytene chromosome 4 and 7 of Rosa. BMC Genet 16:1–10

Kirov I, Van Laere K, Van Roy N, Khrustaleva L (2016) Towards a FISh-based karyotype of *Rosa* L. Comp Cytogenet 10:543

Koning-Boucoiran CFS, Gitonga VW, Yan Z, Dolstra O, van der Linden CG, van der Schoot J, Uenk GE, Verlinden K, Smulders MJM, Krens FA, Maliepaard C (2012) The mode of inheritance in tetraploid cut roses. Theor Appl Genet 125:591–607

Koning-Boucoiran CFS, Esselink GD, Vukosavljev M, van 't Westende WP, Gitonga VW, Krens FA, Voorrips RE, van de Weg WE, Schulz D, Debener T, Maliepaard C, Arens P, Smulders MJ (2014) Using RNA-Seq to assemble a rose transcriptome with more than 13,000 full-length expressed genes and to develop the WagRhSNP 68k axiom SNP array for rose (*Rosa* L.). Front Plant Sci 6:249

Koopman WJM, Wissemann V, De Cock K, Van Huylenbroeck J, De Riek J, Sabatino GJH, Visser D, Vosman B, Ritz CM, Maes B, Werlemark G, Nybom H, Debener T, Linde M, Smulders MJM (2008) AFLP markers as a tool to reconstruct complex relationships: a case study in *Rosa* (Rosaceae). Am J Bot 95:353–366

Kunitake H, Imamizo H, Mii M (1993) Somatic embryogenesis and plant regeneration from immature seed-derived calli of rugosa rose (*Rosa rugosa* Thunb.). Plant Sci 90:187–194

Lammerts WE (1945) The scientific basis of rose breeding. Am Rose Ann 30:71–79

Lee SY, Han BH, Kim YS (2010a) Somatic embryogenesis and shoot development in *Rosa hybrid*a L. Acta Hortic 870:219–225

Lee SY, Jung JH, Kim WH, Kim ST, Lee EK (2010b) Acquirement of transgenic rose plants from embryogenic calluses via *Agrobacterium tumefaciens*. J Plant Biotechnol 37:511–516

Leus L (2005) Resistance breeding for powdery mildew (*Podosphaera pannosa*) and black spot (*Diplocarpon rosae*) in roses. PhD Dissertation, Ghent University

Leus L (2017) Selection strategies for disease resistance in roses. Reference Module in Life Sciences, Elsevier. https://doi.org/10.1016/B978-0-12-809633-8.05008-1

Leus L, Van Huylenbroeck J (2009) Developing resistance to powdery mildew (*Podosphaera pannosa* (Wallr.: Fr.) de Bary): a challenge for rose breeders. In: Zlesak. (ed) RosesFloriculture and ornamental biotechnology. Global Science Books Ltd., UK, Ikenobe, Japan pp 131–138

Leus L, Jeanneteau F, Van Huylenbroeck J, Van Bockstaele E, De Riek J (2004) Molecular evaluation of a collection of rose species and cultivars by AFLP, ITS, rbc L, and mat K. Acta Hortic 651:141–147

Leus L, Dewitte A, Van Huylenbroeck J, Vanhoutte N, Van Bockstaele E, Höfte M (2006) *Podosphaera pannosa* (syn. *Sphaerotheca pannosa*) on *Rosa* and *Prunus* spp.: characterization of pathotypes by differential plant reactions and ITS-sequences. J Phytopathol 154:23–28

Li XQ, Krasnyanski SF, Korban SS (2002a) Somatic embryogenesis, secondary somatic embryogenesis, and shoot organogenesis in *Rosa*. J Plant Physiol 159:313–319

Li XQ, Krasnyanski SF, Korban SS (2002b) Optimization of the uidA gene transfer into somatic embryos of rose via *Agrobacterium tumefaciens*. Plant Physiol Biochem 40:453–459
Li X, Gasic K, Cammue B, Broekaert W, Korban SS (2003) Transgenic rose lines harboring an antimicrobial gene, ace-AMP1, demonstrate enhanced resistance to powdery mildew (*Sphaerotheca pannosa*). Planta 218:226–232
Liang S, Wu X, Byrne D (2017) Flower-size heritability and floral heat-shock tolerance in diploid roses. HortSci 52:682–685
Lim KY, Werlemark G, Matyasek R, Bringloe JB, Sieber V, El Mokadem H, Roberts AV (2005) Evolutionary implications of permanent odd polyploidy in the stable sexual, pentaploid of *Rosa canina* L. Heredity 94:501–506
Li-Marchetti C, Le Bras C, Relion D, Citerne S, Huché-Thélier L, Sakr S, Morel P, Crespel L (2015) Genotypic differences in architectural and physiological responses to water restriction in rose bush. Front Plant Sci 6:355
Linde M, Debener T (2003) Isolation and identification of eight races of powdery mildew of roses (*Podosphaera pannosa*) (Wallr:Fr.) de Bary and the genetic analysis of the resistance gene Rpp1. Theor Appl Genet 107:256–262
Linde M, Shishkoff N (2003) Disease / powdery mildew. In: Roberts, Debener, Gudin (eds) Encyclopedia of rose science. Elsevier, Academic Press, Oxford
Liorzou M, Pernet A, Li S, Chastellier A, Thouroude T, Michel G, Malécot V, Gaillard S, Brieé C, Foucher F, Oghina-Pavie C, Clotault J, Grapin A (2016) Nineteenth century French rose (*Rosa* sp.) germplasm shows a shift over time from a European to an Asian genetic background. J Exp Bot 67:4711–4725
Liu CY, Wang GL, Xie QL, Jin J, Liu GN (2008) A study on the chromosome karyomorphology of 6 species in *Rosa*. J Jiangsu For Sci Technol 35:5–8
Lloyd D, Roberts AV, Short KC (1988) The induction in vitro of adventitious shoots in *Rosa*. Euphytica 37:31–36
Longhi S, Giongo L, Buti M, Surbanovski N, Viola R, Velasco R, Ward JA, Sargent DJ (2014) Molecular genetics and genomics of the Rosoideae: state of the art and future perspectives. Hortic Res 1:1
Ma Y, Chen JY (1991) Chromosome studies of seven roses. J Fujian Coll For 11:215–218
Ma Y, Chen JY (1992) Chromosome studies of six species of *Rosa* in China. Guihaia 12:333–336
Ma Y, Islam-Faridl MN, Crane CF, Stelly DM, Price HJ, Byrne DH (1996) A new procedure to prepare slides of metaphase chromosomes of roses. HortSci 31:855–857
Ma Y, Crana CF, Byrne DH (1997a) Karyotypic relationships among some *Rosa* species. Caryologia 50:317–326
Ma Y, Islam-Faridi MN, Crana CF, Ji Y, Stelly DM, Price HJ, Byrne DH (1997b) In situ hybridization of ribosomal DNA to rose chromosomes. J Hered 88:158–161
Macovei A, Caser M, Donà M, Valassi A, Giovannini A, Carbonera D, Scariot V, Balestrazzi A (2016) Prolonged cold storage affects pollen viability and germination along with hydrogen peroxide and nitric oxide content in *Rosa hybrida*. Not Bot Hortic Agrobo 44:6–10
MacPhail VJ, Kevan PG (2009) Review of the breeding systems of wild roses (*Rosa* spp.). In: Zlesak. (ed) RosesFloriculture and ornamental biotechnology. Global Science Books Ltd., UK, Ikenobe, Japan pp 1–13
Magnard JL, Roccia A, Caissard JC, Vergne Pn Sun P, Hecquet R, Dubois A, Hibrand-Saint Oyant L, Julien F, Nicolè F, Raymond O, Huguet S, Baltenweck R, Meyer S, Claudel P, Jeauffre Jn Rohmer M, Foucher F, Hugueney P, Bendahmane M, Baudino S (2015) Biosynthesis of monoterpene scent compounds in roses. Science 349:81–83
Mandakova T, Singh V, Krämer U, Lysak MA (2015) Genome structure of the heavy metal hyperaccumulator *Noccaea caerulescens* and its stability on metalliferous and nonmetalliferous soils. Plant Physiol 169:674–689
Marchant R, Davey MR, Lucas JA, Lamb CJ, Dixon RA, Power JB (1998a) Expression of a chitinase transgene in rose (*Rosa hybrida* L.) reduces development of blackspot disease (*Diplocarpon rosa* wolf). Mol Breed 4:187–194
Marchant R, Power JB, Lucas JA, Davey MR (1998b) Biolistic transformation of rose (*Rosa hybrida* L.). Ann Bot 81:109–114
Marriott M (2017) Modern (Post-1800). Reference Module in Life Sciences, Elsevier. https://doi.org/10.1016/B978-0-12-809633-8.05063-9
Matsumoto S, Fukui H (1996) Identification of rose cultivars and clonal plants by random amplified polymorphic DNA. Sci Hortic 67:49–54
Matsumoto S, Kouchi M, Yabuki J, Kusunoki M, Ueda Y, Fukui H (1998) Phylogenetic analysis of

the genus *Rosa* using MatK sequence: molecular evidence for the narrow genetic background of modern roses. Sci Hortic 77:73–82
Matsumoto S, Kouchi M, Fukui H (2000) Phylogenetic analysis of the subgenus *Eurosa* using the ITS nrDNA sequence. Acta Hortic 521:193–202
Matthews D, Mottley J, Horan I, Roberts AV (1991) A protoplast to plant system in rose. Plant Cell Tissue Organ Cult 24:173–180
Meneve I (1995) Breeding for disease resistance in roses by means of *Rosa rugosa* and *Rosa fedtschenkoana*. Can Rose Annu 1995:55–57
Meynet J, Barrade R, Duclos A, Siadous R (1994) Dihaploid plants of roses (*Rosa* x *hybrida*, cv. 'Sonia') obtained by parthenogenesis induced using irradiated pollen and in vitro culture of immature seeds. Agronomie 2:169–175
Morpeth DR, Hall AM (2000) Microbial enhancement of seed germination in *Rosa corymbifera* 'Laxa'. Seed Sci Res 10:489–494
Mottley J, Yokoya K, Matthews D, Squirrel J, Wentworth JE (1996) Protoplast fusion and its potential role in the genetic improvement of roses. Acta Hortic 424:393–397
Murali S, Sreedhar D, Lokeswari TS (1996) Regeneration through somatic embryogenesis from petal-derived calli of *Rosa hybrida* L. cv Arizona (hybrid tea). Euphytica 91:271–275
Nakamura N, Hirakawa H, Sato S, Otagaki S, Matsumoto S, Tabata S, Tanaka Y (2017) Genome structure of *Rosa multiflora*, a wild ancestor of cultivated roses, DNA Research, dsx042, https://doi.org/10.1093/dnares/dsx042
Nguyen HN, Schulz D, Winkelmann T, Debener T (2017) Genetic dissection of adventitious shoot regeneration in roses by employing genome-wide association studies. Plant Cell Rep 36:1493–1505
Noriega C, Sondahl MR (1991) Somatic embryogenesis in hybrid tea roses. Biotechnology 9:991–993
Novak P, Neumann P, Pech J, Steinhaisl J, Macas J (2013) RepeatExplorer: a galaxy-based web server for genome-wide characterization of eukaryotic repetitive elements from next-generation sequence reads. Bioinformatics 29:792
Nybom H (2017) DNA fingerprinting. Reference Module in Life Sciences, Elsevier. https://doi.org/10.1016/B978-0-12-809633-8.05044-5
Nybom H, Esselink GD, Werlemark G, Vosman B (2004) Microsatellite DNA marker inheritance indicates preferential pairing between two highly homologous genomes in polyploid and hemisexual dog-roses, *Rosa* L. sect. Caninae DC. Heredity 92:139–150
O'Neill CM, Bancroft I (2000) Comparative physical mapping of segments of the genome of *Brassica oleracea* var. alboglabra that are homoeologous to resequenced regions of chromosomes 4 and 5 of Arabidopsis thaliana. Plant J 23:233–243
Ogilvie I, Cloutier D, Arnold N, Jui PY (1991) The effect of gibberellic acid on fruit and seed set in crosses of garden and winter hardy *Rosa* accessions. Euphytica 52:119–123
Ouyang L, Leus L, Van Labeke MC (2017) Seasonal changes in cold hardiness of garden roses. Paper presented at the 7th international symposium on rose research and cultivation, Angers, France, 2–7 July 2017
Pati PK, Sharma M, Sood A, Sood A, Ahuja PS (2004) Direct shoot regeneration from leaf explants of *Rosa damascena* mill. In Vitro Cell Dev Biol Plant 40:192–195
Pati PK, Rath SP, Sharma M, Sood A, Ahuja PS (2006) In vitro propagation of rose - a review. Biotechnol Adv 24:94–114
Pécrix Y, Rallo G, Folzer H, Cigna M, Gudin S, Le Bris M (2011) Polyploidization mechanisms: temperature environment can induce diploid gamete formation in *Rosa* sp. J Exp Bot 62:3587–3597
Peng T, Chen W, Moens M (2003) Resistance of *Rosa* species and cultivars to *Pratylenchus penetrans*. HortSci 38:560–564
Pertwee J (1995) The production and marketing of roses. Pathfast Publishing, Essex
Pipino L (2011) Improving seed production efficiency for hybrid rose breeding. PhD Dissertation, Ghent University
Pipino L, Van Labeke MC, Mansuino A, Scariot V, Giovannini A, Leus L (2011) Pollen morphology as fertility predictor in hybrid tea roses. Euphytica 178:203–214
Pipino L, Leus L, Scariot V, Van Labeke MC (2013) Embryo and hip development in hybrid roses. Plant Growth Reg 69:107–116
Pourhosseini L, Kermani MJ, Habashi AA, Khalighi A (2013) Efficiency of direct and indirect shoot organogenesis in different genotypes of *Rosa hybrida*. Plant Cell Tissue Organ Cult 112:101–108

Rajapakse S, Hubbard M, Kelly JW, Abbott AG, Ballard RE (1992) Identification of rose cultivars by restriction fragment polymorphism. Sci Hortic 52:237–245

Raymond O, Gouzy J, Just J, Badouin H, Verdenaud M, Lemainque A, Vergne P, Moja S, Choisne N, Pont C, Carrère S, Caissard J, Couloux A, Cottret L, Aury J, Szécsi J, Latrasse D, Madoui M, François L, Fu X, Yang S, Dubois A, Piola F, Larrieu A, Perez M, Labadie K, Perrier L, Govetto B, Labrousse Y, Villand P, Bardoux C, Boltz V, Lopez-Roques C, Heitzler P, Vernoux T, Vandenbussche M, Quesneville H, Boualem A, Bendahmane A, Liu C, Le Bris M, Salse J, Baudino S, Benhamed M, Winckler P, Bendahmane M (2018) The Rosa genome provides new insights into the domestication of modern roses. Nature Genetics. https://doi.org/10.1038/s41588-018-0110-3

Rehder A (1940) Manual of cultivated trees and shrubs hardy in North America. Collier Macmillan Ltd., New York

Reitsma TJ (1966) Pollen morphology of some European Rosaceae. Acta Botanica Neerlandica 15:290–307

Reynders-Aloisi S, Bollereau P (1996) Characterization of genetic diversity in genus *Rosa* by random amplified length polymorphic DNA. Acta Hortic 424:253–259

Ritz CM, Wissemann V (2011) Microsatellite analyses of artificial and apontaneous dogrose hybrids reveal the hybridogenic origin of *Rosa micrantha* by the contribution of unreduced gametes. J Hered 102:217–227

Roberts AV (2007) The use of bead beating to prepare suspensions of nuclei for flow cytometry from fresh leaves, herbarium leaves, petals and pollen. Cytometry A 71A:1039–1044

Roberts AV, Debener T, Gudin S (2003) Introduction. In: Roberts, Debener, Gudin (eds) Encyclopedia of rose science. Elsevier, Academic Press, Oxford

Roberts AV, Gladis T, Brumme H (2009) DNA amounts of roses (*Rosa* L.) and their use in attributing ploidy levels. Plant Cell Rep 28:61–71

Roundey E, Anderson N, Bedard C, Scheiber M, Zlesak D, Byrne D (2017) *Rosa palustris* and *Rosa setigera*: Breeding Challenges. Paper presented at the 7th international symposium on rose research and cultivation, Angers, France, 2–7 July 2017

Rout GR, Debata BK, Das P (1991) Somatic embryogenesis in callus cultures of *Rosa hybrida* L. cv. Landora. Plant Cell Tissue Organ Cult 27:65–69

Rout GR, Samantaray S, Mottley J, Das P (1999) Biotechnology of the rose: a review of recent progress. Sci Hortic 81:201–228

Rowley GD (1956) Germination in *Rosa canina*. Am Rose Ann 41:70–73

Royal FloraHolland (2016) https://www-sys.royalfloraholland.com/en/speciale-paginas/search-in-news/v42149/roses-remain-very-popular. Accessed 4 March 2018

Sarasan V, Roberts AV, Rout GR (2001) Methyl laurate and 6-benzyladenine promote the germination of somatic embryos of a hybrid rose. Plant Cell Rep 20:183–186

Scariot V, Akkak A, Botta R (2006) Characterization and genetic relationships of wild species and old garden roses based on microsatellite analysis. J Am Soc Hortic Sci 131:66–73

Schum A, Hofman K, Ghalib N, Tawfik A (2001) Factors affecting protoplast isolation and plant regeneration in *Rosa* spp. Gartenbauwissenschaft 66:115–122

Semeniuk P (1971a) Inheritance of recurrent blooming in *Rosa wichuraiana*. J Hered 62:203–204

Semeniuk P (1971b) Inheritance of recurrent and non-recurrent blooming in 'goldilocks' x *Rosa wichuraiana* progeny. J Hered 62:319–320

Semeniuk P, Arisumi T (1968) Colchicine-induced tetraploid and cytomerical roses. Bot Gaz 129:190–193

Semeniuk P, Stewart RN (1962) Temperature reversal of after-ripening of rose seeds. Proc Amer Soc Hortic Sci 80:615–621

Senapathi SK, Rout GR (2008) In vitro mutagenesis of rose with ethylmethanesulphonate (EMS) and early selection using RAPD markers. Adv Hortic Sci 3:218–222

Smulders MJM, Arens P, Koning-Boucoiran CFS, Gitonga VW, Krens FA, Atanassov A, Atanassov I, Rusanov KE, Bendahmane M, Dubois A, Raymond O, Caissard JC, Baudino S, Crespel L, Gudin S, Ricci SC, Kovatcheva N, Van Huylenbroeck J, Leus L, Wisseman V, Zimmermann H, Hensen I, Werlemark G, Nybom H (2011) Rosa. In: Kole C (ed) Wild crop relatives: genomic and breeding resources. Springer, Berlin, pp 243–275

Solo K, Collins S, Cheng Q, England B, Hale F, Windham AS, Byrne D, Anderson N, Windham MT (2017) *Rosa* species resistance to Eriophyid mite populations. Conference abstract APS. Phytopathology 107:139

Souq F, Coutos-Thevenot P, Yean H, Delbard G, Maziere Y, Barbe JP, Boulay M (1996) Genetic transformation of roses, 2 examples: one on morphogenesis, the other on anthocyanin biosyn-

thetic pathway. Acta Hortic 424:381–388
Spethmann W, Feuerhahn B (2003) Genetics / species crosses. In: Roberts, Debener, Gudin (eds) Encyclopedia of rose science. Elsevier, Academic Press, Oxford
Spiller M, Linde M, Hibrand-Saint Oyant L, Tsai C-J, Byrne DH, Smulders MJ, Foucher F, Debener T (2011) Towards a unified genetic map for diploid roses. Theor Appl Genet 122:489–500
Squirrell J, Mandegaran Z, Yokoya K, Robets AV, Mottley J (2005) Cell lines and plants obtained after protoplast fusions of Rosa+Rosa, Rosa+Prunus and Rosa+Rubus. Euphytica 146:223–231
Stewart RN, Semeniuk P (1965) The effect of the interaction of temperature with after-ripening requirements and compensating temperature on germination of seeds of 5 species of Rosa. Am J Bot 52:755–760
Svejda (1968) Effect of temperature and seed coat treatment on the germination of rose seeds. Hortscience 3:184–185
Svejda F (1977) Breeding for improvement of flowering attributes of winterhardy *Rosa kordesii* Wuiff hybrids. Euphytica 26:703–708
Svejda F (1979) Inheritance of winterhardiness in roses. Euphytica 28:309–314
Szinay D, Bai Y, Visser R, de Jong H (2010) FISH applications for genomics and plant breeding strategies in tomato and other solanaceous crops. Cytogenet Genome Res 129:199–210
Tanaka Y, Katsumoto Y, Demelis L, Fukuchi-Mizutani M, Fukui Y, Brugliera F, Togami T, Nakamura N, Tsuda S, Mason J (2007) Flower colour modification of roses by expression of a torenian anthocyanin methyltransferase gene. Plant Cell Physiol 48:S221–S221
Torres AM, Milan T, Cubero JI (1993) Identifying rose cultivars using random amplified polymorphic DNA markers. HortSci 28:333–334
Ueda Y (2003) Seed maturation and germination. In: Roberts, Debener, Gudin (eds) Encyclopedia of rose science. Elsevier Academic Press, Oxford
UPOV (2018) website on: Essentially Derived Varieties legi http://www.upov.int/meetings/en/doc_details.jsp?meeting_id=24135&doc_id=186123. Accessed 04 Mar 2018
van der Salm TPM, van der Toorn CJG, Hänisch ten Cate CHH, Dons HJM (1996) Somatic embryogenesis and shoot regeneration from excised adventious roots of the rootstock *Rosa hybrida* L. 'Moneyway'. Plant Cell Rep 15:522–526
van der Salm TPM, van der Toorn CJG, Bouwer R, Hanish ten Cate CH, Don HJM (1997) Production of ROL gene transformed plants of *Rosa hybrida* L. and characterisation of their rooting ability. Mol Breed 3:39–47
van der Salm TPM, Bouwer R, van Dijk AJ, Keizer LCP, Hänisch ten Cate CH, van der Plas LWH, Dons JJM (1998) Stimulation of scion bud release by rol gene transformed rootstocks of *Rosa hybrida* L. J Exp Bot 49:847–852
Van Huylenbroeck J, Leus L, Van Bockstaele E (2005) Interploidy crosses in roses: use of triploids. Acta Hortic 690:109–112
Van Huylenbroeck J, Eeckhaut T, Leus L, Werlemark G, De Riek J (2007) Introduction of wild germplasm in modern roses. Acta Hortic 751:285–290
Vergne P, Maene M, Gabant G, Chauvet A, Debener T, Bendahmane M (2010) Somatic embryogenesis and transformation of the diploid *Rosa chinensis* cv old blush. Plant Cell Tissue Organ Cult 100:73
Visessuwan R, Kawai T, Mii M (1997) Plant regeneration systems from leaf segment culture through embryogenic callus formation of *Rosa hybrida* and *R. canina*. Breed Sci 47:217–222
von Malek B, Debener T (1998) Genetic analysis of resistance to black spot (*Diplocarpon rosae*) in tetraploid roses. Theor Appl Genet 96:228–231
Vosman B, Visser D, van der Voort JR, Smulders MJM, van Eeuwijk F (2004) The establishment of 'essential derivation' among rose varieties, using AFLP. Theor Appl Genet 109:1718–1725
Vukosavljev M (2014) Towards marker assisted breeding in garden roses: from marker development to QTL detection. PhD Dissertation, Wageningen University
Vukosavljev M, Arens P, Voorrips RE, Westende v 't, WP EG, Bourke PM, Cox P, van de Weg WE, Visser RGF, Maliepaard C, Smulders MJ (2016) High-density SNP-based genetic maps for the parents of an outcrossed and a selfed tetraploid garden rose cross, inferred from admixed progeny using the 68k rose SNP array. Hortic Res 3:16052
Walther F, Sauer A (1986) In vitro mutagenesis in roses. Acta Hortic 189:37–46
Wang X, Jacob Y, Mastrantuono S, Bazzano J, Voisin R, Esmenjaud D (2004) Spectrum and inheritance of resistance to the root-knot nematode *Meloidogyne hapla* in *Rosa multiflora* and *R. indica*. Plant Breed 123:79–83
Werlemark G, Carlson-Nilsson U, Uggla M, Nybom H (1995) Effects of temperature treatments on seedling emergence in dogroses, *Rosa* sect. Caninae (L). Acta Agric Scand 45:278–282

Werlemark G, Carlson-Nilsson U, Esselink GD, Nybom H (2009) Studies of intersectional crosses between pentaploid dogrose species (*Rosa* sect. Caninae L.) as seed parents and tetraploid garden roses as pollen donors. In: Zlesak. (ed) RosesFloriculture and ornamental biotechnology. Global Science Books Ltd., UK, Ikenobe, Japan pp 131–138
Whitaker VM, Zuzek K, Hokanson SC (2007) Resistance of 12 rose genotypes to14 isolates of *Diplocarpon rosae* wolf (rose blackspot) collected from eastern North America. Plant Breed 126:83–88
Whitaker VM, Debener T, Roberts AV, Hokanson SC (2010) A standard set ofhost differentials and unified nomenclature for an international collection of *Diplocarpon rosae* races. Plant Pathol 59:745–752
Wissemann V (2003) Conventional taxonomy of wild roses. In: Roberts A, Debener T, Gudin S (eds) Encyclopedia of rose science. Elsevier, Academic Press, Oxford, pp 111–117
Wissemann V, Ritz CM (2005) The genus *Rosa* (Rosoideae, Rosaceae) revisited: molecular analysis of nrITS-1 and *atp*B-*rbc*L intergenic spacer (IGS) versus conventional taxonomy. Bot J Linn Soc 147:275–290
Wylie AP (1954) The history of garden roses – part I. J R Hortic Soc 79:555–571
Yambe Y, Takeno K (1992) Improvement of rose achene germination by treatment with macerating enzymes. HortSci 27:1018–1020
Yambe Y, Hori Y, Takeno K (1992) Levels of endogenous abscisic acid in rose achenes and leaching with activated charcoal to improve seed germination. J Japan Soc Hortic Sci 61:383–387
Yan Z, Denneboom C, Hattendorf A, Dolstra O, Debener T, Stam P, Visser PB (2005) Construction of an integrated map of rose with AFLP, SSR, PK, RGA, RFLP, SCAR and morphological markers. Theor Appl Genet 110:766–777
Yokoya K, Walker S, Sarasan V (1996) Regeneration of rose plants from cell and tissue culture. Acta Hortic 424:333–337
Yokoya K, Kandasamy KI, Walker S, Mandegaran Z, Roberts AV (2000) Resistance of roses to pathotypes of *Diplocarpon rosae*. Ann Appl Biol 136:15–20
Young MA, Schorr P, Baer R (2007) Modern roses 12. American Rose Society, Shreveport
Zakizadeh H, Lutken H, Sriskandarajah S, Serek M, Muller R (2013) Transformation of miniature potted rose (*Rosa hybrida* cv. Linda) with PSAG12-ipt gene delays leaf senescence and enhances resistance to exogenous ethylene. Plant Cell Rep 32:195–205
Zhang D, Germain E, Reynders-Aloisi S, Gandelin MH (2000) Development of amplified fragment length polymorphism markers for variety identification in rose. Acta Hortic 508:113–120
Zhang LH, Byrne DH, Ballard RE, Rajapakse S (2006) Microsatellite marker development in rose and its application in tetraploid mapping. J Am Soc Hortic Sci 131:380–387
Zhang X, Zhang J, Zhang W, Yang T, Xiong Y, Che D (2016) Transcriptome sequencing and de novo analysis of Rosa multiflora under cold stress. Acta Physiol Plant 38:164
Zhou Z, Bao W, Wu N (2009) Effects of scarification, stratification and chemical treatments on the germination of *Rosa soulieana* Crépin achenes. In: da Silva T (ed) Floriculture and ornamental biotechnology. Global Science Books, London, pp 75–80
Zieslin N (1996) Influence of climatic and socio economical factors on mode of cultivation and research of rose plants. Acta Hortic 424:21–22
Zlesak DC (2005) The effects of short-term drying on seed germination in *Rosa*. HortSci 40:1931–1932
Zlesak DC (2006) Rose – *Rosa* x *hybrida*. In: Anderson NO (ed) Flower breeding and genetics: issue, challenges and opportunities for the 21st century. Springer, Dordrecht
Zlesak DC (2008) Warm stratification enhances germination of *Rosa* section Caninae species. HortSci 43:1268
Zlesak DC (2009) Pollen diameter and guard cell length as predictors of ploidy in diverse rose cultivars, species and breeding lines. In: da Silva T (ed) Floriculture and ornamental biotechnology. Global Science Books, London, pp 53–70
Zlesak DC (2010) The effect of gibberellins (GA3 and GA4+7) and ethanol on seed germination of *Rosa eglanteria* and *R. glauca*. Suppl Rose Hybrid Assoc Newslett 41:1–10
Zlesak DC, Thill CA, Anderson NO (2005) Trifluralin-mediated polyploidization of *Rosa chinensis minima* (Sims) Voss seedlings. Euphytica 141:281–290
Zlesak DC, Nelson R, Harp D, Villarreal B, Howell N, Griffin J, Hammond G, George S (2017) Performance of landscape roses grown with minimal input in the north-central, central, and south-Central United States. HortTechnology 27:718–730
Zvi MMB, Shklarman E, Masci T, Kalev H, Debener T, Shafir S, Ovadis M, Vainstein A (2012) PAP1 transcription factor enhances production of phenylpropanoid and terpenoid scent compounds in rose flowers. New Phytol 195:335–345

第28章 郁 金 香

Teresa Orlikowska, Małgorzata Podwyszyńska, Agnieszka Marasek-Ciołakowska, Dariusz Sochacki, and Roman Szymański

摘要 郁金香的高度流行使人们不断创造新的品种来满足市场的需求，包括新的花型、新的花色和生产可持续性。本章介绍了郁金香的种植、育种历史以及目前的园艺现状。基于郁金香生物学基础，我们对郁金香的育种目标和育种方法进行了详细论述。除观赏性外，目前的育种趋势主要集中在生态方向，包括抗病原真菌和抗病毒育种。重点关注的缩短育种进程的方法有远缘杂交、突变、细胞学和分子标记。由于新品种培育周期很长，人们提出了通过组织培养方法繁殖郁金香的可能性。

关键词 亲和杂交与远缘杂交；育种目标；细胞学与分子标记；微繁殖；诱变；多倍体；体细胞克隆；变异源；郁金香

28.1 郁金香在花卉栽培中的历史和现状

郁金香原产于中亚、高加索和阿尔泰山脉，16世纪中叶从土耳其传入欧洲。“tulip”（郁金香）这个名字可能来源于“turban”（土耳其语 tulpana）一词。郁金香鳞茎由圣罗马皇帝 Ferdinand 一世驻奥斯曼帝国的苏莱曼宫廷著名大使 Augerius Gislenius Busbequius 带到维也纳皇家花园，然后经由 Carolus Clusius，传到荷兰莱顿大学的花园，之后开启了郁金香欧洲种植和繁育历程（Pavord 1999）。从那时起，郁金香就引起了收藏家、园丁、园艺家和银行家的兴趣。到今天都很难说是什么样的社会心理机制导致了 17 世纪郁金香在荷兰大受欢迎，当时出现了“郁金香狂热”现象。这很可能是一种人为创造的郁金香需求，需要花色不断改变的郁金香，现在我们已经知道这是病毒感染的结果。最好和最贵的鳞茎是带有碎色花纹的品种；但也正是由于病毒的感染，鳞茎容易退化、消失，或突然获得抗性，最终失去吸引力（Doorenbos 1954；Lesnaw and Ghabrial 2000）。可以说导致了一种通过账号进行资金流动的金融泡沫，泡沫破灭后，荷兰人对郁金香的兴趣依然存在，第一个在欧洲描述郁金香出版插图的人是 Conrad Gesner（Pavord 1999）。尽管大多数郁金香品种的拉丁名都是用 *Tulipa gesneriana*，但该物种的野生型尚未被发

T. Orlikowska (✉) · M. Podwyszyńska · A. Marasek-Ciołakowska
Research Institute of Horticulture, Skierniewice, Poland
e-mail: Teresa.Orlikowska@inhort.pl
D. Sochacki
Warsaw University of Life Sciences – SGGW, Departments of Ornamental Plants, Warsaw, Poland
R. Szymański
Horticulture Farm Roman Szymański, Poznań, Poland

现，可能是郁金香（后面写成 *T. gesneriana*）在被引入欧洲时就是天然杂交品种（Okubo and Sochacki 2013）。

目前，全世界郁金香鳞茎栽培面积约为 13 000hm^2，其中 88%的地区在荷兰，其他国家日本、法国、波兰、德国和新西兰的栽培面积为 122～300hm^2（Buschman 2005）。2016～2017 年，荷兰商业栽培品种达 1800 多个，占地 11 843.91hm^2，但只有 20 个品种栽培面积在 100hm^2 以上（BKD 2017）。5 个最受欢迎的品种（'Strong Gold'、'Leen van der Mark'、'Purple Prince'、'Purple Flag' 和 'Ile de France'）栽培面积占总面积的 19%（2257.5hm^2）（BKD 2017）。荷兰郁金香鳞茎产量超过 40 亿粒，其中 23 亿粒（53%）用于荷兰和其他国家的切花生产（Benschop et al. 2010）。

园丁和收藏家一直在寻找具有新形状、新颜色以及功能特征改变的基因型，这能使郁金香在景观和植物区系组成中得到广泛应用。过去和现在荷兰都是郁金香育种的中心，郁金香是荷兰的象征之一，荷兰的生产商和科研人员共同致力于新品种的培育。郁金香在国际市场上的影响力，吸引了世界上其他育种家积极培育新的种类，这在拥有丰富野生郁金香资源的国家尤其具有前景。例如，最近第一批郁金香品种在中国上市（Qu et al. 2017）。自 20 世纪初，日本培育了约 100 个郁金香品种，其中有一些来自双单倍体，'Kikomachi' 是日本的主要品种（Marasek and Okazaki 2008）。

28.2 郁金香生物学特性

郁金香属属于单子叶植物纲中的一个大科——百合科。根据 Van Raamsdonk 和 De Vries（1995）以及 Van Raamsdonk 等（1997）的分类系统，郁金香属分为两个亚属：郁金香亚属和毛蕊亚属。一般认为该属含 45～100 个代表种，Hall（1940）和 Botschantzeva（1962）报道了约 100 个种，Stork（1984）报道了 45 个种[在 Van Eijk 等（1991）之后]，Van Raamsdonk 和 De Vries（1992，1995）认为有 50～60 个种，Zonneveld（2009）报道了 87 个种，最近 Christenhusz 等（2013）认为是 76 个种。在世界植物名称检索名录中收录了 554 个郁金香属植物，但只有 99 个被接受（Govaerts 2017）。大多数郁金香野生种和品种为二倍体（$2n = 2x = 24$）（Kroon and Jongeriu 1986），体细胞 DNA 2C 值在 32～69pg（Zonneveld 2009）。然而，在 *T. clusiana* 和 *T. kaufmanniana* 的不同基因型中发现了三倍体（$2n = 3x = 36$）（Zonneveld 2009）；在 *T. bifloriformis*、*T. sylvestris*、*T. kolpakowskiana* 和 *T. tetraphylla* 中发现了四倍体（$2n = 4x = 48$）；在 *T. clusiana* 中发现了五倍体（$2n = 5x = 60$）；在 *T. polychrome* 中发现了六倍体（Kroon and Jongerius 1986），这是迄今为止在郁金香属野生种或园艺品种中发现的染色体数目最多的一个种类。

郁金香株高 7～75cm，茎直立坚硬，光滑或被毛，花单朵顶生，只有少数几个种类 *T. biflora*、*T. tarda*、*T. praestans* 和 *T. turkestanica* 为多花头。叶光滑，条状披针形至卵状披针形。一些野生种和栽培种的叶片叶脉间具暗紫色条纹或呈间断的不规则矩形，绝大多数种类叶片灰绿色，覆蜡质层，叶 3～5 枚，其中 2～3 枚宽广、基生。蒴果，内含 100 多粒种子。果实完全成熟后种子遇适宜条件易萌发并长出新的植株。由于杂合度高，子代性状与母本不同，种子繁殖只用于新品种培育。郁金香营养繁殖主要是通过鳞茎增殖。鳞茎肉质、肥厚，为短缩的地下茎，由基部的鳞茎盘和上部 2～6 个同心环状排列的肉质鳞片组成，外被棕褐色膜质鳞茎皮。鳞片的数量和鳞茎的大小决定开花能力。通常一个鳞茎包含 4～5 个鳞片，周茎 6～9cm 就可开花。秋天种植鳞茎时，中心芽已完成微型叶、茎和花的发育，其中花带有三裂片柱头的雌蕊，即完成了花芽分化的最后阶段，也就是所谓的 G 阶段。从生产实用性的角度来说，郁金香鳞茎的理想生命周期为 1 年，但实际上，一个鳞茎从发芽到开花的整个生命周期约 29 个月，若鳞茎太小，第二年不能开花，需再培养 12 个月（Rees 1992）。一个成花球在种植过程中通常包含三代球：母球、子球和新球，子球是替代母球可于第二年开花的球，新球是在子球鳞片间产生的球。在郁金香生产中，叶子枯萎后即可挖球，晾干储存，第二年秋天再种植。

郁金香花被片由三个内片和三个外片组成，未分化成花萼和花冠。花被片长 1.5cm 到几厘米不等，内侧通常有光泽，外侧无。雄蕊 6 枚，3 枚一轮，整齐排列成两个圆圈，围绕着突出的雌蕊，雌蕊 1 枚，柱头三裂，子房细长，呈圆柱形，3 心室，无花柱。花单瓣或复瓣，花型多样，杯型、碗型、星型、钟型、百合花型，或者扭曲的、圆形的花被片。花色极为丰富，除纯蓝色或纯黑色外，有白、黄、粉、红和紫色等，且有不同程度的花色变化，从单色到花被片内外侧不同色，或者边缘不同色。不同品种的花色、花型和花被片内侧基部显著不同。Petrová 和 Faberová（2000）描述了郁金香 10 种花型与 9 种花被片内侧基部特征，依据 UPOV《郁金香特异性、一致性和稳定性测试指南》（UPOV 2006），郁金香主要有 16 种花色（括号中的为代表品种）：①白色（'Snowparrot'），②灰白色（无代表品种），③浅黄色（'Yellow Purissima'），④中黄色（'Yellow Flight'），⑤深黄色（'Yellow Flight'），⑥橙色（'Orange Monarch'），⑦橙红色（'Temple of Beauty'），⑧中红色（'Lefeber's Memory'），⑨深红色（'Prominence'），⑩紫红色（'Blenda'），⑪浅粉色（'Bright Pink Lady'），⑫中粉红色（'Angélique'），⑬深粉色（'Pink Impression'），⑭中紫色（'Attila'），⑮深紫色（'Queen of Night'），⑯棕色（'Cairo'）。郁金香种类繁多，国际园艺分类标准见表 28.1。

表 28.1　郁金香的园艺分类

序号	分组（简写）	描述	代表品种	备注
		早花类		
1	单瓣早花群（SE）	花杯型，单瓣，花茎长 15～45cm	'Christmas Marvel' 'Flair'，'Purple Prince'	
2	重瓣早花群（DE）	花碗型，重瓣，花茎长 40cm	'Monte Carlo' 'Viking'	
		中花类		
3	凯旋系（T）	花杯型，单瓣，花茎长 45～60cm，直立坚实；适宜促成栽培；来源于单瓣早花群与单瓣晚花群的杂交品种	'Apricot Beauty' 'Leen Van Der Mark' 'Strong Gold'	
4	达尔文杂种群（DH）	花卵球形，单瓣，花茎长 70cm；由 *T. gesneriana* 和 *T. fosteriana* 以及其他种杂交而成	'Ad Rem' 'Apeldoorn' 'Van Eijk'	
		晚花类		
5	单瓣晚花群（SL）	花杯型或高脚杯型，花茎长 45～60cm；包括原分类中的达尔文系（Darwin）和卡特吉系（Cottege）	'Menton' 'Queen of Night'	

续表

序号	分组（简写）	描述	代表品种	备注
		晚花类		
6	百合花型群（L）	花瓣顶端渐尖，向后翻卷，类似百合花的花型，花茎长度类型多	‘Claudia’‘Pretty Woman’‘West Point’	
7	流苏花型（Crispa）	单瓣花品种，花瓣边缘有清晰的毛刺或针状突起物	‘Arma’‘Fabio’‘Davenport’	
8	绿瓣群（V）	单瓣花品种，花瓣上带有部分绿色，花茎长36～60cm	‘Greenland’	
9	伦勃朗型（R）	“碎色郁金香”，该类型品种花被片有花纹。原始伦勃朗型郁金香已不再应用于商业化生产，因为这些品种中不常见的大理石色是由病毒感染引起的，常作为收藏品	‘Princess Irene’‘Flaming Parrot’	无病毒并能稳定遗传的品种
10	鹦鹉群（P）	单瓣花品种，花被片卷曲扭转，花茎长50～60cm	‘Bright Parrot’‘Rococco’‘Topparrot’	
11	重瓣晚花群（DL）	芍药、牡丹花型，花茎长40～60cm	‘Angelique’‘Wirosa’‘Yellow Pompenette’	
		变种及杂种		
12	考夫曼群（K）	包括考夫曼郁金香原种、亚种、变种和杂交种；花期早，叶有斑纹。通常基部有鲜艳对比色，株高约30cm	‘Showwinner’‘Giuseppe Verdi’‘The First’	盆栽或园林布置应用
13	福氏群（F）	包括福氏郁金香和福氏郁金香与其他原种的杂交品种；花大，叶片偶有杂色或斑纹	‘Purissima’‘Red Emperor’‘Yellow Emperor’	盆栽或园林布置应用
14	格里氏群（G）	包括格里氏郁金香和格里氏郁金香与其他原种的杂交品种；叶杂色或有条纹，花期比考夫曼群郁金香晚	‘Red Riding Hood’‘Toronto’	盆栽或园林布置应用
15	其他混杂群（M）	包括除以上之外的原种、变种及具有野生种习性的品种		园林布置应用，主要用于自然园或岩石园

改编自 Van Scheepen（1996）、Okubo 和 Sochacki（2013）

母球在秋季种植后只产生不分枝的根（Kawa and De Hertogh 1992）。冬季气温降低，植物体内激素发生变化，这对于郁金香的正常生长和开花至关重要，主要是脱落酸含量降低，促进植物生长的内源激素增加（Saniewski and Kawa-Miszczak 1992）。春季气温升高，地上部迅速生长、开花，和其他百合科植物一样，郁金香会产生一些次级代谢物（生物碱、类黄酮化合物、糖苷、皂苷、单宁），因此，郁金香可以作为一种抗菌剂被当地人用来治疗各种疾病（Ibrahim et al. 2016）。

28.3 郁金香育种目标

郁金香有 14 000 多个品种，郁金香品种分类名录和国际登录（Classified List and International Register of Tulip Names）最新版本记载有 8000 多个品种（Van Scheepen 1996）。每年有 100～150 个新品种在荷兰皇家球根种植者协会（KAVB）注册登记，记录的性状主要有花色、鳞茎大小、花型、生长模式以及花卉园艺中的各类性状。那么当代育种家在寻找什么，答案是新颖。新颖的前提条件是新品种高产（鳞茎大小适合人工调控）、易于促成栽培（特别是在一个短的缺货期）、耐运输、观赏性好。前两个特征主要是针对鳞茎生产商，后两个是针对看重视觉效果的花店和消费者。这样的新品种在花卉业和世界市场都有稳固的潜力，除这些特征外，其他特性也越来越重要，如抗生物和非生物胁迫、瓶插期和盆栽寿命等，而像暖冬耐受性和更低的休眠水平这些特性在一些地区很重要（Schmidt 2015）。全球气温不断上升，成花过程中，温度过高常常会使花芽脱水而造成消蕾。Leeggangers 等（2017）通过分析郁金香成花抑制因子和激活因子基因的表达，部分地回答了环境温度如何影响形态改变，研究表明早在茎尖分生组织发生形态学变化之前，随着环境温度升高，成花抑制因子的表达被抑制，并且在形态学变化结束之前成花激活因子快速表达，这些基因的鉴定将来可用于期望基因型品种的筛选或用于遗传转化。为使绿色植物更有吸引力，不同花期品种的范围应该更广，包括晚花型和早花型。另外一个育种目标是多年生，即不需要每年挖种球和种植，这一育种目标必须与抗病毒和抗真菌感染以及落叶更晚、鳞茎繁育周期缩短相结合（Schmidt 2015）。

郁金香育种家和爱好者的梦想是拥有黑色与蓝色郁金香，Shoj 等（2007，2010）、Momonoi 等（2009）有关蓝色郁金香的研究表明一些品种花基部的蓝色可能是花青素与 Fe^{3+}复合作用的结果，该处 Fe^{3+}浓度比红色部分高 25 倍，而花青素含量在整个花被片中相差不大。最后作者认为，与 Fe^{3+}含量增加相关的空泡铁转运蛋白基因（*TgVit1*）在蓝色细胞中活跃，而与铁储存相关的蛋白质基因 *TgFer1* 只在红色细胞中表达（Shoji et al. 2010）。因此，*TgVit1* 基因表达和 *TgFer1* 基因抑制是郁金香花被片底部呈现蓝色的关键。

对波兰郁金香种植者和育种者 Roman Szymański 来说，他的主要育种目标是观赏性，获得花色更深或绿色的郁金香，优美的外形和好闻的香味也是他选择的重要性状。育种者还将开花很晚如使郁金香 5 月底开花作为育种目标。在 Roman Szymański 的资源收集工作中，除流行品种外，还包括非常古老、稀有、未培育和野生种类的收集，这都是杂交选育工作中的亲本。选择的依据是能高效无性繁殖和具有高抗性，特别是抗病毒感染能力强。优株通常采用无性繁殖，扩大生产中，每年通过酶联免疫吸附测定法（ELISA）检测病毒。一些最重要的自育品种有流苏花型的‘Fringed Black’，它是迄今为止全世界上花色最黑的品种之一，被育种家认为具有茄子颜色，其他流苏花型种类，有橙色和奶油色的‘Frosty Sunrise’、全绿的‘Green Surprise’（属绿瓣群），还有单瓣晚花群中开花最晚的一个红色种类‘The Last’（图 28.1），它们都是 NMG 克隆体。

28.3.1　抗尖孢镰刀菌育种

郁金香尖孢镰刀菌（*Fusarium oxysporum* f. sp. *tulipae* Apt. (Fot)）是一种能引起鳞茎或其基部腐烂的真菌，基腐病最初的症状是鳞茎上形成褐斑，后逐渐扩大，最终基部鳞茎在贮藏过程中腐烂。此外，腐烂的病原体会产生大量乙烯导致次生生理障碍，如呼吸增加、芽减少、根伸长、花芽败育（Cerveny and Miller 2010）。在胁迫过程中，症状表现为植物生长迟缓、叶片变黄，最终死亡，田间症状主要为生长迟缓、花败育、叶片常变红、鳞茎产量下降，基腐病是郁金香生产中最严重的真菌病（Nan Tang et al. 2015）。

Van Eijk 等（1978，1979）首次报道了郁金香抗尖孢镰刀菌的遗传特性，他设计了幼年和成年鳞茎筛选实验，并指出不同品种抗性不同。另一个非常重要的发现是抗性水平的可预见性，对 2 年生幼龄鳞茎和 5 年生成年鳞茎进行抗性实验检测，结果表明二者抗性水平相一致，测试幼年鳞茎只具有初步意义，成年鳞茎对尖孢镰刀菌更敏感（Van Eijk and Eikelboom 1983）。Van Eijk 等（1978，1979）认为尖孢镰刀菌抗性主要是基于基因加性效应。Nan Tang 等（2015）则进行了深入研究，他们以‘Kees Nelis’、‘Cantata’亲本和它们后代不同抗性水平植株为材料进行抗性评估，具体方法是在田间鳞茎上接种三种尖孢镰刀菌菌株，鳞茎收获后评估发病症状，并用特殊检测方法使评估量化，对转绿色荧光蛋白尖孢镰刀菌菌株进行检测，用得到的数据构建遗传图谱，鉴定出 6 个与抗性相关的数量性状位点，以便以后用于分子标记辅助选择。

图 28.1 波兰郁金香育种者 Roman Szymański 培育的郁金香品种：（a）‘Frosty Sunrise’，（b）‘Fringed Black’，（c）breeding clone NMG，（d）‘Green Surprise’

28.3.2 抗灰霉病育种

灰霉病是郁金香的另一种严重真菌病，由于其在种植园中突然蔓延，引起植物广泛腐烂，因此又称为“火”感染，为预防这种病害，需要经常喷洒预防性杀菌剂。Straathof 等（2002）详述了一种关于品种和育种材料抗性水平的温室筛选实验，蜡质层是保护细胞的第一道屏障，事实上，*T. fosteriana* 的易感品种蜡质层薄，*T. gesneriana* 的抗性品种蜡质层则厚（Reyes et al. 2005），*T. tarda* 的抗性最好，但目前为止仍未获得该种与其他品种杂交的后代（Van Tuyl and Van Creij 2007）。

28.3.3 抗郁金香碎色花瓣病（TBV）育种

与病毒相关的疾病可能一直伴随着郁金香的生长，早在 13 世纪就有人观察到这种疾病，当时波斯诗歌中有关于多色郁金香的诗句（Asjes 1994）。在 17 世纪“郁金香狂热”时期，荷兰高度重视能够开五颜六色花朵的种球，弗拉芒派画家描绘了带有病毒症状的郁金香花朵（Lesnaw and Ghabrial 2000）。现在，人们普遍认为碎色是由病毒诱导的与颜色相关的基因沉默引起的，这是 TBV 的症状。在种植园中，避免 TBV 感染几乎是不可能的，因为该病毒通过蚜虫传播 TBV 病毒，即使很短时间的蚜虫接触也会导致植物感染。感染郁金香的病毒种类有很多，但 TBV 是危害最大、分布最广的一种。这种病毒不仅导致感病植株花瓣颜色改变，叶片出现缺绿或条斑，而且会造成种球退化。培育抗性品种是克服这种病毒病的有效方法。根据 Van Tuyl 和 Van Creij（2007）的研究，几个 *T. fosteriana* 品种，如‘Cantata’和‘Princeps’，对 TBV 具有很好的抗性，但没有发现任何一个 *T. gesneriana* 品种具有类似的特性。因此，可用 *T. fosteriana* 品种进行杂交，获得可育后代，并将相关抗性基因导入最佳品种中（Straathof and Eikelboom 1997）。一些品种抗蚜虫性好，或一些品种感病症状明显，易于肉眼观察，而对于白色系或奶油色系的郁金香，花瓣没有明显症状，无法通过肉眼观察识别病株。荷兰开发了一种配备多光谱相机的移动平台，能更准确地检测出花瓣颜色变化或色素沉着（Polder et al. 2014）。抗 TBV 品种选育特别困难是因为植株在生长期间的病害症状强度不同，有些年份可以检测到，而有些年份检测不到，因此必须每年通过酶联免疫吸附测定法（ELISA）筛查（Romanow et al. 1991；Sparnaaij 1991；Eikelboom et al. 1992）。

现代郁金香的育种目标是通过远缘杂交手段将对这三种最重要病害的抗性结合起来，为实现这一目标，有必要对野生种群进行进一步评估，以期找到所需的功能特性，并深入研究如何克服远缘杂交障碍。

28.3.4 郁金香较长切花寿命育种

对于花卉生产商、股东和零售商以及消费者来说，切花寿命是一个非常重要的特征，这也是 Van Eijk 和 Eikelboom（1976，1986）、Van Eijk 等（1977）、Van der Meulen-Muisers 等（1997）的研究课题。他们发现郁金香品种切花寿命是 8～16 天，而杂交后代的是 6～22 天，这说明切花寿命是基因加性效应的结果（Van der Meulen-Muisers et al. 1997）。知道切花寿命遗传特性并进行有效选择，可以有效促进切花寿命延长，他们还发现容器中切花瓶插寿命与田间未采收的花的寿命有关，这一发现使人们在第一次开花时就可以筛选瓶插寿命长的植株，最有效的选择标准是花被片颜色的改变，一个重要的标志是最后一个节间的长度，节间越长，寿命越短；另一个指标是刈割后的吸水强度。此外，促成栽培中周期长的品种采收后瓶插寿命通常更长（Van Eijk and Eikelboom 1976），这有悖于促成栽培中选择周期短的品种。Van der Meulen-Muisers 等（1997）认为父本对枝条伸长的影响大于母本。

28.4 郁金香育种方法

28.4.1 遗传变异来源

自植物栽培思想产生以来，获得改良株型的基础一直是播种最好的单株种子，然后杂交最有利的表型，很可能是通过这种方式在 13 世纪的奥斯曼帝国波斯花园中出现了郁金香的新变种（Pavord 1999）。许多年后，随着郁金香生物学知识的完善和实验室新技术的发展，诱变、多倍体化、胚挽救、不定芽再生和体细胞胚胎发生、遗传转化和分子标记被应用到育种过程中，新品种的离体繁育加速其市场化。

遗传变异的基本来源是目前的栽培品种和收集的野生基因型种类。一些国家特别是那些有自然基因资源或有育种项目的地方，对郁金香实施了原生地和迁地保护。基因库和迁地保护机构主要是一些研究所、大学、植物园、育种公司、球茎生产农场。16～19 世纪，收藏郁金香古老品种最丰富的是荷兰的霍图斯布布尔·玻如姆公园，保存的郁金香品种总数约为 2400 份，其中约 90%已不再用于商业化种植。多亏了前校长 Pieter Boschman，他于 1924 年开始收藏郁金香，一些古老而富有传奇色彩的品种如‘Duc van Tol Yellow’（1595 年）、‘Zomerschoon’（1620 年）或鹦鹉群的‘Perfecta’（1750 年）被保存了下来（Hortus Bulborum 2017）。土耳其是郁金香属植物最重要的基因多样性中心之一，在 Yalova 的 Atatürk Horticultural Central Research Institute 和 Samsun 的 Black Sea Agricultural Research Institute 收藏了大量的郁金香，后者有一个国家郁金香育种项目于 2006 年启动，该项目主要是在郁金香的基因中心 Anatolia 收集和保存郁金香材料，在国家植物基因库做材料的适应性研究鉴定（BSARI 2017）。

英国皇家植物园——邱园的千年种子库储存了 13 个郁金香野生种的 19 份种质资源（MSB 2017），以色列农业研究组织的基因库储存了 5 个种类的 74 份种质资源（IGB 2017）。大型郁金香收集地（包括政府的和私人的）主要在阿塞拜疆、哈萨克斯坦、俄罗斯、捷克、拉脱维亚、立陶宛、波兰、日本和其他郁金香种植国。然而基因库中和迁地保护机构收集的郁金香种质在线数据库没有提供信息或提供的信息非常有限，唯一例外的是捷克的 EVIGEZ 植物遗传资源文件，其中显示了 289 份郁金香属种质，包括一些捷克的品种和无性育种材料（EVIGEZ 2014）。

基因库和收藏的作用不可小觑，主要包括收集、记录、鉴定、评估和保存，为其他基因库和育种提供可能。例如，波兰 Skierniewice 园艺研究所收集了 450 种基因型的郁金香（Sochacki 2014）并进行了如下 3 方面研究：①利用 Fot 代谢物进行基部根的抗性水平研究（Podwyszyńska et al. 2001），②用无症状植物材料进行 TBV 抗性研究（Sochacki 2007），③评价属下材料系统状况以及物种与当前栽培品种之间的亲缘关系和系统发育关系（Yanagisawa et al. 2012）。郁金香在起源地作为野花种植也一直是研究的重点。Qu 等（2016）描述了 17 个原产中国的郁金香，而 Xing 等（2017）评估了 8 种野生郁金香，Trias Blazi 等（2016）介绍了 *T. cypia* 野生种，Eker 等（2014）报道了土耳其的 17 个野生种和 2 个亚种。

28.4.2　兼容形式的杂交

在郁金香育种中，物种内的品种杂交是最重要的育种方式。明确育种目标，选择合适的亲本进行异花授粉。育种公司通常会建立育种圃，大量品种可用于杂交。还有一种方法是采集收集到的种类或基因库中的郁金香花粉进行杂交。杂交时最常用的亲本是 *T. gesneriana* 的栽培品种。这一类群的现代品种遗传成分复杂，既有来自该类群的，也有来自其他类群的，通过品种间杂交可以获得理想的杂交后代。育种的流程包括人工授粉、优株筛选、新品种培育，优株筛选的比例是 0.1%～1%（Van Tuyl and Van Creij 2007）。

在 Roman Szymański（波兰）农场，杂交时最常用的方法是，花开放花粉成熟前去除母本花药，将父本花粉涂抹于母本黏性柱头上，用棉织物包裹柱头。一般花粉成熟比雌蕊早，可提前采集贮藏于冰箱，保存期为一个月左右。种子成熟时（一般从 7 月中旬到 7 月底），采集蒴果，20～25℃下储存于纸袋中，种子可自行萌发。同年 10 月，在混有泥炭基质和肥沃复合肥料的盒子或罐子中播种，覆沙 0.5cm，冬季在有坡度的苗床上过冬，春季转到铝箔温室中，种子萌发，发芽率（0～100%）高低与亲本有关。第一年，幼苗只产生一个 3cm 长圆柱状的小子叶，5 月底叶片干枯变黄，产生直径 2～3mm、重量 10～200mg 的小鳞茎。夏季挖取鳞茎，储存于沙中，秋季的操作流程同第一年，将鳞茎放入盒子或罐中栽培。第三年，将鳞茎直接种植于土壤中。4～6 年后鳞茎长到足够大，内含丰富的贮藏物质，有发育成熟的生殖器官，幼苗可开花。通常鳞茎周径为 6～8cm 时就可以进行生殖生长。鳞茎鳞片之间的芽发育成子球，每年可以产生两到三个新球。这就是为什么一个候选品种需要繁殖 10 年或 10 年以上，才能达到一定数量可以在市场上流通。

现代郁金香育种目标除花色、花型、叶形等主要性状外，还包括不同季节促成栽培适应性、抗病性、运输和贸易中花品质良好、较长的切花寿命。这些性状的测试需要很多年。初筛采用负向选择，如淘汰白化苗、种子不萌发或畸形苗。植株为纯合体时，致命基因控制的性状表达，可能会出现有缺陷的幼苗。第一次开花时，通过花的观赏性和花期长短进行正向筛选，同时可以评价种球产量、Fot 抗性、促成栽培适应性和切花寿命。如果选择的种类产生的小种球过多，那么可以预测该品种不适宜促成栽培。

28.4.3　远缘杂交

通过远缘杂交可以获得全新的种质，如不同原生种间，包括尚未驯化的原生种杂交。创造的新种质具备一些功能特征，特别是一些不常见的特性。尽管远缘杂交困难重重，但从 *T. gesneriana* × *T. fosteriana*、*T. gesneriana* × *T. eichleri*、*T. gesneriana* × *T. kaufmanniana*、*T. gesneriana* × *T. albertii*（正反交）和 *T. greigii* × *T. gesneriana*、*T. praestans* × *T. gesneriana*、*T. tubergeniana* × *T. gesneriana* 和 *T. ingens* × *T. gesneriana* 中获得了部分远缘杂交种（表 28.2）（Custers et al. 1995；Van Creij 1997；Van Creij et al. 1999；Okazaki 2005；Marasek-Ciolakowska et al. 2018）。

表 28.2　参照 Van Raamsdonk 等（1995）和 Marasek-Ciolakowska 等（2018）*Tulipa* 组与 *Eichleres* 组种间杂交的一些改良组合

亚属或组	*Tulipa*						*Eichleres*									
母本 \ 父本	*T. gesneriana*	*T. didieri*	*T. suaveolens*	*T. rhodopea*	*T. marjolettii*	*T. planifolia*	*T. praestans*	*T. micheliana*	*T. kaufmanniana*	*T. vvedenskyi*	*T. tubergeniana*	*T. greigii*	*T. hoogiana*	*T. fosteriana*	*T. ingens*	*T. eichleri*
T. gesneriana	a	a	a	a	a	—	—	—	c	c	c	c	b	c	c	c
T. fosteriana	c	c	—	—	—	—	d	d	d	b	a	a	—	a	a	a
T. kaufmanniana	c	—	d	—	c	—	b	d	a	a	d	—	b	c	—	c
T. greigii	c	c	d	—	d	—	—	a	a	a	c	a	—	a	—	a

a. 几个杂交组合成功，同一组内杂交结实率高（组内杂交）
b. 一个杂交组合成功，杂交结实率高
c. 几个杂交组合成功，杂交结实率低
d. 未获得杂交种子
—. 未进行杂交

由于亲缘关系较远，远缘杂交存在杂交障碍，主要分为受精前和受精后两种障碍。受精前障碍主要包括：花粉在柱头上不萌发，或即使萌发，花粉管在柱头或花柱内的生长长度不能到达胚珠，或即使到达胚珠，胚珠中花粉管生长受阻。为克服这些杂交障碍，可以采用切割柱头或热水处理的方法使柱头上的酶失活（50℃持续 1～3min）（Okazaki and Murakami 1992）。受精后障碍与胚乳降解、胚乳发育异常、幼苗变异（黄化或白化）、不能形成花或生殖器官发育不全有关。表 28.3 列出了一些克服远缘杂交障碍的措施。

表 28.3　郁金香远缘杂交障碍克服方法

受精前杂交障碍	受精后杂交障碍
高温处理柱头（Okazaki and Murakami 1992）	氨基酸或植物生长调节剂处理子房（Van Creij et al. 1999）
重复授粉（Custers et al. 1995）	胚培养、胚珠培养或子房培养（Van Tuyl et al. 1990；Custers et al. 1995；Van Creij et al. 1999，2000；Okazaki 2005）
切割柱头（Van Creij et al. 2000）	
胎座授粉，子房嫁接（Van Creij et al. 1997a，1997b）	

胚挽救是一种离体培养技术，在植物育种中发挥着重要作用。通过胚挽救，可以培育出许多种间和属间杂交种，包括二倍体杂交种（Sharma et al. 1996；Reed 2004）。造成植物种间或属间杂交不亲和性的原因很多，最常见的是胚败育，因此胚挽救技术被广泛应用于克服郁金香种内和种间杂交障碍（Okazaki 2005；Van Tuyl and Van Creij 2007）。胚挽救技术主要包括胚培养、胚珠培养和子房培养，挽救的胚在体外发育成完整植株受多种因素影响，如杂交亲本的基因型、培养器官（胚、胚珠或子房）的类型、授粉后蒴果采收时间以及体外培养条件（培养基成分、光照和温度）。Custers 等（1992，1995）报道了种间杂交 *T. gesneriana* × *T. kaufmanniana* 因胚败育而进行的成功胚挽救。虽然有些杂交可以获得成熟蒴果并能收集到少量种子，但采用胚珠培养或胚培养的方法，能有效提高种间杂交苗的生产效率。Custers 等（1995）认为胚挽救的最佳时间是授粉后的 7～

9周，采用1/2MS（1962年）+4%蔗糖+500mg/L色氨酸+4nmol/L萘乙酸（NAA）+0.75%琼脂的培养基进行胚培养或子房培养，培养初始条件为15℃，暗培养15周。据Niimi（1978）描述，为诱导萌发，使郁金香幼苗发育正常，接种的胚需要进行特定温度处理。通常5℃下培养12周，胚萌发，随后15℃培养12～18周，幼苗产生鳞茎。Okazaki（2005）采用胚培养或胚珠培养法获得了3个组合 *T. gesneriana* × *T. fosteriana*、*T. gesneriana* × *T. eichleri* 和 *T. gesneriana* × *T. greigii* 较多的种间杂交后代。授粉后6～8周取胚和胚珠，采用1/2MS+3%蔗糖+0.2%结冷胶培养基进行胚培养和子房培养，经一系列变温处理，幼苗产生的小鳞茎数量比通过常规杂交获得的多。另外，通过胚珠和子房切片培养，*T. gesneriana* × *T. praestans* 和远缘杂交组合 *T. gesneriana* × *T. agenensis* 也获得了独特的杂交种（Van Creij et al. 1999，2000）。研究人员还比较了授粉后直接进行胚珠培养和授粉后2～9周进行子房切片培养与胚珠培养的效果。在大多数情况下，胚的萌发率随胚发育进程提高，与采用的方法无关。在子房切片培养中，培养基中蔗糖浓度提高到9%可提高胚挽救率。

有时当两个物种杂交不亲和时，可采用与两个亲本均具有兼容性的第三个物种进行杂交来克服杂交不亲和性（Van de Wiel et al. 2010）。用 *T. greigii* 作为第三亲本进行桥梁杂交，成功克服了 *T. gesneriana* 和 *T. kaufmanniana* 两个亲本之间的杂交不亲和性，杂交顺序为[*T. gesneriana* ×（*T. kaufmanniana* × *T. greigii*）]（Van Eijk et al. 1991）。

综上所述，种间杂交可以将不同品种的优良性状整合到新的品种上，包括抗郁金香碎色花瓣病（TBV）、抗郁金香灰霉病（*Botrytis tulipae*）和抗郁金香尖孢镰刀菌（*Fusarium oxysporum* f. sp. *tulipae*）、缩短促成栽培周期、延长切花寿命以及新花型、新花色等（Van Eijk et al. 1991；Van Creij et al. 1997b，1999；Van Raamsdonk et al. 1995；Van Tuyl and Van Creij 2007；Van Tuyl et al. 2012）。

28.4.4 单倍体和双单倍体

据文献记载，已成功报道了25种植物的双单倍体培养体系，并且近几年来双单倍体育种已成为趋势（Małuszyński et al. 2003）。株系一致是杂交品种的基本特性要求，通过双单倍体培养可以很容易获得该特性。

郁金香的几个野生种和栽培品种可以通过体外小孢子培养获得单倍体（Van den Bulk et al. 1994；Van den Bulk and Van Tuyl 1997；Custers et al. 1997）。14个 *T. gesneriana* 品种和13个野生种通过小孢子培养获得了胚状体结构，并培育出两个新品种：'Leen van der Mark' 和 'Rosario'（Van den Bulk and Van Tuyl 1997）。鳞茎在−2℃下贮藏4周后，9℃下生根并长出花芽，从花芽中分离小孢子，一个花芽可分离出约800 000个小孢子。小孢子培养中，1/4MS + 13%蔗糖培养基最适合小孢子发育成多细胞结构。单核早期和中期分离的小孢子形成的多细胞结构数量最多（0.13%），这种结构在含3%蔗糖的固体培养基上能继续生长发育。与合子胚一样，小孢子培养形成的胚也具有休眠特性，低温处理可以打破休眠。低温处理后，胚萌发，形成子叶状结构和匍匐茎，最后形成小鳞茎。'Rosario' 植株结构完整，生长健壮，尽管少数植株为单倍体，但大多数植株都表现出二倍体水平植株的特性，‘Leen van der Mark’的大多数再生植株是单倍体（Van den Bulk and Van Tuyl 1997），作者认为染色体数目自发加倍的能力可能与植株旺盛的生长有关。Custers 等（1997）介绍了郁金香通过小孢子培养产生胚的研究进展，高温预处理（32℃）鳞茎能促进小孢子形成胚，采用改良体系可以产生大量株系，经筛选可用于 F_1 杂交生产。

28.4.5 突变、体细胞无性系变异和多倍体化

突变，无论是自发还是诱发，都是作物改良的重要手段。目前来自 210 种植物的已登记的 3220 多个品种属于突变类型（Bado et al. 2015）。在传统的郁金香培育中，自发突变（也称变异）相当普遍（Van Tuyl and Van Creij 2007）。突变体的产生与品种有关，如 1860 年培育的‘Murillo’（重瓣早花群）产生的突变体最多（Van Scheepen 1996）。目前仍在市场上流通的浅粉-白色花的原始栽培品种，至少产生了 139 种不同花色和不同花型的突变体。另一个 1894 年的老品种‘Bartigon’（单瓣晚花群）也有许多自发突变体，如‘All Bright’、‘Bartigon Duplex’和‘Cordell’。1951 年育成的深受消费者喜爱的红色花三倍体品种‘Apeldoorn’（达尔文杂种群或其他杂交种），变异类型很多，包括常用栽培品种‘Golden Apeldoorn’。尽管有几个鹦鹉群品种是从近交系幼苗中筛选而来，但大多数是从非鹦鹉群植物中获得的突变体（Straathof and Eikelboom 1997）。

郁金香诱变育种历史悠久，20 世纪 30 年代，荷兰育种家 De Mol（1949，1953）就开始了开创性的辐射育种工作，并由此育成了一系列新品种，如从‘Fantasy’中获得的‘Faraday’，由‘Red Champion’育成的‘Estelle Rijnveld’。Grabowska 和 Mynett（1970）用 γ 射线 ^{60}Co（340～440rad）辐射‘Bartigon’鳞茎获得了一个鹦鹉群的品种和一个完全花变异品种。此外，Gaul 和 Mittelstenscheid（1960）证明辐射可以诱导郁金香变异。在日本，Nezu（1965）和 Matsubara 等（1965）进行了大规模的郁金香辐射诱变育种。荷兰 CPRO-DLO 公布了‘Preludium’和‘Lustige Witwe’的几个辐射突变体，叶片有斑纹的‘Santana’、深紫红色花的‘Yvonne’以及‘Rimo’和‘Ivette’，均来自‘Lustige Witwe’（Straathof and Eikelboom 1997）。Broertjes 和 Van Harten（2013）评述了郁金香突变育种进展，影响突变率的因素有辐射剂量、种球发育阶段、种球大小以及基因型，辐射的最佳条件是 6 月采收后或 11 月的大种球顶端分生组织，X 射线辐射剂量 3.5～8Gy。最好采用对辐射不太敏感且营养分生能力强的大种球。育种家观察发现开紫色和粉色花的品种能够产生优雅红色系和艳丽粉色系的变异，甚至是白色变异，这取决于花色苷的浓度。最近，以郁金香品种 *T. fosteriana*‘Albert Heijn’（花瓣粉色）和其自发突变株系‘上农 09’（花瓣红色）为材料揭示了花色突变的分子机制（Yuan et al. 2014）。从花青素成分、花青素转录谱、花色苷合成结构基因（最初是从 *T. fosteriana* 中分离）和类黄酮 3′-羟化酶基因（*TfF3′H1*）（花色苷衍生物合成的关键基因）等方面比较了这两种基因型的差异。*TfF3′H1* 基因启动子的插入突变导致了突变株‘上农 09’花色变化，这种插入抑制 *TfF3′H1* 基因的转录，极大地降低了突变体花被片中花色苷衍生物的积累。此外，与原品种‘Albert Heijn’相比，突变体中类黄酮和黄酮醇含量更高，这与查耳酮合酶基因

（*TfCHS1*）、查耳酮异构酶基因（*TfCHI2*）和黄烷酮 3-羟化酶基因（*TfF3H1*）的高表达水平有关，也可能与突变株花色的改变相关。由于诱变可以改善或稍微改变现有品种的性状，且不需要长期育种过程，因此诱变育种具有优于杂交育种的优势。

Larkin 和 Scowcroft（1981）定义了体细胞无性系变异（SV）是在营养繁殖过程中，尤其是离体培养过程中自发发生的。这种现象是遗传（突变）或表观遗传（非遗传）变化的结果（Karp 1995；Jain and De Klerk 1998；Bairu et al. 2011；Smulders and De Klerk 2011）。典型的遗传变异是染色体数目（整倍体和非整倍体）、染色体结构（易位、缺失、插入和复制）和 DNA 序列（碱基突变）的改变。表观遗传的改变，通常是过渡性的，与基因甲基化有关。一方面，营养繁殖的目的是克隆、大量生产，不希望出现具有不同表型的变种（无性系、体细胞克隆、突变体）；另一方面，体细胞无性系变异有利于作物改良，因为 SV 可能会有适于新品种选育的遗传变异。

Podwyszyńska 等（2006，2010a，2010b）报道了在郁金香再生体系中，不定芽在含噻苯隆（TDZ）的培养基上进行长期培养过程中产生了体细胞无性系变异（图 28.2）。变异率与离体培养时间成正比，离体培养 4～6 年的植株，SV 频率最高，表型变化最大。在离体扩繁了 1.5～3 年的二倍体品种‘Blue Parrot’和‘Prominence’中发现了体细胞无性系变异，三倍体品种‘Giewont’（达尔文杂种群）中有 15.3%的植株会发生变异。离体培养 4 年或 6 年的‘Blue Parrot’和 50%的‘Prominence’植株均会有体细胞无性系变异。据记载，‘Blue Parrot’的变异与花色有关，‘Prominence’和‘Giewont’的变异与花型有关，如百合花型或红色花瓣上有白色条纹，花严重畸形。在长期离体培养的‘Blue Parrot’植株中，有 1%～1.7%杂色叶的体细胞无性系变异。有时在幼嫩植物中也能观察到变异，叶变窄、纵向卷曲或加厚。所有这些变化都是利用基于聚合酶链反应（PCR）的分子标记分析所证实的突变结果。常用的方法有随机扩增多态性 DNA（RAPD）、简单重复序列间区（ISSR）、扩增片段长度多态性（AFLP）以及 5S rDNA 和 25S rDNA 荧光原位杂交（FISH）技术进行核型分析等（Podwyszyńska et al. 2006，2010a；Marasek-Ciolakowska and Podwyszyńska 2008）。此外，在检测郁金香新品种优良材料稳定性的离体培养过程中，ISSR 标记还可用于早期体细胞克隆变异的筛选（Podwyszyńska et al. 2010b）。

图 28.2 从长期微繁殖体系中筛选的体细胞克隆突变体品种‘Agalia’（左侧第一张图）

在目前栽培的多倍体郁金香中，最重要的是达尔文杂种群中三倍体品种。在育种上，四倍体植株可育，比通常不育的三倍体更有价值，四倍体的配子既可以作为新品种来源，也可以用于进一步杂交。Okazaki 等（2005）在花药减数分裂时期利用一氧化二氮（N_2O）诱导产生 2*n* 花粉，并用获得的 2*n* 配子与二倍体品种杂交，从而获得三倍体郁金香品种。授粉一周的植株，在 5～6 个大气压的 N_2O 气缸中放置一天，可获得一批四倍体后代（Straathof and Eikelboom 1997）。另一种方法是有丝分裂多倍体化，即对体细胞使用化学试剂使染色体数目加倍，将细胞阻断在有丝分裂中期，这种多倍体化可用于恢复种间杂交种的育性。Chauvin 等（2005）通过将外植体浸泡在氨磺乐灵溶液中（浓度 5～10mg/L 浸泡 24～48h 或 0.5～1mg/L 浸泡 2 周）进行体外染色体加倍，获得了 30 个四倍体材料，多倍体效率最高可达 20%。Podwyszyńska（2011）用抗有丝分裂剂、稻瘟灵和甲基胺草磷（APM）处理花茎段 7～14 天，通过体外诱导使体细胞染色体加倍，从品种‘Fringed Black’和三个优株的克隆体中获得了几个四倍体植株，多倍体效率达 30%。

28.4.6 离体扩繁

通过分球繁殖生产 10 000 粒郁金香新品种种球，周期长，成本高。而通过离体培养的方法，可以有效缩短繁育周期。组织培养技术广泛应用于许多作物最有价值的育种株系的快速繁殖，20 世纪 90 年代已有文献报道郁金香的离体扩繁，以鳞片、花茎段为外植体，直接再生芽或小鳞茎，但该扩繁方式效率中等（Nishiuchi 1980；Rice et al. 1983；Le Nard et al. 1987；Alderson and Taeb 1990；Baker et al. 1990；Hulscher et al. 1992；Famelaer et al. 1996；Kuijpers and Langens-Gerrits 1997）。有效的二次再生是再生芽或鳞茎无限繁殖的前提，郁金香离体扩繁的最终产物是由试管苗诱导产生的小鳞茎。将离体培养的小鳞茎转移到室温条件下贮藏 5～6 周，之后继续离体培养，9℃下生根，最后将小鳞茎移栽到标准栽培环境中（15～20℃）促使郁金香正常生长。采用该方法，一个母球可以扩繁 200 粒左右的小鳞茎。最近，已证明 *Tulipa tarda* 幼苗的外植体具有很高的再生潜力（Maślanka and Bach 2014）。或许这种方法可以成功地用于野生郁金香的繁殖，如濒危物种的扩繁。现有的高效离体繁育体系中，一个种球 2～3 年能够生产 400～1000 粒小鳞茎（Podwyszyńska and Marasek 2003；Podwyszyńska and Sochacki 2010；Podwyszyńska et al. 2014）。离体扩繁中，不定芽增殖采用含 TDZ、2ip/N6-异戊烯基腺嘌呤和 NAA 的 MS 培养基，经 2 年培养，诱导生成小鳞茎。鳞茎发育周期一般为 8～10 个月，包括三个阶段：①18℃培养 10 周诱导芽休眠；②5℃下冷藏 12～14 周，解除休眠；③18℃在富含蔗糖的培养基上培养 12～14 周形成鳞茎（Podwyszyńska 2006）。不定芽循环增殖已成功地用于维持育种、生产无病毒植物和诱导有丝分裂四倍体（Podwyszyńska 2011；Sochacki and Podwyszyńska 2012）。但是，由于有些品种会发生体细胞无性系变异或者培养基成分发生变化会导致体细胞无性系变异，因此周期性的芽增殖培养时间不宜过长（Podwyszyńska et al. 2006，2010a，2010b）。

郁金香离体培养是通过外植体诱导直接产生体细胞胚或由外植体产生愈伤组织，愈伤组织进一步分化形成体细胞胚。Gude 和 Dijkema（1997）采用郁金香花茎为外植体，在含有 5μmol/L 或 50μmol/L 2,4-D 或毒莠定的培养基上成功诱导出体细胞胚。Podwyszyńska 和 Marasek（1999）报道，通过郁金香胚性愈伤组织诱导产生体细胞胚，可以连续扩繁 5 年。从冷处理的鳞茎中分离得到的花茎段可诱导产生愈伤组织，培养基采用只含生长素（毒莠定、2,4-D、NAA）或含有 TDZ 的改良 MS 培养基。在含 2.5mg/L 2,4-D 与 0.1～0.5mg/L TDZ 的培养基上可以诱导产生胚性愈伤组织。胚状体结构产生一般需要含低浓度生长调节剂的培养基。郁金香胚性化的比例很高（100mg 愈伤组织可以产生 90 个体细胞胚），但最终只有 30%的胚形成芽，4%的胚发育成鳞茎，因此，通过体细胞胚扩繁郁金香的工作还有待进一步深入研究。Bach 和 Ptak（2001）以及 Ptak 和 Bach（2007）报道了利用体细胞胚进行郁金香扩繁的可能性，方法是从鳞茎中分离子房，将子房切段培养，培养基为 MS+25μmol/L 毒莠定+0.5μmol/L 6-苄基腺嘌呤（BAP），之后在含 5μmol/L BAP 和 0.5μmol/L NAA 的培养基中进行继代培养，形成大量体细胞胚，体细胞胚在高蔗糖浓度的培养基上可以产生小鳞茎，因此，深入研究郁金香离体培养各个阶段条件的优化，能有效提高生产效率。

28.4.7　遗传转化

通过遗传转化体系将单一外源基因导入已知品种，并通过单细胞培养获得新的郁金香种类为育种提供了可能。最理想的目标基因是抗病基因，如抗真菌、抗病毒，这些病原菌引起的病害严重影响了郁金香生产。许多物种通过各种方法，如电穿孔、微注射和农杆菌介导，导入目的基因或激活、抑制目的基因表达（Moose and Mumm 2008）。目前 CRISPR（规律间隔成簇短回文重复序列）和 TALEN（类转录激活因子样效应物核酸酶）技术准确率高（Gaj et al. 2013）。单子叶植物中最常用的转化方法是粒子轰击法，如玉米，而通过根癌农杆菌介导，成功率低（Wilmink and Dons 1993）。迄今为止，郁金香遗传转化体系的研究还处于初步阶段，采用的方法有粒子轰击和农杆菌介导（Wilmink et al. 1995a；Wilmink 1996；Chauvin et al. 1997，1999）。遗传转化的关键是要有合适的筛选试剂、报告基因和启动子。过去大多数单子叶植物常用的筛选试剂是卡那霉素，但郁金香对卡那霉素敏感性很低。Wilmink 等（1995b）研究表明，在采用粒子轰击花茎段构建的遗传转化体系中，花茎具有较高的不定芽再生能力，除草剂 Basta（PPT）筛选效率高，水稻肌动蛋白启动子能有效促进 GUS 报告基因的表达，分子水平上检测遗传转化 12～15 个月的植株，证实了再生植株中存在 *bar* 和 *gus* 基因，但蛋白质表达水平较低或不表达。Wilmink（1996）认为，导入的基因只存在于植物组织中的一部分，转化植株中的嵌合结构说明了再生是多细胞起源的。为避免形成嵌合体，植物培养过程中应使用高浓度除草剂 Basta 筛选体系，并将培养时间延长。由于郁金香外植体再生能力低，该属植物发育周期长且复杂，基因遗传转化在郁金香育种中的应用并不十分理想，从转化、转化植株筛选到开花至少需要 8 年时间。

28.4.8　分子标记辅助选择

利用与目标性状连锁的分子标记可以极大地促进和加速育种进程。通过分子标记可对 1～2 年的幼苗进行早期快速筛选。这种手段对于幼年期很长的郁金香来说很重要，在此期间无性系的空间隔离是必需的。由于郁金香功能特征复杂，依据形态和生理指标开发的分子标记数量很少。然而，通过生理、形态和分子方面的研究，开发了一些有用的分子标记，与形态和生理指标相关的分子标记主要与植物的遗传研究有关（Van Eijk and Toxopeus 1968；Van Eijk and Eikelboom 1975，1976），包括1年生幼苗抗尖孢镰刀菌的能力（Van Eijk and Eikelboom 1990；Romanow et al. 1991；Straathof et al. 1997），适宜促成栽培和种球产量高（Van Eijk and Toxopeus 1968）三个方面。切花寿命可以在第一次开花时进行评估（Van der Meulen-Muisers et al. 1997），叶片蜡质层的厚度与灰霉病易感性有关（Reyes et al. 2005）。

基于染色体构建的细胞学标记可用于植物倍性观察（Abedi et al. 2015），鉴别近缘物种的基因组，设计杂交方案或鉴定杂交植物的亲本或异源多倍体物种的祖先，这在不亲和杂交和远缘杂交中尤其重要（Marasek et al. 2006），并证明了渗透性片段的存在（Marasek et al. 2006）。

在郁金香中，关于染色体数目、染色体鉴定和核型分析的研究陆续发表（Sayama et al. 1982；Kroon and Jongerius 1986；Van Raamsdonk and De Vries 1995；Marasek et al. 2006；Abedi et al. 2015）。郁金香属植物染色体基数为 $x = 12$，染色体比较大，但最重要的两个种 *T. gesneriana* 和 *T. fosteriana* 的染色体形态类似，均由中部着丝粒染色体、亚中部着丝粒染色体和亚端部着丝粒染色体组成，没有次缢痕等特殊结构（Marasek et al. 2006）。利用异染色质切片的 Giemsa 染色（C 显带）能有效进行染色体鉴定（Filion 1974；Blakey and Vosa 1982；Van Raamsdonk and De Vries 1995）。Blakey 和 Vosa（1981，1982）利用常规染色和 C 显带技术，建立了 *Eriostemones* 和 *Leiostemones* 亚属郁金香植物的系统发育关系，并根据其形态和异染色质的分布鉴别出几种染色体类型，描述了不同组物种的染色体特征。

以 5S rDNA 和 45S rDNA 为探针的荧光原位杂交（FISH）是一种分子细胞遗传学标记，可以进行物种鉴定。根据 FISH 信号的大小、数量和染色体的形态可以区分 *T. gesneriana* 和 *T. fosteriana* 两个种的品种（Mizuochi et al. 2007）。通过 FISH 的核型分析可以揭示'Prominence'品种和无性系染色体重排、杂交位点数量和杂交信号大小的变化，也能有效地检测 *T. gesneriana* 郁金香品种'Prominence'微繁殖群体的体细胞无性系变异（Marasek-Ciolakowska and Podwyszyńska 2008）（图 28.3）。

在郁金香中，利用基因组原位杂交（GISH）技术有助于鉴定二倍体、三倍体和四倍体达尔文杂种群的基因组构成。达尔文杂种群是由二倍体 *T. gesneriana* 和 *T. fosteriana* 杂交而成的（Marasek et al. 2006；Marasek and Okazaki 2007），结合了 *Tulipa* 和 *Eichleres* 两个属的优良性状，如适宜促成栽培、抗尖孢镰刀菌和抗或部分抗 TBV 病毒，因此越来越受欢迎（Van Tuyl and Van Creij 2007）。对达尔文杂种群三倍体栽培品种的基因组组成分析证实了它们起源于二倍体栽培种产生的二倍体配子（Kroon and Van Eijk 1977；Marasek et al. 2006），该二倍体栽培种具有两个基因组：*T. gesneriana*（G）的基因组和 *T. fosteriana*（F）的基因组，没有重组染色体（Marasek et al. 2006；Marasek and Okazaki 2007）。同样未减数分裂的配子参与了四倍体品种'Ollioules'的形成，它是二倍体品种'Caravelle'和达尔文杂种群品种（Marasek and Okazaki 2007）杂交的结果。GISH 鉴定结果表明，该品种有 3 个 *T. gesneriana* 染色体组和 1 个 *T. fosteriana* 染色体组。根据

Marasek- Ciolakowska 等（2012）的研究，二倍体达尔文杂种群可以产生功能性的 2*n* 配子，也可以产生 *n* 配子，这为从回交 FG 杂种到 *T. gesneriana* 亲本产生不同倍性水平的后代提供了机会（Marasek-Ciolakowska et al. 2014）。

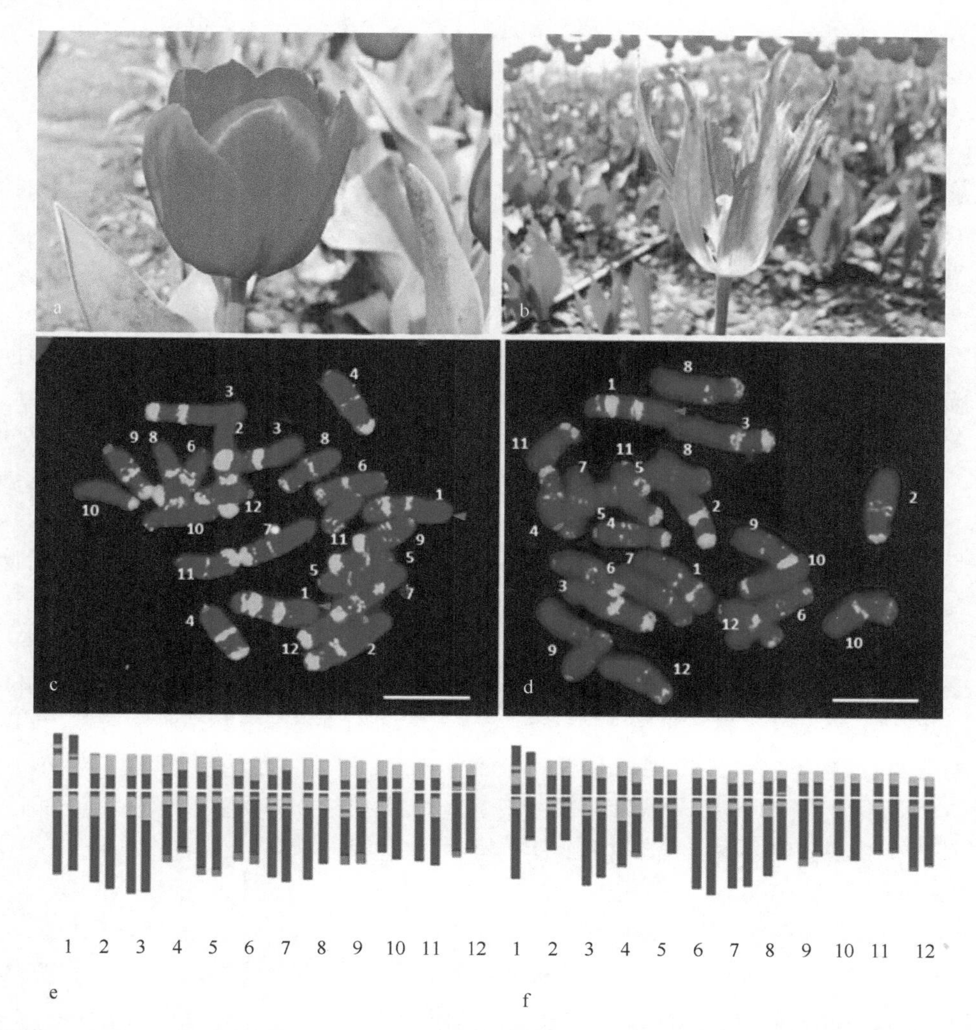

图 28.3 a.‘Prominence’常规繁育；b. 由不定芽长期离体培养获得的体细胞变异体；c、d.‘Prominence’（c）和体细胞无性系（d）染色体的 5S rDNA（绿色）与 45S rDNA（红色）的双色荧光原位杂交图；e、f.‘Prominence’（e）和体细胞无性系（f）染色体标准图。箭头指示微弱的 45S rDNA（红色）和 5S rDNA（绿色）信号，已被检测到但在该图中显示不出来（引自 Marasek-Ciolakowska and Podwyszyńska 2008）

结合 Southern 杂交，用 GISH 和以 45S rDNA、5S rDNA 为探针的 FISH 分子细胞遗传学分析方法，揭示了二倍体品种‘Purissima’的基因组组成，根据其园艺性状，该品种属于福氏群（Van Scheepen 1996），与 *T. gesneriana* 具有高度杂交性。Marasek 和 Okazaki（2008）发现‘Purissima’是一个种间杂交种，其基因组由一个 *T. gesneriana* 基因组和一个 *T. fosteriana* 基因组组成。以‘Purissima’为父本，除获得一个三倍体品种‘Kouki’外，其余杂交后代均为二倍体，在‘Purissima’的二倍体后代中，存在重组染色体，而在代表GGF基因组的三倍体‘Kouki’中没有发现重组染色体。一般染色体易位的总数从1到6不等。在达尔文杂种群二倍体的BC1和BC2后代中，记录着两个亲本染色体间存在大量基因间重组（Marasek- Ciolakowska et al. 2009，2011，2012），这是郁金香渐渗育种的突破。

分子标记在郁金香育种中的作用越来越重要，但大规模应用还不成熟（Arens et al. 2012）。分子标记是基因水平上的遗传标记，在植物育种中利用分子标记可以了解性状的起源与遗传，追踪亲代向后代传递染色体片段的过程，研究基因的连锁，有助于选择合适的杂交亲本，并使早期筛选成为可能（Moose and Mumm 2008）。例如，分子标记用于郁金香品种（Bondrea et al. 2007）和野生种的鉴定（Kiani et al. 2012）以及物种间系统发育分析（Nan Tang et al. 2013；Turktas et al. 2013）。

有些植物物种的基因组已经测序，但基因组大小以及基因中含有大量非编码内含子和重复序列，使像郁金香这种大基因组物种的基因研究非常困难（Shahin et al. 2012）。Moreno Pachon等（2016）采集郁金香不同组织部位进行转录组测序，研究调节植物发育的TCP转录因子家族的功能。Miao等（2016）利用转录组分析揭示了*T. edulis*鳞茎形成的分子机制，旨在对这一濒危物种的繁殖进行遗传学研究。为研究红花郁金香花色苷的生物合成，Yuan等（2013）从郁金香花瓣中克隆了6个与花色苷合成相关的转录因子cDNA全长，研究基因在不同郁金香品种中的表达，比较不同品种间的花色苷含量，旨在为郁金香花色育种提供指导。

郁金香中已克隆出一些重要基因。Momonoi等（2007）分离5个编码1-氨基环丙烷-1-羧酸氧化酶（ACO）的cDNA克隆，该基因在凋萎的花瓣、叶和茎中均有表达。ACO是催化乙烯生物合成的最后一个关键酶，乙烯参与叶片和花的衰老等生理过程。郁金香衰老不具有明显特征，且与乙烯无直接关系，郁金香的ACO由基因家族编码，包含5个基因，这5个基因在郁金香营养器官和生殖器官中所受的调节不同。

Yamagami等（2000）从*T. bakeri*郁金香鳞茎中分离、测定了几丁质酶，并验证了其水解胶质甲壳素的活性，这些研究可用于郁金香抗性标记设计或遗传转化体系中。

Kanno等（2003）研究了*T. gesneriana*郁金香花同源基因的表达，Van Tunen等（1993）通过对B类基因的研究证实郁金香中存在改良的ABC模型，并提出了*T. gesneriana*改良的花器官发育ABC模型，这一结果对于植物花器官形态转化研究具有重要意义。

28.5 持续筛选

持续对高质量鳞茎生产有利的筛选，是保证品种在市场上流通的先决条件。筛选通常是在田间幼苗生长阶段或种球采收后的贮藏阶段进行，这对于所有基因型和表现型品种的保存与选择最健康的种球进行繁殖都是必不可少的。田间筛选主要是去除不符合要求和有病状的植株，种球采收后筛选主要有三种方法：①人工分离大而圆的A级和B级种球；②8月中旬采用水浴漂浮法去除质量轻的种球；③槽上分离扁种球和圆种球（Hekstra 1968；Timmer 1975；Fatel 1997）。在大规模的栽培生产中，上述方法并不适用，但田间清除病毒感染株、选择圆种球的操作可行。对于花色较浅（如白色、奶油色或黄色）的品种，肉眼不易观察，很有可能会保留未检测到的病毒植株，对于这类品种，利用酶联免疫吸附测定法（DAS-ELISA）检测郁金香碎色花瓣病（TBV）病毒是荷兰育种公司的常规操作流程（Knippels 2005，2011）。另外，很多试验，包括田间试验都可以进行自动化检测（Polder et al. 2014）。荷兰哈瓦特克公司（Havatec）已经开发出一种利用X光相机自动选择感染了尖孢镰刀菌的郁金香种球的机器（HAVATEC 2017）。

28.6 新 品 种

每年世界各地培育出大量郁金香新品种。2014～2016年的3年，有140～150个新品种在荷兰皇家球根种植者协会（KAVB）登录，绝大多数由荷兰育种公司培育。2014年澳大利亚和波兰各登录1个新品种，2015年法国登录6个新品种，拉脱维亚登录2个新品种，2016年法国登录4个新品种，拉脱维亚登录2个新品种，中国登录2个新品种，日本登录1个新品种。新登录的品种主要属于凯旋系，该系品种适宜促成栽培、切花寿命长。重瓣早花群、重瓣晚花群（分别有31个和43个）和流苏花型新品种的数量也在不断增加，最受欢迎的是重瓣流苏花型和绿辨群的品种（Bodegom and Van Scheepen 2015，2016，2017）。‘Yellow Crown’和‘Liberstar’两个品种通常被归为凯旋系，花冠独特。‘Ice Cream’属于重瓣晚花群，完全重瓣，花瓣颜色丰富，叶色多样。采用这些亲本进行杂交可以培育一些花期更长、花型与洋蓟花类似的新品种，如‘Harborlight’、‘Purple Tower’和‘Brooklyn’。其中‘Brooklyn’的母本是‘Helene van Dam’×‘Inside Double’的杂交后代，父本是（‘Gander’× *T. schrenkii*）×‘Inside Double’的杂交后代（Vertuco 2017）。除荷兰皇家球根种植者协会外，育种公司或育种人还可以直接向欧盟植物品种局（CPVO）申请新品种权保护，CPVO总部设在法国安格斯，共有431个郁金香栽培品种受保护，其中16个新品种刚投放到市场（CPVO 2017）。

28.7 展 望

郁金香的持续流行是创造新品种的前提，除观赏性外，新品种还应具备种球产量高和促成栽培时间长（从初秋到晚春）的特性。为保护生产环境，作物产品使用的规定越来越严格，这就增加了对抗病和抗虫品种的需求，而对田间和促成栽培管理的要求较低。随着对郁金香遗传知识和重要功能特征机制的深入研究，郁金香育种会越来越高效。毫无疑问，更多的分子生物学工具将会被用于育种工作，新的有效缩短郁金香育种周期的分子标记也将会产生（Krens and Kamo 2013）。由于郁金香栽培过程中会产生大量废弃花朵，因此，使用富含花青素的提取物作为食品染色剂（Arici et al. 2016）或作为活性化合物来源（Sagdic et al. 2013）将成为一个新的思路。

本章译者：

杜文文

云南省农业科学院花卉研究所，昆明 650200

参 考 文 献

Abedi R, Babaci A, Karimzadeh G (2015) Karyological and flow cytometric studies of *Tulipa* (Liliaceae) species from Iran. Plant Syst Evol 301:1473–1484

Alderson PG, Taeb AG (1990) Effect of bulb storage on shoot regeneration from floral stems of tulip *in vitro*. J Hortic Sci 65:71–74

Arens P, Bijman B, Shahin A, van Tuyl JM (2012) Mapping of disease resistance in ornamentals: a long haul. Acta Hortic 953:231–237

Arici M, Karasu S, Baslar M, Toker OS, Sagdic O (2016) Tulip petal as a novel natural food colorant source: extraction optimization and stability studies. Ind Crop Prod 91:215–222

Asjes GJ (1994) Viruses in tulip in the Netherland. Acta Hortic 377:289–300

Bach A, Ptak A (2001) Somatic embryogenesis and plant regeneration from ovaries of *Tulipa gesneriana* L. *in vitro* cultures. Acta Hortic 560:391–394

Bado S, Forster BP, Nielen S, Ghanim A, Lagoda PJ, Till BJ, Laimer M (2015) Plant mutation breeding: current progress and future assessment. Plant Breed Rev 39:23–88

Bairu MW, Aremu AO, Van Staden J (2011) Somaclonal variation in plants: causes and detection methods. Plant Growth Regul 63:147–173

Baker CM, Wilkins HF, Asher PD (1990) Comparisons of precultural treatments and cultural conditions on *in vitro* response of tulip. Acta Hortic 66:83–90

Benschop M, Kamenetsky R, Le Nard M, Okubo H, De Hertogh A (2010) The global flower industry: production, utilization, research. Hortic Rev 36:1–115

BKD (2017) Voorlipige statistiek voorjaarsbloeiers 2016–2017. Bloembollenkeuringsdienst, Lisse

Blakey DH, Vosa CG (1981) Heterochromatin and chromosome variation in cultivated species of *Tulipa,* subg. *Eriostemones* (Liliaceae). Plant Syst Evol 139:47–55

Blakey DH, Vosa CG (1982) Heterochromatin and chromosome variation in cultivated species of *Tulipa* subg. *Leiostemones* (Liliaceae). Plant Syst Evol 139:163–178

Bodegom S, van Scheepen J (2015) KAVB Registraties 2014. Bijlage Bloembollenvisie, 6 February 2015

Bodegom S, van Scheepen J (2016) KAVB Registraties 2015. Bijlage Bloembollenvisie, 18 February 2016

Bodegom S, van Scheepen J (2017) KAVB Registraties 2016. Bijlage Bloembollenvisie, 2 March 2017

Bondrea OI, Pamfil D, van Heusden S, van Tuyl J, Meijer-Dekens F, Bondrea M, Sestras R, Rusu AR, Lucaci M, Patrascu BI, Balteanu VA (2007) AFLP as a modern technique for DNA fingerprinting and identification *Tulipa* cultivars. Bull USAMV-CN 63–64: 317–319

Botschantzeva ZP (1962) Tulips. Taxonomy, morphology, cytology, phytogeography and physiology (Russian edn), English translation: Varekamp HQ. Balkema, Rotterdam

Broertjes C, Van Harten AM (2013) Applied mutation breeding for vegetatively propagated crops, Development in plant science, vol 12. Elsevier, Amsterdam

BSARI (2017) Black Sea Agricultural Research Institute. Horticulture Crops Departments. http://arastirma.tarim.gov.tr/ktae/Sayfalar/EN/Detay.aspx?SayfaId=35, [access date=2017-05-19]

Buschman JCM (2005) Globalisation – flower – flower bulbs – bulb flowers. Acta Hortic 673:27–33

Cerveny CB, Miller WB (2010) Residual effects of ethylene on tulip growth and flowering. Hortscience 45:1164–1166

Chauvin JE, Hamann H, Cohat J, Le Nard M (1997) Selective agents and marker genes for use in genetic transformation of *Gladiolus grandiflorus* and *Tulipa gesneriana*. Acta Hortic 430:291–298

Chauvin JE, Marhadour S, Cohat J, Le Nard M (1999) Effects of gelling agents on *in vitro* regeneration and kanamycin efficiency as a selective agent in plant transformation procedures. Plant Cell Tissue Organ Cult 58:213–217

Chauvin JE, Label A, Kermarrec MP (2005) *In vitro* chromosome doubling in tulip (*Tulipa gesneriana* L.). J Hortic Sci Biotechnol 83:179–186

Christenhusz MJM, Govaerts R, David JC, Hall T, Borland K, Roberts PS, Tuomisto A, Buerki S, Chase MW, Fay MF (2013) Tiptoe through the tulips – cultural history, molecular phylogenetics and classification of *Tulipa* (*Liliaceae*). Bot J Linn Soc 172:280–328

CPVO (2017) Community Plant Variety Office. www.cpvoextranet.cpvo.europa.eu. [access date=2017-06-19]

Custers JBM, Eikelboom W, Bergervoet JHW, Van Eijk JP (1992) *In ovulo* embryo culture of tulip (*Tulipa* L.); effects of culture conditions on seedling and bulblet formation. Sci Hortic 51:111–122

Custers JBM, Eikelboom W, Bergervoet JHW, Eijk JV (1995) Embryo-rescue in the genus *Tulipa* L.; successful direct transfer of *T. kaufmanniana* regel germplasm into *T. gesneriana* L. Euphytica 82:253–261

Custers JBM, Ennik E, Eikelboom W, Dons JJM, Van Lookeren Campagne MM (1997) Embryogenesis from isolated microspores of tulip; towards developing F_1 hybrid varieties. Acta Hortic 430:259–266

De Mol WE (1949) Twenty five years of tulip improvement by X-rays. Pap Mich Acad Sci Arts Lett 35:9–14

De Mol Van Oud L (1953) Röntgen-bestrahlung von Tulpenzwiebeln und ihr modifizierender Einfluss auf die schon vorhandenen Organe sowie auf diejenigen, welche noch nicht geformt sind. Angew Bot 27:143–144

Doorenbos J (1954) Notes on the history of bulb breeding in the Netherlands. Euphytica 3:1–11

Eikelboom W, van Eijk JP, Peters D, van Tuyl JM (1992) Resistance to tulip breaking virus (TBV) in tulip. Acta Hortic 325:631–636

Eker I, Tekin Babac M, Koyuncu M (2014) Revision of the genus *Tulipa* L. (*Liliaceae*) in Turkey. Phytotaxa 157:001–112

EVIGEZ (2014) Evidence genetických zdrojů rostlin v ČR. Plant Genetic Resources in the Czech Republic. http://genbank.vurv.cz/genetic/resources/asp2/default_a.htm [access date=2017-06-17]

Famelaer I, Ennik E, Eikelboom W, Van Tuyl JM, Creemers-Molenaar J (1996) The initiation of callus and regeneration from callus culture of *Tulipa gesneriana*. Plant Cell Tissue Organ Cult 47:51–58

Fatel K (1997) Post-harvest selection of tulip bulbs. Acta Hortic 430:569–575

Filion WG (1974) Differential Giemsa staining in plants. I Banding patterns in three cultivars of *Tulipa*. Chromosome 49:51–60

Gaj T, Gersbach CA, Barbas CF (2013) ZFN, TALEN and CRISPR/Cas-based methods for genome engineering. Trends Biotechnol 31:397–405

Gaul H, Mittelstenscheid L (1960) Hinweise zur Herstellung von Mutationen durch ionsierende strahlen in der Pflanzenzüchtung. Zeitschrift für Pflanzenzüchtung 43:4

Govaerts R (2017) World checklist of *Liliaceae*. Facilitated by the Royal Botanic Gardens, Kew. Published on the Internet; http://apps.kew.org/wcsp/ [access date=2017-05-16]

Grabowska B, Mynett K (1970) Induction of changes in garden tulip (*Tulipa* hybr. Hort.) under the influence of gamma rays ^{60}Co. Biul Inst Hodowli Aklim Rośl 1-2:89–92. (in Polish with English Summary)

Gude H, Dijkema MHGE (1997) Somatic embryogenesis in tulip. Acta Hortic 430:275–280

Hall AD (1940) The genus *Tulipa*. The Royal Horticultural Society, London

HAVATEC (2017) http://www.havatec.com/machines/bulbstar/?lang=en [access date=2017-06-27]

Hekstra G (1968) Selectieve teelt van tulpen gebaseerd op produktie-analyse. Verslagen van de Landbouwhoogschool Onderzoek, m Wageningen, Publikatie Nummer 702:1–83

Hortus Bulborum (2017) Stichting Hortus Bulborum. http://www.hortusbulborum.nl/english [access date=2017–06-17]

Hulscher M, Krijksheld HT, Van der Linde PCG (1992) Propagation of shoots and bulb growth of tulip *in vitro*. Acta Hortic 325:441–446

Ibrahim MF, Hussain FHS, Zanoni G, Vidari G (2016) Phytochemical screening and antibacterial activity of *Tulipa systole* stapf, collected in Kurdistan region-Iraq. ZANCO J Pure Appl Sci 28:44–49

IGB (2017) Israeli GeneBank. Agricultural Research Organisation, Volcani Center, http://www.igb.agri.gov.il/catalog/index.php [access date=2017-06-18]

Jain SM, De Klerk G-J (1998) Somaclonal variation in breeding and propagation of ornamental crops. Plant Tissue Cult Biotechnol 4:63–75

Kanno A, Saeki H, Kameya T, Saedler H, Theissen G (2003) Heterotopic expression of class B floral homeotic genes supports a modified ABC model for tulip (*Tulipa gesneriana*). Plan Mol Biol 52:831–841

Karp A (1995) Somaclonal variation as a tool for crop improvement. Euphytica 85:295–302
Kawa L, De Hertogh AA (1992) Root physiology of ornamental flowering bulbs. Hortic Rev 14:57–88
Kiani M, Memariani F, Zarghami H (2012) Molecular analysis of species of *Tulipa* L. from Iran based on ISSR markers. Plant Syst Evol 298:1515–1522
Knippels PJM (2005) The contribution of quality inspections to the improvement of the quality of the Dutch flowerbulbs and access to export markets. Acta Hortic 673:79–84
Knippels PJM (2011) Recent developments in the inspection schemes of flower bulbs. Acta Hortic 886:147–151
Krens FA, Kamo K (2013) Genomic tools and prospects for new breeding techniques in flower bulb crops. Acta Hortic 974:139–147
Kroon GH, Jongerius MC (1986) Chromosome numbers of *Tulipa* species and the occurrence of hexaploidy. Euphytica 35:73–76
Kroon GH, Van Eijk JP (1977) Polyploidy in tulips (*Tulipa* L.). The occurrence of diploid gametes. Euphytica 26:63–66
Kuijpers AM, Langens-Gerrits M (1997) Propagation of tulip *in vitro*. Acta Hortic 430:321–324
Larkin PJ, Scowcroft WR (1981) Somaclonal variation – novel source of variability from cell cultures for plant improvement. Theor Appl Genet 60:197–214
Le Nard M, Ducommun C, Weber G, Dorion N, Bigot C (1987) Obsevations sur la multiplication *in vitro* de la tulipe (*Tulipa gesneriana* L.) à partir de hampes florales prélevées chez des bulbes en cours de conservation. Agronomie 7:321–329
Leeggangers HA, Nijveen H, Bigas JN, Hilhorst HWM, Immink RGH (2017) Molecular regulation of temperature-dependent floral induction in Tulipa gesneriana. Am Soc Plant Biol 173:1904–1919
Lesnaw JA, Ghabrial SA (2000) Tulip breaking: past, present, and future. Plant Dis 84:1052–1060
Małuszyński M, Kasha KJ, Forster BP, Szarejko I (2003) Doubled haploid production in crop plants: a manual. Kluwer Academic Publ., Dordrecht/Boston/London
Marasek A, Okazaki K (2007) GISH analysis of hybrids produced by interspecific hybridization between *Tulipa gesneriana* and *T. fosteriana*. Acta Hortic 743:133–137
Marasek A, Okazaki K (2008) Analysis of introgression of the *Tulipa fosteriana* genome into *Tulipa gesneriana* using GISH and FISH. Euphytica 160:270–230
Marasek A, Mizuochi H, Okazaki K (2006) The origin of Darwin hybrid tulips analyzed by flow cytometry, karyotype analyses and genomic in situ hybridization. Euphytica 151:279–290
Marasek-Ciolakowska A, Podwyszyńska M (2008) Somaclonal variation in long-term micropropagated tulips (*Tulipa gesneriana* L.) determined by FISH analysis. Floricultural and ornamental biotechnology. Global Sci Books 2:65–72
Marasek-Ciolakowska A, Ramanna MS, Van Tuyl JM (2009) Introgression breeding in genus *Tulipa* analysed by GISH. Acta Hortic 836:105–110
Marasek-Ciolakowska A, Ramanna MS, Van Tuyl JM (2011) Introgression of chromosome segments of *Tulipa forsteriana* into *T. gesneriana* detected through GISH and its implications for breeding virus resistant tulips. Acta Hortic 885:175–182
Marasek-Ciolakowska A, He H, Bijman P, Ramanna MS, Arens P, van Tuyl JM (2012) Assessment of intergenomic recombination through GISH analysis of F1, BC1 and BC2 progenies of *Tulipa gesneriana* and *T. fosteriana*. Plant Syst Evol 298:887–899
Marasek-Ciolakowska A, Xie S, Arens P, Van Tuyl JM (2014) Ploidy manipulation and introgression breeding in Darwin hybrid tulips. Euphytica 189:389–400
Marasek-Ciolakowska A, Nishikawa T, Shea DJ, Okazaki K (2018) Breeding of lilies and tulips – Interspecific hybridization and genetic background. Breeding Sci 68:25–52
Maślanka M, Bach A (2014) Induction of bulb organogenesis in *in vitro* cultures of Tarda tulip (*Tulipa tarda* Stapf.) from seed-derived explants. In Vitro Cell Dev Biol Plant 50:712–721
Matsubara H, Iba S, Oka M, Meshitsuka G (1965) Effects of gamma-irradiation on tulip. II. Effects on various stages of development. Tokyo Metrop Isot Centr Annu Rep 32:157–162
Miao Y, Zhu Z, Guo Q, Zhu Y, Yang X, Sun Y (2016) Transcriptome analysis of differentially expressed genes provides insight into stolon formation in *Tulipa edulis*. Front Plant Sci 7:409
Mizuochi H, Marasek A, Okazaki K (2007) Molecular cloning of *Tulipa fosteriana* rDNA and subsequent FISH analysis yields cytogenetic organization of 5S rDNA and 45S rDNA in *T. gesneriana* and *T. fosteriana*. Euphytica 155:235–248

Momonoi K, Shoji K, Yoshida K (2007) Cloning and characterization of ACC oxidase genes from tulip. Plant Biotechnol 24:241–246

Momonoi K, Yoshida K, Shoji M, Takahashi H, Nakaqmori C, Shoji K, Nitta A, Nishimura M (2009) A vacuolar iron transporter in tulip, TgVit1, is responsible for blue coloration in petal cells through iron accumulation. Plant J 59:437–447

Moose SP, Mumm RH (2008) Molecular plant breeding as the foundation for 21st century crop improvement. Plant Physiol 147:969–977

Moreno-Pachon NM, Leeggangers HACF, Nijven H, Severing E, Hilhorst H, Immink RGH (2016) Elucidating and mining the *Tulipa* and *Lilium* transcriptomes. Plant Mol Biol 92:249–261

MSB (2017) Millenium Seed Bank. Royal Botanic Garden Kew, UK http://apps.kew.org/seedlist/SeedlistServlet [access date=2017-06-17]

Murashige T, Skoog F (1962) A revised medium for rapid growth and bioassays with tobacco tissue cultures. Physiol Plant 15:473–497

Nezu M (1965) Study on the production of bud sports in tulip by ionizing radiation VI. Selection and observation of the mutant progeny. Jpn J Breed 15:113–118

Niimi Y (1978) Influence of low and high temperatures on the initiation and the development of a bulb primordium in isolated tulip embryos. Sci Hortic 9:61–69

Nishiuchi Y (1980) Studies on vegetative propagation of tulip. 4. Regeneration of bulblets in bulb scale segments cultured *in vitro*. J Jpn Soc Hortic Sci 49:235–239

Okazaki K, Murakami K (1992). Effects of flowering time (in forcing culture), stigma excision, and high temperature on overcoming of self incompatibility in tulip. J Jap Soc Hortic Sci 61:405–411

Okazaki K (2005) New aspects of tulip breeding: embryo culture and polyploidy. Acta Hortic 673:127–140

Okazaki K, Kurimoto K, Miyajima I, Enami A, Mizuochi H, Matsumoto Y, Ohya H (2005) Inductionof 2n pollen in tulips by arresting meiotic process with nitrous oxide gas. Euphytica 143:101–114

Okubo H, Sochacki D (2013) Botanical and horticultural aspects of major ornamental geophytes. In: Kamenetsky R, Okubo H (eds) Ornamental geophytes: from basic science to sustainable production. CRC Press, Taylor & Francis Group, Boca Raton, pp 79–121

Pavord A (1999) The tulip. Bloomsbury Publ. Plc, London

Petrová E, Faberová I (2000) Klasifikátor. Descriptor list. Genus *Tulipa* L. Výzkumný ústav okrasného zahradnictví, Prùhonice

Podwyszyńska M (2006) Improvement of bulb formation in micropropagated tulips by treatment with NAA and paclobutrazol or ancymidol. Acta Hortic 725:679–684

Podwyszyńska M (2011) *In vitro* tetraploid induction in tulip. Acta Hortic 961:391–396

Podwyszyńska M, Marasek A (1999) Somatic embryogenesis in tulip. 3rd international symposium in the series recent advances in plant biotechnology, from cells to crops, September 4–10, Stara Lesna. Slovakia Biologia 54(Suppl 7):25–26

Podwyszyńska M, Marasek A (2003) Effect of thidiazuron and paclobutrazol on regeneration potential of flower stalk explants *in vitro* and subsequent shoot multiplication. Acta Soc Bot Pol 72:181–190

Podwyszyńska M, Sochacki D (2010) Micropropagation of tulip: production of virus-free stock plants. In: Jain SM, Ochatt SJ (eds) Protocols for *In Vitro* propagation of ornamental plants, methods in molecular biology (springer protocols) 589. Humana Press/Springer, New York, pp 243–256

Podwyszyńska M, Skrzypczak C, Fatel K, Michalczuk L (2001) Study on usability of *Fusarium oxysporum* Schlecht. f. sp. *tulipae* Apt. metabolite for screening for basal rot resistance in tulip. Acta Agrobotanica 54:71–82

Podwyszyńska M, Niedoba K, Korbin M, Marasek A (2006) Somaclonal variation in micropropagated tulips determined by phenotype and DNA markers. Acta Hortic 714:211–219

Podwyszyńska M, Kuras A, Korbin M (2010a) Somaclonal variation in micropropagated tulips as a source of novel genotypes – field and molecular characteristic. Acta Hortic 855:225–231

Podwyszyńska M, Kuras A, Korbin M (2010b) ISSR evaluation of genetic stability of Polish tulip cultivars propagated in vitro. Biotechnologia 2(89):105–113. (In Polish, English abstract)

Podwyszyńska M, Novák O, Doležal K, Strnad M (2014) Endogenous cytokinin dynamics in micropropagated tulips during bulb formation process influenced by TDZ and iP pre-treatment. Plant Cell Tissue Organ Cult 119:331–346

Polder G, van der Heijden GWAM, Van Doorn J, Baltissen TAHMC (2014) Automatic detection of tulip breaking virus (TBV) in tulip fields using machine vision. Buiosyst Eng 117:35–42
Ptak A, Bach A (2007) Somatic embryogenesis in tulip (*Tulipa gesneriana* L.) flower stem cultures. In Vitro Cell Dev Biol Plant 43:35–39
Qu LW, Xing GM, Zhang YQ, Su JW, Zhao Z, Wang WD, Lei JJ (2016) Native species of the genus *Tulipa* and tulip breeding in China. XII international symposium on Flower Bulbs and Herbaceous Perennials, Kunming, China, June 28 – July 2, 2016. Book of Abstract 109
Qu L-W, Xing G-M, Chen J-J, Lei J-J, Zhang Y-Q (2017) 'Purple Jade': the first tulip cultivar released in China. Hortscience 52:465–466
Reed S (2004) Embryo rescue. Plant development and biotechnology. CRC Press, Boca Raton, pp 235–239
Rees AR (1992) Ornamental bulbs, corms and tubers. CAB International, Wallingford, p 68
Reyes AL, Prins TP, van Empel J-P, van Tuyl JM (2005) Differences in epicuticular wax layer in tulip can influence resistance to *Botrytis tulipae*. Acta Hortic 673:457–461
Rice RD, Alderson PG, Wright NA (1983) Induction of bulbing of tulip shoots *in vitro*. Sci Hortic 20:377–390
Romanow LR, van Eijk JP, Eikelboom W, van Schadewijk AR, Peters D (1991) Determining levels of resistance to *Tulip breaking virus* (TBV) in tulip (*Tulipa* L.) cultivars. Euphytica 51:273–280
Sagdic O, Ekici L, Ozturk I, Tekinay T, Polat B, Tastemur B, Bayram O, Senturk B (2013) Cytotoxic and bioactive properties of different color tulip flowers and degradatioin kinetic of tulip flower anthocyanins. Food Chem Toxicol 58:432–439
Saniewski M, Kawa-Miszczak L (1992) Hormonal control of growth and development of tulips. Acta Hortic 325:43–54
Sayama H, Moue T, Nishimura Y (1982) Cytological study in *Tulipa gesneriana* and *T. fosteriana*. Jpn J Breeding 32:26–34
Schmidt T (2015) Interspecific breeding for warm-winter tolerance in Tulipa gesneriana L. Univ of Minnesota Aquaponics Report, 1–22, http://hdl.handle.net/11299/17843
Shahin A, van Kaauwen M, Esselink D, Bargsten JW, van Tuyl JM, Visser RGF, Arens P (2012) Generation and analysis of expressed sequence tags in the extreme large genomes *Lilium* and *Tulipa*. BMC Genomics 13:640
Sharma DR, Kaur R, Kumar K (1996) Embryo rescue in plants-a review. Euphytica 89:325–337
Shoji K, Miki N, Nakajima N, Momonoi K, Kato C, Yoshida K (2007) Perianth bottom-specific blue color development in tulip cv. Murasakizuisho requires ferric ions. Plant Cell Physiol 48:243–251
Shoji K, Momonoi K, Tsuji T (2010) Alternative expression of vacuolar iron transporter and ferritin genes leads to blue/purple coloration of flowers in tulip cv. Murasakizuisho. Plant Cell Physiol 51:215–224
Smulders MJM, de Klerk GJ (2011) Epigenetics in plant tissue culture. Plant Growth Regul 63:137–146
Sochacki D (2007) Wykrywanie wirusa TBV w gatunkach i odmianach tulipanów. [Detection of *Tulip breaking virus* (TBV) in tulip species and their cultivars] [in Polish with English Abstract]. Zeszyty Problemowe Postępów Nauk Rolniczych 517:705–710
Sochacki D (2014) Kolekcje roślin ozdobnych w Instytucie Ogrodnictwa i ich wykorzystanie. Ogólnopolska Konferencja „Różnorodność biologiczna Polski a Światowy Strategiczny Plan dla Bioróżnorodności 2011–2020 nowe wyzwania i zadania dla ogrodów botanicznych oraz banków genów' [in Polish], June 30 – July 4, 2014, PAS Botanical Garden Center for Biological Diversity Conservation in Warsaw - Powsin. Abstracts 92
Sochacki D, Podwyszyńska M (2012) Virus eradication in narcissus and tulip by chemotherapy. In: Van Tuyl J, Arens P (eds) Floriculture and ornamental biotechnology, Special issue Bulbous Ornamentals, 2, vol 6. Global Science Book, Isleworth, pp 114–121
Sparnaaij LD (1991) Breeding for disease and insect resistance in flower crops. In: Harding J, Singh F, Mol JNM (eds) Genetics and breeding of ornamental species. Kluwer Academic Publishers, Dordrecht, pp 179–211
Straathof TP, Eikelboom W (1997) Tulip breeding at CPRO-DLO. Daffodil and Tulip yearbook 8. Royal Horticultural Society, London. pp 27–33
Straathof TP, Eikelboom W, Van Tuyl JM, Peters D (1997) Screening for TBV-resistance in seedling populations *Tulipa* L. Acta Hortic 430:487–494
Straathof TP, Mes J, Eikelboom W, Van Tuyl JM (2002) A greenhouse screening assay for *Botrytis tulipae* resistance in tulips. Acta Hortic 570:415–421

Tang N, Shahin A, Bijman P, Liu J, van Tuyl JM, Arens P (2013) Genetic diversity and structure in a collection of tulip cultivars assessed by SNP markers. Sci. Hort. 161:286–292

Tang N, van der Lee T, Shahin A, Holdinga M, Bijman P, Caser M, Visser RGF, van Tuyl JM, Arens P (2015) Genetic mapping of resistance to *Fusarium oxysporum* f.sp. *tulipae* in tulip. Mol Breeding 35:122

Timmer MJG (1975) The influence of the choice of planting material on tulip-bulb production. Acta Hortic 47:407–412

Trias-Blazi A, Gücel S, Özden Ö (2016) Current distribution and conservation status reassessment of the Cyprus tulip (*Tulipa cypria*: Liliaceae), new data from northern Cyprus. Plant Biosyst. https://doi.org/10.1080/11263504.2015.1174177

Turktas M, Metin ÖK, Baştuğ B, Ertuğrul F, Saraç YI, Kaya E (2013) Molecular phytogenetic analysis of *Tulipa* (Liliaceae) based on noncoding plastid and nuclear DNA sequences with an emphasis on Turkey. Bot J Linnean Soc 172:270–279

UPOV (2006) Tulip. Guidelines for the conduct of tests for distinctness, uniformity and stability. TG/115/4. International Union for the Protection of New Varieties of Plants, Geneva

Van Creij MGM (1997) Interspecific hybridization in the genus *Tulipa* L. PhD thesis, Wageningen University, The Netherlands

Van Creij MGM, Kerckhoffs DMFJ, Van Tuyl JM (1997a) Interspecific crosses in the genus *Tulipa* L.: identification of pre-fertilization barriers. Sex Plant Reprod 10:116–123

Van Creij MV, Van Went JL, Kerckhoffs DMFJ (1997b) The progamic phase, embryo and endosperm development in an intraspecific *Tulipa gesneriana* L. cross and in the incongruent interspecific cross *T. gesneriana* × *T. agenensis* DC. Sex Plant Reprod 10:241–249

Van Creij MGM, Kerckhoffs DMF, Van Tuyl JM (1999) The effect of ovule age on ovary-slice culture and ovule culture in intraspecific and interspecific crosses with *Tulipa gesneriana* L. Euphytica 108:21–28

Van Creij MGM, Kerckhoffs DMFJ, De Bruijn SM, Vreugdenhil D, Van Tuyl JM (2000) The effect of medium composition on ovary-slice culture and ovule culture in intraspecific *Tulipa gesneriana* crosses. Plant Cell Tissue Org Cult 60:61–67

Van de Wiel C, Schaart J, Niks R, Visser R (2010) Traditional plant breeding methods. Wageningen UR Plant Breeding, Wageningen, May 2010. Report 338

Van den Bulk RW, Van Tuyl JM (1997) *In vitro* induction of haploid plants from the gametophytes of lily and tulip. In: Jain SM, Sopory SK, Veilleux RE (eds) *In vitro* haploid production in higher plants, vol 5. Kluwer Academic Publishers, Dordrecht, pp 73–88

Van den Bulk RW, De Vries-Van Hulten HPJ, Custers JBM, Dons JJM (1994) Induction of embryogenesis in isolated microspores of tulip. Plant Sci 104:101–111

Van Eijk JP, Eikelboom W (1975) Criteria for early selection in tulip breeding. Acta Hortic 47:179–186

Van Eijk JP, Eikelboom W (1976) Possibilities of selection for keeping quality in tulip breeding. Euphytica 25:353–359

Van Eijk JP, Eikelboom W (1983) Breeding for resistance to *Fusarium oxysporum* f.sp. *tulipae* in tulip (*Tulipa* L.). 3. Genotypic evaluation of cultivars and effectiveness of pre-selection. Euphytica 32:505–510

Van Eijk JP, Eikelboom W (1986) Aspects of breeding for keeping quality in *Tulipa*. Acta Hortic 181:237–243

Van Eijk JP, Eikelboom W (1990) Evaluation of breeding research on resistance to *Fusarium oxysporum* in tulip. Acta Hortic 266:357–364

Van Eijk JP, Toxopeus SJ (1968) The possibilities of early selection for forcing ability and productivity in tulips. Euphytica 17:277–283

Van Eijk JP, Eikelboom W, Sparnaij LD (1977) Possibilities of selection for keeping quality in tulip breeding. 2. Euphytica 26:825–828

Van Eijk JP, Bergman BHH, Eikelboom W (1978) Breeding for resistance to *Fusarium oxysporum* f. Sp. *tulipae* in tulip (*Tulipa* L.). 1. Development and screening test for selection. Euphytica 27:441–446

Van Eijk JP, Garretsen F, Eikelboom W (1979) Breeding for resistance to *Fusarium oxysporum* f.sp. *tulipae* in tulip (*Tulipa* L.).2. Phenotypic and genotypic evaluation of cultivars. Euphytica 28:67–71

Van Eijk JP, van Raamsdonk LWD, Eikelboom W, Bino RJ (1991) Interspecific crosses between *Tulipa gesneriana* cultivars and wild *Tulipa* species: a survey. Sex Plant Reprod 4:1–5

Van der Meulen-Muisers JJM, Van Oeveren JC, Van Tuyl JM (1997) Breeding as a tool for improving postharvest quality characters of lily and tulip flowers. Acta Hortic 430:569–575

Van Raamsdonk LWD, De Vries T (1992) Biosystematic studies in *Tulipa* L. sect. *Eriostemones* (*Liliaceae*). Plant Syst Evol 179:27–41
Van Raamsdonk LWD, De Vries T (1995) Species relationships and taxonomy in *Tulipa* subg. *Tulipa* (*Liliaceae*). Plant Syst Evol 195:13–44
Van Raamsdonk LWD, Van Eijk JP, Eikelboom W (1995) Crossability analysis in subgenus Tulipa of the genus Tulipa L. Bot J Linn Soc 117:147–158
Van Raamsdonk LWD, Eikelboom W, De Vries T, Straathof TP (1997) The systematics of the genus *Tulipa* L. Acta Hortic 430:821–827
Van Scheepen J (1996) Classified list and international register of tulip names. Royal General Bulbgrowers' Association, KAVB, Hillegom
Van Tunen AJ, Eikelboom W, Angenent GC (1993) Floral organogenesis in *Tulipa*. Flowering Newslett 16:33–38
Van Tuyl JM, van Creij MG (2007) Tulip. In: Anderson NO (ed) Flower breeding and genetics. Springer, Dordrecht, pp 623–641
Van Tuyl JM, Bino RJ, Custers JBM (1990) Application of *in vitro* pollination, ovary culture, ovule culture and embryo rescue in breeding of *Lilium*, *Tulipa* and *Nerine*, In: De Jong J (ed) Integration of *in vitro* techniques in ornamental plant breeding, Proceedings of the Eucarpia Symposium, Section Ornamentals, November 10–14, 1990, Wageningen, pp 86–97
Van Tuyl JM, Arens P, Marasek-Ciolakowska A (2012) Breeding and genetics of ornamental geophytes. In: Kamenetsky R, Okubo H (eds) Ornamental geophytes: from basic science to sustainable production. CRS Press Inc, London, pp 131–158
Vertuco (2017) Vertuco, Oude Niedorp, The Netherlands. www.vertuco.nl [access date 2017-06-26]
Wilmink A, Dons JJM (1993) Selective agents and marker genes for use in transformation of monocotyledonous plants. Plant Mol Biol Rep 11:165–185
Wilmink A (1996) Genetic modification of tulip by means of particle bombardment. PhD thesis, University Nijmegen, The Netherlands
Wilmink A, Van de Ven BCE, Custers JBM, Van Tuyl JM, Eikelboom W, Dons JJM (1995a) Genetic transformation in *Tulipa* species (tulips). In: Bajaj YPS (ed) Plant protoplasts and genetic engineering VI. Springer, Berlin/Heidelberg, pp 289–298
Wilmink A., Van de Ven BCE, Custers JBM, Nollen Y, Dons JJM (1995b) Adventitious shoot formation in tulip: histological analysis and response to selective agents. Plant Sci 110: 155-164
Xing G, Qu L, Zhang Y, Xue L, Su J, Lei J (2017) Collection and evaluation of wild tulip (*Tulipa* spp.) resources in China. Genet Resour Crop Evol 64:641–652
Yamagami T, Tsutsumi K, Ishiguro M (2000) Cloning, sequencin, and expression of the tulip bulb chitinase-1 cDNA. Biosci Biotechnol Biochem 64:1394–1401
Yanagisawa R, Kuhara T, Nishikawa T, Sochacki D, Marasek-Ciołakowska A, Okazaki K (2012) Phylogenetic analysis of wild and garden tulips using sequences of chloroplast DNA. Acta Hort. 953:103–110
Yuan Y, Ma X, Tang D (2013) Isolation and expression analysis of six putative structural genes involved in anthocyanin biosynthesis in *Tulipa fosteriana*. Sci Hortic 153:93–102
Yuan Y, Ma X, Tang D, Shi Y (2014) Comparison of anthocyanin components, expression of anthocyanin biosynthetic structural genes, and *TfF3'H1* sequences between *Tulipa fosteriana* 'Albert heijn'and its reddish sport. Sci Hortic 175:16–26
Zonneveld BJM (2009) The systematic value of nuclear genome size for 'all' species of *Tulipa* L. (Liliaceae). Plant Syst Evol 281:217–245

第 29 章　温带木本观赏植物

Katrijn Van Laere, Stan C. Hokanson, Ryan Contreras, and Johan Van Huylenbroeck

摘要　木本观赏植物包括大量的属、种和品种，表现出巨大的表型多样性。新引进的木本观赏植物主要来源于自然存在的变异或开放授粉群体的选择。系统的商业育种主要发生在几个重要的属。更先进的技术，如种间杂交和多倍体化已经成功应用于扩大木本植物的改良。由于市场营销和品牌推广的日益成熟，以视觉吸引力为销售点已成为主要的育种目标之一。但真正的挑战仍然是生物（病虫害）和非生物胁迫抗性（抗寒性、耐热性、耐旱性、pH 和抗洪性）。此外，还发现了一些新的发展趋势，如不育育种、多用途育种和具有生态功能的乡土植物的选育。本章还综述了木本植物基因组学和基因工程的研究现状，以及基因组编辑技术应用于快速、精确育种的潜力。

关键词　经典育种；种间育种；多倍体化；生物与非生物抗性；分子育种

29.1　引　　言

木本观赏植物具有丰富的科、属、种和品种多样性。全球总生产面积估计为 1 069 000hm^2，相当于 77 000 个苗圃（AIPH 2015）。中国的生产面积最大，为 740 954hm^2，其次是美国（166 050hm^2）和欧洲（115 000hm^2）。总产值估计为 23 亿欧元，其中欧洲为 9 亿欧元，中国为 80.51 亿欧元，美国为 38.44 亿欧元（AIPH 2015）。国际知名的 *List of Names of Woody Ornamentals* 中，列出了欧洲和美国近 45 000 种首选的木本观赏植物的植物名与商品名（Hoffman 2016），商业种植的乔木和灌木属于 156 科 898 属（Hoffman 2016），同时列出了 9 科 62 属的针叶树种。尽管世界上商业种植和销售的木本观赏植物种类繁多，但大多数栽培植物属于不到 10%的属。

与栽培属和种的高度变异不同，木本观赏植物的商业育种仅限于少数几个重要种类。通过植物育种者权（PBR）的应用，可以发现遗传改良范围相对狭窄的证据。在欧盟（EU），植物品种可以在国家或欧盟层面上得到保护。欧盟植物品种办公室（CPVO）负责在欧盟范围内管理 PBR 系统。据估计，欧洲一半以上的受保护品种通过欧盟体系得到保护（Wegner 2013）。自 1995 年实施 PBR 系统以来，已为 160 个属的木本观赏植

K. Van Laere (✉) · J. Van Huylenbroeck
Flanders Research Institute for Agriculture, Fisheries and Food (ILVO), Plant Sciences Unit, Applied Genetics and Breeding, Melle, Belgium
e-mail: katrijn.vanlaere@ilvo.vlaanderen.be
S. C. Hokanson
University of Minnesota, Department of Horticultural Science, St. Paul, MN, USA
R. Contreras
Oregon State University, Department of Horticulture, Corvallis, OR, USA

物提出了申请（表 29.1）。共提交 7800 多份文件，其中批准 6168 份（78.8%），目前正在审查 790 份（10.1%）。在批准的 PBR 中，3737 个品种权有效，2431 个品种权已终止。当详细研究不同的属时，很明显蔷薇属（*Rosa*）是迄今为止最重要的属，其应用超过 50%，从而揭示了月季育种的影响力。分析了除月季外的其他 3764 份申请资料，其中 2704 个（71.8%）已授权，496 个（13.1%）仍在审查中，只有 15 个属的 50 多个新品种申请（包括有效的、终止的和有效的申请）（表 29.1）。这些观赏植物可分为开花灌木类（绣球、铁线莲、木槿、帚石楠、薰衣草、杜鹃花、醉鱼草）、切花和切叶生产（绣球、金丝桃）、盆栽植物（倒挂金钟、长阶花、欧石楠、马缨丹、茵芋）或木材生产（杨、柳）。除这 15 个属外，紫薇、桉、榆、冬青、六道木、李、槭、合欢和紫金牛等属中至少有 5 个属得到了充分应用，表明近年来的选择或育种工作进展明显。在针叶树中，崖柏属保护品种最多，其次是红豆杉、扁柏、冷杉、云杉和松属。一般来说，在查阅美国植物专利文件时，也可以得出类似的结论和趋势。

表 29.1 欧盟植物新品种系统排名前 100 的树木、灌木及针叶树（1995～2017 年）

属名	收到的申请总数[a]	仍在保护权利	授予和终止的权利	在公示权利[b]
蔷薇属 *Rosa*	4.065	1.722	1.742	294
绣球属 *Hydrangea*	486	241	50	115
金丝桃属 *Hypericum*	283	92	130	21
铁线莲属 *Clematis*	268	190	23	27
木槿属 *Hibiscus*	253	92	85	23
薰衣草属 *Lavandula*	182	86	38	28
帚石楠属 *Calluna*	181	130	23	19
杜鹃花属 *Rhododendron*	155	75	47	18
倒挂金钟属 *Fuchsia*	128	57	44	14
长阶花属 *Hebe*	112	66	14	14
杨属 *Populus*	81	61	12	5
柳属 *Salix*	79	51	0	2
醉鱼草属 *Buddleja*	70	40	3	17
欧石楠属 *Erica*	68	32	14	16
马缨丹属 *Lantana*	61	21	34	3
茵芋属 *Skimmia*	54	38	5	7
叶子花属 *Bougainvillea*	45	13	6	0
朱蕉属 *Cordyline*	43	21	6	1
冬青属 *Ilex*	40	23	3	7
李属 *Prunus*	34	22	4	5
小檗属 *Berberis*	33	28	0	3
槭属 *Acer*	30	20	3	5
六道木属 *Abelia*	28	9	2	7
桉属 *Eucalyptus*	28	5	0	13
榆属 *Ulmus*	27	13	4	8
萼距花属 *Cuphea*	27	12	8	1

续表

属名	收到的申请总数[a]	仍在保护权利	授予和终止的权利	在公示权利[b]
紫薇属 *Lagerstroemia*	26	11	1	14
箭竹属 *Fargesia*	26	19	1	4
崖柏属 *Thuja*	26	19	1	4
避日花属 *Phygelius*	25	11	14	0
山茱萸属 *Cornus*	24	14	4	1
锦带花属 *Weigela*	22	16	3	2
卫矛属 *Euonymus*	22	13	0	4
忍冬属 *Lonicera*	22	13	2	2
毛核木属 *Symphoricarpos*	21	13	3	5
莸属 *Caryopteris*	20	14	1	2
石楠属 *Photinia*	18	12	3	1
常春藤属 *Hedera*	18	6	7	0
风箱果属 *Physocarpus*	17	13	0	2
扁柏属 *Chamaecyparis*	17	7	4	4
麻兰属 *Phormium*	16	11	4	1
绣线菊属 *Spiraea*	16	11	4	0
蜡菊属 *Helichrysum*	15	8	3	2
臭叶木属 *Coprosma*	15	9	1	1
蒂牡花属 *Tibouchina*	15	2	6	0
木兰属 *Magnolia*	14	10	2	1
美洲茶属 *Ceanothus*	14	6	1	1
委陵菜属 *Potentilla*	13	9	1	2
女贞属 *Ligustrum*	13	7	0	1
花葵属 *Lavatera*	12	9	0	1
荚蒾属 *Viburnum*	12	7	1	2
泡桐属 *Paulownia*	12	5	3	1
南天竹属 *Nandina*	11	8	1	1
紫金牛属 *Ardisia*	11	4	1	5
苹果属 *Malus*	11	3	3	2
丁香属 *Syringa*	10	10	0	0
红豆杉属 *Taxus*	10	8	0	2
屈曲花属 *Iberis*	10	7	2	1
南非葵属 *Anisodontea*	10	5	4	0
马醉木属 *Pieris*	10	4	0	3
芍药属 *Paeonia*	10	3	2	2
瑞香属 *Daphne*	10	3	1	2
木藜芦属 *Leucothoe*	9	8	0	1
桦木属 *Betula*	9	7	2	0
大宝石楠属 *Daboecia*	9	6	0	3
云杉属 *Picea*	9	6	1	2

续表

属名	收到的申请总数[a]	仍在保护权利	授予和终止的权利	在公示权利[b]
接骨木属 *Sambucus*	9	5	0	4
木瓜海棠属 *Chaenomeles*	9	5	1	2
迷迭香属 *Rosmarinus*	9	5	1	2
西番莲属 *Passiflora*	9	5	2	1
鱼柳梅属 *Leptospermum*	9	1	7	0
墨西哥橘属 *Choisya*	8	6	1	1
黄栌属 *Cotinus*	8	6	1	1
山茶属 *Camellia*	8	3	4	1
凌霄属 *Campsis*	8	7	0	0
松属 *Pinus*	8	6	0	1
冷杉属 *Abies*	8	6	1	0
海桐属 *Pittosporum*	8	3	1	3
白珠属 *Gaultheria*	8	5	0	1
十大功劳属 *Mahonia*	7	7	0	0
连翘属 *Forsythia*	7	4	1	1
紫荆属 *Cercis*	7	4	2	0
岩蔷薇属 *Cistus*	7	4	2	0
栎属 *Quercus*	7	4	2	0
檵木属 *Loropetalum*	7	2	0	4
旋花属 *Convolvulus*	6	4	0	1
黄杨属 *Buxus*	5	5	0	0
帝王花属 *Protea*	5	3	1	1
枫香树属 *Liquidambar*	5	4	0	0
胡颓子属 *Elaeagnus*	5	3	0	1
木紫草属 *Lithodora*	5	3	1	0
合欢属 *Albizia*	5	2	0	2
南鼠刺属 *Escallonia*	4	4	0	0
白鹃梅属 *Exochorda*	4	4	0	0
分药花属 *Perovskia*	4	4	0	0
悬铃木属 *Platanus*	4	4	0	0
榛属 *Corylus*	4	3	0	1
刺柏属 *Juniperus*	4	3	0	1
火棘属 *Pyracantha*	4	3	1	0
椴属 *Tilia*	4	3	1	0
其他 60 个属[c]	139	80	14	6
合计	7.829	3.737	2.431	790

数据来源于 CPVO 公共检索数据库（截至 2017 年 8 月 1 日有效的应用和标题）

a. 申请总数（1995～2017 年），包括撤回或拒绝的文件

b. 通过技术审查仍在公示的植物育种者权

c. 对于其他 60 个木本观赏植物属，在 EU-PBR 数据库中至少有一个授权的品种，这些属归集在一起

然而，许多木本观赏植物的新品种，特别是树木，在没有植物育种者权保护的情况下被推向市场，由于每个品种生产的数量相对较低，不足以证明植物品种权应用的价值。此外，小的生产量也难以控制，因此妨碍专利费的收取。

本章介绍了温带木本观赏植物的主要育种技术、发展趋势和选择标准，重点介绍了其育种实践和最近的遗传育种研究工作，我们对蔷薇和杜鹃花的育种不做重点介绍，因为在本书的其他独立章节中有所涉及。

29.2　育 种 技 术

29.2.1　传统的鉴定和筛选

自然杂交与多倍体化和无融合生殖一直是植物进化及物种形成的基础（Soltis and Soltis 2009；Soistet et al. 2014），它们对于选育、鉴定和筛选所需的创造性品种极为重要，这种不同的再生相互作用过程产生了许多物种和品种，如栒子（Nybom and Bartish 2007；Li et al. 2017）、石楠（Lo et al. 2009）、悬钩子（Wang et al. 2013a；Clark and Jasieniuk 2012；Socher et al. 2015）、花楸（*Sorbus*）（Ludwig et al. 2013）和委陵菜（*Potentilla* sensu lato）（Morgan et al. 1994）。对欧石楠种子微观形态的研究表明存在自然杂交的证据（Fagundez 2012）。许多例子也描述了自然种间杂交，如小檗（Lubell et al. 2008）[如 *B.* × *ottawensis* 是 *B. vulgaris* × *B. thunbergii* 之间的天然杂交种（Connolly et al. 2013）]；冷杉（Isoda et al. 2000；Krajmerova et al. 2016）；桦树、毛白桦（*B. pubescens*）和矮白桦（*B. nano*）杂交（Karlsdottir et al. 2009；Anamthawat Jonsson and Thorsson 2003）；以及槭属的红枫（*A. rubrum*）和银枫（*A. saccharinum*）（Freeman 1941）之间的天然杂交常有发生，可有目的地获得更多遮阴性强和街道枫树品种（Bachtell 1988）。今天普遍种植的观赏植物中大部分来源于这种自然变异。植物最初是在野外收集并被加入文化中，在欧洲新植物的收集始于几个世纪前，但在 18～19 世纪急剧增加，那时欧洲人开始更频繁地到中国、日本和北美洲旅行（Spongberg 1990）。众所周知，一个重要的绣球品种‘Annabelle’，至今在苗木贸易中还非常流行，据报道它是 20 世纪初在美国伊利诺伊州南部森林中发现的一种天然植物（Homoya 2012）。树木园的植物开发活动对新引进植物品种的鉴定和测试发挥了主导作用（Bean 1989）。例如，野扇花品种‘Dragon Gate’来源于 1980 年在中国云南龙门发现的一种天然植物（Dirr 2011）。种质资源收集仍值得探索，植物园之间的种子交换仍然是获得新种质（如指数种子）的有效途径。为了在自然群体中检测具有独特观赏特性的植物个体，利用现代工具如航空摄影和 GPS 跟踪可以提供帮助。Richards 等（2012）利用该方法，在野外获得了具有特殊秋色和其他目标特性的巨齿红枫基因型，收集并随后无性繁殖，以供进一步评估。而今，遗传资源的交换与利用是在《生物多样性公约》（CBD）和《名古屋议定书》的规定下开展的，使得人们可以获取遗传资源以及公平公正地分享资源。

除了从野外直接引种外，许多品种来源于非人为控制的开放授粉实生苗群体。在木本观赏作物的新品种选育中，这种传统的开放授粉种子采集和播种方法仍经常采用。与切花或盆栽植物相比，木本观赏植物选择具有商业价值品种的要求更为多变。因此，只需种植开放授粉的实生苗，就可能在杂合植株群体中发现新的有价值的变异，这些群体中有趣的变异包括新颖的花朵、特殊的叶型和生长习性。ILVO 就在开放授粉的绣球实生苗群体中成功选择出一些商业品种，如绣球品种‘DVP Pinky’（商品名“Pinky Winky”）和‘ILVOBo’（商品名“BOBO”）。

在杂交率高的物种中，植物间的异倍体区块可以用来促进开放授粉杂交，这项技术利用特定的生殖生物学促进植物间的杂交。例如，挪威枫（*Acer platanoides*）和茶条槭（*A. tataricum* ssp. *ginnala*）的异倍体隔离区块（2*x* 和 4*x*）间开放授粉后分别产生 89% 和 84%的三倍体，而后让这些植株相互杂交后从四倍体中收集种子（Contreras，未发表）。此外，群体间自由杂交经常发生，生长在自然群体附近的苗木可能发生无意杂交。在美国俄勒冈州，发现生长在隔离区块中的血皮枫（*Acer griseum*）与自然生长的大叶枫（*Acer macrophyllum*）杂交，根据形态学和系统发育学很难预测一个杂交种，这些自然杂交种可用随机扩增多态性 DNA（RAPD）标记来证实（Conteras，未发表），虽然它们不具有商业价值，但它说明了在没有人为干预的情况下可由开放授粉恢复自然杂交的可能。

自然突变是另一种重要的新变异来源，很大程度上突变或芽变对新品种的形成做出重要贡献。虽然私营部门引进的植物相对较少，同时附有同行审查的说明事项，然而这种形式的植物引种可以说是有重要意义的。由于大多数木本植物是营养繁殖，其突变产生的新品种通常是稳定的。这些突变的典型例子是具有杂色叶的品种，如红枫、小檗、山茱萸、瑞香、常春藤、绣球、冬青、忍冬和海桐花等，其杂色可能是由差异基因表达、病毒或遗传嵌合体（即嵌合体）引起的（Marcotrigiano 1997）。在针叶树中，许多矮生品种是自然突变的结果，通常称为“丛枝病”（Welch 2012）。值得注意的是，在被子植物和裸子植物中，一些（可能很多）在园林植物中被视为有价值的表型变化和繁殖改良，如杂色植物和矮生植物主要是由病毒感染而非遗传原因所致。

在很多属中更多是采用系统选育，如对枣树（Asatryan and Tel Zur 2014）和梨树（Hardiman and Culley 2010）进行控制性种内杂交，播种成熟的种内杂交果实中的种子，在 F_1 群体中进行选择。在第一轮筛选之后，剩余的实生苗在上市前的几年内（取决于物种）被克隆并再次测试。选择压力是高度可变的，取决于育种者能够种植的种群数量、感兴趣性状的遗传力以及评价性状所需的观察时间。例如，木槿通常可以大量种植，并在一两年内进行评估，而枫树需要更多的空间，必须在接近 30 年的时间范围内进行评估（Hokanson，个人交流）。灌木如普通丁香（*Syringa vulgaris*）已通过种内杂交进行了密集育种，可以认为是中等时间要求，到开花需要花几年的时间，既不需要长时间的观察，也不需要大量的土地。例如，2013 年生产的普通紫丁香杂交 F_1 后代于 2017 年开始开花（Conteras，未发表）。

木本观赏植物通常是营养繁殖。然而，在基于开放授粉或控制杂交的育种计划中，播种是必要的一步（Gosling 2007），许多木本观赏植物的种子存在休眠现象，被视为一个延迟因素。种子休眠的主要类型有三种：物理、生理和形态，以及三种类型的组合。例如，形态生理休眠（MPD）的种子具有发育不全的生理休眠胚胎（Baskin 2009），在种子发芽之前，必须首先克服这些生理限制（Chien 2011）。形态生理休眠的种子需要在长时间温暖的环境下刺激根的发生，然后在长时间的低温下诱导芽的生长（Baskin et al. 2009）。据报道，荚蒾（*Viburnum*）种子具有非深度的单胚轴形态生理休眠特性（Karlsson et al. 2005；Sandiagao et al. 2015），需要在一定时间的温暖环境（15/5～25/15℃）下进行层积处理来打破胚根休眠，然后在一段时间的低温环境（5～20/10℃）下来打破芽休眠。油茶种子也具有非深度的中级生理休眠的特性（Song et al. 2017），种子被埋于容器的 0cm、1cm 和 5cm 土壤中 2 个月，结果表明，埋藏 1cm 深的种子比其他 2 个深度的种子的幼苗数量显著增加。在山茱萸（*Cornus*）种子中，由于抑制物质的存在，内果皮和胚乳负责种子休眠（Fu et al. 2013；Liu et al. 2015），打破这种生理休眠可以通过层埋和硫酸浸泡来实现。悬钩子种子忍受着深度的双重初级休眠，可以通过硫酸处理缩短种子休眠时间（Wada and Reed 2011）。无论是否在冷处理之前用 GA_3 预泡，均被认为可以缩短小檗（Belwal et al. 2015）、十大功劳（Pipinis et al. 2015）和鹅耳枥（Pipinis et al. 2012）等子代的恢复时间。最后，播种绿色种子可以缩短种子休眠期，可能是避免种子休眠的有效途径，West 等（2014）证明，日本丁香（*Syringa reticulata*）的种子在成熟前和干燥前收集，在没有层埋情况下播种，发芽率接近 90%。类似的方法已经成功地缩短了普通丁香种子的发芽时间，并在第一个生长季末产生了更多的植株。紫丁香绿籽播种是否会缩短生育期尚待观察，但它已经用于促进普通紫丁香的基于生长形态的早期筛选（Conteras，未发表）。

目前，单性生殖对于一些木本观赏植物类群的育种计划被证明是一个挑战，大约 80%的栒子（*Cotoneaster*）是单性生殖四倍体（Fryle and Hymo 2009；Sax 1954），使得它们不能作为母本。然而，细尖栒子（*C. apiculatus*）和栒子（*C. splendens*）作为父本与二倍体栒子品种 *C.* × *suecicus*‘Coral Beauty’成功杂交，但仅产生有性后代。

29.2.2　高级育种

在不同的木本观赏植物种类中，采用了胚挽救结合种间杂交、多倍体、突变等先进的育种技术，主要是在那些具有很高商业价值的属中进行的，或者在已经进行了生殖生物学特征描述的属中（表 29.2）。只有少数物种尝试了单倍体诱导、农杆菌诱导须根或原生质体融合，但尚未获得商业品种。

表 29.2　利用先进育种技术的属的概述

［2001 年以来的文献；仅报道具有植物育种者权的属（Hoffman 2016）］

属名	先进的育种技术	参考文献
	灌木和乔木	
六道木属 *Abelia*	种间杂交	Scheiber and Robacker 2003
相思属 *Acacia*	种间杂交	Kato et al. 2014，2012；Yuskianti et al. 2011
	倍性育种，多倍体化	Griffin et al. 2015；Nghiem et al. 2011；Lam et al. 2014；Harbard et al. 2012；Beck et al. 2003

续表

属名	先进的育种技术	参考文献
		灌木和乔木
槭属 *Acer*	种间杂交	Liao et al. 2010；Joung et al. 2001；McNamura et al. 2005
	种质资源鉴定，倍性及基因组大小分析	Wooster et al. 2009；Contreras，未发表
	突变育种（太空飞行）	Sun et al. 2015
	倍性育种，多倍体化	Contreras，未发表
小檗属 *Berberis*	种间杂交	Van Laere，未发表
	倍性育种，多倍体化	Lehrer et al. 2008；Rounsaville and Ranney 2010
	发根基因毛状根诱导	Brijwal and Tamta 2015
桦木属 *Betula*	种间杂交	Czernicka et al. 2014；Hoch et al. 2002
	发根基因毛状根诱导	Piispanen et al. 2003
叶子花属 *Bougainvillea*	突变育种（辐射）	Swaroop et al. 2015
醉鱼草属 *Buddleja*	种间杂交	Van Laere et al. 2011a，2010，2009b；Lindstrom et al. 2004；Elliott et al. 2004
	倍性育种，多倍体化	Dunn and Lindstrom 2007
	突变育种（化学）	Smith and Brand 2012
黄杨属 *Buxus*	种间杂交	Van Laere et al. 2015
	抗性鉴定	Shishikoff et al. 2015
帚石楠属 *Calluna*	种间杂交	Behrend et al. 2015
	倍性育种，多倍体化	Behrend et al. 2015
山茶属 *Camellia*	种间杂交	Ogino et al. 2009；Nishimoto et al. 2004，2003
	倍性育种，多倍体化	Das et al. 2013
	单倍体化	Mishra et al. 2017
	资源鉴定	Singh et al. 2013
	发根基因毛状根诱导	Rana et al. 2016；Zhang et al. 2007
美洲茶属 *Ceanothus*	种间杂交	Van Laere，未发表
	资源鉴定	Bell 2009
连香树属 *Cercidiphyllum*	性状遗传规律分析	Contreras，未发表
紫荆属 *Cercis*	种间杂交	Roberts et al. 2015
	倍性育种，多倍体化	Nadler et al. 2012
铁线莲属 *Clematis*	种间杂交	Wu et al. 2013a
	倍性育种，多倍体化	Tao et al. 2010
桤叶树属 *Clethra*	种间杂交，细胞学分析	Reed et al. 2002
山茱萸属 *Cornus*	种间杂交	Mattera et al. 2015；Shearer and Ranney 2013；Molnar and Capik 2013；Lattier et al. 2014；Morozowska et al. 2013；Wadl et al. 2007，2010；Wang et al. 2007
	多倍体育种	Contreras，未发表

续表

属名	先进的育种技术	参考文献
灌木和乔木		
榛属 *Corylus*	种间杂交	Molnar and Capik 2013；Xie et al. 2014；Mehlenbacher and Thompson 2004
	资源鉴定	Catarcione et al. 2013；Muehlbauer et al. 2014；Rovira et al. 2011；Mehlenbacher and Smith 2009；Molnar et al. 2007
	突变育种（辐射）	Dogan et al. 2007
栒子属 *Cotoneaster*	种间杂交	Contreras et al. 2014
	抗性鉴定	Rothleutner et al. 2014
	倍性育种	Contreras and Friddle 2015a；Rothleutner et al. 2016
南鼠刺属 *Escallonia*	种间杂交	Van Laere，未发表
	倍性育种，多倍体化	Denaeghel et al. 2015；Denaeghel et al. 2018
桉属 *Eucalyptus*	种间杂交	Tan et al. 2017；Weng et al. 2014；Madhibha et al. 2013；Dickinson et al. 2012；Randall et al. 2012；Toloza et al. 2008；Potts and Dungey 2004；Delaporte et al. 2001a，2001b
	资源鉴定	Listyanto et al. 2010
	发根基因毛状根诱导	Plasencia et al. 2016
连翘属 *Forsythia*	种间杂交	Shen et al. 2017
梣属 *Fraxinus*	种间杂交	He et al. 2016；Zeng et al. 2015，2014
倒挂金钟属 *Fuchsia*	种间杂交	Talluri and Murray 2014；Talluri 2012a
	倍性育种，2*n* 配子	Talluri 2012b，2011
长阶花属 *Hebe*	倍性育种，多倍体化	Gallone et al. 2014
	诱变育种（辐射）	Gallone et al. 2012
木槿属 *Hibiscus*	种间杂交	Van Laere et al. 2007，2010；Pounders and Sakhanokho 2016；Kuligowska et al. 2016a；Satya 2012；Magdalita et al. 2016
	倍性育种，多倍体化	Contreras et al. 2009；Contreras and Ruter 2009；Li and Ruter 2017
	诱变育种（辐射）	Van Laere et al. 2009a
	种间杂交	Ma'arup et al. 2012
	发根基因毛状根诱导	Christensen et al. 2009；Mercuri et al. 2010
绣球属 *Hydrangea*	种间杂交	Kudo et al. 2008；Kudo 2016；Cai et al. 2015；Granados-Mendoza et al. 2013；Kardos et al. 2009；Van Laere et al. 2008；Jones and Reed 2006；Reed 2004
	种间杂交	Wiedemann et al. 2015；Reed et al. 2008
	倍性育种，2*n* 配子	Alexander 2017；Crespel and Morel 2014
	原生质体融合	Kaestner et al. 2017
	突变育种	Greer et al. 2008
金丝桃属 *Hypericum*	倍性育种，多倍体化	Meyer et al. 2009
	种间杂交	Olsen et al. 2006
紫薇属 *Lagerstroemia*	种间杂交	Wang et al. 2010；Pounders et al. 2007
	倍性育种，多倍体化	Wang et al. 2012；Zhang et al. 2010；Ye et al. 2010
	突变育种（辐射）	Knauft and Dirr 2007

续表

属名	先进的育种技术	参考文献
	灌木和乔木	
马缨丹属 *Lantana*	种间杂交	Czarnecki and Deng 2009，2008；Czarnecki et al. 2008
女贞属 *Ligustrum*	种间杂交	Van Laere，未发表
	倍性育种，多倍体化	Fetouh et al. 2016
枫香树属 *Liquidambar*	倍性育种，多倍体化	Zhang et al. 2017a
鹅掌楸属 *Liriodendron*	种间杂交	Yao et al. 2016
忍冬属 *Lonicera*	种间杂交	Miyashita and Hoshino 2010
	倍性育种	Miyashita and Hoshino 2015；Li et al. 2011
木兰属 *Magnolia*	种间杂交，基因组大小及倍性分析	Parris et al. 2010
苹果属 *Malus*	种间杂交	Sestras et al. 2009；de Mesquita Dantas et al. 2006；Abe et al. 2011；Bisognin et al. 2009；Baumgartner et al. 2014
	属间杂交	Fischer et al. 2014
	倍性育种，多倍体化	Podwyszynska et al. 2017；Li et al. 2004；Hias et al. 2017；Zhang et al. 2015
	双单倍体	Poisson et al. 2016；Kadota et al. 2002
	种质资源鉴定	Harshman et al. 2017；Harshman and Evans 2015；Yin et al. 2013
	发根基因毛状根诱导	Wu et al. 2012；Zhu et al. 2009；Zhang et al. 2006；Sedira et al. 2005；Yamashita et al. 2004
	突变育种（辐射）	Salvi et al. 2015
芍药属 *Paeonia*	种间杂交	Hao et al. 2013；Cheng and Aoki 2008
西番莲属 *Passiflora*	种间杂交	de Melo et al. 2017，2016a，2016b，2015；Coelho et al. 2016；Ocampo et al. 2016；Soares et al. 2015；Santos et al. 2015，2012；Barbosa et al. 2007；Conceicao et al. 2011
	倍性育种，多倍体化	Rego et al. 2011
山梅花属 *Philadelphus*	种间杂交	Contreras，未发表
悬铃木属 *Platanus*	倍性育种，多倍体化	Liu et al. 2007a
杨属 *Populus*	种间杂交	Hoenicka et al. 2014；Koivuranta et al. 2012；Chauhan et al. 2004
	属间杂交	Bagniewska-Zadworna et al. 2011，2010；Ahmadi et al. 2010；Zenkteler et al. 2005
	倍性育种，2*n* 配子	Tian et al. 2015；Wang et al. 2015，2017；Xi et al. 2012；Guo et al. 2017；Zhang and Kang 2013
	双单倍体	Li et al. 2013a
	种质资源鉴定	Benetka et al. 2012
	原生质体融合	Hennig et al. 2015；Wakita et al. 2005
	发根基因毛状根诱导	Fladung et al. 2013
李属 *Prunus*	种间杂交	Donoso et al. 2015，2016；Zhang and Gu 2015；Komar-Tyomnaya 2015；Szymajda et al. 2015；Pooler and Ma 2013；Schuster et al. 2013；Salava et al. 2013；Esmenjaud et al. 2009；Liu et al. 2007b
	倍性育种，多倍体化	Contreras and Meneghelli 2016；Schulze and Contreras 2017

续表

属名	先进的育种技术	参考文献
		灌木和乔木
李属 *Prunus*	单倍体诱导	Peixe et al. 2004
	种质资源鉴定	Spak et al. 2017；Khadivi-Khub et al. 2016
	原生质体融合	Squirrell et al. 2005
栎属 *Quercus*	单倍体及双单倍体	Pintos et al. 2007，2013
茶藨子属 *Ribes*	突变育种（EMS）	Contreras and Friddle 2015b
悬钩子属 *Rubus*	种间杂交	Dai et al. 2016；Ballington and Fernandez 2008；Clark 2016；Zurawicz 2016
	种质资源鉴定	Giongo et al. 2012；Knight and Fenandez 2008
	原生质体融合	Geerts et al. 2009；Squirrell et al. 2005；Mezzetti et al. 2001
	突变（辐射，化学）	Basaran and Kepenek 2011；Ryu et al. 2017；Tezotto-Uliana et al. 2013
柳属 *Salix*	种间杂交	Fabio et al. 2017；Berlin et al. 2017；Pei et al. 2010
	属间杂交	Bagniewska-Zadworna et al. 2011，2010；Ahmadi et al. 2010；Zenkteler et al. 2005
	单倍体诱导	Wojciechowicz and Kikowska 2009
接骨木属 *Sambucus*	种间杂交	Mikulic-Petkovsek et al. 2014；Simonovik et al. 2007
野扇花属 *Sarcococca*	种间杂交	Denaeghel et al. 2017
	突变育种（EMS）	Contreras，未发表
苦参属 *Sophora*	突变育种（辐射）	Wang et al. 2017a
安息香属 *Styrax*	种质资源鉴定	Lenahan et al. 2010
丁香属 *Syringa*	基因组大小和倍性分析	Lattier and Contreras 2017b
	种间杂交	Lattier and Contreras 2017a
百里香属 *Thymus*	倍性育种，多倍体化	Tavan et al. 2015
蒂牡花属 *Tibouchina*	种间和属间杂交	Hawkins et al. 2016
榆属 *Ulmus*	种间杂交	Solla et al. 2015；Santini et al. 2008
	种质资源鉴定	Bosu et al. 2007；Pinon et al. 2005；Santini et al. 2005；Solla et al. 2005；Buiteveld et al. 2015；Kim 2003
	原生质体融合	Jones et al. 2015
越橘属 *Vaccinium*	种间杂交	Ehlenfeldt and Polashock 2014；Ballington 2009；Wenslaff and Lyrene 2003a，2003b；Vorsa and Polashock 2005；Pathirana et al. 2015；Miyashita et al. 2012
	倍性育种，多倍化	Lyrene 2011，2013，2016；Tsuda et al. 2013；Chavez and Lyrene 2009；Miyashita et al. 2009
	倍性育种，2*n* 配子	Lyrene et al. 2003
	种质资源鉴定	Ehlenfeldt et al. 2009
	突变育种（辐射）	Wang et al. 2017b
荚蒾属 *Viburnum*	种间杂交	Xie et al. 2017；Al-Niemi et al. 2012；Hoch et al. 2004
牡荆属 *Vitex*	倍性育种，多倍体化	Ari et al. 2015
	突变育种（辐射）	Ari et al. 2015

续表

属名	先进的育种技术	参考文献
		灌木和乔木
锦带花属 *Weigela* 紫藤属 *Wisteria*	种间和属间杂交，种质资源鉴定	Benetka and Hyhlikova 2010；Touchell et al. 2006；Yokoyama et al. 2002；Hokanson et al. 2015
		针叶树
冷杉属 *Abies*	种间杂交	Kormutak et al. 2013，2015
云杉属 *Picea*	种间杂交	Major et al. 2005，2015
松属 *Pinus*	种间杂交	Kormutak et al. 2017；Blada et al. 2013；Blada and Tanasie 2013；Lu et al. 2007，2005；Fernando et al. 2005
侧柏属 *Platycladus*	倍性育种，多倍体化	Contreras 2012
崖柏属 *Thuja*	倍性育种，多倍体化	Contreras 2012

29.2.2.1 种间和属间育种

种间杂交是产生新的遗传变异的主要手段之一。在木本观赏植物的几个属中，种间杂交品种已商品化（表 29.2）（Van Huylenbroeck and Van Laere 2010）。在实际的育种计划中，种间杂交是为了将感兴趣的基因从一个物种导入另一个物种。这样，一些特殊的有利性状，如野生种质的新花色和生物或非生物抗性，就可以转化为具有商业价值的性状。在受精前或受精后，种间育种可能会有一些生殖障碍。因此，有计划、循序渐进的方法对于提高成功率是必要的（Eeckhaut et al. 2006）。

种间育种计划的第一步是种质收集及种间和个体间遗传关系的测定。育种群体的基本染色体数目、倍性水平和单倍体基因组大小等信息支持有关杂交组合的直接决策，从而有力地提高杂交计划的效率（Kuligowska et al. 2016b）。在许多属中，基本染色体数目、倍性水平或基因组大小差异已被描述，如醉鱼草（*Buddleja*）（Van Laere et al. 2011a）、黄杨（*Buxus*）（Van Laere et al. 2011b，2015）、绣球（*Hydrangea*）（Van Laere et al. 2008）、木兰（*Magnolia*）（Parris et al. 2010）、野扇花（*Sarcococca*）（Denaeghel et al. 2017）、黄花蒿（Tabur et al. 2012；Pellicer et al. 2010）以及荚蒾（*Viburnum*）（Zhang et al. 2016）。对于醉鱼草，球花醉鱼草（*B. globosa*）（$2n = 2x$）只能四倍化后才能与大叶醉鱼草（*B. davidii*）（$2n = 4x$）杂交成功（Van Laere et al. 2011a）。在绣球中，只有当绣球（*H. macrophylla*）和泽八绣球（*H. serrata*）亲本具有相似的倍性水平与基因组大小时，种间杂交才能成功（Van Laere et al. 2008）。然而，在一些属中，如黄杨，异倍体间的杂交已取得成功，细叶黄杨（*Buxus harlandii*）四倍体和锦熟黄杨（*Buxus sempervirens*）二倍体杂交后产生了大量适合选择的 F_1 后代（Van Laere et al. 2015）。野扇花有 3 个组的基因组大小被鉴定：基因组大小范围在 4.11～4.20pg/2C 的二倍体（$2n = 2x = 24$），基因组大小在 7.25～9.63pg/2C 的二倍体（$2n = 2x = 24$），基因组大小在 8.17～8.33pg/2C 的四倍体（$2n = 4x = 48$）。与不同基因组大小或倍性水平的基因型间杂交相比，同一倍性水平和基因组大小的基因型间的杂交会产生更多的杂交后代。然而，在任何杂交组合中，倍性水平或基因组大小不同完全不是一个障碍（Denaeghel et al. 2017）。同时，在许多属内的种间杂交中，如小檗（Rounsaville and

Ranney 2010）、紫荆（Roberts and Werner 2016）、栒子（Rothleutner et al. 2016），即便是具有相同倍性和基因组大小的物种，属内种间杂交依然很难成功。栒子种间异倍体杂交育性得到恢复，但杂交成功率较低（Contreras and Friddle 2015a）。葡萄牙月桂树（*Prunus lusitanica*）（$2n = 8x = 64$）和桂樱（*P. laurocerasus*）（$2n = 22x = 126$）之间开展过成千上万次的杂交实验仍然失败，若能培育出葡萄牙月桂树（*P. lusitanica*）十六倍体便能使杂交变得更容易，能否解决障碍有待进一步观察（Schulze and Contreras 2017）。

除了倍性和基因组大小的相关信息外，通过分子标记（AFLP、SSR、nrITS、叶绿体 DNA 标记）对育种者收集的基因型分析所得的遗传关系信息对于选择合适的杂交组合是有价值的。这些信息不仅有助于杂交组合的设计，借助分子标记分析也有助于解决不同来源植物的命名和分类错误问题。这一点在一个育种者收集的野扇花属植物（Denaeghel et al. 2017）和黄杨属植物（Van Laere et al. 2011b）中得到了证实。在几种木本观赏植物中进行了遗传变异研究，如六道木（Landrein et al. 2017）、紫荆（Fritsch and Cruz 2012）、美洲茶（Jeong et al. 1997）、金缕梅（Li 2008）、木槿（Van Huylenbroeck et al. 2000）、锦带花和黄锦带（Kim and Kim 1999）、冬青（Cuenoud et al. 2000）、黄杨（Van Laere et al. 2011b；Thammina et al. 2017）、栒子（Li et al. 2014）、绣球（Granados Mendoza et al. 2013）、荚蒾（Clement et al. 2014）、紫薇（He et al. 2012）和铁线莲（Gardner and Hokanson 2005）。种间杂交能力随种间系统发育距离的增加而降低（Sharma 1995）。在绣球属植物中，通过整合可能的亲本之间的亲缘关系以及根据广泛的质体标记计算的遗传距离，可预测育种中最佳备选组合。杂交成功的两种绣球之间，平均遗传距离为 0.010 65，而平均遗传距离等于或大于 0.013 85 时则可能失败（Granados Mendoza et al. 2013）。在野扇花属植物中，杂交亲本之间的遗传距离并不妨碍杂交，但杂交效率较低（Denaeghel et al. 2017）。在丁香属植物中，对亲缘关系较远的物种尝试杂交并没有成功（Lattier and Contreras 2017a）。根据 Li 等（2012）的系统发育实验发现，与遗传距离较远的亲本杂交相比，种间遗传距离较近的亲本杂交获得了相当数量的种子。

此外，亲本间开花物候的差异也会引起实际问题，可以通过培养技术（如冷藏）来延迟开花，或在可加热的温室中培养结合光照同化迫使其提前开花。另一种策略是不同开花期的花粉贮藏采用不同的方法，可在干燥条件或于−4℃、−20℃和−80℃环境下冷藏贮存花粉，某些情况下甚至在液氮中冷藏。荚蒾属植物的花粉在−20℃下保存 12～16 周，萌发率也能保持约 10%（Xie et al. 2017）。在木犀科植物中，茉莉和 *Nycanthes* 的花粉在−60℃下贮藏 48 周后，花粉发芽率分别达到 40.6%和 37%（Perveen and Sanwar 2011）。一般来说，在大多数被子植物中发现的双细胞花粉（Brewbaker 1967）比三细胞花粉寿命长（Hanna and Towill 1995）。

受精成功与否取决于花粉质量和花粉与雌蕊的相互作用。花粉管与胚珠之间的串扰问题可能是杂交失败的原因（Dresselhaus and Franklin-Tong 2013）。更多地了解花粉管引诱剂的诱导机制和识别机制，有助于找到更有效的克服种间杂交障碍的方法（Dresselhaus and Franklin-Tong 2013；Higashiyama and Takeuchi 2015）。测定花粉活力和萌发能力，通过花柱观察花粉生长情况，可以检测到受精前障碍。Xie 等（2017）对地中海荚蒾与荚蒾种间杂交的育性障碍进行了详细研究。不同种间花柱长度和花粉管长度的差异表明可能存在受精前障碍。然而，苯胺蓝染色显示花粉管在授粉后 2～7 天到达子房。在授粉前，植物生长素在柱头富集也说明其对地中海荚蒾和荚蒾杂交结实有一定的促进作用（Xie et al. 2017）。此外，还测定了红锦葵（*H. coccineus*）、戟叶木槿（*H. laevis*）和芙蓉葵（*H. moscheutos*）不同木槿品种间的杂交障碍，通过对花粉管生长的分析，发现花粉管生长受到抑制且生长异常，但这种受精前障碍并没有造成杂交亲本之间的完全生殖隔离（Kuligowska et al. 2016a）。Soaris 等（2013）分析了西番莲不同发育阶段（花期前、花期和花期后）的花粉萌发情况，结果表明，花期是收获花粉用于杂交的最佳时期，因为此时花粉活力和萌发率最高，花粉活力也因物种不同而异。此外，Soaris 等（2015）研究了 11 个西番莲品种的花粉与雌蕊相互作用，发现自交不亲和品种的花粉管抑制是由于胼胝质的不规则沉积，而自交亲和品种的花粉管抑制是由于胼胝质的规则沉积。在茉莉属植物中，6 个互交组合的研究表明花粉活力和柱头易授性随花瓣类型（单瓣、双瓣或多瓣）与花期的不同而不同（Deng et al. 2017a）。尽管花粉粒在母体柱头上正常萌发，但花粉管在雌蕊中受阻，不能到达子房，因此，胚囊仍未受精，这些结果表明，茉莉属植物杂交受到受精前障碍的影响（Deng et al. 2017a）。在一些木本观赏属中，正在开展造成花粉育性降低的减数分裂时花粉母细胞异常的研究。在野扇花属植物中，在柳叶野扇花和云南野扇花中观察到减数分裂异常，出现桥状、迟滞和退化现象。这些导致了具有微核的二元体、三元体和四元体的形成。由于减数分裂异常，花粉育性降低，产生了不同大小的花粉粒（Saggo et al. 2011）。在铁线莲中也发现了由减数分裂异常导致的花粉畸形（Singhal and Kaur 2011；Kumar et al. 2010）。

种间杂交经常出现受精后障碍，主要是由胚乳发育失败导致胚胎败育、杂种败育或缺乏生长活力（Sharma et al. 1996；Orr and Presgraves 2000）。为了克服体外受精后障碍，可以进行胚拯救，这项技术已在几个木本观赏植物属中获得成功。在木槿属植物授粉 11 周后进行离体胚拯救，提高了木槿和华木槿杂交后其胚胎的萌发能力。然而，杂叶色和完全白化造成了许多形成的实生苗丢失（Van Laere et al. 2007）。戟叶木槿（*H. laevis*）、芙蓉葵（*H. moscheutos*）、红锦葵（*H. coccineus*）品种间杂交种表现出无活力、褪绿、坏死、生长发育迟缓和白化现象，从而限制了杂交种的生产（Kuligowska et al. 2016a）。在醉鱼草属植物中，通过离体胚珠培养可获得种间杂种，成功率在很大程度上取决于杂交组合，且观察到单侧不一致性（Van Laere et al. 2009b）。大花六道木（*Abeliax grandiflora* ‘Francis Mason’）和（*A. schumanii*）的种间杂交种子未能萌发。然而，在授粉后的第 4、5、6 周，从瘦果中分离出的胚珠经过培养，胚胎得以恢复，种间杂种得以再生（Scheiber and Robacker 2003）。在一个金雀儿属植物的种间育种计划中，为了克服合子后不亲和胚胎败育，进行了离体胚拯救（Le Gloanic et al. 2015）。共进行了 13 个种间杂交组合，其中，金雀儿、大花金雀儿和红花金儿（*Cytisus* × *boskoopii*）作为母本，紫色金雀儿（*Cytisus purpureus*）为父本。在授粉后 4～6 天收集未成熟果实并进行胚珠的离体培养，胚珠在 1～9 个月后萌发，共获得 38 株 F_1 代幼苗，每个杂交组合中至少有一株。在绣球属植物中，中国绣球（*H. scandens* ssp. *chinensis*）和绣球（*H. macrophylla*）异花授粉后进行胚珠培养产生了种间杂交植株（Kudo et al. 2008；Kudo 2016）。八仙花（*H. macrophylla*）和乔木绣球（*H. arborescens*）种间杂交后进行胚珠培养也会产生种间杂种

（Cai et al. 2015）。此外，通过胚胎培养，获得一个常山（*Dichroa febrifuga*）和八仙花（*H. macrophylla*）的属间杂种（Reed et al. 2008）。其他种间杂交后进行胚拯救的例子可以在荚蒾属（*Viburnum*）（Hoch et al. 1995）、李属（*Prunus*）（Sundouri et al. 2014；Szymajda et al. 2015；Liu et al. 2007b）中发现，属间杂交在柳属和杨属植物（*Salix* × *Populus*）间也有发现（Bagniewska-Zadworna et al. 2011）。

通过形态学描述可以检测所获后代的杂种特性。分析欧洲白蜡（*Fraxinus excelsior*）和狭叶白蜡（*Fraxinus angustifolia*）杂交得到的 F_1 杂种性状，发现大多数表现为中间形态，但由于亲本效应而表现出差异（Thomass et al. 2011）。在石榴属植物中，Yazici 和 Sahin（2016）对来自不同杂交组合的 67 个杂交基因型的 21 个形态特征进行了分析，并将它们与其亲本和具有重要商业价值的石榴属基因型的植株及果实性状进行了比较。与其亲本植株相比，约 95%的木槿（*Hibiscus syriacus*）和庐山木芙蓉（*H. paramutabilis*）杂交种以及 50%的木槿（*H. syriacus*）和华木槿（*H. sinosyriacus*）杂交种具有中间型的花与叶形态（Van Laere et al. 2007）。美国正在为山茱萸属（*Cornus*）开展一项广泛的杂交计划，部分种间 F_1 代杂种表现出较高的观赏品质和抗病性（Mattera et al. 2015）。流式细胞术可用于验证异倍体间杂交种或双亲间基因组大小存在显著差异时的杂交种性质，如黄杨属（*Buxus*）（Van Laere et al. 2015）和野扇花属（*Sarcococca*）（Denaeghel et al. 2017）植物。然而，通过基于聚合酶链反应（PCR）的 DNA 标记技术，如利用 AFLP 和微卫星标记进行的杂交验证是最可靠的，AFLP 对于评估杂交种的性质是有价值的，因为不需要之前的序列信息（Meudt and Clarke 2007），对基因组信息知之甚少的大多数木本植物是有益的。分子标记在许多木本观赏属中已被证实用于杂交验证，如榛属（*Corylus*）（Sathuvalli and Mehlenbacher 2012）、梣属（*Fraxinus*）（Thomasset et al. 2011）、苹果属（*Malus*）和梨属（*Pyrus*）的杂交种（Fischer et al. 2014）、*Passiflora*（Santos et al. 2012）、野扇花属（*Sarcococca*）（Denaeghel et al. 2017）、黄杨属（*Buxus*）（Van Laere et al. 2015）、木槿属（*Hibiscus*）（Van Laere et al. 2007）以及山茱萸属（*Cornus*）（Wang et al. 2007）。在 AFLP 检测的野扇花属杂交子代里，18%的子代中含有 10%～25%的独特雄性亲本标记（Denaeghel et al. 2017）。在黄杨属（Van Laere et al. 2015）和绣球属（Van Laere 2008）植物中也观察到其部分杂种中出现类似情况。部分杂种出现这种情况的假设有两种：一种是只有少数雄性染色体替代雌性染色体的代换系，另一种是增加少数雄性染色体的附加系。种间受精后，两个不同的亲本基因组在一个细胞核内结合。这种新的基因组构成可能导致基因组间的冲突，从而导致遗传和表观遗传的重组与改变，如亲本缺失和染色体消失（Riddle and Birchler 2003；Han et al. 2003）。因此，可以结合分子标记和细胞遗传学技术，如基因组原位杂交（GISH），对这些杂交种进行检测，以监测是否存在外来染色体，阐明两个基因组之间的重组和基因组重排，并证明单亲染色体的消失。尽管 GISH 对于木本植物来说是非常具有挑战性的，因为它们通常具有小的基因组和/或许多小尺寸的染色体，但是对一些木本植物属进行了细胞遗传学鉴定。利用 GISH 在 F_1 和 BC1 植株中明确区分了西番莲亚种与毛西番莲的基因组，从而揭示了重组染色体起源的差异。BC1 杂种中重组染色体数量和性质的差异表明，前体基因组之间具有高度同源性，但也表明 F_1 植物减数分裂重组存在差异（de Melo et al. 2017）。用 GISH 方法对山莓源性染色体和黑莓源性染色体进行了异源多倍体杂交鉴定（Lim et al. 1998a），并对醉鱼草属和木槿属植物 F_1 与 F_2 幼苗群体的杂种特性进行了验证（Van Laere et al. 2010）。在黄钟花属植物中，Contreras 等（2012）用 GISH 和形态学比较的方法确定了肖黄钟花与黄钟花的 F_1 杂种，同时用荧光原位杂交（FISH）记录了 18S DNA 区的 4 个拷贝。

29.2.2.2　多倍体化

许多木本观赏物种使用抗有丝分裂剂使染色体倍增（表 29.2）（Dhooghe et al. 2011）。进行有丝分裂多倍体化实验的主要目的是①诱导新变异（形态学和生理学），②克服物种间由倍性差异造成的杂交障碍，③产生三倍体，④恢复 F_1 杂种的育性，以便将其用于进一步育种。

为了增加南鼠刺属植物的形态变异，在杨梅（*E. rubra*）和玫瑰糠疹（*E. rosea*）上进行了体外多倍体化。可以有效地产生四倍体（Denaegel et al. 2017），并且获得的四倍体表现出变异的表型（Denaegel et al. 2018）。在相思树中，多倍体化实验的目的是将多样性引入育种种群并改善木纤维（Griffin et al. 2015）。据报道，二倍体和四倍体海棠属（*Malus* × *domestica*）叶片的形态学差异明显（Hias et al. 2017；Xue et al. 2015）。据报道，株型紧凑的多倍体植物有醉鱼草（Rose et al. 2000）、木槿（Contreras et al. 2009）和悬铃木（Liu et al. 2007）。除了形态变化和生理变化外，其他性状也发生了变异，如抗逆性和开花（如更长的开花期）。四倍体柑橘幼苗比二倍体基因型更耐盐胁迫。这种变异可能与 $4x$ 基因型的蒸腾速率降低有关（Ruiz et al. 2016）。在忍冬（*Lonicera japonica*）中，四倍体对水分胁迫的抵抗力高于二倍体，与二倍体相比，复水后它们恢复得更快（Li et al. 2009a）。结合生理和分子数据，发现多倍体化提高了海棠的抗旱性（Zhang et al. 2015）。

为了减少葡萄牙月桂树（*Prunus lusitanica*）（$2n = 8x = 64$）和桂樱（*Prunus laurocerasus*）（$2n = 22x = 176$）之间的倍性差异，并期望成功杂交，已对葡萄牙月桂树进行了染色体加倍；最近，开发出了十六倍体葡萄牙月桂树。表型鉴定发现，与未经处理的对照相比，染色体加倍的植物显示出较短的茎、较厚较少的叶子以及较大的保卫细胞（Schulze and Contreras 2017）。然而，目前还没有开花，整个杂交还没有完成。为了将黄色花引入商业化的流苏醉鱼草（*Buddleja davidii*）（$2n = 4x = 76$）品种中，诱导形成了四倍体（*B. globosa*）（$2n = 2x = 38$），随后与流苏醉鱼草成功杂交（Van Laere et al. 2011a）。*B. globosa* 的染色体加倍后花粉萌发能力和种间杂交成功率提高了。F_1 幼苗的植物形态表现出杂种特性，并且存在来自 *B. globosa* 的淡黄色花色。然而，没有一个杂种后代表现得足够好以作为商业品种。在醉鱼草属（*Buddleja*）的另一项研究中，优化了生产可育四倍体形式的杂种浆果醉鱼草（*Buddleja madagascariensis*）× 皱叶醉鱼草（*B. crispa*）的方案，以使橙色花色、叶毛和不裂果特性渗入 *B. davidii* 中。杂交后代产生了四倍体植株，并在多倍体诱导后观察到育性的显著增加。随后，新形成的四倍体与 *B. davidii* 品种之间杂交产生了可育植物（Dunn and Lindstrom 2007）。将种子浸入秋水仙碱水溶液后，可以得到 *Vaccinium darrowii* 四倍体。经秋水仙碱处理的 *V. darrowii*（$4x$）植物和 *V. corymbosum*（$4x$）之间很容易杂交。在这些杂交中，与来自二倍体 *V. darrowii* 的花粉相比，使用四倍体 *V. darrowii* 的花粉产生的植株在坐果率、产量及种子数量方面均有所提高（Chavez and Lyrene 2009）。

诱导四倍体植物也可能是由二倍体和染色体加倍植物杂交产生不育三倍体栽培品种的一个步骤，如锦带花（*Weigela*）（Arène et al. 2007）、帚石楠（*Calluna*）（Behrend et al. 2015）和木槿（*Hibiscus*）（Van Laere et al. 2006；Li and Ruter 2017）就是成功的例子，并在相思树属中尝试过（Harbard et al. 2012）。通过将秋水仙碱溶液滴到新出现的幼苗上可实现女贞加倍，四倍体女贞植物生长和大小、枝条生长、叶片形态以及插条生根表现出显著变化（Fetouh et al. 2016）。四倍体有望在女贞属植物的选择或新品种的选育中发挥重要作用。在栒子属（*Cotoneaster*）中，使用抗火疫病的四倍体 *C. spledens*（$2n = 4x = 68$）作为花粉亲本和一种易感火疫病的二倍体 *C.* × *suecicus*‘Coral Beauty’（$2n = 2x = 34$）杂交（Contreras et al. 2014）。结果显示三倍体杂种对火疫病具有抗性（Contreras and Friddle 2015a），并正在对其生育能力进行筛选。

在育种过程中使用 $2n$ 配子，一种自然的多倍体产生方法，有助于克服种间杂交中的倍性问题。木本观赏植物中有几个例子可以说明未减数配子是多倍体的来源。在一些 *H. syriacus* × *Paramuta bilis* 的 F_1 和 F_2 杂种中，存在未减数的花粉或卵细胞（Van Laere et al. 2009a）。在醉鱼草属（*Buddleja*）中，由 *B. davidii*（$2n = 4x = 76$）和 *B. lindleyana*（$2n = 2x = 38$）杂交获得的杂种有 76 条染色体，这表明 *B. lindleyana* 产生了未减数配子（Van Laere et al. 2011a）。Elliott 等（2004）也报道了类似的观察结果。此外，所描述的 *Buddleja* × *weyeriana*（$2n = 4x = 76$）（Van de Weyer 1920），一个起源于 *B. davidii*（$2n = 4x = 76$）× *B. globosa*（$2n = 2x = 28$）之间杂交 F_2，表明 *Buddleja* 中存在未减数配子。在越橘中，四倍体 *V. corymbosum* L.和二倍体 *V. darrowii* 之间的大多数四倍体杂交种是利用 *V. darrowii* 产生 $2n$ 配子的倾向与强大的三倍体障碍产生的，这大大减少了三倍体杂交种的数量（Chavez and Lyrene 2009）。三倍体紫丁香品种‘President Grévy’（$2n = 3x = 69$）× 二倍体‘Sensation’（$2n = 2x = 46$）杂交产生了一个杂交种，其基因组大于三倍体，但与四倍体不一致（Lattier and Contreras 2017b）。对花粉亲本的进一步研究表明，‘Sensation’产生了 8.5%的未减数配子。在绣球中，存在二倍体和三倍体品种，有证据表明三倍体的植株和花更大。二倍体品种 *H. macrophylla*‘Trophee’先前被证明具有双峰花粉量分布，这可能表明有未减数配子存在（Alexander 2017）。因此，*H. macrophylla*‘Trophee’在与其他一系列二倍体 *H. macrophylla* 品种杂交时被用作亲本。在以‘Trophee’作为雄性亲本的后代中，F_1 代中 94%是三倍体，‘Trophee’的双峰花粉量分布反映了未减数雄性配子存在。F_1 代中三倍体的花序比二倍体更少、更宽。与二倍体相比，三倍体的茎增粗了 16%，其中比二倍体全部数和半数的茎粗 20%（Alexander 2017）。在绣球及其亚种 F_1 代杂种中也观察到了未减数配子的存在（Crespel and Morel 2014）。在杨树中，短期高温可诱导 $2n$ 卵子形成（Wang et al. 2017c；Guo et al. 2017）。通过与未减数配子杂交和杂合子胚胎染色体加倍诱导的多倍体结合了多倍体及杂种的优势，是体现育种优势的两种主要方法。在杨树中，通过高温处理可以诱导胚囊和合子胚染色体加倍，以产生三倍体和四倍体（Guo et al. 2017）。同时，建立了花芽形态特征（授粉后时间）与雌性减数分裂期（胚囊和合子胚发育）的关系，以指导确定花芽处理期。分别在 38℃和 41℃下处理胚囊与合子胚被认为是诱导染色体加倍的最佳条件。此外，细胞遗传学分析表明，授粉后 66～72h（HAP）是染色体加倍的最佳时间，这一时期的特征是单核胚囊和双核胚囊比例较高。在合子胚染色体加倍中，168HAP 为最佳处理时间，此阶段二细胞和四细胞原胚的比例较高（Guo et al. 2017）。

29.2.2.3 突变育种

由于木本植物的营养周期长，突变对于木本植物的育种尤其有利。在木本观赏植物中发现并商业化了叶色与矮化突变体（Arène et al. 2007）。在马缨丹属（*Lantana depressa*）中，已经上市了两种黄色和白色突变体（Datta 1995）。另外，经诱变处理，选育出了红花鲜艳的 *Weigela* 品种（Duron 1992）。一种突变莸属（*Caryopteris*）植物也已商业化（Arène et al. 2007）。用 100～400Gy 剂量的钴（60）源 γ 射线处理荆条（*Vitex*）种子（Ari et al. 2015），种子萌发的 LD_{50} 为 55Gy，幼苗存活的 LD_{50} 为 41.3Gy。在 50Gy 的辐射剂量下，获得了理想的单茎植株类型。突变育种对于荆条（*Vitex*）具有巨大的潜力，可以产生用于景观美化的植物所需的变异（Ari et al. 2015）。通过化学诱变产生了醉鱼草（*Buddleja davidii*）的花斑品种‘Summer Skies’（Smith and Brand 2012）。为了获得这种突变，用化学诱变剂甲基磺酸乙酯（EMS）处理了不同花色的醉鱼草（*Buddleja davidii*）植物种子。‘Summer Skies’作为商业品种已经公布，该品种具有花斑，在其他植物开花之前即春季和初夏具有很强的观赏效果（Smith and Brand 2012）。*Ribes sanguineum* 种子暴露在电磁辐射中，产生了品种‘Oregon Snowflake’，其株型紧凑、分枝密集和叶片多裂（Contreras and Friddle 2015b）。采用 0～1.2%的 EMS 处理马尾松种子，随着剂量的增加，发芽率降低，LD_{50} 约为 0.6%（Contreras，未发表）。存活植株中，0、0.4%和 0.8%处理后株高没有差异，但用 1.2%时株高降低（Contreras，未发表）。

29.2.2.4 先进的体外育种技术

其他先进的育种技术并不常用于木本观赏植物。由原生质体融合产生杂交种的例子见于蔷薇属（*Rosa*）、悬钩子属（*Rusus*）和李属（*Prunus*）（Squirrell et al. 2005）以及柑橘属（*Citrus*）和枸橘属（*Poncirus*）之间（Cheng et al. 2007）。对于绣球来说，体细胞杂交也是获得新的育种基础材料的有力工具（Kaestner et al. 2017）。活的叶肉原生质体从 *H. macrophylla*、*H. paniculata*、*H. arborescens*、*H. quercifolia* 和 *H. febrifuga* 品种中分离得到。在通过聚乙二醇进行电操纵和融合之后，诱导细胞分裂。然而，芽的诱导和植株的再生只能从 *H. macrophylla*‘Schneeball’和 *H. macrophylla*‘Nachtigall’的愈伤组织中获得（Kaestner et al. 2017）。由原生质体融合产生的四倍体毛白杨（*Populus tremula* × *P. tremuloides*）杂交种，与二倍体相比，高度、茎生物量和总叶面积相当或减少（Hennig et al. 2015）。这些四倍体杂交种的耐旱性提高，使它们成为干旱地区短轮伐期生物量种植物种有意义的候选者（Hennig et al. 2015）。在白杨、白桦和桤木之间也尝试了家族间细胞融合（Wakita et al. 2005）。在 *P. alba* 和 *B. platyphylla* 之间进行细胞融合后，获得了两个芽，并且在 *P. alba* 和 *A. firma* 之间融合后再生了 12 株。在这 12 株中，有 7 株具有不同于白杨的锯齿状叶片（Wakita et al. 2005）。

木本观赏作物单倍体（加倍）的研究尚不多见。仅发表了一些研究，如在苹果（Höfer

2003)、杨（Andersen 2003)、栎（Bueno and Manzanera 2003）和柑橘（Germana 2003）以及其他一些植物（Maluszynski et al. 2003)。单倍体在柑橘育种中最有趣的应用之一是通过单倍体和二倍体原生质体融合获得三倍体体细胞杂种。利用流式细胞术对 94 种柑橘再生株系进行倍性分析，结果显示，其中多达 82%是三单倍体，而不是预期的单倍体或双单倍体（Germana et al. 2005)。因此，在培养物中进行再生被认为是在柑橘中获得新三倍体品种的一种快速而有吸引力的方法，这对于现在需要无核水果的水果市场来说可能是非常有吸引力的。同样，在李属植物中，经花药离体培养后产生了纯合子胚（Cimo et al. 2017)。在苹果中，来源花粉的基因型可通过花药离体培养获得（Vanwynsberghe and Keulemans 2004)。然而，在这里，效率很低，在田间，这些纯合基因型表现出不同的生长和生育能力，妨碍了它们在育种项目中的应用。Thammina 等（2011）从胚乳中再生卫矛（*Euonymus alatus* ‘Compactus’）的三倍体，是开发繁殖力降低的三倍体的快速方法。

植物的变异也可以通过插入自然存在的土壤发根农杆菌的 *rol* 基因来诱导。这种细菌通过插入 *rol* 基因的自然、稳定转化引起感染部位毛状根的生长（Christey 2001；Tepfer 1984)。由这些转化根再生的植株 Ri 表型通常具有株高降低、高度分枝的地上部分、花大小改变和生根能力提高的特征（Christey 2001；Tepfer 1984)。不同 Ri 系间表型的差异与 *rol* 基因和内源植物基因的相互作用以及插入基因的拷贝数有关。因此，发根农杆菌创造的有价值新品种不作为转基因品种来评判。在木槿（*Hibiscus rosa-sinensis*）中，以叶段为起始材料，根的转化效率最高（Christensen et al. 2009)，但未实现芽再生。在小檗（Brijwal and Tamta 2015）中也建立了有效的毛状根诱导方案。转基因材料鉴定是通过分析 *rolA* 和 *rolB* 存在于毛状根与转化愈伤组织中。苹果作为一种广泛应用的砧木，成功地与发根农杆菌共培养（Wu et al. 2012)。毛状根由 37%的感染外植体产生。培养 4 周可以观察到芽再生。毛状根再生剂具有较高的生根能力，导致茎短、叶片黄化、皱褶、丛生等形态变异（Wu et al. 2012)。Tsuro 和 Ikedo（2011）利用发根细菌转化了宽窄叶杂交薰衣草 *Lavandula* × *intermedia*，结果发现一半以上的植株在初夏时死于田间，这可能是由于其对季节变化的敏感性增加，而存活下来的植株表现出矮小和广泛的侧枝。与未转化的植物相比，花延迟一个月，并且表现出花梗短（Tsuro and Ikedo 2011)。垂桦（*Betula pendula*）发根细菌转化系（Ri）表现出生长缓慢、株型短及紧缩、叶片较小及较大的根系（Piispanen et al. 2003)。另外，补血草（*Limonium*）的转化植株表现出开花早，花密度高，植株紧凑，叶面积减少，根系发达，花较小且多（Mercuri et al. 2001)。同时，发现了超紧凑、紧凑和半紧凑类型。然而，矮小的程度不能用 T-DNA 拷贝数来解释，因为每种类型都有两个拷贝插入（Mercuri et al. 2003)。

29.2.2.5　种质和品种评估

与现有品种相比，越来越多的新木本观赏植物引入市场，重点转移到了“新颖性”上，而不是提高性能。越来越多的新品种被引进，但是收集和分享这些新品种数据的努力是有限的（Widrechner 2007)。美国建立了几个评价苗圃作物的长期项目。这些实验园，无论是公共的还是私人的，都提供了关于不同气候带和土壤类型中植物品种实际表现的基本信息。NC-7 区域观赏植物实验是美国运行时间最长的景观植物评价网络之一，在 18 个州设有 30 个实验站点（NC-7 2000）。其目的是在广泛的环境和气候极端条件下评估植物的性能，扩大有用的景观植物的范围，同时强调详细的长期（10 年）评估数据。用于实验的植物可以通过植物开发、种子或其他繁殖体交换获得，也可以通过对引进该植物感兴趣的其他机构或育种者直接捐赠获得。收集和报道的数据包括存活率、冬季伤害、观察到的疾病和昆虫问题、观赏品质、场地条件和提供维护。在美国南部地区，一个名为 SERA27：苗圃作物和景观系统的项目或多或少有着相同的目标（SERA27 2017）。芝加哥植物园（Chicago Botanic Garden）等公共花园多年来一直在进行灌木和树木的实验。一些公共花园参与多机构植物种质收集和实验项目，如 NACPEC（Dosmann and Del Tredici 2003）和 PCN（APGA 2018）。这些项目的目的是研究和推动地区苗圃中植物的利用。

在欧洲国家，一些植物实验已经进行了一个多世纪，以告知种植者、批发商、零售商和消费者观赏植物的价值。在荷兰，皇家博斯科园艺学会（Houtman 2013）已经在 19 世纪组织了此类实验。德国和英国等其他国家也有自己的实验系统。2002 年，几个国家决定加强其力量，并设立了欧元区实验（Houtman 2013）。这些工厂实验室分别设在 8 个国家：荷兰、德国、法国、英国、奥地利、比利时、芬兰和爱尔兰。每个国家都有一个独立的机构负责种植和协调这些植物的评估。通过在广阔的地理区域种植和评估植物，可以评估植物的性能和对当地气候的适应性。这些植物是在第二年到第四年之间定级。评审团由种植者、商人和园丁组成；评审团每年举行几次会议。根据当地的喜好，评估标准各不相同，但开花、耐寒和抗病性是一些最重要的特征。关于绣球属（*Hydrangea*）、醉鱼草属（*Buddleja*）、锦带花属（*Weigela*）、长春花属（*Vinca*）及木槿属（*Hibiscus syriacus*）的实验在过去的 10 年中已经建立（Houtman 2013）。

对于榛子（hazelnut），收集了欧洲不同种质的信息（Rovira et al. 2011）。汇编了欧洲 13 个不同藏馆中约 290 种榛子种质的清单，并对其形态和遗传特性进行了评价。另一个例子是悬钩子属（*Rubus*）（Giongo et al. 2012），多年来，在意大利北部对 250 多份悬钩子材料进行了基因型鉴定，以进行品种比较和改进育种方案。研究关键性状与形态、开花和园艺性状以及抗病性。这种种质的筛选为新鲜加工市场育种计划提供了选择，推动了育种市场的发展（Giongo et al. 2012）。

29.3 育 种 目 标

木本观赏植物的育种目标是多样的，高度依赖于特定的属和种。一般来说，植物形态特征如花颜色、形状、花期延长、早开花、重复开花、花瓣、叶（颜色、杂色、形状、亮度）、植物习性和分枝（紧密生长、地被、柱状、球状、非吸盘）、果实（颜色、形状、无核，或者没有水果）是最传统的选择标准。在过去的 10 年中，我们注意到视觉吸引力已经成为零售销售的一个重要标准。零售与品牌推广和市场营销项目的增加齐头并进。除了这些标准外，抗病虫害和非生物胁迫抗性（抗寒性、耐热性、抗旱性、耐盐性、耐高 pH 和抗洪性）也是育种家面临的主要挑战。对于某些属来说，不育性和木材质量是重要的特征。新趋势也正在出现：多用途植物育种，如具有可食用果实的引人入胜的园林植物，可能用作切花，以及繁殖和选择具有生态系统功能的本地植物，如提供野生动植物栖息地、食物、减缓地表径流和水土流失、提供温和的极端温度和/或二氧化碳封存等，仅举几例。一些重要的育种目标和挑战描述如下。

29.3.1 病虫害抗性育种

病虫害是木本植物产业面临的重大挑战。全球贸易的增加导致外来病虫害在以前不存在的地区出现或移动。此外，由于气候条件的变化，病虫或病原在一年内能够完成更多的生殖周期，可能会成为一个突发的问题。其中一些病虫害可能会产生巨大的影响，甚至根除某些物种的商业生产。木本观赏植物的育种面临着巨大的挑战，尤其是考虑到它们的生命周期长、基因组复杂以及缺乏遗传和基因组信息。在开始育种计划之前，在自然种群中检测可能的抗源是很重要的。人工接种方法可以加快幼苗群体的选择。分子标记的应用，特别是在树木上的应用，可以缩短育种周期，有利于新种质的有效选择。

一个历史上著名的例子是荷兰榆病（Dutch elm），它是由 *Ophiostoma ulmi*（Buisman）和后来的 *Ophiostoma novo-ulmi*（Brasier）引起的。这种疾病在 20 世纪蔓延到欧洲和北美洲，杀死了大部分的天然榆树种群以及种植的街道和景观榆树。大型和长期的选育项目始于 1930 年左右的荷兰和美国，至今仍在继续。最近，在意大利、西班牙和法国（Mittempergher and Santini 2004）建立了其他欧洲榆树育种项目。目前市场上有相当数量的抗病品种，这为榆树在不久的将来再次被用作街道树木提供了很好的机会。此外，不同的育种计划产生的这些抗性品种引入了广泛不同的遗传背景，这是积极的方面，表现为新形式的疾病对病原菌抗性的稳定性（Buiteveld et al. 2015）。在杨树中，由杨栅锈菌引起的叶锈病是最具破坏性、广泛传播的病原菌，可侵染商业上重要的杨树品种（Steenackers et al. 1996）。它可能导致年生长量减少 50%以上，连续几年的反复感染会导致杨树人工林的完全丧失。尽管杂交杨树在培育抗病品种方面付出了相当大的努力，但由于新的锈菌不断发展，人们普遍观察到抗性的迅速衰退。因此，主要挑战是选择具有持久的抗锈能力，以防锈菌由多种多样且不断变化的种群组成（Pinon and Frey 2005）。丁香假单胞菌（*Pseudomonas syringae*）在欧洲七叶树引起的欧洲七叶树出血病（*Aesculus hippocastanum*）是另一种正在出现的破坏性疾病，最早于 2002～2003 年发现。该病影响了欧洲西北部多个国家的欧洲七叶树，对景观造成了严重破坏。初步筛选结果表明，马蹄栗自然种群中存在一定的抗性（Pánková et al. 2015）。对于黄杨来说，在过去的 10 年中，真菌盒状枯萎病（*Calonectria pseudonaviculata*）已经成为欧洲和美国的主要疾病，造成了巨大的经济损失。黄杨树种之间的抗病性存在差异（Henricot et al. 2008；Gehesquière 2014），开始了一项旨在选择抗病品种的种间育种计划（Van Laere et al. 2015）。在控制条件下，通过人工接种对获得的幼苗群体进行抗病性筛选，随后在自然感染压力下进行田间试验。今天，第一批耐药基因型正在大规模评估和繁殖，预计到 2019 年将推向市场。栒子属植物易受火疫病侵害。为了确定潜在的抗性来源，采用人工接种的方法，将不同分类群的科托紫菀与欧文氏菌（*Erwinia amylovora*）进行了比较。疾病筛选揭示了物种间易感性的巨大差异，筛选结果被用于设计杂交组合（Contreras et al. 2014）。在山茱萸中，山茱萸炭疽病（*Discula destructive*）是美国佛罗里达州野生山茱萸种群数量减少的原因之一。美国罗格斯大学（Rutgers University）进行了密集的种间育种和选育，培育出了几个抗性品种（Molnar and Cpik 2013）。花园绣球可被叶斑病破坏。对大叶绣球白粉病的发生进行了显微镜观察（Li et al. 2009b）。结果表明，绣球属植物对白粉病的抗性存在变异，说明常规育种可以成功地培育出抗白粉病的品种（Windham et al. 2011）。在筛选绣球对白粉病的抗性及建立测定方法时，了解其抗性机制具有重要的参考价值。欧洲产水曲柳（*Fraxinus*）的一个普遍问题是由病原真菌 *Hymenoscyphus fraxineus* 引起的灰枯病。对来自优良水曲柳的全同胞和半同胞子代进行人工接种鉴定，受控接种的结果表明，与易感克隆的后代相比，低易感克隆的后代发生了较小的坏死。自然感染导致的顶部损伤也明显减少。亲本与后代树冠损伤/坏死发生的相关性证实了遗传抗性的存在，并表明基于控制接种的生物测定法有可能成为一种快速、经济有效的方法，用于评估育种计划中的枯萎病的抗性机制研究（Lobo et al. 2015）。针对苹果赤星病（*Erwinia amylovora*），正在进行苹果抗性育种。在该项目中，分子标记被用于筛选抗性基因型（Cusin et al. 2017）。还有更多的例子描述了木本观赏植物对新病害的抗性，这些新病害在木本观赏植物中出现并引起了巨大的问题，如具有非常广泛的寄主的 *Phytophthora ramorum*（Davidson et al. 2003）、*Ceratocystis platani*（*Pertanus* 中的溃疡斑）（Pilotti et al. 2012）、槭属（*Acer*）的叶斑病和松树（*Pinus*）的泡锈病等（Sniezko and Koch 2017）。

29.3.2 非生物抗性

29.3.2.1 耐寒性

冬季的低温事件是温带气候区自然系统中木本植物物种分布的主要决定因素之一（Larcher 2005）。可以说，在温带地区成功培养木本园林植物品种也是如此。木本植物对寒冷的反应大致可分为三个阶段：①适应，随着时间的推移，对环境刺激（包括较低的温度和减少的日光）的适应性增加；②仲冬抗寒性，即植物被完全驯化并达到其最大抗寒性的过程；③去适应，这是一个相对较快的过程，在该过程中，植物会因温度升高和日照增长而失去抗寒性。植物的基因表达和生理变化会使其适应环境（Levitt 1980；Xin and Browse 2000；Wisniewski et al. 2003）。已知冷积累是数量性状（Guy 1990；Lim et al. 1998b；Pellett 1998）；然而，与该反应相关的基因数目尚不清楚（Bassett and Wisniewski 2009；Fowler and Thomashow 2002；Hannah et al. 2005）。此外，抗寒性还包括一系列形态和结构特征（何时在细胞中形成冰）：细胞的生化特性（蛋白质、碳水化合物、激素）以及环境、肥力、水分、光照、气候对植物的影响（Wisniewski and Gusta 2014）。此外，同一植物内的不同组织对所有这些环境的反应可能不同（Gusta and Wisniewski 2013）。因此，对植物抗寒性的研究需要考虑所讨论植物的特殊形态、生理、物候、生命历史特征和环境要求。测试必须设计为尽可能接近植物将经历寒冷事件的实际生长环境。

从历史上看，植物对冬季寒冷的适应很大程度上被认为是一种植物及时的能力，以及它对特定地区冬季极端低温的反应（Sakai 1966；Sakai and Larcher 1987）。最近，面对全球气候变化引发的环境变化，植物科学家开始考虑其对植物的影响（Gu et al. 2008）。这些变化之一是适应的植物所经历的温度波动（特别是高于冰点），以及其对植物抗寒性的影响。抗寒性的丧失称为去适应，在随后的低温天气中抗寒性恢复称为再适应（Kalberrer et al. 2006）。木本植物会产生两种去适应形式："主动"和"被动"。植物快速主动适应发生在环境温度＞5℃的大幅度增加时，伴随着许多生理变化和生长的恢复。当驯化后的植物感受到低于 5℃的温度升高，且不伴有大规模变化时，被动适应发生（Kalberrer et al. 2006）。植物对温度变化的响应表现出基因型特性。尽管大多数木本植物类群可以重新适应，但重新适应的速率和程度是可变的。此外，重复的去适应/再适应周期似乎限制了最终适应潜力的实现（Kalberer et al. 2006）。

由于木本植物的生长速度较慢且性状复杂，木本植物的耐寒性育种和测试显得较为困难。木本植物抗寒性的测定有两种方法，第一种方法是在室外寒冷的环境中越冬种植植物，并在次年春天检查它们的再生情况和受损率。这些抗寒性的田间试验通常提供植物整体抗寒性的可靠评估，特别是当试验包括已知抗寒性值的标准基因型时。在田间，植物要承受所有可能影响其耐寒性的因素。实地评估的一大缺点是，虽然可以检测和量化冷害，但无法确定冷害是在植物完全适应之前的冬季早期发生，还是在植物完全适应之后的冬季后期发生，或是在冬季寒冷的高峰期发生。因此，评估抗寒性的另一种方法是在实验室中将整个植物或植物部分暴露在冷冻温度下。整个植物的实验室冷冻试验通常使用小型植物，进行一夜的冷冻循环，但根部不受冰冻处理的影响（Hokanson and McNamara 2013）。经过冷冻试验的植物缓慢地解冻、种植，然后在第二年春天生长出来，以评估每一株植物受到的伤害程度。植物部分的实验室冷冻试验一般采用在水浴中受到冷冻温度的组织的电解质泄漏情况而定（Wilner 1959，1960）或在可编程冷冻器中对花芽或血管/形成层组织进行死亡或坏死检查（Hokanson 2010）。实验室冷冻试验比田间试验有一个优势，可以确定在冬季的什么时候以及在什么温度下发生了损伤。此类试验通常在整个冬季定期采用已知标准进行，随着温度下降，每隔一段时间从冰箱中取出重复样品（McNamara and Hokanson 2010）。这种实验室冷冻试验检测的结果通常被理解为相对抗寒性。然而，实验室试验通常被发现高估了植物材料的抗寒性。假定在自然条件下，植物在更长的时间范围内同时经历多种胁迫和冷胁迫，这导致植物在相关野外环境中的耐寒性低于在实验室环境中测试的耐寒性（Strimbeck et al. 2015）。重要的是，当直接比较不同实验室的评估结果时，通常会得出相似的低温耐受性估算值（Burr et al. 1990）。

植物年龄是另一个已经发现但还不完全清楚的对抗寒性有影响的变量。已经注意到，与同种的老植物相比，木本造景植物种的幼苗在更高的温度下会遭受冷害（McNamara and Pellett 2000）。同时已经证明，年龄对耐寒性的影响因物种而异（Cochran and Bernsten 1973；Brown and Bixby 1976；Hummel 1981；Hubackova 1994）。这一现象减缓了寒冷气候育种计划中的育种过程，因为幼苗群体在萌发后的第三年或第四年才被种植到田间进行观察和选择（Hokanson，个人观察）。McNamara 和 Pellett（2000）对一棵 35 岁的黄檗（*Phellodendron amurense*）进行了茎抗逆性试验，并与 10、22 和 34 月龄的幼苗进行了比较。10 月龄的幼苗在 10 月以后没有进行驯化，直到第四个冬天，幼苗才达到成年树的驯化水平。在抗寒园林植物育种计划中，通常采用田间和实验室抗寒性试验进行种质评价。在明尼苏达大学，已经测试了各种分类单元，包括连翘（McNamara and Pellett 1993a）、山梨（McNamara and Pellett 1994a）、蟹苹果（McNamara and Pellett 1995）、锦带花（McNamara and Pellett 1998；McNamara and Hokanson 2010a）、亚洲枫树（McNamara and Hokanson 2010b）、榛（Aiello et al. 2014）和黄杨（Hokanson et al. 2016）以及选择亲本进行育种，育种计划中，行业和消费者对抗寒性的评价来决定品种的选择。该计划自启动以来，已将田间和实验室抗寒性筛选数据纳入其品种释放数据中，如枫树（McNamara and Pellet 1993b；McNamara et al. 2005）和紫藤（Hokanson et al. 2015）。加拿大农业和农业食品部的研究人员对加拿大各地（Z2-5）的种质、栽培品种以及包括崖柏（Rioux et al. 2000；Richer and Rioux 2002）、蟹苹果（Richer et al. 2003a）、绣线菊（Richer et al. 2004）、委陵菜（Richer et al. 2004b）、杜松（Rioux et al. 2004b）在内的加拿大育种计划中的新品种以及各种本地种质（Richer et al. 2003b）进行了广泛的耐寒性测试。

29.3.2.2　耐涝性

无论是由于极端水文事件（USGS 2016），还是由于建筑导致径流在植物周围积聚，或仅仅是由于植物位置不佳，木质景观植物都可能受到景观洪水的影响。例如，大约16%的美国陆地遭受涝灾（Boyer 1982）。根区淹水导致完全（缺氧）或部分缺氧（低氧）。在这些情况下，二氧化碳和乙烯浓度迅速增加，伴随着其他有毒化合物的增加，导致根系系统死亡，最终导致整个植物死亡（Paulle et al. 2010）。在人为设计的景观中有防洪植物是比较理想的，有关农作物与浸水土壤之间关系的许多知识是农作物研究的结果。但是，也有关于天然存在的草本和林木物种种群的研究（Mancusco and Shabala 2010）。人们注意到，利用分类内和种群之间的遗传多样性可以筛选到适应洪水的植物。通常与耐洪性相关的形态和/或生理特征包括肥大的皮孔，根或茎组织中可形成不定根和气孔。本质上，遗传方面似乎是多基因控制，并且已经报道了一些QTL研究（Parelle et al. 2010）。未来的工作在很大程度上取决于与更大群体规模和开发高通量表型鉴定，以准确地响应表征。

关于木本园林植物，耐涝植物的研究主要集中在易受洪水影响或湿地生境上的植物。对木本植物的抗洪能力进行了对照试验。Nash 和 Graves（1993）评估了5种原产于河底、河岸环境的容器栽培分类群[‘Franksred’ red maple（*Acer rubrum* L.‘Franksred’，商标名为 Red Sunset）、甜瓜木兰 *Magnolia virginiana* L.、黑元宝兰 *Nysa sylvatica* Marsh.、秃柏树 *Taxodium distichum* (L.)和木瓜 *Asimina triloba* (L.) Dunal.]遭受水淹和中、重度干旱处理后的发育与形态。作者得出的结论是，甜海湾木兰和秃柏耐受以干旱和洪涝为特征的种植区。将 *Alnus maritima* subsp. *maritima*、*A. maritime* subsp. *georgiensis* 和 *A. maritima* subsp. *oklahomensis*、在容器生长的幼苗、在北美洲东部水分饱和土壤中自然生长的桤木，与其他4种桤木进行了比较，分别是 *A. glutinosa* (L.) Gaertn.、*A. serrulata* (Ait.) Wild.、*A. nitida* (Spach) Endl.和 *A. nepalensis* D. Don，以了解它们对不同土壤湿度水平的反应（Schrader and Graves 2002，2005）。结果表明，在水分饱和土壤上 *A. maritima* 是所有物种中表现最好的，并且在干旱条件下的生长要比除 *A. nepalensis* 以外的其他物种更好。而 subsp. *georgiensis* 是所有亚种中最适应干旱的（Schrader et al. 2005），subsp. *oklahomensis* 可很好地适应干旱，从观赏角度上更可取（Schrader and Graves 2000）。这些结论最终导致决定释放 *A. maritima* 亚种。subsp. *oklahomensis* 被称为“九月的太阳”（Graves and Schrader 2004）。Stewart 等（2007）测定了容器土壤水分对自然发生在湿地边缘的三种植物 *Calycanthus occidentalis*、*Fraxinus anomala* 和 *Pinckneya pubens* 叶片气体交换和碳分配模式的影响。他们得出结论，*Fraxinus anomala* 和 *Pinckneya pubens* 能适应相当宽的土壤水分条件，而在其自然分布范围内出现在湿地边缘的 *Calycanthus occidentalis* 不太耐受。Peterson 和 Graves（2013）比较了三级土壤淹水和三级干旱条件下容器栽培的 *Sambucus* sp. L.与 *Ptelea* sp. 幼苗的相对生长速率（RTR）、光合作用和生物量增加。这两个类群的西部物种对洪水的耐受性都不如中西部物种，后者准确地反映了它们的原生栖息地条件。两种 *Ptelea* 植物都在完全淹水时死亡。然而，*Sambucus nigra* spp. *canadensis* 对完全淹水的耐受性比其西部同类强得多，是唯一一个能在洪水易发地区茁壮成长的物种。两种 *Ptelea* 对干旱的反应相似，都表现出中等抗性。Iwanaga 等（2015）通过将 2 年生容器育苗的幼树完全淹没 4～24 周和 104～164 周，评估了三种在湿地栖息的木本物种，即日本桤木、水杉和落羽杉（落羽杉）。虽然完全淹没并不能反映“典型”的洪水事件，但其结果是有益的，*M. glyptostroboides* 和 *A. japonica* 在淹水条件下存活时间均不超过 8 周。然而，*T. distichum* 在水下存活了 2 年多。值得注意的是，尽管它们来自相似的河岸洪泛区，但它们对洪水的反应大不相同。

29.3.2.3　耐旱性

缺水最常发生在干旱或半干旱地区。但即使在非干旱地区，高效用水的重要性也日益凸显。气候变化模型预测显示，干旱的发生将遍及全球，包括现在湿润的生境（Dai 2011）。随着需水量持续增加，景观用水将受到限制。因此，不久的将来，对干旱敏感的、原本生活在湿润环境中的植物可能会出现缺水现象。同样，城市地区也要种植那些在贫瘠、恶劣环境（如高温、缺水、高盐）中生长的树木或灌木。植物可以采取多种策略来应对干旱胁迫以维持正常生长或保证在干旱时期生存。植物可以通过气孔对蒸腾速率和叶水势调控以及扩展根系从深层土壤中吸收水分来实现避旱或耐旱机制（Bsoul et al. 2006；Lemcoff et al. 2002）。已有研究表明植物应对干旱的机制存在种间和种内差异（Gindaba et al. 2004；Lemcoff et al. 2002；Merchant et al. 2007）。Leksungnoen（2012）发现，在缺水的情况下，桉树（*Eucalyptus*）和大叶枫（*Acer macrophyllum*）植物保持水分吸收能力并正常生长，直到没有水可吸收为止。相反，从外观看，大齿枫（*A. grandidentatum*）组织中仍保存着水分，但在干旱情况下不会进一步生长。Sjörman 等（2015）通过测量 27 个枫树因膨压失水时的叶水势（Ψ_{P0}）研究了其耐旱性基因型，发现了季节性渗透调节与夏季 Ψ_{P0} 之间存在极显著的相关性，表明渗透调节是枫树夏季 Ψ_{P0} 的驱动力。这些数据证实了枫树对水缺乏的耐受范围很广，并显示了将该技术用作筛选新型和传统植物材料抗旱性的工具的可能性。筛选树木或灌木以提高抗旱性需要同时考虑实施干旱胁迫的方法和用于评估干旱响应的标准。由于降雨量常发生变化，在野外研究树木的耐旱性困难重重，在可控的环境中，基于易测量的特征（如生存和生长），底应力筛查对相当大数量不同基因型的初步评估是有效的（Cregg 2004）。叶绿素荧光测量也是评价木本观赏植物耐旱性的一种快速可靠的技术（Percival and Sheriffs 2002）。

除了生理性适应措施，相应的环境条件和栽培技术（如缺水灌溉、适宜基质的应用等）还有助于生产经过干旱胁迫预处理的、健壮的苗木。此外，这些措施可能会加速原生植物的利用，因为它们能适应不利的环境条件：白天的炎热和夜晚的低温、干旱和盐碱化（Franco et al. 2006）。

29.3.2.4 pH 耐受性提高

根据从土壤浆液中检测到的 $H_3O^+_{aq}$ 的量，可以将土壤分类为酸性、中性或碱性。大多数土壤的 pH 在 3～10，pH 小于 3.5（超酸性）或大于 9（强碱性）是不常见的。大多数园林植物在 pH 范围为 6.0～7.0 时生长最佳。但通过进化，有些植物在 pH 超出此范围也可以正常生长。土壤 pH 是决定景观植物种植能否成功的关键因素，因为 pH 通常决定土壤中养分的吸收形式。矿质元素在土壤中的存在形式会影响植物对它们的吸收速度。虽然土壤的 pH 很大程度上取决于矿物质的组成，但认为经营活动也会改变 pH。在建造的景观中 pH 改变的一个主要因素是包含硅酸盐或碳酸盐的建筑材料发生风化作用，如煤渣砌块、砖混凝土和岩石覆盖物，这些物质的风化会导致土壤 pH 升高（DeSutter et al. 2011）。另一个影响因素涉及施工过程中土壤的运动。例如，通过挖掘去除更多的表层有机土（通常 pH 较低）以及将深层土壤（通常 pH 较高）翻到地面。

前面提到的两个过程都会提高土壤 pH。但很多植物对土壤 pH 升高不能很好适应，经典的例子包括红枫（*Acer rubrum* L.），在土壤 pH 高于 5.8 时，会出现由 pH 升高引起的褪绿病（Altland 2006）；桦木（*Betula* spp.）在土壤 pH 高于 6.5 时会出现缺铁黄化（Zhang and Zwiazek 2016a，2016b）；以及针叶栎（*Quercus palustris* Muench.）在土壤 pH 高于 7.0 时，会出现由 pH 升高引起的铁绿化（Neely 1973）。另一个经典的例子是杜鹃花属（*Rhododendron*），本章对此进行了详细介绍（Krebs）。

建造的景观中，除杜鹃花外，很多植物都不能适应土壤 pH 升高的情况，但目前针对木本景观植物 pH 耐受性的育种很少见到报道。目前的研究主要集中在盆栽植物无土培养基质 pH 的控制上。但是，正如 Krebs（本书的单独章节）提到的，这种 pH 操纵研究是在“黑匣子”中进行的。在 pH 升高时，植物是否会对碳酸氢根离子浓度升高和/或低养分利用率做出反应？

Peterson 和 Graves（2009）对美国 *Dirca palustris* L.三个种的幼苗进行了评价，这些幼苗生长在添加了硫酸或试剂级碳酸钙的土壤里，其 pH 分为 5 个梯度：4.5、5.3、5.9、6.5 和 7.3，幼苗对这 5 种 pH 处理的生长响应不是通过这些土壤的 pH 来预测的。研究者发现 *D. palustris* 在酸性处理中生长更好，且在其他无土培养基的研究中也发现了这种现象（Handreck and Black 2002；Symonds et al. 2001）。在矿物质上生长的植物通常在较高的 pH 条件下生长更好，这是因为这类土壤具有更大的阳离子交换能力，并且可以提供更多的微量营养素。

除了上面提到的无土培养基的 pH 问题外，在测定全株高 pH 耐受性时还会出现其他问题。为培养足够大的植株，需要几年时间进行筛选。此外，植物对升高的 pH 的反应可能难以测量。褪绿病很细微，很难辨别，通常在新枝生长过程中表现得最为明显。最典型的，对升高的 pH 处理的响应的检测本质上是破坏性的，涉及在处理期间测量根和茎生物量的积累。为了突破这些限制，Susk 等（2018）利用体外比色测定法来测量落叶杜鹃花根际酸化的能力。该方法基于以下事实：双子叶植物采用策略 1 铁获取机制，该机制是将质子排出根际以降低局部 pH 并创造有利于铁溶解和吸收的条件（Brady and Weil 2004；Briat and Lobréaux 1997）。该方法是指将酚磺酞（酚红）掺入培养基中，并在装有培养基的试管中种植植株。随着介质溶液 pH 由 7.5 逐渐变为 5.5，酚红由红色变为黄色。每个试管在播种时和播种后 3 周分别拍照。使用 MATLAB 脚本分析了每张图片的 2cm × 2cm 区域，以确定种植后的色相、饱和度变化。分别记录根际酸化能力有显著差异的植株。Susko 等（2018）为提高 pH 耐受表型的问题而进行了另一尝试，使用图像分析来测量上述体外研究测试的相同植株的温室种植幼苗的叶面积变化。幼苗在装有无土培养基的塑料容器中生长，该培养基用干重比分别为 0、10%或 20%的粉末状 $CaCO_3$ 处理，以使最终生长培养基的 pH 达到 5.7、7.0 和 7.5。在播种时和播种后 4 周对幼苗拍照，以评估施石灰处理对幼苗叶面积的影响。使用 MATLAB 脚本检测与绿叶表面积相等的 pH 和色相测量值。比较每种幼苗在种植之前和种植后 4 周之间绿叶表面积的变化。$CaCO_3$ 含量增加导致叶面积减少。值得一提的是，植株的平均体外颜色变化率与温室中测得的平均绿叶表面积呈正相关（$R^2 = 0.49$）。

29.3.3　育性

在许多国家，外来（非本地）木本观赏物种被广泛生产与销售。其中一些物种可以建立自给自足的扩散种群，这可能会使它们成为入侵物种并产生巨大的不良生态影响。植物入侵物种清单是由国家、地区甚至是地市制定的。常见的几种重要经济木本观赏植物，正成为入侵物种或具有入侵潜力，如小檗属、醉鱼草属、槭属、梨属、卫矛属等。在某些管辖区，已经制定了禁止出售和使用这类植物的法规。由于种子是快速传播的主要途径，因此培育不育品种是防止入侵的最有效手段之一。此外，植物果实发育缩短了开花期，也因此降低了许多植物属的观赏价值。因此，对木槿属芙蓉葵（Li and Ruter 2017）和木槿（Van Laere et al. 2006）进行了倍性水平的育性育种。通过四倍体芙蓉葵与二倍体木槿杂交获得三倍体。获得的三倍体花期更长，而花比二倍体小，对疫霉的耐受性更高。三倍体花只产生无活力的花粉粒，授粉后果实败育，最终导致三倍体不育（Li and Ruter 2017）。Contreras 等（2009）发现利用异源四倍体红叶槿进行多倍体化培育出一种异同源四倍体，可以使育性降为零，这可能是由减数分裂过程中多价态的形成导致的。多倍体化后育性降低水平存在很大差异，并受到诸如原始植物的染色体数及其基因组特性等因素的影响。在木槿中，可以从秋水仙碱染色体倍增苗与现有品种回交中选育不育株（Van Laere et al. 2006）。木槿中‘DVP azurri’（商品名“Azurri”或“Azurri Blue Satin”）、‘ILVOPS’（商品名“Purple Satin”）和‘ILVO37’（商品名“Orchid Satin”）就是通过该育种程序选育和销售的三种无性系六倍体品种。这些植株的花较大，没有种子，花期延长。利用类似的方法，从二倍体和染色体加倍植株杂交后代中获得了锦带花不育三倍体栽培种（Arène et al. 2007）。此策略也被用于贯叶连翘（*Hypericum androsaemum*）并获得了不育三倍体植株（Trueblood et al. 2010）。在挪威枫（*Acer platanoides*）育种中，也开始了不育三倍体培育工作。挪威枫是在美国不同地区被确定为入侵物种的枫树物种之一。氨磺乐灵（oryzalin）处理后，获得稳定的四倍体植物，之后与二倍体回交（Contreras 2017）。为了促进杂交，将四倍体与二倍体 *Acer platanoides* 和 *A. ginnala* 间作种植并进行杂交，分别获得了 89%和 84%的三倍体。在栒子属（*Cotoneaster*）中，用四倍体（$2n = 4x = 68$）金缕梅作为父本提供花粉和一个二倍体（$2n = 2x = 34$）品种 *C.* × *suecicus*‘珊瑚美人’（‘Coral Beauty’）作为母本杂交产生了不育的三倍体杂种（Contreras et al. 2014）。马缨丹属的许多品种用于商业化生产和景观使用。然而，马缨丹已在佛罗里达州和其他地方被列为一类入侵物种。因此，一个开发不育三倍体马缨丹品种的育种项目被启动（Deng et al. 2017b）。最终选育并释放了 2 个可能的三倍体品系作为新不育马缨丹品种。另一个例子是澳大利亚金合欢，其作为外来物种被广泛种植并且在某些情况下成为入侵物种。如果能够开发出不育的三倍体种植，这一影响可以减小。Harbardt 等（2012）报道一个旨在通过二倍体杂交产生四倍体品系的育种项目正在启动，并获得了具有更重、更厚、更宽以及杯状叶柄的四倍体（Harbardt et al. 2012 ）。

虽说三倍体育性显著降低，但这不是绝对的，很多三倍体的育性存在很大变异。例如，Phillips 等（2016）培育了许多梨的三倍体，与可育的二倍体相比，这些三倍体的育性变异为 0～33.6%，这表明在引种前需要对其育性进行系统的评价。Ramsey 和 Schemske（1998）综述分析了 26 项三倍体研究，这些研究表明三倍体产生的非整倍体花粉最多，但也能产生单倍体（3%）、二倍体（2%）和三倍体（5%）花粉。最有趣的是，在这 26 项研究中，同源三倍体平均育性为 39%，异源三倍体的平均育性为 24%，两者均比非整倍体的育性要高（Ramsey and Schemske 1998）。大部分三倍体至少是半可育的并且在发表前需在单个植株上进行验证。

除加倍外，还可以通过种间杂交来获得不育植株。由于减数分裂过程中父母本染色体配对不正确，杂种可能是不育的。包括山茶（Ackerman and Dermen 1972）和紫薇（Pounders et al. 2007）在内的几个属的木本观赏植物的种间杂种不育已经报道。醉鱼草，被公认为是一种入侵物种，可通过种间杂交培育不育植株。近几年陆续有一些商业品种，如‘Blue Chip’（Werner and Snelling 2009）、‘ILVOargus1’（商品名“Argus White”或“Inspired White”）和‘ILVOargus2’（商品名“Argus Violet”或“Inspired Violet”）。后两种是通过 *Buddleja davidii*（$4x$）和 *B. lindleyana*（$2x$）进行种间杂交产生的（Van Laere et al. 2009b）。

利用化学（如甲基磺酸乙酯）或物理方法（如 γ 射线辐射）进行非定向诱变是育种者降低许多木本植物育性的第三个方法（Lapins 1983）。虽然这种方法可以降低木本景观植物的育性如樱桃（Lapins 1975），但鲜少见到相关报道。其他研究发现，γ-辐射在其他属中也能明显降低育性，如醉鱼草、李（J. Ruter，个人交流）、柑橘和松属（W. Hanna，个人交流）。Smith 和 Noyszewski（待发表）将茶条槭 *Acer ginnala* 与日本小檗 *Berberis thunbergii* 的种子暴露于各种 γ 射线及快速中子辐射下，而将这两个物种的无根软木插枝暴露于较低的 γ 射线辐射范围内。他们发现，大多数茶条槭在 6 年后开始开花，但是有些植物在 8 年后还没有开花。他们的假设是非定向诱变可能破坏了诱导开花的基因并使这些突变体变得不育。作者正在继续观察这些潜在的可用于育性选择的多克隆位点。另外，这项工作中有一些日本小檗（*B. thunbergii*）已经开花，但未能结出可育的种子。作者报道这些突变体正常开花并产生核果，但经检查仅含有异常种子（Smith and Noyszewski，待发表）。

29.3.4　多用途植物

在培育一些木本植物新品种时常常可以达到多重目标，如适合切叶或切花的园林植物，或者是能结果的园林植物。例如，被选作切花或切叶植物的绣球属、金丝桃属、冬青属、毛核木属（*Symphoricarpos*）和紫珠属（*Callicarpa*）。几种常绿植物或秋天色泽漂亮的植物在秋天以切叶植物出售，而春季的灌木会被迫开花分枝，如连翘属植物和欧洲绣球。在某些情况下，如绣球和金丝桃，以切花为目的，开发了完全独立的育种程序。

最近，人们对“可食用景观植物”的兴趣增加，出现了许多交叉植物。在蓝莓（*Vaccinium* sp.）中，具有观赏价值性状的育种多年来一直受人青睐。基于植物的生长习性、颜色、吸引人的浆果颜色、较大的果实等符合家庭园艺审美情趣的特征，培育出了与观赏工业相关的蓝莓品种，如‘Summer Sunset’（NeSmith and Ehlenfeldt 2011）。另一个例子是蓝莓‘Perpetua’品种，具有复壮性和紧凑性，并具有良好的水果品质，这是由商业水果生产商依据育种计划初次培育的标准蓝莓品种。该计划体现了杂交品种的重要性（Finn et al. 2015）。在树莓等物种中还可以找到专门培育具有观赏价值的小果实的其他例子，如矮化品种‘矮蛋糕’（Stephens et al. 2016）。在苹果中，已发现具有柱状生长习性的突变树。这开辟了培育多用途树木的可能性，既可以被业余爱好者取果，也可以为私人或公共花园提供装饰价值（Tobutt 1985）。在过去的 20 年中，各种新的柱状苹果栽培品种已经投放市场。一个育种计划培育了新的观赏榛子品种，该品种的特点是兼具食用和观赏价值。‘Red Dragon’榛子（Mehlenbacher and Smith 2009）抗东部欧洲榛树枯萎病，并有红叶和呈弯曲生长。‘Burgundy Lace’是另一个榛子品种，具有东部欧洲榛树枯萎病抗性和红色的叶子，且更易剥开。

具有多种用途的另一个属是西番莲属，其可作为观赏植物和药用植物，也可作为化妆品工业及其水果和水果衍生物的油料来源。西番莲的叶子、花朵和果实多样而美丽，使其成为观赏植物。西番莲的花具有独有的特征：多类型且色彩亮丽的花丝，雌雄同体，丰富多彩的花瓣和萼片。全世界有 400 多个观赏西番莲杂种。近几年，西番莲的育种工作和分子遗传学研究越来越深入（Cerqueira-Silva et al. 2014；Giovannini et al. 2012）。

29.4　组学时代的木本植物育种

近年来，全球包括基因组学、转录组学、蛋白质组学、代谢组学、RNA 测序（RNA-Seq）等在内的新技术正应用于研究植物对实验条件的反应。第二代测序（NGS）技术和强大的生物信息学使全基因组测序爆炸式增长。这些信息正对复杂基因组进行研究，包括（在较小程度上）木本观赏植物，这将有助于深入研究物种之间的关系（Borem and Fritsche-Neto 2014）。这些“组学”工具彻底改变了植物育种，并刺激了新的育种策略发展。即将到来的挑战是如何将大量的基因组数据和技术转化为可以促进杂交过程并为植物育种者利用的知识。用于木本观赏植物的组学工具的实施非常复杂。木本植物由于童期和育种周期长，种质高度杂合，在回交中很大程度上自交不亲和，诸多育种目标有待商榷。然而，未来的前景大于挑战（Borem and Fritsche-Neto 2014）。

29.4.1 第二代测序

现在有超过 20 种木本植物基因组可供参考（主要是果树和林木）（表 29.3）。

表 29.3 木本植物的参考基因组（www.ncbi.nlm.nil.gov/genome/brows/e/）

物种	参考文献
Actinidia chinensis	Huang et al. 2013
Betula nana	Wang et al. 2013b
Citrus maxima	Wang et al. 2017d
Citrus medica	Wang et al. 2017d
Citrus sinensis	Xu et al. 2013
Eucalyptus grandis	Myburg et al. 2014
Ficus carica	Mori et al. 2017
Hevea brasiliensis	Lau et al. 2016
Malus domestica	Velasco et al. 2010；Daccord et al. 2017
Olea europaea	Unver et al. 2017
Phoenix dactylifera	Al-Dous et al. 2011
Pinus taeda	Neale et al. 2014
Picea glauca	Birol et al. 2013
Picea abies	Nystedt et al. 2013
Populus trichocarpa	Tuskan et al. 2006
Populus pruinosa	Yang et al. 2017
Prunus avium	Shirasawa et al. 2017
Prunus mume	Zhang et al. 2012
Prunus persica	Verde et al. 2013
Pyrus × *bretschneideri*	Wu et al. 2013b
Pyrus communis	Chagné et al. 2014
Quercus robur	Plomion et al. 2016
Vaccinium macrocarpon	Polashock et al. 2014
Vitis vinifera	Jaillon et al. 2007
Ziziphus jujuba	Liu et al. 2014

黑杨木（*Populus trichocarpa*）是第一个具有参考基因组序列的多年生树木，使用

Sanger 技术获得（Tuskan et al. 2006）。该技术还被用于组装第一个桃基因组（Verde et al. 2013）。桃是果树遗传学研究上最透彻的物种之一，其基因组学研究增加部分归因于基因组相对简单：基因组较小（230Mb）、二倍体、有 8 对染色体。其异常短的世代周期（2～4 年）和自我适应能力高，加上经常合作开发必要研究工具的科研单位，为构建其完整基因组的准确序列提供了便利（Arus et al. 2012）。在引入 NGS 之后，就可以使用更多物种的全基因组序列，如 NCBI 网站上所列（表 29.3）。

木本植物的基因组大和复杂阻碍了许多属的完整基因组测序与组装。因此，必须开发新的策略。在火炬松（*Pinus taeda*）中，对基因组进行测序时结合了其生殖生物学和基因组独特的组装方法。火炬松的全基因组鸟枪序列是由一个单倍体的雌配子体获得的（Neale et al. 2014）。

得到拥有大量简单序列重复（SSR）、单核苷酸多态性（SNP）、插入和缺失（indel）或裂解的扩增多态序列（CAPS）的木本观赏植物的序列数据，会花更少成本，更便于开发特异序列的分子标记。这对研究系统发育和多样性、寻找杂交亲本、建立新的育种技术和杂交验证策略具有重要意义。

29.4.2 全基因组遗传多样性研究

育种的主要挑战是获取和利用种质资源与野生近缘种中广泛的遗传变异。大多数高通量基因分型平台用于研究多样性和种群结构（Badenes et al. 2016）。在多样性研究中，开发出来的序列用于鉴定分子标记，如 SSR、CAPS、插入缺失、转录间隔区（ITS）和 SNP 标记，它们在群体和群体之间具有高度多态性。ITS 标记用于研究丁香种质之间的遗传关系（Smolik et al. 2010）。在桃中，II 9 K SNP 芯片被开发设计出来（Verde et al. 2012；International Peach SNP consorium），用于鉴别桃、樱桃和杏等品种的基因分型（Romeu et al. 2014；Sanchez et al. 2014）。在甜樱桃里，开发了 CAPS 和插入缺失标记（Shirasawa et al. 2017）。这些标记有助于李属的育种计划。在苹果属里，9 K SNP 芯片也被开发出来，用于研究苹果属的等位基因多样性育种（Chagné et al. 2012；RosBREED SNP consortium）。而在木本物种多样性研究中，最常用的标记仍然是简单序列重复（SSR），如在李属（Aranzana et al. 2003）、苹果属（Silfverberg-Dilworth et al. 2006；Moriya et al. 2012a；Van Dyk et al. 2010）、梨属（Yamamoto et al. 2002；Chen et al. 2015b）、山茱萸属（Wang et al. 2008a）、柑橘属（Barkley et al. 2006）、木犀榄属（Fernandez et al. 2015）、榛属（Martins et al. 2014；Akin et al. 2016；Sathuvalli et al. 2012）、栲属（Sutherland et al. 2010）、山茶属（Caser et al. 2010）等。SSR 标记的应用证明了金合欢和马占相思杂交的 F_1 与回交后代的存在（Le et al. 2017）。基因间遗传距离的估计显示出与谱系记录一致的相关性模式。马占相思（*Acacia mangium*）中的 SSR 标记也可用于确定二倍体马占相思和秋水仙碱诱导的新四倍体基因型的繁育系统，其特征是幼苗后代中杂交种的比例较高（Griffin et al. 2012）。二倍体繁殖后代的方式主要是杂交，而四倍体产生后代的方式几乎是完全自交。后合子因素是育种系统存在差异的主要原因，但细胞遗传学原因尚待研究（Griffin et al. 2012）。此外，RAPD 标记被证明有助于揭示相思树之间的遗传多样性

（Nanda et al. 2004）。对于绣球，最近的研究大大加深了对观赏绣球物种之间繁衍进化知识的认识。RAPD 和叶绿体 DNA 标记（Uemaehi et al. 2014）、DNA 含量测定（Cerbah et al. 2001）和 SSR 标记（Reed and Rinehart 2009；Rinehart et al. 2006）显示了大多数广泛种植的绣球的遗传多样性。对于小范围种系杂交情况，开发全基因组标记用于发现潜在直系同源的单拷贝标记（Granados Mendoza et al. 2015）。潜在亲本的系统发育相关性以及由广泛的质体标记计算出的遗传距离的整合可预测杂交效率（Granados Mendoza et al. 2013）。平均遗传距离为 0.010 65 时，两物种间可以杂交成功，而在平均遗传距离为 0.013 85 或更高的情况下两物种间往往不能成功杂交。作者建议使用 *H. arborescence*、*H. sargentiana*、*H. integrifolia* 和 *H. seemannii* 来检测杂交亲本遗传距离，用这些标记检测杂交育种成功性往往效果更好（Granados Mendoza et al. 2013）。

染色体数目和基因组大小也可以作为属内系统发育成功重建的依据，如七叶树属（Krahulcova et al. 2017）、黄杨属（Van Laere et al. 2011b）、野扇花属（Denaeghel et al. 2017）、醉鱼草属（Van Laere et al. 2009b）、绣球属（Van Laere et al. 2008）、荚蒾属（Zhang et al. 2016）和栒子属（Rothleutner et al. 2016）等。此外，染色体生物学在植物基因组结构和功能领域的应用推动了植物细胞基因组学的发展。DNA 碱基特异性基因组荧光原位杂交 GISH 和基于荧光原位杂交 FISH 的染色体绘图等先进技术的发展极大地促进了植物中染色体特异性标记的鉴定、定位和作图，这对于检测植物染色体、杂种性质、外来染色体和染色体畸变、体细胞克隆变异分析、物种鉴定以及多样性分析非常重要（Talukdar and Sinjushin 2015）。流式细胞仪检测核 DNA 含量和碱基组成（GC%），荧光色带检测富含 GC 和 AT 的 NDA 区，以及荧光原位杂交（FISH）进行 5S 和 18S-5.8S-26S 染色体作图，分别用于卷柏与米曲霉以及大叶卷柏（Mortreau et al. 2010），槲寄生叶和潘氏嗜血杆菌的研究（Van Laere et al. 2008）。这些细胞遗传学特征的种内和种间变异为物种鉴别提供了依据。28 个悬钩子比较研究应用于核型分析。此外，在 6 个悬钩子类群（$2n = 14$ 至 $2n = 56$）中评估 45S 和 5S 核糖体 DNA（FISH）的减数分裂配对行为和分布模式的结果支持了悬钩子多倍体可能是异源多倍体的观点（Chen et al. 2015c）。在四倍体和八倍体样品中仅观察到很少的多价体。FISH 分析显示在二倍体类群中有两个 45S rDNA 位点，除了四倍体濑户内罗非鱼（*R. setchuenensis*）只有三个位点，其他每个研究的多倍体预期倍数均为该数的倍数。二倍体和四倍体类群都有两个 5S rDNA 位点，而八倍体布氏杆菌（*R. buergeri*）携带三个位点。

29.4.3 性状定位和分子标记辅助选择育种

基因组数据用于鉴定和表征与重要性状相关的基因与基因家族，为分子标记辅助选择和植物育种提供了新工具。鉴定和标记重要的特性，如生物抗性和非生物抗性，对于木本植物育种具有重要意义。然而，MAS 在木本物种上成功的实例数量仍然很少。

筛选与性状相关的标记是先决条件。天然遗传变异连锁作图（基于图的克隆）的使用可以识别潜在的基因和数量性状基因座（QTL）。在柳属（*Salix*）中，鉴定了在盐胁迫下表达的生长假单胞菌（*S. matsudana*）的生长 QTL，提供了耐盐蒿柳（*Salix*）品种分子标记辅助选择育种的重要信息(Zhang et al. 2017b)。在绣球(*Hydrangea macrophylla*)中，正在进行 QTL 分析和全基因组标记分析，以绘制负责带帽或拖把花序的基因图谱为主要筛选目标（Traenkner and Engel 2017）。花序类型以单基因、显性-隐性方式遗传，拖把型是隐性的，而带帽型是显性的。头带花序的绣球植物更好看些，因此受到消费者的青睐。如果将带花序植株与带帽（lacecap）植株杂交，则所有子代都会形成带花序。将此类 F_1 与另一株带花序植株回交，后代将分离出带花序和带帽（lacecap）植株，从而可以筛选出带花序植株（Traenkner and Engel 2017）。但是绣球植物需要大约 13 个月才能发育出花序，这延迟了花序类型的确定。因此，使用与花序类型相关的分子标记将能够在复杂的育种计划中找到带花序的等位基因，并在苗期就通过花序基因进行分子标记辅助选择（Traenkner and Engel 2017）。石楠有美化环境的功能，其突变而形成的延长的类似灯笼的花形态有着巨大的经济价值。目前已经构建了石楠的遗传连锁图谱，在该图谱上绘制了“花朵类型”的基因座。确认了最重要的园艺性状“花型”、“花色”和“叶色”的独立遗传性，并且确定了“花型”的分子标记辅助选择可用于分子标记辅助选择育种和进一步筛选灯笼花序结构（Behrend et al. 2013）。为了增进对山茱叶子颜色遗传的理解，并在育种过程中尽早选择红色叶子基因型，鉴定了假定的叶子颜色 QTL（Wadl et al. 2011）。在遗传群体中，叶子的颜色分为绿色和红色表型，并在三年内的 5 个春季对颜色进行视觉评估。但是，随着时间的推移，检测到 QTL 不稳定，这为开花山茱中红色色素表达的复杂遗传学提供了证据（Wadl et al. 2011）。在栗树、橡树和山毛榉中，还发现了许多种内和种间全基因家族，用于鉴定与栗枯病（*C. parasitica*）、法格斯树皮病和橡树猝死有关的基因和 QTL（Koch et al. 2010；Kremer et al. 2012），以及栎属（*Quercus*）中有关涝渍和水利用效率的特征基因（Parelle et al. 2007；Brendel et al. 2008；Kremer et al. 2012）。日本梨的 QTL 分析产生了一批标记，这些标记与由纳氏菌（*Venturia nashicola*）引起的结病的抗性基因相关（Gonai et al. 2012），并与单个等位基因 S-基因座（Nashima et al. 2015）控制的自交不亲和性相关。可以建立用于快速和可靠测定 S-基因型的分子测定方法。

在有参考基因组序列作图的种群（表 29.3）中，与目标性状相关的主要 QTL 被鉴定出，对具有极端表型的单个基因组进行重新测序，然后与参考基因组进行短 reads 序列比对，这有助于鉴定与此类性状相关的多态性性状（Badenes et al. 2016）。使用此程序，Zhebentyayeva 等（2014）确定了两个桃中休眠相关基因 *PpeDAM5*、*PpeDAM6*，这 2 个标记可作为与低温和芽休眠基因关联的候选基因。在苹果中，参与柱状生长习性表型的 *Co* 基因被精细定位在连锁群 10 号染色体的 200kb 区域中（Bai et al. 2012；Moriya et al. 2012b）。来自具有柱状性状的‘Wijcik’突变体和具有正常表型的野生型‘McIntosh’高通量基因组测序的证据表明，插入的 1956bp 移动 DNA 元件可能是 *Co* 突变的起源（Wolters et al. 2013；Otto et al. 2014）。插入基因位于一个基因间区域，显著影响 2OG-Fe（II）加氧酶基因 *MdCo31* 的表达。通过组成型表达，该基因在拟南芥中有 *Co* 样表型的复制，从功能上证实了 *MdCo31* 参与柱状生长习性。

然而，通过 QTL 分析开发标记取决于是否有连锁不平衡（如单个全同胞家族）和影响目标性状基因分离变异的天然或合成种群的可用性。也就存在两个重大障碍：首先也是最重要的难题是需要大量的试验群体，这对于苗木育种有很大困难，需要进行筛选以实现足够的遗传精度及进行物理隔离。其次，难以用数量性状基因座（QTL）进行精细定位。此外，QTL 测绘还难以获得遗传更为复杂和更具有经济价值的性状基因，如木材质量（Strauss and Bradshaw 2004）。因此，基于遗传图谱的克隆技术对于许多树种而言通常是不可行的。作为替代方案，可以通过基因组和 EST 测序来鉴定新基因。例如，对于重要的开花性状基因可以通过其表达随年龄或繁殖阶段的变化而鉴定（Strauss and Bradshaw 2004）。RNA 测序（RNA-Seq）是一种强大的分析转录组的工具，因为可以通过量化由 NGS 技术获得的短 cDNA 读数来精确测量样品中每个基因的表达水平，从而可以比较两个转录样本之间的基因型和转录条件（Martin and Wang 2011）。因此，可以通过转录组提供的基因表达数据确认基因组与细胞表征之间的动态联系。RNA 读数可定位在参考基因组上，或在基因组序列不可用时简单地比对和组装，这种方法适用于任何植物。例如，对水曲柳应用了相关的转录组学分析，发现了中灰烬枯死病（ash dieback）症状评分的基因序列和基因表达的遗传多样性（Harper et al. 2016）。鉴定出与受感染树木的树冠破坏密切相关的标记。使用这些标记，在其他种树木的测试小组中预测了表型，成功地鉴定了对该病易感性较低的植株。对于桃李，在对照和洪涝胁迫植物上进行了转录组分析（Klumb et al. 2017）。该转录组分析支持开发特定的分子标记，以研究淹没条件下桃李反应的分子机制。这对于协助育种计划选择更具耐受性的基因型至关重要（Klumb et al. 2017）。另一个例子来自泡桐（*Paulownia tomentosa*），这是一棵生长在中国盐渍地的树木（Zhao et al. 2017）。RNA-Seq 用于分析盐胁迫对二倍体和四倍体白杨植株的影响。差异表达基因的功能注释表明，植物激素信号转导和光合活性对于植物对高盐条件的反应至关重要。利用实时定量 PCR 分析验证了 8 个差异表达基因的表达模式。该结果可能有助于加速栽培泡桐和其他物种的遗传改良，以增强其适应盐渍土的能力（Zhao et al. 2017）。桂花的转录组分析可以鉴定几种与代谢途径相关的主要基因，并为桂花进一步的功能基因组研究提供有用的数据（Mu et al. 2014）。转录组研究涉及的其他木质观赏器官包括花序和花的形态，如山茱和芙蓉。绣球中的铝耐受性和积累、木材质量特征、次生细胞壁形成和木质素生物合成等研究，这些主要在树木中进行，如杨（Chen et al. 2015）、桉（Thavamanikumar et al. 2014）、相思树（Wong et al. 2011）和栎属（Soler et al. 2007）。对缺水的转录组相应的研究（Dong et al. 2014a，2014b；Behringer et al. 2015；Dong et al. 2017），包括营养缺乏症（Fan et al. 2014）、热休克（Chen et al. 2014）、耐寒性（Chen et al. 2014）和疾病抵抗力等相关研究。气候变化和病虫害综合治理措施，以后也可用于分子标记辅助选择抗性等位基因（Liu et al. 2013；Barakat et al. 2009；Harper et al. 2016；Perdiguero et al. 2015；Sathuvalli et al. 2011）。

在其他方面，基于基因组学的更先进方法，如全基因组关联研究（GWAS）和基因组选择（GS），有望在果树育种中应用，但这需要很长时间来培育一个品种（Minamikawa et al. 2017）。例如，在柑橘中，对一个亲本（111 个变种）和一个后代群体（来自 35 个全同胞家族的 676 个个体）进行了 1841 个 SNP 的基因分型，并对 17 个水果品质性状进行了表型分析，结果表明 GWAS 和 GS 均可作为柑橘果实性状遗传的有效工具

（Minamikawa et al. 2017）。GWAS 和 GS 也正在成为梨、苹果与森林育种计划的有力工具，如桉（Tan et al. 2017），可作为长寿植物双亲 QTL 作图的替代方法进行植物研究（Yamamoto and Terakami 2016）。

29.4.4　基因工程和突变基因组学

基因工程可以产生新的特征，而这些特征不会引起植物本身基因组的自然变异。此外，基因工程技术可以缩短木本植物的育种周期，而即使用分子标记辅助选择技术也需要很长时间。但是，迄今为止，木本观赏作物的市场化程度很低，公众和零售商对转基因生物的接受度有限，这限制了木本植物的基因工程发展。

尽管有多种木本植物基因工程技术，但农杆菌介导的转化是木本植物物种最常用的方法（Osakabe et al. 2016）。尽管木本植物采用农杆菌介导的转化方法较为困难，但组织培养系统的改进已允许生产转基因木本植物（Osakabe et al. 2016）。杨属是最早进行遗传转化的硬木树种之一，已经产生了农杆菌介导转化的胡杨，改变了除草剂抗性（Fillatti et al. 1987）、生长和木材特性（Tuominen et al. 1995）、氮素代谢（Gallardo et al. 1999）和木质素含量（Hu et al. 1998；Li et al. 2013b），并提高了耐盐性（Hu et al. 2005；Li et al. 2009c）。已经获得了遗传转化桉树，其引入了内源和外源基因用于生产具有改良次生细胞壁的植株（Hussey et al. 2011），并改善了耐盐性（Matsunaga et al. 2012；Yu et al. 2013）。还获得了辐射松（Grant et al. 2004）、针叶松（Wenck et al. 1999；Tang et al. 2001；Tang et al. 2005）和云杉（Wenck et al. 1999）的转基因植物。通过农杆菌介导转化豇豆蚜传花叶病毒（CABMV）基因片段产生了对 CABMV 产生抗性的西番莲（Correa et al. 2013，2015；Monteiro-Hara et al. 2011）。APETALA（AP1）参与开花，一直是许多研究的重点，包括木本植物。Huang 等（2014）利用农杆菌介导的转化产生了过表达 BpAP1 的白桦 × 桦木系，已证明 BpAP1 可以通过有性繁殖而遗传，BpAP1 的过表达可引起提早开花、矮化和幼年期缩短。这些结果加深了对花芽分化机制的理解，并为桦木和其他木本植物的育种开辟了新的可能性（Huang et al. 2014；Kost et al. 2015）。来自野苹果 *Malus* × *robusta* 的顺式基因，从易受火疫病侵害的栽培品种‘Gala Galaxy’开发了一种苹果变异系（cisgenic），赋予了其对火疫病的抗性，转化的品系仅携带顺式基因及其天然调控序列（Kost et al. 2015）。纯天然葡萄植株显示白粉病的发生率降低且黑腐病的严重程度降低（Dhekney et al. 2011）。开发无外源 DNA 的内源/同源柑橘栽培品种，其转化效率约为 0.67%（An et al. 2013）。另一种转化技术即 RNA 干扰，这项技术使用复杂的分子机制，其主要功能是抑制基因表达，可被双链 RNA 分子激活。这个基因作用过程中称为“敲除”（Limera et al. 2017）。RNAi 已应用于木本水果物种中，以引入病原体抗性，如在李属中引入李子痘病毒抗性（Scorza et al. 2013）和在苹果属中引入白粉病抗性（Pessina et al. 2016），并改善收获质量，如减少了家蝇的繁殖力并增加了花的吸引力（Klocko et al. 2016）。

根癌农杆菌介导的基因修饰方法是使用未经修饰的细菌（如根癌农杆菌）进行自然转化，该菌株缺少重组核酸，被认为可产生非转基因植物（Lütken et al. 2012）。综上所述，观赏开花植物与根癌农杆菌的共培养以及随之而来的 *rol* 基因在植物基因组中的整合可能会诱导出所需的植物性状，如矮化和浓密的表型。另外，获得改善的生根能力对于那些不易生根的木质植物来说是有用的。尽管整合了 *rol* 基因的植物数量正在增加（Casanova et al. 2005）（表 29.2），但该技术对于木质观赏植物仍然具有挑战性。

突变基因组学被定义为可应用的突变育种，其中基因组信息和软件工具用于育种设计、突变体的筛选、选择和验证（Talukdar and Sinjushin 2015）。基因组编辑工具，如ZFNs、TALENs 和 CRISPR/Cas9，已被开发出来为位点特异性内切核酸引入特定位点的突变。Peer 等（2015）使用 *uidA* 基因的 ZFN 靶向，*uidA* 基因表达 β-葡萄糖醛酸酶（GUS），并由此证明了基于 ZFN 的定点诱变在家蝇和无花果中成功应用。自首次报道 CRISPR/Cas9 的基因编辑潜力以来（Sander and Joung 2014），这项技术正在改变从医学到农业生物学的各个方面，并为改善农作物提供了许多可能。对于具有高度异源基因组和较长生成周期的木本观赏作物，CRISPR/Cas9 提供了加速遗传改良的简便方法。CRISPR/Cas9在木本植物中的首次应用是在白杨（*Populus tremula*）× 白杨假单胞菌（*P. alba*）杂种中，编辑参与苯丙烷代谢的 4-香豆酸酯：CoA 连接酶（4CL）基因家族（Zhou et al. 2015）。其中一个基因 *4CL1* 具有参与木质素生物合成的特征（Boerjan et al. 2003）。使用 CRISPR/Cas9 进行基因组编辑，该基因中产生了小插入缺失，木质素含量降低了（Zhou et al. 2015）。在同一研究中，靶向 4CL2 被认为与类黄酮生物合成有关。CRISPR/Cas9 诱导的 4CL2 突变导致根中单宁水平降低，为支持 4CL 在类黄酮生物合成中的功能提供了证据（Zhou et al. 2015；Tsai and Xue 2015）。还报道了另一物种杨属毛白杨（*P. tomentosa*），以八氢番茄红素脱氢酶（PDS）为靶向，CRISPR/Cas9 成功实现基因编辑。该研究中 4 个 gRNA 被多重化，大约 50%处理过的植物表现出白化病表型（Fan et al. 2015）。在柑橘中，CRISPR/Cas9 已成功用于修饰溃疡病易感基因 CsLOB1，为产生抗病柑橘品种提供了可行途径（Jia et al. 2017；Peng et al. 2017）。在苹果（Nishitani et al. 2016）和葡萄（*Vitis vinifera*）（Nakajima et al. 2017）中，番茄红素脱氢酶（PDS）发生了突变。在 31.8%的再生苗中观察到清晰和部分苹果白化病表型，并且通过 DNA 测序确认了苹果中 PDS 的双等位基因突变（Nishitani et al. 2016）。使用核糖核蛋白可以对葡萄 MLO-7 与苹果 DIPM-1、2 和 4 的原生质体进行有效的定向诱变，从而在葡萄中产生白粉病抗性，并在苹果中产生白叶枯病抗性（Malnoy et al. 2016）。

基因组编辑中的一个主要问题是 DNA 修饰的特异性，因为非特异性 CRISPR/Cas9 活性引起的脱靶切割可能引起意想不到的突变。对于以全基因组、节段和串联重复作为特征的植物基因组来说尤其如此（Tsai and Xue 2015）。在具有高度异源多年生木本杨树的研究结果表明，CRISPR/Cas9 的脱靶切割可能性很低（Tsai and Xue 2015）。木本植物特有的其他一些限制可能会妨碍 CRISPR/Cas9 的使用。许多森林和园艺树木物种与基因型对组织培养的都是有一定困难（Tsai and Xue 2015）。有效的体外再生方案的开发取决于基因型和所用起始植物组织的类型。此外，序列多态性在大多数开花植物，特别是自然界中杂种的木本植物的基因组编辑中值得关注。例如，据报道，杨属（60bp 为一个片段）（Ingvarsson 2005）和桉属（16～33bp 为一个片段）（Külheim et al. 2009）基因的 SNP 频率很高。几种基于 Web 的 gRNA 设计程序可用于植物，但是它们在异型物种中应用受到限制，因为预加载的基因组序列缺乏双等位基因（或多倍体）覆盖（Tsai and Xue

2015）。可以从 T1 代中分离获得 DNA 编辑但未转基因的子代。但是，大多数木本物种的漫长童期阻碍了这一进程。最后，异种后代将不同于亲本，这需要额外的筛选和选择。然而，将突变引入优良基因型已经应用于农作物和树木育种中，并非是由 CRISPR/Cas9 编辑的突变体所独有。一个例子是火炬松（*Pinus taeda*），它带有天然的、参与木质素生物合成的肉桂醇脱氢酶无效等位基因，在美国的育种计划中被用作亲本（Gill et al. 2003）。另一个例子是早熟的 FT 李子，该李子在快速育种计划中被用作亲本，以开发改良的品种（Callahan et al. 2015）。基于病毒系统的瞬时表达系统可产生杂交非转基因植物。开发和优化用于去除 T-DNA 而不交叉的创新方法或其他转染方法，如使用核糖核蛋白（RNP）的开发和优化将是必要的。

通过基因组编辑的农作物品种与天然存在的或化学诱导的突变体没有区别，因此对于遵守产品法规的国家，基因编辑作物可能不受到转基因植物（GMO）法规的约束（Voytas and Gao 2014）。预期将鼓励基于 CRISPR/Cas9 的基因组编辑应用于林业、水果和坚果以及木本观赏植物产业（Tsai and Xue 2015）。

29.4.5　表型组学

为鉴定目标基因和种质，现代作物改良技术既依赖于 DNA 测序，又依赖于对植物性状的准确定量（Fahlgren et al. 2015）。随着 DNA 测序技术的飞速发展，植物表型鉴定逐渐成为瓶颈，因为需要强大的表型分析工具来选择具有理想性状的植物。近年来，越来越多的图像捕获和分析工具出现以及对硬件与软件的改善使得高通量植物形态学领域飞速发展。这些进步为更深入地分析植物特征创造了条件（Fahlgren et al. 2015；Araus and Cairns 2014）。高通量表型分析平台的一项关键进步是能够以突破性方式捕获植株性状。由于减少了重复采样集而提高了基因型、处理和生物学重复的实验容量。除了提高容量的潜力，Lattier 和 Contreras（2014）还描述了使用比色法对花色进行更准确的表型分析，作为描述花性状遗传力研究的一部分。该工具以简化诸如花色的性状表型依据，然后将比色数据转换为更常用的值，如与 RHS 色卡相对应的值。

在农作物中，技术创新和专业化如无人机和在农场范围内安装传感器等可以提高农作物生长发育监控的分辨率、精度和规模（Bevan et al. 2017）。随着个体数量的增加，会得到越来越多的数据，并有望应用于育种。但由于木本观赏植物市场份额通常太低，经常无法开发和使用这些新技术。大多数木本观赏植物的表型研究是在少量的植株上手动测量一些性状（如节间长度、叶长和宽度、花直径、枝长等），这是一种劳动密集型工作且对环境变化敏感。现在，低成本的高通量表型分析方法已经开发，并且对木本观赏植物的图像捕获和分析进行了描述（Susko et al. 2018）。对于较大的植株，可以为每个单株收集图像数据，包括顶视图和侧视图，然后采用图像处理算法识别植物的像素，并采用所识别的图像测量每株植物的形态特征（如形状、结构、叶面积、体积等）（Yang et al. 2014）。目前已经用图像分析比较了南美鼠刺二倍体和四倍体基因型之间的植株结构性状（Denaeghel et al. 2018），并用 ImageJ 分析了顶视图和侧视图图像（Abramoff et al. 2004）。采用诸如边框、凸包和最小边界圆之类的参数可以计算每种植物的循环性、密度和表面积（Denaeghel et al. 2018）。Dudits 等（2016）使用基于图像分析的半自动平台将蒿柳（*Salix viminalis*）四倍体茎和根的生长与二倍体进行了比较。丁香假单胞菌可导致多种木本物种发生茎坏死和溃疡病，对病灶随着时间的推移在木质组织中的进展进行定性定量检测是育种者选择抗药性的关键（Li et al. 2015）。Li 等（2015）使用自动图像分析开发了一种快速可靠的病变定量方法，该方法比用眼睛评分更客观。图像分析也可用于估算叶面积指数，如在甜樱桃（*Prunus avium*）（Mora et al. 2016）、土耳其栎（*Quercus cerris*）、欧洲鹅耳枥（*Carpinus betulus*）和欧洲山毛榉（*Fagus sylvatica*）中所用的方法（Chianucci et al. 2014）。

附属相机系统可以完成彩色成像。红外光与植物组织相互作用的特性使对表型进行红外分析有潜在用途（Seelig et al. 2008）。耐旱性和耐盐碱性是两个重要的非生物胁迫耐受性，干旱和盐碱等胁迫会使植株产生非常相似的表型效应，如气孔关闭。气孔关闭的作用是降低光合作用和呼吸作用，这意味着在土壤水分限制下生长的植物的冠层温度会升高。红外热成像技术可根据温度检测物体发出的长波红外辐射（Sirault et al. 2009），可用于选择能够在胁迫下保持气孔开放的基因型，以快速、低成本地进行筛选（Furbank and Tester 2011）。在柑橘中，开发了一种方法，通过热成像和热像仪同时分析温度与可见表型来分析自动拍摄的单株树的正面图像（Jimenez-Bello et al. 2011；Ballester et al. 2013）。该工具可以检测具有不连续冠层的多年生木本植物的水分状况。但研究结果显示，评估的农作物类型对结果有重要影响，因为柿子树的冠层温度与植物水分状况之间具有很好的相关性，而在柑柚（*Citrus clementina*）和柑橘（*Citrus sinensis*）中，树种之间的温差与植物水势不一定总是相关（Jimenez-Bello et al. 2011；Ballester et al. 2013）。

除气孔导度外，出芽（表明生态休眠的结束和生长期的开始）也受到外界胁迫和气候变化的影响。营养芽受到的寒冷和胁迫水平会影响发芽时间（Kleinknecht et al. 2015；Pope et al. 2013）。Kleinknecht 等（2015）设计了一种传感器，用于测量温带木本植物芽的发生时间，连同关键环境驱动因素（即温度和光周期）的数据，在松果松（*Pinus pinea*）、花旗松（*Pseudotsuga menziesii*）和毛白杨 × 银白杨（*Populus tremula* × *Populus alba*）杂交种上验证了该系统。

叶绿素荧光成像是检测植物胁迫的另一种方法（Jansen et al. 2009）。在胁迫研究中，最容易测量且最常用的荧光参数是暗适应 F_v/F_m，它是光系统 II 中光收集的固有光化学效率的量度（Furbank and Tester 2011）。对橄榄树的一项研究（Faraloni et al. 2011）表明，叶绿素荧光成像技术可用于快速量化高等植物对生理胁迫的响应。此外，植株内测量结果表明，整株植物对干旱的响应与体外观察到的叶片耐脱水水平有关（Faraloni et al. 2011）。Salvatori 等（2014）用多通道荧光计分析了几种木本和草本植物的胁迫反应，其中包括杨梅（*Arbutus unedo*）、刺叶栎（*Quercus ilex*）、黑果绣球（*Viburnum lantana*）和酿酒葡萄（*Vitis vinifera*）。结果表明，基于叶绿素荧光数据，使用这种多通道荧光计可以很好地了解植物的胁迫反应，同时综合分析快速荧光（PF）和调制反射率（MR）在不同条件下植物体对胁迫反应将是最有效的途径（Salvatori et al. 2014）。

尽管可以使用二维（2D）图像数据有效地估计三维（3D）特性，但是 3D 植物建模对于测量建筑特征可能特别有帮助。可以使用激光扫描技术和深度传感器直接测量三维结构（Paulus et al. 2013）。在 *Salix* 的一项研究中，通过确定分支直径的变化检测极值和测量精度，探索了 3D 相机的实用程序以测量植物冠层结构，用于收集有关木本植物分支架构的 3D 数据（Nock et al. 2013）。结合 3D 建模和红外热成像技术，可以研究云杉芽原基的超低温耐受机制（Kuprian et al. 2017）。X 射线计算机断层扫描（microCT）是一种可以对扫描对象进行 3D 重建的成像方法（Larabell and Nugent 2010）。这项技术用于研究葡萄移植区的内部结构和 3D 组织（Milien et al. 2012）。在桉树中，分析了易脱水植物的缺水脆弱性（Nolf et al. 2017）。X 射线显微照相分析的结果与对同一样品液压测量获得的结果一致。microCT 还用于鉴定红枫根段内的木质部血管末端（Wason et al. 2017）。

植物表型组学最大、最耗时的挑战是确定图像提取的性状和生物学性状的相关性。当前基于图像的表型试验通常针对特异的问题。但也可以使用新算法从这些图像数据集中提取更多数据，类似于重新分析测序数据（Fahlgren et al. 2015）。因此，植物表型组学领域是跨学科的，是植物科学、工程学和计算机科学之间相互交叉的一门学科。

29.5　结　　论

本章提出了有关木本观赏植物育种所有方面的观点。与高度变异的栽培属、种相比，木本观赏植物的商业育种仅在一些重要的树种中进行。所以，可以得出结论，虽然诸多先进的育种技术已经得到应用，但传统的选择育种方法仍然非常重要。许多品种是通过种间杂交、有丝分裂和减数分裂多倍体化或突变产生的。其他技术，如根部与细菌共培养、原生质体融合、单倍体诱导等，在木本观赏植物中仍然处于研究试验阶段，但在未来可以期待其实际应用。

不同的种属其育种目标会有很大差异。随着品牌和市场化的发展，视觉吸引力（在销售时）是木本观赏植物的重要指标。由于某些物种的入侵性，通常需要培育一些不育品种。一种新兴趋势是多用途植物的育种。然而，总体上主要的挑战仍然是获得对生物和非生物胁迫的抵抗力。

测序技术的发展以及木本植物基因组和转录组的实用性提高，为植物育种者提供了有用的工具，对重要植物过程背后遗传学的了解越来越深入，并且能以低成本开发生产序列特异性标记。虽然目前用于木本植物的辅助基因组选择的标记数目是有限的，但这可能会对木本植物属的杂交验证和分子标记辅助选择产生积极作用。此外，针对木本植物育种周期长、童期长、基因组高度杂合的特征，一些新的育种技术，如 CRISPR/Cas9，可以提供一种简便的方法来加速木本植物的基因组改良。这将加快育种和选择进程，新品种的开发将更高效而精确。

本章译者：

晏慧君，李绅崇

云南省农业科学院花卉研究所，国家观赏园艺工程技术研究中心，云南省花卉育种重点实验室，昆明 650200

参考文献

Abe K, Kotoda N, Kato H, Soejima JI (2011) Genetic studies on resistance to Valsa canker in apple: genetic variance and breeding values estimated from intra- and inter-specific hybrid progeny populations. Tree Genet Genomes 7(2):363–372

Abramoff MD, Magalhaes PJ, Ram SJ (2004) Image processing with ImageJ. Biophoton Int 11(7):36–42

Ackerman WL, Dermen H (1972) Fertile colchiploid from a sterile interspecific *Camellia* hybrid. J Hered 63(2):55

Ahmadi A, Azadfar D, Mofidabadi AJ (2010) Study of inter-generic hybridization possibility between *Salix aegyptiaca* and *Populus caspica* to achieve new hybrids. International Journal of Plant Production 4(2):143–147

Aiello AS, Rothleutner J, McNamara S, Hokanson SC (2014) Mid-winter cold hardiness of *Corylus fargesii* germplasm as determined in laboratory freezing tests. Hortscience 48:S339

AIPH (2015) International statistics flowers and plants 2015. UK, Reading

Akin M, Nyberg A, Postman J, Mehlenbacher S, Bassil NV (2016) A multiplexed microsatellite fingerprinting set for hazelnut cultivar identification. Eur J Hortic Sci 81(6):327–338

Al-Dous EK, George B, Al-Mahmoud ME, Al-Jaber MY, Wang H, Salameh YM, Al-Azwani EK, Chaluvadi S, Pontaroli AC, DeBarry J, Arondel V, Ohlrogge J, Saie IJ, Suliman-Elmeer KM, Bennetzen JL, Kruegger RR, Malek JA (2011) De novo genome sequencing and comparative genomics of date palm (*Phoenix dactylifera*). Nat Biotechnol 29(6):521–527

Alexander L (2017) Production of triploid *Hydrangea macrophylla* via unreduced gamete breeding. Hortscience 52(2):221–224

Al-Niemi T, Weeden NF, McCown BH, Hoch WA (2012) Genetic analysis of an interspecific cross in ornamental *Viburnum* (*Viburnum*). J Hered 103(1):2–12

Altland J (2006) Foliar chlorosis in field-grown red maples. Hortscience 41:1347–1350

American Public Garden Association (APGA) (2018) About the plant collections network. https://publicgardens.org/programs/about-plant-collections-network. Accessed 1/18

An C, Orbovic V, Mou Z (2013) An efficient introgenic vector for generating introgenic and cisgenic plants in *Citrus*. Am J Plant Sci 4:2131–2137

Anamthawat-Jonsson K, Thorsson AT (2003) Natural hybridisation in birch: triploid hybrids between *Betula nana* and *B. pubescens*. Plant Cell Tissue Org Cult 75(2):99–107

Andersen SB (2003) Doubled haploid production in poplar. In: Maluszynski M, Kasha KJ, Forster BP, Szarejko I (eds) Doubled haploid production in crop plants – a manual. Kluwer Academic Publishers, Dordrecht

Aranzana MJ, Pineda A, Cosson P, Dirlewanger E, Ascasibar J, Cipriani G et al (2003) A set of simple-sequence repeat (SSR) markers covering the *Prunus* genome. Theor Appl Genet 106:819–825

Araus JL, Cairns JE (2014) Field high-throughput phenotyping: the new crop breeding frontier. Trends Plant Sci 19(1):52–61

Arène L, Bellenot-Kapusta V, Belin J, Cadic A, Clérac M, Decourtye L, Duron M (2007) Breeding program on woody ornamental plants in Angers – France: a collaboration of 32 years between INRA and SAPHO. Acta Hort 743:35–38

Ari E, Djapo H, Mutlu N, Gurbuz E, Karaguzel O (2015) Creation of variation through gamma irradiation and polyploidization in *Vitex agnus-castus* L. Sci Hortic 195:74–81

Arús P, Verde I, Sosinski B, Zhebentyayeva T, Abbott AG (2012) The peach genome. Tree Genet Genomes 8(3):531–547

Asatryan A, Tel-Zur N (2014) Intraspecific and interspecific crossability in three *Ziziphus* species (*Rhamnaceae*). Genet Resour Crop Evol 61(1):215–233

Bachtell KR (1988) *Acer* x *freemanii* – a source for new shade tree selections. IPPS-Eastern Reg 38:509–514

Badenes ML, Fernandez I Marti A, Rios G, Rubio-Cabetas MJ (2016) Application of genomic technologies to the breeding of trees. Frontiers in Genetics 7, article 198

Bagniewska-Zadworna A, Wojciechowicz MK, Zenkteler M, Jezowski S, Zenkteler E (2010) Cytological analysis of hybrid embryos of intergeneric crosses between *Salix viminalis* and *Populus* species. Aust J Bot 58(1):42–48

Bagniewska-Zadworna A, Zenkteler M, Zenkteler E, Wojciechowicz MK, Barakat A, Carlson JE (2011) A successful application of the embryo rescue technique as a model for studying crosses

between *Salix viminalis* and *Populus* species. Aust J Bot 59(4):382–392
Bai T, Zhu Y, Fernandez-Fernandez F, Keulemans J, Brown S, Xu K (2012) Fine genetic mapping of the Co locus controlling columnar growth habit in apple. Mol Gen Genomics 287:437–450
Ballester C, Jimenez-Bello MA, Castel JR, Intrigliolo DS (2013) Usefulness of thermography for plant water stress detection in *Citrus* and persimmon trees. Agric For Meteorol 168:120–129
Ballington JR (2009) The role of interspecific hybridization in blueberry improvement. Acta Hort 810:49–59
Ballington JR, Fernandez GE (2008) Breeding raspberries adapted to warm humid climates with fluctuating temperatures in winter. Acta Hort 777:87–90
Barakat A, DiLoreto DS, Zhang Y, Smith C, Baier K, Powell WA, Wheeler N, Sederoff R, Carlson JE (2009) Comparison of the transcriptomes of American chestnut (*Castanea dentata*) and Chinese chestnut (*Castanea mollissima*) in response to the chestnut blight infection. BMC Plant Biol 9:51
Barbosa LV, Mondin M, Oliveira CA, Souza AP, Vieira MLC (2007) Cytological behaviour of the somatic hybrids *Passiflora edulis* f. flavicarpa plus *P-cincinnata*. Plant Breed 126(3):323–328
Barkley NA, Roosse ML, Krueger RR, Federici CT (2006) Assessing genetic diversity and population structure in a *Citrus* germplasm collection utilizing simple sequence repeat markers (SSRs). Theor Appl Genet 112:1519–1531
Basaran P, Kepenek K (2011) Fruit quality attributes of blackberry (*Rubus sanctus*) mutants obtained by Co-60 gamma irradiation. Biotechnol Bioprocess Eng 16(3):587–592
Baskin CC, Chen SY, Chien CT, Baskin JM (2009) Overview of seed dormancy in *Viburnum* (*Caprifoliaceae*). Propag Ornam Plants 9(3):115–121
Bassett CL, Wisniewski M (2009) Global expression of cold-responsive genes in fruit trees. In: Gusta L, Wisniewski M, Tanino K (eds) Plant cold hardiness: from the laboratory to the field. CABI, Oxford, UK, pp 72–79
Baumgartner IO, Patocchi A, Lussi L, Kellerhals M, Peil A (2014) Accelerated introgression of fire blight resistance from *Malus* x *robusta* 5 and other wild Germplasm into elite apples. Acta Hort 1056:281–287
Bean WJ (1989) Trees and shrubs hardy in the British isles: volume I. John Murray Publishers, London
Beck SL, Dunlop RW, Fossey A (2003) Evaluation of induced polyploidy in *Acacia mearnsii* through stomatal counts and guard cell measurements. S Afr J Bot 69(4):563–567
Behrend A, Borchert T, Spiller M, Hohe A (2013) AFLP-based genetic mapping of the "bud-flowering" trait in heather (*Calluna vulgaris*). BMC Genet 14:64
Behrend A, Gluschak A, Przybyla A, Hohe A (2015) Interploid crosses in heather (*Calluna vulgaris*). Sci Hortic 181:162–167
Behringer D, Zimmermann H, Ziegenhagen B, Liepelt S (2015) Differential gene expression reveals candidate genes for drought stress response in *Abies alba* (*Pinaceae*). PLoS One 10:e0124564
Bell NC (2009) Evaluation of growth, flowering, and cold hardiness of *Ceanothus* in western Oregon. HortTechnology 19(2):411–417
Belwal T, Bisht A, Bhatt ID, Rawal RS (2015) Influence of seed priming and storage time on germination and enzymatic activity of selected *Berberis* species. Plant Growth Regul 77(2):189–199
Benetka V, Hyhlikova M (2010) Low and compact varieties of *Weigela* and *Potentilla* cultivated in Pruhonice. Acta Hort 885:61–63
Benetka V, Novotna K, Stochlova P (2012) Wild populations as a source of germplasm for black poplar (*Populus nigra* L.) breeding programs. Tree Genet Genomes 8(5):1073–1084
Berlin S, Hallingback HR, Beyer F, Nordh NE, Weih M, Ronnberg-Wastljung AC (2017) Genetics of phenotypic plasticity and biomass traits in hybrid willows across contrasting environments and years. Ann Bot 120(1):87–100
Bevan MW, Uauy C, wulff BBH, Zhou J, Krasileva K, Clark MD (2017) Genomic innovation for crop improvement. Nature 543(7645):346–354
Birol I, Raymond A, Jackman SD, Pleasance S, Coope R, Taylor GA, Yuen MM, Keeling CI, Brand D, Vandervalk BP, Kirk H, Pandoh P, Moore RA, Zhao Y, Mungall AJ, Jaquish B, Yanchuk A, Ritland C, Boyle B, Bousquet J, Ritland K, Mackay J, Bohlmann J, Jones SJ (2013) Assembling the 20 Gb white spruce (*Picea glauca*) genome from whole-genome shotgun sequencing data. Bioinformatics 29(12):1492–1497
Bisognin C, Seemueller E, Citterio S, Velasco R, Grando MS, Jarausch W (2009) Use of SSR markers to assess sexual vs. apomictic origin and ploidy level of breeding progeny derived from crosses of apple proliferation-resistant *Malus sieboldii* and its hybrids with *Malus* x

domestica cultivars. Plant Breed 128(5):507–513
Blada I, Tanasie S (2013) Growth, straightness and survival at age 32 in a *Pinus strobus* x *P. wallichiana* F-1 hybrid population (experiment 2). Ann For Res 56(1):15–30
Blada I, Tanasie S, Dinu C, Bratu I (2013) Growth, straightness and survival at age 32 in a *Pinus strobus* x *P. wallichiana* F-1 hybrid population (experiment 1). Ann For Res 56(2):269–282
Boerjan W, Ralph J, Baucher M (2003) Lignin biosynthesis. Annu Rev Plant Biol 54:519–546
Borem A, Fritsche-Neeto R (2014) Omics in plant breeding. John Wiley and Sons, Inc
Bosu PP, Miller F, Wagner MR (2007) Susceptibility of 32 elm species and hybrids (*Ulmus* spp.) to the elm leaf beetle (Coleoptera: chrysomelidae) under field conditions in Arizona. J Econ Entomol 100(6):1808–1814
Boyer JS (1982) Plant productivity and environment. Science 218:443–448
Brady NC, Weil R (2004) Elements of the nature and properties of soils, 2nd edn. Pearson Prentice Hall, Upper Saddle River, NJ
Brendel O, Le Thiec D, Saintagne C, Bodénès C, Kremer A, Guehl JM (2008) Detection of quantitative trait loci controlling water use efficiency and related traits in *Quercus robur* L. Tree Genet Genome 4:263–278
Brewbaker JL (1967) The distribution and phylogenetic significance of binucleate and trinucleate pollen grains in the angiosperms. Am J Bot 54(9):1069–1083
Briat JF, Lobréaux S (1997) Iron transport and storage in plants. Trends Plant Sci 2:187–193
Brijwal L, Tamta S (2015) *Agrobacterium rhizogenes* mediated hairy root induction in endangered *Berberis aristata* DC. Spring 4:443
Brown GN, Bixby JA (1976) Relationship between black locust seedling age and induction of cold hardiness. For Sci 22:208–210
Bsoul E, St. Hilaire R, Van Leeuwen DM (2006) Bigtooth maples exposed to asynchronous cyclic irrigation show provenance differences in drought adaptation mechanisms. J Am Soc Hortic Sci 131(4):459–468
Bueno MA, Manzanera JA (2003) Oak anther culture. In: Maluszynski M, Kasha KJ, Forster BP, Szarejko I (eds) Doubled haploid production in crop plants – a manual. Kluwer Academic Publishers, Dordrecht
Buiteveld J, Van Der Werf B, Hiemstra JA (2015) Comparison of commercial elm cultivars and promising unreleased Dutch clones for resistance to *Ophiostoma novo-ulmi*. iForest 8:158–164
Burr KE, Tinus RW, King RM (1990) Comparison of three cold hardiness tests for conifer seedlings. Tree Physiol 6:351–369
Cai M, Wang K, Luo L, Pan HT, Zhang QX (2015) Production of interspecific hybrids between *Hydrangea macrophylla* and *Hydrangea arborescens* via ovary culture. Hortscience 50(12):1765–1769
Callahan A, Dardick C, Tosetti R, Lalli D, Scorza R (2015) 21st century approach to improving Burbank's Stoneless' plum. Hortscience 50:195–200
Casanova E, Trillas MI, Moysset L, Vainstein A (2005) Influences of rol genes in floriculture. Biotechnol Adv 23:3–39
Caser M, Marinoni DT, Scariot V (2010) Microsatellite-based genetic relationships in the genus *Camellia*: potential for improving cultivars. Genome 53(5):384–399
Catarcione G, Vittori D, Bizzarri S, Rugini E, De Pace C (2013) Hazelnut (*Corylus avellana*) genetic resources for the improvement of phenology, pest resistance and seed size traits. Acta Hort 976:91–97
Cerbah M, Mortreau E, Brown S, Siljak-Yakovlev S, Bertrand H, Lambert C (2001) Genome size variation and species relationships in the genus *Hydrangea*. Theor Appl Genet 103(1):45–51
Cerqueira-Silva CBM, Jesus ON, Santos ESL, Corrêa RX, Souza AP (2014) Genetic breeding and diversity of the genus *Passiflora*: progress and perspectives in molecular and genetic studies. Int J Mol Sci 15(8):14122–14152
Chagné D, Crowhurst RN, Troggio M, Davey MW, Gilmore B, Lawley C, Vanderzande S, Hellens RP, Kumar S, Cestaro A, Velasco R, Main D, Rees JD, Iezzoni A, Mockler T, Wilhelm L, Van de Weg E, Gardiner SE, Bassil N, Peace C (2012) Genome-wide SNP detection, validation, and development of an 8K SNP array for apple. PLoS One 7:e31745
Chagné D, Crowhurst RN, Pindo M, Thirimawithana A, Deng C, Ireland H, Fiers M, Dzierzon H, Cestaro A, Fontana P, Bianco L, Lu A, Storey R, Knäbel M, Saeed M, Montanari S, Kim YK, Nicolini D, Larger S, Stefani E, Allan AC, Bowen J, Harvey I, Johnston J, Malnoy M, Troggio M, Perchepied L, Sawyer G, Wiedow C, Won K, Viola R, Hellens RP, Brewer L, Bus VG, Schaffer RJ, Gardiner SE, Velasco R (2014) The draft genome sequence of European pear (*Pyrus communis* L. Bartlett). PLoS One 9:e92644

Chauhan N, Negi MS, Sabharwal V, Khurana DK, Lakshmikumaran M (2004) Screening interspecific hybrids of *Populus* (*P-ciliata* x *maximowiczii*) using AFLP markers. Theor Appl Genet 108(5):951–957

Chavez DJ, Lyrene PM (2009) Production and identification of colchicine-derived tetraploid *Vaccinium darrowii* and its use in breeding. J Am Soc Hortic Sci 134(3):356–363

Chen X, Zhang J, Liu Q, Guo W, Zhao T, Ma Q, Wang G (2014) Transcriptome sequencing and identification of cold tolerance genes in hardy *Corylus* species(*C. heterophylla* Fisch) floral buds. PLoS One 9(9):e108604

Chen H, Lu C, Jiang H, Peng J (2015a) Global Transcriptome analysis reveals distinct aluminum-tolerance pathways in the Al-accumulating species *Hydrangea macrophylla* and marker identification. PLoS One 10(12):e0144927

Chen H, Song Y, Li LT, Khan MA, Li XG, Korban SS, Wu J, Zhang SL (2015b) Construction of a high-density simple sequence repeat consensus genetic map for pear (*Pyrus* spp.). Plant Mol Biol Reprod 33:316–325

Chen Q, Wang Y, Nan H, Zhang L, Tang H, Wang X (2015c) Meiotic configuration and rDNA distribution patterns in six *Rubus* taxa. Indian Journal of Genetics and Plant Breeding 75(2):242–249

Cheng F, Aoki N (2008) Crosses of Chinese and Japanese tree peony and a primary study on in vitro culture of hybrid embryos. Acta Hort 766:367

Cheng YJ, Guo WW, Deng XX (2007) Inter-generic mitochondrial genome recombination in somatic hybrids between *Citrus* and *Poncirus*. J Hort Science and Biotechnology 82:849–854

Chianucci F, Cutini A, Corona P, Puletti N (2014) Estimation of leaf area index in understory deciduous trees using digital photography. Agric For Meteorol 198-199:259–264

Chien CT, Chen SY, Tsai CC, Baskin JM, Baskin CC, KuoHuang LL (2011) Deep simple epicotyl morphophysiological dormancy in seeds of two *Viburnum* species, with special reference to shoot growth and development inside seed. Ann Bot 108(1):13–22

Christensen B, Sriskandarajah S, Müller R (2009) Transformation of *Hibiscus rosa-sinensis* L. by *Agrobacterium rhizogenes*. J Hortic Sci Biotechnol 84(2):204–208

Christey MC (2001) Invited review: use of Ri-mediated transformation for production of transgenic plants. In Vitro Cell Dev Biol Plant 37:687–700

Cimo G, Marchese A, Germana MA (2017) Microspore embryogenesis induced through in vitro anther culture of almond (*Prunus dulcis* Mill.). Plant Cell Tissue Org Cult 128(1):85–95

Clark JR (2016) Breeding southern US blackberries, idea to industry. Acta Hort 1133:3–11

Clark LV, Jasieniuk M (2012) Spontaneous hybrids between native and exotic *Rubus* in the Western United States produce offspring both by apomixis and by sexual recombination. Heredity 109(5):320–328

Clement WL, Arakaki M, Sweeney PW, Edwards EJ, Donoghue MJ (2014) A chloroplast tree for *Viburnum* (*Adoxaceae*) and its implications for phylogenetic classification and character evolution. Am J Bot 101(6):1029–1049

Cochran PH, Berntsen CM (1973) Tolerance of lodgepole and ponderosa pine seedlings to low night temperatures. For Sci 19:272–280

Coelho MSE, Bortoleti KCD, de Araujo FP, de Melo NF (2016) Cytogenetic characterization of the *Passiflora edulis* Sims x *Passiflora cincinnata* mast. interspecific hybrid and its parents. Euphytica 210(1):93–104

Conceicao LDHCS, Belo GO, Souza MM, Santos SF, Cerqueira-Silva CBM, Correa RX (2011) Confirmation of cross-fertilization using molecular markers in ornamental passion flower hybrids. Genet Mol Res 10(1):47–52

Connolly BA, Anderson GJ, Brand MH (2013) Occurrence and fertility of feral hybrid barberry *Berberis* X *ottawensis* (Berberidaceae) in Connecticut and Massachusetts. Rhodora 115(962):121–132

Contreras RN (2012) A simple chromosome doubling technique is effective for three species of Cupressaceae. Hortscience 47(6):712–714

Contreras RN (2017) The struggle is real (and fun!): long term breeding at a public university. Combined Proceedings of the International Plant Propagators' Society, Wilsonville, Oregon, USA. 67 (In press)

Contreras RN, Friddle MW (2015a) Fire blight resistance among interspecific and interploidy F1 hybrids of *Cotoneaster*. Proc. 60th Ann. SNA Res. Conf 60:186–190

Contreras RN, Friddle MW (2015b) 'Oregon Snowflake' flowering currant. Hortscience 52(2):320–321

Contreras RN, Meneghelli L (2016) In vitro chromosome doubling of Prunus laurocerasus 'Otto Luyken' and 'Schipkaensis'. Hortscience 51(12):1463–1466

Contreras RN, Ruter JM (2009) An Oryzalin induced Polyploid from a hybrid of *Hibiscus acetosella* x *H. radiatus* (Malvaceae) exhibits reduced fertility and altered morphology. Hortscience 44(4):1177

Contreras RN, Ruter JM, Hanna WW (2009) An Oryzalin-induced Autoallooctoploid of *Hibiscus acetosella* 'Panama Red'. J Am Soc Hortic Sci 134(5):553–559

Contreras R, Rothleutner J, Stockwell VO (2014) Breeding for fire blight resistance and sterility in *Cotoneaster*. Acta Hort 1056:221–224

Correa MF, Monteiro-Hara ACBA, Mello APOA, Harakava R, Rezende JAM, Mendes BMJ (2013) Progress towards genetic transformation of Passiflora edulis f. Flavicarpa for resistance to cowpea-aphid borne mosaic virus (CABMV). In Vitro Cellular & Developmental Biology-Animal 49:S59

Correa MF, Pinto APC, Rezende JAM, Harakava R, Mendes BMJ (2015) Genetic transformation of sweet passion fruit (*Passiflora alata*) and reactions of the transgenic plants to. Cowpea aphid borne mosaic virus European Journal of Plant Pathology 143(4):813–821

Cregg BM (2004) Improving drought tolerance of trees: theoretical and practical considerations. Acta Hort 630:147–158

Crespel L, Morel P (2014) Pollen viability and meiotic behaviour in intraspecific hybrids of *Hydrangea aspera* subsp. aspera Kawakami group x subsp. sargentiana. Plant Breed 133(4):536–541

Cuenoud P, Martinez MAD, Loizeau PA, Spichiger R, Andrews S, Manen JF (2000) Molecular phylogeny and biogeography of the genus *Ilex* L. (aquifoliaceae). Ann Bot 85:111–122

Cusin R, Revers LF, Maraschin FD (2017) New biotechnological tools to accelerate scab-resistance trait transfer to apple. Genet Mol Biol 40(1):305–311

Czarnecki D, Deng Z (2008) The effects of cultivar, ploidy level, direction of pollination, and temperature on seed set and production of triploids in *Lantana camara*. Hortscience 43(4):1093–1094

Czarnecki DM, Deng Z (2009) Occurrence of unreduced female gametes leads to sexual polyploidization in *Lantana*. J Am Soc Hortic Sci 134(5):560–566

Czarnecki D, Deng Z, Clark DG (2008) Assessment of ploidy levels, pollen viability, and seed production of Lantana camara cultivars and breeding lines. Hortscience 43(4):1195–1196

Czernicka M, Plawiak J, Muras P (2014) Genetic diversity of F1 and F2 interspecific hybrids between dwarf birch (Betula nana L.) and Himalayan birch (B. utilis var. jacquemontii (Spach) Winkl. 'Doorenbos') using RAPD-PCR markers and ploidy analysis. Acta Biochemica Polonica 61(2):195–199

Daccord N, Celton JM, Linsmith G, Becker C, Choisne N, Schijlen E, Van de Geest H, Bianco L, Micheletti D, Velasco R, Di Pierro EA, Gouzy J, Rees DJG, Guérif P, Muranty H, Durel CE, Laurens F, Lespinasse Y, Gaillard S, Aubourg S, Quesneville H, Weigel D, Van De Weg E, Troggio M, Bucher E (2017) High-quality de novo assembly of the apple genome and methylome dynamics of early fruit development. Nat Genet 49:1099–1106

Dai A (2011) Drought under global warming: a review. Adv Rev 2:45–65

Dai H, Liu S, Xiao D (2016) Botanical traits and cold hardiness of interspecific hybrids between European and Chinese raspberries. Acta Hort 1133:61–65

Das SK, Sabhapondit S, Ahmed G, Das S (2013) Biochemical evaluation of triploid progenies of diploid x Tetraploid breeding populations of *Camellia* for genotypes rich in Catechin and Caffeine. Biochem Genet 51(5–6):358–376

Datta SK (1995) Induced mutation for plant domestication: *Lantana depressa* Naud. Proc of the Indian National Science Academy Part B Biological Sciences 61:73–78

Davidson J, Werres S, Garbelotto M, Hansen E, Rizzo D (2003) Sudden oak death and associated diseases caused by *Phytophthora ramorum*. Plant Health Progress 23:372–379

de Melo CAF, Silva GS, Souza MM (2015) Establishment of the genomic in situ hybridization (GISH) technique for analysis in interspecific hybrids of *Passiflora*. Genet Mol Res 14(1):2176–2188

de Melo CAF, Silva GS, Souza MM (2016a) Using FISH and GISH techniques for recombination analysis on BC1 interspecific hybrids involving *Passiflora sublanceolata*. Cytogenet Genome Res 148(2–3):106–107

de Melo CAF, Souza MM, Viana AP, Santos EA, Souza VDO, Correa RX (2016b) Morphological characterization and genetic parameter estimation in backcrossed progenies of *Passiflora* L. for ornamental use. Sci Hortic 212:91–103

de Melo CAF, Souza MM, Silva GS (2017) Karyotype analysis by FISH and GISH techniques on artificial backcrossed interspecific hybrids involving *Passiflora sublanceolata* (Killip)

MacDougal (Passifloraceae). Euphytica 213(8):161
de Mesquita Dantas AC, Boneti JI, Nodari RO, Guerra MP (2006) Embryo rescue from interspecific crosses in apple rootstocks. Pesquisa Agropecuaria Brasileira 41(6):969–973
Delaporte KL, Conran JG, Sedgley M (2001a) Interspecific hybridization between three closely related ornamental *Eucalyptus* species: *E. macrocarpa*, *E. youngiana* and *E. pyriformis*. Journal of Horticultural Science & Biotechnology 76(4):384–391
Delaporte KL, Conran JG, Sedgley M (2001b) Interspecific hybridization within *Eucalyptus* (Myrtaceae): Subgenus Symphyomyrtus, sections Bisectae and Adnataria. Int J Plant Sci 162(6):1317–1326
Denaeghel H, Van Laere K, Leus L, Van Huylenbroeck J, Van Labeke MC (2015) Induction of tetraploids in *Escallonia* spp. Acta Hort 1087:453–458
Denaeghel H, Van Laere K, Leus L, Van Huylenbroeck J, Van Labeke MC (2017) Interspecific hybridization in *Sarcococca* supported by analysis of ploidy level, genome size and genetic relationships. Euphytica 213(7):149
Denaeghel H, Van Laere K, Leus L, Lootens P, Van Huylenbroeck J, van Labeke MC (2018) The variable effect of polyploidization on the phenotype in *Escallonia*. Accepted for publication in Frontiers, Frontiers in Plant Science, 9(354)
Deng YM, Sun XB, Gu GS, Jia XP, Liang LJ, Su JL (2017a) Identification of pre-fertilization reproductive barriers and the underlying cytological mechanism in crosses among three petal-types of *Jasminum sambac* and their relevance to phylogenetic relationships. PLoS One 12(4):e0176026
Deng ZN, Wilson SB, Ying XB, Czarnecki DM (2017b) Infertile *Lantana camara* cultivars UF-1011-2 and UF-1013A-2A. Hortscience 52(4):652–657
DeSutter T, Prunty L, Bell J (2011) Concrete grinding residue characterization and influence on infiltration. J Environ Qual 40:1–6
Dhekney SA, Li ZT, Gray DJ (2011) Grapevines engineered to express cisgenic *Vitis vinifera* thaumatin-like protein exhibit fungal disease resistance in vitro cell. Dev Biol Plant 47:458–466
Dhooghe E, Van Laere K, Eeckhaut T, Leus L, Van Huylenbroeck J (2011) Mitotic chromosome doubling of plant tissues in vitro. Plant Cell Tissue Organ Cult 104(3):359–373
Dickinson GR, Lee DJ, Wallace HM (2012) The influence of pre- and post-zygotic barriers on interspecific *Corymbia* hybridization. Ann Bot 109(7):1215–1226
Dirr MA (2011) Dirr's encyclopedia of trees and shrubs. Timber Press, Portland
Dogan A, Siyakus G, Severcan F (2007) FTIR spectroscopic characterization of irradiated hazelnut (*Corylus avellana* L.). Food Chem 100(3):1106–1114
Dong Y, Fan G, Deng M, Xu E, Zhao Z (2014a) Genome-wide expression profiling of the transcriptomes of four *Paulownia tomentosa* accessions in response to drought. Genomics 104:295–305
Dong Y, Fan G, Zhao Z, Deng M (2014b) Transcriptome expression profiling in response to drought stress in *Paulownia australis*. Int J Mol Sci 15(3):4583–4607
Dong B, Wu B, Hong W, Li X, Li Z, Xue L, Huang Y (2017) Transcriptome analysis of the tea oil Camellia (*Camellia oleifera*) reveals candidate drought stress genes. PLoS One 12(7):e0181835
Donoso JM, Eduardo I, Picanol R, Batlle I, Howad W, Aranzana MJ, Arus P (2015) High-density mapping suggests cytoplasmic male sterility with two restorer genes in almond x peach progenies. Horticult Res 2:15016
Donoso JM, Picanol R, Serra O, Howad W, Alegre S, Arus P, Eduardo I (2016) Exploring almond genetic variability useful for peach improvement: mapping major genes and QTLs in two interspecific almond x peach populations. Mol Breed 36(2):16
Dosmann M, Del Tredici P (2003) Plant introduction, distribution, and survival: a case study of the 1980 Sino-American botanical expedition. Bioscience 53:588–597
Dresselhaus T, Franklin-Tong N (2013) Male-female crosstalk during pollen germination, tube growth and guidance, and double fertilization. Mol Plant 6:1018–1036
Dudits D, Török K, Cseri A, Paul K, Nagy AV, Nagy B, Sass L, Ferenc G, Vankova R, Dobrev P, Vass I, Ayaydin F (2016) Response of organ structure and physiology to autotetraploidisation in early development of energy Willow *Salix viminalis*. Plant Physiol 170:1504–1523
Dunn BL, Lindstrom JT (2007) Oryzalin-induced chromosome doubling in *Buddleja* to facilitate interspecific hybridization. Hortscience 42:1326–1328
Duron M (1992) Induced mutations through EMS treatment after adventitious bud formation on shoot internodes of *Weigela* cv. Bristol Ruby Acta Hort 320:113–118
Eeckhaut T, Van Laere K, De Riek J, Van Huylenbroeck J (2006) Overcoming interspecific barriers in ornamental plant breeding. In: Teixeira Da Silva JA (ed) Floriculture, ornamental and plant

biotechnology: advances and topical issues. Global Science Books, London, UK

Ehlenfeldt MK, Polashock JJ (2014) Highly fertile intersectional blueberry hybrids of *Vaccinium padifolium* section Hemimyrtillus and *V. corymbosum* section Cyanococcus. J Am Soc Hortic Sci 139(1):30–38

Ehlenfeldt MK, Rowland LJ, Ogden EL, Vinyard BT (2009) Cold hardiness of southern-adapted blueberry (*Vaccinium* x hybrid) genotypes and the potential for their use in northern-adapted blueberry breeding. Plant Breed 128(4):393–396

Elliott W, Werner DJ, Fantz PR (2004) A hybrid of *Buddleja davidii* var. nanhoensis 'Nanho Purple' and *B-lindleyana*. Hortscience 39(7):1581–1583

Esmenjaud D, Voisin R, Van Ghelder C, Bosselut N, Lafargue B, Di Vito M, Dirlewanger E, Poessel JL, Kleinhentz M (2009) Genetic dissection of resistance to root-knot nematodes *Meloidogyne* spp. in plum, peach, almond, and apricot from various segregating interspecific *Prunus* progenies. Tree Genet Genomes 5(2):279–289

Fabio ES, Kemanian AR, Montes F, Miller RO, Smart LB (2017) A mixed model approach for evaluating yield improvements in interspecific hybrids of shrub willow, a dedicated bioenergy crop. Ind Crop Prod 96:57–70

Fagundez J (2012) Study of some European wild hybrids of *Erica* L. (Ericaceae), with descriptions of a new nothospecies: *Erica x nelsonii* Fagundez and a new nothosubspecies: *Erica* x *veitchii* nothosubsp asturica Fagundez. Candollea 67(1):51–57

Fahlgren N, Gehan MA, Baxter I (2015) Lights, camera, action: high-throughput plant phenotyping is ready for close-up. Curr Opin Plant Biol 24:93–99

Fan F, Cui B, Zhang T, Qiao G, Ding G, Wen X (2014) The temporal transcriptomic response of *Pinus massoniana* seedlings to phosphorus deficiency. PLoS One 9:e105068

Fan D, Liu T, Li C, Jiao B, Li S, Hou Y, Luo K (2015) Efficient CRISPR/Cas9-mediated targeted mutagenesis in *Populus* in the first generation. Sci Rep 5:12217

Faraloni C, Cutino I, Petruccelli R, Leva AR, Lazzeri S, Torzillo G (2011) Chlorophyll fluorescence technique as a rapid tool for in vitro screening of olive cultivars (*Olea europaea* L.) tolerant to drought stress. Environ Exp Bot 73:49–56

Fernandez IMA, Font IFC, Socias ICR, Rubio-Cabetas MJ (2015) Genetic relationships and population structure of local olive tree accessions from Northeastern Spain revealed by SSR markers. Acta Physiol Plant 37:1726–1735

Fernando DD, Long SM, Sniezko RA (2005) Sexual reproduction and crossing barriers in white pines: the case between *Pinus lambertiana* (sugar pine) and *P-monticola* (western white pine). Tree Genet Genomes 1(4):143–150

Fetouh MI, Kareem A, Knox GW, Wilson SB, Deng Z (2016) Induction, identification, and characterization of Tetraploids in Japanese rivet (*Ligustrum japonicum*). Hortscience 51(11):1371–1377

Fillatti JJ, Kiser J, Rose R, Comai L (1987) Efficient transfer of a glyphosate tolerance gene into tomato using a binary *Agrobacterium tumefaciens* vector. Bio Technol 5:726–730

Finn CE, Strik BC, Mackey TA, Hummer KE, Martin RR (2015) Perpetua ornamental reflowering blueberry. Hortscience 50:1828–1829

Fischer TC, Malnoy M, Hofmann T, Schwab W, Palmieri L, Wehrens R, Schuch LA, Muller M, Schimmelpfeng H, Velasco R, Martens S (2014) F-1 hybrid of cultivated apple (*Malus* x *domestica*) and European pear (*Pyrus communis*) with fertile F-2 offspring. Mol Breed 34(3):817–828

Fladung M, Hoenicka H, Ahuja MR (2013) Genomic stability and long-term transgene expression in poplar. Transgenic Res 22(6):1167–1178

Fowler S, Thomashow MF (2002) *Arabidopsis* transcriptome profiling indicates that multiple regulatory pathways are activated during cold acclimation in addition to the CBF cold responsive pathway. Plant Cell 14:1675–1690

Franco JA, Martínez-Sánchez JJ, Fernández JA, Bañón S (2006) Selection and nursery production of ornamental plants for landscaping and xerogardening in semi-arid environments. J Hort Sci Biotech 81(1):3–17

Freeman OM (1941) A red maple, silver maple hybrid. J Hered 32:11–14

Fritsch PW, Cruz BC (2012) Phylogeny of *Cercis* based on DNA sequences of nuclear ITS and four plastid regions: implications for transatlantic historical biogeography. Mol Phylogenet Evol 62(3):816–825

Fryer J, Hylmö B (2009) *Cotoneaster*s: a comprehensive guide to shrubs for flowers, fruit, and foliage. Timber Press, Portland

Fu XX, Liu HN, Zhou XD, Shang XL (2013) Seed dormancy mechanism and dormancy breaking techniques for *Cornus kousa* var. chinensis. Seed Sci Technol 41(3):458–463

Furbank RT, Tester M (2011) Phenomics – technologies to relieve the phenotyping bottleneck. Trends Plant Sci 16(12):635–644

Gallardo F, Fu J, Canton FR, Garcia-Gutierrez A, Canovas FM, Kirby EG (1999) Expression of a conifer glutamine synthetase gene in transgenic poplar. Planta 210:19–26

Gallone A, Hunter A, Douglas GC (2012) Radiosensitivity of *Hebe* 'Oratia Beauty' and 'Wiri Mist' irradiated in vitro with gamma-rays from Co-60. Sci Hortic 138:36–42

Gallone A, Hunter A, Douglas GC (2014) Polyploid induction in vitro using colchicine and oryzalin on *Hebe* 'Oratia Beauty': production and characterization of the vegetative traits. Sci Hortic 179:59–66

Gardner N, Hokanson SC (2005) Intersimple sequence repeat fingerprinting and genetic variation in a collection of *Clematis* cultivars and commercial germplasm. Hortscience 40:1982–1987

Geerts P, Henneguez A, Druart P, Watillon B (2009) Protoplast electro-fusion technology as a tool for somatic hybridisation between strawberries and raspberries. Acta Hort 842:495–498

Gehesquière B (2014) *Cylindrocladium buxicola* nom. Cons; prop. (syn. *Calonectria pseudonaviculata*) on *Buxus*: molecular characterisation, epidemiology, host resistance and fungicide control. In: Phd thesis. Ghent University, Belgium

Germana MA (2003) Haploids and doubled haploids in *Citrus* ssp. In: Maluszynski M, Kasha KJ, Forster BP, Szarejko I (eds) Doubled haploid production in crop plants – a manual. Kluwer Academic Publishers, Dordrecht

Germana MA, Chiancone B, Lain O, Testolin R (2005) Anther culture in *Citrus clementina*: a way to regenerate tri-haploids. Aust J Agric Res 56(8):839–845

Gill GP, Brown GR, Neale DB (2003) A sequence mutation in the cinnamyl alcohol dehydrogenase gene associated with altered lignification in loblolly pine. Plant Biotechnol J 1:253–258

Gindaba J, Rozanov A, Negash L (2004) Response of seedlings of two *Eucalyptus* and three deciduous tree species from Ethiopia to severe water stress. For Ecol Manag 201(1):121–131

Giongo L, Palmieri L, Grassi A, Grisenti M, Poncetta P, Velasco R (2012) Phenotyping and genotyping of *Rubus* Germplasm for the improvement of quality traits in the raspberry breeding program. Acta Hort 946:77–81

Giovannini A, Dente F, De Benedetti L, Nicoletti F, Braglia L, Gavazzi F, Mercuri A (2012) Interspecific hybridization in ornamental passion flowers. Acta Hort 953:111–118

Gonai T, Terakami S, Nishitani C, Yamamoto T, Kasumi M (2012) Fine mapping of the scab resistance gene of Japanese pear 'Kinchaku' for efficient marker-assisted selection. Bull Ibaraki Plant Biotech Inst 12:27–33

Gosling P (2007) Raising trees and shrubs from seed – practice guide. Forestry Commission, Edinburgh, UK

Granados Mendoza C, Wanke S, Goetghebeur P, Samain MS (2013) Facilitating wide hybridization in hydrangeas. l. Cultivars: a phylogenetic and marker-assisted breeding approach. Mol Breed 32(1):233–239

Granados Mendoza C, Naumann J, Samain MS, Goetghebeur P, De Smet Y, Wanke S (2015) A genome-scale mining strategy for recovering novel rapidly-evolving nuclear single-copy genes for addressing shallow-scale phylogenetics in *Hydrangea*. BMC Evol Biol 15:132

Grant JE, Cooper PA, Dale TM (2004) Transgenic *Pinus radiata* from *Agrobacterium tumefaciens*-mediated transformation of cotyledons. Plant Cell Rep 22:894–902

Graves WR, Schrader JA (2004) 'September Sun' seaside alder: an autumn-blooming shrub native to North America. Hortscience 39:438–439

Greer S, Adkins J, Reed S (2008) Increasing phenotypic diversity in *Hydrangea macrophylla* using targeted and random mutation. Hortscience 43(4):1101

Griffin AR, Vuong TD, Vaillancourt RE, Harbard JL, Harwood CE, Nghiem CQ, Ha TH (2012) The breeding systems of diploid and neoautotetraploid clones of *Acacia mangium* Willd. In a synthetic sympatric population in Vietnam. Sex Plant Reprod 25(4):257–265

Griffin AR, Chi NQ, Harbard JL, Son DH, Harwood CE, Price A, Vuong TD, Koutoulis A, Thinh NH (2015) Breeding polyploid varieties of tropical acacias: progress and prospects. Southern forests 77(1):41–50

Gu L, Hanson PJ, Post WM, Kaiser DP, Yang B, Nemani R, Pallardy SG, Meyers T (2008) The 2007 Eastern US spring freeze: increased cold damage in a warming world? Bioscience 58:253–262

Guo L, Xu W, Zhang Y, Zhang J, Wei Z (2017) Inducing triploids and tetraploids with high temperatures in *Populus* sect. Tacamahaca Plant Cell Reports 36(2):313–326

Gusta LV, Wisniewski M (2013) Understanding plant cold hardiness: an opinion. Physiol Plant 147:3–14

Guy CL (1990) Cold acclimation and freezing stress tolerance: role of protein metabolism. Annu Rev Plant Physiol Plant Mol Biol 41:187–223

Han FP, Fedak G, Quellet T, Liu B (2003) Rapid genomic changes in interspecific and intergeneric hybrids and allopolyploids of Triticeae. Genome 46:716–723

Handreck K, Black N (2002) Growing media for ornamental plants and turf. Univ. of New South Wales Press, Sydney, Australia

Hanna WW, Towill LE (1995) Long-term pollen storage. In: Janick J (ed) Plant breeding reviews, vol 13. Wiley and Sons, Inc., Oxford, UK

Hannah MA, Heyer AG, Hincha DK (2005) A global survey of gene regulation during cold acclimation in *Arabidopsis thaliana*. PLoS Genet 1:e26

Hao Q, Aoki N, Katayama J, Kako T, Cheon KS, Akazawa Y, Kobayashi N (2013) Crossability of American tree peony 'High Noon' as seed parent with Japanese cultivars to breed superior cultivars. Euphytica 191(1):35–44

Harbard JL, Griffin AR, Foster S, Brooker C, Kha LD, Koutoulis A (2012) Production of colchicine-induced autotetraploids as a basis for sterility breeding in *Acacia mangium* Willd. Forestry 85(3):427–436

Hardiman NA, Culley TM (2010) Reproductive success of cultivated *Pyrus Calleryana* (Rosaceae) and establishment ability of invasive, hybrid progeny. Am J Bot 97(10):1698–1706

Harper AL, McKinney LV, Nielsen LR, Havlickova L, Li Y, Trick M, Sollars ESA, Janacek SH, Downie JA, Buggs RJA, Kjaer D, Bancroft I (2016) Molecular markers for tolerance of European ash (*Fraxinus excelsior*) to dieback disease identified using associative transcriptomics. Sci Rep 6:19335

Harshman JM, Evans K (2015) Survey of moldy Core incidence in Germplasm from the three US apple breeding programs. Journal of the American Pomological Society 69(1):51–57

Harshman JM, Evans KM, Allen H, Potts R, Flamenco J, Aldwinckle HS, Wisniewski ME, Norelli JL (2017) Fire blight resistance in wild accessions of *Malus sieversii*. Plant Dis 101(10):1738–1745

Hawkins SM, Ruter JM, Robacker CD (2016) Interspecific and Intergeneric hybridization in *Dissotis* and *Tibouchina*. Hortscience 51(4):325–329

He D, Liu Y, Cai M, Pan HT, Ahang QX, Wang XY, Wang XJ (2012) Genetic diversity of *Lagerstroemia* (Lythraceae) species assessed by simple sequence repeat markers. Genet Mol Res 11(3):3522–3533

He Z, Zhan Y, Zeng F, Zhao X, Wang X (2016) Drought physiology and gene expression characteristics of *Fraxinus* interspecific hybrids. Plant Growth Regul 78(2):179–193

Hennig A, Kleinschmit JRG, Schoneberg S, Loeffler S, Janssen A, Polle A (2015) Water consumption and biomass production of protoplast fusion lines of poplar hybrids under drought stress. Front Plant Sci 6:330

Henricot B, Gorton C, Denton G, denton J (2008) Studies on the control of *Cylindrocladium buxicola* using fungicides and host resistance. Plant Dis 92:1273–1279

Hias N, Leus L, Davey MW, Vanderzande S, Van Huylenbroeck J, Keulemans J (2017) Effect of polyploidization on morphology in two apple (*Malus* x *domestica*) genotypes. Hortic Sci 44(2):55–63

Higashiyama T, Takeuchi H (2015) The mechanism and key molecules involved in pollen tube guidance. Annu Rev Plant Biol 66:393–413

Hoch WA, Zeldin EL, Nienhuis J, McCown BH (1995) Generation and identification of new *Viburnum* hybrids. J Environ Hortic 13:193–195

Hoch WA, Jung G, McCown BH (2002) Effectiveness of interspecific hybridization for incorporating birch leafminer (*Fenusa pusilla*) resistance into white-barked *Betula*. J Am Soc Hortic Sci 127(6):957–962

Hoch WA, McCown BH, Weston PA (2004) The potential of breeding for resistance to the introduced pest, viburnum leaf beetle (*Pyrrhalta viburni*). Acta Hort 630:53–55

Hoenicka H, Lehnhardt D, Nilsson O, Hanelt D, Fladung M (2014) Successful crossings with early flowering transgenic poplar: interspecific crossings, but not transgenesis, promoted aberrant phenotypes in offspring. Plant Biotechnol J 12(8):1066–1074

Höfer M (2003) In vitro androgenesis in apple. In: Maluszynski M, Kasha KJ, Forster BP, Szarejko I (eds) Doubled haploid production in crop plants – a manual. Kluwer Academic Publishers, Dordrecht

Hoffman MHA (2016) List of Names of Woody Plants/Naamlijst van Houtige Gewassen/Namenliste Gehölze/Liste des Noms des Plantes Ligneuses: International Standard ENA 2016–2020. 1080p

Hokanson SC (2010) Lights in the land of 10,000 lakes. In: Rhododendrons, Camellias, and Magnolias. Royal Horticultural Society Press, London, UK, pp 22–332

Hokanson SC, McNamara S (2013) Can't always get what we want! Finding and creating cold for hardiness screening at the University of Minnesota. Acta Hort 990:193–202

Hokanson SC, McNamara S, Zuzek K, Zins M, Rose N (2015) *Wisteria frutescens*, American wisteria, 'Betty Matthews', first editions®. Hortscience 50:317–319

Hokanson SC, Susko AQ, McNamara S (2016) A field evaluation of *Buxus* cultivars and species germplasm. Hortscience 51:S101

Homoya MA (2012) Wildflowers and ferns of Indiana forests – a field guide. Indiana University Press, Bloomington

Houtman R (2013) Euro-trials: the first example of international co-operation in plant trials. Acta Hort 980:23–27

Hu WJ, Kawaoka A, Tsai CJ, Lung JH, Osakabe K, Ebinuma H, Chiang VL (1998) Compartmentalised expression of two structurally and functionally distinct 4-coumarate:CoA ligase genes in aspen (*Populus tremuloides*). Proc Natl Acad Sci U S A 95:5407–5412

Hu L, Lu H, Liu Q, Chen X, Jiang X (2005) Overexpression of *mtlD* gene in transgenic *Populus tomentosa* improves salt tolerance through accumulation of mannitol. Tree Physiol 25:1273–1281

Huang S, Ding J, Deng D, Tang W, Sun H, Liu D, Zhang L, Niu X, Zhang X, Meng M, Yu J, Liu J, Han Y, Shi W, Zhang D, Cao S, Wei Z, Cui Y, Xia Y, Zeng H, Bao K, Lin L, Min Y, Zhang H, Miao M, Tang X, Zhu Y, Sui Y, Li G, Sun H, Yue J, Sun J, Liu F, Zhou L, Lei L, Zheng X, Liu M, Huang L, Song J, Xu C, Li J, Ye K, Zhong S, Lu BR, He G, Xiao F, Wang HL, Zheng H, Fei Z, Liu Y (2013) Draft genome of the kiwifruit *Actinidia chinensis*. Nat Commun 4:2640

Huang H, Wang S, Jiang J, Liu G, Li H, Chen S, Xu H (2014) Overexpression of BpAP1 induces early flowering and produces dwarfism in *Betula platyphylla* x *Betula pendula*. Physiol Plant 151(4):495–506

Hubackova M (1994) The grapevine buds cold hardiness in juvenile plants. Ochr Rostl 30:305–309

Hummel R (1981) Temperature, prolonged short day, environmental preconditioning, and plantage effects on the acclimation response in *Cornus sericea* L. red-osier dogwood. PhD Diss., Univ. of Minnesota, St. Paul

Hussey SG, Mizrachi E, Spokevicius AV, Bossinger G, Berger DK, Myburg AA (2011) SND2, a NAC transcription factor gene, regulates genes involved in secondary cell wall development in *Arabidopsis* fibres and increases fibre cell area in *Eucalyptus*. BMC Plant Biol 11:173

Ingvarsson PK (2005) Nucleotide polymorphism and linkage disequilibrium within and among natural populations of European aspen (*Populus tremula* L. Salicaceae). Genetics 169:945–953

Isoda K, Shiraishi S, Watanabe S, Kitamura K (2000) Molecular evidence of natural hybridization between *Abies veitchii* and *A-homplepis* (Pinaceae) revealed by chloroplast, mitochondrial and nuclear DNA markers. Mol Ecol 9(12):1965–1974

Iwanaga F, Tanaka K, Nakazato I, Yamamoto F (2015) Effects of submergence on growth and survival of saplings of three wetland trees differing in adaptive mechanisms for flood tolerance. Forest Systems 24(1):e-001

Jaillon O, Aury JM, Noel B, Policriti A, Clepet C, Casagrande A, Choisne N, Aubourg S, Vitulo N, Jubin C, Vezzi A, Legeai F, Hugueney P, Dasilva C, Horner D, Mica E, Jublot D, Poulain J, Bruyère C, Billault A, Segurens B, Gouyvenoux M, Ugarte E, Cattonaro F, Anthouard V, Vico V, Del Fabbro C, Alaux M, Di Gaspero G, Dumas V, Felice N, Paillard S, Juman I, Moroldo M, Scalabrin S, Canaguier A, Le Clainche I, Malacrida G, Durand E, Pesole G, Laucou V, Chatelet P, Merdinoglu D, Delledonne M, Pezzotti M, Lecharny A, Scarpelli C, Artiguenave F, Pè ME, Valle G, Morgante M, Caboche M, Adam-Blondon AF, Weissenbach J, Quétier F, Wincker P (2007) The grapevine genome sequence suggests ancestral hexaploidization in major angiosperm phyla. Nature 447(7161):463–467

Jansen M, Gilmer F, Biskup B, Nagel KA, Fischbach A, Briem S, Dreissen G, Tittmann S, Braun S, De Jaeger I, Metzlaff M, Schurr U, Scharr H, Walter A (2009) Simultaneous phenotyping of leaf growth and chlorophyll fluorescence via GROWSCREEN FLUORO allows detection of stress tolerance in *Arabidopsis thaliana* and other rosette plants. Funct Plant Biol 36(11):902–914

Jeong SC, Liston A, Myrold DD (1997) Molecular phylogeny of the genus Ceanothus (Rhamnaceae) using rbcL and ndhF sequences. Theor Appl Gen 94:852–857

Jia H, Zhang Y, Orbovic V, Xu J, White F, Jones J, Wang N (2017) Genome editing of the disease susceptibility gene CsLOB1 in *Citrus* confers resistance to citrus canker. Plant Biotechnol J 15(7):817–823

Jimenez-Bello MA, Ballester C, Castel JR, Intrigliolo DS (2011) Development and validation of an automatic thermal imaging process for assessing plant water status. Agric Water Manag 98:1497–1504

Jones KA, Reed SM (2006) Production and verification of *Hydrangea arborescens* 'Dardom' x *H-involucrata* hybrids. Hortscience 41(3):564–566

Jones AMP, Shukla MR, Biswas GCG, Saxena PK (2015) Protoplast-to-plant regeneration of American elm (*Ulmus americana*). Protoplasma 252(3):925–931

Joung YH, Roh MS, Bentz SE (2001) Characterization of *Acer griseum* and its putative interspecific hybrids. Acta Hort 546:217–220

Kadota M, Han DS, Niimi Y (2002) Plant regeneration from anther-derived embryos of apple and pear. Hortscience 37(6):962–965

Kaestner U, Klocke E, Abel S (2017) Regeneration of protoplasts after somatic hybridisation of *Hydrangea*. Plant Cell Tissue Org Cult 129(3):359–373

Kalberer SR, Wisniewski M, Arora R (2006) Deacclimation and reacclimation of cold-hardy plants: current understanding and emerging concepts. Plant Sci 171:3–16

Kardos JH, Robacker CD, Dirr MA, Rinehart TA (2009) Production and verification of *Hydrangea macrophylla* x *H. angustipetala* hybrids. Hortscience 44(6):1534–1537

Karlsdottir L, Hallsdottir M, Thorsson AT, Anamthawat-Jonsson K (2009) Evidence of hybridisation between *Betula pubescens* and *B. nana* in Iceland during the early Holocene. Rev Palaeobot Palynol 156(3–4):350–357

Karlsson LM, Hidayati SN, Walck JL, Milberg P (2005) Complex combination of seed dormancy and seedling development determine emergence of *Viburnum tinus* (Caprifoliaceae). Ann Bot 95(2):323–330

Kato K, Yamaguchi S, Chigira O, Ogawa Y, Isoda K (2012) Tube pollination using stored pollen for creating *Acacia auriculiformis* hybrids. J Trop For Sci 24(2):209–216

Kato K, Yamaguchi S, Chigira O, Hanaoka S (2014) Comparative study of reciprocal crossing for establishment of *Acacia* hybrids. J Trop For Sci 26(4):469–483

Khadivi-Khub A, Sarooghi F, Abbasi F (2016) Phenotypic variation of *Prunus scoparia* germplasm: implications for breeding. Sci Hortic 207:193–202

Kim K (2003) An integrated approach to realizing the potential of Asian plant germplasm. Acta Hort 620:383–387

Kim YD, Kim SH (1999) Phylogeny of *Weigela* and *Diervilla* (Caprifoliaceae) based on nuclear rDNA ITS sequences: biogeographic and taxonomic implications. J Plant Res 112:331–341

Kim YM, Kim S, Koo N, Shin AY, Yeom SI, Seo E, Park SJ, Kang WH, Kim MS, Park J, Jang I, Kim PG, Byeon I, Kim MS, Choi J, Ko G, Hwang JH, Yang TJ, Choi SB, Lee JM, Lim KB, Lee J, Choi IY, Park BS, Kwon SY, Choi D, Kim RW (2017) Genome analysis of *Hibiscus syriacus* provides insights of polyploidization and indeterminate flowering in woody plants. DNA Res 24(1):71–80

Kleinknecht GJ, Lintz HE, Kruger A, Niemeier JJ, Salino-Hugg MJ, Thomas CK, Still CJ, Kim Y (2015) Introducing a sensor to measure budburst and its environmental drivers. Front Plant Sci 6, article 123:1–11

Klocko AL, Borejsza-Wysocka E, Brunner AM, Shevchenko O, Aldwinckle H, Strauss SH (2016) Transgenic suppression of AGAMOUS genes in apple reduces fertility and increases floral attractiveness. PLoS One 11:e015421

Klumb EK, Arge LWP, do Amaral MN, Rickes LN, Benitez LC, Braga EJB, Bianchi VJ (2017) Transcriptome profiling of *Prunus persica* plants under flooding. Trees-Structure and Function 31(4):1127–1135

Knauft D, Dirr M (2007) Irradiation of *Lagerstroemia* to induce sterility. Hortscience 42(3):445

Knight VH, Fenandez F (2008) Screening for resistance to *Phytophthora fragariae* var. rubi in Rubus germplasm at East Malling. Acta Hort 777:353

Koch JL, Carey DW, Mason ME, Nelson CD (2010) Assessment of beech scale resistance in full- and half-sibling American beech families. Can J For Res 40:265–272

Koivuranta L, Latva-Karjanmaa T, Pulkkinen P (2012) The effect of temperature on seed quality and quantity in crosses between European (*Populus tremula*) and hybrid aspens (*P. tremula* x *P. tremuloides*). Silva Fennica 46(1):17–26

Komar-Tyomnaya L (2015) Use of wild species in ornamental peach breeding. Acta Hort 1087:415–421

Kormutak A, Vookova B, Camek V, Salaj T, Galgoci M, Manka P, Bolecek P, Kuna R, Kobliha J, Lukacik I, Goemoery D (2013) Artificial hybridization of some *Abies* species. Plant Syst Evol 299(6):1175–1184

Kormutak A, Galgoci M, Manka P, Bolecek P, Camek V, Vookova B, Goemoery D (2015) Growth characteristics and needle structure in some interspecific hybrids of *Abies cephalonica* Loud. Dendrobiology 73:47–53

Kormutak A, Galgoci M, Manka P, Koubova M, Jopcik M, Sukenikova D, Bolecek P, Gomory D (2017) Field-based artificial crossings indicate partial compatibility of reciprocal crosses between *Pinus sylvestris* and *Pinus mugo* and unexpected chloroplast DNA inheritance. Tree Genet Genomes 13(3):68

Kost TD, Gessler C, Jänsch M, Flachowsky H, Patocchi A, Broggnini GAL (2015) Development of the first cisgenic apple with increased resistance to fire blight. PLoS One 10:e0143980

Krahulcova A, Travnicek P, Krahulec F, Rejmanek M (2017) Small genomes and large seeds: chromosome numbers, genome size and seed mass in diploid *Aesculus* species (Sapindaceae). Ann Bot 119(6):957–964

Krajmerova D, Paule L, Zhelev P, Volekova M, Evtimov I, Gagov V, Gomory D (2016) Natural hybridization in eastern-Mediterranean firs: the case of *Abies borisii*-regis. Plant Biosystems 150(6):1189–1199

Kremer A, Abbott AG, Carlson JE, Manon PS, Plomion C, Sisco P, Staton ME, Ueno S, Vendramin GG (2012) Genomics of Fagaceae. Tree Genet Genomes 8(3):583–610

Kudo N (2016) Production of interspecific hybrid plants between *Hydrangea scandens* subsp chinensis and *Hydrangea macrophylla* via ovule culture. Acta Hort 1140:117–118

Kudo N, Matsui T, Okada T (2008) A novel interspecific hybrid plant between *Hydrangea scandens* ssp. chinensis and *H. macrophylla* via ovule culture. Plant Biotechnology 25(6):529–533

Külheim C, Hui Yeoh S, Maintz J, Foley WJ, Moran GF (2009) Comparative SNP diversity among four *Eucalyptus* species for genes from secondary metabolite biosynthetic pathways. BMC Genomics 10:452

Kuligowska K, Lutken H, Christensen B, Muller R (2016a) Interspecific hybridization among cultivars of hardy *Hibiscus* species section Muenchhusia. Breed Sci 66(2):300–308

Kuligowska K, Lutken H, Mueller R (2016b) Towards development of new ornamental plants: status and progress in wide hybridisation. Planta 244(1):1–17

Kumar P, Singhal VK, Kaur D, Kaur S (2010) Cytomixis and associated meiotic abnormalities affecting pollen fertility in *Clematis orientalis*. Biol Plant 54(1):181–184

Kuprian E, Munkler C, Resnyak A, Zimmermann S, Tuong TD, Gierlinger N, Müller T, Livingston DP, Neuner G (2017) Complex bud architecture and cell-specific chemical patterns enable supercooling of *Picea abies* bud primordia. Plant Cell Environ 40:3101–3112

Lam HK, Harbard JL, Koutoulis A (2014) Tetraploid induction of *Acacia crassicarpa* using colchicine and oryzalin. J Trop For Sci 26(3):347–354

Landrein S, Buerki S, Wang HF, Clarkson JJ (2017) Untangling the reticulate history of species complexes and horticultural breeds in *Abelia* (Caprifoliaceae). Ann Bot 120(2):257–269

Lapins KO (1975) Polyploidy and mutations induced in apricot by colchicine treatment. Can J Genet Cytol 17:591–599

Lapins KO (1983) Mutation breeding. In: Moore JN, Janick J (eds) Methods in fruit breeding. Purdue Univ. Press, West Lafayette, IN, pp 74–99

Larabell CA, Nugent KA (2010) Imaging cellular architecture with X-rays. Curr Opin Struct Bio 20:623–631

Larcher W (2005) Climatic constraints drive the evolution of low temperature resistance in woody plants. Journal of Agricultural Meteorology 61:189–202

Lattier JD, Contreras RN (2014) Colorimetric phenotyping of tetraploid progeny exhibiting incomplete dominance for flower color. Independent Plant Breeders Conference, Grand Rapid, MI

Lattier JD, Contreras RN (2017a) Intraspecific, interspecific, and Interseries cross-compatibility in Lilac. J Amer Soc Hort Sci 142(4):279–288

Lattier JD, Contreras RN (2017b) Ploidy and genome size in lilac species, cultivars, and interploid hybrids. J. Amer. Soc. Hort. Sci. 142(5):355–366

Lattier JD, Touchell DH, Ranney TG (2014) Micropropagation of an interspecific hybrid dogwood (*Cornus* 'ncch1'). Propagation of Ornamental Plants 14(4):184–190

Lau NS, Makita Y, Kawashima M, Taylor TD, Kondo S, Othman AS, Shu-Chien AC, Matsui M (2016) The rubber tree genome shows expansion of gene family associated with rubber biosynthesis. Sci Rep 6:28594

Le Gloanic A, Malécot V, Belin J, Heinry E, Kapusta V (2015) Interspecific hybridization in the tribe Genisteae using in vitro embryo rescue. Acta Hort 1087:315–320

Le S, Harwood CE, Griffin AR, Do SH, Ha TH, Ratnam W, Vaillancourt RE (2017) Using SSR markers for hybrid identification and resource management in Vietnamese *Acacia* breeding.

Tree Genet Genomes 13(5):102

Lehrer JM, Brand MH, Lubell JD (2008) Induction of tetraploidy in meristematically active seeds of Japanese barberry (*Berberis thunbergii* var. atropurpurea) through exposure to colchicine and oryzalin. Sci Hortic 119(1):67–71

Leksungnoen N (2012) The relationship between salinity and drought tolerance in turfgrasses and woody species. All Graduate Theses and Dissertations. Utah State University, Paper 1196

Lemcoff JH, Guarnaschelli AB, Garau AM, Prystupa P (2002) Elastic and osmotic adjustments in rooted cuttings of several clones of *Eucalyptus camaldulensis* Dehnh. From southeastern Australia after a drought. Flora 197:134–142

Lenahan OM, Graves WR, Arora R (2010) Cold-hardiness and Deacclimation of *Styrax americanus* from three provenances. Hortscience 45(12):1819–1823

Levitt J (1980) Response of plant to environmental stresses. In: I. Chilling, freezing, and high temperature stresses, 2nd edn. Academic Press, New York

Li J (2008) Molecular phylogenetics of Hamamelidaceae: evidence from DNA sequences of nuclear and chloroplast genomes. In: Sharma AK, Sharma A (eds) Plant genome: biodiversity and evolution, volI, part E: phanerograms-angiosperm. Science Publishers, pp 227–250

Li ZT, Ruter JM (2017) Development and evaluation of diploid and polyploid *Hibiscus moscheutos*. Hortscience 52(5):676–681

Li YH, Han ZH, Xu X (2004) Segregation patterns of AFLP markers in F-1 hybrids of a cross between tetraploid and diploid species in the genus *Malus*. Plant Breed 123(4):316–320

Li WD, Biswas DK, Xu H, Xu CQ, Wang XZ, Liu JK, Jiang GM (2009a) Photosynthetic responses to chromosome doubling in relation to leaf anatomy in *Lonicera japonica* subjected to water stress. Funct Plant Biol 36:783–792

Li YH, Windham MT, Trigiano RN, Reed SM, Spiers JM, Rinehart TA (2009b) Bright-field and fluorescence microscopic study of development of *Erysiphe polygoni* in susceptible and resistant bigleaf *Hydrangea*. Plant Dis 93(2):130–134

Li Y, Su X, Zhang B, Huang Q, Zhang X, Huang R (2009c) Expression of jasmonic ethylene responsive factor gene in transgenic poplar tree leads to increased salt tolerance. Tree Physiol 29:273–279

Li WD, Hu X, Liu JK, Jiang GM, Li O, Xing D (2011) Chromosome doubling can increase heat tolerance in *Lonicera japonica* as indicated by chlorophyll fluorescence imaging. Biol Plant 55(2):279–284

Li J, Goldman-Huertas B, DeYoung J, Alexander J (2012) Phylogenetics and diversification of Syringa inferred from nuclear and plastid DNA sequences. Castanea 77(1):82–88

Li Y, Li H, Chen Z, Ji LX, Ye MX, Wang J, Wang L, An XM (2013a) Haploid plants from anther cultures of poplar (*Populus* x *beijingensis*). Plant Cell Tissue Org Cult 114(1):39–48

Li L, Zhou Y, Cheng X, Sun J, Marita JM, Ralph J, Chinag VL (2013b) Combinatorial modification of multiple lignin traits in trees through multigene cotransformation. Proc Natl Acad Sci U S A 100:4939–4944

Li FF, Fan Q, Li QY, Chen SF, Guo W, Cui DF, Liao WB (2014) Molecular phylogeny of *Cotoneaster* (Rosaceae) inferred from nuclear ITS and multiple chloroplast sequences. Plant Syst Evol 300(6):1533–1546

Li B, Hulin MT, Brain P, Mansfield JW, Jackson RW, Harrison RJ (2015) Rapid, automated detection of stem canker symptoms in woody perennials using artificial neural network analysis. Plant Methods 11:57–66

Li MW, Chen SF, Zhou RC, Fan Q, Li FF, Liao WB (2017) Molecular evidence for natural hybridization between *Cotoneaster dielsianus* and *C. glaucophyllus*. Front Plant Sci 8:art. 704

Liao PC, Shih HC, Yen TB, LU SY, Cheng YP, Chiang YC (2010) Molecular evaluation of interspecific hybrids between *Acer albopurpurascens* and *A. buergerianum* var. formosanum. Bot Stud 51(4):413–420

Lim KY, Leitch IJ, Leitch AR (1998a) Genomic characterisation and the detection of raspberry chromatin in polyploid *Rubus*. Theor Appl Genet 97(7):1027–1033

Lim CC, Krebs SL, Arora R (1998b) Genetic study of freeze-tolerance in *Rhododendron* populations: implications for cold hardiness breeding. J Am Rhododendron Soc 52:143–148

Limera C, Sabbadini S, Sweet JB, Mezzetti B (2017) New biotechnological tools for the genetic improvement of major woody fruit species. Front Plant Sci 8, 1418

Lindstrom JT, Bujarski GT, Burkett BM (2004) A novel intersectional *Buddleja* hybrid. Hortscience 39:642–643

Listyanto T, Glencross K, Nichols JD, Schoer L, Harwood C (2010) Performance of eight eucalypt species and interspecific hybrid combinations at three sites in northern New South Wales,

Australia. Aust For 73(1):47–52
Liu G, Li Z, Bao M (2007a) Colchicine-induced chromosome doubling in *Platanus acerifolia* and its effect on plant morphology. Euphytica 157:145–154
Liu W, Chen X, Liu G, Liang Q, He T, Feng J (2007b) Interspecific hybridization of *Prunus persica* with *P-armeniaca* and *P-salicina* using embryo rescue. Plant Cell Tissue Org Cult 88(3):289–299
Liu R, Dong Y, Fan G, Zhao Z, Deng M, Cao X, Niu S (2013) Discovery of genes related to witches broom disease in *Paulownia tomentosa* x *Paulownia fortunei* by a De novo assembled transcriptome. PLoS One 8(11):e80238
Liu MJ, Zhao J, Cai QL, Liu GC, Wang JR, Zhao ZH, Liu P, Dai L, Yan G, Wang WJ, Li XS, Chen Y, Sun YD, Liu ZG, Lin MJ, Xiao J, Chen YY, Li XF, Wu B, Ma Y, Jian JB, Yang W, Yuan Z, Sun XC, Wei YL, Yu LL, Zhang C, Liao SG, He RJ, Guang XM, Wang Z, Zhang YY, Luo LH (2014) The complex jujube genome provides insights into fruit tree biology. Nat Commun 5:5315
Liu H, Qian C, Zhou J, Zhang X, Ma Q, Li S (2015) Causes and breaking of seed dormancy in flowering dogwood (*Cornus florida* L.). Hortscience 50(7):1041–1044
Lo EY, Stefanovic S, Dickinson TA (2009) Population genetic structure of diploid sexual and polyploid apomictic hawthorns (*Crataegus*; Rosaceae) in the Pacific Northwest. Mol Ecol 18:1145–1160
Lobo A, McKinney LV, Hansen JK, Kjaer ED, Nielsen LR (2015) Genetic variation in die-back resistance in *Fraxinus excelsior* confirmed by progeny inoculation assay. For Pathol 45(5):379–387
Lu PX, Sinclair RW, Boult TJ, Blake SG (2005) Seedling survival of *Pinus strobus* and its interspecific hybrids after artificial inoculation of *Cronartium ribicola*. For Ecol Manag 214(1–3):344–357
Lu P, Colombo SJ, Sinclair RW (2007) Cold hardiness of interspecific hybrids between *Pinus strobus* and *P-wallichiana* measured by post-freezing needle electrolyte leakage. Tree Physiol 27(2):243–250
Lubell JD, Brand MH, Lehrer JM (2008) AFLP identification of *Berberis thunbergii* cultivars, inter-specific hybrids, and their parental species. Journal of horticultural science and biotechnology 82(1):55–63
Ludwig S, Robertson A, Rich TC, Djordjević M, Cerović R, Houston L, Harris SA, Hiscock SJ (2013) Breeding systems, hybridization and continuing evolution in Avon gorge *Sorbus*. Ann Bot 111:563–575
Lütken H, Clarke JL, Müller R (2012) Genetic engineering and sustainable production of ornamentals: current status and future directions. Plant Cell Rep 31:1141–1157
Lyrene PM (2011) First report of *Vaccinium arboretum* hybrids with cultivated highbush blueberry. Hortscience 46(4):563–566
Lyrene PM (2013) Fertility and other characteristics of F-1 and backcross(1) progeny from an intersectional blueberry cross [(highbush cultivar x *Vaccinium arboreum*) x highbush cultivar]. Hortscience 48(2):146–149
Lyrene PM (2016) Phenotype and fertility of intersectional hybrids between tetraploid highbush blueberry and colchicine-treated *Vaccinium stamineum*. Hortscience 51(1):15–22
Lyrene PM, Vorsa N, Ballington JR (2003) Polyploidy and sexual polyploidization in the genus Vaccinium. Euphytica 133(1):27–36
Ma'arup R, Abd Aziz M, Osman M (2012) Development of a procedure for production of haploid plants through microspore culture of roselle (*Hibiscus sabdariffa* L.). Sci Hortic 145:52–61
Madhibha T, Murepa R, Musokonyi C, Gapare W (2013) Genetic parameter estimates for interspecific *Eucalyptus* hybrids and implications for hybrid breeding strategy. New For 44(1):63–84
Magdalita PM, Cayaban MFH, Gregorio MT, Silverio JV (2016) Development and characterization of nine new *Hibiscus* hybrids. Philippine Journal of Crop Science 41(2):31–45
Major JE, Mosseler A, Johnsen KH, Rajora OP, Barsi DC, Kim KH, Park JM, Campbell M (2005) Reproductive barriers and hybridity in two spruces, *Picea rubens* and *Picea mariana*, sympatric in eastern North America. Canadian Journal of Botany-Revue Canadienne de Botanique 83(2):163–175
Major JE, Mosseler A, Johnsen KH, Campbell M, Malcolm J (2015) Growth and allocation of Picea rubens, Picea mariana, and their hybrids under ambient and elevated CO2. Can J For Res 45(7):877–887
Malnoy M, Viola R, Jung MH, Koo OJ, Kim S, Kim JS, Velasco R, Kanchiswamy CN (2016) DNA-free genetically edited grapevine and apple protoplast using CRISPR/Cas9 Ribonucleoproteins.

Front Plant Sci 7:1904
Maluszynski M, Kasha KJ, Szarejko I (2003) Published doubled haploid protocols in plant species. In: Maluszynski M, Kasha KJ, Forster BP, Szarejko I (eds) Doubled haploid production in crop plants – a manual. Kluwer Academic Publishers, Dordrecht
Mancuso S, Shabala S (2010) Waterlogging signaling and tolerance in plants. Springer, Berlin Heidelberg, Germany
Marcotrigiano M (1997) Chimeras and variegation: patterns of deceit. Hortscience 32(5):773–784
Martin JA, Wang Z (2011) Next-generation transcriptome assembly. Nat Rev Gent 12:671–682
Martins S, Simoes F, Matos J, Silva AP, Carnide V (2014) Genetic relationship among wild, landraces and cultivars of hazelnut (*Corylus avellana*) from Portugal revealed through ISSR and AFLP markers. Plant Syst Evol 300(5):1035–1046
Matsunaga E, Nanto K, Oishi M, Ebinuma H, Morishita Y, Sakurai N, Suzuki H, Shibata D, Shimada T (2012) *Agrobacterium*-mediated transformation of *Eucalyptus globulus* using explants with shoot apex with introduction of bacterial choline oxidase gene to enhance salt tolerance. Plant Cell Rep 31:225–235
Mattera R, Molnar T, Struwe L (2015) *Cornus* x *elwinortonii* and *Cornus* x *rutgersensis* (Cornaceae), new names for two artificially produced hybrids of big-bracted dogwoods. Phytokeys 55:93–111
McNamara S, Hokanson SC (2010a) Cold hardiness of *Weigela* (*Weigela florida* Bunge) cultivars. J Environ Hort 28:35–40
McNamara S, Hokanson SC (2010b) A preliminary evaluation of cold hardiness in six Asian maple taxa. ASHS Annual Meeting Abstract Program, p124
McNamara S, Pellett H (1993a) Flower bud hardiness of *Forsythia* cultivars. J Environ Hort 11:35–38
McNamara S, Pellett H (1993b) *Acer rubrum* 'Autumn Spire'. J Environ Hort 11:147–148
McNamara S, Pellett H (1994) Cold hardiness of landscape pear taxa. J Environ Hort 12:227–230
McNamara S, Pellett H (1995) Cold hardiness of flowering crabapple cultivars. J Environ Hort 14:111–114
McNamara S, Pellett H (1998) Cold hardiness of *Weigela* cultivars. J Environ Hort 16:238–242
McNamara S, Pellett H (2000) Cold hardiness of *Phellodendron sachalinense* Friedr Schmidt seedlings increase with age. Hortscience 35:304–305
McNamara S, Zuzek K, Rose N, Pellett H, Hokanson SC (2005) 'Firefall' freeman maple. Hortscience 40:269–271
Mehlenbacher SA, Smith DC (2009) 'Red Dragon' ornamental hazelnut. Hortscience 44(3):843–844
Mehlenbacher SA, Thompson MM (2004) Inheritance of style color in hazelnut. Hortscience 39(3):475–476
Merchant A, Callister A, Arndt S, Tausz M, Adams M (2007) Contrasting physiological responses of six *Eucalyptus* species to water deficit. Ann Bot 100:1507–1515
Mercuri A, Bruna S, De Benedetti L, Burchi G, Schiva T (2001) Modification of plant architecture in *Limonium* spp. induced by *rol* genes. Plant cell. Tissue and Organ Culture 65:247–253
Mercuri A, Anfosso L, Burchi G, Bruna S, De Benedetti L, Schiva T (2003) *Rol* genes and new genotypes of *Limonium gmelinii* through *Agrobacterium*-mediated transformation. Acta Hort 624:455–462
Mercuri A, Braglia L, De Benedetti L, Ballardini M, Nicoletti F, Bianchini C (2010) New genotypes of *Hibiscus* x *rosa-sinensis* through classical breeding and genetic transformation. Acta Hort 855:201–207
Meudt HM, Clarke AC (2007) Almost forgotten or latest practice? AFLP applications, analyses and advances. Trends Plant Sci 12:106–117
Meyer E, Touchell DH, Ranney TG (2009) In vitro shoot regeneration and polyploid induction from leaves of *Hypericum* species. Hortscience 44(7):1957–1961
Mezzetti B, Landi L, Phan BH, Taruschio L, Lim KY (2001) Peg-mediated fusion of *Rubus idaeus* (raspberry) and *R-fruticosus* (blackberry) protoplasts, selection and characterisation of callus lines. Plant Biosystems 135(1):63–69
Mikulic-Petkovsek M, Schmitzer V, Slatnar A, Todorovic B, Veberic R, Stampar F, Ivancic A (2014) Investigation of anthocyanin profile of four elderberry species and interspecific hybrids. J Agric Food Chem 62(24):5573–5580
Milien M, Renault-Spilmont AS, Cookson SJ, Sarrazin A, Verdeil JL (2012) Visualisation of the 3D structure of the graft union of grapevine using X-ray tomography. Sci Hortic 144:130–140
Minamikawa MF, Nonaka K, Kaminuma E, Kajiya-Kanegae H, Onogi A, Goto S, Yoshioka T,

Imai A, Hamada H, Hayashi T, Matsumoto S, Katayose Y, Toyoda A, Fujiyama A, Nakamura Y, Shimizu T, Iwata H (2017) Genome-wide association study and genomic prediction in *Citrus*: potential of genomics-assisted breeding for fruit quality traits. Sci Rep 7:4721

Mishra VK, Bajpai R, Chaturvedi R (2017) An efficient and reproducible method for development of androgenic haploid plants from in vitro anther cultures of *Camellia assamica* ssp assamica (Masters). In Vitro Cell Dev Biol Plant 53(3):239–248

Mittempergher L, Santini A (2004) The history of elm breeding. Investigación Agraria Sistemas y Recursos Forestales 13:161–177

Miyashita T, Hoshino Y (2010) Interspecific hybridization in *Lonicera caerulea* and *Lonicera gracilipes*: the occurrence of green/albino plants by reciprocal crossing. Sci Hortic 125(4):692–699

Miyashita T, Hoshino Y (2015) Interploid and intraploid hybridizations to produce polyploid Haskap (*Lonicera caerulea* var. emphyllocalyx) plants. Euphytica 201(1):15–27

Miyashita C, Ishikawa S, Mii M (2009) In vitro induction of the amphiploid in interspecific hybrid of blueberry (*Vaccinium corymbosum* x *Vaccinium ashei*) with colchicine treatment. Sci Hortic 122(3):375–379

Miyashita C, Mii M, Aung T, Ogiwara I (2012) Effect of cross direction and cultivars on cross-ability of interspecific hybridization between *Vaccinium corymbosum* and *Vaccinium virgatum*. Sci Hortic 142:1–6

Molnar TJ, Capik JM (2013) The Rutgers University woody ornamentals breeding program: past, present and future. Acta Hort 990:271–280

Molnar TJ, Zaurov DE, Goffreda JC, Mehlenbacher SA (2007) Survey of hazelnut germplasm from Russia and Crimea for response to eastern filbert blight. Hortscience 42(1):51–56

Monteiro-Hara ACBA, Jadao AS, Mendes BMJ, Rezende JAM, Trevisan F, Mello APOA, Vieira MLC, Meletti LMM, Piedade SMDS (2011) Genetic transformation of passionflower and evaluation of R-1 and R-2 generations for resistance to Cowpea aphid borne mosaic virus. Plant Dis 95(8):1021–1025

Mora M, Avila F, Carrasco-Benavides M, Maldonado G, Olguin-Caceres J, Fuentes S (2016) Automated computation of leaf area index from fruit trees using improved image processing algorithms applied to canopy cover digital photographies. Comput Electron Agric 123:195–202

Morgan DR, Soltis DE, Robertson KR (1994) Systematic and evolutionary implications of *rbcL* sequence variation in Rosaceae. Am J Bot 81:890–903

Mori K, Shirasawa K, Nogata H, Hirata C, Tashiro K, Habu T, Kim S, Himeno S, Kuhara S, Ikegami H (2017) Identification of RAN1 orthologue associated with sex determination through whole genome sequencing analysis in fig (*Ficus carica* L.). Sci Rep 25(7):41124

Moriya S, Iwanami H, Kotoda N, Haji T, Okada K, Terakami S, Mimida N, Yamamoto T, Abe K (2012a) Aligned genetic linkage maps of apple rootstock cultivar ‘JM7’ and *Malus sieboldii* ‘Sanashi 63’ constructed with novel EST-SSRs. Tree Genet Genomes 8:709–723

Moriya S, Okada K, Haji T, Yamamoto T, Abe K (2012b) Fine mapping of Co, a gene controlling columnar growth habit located on apple (*Malus domestica* Borkh.) linkage group 10. Plant Breed 131:437–450

Morozowska M, Gawronska B, Woznicka A (2013) Morphological, anatomical and genetic differentiation of *Cornus mas*, *Cornus officinalis* and their interspecific hybrid. Dendrobiology 70:45–57

Mortreau E, Siljak-Yakovlev S, Cerbah M, Brown SC, Bertrand H, Lambert C (2010) Cytogenetic characterization of *Hydrangea involucrata* Sieb. and *H. aspera* D. Don complex (Hydrangeaceae): genetic, evolutional, and taxonomic implications. Tree Genet Genomes 6(1):137–148

Mu HN, Li HG, Wang LG, Yang XL, Sun TZ, Xu C (2014) Transcriptome sequencing and analysis of sweet *Osmanthus* (*Osmanthus fragrans* Lour.). Genes & Genomics 36(6):777–788

Muehlbauer MF, Honig JA, Capik JM, Vaiciunas JN, Molnar TJ (2014) Characterization of eastern filbert blight-resistant hazelnut Germplasm using microsatellite markers. J Am Soc Hortic Sci 139(4):399–432

Myburg AA, Grattapaglia D, Tuskan GA, Hellsten U, Hayes RD, Grimwood J, Jenkins J, Lindquist E, Tice H, Bauer D, Goodstein DM, Dubchak I, Poliakov A, Mizrachi E, Kullan AR, Hussey SG, Pinard D, van der Merwe K, Singh P, van Jaarsveld I, Silva-Junior OB, Togawa RC, Pappas MR, Faria DA, Sansaloni CP, Petroli CD, Yang X, Ranjan P, Tschaplinski TJ, Ye CY, Li T, Sterck L, Vanneste K, Murat F, Soler M, Clemente HS, Saidi N, Cassan-Wang H, Dunand C, Hefer CA, Bornberg-Bauer E, Kersting AR, Vining K, Amarasinghe V, Ranik M, Naithani S, Elser J, Boyd AE, Liston A, Spatafora JW, Dharmwardhana P, Raja R, Sullivan C, Romanel E, Alves-Ferreira M, Külheim C, Foley W, Carocha V, Paiva J, Kudrna D, Brommonschenkel SH,

Pasquali G, Byrne M, Rigault P, Tibbits J, Spokevicius A, Jones RC, Steane DA, Vaillancourt RE, Potts BM, Joubert F, K B, Pappas GJ, Strauss SH, Jaiswal P, Grima-Pettenati J, Salse J, Van de Peer Y, Rokhsar DS, Schmutz J (2014) The genome of *Eucalyptus grandis*. Nature 510(7505):356–362

Nadler JD, Pooler M, Olsen RT, Coleman GD (2012) In vitro induction of polyploidy in *Cercis glabra* Pamp. Sci Hortic 148:126–130

Nakajima I, Ban Y, Azuma A, Onoue N, Moriguchi T, Yamamoto T, Toki S, Endo M (2017) CRISPR/Cas9-mediated targeted mutagenesis in grape. PLoS One 12:e0177966

Nanda RM, Nayak S, Rout GR (2004) Studies on genetic relatedness of *Acacia* tree species using RAPD markers. Biologia 59(1):115–120

Nash LJ, Graves WR (1993) Drought and flood stress effects on plant development and leaf water relations of 5 taxa of trees native to bottomland habitats. J Amer Soc Hort Sci 118:845–850

Nashima K, Terakami S, Nishio S, Kunihisa M, Nishitani C, Saito T, Yamamoto T (2015) S-genotype identification based on allele-specific PCR in Japanese pear. Breed Sci 65:208–215

NC-7 (2000) The NC-7 website: a new way to find woody ornamental. https://www.ars.usda.gov/news-events/news/research-news/2000/the-nc-7-website-a-new-way-to-find-woody-ornamentals

Neale DB, Wegrzyn JL, Stevens KA, Zimin AV, Puiu D, Crepeau MW, Cardeno C, Koriabine M, Holtz-Morris AE, Liechty JD, Martínez-García PJ, Vasquez-Gross HA, Lin BY, Zieve JJ, Dougherty WM, Fuentes-Soriano S, Wu LS, Gilbert D, Marçais G, Roberts M, Holt C, Yandell M, Davis JM, Smith KE, Dean JF, Lorenz WW, Whetten RW, Sederoff R, Wheeler N, McGuire PE, Main D, Loopstra CA, Mockaitis K, de Jong PJ, Yorke JA, Salzberg SL, Langley CH (2014) Decoding the massive genome of loblolly pine using haploid DNA and novel assembly strategies. Genome Biol 15(3):R59

Neely D (1973) Pin oak chlorosis. J For 71:340–342

NeSmith DS, Ehlenfeldt MK (2011) ‘Summer Sunset’: a new ornamental blueberry. Hortscience 46(11):1560–1561

Nghiem CQ, Harwoon CE, Harbard JL, Griffin AR, Ha TH, Koutoulis A (2011) Floral phenology and morphology of colchicine-induced tetraploid *Acacia mangium* compared with diploid *A. mangium* and *A. auriculiformis*: implications for interploidy pollination. Aust J Bot 59(6):582–592

Nishimoto S, Shimizu K, Hashimoto F, Sakata Y (2003) Interspecific hybrids of *Camellia chrysantha* x *C. japonica* by ovule culture. Journal of the Japanese Society for Horticultural Science 72(3):236–242

Nishimoto S, Hashimoto F, Shimizu K, Sakata Y (2004) Petal coloration of interspecific hybrids between *Camellia chrysantha* x *C. japonica*. Journal of the Japanese Society for Horticultural Science 73(2):189–191

Nishitani C, Hirai SK, Masato W, Kazuma O, Keishi O, Toshiya Y et al (2016) Efficient genome editing in apple using a CRISPR/Cas9 system. Sci Rep 6:31481

Nock CA, Taugourdeau O, Delagrange S, Messier C (2013) Assessing the potential of low-cost 3D cameras for the rapid measurement of plant woody structure. Sensors 13:16216–16233

Nolf M, Lopez R, Peters JMR, Flavel RJ, Koloadin LS, Young IM, Choat B (2017) Visualization of xylem embolism by X-ray microtomography: a direct test against hydraulic measurements. New Phytol 214:890–898

Nybom H, Bartish IV (2007) DNA markers and morphometry reveal multiclonal and poorly defined taxa in an apomict *Cotoneaster* species complex. Taxon 56:119–128

Nystedt B, Street NR, Wetterbom A, Zuccolo A, Lin YC, Scofield DG, Vezzi F, Delhomme N, Giacomello S, Alexeyenko A, Vicedomini R, Sahlin K, Sherwood E, Elfstrand M, Gramzow L, Holmberg K, Hallman J, Keech O, Klasson L, Koriabine M, Kucukoglu M, Kaller M, Luthman J, Lysholm F, Niittyla T, Olson A, Rilakovic N, Ritland C, Rossello JA, Sena J, Svensson T, Talavera-Lopez C, Theissen G, Tuominen H, Vanneste K, Wu ZQ, Zhang B, Zerbe P, Arvestad L, Bhalerao R, Bohlmann J, Bousquet J, Gil RG, Hvidsten TR, de Jong P, MacKay J, Morgante M, Ritland K, Sundberg B, Thompson SL, Van de Peer Y, Andersson B, Nilsson O, Ingvarsson PK, Lundeberg J, Jansson S (2013) The Norway spruce genome sequence and conifer genome evolution. Nature 497(7451):579–584

Ocampo J, Arias JC, Urrea R (2016) Interspecific hybridization between cultivated and wild species of genus *Passiflora* L. Euphytica 209(2):395–408

Ogino A, Tanaka J, Taniguchi F, Yamamoto M, Yamada K (2009) Detection and characterization of caffeine-less tea plants originated from interspecific hybridization. Breed Sci 59(3):277–283

Olsen RT, Ranney TG, Werner DJ (2006) Fertility and inheritance of variegated and purple foliage across a polyploid series in *Hypericum androsaemum* L. J Am Soc Hortic Sci 131(6):725–730

Orr HA, Presgraves DC (2000) Speciation by postzygotic isolation: forces, genes and molecules. Bioassays 22:1085–1094

Osakabe Y, Sugano SS, Osakabe K (2016) Genome engineering of woody plants: past, present and future. J Wood Sci 62:217–225

Otto D, Peterson R, Braukseipe B, Braun P, Schmidt ER (2014) The columnar mutation ("Co gene") of apple (*Malus domestica*) is associated with an integration of a gypsy-like retrotransposon. Mol Breed 33:863–880

Pánková I, Krejzar V, Mertelík J, Kloudová K (2015) The occurrence of lines tolerant to the causal agent of bleeding canker, *Pseudomonas syringae v.* aesculi, in a natural horse chestnut population in Central Europe. Eur J Plant Pathol 142:37–47

Parelle J, Zapater M, Scotti-Saintagne C, Kremer A, Jolivet Y, Dreyer E, Brendel O (2007) Quantitative trait loci of tolerance to water-logging in an European oak (*Quercus robur* L.): physiological relevance and temporal effect patterns. Plant Cell Environ 30:422–434

Parelle J, Dreyer E, Brendel O (2010) Genetic variability and determinism of adaptation of plants to soil waterlogging. In: Mancuso S, Shabala S (eds) Waterlogging signaling and tolerance in plants. Springer-Verlag, Berlin Heidelberg, Germany

Parris JK, Ranney TG, Knap HT, Baird WV (2010) Ploidy levels, relative genome sizes, and base pair composition in *Magnolia*. J Amer Soc Hort Sci 135(6):533–547

Pathirana R, Wiedow C, Pathirana S, Hedderley D, Morgan E, Scalzo J, Frew T, Timmerman-Vaughan G (2015) Ovule culture and embryo rescue facilitate interspecific hybridisation in blueberry (*Vaccinium* spp.). Acta Hort 1083:123–132

Paulus S, Dupuis J, Mahlein AK, Kuhlmann H (2013) Surface feature based classification of plant organs from 3D laser scanned point clauds for plant phenotyping. BMC bioinformatics 14:238

Peer R, Rivlin G, Golobovitch S, Lapidot M, Gal-On A, Vainstein A, Tzfira T, Flaishman MA (2015) Targeted mutagenesis using zinc-finger nucleases in perennial fruit trees. Planta 241:941–951

Pei MH, Ruiz C, Shield I, Macalpine W, Lindegaard K, Bayon C, Karp A (2010) Mendelian inheritance of rust resistance to *Melampsora larici-epitea* in crosses between *Salix sachalinensis* and *S. viminalis*. Plant Pathol 59(5):862–872

Peixe A, Barroso J, Potes A, Pais MS (2004) Induction of haploid morphogenic calluses from in vitro cultured anthers of *Prunus armeniaca* cv. 'Harcot'. Plant Cell Tissue Org Cult 77(1):35–41

Pellet H (1998) Breeding of cold hardy woody landscape plants. In: Li PH, Chen THH (eds) Plant cold hardiness: molecular biology, biochemistry, and physiology. Plenum Press, New York, USA, pp p317–p324

Pellicer J, Garcia S, Canela MA, Garnatje T, Korobkov AA, Twibell JD, Valles J (2010) Genome size dynamics in *Artemisia* L. (Asteraceae): following the track of polyploidy. Plant Biol 12(5):820–830

Peng A, Chen S, Lei T, Xu L, He Y, Wu L et al (2017) Engineering canker-resistant plants through CRISPR/Cas9-targeted editing of the susceptibility gene CsLOB1 promotor in *Citrus*. Plant Biotech J 15(12):1509–1519

Percival GC, Sheriffs CN (2002) Identification of drought-tolerant woody perennials using chlorophyll fluorescence. J Arboric 28(5):215–223

Perdiguero P, Venturas M, Cervera MT, Gil CC (2015) Massive sequencing of *Ulmus* minor's transcriptome provides new molecular tools for a genus under the constant threat of Dutch elm disease. Front Plant Sci 6:541

Perveen A, Sarwar GR (2011) Pollen germination capacity of two cultivated species (*Jasminum sambac* (L.) Ait. and *Nyctanthes arbor-tristis* L. of family Oleaceae). Pak J Bot 43(4):2109–2112

Pessina S, Angeli D, Martens S, Visser R, Bai Y, Salamini F, Velasco R, Schouten HJ, Malnoy M (2016) The knock-down of the expression of MdMLO19 reduces susceptibility to powdery mildew (*Podosphaera leucotricha*) in apple (*Malus domestica*). Plant Biotechnol J 14:2033–2044

Peterson BJ, Graves WR (2009) Variation in development and response to root-zone pH among seedlings of *Dirca palustris* (Thymelaeaceae) from three provenances. Hortscience 44:1319–1322

Peterson BJ, Graves WR (2013) Responses to root-zone water content of shrub congeners from eastern North America and Mediterranean California. Hortscience 48:715–719

Phillips WD, Ranney TG, Touchell DH, Eaker TA (2016) Fertility and reproductive pathways of triploid flowering pears (Pyrus sp.). Hortscience 51(8):968–971

Piispanen R, Aronen T, Chen XW, Saranpaa P, Haggman H (2003) Silver birch (*Betula pendula*) plants with *aux* and *rol* genes show consistent changes in morphology, xylem structure and

chemistry. Tree Physiol 23(11):721–733
Pilotti M, Lumia V, Di Lernia G, Brunetti A (2012) Development of real-time PCR for in wood-detection of *Ceratocystis platani*, the agent of canker stain of *Platanus* spp. Eur J Plant Pathol 135(1):61–69
Pinon J, Frey P (2005) Interactions between poplar clones and *Melampsora* populations and their implications for breeding for durable resistance. In: Pei MH, McCracken AR (eds) Rust diseases of willow and poplar. CAB International, Wallingford, UK, pp p139–p154
Pinon J, Husson C, Collin E (2005) Susceptibility of native French elm clones to *Ophiostoma novo-ulmi*. Ann For Sci 62(7):689–696
Pintos B, Manzanera JA, Bueno MA (2007) Antimitotic agents increase the production of doubled-haploid embryos from cork oak anther culture. J Plant Physiol 164(12):1595–1604
Pintos B, Sanchez N, Bueno MA, Navarro MR, Jorrin J, Manzanera JA, Gomez-Garay A (2013) Induction of *Quercus ilex* L. haploid and doubled-haploid embryos from anther cultures by temperature-stress. Silvae Genetica 62(4–5):210–218
Pipinis E, Milios E, Kiamos N, Mavrokordopoulou O, Smiris P (2012) Effects of stratification and pre-treatment with gibberellic acid on seed germination of two *Carpinus* species. Seed Sci Technol 40(1):21–31
Pipinis E, Mavrokordopoulou O, Milios E, Diamanta A, Kotili I, Smiris P (2015) Effects of dormancy-breaking treatments on seed germination of *Koelreuteria paniculata* and *Mahonia aquifolium*. Dendrobiology 74:149–155
Plasencia A, Soler M, Dupas A, Ladouce N, Silva-Martins G, Martinez Y, Lapierre C, Franche C, Truchet I, Grima-Pettenati J (2016) *Eucalyptus* hairy roots, a fast, efficient and versatile tool to explore function and expression of genes involved in wood formation. Plant Biotechnol J 14(6):1381–1393
Plomion C, Aury JM, Amselem J, Alaeitabar T, Barbe V, Belser V, Berges H, Bodenes C, Boudet N, Boury C, Canaguier A, Couloux A, Da Silva C, Duplessis S, Ehrenmann F, Estrada-Mairey B, Fouteau S, Francillonne N, Gaspin C, Guichard C, Klopp C, Labadie K, Lalanne C, Le Clainche I, Leple JC, Le Provost G, Leroy T, Lesur I, Martin F, Mercier J, Michotey C, Murat F, Salin F, Steinbach D, Faivre-Rampant P, Wincker P, Salse J, Quesneville H, Kremer A (2016) Decoding the oak genome: public release of sequence data, assembly, annotation and publication strategies. Mol Ecol Resour 16:254–265
Podwyszynska M, Sowik I, Machlanska A, Kruczynska D, Dyki B (2017) In vitro tetraploid induction of *Malus* x *domestica* Borkh. Using leaf or shoot explants. Sci Hortic 226:379–388
Poisson AS, Berthelot P, Le Bras C, Grapin A, Vergne E, Chevreau E (2016) A droplet-vitrification protocol enabled cryopreservation of doubled haploid explants of *Malus* x *domestica* Borkh. 'Golden Delicious'. Sci Hortic 209:187–191
Polashock J, Zelzion E, Fajardo D, Zalapa J, Georgi L, Bhattacharya D, Vorsa N (2014) The American cranberry: first insights into the whole genome of a species adapted to bog habitat. BMC Plant Biol 14:165
Pooler M, Ma H (2013) Interspecific hybridizations in ornamental flowering cherries validated by simple sequence repeat analysis. J Am Soc Hortic Sci 138(3):198–204
Pope KS, Dose V, Da Silva D, Brown PH, Leslie CA, Dejong TM (2013) Detecting nonlinear response of spring phenology to climate change by Bayesian analysis. Glob Change Biol 19:1518–1525
Potts BM, Dungey HS (2004) Interspecific hybridization of *Eucalyptus*: key issues for breeders and geneticists. New For 27(2):115–138
Pounders CT, Sakhanokho HF (2016) 'Hapa White', 'Hapa Pink', and 'Hapa Red' interspecific hybrid *Hibiscus* cultivars. Hortscience 51(12):1616–1617
Pounders C, Rinehart T, Sakhanokho H (2007) Evaluation of interspecific hybrids between *Lagerstroemia indica* and *L-speciosa*. Hortscience 42(6):1317–1322
Ramsey J, Schemske DW (1998) Pathways, mechanisms, and rates of polyploid formation in flowering plants. Ann Rev of Ecology and Systematics 29:467–501
Rana MM, Han ZX, Song DP, Liu GF, Li DX, Wan XC, Karthikeyan A, Wei S (2016) Effect of medium supplements on *Agrobacterium rhizogenes* mediated hairy root induction from the callus tissues of *Camellia sinensis* var. sinensis. Int J Mol Sci 17(7):1132
Randall BW, Walton DA, Lee DJ, Wallace HM (2012) Pollination methods, stigma receptivity and pollen tube growth in *Eucalyptus argophloia*. Silvae Genetica 61(3):121–126
Reed SM (2004) Self-incompatibility and time of stigma receptivity in two species of *Hydrangea*. Hortscience 39(2):312–315
Reed SM, Rinehart TA (2009) Simple-sequence repeat marker analysis of genetic relationships

within *Hydrangea paniculata*. Hortscience 44(1):27–31
Reed SM, Joung Y, Roh MS (2002) Interspecific hybridization in *Clethra*. Hortscience 37:393–397
Reed SM, Jones KD, Rinehart TA (2008) Production and characterization of intergeneric hybrids between *Dichroa febrifuga* and *Hydrangea macrophylla*. J Am Soc Hortic Sci 133(1):84–91
Rego MM, Rego ER, Bruckner CH, Finger FL, Otoni WC (2011) In vitro induction of autotetraploids from diploid yellow passion fruit mediated by colchicine and oryzalin. Plant Cell Tissue Org Cult 107(3):451–459
Renfro SE, Burkett BM, Dunn BL, Lindstrom JT (2007) 'Asian Moon' *Buddleja*. Hortscience 42(6):1486–1487
Richards MR, Rupp LA, Kjelgren R, Rasmussen VP (2012) Selection and budding propagation of native Bigtooth maple for water-conserving landscapes. HortTechnology 22:669–676
Richer C, Rioux JA (2002) Evaluation of the growth and winter hardiness of *Thuja occidentalis* L. and five cultivars under North-East Canadian climatic conditions. Canadian J of Plant Science 82:169–175
Richer C, Rioux JA, Lamy MP (2003a) Winter damage and growth evaluation of seven ornamental crab apple cultivars in the diverse weather conditions of the Canadian north-east. Canadian J of Plant Science 83:835–849
Richer C, Rioux JA, Lamy MP (2003b) Tolerance evaluation of six indigenous or naturalized species under different northeastern Canadian climatic conditions. Canadian J of Plant Science 83:825–833
Richer C, Rioux JA, Lamy MP (2004a) Evaluation of the growth and winter hardiness of six cultivars of *Spiraea* under northeastern Canadian climatic conditions. Canadian J of Plant Science 84:265–277
Richer C, Rioux JA, Lamy MP (2004b) Interpretation of the response of ten cultivars of *Potentilla fruticosa* evaluated under North-East Canadian climatic conditions. Canadian J of Plant Science 84:265–277
Riddle NC, Birchler JA (2003) Effects of reunited diverged regulatory hierarchies in allopolyploids and species hybrids. Trends Genet 19:597–600
Rinehart TA, Scheffler BE, Reed SM (2006) Genetic diversity estimates for the genus *Hydrangea* and development of a molecular key based on SSR. J Am Soc Hortic Sci 131(6):787–797
Rioux JA, Marquis P, Richer C, Lamy MP (2000) Evaluation of the winter hardiness of *Thuja occidentalis* L. and eight cultivars under North-East Canadian climatic conditions. Canadian J of Plant Science 80:631–637
Rioux JA, Richer C, Lamy MP (2004) Tolerance evaluation of eleven junipers (*Juniperus sp.*) under North-Eastern Canadian climatic conditions. Canadian J of Plant Sciences 84:1135–1153
Roberts DJ, Werner DJ (2016) Genome size and ploidy levels of *Cercis* (Redbud) species, cultivars, and botanical varieties. Hortscience 51(4):330–333
Roberts DJ, Werner DJ, Wadl PA, Trigiano RN (2015) Inheritance and allelism of morphological traits in eastern redbud (*Cercis canadensis* L.). Horticulture Research 2:15049
Romeu JF, Monforte AJ, Sanchez G, Granell A, Garcia-Brunton J, Badenes ML, Rios G (2014) Quantitative trait loci affecting reproductive phenology in peach. BMC Plant Biol 14:52
Rose JB, Kubba J, Tobutt KR (2000) Induction of tetraploidy in *Buddleia globosa*. Plant Cell Tissue Org Cult 63(2):121–125
Rothleutner JJ, Contreras RN, Stockwell VO, Owen JS Jr (2014) Screening *Cotoneaster* for resistance to fire blight by artificial inoculation. Hortscience 49(12):1480–1485
Rothleutner JJ, Friddle MW, Contreras RN (2016) Ploidy levels, relative genome sizes, and base pair composition in *Cotoneaster*. J Am Soc Hortic Sci 141(5):457–466
Rounsaville TJ, Ranney TG (2010) Ploidy levels and genome sizes of *Berberis* L. and *Mahonia* Nutt. species, hybrids, and cultivars. Hortscience 45(7):1029–1033
Rovira M, Avanzato D, Bacchetta L, Botta R, Drogoudi P, Ferreira JJ, Sarraquigne JP, Silva AP, Solar A (2011) European *Corylus avellana* L. germplasm collections. Acta Hort 918:871–876
Ruiz M, Quinones A, Martinez-Alcantara B, Aleza P, Morillon R, Navarro L, Primo-Millo E, Martinez-Cuenca MR (2016) Effects of salinity on diploid (2x) and doubled diploid (4x) *Citrus macrophylla* genotypes. Sci Hortic 207:33–40
Ryu HW, Cho BO, Ryu J, Jin CH, Kim JB, Kang SY, Han AR (2017) Anthocyanin contents enhancement with gamma irradiated mutagenesis in blackberry (*Rubus fruticosus*). Nat Prod Commun 12(9):1451–1454
Saggoo MIS, Faroog U, Lovleen (2011) Meiotic studies in *Sarcococca* species (Buxaceae) from western Himalayas. Cytologia 76(3):329–335
Sakai A (1966) Studies of frost hardiness in woody plants. II. Effect of temperature on hardening.

Plant Physiol 41:353–359
Sakai A, Larcher W (1987) Frost survival of plants: responses and adaptation to freezing stress. Springer, Berlin
Salava J, Polak J, Oukropec I (2013) Evaluation of the *Prunus* interspecific progenies for resistance to plum pox virus. Czech J Genet Plant Breed 49(2):65–69
Salvatori E, Fusaro L, Gottardini E, Pollastrini M, Goltsev V, Strasser RJ, Bussotti F (2014) Plant stress analysis; application of prompt, delayed chlorophyll fluorescence and 820 nm modulated reflectance. Insights from independent experiments. Plant Physiol Biochem 85:105–113
Salvi S, Piazza S, Predieri S, Fuochi P, Velasco R, Malnoy M (2015) High frequency of chromosome deletions in regenerated and mutagenized apple (*Malus* x *domestica* Borkh.) seedlings. Mol Breed 35(1):4
Sanchez G, Martinez J, Romeu J, Garcia J, Monforte AJ, Badenes ML et al (2014) The peach volatilome modularity is reflected at the genetic and environmental response levels in a QTL mapping population. BMC Plant Biol 14:137
Sander JD, Joung JK (2014) CRISPR-Cas systems for editing, regulating and targeting genomes. Nat Biotechnol 32(4):347–355
Santiago A, Ferrandis P, Herranz JM (2015) Non-deep simple morphophysiological dormancy in seeds of *Viburnum lantana* (Caprifoliaceae), a new dormancy level in the genus *Viburnum*. Seed Sci Res 25(1):46–56
Santini A, Fagnani A, Ferrini F, Ghelardini L, Mittempergher L (2005) Variation among Italian and French elm clones in their response to *Ophiostoma novo-ulmi* inoculation. For Pathol 35(3):183–193
Santini A, La Porta N, Ghelardini L, Mittempergher L (2008) Breeding against Dutch elm disease adapted to the Mediterranean climate. Euphytica 163(1):45–56
Santos EA, Souza MM, Abreu PP, da Conceicao LDHCS, Araujo IS, Vianna AP, de Almeirda AAF, Freitas JCD (2012) Confirmation and characterization of interspecific hybrids of *Passiflora* L. (Passifloraceae) for ornamental use. Euphytica 184(3):389–399
Santos EA, Viana AP, Freitas JCO, Rodrigues DL, Tavares RF, Paiva CL, Souza MM (2015) Genotype selection by REML/BLUP methodology in a segregating population from an interspecific *Passiflora* spp. crossing. Euphytica 204(1):1–11
Sathuvalli VR, Mehlenbacher SA (2012) Characterization of American hazelnut (*Corylus americana*) accessions and *Corylus americana* x *Corylus avellana* hybrids using microsatellite markers. Genet Resour Crop Evol 59(6):1055–1075
Sathuvalli VR, Mehlenbacher SA, Smith DC (2011) DNA markers linked to eastern filbert blight resistance from a hazelnut selection from the Republic of Georgia. J Am Soc Hortic Sci 136(5):350–357
Satya P (2012) Prezygotic interspecific hybridization barriers between kenaf (*Hibiscus cannabinus* L.) and four wild relatives. Plant Breed 131(5):648–655
Sax HJ (1954) Polyploidy and apomixis in *Cotoneaster*. J Arnold Arb 35(4):334–365
Scheiber SM, Robacker CD (2003) Interspecific hybridization between *Abelia* x *grandiflora* ' Francis Mason and *A. schumannii* via ovule culture. Euphytica 132:1–6
Schrader JA, Graves WR (2000) *Alnus maritima:* A rare woody species from the new world. New Plantsman 7:74–82
Schrader JA, Graves WR (2002) Intraspecific systematics of *Alnus maritima* (Betulaceae) from three widely disjunct provenances. Castanea 67:380–401
Schrader JA, Gardner SJ, Graves WR (2005) Resistance to water stress of *Alnus maritima*: intraspecific variation and comparisons to other alders. Environ Exp Bot 53:281–298
Schulze JA, Contreras RN (2017) In vivo chromosome doubling of *Prunus lusitanica* and preliminary morphological observations. Hortscience 52:332–337
Schuster M, Grafe C, Hoberg E, Schuetze W (2013) Interspecific hybridization in sweet and sour cherry breeding. Acta Hort 976:79–86
Scorza R, Callahan A, Dardick C, Ravelonandro M, Polak J, Malinowski T, Zagrai I, Cambra M, Kamenova I (2013) Genetic engineering of plum pox virus resistance: 'HoneySweet' plum – from concept to product. Plant Cell Tissue Organ Cult 115:1–12
Sedira M, Butler E, Gallagher T, Welander M (2005) Verification of auxin-induced gene expression during adventitious rooting in rolB-transformed and untransformed apple Jork 9. Plant Sci 168(5):1193–1198
Seelig HD, Hoehn A, Stodieck LS, Klaus DM, Adams WW, Emery WJ (2008) The assessment of leaf water content using leaf reflectance ratios in the visible, near-, and short-wave-infrared. Int J Remote Sens 29:3701–3713

SERA27: Nursery Crop and Landscape (2017) systemshttps://www.nimss.org/projects/view/mrp/outline/18468

Sestras R, Pamfil D, Sestras A, Jaentschi L, Bolboaca S, Dan C (2009) Inheritance of vigour tree in F1 apple interspecific hybrids. Notulae Botanicae Horti Agrobotanici Cluj-Napoca 37(1):70–73

Sharma H (1995) How wide can a wide cross be? Euphytica 82:43–64

Sharma D, Kaur R, Kumar K (1996) Embryo rescue in plants – a review. Euphytica 89:325–337

Shearer K, Ranney TG (2013) Ploidy levels and relative genome sizes of species, hybrids, and cultivars of dogwood (*Cornus* spp.). Hortscience 48(7):825–830

Shen JS, Xu TL, Shi C, Cheng TR, Wang J, Pan HT, Zhang QX (2017) Obtainment of an intergeneric hybrid between *Forsythia* and *Abeliophyllum*. Euphytica 213(4):95

Shirasawa K, Isuzugawa K, Ikenaga M, Saito Y, Yamamoto T, Hirakawa H, Isobe S (2017) The genome sequence of sweet cherry (*Prunus avium*) for use in genomics-assisted breeding. DNA Res 24(5):499–508

Shishikoff N, Daughtrey M, Aker S, Olsen RT (2015) Evaluating boxwood susceptibility to *Calonectria pseudonaviculata* using cuttings from the National Boxwood Collection. Plant Health Progress 16(1):11–15

Silfverberg-Dilworth E, Matasci CL, Van de Weg WE, Van Kaauwen MPW, Walser M, Kodde LP et al (2006) Microsatellite markers spanning the apple (*Malus* x *domestica* Borkh.) genome. Tree Genet Genomes 2:202–224

Simonovik B, Ivancic A, Jakse J, Bohanec B (2007) Production and genetic evaluation of interspecific hybrids within the genus *Sambucus*. Plant Breed 126(6):628–633

Singh S, Sud RK, Gulati A, Joshi R, Yadav AK, Sharma RK (2013) Germplasm appraisal of western Himalayan tea: a breeding strategy for yield and quality improvement. Genet Resour Crop Evol 60(4):1501–1513

Singhal VK, Kaur D (2011) Cytomixis induced meiotic irregularities and pollen malformation in *Clematis graveolens* Lindley from the cold deserts of Kinnaur District of Himachal Pradesh (India). Cytologia 76(3):319–327

Sirault XRR, James RA, Furbank RT (2009) A new screening method for osmotic component of salinity tolerance in cereals using infrared thermography. Funct Plant Biol 36:970

Sjöman H, Hirons AD, Bassuk NL (2015) Urban forest resilience through tree selection—variation in drought tolerance in *Acer*. Urban Forestry and Urban Greening 14:858–865

Smith WA, Brand MH (2012) 'Summer Skies' *Buddleja davidii*. Hortscience 47(1):126–127

Smith AG, Noyszewski AK (in press) Mutagenesis breeding for seedless varieties of popular landscape plants. Accepted for publication in Acta Hort Acta Hort 1191: 43–51

Smolik M, Andrys D, Franas A, Krupa-Malkiewics M, Malinowska K (2010) Polymorphism in *Syringa* rDNA regions assessed by PCR technique. Dendrobiology 64:55–64

Sniezko RA, Koch J (2017) Breeding trees resistant to insects and diseases: putting theory into application. Biol Invasions 89:1–24

Soares TL, de Jesus ON, dos Santos-Serejo JA, de Oliveira EJ (2013) In vitro pollen germination and pollen viability in passion fruit (*Passiflora* spp.). Rev Bras Frutic 35(4):1116–1126

Soares TL, de Jesus ON, de Souza EH, de Oliveira EJ (2015) Reproductive biology and pollen-pistil interactions in *Passiflora* species with ornamental potential. Sci Hortic 197:339–349

Sochor M, Vašut RJ, Sharbel TF, Trávnícek B (2015) How just a few makes a lot: speciation via reticulation and apomixis on example of European brambles (*Rubus* subgen. Rubus, Rosaceae). Mol Phylogenet Evol 89:13–27

Soler M, Serra O, Molinas M, Huguet G, Fluch S, Figueras M (2007) A genomic approach to suberin biosynthesis and cork differentiation. Plant Physiol 144:419–431

Solla A, Bohnens J, Collin E, Diamandis S, Franke A, Gil L, Buron M, Santini A, Mittempergher L, Pinon J et al (2005) Screening European elms for resistance to *Ophiostoma novo-ulmi*. For Sci 51(2):134–141

Solla A, Lopez-Almansa JC, Martin JA, Gil L (2015) Genetic variation and heritability estimates of *Ulmus minor* and *Ulmus pumila* hybrids for budburst, growth and tolerance to Ophiostoma novo-ulmi. Iforest-Biogeosciences and Forestry 8:422–430

Soltis PS, Soltis DE (2009) The role of hybridization in plant speciation. Annu Rev Plant Biol 60:561–588

Soltis PS, Liu X, Marchant DB, Visger CJ, Soltis DE (2014) Polyploidy and novelty: gottlieb's legacy. Phil Trans R Soc B 369:20130351

Song D, Jaganathan GK, Han Y, Liu B (2017) Seed dormancy in *Camellia sinensis* L. (Theaceae): effects of cold-stratification and exogenous gibberellic acid application on germination. Botany 95(2):147–152

Spak J, Pribylova J, Safarova D, Lenz O, Koloniuk I, Navratil M, Franova J, Spakova V, Paprstein F (2017) Cherry necrotic rusty mottle and cherry green ring mottle viruses in Czech cherry Germplasm. Plant Prot Sci 53(4):195–200

Spongberg SA (1990) A Reunion of trees: the discovery of exotic plants and their introduction into North American and European landscapes. Harvard University Press, Cambridge, Mass

Squirrell J, Mandegaran Z, Yokoya K, Roberts AV, Mottley J (2005) Cell lines and plants obtained after protoplast fusions of *Rosa* plus *Rosa*, *Rosa* plus *Prunus* and *Rosa* plus *Rubus*. Euphytica 146(3):223–231

Steenackers J, Steenackers M, Steenackers V, Stevens M (1996) Poplar diseases, consequences on growth and wood quality. Biomass Bioenergy 10:267–274

Stephens MJ, Gaudion J, Hall HK, Enfield JR (2016) 'NR7' (marketed as 'Raspberry shortcake (R)' and 'Ruby beauty (R)') red raspberry. Hortscience 51:112–115

Stewart JR, Landes RD, Koeser AK, Pettay AL (2007) Net photosynthesis and growth of three novel woody species under water stress: *Calycanthus occidentalis*, *Fraxinus anomala*, and *Pinckneya pubens*. Hortscience 42:1341–1345

Strauss SH, Bradshaw HD (2004) The bioengineered Forest: challenges for science and technology. Resources for future, Washington DC. 245 pp

Strimbeck GR, Schaberg PG, Fossdal CG, Schröder WP, Kjellsen TD (2015) Extreme low temperature tolerance in woody plants. Front Plant Sci 6:884

Sun YH, Zhang YY, Yuan CQ, Yang Q, Long C, Li Y, Yang MS (2015) Assessment of genetic diversity and variation of *Acer* mono max seedlings after spaceflight. Pak J Bot 47(1):197–202

Sundouri AS, Singh H, Gill MIS, Thakur A, Sangwan AK (2014) In-vitro germination of hybrid embryo rescued from low chill peaches as affected by stratification period and embryo age. Indian Journal of Horticulture 71(2):151–155

Susko AQ, Rinehart TA, Bradeen JM, Hokanson SC (2018) An evaluation of two seedling phenotyping protocols to assess pH adaptability in deciduous azalea (*Rhododendron* sect. *Pentanthera* G. Don). Hortscience. in press

Sutherland BG, Belaj A, Nier S, Cottrell JE, Vaughan SP, Hubert J, Russell K (2010) Molecular biodiversity and population structure in common ash (*Fraxinus excelsior* L.) in Britain: implications for conservation. Mol Ecol 19(11):2196–2211

Swaroop K, Jain R, Janakiram T (2015) Effect of different doses of gamma rays for induction of mutation in *Bougainvillea* cv Mahatma Gandhi. Indian Journal of Agricultural Sciences 85(9):1245–1247

Symonds WL, Campbell LC, Clemens J (2001) Response of ornamental *Eucalyptus* from acidic and alkaline habitats to potting medium pH. Scientia Hort 88:121–131

Szymajda M, Napriorkowska B, Korbin M, Zurawicz E (2015) Studies on the interspecific crossing compatibility among three *Prunus* species and their hybrids. Hortic Sci 42(2):70–82

Tabur S, Civelek S, Oney S, Ergun SBY, Kursat M, Turkoglu I (2012) Chromosome counts and karyomorphology of some species of *Artemisia* (Asteraceae) from Turkey. Turk J Bot 36(3):235–246

Talluri RS (2011) Gametes with somatic chromosome number and their significance in interspecific hybridization in *Fuchsia*. Biol Plant 55(3):596–600

Talluri RS (2012a) Barriers to gene flow in interspecific hybridization in *Fuchsia* L. (Onagraceae). J Genet 91(1):81–85

Talluri RS (2012b) Interploidy interspecific hybridization in *Fuchsia*. J Genet 91(1):71–74

Talluri RS, Murray BG (2014) Variability in interspecific hybrids of *Fuchsia*. Indian Journal of Genetics and Plant Breeding 74(1):119–121

Talukdar D, Sinjushin A (2015) Cytogenomics and mutagenomics in plant functional biology and breeding. In: Barh D, Khan MS, Davies E (eds) Plantomics: the omics of plant science. Springer, India, pp 113–156

Tan B, Grattapaglia D, Martins GS, Ferreira KZ, Sundberg B, Ingvarsson PK (2017) Evaluating the accuracy of genomic prediction of growth and wood traits in two *Eucalyptus* species and their F-1 hybrids. BMC Plant Biol 17:110

Tang W, Sederoff R, Whetten R (2001) Regeneration of transgenic loblolly pine (*Pinus taeda* L.) from zygotic embryos transformed with *Agrobacterium tumefaciens*. Plant 213:981–989

Tang W, Peng X, Newton RJ (2005) Enhanced tolerance to salt stress in transgenic loblolly pine simultaneously expressing two genes encoding mannitol-1-phosphate dehydrogenase and glucitol-6-phosphate dehydrogenase. Plant Physiol Biochem 43:139–146

Tao Y, Wang LY, Roh MS (2010) Confirmation of *Clematis* hybrids using molecular markers. Sci Hortic 125(2):136–145

Tavan M, Mirjalili MH, Karimzadeh G (2015) In vitro polyploidy induction: changes in morphological, anatomical and phytochemical characteristics of *Thymus persicus* (Lamiaceae). Plant Cell Tissue Org Cult 122(3):573–583

Tepfer D (1984) Transformation of several species of higher plants by *Agrobacterium rhizogenes*: sexual transmission of the transformed genotype and phenotype. Cell 37:959–967

Tezotto-Uliana JV, Berno ND, Saji FRQ, Kluge RA (2013) Gamma radiation: an efficient technology to conserve the quality of fresh raspberries. Sci Hortic 164:348–352

Thammina C, He M, Lu L, Cao K, Yu H, Chen Y, Tian L, Chen J, McAvoy R, Ellis D, Zhao D, Wang Y, Zhang X, Li Y (2011) In vitro regeneration of triploid plants of *Euonymus alatus* 'Compactus' (burning bush) from endosperm tissues. Hortscience 46(8):1141–1147

Thammina CS, Olsen RT, Kramer M, Pooler MR (2017) Genetic relationships of boxwood (*Buxus* L.) accessions based on genic simple sequence repeat markers. Genet Resour Crop Evol 64(6):1281–1293

Thavamanikumar S, Southerton S, Thumma B (2014) RNA-seq using two populations reveals genes and alleles controlling wood traits and growth in Eucalyptus nitens. PLoS One 9:e101104

Thomasset M, Fernandez-Manjarres JF, Douglas GC, Frascaria-Lacoste N, Raquin C, Hodkinson TR (2011) Molecular and morphological characterization of reciprocal F1 hybrid Ash (*Fraxinus excelsior* x *Fraxinus angustifolia*, Oleaceae) and parental species. Int J Plant Sci 172(3):423–433

Tian J, Wang J, Dong L, Dai F, Wang J (2015) Pollen variation as a response to hybridisation in *Populus* L. section Aigeiros Duby. Euphytica 206(2):433–443

Tobutt KR (1985) Breeding columnar apples at East Malling. Acta Hort 159:63–68

Toloza AC, Lucia A, Zerba E, Masuh H, Picollo MI (2008) Interspecific hybridization of *Eucalyptus* as a potential tool to improve the bioactivity of essential oils against permethrin-resistant head lice from Argentina. Bioresour Technol 99(15):7341–7347

Touchell D, Vitoria Z, Ranney T (2006) Intergeneric hybrids between *Weigela* and *Diervilla* (Caprifoliaceae). Hortscience 41(4):1008

Traenkner C, Engel F (2017) Genotyping-by-sequencing facilitates genetic mapping of the inflorescence type in *Hydrangea*. Julius-Kuhn-Archiv 457:46–47

Trueblood CE, Ranney TG, Lynch NP, Neal JC, Olsen RT (2010) Evaluating fertility of triploid clones of *Hypericum androsaemum* L. for use as non-invasive landscape plants. Hortscience 45:1026–1028

Tsai CJ, Xue LJ (2015) CRISPRing into the woods. GM crops and Food 6:206–215

Tsuda H, Kunitake H, Yamasaki M, Komatsu H, Yoshioka K (2013) Production of intersectional hybrids between colchicine-induced Tetraploid Shashanbo (*Vaccinium bracteatum*) and Highbush blueberry 'Spartan'. J Am Soc Hortic Sci 138(4):317–324

Tsuro M, Ikedo H (2011) Changes in morphological phenotypes and essential oil components in lavandin (*Lavandula×intermedia* Emeric ex Loisel.) transformed with wild-type strains of *Agrobacterium rhizogenes*. Sci Hortic 130(3):647–652

Tuominen H, Sitbon F, Jacobsson C, Sandberg G, Olsson O, Sundberg B (1995) Altered growth and wood characteristics in transgenic hybrid aspen expressing *Agrobacterium tumefaciens* T-DNA indole acetic acid-biosynthetic genes. Plant Physiol 109:1179–1189

Tuskan GA, Difazio S, Jansson S, Bohlmann J, Grigoriev I, Hellsten U, Putnam N, Ralph S, Rombauts S, Salamov A, Schein J, Sterck L, Aerts A, Bhalerao RR, Bhalerao RP, Blaudez D, Boerjan W, Brun A, Brunner A, Busov V, Campbell M, Carlson J, Chalot M, Chapman J, Chen GL, Cooper D, Coutinho PM, Couturier J, Covert S, Cronk Q, Cunningham R, Davis J, Degroeve S, Déjardin A, Depamphilis C, Detter J, Dirks B, Dubchak I, Duplessis S, Ehlting J, Ellis B, Gendler K, Goodstein D, Gribskov M, Grimwood J, Groover A, Gunter L, Hamberger B, Heinze B, Helariutta Y, Henrissat B, Holligan D, Holt R, Huang W, Islam-Faridi N, Jones S, Jones-Rhoades M, Jorgensen R, Joshi C, Kangasjärvi J, Karlsson J, Kelleher C, Kirkpatrick R, Kirst M, Kohler A, Kalluri U, Larimer F, Leebens-Mack J, Leplé JC, Locascio P, Lou Y, Lucas S, Martin F, Montanini B, Napoli C, Nelson DR, Nelson C, Nieminen K, Nilsson O, Pereda V, Peter G, Philippe R, Pilate G, Poliakov A, Razumovskaya J, Richardson P, Rinaldi C, Ritland K, Rouzé P, Ryaboy D, Schmutz J, Schrader J, Segerman B, Shin H, Siddiqui A, Sterky F, Terry A, Tsai CJ, Uberbacher E, Unneberg P, Vahala J, Wall K, Wessler S, Yang G, Yin T, Douglas C, Marra M, Sandberg G, Van de Peer Y, Rokhsar D (2006) The genome of black cottonwood, *Populus trichocarpa* (Torr. & Gray). Science 313(5793):1596–1604

Uemaehi T, Mizuhara Y, Deguchi K, Shinjo Y, Kajino E, Ohba H (2014) Phylogenetic relationship of *Hydrangea macrophylla* (Thunb.) Ser. and *H. serrata* (Thunb.) Ser. Evaluated using RAPD markers and plastid DNA sequences. Journal of the Japanese Society for Horticultural Science

83(2):163–171
United States Geological Survey (USGS) (2016) USGS Water Science School. https://water.usgs.gov/edu/100yearflood.html
Unver T, Wu Z, Sterck L, Turktas M, Lohaus R, Li Z, Yang M, He L, Deng T, Escalante FJ, Llorens C, Roig FJ, Parmaksiz I, Dundar E, Xie F, Zhang B, Ipek A, Uranbey S, Erayman M, Ilhan E, Badad O, Ghazal H, Lightfoot DA, Kasarla P, Colantonio V, Tombuloglu H, Hernandez P, Mete N, Cetin O, Van Montagu M, Yang H, Gao Q, Dorado G, Van de Peer Y (2017) Genome of wild olive and the evolution of oil biosynthesis. Proc Natl Acad Sci U S A 114(44):E9413–E9422
Van de Weyer W (1920) Hybrid buddleias. The Gardeners' Chronicle, ser 3 68:181
Van Dyk MM, Soeker MK, Labuschagne IF, Rees DJG (2010) Identification of a major QTL for time of initial vegetative bud-break in apple (*Malus* x *domestica* Borkh.). Tree Genet Genomes 6:489–502
Van Huylenbroeck J, Van Laere K (2010) Breeding strategies for woody ornamentals. Acta Hort 885:391–401
Van Huylenbroeck JM, De Riek J, De Loose M (2000) Genetic relationships among *Hibiscus syriacus*, *Hibiscus sinosyriacus* and *Hibiscus paramutabilis* revealed by AFLP, morphology and ploidy analysis. Genet Resour Crop Evol 47:335–343
Van Laere K (2008) Interspecific hybridization in woody ornamentals PhD thesis, Ghent University 163p
Van Laere K, Van Huylenbroeck J, Eeckhaut T, Van Bockstaele E (2006) Breeding strategies to increase genetic variability within *Hibiscus syriacus*. Acta Hort 714:75–81
Van Laere K, Van Huylenbroeck JM, Van Bockstaele E (2007) Interspecific hybridisation between *Hibiscus syriacus*, *Hibiscus sinosyriacus* and *Hibiscus paramutabilis*. Euphytica 155(1–2):271–283
Van Laere K, Van Huylenbroeck J, Van Bockstaele E (2008) Karyotype analysis and physical mapping of 45S rRNA genes in *Hydrangea* species by fluorescence in situ hybridization. Plant Breed 127(3):301–307
Van Laere K, Dewitte A, Van Huylenbroeck J, Van Bockstaele E (2009a) Evidence for the occurrence of unreduced gametes in interspecific hybrids of *Hibiscus*. Journal of Horticultural Science & Biotechnology 84(2):240–247
Van Laere K, Leus L, Van Huylenbroeck J, Van Bockstaele E (2009b) Interspecific hybridisation and genome size analysis in *Buddleja*. Euphytica 166(3):445–456
Van Laere K, Khrustaleva L, Van Huylenbroeck J, Van Bockstaele E (2010) Application of GISH to characterize woody ornamental hybrids with small genomes and chromosomes. Plant Breed 129(4):442–447
Van Laere K, Van Huylenbroeck J, Van Bockstaele E (2011a) Introgression of yellow flower colour in *Buddleja davidii* by means of polyploidisation and interspecific hybridisation. Eur J Hortic Sci 38(3):96–103
Van Laere K, Hermans D, Leus L, Van Huylenbroeck J (2011b) Genetic relationships in European and Asiatic *Buxus* species based on AFLP markers, genome sizes and chromosome numbers. Plant Syst Evol 293(1–4):1–11
Van Laere K, Hermans D, Leus L, Van Huylenbroeck J (2015) Interspecific hybridisation within *Buxus* spp. Sci Hortic 185:139–144
Vanwynsberghe L, Keulemans J (2004) Homozygous genotypes for apple breeding: a feasible approach? Acta Hort 663:803–807
Velasco R, Zharkikh A, Affourtit J, Dhingra A, Cestaro A, Kalyanaraman A, Fontana P, Bhatnagar SK, Troggio M, Pruss D, Salvi S, Pindo M, Baldi P, Castelletti S, Cavaiuolo M, Coppola G, Costa F, Cova V, Dal Ri A, Goremykin V, Komjanc M, Longhi S, Magnago P, Malacarne G, Malnoy M, Micheletti D, Moretto M, Perazzolli M, Si-Ammou A, Vezzulli S, Zini E, Eldredge G, Fitzgerald LM, Gutin N, Lanchbury J, Macalma T, Mitchell JT, Reid J, Wardell B, Kodira C, Chen Z, Desany B, Niazi F, Palmer M, Koepke T, Jiwan D, Schaeffer S, Krishnan V, Wu C, Chu VT, King ST, Vick J, Tao Q, Mraz A, Stormo A, Stormo K, Bogden R, Ederle D, Stella A, Vecchietti A, Kater MM, Masiero S, Lasserre P, Lespinasse Y, Allan AC, Bus V, Chagne D, Crowhurst RN, Gleave AP, Lavezzo E, Fawcett JA, Proost S, Rouze P, Sterck L, Toppo S, Lazzari B, Hellens RP, Durel CE, Gutin A, Bumgarner RE, Gardiner SE, Skolnick M, Egholm M, Van de Peer Y, Salamini F, Viola R (2010) The genome of the domesticated apple (*Malus* x *domestica* Borkh). Nat Genet 42:833–839
Verde I, Bassil N, Scalabrin S, Gilmore B, Lawley CT, Gasic K Micheletti D, Rosyara UR, Cattonaro F, Vendramin E, Main D, Aramini V, Blas AL, Mockler TC, Bryant DW, Wilhelm L, Troggio M, Sosinski B, Jose Aranzana M, Arus P, Iezzoni A, Morgante M, Peace C (2012)

Development and evaluation of a 9K SNP array for peach by internationally coordinated SNP detection and validation in breeding germplasm. PLoS One 7:e35668

Verde I, Abbott AG, Scalabrin S, Jung S, Shu S, Marroni F, Zhebentyayeva T, Dettori MT, Grimwood J, Cattonaro F, Zuccolo A, Rossini L, Jenkins J, Vendramin E, Meisel LA, Decroocq V, Sosinski B, Prochnik S, Mitros T, Policriti A, Cipriani G, Dondini L, Ficklin S, Goodstein DM, Xuan P, Del Fabbro C, Aramini V, Copetti D, Gonzalez S, Horner DS, Falchi R, Lucas S, Mica E, Maldonado J, Lazzari B, Bielenberg D, Pirona R, Miculan M, Barakat A, Testolin R, Stella A, Tartarini S, Tonutti P, Arús P, Orellana A, Wells C, Main D, Vizzotto G, Silva H, Salamini F, Schmutz J, Morgante M, Rokhsar DS (2013) The high-quality draft genome of peach (*Prunus persica*) identifies unique patterns of genetic diversity, domestication and genome evolution. Nat Genet 45(5):487–494

Vorsa N, Polashock JJ (2005) Alteration of anthocyanin glycosylation in cranberry through interspecific hybridization. J Am Soc Hortic Sci 130(5):711–715

Voytas DF, Gao C (2014) Precision genome engineering and agriculture: opportunities and regulatory challenges. PLoS Biol 12:e1001877

Wada S, Reed BM (2011) Standardizing germination protocols for diverse raspberry and blackberry species. Sci Hortic 132:42–49

Wadl PA, Skinner JA, Wang X, Rinehart TA, Reed SM, Pantalone VR, Windham MT, Trigiano RN (2007) Breeding intra- and interspecific *Cornus* species. Hortscience 42(4):900–901

Wadl PA, Wang X, Pantalone VR, Trigiano RN (2010) Inheritance of red foliage in flowering dogwood (*Cornus florida* L.). Euphytica 176(1):99–104

Wadl PA, Saxton AM, Wang X, Pantalone VR, Rinehart TA, Trigiano RN (2011) Quantitative trait loci associated with red foliage in *Cornus florida* L. Mol Breed 27(3):409–416

Wakita Y, Yokota S, Yoshizawa N, Katsuki T, Nishiyama Y, Yokoyama T, Fukui M, Sasamoto H (2005) Interfamilial cell fusion among leaf protoplasts of *Populus alba*, *Betula platyphylla* and *Alnus firma*: assessment of electric treatment and in vitro culture conditions. Plant Cell Tissue Org Cult 83(3):319–326

Wang X, Gann A, Reed S, Windham M, Trigiano R (2007) Identification of interspecific hybrids between *Cornus kousa* and *C. florida* L. using SSR markers. Hortscience 42(4):964

Wang X, Trigiano R, Windham M, Scheffler B, Rinehart T, Spiers J (2008a) Development and characterization of simple sequence repeats for flowering dogwood (*Cornus florida* L.). Tree Genet Genomes 4:461–468

Wang XR, Tang HR, Duan J, Li L (2008b) A comparative study on karyotypes of 28 taxa in Rubus sect. Idaeobatus and sect. Malachobatus (Rosaceae) from China. J Syst Evol 46(4):505–515

Wang X, Genovesi A, Pounders C, Cabrera RI (2010) A preliminary report on the use of embryo rescue techniques with intra- and interspecific hybrids in crape Myrtle (*Lagerstroemia* L.). Hortscience 45(8):122–123

Wang XJ, Wang XF, Cai M, He D, Pan HT, Zhang QX (2012) In vitro chromosome doubling and tetraploid identification in *Lagerstroemia indica*. Journal of Food Agriculture & Environment 10(3–4):1364–1367

Wang Y, Chen Q, He W, Chen T, Nan H, Tang HR, Wang XR (2013a) Genetic relationships between Rubus parvifolius and R. coreanus (Rosaceae), and preliminary identification one of their putative hybrids. Indian journal of genetics and plant breeding 73(1):72–81

Wang N, Thomson M, Bodles WJ, Crawford RM, Hunt HV, Featherstone AW, Pellicer J, Buggs RJ (2013b) Genome sequence of dwarf birch (Betula nana) and cross-species RAD markers. Mol Ecol 22(11):3098–3111

Wang J, You H, Tian J, Wang Y, Liu M, Duan W (2015) Abnormal meiotic chromosome behavior and gametic variation induced by intersectional hybridization in *Populus* L. Tree Genet Genomes 11(3):61

Wang P, Zhang Y, Zhao L, Mo B, Luo T (2017a) Effect of gamma rays on Sophora davidii and detection of DNA polymorphism through ISSR marker. Biomed Research International X:8576404

Wang C, Gao Y, Tao Y, Wu X, Cui Z (2017b) Gamma-irradiation treatment decreased degradation of cell-wall polysaccharides in blueberry fruit during cold storage. Postharvest Biol Technol 131:31–38

Wang J, Huo B, Liu W, Li D, Liao L (2017c) Abnormal meiosis in an intersectional allotriploid of *Populus* L. and segregation of ploidy levels in 2X x 3Xprogeny. PLoS One 12(7):e018176 7

Wang X, Xu Y, Zhang S, Cao L, Huang Y, Cheng J, Wu G, Tian S, Chen C, Liu Y, Yu H, Yang X, Lan H, Wang N, Wang L, Xu J, Jiang X, Xie Z, Tan M, Larkin RM, Chen LL, Ma BG, Ruan Y, Deng X, Xu Q (2017d) Genomic analyses of primitive, wild and cultivated Citrus provide

insights into asexual reproduction. Nat Genet 49(5):765–772
Wason JW, Huggett BA, Brodersen CR (2017) MicroCT imaging as a tool to study vessel endings in situ. Am J Bot 104(9):1424–1430
Wegner J (2013) Intellectual property on woody ornamental plant varieties. Acta Hort:990, 221–228
Welch HJ (2012) The conifer manual, volume 1. Springer Science + Business media B.V, Dordrecht
Wenck AR, Quinn M, Whetten RW, Pullman G, Sederoff R (1999) High efficiency *Agrobacterium*-mediated transformation of Norway spruce (*Picea abies*) and loblolly pine (*Pinus taeda*). Plant Mol Bio 39:407–4016
Weng Q, He X, Li F, Li M, Yu X, Shi J, Gan S (2014) Hybridizing ability and heterosis between *Eucalyptus urophylla* and *E. tereticornis* for growth and wood density over two environments. Silvae Genetica 63(1–2):15–24
Wenslaff TF, Lyrene PM (2003a) Chromosome homology in tetraploid southern highbush x *Vaccinium elliottii* hybrids. Hortscience 38(2):263–265
Wenslaff TF, Lyrene PM (2003b) Unilateral cross compatibility in *Vaccinium elliottii* x *V. arboreum*, an intersectional blueberry hybrid. Euphytica 131(3):255–258
Werner DJ, Snelling LK (2009) 'Blue Chip' and 'Miss Ruby' *Buddleja*. Hortscience 44(3):841–842
West TP, DeMarais SL, Lee CW (2014) Germination of non-stratified Japanese tree lilac seeds as influenced by seed capsule maturity and moisture content. HortTechnology 24(2):177–180
Widrlechner MP (2007) Old and new trends influencing the introduction of new nursery crops. In: Janick J, Whipkey A (eds) Issues in new crops and new uses. ASHS Press, Alexandria, pp 237–245
Wiedemann M, Meinl K, Samain MS, Klocke E, Abel S, Wanke S (2015) Intergeneric hybrids between species of *Hydrangea* and *Dichroa* – their germination in vivo and in vitro and molecular verification by RAPD analysis. Acta Hort 1087:333–338
Wilner J (1959) Note on an electrolyte procedure for differentiating between frost injury of roots and shoots in woody plants. Can J Plant Sci 39:512–513
Wilner J (1960) Relative and absolute electrolytic conductance tests for frost hardiness of apple varieties. Can J Plant Sci 40:630–637
Windham MT, Reed SM, Mmbaga MT, Windham AS, Li Y, Rinehart T (2011) Evaluation of powdery mildew resistance in *Hydrangea macrophylla*. J Environ Hort 29:60–64
Wisniewski M, Gusta LV (2014) The biology of cold hardiness: adaptive strategies (preface). Environ Exp Bot 106:1–3
Wisniewski M, Bassett C, Gusta LV (2003) An overview of cold hardiness in woody plants: seeing the forest through the trees. Hortscience 38:952–959
Wojciechowicz MK, Kikowska MA (2009) Induction of multi-nucleate microspores in anther culture of *Salix viminalis* L. Dendrobiology 61:55–64
Wolters PJ, Schouten HJ, Velasco R, Si-Ammour A, Baldi P (2013) Evidence for regulation of columnar habit by a putative 2OG-Fe(II) oxygenase. New Phytol 200:993–999
Wong MML, Cannon CH, Wickneswari R (2011) Identification of lignin genes and regulatory sequences involved in secondary call wall formation in Acacia auriculiformis and Acacia mangium via de novo transcriptome sequencing. BMC Genomics 12:342
Wooster LD, Bassuk N (2009) Evaluation of *Acer truncatum* germplasm for use in urban landscape plantings. Hortscience 44(4):1078–1079
Wu J, Wang Y, Zhang LX, Zhang XZ, Kong J, Lu J, Han ZH (2012) High-efficiency regeneration of agrobacterium rhizogenes-induced hairy root in apple rootstock *Malus baccata* (L.) Borkh. Plant Cell Tissue Org Cult 111(2):183–189
Wu Y, Li W, Dong J, Yang N, Zhao X, Yang W (2013a) Tetraploid induction and cytogenetic characterization for *Clematis heracleifolia*. Caryologia 66(3):215–220
Wu J, Wang Z, Shi Z, Zhang S, Ming R, Zhu S, Khan MA, Tao S, Korban SS, Wang H, Chen NJ, Nishio T, Xu X, Cong L, Qi K, Huang X, Wang Y, Zhao X, Wu J, Deng C, Gou C, Zhou W, Yin H, Qin G, Sha Y, Tao Y, Chen H, Yang Y, Song Y, Zhan D, Wang J, Li L, Dai M, Gu C, Wang Y, Shi D, Wang X, Zhang H, Zeng L, Zheng D, Wang C, Chen M, Wang G, Xie L, Sovero V, Sha S, Huang W, Zhang S, Zhang M, Sun J, Xu L, Li Y, Liu X, Li Q, Shen J, Wang J, Paull RE, Bennetzen JL, Wang J, Zhang S (2013b) The genome of the pear (*Pyrus bretschneideri* Rehd.). Genome Res 23(2):396–408
Xi X, Li D, Xu W, Guo L, Zhang J, Li B (2012) 2n egg formation in *Populus* x *euramericana* (Dode) Guinier. Tree Genet Genomes 8(6):1237–1245
Xie M, Zheng JL, Wang DM (2014) Achievements and perspective in hazelnut breeding in China. Acta Hortic 1052:41–43

Xie W, Leus L, Wang JH, Van Laere K (2017) Fertility barriers in interspecific crosses within *Viburnum*. Euphytica 213:34

Xin Z, Browse J (2000) Cold comfort farm: the acclimation of plants to freezing temperatures. Plant Cell and Environment 23:893–902

Xu Q, Chen LL, Ruan X, Chen D, Zhu A, Chen C, Bertrand D, Jiao WB, Hao BH, Lyon MP, Chen J, Gao S, Xing F, Lan H, Chang JW, Ge X, Lei Y, Hu Q, Miao Y, Wang L, Xiao S, Biswas MK, Zeng W, Guo F, Cao H, Yang X, Xu XW, Cheng YJ, Xu J, Liu JH, Luo OJ, Tang Z, Guo WW, Kuang H, Zhang HY, Roose ML, Nagarajan N, Deng XX, Ruan Y (2013) The draft genome of sweet orange (*Citrus sinensis*). Nat Genet 45(1):59–66

Xue H, Zhang F, Zhang ZH, Fu JF, Wang F, Zhang B, Ma Y (2015) Differences in salt tolerance between diploid and autotetraploid apple seedlings exposed to salt stress. Sci Hortic 190:24–30

Yamamoto T, Terakami S (2016) Genomics of pear and other Rosaceae fruit trees. Breed Sci 66(1):148–159

Yamamoto T, Kimura T, Shoda M, Imai T, Saito T, Sawamura Y, Kotobuki K, Hayashi T, Matsuta N (2002) Genetic linkage maps constructed by using an interspecific cross between Japanese and European pears. Theor Appl Gent 106:9–18

Yamashita H, Daimon H, Akasaka-Kennedy Y, Masuda T (2004) Plant regeneration from hairy roots of apple rootstock, *Malus prunifolia* Borkh. var. ringo Asami, strain Nagano no. 1, transformed by *Agrobacterium rhizogenes*. Journal of the Japanese Society for Horticultural Science 73(6):505–510

Yang W, Guo Z, Huang C, Duan L, Chen G, Jiang N, Fang W, Feng H, Xie W, Lian X, Wang G, Luo Q, Zhang Q, Liu Q, Xiong L (2014) Combining high-throughput phenotyping and genome-wide association studies to reveal natural genetic variation in rice. Nat Commun 5:5087

Yang W, Wang K, Zhang J, Ma J, Liu J, Ma T (2017) The draft genome sequence of a desert tree-*Populus pruinosa*. Gigascience 6(9):6

Yao J, Li H, Ye J, Shi L (2016) Relationship between parental genetic distance and offspring's heterosis for early growth traits in *Liriodendron*: implication for parent pair selection in cross breeding. New For 47(1):163–177

Yazici K, Sahin A (2016) Characterization of pomegranate (*Punica granatum* L.) hybrids and their potential use in further breeding. Turk J Agric For 40(6):813–824

Ye YM, Tong J, Shi XP, Yuan W, Li GR (2010) Morphological and cytological studies of diploid and colchicine-induced tetraploid lines of crape myrtle (*Lagerstroemia indica* L.). Sci Hortic 124(1):95–101

Yin L, Li M, Ke X, Li C, Zou Y, Liang D, Ma F (2013) Evaluation of *Malus* germplasm resistance to marssonina apple blotch. Eur J Plant Pathol 136(3):597–602

Yokoyama J, Fukuda T, Yokoyama A, Maki M (2002) The intersectional hybrid between *Weigela hortensis* and *W-maximowiczii* (Caprifoliaceae). Bot J Linn Soc 138(3):369–380

Yu X, Kikuchi A, Shimazaki T, Yamada A, Ozeki Y, Matsunaga E, Ebinuma H, Watanabe KN (2013) Assessment of the salt tolerance and environmental biosafety of *Eucalyptus camaldulensis* harboring a mangrin transgene. J Plant Res 126:141–150

Yuskianti V, Xin HF, Xiang ZB, Shiraishi S (2011) Diagnosis of interspecific hybrids between *Acacia mangium* and *A. auriculiformis* using single nucleotide polymorphism (SNP) markers. Silvae genetica 60(3–4):85–92

Zeng FS, Zhou S, Zhan YG, Dong J (2014) Drought resistance and DNA methylation of interspecific hybrids between *Fraxinus mandshurica* and *Fraxinus Americana*. Trees Structure and Function 28(6):1679–1692

Zeng FS, Li LL, Liang NS, Wang X, Li X, Zhan YG (2015) Salt tolerance and alterations in cytosine methylation in the interspecific hybrids of *Fraxinus velutina* and *Fraxinus mandshurica*. Euphytica 205(3):721–737

Zenkteler M, Wojciechowicz M, Bagniewska-Zadworna A, Zenkteler E, Jezowski S (2005) Intergeneric crossability studies on obtaining hybrids between *Salix viminalis* and four *Populus* species – in vivo and in vitro pollination of pistils and the formation of embryos and plantlets. Trees-Structure and Function 19(6):638–643

Zhang Q, Gu D (2015) Development of a new hybrid between *Prunus tomentosa* Thunb. and *Prunus salicina* Lindl. Hortscience 50(4):517–519

Zhang P, Kang X (2013) Occurrence and cytological mechanism of numerically unreduced pollen in diploid *Populus euphratica*. Silvae Genetica 62(6):285–291

Zhang W, Zwiazek JJ (2016a) Effects of root medium pH on root water transport and apoplastic pH in red-osier dogwood (*Cornus sericea*) and paper birch (*Betula papyrifera*) seedlings. Plant Biol J 18:1001–1007

Zhang W, Zwiazek JJ (2016b) Responses of reclamation plants to high root zone pH: effects of phosphorus and calcium availability. J Environ Qual. in press

Zhang Z, Sun A, Cong Y, Sheng B, Yao Q, Cheng ZM (2006) *Agrobacterium*-mediated transformation of the apple rootstock *Malus micromalus* Makino with the ROLC gene. In Vitro Cellular & Developmental Biology-Plant 42(6):491–497

Zhang GH, Liang YR, Jin J, Lu JL, Borthakur D, Dong JJ, Zheng XQ (2007) Induction of hairy roots by *Agrobacterium rhizogenes* in relation to L-theanine production in *Camellia sinensis*. Journal of Horticultural Science & Biotechnology 82(4):636–640

Zhang QY, Luo FX, Liu L, Guo FC (2010) In vitro induction of tetraploids in crape myrtle (*Lagerstroemia indica* L.). Plant Cell Tissue Org Cult 101(1):41–47

Zhang Q, Chen W, Sun L, Zhao F, Huang B, Yang W, Tao Y, Wang J, Yuan Z, Fan G, Xing Z, Han C, Pan H, Zhong X, Shi W, Liang X, Du D, Sun F, Xu Z, Hao R, Lv T, Lv Y, Zheng Z, Sun M, Luo L, Cai M, Gao Y, Wang J, Yin Y, Xu X, Cheng T, Wang J (2012) The genome of *Prunus mume*. Nat Commun 3:1318

Zhang J, Franks RG, Liu X, Kang M, Keebler JEM, Schaff JE, Huang HW, Xiang QY (2013) De novo sequencing, characterisation, and comparison of inflorescence transcriptomes of *Cornus canadensis* and *C. florida* (Cornaceae). PLoS One 8(12):e82674

Zhang F, Xue H, Lu X, Zhang B, Wang F, Ma Y, Zhang Z (2015) Autotetraploidization enhances drought stress tolerance in two apple cultivars. Trees-Structure and Function 29(6):1773–1780

Zhang NN, Sun WB, Yang J (2016) Chromosome counts and karyotype analysis of *Viburnum* taxa (Adoxaceae). Caryologia 69(1):12–19

Zhang Y, Wang Z, Qi S, Wang X, Zhao J, Zhang J, Li B, Zhang Y, Liu X, Yuan W (2017a) In vitro tetraploid induction from leaf and petiole explants of hybrid Sweetgum (*Liquidambar styraciflua* x *Liquidambar formosana*). Forests 8(8):264

Zhang J, Yuan H, Yang Q, Li M, Wang Y, Li Y, Ma X, Tan F, Wu R (2017b) The genetic architecture of growth traits in *Salix matsudana* under salt stress. Horticulture Research 4:17024

Zhao Z, Li Y, Liu H, Zhai X, Deng M, Dong Y, Fan G (2017) Genome-wide expression analysis of salt-stressed diploid and autotetraploid *Paulownia tomentosa*. PLoS One 12(10):e0185455

Zhebentyayeva TN, Fan S, Chandra A, Bielenberg DG, Reighard GL, Okie WR, Abbott AG (2014) Dissection of chilling requirement and bloom date QTLs in peach using a whole genome sequencing of sibling trees from an F2 mapping population. Tree Genet Genomes 10:35–51

Zhou X, Jacobs TB, Xue LJ, Harding SA, Tsai CJ (2015) Exploiting SNPs for biallelic CRISPR mutations in the outcrossing woody perennial *Populus* reveals 4-coumarate:CoA ligase specificity and redundancy. New Phytol 208:298–301

Zhu LH, Li XY, Kale L, Welander M (2009) Plant size control in apple through genetic engineering. Acta Hort 839:689–694

Zurawicz E (2016) Cross-pollination increases the number of drupelets in the fruits of red raspberry (*Rubus idaeus* L.). Acta Hort 1133:145–151